Pseudo-Differential Operators
Theory and Applications
Vol. 2

Managing Editor

M.W. Wong (York University, Canada)

Editorial Board

Luigi Rodino (Università di Torino, Italy)
Bert-Wolfgang Schulze (Universität Potsdam, Germany)
Johannes Sjöstrand (École Polytechnique, Palaiseau, France)
Sundaram Thangavelu (Indian Institute of Science at Bangalore, India)
Marciej Zworski (University of California at Berkeley, USA)

Pseudo-Differential Operators: Theory and Applications is a series of moderately priced graduate-level textbooks and monographs appealing to students and experts alike. Pseudo-differential operators are understood in a very broad sense and include such topics as harmonic analysis, PDE, geometry, mathematical physics, microlocal analysis, time-frequency analysis, imaging and computations. Modern trends and novel applications in mathematics, natural sciences, medicine, scientific computing, and engineering are highlighted.

Michael Ruzhansky | Ville Turunen

Pseudo-Differential Operators and Symmetries

Background Analysis and Advanced Topics

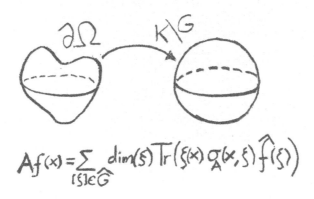

$$Af(x) = \sum_{[\xi]\in\widehat{G}} \dim(\xi) \operatorname{Tr}\left(\xi(x)\, \sigma_A(x,\xi)\, \widehat{f}(\xi)\right)$$

Birkhäuser
Basel · Boston · Berlin

Authors:

Michael Ruzhansky
Department of Mathematics
Imperial College London
180 Queen's Gate
London SW7 2AZ
United Kingdom
e-mail: m.ruzhansky@imperial.ac.uk

Ville Turunen
Institute of Mathematics
Helsinki University of Technology
P.O. Box 1100
FI-02015 TKK
Finland
e-mail: ville.turunen@hut.fi

2000 Mathematics Subject Classification: 35Sxx, 58J40; 43A77, 43A80, 43A85

Library of Congress Control Number: 2009929498

Bibliographic information published by Die Deutsche Bibliothek
Die Deutsche Bibliothek lists this publication in the Deutsche Nationalbibliografie;
detailed bibliographic data is available in the Internet at <http://dnb.ddb.de>.

ISBN 978-3-7643-8513-2 Birkhäuser Verlag AG, Basel · Boston · Berlin

© 2010 Birkhäuser Verlag AG
Basel · Boston · Berlin
P.O. Box 133, CH-4010 Basel, Switzerland
Part of Springer Science+Business Media
Printed on acid-free paper produced of chlorine-free pulp. TCF∞
Printed in Germany

ISBN 978-3-7643-8513-2   e-ISBN 978-3-7643-8514-9

9 8 7 6 5 4 3 2 1   www.birkhauser.ch

Contents

Part II   Commutative Symmetries

Part III   Representation Theory of Compact Groups

Part IV   Non-commutative Symmetries

11 Fourier Analysis on SU(2)

12 Pseudo-differential Operators on SU(2)

Preface

This monograph is devoted to the development of the theory of pseudo-differential operators on spaces with symmetries. Such spaces are the Euclidean space $\mathbb{R}^n$, the torus $\mathbb{T}^n$, compact Lie groups and compact homogeneous spaces.

The book consists of several parts. One of our aims has been not only to present new results on pseudo-differential operators but also to show parallels between different approaches to pseudo-differential operators on different spaces. Moreover, we tried to present the material in a self-contained way to make it accessible for readers approaching the material for the first time.

However, different spaces on which we develop the theory of pseudo-differential operators require different backgrounds. Thus, while operators on the Euclidean space in Chapter 2 rely on the well-known Euclidean Fourier analysis, pseudo-differential operators on the torus and more general Lie groups in Chapters 4 and 10 require certain backgrounds in discrete analysis and in the representation theory of compact Lie groups, which we therefore present in Chapter 3 and in Part III, respectively. Moreover, anyone who wishes to work with pseudo-differential operators on Lie groups will certainly benefit from a good grasp of certain aspects of representation theory. That is why we present the main elements of this theory in Part III, thus eliminating the necessity for the reader to consult other sources for most of the time. Similarly, the backgrounds for the theory of pseudo-differential operators on $\mathbb{S}^3$ and $SU(2)$ developed in Chapter 12 can be found in Chapter 11 presented in a self-contained way suitable for immediate use.

However, it was still not a simple matter to make a self-contained presentation of these theories without referring to basics of the more general analysis. Thus, in hoping that this monograph may serve as a guide to different aspects of pseudo-differential operators, we decided to include the basics of analysis that are certainly useful for anyone working with pseudo-differential operators.

Overall, we tried to supplement all the material with exercises for learning the ideas and practicing the techniques. They range from elementary problems to more challenging ones. In fact, on many occasions where other authors could say "it is easy to see" or "one can check", we prefer to present it as an exercise. At the same time, more challenging exercises also serve as an excellent way to present more aspects of the discussed material.

We would like to thank Professor G. Vainikko, who introduced V. Turunen to pseudo-differential equations on circles [137], leading naturally to the non-commutative setting of the doctoral thesis. The thesis work was crucially influenced by a visit to M.E. Taylor in spring 2000. We are grateful to Professor M.W. Wong for suggesting that we write this monograph, to our students for giving us useful feedback on the background material of the book, and to Dr. J. Wirth for reading the manuscript and for his useful feedback and numerous comments, which led to clarifications of the presentation, especially of the material from Section 10.3. Most of the work was carried out at the pleasant atmospheres provided by Helsinki University of Technology and Imperial College London. Moreover, over the years, we have outlined substantial parts of the monograph elsewhere: particularly, we appreciate the hospitality of University of North Carolina at Chapel Hill, University of Torino and Osaka University. The work of M. Ruzhansky was supported in part by EPSRC grants EP/E062873/01 and EP/G007233/1. The travels of V. Turunen were financed by the Magnus Ehrnrooth Foundation, by the Vilho, Yrjö and Kalle Väisälä Foundation of the Finnish Academy of Science and Letters, and by the Finnish Cultural Foundation. Finally, our loving thanks go to our families for all the encouragement and understanding that we received while working on this monograph.

March 2009   Michael Ruzhansky, London
                                                    Ville Turunen, Helsinki

Introduction

Historical notes

Pseudo-differential operators (ΨDO) can be considered as natural extensions of linear partial differential operators, with which they share many essential properties. The study of pseudo-differential operators grew out of research in the 1960s on singular integral operators; being a relatively young subject, the theory is only now reaching a stable form.

Pseudo-differential operators are generalisations of linear partial differential operators, with roots entwined deep down in solving differential equations.

Among the most influential predecessors of the theory of pseudo-differential operators one must mention the works of Solomon Grigorievich Mikhlin, Alberto Calderón and Antoni Szczepan Zygmund. Around 1957, anticipating novel methods, Alberto Calderón proved the local uniqueness theorem of the Cauchy problem of a partial differential equation. This proof involved the idea of studying the algebraic theory of characteristic polynomials of differential equations.

Another landmark was set in ca. 1963, when Michael Atiyah and Isadore Singer presented their celebrated index theorem. Applying operators, which nowadays are recognised as pseudo-differential operators, it was shown that the geometric and analytical indices of Erik Ivar Fredholm's "Fredholm operator" on a compact manifold are equal. In particular, these successes by Calderón and Atiyah–Singer motivated developing a comprehensive theory for these newly found tools. The Atiyah–Singer index theorem is also tied to the advent of K-theory, a significant field of study in itself.

The evolution of the pseudo-differential theory was then rapid. In 1963, Peter Lax proposed some singular integral representations using Jean Baptiste Joseph Fourier's "Fourier series". A little later, Joseph Kohn and Louis Nirenberg presented a more useful approach with the aid of Fourier integral operators and named their representations pseudo-differential operators. Showing that these operators form an algebra, they derived a broad theory, and their results were applied by Peter Lax and Kurt Otto Friedrichs in boundary problems of linear partial differential equations. Other related studies were conducted by Agranovich, Bokobza, Kumano-go, Schwartz, Seeley, Unterberger, and foremost, by Lars Hörmander, who coined the modern pseudo-differential theory in 1965, leading into a vast range of methods and results. The efforts of Kohn, Nirenberg, and Hörmander gave birth to symbol analysis, which is the basis of the theory of pseudo-differential operators.

It is interesting how the ideas of symbol analysis have matured over about 200 years. Already Joseph Lagrange and Augustin Cauchy studied the assignment of a characteristic polynomial to the corresponding differential operator. In the 1880s, Oliver Heaviside developed an operational calculus for the solution of ordinary differential equations met in the theory of electrical circuits. A more sophisticated problem of this kind, related to quantum mechanics, was solved by Hermann Weyl in 1927, and eventually the concept of the symbol of an operator was introduced by Solomon Grigorievich Mikhlin in 1936. After all, there is nothing new under the sun.

Since the mid-1960s, pseudo-differential operators have been widely applied in research on partial differential equations: along with new theorems, they have provided a better understanding of parts of classical analysis including, for instance, Sergei Lvovich Sobolev's "Sobolev spaces", potentials, George Green's "Green functions", fundamental solutions, and the index theory of elliptic operators. Furthermore, they appear naturally when reducing elliptic boundary value problems to the boundary. Briefly, modern mathematical analysis has gained valuable clarity with the unifying aid of pseudo-differential operators. Fourier integral operators are more general than pseudo-differential operators, having the same status in the study of hyperbolic equations as pseudo-differential operators have with respect to elliptic equations.

A natural approach to treat pseudo-differential operators on n-dimensional C^∞-manifolds is to use the theory of $\mathbb{R}^n$ locally: this can be done, since the classes of pseudo-differential operators are invariant under smooth changes of coordinates. However, on periodic spaces (tori) $\mathbb{T}^n$, this could be a clumsy way of thinking, as the local theory is plagued with rather technical convergence and local coordinate questions. The compact group structure of the torus is important from the harmonic analysis point of view.

In 1979 (and 1985) Mikhail Semenovich Agranovich (see [3]) presented an appealing formulation of pseudo-differential operators on the unit circle $\mathbb{S}^1$ using Fourier series. Hence, the independent study of periodic pseudo-differential operators was initiated. The equivalence of local and global definitions of periodic pseudo-differential operators was completely proven by William McLean in 1989. By then, the global definition was widely adopted and used by Agranovich, Amosov, D.N. Arnold, Elschner, McLean, Saranen, Schmidt, Sloan, and Wendland among others. Its effectiveness has been recognised particularly in the numerical analysis of boundary integral equations.

The literature on pseudo-differential operators is extensive. At the time of writing of this paragraph (28 January 2009), a search on MathSciNet showed 1107 entries with words "pseudodifferential operator" in the title (among which 33 are books), 436 entries with words "pseudo-differential operator" in the title (among which 37 are books), 3971 entries with words "pseudodifferential operator" anywhere (among which 417 are books), and 1509 entries with words "pseudo-differential operator" anywhere (among which 151 are books). Most of these works are devoted to the analysis on $\mathbb{R}^n$ and thus we have no means to give a comprehensive overview there. Thus, the emphasis of this monograph is on pseudo-differential operators on the torus, on Lie groups, and on spaces with symmetries, in which cases the literature is much more limited.

Periodic pseudo-differential operators

It turns out that the pseudo-differential and periodic pseudo-differential theories are analogous, the periodic case actually being more discernible.

Despite the intense research on periodic integral equations, the theory of periodic pseudo-differential operators has been difficult to find in the literature. On the other hand, the wealth of publications on general pseudo-differential operators is cumbersome for the periodic case, and it is too easy to get lost in the midst of irrelevant technical details.

In the sequel the elementary properties of periodic pseudo-differential operators are studied. The prerequisites for understanding the theory are more modest than one might expect. Of course, a basic knowledge of functional analysis is necessary, but the simple central tools are Gottfried Wilhelm von Leibniz' "Leibniz formula", Brook Taylor's "Taylor expansion", and Jean Baptiste Joseph Fourier's "Fourier transform". In the periodic case, these familiar concepts of the classical calculus are to be expressed in discrete forms using differences and summation instead of derivatives and integration.

Our working spaces will be the Sobolev spaces $H^s(\mathbb{T}^n)$ on the compact torus group $\mathbb{T}^n$. These spaces ideally reflect smoothness properties, which are of fundamental significance for pseudo-differential operators, as the traditional operator theoretic methods fail to be satisfactory – pseudo-differential operators and periodic pseudo-differential operators do not form any reasonable normed algebra.

The structure of the treatment of periodic pseudo-differential operators is the following: first, introduction of necessary functional analytic prerequisites, then development of useful tools for analysis of series and periodic functions, and after that the presentation of the theory of periodic pseudo-differential operators. The focus of the study is on symbolic analysis.

The techniques of the extension of symbols and the periodisation of operators allow one effectively to relate the Euclidean and the periodic theories, and to use one to derive results in the other. However, we tried to reduce a reliance on such ideas, keeping in mind the development of the subject on Lie groups where such a relation is not readily available. From this point of view, analysis on the torus can be viewed rather as a special case of analysis on a Lie group than the periodic Euclidean case.

The main justification of this work on the torus, from the authors' point of view, is the unification and development of the global theory of periodic pseudo-differential operators. It becomes evident how elegant this theory is, especially when compared to the theory on $\mathbb{R}^n$; and as such, periodic pseudo-differential operators may actually serve as a nice first introduction to the general theory of pseudo-differential operators. For those who have already acquainted themselves with pseudo-differential operators this work may still offer another aspect of the analysis. Thus, there is a hope that these tools will find various uses.

Although we decided not to discuss Fourier integral operators on $\mathbb{R}^n$, we devote some efforts to analysing operators that we call Fourier series operators. These are analogues of Fourier integral operators on the torus and we study them in terms of toroidal quantization. The main new difficulty here is that while pseudo-differential operators do not move the wave front sets of distributions, this is no longer the case for Fourier series operators. Thus, we are quickly forced to make extensions of functions from an integer lattice to Euclidean space on the frequency side. The analysis presented here shows certain limitation of the use of Fourier series operators; however, we succeed in establishing elements of calculus for them and discuss an application to hyperbolic partial differential equations.

Pseudo-differential operators on Lie groups

Non-commutative Lie groups and homogeneous spaces play important roles in different areas of mathematics. Some fundamental examples include spheres $\mathbb{S}^n$, which are homogeneous spaces under the action of the orthogonal groups. The important special case is the three-dimensional sphere $\mathbb{S}^3$ which happens to be also a group. However, while the general theory of pseudo-differential operators is available on such spaces, it presents certain limitations. First, working in local coordinates often makes it very complicated to keep track of the global geometric features. For example, a fundamental property that spheres are fixed by rotations becomes almost untraceable when looking at it in local coordinates. Another limitation is that while the local approach yields an invariant notion of the principal symbol, the full symbol is not readily available. This presents profound complications in applying the theory of pseudo-differential operators to problems on manifolds that depend on knowledge of the full symbol of an operator.

In general, it is a natural idea to build pseudo-differential operators out of smooth families of convolution operators on Lie groups. There have been many works aiming at the understanding of pseudo-differential operators on Lie groups from this point of view, e.g., the works on left-invariant operators [121, 78, 40], convolution calculus on nilpotent Lie groups [77], L^2-boundedness of convolution operators related to Howe's conjecture [57, 41], and many others.

However, in this work, we strive to develop the convolution approach into a symbolic quantization, which always provides a much more convenient framework for the analysis of operators. For this, our analysis of operators and their symbols is based on the representation theory of Lie groups. This leads to a description of the full symbol of a pseudo-differential operator on a Lie group as a sequence of matrices of growing sizes equal to the dimensions of the corresponding representations of the group. We also characterise, in terms of the introduced quantizations, standard Hörmander classes Ψ^m on Lie groups. One of the advantages of the presented approach is that we obtain a notion of full (global) symbols which matches the underlying Fourier analysis on the group in a perfect way. For a group G, such a symbol can be interpreted as a mapping defined on the space $G \times \widehat{G}$, where $\widehat{G}$ is the unitary dual of a compact Lie group G. In a nutshell, this analysis can be regarded as a non-commutative analogue of the Kohn–Nirenberg quantization of pseudo-differential operators that was proposed by Joseph Kohn and Louis Nirenberg in [68] in the Euclidean setting. As such, the present research is perhaps most closely related to the work of Michael Taylor [128], who, however, in his analysis used an exponential mapping to rely on pseudo-differential operators on a Lie algebra which can be viewed as a Euclidean space with the corresponding standard theory of pseudo-differential operators. However, the approach developed in this work is different from that of [128, 129] in the sense that we rely on the group structure directly and thus are not restricted to neighbourhoods of the neutral element, thus being able to approach global symbol classes directly. Some aspects of the analysis presented in this part appeared in [99].

As an important example, the approach developed here gives us quite detailed information on the global quantization of operators on the three-dimensional sphere $\mathbb{S}^3$. More generally, we note that if we have a closed simply-connected three-dimensional manifold M, then by the recently resolved Poincaré conjecture there exists a global diffeomorphism $M \simeq \mathbb{S}^3 \simeq \mathrm{SU}(2)$ that turns M into a Lie group with a group structure induced by $\mathbb{S}^3$ (or by $\mathrm{SU}(2)$). Thus, we can use the approach developed for $\mathrm{SU}(2)$ to immediately obtain the corresponding global quantization of operators on M with respect to this induced group product. In fact, all the formulae remain completely the same since the unitary dual of $\mathrm{SU}(2)$ (or $\mathbb{S}^3$ in the quaternionic $\mathbb{R}^4$) is mapped by this diffeomorphism as well. An interesting feature of the pseudo-differential operators from Hörmander's classes Ψ^m on these spaces is that they have matrix-valued full symbols with a remarkable rapid off-diagonal decay property.

We also introduce a general machinery with which we obtain global quantization on homogeneous spaces using the one on the Lie group that acts on the space. Although we do not yet have general analogues of the diffeomorphic Poincaré conjecture in higher dimensions, this already covers cases when M is a convex surface or a surface with positive curvature tensor, as well as more general manifolds in terms of their Pontryagin class, etc.

Conventions

Each part or a chapter of the book is preceded by a short introduction explaining the layout and conventions. However, let us mention now several conventions that hold throughout the book.

Constants will be usually denoted by C (sometimes with subscripts), and their values may differ on different occasions, even when appearing in subsequent estimates. Throughout the book, the notation for the Laplace operator is $\mathcal{L}$ in order not to confuse it with difference operators which are denoted by $\triangle$.

In Chapters 3 and 4 we encounter a notational difficulty that both frequencies and multi-indices are integers with different conventions for norms than are normally used in the literature. To address this issue, there we let $|\alpha| = |\alpha|_{\ell^1}$ be the ℓ^1-norm (of the multi-index α) and $\|\xi\| = \|\xi\|_{\ell^2}$ be the Euclidean ℓ^2 norm (of the frequency $\xi \in \mathbb{Z}^n$). However, in other chapters we write a more traditional $|\xi|$ for the length of the vector ξ in $\mathbb{R}^n$, and reserve the notation $\|\cdot\|_X$ for a norm in a normed space X. However, there should be no confusion with this notation since we usually make it clear which norm we use. In Part IV, $\xi = \xi(x)$ stands for a representation, so that we can still use the usual notation $\sigma(x, \xi)$ for symbols.

Part I

Foundations of Analysis

Part I of the monograph contains preliminary material that could be useful for anyone working in the theory of pseudo-differential operators.

The material of the book is on the intersection of classical analysis with the representation theory of Lie groups. Aiming at making the presentation self-sufficient we include preliminary material that may be used as a reference for concepts developed later. In any case, the material presented in this part may be used either as a reference or as an independent textbook on the foundations of analysis.

Throughout the book, we assume that the reader has survived undergraduate calculus courses, so that concepts like *partial derivatives* and the *Riemann integral* are familiar. Otherwise, the prerequisites for understanding the material in this book are quite modest. We shall start with a naive version of a set theory, metric spaces, topology, functional analysis, measure theory and integration in Lebesgue's sense.

Chapter A

Sets, Topology and Metrics

First, we present the basic notations and properties of sets, used elsewhere in the book. The set theory involved is "naive", sufficient for our purposes; for a thorough treatment, see, e.g., [46]. The sets of integer, rational, real or complex numbers will be taken for granted, we shall not construct them.

Let us first list some abbreviations that we are going to use:

- "*P and Q*" means that both properties P and Q are true.
- "*P or Q*" means that at least one of the properties P and Q is true.
- "$P \Rightarrow Q$" reads "*If P then Q*", meaning that "P is false or Q is true". Equivalently "$Q \Leftarrow P$", i.e., "*Q only if P*".
- "$P \Longleftrightarrow Q$" is "$P \Rightarrow Q$ and $P \Leftarrow Q$", reading "*P if and only if Q*".
- "$\exists x$" reads "*There exists x*".
- "$\exists! x$" reads "*There exists a unique x*".
- "$\forall x$" reads "*For every x*".
- "$P := Q$" or "$Q =: P$" reads "*P is defined to be Q*".

A.1 Sets, collections, families

Naively, a *set* (or a *collection* or a *family*) A consists of *points* (or *elements* or *members*) x.

Example. Sets of points, like a *collection of coins*, a *family of two parents and three children*, a *flock of sheep*, a *pack of wolves*, or a *crowd of protesters*.

Example. Points in a set, like the *members of a parliament*, the *flowers in a bundle*, or the *stars in a constellation*.

We denote $x \in A$ if the element x *belongs to* the set A, and $x \notin A$ if x does not belong to A. A set A is a *subset* of a set B, denoted by $A \subset B$ or $B \supset A$, if

$$\forall x : \ x \in A \Rightarrow x \in B.$$

Sets A, B are *equal*, denoted by $A = B$, if $A \subset B$ and $B \subset A$, i.e.,

$$\forall x : \ x \in A \iff x \in B.$$

If $A \subset B$ and $A \neq B$ then A is called a *proper subset* of B.

Remark A.1.1 (**Notation for numbers**). The sets of *integer, rational, real and complex numbers* are respectively $\mathbb{Z}$, $\mathbb{Q}$, $\mathbb{R}$ and $\mathbb{C}$; let $\mathbb{N} = \mathbb{Z}^+$ and $\mathbb{R}^+$ stand for the corresponding subsets of (strictly) positive numbers. Then

$$\mathbb{Z}^+ \subset \mathbb{Z} \subset \mathbb{Q} \subset \mathbb{R} \subset \mathbb{C}.$$

We also write $\mathbb{N}_0 = \mathbb{N} \cup \{0\}$.

There are various ways for expressing sets. Sometimes all the elements can be listed:

- The *empty set* $\emptyset = \{\}$ is the unique set without elements: $\forall x : x \notin \emptyset$.
- Set $\{x\}$ consists of a single element $x \in \{x\}$.
- Set $\{x, y\} = \{y, x\}$ consists of elements x and y. And so on. Yet $\{x\} = \{x, x\} = \{x, x, x\}$ etc.

A set consisting of those elements for which property P holds can be denoted by

$$\{x : \ P(x)\} = \{x \mid P(x)\} .$$

A set consisting of finitely many elements $x_1, \ldots, x_n$ could be denoted by

$$
\begin{aligned}
\{x_1, \ldots, x_n\} \ &= \ \{x_k : \ k \in \{1, \ldots, n\}\} \\
&= \ \{x_k \mid k \in \mathbb{Z}^+ : \ k \leq n\} \\
&= \ \{x_k\}_{k=1}^n ,
\end{aligned}
$$

and the infinite set of positive integers by

$$\mathbb{Z}^+ \ = \ \{1, 2, 3, 4, 5, \cdots\} .$$

The *power set* $\mathcal{P}(X)$ consists of all the subsets of X,

$$\mathcal{P}(X) = \{A : \ A \subset X\}$$

Example. For the set $X = \{1\}$, we have

$$
\begin{aligned}
\mathcal{P}(X) \ &= \ \{\emptyset, \{1\}\}, \\
\mathcal{P}(\mathcal{P}(X)) \ &= \ \{\emptyset, \{\emptyset\}, \{1\}, \{\emptyset, \{1\}\}\},
\end{aligned}
$$

and we leave it as an exercise to find $\mathcal{P}(\mathcal{P}(\mathcal{P}(X)))$, which contains $2^4 = 16$ elements in this case.

Example. Always at least $\emptyset, X \in \mathcal{P}(X)$. If $x \in X$, then $\{x\} \in \mathcal{P}(X)$ and $\{\{x\}\} \in \mathcal{P}(\mathcal{P}(X))$,

$$x \neq \{x\} \neq \{\{x\}\} \neq \cdots,$$
$$x \in \{x\} \in \{\{x\}\} \in \cdots.$$

However, we shall allow neither $x \in x$ nor $x \notin x$; consider *Russell's paradox*: given $x = \{a : a \notin a\}$, is $x \in x$?

For $A, B \subset X$, let us define the *union* $A \cup B$, the *intersection* $A \cap B$ and the *difference* $A \setminus B$ by

$$
\begin{aligned}
A \cup B &:= \{x : x \in A \text{ or } x \in B\}, \\
A \cap B &:= \{x : x \in A \text{ and } x \in B\}, \\
A \setminus B &:= \{x : x \in A \text{ and } x \notin B\}.
\end{aligned}
$$

The *complement* A^c of A in X is defined by $A^c := X \setminus A$.

Example. If $A = \{1, 2\}$ and $B = \{2, 3\}$ then $A \cup B = \{1, 2, 3\}$, $A \cap B = \{2\}$ and $A \setminus B = \{1\}$.

Example. $\mathbb{R} \setminus \mathbb{Q}$ is the set of irrational numbers.

Exercise A.1.2. Show that

$$
\begin{aligned}
(A \cup B) \cup C &= A \cup (B \cup C), \\
(A \cap B) \cap C &= A \cap (B \cap C), \\
(A \cup B) \cap C &= (A \cap C) \cup (B \cap C), \\
(A \cap B) \cup C &= (A \cup C) \cap (B \cup C).
\end{aligned}
$$

Notice that in the latter two cases above, the order of the parentheses is essential. On the other hand, the associativity in the first two equalities allows us to abbreviate $A \cup B \cup C := (A \cup B) \cup C$ and $A \cap B \cap C := (A \cap B) \cap C$ and so on.

Definition A.1.3 (Index sets). Let I be any set and assume that for every $i \in I$ we are given a set A_i. Then I is an *index set* for the collection of sets A_i.

Definition A.1.4 (Unions and intersections of families). For a family $\mathcal{A} \subset \mathcal{P}(X)$, the *union* $\bigcup \mathcal{A}$ and the *intersection* $\bigcap \mathcal{A}$ are defined by

$$\bigcup \mathcal{A} = \bigcup_{B \in \mathcal{A}} B := \{x \mid \exists B \in \mathcal{A} : x \in B\},$$

$$\bigcap \mathcal{A} = \bigcap_{B \in \mathcal{A}} B := \{x \mid \forall B \in \mathcal{A} : x \in B\}.$$

Example. If $\mathcal{A} = \{B, C\}$ then $\bigcup \mathcal{A} = B \cup C$ and $\bigcap \mathcal{A} = B \cap C$.

Notice that if $\mathcal{A} \subset \mathcal{B} \subset \mathcal{P}(X)$ then

$$\emptyset \subset \bigcup \mathcal{A} \subset \bigcup \mathcal{B} \subset X \quad \text{and} \quad \emptyset \subset \bigcap \mathcal{B} \subset \bigcap \mathcal{A} \subset X.$$

Especially, for $\emptyset \subset \mathcal{P}(X)$ we have

$$\bigcup \emptyset = \emptyset \quad \text{and} \quad \bigcap \emptyset = X. \tag{A.1}$$

Notice that $A \cup B = \bigcup \{A, B\}$ and $A \cap B = \bigcap \{A, B\}$. For unions (and similarly for intersections), the following notations are also commonplace:

$$\bigcup_{j \in K} A_j \ := \ \bigcup \{A_j \mid j \in K\},$$

$$\bigcup_{k=1}^{n} A_k \ := \ \bigcup \{A_k \mid k \in \mathbb{Z}^+ : 1 \le k \le n\},$$

$$\bigcup_{k=1}^{\infty} A_k \ := \ \bigcup \{A_k \mid k \in \mathbb{Z}^+\}.$$

Example. $\displaystyle\bigcap_{k=1}^{3} A_k = A_1 \cap A_2 \cap A_3.$

Exercise A.1.5 (de Morgan's rules). Prove *de Morgan's rules*:

$$X \setminus \bigcup_{j \in K} A_j \ = \ \bigcap_{j \in K} (X \setminus A_j),$$

$$X \setminus \bigcap_{j \in K} A_j \ = \ \bigcup_{j \in K} (X \setminus A_j).$$

A.2   Relations, functions, equivalences and orders

The *Cartesian product* of sets A and B is

$$A \times B = \{(x, y) : x \in A, \ y \in B\},$$

where the elements $(x, y) := \{x, \{x, y\}\}$ are ordered pairs: if $x \ne y$ then $(x, y) \ne (y, x)$, whereas $\{x, y\} = \{y, x\}$. A *relation from A to B* is a subset $R \subset A \times B$. We write xRy if $(x, y) \in R$, saying "x is in relation R to y"; analogously, $x \not R y$ means $(x, y) \notin R$ ("x is not in relation R to y").

Functions. A relation $f \subset X \times Y$ is called a *function* (or a *mapping*) from X to Y, denoted by

$$f : X \to Y \quad \text{or} \quad X \overset{f}{\to} Y,$$

if for each $x \in X$ there exists a unique $y \in Y$ such that $(x, y) \in f$:

$$\forall x \in X \; \exists! y \in Y : \quad (x, y) \in f;$$

in this case, we write

$$y := f(x) \quad \text{or} \quad x \mapsto f(x) = y.$$

Intuitively, a function $f : X \to Y$ is a rule taking $x \in X$ to $f(x) \in Y$. Functions $X \xrightarrow{f} Y$ and $Y \xrightarrow{g} Z$ yield a *composition* $X \xrightarrow{g \circ f} Z$ by $g \circ f(x) := g(f(x))$. The *restriction* of $f : X \to Y$ to $A \subset X$ is $f|_A : A \to Y$ defined by $f|_A(x) := f(x)$.

Example. The *characteristic function* of a set $E \in \mathcal{P}(X)$ is $\chi_E : X \to \mathbb{R}$ defined by

$$\chi_E(x) := \begin{cases} 1, & \text{if } x \in E, \\ 0, & \text{if } x \notin E. \end{cases}$$

Definition A.2.1 (Injections, surjections, bijections). A function $f : X \to Y$ is

- an *injection* if $f(x_1) = f(x_2)$ implies $x_1 = x_2$,
- a *surjection* if for every $y \in Y$ there exists $x \in X$ such that $f(x) = y$,
- and a *bijection* if it is both injective and surjective, and in this case we may define the *inverse function* $f^{-1} : Y \to X$ such that $f(x) = y$ if and only if $x = f^{-1}(y)$.

Definition A.2.2. (Image and preimage) A function $f : X \to Y$ begets functions

$$f^+ : \mathcal{P}(X) \to \mathcal{P}(Y), \quad f^+(A) = f(A) := \{f(x) \in Y : x \in A\},$$
$$f^- : \mathcal{P}(Y) \to \mathcal{P}(X), \quad f^-(B) = f^{-1}(B) := \{x \in Y : f(x) \in B\}.$$

Sets $f(A)$ and $f^{-1}(B)$ are called the *image* of $A \subset X$ and the *preimage* of $B \subset Y$, respectively.

Exercise A.2.3. Let $f : X \to Y$, $A \subset X$ and $B \subset Y$. Show that

$$A \subset f^{-1}(f(A)) \quad \text{and} \quad f(f^{-1}(B)) \subset B.$$

Give examples showing that these subsets can be proper.

Exercise A.2.4. Let $f : X \to Y$, $A_0 \subset X$, $B_0 \subset Y$, $\mathcal{A} \subset \mathcal{P}(X)$ and $\mathcal{B} \subset \mathcal{P}(Y)$. Show that

$$\begin{cases} f(\bigcup \mathcal{A}) & = & \bigcup_{A \in \mathcal{A}} f(A), \\ f(\bigcap \mathcal{A}) & \subset & \bigcap_{A \in \mathcal{A}} f(A), \\ f(X \setminus A_0) & \supset & Y \setminus f(A_0), \end{cases}$$

where the subsets can be proper, while

$$\begin{cases} f^{-1}(\bigcup \mathcal{B}) & = & \bigcup_{B \in \mathcal{B}} f^{-1}(B), \\ f^{-1}(\bigcap \mathcal{B}) & = & \bigcap_{B \in \mathcal{B}} f^{-1}(B), \\ f^{-1}(Y \setminus B_0) & = & X \setminus f^{-1}(B_0). \end{cases}$$

These set-operation-friendly properties of $f^{-1} : \mathcal{P}(Y) \to \mathcal{P}(X)$ will be encountered later in topology and measure theory.

Definition A.2.5 (Induced and co-induced families). Let $f : X \to Y$, $\mathcal{A} \subset \mathcal{P}(X)$ and $\mathcal{B} \subset \mathcal{P}(Y)$. Then f is said to *induce* the family $f^{-1}(\mathcal{B}) \subset \mathcal{P}(X)$ and to *co-induce* the family $\mathcal{D} \subset \mathcal{P}(Y)$, where

$$f^{-1}(\mathcal{B}) := \{f^{-1}(B) \mid B \in \mathcal{B}\},$$
$$\mathcal{D} := \{B \subset Y \mid f^{-1}(B) \in \mathcal{A}\}.$$

Equivalences

Definition A.2.6 (Equivalence relation). A subset $\sim$ of $X \times X$ is an *equivalence relation* on X if it is

1. *reflexive*: $x \sim x$ (for all $x \in X$);
2. *symmetric*: if $x \sim y$ then $y \sim x$ (for all $x, y \in X$);
3. *transitive*: if $x \sim y$ and $y \sim z$ then $x \sim z$ (for all $x, y, z \in X$).

The *equivalence class* of $x \in X$ is

$$[x] := \{y \in X \mid x \sim y\},$$

and the equivalence classes form the *quotient space*

$$X/\sim := \{[x] \mid x \in X\}.$$

Notice that $x \in [x] \subset X$, that $[x] \cap [y] = \emptyset$ if $[x] \neq [y]$, and that $X = \bigcup_{x \in X}[x]$. *Example.* Clearly, the identity relation $=$ is an equivalence relation on X, and $f(x) := \{x\}$ defines a natural bijection $f : X \to X/=$.

Example. Let X and Y denote the sets of all women and men, respectively. For simplicity, we may assume the disjointness $X \cap Y = \emptyset$. Let $Isolde, Juliet \in X$ and $Romeo, Tristan \in Y$. For $a, b \in X \cup Y$, let $x \sim y$ if and only if a and b are of the same gender. Then

$$Y = [Tristan] = [Romeo] \quad \neq \quad [Juliet] = [Isolde] = X,$$
$$X \cup Y = [Romeo] \cup [Juliet],$$
$$(X \cup Y)/\sim = \{[Romeo], [Juliet]\}.$$

Exercise A.2.7. Let us define a relation $\sim$ in the Euclidean plane $\mathbb{R}^2$ by setting $(x_1, x_2) \sim (y_1, y_2)$ if and only if $x_1 - y_1, x_2 - y_2 \in \mathbb{Z}$. Show that $\sim$ is an equivalence relation. What is the equivalence class of the origin $(0, 0) \in \mathbb{R}^2$? What is common between a doughnut and the quotient space here?

Exercise A.2.8. Let us define a relation $\sim$ in the punctured Euclidean space $\mathbb{R}^3 \setminus \{(0, 0, 0)\}$ by setting $(x_1, x_2, x_3) \sim (y_1, y_2, y_3)$ if and only if $(x_1, x_2, x_3) = (ty_1, ty_2, ty_3)$ for some $t \in \mathbb{R}^+$. Prove that $\sim$ is an equivalence relation. What is common between a sphere and the quotient space here?

Orders

Definition A.2.9 (Partial order). A non-empty set X is *partially ordered* if there is a *partial order* $\leq$ on X. That is, $\leq$ is a relation from X to X, such that it is

1. *reflexive*: $x \leq x$ (for all $x \in X$);
2. *anti-symmetric*: if $x \leq y$ and $y \leq x$ then $x = y$ (for all $x, y \in X$);
3. *transitive*: if $x \leq y$ and $y \leq z$ then $x \leq z$ (for all $x, y, z \in X$).

We say that y *is greater than* x (or x *is less than* y), denoted by $x < y$, if $x \leq y$ and $x \neq y$.

Example. The set $\mathbb{R}$ of real numbers has the usual order $\leq$. Naturally, any of its non-empty subsets, e.g., $\mathbb{Z}^+ \subset \mathbb{R}$, inherits the order. The set $[-\infty, +\infty] = \mathbb{R} \cup \{-\infty, +\infty\}$ has the order $\leq$ extended from $\mathbb{R}$, with conventions $-\infty \leq x$ and $x \leq +\infty$ for every $x \in [-\infty, +\infty]$.

Example. Let us order $X = \mathcal{P}(S)$ by inclusion. That is, for $A, B \subset S$, let $A \leq B$ if and only if $A \subset B$.

Example. Let X, Y be sets, where Y has a partial order $\leq$. We may introduce a new partial order for all functions $f, g : X \to Y$ by setting

$$ f \leq g \quad \overset{\text{definition}}{\iff} \quad \forall x \in X : f(x) \leq g(x). $$

This partial order is commonplace especially when $Y = \mathbb{R}$ or $Y = [-\infty, \infty]$.

Definition A.2.10 (Chains and total order). A non-empty subset $K \subset X$ is a *chain* if $x \leq y$ or $y \leq x$ for all $x, y \in K$. The partial order is *total* (or *linear*) if the whole set X is a chain.

Example. $[-\infty, +\infty]$ is a chain with the usual partial order. Thereby also its subsets are chains, e.g., $\mathbb{R}$ and $\mathbb{Z}^+$. If $\{A_j : j \in J\} \subset \mathcal{P}(S)$ is a chain then $A_j \subset A_k$ or $A_k \subset A_j$ for each $j, k \in J$. Moreover, $\mathcal{P}(S)$ is not a chain if S has more than one element.

Definition A.2.11 (Bounds). Let $\leq$ be a partial order on X. The sets of *upper and lower bounds* of $A \subset X$ are defined, respectively, by

$$ \uparrow A \; := \; \{x \in X \mid \forall a \in A : a \leq x\}, $$
$$ \downarrow A \; := \; \{x \in X \mid \forall a \in A : x \leq a\}. $$

If $x \in A \cap \uparrow A$ then it is the *maximum of A*, denoted by $x = \max(A)$. If $x \in A \cap \downarrow A$ then it is the *minimum of A*, denoted by $x = \min(A)$. If $A \cap \uparrow \{z\} = \{z\}$ then the element $z \in A$ is called *maximal in A*. Similarly, if $A \cap \downarrow \{z\} = \{z\}$ then the element $z \in A$ is called *minimal in A*. If $\sup(A) := \min(\uparrow A) \in X$ exists, it is called the *supremum of A*, and if $\inf(A) := \max(\downarrow A) \in X$ exists, it is the *infimum of A*.

Remark A.2.12. Notations like

$$\sup_{k \geq 1} x_k = \sup_{k \in \mathbb{Z}^+} x_k = \sup\{x_k : k \in \mathbb{Z}^+\}$$

are quite common.

Example. The minimum in $\mathbb{Z}^+$ is 1, but there is no maximal element. For each $A \subset [-\infty, \infty]$, the infimum and the supremum exist.

Example. Let $X = \mathcal{P}(S)$. Then $\max(X) = S$ and $\min(X) = \emptyset$. If $\mathcal{A} \subset X$ then $\sup(\mathcal{A}) = \bigcup \mathcal{A}$ and $\inf(\mathcal{A}) = \bigcap \mathcal{A}$. For each $x \in S$, element $S \setminus \{x\} \in X$ is maximal in the subset $X \setminus \{S\}$.

Definition A.2.13 (lim sup and lim inf). Let $x_k \in X$ for each $k \in \mathbb{Z}^+$. If the following supremums and infimums exist, let

$$\limsup_{k \to \infty} x_k \quad := \quad \inf\left\{\sup\{x_k : j \leq k\} \mid j \in \mathbb{Z}^+\right\},$$

$$\liminf_{k \to \infty} x_k \quad := \quad \sup\left\{\inf\{x_k : j \leq k\} \mid j \in \mathbb{Z}^+\right\}.$$

Example. Let $E_k \in \mathcal{P}(X)$ for each $k \in \mathbb{Z}^+$. Then

$$\limsup_{k \to \infty} E_k \quad = \quad \bigcap_{j=1}^{\infty} \bigcup_{k=j}^{\infty} E_k,$$

$$\liminf_{k \to \infty} E_k \quad = \quad \bigcup_{j=1}^{\infty} \bigcap_{k=j}^{\infty} E_k.$$

Exercise A.2.14. Let $A = \limsup_{k \to \infty} E_k$ and $B = \liminf_{k \to \infty} E_k$ as in the example above. Show that

$$\chi_A = \limsup_{k \to \infty} \chi_{E_k} \quad \text{and} \quad \chi_B = \liminf_{k \to \infty} \chi_{E_k},$$

where $\chi_E : X \to \mathbb{R}$ is the characteristic function of $E \subset X$.

A.3   Dominoes tumbling and transfinite induction

The principle of mathematical induction can be compared to a sequence of dominoes, falling over one after another when the first tumbles down. More precisely,

if $1 \in S \subset \mathbb{Z}^+$ and $n \in S \Rightarrow n+1 \in S$ for every $n \in \mathbb{Z}^+$, then $S = \mathbb{Z}^+$.

The Transfinite Induction Principle generalises this, working on any well-ordered set.

Definition A.3.1 (Well-ordered sets). A partially ordered set X is said to be *well ordered*, if $\min(A)$ exists whenever $\emptyset \neq A \subset X$.

Example. With its usual order, $\mathbb{Z}^+$ is well ordered. With their usual orders, $\mathbb{Z}$, $\mathbb{R}$ and $[-\infty, +\infty]$ are not well ordered. With the inclusion order, $\mathcal{P}(S)$ is not well ordered, if there is more than one element in S.

Theorem A.3.2 (Transfinite Induction Principle). *Let X be well ordered and $S \subset X$. Assume that for each $x \in X$ it holds that $x \in S$ if*

$$\{y \in X : y < x\} \subset S.$$

Then $S = X$.

Exercise A.3.3. Prove the Transfinite Induction Principle.

Exercise A.3.4 (Transfinite $\Longrightarrow$ mathematical induction). Check that in the case $X = \mathbb{Z}^+$, the Transfinite Induction Principle is the usual *mathematical induction.*

The value of the Transfinite Induction Principle might be limited, as we have to assume the well-ordering of the underlying set. Actually, many (but not all) working mathematicians assume that

every non-empty set can be well ordered,

which is the so-called **Well-Ordering Principle**. Is such a principle likely to be true? After all, for example on sets $\mathbb{R}$ or $\mathcal{P}(\mathbb{Z}^+)$, can we imagine what well-orderings might look like? All the elementary tools which we use in our mathematical reasoning should be at least believable, so maybe the Well-Ordering Principle does not appear as a satisfying set theoretic axiom. Could we perhaps prove or disprove it from other, intuitively more reliable principles? We shall return to this question later.

A.4   Axiom of Choice: equivalent formulations

In this section we shall consider how to calculate the number of points in a set, and what infinity might mean in general.

Choosing. We may always choose one point out of a non-empty set, no matter how many points there are around. But sometimes we need infinitely many tasks done at once. For instance, we might want to choose a point from each of the non-empty subsets $A \subset X$ in no time at all: as a tool, we need the Axiom of Choice for X.

Definition A.4.1 (Choice function). Let $X \neq \emptyset$. A mapping $f : \mathcal{P}(X) \to X$ is called a *choice function* on X if $f(A) \in A$ whenever $\emptyset \neq A \subset X$.

Example. Let $X = \{p, q\}$ where $p \neq q$. Let $f : \mathcal{P}(X) \to X$ such that $f(X) = p = f(\{p\})$ and $f(\{q\}) = q$. Then f is a choice function on X.

The following Axiom of Choice should be considered as an axiom or a fundamental principle. In this section we discuss its implications.

Axiom A.4.2 (Axiom of Choice). For every non-empty set there exists a choice function.

Exercise A.4.3. Prove that a choice function exists on a well-ordered set. Thus the Well-Ordering Principle implies the Axiom of Choice.

The Axiom of Choice might look more convincing than the Well-Ordering Principle. Yet, we should be careful, as we are dealing with all kinds of sets, about which our intuition might be deficient. The Axiom of Choice might be plausible for $X = \mathbb{Z}^+$, or maybe even for $X = \mathbb{R}$, but can we be sure whether it is true in general? Nevertheless, let us add Axiom A.4.2 to our set-theoretic tool box.

There are plenty of equivalent formulations for the Axiom of Choice. In the sequel, we present some variants, starting with the "Axiom of Choice for Cartesian Products", to be presented soon.

Definition A.4.4 (Cartesian product). Let X_j be a set for each $j \in J$. The *Cartesian product* is defined to be

$$\prod_{j \in J} X_j := \left\{ f \ \middle| \ f : J \to \bigcup_{j \in J} X_j \quad \text{and} \quad \forall j \in J : \ f(j) \in X_j \right\}.$$

If $X_j = X$ for each $j \in J$, we write $X^J := \prod_{j \in J} X_j$. The elements $f \in X^J$ are then functions $f : J \to X$. Moreover, let $X^n := X^{\mathbb{Z}_n}$, where

$$\mathbb{Z}_n := \{k \in \mathbb{Z}^+ \mid k \leq n\}.$$

Exercise A.4.5. Give an example of a bijection

$$g : X_1 \times X_2 \to \prod_{j \in \{1,2\}} X_j,$$

especially in the case $X \times X \to X^2$. Thereby $X_1 \times X_2$ can be identified with the Cartesian product $\prod_{j \in \{1,2\}} X_j$.

Exercise A.4.6. Give a bijection $g : \mathcal{P}(X) \to \{0,1\}^X$.

The following Theorem A.4.7 is a consequence of the Axiom of Choice A.4.2. However, by Exercise A.4.8, Theorem A.4.7 also implies the Axiom of Choice, and thus it could have been taken as an axiom itself.

Theorem A.4.7 (Axiom of Choice for Cartesian Products). *The Cartesian product of non-empty sets is non-empty.*

Exercise A.4.8. Show that the Axiom of Choice is equivalent to the Axiom of Choice for Cartesian Products.

Theorem A.4.9 (Hausdorff Maximal Principle). *Any chain is contained in a maximal chain.*

Proof. Let $(X, \leq)$ be a partially ordered set with a chain $C_0 \subset X$. Let

$$\mathcal{T} := \{C \mid C \subset X \text{ is a chain such that } C_0 \subset C\}.$$

Now $C_0 \in \mathcal{T}$, so $\mathcal{T} \neq \emptyset$. Let $f : \mathcal{P}(X) \to X$ be a choice function for X. Let us define $s : \mathcal{T} \to \mathcal{T}$ such that $s(C) = C$ if $C \in \mathcal{T}$ is maximal, and if $C \in \mathcal{T}$ is not maximal then

$$s(C) := C \cup \{f(\{x \in X \setminus C : C \cup \{x\} \in \mathcal{T}\})\};$$

in this latter case, the chain $s(C)$ is obtained by adding one element to the chain C. The claim follows if we can show that $C = s(C)$ for some $C \in \mathcal{T}$. Let $\mathcal{U} \subset \mathcal{T}$ be a *tower* if

- $C_0 \in \mathcal{U}$,
- $\bigcup \mathcal{K} \in \mathcal{U}$ for any chain $\mathcal{K} \subset \mathcal{U}$,
- $s(\mathcal{U}) \subset \mathcal{U}$. In other words: if $A \in \mathcal{U}$ then $s(A) \in \mathcal{U}$.

For instance, $\mathcal{T}$ is a tower. Let $\mathcal{V}$ be the intersection of all towers. Clearly, $\mathcal{V}$ is a tower, in fact the minimal tower. It will turn out that $\bigcup \mathcal{V} \in \mathcal{T}$ is a maximal chain. This follows if we can show that $\mathcal{V}' \subset \mathcal{V}$ is a tower, where

$$\mathcal{V}' := \{C \in \mathcal{V} \mid \forall B \in \mathcal{V} : B \subset C \text{ or } C \subset B\},$$

since the minimality would imply $\mathcal{V} = \mathcal{V}'$. Clearly, $C_0 \in \mathcal{V}'$, and if $\mathcal{K} \subset \mathcal{V}'$ is a chain then $\bigcup \mathcal{K} \in \mathcal{V}'$. Let $C \in \mathcal{V}'$; we have to show that $s(C) \in \mathcal{V}'$. This follows, if we can show that $\langle C \rangle \subset \mathcal{V}$ is a tower, where

$$\langle C \rangle := \{A \in \mathcal{V} \mid A \subset C \text{ or } s(C) \subset A\}.$$

Clearly, $C_0 \in \langle C \rangle$, and if $\mathcal{K} \subset \langle C \rangle$ is a chain then $\bigcup \mathcal{K} \in \langle C \rangle$. Let $A \in \langle C \rangle$; we have to show that $s(A) \in \langle C \rangle$, i.e., show that $s(A) \subset C$ or $s(C) \subset s(A)$. Since $C \in \mathcal{V}'$, we have $s(A) \subset C$ or $C \subset s(A)$. Suppose the non-trivial case "$C \subset s(A)$ and $A \subset C$". Since $s(A) = A \cup \{x\}$ for some $x \in X$, we must have $s(A) = C$ or $C = A$. The proof is complete. □

Theorem A.4.10 (Zorn's Lemma). *A partially ordered set where every chain has an upper bound has a maximal element.*

Exercise A.4.11 (Hausdorff Maximal Principle $\iff$ Zorn's Lemma). Show that the Hausdorff Maximal Principle is equivalent to Zorn's Lemma.

Theorem A.4.12 (Zorn's Lemma $\implies$ Axiom of Choice). *Zorn's Lemma implies the Axiom of Choice.*

Proof. Let X be a non-empty set. Let

$$P := \{f \mid f : \mathcal{P}(A) \to A \text{ is a choice function for some } A \subset X\}.$$

Now $P \neq \emptyset$, because $(\{x\} \mapsto x) : \mathcal{P}(\{x\}) \to \{x\}$ belongs to P for any $x \in X$. Let us endow P with the partial order $\leq$ by inclusion:

$$f \leq g \quad \overset{\text{definition}}{\iff} \quad f \subset g.$$

(here recall that $f \in P$ is a subset $f \subset \mathcal{P}(A) \times A$ for some $A \subset X$). Suppose $C = \{f_j : j \in J\} \subset P$ is a chain. Then it is easy to verify that

$$\bigcup C = \bigcup_{j \in J} f_j \in P$$

is an upper bound for C, so according to **Zorn's Lemma** there exists a maximal element $f \in P$, which is a choice function for some $A \subset X$. We have to show that $A = X$. On the contrary, suppose $B \subset X$ such that $B \notin \mathcal{P}(A)$. Take $x \in B$. Then $f \subset f \cup \{(B, x)\} \in P$, which would contradict the maximality of f. Hence f must be a choice function for $A = X$. $\qquad\qquad\square$

How many points? Intuitively, cardinality measures the number of the elements in a set. Cardinality is a relative concept: sets A, B are compared by whether there is an injection, a surjection or a bijection from one to another. The most interesting results concern infinite sets.

Definition A.4.13 (Cardinality). Sets A, B have the same *cardinality*, denoted by

$$|A| = |B| \quad (\text{or } A \sim B),$$

if there exists a bijection $f : A \to B$. If there exists $C \subset B$ such that $|A| = |C|$, we write

$$|A| \leq |B|.$$

Moreover, $|A| \leq |B| \neq |A|$ is abbreviated by

$$|A| < |B|.$$

The cardinality of a set A is often also denoted by $\mathrm{card}(A)$.

Exercise A.4.14. Let $|A| = |B|$. Show that $|\mathcal{P}(A)| = |\mathcal{P}(B)|$.

Exercise A.4.15. Show that $|\mathbb{Z}^+| = |\mathbb{Z}|$.

Remark A.4.16. Clearly for every set A, B, C we have

$$A \sim A,$$
$$A \sim B \iff B \sim A,$$
$$A \sim B \text{ and } B \sim C \implies A \sim C;$$

formally this is an equivalence relation, though we may have difficulties when discussing the "set of all sets". Notice that $|A| \leq |B|$ means that there is an injection $f : A \to B$, and in this case we may identify set A with $f(A) \subset B$. Obviously,

$$|A| \leq |B| \leq |C| \implies |A| \leq |C|.$$

It is less obvious whether $|A| = |B|$ when $|A| \leq |B| \leq |A|$:

Theorem A.4.17 (Schröder–Bernstein). *Let* $|X| \le |Y|$ *and* $|Y| \le |X|$. *Then* $|X| = |Y|$.

Proof. Let $f : X \to Y$ and $g : Y \to X$ be injections. Let $X_0 := X$ and $X_1 := g(Y)$. Define inductively $\{X_k : k \in \mathbb{Z}^+\} \subset \mathcal{P}(X)$ by

$$X_{k+2} := g(f(X_k)).$$

Let $X_\infty := \bigcap_{k=0}^{\infty} X_k$. Now $X_\infty \subset X_{k+1} \subset X_k$ for each $k \ge 0$. Moreover,

$$X_k \setminus X_{k+1} \sim \begin{cases} X_0 \setminus X_1, & \text{if } k \text{ is odd}, \\ X_1 \setminus X_2, & \text{if } k \text{ is even}, \end{cases}$$

so that

$$
\begin{aligned}
X &= X_\infty \cup \bigcup_{k=0}^{\infty} (X_k \setminus X_{k+1}) \\
&\sim X_\infty \cup \bigcup_{k=0}^{\infty} (X_{k+1} \setminus X_{k+2}) \\
&= X_1 \\
&\sim Y.
\end{aligned}
$$

Thus $X \sim Y$. $\qquad\qquad\square$

The following Law of Trichotomy is equivalent to the Axiom of Choice, though we derive it only as a corollary to Zorn's Lemma:

Theorem A.4.18 (The Law of Trichotomy). *Let* X, Y *be sets. Then exactly one of the following holds:*

$$|X| < |Y|, \quad |X| = |Y|, \quad |Y| < |X|.$$

Proof. Assume the non-trivial case $X, Y \ne \emptyset$. Let us define

$$\mathcal{J} := \{f \mid A \subset X, \ f : A \to Y \text{ injective}\}.$$

Clearly, $\mathcal{J} \ne \emptyset$. Thus we may define a partial order $\le$ on $\mathcal{J}$ by

$$g \le h \iff g \subset h;$$

notice here that $g, h \subset X \times Y$. Let $\mathcal{K} \subset \mathcal{J}$ be a *chain*. Then it has an upper bound $\bigcup \mathcal{K} \in \mathcal{J}$. Hence by **Zorn's Lemma**, there exists a maximal element $f \in \mathcal{J}$. Now $f : A \to Y$ is injective, where $A \subset X$. If $A = X$ then

$$|X| \le |Y|.$$

If $f(A) = Y$ then
$$|Y| = |A| \leq |X|.$$

So let us suppose that $A \neq X$ and $f(A) \neq Y$. Then take $x_0 \in X \setminus A$ and $y_0 \in Y \setminus f(A)$. Define

$$g : A \cup \{x_0\} \to Y, \quad g(x) := \begin{cases} f(x), & \text{if } x \in A, \\ y_0, & \text{if } x = x_0. \end{cases}$$

Then $g \in \mathcal{J}$ and $f \leq g \neq f$, which contradicts the maximality of f. Thereby $A = X$ or $f(A) = Y$, meaning

$$|X| \leq |Y| \quad \text{or} \quad |Y| \leq |X|.$$

Finally, if $|X| \leq |Y|$ and $|Y| \leq |X|$ then $|X| = |Y|$ by Theorem A.4.17.   □

There is no greatest cardinality:

Theorem A.4.19 (No greatest cardinality). *Let X be a set. Then $|X| < |\mathcal{P}(X)|$.*

Proof. If $X = \emptyset$ then $\mathcal{P}(X) = \{\emptyset\}$, and the only injection from X to $\mathcal{P}(X)$ is then the empty relation, which is not a bijection. Assume that $X \neq \emptyset$. Then function

$$f : X \to \mathcal{P}(X), \quad f(x) := \{x\}$$

is an injection, establishing $|X| \leq |\mathcal{P}(X)|$. To get a contradiction, assume that $X \sim \mathcal{P}(X)$, so that there exists a bijection $g : X \to \mathcal{P}(X)$. Let

$$A := \{x \in X : x \notin g(x)\}.$$

Let $x_0 := g^{-1}(A)$. Now $x_0 \in A$ if and only if

$$x_0 \notin g(x_0) = A,$$

which is a contradiction.   □

Definition A.4.20 (Counting). Let A, B, C, D be sets. For $n \in \mathbb{Z}^+$, let

$$\mathbb{Z}_n := \{k \in \mathbb{Z}^+ \mid k \leq n\} = \{1, \ldots, n\}.$$

We say that $|\emptyset| = 0$, $|\mathbb{Z}_n| = n$,
A is *finite* if $|A| = n$ for some $n \in \mathbb{Z}^+ \cup \{0\}$.
B is *infinite* if it is not finite.
C is *countable* if $|C| \leq |\mathbb{Z}^+|$.
D is *uncountable* if it is not countable.

Remark A.4.21. To strive for transparency in the proofs in this section, let us forget the Law of Trichotomy, which would provide short-cuts like

$$|X| < |Y| \iff |Y| \nleq |X|.$$

The reader may easily simplify parts of the reasoning using this. The reader is also encouraged to find out where we use the Axiom of Choice or some other nontrivial tools.

Proposition A.4.22. *Let A, B be sets. Then $|A| < |\mathbb{Z}^+| \le |B|$ if and only if A is finite and B is infinite.*

Proof. Let $A \ne \emptyset$ be finite, so $A \sim \mathbb{Z}_n \subset \mathbb{Z}^+$ for some $n \in \mathbb{Z}^+$. Hence $|A| \le |\mathbb{Z}^+|$. If $f : \mathbb{Z}^+ \to A$ then $f(n+1) \in f(\mathbb{Z}_n)$, so f is not injective, especially not bijective. Thus $|\mathbb{Z}^+| \nleq |A|$ and $|A| < |\mathbb{Z}^+|$. Consequently, if $|\mathbb{Z}^+| \le |B|$ then B is infinite.

Let B be infinite. Take $x_1 \in B \ne \emptyset$. Let $A_n = \{x_1, \ldots, x_n\} \subset B$ be a finite set. Inductively, take $x_{n+1} \in B \setminus A_n \ne \emptyset$. Define

$$\begin{cases} g : \mathbb{Z}^+ \to B, \\ g(n) := x_n. \end{cases}$$

Now g is injective. Hence $|\mathbb{Z}^+| \le |B|$.

Let $B \subset \mathbb{Z}^+$ be infinite. Define $h : \mathbb{Z}^+ \to B$ inductively by

$$\begin{cases} h(1) := \min(B), \\ h(n+1) := \min\left(B \setminus \{h(1), \ldots, h(n)\}\right). \end{cases}$$

Now h is a bijection: $|B| = |\mathbb{Z}^+|$. So if $|A| < |\mathbb{Z}^+|$ then A is finite. $\qquad\square$

Proposition A.4.23. *Let C, D be sets. Then $|C| \le |\mathbb{Z}^+| < |D|$ if and only if C is countable and D is uncountable.*

Proof. Property $|C| \le |\mathbb{Z}^+|$ is just the definition of countability. Let D be uncountable, i.e., $|D| \nleq |\mathbb{Z}^+|$. By Proposition A.4.22, D is not finite, i.e., it is infinite, i.e., $|\mathbb{Z}^+| \le |D|$. Because of $|\mathbb{Z}^+| \ne |D|$, we have $|\mathbb{Z}^+| < |D|$.

Let $|\mathbb{Z}^+| < |D|$. By Proposition A.4.22, D is infinite, i.e., $|D| \nleq |\mathbb{Z}^+|$. Because of $|\mathbb{Z}^+| \ne |D|$, we have even $|D| \nleq |\mathbb{Z}^+|$, i.e., D is uncountable. $\qquad\square$

Remark A.4.24. Let us collect the results from Propositions A.4.22 and A.4.23: For sets A, B, C, D,

$$\begin{cases} |A| < |\mathbb{Z}^+| \le |B|, \\ |C| \le |\mathbb{Z}^+| < |D| \end{cases}$$

if and only if A is finite, B is infinite, C is countable, and D is uncountable. In the proofs, we used induction, i.e., well-ordering for $\mathbb{Z}^+$.

Proposition A.4.25 (Cantor). *Let $A_k \subset X$ be a countable subset for each $k \in \mathbb{Z}^+$. Then $\bigcup_{k=1}^{\infty} A_k$ is countable.*

Proof. We may enumerate the elements of each countable A_k:

$$A_k := \{a_{kj} : j \in \mathbb{Z}^+\},$$

$$A_1 = \{a_{11}, a_{12}, a_{13}, a_{14}, \cdots\},$$
$$A_2 = \{a_{21}, a_{22}, a_{23}, a_{24}, \cdots\},$$
$$A_3 = \{a_{31}, a_{32}, a_{33}, a_{34}, \cdots\},$$
$$A_4 = \{a_{41}, a_{42}, a_{43}, a_{44}, \cdots\},$$

$$\vdots$$

Their union is enumerated by

$$\cup_{k=1}^{\infty} A_k = \{a_{11},$$
$$a_{21}, a_{12},$$
$$a_{31}, a_{22}, a_{13},$$
$$a_{41}, a_{32}, a_{23}, a_{14}, \cdots\}$$
$$= \{a_{k-j+1,j} : 1 \leq j \leq k, \ k \in \mathbb{Z}^+\}. \qquad \square$$

Exercise A.4.26. Show that the set $\mathbb{Q}$ of rational numbers is countably infinite.

Exercise A.4.27 (Algebraic numbers). A number $\lambda \in \mathbb{C}$ is called *algebraic* if $p(\lambda) = 0$ for some non-zero polynomial p with integer coefficients, i.e., if some polynomial

$$p(z) = \sum_{k=0}^{n} a_k z^k,$$

where $n \in \mathbb{Z}^+$, $\{a_k\}_{k=0}^{n} \subset \mathbb{Z}$ and $a_n \neq 0$. Let $\mathbb{A} \subset \mathbb{C}$ be the set of algebraic numbers. Show that $\mathbb{Q} \subset \mathbb{A}$, that $\mathbb{A}$ is countable, and give an example of a number $\lambda \in (\mathbb{R} \cap \mathbb{A}) \setminus \mathbb{Q}$.

Proposition A.4.28. $|\mathbb{R}| = |\mathcal{P}(\mathbb{Z}^+)|.$

Proof. Let us define

$$f : \mathbb{R} \to \mathcal{P}(\mathbb{Q}), \quad f(x) := \{r \in \mathbb{Q} : r < x\}.$$

Obviously f is injective, hence $|\mathbb{R}| \leq |\mathcal{P}(\mathbb{Q})|$. By Exercise A.4.26, $|\mathbb{Q}| = |\mathbb{Z}^+|$, implying $|\mathcal{P}(\mathbb{Q})| = |\mathcal{P}(\mathbb{Z}^+)|$. On the other hand, let us define

$$g : \mathcal{P}(\mathbb{Z}^+) \to \mathbb{R}, \quad g(A) := \sum_{k \in A} 10^{-k}.$$

For instance, $0 = g(\emptyset) \leq g(A) \leq g(\mathbb{Z}^+) = 1/9$. Nevertheless, g is injective, implying $|\mathcal{P}(\mathbb{Z}^+)| \leq |\mathbb{R}|$. This completes the proof. $\qquad \square$

Exercise A.4.29. Let X be an uncountable set. Show that there exists an uncountable subset $S \subset X$ such that $X \setminus S$ is also uncountable.

A.5 Well-Ordering Principle revisited

Trivially, the Well-Ordering Principle implies the Axiom of Choice. Actually, there is the reverse implication, too:

Theorem A.5.1 (Well-Ordering Principle). *Every non-empty set can be well ordered.*

Proof. Let $X \neq \emptyset$. Let

$$P := \{(A_j, \leq_j) \mid j \in J, \ A_j \subset X, \ (A_j, \leq_j) \text{ well-ordered}\}.$$

Clearly, $P \neq \emptyset$. Define a partial order $\leq$ on P by inclusion:

$$(A_j, \leq_j) \leq (A_k, \leq_k) \overset{\text{definition}}{\Longleftrightarrow} \leq_j \subset \leq_k .$$

Take a chain $C \subset P$. Let

$$B := \bigcup_{(A_j, \leq_j) \in C} A_j, \quad \leq_B := \bigcup_{(A_j, \leq_j) \in C} \leq_j .$$

Then $(B, \leq_B) \in P$ is an upper bound for the chain $C \subset P$, so there exists a maximal element $(A, \leq_A) \in P$ by **Zorn's Lemma** A.4.10. Now, if there was $x \in X \setminus A$, then we easily see that $A \cup \{x\}$ could be well ordered by $\leq_x$ for which $\leq_A \subset \leq_x$, which would contradict the maximality of $(A, \leq_A)$. Therefore $A = X$ has been well ordered. □

Although we already know that the Well-Ordering Principle and the Hausdorff Maximal Principle are equivalent, let us demonstrate how to use transfinite induction in a related proof:

Proposition A.5.2 (Well-Ordering Principle $\Longrightarrow$ Hausdorff Maximal Principle). *The Well-Ordering Principle implies the Hausdorff Maximal Principle.*

Proof. Let $(X, \leq)$ be well ordered, i.e., there exists $\min(A) \in A$ whenever $\emptyset \neq A \subset X$. Let $\leq_0$ be a partial order on X. Let us define $f : X \to \mathcal{P}(X)$ by transfinite induction in the following way:

$$f(x) := \begin{cases} \{x\}, & \text{if } \{x\} \cup f(\{y : y < x\}) \text{ is a chain with respect to } \leq_0, \\ \emptyset & \text{otherwise.} \end{cases}$$

Then $f(X) \subset \mathcal{P}(X)$ is a maximal chain. □

Exercise A.5.3. Fill in the details in the proof of Proposition A.5.2.

Remark A.5.4 (**Formulations of the Axiom of Choice**). Collecting earlier results and exercises, we see that the following claims are equivalent: the Axiom of Choice, the Axiom of Choice for Cartesian Products, the Hausdorff Maximal Principle, Zorn's Lemma, and the Well-Ordering Principle. The Law of Trichotomy was derived as a corollary to these, but it is actually another equivalent formulation for the Axiom of Choice (see, e.g., [124]).

Remark A.5.5 (**Continuum Hypothesis**). When working in analysis, one does not often pay much attention to the underlying set theoretic foundations. Yet, there are many deep problems involved. For instance, it can be shown that there is the smallest uncountable cardinality $|\Omega|$, i.e., whenever S is uncountable then

$$|\mathbb{Z}^+| < |\Omega| \le |S|.$$

So $|\Omega| \le |\mathbb{R}|$. A natural question is whether $|\Omega| = |\mathbb{R}|$? Actually, in year 1900, David Hilbert proposed the so-called *Continuum Hypothesis*

$$|\Omega| = |\mathbb{R}|.$$

The *Generalised Continuum Hypothesis* is that if X, Y are infinite sets and $|X| \le |Y| \le |\mathcal{P}(X)|$ then $|X| = |Y|$ or $|Y| = |\mathcal{P}(X)|$. Without going into details, let (ZF) denote the *Zermelo–Fraenkel* axioms for set theory, (AC) the Axiom of Choice, and (CH) the Generalised Continuum Hypothesis. From 1930s to 1960s, Kurt Gödel and Paul Cohen discovered that:

1. Within (ZF) one cannot prove whether (ZF) is consistent.
2. (ZF+AC+CH) is consistent if (ZF) is consistent.
3. (AC) is independent of (ZF).
4. (CH) is independent of (ZF+AC).

The reader will be notified, whenever we apply (AC) or its equivalents (which is not that often); in this book, we shall not need (CH) at all.

A.6   Metric spaces

Definition A.6.1 (Metric space). A function $d : X \times X \to [0, \infty)$ is called a *metric* on the set X if for every $x, y, z \in X$ we have

$$
\begin{aligned}
d(x, y) = 0 &\iff x = y && \text{(non-degeneracy)};\\
d(x, y) &= d(y, x) && \text{(symmetry)};\\
d(x, z) &\le d(x, y) + d(y, z) && \text{(triangle inequality)}.
\end{aligned}
$$

Then (X, d) (or simply X when d is evident) is called a *metric space*. Sometimes a metric is called a *distance function*. When $x \in X$ and $r > 0$,

$$B_r(x) := \{y \in X \mid d(x, y) < r\}$$

is called the *x-centered open ball of radius r*. If we want to emphasise that the ball is taken with respect to metric d, we will write $B_d(x, r)$.

Remark A.6.2. In a metric space (X, d),

$$\bigcup_{k=1}^{\infty} B_k(x) = X \quad \text{and} \quad \bigcap_{k=1}^{\infty} B_{1/k}(x) = \{x\}.$$

Example (**Discrete metric**). The mapping $d : X \times X \to [0, \infty)$ defined by

$$d(x, y) := \begin{cases} 1, & \text{if } x \neq y, \\ 0, & \text{if } x = y \end{cases}$$

is called the *discrete metric* on X. Here

$$B_r(x) = \begin{cases} X, & \text{if } 1 < r, \\ \{x\}, & \text{if } 0 < r \leq 1. \end{cases}$$

Example. Normed vector spaces form a very important class of metric spaces, see Definition B.4.1.

Exercise A.6.3. For $1 \leq p < \infty$,

$$d_p(x, y) = \|x - y\|_p := \left(\sum_{j=1}^{n} |x_j - y_j|^p \right)^{1/p}$$

defines a metric $d_p : \mathbb{R}^n \times \mathbb{R}^n \to [0, \infty)$. Function

$$d_\infty(x, y) = \max_{1 \leq j \leq n} |x_j - y_j|$$

also turns $\mathbb{R}^n$ into a metric space. Unless otherwise mentioned, the space $\mathbb{R}^n$ is endowed with the *Euclidean metric* d_2 (distance "as the crow flies").

Exercise A.6.4 (Sup-metric). Let $a < b$ and let $B([a, b])$ be the space of all bounded functions $f : [a, b] \to \mathbb{R}$. Show that function

$$d_\infty(f, g) = \sup_{y \in [a,b]} |f(y) - g(y)|$$

turns $B([a, b])$ into a metric space. It is called the *sup-metric*.

Remark A.6.5 (**Metric subspaces**). If $A \subset X$ and $d : X \times X \to [0, \infty)$ is a metric then the restriction

$$d|_{A \times A} : A \times A \to [0, \infty)$$

is a metric on A, with $B_{d|_{A \times A}}(x, r) = A \cap B_d(x, r)$.

Exercise A.6.6. Let $a < b$ and let $C([a, b])$ be the space of all continuous functions $f : [a, b] \to \mathbb{R}$. Show the following statements: The function $d_\infty(f, g) = \sup_{y \in [a,b]} |f(y) - g(y)|$ turns $(C([a, b]), d_\infty)$ into a metric subspace of $(B([a, b]), d_\infty)$. The space $C([a, b])$ also becomes a metric space with metric

$$d_p(f, g) = \left(\int_a^b |f(y) - g(y)|^p \, dy \right)^{1/p},$$

for any $1 \leq p < \infty$. However, $B([a, b])$ with these d_p is not a metric space.

Definition A.6.7 (Diameter and bounded sets). The *diameter* of a set $A \subset X$ in a metric space (X, d) is

$$\operatorname{diam}(A) := \sup \{d(x, y) \mid x, y \in A\},$$

with convention $\operatorname{diam}(\emptyset) = 0$. A set $A \subset X$ is said to be *bounded*, if $\operatorname{diam}(A) < \infty$.

Example. $\operatorname{diam}(\{x\}) = 0$, $\operatorname{diam}(\{x, y\}) = d(x, y)$, and

$$\operatorname{diam}(\{x, y, z\}) = \max \{d(x, y), d(y, z), d(x, z)\}.$$

Exercise A.6.8. Show that $\operatorname{diam}(B_r(x)) \leq 2r$, so that balls are bounded.

Definition A.6.9 (Distance between sets). The *distance* between sets $A, B \subset X$ is

$$\operatorname{dist}(A, B) := \inf \{d(x, y) \mid x \in A, \ y \in B\},$$

with the convention that $\operatorname{dist}(A, \emptyset) = \infty$.

Exercise A.6.10. Show that $A \cap B_r(x) \neq \emptyset$ if and only if $\operatorname{dist}(\{x\}, A) < r$.

We note that the function $\operatorname{dist}(A, B)$ does not define a metric on subsets of X. For example:

Exercise A.6.11. Give an example of sets $A, B \subset \mathbb{R}^2$ for which $\operatorname{dist}(A, B) = 0$ even though $A \cap B = \emptyset$. Here we consider naturally the Euclidean metric.

Exercise A.6.12. Show that set S in a metric space (X, d) is bounded if and only if there exist some $a \in X$ and $r > 0$ such that $S \subset B_r(a)$.

Lemma A.6.13. *Let S be a bounded set in a metric space (X, d) and let $c \in X$. Then $S \subset B_R(c)$ for some $R > 0$.*

Proof. Since S is a bounded set, there exist some $a \in X$ and $r > 0$ such that $S \subset B_r(a)$. Consequently, for all $x \in S$ we have

$$d(x, c) \leq d(x, a) + d(a, c) < r + d(a, c),$$

so the statement follows with $R = r + d(a, c)$. $\square$

Proposition A.6.14. *The union of finitely many bounded sets in a non-empty metric space is bounded.*

Proof. Let $S_1, \ldots, S_n$ be bounded sets in a non-empty metric space (X, d). Let us take some $c \in X$. Then by Lemma A.6.13 there exists some R_i, $i = 1, \ldots, n$, such that $S_i \subset B_{R_i}(c)$. If we take $R = \max\{R_1, \ldots, R_n\}$, then we have $S_i \subset B_{R_i}(c) \subset B_R(c)$, which implies that $\cup_{i=1}^n S_i \subset B_R(c)$ is bounded. $\square$

Remark A.6.15. We note that the union of infinitely many bounded sets does not have to be bounded. For example, the union of sets $S_i = (0, i) \subset \mathbb{R}$, $i \in \mathbb{N}$, is not bounded in $(\mathbb{R}, d_\infty)$.

Usually, the topological properties can be characterised with *generalised sequences* (or *nets*). Now, we briefly study this phenomenon in metric topology, where ordinary sequences suffice.

Definition A.6.16 (Sequences). A *sequence* in a set A is a mapping $x : \mathbb{Z}^+ \to A$. We write $x_k := x(k)$ and

$$x = (x_k)_{k \in \mathbb{Z}^+} = (x_k)_{k=1}^{\infty} = (x_1, x_2, x_3, \ldots).$$

Notice that $x \neq \{x_1, x_2, x_3, \cdots\} = \{x_k : k \in \mathbb{Z}^+\}$.

Definition A.6.17 (Convergence). Let (X, d) be a metric space. A sequence $x : \mathbb{Z}^+ \to X$ *converges* to a point $p \in X$, if $\lim_{k \to \infty} d(x_k, p) = 0$, i.e.,

$$\forall \varepsilon > 0 \; \exists k_\varepsilon \in \mathbb{Z}^+ : \quad k \geq k_\varepsilon \Rightarrow d(x_k, p) < \varepsilon.$$

In such a case, we write $\lim_{k \to \infty} x_k = p$ or $x_k \to p$ or $x_k \xrightarrow[k \to \infty]{d} p$ etc.

Clearly, $x_k \to p$ as $k \to \infty$ if and only if

$$\forall \varepsilon > 0 \quad \exists N : \quad k \geq N \Rightarrow x_k \in B_\varepsilon(p).$$

We now collect some properties of limits.

Proposition A.6.18 (Uniqueness of limits in metric spaces). *Let (X, d) be a metric space. If $x_k \to p$ and $x_k \to q$ as $k \to \infty$, then $p = q$.*

Proof. Let $\varepsilon > 0$. Since $x_k \to p$ and $x_k \to q$ as $k \to \infty$, it follows that there are some numbers N_1, N_2 such that $d(x_k, p) < \varepsilon$ for all $k > N_1$ and such that $d(x_k, q) < \varepsilon$ for all $k > N_2$. Hence by the triangle inequality for all $k > \max\{N_1, N_2\}$ we have $d(p, q) \leq d(p, x_k) + d(x_k, q) < 2\varepsilon$. Since this conclusion is true for any $\varepsilon > 0$, it follows that $d(p, q) = 0$ and hence $p = q$. $\square$

A.7 Topological spaces

Previously, a metric provided a way of measuring distances between sets. The branch of mathematics called topology can be thought as a way to describe "qualitative geography of a set" without referring to specific numerical distance values. We begin by considering properties of metric spaces that motivate the definition of topology which follows after them.

Definition A.7.1 (Open sets and neighbourhoods). A set $U \subset X$ in a metric space X is said to be *open* if for every $x \in U$ there is some $\varepsilon > 0$ such that $B_\varepsilon(x) \subset U$. For a point $x \in X$, any open set containing x is called an *open neighbourhood* of x.

Proposition A.7.2. *Every ball $B_r(a)$ in a metric space (X, d) is open.*

Proof. Let $x \in B_r(a)$. Then the number $\varepsilon = r - d(x,a) > 0$ is positive, and $B_\varepsilon(x) \subset B_r(a)$. Indeed, for any $y \in B_\varepsilon(x)$ we have $d(y,a) \leq d(y,x) + d(x,a) < \varepsilon + d(x,a) = r$. □

Proposition A.7.3. *Let (X,d) be a metric space. Then $x_k \to p$ as $k \to \infty$ if and only if every open neighbourhood of x contains all but finitely many of the points x_k.*

Proof. "If" implication is immediate because balls are open. On the other hand, let $p \in U$ where U is an open set. Then there is some $\varepsilon > 0$ such that $B_\varepsilon(p) \subset U$. Now, if $x_k \to p$ as $k \to \infty$, there is some N such that for all $k > N$ we have $x_k \in B_\varepsilon(p) \subset U$, implying the statement. □

Definition A.7.4 (Continuous mappings in metric spaces). Let (X_1, d_1) and (X_2, d_2) be two metric spaces, let $f : X_1 \to X_2$, and let $a \in X_1$. Then f is said to be *continuous at a* if for every $\varepsilon > 0$ there is some $\delta > 0$ such that $d_1(x,a) < \delta$ implies $d_2(f(x), f(a)) < \varepsilon$. The mapping f is said to be *continuous (on X_1)* if it is continuous at all points of X_1.

Example. Let $X_1 = C([a,b])$ and $X_2 = \mathbb{R}$ be equipped with the sup-metrics d_1 and d_2, respectively. Then mapping $\Phi : X_1 \to X_2$ defined by $\Phi(h) = \int_a^b h(y)dy$ is continuous.

Definition A.7.5 (Preimage). Let $f : X_1 \to X_2$ be a mapping and let $S \subset X_2$ be any subset of X_2. Then the preimage of S under f is defined by

$$f^{-1}(S) = \{x \in X_1 : f(x) \in S\}.$$

Theorem A.7.6. *Let $(X_1, d_1), (X_2, d_2)$ be metric spaces and let $f : X_1 \to X_2$. Then the following statements are equivalent:*

(i) *f is continuous on X_1;*
(ii) *for every $a \in X_1$ and every ball $B_\varepsilon(f(a)) \subset X_2$ there is a ball $B_\delta(a) \subset X_1$ such that $B_\delta(a) \subset f^{-1}(B_\varepsilon(f(a)))$;*
(iii) *for every open set $U \subset X_2$ its preimage $f^{-1}(U)$ is open in X_1.*

Proof. First, let us show the equivalence of (i) and (ii). Condition (i) is equivalent to saying that for every $\varepsilon > 0$ there is $\delta > 0$ such that $d_1(x,a) < \delta$ implies $d_2(f(a), f(x)) < \varepsilon$. In turn this is equivalent to saying that for every $\varepsilon > 0$ there is $\delta > 0$ such that $x \in B_\delta(a)$ implies $f(x) \in B_\varepsilon(f(a))$, which means that $B_\delta(a) \subset f^{-1}(B_\varepsilon(f(a)))$.

To show that (ii) implies (iii), let us assume that f is continuous and that $U \subset X_2$ is open. Take $x \in f^{-1}(U)$. Then $f(x) \in U$ and since U is open there is some $\varepsilon > 0$ such that $B_\varepsilon(f(a)) \subset U$. Consequently, by (ii), there exists $\delta > 0$ such that $B_\delta(a) \subset f^{-1}(B_\varepsilon(f(a))) \subset f^{-1}(U)$, implying that $f^{-1}(U)$ is open.

Finally, let us show that (iii) implies (ii). We observe that by (iii) for every $a \in X_1$ and every $\varepsilon > 0$ the set $f^{-1}(B_\varepsilon(f(a)))$ is an open set containing a. Hence there is some $\delta > 0$ such that $B_\delta(a) \subset f^{-1}(B_\varepsilon(f(a)))$, completing the proof. □

Theorem A.7.7. *Let X be a metric space. We have the following properties of open sets in X:*

(T1) $\emptyset$ *and X are open sets in X.*

(T2)   *The union of any collection of open subsets of X is open.*

(T3)   *The intersection of a finite collection of open subsets of X is open.*

Proof. It is obvious that the empty set $\emptyset$ is open. Moreover, for any $x \in X$ and any $\varepsilon > 0$ we have $B_\varepsilon(x) \subset X$, implying that X is also open.

To show (T2), suppose that we have a collection $\{A_i\}_{i \in I}$ of open sets in X, for an index set I. Let $a \in \cup_{i \in I} A_i$. Then there is some $j \in I$ such that $a \in A_j$ and since A_j is open there is some $\varepsilon > 0$ such that $B_\varepsilon(a) \subset A_j \subset \cup_{i \in I} A_i$, implying (T2).

To show (T3), assume that $A_1, \ldots, A_n$ is a finite collection of open sets and let $a \in \cap_{i=1}^n A_i$. It follows that for every $i = 1, \ldots, n$ we have $a \in A_i$ and hence there is $\varepsilon_i > 0$ such that $B_{\varepsilon_i}(a) \subset A_i$. Let now $\varepsilon = \min\{\varepsilon_1, \ldots, \varepsilon_n\}$. Then $B_\varepsilon(a) \subset A_i$ for all i and hence $B_\varepsilon(a) \subset \cap_{i=1}^n A_i$ implying that the intersection of A_i's is open. $\square$

Definition A.7.8 (Topology). A family of sets $\tau \subset \mathcal{P}(X)$ is called a *topology* on the set X if

1. $\bigcup \mathcal{U} \in \tau$ for every collection $\mathcal{U} \subset \tau$, and

2. $\bigcap \mathcal{U} \in \tau$ for every finite collection $\mathcal{U} \subset \tau$.

Then (X, τ) (or simply X when τ is evident) is called a *topological space*; a set $A \subset X$ is called *open* (or *τ-open*) if $A \in \tau$, and *closed* (or *τ-closed*) if $X \setminus A \in \tau$. Let the collection of τ-closed sets be denoted by

$$\tau^* = \{X \setminus U : U \in \tau\}.$$

Then the axioms of the topology become naturally complemented:

1. $\bigcap \mathcal{A} \in \tau^*$ for every collection $\mathcal{A} \subset \tau^*$, and

2. $\bigcup \mathcal{A} \in \tau^*$ for every finite collection $\mathcal{A} \subset \tau^*$.

Remark A.7.9. Recall our natural conventions (A.1) for the union and the intersection of the empty family. Thereby $\tau \subset \mathcal{P}(X)$ is a topology if and only if the following conditions hold:

(T1) $\emptyset, X \in \tau$,

(T2) $\bigcup \mathcal{U} \in \tau$ for every non-empty collection $\mathcal{U} \subset \tau$, and

(T3) $U \cap V \in \tau$ for every $U, V \in \tau$.

Consequently, for any topology of X, the subsets $\emptyset \subset X$ and $X \subset X$ are always both open and closed.

Proposition A.7.3 motivates the following notion of convergence in topological spaces.

Definition A.7.10 (Convergence in topological spaces). Let (X, τ) be a topological space. We say that a sequence x_k *converges* to p as $k \to \infty$, and write $x_k \to p$ as $k \to \infty$, if every open neighbourhood of p contains all but finitely many of points x_k.

Proposition A.7.11. *Let X and Y be topological spaces and let $f : X \to Y$ be continuous. If $x_k \to p$ in X as $k \to \infty$ then $f(x_k) \to f(p)$ in Y as $k \to \infty$.*

Proof. Let U be an open set in Y containing $f(p)$. Then $p \in f^{-1}(U)$ and $f^{-1}(U)$ is open in X, implying that there is N such that $x_k \in f^{-1}(U)$ for all $k > N$. Consequently, $f(x_k) \in U$ for all $k > N$ implying that $f(x_k) \to f(p)$ in Y as $k \to \infty$.   □

Corollary A.7.12. *Any metric space is a topological space by Theorem A.7.7. The canonical topology of a metric space (X, d) is the family τ consisting of all sets in (X, d) which are open according to Definition A.7.1. This canonical metric topology will be denoted by τ_d or by $\tau(d)$. Metric convergence in (X, d) is equivalent to the topological convergence in the canonical metric topology (X, τ_d).*

Remark A.7.13. Notice that the intersection of any finite collection of τ-open sets is τ-open. On the other hand, it may well be that a countably infinite intersection of open sets is not open. In a metric space (X, d),

$$\bigcap_{k=1}^{\infty} B_{1/k}(x) = \{x\}.$$

Now $\{x\} \in \tau_d$ if and only if $\{x\} = B_r(x)$ for some $r > 0$.

Corollary A.7.14 (Properties of closed sets). *Let X be a topological space. We have the following properties of closed sets in X:*

(C1) $\emptyset$ *and X are closed in X.*
(C2)   *The intersection of any collection of closed subsets of X is closed.*
(C3)   *The union of a finite collection of closed subsets of X is closed.*

Proof. Let A_i, $i \in I$, be any collection of subsets of X. The corollary follows immediately from Remark A.7.9 and de Morgan's rules

$$X \backslash \bigcup_{i \in I} A_i = \bigcap_{i \in I} (X \backslash A_i), \quad X \backslash \bigcap_{i \in I} A_i = \bigcup_{i \in I} (X \backslash A_i),$$

see Exercise A.1.5.   □

Definition A.7.15 (Comparing metric topologies). Let d_1, d_2 be two metrics on a set X. The topology $\tau(d_1)$ defined by d_1 is said to be *stronger* than topology $\tau(d_2)$ defined by d_2 if $\tau(d_1) \supset \tau(d_2)$. In this case the topology $\tau(d_2)$ is also said to be *weaker* than $\tau(d_1)$. Metrics d_1, d_2 on a set X are said to be *equivalent* if they define the same topology $\tau(d_1) = \tau(d_2)$.

Proposition A.7.16 (Criterion for comparing metric topologies). *Let d_1, d_2 be two metrics on a set X such that there is a constant $C > 0$ such that $d_2(x, y) \leq Cd_1(x, y)$ for all $x, y \in X$. Then $\tau(d_2) \subset \tau(d_1)$, i.e., every d_2-open set is also d_1-open.*

Consequently, if there is a constant $C > 0$ such that

$$C^{-1}d_1(x, y) \leq d_2(x, y) \leq Cd_1(x, y), \tag{A.2}$$

for all $x, y \in X$, then metrics d_1 and d_2 are equivalent. Such metrics are called Lipschitz equivalent.

Sometimes such metrics are called just *equivalent*, however we use the term "Lipschitz" to distinguish this equivalence from the one in Definition A.7.15.

Proof. Fixing the constant $C > 0$ from (A.2), we observe that $d_1(x, y) < r$ implies $d_2(x, y) < Cr$, which means that $B_{d_1}(x, r) \subset B_{d_2}(x, Cr)$. Let now $U \in \tau(d_2)$ and let $x \in U$. Then there is some $\varepsilon > 0$ such that $B_{d_2}(x, \varepsilon) \subset U$ implying that $B_{d_1}(x, \varepsilon/C) \subset U$. Hence $U \in \tau(d_1)$. $\qquad\square$

Exercise A.7.17. Prove that the metrics d_p, $1 \leq p \leq \infty$, from Exercise A.6.3, are all Lipschitz equivalent. The corresponding topology is called the Euclidean metric topology on $\mathbb{R}^n$.

Definition A.7.18 (Relative topology). Let (X, τ) be a topological space and let $A \subset X$. Then we define the *relative topology* on A by

$$\tau_A = \{U \cap A : U \in \tau\}.$$

Proposition A.7.19 (Relative topology is a topology). *Any subset A of a topological space (X, τ) when equipped with the relative topology τ_A is a topological space.*

Proof. We have to check the properties (T1)–(T3) of Remark A.7.9. It is easy to see that $\emptyset = \emptyset \cap A \in \tau_A$ and that $A = X \cap A \in \tau_A$. To show (T2), let $V_i \in \tau_A$, $i \in I$, be a family of sets from τ_A. Then there exist sets $U_i \in \tau$ such that $V_i = U_i \cap A$. Consequently, we have

$$\bigcup_{i \in I} V_i = \bigcup_{i \in I}(U_i \cap A) = \left(\bigcup_{i \in I} U_i\right) \cap A \in \tau_A.$$

To show (T3), let $V_1, \ldots, V_n$ be a family of sets from τ_A. It follows that there exist sets $U_i \in \tau$ such that $V_i = U_i \cap A$. Consequently, we have

$$\bigcap_{i=1}^{n} V_i = \bigcap_{i=1}^{n}(U_i \cap A) = \left(\bigcap_{i=1}^{n} U_i\right) \cap A \in \tau_A,$$

completing the proof. $\qquad\square$

Remark A.7.20 (**Metric subspaces**). Let (X, d) be a metric space with canonical topology $\tau(d)$. Let $Y \subset X$ be a subset of X and let us define $d_Y = d|_{Y \times Y}$. Then $\tau(d_Y) = \tau(d)_Y$, i.e., the canonical topology of the metric subspace coincides with the relative topology of the metric space.

Definition A.7.21 (Product topology). Let (X_1, τ_1) and (X_2, τ_2) be topological spaces. A subset of $X_1 \times X_2$ is said to be *open in the product topology* if it is a union of sets of the form $U_1 \times U_2$, where $U_1 \in \tau_1$, $U_2 \in \tau_2$. The collection of all such open sets is denoted by $\tau_1 \otimes \tau_2$.

Proposition A.7.22 (Product topology is a topology). *The set $X_1 \times X_2$ with the collection $\tau_1 \otimes \tau_2$ is a topological space.*

Proof. We have to check properties (T1)–(T3) of Remark A.7.9. It is easy to see that $\emptyset = \emptyset \times \emptyset \in \tau_1 \otimes \tau_2$ and that $X_1 \times X_2 \in \tau_1 \otimes \tau_2$.

To show (T2), assume that $A_\alpha \in \tau_1 \otimes \tau_2$ for all $\alpha \in I$. Then each A_α is a union of sets of the form $U_1 \times U_2$ with $U_1 \in \tau_1$, $U_2 \in \tau_2$. Consequently, the union $\cup_{\alpha \in I} A_\alpha$ is a union of sets of the same form and does, therefore, also belong to $\tau_1 \otimes \tau_2$.

To show (T3), even for n sets, assume that $A_i \in \tau_1 \otimes \tau_2$, for all $i = 1, \ldots, n$. By definition there exist collections $U_{\alpha_i}^i \in \tau_1$, $V_{\alpha_i}^i \in \tau_2$, $\alpha_i \in I_i$, $i = 1, \ldots, n$, such that

$$A_i = \bigcup_{\alpha_i \in I_i} (U_{\alpha_i}^i \times V_{\alpha_i}^i), \ i = 1, \ldots, n.$$

Consequently,

$$\bigcap_{i=1}^n A_i = \bigcup_{\alpha_i \in I_i, \, 1 \leq i \leq n} \left((\cap_{j=1}^n U_{\alpha_i}^j) \times (\cap_{j=1}^n V_{\alpha_i}^j) \right) \in \tau_1 \otimes \tau_2,$$

completing the proof. $\square$

Theorem A.7.23 (Topologies on $\mathbb{R}^2$). *The product topology on $\mathbb{R} \times \mathbb{R}$ is the Euclidean metric topology of $\mathbb{R}^2$.*

Proof. We start by proving that every set open in the product topology of $\mathbb{R}^2$ is also open in the Euclidean topology of $\mathbb{R}^2$. First we note that any open set in $\mathbb{R}$ in the Euclidean topology is a union of open intervals, i.e., every open set U can be written as $U = \cup_{x \in U} B_{\varepsilon_x}(x)$, where $B_{\varepsilon_x}(x)$ is an open ball centred at x with some $\varepsilon_x > 0$. Then we note that every open rectangle in $\mathbb{R}^2$ is open in the Euclidean topology. Indeed, any rectangle $R = (a, b) \times (c, d)$ in $\mathbb{R}^2$ can be written as a union of balls, i.e.,

$$R = \cup_{x \in R} B_{\varepsilon_x}(x),$$

with balls $B_{\varepsilon_x}(x)$ taken with respect to d_2, with some $\varepsilon_x > 0$, implying that R is open in the Euclidean topology of $\mathbb{R}^2$. Finally, we note that any open set A in the product topology is a union of sets of the form $U_1 \times U_2$, where U_1, U_2 are open in

$\mathbb{R}$. Consequently, writing both U_1 and U_2 as unions of open intervals, we obtain A is a union of open rectangles in $\mathbb{R}^2$, which we showed to be open in the Euclidean topology, implying in turn that A is also open in the Euclidean topology of $\mathbb{R}^2$.

Conversely, let us prove that every set open in the Euclidean topology of $\mathbb{R}^2$ is also open in the product topology of $\mathbb{R}^2$. First we note that clearly every disc $B_\varepsilon(x)$ in $\mathbb{R}^2$ can be written as a union of open rectangles and is, therefore, open in the product topology of $\mathbb{R}^2$. Consequently, every open set U in the Euclidean topology can be written as $U = \bigcup_{x \in U} B_{\varepsilon_x}(x)$ for some $\varepsilon_x > 0$, so that it is also open in the product topology as a union of open sets. $\qquad\square$

A.8   Kuratowski's closure

In this section we describe another approach to topology based on Kuratowski's closure operator. This provides another (and perhaps more intuitive) approach to some notions of the previous section.

Definition A.8.1 (Metric interior, closure, boundary, etc.). In a metric space (X, d), the *metric closure* of $A \subset X$ is

$$\overline{A} = \mathrm{cl}_d(A) := \{x \in X \mid \forall r > 0 : \ A \cap B_r(x) \neq \emptyset\}.$$

In other words, $x \in \mathrm{cl}_d(A) \iff \mathrm{dist}(\{x\}, A) = 0$ (i.e., "x is close to A"). This is also equivalent to saying that every ball around x contains point(s) of A.

The *metric interior* $\mathrm{int}_d(A)$, the *metric exterior* $\mathrm{ext}_d(A)$ and the *metric boundary* $\partial_d(A)$ are defined by

$$
\begin{aligned}
\mathrm{int}_d(A) \quad &:= \quad X \setminus \mathrm{cl}_d(X \setminus A), \\
\mathrm{ext}_d(A) \quad &:= \quad X \setminus \mathrm{cl}_d(A), \\
\partial_d(A) \quad &:= \quad \mathrm{cl}_d(A) \cap \mathrm{cl}_d(X \setminus A).
\end{aligned}
$$

Notice that in this way, we have defined mappings

$$\mathrm{cl}_d, \mathrm{int}_d, \mathrm{ext}_d, \partial_d : \mathcal{P}(X) \to \mathcal{P}(X).$$

Exercise A.8.2. Let (X, d) be a metric space and $A \subset X$. Prove the following claims:

$$
\begin{aligned}
\mathrm{int}_d(A) \quad &= \quad \{x \in X \mid \exists r > 0 : \ B_r(x) \subset A\}, \\
\partial_d(A) \quad &= \quad \mathrm{cl}_d(A) \setminus \mathrm{int}_d(A), \\
X \quad &= \quad \mathrm{int}_d(A) \cup \partial_d(A) \cup \mathrm{ext}_d(A).
\end{aligned}
$$

Consequently, prove that $\mathrm{cl}_d(A)$ is closed for any set $A \subset X$.

Definition A.8.3 (Metric topology). Let (X, d) be a metric space. Then

$$\tau_d := \mathrm{int}_d(\mathcal{P}(X)) = \{\mathrm{int}_d(A) \mid A \subset X\}$$

is called the *metric topology* or the family of *metrically open sets*. The corresponding family of *metrically closed sets* is

$$\tau_d^* := \mathrm{cl}_d(\mathcal{P}(X)) = \{\mathrm{cl}_d(A) \mid A \subset X\}.$$

By the following Lemma A.8.4, we have

- a set $C \subset X$ is metrically closed if and only if $C = \mathrm{cl}_d(C)$,
- a set $U \subset X$ is metrically open if and only if $U = \mathrm{int}_d(U)$.

Lemma A.8.4. *Let (X, d) be a metric space and $A \subset X$. Then*

$$\begin{aligned}
\mathrm{cl}_d\left(\mathrm{cl}_d(A)\right) &= \mathrm{cl}_d(A), & (A.3)\\
\mathrm{int}_d\left(\mathrm{int}_d(A)\right) &= \mathrm{int}_d(A). & (A.4)
\end{aligned}$$

Proof. Let $C = \mathrm{cl}_d(A)$. Trivially, $C \subset \mathrm{cl}_d(C)$. Let $x \in \mathrm{cl}_d(C)$. Let $r > 0$. Take $y \in C \cap B_r(x)$, and then $z \in A \cap B_r(y)$. Hence

$$d(x, z) \leq d(x, y) + d(y, z) < 2r,$$

so $x \in C$. Thus (A.3) is obtained. By the definition of the metric interior, (A.3) implies (A.4). $\qquad\square$

Definition A.8.5 (Topological interior, closure, boundary, etc.). Let τ be a topology on X. For $A \subset X$, the *interior* $\mathrm{int}_\tau(A)$ is the largest open subset of A, and the *closure* $\overline{A} = \mathrm{cl}_\tau(A)$ is the smallest closed set containing A. That is,

$$\begin{aligned}
\overline{A} = \mathrm{int}_\tau(A) &:= \bigcup\{U \in \tau \mid U \subset A\},\\
\mathrm{cl}_\tau(A) &:= \bigcap\{S \in \tau^* \mid A \subset S\}.
\end{aligned}$$

These define mappings $\mathrm{int}_\tau, \mathrm{cl}_\tau : \mathcal{P}(X) \to \mathcal{P}(X)$. The *boundary* $\partial_\tau(A)$ of a set $A \subset X$ is defined by

$$\partial_\tau(A) := \mathrm{cl}_\tau(A) \cap \mathrm{cl}_\tau(X \setminus A).$$

A set $A \subset X$ is *dense* if $\mathrm{cl}_\tau(A) = X$. The topological space (X, τ) is *separable* if it has a countable dense subset. A point $x \in X$ is an *isolated point* of a set $A \subset X$ if $A \cap U = \{x\}$ for some $U \in \tau$. A point $y \in X$ is an *accumulation point* of a set $B \subset X$ if $(B \cap V) \setminus \{y\} \neq \emptyset$ for every $V \in \tau$. A *neighbourhood of* $x \in X$ is any open set $U \subset X$ containing x. The family of neighbourhoods of $x \in X$ is denoted by

$$\mathcal{V}_\tau(x) := \{U \in \tau \mid x \in U\}$$

(or simply $\mathcal{V}(x)$, when τ is evident).

Remark A.8.6. Intuitively, the closure $\mathrm{cl}_\tau(A) \subset X$ contains those points that are *close to* A. Clearly,

$$\begin{aligned} \tau &= \{\mathrm{int}_\tau(A) \mid A \subset X\}, \\ \tau^* &= \{\mathrm{cl}_\tau(A) \mid A \subset X\}. \end{aligned}$$

Moreover, $U \in \tau$ if and only if $U = \mathrm{int}_\tau(U)$, and $C \in \tau^*$ if and only if $C = \mathrm{cl}_\tau(C)$.

Exercise A.8.7. Prove that

$$\partial_\tau(A) = \mathrm{cl}_\tau(A) \setminus \mathrm{int}_\tau(A).$$

Exercise A.8.8. Let τ_d be the metric topology of a metric space (X, d). Show that $\mathrm{int}_d = \mathrm{int}_{\tau_d}$ and that $\mathrm{cl}_d = \mathrm{cl}_{\tau_d}$.

Proposition A.8.9 (A characterisation of open sets). *Let A be a subset of a topological space X. Then A is open if and only if for every $x \in A$ there is an open set U_x containing x such that $U_x \subset S$.*

Proof. If A is open we can take $U_x = A$ for every $x \in A$. Conversely, writing $A = \cup_{x \in A} U_x$ by property (T2) of open sets we get that A is open if all U_x are open. $\qquad\square$

Proposition A.8.10 (A characterisation of closures). *Let A be a subset of a topological space X. Then $x \in \overline{A}$ if and only if every open set containing x contains a point of A.*

Proof. We will prove that $x \notin \overline{A}$ if and only if there is an open set U such that $x \in U$ but $A \cap U = \emptyset$. Since $\overline{A}$ is defined as the intersection of all closed sets containing A, it follows that $x \notin \overline{A}$ means that there is a closed set C such that $A \subset C$ and $x \notin C$. Set $U = X \setminus U$ is then the required set. $\qquad\square$

Definition A.8.11 (Closure operator). Let X be a set. A *closure operator* on X is a mapping $c : \mathcal{P}(X) \to \mathcal{P}(X)$ satisfying *Kuratowski's closure axioms*:

1. $c(\emptyset) = \emptyset$,
2. $A \subset c(A)$,
3. $c(c(A)) = c(A)$,
4. $c(A \cup B) = c(A) \cup c(B)$.

Instead of a closure operator $c : \mathcal{P}(X) \to \mathcal{P}(X)$, we could study an *interior operator* $i : \mathcal{P}(X) \to \mathcal{P}(X)$, related to each other by

$$\begin{aligned} i(S) &= X \setminus c(X \setminus S), \\ c(A) &= X \setminus i(X \setminus A). \end{aligned}$$

Kuratowski's closure axioms become *interior axioms*:

1. $i(X) = X$,
2. $i(S) \subset S$,
3. $i(i(S)) = i(S)$,
4. $i(S \cap T) = i(S) \cap i(T)$.

Theorem A.8.12. *Let* (X, τ) *be a topological space. Then the mappings* $\mathrm{int}_\tau, \mathrm{cl}_\tau :$ $\mathcal{P}(X) \to \mathcal{P}(X)$ *are interior and closure operators, respectively.*

Proof. Obviously, $\mathrm{int}_\tau(X) = X$ and $\mathrm{int}_\tau(A) \subset A$. Moreover, $\mathrm{int}_\tau(U) = U$ for $U \in \tau$, and $\mathrm{int}_\tau(A) \in \tau$, because τ is a topology. Hence $\mathrm{int}_\tau(\mathrm{int}_\tau(A)) = \mathrm{int}_\tau(A)$. Finally,

$$\begin{cases} \mathrm{int}_\tau(A \cap B) \subset \mathrm{int}_\tau(A) \subset A, \\ \mathrm{int}_\tau(A \cap B) \subset \mathrm{int}_\tau(B) \subset B, \end{cases}$$

yielding

$$\mathrm{int}_\tau(\mathrm{int}_\tau(A \cap B)) \subset \mathrm{int}_\tau(\mathrm{int}_\tau(A) \cap \mathrm{int}_\tau(B)) \subset \mathrm{int}_\tau(A \cap B),$$

where $\mathrm{int}_\tau(A) \cap \mathrm{int}_\tau(B) \in \tau$, so that $\mathrm{int}_\tau(A \cap B) = \mathrm{int}_\tau(A) \cap \mathrm{int}_\tau(B)$. □

Theorem A.8.13. *Let* $i : \mathcal{P}(X) \to \mathcal{P}(X)$ *be an interior operator. Then the family* $\tau_i = i(\mathcal{P}(X)) = \{i(A) : A \subset X\}$ *is a topology. Moreover,* $i = \mathrm{int}_{\tau_i}$.

Proof. First,

$$\emptyset = i(\emptyset) \in \tau_i, \quad X = i(X) \in \tau_i.$$

Second, if $A, B \in \tau_i$ then $A \cap B = i(A) \cap i(B) = i(A \cap B) \in \tau_i$. Third, let $\mathcal{A} = \{A_j : j \in J\} \subset \tau_i$. Now

$$\bigcup \mathcal{A} \overset{A_j = i(A_j)}{=\!=} \bigcup_{j \in J} i(A_j) \overset{i(A_j) \subset i(\bigcup \mathcal{A})}{\subset} i\left(\bigcup \mathcal{A}\right) \subset \bigcup \mathcal{A}.$$

Thus $\bigcup \mathcal{A} = i(\bigcup \mathcal{A}) \in \tau_i$. Next,

$$\begin{aligned} \mathrm{int}_{\tau_i}(A) &= \bigcup \{U \in \tau_i \mid U \subset A\} \\ &= \bigcup \{i(B) \mid i(B) \subset A, \ B \subset X\}. \end{aligned}$$

Here we see that $i(A) \subset \mathrm{int}_{\tau_i}(A)$. Moreover, if $i(A) \subset i(B) \subset A$ then $i(A) = i(i(A)) \subset i(i(B)) = i(B) \subset i(A)$. Hence $i(A) = \mathrm{int}_{\tau_i}(A)$. □

Remark A.8.14. Above we have seen how topologies and closure operators (or interior operators) on a set are in bijective correspondence.

Exercise A.8.15. For each $j \in J$, let τ_j be a topology on X. Prove that $\tau = \bigcap_{j \in J} \tau_j$ is a topology. Give an example, where $\bigcup_{j \in J} \tau_j$ is not a topology.

Definition A.8.16 (Base of topology). Let (X, τ) be a topological space. A family $\mathcal{B} \subset \mathcal{P}(X)$ is called a *base* (or *basis*) for the topology τ if any open set is a union of some members of $\mathcal{B}$, i.e.,

$$\tau = \left\{ \bigcup \mathcal{B}' : \ \mathcal{B}' \subset \mathcal{B} \right\}.$$

A family $\mathcal{A} \subset \mathcal{P}(X)$ is called a *subbase* (or *subbasis*) for the topology τ if

$$\left\{ \bigcap \mathcal{A}' : \ \mathcal{A}' \subset \mathcal{A} \text{ is finite} \right\}$$

is a base for the topology. A topology is called *second countable* if it has a countable base.

Example. Trivially a topology τ is a base for itself, as $U = \bigcup \{U\}$ for every $U \in \tau$. If (X, d) is a metric space then

$$\mathcal{B} := \{ B_r(x) \mid x \in X, \ r > 0 \}$$

constitutes a base for τ_d.

Exercise A.8.17. Let $\mathcal{A} \subset \mathcal{P}(X)$. Show that there is the minimal topology $\tau_{\mathcal{A}}$ on X such that $\mathcal{A} \subset \tau_{\mathcal{A}}$: more precisely, if σ is a topology on X for which $\mathcal{A} \subset \sigma$, then $\tau_{\mathcal{A}} \subset \sigma$.

Exercise A.8.18. Let $\tau_{\mathcal{A}}$ be as in the previous exercise. Prove that a base for this topology is provided by

$$\mathcal{B} = \left\{ \bigcap \mathcal{A}' : \ \mathcal{A}' \subset \mathcal{A} \cup \{X\} \text{ is finite} \right\}.$$

Finally, we give another proof of Corollary A.7.12 that metric spaces are topological spaces using the introduced notions of interior and closure.

Theorem A.8.19 (Metric topology is a topology). *Any metric topology is a topology.*

Proof. Let τ_d be the metric topology of (X, d). By Lemma A.8.4, $U \in \tau_d$ if and only if $U = \mathrm{int}_d(U)$. Now $\emptyset, X \in \tau_d$, because

$$\begin{cases} \mathrm{int}_d(\emptyset) = \{ x \in X \mid \exists r > 0 : \ B_r(x) \subset \emptyset \} = \emptyset, \\ \mathrm{int}_d(X) = \{ x \in X \mid \exists r > 0 : \ B_r(x) \subset X \} = X. \end{cases}$$

Next, if $B_r(x) \subset U$ and $B_s(x) \subset V$ then $B_{\min\{r,s\}}(x) \subset U \cap V$. Thus if $U, V \in \tau_d$ then $U \cap V \in \tau_d$. Finally, if $B_r(x) \subset U_k$ for some $k \in J$ then $B_r(x) \subset \bigcup_{j \in J} U_j$. Thus if $\{U_j : \ j \in J\} \subset \tau_d$ then $\bigcup_{j \in J} U_j \in \tau_d$. $\square$

Exercise A.8.20 (Product topology). Let X, Y be topological spaces with bases $\mathcal{B}_X, \mathcal{B}_Y$, respectively. Show that sets

$$\{U \times V \mid U \in \mathcal{B}_X, \ V \in \mathcal{B}_Y\}$$

form a base for the product topology of $X \times Y = \{(x, y) \mid x \in X, \ y \in Y\}$ from Definition A.7.21.

The metric topology (but not only, cf. topological spaces with countable topology bases) can be characterised by the limits of sequences:

Theorem A.8.21. Let (X, d) be a metric space, $p \in X$ and $A \subset X$. Then $p \in c_d(A)$ if and only if some sequence $x : \mathbb{Z}^+ \to A$ converges to p.

Proof. Let $x_k \to p$, where $x_k \in A$ for each $k \in \mathbb{Z}^+$. That is,

$$\forall \varepsilon > 0 \ \exists k_\varepsilon \in \mathbb{Z}^+ : \quad k \geq k_\varepsilon \ \Rightarrow \ x_k \in B_d(p, \varepsilon).$$

Thus $A \cap B_d(p, \varepsilon) \neq \emptyset$ for every $\varepsilon > 0$. Thereby $p \in c_d(A)$.

Let $p \in c_d(A)$, that is $A \cap B_r(x) \neq \emptyset$ for all $r > 0$. For each $k \in \mathbb{Z}^+$, take $x_k \in A \cap B_d(p, 1/k)$. Now $(x_k)_{k=1}^\infty$ is a sequence in A, converging to p, because $d(x_k, p) < 1/k$. $\square$

A.9 Complete metric spaces

In this section we discuss complete metric spaces, give a sample application to Fredholm integral equations using Banach's Fixed Point Theorem, and show that every metric space can be "completed" and such a completion is essentially unique. Later, we will revisit this topic again to show completeness of $\mathbb{R}$ and $\mathbb{R}^n$ in Theorem A.13.10 and Corollary A.13.11. Completeness in topological vector spaces will be discussed in Section B.2.

Definition A.9.1 (Cauchy sequences and completeness). Let (X, d) be a metric space. A sequence $x : \mathbb{Z}^+ \to X$ is a *Cauchy sequence* if

$$\forall \varepsilon > 0 \ \exists k_\varepsilon \in \mathbb{Z}^+ : \ i, j \geq k_\varepsilon \ \Rightarrow \ d(x_i, x_j) < \varepsilon.$$

A metric space is called *complete* if all Cauchy sequences converge.

Example. The Euclidean metric space $(\mathbb{R}^n, d)$ is complete (see Corollary A.13.11), but its dense subset $\mathbb{Q}^n$ is not (metric of course inherited from d). For instance, *Napier's constant* $e \in \mathbb{R} \setminus \mathbb{Q}$ is obtained as the limit of numbers $\sum_{j=0}^k 1/j! \in \mathbb{Q}$.

Lemma A.9.2 (Properties of Cauchy sequences). *We have the following properties:*

(1) *Every convergent sequence is a Cauchy sequence.*

(2) *Every Cauchy sequence is bounded.*

(3) *If a Cauchy sequence has a convergent subsequence, it converges to the same limit.*

Proof. We assume that a metric space (X, d) is non-empty. To prove (1), let $x_k \to p$. We want to show that $(x_k)_{k=1}^\infty$ is a Cauchy sequence. Let $\varepsilon > 0$. Take $k_\varepsilon \in \mathbb{Z}^+$ such that $d(x_k, p) < \varepsilon$ if $k \geq k_\varepsilon$. Let $i, j \geq k_\varepsilon$. Then

$$d(x_i, x_j) \leq d(x_i, p) + d(p, x_j) < 2\varepsilon.$$

To prove (2), let $(x_k)_{k=1}^\infty$ be a Cauchy sequence. Take $\varepsilon = 1$. Then there is some k such that for $i, j \geq k$ we have $d(x_i, x_j) < 1$. Let us now fix some $a \in X$. Then for $i > k$ we have

$$d(a, x_i) \leq d(a, x_{k+1}) + d(x_{k+1}, x_i) < \rho + 1,$$

with $\rho = d(a, x_{k+1})$. Setting $R := \max\{d(a, x_1), \ldots, d(a, x_k), \rho\}$, we get that $x_i \in B_{R+1}(a)$ for all i.

To prove (3), let $(x_n)_{n=1}^\infty$ be a Cauchy sequence, with a convergent subsequence $x_{n_i} \to p \in X$. Fix some $\varepsilon > 0$. Then there is some k such that for all $n, m \geq k$ we have $d(x_n, x_m) < \varepsilon$. At the same time, there is some N such that for $n_i > N$, we have $d(x_{n_i}, p) < \varepsilon$. Consequently, for $n \geq \max\{k, N\}$, we get

$$d(x_n, p) \leq d(x_n, x_{n_i}) + d(x_{n_i}, p) < 2\varepsilon,$$

which means that $x_n \to p$ as $n \to \infty$. $\qquad\square$

Theorem A.9.3. *Let (X, d) be a complete metric space, and $A \subset X$. Then $(A, d|_{A \times A})$ is complete if and only if $A \subset X$ is closed.*

Proof. Let $A \subset X$ be closed. Take a Cauchy sequence $x : \mathbb{Z}^+ \to A$. Due to the completeness of (X, d), x converges to a point $p \in X$. Now $p \in A$, because A is closed. Thus $(A, d|_{A \times A})$ is complete.

Suppose $(A, d|_{A \times A})$ is complete. We have to show that $c_d(A) = A$. Take $p \in c_d(A)$. For each $k \in \mathbb{Z}^+$, take $x_k \in A \cap B_d(p, 1/k)$. Clearly, $x_k \to p$, so $(x_k)_{k=1}^\infty$ is a Cauchy sequence in A. Due to the completeness of $(A, d|_{A \times A})$, $x_k \to a$ for some $a \in A$. Because the limits in X are unique, $p = a \in A$. Thus $A = c_d(A)$ is closed. $\qquad\square$

We now show one application of the notion of completeness to solving integral equations.

Definition A.9.4 (Pointwise convergence of functions). Let $f_n : [a, b] \to \mathbb{R}$ be a sequence of functions and let $f : [a, b] \to \mathbb{R}$. Then we say that f_n converges to f *pointwise* on $[a, b]$ if $f_n(x) \to f(x)$ as $n \to \infty$ for all $x \in [a, b]$. In other words, this means that

$$\forall x \in [a, b] \quad \forall \varepsilon > 0 \quad \exists N = N(\varepsilon, x): \quad n > N \implies |f_n(x) - f(x)| < \varepsilon.$$

As before, by $C([a, b])$ we denote the space of all continuous functions $f : [a, b] \to \mathbb{R}$. By default we always equip it with the sup-metric d_∞.

Exercise A.9.5. Find a sequence of continuous functions $f_n \in C([0,1])$ such that $f_n \to f$ pointwise on $[0,1]$, but $f : [0,1] \to \mathbb{R}$ is not continuous on $[0,1]$.

To remedy this situation, we introduce another notion of convergence of functions:

Definition A.9.6 (Uniform convergence of functions). Let $f_n : [a,b] \to \mathbb{R}$ be a sequence of functions and let $f : [a,b] \to \mathbb{R}$. Then we say that f_n converges to f *uniformly* on $[a,b]$ if

$$\forall \varepsilon > 0 \quad \exists N = N(\varepsilon): \quad \forall n > N \quad x \in [a,b] \implies |f_n(x) - f(x)| < \varepsilon.$$

The difference with the pointwise convergence here is that the same index N works for all $x \in [a,b]$.

Theorem A.9.7. *Let $f_n \in C([a,b])$ be a sequence of continuous functions, let $f : [a,b] \to \mathbb{R}$, and suppose that f_n converges to f uniformly on $[a,b]$. Then f is continuous on $[a,b]$.*

Proof. Fix $\varepsilon > 0$. Since $f_n \to f$ uniformly, there is some $N = N(\varepsilon)$ such that for all $n > N$ and all $x \in [a,b]$ we have $|f_n(x) - f(x)| < \varepsilon$. Let $c \in [a,b]$. We will show that f is continuous at c. Since every function f_n is continuous at c, there is some $\delta = \delta(n) > 0$ such that $|x - c| < \delta$ implies $|f_n(x) - f_n(c)| < \varepsilon$. Taking some $n > N$, we get

$$|f(x) - f(c)| \leq |f(x) - f_n(x)| + |f_n(x) - f_n(c)| + |f_n(c) - f(c)| < 3\varepsilon$$

for all $|x - c| < \delta$, implying that f is continuous at c. $\qquad\qquad\square$

This result extends to uniform limits of continuous functions on general topological spaces, see Exercise C.2.18.

Proposition A.9.8 (Metric uniform convergence). *We have $f_n \to f$ in metric space $(C([a,b]), d_\infty)$ if and only if $f_n \to f$ uniformly on $[a,b]$.*

Proof. Convergence $f_n \to f$ in metric space $(C([a,b]), d_\infty)$ means that for every $\varepsilon > 0$ there is N such that for all $n > N$ we have $\sup_{x \in [a,b]} |f_n(x) - f(x)| < \varepsilon$. But this means that $|f_n(x) - f(x)| < \varepsilon$ for all $x \in [a,b]$, which is the uniform convergence. $\qquad\qquad\square$

Theorem A.9.9 (Completeness of continuous functions). *Space $C([a,b])$ with sup-metric d_∞ is complete.*

Proof. Let $f_n \in C([a,b])$ be a Cauchy sequence. Fix $\varepsilon > 0$. Then there is some N such that for all $m, n > N$ we have

$$\sup_{x \in [a,b]} |f_n(x) - f_m(x)| < \varepsilon. \tag{A.5}$$

Therefore, for each $x \in [a, b]$ the sequence $(f_n(x))_{n=1}^{\infty}$ is a Cauchy sequence in $\mathbb{R}$. If we use that $\mathbb{R}$ is complete (see Theorem A.13.10), it converges to some point in $\mathbb{R}$, which we call $f(x)$. Thus, for every $x \in [a, b]$ we have $f_n(x) \to f(x)$ as $n \to \infty$. Passing to the limit as $n \to \infty$ in (A.5), we obtain $\sup_{x \in [a,b]} |f(x) - f_m(x)| \leq \varepsilon$, which means that $d_\infty(f, f_m) \leq \varepsilon$, completing the proof. □

Theorem A.9.10 (Banach's Fixed Point Theorem). *Let (X, d) be a non-empty complete metric space, let $k < 1$ be a constant, and let $f : X \to X$ be such that*

$$d(f(x), f(y)) \leq k \, d(x, y) \tag{A.6}$$

for all $x, y \in X$. Then there exists a unique point $a \in X$ such that $a = f(a)$.

A mapping f satisfying (A.6) with some constant $k < 1$ is called a *contraction*. A point a such that $a = f(a)$ is called a *fixed point* of f.

Exercise A.9.11. Show that the conditions of Theorem A.9.10 are indispensable. For example, the conclusion of Theorem A.9.10 fails if X is not complete. Show that it also fails if $k \geq 1$. Finally, give an example of a function $f : X \to X$ satisfying

$$d(f(x), f(y)) < d(x, y)$$

instead of (A.6) on a complete metric space $X \neq \emptyset$ such that f does not have fixed points.

Proof of Theorem A.9.10. First we observe that f is continuous. Indeed, if $d(x, y) < \varepsilon$, it follows that $d(f(x), f(y)) \leq kd(x, y) < k\varepsilon < \epsilon$. We now construct a certain Cauchy sequence, whose limit will be the required fixed point of f. Take any $x_0 \in X$. For all $n \geq 0$, define $x_{n+1} = f(x_n)$. Then for all $n \geq 1$ we have

$$d(x_{n+1}, x_n) = d(f(x_{n+1}), f(x_n)) \leq kd(x_n, x_{n-1}),$$

implying that $d(x_{n+1}, x_n) \leq k^n d(x_1, x_0)$. Consequently, for $n > m \geq 1$, we have

$$
\begin{aligned}
d(x_n, x_m) &\leq d(x_n, x_{n-1}) + \cdots + d(x_{m+1}, x_m) \\
&\leq (k^{n-1} + \cdots + k^m)d(x_1, x_0) \\
&\leq k^m \sum_{i=0}^{\infty} k^i \, d(x_1, x_0) \\
&= \frac{k^m}{1 - k} d(x_1, x_0).
\end{aligned}
$$

Since $k < 1$ it follows that $d(x_n, x_m) \to 0$ as $n, m \to \infty$ which means that x_n is a Cauchy sequence. Since X is complete, $x_n \to a$ for some $a \in X$. We claim that a is a fixed point of f. Indeed, since $x_n \to a$ and since f is continuous, we have $f(x_n) \to f(a)$ by Proposition A.7.11. Therefore, $x_{n+1} \to f(a)$ as $n \to \infty$, and by the uniqueness of limits in metric spaces Proposition A.6.18 we have $f(a) = a$.

Finally, let us show that there is only one fixed point. Suppose that $f(a) = a$ and $f(b) = b$. It follows that $d(a, b) = d(f(a), f(b)) \leq kd(a, b)$ and since $k < 1$, we must have $d(a, b) = 0$ and hence $a = b$. □

Corollary A.9.12 (Fredholm integral equations). *Let $p : [0, 1] \to \mathbb{R}$ be continuous, $p \geq 0$, and such that $\int_0^1 p(t)\, dt < 1$. Let $g \in C([0, 1])$. Then there exists a unique function $f \in C([0, 1])$ such that*

$$f(x) = g(x) - \int_0^x f(t)\, p(t)\, dt.$$

Proof. As usual, let us equip $C([0, 1])$ with the sup-metric d_∞, and let us define $T : C([0, 1]) \to C([0, 1])$ by

$$(Tf)(x) = g(x) - \int_0^x f(t)\, p(t)\, dt.$$

We claim that Y is a contraction, which together with the completeness of $C([0, 1])$ in Theorem A.9.9 and Banach's Fixed Point Theorem A.9.10 would imply the statement. We have

$$d_\infty(Tf, Tg) = \sup_{x \in [0,1]} \left| \int_0^x (f(t) - g(t))\, p(t)\, dt \right| \leq \sup_{x \in [0,1]} \int_0^x |f(t) - g(t)|\, p(t)\, dt$$

$$= \int_0^1 |f(t) - g(t)|\, p(t)\, dt \leq \sup_{x \in [0,1]} |f(x) - g(x)| \int_0^1 p(t)\, dt$$

$$\leq kd(f, g),$$

where $k = \int_0^1 p(t)\, dt < 1$. $\qquad\qquad\square$

Finally we will show that every metric space can be "completed" to become a complete metric space and such "completion" is essentially unique.

Definition A.9.13 (Completion). Let (X, d) be a metric space. A complete metric space X^* is said to be a *completion* of X if X is a topological subspace of X^* and if $\overline{X} = X^*$ (i.e., if X is dense in X^*).

Remark A.9.14. Completion of a metric space can be defined in another way: a complete metric space (X^*, d^*) is a completion of (X, d) if there exists an isometry $\iota : X \to X^*$ such that the image $\iota(X)$ is dense in X^*. In the proof of Theorem A.9.15, we are actually using this idea: there X^* is the family of Cauchy sequences in X, and the points of X are naturally identified with the constant sequences.

Theorem A.9.15 (Completions of metric spaces). *Every metric space (X, d) has a completion. This completion is unique up to an isometry leaving X fixed.*

Proof. **Existence.** We will construct a completion as a space of equivalence classes of Cauchy sequences in X. Thus, we will call Cauchy sequences $(x_n)_{n=1}^\infty$ and $(x_n')_{n=1}^\infty$ equivalent if $d(x_n, x_n') \to 0$ as $n \to \infty$. One can readily see that this is an equivalence relation as in Definition A.2.6, and we define X^* to be the space of equivalence classes of such Cauchy sequences. Space X^* has a metric d^* defined as follows. For $x^*, y^* \in X^*$, pick some representatives $(x_n)_{n=1}^\infty \in x^*$ and

$(y_n)_{n=1}^{\infty} \in y^*$, and set

$$d^*(x^*, y^*) := \lim_{n \to \infty} d(x_n, y_n). \tag{A.7}$$

We first check that d^* is a well-defined function on X^*, namely that the limit in (A.7) exists and that it is independent of the choice of representatives on equivalence classes x^* and y^*. To check that the limit exists, we use the fact that x_n and y_n are Cauchy sequences, so for n and m sufficiently large we can estimate

$$\begin{aligned}
&|d(x_n, y_n) - d(x_m, y_m)| \\
=\quad &|d(x_n, y_n) - d(x_n, y_m) + d(x_n, y_m) - d(x_m, y_m)| \\
\leq\quad &|d(x_n, y_n) - d(x_n, y_m)| + |d(x_n, y_m) - d(x_m, y_m)| \\
\leq\quad &d(y_n, y_m) + d(x_n, x_m),
\end{aligned}$$

and the latter goes to zero as $n, m \to \infty$. It follows that the sequence of real numbers $(d(x_n, y_n))_{n=1}^{\infty}$ is a Cauchy sequence in $\mathbb{R}$, and hence converges because $\mathbb{R}$ is complete by Theorem A.13.10 (which will be proved later).

Let us now show that $d^*(x^*, y^*)$ is independent of the choice of representatives from x^* and y^*. Let us take $(x_n)_{n=1}^{\infty}, (x_n')_{n=1}^{\infty} \in x^*$ and $(y_n)_{n=1}^{\infty}, (y_n')_{n=1}^{\infty} \in y^*$. Then by a calculation similar to the one before we can show that

$$|d(x_n, y_n) - d(x_n', y_n')| \leq d(x_n, x_n') + d(y_n, y_n'),$$

which implies that $\lim_{n \to \infty} d(x_n, y_n) = \lim_{n \to \infty} d(x_n', y_n')$.

We now claim that (X^*, d^*) is a metric space. Non-degeneracy and symmetry in Definition A.6.1 are straightforward. The triangle inequality for d^* follows from that for d. Indeed, passing to the limit as $n \to \infty$ in the inequality $d(x_n, z_n) \leq d(x_n, y_n) + d(y_n, z_n)$, we get $d^*(x^*, z^*) \leq d^*(x^*, y^*) + d^*(y^*, z^*)$.

Next we will verify that X^* is a completion of X. We first have to check that (X, d) is a topological subspace of (X^*, d^*). We observe that for every $x \in X$ its equivalence class contains the convergent constant sequence $(x_n = x)_{n=1}^{\infty}$, and hence any equivalent Cauchy sequence must be also convergent. Thus, the class x^* consists of all sequences $(x_n)_{n=1}^{\infty}$ convergent to x. Now, if $x, y \in X$ and $(x_n)_{n=1}^{\infty} \in x^*, (y_n)_{n=1}^{\infty} \in y^*$, we have $x_n \to x$ and $y_n \to y$ as $n \to \infty$, and hence $d(x, y) = \lim_{n \to \infty} d(x_n, y_n) = d^*(x^*, y^*)$. Therefore, the mapping $x \mapsto x^*$ is an isometry from X to X^* and hence X is a topological subspace of X^* if we identify it with its image under this isometry. Thus, in the sequel we will no longer distinguish between X and its image in X^*.

We next show that X is dense in X^*. Let $x^* \in X^*$, let $\varepsilon > 0$, and let $(x_n)_{n=1}^{\infty} \in x^*$. Since x_n is a Cauchy sequence, there is some N such that for all $n, m > N$ we have $d(x_n, x_m) < \varepsilon$. Letting $m \to \infty$, we get $d^*(x_n, x^*) = \lim_{m \to \infty} d(x_n, x_m) \leq \varepsilon$. Therefore, any neighbourhood of x^* contains a point of X, which means that $\overline{X} = X^*$ by Proposition A.8.10.

Finally, we show that (X^*, d^*) is complete. First we observe that by the construction of X^* any Cauchy sequence $(x_n)_{n=1}^{\infty}$ of points of X converges to $x^* \in X^*$, for $x^* \ni (x_n)_{n=1}^{\infty}$. Second, for any Cauchy sequence x_n^* of points in

X^* there is an equivalent sequence x_n of points of X because $\overline{X} = X^*$. Indeed, for every n there is a point $x_n \in X$ such that $d^*(x_n, x_n^*) < \frac{1}{n}$. Sequence x_n is then a Cauchy sequence and by the first part of this argument it converges to its equivalence class x^* in X^*. Therefore, x_n^* also converges to x^* in (X^*, d^*).

Uniqueness. We want to show that if (X^*, d^*) and (X^{**}, d^{**}) are two completions of X then there is a bijection $f : X^* \to X^{**}$ such that $f(x) = x$ for all $x \in X$, and such that $f(x^*) = x^{**}$, $f(y^*) = y^{**}$ implies that $d^*(x^*, y^*) = d^{**}(x^{**}, y^{**})$. We define f in the following way. For $x^* \in X^*$, in view of the density of X in X^*, there exists a sequence $x_n \in X$ such that $x_n \to x^*$ in (X^*, d^*). Therefore, x_n is a Cauchy sequence in X, and since X^{**} is also a completion of X and is complete, it has some limit in X^{**}, so that $x_n \to x^{**}$ in (X^{**}, d^{**}). One can readily see that this x^{**} is independent of the choice of sequence x_n convergent to x^*. We define f by setting $x^{**} = f(x^*)$.

By construction it is clear that $f(x) = x$ for all $x \in X$. Moreover, let $x_n \to x^*$ and $y_n \to y^*$ in (X^*, d^*) and let $x_n \to x^{**}$ and $y_n \to y^{**}$ in (X^{**}, d^{**}). Consequently,

$$d^*(x^*, y^*) = \lim_{n \to \infty} d^*(x_n, y_n) = \lim_{n \to \infty} d(x_n, y_n)$$
$$= \lim_{n \to \infty} d^{**}(x_n, y_n) = d^{**}(x^{**}, y^{**}),$$

completing the proof. $\square$

A.10   Continuity and homeomorphisms

Recall that an expression like "(X, τ) is a topological space" is often abbreviated by "X is a topological space". In the sequel, to simplify notation, we may use the same letter c for the closure operators of different topological spaces: that is, if $A \subset X$ and $B \subset Y$, $c(A)$ is the closure in the topology of X, and $c(B)$ is the closure in the topology of Y. If needed, we shall express which topologies are meant. In reading the following definition, recall how we have interpreted $x \in c(A)$ as "$x \in X$ is close to $A \subset X$":

Definition A.10.1 (Continuous mappings). A mapping $f : X \to Y$ is *continuous at point* $x \in X$ if

$$x \in c(A) \quad \implies \quad f(x) \in c(f(A))$$

for every $A \subset X$. A mapping $f : X \to Y$ is *continuous* if it is continuous at every point $x \in X$, i.e.,

$$f(c(A)) \subset c(f(A))$$

for every $A \subset X$. If precision is needed, we may emphasize the topologies involved and, instead of mere *continuity*, speak specifically about (τ_X, τ_Y)-*continuity*. The set of continuous functions from X to Y is often denoted by $C(X, Y)$, with convention $C(X) = C(X, \mathbb{R})$ (or $C(X) = C(X, \mathbb{C})$).

Exercise A.10.2. Let $c \in \mathbb{R}$. Let $f, g : X \to \mathbb{R}$ be continuous, where we use the Euclidean metric topology on $\mathbb{R}$. Show that the following functions $X \to \mathbb{R}$ are then continuous: cf, $f + g$, fg, $|f|$, $\max\{f, g\}$, $\min\{f, g\}$ (here, e.g., $\max\{f, g\}(x) := \max\{f(x), g(x)\}$ etc.). Moreover show that if $g(x) \neq 0$ then f/g is continuous at $x \in X$.

Exercise A.10.3. Let (X_1, τ_1) and (X_2, τ_2) be topological spaces. Show that a mapping $f : X_1 \to X_2$ is continuous at $x \in X_1$ if and only if

$$\forall V \in \mathcal{V}_{\tau_2}(f(x)) \quad \exists U \in \mathcal{V}_{\tau_1}(x) : \quad f(U) \subset V.$$

Exercise A.10.4. Let (X, d_X) and (Y, d_Y) be metric spaces, $p \in X$ and $f : X \to Y$. Show that the following conditions are equivalent:

1. f is continuous at $p \in X$ (with respect to the metric topologies).
2. $\forall \varepsilon > 0 \; \exists \delta > 0 \; \forall w \in X : \; d_X(p, w) < \delta \Rightarrow d_Y(f(p), f(w)) < \varepsilon$.
3. $f(x_k) \to f(p)$ whenever $x_k \to p$.

Theorem A.10.5. *Let $f : X \to Y$. Then f is continuous if and only if $f^{-1}(V) \in \tau_X$ for every $V \in \tau_Y$.*

Remark A.10.6. The continuity criterion here might be read as: *"preimages of open sets are open"*. Sometimes this condition is taken as the definition of continuity of f. Equivalently, by taking complements, this means *"preimages of closed sets are closed"*.

Proof. Let us assume that "preimages of closed sets are closed". Then $A' = f^{-1}(c(f(A)))$ is closed, and $A \subset A'$, so $c(A) \subset c(A') = A'$. Hence

$$f(c(A)) \subset f(A') \subset c(f(A)).$$

Property $f(c(A)) \subset c(f(A))$ means the continuity of $f : X \to Y$.

Conversely, let $f : X \to Y$ be continuous. Let $A = f^{-1}(c(B))$, where $B \subset Y$. Then $f(c(A)) \subset c(f(A)) \subset c(c(B)) = c(B)$, so

$$c(A) \subset f^{-1}(f(c(A))) \subset f^{-1}(c(B)) = A.$$

Therefore $c(A) = A$, i.e., A is closed. $\qquad\qquad\qquad\qquad\qquad\qquad \square$

Corollary A.10.7. *Let $f : X \to Y$, and let τ_Y be a topology on Y. Then $f^{-1}(\tau_Y) = \{f^{-1}(V) \mid V \in \tau_Y\}$ is a topology on X. Moreover, f is (τ_X, τ_Y)-continuous if and only if $f^{-1}(\tau_Y) \subset \tau_X$.*

Exercise A.10.8. Prove Corollary A.10.7. The topology $f^{-1}(\tau_Y)$ is called the topology *induced* from τ_Y by f. Show that the relative topology on a subset $A \subset X$ of a topological space X in Definition A.7.18 is induced by the identity mapping $A \to X$.

Definition A.10.9 (Induced topology). Let $\mathcal{F}$ be a family of mappings $f : X \to Y$, where (Y, τ_Y) is a topological space. Then

$$\bigcap_{f \in \mathcal{F}} f^{-1}(\tau_Y) \subset \mathcal{P}(X)$$

is the topology *induced* from τ_Y by $\mathcal{F}$.

Proposition A.10.10. *Let X, Y, Z be topological spaces and let $f : X \to Y$ and $g : Y \to Z$ be continuous. Then $g \circ f : X \to Z$ is continuous.*

Proof. We will use Theorem A.10.5. Let U be open in Z. Then $g^{-1}(U)$ is open in Y and hence $(g \circ f)^{-1}(U) = f^{-1}(g^{-1}(U))$ is open in X, implying that $g \circ f$ is continuous. $\square$

Exercise A.10.11. Prove Proposition A.10.10 directly from Definition A.10.1.

Definition A.10.12 (Homeomorphisms and topological equivalence). A bijective mapping $f : X \to Y$ is a *homeomorphism* if both f and f^{-1} are continuous. In this case we say that the corresponding topological spaces (X, τ_X) and (Y, τ_Y) are *homeomorphic*. Homeomorphic spaces are also called *topologically equivalent*. A property which holds in all topologically equivalent spaces is called a *topological property*.

Example. Any two open intervals in $\mathbb{R}$ are topologically equivalent. For a set X, properties "X has five elements" or "all subsets of X are open" are topological properties.

Remark A.10.13. A homeomorphism is a topological isomorphism: homeomorphic spaces are topologically the same. As the saying goes, a topologist is a person who does not know the difference between a doughnut and a coffee cup. Let us denote briefly $X \approx Y$ when (X, τ_X) and (Y, τ_Y) are homeomorphic. It is easy to see that we have an equivalence

$$X \approx X,$$
$$X \approx Y \implies Y \approx X,$$
$$X \approx Y \text{ and } Y \approx Z \implies X \approx Z.$$

Analogously, there is a concept of metric space isomorphisms: a bijective mapping $f : X \to Y$ between metric spaces $(X, d_X), (Y, d_Y)$ is called an *isometric isomorphism* if $d_Y(f(a), f(b)) = d_X(a, b)$ for every $a \in X$ and $b \in Y$.

Example. The reader may check that $(x \mapsto x/(1 + |x|)) : \mathbb{R} \approx (-1, 1)$. Using algebraic topology, one can prove that $\mathbb{R}^m \approx \mathbb{R}^n$ if and only if $m = n$ (this is not trivial!).

Example. Any isometric isomorphism is a homeomorphism. Clearly the unbounded $\mathbb{R}$ and the bounded $(-1, 1)$ are not isometrically isomorphic. An orthogonal linear operator $A : \mathbb{R}^n \to \mathbb{R}^n$ is an isometric isomorphism, when $\mathbb{R}^n$ is endowed with the

Euclidean norm. The forward shift operator on $\ell^p(\mathbb{Z})$ is an isometric isomorphism, but the forward shift operator on $\ell^p(\mathbb{N})$ is only a non-surjective isometry.

Exercise A.10.14. Let (X, d_X) and (Y, d_Y) be metric spaces. Recall that $f : X \to Y$ is continuous if and only if

$$\forall a \in X \; \forall \varepsilon > 0 \; \exists \delta > 0 \; \forall b \in X : \quad d_X(a, b) < \delta \implies d_Y(f(a), f(b)) < \varepsilon.$$

A function $f : X \to Y$ is *uniformly continuous* if

$$\forall \varepsilon > 0 \; \exists \delta > 0 \; \forall a, b \in X : \quad d_X(a, b) < \delta \implies d_Y(f(a), f(b)) < \varepsilon,$$

and *Lipschitz-continuous* if

$$\exists C < \infty \; \forall a, b \in X : \quad d_Y(f(a), f(b)) \leq C \, d_X(a, b).$$

Prove that Lipschitz-continuity implies uniform continuity, and that uniform continuity implies continuity; give examples showing that these implications cannot be reversed.

Theorem A.10.15. *Two metrics d_1, d_2 on a set X are equivalent if and only if the identity mapping from (X, d_1) to (X, d_2) is a homeomorphism.*

Proof. Let $id(x) = x$ be the identity mapping from (X, d_1) to (X, d_2). Since $id^{-1}(U) = U$ for any set U, the forward implication follows from the definition of a continuous mapping and that of equivalent metrics. On the other hand, suppose the identity map is a homeomorphism. Again, since $id^{-1}(U) = U$ we get that every set open in (X, d_2) is open in (X, d_1) since id is continuous. The converse is true since id^{-1} is also continuous. $\qquad\square$

A.11 Compact topological spaces

Eventually, we will mainly concentrate on compact Hausdorff spaces, but in this section we deal with more general classes of topological spaces.

Definition A.11.1 (Coverings). Let X be a set and $K \subset X$. A family $\mathcal{U} \subset \mathcal{P}(X)$ is called a *cover of K* if

$$K \subset \bigcup \mathcal{U};$$

if the cover $\mathcal{U}$ is a finite set, it is called a *finite cover*. A cover $\mathcal{U}$ of $K \subset X$ has a *subcover* $\mathcal{U}' \subset \mathcal{U}$ if $\mathcal{U}'$ itself is a cover of K. In a topological space, an *open cover* refers to a cover consisting of open sets.

Definition A.11.2 (Compact sets). Let (X, τ) be a topological space. A subset $K \subset X$ is *compact* (more precisely τ-*compact*) if every open cover of K has a finite subcover. We say that (X, τ) is a *compact space* if X itself is τ-compact. A topological space is *locally compact* if each of its points has a neighbourhood whose closure is compact.

Remark A.11.3. Briefly, in a topological space (X, τ), $K \subset X$ is compact if and only if the following holds: given any family $\mathcal{U} \subset \tau$ such that $K \subset \bigcup \mathcal{U}$, there exists a finite subfamily $\mathcal{U}' \subset \mathcal{U}$ such that $K \subset \bigcup \mathcal{U}'$.

Remark A.11.4. Let us consider a tongue-in-cheek geographical-zoological analogue for compactness: In a space or universe (X, τ), let non-empty open sets correspond to territories of angry animals; recall the metaphor that a point $x \in U \in \tau$ is "far away from (i.e., not close to) the set $X \setminus U$". Compactness of an island $K \subset X$ means that any given territorial cover $\mathcal{U}$ has a finite subcover $\mathcal{U}'$: already a finite number of beasts governs the whole island.

Example.

1. If τ_1 and τ_2 are topologies of X, $\tau_1 \subset \tau_2$, and (X, τ_2) is a compact space then (X, τ_1) is a compact space.

2. $(X, \{\emptyset, X\})$ is a compact space.

3. If $|X| = \infty$ then $(X, \mathcal{P}(X))$ is not a compact space, but it is locally compact. Clearly any space with a finite topology is compact. Even though a compact topology can be of *any* cardinality, it is in a sense "not far away from being finite".

4. A metric space is compact if and only if it is sequentially compact (i.e., every sequence contains a converging subsequence, see Theorem A.13.4).

5. A subset $X \subset \mathbb{R}^n$ is compact if and only if it is closed and bounded (Heine–Borel Theorem A.13.7).

6. Theorem B.4.21 due to Frigyes Riesz asserts that a closed ball in a normed vector space over $\mathbb{C}$ (or $\mathbb{R}$) is compact (i.e., the space is locally compact) if and only if the vector space is finite-dimensional.

Of course, we may work with a complemented version of the compactness criterion in terms of closed sets:

Proposition A.11.5 (Finite intersection property). *A topological space X is compact if and only if the closed sets in X have the finite intersection property, which means that any collection $\{F_\alpha\}_\alpha$ of closed sets in X with $\cap_\alpha F_\alpha = \emptyset$ has a finite subcollection $\{F_i\}_{i=1}^n \subset \{F_\alpha\}_\alpha$ such that $\cap_{i=1}^n F_i = \emptyset$.*

Proof. Defining $U_\alpha = X \setminus F_\alpha$, we observe that condition $\cap_\alpha F_\alpha = \emptyset$ means that $\{U_\alpha\}_\alpha$ is an open covering of X. The condition that X is compact means that any such covering has some finite subcollection $\{U_i\}_{i=1}^n$ with $\cup_{i=1}^n U_i = X$, which in turn means that $\cap_{i=1}^n F_i = \emptyset$. $\qquad \square$

Proposition A.11.6 (Characterisation of compact subspaces). *Let (X, τ) be a topological space and let $Y \subset X$. Topological subspace (Y, τ_Y) is compact if and only if every collection $\{U_\alpha\}_{\alpha \in I}$ of sets $U_\alpha \in \tau$ with $\bigcup_{\alpha \in I} U_\alpha \supset Y$ has a finite subcollection that covers Y.*

Proof. Assume that (Y, τ_Y) is compact and let $\{U_\alpha\}_{\alpha \in I}$ be a collection of sets $U_\alpha \in \tau$ with $\bigcup_{\alpha \in I} U_\alpha \supset Y$. Then the collection $\{U_\alpha \cap Y\}_{\alpha \in I}$ is an open cover of (Y, τ_Y) and hence has a finite subcover $\{U_i \cap Y\}_{i=1}^n$. The corresponding collection $\{U_i\}_{i=1}^n$ is a finite subcollection of $\{U_\alpha\}_{\alpha \in I}$ that covers Y.

Conversely, let $\{V_\alpha\}_{\alpha \in I} \subset \tau_Y$ be an open cover of Y. Then there exist sets $U_\alpha \in \tau$ such that $V_\alpha = U_\alpha \cap Y$. Consequently, $\{U_\alpha\}_{\alpha \in I} \subset \tau$ is a cover of Y, and by assumption it has a finite subcollection $\{U_i\}_{i=1}^n$ that covers Y. The corresponding collection $\{V_i\}_{i=1}^n$ is then a finite open cover of Y. $\qquad\square$

Exercise A.11.7. Show that a finite set in a topological space is compact.

Exercise A.11.8. Let $x \in \mathbb{R}^n$ and $r > 0$. Show that the open ball $B_r(x) \subset \mathbb{R}^n$ is not compact in the Euclidean metric topology.

Exercise A.11.9. Prove that a union of two compact sets is compact.

Proposition A.11.10. *Let (X, τ) be a topological space, $K \subset X$ compact and $S \subset X$ closed. Then $K \cap S$ is compact.*

Proof. Let $\mathcal{U}$ be an open cover of $K \cap S$. Then $\mathcal{U} \cup \{X \setminus S\}$ is an open cover of K, thus having a finite subcover $\mathcal{U}'$. Then $\mathcal{U}' \cap \mathcal{U} \subset \mathcal{U}$ is a finite subcover of $K \cap S$. $\qquad\square$

Proposition A.11.11 (Some properties of compact sets). *We have the following properties:*

(1) *A closed subset of a compact topological space is compact.*

(2) *A compact subset of a metric space is bounded (and closed).*

Proof. To prove (1), let Y be a closed subset of a compact topological space (X, τ). Let $\{U_\alpha\}_{\alpha \in I} \subset \tau$ be an open cover of Y. Since Y is closed, its complement $X \setminus Y$ is open, and collection $\{X \setminus Y, U_\alpha\}_{\alpha \in I}$ is an open cover of X. Since X is compact, it has a finite subcover and since $X \setminus Y$ is disjoint from Y, removing $X \setminus Y$ (if necessary) from this subcover we obtain a finite subcover of Y.

To prove (2), let Y be a compact subspace of a metric space (X, d). A collection of unit balls $\{B_1(y)\}_{y \in Y}$ is an open cover of Y, and hence it has a finite subcover, say $\{B_1(y_i)\}_{i=1}^n$. Applying Proposition A.6.14 we obtain that Y must be bounded. $\qquad\square$

Proposition A.11.12. *Let X be a compact space and $f : X \to Y$ continuous. Then $f(X) \subset Y$ is compact.*

Proof. Let $\mathcal{V}$ be an open cover of $f(X)$. Then $\mathcal{U} := \{f^{-1}(V) \mid V \in \mathcal{V}\}$ is an open cover of X, thus having a finite subcover $\mathcal{U}' = \{f^{-1}(V) \mid V \in \mathcal{V}'\}$, where $\mathcal{V}' \subset \mathcal{V}$ is a finite collection. Then $f(X)$ is covered by $\mathcal{V}' \subset \mathcal{V}$: if $y \in f(X)$ then $y = f(x)$ for some $x \in X$, so $x \in f^{-1}(V_0)$ for some $V_0 \in \mathcal{V}'$, so $y = f(x) \in f(f^{-1}(V_0)) \subset V_0$. $\qquad\square$

Corollary A.11.13. *Let $f : X \to \mathbb{R}$ be a continuous mapping from a compact topological space X to $\mathbb{R}$ equipped with the Euclidean topology. Then $f(X)$ is a bounded subset of $\mathbb{R}$.*

Theorem A.11.14 (Product of compact spaces is compact). *Let X, Y be compact topological spaces. Then $X \times Y$ in the product topology is compact.*

Proof. Let $\mathcal{C} = \{W_\alpha\}_{\alpha \in I}$ be an open cover of $X \times Y$ in the product topology. In particular, it means that each W_α is a union of "rectangles" of the form $U \times Y$ where U and V are open in X and Y, respectively. For every (x, y) there is a "rectangle" $U_x^y \times V_x^y$ and the corresponding set W_x^y such that

$$(x, y) \in U_x^y \times V_x^y \subset W_x^y \in \mathcal{C}.$$

For every $x \in y$, collection $\{V_x^y\}_{y \in Y}$ is an open covering of Y which then must have some finite subcover, which we denote by $\{V_x^{y_i(x)}\}_{i=1}^{n(x)}$. Set $U_x = \cap_{i=1}^{n(x)} U_x^{y_i(x)}$ is open in X and collection $\{W_x^{y_i(x)}\}_{i=1}^{n(x)}$ is a cover of $U_x \times Y$.

In turn, the collection $\{U_x\}_{x \in X}$ is an open cover of X which then must have some finite subcover, which we denote by $\{U_{x_j}\}_{j=1}^{m}$. We now claim that the collection $\{W_{x_j}^{y_i(x_j)}\}_{ij} \subset \mathcal{C}$ is a finite cover of $X \times Y$. Indeed, for every $(x, y) \in X \times Y$ there is some U_{x_j} that contains x, and then there is some $V_{x_j}^{y_j(x_j)}$ that contains y, implying that $(x, y) \in W_{x_j}^{y_j(x_j)}$. $\square$

Lemma A.11.15. *Let (X, τ) be a compact space and $S \subset X$ infinite. Then S has an accumulation point.*

Proof. Recall that $x \in X$ is an accumulation point of $S \subset X$ if

$$\forall U \in \tau : \quad x \in U \implies (S \cap U) \setminus \{x\} \neq \emptyset.$$

Suppose $S \subset X$ has no accumulation points, i.e.,

$$\forall x \in X \ \exists U_x \in \tau : \quad x \in U_x \text{ and } S \cap U_x \subset \{x\}.$$

Now $\mathcal{U} = \{U_x : x \in X\}$ is an open cover of X, having a finite subcover $\mathcal{U}' \subset \mathcal{U}$ by compactness. Then

$$S = S \cap \left(\bigcup \mathcal{U}'\right) = \bigcup_{U_x \in \mathcal{U}'} (S \cap U_x).$$

Here the union is finite, and $S \cap U_x \subset \{x\}$ in each case. Thus S is finite. $\square$

A.12 Compact Hausdorff spaces

Next we are going to witness how beautiful compact Hausdorff topologies are. Among topological spaces, Hausdorff spaces are those where points are distinctively separated by open neighbourhoods; this happens especially in metric topology. Roughly, Hausdorff spaces have enough open sets to distinguish between any two points, while compact spaces "do not have too many open sets". Combining these two properties, compact Hausdorff spaces form a useful class of topological spaces.

Definition A.12.1 (Hausdorff spaces). A topological space (X, τ) is called a *Hausdorff space* if for each $a, b \in X$, where $a \neq b$, there exists $U, V \in \tau$ such that $a \in U$, $b \in V$ and $U \cap V = \emptyset$.

Example.

1. If τ_1 and τ_2 are topologies of X, $\tau_1 \subset \tau_2$, and (X, τ_1) is a Hausdorff space then (X, τ_2) is a Hausdorff space.
2. $(X, \mathcal{P}(X))$ is a Hausdorff space.
3. If X has more than one point and $\tau = \{\emptyset, X\}$ then (X, τ) is not Hausdorff.
4. Clearly any metric space (X, d) is a Hausdorff space; if $x, y \in X$, $x \neq y$, then $B_r(x) \cap B_r(y) = \emptyset$, when $r \leq d(x, y)/2$.
5. The distribution spaces $\mathcal{D}'(\mathbb{R}^n)$, $\mathcal{S}'(\mathbb{R}^n)$ and $\mathcal{E}'(\mathbb{R}^n)$ are non-metrisable Hausdorff spaces.

Theorem A.12.2. *In Hausdorff spaces, we have the following properties:*

(1) *Every convergent sequence has a unique limit.*

(2) *All finite sets are closed.*

(3) *Every topological subspace is also Hausdorff.*

(4) *A compact subspace of a Hausdorff space is closed.*

(5) *A subset of a compact Hausdorff space is compact if and only if it is closed.*

Proof. To prove (1), let x_n be a sequence such that $x_n \to p$ and $x_n \to q$ as $n \to \infty$. Assume $p \neq q$. Then there exist open sets U, V such that $p \in U, q \in V$ and $U \cap V = \emptyset$. Consequently, there are numbers N and M such that for all $n > N$ we have $x_n \in U$ and for all $n > M$ we have $x_n \in V$, which yields a contradiction.

To prove (2), in view of property (C3) of Corollary A.7.14 it is enough to show that one-point sets $\{x\}$ in a Hausdorff topological space X are closed. For every $y \in X \backslash \{x\}$ there exist open disjoint sets $U_y \ni x$ and $V_y \ni y$. Since $x \notin V_y$ it follows that $V_y \subset X \backslash \{x\}$ and hence $X \backslash \{x\} = \bigcup_{y \in X \backslash \{x\}} V_y$, implying that $X \backslash \{x\}$ is open.

To prove (3), let Y be a subset of a Hausdorff topological space (X, τ) and let τ_Y be the relative topology on Y. Let $a, b \in Y$ be such that $a \neq b$. Since (X, τ) is Hausdorff there exist open disjoint sets $U, V \in \tau$ such that $a \in U$ and $b \in V$. Consequently, $a \in U \cap Y \in \tau_Y$ and $b \in V \cap Y \in \tau_Y$, and $U \cap Y$ and $V \cap Y$ are disjoint, implying that (Y, τ_Y) is Hausdorff.

To prove (4), let Y be a compact subspace of a topological space X. If $Y = X$ the statement is trivial. Assuming that $Y \neq X$, let us take some $x \in X \backslash Y$. Then for every $y \in Y$ there are open disjoint sets $U_y \ni x$ and $V_y \ni y$. The collection $\{V_y\}_{y \in Y}$ is a covering of Y, and hence by Proposition A.11.6 there is a finite collection $V_{y_1}, \ldots, V_{y_n}$ still covering Y. Then set $U_x = \cap_{i=1}^n U_{y_i}$ is open, $x \in U_x$, and $U_x \cap Y = \emptyset$. Therefore, $X \backslash Y = \cup_{x \in X \backslash Y} U_x$ is open and hence Y is closed.

Statement (5) follows immediately from (4) and property (1) of Proposition A.11.11. $\qquad\qquad\square$

Theorem A.12.3 (Hausdorff property is a topological property). *Let $f : X_1 \to X_2$ be an injective and continuous mapping between topological spaces (X_1, τ_1) and (X_2, τ_2). If (X_2, τ_2) is Hausdorff then (X_1, τ_1) is also Hausdorff. Consequently, the Hausdorff property is a topological property.*

Proof. Let $x, y \in X_1$ be such that $x \neq y$. Since f is injective, we have $f(x) \neq f(y)$ and since (X_2, τ_2) is Hausdorff there exist open disjoint sets $U, V \in \tau_2$ such that $f(x) \in U$ and $f(y) \in V$. Since f is continuous, sets $f^{-1}(U)$ and $f^{-1}(V)$ are open disjoint neighbourhoods of x and y in X_1, respectively, implying that (X_1, τ_1) is also Hausdorff. That the Hausdorff property is a topological property follows immediately from this. $\qquad\square$

Exercise A.12.4 (Product of Hausdorff spaces). If (X_1, τ_1) and (X_1, τ_1) are Hausdorff topological spaces, show that $(X_1 \times X_2, \tau_1 \otimes \tau_2)$ is a Hausdorff topological space.

Theorem A.12.5. *Let X be a Hausdorff space, $A, B \subset X$ compact subsets, and $A \cap B = \emptyset$. Then there exist open sets $U, V \subset X$ such that $A \subset U$, $B \subset V$, and $U \cap V = \emptyset$.*

Proof. The proof is trivial if $A = \emptyset$ or $B = \emptyset$. So assume $x \in A$ and $y \in B$. Since X is a Hausdorff space and $x \neq y$, we can choose neighbourhoods $U_{xy} \in \mathcal{V}(x)$ and $V_{xy} \in \mathcal{V}(y)$ such that $U_{xy} \cap V_{xy} = \emptyset$. The collection $\mathcal{P} = \{V_{xy} \mid y \in B\}$ is an open cover of the compact set B, so that it has a finite subcover

$$\mathcal{P}_x = \{V_{xy_j} \mid 1 \leq j \leq n_x\} \subset \mathcal{P}$$

for some $n_x \in \mathbb{N}$. Let

$$U_x := \bigcap_{j=1}^{n_x} U_{xy_j}.$$

Now $\mathcal{O} = \{U_x \mid x \in A\}$ is an open cover of the compact set A, so that it has a finite subcover

$$\mathcal{O}' = \{U_{x_i} \mid 1 \leq i \leq m\} \subset \mathcal{O}.$$

Then define

$$U := \bigcup \mathcal{O}', \quad V := \bigcap_{i=1}^{m} \bigcup \mathcal{P}_{x_i}.$$

It is an easy task to check that U and V have the desired properties. $\qquad\square$

Corollary A.12.6. *Let X be a compact Hausdorff space, $x \in X$, and $W \in \mathcal{V}(x)$. Then there exists $U \in \mathcal{V}(x)$ such that $\overline{U} \subset W$.*

Proof. Now $\{x\}$ and $X \setminus W$ are closed sets in a compact space, thus they are compact. Since these sets are disjoint, there exist open disjoint sets $U, V \subset X$ such that $x \in U$ and $X \setminus W \subset V$; i.e., $x \in U \subset X \setminus V \subset W$. Hence $x \in U \subset \overline{U} \subset X \setminus V \subset W$. $\qquad\square$

Proposition A.12.7. *Let (X, τ_X) be a compact space and (Y, τ_Y) a Hausdorff space. Any bijective continuous mapping $f : X \to Y$ is a homeomorphism.*

Proof. Let $U \in \tau_X$. Then $X \setminus U$ is closed, hence compact. Consequently, $f(X \setminus U)$ is compact, and due to the Hausdorff property $f(X \setminus U)$ is closed. Therefore $(f^{-1})^{-1}(U) = f(U)$ is open. $\qquad\square$

Corollary A.12.8. *Let X be a set with a compact topology τ_2 and a Hausdorff topology τ_1. If $\tau_1 \subset \tau_2$ then $\tau_1 = \tau_2$.*

Proof. The identity mapping $(x \mapsto x) : X \to X$ is a continuous bijection from (X, τ_2) to (X, τ_1). $\qquad\square$

A more direct proof of the corollary. Let $U \in \tau_2$. Since (X, τ_2) is compact and $X \setminus U$ is τ_2-closed, $X \setminus U$ must be τ_2-compact. Now $\tau_1 \subset \tau_2$, so that $X \setminus U$ is τ_1-compact. (X, τ_1) is Hausdorff, implying that $X \setminus U$ is τ_1-closed, thus $U \in \tau_1$; this yields $\tau_2 \subset \tau_1$. $\qquad\square$

Definition A.12.9 (Separating points). A family $\mathcal{F}$ of mappings $X \to \mathbb{C}$ is said to *separate the points of the set X* if there exists $f \in \mathcal{F}$ such that $f(x) \neq f(y)$ whenever $x \neq y$.

Definition A.12.10 (Support). The *support* of a function $f \in C(X)$ is the set
$$\operatorname{supp}(f) := \overline{\{x \in X \mid f(x) \neq 0\}}.$$

Let $f \in C(X)$ such that $0 \leq f \leq 1$. Notations
$$K \prec f, \quad f \prec U$$
mean, respectively, that $K \subset X$ is compact and $\chi_K \leq f$, and that $U \subset X$ is open and $\operatorname{supp}(f) \subset U$.

Theorem A.12.11 (Urysohn's Lemma). *Let X be a compact Hausdorff space, A, $B \subset X$ closed non-empty sets, $A \cap B = \emptyset$. Then there exists $f \in C(X)$ and $U \subset X \setminus A$ such that $B \prec f \prec U$. Especially, we find f such that*
$$0 \leq f \leq 1, \quad f(A) = \{0\}, \quad f(B) = \{1\}.$$

Proof. The set $\mathbb{Q} \cap [0, 1]$ is countably infinite; let $\phi : \mathbb{N} \to \mathbb{Q} \cap [0, 1]$ be a bijection satisfying $\phi(0) = 0$ and $\phi(1) = 1$. Choose open sets $U_0, U_1 \subset X$ such that
$$A \subset U_0 \subset \overline{U_0} \subset U_1 \subset \overline{U_1} \subset X \setminus B.$$

Then we proceed inductively as follows: Suppose we have chosen open sets $U_{\phi(0)}$, $U_{\phi(1)}, \ldots, U_{\phi(n)}$ such that
$$\phi(i) < \phi(j) \Rightarrow \overline{U_{\phi(i)}} \subset U_{\phi(j)}.$$

Let us choose an open set $U_{\phi(n+1)} \subset X$ such that

$$\phi(i) < \phi(n+1) < \phi(j) \Rightarrow \overline{U_{\phi(i)}} \subset U_{\phi(n+1)} \subset \overline{U_{\phi(n+1)}} \subset U_{\phi(j)}$$

whenever $0 \leq i, j \leq n$. Let us define

$$r < 0 \Rightarrow U_r := \emptyset, \quad s > 1 \Rightarrow U_s := X.$$

Hence for each $q \in \mathbb{Q}$ we get an open set $U_q \subset X$ such that

$$\forall r, s \in \mathbb{Q} : r < s \Rightarrow \overline{U_r} \subset U_s.$$

Let us define a function $f : X \rightarrow [0, 1]$ by

$$f(x) := \inf\{r : x \in U_r\}.$$

Clearly $0 \leq f \leq 1$, $f(A) = \{0\}$ and $f(B) = \{1\}$.

Let us prove that f is continuous. Take $x \in X$ and $\varepsilon > 0$. Take $r, s \in \mathbb{Q}$ such that

$$f(x) - \varepsilon < r < f(x) < s < f(x) + \varepsilon;$$

then f is continuous at x, since $x \in U_s \setminus \overline{U_r}$ and for every $y \in U_s \setminus \overline{U_r}$ we have $|f(y) - f(x)| < \varepsilon$. Thus $f \in C(X)$. □

Corollary A.12.12. Let X be a compact space. Then $C(X)$ separates the points of X if and only if X is Hausdorff.

Exercise A.12.13. Prove the previous corollary.

Definition A.12.14 (Partition of unity). A *partition of unity* on $K \subset X$ in a topological space (X, τ) is a family $\mathcal{F} = \{\phi_j : X \rightarrow [0, 1] \mid j \in J\}$ of continuous functions such that

$$\chi_K \leq \sum_{j \in J} \phi_j \leq 1,$$

where the sum is required to be *locally finite*: for each $x \in X$ there exists $U \in \mathcal{V}(x)$ such that $\mathrm{supp}(\phi_j) \subset X \setminus U$ for all but finitely many $\phi_j \in \mathcal{F}$. Moreover, if now $\phi_j \prec U_j$ for all $j \in J$, where $\mathcal{U} = \{U_j : j \in J\}$ is an open cover of X, then $\mathcal{F}$ is called a *partition of unity on K subordinate to $\mathcal{U}$*.

Corollary A.12.15 (Partition of unity). Let $\mathcal{U}$ be an open cover of a compact set $K \subset X$ in a Hausdorff space (X, τ). Then there exists a partition of unity on K subordinate to $\mathcal{U}$.

Proof. Assume the non-trivial case $K \neq \emptyset$. Take a finite subcover $\mathcal{U}' = \{U_j \mid 1 \leq j \leq n\} \subset \mathcal{U}$. For $x \in K$, take $j \in \{1, \ldots, n\}$ such that $x \in U_j$; then choose $V_x \in \mathcal{V}(x)$ such that $\overline{V_x} \subset U_j$. Then $\mathcal{O} = \{V_x \mid x \in K\}$ is an open cover of K, thus having a finite subcover $\mathcal{O}' \subset \mathcal{O}$. Let

$$K_j := \bigcup \{\overline{V} \in \mathcal{O}' : \overline{V} \subset U_j\}.$$

Urysohn's Lemma provides functions $f_j \in C(X)$ satisfying $K_j \prec f_j \prec U_j$. Again by Urysohn's Lemma, there exists $g \in C(X)$ such that

$$\bigcup_{j=1}^{n} K_j \prec g \prec \left\{ x \in X : \sum_{k=1}^{n} f_k(x) > 0 \right\}.$$

Notice that $K \subset \bigcup_{j=1}^{n} K_j$. Let

$$\phi_j := f_j / (1 - g + \sum_{k=1}^{n} f_k).$$

Then $\{\phi_j \in C(X)\}_{j=1}^{n}$ provides a desired partition of unity. □

Exercise A.12.16. In a compact metric space (X, d), Urysohn's Lemma is much easier to obtain: When $A, B \subset X$ are closed and non-empty such that $A \cap B = \emptyset$, define $f : X \to \mathbb{R}$ by

$$f(x) := \min \left\{ 1, \frac{\mathrm{dist}(A, \{x\})}{\mathrm{dist}(A, B)} \right\}.$$

Show that f is continuous, $0 \leq f \leq 1$, $f(A) = \{0\}$ and $f(B) = 1$.

Definition A.12.17 (Equicontinuity). Let X be a topological space. A family $\mathcal{F}$ of mappings $f : X \to \mathbb{C}$ is called *equicontinuous* at $p \in X$ if for every $\varepsilon > 0$ there exists a neighbourhood $U \subset X$ of p such that $|f(x) - f(p)| < \varepsilon$ whenever $f \in \mathcal{F}$ and $x \in U$.

Exercise A.12.18. Prove the following Theorem A.12.19. (Hint: a bounded sequence of numbers has a convergent subsequence...)

Theorem A.12.19 (Arzelà-Ascoli Theorem). *Let $K \subset \mathbb{R}^n$ be compact. For each $j \in \mathbb{Z}^+$, let $f_j : K \to \mathbb{C}$ be continuous, and assume that $\mathcal{F} = \{f_j \mid j \in \mathbb{Z}^+\}$ is equicontinuous on K. If $\mathcal{F}$ is bounded, i.e.,*

$$\sup_{x \in K, j \in \mathbb{Z}^+} |f_j(x)| < \infty,$$

then there is a subsequence $\{f_{j_k} \mid k \in \mathbb{Z}^+\}$ that converges uniformly on K.

A.13 Sequential compactness

In this section, a metric space (X, d) is endowed with its canonical metric topology τ_d.

Proposition A.13.1 (Closed and bounded if compact in metric). *Let (X, d) be a metric space, and let $K \subset X$ be compact. Then K is closed and bounded.*

Proof. Let us assume $K \neq \emptyset$, to avoid a triviality. Let $x_0 \in X$. Then $\mathcal{U} = \{B_k(x_0) \mid k \in \mathbb{Z}^+\}$ is an open cover of K. Due to compactness of K, there is a subcover $\mathcal{U}' = \{B_k(x_0) \mid k \in S\}$, where $S \subset \mathbb{Z}^+$ is finite. Now

$$K \subset \bigcup \mathcal{U}' = \bigcup_{k \in S} B_k(x_0) = B_{\max(S)}(x_0).$$

Therefore $\operatorname{diam}(K) \leq 2\max(S) < \infty$, so K is bounded.

We have to prove that K is closed. Let $x \in X \setminus K$. Then

$$\mathcal{V} := \{B_{d(x,y)/2}(y) \mid y \in K\}$$

is an open cover of K. By compactness, there is a finite subcover

$$\mathcal{V}' = \{B_{d(x,y_j)/2}(y_j)\}_{j=1}^n.$$

Let $r := \min\{d(x,y_j)/2\}_{j=1}^n$. Then

$$B_r(x) \cap K \subset \bigcup_{j=1}^n \left(B_r(x) \cap B_{d(x,y_j)/2}(y_j)\right) = \emptyset,$$

so $x \notin c_d(K)$. Thereby $K = c_d(K)$ is closed. $\qquad\square$

Exercise A.13.2. Give an example of a bounded non-compact metric space.

Definition A.13.3 (Sequential compactness). A metric space is *sequentially compact* if each of its sequences has a converging subsequence. That is, given a sequence $(x_k)_{k=1}^\infty$ in a sequentially compact metric space (X,d), there is a converging sequence $(x_{k_j})_{j=1}^\infty$, where $k_{j+1} > k_j \in \mathbb{Z}^+$ for each $j \in \mathbb{Z}^+$.

Theorem A.13.4 (Compact $\Leftrightarrow$ sequentially compact in metric spaces). *A metric space (X,d) is compact if and only if it is sequentially compact.*

Proof. Let us assume that $X \neq \emptyset$ is compact. Take a sequence $(x_k)_{k=1}^\infty$ in X. If the set $\{x_k : k \in \mathbb{Z}^+\}$ is finite, there exists $y \in X$ such that $y = x_k$ for infinitely many $k \in \mathbb{Z}^+$. Then a desired convergent subsequence is given by $(y, y, y, \ldots)$. Now assume that the set $S := \{x_k : k \in \mathbb{Z}^+\}$ is infinite, so it has an accumulation point $p \in X$ by Lemma A.11.15. Take $k_1 \in \mathbb{Z}^+$ such that $x_{k_1} \in S \cap B_1(p)$. Inductively, take $k_{j+1} > k_j \in \mathbb{Z}^+$ such that $x_{k_{j+1}} \in S \cap B_{1/j}(p)$. Then $d(p, x_{k_{j+1}}) < 1/j \to_{j \to \infty} 0$, so $x_{k_j} \to_{j \to \infty} p$. We have proven that a compact metric space is sequentially compact.

Now let (X,d) be sequentially compact. We want to show that its metric topology is compact. Take an open cover $\mathcal{U} = \{U_\alpha : \alpha \in A\}$ of X. We claim that

$$\exists \varepsilon_0 > 0 \ \forall x \in X \ \exists \alpha \in A : \quad B_{\varepsilon_0}(x) \subset U_\alpha. \tag{A.8}$$

Let us prove this by deducing a contradiction from the logically negated assumption

$$\forall \varepsilon > 0 \ \exists x \in X \ \forall \alpha \in A : \quad B_\varepsilon(x) \not\subset U_\alpha.$$

This would especially imply

$$\forall k \in \mathbb{Z}^+ \ \exists x_k \in X \ \forall \alpha \in A : \quad B_{1/k}(x_k) \not\subset U_\alpha.$$

This gives us a sequence $(x_k)_{k=1}^\infty$, which by sequential compactness has a subsequence $(x_{k_j})_{j=1}^\infty$ converging to a point $p \in X$. Since $\mathcal{U}$ covers X, we have $p \in U_{\alpha_p}$ for some $\alpha_p \in A$. Since U_{α_p} is open, $B_\varepsilon(p) \subset U_{\alpha_p}$ for some $\varepsilon > 0$. But for large enough j,

$$B_{1/k_j}(x_{k_j}) \subset B_\varepsilon(p) \subset U_{\alpha_p}.$$

This is a contradiction, so (A.8) must be true. Now we claim that

$$X \text{ can be covered with finitely many open balls of radius } \varepsilon_0. \tag{A.9}$$

What happens if (A.9) is not true? Then take $x_1 \in X$, and inductively

$$x_{k+1} \in X \setminus \bigcup_{j=1}^k B_{\varepsilon_0}(x_j) \neq \emptyset,$$

where the non-emptiness of the set is due to the counter-assumption.

Now $d(x_j, x_k) \geq \varepsilon_0 > 0$ if $j \neq k$, so the sequence $(x_k)_{k=1}^\infty$ does not have a convergent subsequence. But this would contradict the sequential compactness. Hence (A.9) must be true. □

Exercise A.13.5. Think why the compactness of X follows from (A.8) and (A.9).

Exercise A.13.6. Show that a compact metric space is complete.

Corollary A.13.7 (Heine–Borel Theorem). *Let* $\mathbb{R}^n$ *be endowed with its Euclidean topology. Then* $K \subset \mathbb{R}^n$ *is compact if and only if it is closed and bounded.*

Proof. In any metric topology, compactness implies closedness and boundedness, see Proposition A.13.1. So let $S \subset \mathbb{R}^n$ be non-empty, closed and bounded. We shall prove that it is sequentially compact. Take a sequence $(x_k)_{k=1}^\infty$ in S. By boundedness, there exist $a, b \in \mathbb{R}$ such that $S \subset [a,b]^n =: Q_1$. That is, Q_1 is a closed cube of sidelength $b - a$.

Now we chop Q_1 inductively into pieces. When the cube Q_j has been chosen, we decompose Q_j "dyadically" to a union of 2^n cubes $Q_{j+1,m}$ (here $m \in \{1, \ldots, 2^n\}$), whose interiors are disjoint and whose sidelengths are $2^{-j}(b-a)$. Choose $Q_j \in \{Q_{j+1,m} : j \in \{1, \ldots, 2^n\}\}$ such that $x_k \in Q_{j+1}$ for infinitely many $k \in \mathbb{Z}^+$.

We construct the convergent subsequence $(x_{k_j})_{j=1}^\infty$ inductively. Let $k_1 := 1$. Take $k_{j+1} > k_j \in \mathbb{Z}^+$ such that $x_{k_{j+1}} \in Q_{j+1}$. Now $(x_{k_j})_{j=1}^\infty$ is a Cauchy sequence, because

$$\begin{cases} Q_1 \supset Q_2 \supset Q_3 \supset \cdots \supset Q_j \supset Q_{j+1} \supset \cdots, \\ \mathrm{diam}(Q_{j+1}) = \sqrt{n}\, 2^{-j}(b-a) \to_{j\to\infty} 0. \end{cases}$$

Due to the completeness of $\mathbb{R}^n$, the Cauchy sequence $(x_{k_j})_{j=1}^\infty$ of $S \subset \mathbb{R}^n$ converges to a point $p \in \mathbb{R}^n$. But $p \in S$, because S is closed. Thus S is sequentially compact. $\qquad \square$

Corollary A.13.8. *Let (X,τ) be a compact topological space and $f : X \to \mathbb{R}$ continuous. Then there exist $\max(f(X)), \min(f(X)) \in \mathbb{R}$.*

Proof. Assume that $X \neq \emptyset$. By Proposition A.11.12, $f(X) \subset \mathbb{R}$ is compact. By the Heine–Borel Theorem A.13.7, equivalently $f(X) \subset \mathbb{R}$ is closed and bounded. Thereby $\sup(f(X)), \inf(f(X)) \in f(X)$. $\qquad \square$

We note that the Heine–Borel theorem can also be proved without referring to the sequential compactness. For simplicity, we show this in the one-dimensional case.

Theorem A.13.9 (Heine-Borel Theorem in 1D). *Closed intervals $[a,b]$ are compact in $\mathbb{R}$ in the Euclidean topology.*

Proof. We will assume $a < b$ since otherwise the statement is trivial. For an open covering $\mathcal{C} = \{U_\alpha\}_{\alpha \in I}$ of $[a,b]$ let $S \subset [a,b]$ be defined by

$$S = \{x \in [a,b] : [a,x] \text{ can be covered by finitely many sets from } \mathcal{C}\}.$$

The statement of the theorem will follow if we show that $b \in S$. Since $S \neq 0$ in view of $a \in S$ and since $S \subset [a,b]$ is bounded, we can define $c = \sup S$ so that $c \in [a,b]$. The statement of the theorem will follow if we show that $c \in S$ and that $c = b$.

To show that $c \in S$, we observe that since $c \in [a,b]$, there is some set $U_c \in \mathcal{C}$ such that $c \in U_c$. Since U_c is open, there is some $\varepsilon > 0$ such that $(c-\varepsilon, c] \subset U_c$. At the same time, since $c - \varepsilon < c = \sup S$, the closed interval $[a, c-\varepsilon]$ can be covered by finitely many sets from $\mathcal{C}$ by the definition of S and c. Consequently, adding U_c to this finite collection of sets from $\mathcal{C}$ we obtain a finite covering of $[a,c]$, implying that $c \in S$.

To show that $c = b$, let us assume that $c < b$. As before, let U_c be such that $c \in U_c \in \mathcal{C}$. Since U_c is open and $c < b$, there is $\varepsilon > 0$ such that $[c, c+2\varepsilon) \subset U_c$. Since $c \in S$, closed interval $[a,c]$ can be covered by finitely many sets from $\mathcal{C}$, and adding U_c to this finite collection we obtain a finite covering of $[a, c+\varepsilon]$ which is a contradiction with $c = \sup S$. $\qquad \square$

Theorem A.13.10 ($\mathbb{R}$ is complete). *The real line $\mathbb{R}$ with the Euclidean metric is complete.*

Proof. Let x_n be a Cauchy sequence in $\mathbb{R}$. By Lemma A.9.2, (2), the set $\{x_n\}_{n=1}^{\infty}$ is bounded, i.e., there are some $a, b \in \mathbb{R}$ such that $\{x_n\}_{n=1}^{\infty} \subset [a, b]$. By the Heine–Borel Theorem in Corollary A.13.7 or in Theorem A.13.9 the interval $[a, b]$ is compact, and by Exercise A.13.6 it must be complete. Therefore, x_n must have a convergent subsequence, and since it is a Cauchy sequence, the whole sequence is convergent by Lemma A.9.2, (3). □

Corollary A.13.11 ($\mathbb{R}^n$ is complete). *The space $\mathbb{R}^n$ is complete with respect to any of the Lipschitz equivalent metrics d_p, $1 \leq p \leq \infty$.*

Proof. Since all metrics d_p are Lipschitz equivalent, it is enough to take one, e.g., d_∞. Writing $\bar{x}_k = (x_k^{(1)}, \ldots, x_k^{(n)})$ and $d_\infty(\bar{x}_k, \bar{x}_l) = \max_{1 \leq i \leq n} |x_k^{(i)} - x_l^{(i)}|$, we have that $d_\infty(\bar{x}_k, \bar{x}_l) < \varepsilon$ implies $|x_k^{(i)} - x_l^{(i)}| < \varepsilon$ for all $i = 1, \ldots, n$. Thus, if $\bar{x}_k \in \mathbb{R}^n$ is a Cauchy sequence in $\mathbb{R}^n$, it follows that $x_k^{(i)}$ is a Cauchy sequence in $\mathbb{R}$ for all i, and hence it has a limit, say $x^{(i)}$, for all i, by Theorem A.13.10. Writing $\bar{x} = (x^{(1)}, \ldots, x^{(n)})$, we claim that $\bar{x}_k \to \bar{x}$ as $k \to \infty$. Indeed, let $\varepsilon > 0$. Then for all i there is a number N_i such that $k > N_i$ implies $|x_k^{(i)} - x^{(i)}| < \varepsilon$. Therefore, for $k > \max_{1 \leq i \leq n} N_i$, we have $|x_k^{(i)} - x^{(i)}| < \varepsilon$ for all i, which means that $d_\infty(\bar{x}_k, \bar{x}) < \varepsilon$. □

Alternative proof of Theorem A.13.4. We now state several results of independent importance that will give another proof of Theorem A.13.4.

Lemma A.13.12 (Lebesgue's covering lemma). *Let $\mathcal{C}$ be an open covering of a sequentially compact metric space (X, d). Then there is $\varepsilon > 0$ such that every ball with radius ε is contained in some set from the covering $\mathcal{C}$. Such ε is called a Lebesgue number of the covering $\mathcal{C}$.*

Proof. Suppose that no such $\varepsilon > 0$ exists. It means that for every $n \in \mathbb{N}$ there is a ball $B_{1/n}(x_n)$ which is not contained in any set from $\mathcal{C}$. Let $(x_{n_j})_{j=1}^{\infty}$ be a convergent subsequence of $(x_n)_{n=1}^{\infty}$ with some limit $x \in X$, so that $x_{n_j} \to X$ as $j \to \infty$. Let $U \in \mathcal{C}$ be a set in $\mathcal{C}$ containing x. Since U is open, there is some $\delta > 0$ such that $B_{2\delta}(x) \subset U$. Now let N be one of indices n_j such that $d(x_N, x) < \delta$ and such that $\frac{1}{N} < \delta$. We claim that $B_{1/N}(x_N) \subset B_{2\delta}(x)$ which would be a contradiction with our choice of the sequence x_n and the fact that $B_{2\delta}(x) \subset U$. Indeed, if $y \in B_{1/N}(x_N)$, we have

$$d(y, x) \leq d(y, x_N) + d(x_N, x) < \frac{1}{N} + \delta < 2\delta,$$

which means that $y \in B_{2\delta}(x)$, completing the proof. □

Lemma A.13.13 (Totally bounded metric spaces). *Let X be a sequentially compact metric space. Then X is totally bounded, which means that for every $\varepsilon > 0$ there are finitely many balls in X with radius ε that cover X.*

Proof. Suppose that there is $\varepsilon > 0$ such that no finitely many balls in X with radius ε cover X. We will now construct a sequence of points in X with no convergent subsequence. Let $x_1 \in X$ be an arbitrary point. Let x_2 be any point in $X \backslash B_\varepsilon(x_1)$. Inductively, suppose we have points $x_1, \ldots, x_n \in X$ such that $x_j \in X \backslash \cup_{i=1}^{j-1} B_\varepsilon(x_i)$. Since the collection $\{B_\varepsilon(x_i)\}_{i=1}^n$ does not cover X, we can always choose some $x_n \in X \backslash \cup_{i=1}^{n-1} B_\varepsilon(x_i)$. All points in this sequence have the property that $d(x_n, x_k) \geq \varepsilon$ for any $n, k \in \mathbb{N}$ which means that sequence $(x_n)_{n=1}^\infty$ can not have any convergent subsequence. $\qquad\square$

Exercise A.13.14. In general, a metric space is said to be *totally bounded* if

$$\forall \varepsilon > 0 \; \exists \{x_j \mid j \in \{1, \ldots, n_\varepsilon\}\} \subset X : \quad X = \bigcup_{j=1}^{n_\varepsilon} B_\varepsilon(x_j).$$

Show that a metric space (X, d) is compact if and only if it is bounded and totally bounded.

Alternative proof of Theorem A.13.4. Let (X, d) be a metric space. First we will prove that if X is compact it is sequentially compact. Let $(x_n)_{n=1}^\infty$ be a sequence of points in x. Define $A_n = \{x_n, x_{n+1}, x_{n+2}, \ldots\}$, so that $A_1 = \{x_n\}_{n=1}^\infty$. Let $F_n = \overline{A_n}$. Clearly F_n is a closed set and the intersection of any finite number of sets F_n is non-empty since it contains A_N for some N. Since X is compact, by the finite intersection property in Proposition A.11.5 we have $\cap_{n=1}^\infty F_n \neq \emptyset$. Let now $x \in \cap_{n=1}^\infty F_n$, so that $x \in F_n = \overline{A_n}$ for all n. Using a characterisation of closures in Proposition A.8.10, it follows that every open ball $B_{1/j}(x)$ contains a point $x_{n_j} \in A_{n_j}$ with n_j as large as we want. Therefore, we have a subsequence $\{x_{n_j}\}$ of $\{x_n\}$ such that $d(x_{n_j}, x) < 1/j$, which means that it is a convergent subsequence of $\{x_n\}$.

Let us now prove that if X is sequentially compact it is compact. Let $\mathcal{C}$ be an open cover of X and let $\varepsilon > 0$ be its Lebesgue number according to Lemma A.13.12. By Lemma A.13.13, X is totally bounded, so that it is covered by finitely many balls $\{B_\varepsilon(x_i)\}_{i=1}^n$. Since ε is a Lebesgue number, for every $i = 1, \ldots, n$, there is some $U_i \in \mathcal{C}$ such that $B_\varepsilon(x_i) \subset U_i$. Consequently, $\{U_i\}_{i=1}^n$ must be a cover for X. $\qquad\square$

A.14   Stone–Weierstrass theorem

In the sequel we study densities of subalgebras in $C(X)$. These results will be applied in characterising function algebras among Banach algebras. For material concerning algebras we refer to Chapter D. First we study continuous functions on $[a, b] \subset \mathbb{R}$:

Theorem A.14.1 (Weierstrass Theorem (1885)). *Polynomials are dense in* $C([a,b])$.

Proof. Evidently, it is enough to consider the case $[a, b] = [0, 1]$. Let $f \in C([0, 1])$, and let $g(x) = f(x) - (f(0) + (f(1) - f(0))x)$; then $g \in C(\mathbb{R})$ if we define $g(x) = 0$ for $x \in \mathbb{R} \setminus [0, 1]$. For $n \in \mathbb{N}$ let us define $k_n : \mathbb{R} \to [0, \infty)$ by

$$k_n(x) := \begin{cases} \frac{(1-x^2)^n}{\int_{-1}^{1}(1-t^2)^n \, dt}, & \text{when } |x| < 1, \\ 0, & \text{when } |x| \geq 1. \end{cases}$$

Then define $P_n := g * k_n$ (*convolution of g and k_n*), that is

$$P_n(x) = \int_{-\infty}^{\infty} g(x-t) \, k_n(t) \, dt = \int_{-\infty}^{\infty} g(t) \, k_n(x-t) \, dt = \int_0^1 g(t) \, k_n(x-t) \, dt,$$

and from this last expression we see that P_n is a polynomial on $[0, 1]$. Notice that P_n is real valued if f is real valued. Take any $\varepsilon > 0$. The function g is uniformly continuous, so that there exists $\delta > 0$ such that

$$\forall x, y \in \mathbb{R} : \ |x - y| < \delta \Rightarrow |g(x) - g(y)| < \varepsilon.$$

Let $\|g\| = \max_{t \in [0,1]} |g(t)|$. Take $x \in [0, 1]$. Then

$$|P_n(x) - g(x)| = \left| \int_{-\infty}^{\infty} g(x-t) \, k_n(t) \, dt - g(x) \int_{-\infty}^{\infty} k_n(t) \, dt \right|$$

$$= \left| \int_{-1}^{1} (g(x-t) - g(x)) \, k_n(t) \, dt \right|$$

$$\leq \int_{-1}^{1} |g(x-t) - g(x)| \, k_n(t) \, dt$$

$$\leq \int_{-1}^{-\delta} 2\|g\| \, k_n(t) \, dt + \int_{-\delta}^{\delta} \varepsilon \, k_n(t) \, dt + \int_{\delta}^{1} 2\|g\| \, k_n(t) \, dt$$

$$\leq 4\|g\| \int_{\delta}^{1} k_n(t) \, dt + \varepsilon.$$

The reader may verify that $\int_{\delta}^{1} k_n(t) \, dt \to_{n \to \infty} 0$ for every $\delta > 0$. Hence $\|Q_n - f\| \to_{n \to \infty} 0$, where $Q_n(x) = P_n(x) + f(0) + (f(1) - f(0))x$. □

Exercise A.14.2. Show that $\int_{\delta}^{1} k_n(t) \, dt \to_{n \to \infty} 0$ in the proof of the Weierstrass Theorem A.14.1.

Definition A.14.3 (Involutive subalgebras). For $f : X \to \mathbb{C}$ let us define $f^* : X \to \mathbb{C}$ by $f^*(x) := \overline{f(x)}$, and define $|f| : X \to \mathbb{C}$ by $|f|(x) := |f(x)|$. A subalgebra $\mathcal{A} \subset \mathcal{F}(X)$ is called *involutive* if $f^* \in \mathcal{A}$ whenever $f \in \mathcal{A}$.

Theorem A.14.4 (Stone–Weierstrass Theorem (1937)). *Let X be a compact space. Let $\mathcal{A} \subset C(X)$ be an involutive subalgebra separating the points of X. Then $\mathcal{A}$ is dense in $C(X)$.*

Proof. If $f \in \mathcal{A}$ then $f^* \in \mathcal{A}$, so that the real part $\mathrm{Re}f = \dfrac{f + f^*}{2}$ belongs to $\mathcal{A}$. Let us define

$$\mathcal{A}_{\mathbb{R}} := \{\mathrm{Re}f \mid f \in \mathcal{A}\};$$

this is a $\mathbb{R}$-subalgebra of the $\mathbb{R}$-algebra $C(X, \mathbb{R})$ of continuous real-valued functions on X. Then

$$\mathcal{A} = \{f + ig \mid f, g \in \mathcal{A}_{\mathbb{R}}\},$$

so that $\mathcal{A}_{\mathbb{R}}$ separates the points of X. If we can show that $\mathcal{A}_{\mathbb{R}}$ is dense in $C(X, \mathbb{R})$ then $\mathcal{A}$ would be dense in $C(X)$.

First we have to show that $\overline{\mathcal{A}_{\mathbb{R}}}$ is closed under taking maximums and minimums. For $f, g \in C(X, \mathbb{R})$ we define

$$\max(f, g)(x) := \max(f(x), g(x)), \quad \min(f, g)(x) := \min(f(x), g(x)).$$

Notice that $\overline{\mathcal{A}_{\mathbb{R}}}$ is an algebra over the field $\mathbb{R}$. Since

$$\max(f, g) = \frac{f + g}{2} + \frac{|f - g|}{2}, \quad \min(f, g) = \frac{f + g}{2} - \frac{|f - g|}{2},$$

it is enough to prove that $|h| \in \overline{\mathcal{A}_{\mathbb{R}}}$ whenever $h \in \overline{\mathcal{A}_{\mathbb{R}}}$. Let $h \in \overline{\mathcal{A}_{\mathbb{R}}}$. By the Weierstrass Theorem A.14.1 there is a sequence of polynomials $P_n : \mathbb{R} \to \mathbb{R}$ such that

$$P_n(x) \to_{n \to \infty} |x|$$

uniformly on the interval $[-\|h\|, \|h\|]$. Thereby

$$\||h| - P_n(h)\| \to_{n \to \infty} 0,$$

where $P_n(h)(x) := P_n(h(x))$. Since $P_n(h) \in \overline{\mathcal{A}_{\mathbb{R}}}$ for every n, this implies that $|h| \in \overline{\mathcal{A}_{\mathbb{R}}}$. Now we know that $\max(f, g), \min(f, g) \in \overline{\mathcal{A}_{\mathbb{R}}}$ whenever $f, g \in \overline{\mathcal{A}_{\mathbb{R}}}$.

Now we are ready to prove that $f \in C(X, \mathbb{R})$ can be approximated by elements of $\mathcal{A}_{\mathbb{R}}$. Take $\varepsilon > 0$ and $x, y \in X$, $x \neq y$. Since $\mathcal{A}_{\mathbb{R}}$ separates the points of X, we may pick $h \in \mathcal{A}_{\mathbb{R}}$ such that $h(x) \neq h(y)$. Let $g_{xx} = f(x)\mathbf{1}$, and let

$$g_{xy}(z) := \frac{h(z) - h(y)}{h(x) - h(y)} \, f(x) + \frac{h(z) - h(x)}{h(y) - h(x)} \, f(y).$$

Here $g_{xx}, g_{xy} \in \mathcal{A}_{\mathbb{R}}$, since $\mathcal{A}_{\mathbb{R}}$ is an algebra. Furthermore,

$$g_{xy}(x) = f(x), \quad g_{xy}(y) = f(y).$$

Due to the continuity of g_{xy}, there is an open set $V_{xy} \in \mathcal{V}(y)$ such that

$$z \in V_{xy} \implies f(z) - \varepsilon < g_{xy}(z).$$

Now $\{V_{xy} \mid y \in X\}$ is an open cover of the compact space X, so that there is a finite subcover $\{V_{xy_j} \mid 1 \leq j \leq n\}$. Define

$$g_x := \max_{1 \leq j \leq n} g_{xy_j};$$

$g_x \in \overline{\mathcal{A}_\mathbb{R}}$, because $\overline{\mathcal{A}_\mathbb{R}}$ is closed under taking maximums. Moreover,

$$\forall z \in X : \ f(z) - \varepsilon < g_x(z).$$

Due to the continuity of g_x (and since $g_x(x) = f(x)$), there is an open set $U_x \in \mathcal{V}(x)$ such that

$$z \in U_x \ \Rightarrow \ g_x(z) < f(z) + \varepsilon.$$

Now $\{U_x \mid x \in X\}$ is an open cover of the compact space X, so that there is a finite subcover $\{U_{x_i} \mid 1 \leq i \leq m\}$. Define

$$g := \min_{1 \leq i \leq m} g_{x_i};$$

$g \in \overline{\mathcal{A}_\mathbb{R}}$, because $\overline{\mathcal{A}_\mathbb{R}}$ is closed under taking minimums. Moreover,

$$\forall z \in X : \ g(z) < f(z) + \varepsilon.$$

Thus

$$f(z) - \varepsilon < \min_{1 \leq i \leq m} g_{x_i}(z) = g(z) < f(z) + \varepsilon,$$

that is $|g(z) - f(z)| < \varepsilon$ for every $z \in X$, i.e., $\|g - f\| < \varepsilon$. Hence $\mathcal{A}_\mathbb{R}$ is dense in $C(X, \mathbb{R})$ implying that $\mathcal{A}$ is dense in $C(X)$. □

Remark A.14.5. Notice that under the assumptions of the Stone–Weierstrass Theorem, the compact space is actually a compact Hausdorff space, since continuous functions separate the points.

A.15 Manifolds

We now give an example of Hausdorff spaces which is a starting point of the geometric analysis. We will come back to this topic with more details in Section 5.2.

Definition A.15.1 (Manifold). A topological space (X, τ) is called an *n-dimensional (topological) manifold* if it is second countable, Hausdorff and each of its points has a neighbourhood homeomorphic to an open set of the Euclidean space $\mathbb{R}^n$. If $\phi : U \to U'$ is a homeomorphism, where $U \in \tau$ and $U' \subset \mathbb{R}^n$ is open then the pair (U, ϕ) is called a *chart* on X.

Exercise A.15.2. Show that the sphere $\mathbb{S}^n = \left\{ x \in \mathbb{R}^{n+1} : \ \sum_{j=1}^{n+1} x_j^2 = 1 \right\}$ is an n-dimensional manifold.

Exercise A.15.3. Let X and Y be manifolds of respective dimensions m, n. Show that $X \times Y$ is a manifold of dimension $m + n$.

Definition A.15.4 (Differentiable manifold). Let (X, τ) be an n-dimensional manifold. A collection $\mathcal{A} = \{(U_i, \phi_i) : i \in I\}$ of charts on X is called a C^k-*atlas* if $\{U_i : i \in I\}$ is a cover of X and if the mappings

$$\left(x \mapsto \phi_j(\phi_i^{-1}(x)) \right) : \phi_i(U_i \cap U_j) \to \phi_j(U_i \cap U_j)$$

are C^k-smooth whenever $U_i \cap U_j \neq \emptyset$. If there is a C^k-atlas then X is called a C^k-*manifold* (*differentiable manifold*).

A.16 Connectedness and path-connectedness

In this section we discuss notions of connected and path-connected topological spaces and a relation between them.

Proposition A.16.1. *Let (X, τ) be a topological space. Then the following statements are equivalent*

(i) *There exist non-empty open subsets U, V of X such that $U \cap V = \emptyset$ and $U \cup V = X$.*

(ii) *There exists a non-empty subset U of X such that U is open and closed and such that $U \neq X$.*

(iii) *There exists a continuous surjective mapping from X to the set $\{0, 1\}$ equipped with the discrete topology.*

Proof. Statements (i) and (ii) are equivalent if we take $V = X \backslash U$. Let us show that (i) implies (iii). Define a mapping f by $f(x) = 0$ for $x \in U$ and $f(x) = 1$ for $x \in V$. Since U and V are non-empty, the mapping f is surjective. If W is any subset of $\{0, 1\}$, its preimage $f^{-1}(W)$ is one of the sets $\emptyset, U, V, X$. Since all of them are open, f is continuous.

Finally, to show that (iii) implies (i), we set $U = f^{-1}(0)$ and $V = f^{-1}(1)$. Since f is continuous, both sets are open. Moreover, clearly they are disjoint, $U \cup V = X$, and they are non-empty because f is surjective. $\qquad\square$

Definition A.16.2 (Connected topological space). A topological space (X, τ) is said to be *disconnected* if it satisfies any of the equivalent properties of Proposition A.16.1. Otherwise, it is said to be *connected*.

Proposition A.16.3 ("Connectedness" is a topological property). *Let X and Y be topological spaces and let $f : X \to Y$ be continuous. If X is connected, then $f(X)$ is also connected. Consequently, "connectedness" is a topological property.*

Proof. Suppose that $f(X) = U \cup V$ with U, V as in Proposition A.16.1, (i). Then $X = f^{-1}(U) \cup f^{-1}(V)$ and sets $f^{-1}(U), f^{-1}(V)$ satisfy conditions of Proposition A.16.1, (i), yielding a contradiction. $\qquad\square$

Exercise A.16.4. Prove that a subset A of a topological space X is disconnected (in the relative topology) if and only if there are open sets U, V in X such that $U \cap A \neq \emptyset$, $V \cap A \neq \emptyset$, $A \subset U \cup V$ and $U \cap V \cap A = \emptyset$.

Proposition A.16.5 (Closures are connected). *Let X be a topological space and let $A \subset X$. If A is connected, then its closure $\overline{A}$ is also connected.*

Proof. Let U and V be open sets in X such that $\overline{A} \subset U \cup V$ and $U \cap V \cap \overline{A} = \emptyset$. Since $A \subset \overline{A}$, we have $A \subset U \cup V$ and $U \cap V \cap A = \emptyset$. Since A is connected, by Exercise A.16.4 we must then have either $U \cap A = \emptyset$ or $V \cap A = \emptyset$, which means that either $A \subset X \backslash U$ or $A \subset X \backslash V$. Since the sets $X \backslash U$ and $X \backslash V$ are closed in X, it follows that we have either $\overline{A} \subset X \backslash U$ or $\overline{A} \subset X \backslash V$, which means that either $U \cap \overline{A} = \emptyset$ or $V \cap \overline{A} = \emptyset$. By Exercise A.16.4 again, it means that $\overline{A}$ is connected. $\qquad\square$

Definition A.16.6 (Path-connected topological spaces). A topological space X is said to be *path-connected* if for any two points $a, b \in X$ there is a *path* from a to b, i.e., a continuous mapping $\gamma : [0, 1] \to X$ such that $\gamma(0) = a$ and $\gamma(1) = b$.

Theorem A.16.7 (Path-connected $\Longrightarrow$ connected). *A path-connected topological space is connected.*

Exercise A.16.8. Show that the converse is not true. For example, prove that the set $X = \{(0, t) : -1 \leq t \leq 1\} \cup \{(t, \sin \frac{1}{t}), t > 0\}$ in the relative topology of the Euclidean space $\mathbb{R}^2$ is connected but not path-connected.

We first prove a special case of Theorem A.16.7, namely we show that intervals in $\mathbb{R}$ are connected. We then reduce the general case to this one. By an interval in $\mathbb{R}$ we understand any open or closed or half-open, finite or infinite interval.

Theorem A.16.9 (Interval in $\mathbb{R} \Longrightarrow$ connected). *Every interval I in $\mathbb{R}$ with the Euclidean topology is connected.*

Proof. We will prove it by contradiction. Suppose $I = U \cup V$, where U and V are non-empty, disjoint, and open in the relative topology of I. Let $u \in U$, $v \in V$, and assume $u < v$. Since I is an interval we have $[u, v] \subset I$, and we write

$$A = \{x \in I : u \leq x \text{ and } [u, x] \subset U\}.$$

Since $u \in A$, A is non-empty, and since $v \notin U$, A is bounded above. Thus, we can define $w = \sup A$, and we have $[u, w) \subset U$. Since $w \in [u, v]$, we also have $w \in I = U \cup V$, so that either $w \in U$ or $w \in V$.

We will now show that both choices are impossible. Suppose $w \in U$. Then $w < v$ and since U is open, there is some $\delta > 0$ such that $(w - \delta, w + \delta) \cap A \subset U$. Now, if we take some $z \in (w, w + \delta) \cap A$, we have $[w, z] \subset U$, so that also $[u, z] \in U$, contradicting $w = \sup A$.

Suppose now $w \in V$. Then $u < w$ and since V is open, there is some $\delta > 0$ such that $(w - \delta, w + \delta) \cap A \subset V$. Now, if we take some $z \in (w - \delta, w) \cap A$, we

have $(z, w] \subset V$, so that for all $x \in (z, w]$ we have that $[u, x] \not\subset U$, contradicting
$w = \sup A$ again. $\square$

Proof of Theorem A.16.7. Let X be a path-connected topological space and let f
be a continuous mapping from X to $\{0, 1\}$ equipped with the discrete topology.
By Proposition A.16.1 it is enough to show that f must be constant. Without
loss of generality, suppose that $f(x) = 0$ for some $x \in X$. Let $y \in X$ and let γ
be a path from x to y. Then the composition mapping $f \circ \gamma : [0, 1] \to \{0, 1\}$ is
continuous. Since $[0, 1]$ is connected by Theorem A.16.9 it follows that $f \circ \gamma$ can
not be surjective, so that $f(y) = f(\gamma(1)) = f(\gamma(0)) = f(x) = 0$. Thus, $f(y) = 0$
for all $y \in X$, which means that f can not be surjective. $\square$

Theorem A.16.9 has the converse:

Theorem A.16.10 (Connected in $\mathbb{R} \Longrightarrow$ interval). *If I is a connected subset of $\mathbb{R}$
it must be an interval.*

Proof. First we show that $I \subset \mathbb{R}$ is an interval if and only if for any $a, c \in I$ and
any $b \in \mathbb{R}$ with $a < b < c$ we must have $b \in I$.

If I is an interval the implication is trivial. Conversely, we will prove that
if $a \in I$, then $I \cap [a, \infty)$ is $[a, \infty)$ or $[a, e]$ or $[a, e)$ for some $e \in \mathbb{R}$. If I is not
bounded above, then for any $b > a$ there is $c \in I$ such that $c > b$. Hence $b \in I$ by
assumption. In this case $I \cap [a, \infty) = [a, \infty)$. So we may assume that I is bounded
above and let $e = \sup I$. If $e = a$, then $I \cap [a, \infty) = [a, a]$, so we may assume $e > a$.
Then for any b with $a < b < e$ there is some $c \in I$ such that $a < b < c$, and hence
$b \in I$ by assumption. Therefore, $I \cap [a, \infty)$ is $[a, e]$ or $[a, e)$ depending on whether
$e \in I$ or not. Arguing in a similar way for $I \cap (-\infty, a]$ we get that I must be an
interval.

Now, suppose I is not an interval. By the above claim, there exits some
$a, c \in I$ and $b \notin I$ such that $a < b < c$. But then $U = I \cap (-\infty, b)$ and $V = I \cap (b, \infty)$
is a decomposition of I into a union of non-empty, open disjoint sets with $U \cup V = I$,
contradicting the assumption that I is connected. $\square$

We will now show a converse to Theorem A.16.7, provided that we are dealing
with subsets of $\mathbb{R}^n$.

Theorem A.16.11 (Open connected in $\mathbb{R}^n \Longrightarrow$ path-connected). *Every open connected subset of $\mathbb{R}^n$ with the Euclidean topology is path-connected.*

Proof. First we note that if we have a path γ_1 from a to b and a path γ_2 from
b to c, we can glue them together to obtain a path from a to c, e.g., by setting
$\gamma(t) = \gamma_1(2t)$ for $0 \le t \le \frac{1}{2}$, and $\gamma(t) = \gamma_2(2t - 1)$ for $\frac{1}{2} \le t \le 1$.

Let A be a non-empty open connected set in $\mathbb{R}^n$ (the statement is trivial for
the empty set). Take some $a \in A$ and define

$$U = \{b \in A : \text{ there is a path from } a \text{ to } b \text{ in } A\}.$$

We claim that U is open and closed. Indeed, if $b \in U$, we have $B_\varepsilon(b) \subset A$ for some $\varepsilon > 0$. Consequently, for any $c \in B_\varepsilon(b)$ we have a path from a to b in A by the definition of U, and we obviously also have a path from b to c in $B_\varepsilon(b)$ (e.g., just a straight line). Glueing these paths together, we obtain a path from a to c in A, which means that $B_\varepsilon(b) \subset U$ and hence U is open. To show that U is also closed, take some $b \in A\backslash U$. Then we have $B_\varepsilon(b) \subset A$ for some $\varepsilon > 0$. If now $c \in B_\varepsilon(b)$, there is a path from c to b in $B_\varepsilon(b)$. Consequently, there can be no path from a to c because otherwise there would be a path from a to b in A. Thus, $c \in A\backslash U$, implying that $A\backslash U$ is open.

Finally, writing $A = U \cup (A\backslash U)$ as a union of two disjoint open sets, and observing that U contains a and is, therefore, non-empty, it follows that $A\backslash U = \emptyset$ because A is connected. But this means that $A = U$ and hence A is path-connected. $\qquad\square$

A.17 Co-induction and quotient spaces

Definition A.17.1 (Co-induced topology). Let X and J be sets, (X_j, τ_j) be topological spaces for every $j \in J$, and $\mathcal{F} = \{f_j : X_j \to X \mid j \in J\}$ be a family of mappings. The $\mathcal{F}$-*co-induced topology* of X is the strongest topology τ on X such that the mappings f_j are continuous for every $j \in J$.

Exercise A.17.2. Let τ be the co-induced topology from Definition A.17.1. Show that

$$\tau = \left\{ U \subset X \mid \forall j \in J : \ f_j^{-1}(U) \in \tau_j \right\}.$$

Definition A.17.3 (Quotient topology). Let (X, τ_X) be a topological space, and let $\sim$ be an equivalence relation on X. Let

$$[x] := \{y \in X \mid x \sim y\},$$

$$X/\!\sim := \{[x] \mid x \in X\},$$

and define the *quotient map* $\pi : X \to X/\!\sim$ by $\pi(x) := [x]$. The *quotient topology* of the *quotient space* $X/\!\sim$ is the $\{\pi\}$-co-induced topology on $X/\!\sim$.

Exercise A.17.4. Show that $X/\!\sim$ is compact if X is compact.

Example. Let $\mathcal{A}$ be a topological vector space and $\mathcal{J}$ its subspace. Let us write $[x] := x + \mathcal{J}$ for $x \in \mathcal{A}$. Then the quotient topology of $\mathcal{A}/\mathcal{J} = \{[x] \mid x \in \mathcal{A}\}$ is the topology co-induced by the family $\{(x \mapsto [x]) : \mathcal{A} \to \mathcal{A}/\mathcal{J}\}$.

Remark A.17.5. The message of the following Exercise A.17.6 is that if our compact space X is not Hausdorff, we can factor out the inessential information that the continuous functions $f : X \to \mathbb{C}$ do not see, to obtain a compact Hausdorff space related nicely to X.

Exercise A.17.6. Let X be a topological space, and let us define a relation $R \subset X \times X$ by

$$(x, y) \in R \quad \overset{\text{definition}}{\Longleftrightarrow} \quad \forall f \in C(X): \ f(x) = f(y).$$

Prove:

(a) R is an equivalence relation on X.

(b) There is a natural bijection between the sets $C(X)$ and $C(X/R)$.

(c) X/R is a Hausdorff space.

(d) If X is a compact Hausdorff space then $X \cong X/R$.

Exercise A.17.7. For $A \subset X$, let us define the equivalence relation R_A by

$$(x, y) \in R_A \quad \overset{\text{definition}}{\Longleftrightarrow} \quad x = y \text{ or } \{x, y\} \subset A.$$

Let X be a topological space, and let $\infty \subset X$ be a closed subset. Prove that the mapping

$$X \setminus \infty \to (X/R_\infty) \setminus \{\infty\}, \quad x \mapsto [x],$$

is a homeomorphism.

Finally, let us state a basic property of co-induced topologies:

Proposition A.17.8. *Let X have the $\mathcal{F}$-co-induced topology, and Y be a topological space. A mapping $g : X \to Y$ is continuous if and only if $g \circ f$ is continuous for every $f \in \mathcal{F}$.*

Proof. If g is continuous then the composed mapping $g \circ f$ is continuous for every $f \in \mathcal{F}$.

Conversely, suppose $g \circ f_j$ is continuous for every $f_j \in \mathcal{F}$, $f_j : X_j \to X$. Let $V \subset Y$ be open. Then

$$f_j^{-1}(g^{-1}(V)) = (g \circ f_j)^{-1}(V) \subset X_j \quad \text{is open;}$$

thereby $g^{-1}(V) = f_j(f_j^{-1}(g^{-1}(V))) \subset X$ is open. $\qquad\square$

Corollary A.17.9. *Let X, Y be topological spaces, R be an equivalence relation on X, and endow X/R with the quotient topology. A mapping $f : X/R \to Y$ is continuous if and only if $(x \mapsto f([x])) : X \to Y$ is continuous.* $\qquad\square$

A.18   Induction and product spaces

The main theorem of this section is Tihonov's theorem which is a generalisation of Theorem A.11.14 to infinitely many sets. However, we also discuss other topologies induced by infinite families, and some of their properties.

Definition A.18.1 (Induced topology). Let X and J be sets, (X_j, τ_j) be topological spaces for every $j \in J$ and $\mathcal{F} = \{f_j : X \to X_j \mid j \in J\}$ be a family of mappings. The $\mathcal{F}$-*induced topology* of X is the weakest topology τ on X such that the mappings f_j are continuous for every $j \in J$.

Example. Let (X, τ_X) be a topological space, $A \subset X$, and let $\iota : A \to X$ be defined by $\iota(a) = a$. Then the $\{\iota\}$-induced topology on A is

$$\tau_X|_A := \{U \cap A \mid U \in \tau_X\}.$$

This is called the *relative topology* of A, see Definition A.7.18. Let $f : X \to Y$. The restriction $f|_A = f \circ \iota : A \to Y$ satisfies $f|_A(a) = f(a)$ for every $a \in A \subset X$.

Exercise A.18.2. Prove **Tietze's Extension Theorem**: *Let X be a compact Hausdorff space, $K \subset X$ closed and $f \in C(K)$. Then there exists $F \in C(X)$ such that $F|_K = f$.* (Hint: approximate F by continuous functions that would exist by Urysohn's lemma.)

Example. Let (X, τ) be a topological space. Let σ be the $C(X) = C(X, \tau)$-induced topology, i.e., the weakest topology on X making the all τ-continuous functions continuous. Obviously, $\sigma \subset \tau$, and $C(X, \sigma) = C(X, \tau)$. If (X, τ) is a compact Hausdorff space it is easy to check that $\sigma = \tau$.

Example. Let X, Y be topological spaces with bases $\mathcal{B}_X, \mathcal{B}_Y$, respectively. Recall that the product topology for $X \times Y = \{(x, y) \mid x \in X, \ y \in Y\}$ has a base

$$\{U \times V \mid U \in \mathcal{B}_X, \ V \in \mathcal{B}_Y\}.$$

This topology is actually induced by the family

$$\{p_X : X \times Y \to X, \ p_Y : X \times Y \to Y\},$$

where the *coordinate projections* p_X and p_Y are defined by $p_X((x, y)) = x$ and $p_Y((x, y)) = y$.

Definition A.18.3 (Product topology). Let X_j be a set for every $j \in J$. The *Cartesian product*

$$X = \prod_{j \in J} X_j$$

is the set of the mappings

$$x : J \to \bigcup_{j \in J} X_j \quad \text{such that} \quad \forall j \in J : \ x(j) \in X_j.$$

Due to the Axiom of Choice, X is non-empty if all X_j are non-empty. The mapping

$$p_j : X \to X_j, \quad x \mapsto x_j := x(j),$$

is called the jth *coordinate projection*. Let (X_j, τ_j) be topological spaces. Let $X := \prod_{j \in J} X_j$ be the Cartesian product. Then the $\{p_j \mid j \in J\}$-induced topology on X is called the *product topology* of X.

If $X_j = Y$ for all $j \in J$, it is customary to write

$$\prod_{j \in J} X_j = Y^J = \{f \mid f : J \to Y\}.$$

Let us state a basic property of induced topologies:

Proposition A.18.4. *Let X have the $\mathcal{F}$-induced topology, and Y be a topological space. A mapping $g : Y \to X$ is continuous if and only if $f \circ g$ is continuous for every $f \in \mathcal{F}$.*

Proof. If g is continuous then the composed mapping $f \circ g$ is continuous for every $f \in \mathcal{F}$, by Proposition A.10.10.

Conversely, suppose $f_j \circ g$ is continuous for every $f_j \in \mathcal{F}$, $f : X \to X_j$. Let $y \in Y$, $V \subset X$ be open, $g(y) \in V$. Then there exist $\{f_{j_k}\}_{k=1}^n \subset \mathcal{F}$ and open sets $W_{j_k} \subset X_{j_k}$ such that

$$g(y) \in \bigcap_{k=1}^n f_{j_k}^{-1}(W_{j_k}) \subset V.$$

Let

$$U := \bigcap_{k=1}^n (f_{j_k} \circ g)^{-1}(W_{j_k}).$$

Then $U \subset Y$ is open, $y \in U$, and $g(U) \subset V$; hence $g : Y \to X$ is continuous at an arbitrary point $y \in Y$, i.e., $g \in C(Y, X)$.   □

Remark A.18.5 (**Hausdorff preserved in products**). It is easy to see that a Cartesian product of Hausdorff spaces is always Hausdorff: if $X = \prod_{j \in J} X_j$ and $x, y \in X$, $x \neq y$, then there exists $j \in J$ such that $x_j \neq y_j$. Therefore there are open sets $U_j, V_j \subset X_j$ such that

$$x_j \in U_j, \quad y_j \in V_j, \quad U_j \cap V_j = \emptyset.$$

Let $U := p_j^{-1}(U_j)$ and $V := p^{-1}(V_j)$. Then $U, V \subset X$ are open,

$$x \in U, \quad y \in V, \quad U \cap V = \emptyset.$$

Also compactness is preserved in products; this is stated in Tihonov's Theorem (Tychonoff's Theorem). Before proving this we introduce a tool, which can be compared with Proposition A.11.5:

Definition A.18.6 (Non-Empty Finite InterSection (NEFIS) property). Let X be a set. Let NEFIS(X) be the set of those families $\mathcal{F} \subset \mathcal{P}(X)$ such that every finite subfamily of $\mathcal{F}$ has a non-empty intersection. In other words, a family $\mathcal{F} \subset \mathcal{P}(X)$ belongs to NEFIS(X) if and only if $\bigcap \mathcal{F}' \neq \emptyset$ for every finite subfamily $\mathcal{F}' \subset \mathcal{F}$.

Lemma A.18.7. *A topological space X is compact if and only if $\mathcal{F} \notin \text{NEFIS}(X)$ whenever $\mathcal{F} \subset \mathcal{P}(X)$ is a family of closed sets satisfying $\bigcap \mathcal{F} = \emptyset$.*

Proof. Let X be a **set**, $\mathcal{U} \subset \mathcal{P}(X)$, and $\mathcal{F} := \{X \setminus U \mid U \in \mathcal{U}\}$. Then

$$\bigcap \mathcal{F} = \bigcap_{U \in \mathcal{U}} (X \setminus U) = X \setminus \bigcup \mathcal{U},$$

so that $\mathcal{U}$ is a cover of X if and only if $\bigcap \mathcal{F} = \emptyset$. Now the claim follows the definition of compactness. $\qquad\qquad\square$

Theorem A.18.8 (Tihonov's Theorem (1935)). *Let X_j be a compact space for every $j \in J$. Then $X = \prod\limits_{j \in J} X_j$ is compact.*

Proof. To avoid the trivial case, suppose $X_j \neq \emptyset$ for every $j \in J$. Let $\mathcal{F} \in \text{NEFIS}(X)$ be a family of closed sets. In order to prove the compactness of X we have to show that $\bigcap \mathcal{F} \neq \emptyset$.

Let

$$P := \{\mathcal{G} \in \text{NEFIS}(X) \mid \mathcal{F} \subset \mathcal{G}\}.$$

Let us equip the set P with a partial order relation $\leq$:

$$\mathcal{G} \leq \mathcal{H} \overset{\text{definition}}{\Longleftrightarrow} \mathcal{G} \subset \mathcal{H}.$$

The **Hausdorff Maximal Principle** A.4.9 says that the chain $\{\mathcal{F}\} \subset P$ belongs to a maximal chain $C \subset P$. The reader may verify that $\mathcal{G} := \bigcup C \in P$ is a maximal element of P.

Notice that the maximal element $\mathcal{G}$ may contain non-closed sets. For every $j \in J$ the family

$$\{p_j(G) \mid G \in \mathcal{G}\}$$

belongs to $\text{NEFIS}(X_j)$. Define

$$\mathcal{G}_j := \{\overline{p_j(G)} \mid G \in \mathcal{G}\}.$$

Clearly also $\mathcal{G}_j \in \text{NEFIS}(X_j)$, and the elements of $\mathcal{G}_j$ are closed sets in X_j. Since X_j is compact, we have $\bigcap \mathcal{G}_j \neq \emptyset$. Hence, by the **Axiom of Choice** A.4.2, there is an element $x := (x_j)_{j \in J} \in X$ such that

$$x_j \in \bigcap \mathcal{G}_j.$$

We shall show that $x \in \bigcap \mathcal{F}$, which proves Tihonov's Theorem.

If $V_j \subset X_j$ is a neighbourhood of x_j and $G \in \mathcal{G}$ then

$$p_j(G) \cap V_j \neq \emptyset,$$

because $x_j \in \overline{p_j(G)}$. Thus

$$G \cap p_j^{-1}(V_j) \neq \emptyset$$

for every $G \in \mathcal{G}$, so that $\mathcal{G} \cup \{p_j^{-1}(V_j)\}$ belongs to P; the maximality of $\mathcal{G}$ implies that

$$p_j^{-1}(V_j) \in \mathcal{G}.$$

Let $V \in \tau_X$ be a neighbourhood of x. Due to the definition of the product topology,

$$x \in \bigcap_{k=1}^{n} p_{j_k}^{-1}(V_{j_k}) \subset V$$

for some finite index set $\{j_k\}_{k=1}^{n} \subset J$, where $V_{j_k} \subset X_{j_k}$ is a neighbourhood of x_{j_k}. Due to the maximality of $\mathcal{G}$, any finite intersection of members of $\mathcal{G}$ belongs to $\mathcal{G}$, so that

$$\bigcap_{k=1}^{n} p_{j_k}^{-1}(V_{j_k}) \in \mathcal{G}.$$

Therefore for every $G \in \mathcal{G}$ and $V \in \mathcal{V}_{\tau_X}(x)$ we have

$$G \cap V \neq \emptyset.$$

Hence $x \in \overline{G}$ for every $G \in \mathcal{G}$, yielding

$$x \in \bigcap_{G \in \mathcal{G}} \overline{G} \overset{\mathcal{F} \subset \mathcal{G}}{\subset} \bigcap_{F \in \mathcal{F}} \overline{F} = \bigcap_{F \in \mathcal{F}} F = \bigcap \mathcal{F},$$

so that $\bigcap \mathcal{F} \neq \emptyset$. $\square$

Remark A.18.9. Actually, Tihonov's Theorem A.18.8 is equivalent to the Axiom of Choice A.4.2; we shall not prove this.

A.19 Metrisable topologies

It is often very useful to know whether a topology on a space comes from some metric. Here we try to construct metrics on compact spaces. We shall learn that a compact space X is metrisable if and only if the corresponding normed algebra $C(X)$ is separable. Metrisability is equivalent to the existence of a countable family of continuous functions separating the points of the space. As a vague analogy to manifolds, the reader may view such a countable family as a set of coordinate functions on the space.

Definition A.19.1 (Metrisable topology). A topological space (X, τ) is called *metrisable* if there exists a metric d on X such that the topology τ is the canonical metric topology of (X, d), i.e., if there exists a metric d on X such that $\tau = \tau_d$.

Example (**Discrete topology**). The discrete topology on the set X is the collection τ of all subsets of X. This is a metric topology corresponding to the discrete metric.

Exercise A.19.2. Let X, Y be metrisable. Prove that $X \times Y$ is metrisable, and that

$$(x_n, y_n) \overset{X \times Y}{\to} (x, y) \quad \Leftrightarrow \quad x_n \overset{X}{\to} x \text{ and } y_n \overset{Y}{\to} y.$$

Remark A.19.3. There are plenty of non-metrisable topological spaces, the easiest example being X with more than one point and with $\tau = \{\emptyset, X\}$. If X is an infinite-dimensional Banach space then the weak*-topology[1] of $X' := \mathcal{L}(X, \mathbb{C})$ is not metrisable. The distribution spaces $\mathcal{D}'(\mathbb{R}^n)$, $\mathcal{S}'(\mathbb{R}^n)$ and $\mathcal{E}'(\mathbb{R}^n)$ are non-metrisable topological spaces. We shall later prove that for the compact Hausdorff spaces metrisability is equivalent to the existence of a countable base.

Exercise A.19.4. Show that (X, τ) is a topological space, where

$$\tau = \{U \subset X \mid U = \emptyset, \text{ or } X \setminus U \text{ is finite}\}.$$

When is this topology metrisable?

Theorem A.19.5. *Let (X, τ) be compact. Assume that there exists a countable family $\mathcal{F} \subset C(X)$ separating the points of X. Then (X, τ) is metrisable.*

Proof. Let $\mathcal{F} = \{f_n\}_{n=0}^{\infty} \subset C(X)$ separate the points of X. We can assume that $|f_n| \leq 1$ for every $n \in \mathbb{N}$; otherwise consider for instance functions $x \mapsto f_n(x)/(1 + |f_n(x)|)$. Let us define

$$d(x, y) := \sup_{n \in \mathbb{N}} 2^{-n} |f_n(x) - f_n(y)|$$

for every $x, y \in X$. Next we prove that $d : X \times X \to [0, \infty)$ is a metric: $d(x, y) = 0 \Leftrightarrow x = y$, because $\{f_n\}_{n=0}^{\infty}$ is a separating family. Clearly also $d(x, y) = d(y, x)$ for every $x, y \in X$. Let $x, y, z \in X$. We have the triangle inequality:

$$
\begin{aligned}
d(x, z) &= \sup_{n \in \mathbb{N}} 2^{-n} |f_n(x) - f_n(z)| \\
&\leq \sup_{n \in \mathbb{N}} (2^{-n} |f_n(x) - f_n(y)| + 2^{-n} |f_n(y) - f_n(z)|) \\
&\leq \sup_{m \in \mathbb{N}} 2^{-m} |f_m(x) - f_m(y)| + \sup_{n \in \mathbb{N}} 2^{-n} |f_n(y) - f_n(z)| \\
&= d(x, y) + d(y, z).
\end{aligned}
$$

Hence d is a metric on X.

Finally, let us prove that the metric topology coincides with the original topology, $\tau_d = \tau$. Let $x \in X$, $\varepsilon > 0$. Take $N \in \mathbb{N}$ such that $2^{-N} < \varepsilon$. Define

$$U_n := f_n^{-1}(B_\varepsilon(f_n(x))) \in \mathcal{V}_\tau(x), \quad U := \bigcap_{n=0}^{N} U_n \in \mathcal{V}_\tau(x).$$

---

[1] see Definition B.4.35

If $y \in U$ then
$$d(x, y) = \sup_{n \in \mathbb{N}} 2^{-n} |f_n(x) - f_n(y)| < \varepsilon.$$

Thus $x \in U \subset B_\varepsilon(x) = \{y \in X \mid d(x, y) < \varepsilon\}$. This proves that the original topology τ is finer than the metric topology τ_d, i.e., $\tau_d \subset \tau$. Combined with the facts that (X, τ) is compact and (X, τ_d) is Hausdorff, this implies that we must have $\tau_d = \tau$, by Corollary A.12.8. $\qquad\square$

Corollary A.19.6. *Let X be a compact Hausdorff space. Then X is metrisable if and only if it has a countable basis.*

Proof. Suppose X is a compact space, metrisable with a metric d. Let $r > 0$. Then $\mathcal{B}_r = \{B_d(x, r) \mid x \in X\}$ is an open cover of X, thus having a finite subcover $\mathcal{B}_r' \subset \mathcal{B}_r$. Then $\mathcal{B} := \bigcup_{n=1}^{\infty} \mathcal{B}_{1/n}'$ is a countable basis for X.

Conversely, suppose X is a compact Hausdorff space with a countable basis $\mathcal{B}$. Then the family
$$\mathcal{C} := \{(B_1, B_2) \in \mathcal{B} \times \mathcal{B} \mid \overline{B_1} \subset B_2\}$$

is countable. For each $(B_1, B_2) \in \mathcal{C}$, Urysohn's Lemma (Theorem A.12.11) provides a function $f_{B_1 B_2} \in C(X)$ satisfying
$$f_{B_1 B_2}(\overline{B_1}) = \{0\} \quad \text{and} \quad f_{B_1 B_2}(X \setminus B_2) = \{1\}.$$

Next we show that the countable family
$$\mathcal{F} = \{f_{B_1 B_2} : (B_1, B_2) \in \mathcal{C}\} \subset C(X)$$

separates the points of X. Indeed, Take $x, y \in X$, $x \neq y$. Then $W := X \setminus \{y\} \in \mathcal{V}(x)$. Since X is a compact Hausdorff space, by Corollary A.12.6 there exists $U \in \mathcal{V}(x)$ such that $\overline{U} \subset W$. Take $B', B \in \mathcal{B}$ such that $x \in B' \subset \overline{B'} \subset B \subset U$. Then $f_{B'B}(x) = 0 \neq 1 = f_{B'B}(y)$. Thus X is metrisable. $\qquad\square$

Corollary A.19.7. *Let X be a compact Hausdorff space. Then X is metrisable if and only if $C(X)$ is separable.*

Proof. Suppose X is a metrisable compact space. Let $\mathcal{F} \subset C(X)$ be a countable family separating the points of X (as in the proof of the previous corollary). Let $\mathcal{G}$ be the set of finite products of functions f for which $f \in \mathcal{F} \cup \mathcal{F}^* \cup \{\mathbf{1}\}$; the set $\mathcal{G} = \{g_j\}_{j=0}^{\infty}$ is countable. The linear span $\mathcal{A}$ of $\mathcal{G}$ is the involutive algebra generated by $\mathcal{F}$ (the smallest $*$-algebra containing $\mathcal{F}$, see Definition D.5.1); due to the Stone–Weierstrass Theorem (see Theorem A.14.4), $\mathcal{A}$ is dense in $C(X)$. If $S \subset \mathbb{C}$ is a countable dense set then
$$\{\lambda_0 \mathbf{1} + \sum_{j=1}^{n} \lambda_j g_j \mid n \in \mathbb{Z}^+, \ (\lambda_j)_{j=0}^{n} \subset S\}$$

is a countable dense subset of $\mathcal{A}$, thereby dense in $C(X)$.

Conversely, assume that $\mathcal{F} = \{f_n\}_{n=0}^{\infty} \subset C(X)$ is a dense subset. Take $x, y \in X$, $x \neq y$. By Urysohn's Lemma (Theorem A.12.11) there exists $f \in C(X)$ such that $f(x) = 0 \neq 1 = f(y)$. Take $f_n \in \mathcal{F}$ such that $\|f - f_n\| < 1/2$. Then

$$|f_n(x)| < 1/2 \quad \text{and} \quad |f_n(y)| > 1/2,$$

so that $f_n(x) \neq f_n(y)$; $\mathcal{F}$ separates the points of X.   □

Exercise A.19.8. Prove that a topological space with a countable basis is separable. Prove that a metric space has a countable basis if and only if it is separable.

Exercise A.19.9. There are non-metrisable separable compact Hausdorff spaces! Prove that X,

$$X = \{f : [0, 1] \rightarrow [0, 1] \mid x \leq y \Rightarrow f(x) \leq f(y)\},$$

endowed with a relative topology, is such a space. Hint: Tihonov's Theorem.

A.20   Topology via generalised sequences

Definition A.20.1 (Directed set). A non-empty set J is *directed* if there exists a relation "$\leq$" such that $\leq \subset J \times J$ (where $(x, y) \in \leq$ is usually denoted by $x \leq y$) such that for every $x, y, z \in J$ it holds that

1. $x \leq x$,

2. if $x \leq y$ and $y \leq z$ then $x \leq z$,

3. there exists $w \in J$ such that $x \leq w$ and $y \leq w$.

Definition A.20.2 (Nets and convergence). A *net* (or a *generalised sequence*) in a topological space (X, τ) is a mapping $(j \mapsto x_j) : J \rightarrow X$, denoted also by $(x_j)_{j \in J}$, where J is a directed set. If $K \subset J$ is a directed set (with respect to the natural inherited relation $\leq$) then the net $(x_j)_{j \in K}$ is called a *subnet* of the net $(x_j)_{j \in J}$. A net $(x_j)_{j \in J}$ *converges to a point* $p \in X$, denoted by

$$x_j \rightarrow p \quad \text{or} \quad x_j \xrightarrow[j \in J]{} p \quad \text{or} \quad \lim_{j \in J} x_j = \lim x_j = p,$$

if for every neighbourhood U of p there exists $j_U \in J$ such that $x_j \in U$ whenever $j_U \leq j$.

Example. A *sequence* $(x_j)_{j \in \mathbb{Z}^+}$ is a net, where $\mathbb{Z}^+$ is directed by the usual partial order; sequences characterise topology in spaces of countable local bases, for instance metric spaces. But there are more complicated topologies, where sequences are not enough; for instance, the weak*-topology for the dual of an infinite-dimensional topological vector space.

Exercise A.20.3 (Nets and closure). Let X be a topological space. Show that $p \in X$ belongs to the closure of $S \subset X$ if and only if there exists a net $(x_j)_{j \in J} : J \rightarrow S$ such that $x_j \rightarrow p$.

Exercise A.20.4 (Nets and continuity). Show that a function $f : X \to Y$ is continuous at $p \in X$ if and only if $f(x_j) \to f(p)$ whenever $x_j \to p$ for nets $(x_j)_{j \in J}$ in X.

Exercise A.20.5 (Nets and compactness). Show that a topological space X is compact if and only if its every net has a converging subnet.

Exercise A.20.6. In the spirit of Exercises A.20.3, A.20.4 and A.20.5, express other topological concepts via nets.

Chapter B

Elementary Functional Analysis

We assume that the reader already has knowledge of (complex) matrices, determinants, etc. In this chapter, we shall present basic machinery for dealing with vector spaces, especially Banach and Hilbert spaces. We do not go into depth in this direction as there are plenty of excellent specialised monographs available devoted to various aspects of the subject, see, e.g., [11, 35, 53, 59, 63, 70, 87, 89, 90, 116, 134, 146, 153]. However, we still make an independent presentation of a collection of results which are indispensable for anyone working in analysis, and which are useful for other parts of this book.

B.1 Vector spaces

Definition B.1.1 (Vector space). Let $\mathbb{K} \in \{\mathbb{R}, \mathbb{C}\}$. A $\mathbb{K}$-*vector space* (or a *vector space over the field* $\mathbb{K}$, or a *vector space* if $\mathbb{K}$ is implicitly known) is a set V endowed with mappings

$$((x, y) \mapsto x + y) : V \times V \to V,$$
$$((\lambda, x) \mapsto \lambda x) : \mathbb{K} \times V \to V$$

such that there exists an *origin* $0 \in V$ and such that the following properties hold:

$$(x + y) + z = x + (y + z),$$
$$x + 0 = x,$$
$$x + (-1)x = 0,$$
$$x + y = y + x,$$
$$1x = x,$$
$$\lambda(\mu x) = (\lambda \mu)x,$$
$$\lambda(x + y) = \lambda x + \lambda y,$$
$$(\lambda + \mu)x = \lambda x + \mu x$$

for all $x, y, z \in V$ and $\lambda, \mu \in \mathbb{K}$. We may write $x + y + z := (x + y) + z$ and $-x := (-1)x$. Elements of a vector space are called *vectors*.

Definition B.1.2 (Convex and balanced sets). A subset C of a vector space is *convex* if $tx + (1 - t)y \in C$ for every $x, y \in C$ whenever $0 < t < 1$. A subset B of a vector space is *balanced* if $\lambda x \in B$ for every $x \in B$ whenever $|\lambda| \le 1$.

Example. $\mathbb{K} \in \{\mathbb{R}, \mathbb{C}\}$ is itself a vector space over $\mathbb{K}$, likewise $\mathbb{K}^n$ with operations $(x_k)_{k=1}^n + (y_k)_{k=1}^n := (x_k + y_k)_{k=1}^n$ and $\lambda(x_k)_{k=1}^n := (\lambda x_k)_{k=1}^n$.

Example. Let V be a $\mathbb{K}$-vector space and $X \ne \emptyset$. The set V^X of mappings $f : X \to V$ is a $\mathbb{K}$-vector space with pointwise operations $(f + g)(x) := f(x) + g(x)$ and $(\lambda f)(x) := \lambda\, f(x)$. The vector space $\mathbb{K}^n$ can be naturally identified with $\mathbb{K}^X$, where $X = \{k \in \mathbb{Z}^+ : k \le n\}$.

Example. Let V be a vector space such that its vector operations restricted to $W \subset V$ endow this subset with the vector space structure. Then W is called a *vector subspace*. A vector space V has always *trivial subspaces* $\{0\}$ and V. The vector space V^X has, e.g., the subspace $\{f : X \to V \mid \forall x \in K : f(x) = 0\}$, where $K \subset X$ is a fixed subset.

Definition B.1.3 (Algebraic basis). Let V be a vector space and $S \subset V$. Let us write

$$\sum_{x \in S} \lambda(x)x = \sum_{x \in S:\ \lambda(x) \ne 0} \lambda(x)x,$$

when $\lambda : S \to \mathbb{K}$ is finitely supported, i.e., $\{x \in S : \lambda(x) \ne 0\}$ is finite. The *span* of a subset S of a vector space V is

$$\mathrm{span}(S) := \left\{ \sum_{x \in S} \lambda(x)x \,\middle|\, \lambda : S \to \mathbb{K} \text{ finitely supported} \right\}.$$

Thus $\mathrm{span}(S)$ is the smallest subspace containing $S \subset V$. A subset S of a $\mathbb{K}$-vector space is said to be *linearly independent* if

$$\sum_{x \in S} \lambda(x)x = 0 \quad \Rightarrow \quad \lambda \equiv 0.$$

A subset is *linearly dependent* if it is not linearly independent. A subset $S \subset V$ is called an *algebraic basis* (or a *Hamel basis*) of V if S is linearly independent and $V = \mathrm{span}(S)$.

Remark B.1.4. Let $\mathcal{B}$ be an algebraic basis for V. Then there exists a unique set of functions $(x \mapsto \langle x, b \rangle_\mathcal{B}) : V \to \mathbb{K}$ such that

$$x = \sum_{b \in \mathcal{B}} \langle x, b \rangle_\mathcal{B}\, b$$

for every $x \in V$. Notice that $\langle x, b \rangle_\mathcal{B} \ne 0$ for at most finitely many $b \in \mathcal{B}$. Consider this, e.g., with respect to the vector space $\mathbb{K}^X$ in the example before.

Example. The canonical algebraic basis for $\mathbb{K}^n$ is $\{e_k\}_{k=1}^n$, where $e_k = (\delta_{jk})_{j=1}^n$ and $\delta_{kk} = 1$ and $\delta_{jk} = 0$ otherwise.

Lemma B.1.5. *Every vector space $V \neq \{0\}$ has an algebraic basis. Moreover, any two algebraic bases have the same cardinality*[1].

Proof. Let $\mathcal{F}$ be the family of all linearly independent subsets of V. Now $\mathcal{F} \neq \emptyset$, because $\{x\} \in \mathcal{F}$ for every $x \in V \setminus \{0\}$. Endow $\mathcal{F}$ with a partial order by inclusion. Let $\mathcal{C} \subset \mathcal{F}$ be a chain and let $F := \bigcup \mathcal{C}$. It is easy to check that $F \in \mathcal{F}$ is an upper bound for $\mathcal{C}$. Thereby there is a maximal element $M \in \mathcal{F}$. Obviously, M is an algebraic basis for V.

Let $\mathcal{A}, \mathcal{B}$ be algebraic bases for V. The reader may prove (by induction) that $\mathrm{card}(\mathcal{A}) = \mathrm{card}(\mathcal{B})$ when $\mathcal{A}$ is finite. So suppose $\mathrm{card}(\mathcal{A}) \leq \mathrm{card}(\mathcal{B})$, where $\mathcal{A}$ is infinite. Now $\mathrm{card}(\mathcal{A}) = \mathrm{card}(S)$, where

$$S := \{(a, b) \in \mathcal{A} \times \mathcal{B} : \langle a, b \rangle_\mathcal{B} \neq \emptyset\}.$$

Assume $\mathrm{card}(\mathcal{A}) < \mathrm{card}(\mathcal{B})$. Thus

$$\exists b_0 \in \mathcal{B} \; \forall a \in \mathcal{A} : \; \langle a, b_0 \rangle_\mathcal{B} = 0.$$

But then

$$
\begin{aligned}
b_0 &= \sum_{a \in \mathcal{A}} \langle b_0, a \rangle_\mathcal{A} \, a \\
&= \sum_{a \in \mathcal{A}} \langle b_0, a \rangle_\mathcal{A} \sum_{b \in \mathcal{B}} \langle a, b \rangle_\mathcal{B} \, b \\
&= \sum_{b \in \mathcal{B}} \left(\sum_{a \in \mathcal{A}} \langle b_0, a \rangle_\mathcal{A} \, \langle a, b \rangle_\mathcal{B} \right) b \\
&= \sum_{b \in \mathcal{B} \setminus \{b_0\}} \left(\sum_{a \in \mathcal{A}} \langle b_0, a \rangle_\mathcal{A} \, \langle a, b \rangle_\mathcal{B} \right) b \\
&\in \; \mathrm{span}(\mathcal{B} \setminus \{b_0\}),
\end{aligned}
$$

contradicting the linear independence of $\mathcal{B}$. Thus $\mathrm{card}(A) = \mathrm{card}(B)$. $\square$

Definition B.1.6 (Algebraic dimension). By Lemma B.1.5, we may define the *algebraic dimension* $\dim(V)$ of a vector space V to be the cardinality of any of its algebraic bases. The vector space V is said to be *finite-dimensional* if $\dim(V)$ is finite, and *infinite-dimensional* otherwise.

---

[1] Here we will use the notation $\mathrm{card}(\mathcal{A})$ for the cardinality of $\mathcal{A}$ to avoid any confusion with the notation for norms; see Definition A.4.13.

Definition B.1.7 (Linear operators and functionals). Let V, W be $\mathbb{K}$-vector spaces. A mapping $A : V \to W$ is called a *linear operator* (or a linear mapping), denoted $A \in L(V, W)$, if

$$\begin{cases} A(u + v) = A(u) + A(v), \\ A(\lambda v) = \lambda\, A(v) \end{cases}$$

for every $u, v \in V$ and $\lambda \in \mathbb{K}$. Then it is customary to write $Av := A(v)$, and $L(V) := L(V, V)$. A linear mapping $f : V \to \mathbb{K}$ is called a *linear functional*.

Definition B.1.8 (Kernel and image). The *kernel* $\mathrm{Ker}(A) \subset V$ of a linear operator $A : V \to W$ is defined by

$$\mathrm{Ker}(A) := \{u \in V : Au = 0\},$$

where 0 is the origin of the vector space W. The image $\mathrm{Im}(A) \subset W$ of A is defined by

$$\mathrm{Im}(A) := \{Au : u \in V\}.$$

Exercise B.1.9. Show that $\mathrm{Ker}(A)$ is a vector subspace of V and that $\mathrm{Im}(A)$ is a vector subspace of W.

Exercise B.1.10. Let $C \subset V$ be convex and $A \in L(V, W)$. Show that $A(C) \subset W$ is convex.

Definition B.1.11 (Spectrum of an operator). Let V be a $\mathbb{K}$-vector space. Let $I \in L(V)$ denote the identity operator $(x \mapsto x) : V \to V$. The *spectrum* of $A \in L(V)$ is

$$\sigma(A) := \{\lambda \in \mathbb{K} : \lambda I - A \text{ is not bijective}\}.$$

Exercise B.1.12. Appealing to the Fundamental Theorem of Algebra, show that $\sigma(A) \neq \emptyset$ for $A \in L(\mathbb{C}^n)$.

Exercise B.1.13. Give an example, where $\sigma(A) = \emptyset \neq \sigma(A^2)$.

Exercise B.1.14. Show that $\sigma(A) = \{0\}$ if A is nilpotent, i.e., if $A^k = 0$ for some $k \in \mathbb{Z}^+$.

Exercise B.1.15. Show that $\sigma(AB) \cup \{0\} = \sigma(BA) \cup \{0\}$ in general, and that $\sigma(AB) = \sigma(BA)$ if A is bijective.

Definition B.1.16 (Quotient vector space). Let M be a subspace of a $\mathbb{K}$-vector space V. Let us endow the quotient set $V/M := \{x + M \mid x \in V\}$ with the operations

$$\begin{aligned} ((x + M, y + M) \mapsto x + y + M) &: V/M \times V/M \to V/M, \\ (\lambda, (y + M)) \mapsto \lambda y + M) &: \mathbb{K} \times V/M \to V/M. \end{aligned}$$

Then it is easy to show that with these operations, this so-called *quotient vector space* is indeed a vector space.

Remark B.1.17. In the case of a topological vector space V (see Definition B.2.1), the quotient V/M is endowed with the quotient topology, and then V/M is a topological vector space if and only if the original subspace $M \subset V$ was closed.

B.1.1 Tensor products

The basic idea in multilinear algebra is to linearise multilinear operators. The functional analytic foundation is provided by tensor products that we concisely review here. We also introduce locally convex spaces and Fréchet spaces as well as Montel and nuclear spaces.

Definition B.1.18 (Bilinear mappings). Let X_j $(1 \leq j \leq r)$ and V be $\mathbb{K}$-vector spaces (that is, vector spaces over the field $\mathbb{K}$). A mapping $A : X_1 \times X_2 \to V$ is *2-linear* (or *bilinear*) if $x \mapsto A(x, x_2)$ and $x \mapsto A(x_1, x)$ are linear mappings for each $x_j \in X_j$. The reader may guess what conditions an *r-linear* mapping

$$X_1 \times \cdots \times X_r \to V$$

satisfies.

Definition B.1.19 (Tensor product of spaces). The *algebraic tensor product* of $\mathbb{K}$-vector spaces $X_1, \ldots, X_r$ is a $\mathbb{K}$-vector space V endowed with an r-linear mapping i such that for every $\mathbb{K}$-vector space W and for every r-linear mapping

$$A : X_1 \times \cdots \times X_r \to W$$

there exists a (unique) linear mapping $\tilde{A} : V \to W$ satisfying $\tilde{A}i = A$. (The reader is encouraged to draw a commutative diagram involving the vector spaces and mappings $i, A, \tilde{A}$!) Any two tensor products for $X_1, \ldots, X_r$ can easily be seen isomorphic, so that we may denote *the* tensor product of these vector spaces by

$$X_1 \otimes \cdots \otimes X_r.$$

In fact, such a tensor product always exists. Indeed, let X, Y be $\mathbb{K}$-vector spaces. We may formally define the set $B := \{x \otimes y \mid x \in X, \ y \in Y\}$, where $x \otimes y = a \otimes b$ if and only if $x = a$ and $y = b$. Let Z be the $\mathbb{K}$-vector space with basis B, i.e.,

$$
\begin{aligned}
Z \ &= \ \operatorname{span}\{x \otimes y \mid x \in X, \ y \in Y\} \\
&= \ \left\{\sum_{j=0}^{n} \lambda_j (x_j \otimes y_j) : \ n \in \mathbb{N}, \ \lambda_j \in \mathbb{K}, \ x_j \in X, \ y_j \in Y\right\}.
\end{aligned}
$$

Let

$$
\begin{aligned}
[0 \otimes 0] \ := \ \operatorname{span}\{ &\alpha_1\beta_1(x_1 \otimes y_1) + \alpha_1\beta_2(x_1 \otimes y_2) \\
&+ \alpha_2\beta_1(x_2 \otimes y_1) + \alpha_2\beta_2(x_2 \otimes y_2) \\
&- (\alpha_1 x_1 + \alpha_2 x_2) \otimes (\beta_1 y_1 + \beta_2 y_2) : \\
&\alpha_j, \beta_j \in \mathbb{K}, \ x_j \in X, \ y_j \in Y\}.
\end{aligned}
$$

For $z \in Z$, let $[z] := z + [0 \otimes 0]$. The *tensor product* of X, Y is the $\mathbb{K}$-vector space

$$X \otimes Y := Z/[0 \otimes 0] = \{[z] \mid z \in Z\},$$

where $([z_1], [z_2]) \mapsto [z_1 + z_2]$ and $(\lambda, [z]) \mapsto [\lambda z]$ are well defined mappings $(X \otimes Y) \times (X \otimes Y) \to X \otimes Y$ and $\mathbb{K} \times (X \otimes Y) \to X \otimes Y$, respectively.

Definition B.1.20 (Tensor product of operators). Let X, Y, V, W be $\mathbb{K}$-vector spaces, and let $A : X \to V$ and $B : Y \to W$ be linear operators. The *tensor product* of A, B is the linear operator $A \otimes B : X \otimes Y \to V \otimes W$, which is the unique linear extension of the mapping $x \otimes y \mapsto Ax \otimes By$, where $x \in X$ and $y \in Y$.

Example. Let X and Y be finite-dimensional $\mathbb{K}$-vector spaces with bases

$$\{x_i\}_{i=1}^{\dim(X)} \quad \text{and} \quad \{y_j\}_{j=1}^{\dim(Y)},$$

respectively. Then $X \otimes Y$ has a basis

$$\{x_i \otimes y_j \mid 1 \le i \le \dim(X),\ 1 \le j \le \dim(Y)\}.$$

Let S be a finite set. Let $\mathcal{F}(S)$ be the $\mathbb{K}$-vector space of functions $S \to \mathbb{K}$; it has a basis $\{\delta_x \mid x \in S\}$, where $\delta_x(y) = 1$ if $x = y$, and $\delta_x(y) = 0$ otherwise. Now it is easy to see that for finite sets S_1, S_2 the vector spaces $\mathcal{F}(S_1) \otimes \mathcal{F}(S_2)$ and $\mathcal{F}(S_1 \times S_2)$ are isomorphic; for $f_j \in \mathcal{F}(S_j)$, we may regard $f_1 \otimes f_2 \in \mathcal{F}(S_1) \otimes \mathcal{F}(S_2)$ as a function $f_1 \otimes f_2 \in \mathcal{F}(S_1 \times S_2)$ by

$$(f_1 \otimes f_2)(x_1, x_2) := f_1(x_1)\, f_2(x_2).$$

Definition B.1.21 (Inner product on $V \otimes W$). Suppose V, W are finite-dimensional inner product spaces over $\mathbb{K}$. The natural inner product for $V \otimes W$ is obtained by extending

$$\langle v_1 \otimes w_1, v_2 \otimes w_2 \rangle_{V \otimes W} := \langle v_1, v_2 \rangle_V\, \langle w_1, w_2 \rangle_W.$$

Definition B.1.22 (Duals of tensor product spaces). The dual $(V \otimes W)'$ of a tensor product space $V \otimes W$ is naturally identified with $V' \otimes W'$.

Alternative approach to tensor products. Now we briefly describe another approach to tensor products.

Definition B.1.23 (Algebraic tensor product). Let $\mathbb{K} \in \{\mathbb{R}, \mathbb{C}\}$. Let X, Y be $\mathbb{K}$-vector spaces, and X', Y' their respective algebraic duals, i.e., the spaces of linear functionals $X \to \mathbb{K}$ and $Y \to \mathbb{K}$. For $x \in X$ and $y \in Y$, define the bilinear functional $x \otimes y : X' \times Y' \to \mathbb{K}$ by

$$(x \otimes y)(x', y') := x'(x)\, y'(y).$$

Let $B(X', Y')$ denote the space of all bilinear functionals $X' \times Y' \to \mathbb{K}$. The *algebraic tensor product* (or simply the *tensor product*) $X \otimes Y$ is the vector subspace of $B(X', Y')$ which is spanned by the set $\{x \otimes y :\ x \in X,\ y \in Y\}$.

Exercise B.1.24. Show that $a : (X \otimes Y)' \to B(X, Y)$ is a linear bijection, where $a(f)(x \otimes y) := f(x, y)$ for $f \in (X \otimes Y)'$, $x \in X$ and $y \in Y$.

Exercise B.1.25. Let X, Y, Z be $\mathbb{K}$-vector spaces. Let $B(X, Y; Z)$ denote the vector space of bilinear mappings $X \times Y \to Z$. Find a linear bijection $B(X, Y; Z) \to L(X \otimes Y, Z)$, where $L(V, Z)$ is the vector space of linear mappings $V \to Z$.

B.2 Topological vector spaces

Vector spaces can be combined with topology. For reader's convenience, if one has not encountered Banach and Hilbert spaces yet, we suggest skipping the sections on topological vector spaces and locally convex spaces at this point, and returning here later.

Definition B.2.1 (Topological vector space). A *topological vector space* V over a field $\mathbb{K} \in \{\mathbb{R}, \mathbb{C}\}$ is both a topological space and a vector space over $\mathbb{K}$ such that $\{0\} \subset V$ is closed and such that the mappings

$$((\lambda, x) \mapsto \lambda x) \quad : \quad \mathbb{K} \times V \to V,$$
$$((x, y) \mapsto x + y) \quad : \quad V \times V \to V$$

are continuous. The *dual space* V' of a topological vector space V consists of continuous linear functionals $f : V \to \mathbb{K}$.

Exercise B.2.2. Show that a topological vector space is a Hausdorff space.

Exercise B.2.3. Show that in a topological vector space every neighbourhood of 0 contains a balanced neighbourhood of 0.

Exercise B.2.4. Prove that a topological vector space V is metrisable if and only if it has a countable family $\{U_j\}_{j=1}^{\infty}$ of neighbourhoods of $0 \in V$ such that $\bigcap_{j=1}^{\infty} U_j = \{0\}$. Moreover, show that in this case a compatible metric $d : V \times V \to [0, \infty)$ can be chosen translation-invariant in the sense that $d(x + z, y + z) = d(x, y)$ for every $x, y, z \in V$.

Definition B.2.5 (Equicontinuity in vector space). Let X be a topological space and V a topological vector space. A family $\mathcal{F}$ of mappings $f : X \to V$ is called *equicontinuous at* $p \in X$ if for every neighbourhood $W \subset V$ of $f(p)$ there exists a neighbourhood $U \subset X$ of p such that $f(x) \in W$ whenever $f \in \mathcal{F}$ and $x \in U$.

Remark B.2.6 (**NEFIS property and compactness**). Recall the Non-Empty Finite Intersection property (NEFIS) from Definition A.18.6: that is, we denote by NEFIS(X) the set of those families $\mathcal{F} \subset \mathcal{P}(X)$ such that every finite subfamily of $\mathcal{F}$ has a non-empty intersection. Recall also that a topological space X is compact if and only if $\bigcap \mathcal{F} \neq \emptyset$ whenever $\mathcal{F} \in$ NEFIS(X) consists of closed sets.

Definition B.2.7 (Small sets property). Let X be a topological vector space. A family $\mathcal{F} \subset \mathcal{P}(X)$ is said to *contain small sets* if for every neighbourhood U of $0 \in X$ there exists $x \in X$ and $S \in \mathcal{F}$ such that $S \subset x + U$.

Definition B.2.8 (Completeness of topological vector spaces). A subset S of a topological vector space X is called *complete* if $\bigcap \mathcal{F} \neq \emptyset$ whenever $\mathcal{F} \in \mathrm{NEFIS}(X)$ consists of closed subsets of S and contains small sets.

Exercise B.2.9 (Completeness and Cauchy nets). A net $(x_j)_{j \in J}$ in a topological vector space X is called *a Cauchy net* if for every neighbourhood V of $0 \in X$ there exists $k = k_V \in J$ such that $x_i - x_j \in V$ whenever $k \leq i, j$. Prove that $S \subset X$ is complete if and only if each Cauchy net in S converges to a point in S.

Exercise B.2.10. Show that a complete subset of a topological vector space is closed, and that a closed subset of a complete topological vector space is complete.

Exercise B.2.11 (Completeness and Cartesian product). Let X_j be a topological vector space for each $j \in J$. Show that the product space $X = \prod_{j \in J} X_j$ is complete if and only if X_j is complete for every $j \in J$.

Definition B.2.12 (Total boundedness in topological vector spaces). A subset S of a topological vector space X is *totally bounded* if for every neighbourhood U of $0 \in X$ there exists a finite set $F \subset X$ such that $S \subset F + U$.

Exercise B.2.13 (Hausdorff Total Boundedness Theorem). Prove the following *Hausdorff Total Boundedness Theorem: A subset of a topological vector space is compact if and only if it is totally bounded and complete.*

Definition B.2.14 (Completion of a topological vector space). A *completion of a topological vector space* X is an injective open continuous linear mapping $\iota : X \to \widehat{X}$, where $\iota(X)$ is a dense subset of the complete topological vector space $\widehat{X}$.

Exercise B.2.15 (Existence and uniqueness of completion). Let X be a topological vector space. Show that it has a completion $\iota : X \to \widehat{X}$, and that this completion is unique in the following sense: if $\kappa : X \to Z$ is another completion, then the linear mapping $(\iota(x) \mapsto \kappa(x)) : \iota(X) \to Z$ has a unique continuous extension to an isomorphism $\widehat{X} \to Z$ of topological vector spaces.

Exercise B.2.16 (Extension of continuous linear operators). Let $A : X \to Y$ be continuous and linear, where the topological vector spaces X, Y have respective completions $\iota_X : X \to \widehat{X}$ and $\iota_Y : Y \to \widehat{Y}$. Show that there exists a unique continuous linear mapping $\widehat{A} : \widehat{X} \to \widehat{Y}$ such that $\widehat{A} \circ \iota_X = \iota_Y \circ A$, i.e., that the following diagram is commutative:

$$
\begin{array}{ccc}
X & \xrightarrow{\;A\;} & Y \\
\iota_X \downarrow & & \downarrow \iota_Y \\
\widehat{X} & \xrightarrow{\;\widehat{A}\;} & \widehat{Y}
\end{array}
$$

B.3 Locally convex spaces

A locally convex space is a topological vector space where a local base for the topology can be given by convex neighbourhoods. If the reader is not familiar with Banach and Hilbert spaces yet, we suggest first examining those concepts, and returning to this section only afterwards. This is why in this section we will freely refer to Section B.4 to illustrate the introduced concepts in a simpler setting of Banach spaces. In the sequel, we present some essential results for locally convex spaces in a series of exercises of widely varying difficulty, for which the reader may find help, e.g., from [63], [89] and [134].

Definition B.3.1 (Locally convex spaces). A topological vector space V (over $\mathbb{K}$) is called *locally convex* if for every neighbourhood U of $0 \in V$ there exists a convex neighbourhood C such that $0 \in C \subset U$.

Exercise B.3.2. Show that in a locally convex space each neighbourhood of 0 contains a convex balanced neighbourhood of 0.

Exercise B.3.3. Let $\mathcal{U}$ be the family of all convex balanced neighbourhoods of 0 in a topological vector space V. For $U \in \mathcal{U}$, define a so-called *Minkowski functional* $p_U : V \to [0, \infty)$ by

$$p_U(x) := \inf \left\{ \lambda \in \mathbb{R}^+ : \ x/\lambda \in U \right\}.$$

Show that p_U is a seminorm (see Definition B.4.1). Moreover, prove that V is locally convex if and only if its topology is induced by the family

$$\{p_U : V \to [0, \infty) \mid U \in \mathcal{U}\}.$$

Definition B.3.4 (Fréchet spaces). A locally convex space having a complete (and translation-invariant) metric is called a *Fréchet space*.

Exercise B.3.5. Show that a locally convex space V is metrisable if and only if it has the following property: there exists a countable collection $\{p_k\}_{k=1}^\infty$ of continuous seminorms $p_k : V \to [0, \infty)$ such that for every $x \in V \setminus \{0\}$ there exists $k_x \in \mathbb{Z}^+$ satisfying $p_{k_x}(x) \neq 0$ (i.e., the seminorm family separates the points of V).

Exercise B.3.6. Let $k \in \mathbb{Z}^+ \cup \{0, \infty\}$ and let $U \subset \mathbb{R}^n$ be an open non-empty set. Endow space $C^k(U)$ with a Fréchet space structure.

Exercise B.3.7. Let $\Omega \subset \mathbb{C}$ be open and non-empty. Endow the space $\mathbb{H}(\Omega) \subset C(\Omega)$ of analytic functions $f : \Omega \to \mathbb{C}$ with a structure of a Fréchet space.

Definition B.3.8 (Schwartz space). For $f \in C^\infty(\mathbb{R}^n)$ and $\alpha, \beta \in \mathbb{N}_0^n$, let

$$p_{\alpha\beta}(f) := \sup_{x \in \mathbb{R}^n} \left| x^\beta \partial_x^\alpha f(x) \right|.$$

If $p_{\alpha\beta}(f) < \infty$ for every α, β, then f is called a *rapidly decreasing smooth function*. The collection of such functions is called the *Schwartz space* $\mathcal{S}(\mathbb{R}^n)$.

Exercise B.3.9. Show that the Schwartz space $\mathcal{S}(\mathbb{R}^n)$ is a Fréchet space.

Definition B.3.10 (LF-space). A $\mathbb{C}$-vector space X is called an *LF-space* (or a *limit of Fréchet spaces*) if $X = \bigcup_{j=1}^{\infty} X_j$, where each $X_j \subset X_{j+1}$ is a subspace of X, having a Fréchet space topology τ_j such that $\tau_j = \{U \cap X_j : U \in \tau_{j+1}\}$. The topology τ of the LF-space X is then generated by the set

$$\tau := \{x + V \mid x \in X, \ V \in \mathcal{C}, \ V \cap X_j \in \tau_j \text{ for every } j\},$$

where $\mathcal{C}$ is the family of those convex subsets of X that contain 0.

Exercise B.3.11. Let τ be the topology of an LF-space $X = \bigcup_{j=1}^{\infty} X_j$ as in Definition B.3.10. Prove that
$$\tau_j = \{U \cap X_j \mid U \in \tau\}.$$
Moreover, show that a linear functional $f : X \to \mathbb{C}$ is continuous if and only if the restriction $f|_{X_j} : X_j \to \mathbb{C}$ is continuous for every j.

Exercise B.3.12. Let $U \subset \mathbb{R}^n$ be an open non-empty set. Let $\mathcal{D}(U)$ consist of compactly supported C^∞-smooth functions $f : U \to \mathbb{C}$. Endow $\mathcal{D}(U)$ with an LF-space structure; this is not a Fréchet space anymore.

Definition B.3.13 (Test functions and distributions). The LF-space $\mathcal{D}(U)$ of Exercise B.3.12 is called the space of *test functions*, and a continuous linear functional $f : \mathcal{D}(U) \to \mathbb{C}$ is called a *distribution on $U \subset \mathbb{R}^n$*. The space of distributions on U is denoted by $\mathcal{D}'(U)$.

Exercise B.3.14 (Locally convex Hahn–Banach Theorem). Prove the following analogue of the Hahn–Banach Theorem B.4.25: *Let X be a locally convex space (over $\mathbb{K}$) and $f : M \to \mathbb{K}$ be a continuous linear functional on a vector subspace $M \subset X$. Then there exists a continuous extension $F : X \to \mathbb{K}$ such that $F|_M = f$.*

Exercise B.3.15. Let X be a $\mathbb{K}$-vector space, and suppose V is a vector space of linear functionals $f : X \to \mathbb{K}$ that separates the points of X. Show that V induces a locally convex topology on X, and that then the dual $X' = V$.

Definition B.3.16 (Weak topology). Let X be a topological vector space such that the dual $X' = \mathcal{L}(X, \mathbb{K})$ separates the points of X. The X'-induced topology is called the *weak topology of X*.

Exercise B.3.17 (Closure of convex sets). Let X be locally convex and $C \subset X$ convex. Show that the closure of C is the same in both the original topology and the weak topology.

Definition B.3.18 (Weak*-topology). Let X be a topological vector space. The *weak*-topology* of the dual X' is the topology induced by the family $\{x' \mid x \in X\}$, where $x' : X' \to \mathbb{K}$ is defined by $x'(f) := f(x)$.

Exercise B.3.19. Let $x \in X$. Show that $x' = (f \mapsto f(x)) : X' \to \mathbb{K}$ is linear. Moreover, prove that if a linear functional $f : X' \to \mathbb{K}$ is continuous with respect to the weak*-topology, then $f = x'$ for some $x \in X$.

Exercise B.3.20 (Banach–Alaoglu Theorem in topological vector spaces). Prove the following generalisation of the Banach–Alaoglu Theorem B.4.36: *Let X be a topological vector space. Let $U \subset X$ be a neighbourhood of $0 \in X$, and let*

$$K := \{ f \in X' \mid \forall x \in U : \ |f(x)| \leq 1 \}.$$

Then $K \subset X'$ is compact in the weak*-topology.

Definition B.3.21 (Convex hull). The *convex hull* of a subset S of a vector space X is the intersection of all convex sets that contain S. (Notice that at least X is a convex set containing S.)

Exercise B.3.22. Show that the convex hull of S is the smallest convex set that contains S.

Exercise B.3.23. Show that $x \in X$ belongs to the convex hull of S if and only if $x = \sum_{k=1}^{n} t_k x_k$ for some $n \in \mathbb{Z}^+$, where the vectors $x_k \in S$, and $t_k > 0$ are such that $\sum_{k=1}^{n} t_k = 1$.

Definition B.3.24 (Extreme set). Let K be a subset of a vector space X. A non-empty set $E \subset K$ is called an *extreme set* of K if the conditions

$$\begin{cases} x, y \in K, \\ tx + (1 - t)y \in E \quad \text{for some } t \in (0, 1) \end{cases}$$

imply that $x, y \in E$. A point $z \in K$ is called an *extreme point* of $K \subset X$ if $\{z\}$ is an extreme set of K (alternative characterisation: if $x, y \in K$ and $z = tx + (1-t)y$ for some $0 < t < 1$ then $x = y = z$).

Exercise B.3.25 (Krein–Milman Theorem). Prove the following *Krein–Milman Theorem: Let X be a locally convex space and $K \subset X$ compact. Then K is contained in the closure of the convex hull of the set of the extreme points of K.*
(Hint: The first problem is the very existence of extreme points. The family of compact extreme sets of K can be ordered by inclusion, and by the Hausdorff Maximal Principle there is a maximal chain. Notice that X' separates the points of X...)

Exercise B.3.26. Let K be a compact subset of a Fréchet space X. Show that the closure of the convex hull of K is compact.

Exercise B.3.27. Let $f : G \to X$ be continuous, where X is a Fréchet space and G is a compact Hausdorff space. Let μ be a finite positive Borel measure on G. Show that there exists a unique vector $v \in X$ such that

$$\phi(v) = \int_G \phi(f) \, \mathrm{d}\mu$$

for every $\phi \in X'$.

Definition B.3.28 (Pettis integral). Let $f : G \to X$, μ and v be as in Exercise B.3.27. Then the vector $v \in X$ is called the *Pettis integral* (or the *weak integral*) of f with respect to μ, denoted by

$$v = \int_G f \, d\mu.$$

Exercise B.3.29. Let $f : G \to X$ and μ be as in Definition B.3.28. Assume that X is even a Banach space. Show that

$$\left\| \int_G f \, d\mu \right\| \leq \int_G \|f\| \, d\mu.$$

Definition B.3.30 (Barreled space). A subset B of a topological vector space X is called a *barrel* if it is closed, balanced, convex and $X = \bigcup_{t>0} tB$. A topological vector space is called *barreled* if its every barrel contains a neighbourhood of the origin.

Remark B.3.31. Notice that a barreled space is not necessarily locally convex.

Exercise B.3.32 (LF-spaces are barreled). Show that LF-spaces are barreled.

Definition B.3.33 (Heine–Borel property). A metric space is said to satisfy the *Heine–Borel property* if its closed and bounded sets are compact.

Definition B.3.34 (Montel space). A barreled locally convex space with the Heine–Borel property is called a *Montel space*.

Exercise B.3.35. Prove that $C^\infty(U)$ and $\mathcal{D}(U)$ are Montel spaces, where $U \subset \mathbb{R}^n$ is open and non-empty.

Exercise B.3.36. Let $\Omega \subset \mathbb{C}$ be open and non-empty. Show that the space $\mathbb{H}(\Omega)$ of analytic functions on Ω is a Montel space.

Exercise B.3.37. Prove that the Schwartz space $\mathcal{S}(\mathbb{R}^n)$ is a Montel space.

Exercise B.3.38. Let $U \subset \mathbb{R}^n$ be open and non-empty. Show that $C^k(U)$ is not a Montel space if $k \in \mathbb{N}_0$.

B.3.1   Topological tensor products

In this section we review the topological tensor products. If the reader is interested in more details on this subject we refer to [87] and to [134].

Definition B.3.39 (Projective tensor product). Let $X \otimes Y$ be the algebraic tensor product of locally convex spaces X, Y. The *projective tensor topology* or the *π-topology* of $X \otimes Y$ is the strongest topology for which the bilinear mapping $((x, y) \mapsto x \otimes y) : X \times Y \to X \otimes Y$ is continuous. This topological space is denoted by $X \otimes_\pi Y$, and its completion by $X \widehat{\otimes}_\pi Y$.

Exercise B.3.40. Let X, Y be locally convex spaces over $\mathbb{C}$. Show that the dual of $X \otimes_\pi Y$ (and also of $X \widehat{\otimes}_\pi Y$) is isomorphic to the space of continuous bilinear mappings $X \times Y \to \mathbb{C}$.

Exercise B.3.41. Let X, Y be locally convex metrisable spaces. Show that $X \widehat{\otimes}_\pi Y$ is a Fréchet space. Moreover, if X, Y are barreled, show that $X \otimes Y$ is barreled.

Exercise B.3.42. Let X, Y be locally convex metrisable barreled spaces. Show that $X \otimes Y$ is barreled.

Exercise B.3.43 (Projective Banach tensor product). Let X, Y be Banach spaces. For $f \in X \otimes Y$, define

$$\|f\|_\pi := \inf \left\{ \sum_j \|x_j\| \, \|y_j\| : \ f = \sum_j x_j \otimes y_j \right\}.$$

Show that $f \mapsto \|f\|_\pi$ is a norm on $X \otimes Y$, and that the corresponding norm topology is the projective tensor topology.

Exercise B.3.44. Let X, Y be locally convex spaces over $\mathbb{C}$. Show that the algebraic tensor product $X \otimes Y$ can be identified with the space $B(X', Y')$ of continuous bilinear functionals $X' \times Y' \to \mathbb{C}$, where X' and Y' are the dual spaces with weak topologies.

Definition B.3.45 (Injective tensor product). Let X, Y be locally convex spaces over $\mathbb{C}$. Let $\widetilde{B}(X', Y')$ be the space of those bilinear functionals $X' \times Y' \to \mathbb{C}$ that are continuous separately in each variable. Endow $\widetilde{B}(X', Y')$ with the topology τ of uniform convergence on the products of an equicontinuous subset of X' and an equicontinuous subset of Y'. Interpreting $X \otimes Y \subset \widetilde{B}(X', Y')$ as in Exercise B.3.44, let the *injective tensor topology* be the restriction of τ to $X \otimes Y$. This topological space is denoted by $X \otimes_\varepsilon Y$, and its completion by $X \widehat{\otimes}_\varepsilon Y$.

Exercise B.3.46. Let X, Y be locally convex spaces over $\mathbb{C}$. Show that the bilinear mapping $((x, y) \mapsto x \otimes y) : X \times Y \to X \otimes_\varepsilon Y$ is continuous. From this, deduce that the injective topology of $X \otimes Y$ is coarser than the projective topology (i.e., is a subset of the projective topology).

Exercise B.3.47. Studying the mapping $((x, y) \mapsto x \otimes y) : X \times Y \to X \otimes Y$, explain how the inclusion $X \widehat{\otimes}_\pi Y \subset X \widehat{\otimes}_\varepsilon Y$ should be understood.

Exercise B.3.48 (Injective Banach tensor product). Let X, Y be Banach spaces. For $f \in X \otimes Y$, define

$$\|f\|_\varepsilon := \sup \left\{ |x' \otimes y'(f)| : \ x' \in X', \ y' \in Y', \ \|x'\| = 1 = \|y'\| \right\}.$$

Show that $f \mapsto \|f\|_\varepsilon$ is a norm on $X \otimes Y$, and that the corresponding norm topology is the injective tensor topology.

Definition B.3.49 (Nuclear space). A locally convex space X is called *nuclear* if $X \widehat{\otimes}_\pi Y = X \widehat{\otimes}_\varepsilon Y$ for every locally convex space Y (where the equality of sets is understood in the sense of Exercise B.3.47). In such a case, these completed tensor products are written simply $X \widehat{\otimes} Y$.

Exercise B.3.50. Let X, Y be nuclear spaces, and let $M, N \subset X$ be vector subspaces such that N is closed. Show that M, X/N, $X \times Y$ and $X \widehat{\otimes} Y$ are nuclear spaces.

Exercise B.3.51. Show that $C^\infty(U)$, $\mathcal{D}(U)$, $\mathcal{S}(\mathbb{R}^n)$, $\mathbb{H}(\Omega)$ and their dual spaces (of distributions) are nuclear.

Exercise B.3.52. Let X, Y be Fréchet spaces and X nuclear. Show that $\mathcal{L}(X', Y) \cong X \widehat{\otimes} Y$, $\mathcal{L}(X, Y) \cong X' \widehat{\otimes} Y$, and that $X' \widehat{\otimes} Y' \cong (X \widehat{\otimes} Y)'$.

Exercise B.3.53. Prove the following *Schwartz Kernel Theorem B.3.55*:

Remark B.3.54. In the following Schwartz Kernel Theorem B.3.55, we denote $\langle \psi, A\phi \rangle := (A\phi)(\psi)$, and $\langle \psi \otimes \phi, K_A \rangle := K_A(\psi \otimes \phi)$.

Theorem B.3.55 (Schwartz Kernel Theorem). *Let $U \subset \mathbb{R}^m$, $V \subset \mathbb{R}^n$ be open and non-empty, and let $A : \mathcal{D}(U) \to \mathcal{D}'(V)$ be linear and continuous. Then there exists a unique distribution $K_A \in \mathcal{D}'(V) \widehat{\otimes} \mathcal{D}'(U) \cong \mathcal{D}'(V \times U)$ such that*

$$\langle \psi, A\phi \rangle = \langle \psi \otimes \phi, K_A \rangle$$

for every $\phi \in \mathcal{D}(U)$ and $\psi \in \mathcal{D}(V)$. Moreover, if $A : \mathcal{D}(U) \to C^\infty(V)$ is continuous then it can be interpreted that $K_A \in C^\infty(V, \mathcal{D}'(U))$.

Definition B.3.56 (Schwartz kernel). The distribution K_A in Theorem B.3.55 is called the *Schwartz kernel* of A, written informally as

$$A\phi(x) = \int_V K_A(x, y)\, \phi(y)\, dy.$$

Exercise B.3.57. Let $A : \mathcal{D}(U) \to \mathcal{D}'(V)$ be continuous and linear as in Theorem B.3.55. Give necessary and sufficient conditions for A such that $K_A \in C^\infty(V \times U)$.

Exercise B.3.58. Find variants of the Schwartz Kernel Theorem B.3.55 for Schwartz functions and for tempered distributions.

B.4 Banach spaces

Definition B.4.1 (Seminorm and norm; normed spaces). Let X be a $\mathbb{K}$-vector space. A mapping $p : X \to \mathbb{R}$ is a *seminorm* if

$$\begin{cases} p(x + y) \le p(x) + p(y), \\ p(\lambda x) = |\lambda|\, p(x) \end{cases}$$

for every $x, y \in X$ and $\lambda \in \mathbb{R}$. If $p : X \to \mathbb{R}$ is a seminorm for which $p(x) = 0$ implies $x = 0$, then it is called a *norm*. Typically, a norm on X is written as $x \mapsto \|x\|_X$ or simply $\|x\|$. A vector space with a norm is called a *normed space*.

Example. On the vector space $\mathbb{K}$, the absolute value mapping $x \mapsto |x|$ is a norm.

Exercise B.4.2. Let $p : X \to [0, \infty)$ be a seminorm on a $\mathbb{K}$-vector space X and

$$x \sim y \quad \overset{\text{definition}}{\Longleftrightarrow} \quad p(x - y) = 0.$$

Prove the following claims:

(a) $\sim$ is an equivalence relation on X.
(b) The set $L := \{[x] : x \in X\}$, with $[x] := \{y \in X : x \sim y\}$, is an $\mathbb{R}$-vector space when endowed with operations

$$[x] + [y] := [x + y], \quad \lambda[x] := [\lambda x]$$

and the norm $[x] \mapsto p(x)$.

Exercise B.4.3. Let $w_j \geq 0$ for every $j \in J$. Define

$$\sum_{j \in J} w_j := \sup \left\{ \sum_{k \in K} w_k : K \subset J \text{ finite} \right\}.$$

Show that $\{j \in J : w_j > 0\}$ is at most countable if $\sum_{j \in J} w_j < \infty$.

Exercise B.4.4. For $x \in \mathbb{K}^J$ define

$$\|x\|_{\ell^p} := \begin{cases} \left(\sum_{j \in J} |x_j|^p \right)^{1/p}, & \text{if } 1 \leq p < \infty, \\ \sup_{j \in J} |x_j|, & \text{if } p = \infty, \end{cases}$$

where $x_j := x(j)$. Show that $\ell^p(J) := \{x \in \mathbb{K}^J : \|x\|_{\ell^p} < \infty\}$ is a Banach space with respect to the norm $x \mapsto \|x\|_{\ell^p}$.

Exercise B.4.5. Norms p_1, p_2 on a vector space V are called (Lipschitz) *equivalent* if $a^{-1} p_1(x) \leq p_2(x) \leq a p_1(x)$ for every $x \in V$, where $a \geq 1$ is a constant. Show that any two norms on a finite-dimensional space V are equivalent. Consequently, a finite-dimensional normed space is a Banach space.

Exercise B.4.6. Let K be a compact space. Show that

$$C(K) := \{f : K \to \mathbb{K} \mid f \text{ continuous }\}$$

is a Banach space when endowed with the norm

$$f \mapsto \|f\|_{C(K)} := \sup_{x \in K} |f(x)|.$$

Remark B.4.7. The previous exercise deals with special cases of of $L^p(\mu)$, the *Lebesgue p-spaces*. These Banach spaces are introduced in Definition C.4.6.

Definition B.4.8 (Normed and Banach spaces). Notice that the *norm metric* $((x, y) \mapsto \|x - y\|) : X \times X \to \mathbb{R}$ is a metric on X. Let τ_X denote the corresponding metric topology, called the *norm topology*, where the *open ball centered at $x \in X$ with radius $r > 0$* is

$$\mathbb{B}_X(x, r) = \mathbb{B}(x, r) = \{y \in X : \|x - y\| < r\}.$$

Ball $\mathbb{B}_X(0, 1)$ is called the *open unit ball*. The *closed ball centered at $x \in X$ with radius $r > 0$* is

$$\overline{\mathbb{B}}(x, r) := \{y \in X : \|x - y\| \leq r\}.$$

Notice that here $\overline{\mathbb{B}(x, r)} = \overline{\mathbb{B}}(x, r)$, where $\overline{S}$ refers to the norm closure of a set $S \subset X$. A *Banach space* is a normed space where the norm metric is complete.

Exercise B.4.9. Show that $V := \{x \in \ell^p(J) : \{j \in J : x_j \neq 0\} \text{ finite}\}$ is a dense normed vector subspace of $\ell^p(J)$.

Definition B.4.10 (Bounded linear operators). A linear mapping $A : X \to Y$ between normed spaces X, Y is called *bounded*, denoted $A \in \mathcal{L}(X, Y)$, if

$$\|Ax\| \leq C \|x\|$$

for every $x \in X$, where $C < \infty$ is a constant. The *norm of $A \in \mathcal{L}(X, Y)$* is

$$\|A\| := \sup_{x \in X: \|x\| \leq 1} \|Ax\|.$$

This norm is also called the *operator norm* and is often denoted by $\|A\|_{op}$. We often abbreviate $\mathcal{L}(X) := \mathcal{L}(X, X)$.

Exercise B.4.11. Let $A : X \to Y$ be a linear operator between normed spaces X and Y. Show that A is bounded if and only if it is continuous.

Exercise B.4.12. Show that $\mathcal{L}(X, Y)$ is really a normed space.

Exercise B.4.13. Show that $\|AB\| \leq \|A\|\|B\|$ if $B \in \mathcal{L}(X, Y)$ and $A \in \mathcal{L}(Y, Z)$.

Exercise B.4.14. Show that $\mathcal{L}(X, Y)$ is a Banach space if Y is a Banach space.

Definition B.4.15 (Duals). Let V be a Banach space over $\mathbb{K}$. The dual of V is the space

$$V' = \mathcal{L}(V, \mathbb{K}) := \{A : V \to \mathbb{K} \mid A \text{ bounded and linear}\}$$

endowed with the (operator) norm

$$A \mapsto \|A\| := \sup_{f \in V: \|f\|_V \leq 1} |A(f)|.$$

Exercise B.4.16. Prove that V' is a Banach space.

Definition B.4.17 (Compact linear operator). Let X, Y be normed spaces, and let $\overline{\mathbb{B}} = \overline{\mathbb{B}(0,1)} = \{x \in X : \|x\| \leq 1\}$. A linear mapping $A : X \to Y$ is called *compact*, written $A \in \mathcal{LC}(X, Y)$, if the closure of $A(\overline{\mathbb{B}}) \subset Y$ is compact. We also write $\mathcal{LC}(X) := \mathcal{LC}(X, X)$.

Exercise B.4.18. Show that $\mathcal{LC}(X, Y)$ is a linear subspace of $\mathcal{L}(X, Y)$, and it is closed if Y is complete.

Exercise B.4.19. Let $B_0, C_0 \in \mathcal{L}(X, Y)$ and $B_1, C_1 \in \mathcal{L}(Y, Z)$ such that C_0, C_1 are compact. Show that $C_1 B_0, B_1 C_0$ are compact.

Lemma B.4.20. (Almost Orthogonality Lemma [F. Riesz]) *Let X be a normed space with closed subspace $Y \neq X$. For each $\varepsilon > 0$ there exists $x_\varepsilon \in X$ such that $\|x_\varepsilon\| = 1$ and $\mathrm{dist}(x_\varepsilon, Y) \geq 1 - \varepsilon$.*

Proof. Let $z \in X \setminus Y$ and $r := \mathrm{dist}(z, Y) > 0$. Take $y = y_\varepsilon \in Y$ such that $r \leq \|z - y\| < (1 - \varepsilon)^{-1} r$. Let $x_\varepsilon := (z - y)/\|z - y\|$. If $u \in Y$ then

$$
\begin{aligned}
\|x_\varepsilon - u\| &= \left\| \frac{z - y}{\|z - y\|} - u \right\| \\
&= \frac{\|z - (y + \|z - y\| u)\|}{\|z - y\|} \\
&> \frac{r}{(1 - \varepsilon)^{-1} r} \\
&= 1 - \varepsilon,
\end{aligned}
$$

showing that $\mathrm{dist}(x_\varepsilon, Y) \geq 1 - \varepsilon$. $\qquad\square$

Theorem B.4.21 (Riesz's Compactness Theorem). *Let X be a normed space. Then X is finite-dimensional if and only if $\overline{\mathbb{B}(0,1)}$ is compact.*

Proof. By the Heine–Borel Theorem, a set in a finite-dimensional normed space is compact if and only if it is bounded.

Now let X be infinite-dimensional. Let $0 < \varepsilon < 1$ and $x_1 \in X$ such that $\|x_1\| = 1$. Inductively, let $Y_k := \mathrm{span}\{x_j\}_{j=1}^k \neq X$, and choose $x_{k+1} \in X \setminus Y_k \neq \emptyset$ such that $\|x_{k+1}\| = 1$ and $\mathrm{dist}(x_{k+1}, Y_k) > 1 - \varepsilon$. Then it is clear that the sequence $(x_k)_{k=1}^\infty$ does not have a converging subsequence. Hence by Theorem A.13.4, $\overline{\mathbb{B}(0,1)}$ is not compact. $\qquad\square$

Remark B.4.22 (**Is identity compact?**). Riesz's Compactness Theorem B.4.21 could also be stated: *a normed space X is finite-dimensional if and only if the identity mapping $I = (x \mapsto x) : X \to X$ is compact.* This together with the results of Exercises B.4.18 and B.4.19 proves that $\mathcal{LC}(X)$ is a closed two-sided proper ideal of $\mathcal{L}(X)$, where X is a Banach space and X is not finite-dimensional.

Theorem B.4.23 (Baire's Theorem). *Let (X, d) be a complete metric space and $U_j \subset X$ be dense and open for each $k \in \mathbb{Z}^+$. Then $G = \bigcap_{k=1}^{\infty} U_k$ is dense.*

Proof. We must show that $G \cap \mathbb{B}(x_0, r_0) \neq \emptyset$ for any $x_0 \in X$ and $r_0 > 0$. Assuming $X \neq \emptyset$, take x_1 and r_1 such that

$$\overline{\mathbb{B}(x_1, r_1)} \subset U_1 \cap \mathbb{B}(x_0, r_0).$$

Inductively, we choose x_{k+1} and $r_{k+1} < 1/k$ so that

$$\overline{\mathbb{B}(x_{k+1}, r_{k+1})} \subset U_{k+1} \cap \mathbb{B}(x_k, r_k).$$

Then $(x_k)_{k=1}^{\infty}$ is a Cauchy sequence, thus converging to some $x \in X$ by completeness. By construction, $x \in G \cap \mathbb{B}(x_0, r_0)$. $\square$

Exercise B.4.24 (Baire's Theorem and interior points). Clearly, Baire's Theorem B.4.23 is equivalent to the following: in a complete metric space, a countable union of sets without interior points is without interior points. Use this to prove that an algebraic basis of an infinite-dimensional Banach space must be uncountable.

Theorem B.4.25 (Hahn–Banach Theorem). *Let X be a real normed space and $f : M_f \to \mathbb{R}$ be bounded and linear on a vector subspace $M_f \subset X$. Then there exists extension $F : X \to \mathbb{R}$ such that $F|_{M_f} = f$ and $\|F\| = \|f\|$.*

Proof. Let

$$S := \{h : M_h \to \mathbb{R} \quad | \quad h \text{ linear on vector subspace } M_h \subset X,$$
$$M_f \subset M_h, \ \|h\| = \|f\|\}.$$

Then $f \in S \neq \emptyset$. Endow S with the partial order

$$g \leq h \iff \begin{cases} M_g \subset M_h, \\ g = h|_{M_g}. \end{cases}$$

Take a chain $(f_j)_{j \in J} \subset S$. Then $f_j \leq h$ for each $j \in J$, where $h \in S$ is defined so that $M_h = \bigcup_{j \in J} M_{f_j}$, $h|_{M_{f_j}} = f_j$. Thereby, in view of Zorn's lemma (Theorem A.4.10), there is a maximal element $F : M_F \to \mathbb{R}$ in S. Suppose $M_F \neq X$. Then take $x_0 \in X \setminus M_F$. Given $a \in \mathbb{R}$, define

$$G_a : M_F + \mathbb{R}x_0 \to \mathbb{R}, \quad G_a(u + tx_0) = F(u) - ta.$$

Then G_a is bounded, linear, and $G_a|_{M_F} = F$. Hence $\|G_a\| \geq \|F\| = \|f\|$. Could it be that $\|G_a\| = \|F\|$ (this would contradict the maximality of F)? For any $u, v \in M_F$,

$$\begin{aligned} |F(u) - F(v)| &= |F(u - v)| \\ &\leq \|F\| \, \|u - v\| \\ &\leq \|F\| \, (\|u + x_0\| + \|v + x_0\|). \end{aligned}$$

Hence there exists $a_0 \in \mathbb{R}$ such that

$$F(u) - \|F\| \, \|u + x_0\| \le a_0 \le F(v) + \|F\| \, \|v + x_0\|$$

for every $u, v \in M_F$. Thus

$$|F(w) - a_0| \le \|F\| \, \|w + x_0\|$$

for every $w \in M_F$. From this (assuming that non-trivially $t \ne 0$), we get

$$
\begin{aligned}
|G_{a_0}(u + tx_0)| &= |t| \, |u/t - a_0| \\
&\le |t| \, \|F\| \, \|u/t + x_0\| \\
&= \|F\| \, \|u + tx_0\|;
\end{aligned}
$$

but this means $\|G_{a_0}\| \le \|F\|$, a contradiction. $\qquad \square$

Exercise B.4.26 (Complex version of the Hahn–Banach Theorem). Prove the complex version of the Hahn–Banach Theorem: Let X be a complex normed space and $f : M_f \to \mathbb{C}$ be bounded and linear on a vector subspace $M_f \subset X$. Then there exists an extension $F : X \to \mathbb{C}$ such that $F|_{M_f} = f$ and $\|F\| = \|f\|$.

Corollary B.4.27. *Let X be a normed space and $x \in X$. Then*

$$\|x\| = \max \{|F(x)| : \ F \in L(X, \mathbb{K}), \ \|F\| \le 1\}.$$

Corollary B.4.28 (Hahn–Banach $\Longrightarrow$ Riesz' Compactness Theorem). *Let X be a normed space. Then $\overline{\mathbb{B}}(0,1)$ is compact if and only if it is finite-dimensional.*

Proof. By the Heine–Borel Theorem, a closed set in a finite-dimensional normed space is compact if and only if it is bounded.

The proof for the converse follows [28]: Suppose X is locally compact and let $S^1 := \{x \in X : \ \|x\| = 1\}$. Then $\{S^1 \cap \mathrm{Ker}(f) : \ f \in \mathcal{L}(X, \mathbb{K})\}$ is a family of compact sets, whose intersection is empty by the Hahn–Banach Theorem. Thereby there exists $\{f_k\}_{k=1}^n \subset \mathcal{L}(X, \mathbb{K})$ such that

$$\bigcap_{k=1}^n S^1 \cap \mathrm{Ker}(f_k) = \emptyset, \quad \text{i.e.,} \quad \bigcap_{k=1}^n \mathrm{Ker}(f_k) = \{0\}.$$

Since the co-dimension of $\mathrm{Ker}(f_k) \le 1$, this implies that $\dim(X) \le n$. $\qquad \square$

Theorem B.4.29 (Banach–Steinhaus Theorem, or Uniform Boundedness Principle). *Let X be a Banach space, let Y be a normed space, and let $\{A_j\}_{j \in J} \subset \mathcal{L}(X, Y)$ be such that*

$$\sup_{j \in J} \|A_j x\| < \infty$$

for every $x \in X$. Then $\sup_{j \in J} \|A_j\| < \infty$.

Proof. Let $p_j(x) := \|A_j x\|$ and $p(x) := \sup\{p_j(x) \mid j \in J\}$. Clearly, $p, p_j : X \to \mathbb{R}$ are seminorms. Moreover, p_j is continuous for every $j \in J$, and we must show that also p is continuous. Since p_j is continuous for every $j \in J$, set

$$U_k := \{x \in X : \ p(x) > k\} = \bigcup_{j \in J} \{x \in X : \ p_j(x) > k\}$$

is open. Now $\bigcap_{k=1}^{\infty} U_k = \emptyset$, so that by Baire's Theorem B.4.23 there exists $k_0 \in \mathbb{Z}^+$ for which $\overline{U_{k_0}} \neq X$; actually, here $\overline{U_1} \neq X$, because $U_1 = k^{-1}U_k$. Choose $x_0 \in X$ and $r_0 > 0$ such that

$$\mathbb{B}(x_0, r_0) \subset X \setminus \overline{U_1}.$$

If $z \in \mathbb{B}(0, 1)$ then

$$
\begin{aligned}
r_0 \, p(z) \ &= \ p(r_0 z) \\
&\leq \ p(x_0 + r_0 z) + p(-x_0) \\
&\leq \ 2.
\end{aligned}
$$

Thus $\|A_j\| \leq 2/r_0$ for every $j \in J$. $\square$

Definition B.4.30 (Open mappings). A mapping $f : X \to Y$ between topological spaces X, Y is said to be *open*, if $f(U) \subset Y$ is open for every open $U \subset X$.

Theorem B.4.31 (Open Mapping Theorem). *Let $A \in \mathcal{L}(X, Y)$ be surjective, where X, Y are Banach spaces. Then A is open.*

Proof. It is sufficient to show that $\mathbb{B}_Y(0, r) \subset A(\mathbb{B}_X(0, 1))$ for some $r > 0$. For each $k \in \mathbb{Z}^+$, set $U_k := Y \setminus \overline{A(\mathbb{B}_X(0, k))}$ is open. Now $\bigcap_{k=1}^{\infty} U_k = \emptyset$, because A is surjective. By Baire's Theorem B.4.23, $\overline{U_{k_0}} \neq Y$ for some $k_0 \in \mathbb{Z}^+$; actually, $\overline{U_1} \neq Y$, because $A(\mathbb{B}_X(0, 1)) = k^{-1}A(\mathbb{B}_X(0, k))$. Take $y_0 \in Y$ and $r_0 > 0$ such that

$$\mathbb{B}_Y(y_0, r_0) \subset Y \setminus \overline{U_1}.$$

Now

$$\mathbb{B}_Y(y_0, r_0) \subset Y \setminus U_1 = \overline{A(\mathbb{B}_X(0, 1))}.$$

Let $\varepsilon > 0$ and $y \in \mathbb{B}_Y(0, r_0)$. Take $w_1, w_2 \in \mathbb{B}_X(0, 1)$ such that

$$
\begin{aligned}
\|y_0 - A w_1\| \ &< \ \varepsilon/2, \\
\|(y_0 + y) - A w_2\| \ &< \ \varepsilon/2.
\end{aligned}
$$

Then $w_1 - w_2 \in \mathbb{B}_X(0, 2)$ and $\|y - A(w_1 - w_2)\| < \varepsilon$. By linearity, this yields

$$\forall \varepsilon > 0 \ \forall y \in \mathbb{B}_Y(0, r_0) \ \exists x \in \mathbb{B}_X(0, 2\|y\|/r_0) : \ \|y - Ax\| < \varepsilon.$$

Thus if $z \in \mathbb{B}_Y(0, r_0)$, take $x_0 \in \mathbb{B}_X(0, 2)$ such that $\|z - Ax_0\| < r_0/2$. Inductively, choose $x_k \in \mathbb{B}_X(0, 2^{1-k})$ such that $\|z - A\sum_{j=0}^{k} x_j\| < 2^{1-k}r_0$. Now $\sum_{j=0}^{k} x_j \to_k$

$x \in \overline{\mathbb{B}_X(0,4)} \subset \mathbb{B}_X(0,5)$, because X is complete. We have $z = Ax$ by continuity of A. Thereby

$$\mathbb{B}_Y(0, r_0) \subset A(\mathbb{B}_X(0,5)),$$

implying $\mathbb{B}_Y(0, r_0/5) \subset A(\mathbb{B}_X(0,1))$. $\qquad\qquad\qquad\qquad\qquad\qquad\square$

Corollary B.4.32 (Bounded Inverse Theorem). *Let $B \in \mathcal{L}(X, Y)$ be bijective - between Banach spaces X, Y. Then B^{-1} is continuous.*

Definition B.4.33 (Graph). The *graph* of a mapping $f : X \to Y$ is

$$\begin{aligned} \Gamma(f) \quad &:= \quad \{(x, f(x)) \mid x \in X\} \\ &\subset \quad X \times Y. \end{aligned}$$

Theorem B.4.34 (Closed Graph Theorem). *Let $A : X \to Y$ be a linear mapping between Banach spaces X, Y. Then A is continuous if and only if its graph is closed in $X \times Y$.*

Proof. Suppose A is continuous. Take a Cauchy sequence $((x_j, Ax_j))_{j=1}^{\infty}$ of $\Gamma(A) \subset X \times Y$. Then $(x_j)_{j=1}^{\infty}$ is a Cauchy sequence of X, thereby converging to some $x \in X$ by completeness. Then $Ax_j \to Ax$ by the continuity of A. Hence $(x_j, Ax_j) \to (x, Ax) \in \Gamma(A)$; the graph is closed.

Now assume that $\Gamma(A) \subset X \times Y$ is closed. Thus the graph is a Banach subspace of $X \times Y$. Define a mapping $B := (x \mapsto (x, Ax)) : X \to \Gamma(A)$. It is easy to see that B is a linear bijection. By the Open Mapping Theorem, B is continuous. This implies the continuity of A. $\qquad\qquad\qquad\qquad\qquad\square$

Definition B.4.35 (Weak*-topology). Let $x \mapsto \|x\|$ be the norm of a normed vector space X over a field $\mathbb{K} \in \{\mathbb{R}, \mathbb{C}\}$. The dual space $X' = \mathcal{L}(X, \mathbb{K})$ of X is a set of bounded linear functionals $f : X \to \mathbb{K}$, having a norm

$$\|f\| := \sup_{x \in X: \ \|x\| \leq 1} |f(x)|.$$

This endows X' with a Banach space structure. However, it is often better to use a weaker topology for the dual: let us define $x(f) := f(x)$ for every $x \in X$ and $f \in X'$; this gives the interpretation $X \subset X'' := \mathcal{L}(X', \mathbb{K})$, because

$$|x(f)| = |f(x)| \leq \|f\| \|x\|.$$

So we may treat X as a set of functions $X' \to \mathbb{K}$, and we define the *weak*-topology* of X' to be the X-induced[2] topology of X'.

Theorem B.4.36 (Banach–Alaoglu Theorem). *Let X be a Banach space. Then the closed unit ball*

$$K := \overline{B_{X'}(0,1)} = \{\phi \in X' : \|\phi\|_{X'} \leq 1\}$$

of X' is weak-compact.*

---

[2] see Definition A.18.1

Proof. Due to Tihonov's Theorem A.18.8,

$$P := \prod_{x \in X} \{\lambda \in \mathbb{C} : \ |\lambda| \le \|x\|\} = \overline{\mathbb{D}(0, \|x\|)}^X$$

is compact in the product topology τ_P. Any element $f \in P$ is a mapping

$$f : X \to \mathbb{C} \quad \text{such that} \quad f(x) \le \|x\|.$$

Hence $K = X' \cap P$. Let τ_1 and τ_2 be the relative topologies of K inherited from the weak*-topology $\tau_{X'}$ of X' and the product topology τ_P of P, respectively. We shall prove that $\tau_1 = \tau_2$ and that $K \subset P$ is closed; this would show that K is a compact Hausdorff space.

First, let $\phi \in X'$, $f \in P$, $S \subset X$, and $\delta > 0$. Define

$$\begin{aligned} U(\phi, S, \delta) &:= \{\psi \in X' : \ x \in S \Rightarrow |\psi x - \phi x| < \delta\}, \\ V(f, S, \delta) &:= \{g \in P : \ x \in S \Rightarrow |g(x) - f(x)| < \delta\}. \end{aligned}$$

Then

$$\begin{aligned} \mathcal{U} &:= \{U(\phi, S, \delta) \mid \phi \in X', \ S \subset X \text{ finite}, \ \delta > 0\}, \\ \mathcal{V} &:= \{V(f, S, \delta) \mid f \in P, \ S \subset X \text{ finite}, \ \delta > 0\} \end{aligned}$$

are bases for the topologies $\tau_{X'}$ and τ_P, respectively. Clearly

$$K \cap U(\phi, S, \delta) = K \cap V(\phi, S, \delta),$$

so that the topologies $\tau_{X'}$ and τ_P agree on K, i.e., $\tau_1 = \tau_2$.

Still we have to show that $K \subset P$ is closed. Let $f \in \overline{K} \subset P$. First we show that f is linear. Take $x, y \in X$, $\lambda, \mu \in \mathbb{C}$ and $\delta > 0$. Choose $\phi_\delta \in K$ such that

$$f \in V(\phi_\delta, \{x, y, \lambda x + \mu y\}, \delta).$$

Then

$$\begin{aligned} &|f(\lambda x + \mu y) - (\lambda f(x) + \mu f(y))| \\ \le\ & |f(\lambda x + \mu y) - \phi_\delta(\lambda x + \mu y)| + |\phi_\delta(\lambda x + \mu y) - (\lambda f(x) + \mu f(y))| \\ =\ & |f(\lambda x + \mu y) - \phi_\delta(\lambda x + \mu y)| + |\lambda(\phi_\delta x - f(x)) + \mu(\phi_\delta y - f(y))| \\ \le\ & |f(\lambda x + \mu y) - \phi_\delta(\lambda x + \mu y)| + |\lambda| \, |\phi_\delta x - f(x)| + |\mu| \, |\phi_\delta y - f(y)| \\ \le\ & \delta \, (1 + |\lambda| + |\mu|). \end{aligned}$$

This holds for every $\delta > 0$, so that actually

$$f(\lambda x + \mu y) = \lambda f(x) + \mu f(y),$$

f is linear! Moreover, $\|f\| \le 1$, because

$$|f(x)| \le |f(x) - \phi_\delta x| + |\phi_\delta x| \le \delta + \|x\|.$$

Hence $f \in K$, K is closed. $\qquad\qquad\qquad\qquad\qquad\qquad\qquad\qquad\qquad\square$

Remark B.4.37. The Banach–Alaoglu Theorem B.4.36 implies that a bounded weak*-closed subset of the dual space is a compact Hausdorff space in the relative weak*-topology. However, in a normed space norm-closed balls are compact if and only if the dimension is finite!

B.4.1 Banach space adjoint

We now come back to the adjoints of Banach spaces and of operators introduced in Definition B.4.15. Here we give a condensed treatment to acquaint the reader with the topic.

Definition B.4.38 (Duality). Let X be a Banach space and $X' = \mathcal{L}(X, \mathbb{K})$ its dual. For $x \in X$ and $x' \in X'$ let us write

$$\langle x, x' \rangle := x'(x).$$

We endow X' with the norm $x' \mapsto \|x'\|$ given by

$$\|x'\| := \sup \left\{ |\langle x, x' \rangle| : x \in X, \ \|x\| \leq 1 \right\}.$$

Exercise B.4.39. Let X be a Banach space and $x \in X$. Show that

$$\|x\| = \sup \left\{ |\langle x, x' \rangle| : x' \in X', \ \|x'\| \leq 1 \right\}.$$

Exercise B.4.40. Let X, Y be Banach spaces with respective duals X', Y'. Let $A \in \mathcal{L}(X, Y)$. Show that there exists a unique $A' \in \mathcal{L}(Y', X')$ such that

$$\langle Ax, y' \rangle = \langle x, A'(y') \rangle \tag{B.1}$$

for every $x \in X$ and $y' \in Y'$. Prove also that

$$\|A'\| = \|A\|.$$

Definition B.4.41 (Adjoint operator). Let $A \in \mathcal{L}(X, Y)$ be as in Exercise B.4.40. Then $A' \in \mathcal{L}(Y', X')$ defined by (B.1) is called the (*Banach*) *adjoint* of A.

Exercise B.4.42. Show that $A \in \mathcal{L}(X, Y)$ is compact if and only if $A' \in \mathcal{L}(Y', X')$ is compact.

Definition B.4.43 (Complemented subspace). A closed subspace V of a topological vector space X is said to be *complemented* in X by a subspace $W \subset X$ if

$$\begin{cases} V + W = X & \text{and} \\ V \cap W = \{0\}. \end{cases}$$

Then we write $X = V \oplus W$, saying that X is the *direct sum* of V and W.

Exercise B.4.44. Show that a closed subspace V is complemented in X if X/V is finite-dimensional.

Exercise B.4.45. Show that a finite-dimensional subspace of a locally convex space is complemented. (Hint: Hahn–Banach.)

Exercise B.4.46. Let $A \in \mathcal{L}(X)$ be compact, where X is a Banach space. Let λ be a non-zero scalar. Show that the range set

$$(\lambda I - A)(X) = \{\lambda x - Ax : \ x \in X\}$$

is closed, $\mathrm{Ker}(\lambda I - A) = \{x \in X : Ax = \lambda x\}$ is finite-dimensional, and that

$$
\begin{aligned}
& \dim\left(\mathrm{Ker}(\lambda I - A)\right) \\
= {} & \dim\left(\mathrm{Ker}(\lambda I - A')\right) \\
= {} & \dim\left(X/((\lambda I - A)(X))\right) \\
= {} & \dim\left(X'/((\lambda I - A')(X'))\right).
\end{aligned}
$$

Definition B.4.47 (Reflexive space). Let X be a Banach space and $X' = \mathcal{L}(X, \mathbb{K})$ its dual Banach space. The *second dual* of X is $X'' := (X')' = \mathcal{L}(X', \mathbb{K})$. It is then easy to show that we can define a linear isometry $(x \mapsto x'') : X \to X''$ onto a closed subspace of X'' by

$$x''(f) := f(x).$$

Thus X can be regarded as a subspace of X''. If $X'' = \{x'' : \ x \in X\}$ then X is called *reflexive*.

Exercise B.4.48. Show that $(x \mapsto x'') : X \to X''$ in Definition B.4.47 has the claimed properties.

Exercise B.4.49. Let $1 < p < \infty$. Show that $\ell^p = \ell^p(\mathbb{Z}^+)$ is reflexive. What about ℓ^1 and ℓ^∞?

Exercise B.4.50. Show that $C([0, 1])$ is not reflexive.

Exercise B.4.51. Let X be a Banach space. Prove that X is reflexive if and only if its closed unit ball is compact in the weak topology. (Hint: Hahn–Banach and Banach–Alaoglu).

Exercise B.4.52. Let V be a closed subspace of a reflexive Banach space X. Show that V and X/V are reflexive.

Exercise B.4.53. Show that X is reflexive if and only if X' is reflexive.

B.5 Hilbert spaces

Definition B.5.1 (Inner product and Hilbert spaces). Let $\mathcal{H}$ be a $\mathbb{C}$-vector space. A mapping $((x, y) \mapsto \langle x, y \rangle) : \mathcal{H} \times \mathcal{H} \to \mathbb{C}$ is an *inner product* if

$$
\begin{aligned}
&\langle x + y, z \rangle = \langle x, z \rangle + \langle y, z \rangle, \\
&\langle \lambda x, y \rangle = \lambda \langle x, y \rangle, \\
&\langle y, x \rangle = \overline{\langle x, y \rangle}, \\
&\langle x, x \rangle \geq 0, \\
&\langle x, x \rangle = 0 \quad \Rightarrow \quad x = 0
\end{aligned}
$$

for every $x, y \in \mathcal{H}$ and $\lambda, \mu \in \mathbb{C}$. Then $\mathcal{H}$ endowed with the inner product is called an *inner product space*. An inner product defines the canonical norm

$$
\|x\| := \langle x, x \rangle^{1/2};
$$

we shall soon prove that this is a norm in the usual sense. $\mathcal{H}$ is called a *Hilbert space* (or a *complete inner product space*) if it is a Banach space with respect to the canonical norm.

Exercise B.5.2. Show that $\ell^2(J)$ is a Hilbert space, where

$$
\langle x, y \rangle = \sum_{j \in J} x_j \overline{y_j}.
$$

Definition B.5.3 (Orthogonality). Vectors $x, y \in \mathcal{H}$ are said to be *orthogonal* in an inner product space $\mathcal{H}$, denoted $x \perp y$, if $\langle x, y \rangle = 0$. For $S \subset \mathcal{H}$, let

$$
S^\perp := \{ x \in \mathcal{H} \mid \forall y \in S : x \perp y \}.
$$

Subspaces $M, N \subset \mathcal{H}$ are called *orthogonal*, denoted by $M \perp N$, if $\langle x, y \rangle = 0$ for every $x \in M$ and $y \in N$. A collection $\{x_\alpha\}_{\alpha \in I}$ is called *orthonormal* if $\|x_\alpha\| = 1$ for all $\alpha \in I$ and if $\langle x_\alpha, x_\beta \rangle = 0$ for all $\alpha \neq \beta$, $\alpha, \beta \in I$.

Exercise B.5.4. Show that $S^\perp \subset \mathcal{H}$ is a closed vector subspace, and that $S \subset (S^\perp)^\perp$. Show that if V is a closed vector subspace of $\mathcal{H}$ then $V = (V^\perp)^\perp$.

Exercise B.5.5 (Pythagoras' Theorem). Let $x_1, x_2, \ldots, x_n \in \mathcal{H}$ be mutually orthogonal, i.e., assume that $x_i \perp x_j$ for all $i \neq j$. Prove that $\| \sum_{j=1}^n x_j \|^2 = \sum_{j=1}^n \|x_j\|^2$. (This generalised the famous theorem of Pythagoras of Samos on the triangles in the plane.)

Proposition B.5.6 (Cauchy–Schwarz inequality). *Let $\mathcal{H}$ be an inner product space. Then*

$$
|\langle x, y \rangle| \leq \|x\| \, \|y\| \tag{B.2}
$$

for every $x, y \in \mathcal{H}$.

Proof. We may assume that $x \neq 0$ and $y \neq 0$, otherwise the statement is trivial. For $t \in \mathbb{R}$,

$$
\begin{aligned}
0 \;\leq\;& \|x - ty\|^2 \\
=\;& \langle x - ty, x - ty \rangle \\
=\;& \langle x, x \rangle - t\langle x, y \rangle - t\langle y, x \rangle + t^2 \langle y, y \rangle \\
=\;& \|y\|^2 t^2 - 2t\mathrm{Re}\langle x, y \rangle + \|x\|^2 \\
=\;& \|y\|^2 \left(t - \frac{\mathrm{Re}\langle x, y \rangle}{\|y\|^2} \right)^2 + \left(\|x\|^2 - \left(\frac{\mathrm{Re}\langle x, y \rangle}{\|y\|} \right)^2 \right).
\end{aligned}
$$

Taking $t = \frac{\mathrm{Re}\langle x, y \rangle}{\|y\|^2}$, we get

$$
|\mathrm{Re}\langle x, y \rangle| \leq \|x\| \, \|y\|
$$

for every $x, y \in \mathcal{H}$. Now $\langle x, y \rangle = |\langle x, y \rangle| e^{i\phi}$ for some $\phi \in \mathbb{R}$, and

$$
|\langle x, y \rangle| = \langle e^{-i\phi} x, y \rangle = \left| \mathrm{Re}\langle e^{-i\phi} x, y \rangle \right| \leq \|e^{-i\phi} x\| \, \|y\| = \|x\| \, \|y\|.
$$

This completes the proof.   □

Corollary B.5.7 (Triangle inequality). *Let $\mathcal{H}$ be an inner product space. Then*

$$
\|x + y\| \leq \|x\| + \|y\|.
$$

Consequently, the canonical norm of an inner product space is a norm in the usual sense.

Proof. Now

$$
\begin{aligned}
\|x + y\|^2 \;=\;& \langle x + y, x + y \rangle \\
=\;& \langle x, x \rangle + \langle x, y \rangle + \langle y, x \rangle + \langle y, y \rangle \\
\overset{\text{(B.2)}}{\leq}\;& \|x\|^2 + 2\|x\| \, \|y\| + \|y\|^2 \\
=\;& (\|x\| + \|y\|)^2,
\end{aligned}
$$

completing the proof.   □

Remark B.5.8. One may naturally study $\mathbb{R}$-Hilbert spaces, where the scalar field is $\mathbb{R}$ and the inner product takes real values. Then

$$
\langle x, y \rangle = \frac{\|x\|^2 + \|y\|^2 - \|x - y\|^2}{2}.
$$

Thus the inner product can be recovered from the norm here.

Exercise B.5.9. Prove this remark. In ($\mathbb{C}$-) Hilbert spaces, prove that

$$\langle x, y \rangle = \frac{\|x+y\|^2 - \|x-y\|^2 + i\|x+iy\|^2 - i\|x-iy\|^2}{4}.$$

Exercise B.5.10. Every Hilbert space is canonically a Banach space, but not vice versa: in a real Banach space, $(x, y) \mapsto (\|x\|^2 + \|y\|^2 - \|x-y\|^2)/2$ does not always define an inner product. Present some examples.

Lemma B.5.11. *Let $\mathcal{H}$ be a Hilbert space. Suppose $C \subset \mathcal{H}$ is closed, convex and non-empty. Then there exists unique $z \in C$ such that $\|z\| = \inf\{\|x\| : x \in C\}$.*

Proof. Let $r := \inf\{\|x\| : x \in C\}$. For any $x, y \in \mathcal{H}$, the *parallelogram identity*

$$\|x+y\|^2 + \|x-y\|^2 = 2(\|x\|^2 + \|y\|^2) \tag{B.3}$$

holds. Take a sequence $(x_k)_{k=1}^\infty$ in C such that $\|x_k\| \to_{k\to\infty} r$. Now $(x_j + x_k)/2 \in C$ due to convexity, so that $4r^2 \le \|x_j + x_k\|^2$. Hence

$$
\begin{aligned}
4r^2 + \|x_j - x_k\|^2 \quad &\le \quad \|x_j + x_k\|^2 + \|x_j - x_k\|^2 \\
&\overset{(B.3)}{=} \quad 2(\|x_j\|^2 + \|x_k\|^2) \\
&\xrightarrow[j,k\to\infty]{} \quad 4r^2,
\end{aligned}
$$

implying $\|x_j - x_k\| \to_{j,k\to\infty} 0$. Thus $(x_k)_{k=1}^\infty$ is a Cauchy sequence, converging to some $z \in C$ with $\|z\| = r$ (recall that $\mathcal{H}$ is complete and $C \subset \mathcal{H}$ is closed). If $z' \in C$ satisfies $\|z'\| = d$ then the alternating sequence $(z, z', z, z', \ldots)$ would be a Cauchy sequence, by the reasoning above: hence $z = z'$. $\qquad\square$

Exercise B.5.12 (Parallelogram identity). Show that the parallelogram identity (B.3):

$$\|x+y\|^2 + \|x-y\|^2 = 2(\|x\|^2 + \|y\|^2)$$

holds for all $x, y \in \mathcal{H}$.

Lemma B.5.13. *Let M be a vector subspace in a Hilbert space $\mathcal{H}$. Let $\|z\| \le \|z+u\|$ for every $u \in M$. Then $z \in M^\perp$.*

Proof. To get a contradiction, assume $\langle z, v \rangle \ne 0$ for some $v \in M$. Multiplying v by a scalar, we may assume that $\mathrm{Re}\langle z, v \rangle \ne 0$. If $r \in \mathbb{R}$ then

$$0 \le \|z - rv\|^2 - \|z\|^2 = r^2\|v\|^2 - 2r\mathrm{Re}\langle z, v \rangle = r(r\|v\|^2 - 2\mathrm{Re}\langle z, v \rangle),$$

but this inequality fails when r is between 0 and $2\mathrm{Re}\langle z, v \rangle/\|v\|^2$. $\qquad\square$

Definition B.5.14 (Orthogonal projection). Let M be a closed subspace of a Hilbert space $\mathcal{H}$. Then we may define $P_M : \mathcal{H} \to \mathcal{H}$ so that $P_M(x) \in M$ is the point in M closest to $x \in \mathcal{H}$. Mapping P_M is called the *orthogonal projection onto M*.

Proposition B.5.15. *Operator $P_M : \mathcal{H} \to \mathcal{H}$ defined above is linear, and $\|P_M\| = 1$ (unless $M = \{0\}$). Moreover, $P_{M^\perp} = I - P_M$.*

Proof. Let $x \in \mathcal{H}$, $P := P_M$ and $Q = I - P$. By Definition B.5.14, $P(x) \in M$ and

$$\|Q(x)\| \leq \|Q(x) + u\|$$

for every $u \in M$. This implies $Q(x) \in M^\perp$ by Lemma B.5.13. Let $x, y \in \mathcal{H}$ and $\lambda, \mu \in \mathbb{C}$. Since

$$\begin{aligned} \lambda x &= \lambda \left(P(x) + Q(x) \right), \\ \mu y &= \mu \left(P(y) + Q(y) \right), \\ \lambda x + \mu y &= P(\lambda x + \mu y) + Q(\lambda x + \mu y), \end{aligned}$$

we get

$$M \ni P(\lambda x + \mu y) - \lambda P(x) - \mu P(y) = \lambda Q(x) + \mu Q(y) - Q(\lambda x + \mu y) \in M^\perp.$$

This implies the linearity of P, because $M \cap M^\perp = \{0\}$. Finally,

$$\|x\|^2 = \|Px + Qx\|^2 = \|Px\|^2 + \|Qx\|^2 + 2\Re\langle Px, Qx \rangle = \|Px\|^2 + \|Qx\|^2;$$

in particular, $\|Px\| \leq \|x\|$. $\square$

Remark B.5.16. We have proven that

$$\mathcal{H} = M \oplus M^\perp.$$

This means that $M, M^\perp$ are closed subspaces of the Hilbert space $\mathcal{H}$ such that $M \perp M^\perp$ and that $M + M^\perp = \mathcal{H}$.

Definition B.5.17 (Direct sum). Let $\{\mathcal{H}_j : j \in J\}$ be a family of pair-wise orthogonal closed subspaces of $\mathcal{H}$. If the span of $\bigcup_{j \in J} \mathcal{H}_j$ is dense in $\mathcal{H}$ then H is said to be a *direct sum* of $\{\mathcal{H}_j : j \in J\}$, denoted by

$$\mathcal{H} = \bigoplus_{j \in J} \mathcal{H}_j.$$

If $\mathcal{H}$ is a direct sum of $\{M_j\}_{j=1}^k$, we write $\mathcal{H} = \bigoplus_{j=1}^k M_j$. Especially, $M_1 \oplus M_2 = \bigoplus_{j=1}^2 M_j$.

Remark B.5.18. If $\mathcal{H}$ is a Hilbert space, it is easy to see that $f = (x \mapsto \langle x, y \rangle) : \mathcal{H} \to \mathbb{C}$ is a linear functional, and $\|f\| = \|y\|$ due to the Cauchy–Schwarz inequality and to $f(y) = \|y\|^2$. Actually, there are no other kinds of bounded linear functionals on a Hilbert space:

Theorem B.5.19 (Riesz (Hilbert Space) Representation Theorem). *Let $f : \mathcal{H} \to \mathbb{C}$ be a bounded linear functional on a Hilbert space $\mathcal{H}$. Then there exists a unique $y \in \mathcal{H}$ such that*

$$f(x) = \langle x, y \rangle$$

for every $x \in \mathcal{H}$. Moreover, $\|f\| = \|y\|$.

Sometimes this theorem is also called the Fréchet–Riesz (representation) theorem.

Proof. Assume the non-trivial case $f \neq 0$. Thus we may choose $u \in \mathrm{Ker}(f)^\perp$ for which $\|u\| = 1$. Pursuing for a suitable representative $y \in \mathcal{H}$, we notice that $f(u) = \langle u, f(u)u \rangle$, inspiring an investigation:

$$
\begin{aligned}
\langle x, \overline{f(u)}u \rangle - f(x) &= \langle f(u)x, u \rangle - \langle f(x)u, u \rangle \\
&= \langle f(u)x - f(x)u, u \rangle \\
&= 0,
\end{aligned}
$$

since $f(u)x - f(x)u \in \mathrm{Ker}(f)$. Thus $f(x) = \langle x, \overline{f(u)}u \rangle$ for every $x \in \mathcal{H}$. Furthermore, if $f(x) = \langle x, y \rangle = \langle x, z \rangle$ for every $x \in \mathcal{H}$ then

$$0 = f(x) - f(x) = \langle x, y \rangle - \langle x, z \rangle = \langle x, y - z \rangle \overset{x=y-z}{=} \|y - z\|^2,$$

so that $y = z$. $\qquad\square$

Definition B.5.20 (Adjoint operator). Let $\mathcal{H}$ be a Hilbert space, $z \in \mathcal{H}$ and $A \in \mathcal{L}(\mathcal{H})$. Then a bounded linear functional on $\mathcal{H}$ is defined by $x \mapsto \langle Ax, z \rangle$, so that by Theorem B.5.19 there exists a unique vector $A^*z \in \mathcal{H}$ satisfying

$$\langle Ax, z \rangle = \langle x, A^*z \rangle$$

for every $x \in \mathcal{H}$. This defines a mapping $A^* : \mathcal{H} \to \mathcal{H}$, which is called the *adjoint* of $A \in \mathcal{L}(\mathcal{H})$. If $A^* = A$ then A is called *self-adjoint*.

Exercise B.5.21. Let $\lambda \in \mathbb{C}$ and $A, B \in \mathcal{L}(\mathcal{H})$. Show that $(\lambda A)^* = \overline{\lambda}A^*$, $(A+B)^* = A^* + B^*$ and $(AB)^* = B^*A^*$.

Exercise B.5.22. Show that the adjoint operator $A^* : \mathcal{H} \to \mathcal{H}$ of $A \in \mathcal{L}(\mathcal{H})$ is linear and bounded. Moreover, show that $(A^*)^* = A$, $\|A^*A\| = \|A\|^2$ and $\|A^*\| = \|A\|$.

Lemma B.5.23. *Let $A^* = A \in \mathcal{L}(\mathcal{H})$. Then*

$$\|A\| = \sup_{x:\ \|x\|\leq 1} |\langle Ax, x \rangle|.$$

Proof. Let $r := \sup\{|\langle Ax, x \rangle| : x \in \mathcal{H}, \|x\| \leq 1\}$. Then

$$r \overset{(B.2)}{\leq} \sup_{x:\ \|x\|\leq 1} \|Ax\|\,\|x\| \leq \|A\|.$$

Let us assume that $Ax \neq 0$ for $\|x\| = 1$, and let $y := Ax/\|Ax\|$. Since $A^* = A$, we have $\langle Ax, y \rangle = \langle x, Ay \rangle = \langle Ay, x \rangle \in \mathbb{R}$, so that

$$
\begin{aligned}
\|Ax\| &= \langle Ax, y \rangle \\
&\overset{A^*=A}{=} \frac{1}{4} \left(\langle A(x+y), x+y \rangle - \langle A(x-y), x-y \rangle \right) \\
&\leq \frac{1}{4} \left(|\langle A(x+y), x+y \rangle| + |\langle A(x-y), x-y \rangle| \right) \\
&\leq \frac{r}{4} \left(\|x+y\|^2 + \|x-y\|^2 \right) \\
&\overset{(B.3)}{=} \frac{r}{2} \left(\|x\|^2 + \|y\|^2 \right) \\
&= r.
\end{aligned}
$$

This concludes the proof. $\square$

Lemma B.5.24. *Let $\mathcal{H} \neq \{0\}$. Let $A^* = A \in \mathcal{L}(\mathcal{H})$ be compact. Then there exists a non-zero $x \in \mathcal{H}$ such that $Ax = +\|A\|x$ or $Ax = -\|A\|x$.*

Proof. Assume the non-trivial case $\|A\| > 0$. By Lemma B.5.23, we may choose $\lambda \in \{\pm\|A\|\}$ to be an accumulation point of the set $\{\langle Ax, x \rangle : x \in \mathcal{H}, \|x\| \leq 1\}$. For each $k \in \mathbb{Z}^+$, take $x_k \in \mathcal{H}$ such that $\|x_k\| \leq 1$ and $\langle Ax_k, x_k \rangle \to_{k \to \infty} \lambda$. Since A is compact, by Theorem A.13.4 it follows that the sequence $(Ax_k)_{k=1}^\infty$ has a convergent subsequence; omitting elements from the sequence, we may assume that $z := \lim_k Ax_k \in \mathcal{H}$ exists. Now

$$
\begin{aligned}
0 &\leq \|Ax_k - \lambda x_k\|^2 \\
&= \|Ax_k\|^2 + \lambda^2 \|x_k\|^2 - 2\lambda\langle Ax_k, x_k \rangle \\
&\leq \|A\|^2 + \lambda^2 - 2\lambda\langle Ax_k, x_k \rangle \\
&\xrightarrow[k \to \infty]{} 0,
\end{aligned}
$$

implying that $\lim_k \lambda x_k$ exists and is equal to $\lim_k Ax_k = z$. Finally, let $x := z/\lambda$, so that by continuity $Ax = \lim_k Ax_k = \lambda x$. $\square$

Theorem B.5.25 (Diagonalisation of compact self-adjoint operators). *Let $\mathcal{H}$ be infinite-dimensional and $A^* = A \in \mathcal{L}(\mathcal{H})$ be compact. Then there exist $\{\lambda_k\}_{k=1}^\infty \subset \mathbb{R}$ and an orthonormal set $\{x_k\}_{k=1}^\infty \subset \mathcal{H}$ such that $|\lambda_{k+1}| \leq |\lambda_k|$, $\lim_k \lambda_k = 0$ and*

$$
Ax = \sum_{k=1}^\infty \lambda_k \langle x, x_k \rangle \, x_k
$$

for every $x \in \mathcal{H}$.

Proof. By Lemma B.5.24, take $\lambda_1 \in \mathbb{R}$ and $x_1 \in \mathcal{H}$ such that $\|x_1\| = 1$, $Ax_1 = \lambda_1 x_1$ and $\|A_1\| = |\lambda_1|$. Then we proceed by induction as follows. Let $\mathcal{H}_k := \left(\{x_j\}_{j=1}^{k-1} \right)^\perp$.

Then $A_k^* = A_k := A|_{\mathcal{H}_k} \in \mathcal{L}(\mathcal{H}_k)$ is compact as it is a finite-dimensional operator, so we may apply Lemma B.5.24 to choose $\lambda_k \in \mathbb{R}$ and $x_k \in \mathcal{H}_k$ such that $\|x_k\| = 1$, $Ax_k = \lambda_k x_k$ and $\|A_k\| := |\lambda_k|$. Since $\mathcal{H}$ is infinite-dimensional, we obtain an orthonormal family $\{x_k\}_{k=1}^{\infty} \subset \mathcal{H}$, and $Ax_k = \lambda_k x_k$ for each $k \in \mathbb{Z}^+$, where $|\lambda_{k+1}| \leq |\lambda_k|$.

Since A is compact, $(Ax_k)_{k=1}^{\infty}$ has a converging subsequence. Actually, $(A_k)_{k=1}^{\infty}$ itself must converge and $\lambda_k \to 0$, because

$$\|Ax_j - Ax_k\| = \|\lambda_j x_j - \lambda_k x_k\| = \sqrt{\lambda_j^2 + \lambda_k^2} \geq |\lambda_k|$$

for every $j, k \in \mathbb{Z}^+$. If $x \in \mathcal{H}$ then $z_k := x - \sum_{j=1}^{k-1}\langle x, x_k\rangle\, x_k \in \mathcal{H}_k$, and

$$\|Az_k\| = \|A_k z_k\| \leq \|A_k\|\|z_k\| = |\lambda_k|\,\|z_k\| \leq |\lambda_k|\,\|x\| \xrightarrow[k\to\infty]{} 0,$$

completing the proof. $\qquad\qquad\qquad\qquad\qquad\qquad\qquad\qquad\qquad\qquad\qquad\qquad\square$

Corollary B.5.26 (Hilbert–Schmidt Spectral Theorem). *Let $A^* = A \in \mathcal{L}(\mathcal{H})$ be compact. Then $\sigma(A)$ is at most countable, and $\operatorname{Ker}(\lambda I - A)$ is finite-dimensional if $0 \neq \lambda \in \sigma(A)$. Moreover, $\sigma(A) \setminus \{0\}$ is discrete, and*

$$\mathcal{H} = \bigoplus_{\lambda \in \sigma(A)} \operatorname{Ker}(\lambda I - A).$$

Exercise B.5.27. Prove the Hilbert–Schmidt Spectral Theorem using Theorem B.5.25.

Definition B.5.28 (Weak topology on a Hilbert space). The *weak topology* of a Hilbert space $\mathcal{H}$ is the smallest topology for which mappings

$$(u \mapsto \langle u, v\rangle_{\mathcal{H}}) : \mathcal{H} \to \mathbb{C}$$

are continuous for all $v \in \mathcal{H}$.

Exercise B.5.29 (Weak = weak* in Hilbert spaces). Show that Hilbert spaces are reflexive. Prove that in a Hilbert space the weak topology is the same as the weak*-topology, introduced in Definition B.4.35.

As a consequence of Exercise B.5.29 and the Banach–Alaoglu Theorem B.4.36 we obtain:

Theorem B.5.30 (Banach–Alaoglu Theorem for Hilbert spaces). *Let $\mathcal{H}$ be a Hilbert space. Its closed unit ball*

$$\overline{\mathbb{B}} = \{v \in \mathcal{H} : \|v\|_{\mathcal{H}} \leq 1\}$$

is compact in the weak topology.

Exercise B.5.31. Let $\{e_\alpha\}_{\alpha \in I}$ be an orthonormal collection in $\mathcal{H}$ and let $x \in \mathcal{H}$. Show that

$$\sum_{\alpha \in I} |\langle x, e_\alpha \rangle|^2 \leq \|x\|^2. \tag{B.4}$$

(Hint: Pythagoras' theorem.) Consequently, deduce from Exercise B.4.3 that the set of α such that $\langle x, e_\alpha \rangle \neq 0$ is at most countable.

We finish with the following theorem which is of importance, because it allows one to decompose elements into "simpler ones", which is particularly important in applications.

Theorem B.5.32 (Orthonormal sets in a Hilbert space). *Let $\{e_\alpha\}_{\alpha \in I}$ be an orthonormal set in the Hilbert space $\mathcal{H}$. Then the following conditions are equivalent:*

(i) *For every $x \in \mathcal{H}$ there are only countably many $\alpha \in I$ such that $\langle x, e_\alpha \rangle \neq 0$, and the equality*

$$x = \sum_{\alpha \in I} \langle x, e_\alpha \rangle e_\alpha$$

holds, where the series is converging in norm, independent of any ordering of its terms.

(ii) *If $\langle x, e_\alpha \rangle = 0$ for all $\alpha \in I$, then $x = 0$.*

(iii) *(Plancherel's identity) For every $x \in \mathcal{H}$ it holds that $\|x\|^2 = \sum_{\alpha \in I} |\langle x, e_\alpha \rangle|^2$.*

Proof. (i) $\Rightarrow$ (iii). This follows by enumerating countably many e_α's with $\langle x, e_\alpha \rangle \neq 0$ by $\{e_j\}_{j=1}^\infty$, and using the identity

$$\|x\|^2 - \sum_{j=1}^n |\langle x, e_j \rangle|^2 = \|x - \sum_{j=1}^n \langle x, e_j \rangle e_j\|^2.$$

(iii) $\Rightarrow$ (ii) is automatic. Finally, let us show (ii) $\Rightarrow$ (i). It follows from the last part of Exercise B.5.31 that the collection of e_α with $\langle x, e_\alpha \rangle \neq 0$ is countable, so it can be enumerated by $\{e_j\}_{j=1}^\infty$. Now, the identity

$$\|\sum_{j=j_1}^{j_2} \langle x, e_j \rangle e_j\|^2 = \sum_{j=j_1}^{j_2} \|\langle x, e_j \rangle\|^2$$

and (B.4) imply that the right-hand side $\to 0$ as $j_1, j_2 \to \infty$. This means that the series $\sum_{j=1}^\infty \langle x, e_j \rangle e_j$ converges. Setting $y := x - \sum_{j=1}^\infty \langle x, e_j \rangle e_j$ we see that $\langle y, e_\alpha \rangle = 0$ for all $\alpha \in I$, which implies that $y = 0$. $\square$

Exercise B.5.33. Verify the identities stated in the proof.

Definition B.5.34 (Orthonormal basis). An orthonormal set satisfying conditions of Theorem B.5.32 is called an *orthonormal basis* of the Hilbert space $\mathcal{H}$. Then we have the following properties

Theorem B.5.35 (Every Hilbert space has an orthonormal basis). *Every Hilbert space $\mathcal{H}$ has an orthonormal basis. An orthonormal basis is countable if and only if $\mathcal{H}$ is separable, in which case any other basis is also countable.*

Exercise B.5.36. Prove Theorem B.5.35: the first part follows from Zorn's lemma if we order orthonormal collections by inclusion, since the maximal element would satisfy property (ii) of Theorem B.5.32. The second part follows from the Gram–Schmidt process;

Exercise B.5.37 (Gram–Schmidt orthonormalisation process). Let $\{x_k\}_{k=1}^{\infty}$ be a linearly independent family of vectors in a Hilbert space $\mathcal{H}$. Let

$$\begin{cases} y_1 := x_1, \\[2mm] e_k := y_k/\|y_k\|, \quad \text{and} \\[2mm] y_{k+1} := x_{k+1} - \displaystyle\sum_{j=1}^{k} \langle x_{k+1},\, e_j \rangle e_j \end{cases}$$

for all $k \in \mathbb{Z}^+$. Show that $\{e_k\}_{k=1}^{\infty}$ is an orthonormal set in $\mathcal{H}$, such that

$$\operatorname{span}\{e_k\}_{k=1}^{n} = \operatorname{span}\{x_k\}_{k=1}^{n}$$

for every $n \in \mathbb{Z}^+ \cup \{\infty\}$.

B.5.1 Trace class, Hilbert–Schmidt, and Schatten classes

Definition B.5.38 (Trace class operators). Let $\mathcal{H}$ be a Hilbert space with orthonormal basis $\{e_j \mid j \in J\}$. Let $A \in \mathcal{L}(\mathcal{H})$. Let us write

$$\|A\|_{S_1} := \sum_{j \in J} |\langle Ae_j, e_j \rangle_{\mathcal{H}}|;$$

this is the *trace norm* of A, and the *trace class* is the (Banach) space

$$S_1 = S_1(\mathcal{H}) := \{A \in \mathcal{L}(\mathcal{H}): \ \|A\|_{S_1} < \infty\}.$$

The *trace* is the linear functional $\operatorname{Tr}: S_1(\mathcal{H}) \to \mathbb{C}$, defined by

$$A \mapsto \sum_{j \in J} \langle Ae_j, e_j \rangle_{\mathcal{H}}.$$

Exercise B.5.39. Verify that the definition of the trace is independent of the choice of the orthonormal basis for $\mathcal{H}$. Consequently, if $(a_{ij})_{i,j \in J}$ is the matrix representation of $A \in S_1$ with respect to the chosen basis, then $\operatorname{Tr}(A) = \sum_{j \in J} a_{jj}$.

Exercise B.5.40 (Properties of trace). Prove the following properties of the trace functional:

$$
\begin{aligned}
\mathrm{Tr}(AB) &= \mathrm{Tr}(BA), \\
\mathrm{Tr}(A^*) &= \overline{\mathrm{Tr}(A)}, \\
\mathrm{Tr}(A^*A) &\geq 0, \\
\mathrm{Tr}(A \oplus B) &= \mathrm{Tr}(A) + \mathrm{Tr}(B), \\
\dim(\mathcal{H}) < \infty \;\Rightarrow\; &\begin{cases} \mathrm{Tr}(I_{\mathcal{H}}) = \dim(\mathcal{H}), \\ \mathrm{Tr}(A \otimes B) = \mathrm{Tr}(A)\,\mathrm{Tr}(B). \end{cases}
\end{aligned}
$$

Exercise B.5.41 (Trace on a finite-dimensional space). Show that the trace on a finite-dimensional vector space is independent of the choice of inner product. Thus, the trace of a square matrix is defined to be the sum of its diagonal elements; moreover, the trace is the sum of the eigenvalues (with multiplicities counted).

Exercise B.5.42. Let $\mathcal{H}$ be finite-dimensional. Let $f : \mathcal{L}(\mathcal{H}) \to \mathbb{C}$ be a linear functional satisfying

$$
\begin{cases}
f(AB) = f(BA), \\
f(A^*A) \geq 0, \\
f(I_{\mathcal{H}}) = \dim(\mathcal{H})
\end{cases}
$$

for all $A, B \in \mathcal{L}(\mathcal{H})$. Show that $f = \mathrm{Tr}$.

Definition B.5.43 (Hilbert-Schmidt operators). The space of *Hilbert–Schmidt operators* is

$$
S_2 = S_2(\mathcal{H}) := \{A \in \mathcal{L}(\mathcal{H}) : \; A^*A \in S_1(\mathcal{H})\},
$$

and it can be endowed with a Hilbert space structure via the inner product

$$
\langle A, B \rangle_{S_2} := \mathrm{Tr}(AB^*).
$$

The Hilbert–Schmidt norm is then

$$
\|A\|_{HS} = \|A\|_{S_2} := \langle A, A \rangle_{S_2}^{1/2}.
$$

The case of the Hilbert–Schmidt norm on the finite-dimensional spaces will be discussed in more detail in Section 12.6.

Remark B.5.44. In general, there are inclusions $S_1 \subset S_2 \subset \mathcal{K} \subset S_\infty$, where $S_\infty := \mathcal{L}(\mathcal{H})$ and $\mathcal{K} \subset S_\infty$ is the subspace of compact linear operators. Moreover,

$$
\|A\|_{S_\infty} \leq \|A\|_{S_2} \leq \|A\|_{S_1}
$$

for all $A \in S_\infty$. One can show that the dual $\mathcal{K}' = \mathcal{L}(\mathcal{K}, \mathbb{C})$ is isometrically isomorphic to S_1, and that $(S_1)'$ is isometrically isomorphic to S_∞. In the latter case, it turns out that a bounded linear functional on S_1 is of the form $A \mapsto \mathrm{Tr}(AB)$ for some $B \in S_\infty$. These phenomena are related to properties of the sequence spaces $\ell^p = \ell^p(\mathbb{Z}^+)$. In analogy to the operator spaces, $\ell^1 \subset \ell^2 \subset c_0 \subset \ell^\infty$, where c_0 is the space of sequences converging to 0, playing the counterpart of space $\mathcal{K}$.

Remark B.5.45 (**Schatten classes**). Trace class operators S_1 and Hilbert–Schmidt operators S_2 turn out to be special cases of the *Schatten classes* S_p, $1 \le p < \infty$. These classes can be introduced with the help of the singular values $\mu^2 \in \sigma(A^*A)$. To avoid the technicalities we assume that all the operators below are compact. Thus, for $A \in \mathcal{L}(\mathcal{H})$ we set

$$\|A\|_{S_p} := \left(\sum_{\mu^2 \in \sigma(A^*A)} \mu^p \right)^{1/p}.$$

We note that operators that satisfy $\|A\|_{S_p} < \infty$ must have at most countable spectrum $\sigma(A^*A)$ in view of Exercise B.4.3, but in our case this is automatically satisfied since we assumed that A is compact. Therefore, denoting the sequence of singular values $\mu_j^2 \in \sigma(A^*A)$, counted with multiplicities, we have

$$\|A\|_{S_p} = \|\{\mu_j\}_j\|_{\ell^p}.$$

The Schatten class S_p is then defined as the space

$$S_p = S_p(\mathcal{H}) := \left\{ A \in \mathcal{L}(\mathcal{H}) : \|A\|_{S_p} < \infty \right\}.$$

With this norm, $S_p(\mathcal{H})$ is a Banach space, and $S_2(\mathcal{H})$ is a Hilbert space. In analogy to the trace class and Hilbert–Schmidt operators, one can show that actually $\|A\|_{S_p}^p = \mathrm{Tr}(|A^*A|^{p/2}) = \mathrm{Tr}(|A|^p)$ for a compact operator A.

Exercise B.5.46. Show that the Schatten classes S_1 and S_2 coincide with the previously defined trace class and Hilbert–Schmidt class, respectively.

Exercise B.5.47 (Hölder's inequality for Schatten classes). Show that a Schatten class S_p is an ideal in $\mathcal{L}(\mathcal{H})$. Let $\mathcal{H}$ be separable. Show that if $1 \le p \le \infty$, $\frac{1}{p} + \frac{1}{q} = 1$, $A \in S_p$ and $B \in S_q$, then

$$\|AB\|_{S_1} \le \|A\|_{S_p} \|B\|_{S_q}.$$

(Hint: approximate operators by matrices.)

Chapter C

Measure Theory and Integration

This chapter provides sufficient general information about measures and integration for the purposes of this book. The starting point is the concept of an outer measure, which "measures weights of subsets of a space". We should first consider how to sum such weights, which are either infinite or non-negative real numbers. For a finite set K, notation

$$\sum_{j \in K} a_j$$

abbreviates the usual sum of numbers $a_j \in [0, \infty]$ over the index set K. The conventions here are that $a + \infty = \infty$ for all $a \in [0, \infty]$, and that

$$\sum_{j \in \emptyset} a_j = 0.$$

Infinite summations are defined by limits as follows:

Definition C.0.1. The *sum* of numbers $a_j \in [0, \infty]$ over the index set J is

$$\sum_{j \in J} a_j := \sup \left\{ \sum_{j \in K} a_j : \quad K \subset J \text{ is finite} \right\}.$$

Exercise C.0.2. Let $0 < a_j < \infty$ for each $j \in J$. Suppose

$$\sum_{j \in J} a_j < \infty.$$

Show that J is at most countable.

The message of Exercise C.0.2 is that for positive numbers, only countable summations are interesting. In measure theory, where summations are fundamental, such a "restriction to countability" will be encountered recurrently.

C.1   Measures and outer measures

Recall that for a set X, by $\mathcal{P}(X) := \{E \mid E \subset X\}$ we denote its *power set*, i.e., the family of all subsets of X. Let us write $E^c := X \setminus E = \{x \in X : x \notin E\}$ for the complement set, when the space X is implicitly known from the context.

C.1.1   Measuring sets

Definition C.1.1 (Outer measure). A mapping $\psi : \mathcal{P}(X) \to [0, \infty]$ is an *outer measure* on a set $X \neq \emptyset$ if

$$\psi(\emptyset) = 0,$$

$$E \subset F \quad \Rightarrow \quad \psi(E) \leq \psi(F),$$

$$\psi\left(\bigcup_{j=1}^{\infty} E_j\right) \leq \sum_{j=1}^{\infty} \psi(E_j)$$

for every $E, F \subset X$ and $\{E_j\}_{j=1}^{\infty} \subset \mathcal{P}(X)$.

Intuitively, an outer measure is weighs the subsets of a space.

Example. Define $\psi : \mathcal{P}(X) \to [0, \infty]$ by $\psi(\emptyset) = 0$ and $\psi(E) = 1$, when $\emptyset \neq E \subset X$. This is an outer measure.

Example. Let $\psi : \mathcal{P}(X) \to [0, \infty]$, where $\psi(E)$ is the number of points in the set $E \subset X$. Such an outer measure is called a *counting measure* for obvious reasons.

At first sight, constructing meaningful non-trivial outer measures may appear difficult. However, there is an easy and useful method for generating outer measures out of simpler set functions, which we call the measurelets:

Definition C.1.2 (Measurelets). Let $\mathcal{A} \subset \mathcal{P}(X)$ cover X, i.e., $X = \bigcup \mathcal{A}$. We call a mapping $m : \mathcal{A} \to [0, \infty]$ a *measurelet* on X. Members of the family $\mathcal{A}$ are called the *elementary sets*. A measurelet $m : \mathcal{A} \to [0, \infty]$ on X *generates* a mapping $m^* : \mathcal{P}(X) \to [0, \infty]$ defined by

$$m^*(E) := \inf\left\{\sum_{A \in \mathcal{B}} m(A) : \quad \mathcal{B} \subset \mathcal{A} \text{ is countable, } E \subset \bigcup \mathcal{B}\right\}.$$

Exercise C.1.3. Let

$$\mathcal{A} := \{\emptyset, \mathbb{R}^2\} \cup \{S \subset \mathbb{R}^2 : S \text{ a finite union of polygons}\}.$$

Let us define a measurelet $A : \mathcal{A} \to [0, \infty]$ by the following informal demands:

(1) $A(rectangle) = base \cdot height$.

(2) $A(S_1 \cup S_2) = A(S_1) + A(S_2)$, if the interiors of the sets S_1, S_2 are disjoint.

(3) The measurelet A does not change in translations nor rotations of sets.

Using these rules, calculate the measurelets of a parallelogram and a triangle.

Apparently, there are plenty of measurelets: almost anything goes. Especially, outer measures are measurelets.

Theorem C.1.4. *Let* $m : \mathcal{A} \to [0,\infty]$ *be a measurelet on a set* X. *Then* $m^* : \mathcal{P}(X) \to [0,\infty]$ *is an outer measure for which* $m^*(A) \le m(A)$ *for every* $A \in \mathcal{A}$.

Proof. Clearly, $m^* : \mathcal{P}(X) \to [0,\infty]$ is well defined, and $m^*(A) \le m(A)$ for every $A \in \mathcal{A}$. We see that $m^*(\emptyset) = 0$, because $\sum_{A \in \emptyset} m(A) = 0$, $\emptyset \subset \mathcal{A}$ is countable, and $\emptyset \subset \bigcup \emptyset$. Next, if $E \subset F \subset X$ then $m^*(E) \le m^*(F)$, because any cover $\{A_j\}_{j=1}^{\infty}$ of F is also a cover of E. Lastly, let $\{E_j\}_{j=1}^{\infty} \subset \mathcal{P}(X)$. Take $\varepsilon > 0$. For each $j \ge 1$, choose $\{A_{jk}\}_{k=1}^{\infty} \subset \mathcal{A}$ such that

$$E_j \subset \bigcup_{k=1}^{\infty} A_{jk} \quad \text{and} \quad m^*(E_j) + 2^{-j}\varepsilon \ge \sum_{k=1}^{\infty} m(A_{jk}).$$

Then $\{A_{jk}\}_{j,k=1}^{\infty} \subset \mathcal{A}$ is a countable cover of $\bigcup_{j=1}^{\infty} E_j \subset X$, and

$$m^* \left(\bigcup_{j=1}^{\infty} E_j \right) \le \sum_{j=1}^{\infty} \sum_{k=1}^{\infty} m(A_{jk})$$

$$\le \sum_{j=1}^{\infty} m^*(E_j) + \varepsilon.$$

Thus $m^*(\bigcup_{j=1}^{\infty} E_j) \le \sum_{j=1}^{\infty} m^*(E_j)$; the proof is complete. $\qquad\square$

Definition C.1.5 (Lebesgue's outer measure). On the Euclidean space $X = \mathbb{R}^n$, let us define the partial order $\le$ by

$$a \le b \quad \overset{\text{definition}}{\Longleftrightarrow} \quad \forall j \in \{1,\dots,n\} : \ a_j \le b_j.$$

When $a \le b$, let the n-*interval* be

$$[a,b] := [a_1, b_1] \times \cdots \times [a_n, b_n] = \{x \in \mathbb{R}^n : \ a \le x \le b\}.$$

For $\mathcal{A} = \{[a,b] : \ a,b \in X, \ a \le b\}$ let us define the *Lebesgue measurelet* $m : \mathcal{A} \to [0,\infty]$ by

$$m([a,b]) := \text{volume}([a,b]) = \prod_{j=1}^{n} |a_j - b_j|.$$

Then the generated outer measure $\lambda^* = \lambda_{\mathbb{R}^n}^* := m^* : \mathcal{P}(\mathbb{R}^n) \to [0,\infty]$ is called the *Lebesgue outer measure* of $\mathbb{R}^n$.

Exercise C.1.6. Give an example of an outer measure that cannot be generated by a measurelet.

Definition C.1.7 (Outer measure measurability). Let $\psi : \mathcal{P}(X) \to [0,\infty]$ be an outer measure. A set $E \subset X$ is called ψ-*measurable* if

$$\psi(S) = \psi(E \cap S) + \psi(E^c \cap S)$$

for every $S \subset X$, where $E^c = X \backslash E$. The family of ψ-measurable sets is denoted by

$$\mathcal{M}(\psi) \subset \mathcal{P}(X).$$

Remark C.1.8. Notice that trivially

$$\psi(S) \leq \psi(E \cap S) + \psi(E^c \cap S)$$

by the properties of the outer measure. Intuitively, a measurable set E "sharply cuts" "rough" sets $S \subset X$ into two disjoint pieces, $E \cap S$ and $E^c \cap S$.

Remark C.1.9 (**Non-measurability**). The Axiom of Choice can be used to "construct" a subset $E \subset \mathbb{R}^n$ which is not Lebesgue measurable. We will discuss this topic in Section C.1.4.

Exercise C.1.10. Let $\psi : \mathcal{P}(X) \to [0,\infty]$ be an outer measure and $E \subset X$. Define $\psi_E : \mathcal{P}(X) \to [0,\infty]$ by $\psi_E(S) := \psi(E \cap S)$. Show that ψ_E is an outer measure for which $\mathcal{M}(\psi) \subset \mathcal{M}(\psi_E)$.

Lemma C.1.11. *Let* $\psi : \mathcal{P}(X) \to [0,\infty]$ *be an outer measure and* $\psi(E) = 0$. *Then* $E \in \mathcal{M}(\psi)$.

Proof. Let $S \subset X$. Then

$$\begin{aligned} \psi(S) &\leq \psi(E \cap S) + \psi(E^c \cap S) \\ &\leq \psi(E) + \psi(S) \\ &= \psi(S), \end{aligned}$$

so that $\psi(S) = \psi(E \cap S) + \psi(E^c \cap S)$; set E is ψ-measurable. $\qquad\square$

Lemma C.1.12. *Let* $E, F \in \mathcal{M}(\psi)$. *Then* $E^c, E \cap F, E \cup F \in \mathcal{M}(\psi)$.

Proof. The definition of ψ-measurability is clearly complement symmetric, so that $E \in \mathcal{M}(\psi) \iff E^c \in \mathcal{M}(\psi)$. Next, it is sufficient to deal with $E \cup F$, since $E \cap F = (E^c \cup F^c)^c$. Take $S \subset X$. Then

$$\begin{aligned} \psi(S) \quad &\leq \quad \psi((E \cup F) \cap S) + \psi((E \cup F)^c \cap S) \\ &= \quad \psi((E \cup F) \cap S) + \psi(E^c \cap F^c \cap S) \\ &\overset{E \in \mathcal{M}(\psi)}{=} \quad \psi(E \cap S) + \psi(E^c \cap F \cap S) + \psi(E^c \cap F^c \cap S) \\ &\overset{F \in \mathcal{M}(\psi)}{=} \quad \psi(E \cap S) + \psi(E^c \cap S) \\ &\overset{E \in \mathcal{M}(\psi)}{=} \quad \psi(S). \end{aligned}$$

Hence $\psi(S) = \psi((E \cup F) \cap S) + \psi((E \cup F)^c \cap S)$, so that $E \cup F$ is ψ-measurable. $\quad\square$

Exercise C.1.13. Let $\psi : \mathcal{P}(X) \to [0, \infty]$ be an outer measure. Let $E \subset S \subset X$, $E \in \mathcal{M}(\psi)$ and $\psi(E) < \infty$. Show that $\psi(S \setminus E) = \psi(S) - \psi(E)$.

Definition C.1.14 (σ-algebras). A family $\mathcal{M} \subset \mathcal{P}(X)$ is called a σ-algebra on X (pronounced: *sigma-algebra*) if

1. $\bigcup \mathcal{E} \in \mathcal{M}$ for every countable collection $\mathcal{E} \subset \mathcal{M}$, and
2. $E^c \in \mathcal{M}$ for every $E \in \mathcal{M}$.

Remark C.1.15. Here, recall the conventions for the union and the intersection of the empty family: for $\mathcal{A} = \emptyset \subset \mathcal{P}(X)$, we naturally define $\bigcup \mathcal{A} = \emptyset$, but notice that $\bigcap \mathcal{A} = X$ (this is not as surprising as it might first seem). Thereby $\mathcal{M}$ is a σ-algebra on X if and only if

1. $\bigcup_{j=1}^{\infty} E_j \in \mathcal{M}$ whenever $\{E_j\}_{j=1}^{\infty} \subset \mathcal{M}$,
2. $E^c \in \mathcal{M}$ for every $E \in \mathcal{M}$, and
3. $\emptyset \in \mathcal{M}$.

Thus, a σ-algebra on X contains always at least subsets $\emptyset \subset X$ and $X \subset X$.

Proposition C.1.16. *Let $\mathcal{A} \subset \mathcal{P}(X)$. There exists the smallest σ-algebra $\Sigma(\mathcal{A})$ on X containing $\mathcal{A}$, called the σ-algebra generated by $\mathcal{A}$.*

A word of warning: there is no summation in this σ-algebra business here, even though we have used the capital sigma symbol Σ.

Exercise C.1.17. Prove Proposition C.1.16.

Definition C.1.18 (Borel σ-algebra). The *Borel σ-algebra* of a topological space (X, τ) is $\Sigma(\tau) \subset \mathcal{P}(X)$. The members of $\Sigma(\tau)$ are called *Borel sets*.

Definition C.1.19 (Disjoint family). A family $\mathcal{A} \subset \mathcal{P}(X)$ is called *disjoint* if $A \cap B = \emptyset$ for every $A, B \in \mathcal{A}$ for which $A \neq B$.

Remark C.1.20 **(Disjointisation).** In measure theory, the following "disjointisation" process often comes in handy. Let $\mathcal{M}$ be a σ-algebra and $\{E_j\}_{j=1}^{\infty} \subset \mathcal{M}$. Let $F_1 := E_1$ and

$$F_{k+1} := E_{k+1} \setminus \bigcup_{j=1}^{k} E_j.$$

Now $\{F_k\}_{k=1}^{\infty} \subset \mathcal{M}$ is a disjoint family satisfying $F_k \subset E_k$ and

$$\bigcup_{j=1}^{\infty} E_j = \bigcup_{k=1}^{\infty} F_k.$$

Proposition C.1.21. *Let $\psi : \mathcal{P}(X) \to [0, \infty]$ be an outer measure. Let $\{F_k\}_{k=1}^{\infty} \subset \mathcal{M}(\psi)$ be disjoint. Then $\bigcup_{k=1}^{\infty} F_k \in \mathcal{M}(\psi)$ and*

$$\psi\left(\bigcup_{k=1}^{\infty} F_k \cap S\right) = \sum_{k=1}^{\infty} \psi(F_k \cap S) \tag{C.1}$$

for every $S \subset X$, especially

$$\psi \left(\bigcup_{k=1}^{\infty} F_k \right) = \sum_{k=1}^{\infty} \psi(F_k).$$

Proof. Let $E := \bigcup_{k=1}^{\infty} F_k$. Take $S \subset X$. By Lemma C.1.12, $G_n := \bigcup_{k=1}^{n} F_k \in \mathcal{M}(\psi)$. Now

$$
\begin{array}{rl}
\psi(S) \quad \leq & \psi(E \cap S) + \psi(E^c \cap S) \\[2mm]
\leq & \displaystyle\sum_{k=1}^{\infty} \psi(F_k \cap S) + \psi(E^c \cap S) \\[2mm]
= & \displaystyle\lim_{n\to\infty} \left(\sum_{k=1}^{n} \psi(F_k \cap S) + \psi(E^c \cap S) \right) \\[2mm]
\overset{\{F_k\}_{k=1}^{n} \subset \mathcal{M}(\psi) \text{ disjoint}}{=} & \displaystyle\lim_{n\to\infty} \left(\psi(G_n \cap S) + \psi(E^c \cap S) \right) \\[2mm]
\overset{E^c \subset G_n^c}{\leq} & \displaystyle\lim_{n\to\infty} \left(\psi(G_n \cap S) + \psi(G_n^c \cap S) \right) \\[2mm]
\overset{G_n \in \mathcal{M}(\psi)}{=} & \psi(S).
\end{array}
$$

Hence $\psi(S) = \psi(E \cap S) + \psi(E^c \cap S)$, meaning that $E \in \mathcal{M}(\psi)$. Moreover, (C.1) follows from the above chain of (in)equalities. $\qquad\square$

Corollary C.1.22. *Let $\psi : \mathcal{P}(X) \to [0, \infty]$ be an outer measure. For each $k \geq 1$, let $E_k \in \mathcal{M}(\psi)$ be such that $E_k \subset E_{k+1}$. Then*

$$\psi \left(\bigcup_{k=1}^{\infty} E_k \right) = \lim_{k\to\infty} \psi(E_k). \tag{C.2}$$

For each $k \geq 1$, let $F_k \in \mathcal{M}(\psi)$ such that $F_k \supset F_{k+1}$ and $\psi(F_1) < \infty$. Then

$$\psi \left(\bigcap_{k=1}^{\infty} F_k \right) = \lim_{k\to\infty} \psi(F_k). \tag{C.3}$$

Proof. Let us assume that $\psi(E_k) < \infty$ for every $k \geq 1$, for otherwise the first claim is trivial. Thereby

$$
\begin{array}{rl}
\psi(\bigcup_{k=1}^{\infty} E_k) \quad = & \psi \left(E_1 \cup \bigcup_{k=1}^{\infty} (E_{k+1} \setminus E_k) \right) \\[2mm]
\overset{\text{Prop. C.1.21}}{=} & \psi(E_1) + \displaystyle\sum_{k=1}^{\infty} \psi(E_{k+1} \setminus E_k) \\[2mm]
\overset{\text{Exercise C.1.13}}{=} & \psi(E_1) + \displaystyle\lim_{n\to\infty} \sum_{k=1}^{n} (\psi(E_{k+1}) - \psi(E_k)) \\[2mm]
= & \displaystyle\lim_{n\to\infty} \psi(E_{n+1}).
\end{array}
$$

Now

$$
\psi(F_1) \quad = \quad \psi\left(\left(\bigcap_{k=1}^{\infty} F_k\right) \cup \bigcup_{j=1}^{\infty} (F_1 \setminus F_j)\right)
$$

$$
\overset{\text{Prop. C.1.21}}{=} \quad \psi\left(\bigcap_{k=1}^{\infty} F_k\right) + \psi\left(\bigcup_{j=1}^{\infty} (F_1 \setminus F_j)\right)
$$

$$
\overset{\text{(C.2)}}{=} \quad \psi\left(\bigcap_{k=1}^{\infty} F_k\right) + \lim_{j\to\infty} \psi(F_1 \setminus F_j)
$$

$$
\overset{\text{Exercise C.1.13}}{=} \quad \psi\left(\bigcap_{k=1}^{\infty} F_k\right) + \lim_{j\to\infty} (\psi(F_1) - \psi(F_j)),
$$

from which (C.3) follows, since $\psi(F_1) < \infty$. $\qquad\square$

Exercise C.1.23. Give an example of an outer measure $\psi : \mathcal{P}(X) \to [0,\infty]$ and sets $E_k \subset X$ such that $E_k \subset E_{k+1}$ for all $k \in \mathbb{Z}^+$ and

$$
\psi\left(\bigcup_{k=1}^{\infty} E_k\right) \neq \lim_{k\to\infty} \psi(E_k).
$$

Exercise C.1.24. Give an example that shows the indispensability of the assumption $\psi(F_1) < \infty$ in Corollary C.1.22. For instance, find an outer measure $\varphi : \mathcal{P}(\mathbb{Z}) \to [0,\infty]$ and a family $\{F_k\}_{k=1}^{\infty} \subset \mathcal{M}(\varphi)$ for which

$$
\varphi\left(\bigcap_{k=1}^{\infty} F_k\right) \neq \lim_{k\to\infty} \varphi(F_k),
$$

even though $F_k \supset F_{k+1}$ for all k.

Theorem C.1.25. *Let* $\psi : \mathcal{P}(X) \to [0,\infty]$ *be an outer measure. Then the* ψ-*measurable sets form a* σ-*algebra* $\mathcal{M}(\psi)$.

Proof. $\emptyset \in \mathcal{M}(\psi)$ due to Lemma C.1.11. By Lemma C.1.12, we know that $\mathcal{M}(\psi)$ is closed under taking complements. We must prove that it is closed also under taking countable unions. Let $\{E_j\}_{j=1}^{\infty} \subset \mathcal{M}(\psi)$. Applying the disjointisation process of Remark C.1.20, we obtain a disjoint family $\{F_k\}_{k=1}^{\infty} \subset \mathcal{M}(\psi)$, for which $\bigcup_{j=1}^{\infty} E_j = \bigcup_{k=1}^{\infty} F_k$. Exploiting Proposition C.1.21, the proof is concluded. $\qquad\square$

Definition C.1.26 (Measures and measure spaces). Let $\mathcal{M}$ be a σ-algebra on X. A mapping $\mu : \mathcal{M} \to [0,\infty]$ is called a *measure* on X if

$$
\mu(\emptyset) = 0,
$$

$$
\mu\left(\bigcup_{j=1}^{\infty} E_j\right) = \sum_{j=1}^{\infty} \mu(E_j)
$$

whenever $\{E_j\}_{j=1}^\infty \subset \mathcal{M}$ is a disjoint family. Then the triple $(X, \mathcal{M}, \mu)$ is called a *measure space*; such a measure space and the corresponding measure μ are called:

- *finite*, if $\mu(X) < \infty$;
- *probability*, if μ is a finite measure with $\mu(X) = 1$;
- *complete*, if $F \in \mathcal{M}$ whenever there exists $E \in \mathcal{M}$ such that $F \subset E$ and $\mu(E) = 0$;
- *Borel*, if $\mathcal{M} = \Sigma(\tau)$, σ-algebra of the Borel sets in a topological space (X, τ). However, sometimes the Borel condition may mean $\Sigma(\tau) \subset \mathcal{M}$ (more on this later).

Theorem C.1.27. *Let $\psi : \mathcal{P}(X) \to [0, \infty]$ be an outer measure. Then the restriction $\psi|_{\mathcal{M}(\psi)} : \mathcal{M}(\psi) \to [0, \infty]$ is a complete measure.*

Proof. This follows by Proposition C.1.21 and Lemma C.1.11.   □

Exercise C.1.28. Let $\mu_k : \mathcal{M} \to [0, \infty]$ be measures for which $\mu_k(E) \leq \mu_{k+1}(E)$ for every $E \in \mathcal{M}$ (and all $k \in \mathbb{Z}^+$). Show that $\mu : \mathcal{M} \to [0, \infty]$, where

$$\mu(E) := \lim_{k \to \infty} \mu_k(E).$$

Exercise C.1.29 (Borel–Cantelli Lemma). Let $(X, \mathcal{M}, \mu)$ be a measure space, $\{E_j\}_{j=1}^\infty \subset \mathcal{M}$ and

$$E := \left\{ x \in X \mid \{j \in \mathbb{Z}^+ : x \in E_j\} \text{ is infinite} \right\}.$$

Prove that $\mu(E) = 0$ if

$$\sum_{j=1}^\infty \mu(E_j) < \infty.$$

This is the so-called *Borel–Cantelli Lemma*.

Remark C.1.30. By Theorem C.1.4, any measure μ generates the outer measure μ^*, whose restriction $\mu^*|_{\mathcal{M}(\mu^*)}$ is a complete measure, which generates an outer measure, and so on. Fortunately, this back-and-forth-process terminates, as we shall see in Theorems C.1.35 and C.1.36.

Lemma C.1.31. *Let $\mu : \mathcal{M} \to [0, \infty]$ be a measure on X. Then for every $S \subset X$ there exists $A \in \mathcal{M}$ such that*

$$S \subset A : \ \mu^*(S) = \mu(A).$$

Consequently,

$$\mu^*(S) = \min \{\mu(A) : \ S \subset A \in \mathcal{M}\}.$$

Remark C.1.32. An outer measure $\psi : \mathcal{P}(X) \to [0,\infty]$ is called $\mathcal{M}$-*regular* if $\mathcal{M} \subset \mathcal{M}(\psi)$ and

$$\forall S \subset X \,\exists A \in \mathcal{M} : S \subset A, \; \psi(S) = \psi(A);$$

according to Lemma C.1.31, the outer measure μ^* generated by a measure $\mu : \mathcal{M} \to [0, \infty]$ is $\mathcal{M}$-regular.

Proof. If $S \subset X$ then

$$
\begin{aligned}
\mu^*(S) &= \inf \left\{ \sum_{j=1}^{\infty} \mu(A_j) : S \subset \bigcup_{j=1}^{\infty} A_j, \; \{A_j\}_{j=1}^{\infty} \subset \mathcal{M} \right\} \\
&\geq \inf \left\{ \mu \left(\bigcup_{j=1}^{\infty} A_j \right) : S \subset \bigcup_{j=1}^{\infty} A_j, \; \{A_j\}_{j=1}^{\infty} \subset \mathcal{M} \right\} \\
&= \inf \left\{ \mu(A) : S \subset A, \; A \in \mathcal{M} \right\} \\
&\geq \mu^*(S).
\end{aligned}
$$

Thus $\mu^*(S) = \inf\{\mu(A) : S \subset A \in \mathcal{M}\}$. For $\varepsilon > 0$, choose $A_\varepsilon \in \mathcal{M}$ such that $S \subset A_\varepsilon$ and $\mu^*(S) + \varepsilon \geq \mu(A_\varepsilon)$. Let $A_0 := \bigcap_{k=1}^{\infty} A_{1/k} \in \mathcal{M}$. Then $S \subset A_0$, and

$$\mu^*(S) \leq \mu(A_0) \leq \mu(A_\varepsilon) \leq \mu^*(S) + \varepsilon$$

implies $\mu^*(S) = \mu(A_0)$.   □

Exercise C.1.33. Let $\psi : \mathcal{P}(X) \to [0, \infty]$ be an $\mathcal{M}$-regular outer measure and $E \in \mathcal{M}(\psi)$. Define $\psi_E : \mathcal{P}(X) \to [0, \infty]$ by $\psi_E(S) := \psi(E \cap S)$ as in Exercise C.1.10. Show that ψ_E is an $\mathcal{M}$-regular outer measure.

Exercise C.1.34. Let $(X, \mathcal{M}, \mu)$ be a measure space and $E_k \subset X$ such that $E_k \subset E_{k+1}$ for all $k \in \mathbb{Z}^+$. Show that

$$\mu^* \left(\bigcup_{k=1}^{\infty} E_k \right) = \lim_{k \to \infty} \mu^*(E_k).$$

Notice that this does not violate Exercise C.1.23.

Theorem C.1.35 (Carathéodory–Hahn extension). *Let $\mu : \mathcal{M} \to [0,\infty]$ be a measure. Then $\mathcal{M} \subset \mathcal{M}(\mu^*)$ and $\mu = \mu^*|_{\mathcal{M}}$.*

Proof. Let $E \in \mathcal{M}$. Then $\mu^*(E) = \mu(E)$, because trivially $\mu^*(E) \leq \mu(E)$ and because

$$\mu(E) \leq \mu \left(\bigcup_{j=1}^{\infty} E_j \right) \leq \sum_{j=1}^{\infty} \mu(E_j)$$

for any $\{E_j\}_{j=1}^{\infty} \subset \mathcal{M}$ covering E. To prove $\mathcal{M} \subset \mathcal{M}(\mu^*)$, we must show that

$$\mu^*(S) = \mu^*(E \cap S) + \mu^*(E^c \cap S)$$

for any $S \subset X$. This follows, because

$$
\begin{aligned}
& \mu^*(E \cap S) + \mu^*(E^c \cap S) \\
\geq\ & \mu^*(S) \\
\overset{\text{Lemma C.1.31}}{=}\ & \inf\{\mu(A) : S \subset A \in \mathcal{M}\} \\
=\ & \inf\{\mu(A \cap E) + \mu(A \cap E^c) : S \subset A \in \mathcal{M}\} \\
\geq\ & \mu^*(E \cap S) + \mu^*(E^c \cap S).
\end{aligned}
$$

This concludes the proof. □

Theorem C.1.36. *Let* $\mu : \mathcal{M} \to [0, \infty]$ *be a measure. Then* $\mu^* = \left(\mu^*|_{\mathcal{M}(\mu^*)}\right)^*$.

Proof. Let $\nu := \mu^*|_{\mathcal{M}(\mu^*)}$. We must show that $\nu^* = \mu^*$. Since $\mu = \mu^*|_{\mathcal{M}}$ and $\mathcal{M} \subset \mathcal{M}(\mu^*)$ by Theorem C.1.35, we see that μ is a restriction of ν, and thus the investigation of Definition C.1.2 yields $\nu^* \leq \mu^*$. Moreover,

$$
\begin{aligned}
\mu^*(S) \quad \geq\ & \quad \nu^*(S) \\
\overset{\text{Lemma C.1.31}}{=}\ & \inf\{\nu(A) : S \subset A \in \mathcal{M}(\mu^*)\} \\
\overset{\text{Lemma C.1.31}}{=}\ & \inf\{\mu(B) : S \subset A \in \mathcal{M}(\mu^*),\ A \subset B \in \mathcal{M}\} \\
\geq\ & \inf\{\mu(B) : S \subset B \in \mathcal{M}\} \\
\geq\ & \mu^*(S),
\end{aligned}
$$

so that $\mu^*(S) = \nu^*(S)$. □

Remark C.1.37. In the sequel, measures are often required to be complete. This restriction is not severe, as measures can always be completed, e.g., by the Carathéodory–Hahn extension, whose naturality is proclaimed by Theorems C.1.35 and C.1.36: if $(X, \mathcal{M}, \mu)$ is a measure space, $\mathcal{N} = \mathcal{M}(\mu^*)$ and $\nu = \mu^*|_{\mathcal{N}}$, then $(X, \mathcal{N}, \nu)$ is a complete measure space such that $\mathcal{M} \subset \mathcal{N}$ and $\mu = \nu|_{\mathcal{M}}$, with $\mu^* = \nu^*$. So, from this point onwards, we may assume that a measure $\mu : \mathcal{M} \to [0, \infty]$ is already Carathéodory–Hahn complete, i.e., that $\mathcal{M} = \mathcal{M}(\mu^*)$.

C.1.2   Borel regularity

Borel measures are particularly important, providing a link with topology on the space. We will study such measures in this section.

Definition C.1.38 (Borel regular outer measures). Let (X, τ) be a topological space and $\Sigma(\tau)$ its Borel σ-algebra. An outer measure $\psi : \mathcal{P}(X) \to [0, \infty]$ is *Borel regular* if it is $\Sigma(\tau)$-regular.

Definition C.1.39 (Metric outer measure). An outer measure $\psi : \mathcal{P}(X) \to [0, \infty]$ on a metric space (X, d) is called a *metric outer measure* if it satisfies the following *Carathéodory condition*:

$$\text{dist}(A, B) > 0 \quad \Rightarrow \quad \psi(A \cup B) = \psi(A) + \psi(B). \qquad (C.4)$$

This condition characterises measurability of Borel sets of a metric space:

Theorem C.1.40. *Let τ_d be the metric topology of a metric space (X, d). An outer measure $\psi : \mathcal{P}(X) \to [0, \infty]$ is a metric outer measure if and only if $\tau_d \subset \mathcal{M}(\psi)$.*

Proof. The "if" part of the proof is left for the reader as Exercise C.1.41. Take $U \in \tau_d$. To show that $U \in \mathcal{M}(\psi)$, we need to prove $\psi(A \cup B) = \psi(A) + \psi(B)$ when $A \subset U$ and $B \subset U^c$. We may assume that $\psi(A), \psi(B) < \infty$. For each $k \in \mathbb{Z}^+$, let

$$A_k := \{x \in A \mid \text{dist}(x, U^c) \geq 1/k\}.$$

Then $\text{dist}(A_k, B) \geq 1/k$, enabling the application of the Carathéodory condition (C.4) in

$$
\psi(A) + \psi(B) \overset{\text{trivial}}{\geq} \psi(A \cup B)
$$
$$
\overset{A \supset A_k}{=} \psi(A_k \cup B)
$$
$$
\overset{(C.4)}{=} \psi(A_k) + \psi(B).
$$

Clearly

$$\psi(A_k) \leq \psi(A) \leq \psi(A_k) + \psi(A \setminus A_k),$$

so we have to show that $\psi(A \setminus A_k) \to 0$. Here $A = \bigcup_{k=1}^{\infty} A_k$, since U is open. Consequently

$$
\psi(A \setminus A_k) = \psi\left(\bigcup_{l=k}^{\infty} (A_{l+1} \setminus A_l) \right)
$$
$$
\leq \sum_{l=k}^{\infty} \psi(A_{l+1} \setminus A_l)
$$
$$
\overset{(C.4)}{=} \psi\left(\bigcup_{m=1}^{\infty} (A_{k+2m+1} \setminus A_{k+2m}) \right) + \psi\left(\bigcup_{m=1}^{\infty} (A_{k+2m} \setminus A_{k+2m-1}) \right)
$$
$$
\leq 2\,\psi(A) < \infty.
$$

Thus $\psi(A \setminus A_k) \leq \displaystyle\sum_{l=k}^{\infty} \psi(A_{l+1} \setminus A_l) \xrightarrow{k \to \infty} 0$. $\qquad \square$

Exercise C.1.41. Let (X, d) be a metric space. Complete the proof of Theorem C.1.40 by showing that if $\Sigma(\tau_d) \subset \mathcal{M}(\psi)$ then ψ is a metric outer measure.

Theorem C.1.42 (Topological approximation of measurable sets). *Let (X, d) be a metric space and $\psi : \mathcal{P}(X) \to [0, \infty]$ be a Borel regular outer measure such that $\psi(X) < \infty$. Let $E \subset X$. Then the following statements are equivalent:*

1. *$E \in \mathcal{M}(\psi)$.*

2. *E can be ψ-approximated topologically: more precisely, for each $\varepsilon > 0$ there exist closed $F_\varepsilon \subset X$ and open $G_\varepsilon \subset X$ such that $G_\varepsilon \supset E \supset F_\varepsilon$ and $\psi(G_\varepsilon \setminus F_\varepsilon) < \varepsilon$.*

Proof. Let us assume the second condition. Let $E \subset X$ such that for each $\varepsilon > 0$ there exists a closed set $F_\varepsilon \subset X$ such that

$$F_\varepsilon \subset E \quad \text{and} \quad \psi(E \setminus F_\varepsilon) < \varepsilon.$$

If $F = \bigcup_{k=1}^\infty F_{1/k}$ then $E \supset F \in \Sigma(\tau_d) \subset \mathcal{M}(\psi)$, since we assume the measurability of the Borel sets. Moreover, $E \in \mathcal{M}(\psi)$, because

$$E = F \cup (E \setminus F),$$

where $E \setminus F \in \mathcal{M}(\psi)$ due to

$$0 \leq \psi(E \setminus F) \setminus \psi(E \setminus F_{1/k}) < \frac{1}{k} \xrightarrow[k \to \infty]{} 0 .$$

Thus the second condition of the theorem implies the first one. Notice that here we did not even need the assumption $\psi(X) < \infty$ nor the sets G_ε!

Conversely, we must show that ψ-measurable sets can be ψ-approximated topologically. This can be done by showing that

$$\mathcal{D} := \{A \in \mathcal{M}(\psi) \mid A \text{ can be } \psi\text{-approximated topologically}\}$$

is a σ-algebra containing τ_d; then the Borel regularity will imply $\mathcal{D} = \mathcal{M}(\psi)$. Trivially, $\emptyset \in \mathcal{D}$, and if $A \in \mathcal{D}$ then also $A^c \in \mathcal{D}$. Let $\{A_k\}_{k=1}^\infty \subset \mathcal{D}$; now $\mathcal{D}$ is a σ-algebra if $A := \bigcap_{k=1}^\infty A_k \in \mathcal{D}$. Clearly, $A \in \mathcal{M}(\psi)$, because each $A_k \in \mathcal{M}(\psi)$. By the topological ψ-approximation, for each $k \in \mathbb{Z}^+$ we can take closed $F_k \subset X$ and open $G_k \subset X$ such that

$$G_k \supset A_k \supset F_k \quad \text{and} \quad \begin{cases} \psi(G_k \setminus A_k) \leq 2^{-k}\varepsilon, \\ \psi(A_k \setminus F_k) \leq 2^{-k}\varepsilon. \end{cases}$$

Then the closed set $\bigcap_{k=1}^\infty F_k$ ψ-approximates the set $\bigcap_{k=1}^\infty A_k$ from inside:

$$\psi\left(\bigcap_{k=1}^\infty A_k \setminus \bigcap_{k=1}^\infty F_k \right) \leq \psi\left(\bigcup_{k=1}^\infty (A_k \setminus F_k) \right)$$

$$\leq \sum_{k=1}^\infty \psi(A_k \setminus F_k)$$

$$\leq \varepsilon.$$

On the other hand, for large enough $n \in \mathbb{Z}^+$, the open set $\bigcap_{k=1}^{n} G_k$ ψ-approximates the set $\bigcap_{k=1}^{\infty} A_k$ from outside:

$$\psi\left(\bigcap_{k=1}^{n} G_k \setminus \bigcap_{k=1}^{\infty} A_k\right) \xrightarrow[n\to\infty]{\psi(X)<\infty} \psi\left(\bigcap_{k=1}^{\infty} G_k \setminus \bigcap_{k=1}^{\infty} A_k\right)$$
$$\leq \quad \cdots \quad \leq \quad \varepsilon.$$

Thus we have seen that $\mathcal{D}$ is a σ-algebra, so in proving that $\tau_d \subset \mathcal{D}$, it suffices to show that $F \in \mathcal{D}$ when $F \subset X$ is closed. First, $F \in \mathcal{M}(\psi)$, because the Borel sets are measurable. Clearly, the closed set F ψ-approximates itself from inside, as $\psi(F \setminus F) = \psi(\emptyset) = 0$. Let $U_\varepsilon := \bigcup_{x \in F} B_\varepsilon(x)$, where $B_r(x) = \{y \in X : d(x,y) < r\}$ is an open ball. Thus $U_\varepsilon \subset X$ is open, $F \subset U_\varepsilon$, and

$$\psi(U_{1/k} \setminus F) \xrightarrow[k\to\infty]{\psi(X)<\infty} \psi\left(\bigcap_{k=1}^{\infty}(U_{1/k} \setminus F)\right) \overset{F \text{ closed}}{=} \psi(\emptyset) = 0.$$

Hence F can be ψ-approximated by open sets U_ε from outside.

Now we know that $\mathcal{D} \supset \Sigma(\tau_d)$ is a σ-algebra. Take $E \in \mathcal{M}(\psi)$. By the Borel regularity, there exist Borel sets $F, G \subset X$ such that

$$G \supset E, \qquad \psi(G) = \psi(E),$$
$$F^c \supset E^c, \qquad \psi(F^c) = \psi(E^c).$$

By the topological ψ-approximation, take closed $F_\varepsilon \subset X$ and open $G_\varepsilon \subset X$ such that

$$G_\varepsilon \supset G \supset E \supset F \supset F_\varepsilon,$$
$$\psi(G_\varepsilon \setminus G) < \varepsilon, \qquad \psi(F \setminus F_\varepsilon) < \varepsilon.$$

Then

$$\psi(G_\varepsilon \setminus E) \leq \psi(G_\varepsilon \setminus G) + \psi(G \setminus E) < \varepsilon,$$
$$\psi(E \setminus F_\varepsilon) \leq \psi(E \setminus F) + \psi(F \setminus F_\varepsilon) < \varepsilon,$$

completing the proof. $\qquad \square$

Remark C.1.43. From the proof of Theorem C.1.42 we see that $E \in \mathcal{M}(\psi)$ if and only if $E = B \cup N$, where B is a Borel set and $\psi(N) = 0$.

Exercise C.1.44. In Theorem C.1.42 we assumed that $\psi(X) < \infty$. Prove an analogous result assuming that $\psi(B_r(x)) < \infty$ whenever $0 < r < \infty$; prove also that Remark C.1.43 holds in this generalisation (hint: Exercises C.1.10 and C.1.33). Notice that this new assumption is satisfied by the Lebesgue outer measure $\lambda_{\mathbb{R}^n}^*$.

C.1.3   On Lebesgue measure

Recall the Lebesgue outer measure $\lambda_{\mathbb{R}^n}^* : \mathcal{P}(\mathbb{R}^n) \to [0, \infty]$ from Definition C.1.5: first, we set the volume of an n-interval

$$I_{ab} := [a, b] = [a_1, b_1] \times \cdots \times [a_n, b_n] \subset \mathbb{R}^n$$

to be

$$m(I_{ab}) = \text{volume}(I_{ab})$$
$$= (b_1 - a_1) \cdots (b_n - a_n),$$

and define

$$\lambda_{\mathbb{R}^n}^* (E) = \inf \left\{ \sum_{k=1}^{\infty} m(I_k) : \ E \subset \bigcup_{j=1}^{\infty} I_k, \ I_k \subset \mathbb{R}^n \text{ an } n\text{-interval} \right\}.$$

Definition C.1.45 (Lebesgue measure). The *Lebesgue measure* $\lambda_{\mathbb{R}^n} : \mathcal{M} \to [0, \infty]$ is the restriction of the Lebesgue outer measure $\lambda_{\mathbb{R}^n}^*$ to the σ-algebra $\mathcal{M} = \mathcal{M}(\lambda_{\mathbb{R}^n}^*)$.

For a measurable set $E \subset \mathbb{R}^n$, the number $\lambda_{\mathbb{R}^n}(E) \in [0, \infty]$ can be thought as an "n-dimensional volume". Next we try to justify this claim.

Proposition C.1.46. *For any n-interval $I_{ab} \subset \mathbb{R}^n$,*

$$\lambda_{\mathbb{R}^n}^* (I_{ab}) = \text{volume}(I_{ab}).$$

Proof. Trivially $\lambda_{\mathbb{R}^n}^* (I_{ab}) \leq \text{volume}(I_{ab})$ by the definition of the Lebesgue outer measure. Conversely, take $\varepsilon > 0$. Take a family $\{A_j\}_{j=1}^{\infty}$ of n-intervals such that $I_{ab} \subset \bigcup_{j=1}^{\infty} A_j$ and

$$\sum_{j=1}^{\infty} (A_j) < \lambda_{\mathbb{R}^n}^* (I_{ab}) + \varepsilon.$$

Take a family $\{B_j\}_{j=1}^{\infty}$ of n-intervals such that

$$A_j \subset \text{int}(B_j) \quad \text{and} \quad \text{volume}(B_j) \leq \text{volume}(A_j) + 2^{-j} \varepsilon.$$

Then $\{\text{int}(B_j)\}_{j=1}^{\infty}$ is an open cover of the compact set $I_{ab} \subset \mathbb{R}^n$, thus having a finite subcover $\{\text{int}(B_{j_k})\}_{k=1}^{l}$. Take a family $\{K_i\}_{i=1}^{m}$ of n-intervals such that $I_{ab} = \bigcup_{i=1}^{m} K_i$, $\{\text{int}(K_i)\}_{i=1}^{m}$ is disjoint and that for each i there exists j_k such that $K_i \subset \text{int}(B_{j_k})$; that is, the idea is to chop the n-interval I_{ab} into small enough n-intervals K_i. Then

$$
\begin{aligned}
\text{volume}(I_{ab}) &= \sum_{i=1}^{m} \text{volume}(K_i) \\
&\leq \sum_{k=1}^{l} \text{volume}(B_{j_k}) \\
&\leq \sum_{j=1}^{\infty} \text{volume}(B_j) \\
&\leq \sum_{j=1}^{\infty} \text{volume}(A_j) + \varepsilon \\
&\leq \lambda^*_{\mathbb{R}^n}(I_{ab}) + 2\varepsilon.
\end{aligned}
$$

Hence $\text{volume}(I_{ab}) \leq \lambda^*_{\mathbb{R}^n}(I_{ab})$. $\qquad\square$

Exercise C.1.47. Let $t \in \mathbb{R}$, $x \in \mathbb{R}^n$ $E \subset \mathbb{R}^n$,

$$
\begin{aligned}
tE &:= \{ty \mid y \in E\} \quad \text{and} \\
x + E &:= \{x + y \mid y \in E\}.
\end{aligned}
$$

Show that

$$
\begin{aligned}
\lambda^*_{\mathbb{R}^n}(tE) &= |t|^n \, \lambda^*_{\mathbb{R}^n}(E), \\
\lambda^*_{\mathbb{R}^n}(x + E) &= \lambda^*_{\mathbb{R}^n}(E).
\end{aligned}
$$

Moreover, show that $tE, x + E \in \mathcal{M}$ if $E \in \mathcal{M} = \mathcal{M}(\lambda^*_{\mathbb{R}^n})$.

Remark C.1.48 (**Translation and rotation invariance of Lebesgue measure**). The *translation invariance* of the Lebesgue (outer) measure refers to the invariance under mapping $E \mapsto x + E$ in Exercise C.1.47. The Lebesgue measure behaves well also under linear mappings: for a linear mapping $A : \mathbb{R}^n \to \mathbb{R}^n$ and $E \subset \mathbb{R}^n$, let $AE := \{Ay \mid y \in E\}$. Then

$$
\lambda^*_{\mathbb{R}^n}(AE) = |\det(A)| \, \lambda^*_{\mathbb{R}^n}(E),
$$

where $\det(A) \in \mathbb{R}$ is the determinant of A. Moreover, $AE \in \mathcal{M}$ if $E \in \mathcal{M} = \mathcal{M}(\lambda^*_{\mathbb{R}^n})$. Especially, the invariance under the orthogonal mappings is called the *rotation invariance*.

Lemma C.1.49. *Let us define the half-space* $E = \{x \in \mathbb{R}^n \mid x_i \geq 0\}$, *where* $i \in \{1, \ldots, n\}$. *Then* $E \in \mathcal{M} = \mathcal{M}(\lambda^*_{\mathbb{R}^n})$.

Proof. Clearly, it is sufficient to deal with the case $i = 1$. Take $S \subset \mathbb{R}^n$. Let $\varepsilon > 0$. Take a family $\{A_j\}_{j=1}^{n}$ of n-intervals such that $S \subset \bigcup_{j=1}^{\infty} A_j$ and

$$
\sum_{j=1}^{n} \text{volume}(A_j) \leq \lambda^*_{\mathbb{R}^n}(S) + \varepsilon.
$$

Notice that $E \cap A_j$ and $\overline{E^c} \cap A_j$ are n-intervals, so that

$$
\begin{aligned}
\lambda^*_{\mathbb{R}^n}(S) &\leq \lambda^*_{\mathbb{R}^n}(E \cap S) + \lambda^*_{\mathbb{R}^n}(E^c \cap S) \\
&\leq \lambda^*_{\mathbb{R}^n}(E \cap S) + \lambda^*_{\mathbb{R}^n}(\overline{E^c} \cap S) \\
&\leq \lambda^*_{\mathbb{R}^n}\left(\bigcup_{j=1}^{\infty} (E \cap A_j) \right) + \lambda^*_{\mathbb{R}^n}\left(\bigcup_{j=1}^{\infty} (\overline{E^c} \cap A_j) \right) \\
&\leq \sum_{j=1}^{\infty} \left(\mathrm{volume}(E \cap A_j) + \mathrm{volume}(\overline{E^c} \cap A_j) \right) \\
&= \sum_{j=1}^{\infty} \mathrm{volume}(A_j) \\
&\leq \lambda^*_{\mathbb{R}^n}(S) + \varepsilon.
\end{aligned}
$$

Thus $\lambda^*_{\mathbb{R}^n}(S) = \lambda^*_{\mathbb{R}^n}(E \cap S) + \lambda^*_{\mathbb{R}^n}(E^c \cap S)$. This proves the Lebesgue measurability of the half-space $E \subset \mathbb{R}^n$. $\square$

Corollary C.1.50. *The closed n-interval $[a, b] \subset \mathbb{R}^n$ is Lebesgue measurable, and so is its interior.*

Proof. First,

$$
[a, b] = \bigcap_{k=1}^{n} \left(\{x \in \mathbb{R}^n : a_k \leq x_k\} \cap \{x \in \mathbb{R}^n : x_k \leq b_k\} \right),
$$

so it is measurable, as a finite intersection of measurable sets. Finally, if $c = (1, \ldots, 1) \in \mathbb{R}^n$ then the interior

$$
\mathrm{int}([a, b]) = \bigcup_{k=1}^{\infty} [a + c/k, b - c/k].
$$

Being a countable union of measurable sets, the interior is measurable. $\square$

Definition C.1.51. For $x \in \mathbb{R}^n$ and $r > 0$, let the *open cube* be

$$
\begin{aligned}
Q(x, r) &:= x + (-r, +r)^n \\
&= \{y \in \mathbb{R}^n \mid \forall i \in \{1, \ldots, n\} : |x_i - y_i| < r\}.
\end{aligned}
$$

This is a Lebesgue measurable set, as it is the interior of the closed n-interval $\overline{Q(x, r)} = [a, b]$, where $a = x - (r, \ldots, r)$ and $b = x + (r, \ldots, r)$.

Corollary C.1.52 (Lebesgue outer measure is Borel regular). *Lebesgue outer measure $\lambda^* : \mathcal{P}(\mathbb{R}^n) \to [0, \infty]$ is Borel regular.*

Proof. Let $U \subset \mathbb{R}^n$ be open. It is easy to check that

$$Q(x, r/\sqrt{n}) \subset B_r(x) \subset Q(x, r).$$

Thus $x \in U$ if and only if $Q(x, r) \subset U$ for some $r > 0$. If $Q(x, r) \subset U$, take $z \in \mathbb{Q}^n \cap B_{r/2}(x)$; then $x \in Q(z, r/2) \subset Q(x, r) \subset U$. Thus

$$U = \bigcup \{Q(z, 1/m) : \ z \in \mathbb{Q}^n, \ m \in \mathbb{Z}^+, Q(z, 1/m) \subset U\},$$

which is measurable as a countable union of measurable sets. □

Remark C.1.53. It now turns out that Lebesgue measurable sets are nearly open or closed sets:

Theorem C.1.54 (Topological approximation of Lebesgue measurable sets). *Let $E \subset \mathbb{R}^n$. The following three conditions are equivalent:*

1. $E \in \mathcal{M}(\lambda_{\mathbb{R}^n}^*)$.
2. *For every $\varepsilon > 0$ there exists an open set $U \subset \mathbb{R}^n$ such that $E \subset U$ and $\lambda_{\mathbb{R}^n}^*(U \setminus E) < \varepsilon$.*
3. *For every $\varepsilon > 0$ there exists a closed set $S \subset \mathbb{R}^n$ such that $S \subset E$ and $\lambda_{\mathbb{R}^n}^*(E \setminus S) < \varepsilon$.*

Proof. Let us show that the first condition implies the second one. Suppose $E \subset \mathbb{R}^n$ is Lebesgue measurable. Let $\varepsilon > 0$. For a moment, assume that

$$\lambda_{\mathbb{R}^n}(E) < \infty. \tag{C.5}$$

Take a family $\{A_j\}_{j=1}^\infty$ of n-intervals such $E \subset \bigcup_{j=1}^\infty A_j$ and

$$\sum_{j=1}^\infty \text{volume}(A_j) < \lambda_{\mathbb{R}^n}(E) + \varepsilon. \tag{C.6}$$

We may think that this is an ε-tight cover of E, and we may loosen it a bit by taking a family $\{B_j\}_{j=1}^\infty$ on n-intervals such that $A_j \subset \text{int}(B_j)$ and

$$\lambda_{\mathbb{R}^n}(B_j) \le \lambda_{\mathbb{R}^n}(A_j) + 2^{-j}\varepsilon. \tag{C.7}$$

Let $U := \bigcup_{j=1}^\infty \text{int}(B_j)$. Then $U \subset \mathbb{R}^n$ is open, $E \subset U$ and

$$
\begin{aligned}
\lambda_{\mathbb{R}^n}(U) \quad &\le \quad \sum_{j=1}^\infty \lambda_{\mathbb{R}^n}(B_j) \\
&\overset{(C.7)}{\le} \quad \sum_{j=1}^\infty \lambda_{\mathbb{R}^n}(A_j) + \varepsilon \\
&\overset{(C.6)}{<} \quad \lambda_{\mathbb{R}^n}(E) + 2\varepsilon.
\end{aligned}
$$

From this we get (as E, U are measurable and $E \subset U$)

$$\lambda_{\mathbb{R}^n}^*(U \setminus E) = \lambda_{\mathbb{R}^n}(U \setminus E) \overset{(C.5)}{=} \lambda_{\mathbb{R}^n}(U) - \lambda_{\mathbb{R}^n}(E) < 2\varepsilon.$$

Thus the case of (C.5) is completely solved. Now let us forget the restriction (C.5), and let $E_k := E \cap B_d(0, k)$, where d is the Euclidean distance. Then

$$E_k \in \mathcal{M}(\lambda_{\mathbb{R}^n}^*), \quad \lambda_{\mathbb{R}^n}(E_k) < \infty, \quad E = \bigcup_{k=1}^{\infty} E_k.$$

By the earlier part of the proof, for each k there exists an open set $U_k \subset \mathbb{R}^n$ such that $E_k \subset U_k$ and
$$\lambda_{\mathbb{R}^n}(U_k \setminus E_k) < 2^{-k}\varepsilon.$$
Let $U := \bigcup_{k=1}^{\infty} U_k$. Then U is open, $E \subset U$ and

$$U \setminus E = \left(\bigcup_{k=1}^{\infty} U_k \right) \setminus E = \bigcup_{k=1}^{\infty} (U_k \setminus E) \subset \bigcup_{k=1}^{\infty} (U_k \setminus E_k),$$

implying

$$\lambda_{\mathbb{R}^n}^*(U \setminus E) \leq \sum_{k=1}^{\infty} \lambda_{\mathbb{R}^n}^*(U_k \setminus E_k) < \sum_{k=1}^{\infty} 2^{-k}\varepsilon = \varepsilon.$$

Thus the first condition in Theorem C.1.54 implies the second one.

Let us now assume the second condition, about approximation by open sets from outside: thereby for each $k \in \mathbb{Z}^+$ there exists an open set $U_k \subset \mathbb{R}^n$ such that $E \subset U_k$ and

$$\lambda_{\mathbb{R}^n}^*(U_k \setminus E) < \frac{1}{k}.$$

Let $G := \bigcap_{k=1}^{\infty} U_k$. Then $E \subset G \in \mathcal{M}(\lambda_{\mathbb{R}^n}^*)$, and $G \setminus E \subset U_j \setminus E$ for every $j \in \mathbb{Z}^+$. Hence

$$\lambda_{\mathbb{R}^n}^*(G \setminus E) \leq \lambda_{\mathbb{R}^n}^*(U_j \setminus E) < \frac{1}{j} \xrightarrow[j \to \infty]{} 0,$$

so that $\lambda_{\mathbb{R}^n}^*(G \setminus E) = 0$. Thus $G \setminus E \in \mathcal{M}(\lambda_{\mathbb{R}^n}^*)$ by Lemma C.1.11, so that $E = G \setminus (G \setminus E) \in \mathcal{M}(\lambda_{\mathbb{R}^n}^*)$. This shows that the second condition implies the first one in Theorem C.1.54.

Let us now show that the first and the second conditions imply the third condition. Let $E \in \mathcal{M}(\lambda_{\mathbb{R}^n}^*)$. Take $\varepsilon > 0$. Since $E^c \in \mathcal{M}(\lambda_{\mathbb{R}^n}^*)$, there exists an open set $U \subset \mathbb{R}^n$ such that $E^c \subset U$ and

$$\lambda_{\mathbb{R}^n}(U \setminus E^c) < \varepsilon.$$

Now $S := U^c \subset \mathbb{R}^n$ is closed, $S \subset E$, and $E \setminus S = U \setminus E^c$. This establishes the third condition, about approximation by closed sets from inside.

The rest of the proof is left for the reader as an exercise. Naturally, the reasoning can be made similar to the case where the second condition implied the first one.   □

Exercise C.1.55. Complete the proof of Theorem C.1.54 by showing that the third condition implies the first one.

Remark C.1.56 (**Lebesgue is "almost" Borel**). From the proof of Theorem C.1.54 and from a solution to Exercise C.1.55, we may notice that a set $E \subset \mathbb{R}^n$ is Lebesgue measurable if and only if there exist Borel sets $F, G \subset \mathbb{R}^n$ such that $F \subset E \subset G$ and

$$\lambda_{\mathbb{R}^n}(G \setminus F) = 0.$$

Moreover, closer examination reveals that G can be taken as a countable intersection of open sets, and correspondingly F as a countable union of closed sets. In this sense, a Lebesgue measurable set is almost Borel (up to measure zero), and it looks nearly as if open from outside, and nearly as if closed from inside.

C.1.4 Lebesgue non-measurable sets

The Axiom of Choice (Axiom A.4.2) can be used to "construct" a Lebesgue non-measurable subset $S \subset \mathbb{R}^n$. Let $f : \mathcal{P}(\mathbb{R}^n) \to \mathbb{R}^n$ be a choice function. Let

$$S := \{f(x + \mathbb{Q}^n) \mid x \in \mathbb{R}^n\}.$$

Let us show that this set is non-measurable. Now $\lambda^*_{\mathbb{R}^n}(S) > 0$, because $\mathbb{R}^n = \mathbb{Q}^n + S$ is the union of a countable family $\{q + S \mid q \in \mathbb{Q}^n\}$, where $\lambda^*_{\mathbb{R}^n}(r+S) = \lambda^*_{\mathbb{R}^n}(S)$. Moreover, if $0 \neq q \in \mathbb{Q}^n$ then $S \cap (q + S) = \emptyset$. By the following result, this proves the non-measurability of S:

Proposition C.1.57. *Let $S \subset \mathbb{R}^n$ be Lebesgue measurable and $\lambda_{\mathbb{R}^n}(S) > 0$. Then there exists $\delta > 0$ such that $\lambda_{\mathbb{R}^n}(S \cap (x + S)) > 0$ whenever $\|x\|_{\mathbb{R}^n} < \delta$.*

Proof. Let $0 < \varepsilon < 1$. Since $\lambda(S) > 0$, there exists an n-interval $I = [a, b] \subset \mathbb{R}^n$ such that

$$\lambda(S \cap I) = (1 - \varepsilon)\,\lambda(I) > 0.$$

Let $E = S \cap I$. Then $\lambda(I \setminus E) = \lambda(I) - \lambda(E) = \varepsilon\,\lambda(I)$ due to the measurability of E. For any $x \in \mathbb{R}^n$,

$$I \cap (x + I) = (E \cup (x + E)) \cup (I \setminus E) \cup ((x + I) \setminus (x + E)),$$

so that

$$\begin{aligned}
\lambda(I \cap (x + I)) &\leq \lambda(E \cap (x + E)) + \lambda(I \setminus E) + \lambda((x + I) \setminus (x + E)) \\
&= \lambda(E \cap (x + E)) + 2\varepsilon\,\lambda(I),
\end{aligned}$$

where the last equality follows by the translation invariance of the Lebesgue measure. The reader easily verifies that $\lim_{x \to 0} \lambda(I + (x + I)) = \lambda(I)$. Thus the claim follows if we choose ε small enough. $\qquad\square$

Exercise C.1.58. Let $I = [a, b] \subset \mathbb{R}^n$ be an n-interval. Show that

$$\lambda_{\mathbb{R}^n}\left(I \cap (x + I)\right) \xrightarrow[\|x\|_{\mathbb{R}^n} \to 0]{} \lambda_{\mathbb{R}^n}(I).$$

Actually, it can be shown that in the Zermelo–Fraenkel set theory without the Axiom of Choice, one cannot prove the existence of Lebesgue non-measurable sets: see [114]. In practice, we do not have to worry about non-Lebesgue-measurability much.

C.2   Measurable functions

In topology, continuous functions were essential; in measure theory, the nice functions are the measurable ones. Before going into details, let us sketch the common framework behind both continuity and measurability. Let us say that $f : X \to Y$ *induces* (or *pulls back*) from a family $\mathcal{B} \subset \mathcal{P}(Y)$ a new family $f^*(\mathcal{B}) \subset \mathcal{P}(X)$ defined by

$$f^*(\mathcal{B}) := \left\{ f^{-1}(B) \subset X \mid B \in \mathcal{B} \right\},$$

and $f : X \to Y$ *co-induces* (or *pushes forward*) from a family $\mathcal{A} \subset \mathcal{P}(X)$ a new family $f_*(\mathcal{A}) \subset \mathcal{P}(Y)$ defined by

$$f_*(\mathcal{A}) := \left\{ B \subset Y \mid f^{-1}(B) \in \mathcal{A} \right\}.$$

Here if $\mathcal{A}, \mathcal{B}$ are topologies (or respectively σ-algebras) then $f_*(\mathcal{A}), f^*(\mathcal{B})$ are also topologies (or respectively σ-algebras), since $f^{-1} : \mathcal{P}(Y) \to \mathcal{P}(X)$ preserves unions, intersections and complementations.

Exercise C.2.1. Let $\mathcal{A}, \mathcal{B}$ be σ-algebras. Check that $f_*(\mathcal{A}), f^*(\mathcal{B})$ are indeed σ-algebras.

C.2.1   Well-behaving functions

Definition C.2.2 (Measurability). Let $\mathcal{M}_X, \mathcal{M}_Y$ be σ-algebras on X and Y, respectively. A function $f : X \to Y$ is called $(\mathcal{M}_X, \mathcal{M}_Y)$-*measurable* if

$$f^{-1}(V) \in \mathcal{M}_X$$

for every $V \in \mathcal{M}_Y$; that is, if $f^*(\mathcal{M}_Y) \subset \mathcal{M}_X$.

Remark C.2.3. We see that the measurability behaves well in compositions provided that the involved σ-algebras naturally match: if

$$X \xrightarrow[(\mathcal{M}, \mathcal{N})\text{-measurable}]{f} Y \xrightarrow[(\mathcal{N}, \mathcal{O})\text{-measurable}]{g} Z$$

then $g \circ f : X \to Z$ is $(\mathcal{M}, \mathcal{O})$-measurable. For us, a most important case is $Y = Z = [-\infty, +\infty] = \mathbb{R} \cup \{-\infty, +\infty\}$, for which the canonical σ-algebra will be the collection $\Sigma(\tau_\infty)$ of Borel sets, where $\tau_\infty \subset \mathcal{P}([-\infty, +\infty])$ is the smallest topology for which all the intervals $[a, b] \subset [-\infty, +\infty]$ are closed.

Definition C.2.4 (Borel/Lebesgue mesurability). Let $\mathcal{M}$ be a σ-algebra on X, and let τ_X be a topology of X. A function $f : X \to [-\infty, +\infty]$ is called

- $\mathcal{M}$-*measurable* if it is $(\mathcal{M}, \Sigma(\tau_\infty))$-measurable, and
- *Borel measurable* if it is $\Sigma(\tau_X)$-measurable.

A function $f : \mathbb{R}^n \to [-\infty, +\infty]$ is called *Lebesgue measurable* if it is $\mathcal{M}(\lambda_{\mathbb{R}^n}^*)$-measurable.

Definition C.2.5. The *characteristic function* $\chi_E : X \to \mathbb{R}$ of a subset $E \subset X$ is defined by

$$\chi_E(x) := \begin{cases} 1, & \text{if } x \in E, \\ 0, & \text{if } x \in E^c. \end{cases}$$

Notice that χ_E is $\mathcal{M}$-measurable if and only if $E \in \mathcal{M}$.

Definition C.2.6. Let $a \in \mathbb{R}$ and $f, g : X \to [-\infty, +\infty]$. We write

$$\begin{aligned} \{f > a\} &:= \{x \in X \mid f(x) > a\}, \\ \{f > g\} &:= \{x \in X \mid f(x) > g(x)\}. \end{aligned}$$

In an analogous manner one defines sets

$$\{f < a\}, \{f \geq a\}, \{f \leq a\}, \{f = a\}, \{f \neq a\},$$

$$\{f < g\}, \{f \geq g\}, \{f \leq g\}, \{f = g\}, \{f \neq g\},$$

and so on.

Theorem C.2.7. *Let $\mathcal{M}$ be a σ-algebra on X and $f : X \to [-\infty, +\infty]$. Then the following conditions are equivalent:*

1. *f is $\mathcal{M}$-measurable.*
2. *$\{f > a\}$ is measurable for each $a \in \mathbb{R}$.*
3. *$\{f \geq a\}$ is measurable for each $a \in \mathbb{R}$.*
4. *$\{f < a\}$ is measurable for each $a \in \mathbb{R}$.*
5. *$\{f \leq a\}$ is measurable for each $a \in \mathbb{R}$.*

Proof. If f is $\mathcal{M}$-measurable then $\{f > a\} = f^{-1}((a, +\infty]) \in \mathcal{M}$, because $(a, +\infty] \subset [-\infty, +\infty]$ is a Borel set.

Now suppose $\{f > a\} \in \mathcal{M}$ for every $a \in \mathbb{R}$: we have to show that f is $\mathcal{M}$-measurable. We notice that f is $(\mathcal{M}, \mathcal{D})$-measurable, where

$$\mathcal{D} := f_*(\mathcal{M}) = \left\{ B \subset [-\infty, +\infty] \mid f^{-1}(B) \in \mathcal{M} \right\}.$$

Furthermore, f is $\mathcal{M}$-measurable, because $\Sigma(\tau_\infty) \subset \mathcal{D}$, because for every $[a, b] \subset [-\infty, +\infty]$ we have

$$
\begin{aligned}
f^{-1}([a,b]) &= \{f \geq a\} \cap \{f \leq b\} \\
&= \bigcap_{k=1}^{\infty} \{f > a - 1/k\} \cap \{f > b\}^c \quad \in \quad \mathcal{M};
\end{aligned}
$$

recall that $\Sigma(\tau_\infty)$ is the smallest σ-algebra containing every interval. Thus f is $\mathcal{M}$-measurable. All the other claims have essentially similar proofs. $\square$

Remark C.2.8. Let $f, g : X \to [-\infty, +\infty]$ be $\mathcal{M}$-measurable. By Theorem C.2.7, then $\{f > g\} \in \mathcal{M}$, because

$$
\{f > g\} = \bigcup_{r \in \mathbb{Q}} (\{f > r\} \cap \{g < r\});
$$

notice that here the union is countable! Similarly, also

$$
\{f \geq g\}, \{f < g\}, \{f \leq g\}, \{f = g\}, \{f \neq g\} \in \mathcal{M}.
$$

Example. A continuous function $f : X \to [-\infty, +\infty]$ is Borel measurable, because $\{f \geq a\} \subset X$ is closed for each $a \in \mathbb{R}$. Therefore a continuous function $f : \mathbb{R}^n \to [-\infty, +\infty]$ is Lebesgue measurable, because Borel sets in $\mathbb{R}^n$ are Lebesgue measurable.

Theorem C.2.9. *Let $\lambda \in \mathbb{R}$ and $0 < p < \infty$. Let $f, g : X \to \mathbb{R}$ be $\mathcal{M}$-measurable. Then*

$$
\lambda f, f + g, fg, |f|^p, \min(f, g), \max(f, g) : X \to \mathbb{R}
$$

are $\mathcal{M}$-measurable. Moreover, if $0 \notin f(X) = \{f(x) : x \in X\}$ then $1/f$ is $\mathcal{M}$-measurable.

Proof. The reader may easily show that λf is $\mathcal{M}$-measurable. If $a \in \mathbb{R}$ then

$$
\begin{aligned}
\{f + g > a\} &= \bigcup_{q \in \mathbb{Q}:\ q > a} \{f + g > q\} \\
&= \bigcup_{r, s \in \mathbb{Q}:\ r + s > a} (\{f > r\} \cap \{g > s\}) \quad \in \quad \mathcal{M},
\end{aligned}
$$

showing that $f + g$ is $\mathcal{M}$-measurable. If $a \geq 0$ then

$$
\{f^2 > a\} = \{f > \sqrt{a}\} \cup \{f < -\sqrt{a}\},
$$

so that f^2 is $\mathcal{M}$-measurable. Thereby also

$$
fg = \frac{(f + g)^2 - (f - g)^2}{4}
$$

is $\mathcal{M}$-measurable. If $0 \notin f(X)$ then $1/f : X \to \mathbb{R}$ is $\mathcal{M}$-measurable, since it is a composition of

- the $\mathcal{M}$-measurable mapping $(x \mapsto f(x)) : X \to \mathbb{R} \setminus \{0\}$, and
- the continuous mapping $(t \mapsto 1/t) : \mathbb{R} \setminus \{0\} \to \mathbb{R}$.

The rest of the proof is left as Exercise C.2.10. □

Exercise C.2.10. Complete the proof of Theorem C.2.9.

Remark C.2.11. Notice that f^2 can be $\mathcal{M}$-measurable even if f is not: consider, e.g., $f = \chi_E - 1/2$, where $E \notin \mathcal{M}$.

Definition C.2.12 (μ-almost everywhere). Let $(X, \mathcal{M}, \mu)$ be a complete measure space. We say that a property holds μ-*almost everywhere* (abbreviated μ-*a.e.*) if it holds in a set $N^c = X \setminus N$, where $N \in \mathcal{M}$ and $\mu(N) = 0$.

Theorem C.2.13. *Let* $(X, \mathcal{M}, \mu)$ *be complete and* $f, g : X \to [-\infty, +\infty]$. *Let* f *be* $\mathcal{M}$-*measurable and* $f = g$ μ-*a.e. Then* g *is* $\mathcal{M}$-*measurable.*

Proof. Let $N := \{f \neq g\} \in \mathcal{M}$. We have to show that $\{g > a\} \in \mathcal{M}$ for any $a \in \mathbb{R}$. Notice that

$$\begin{aligned} \{g > a\} &= (N \cap \{g > a\}) \cup (N^c \cap \{g > a\}) \\ &= (N \cap \{g > a\}) \cup (N^c \cap \{f > a\}). \end{aligned}$$

Clearly, $N^c \cap \{f > a\} \in \mathcal{M}$. Moreover, $N \cap \{g > a\} \in \mathcal{M}$, because μ is complete and $\mu^*(N \cap \{g > a\}) \leq \mu(N) = 0$. □

Definition C.2.14 (Distinguishing functions?). Let $(X, \mathcal{M}, \mu)$ be complete. Write $f \sim_\mu g$, if $f = g$ μ-almost everywhere: we may identify those functions that μ "does not distinguish". Especially, if $f : X \to [-\infty, +\infty]$ such that $\mu(\{|f| = \infty\}) = 0$, we may identify f with $g : X \to \mathbb{R}$ defined by

$$g(x) := \begin{cases} f(x), & \text{when } f(x) \in \mathbb{R}, \\ 0, & \text{otherwise.} \end{cases}$$

C.2.2 Sequences of measurable functions

Theorem C.2.15. *Let* $f_j : X \to [-\infty, +\infty]$ *be* $\mathcal{M}$-*measurable for each* $j \in \mathbb{Z}^+$. *Then*

$$\sup_{j \in \mathbb{Z}^+} f_j, \quad \inf_{j \in \mathbb{Z}^+} f_j, \quad \limsup_{j \to \infty} f_j, \quad \liminf_{j \to \infty} f_j$$

are also $\mathcal{M}$-*measurable.*

Proof. First,

$$\left\{ \sup_{j \in \mathbb{Z}^+} f_j > a \right\} = \bigcup_{j=1}^{\infty} \{f_j > a\} \in \mathcal{M}.$$

Second, the case of the infimum is handled analogously. Third, these previous cases imply the results for $\limsup$ and $\liminf$. $\qquad\square$

Definition C.2.16 (Convergences). Let $f_j, f : X \to \mathbb{R}$, where $j \in \mathbb{Z}^+$. Let us define various convergences $f_j \to f$ in the following manner: We say that $f_j \to f$ *pointwise* (word "*pointwise*" often omitted) if

$$\forall x \in X : \ |f_j(x) - f(x)| \xrightarrow[j \to \infty]{} 0.$$

Saying that $f_j \to f$ *uniformly* means

$$\sup_{x \in X} : \ |f_j(x) - f(x)| \xrightarrow[j \to \infty]{} 0.$$

Let $(X, \mathcal{M}, \mu)$ be complete, $f_j : X \to \mathbb{R}$ be $\mathcal{M}$-measurable and $f : X \to [-\infty, +\infty]$. We say that $f_j \to f$ μ-*a.e.* if

$$f_j \to f \text{ pointwise } \mu\text{-a.e. on } X.$$

Saying that $f_j \to f$ μ-*almost uniformly* means that

$$\begin{cases} \forall \varepsilon > 0 \ \exists A_\varepsilon \in \mathcal{M} : \ (f_j - f)|_{A_\varepsilon} \xrightarrow[j \to \infty]{} 0 \text{ uniformly}, \\ \mu(A_\varepsilon^c) < \varepsilon. \end{cases}$$

Saying that $f_j \to f$ *in measure* μ means

$$\forall \varepsilon > 0 : \ \mu^* \left(\{|f_j - f| \geq \varepsilon\} \right) \xrightarrow[j \to \infty]{} 0.$$

Exercise C.2.17. Let functions $f_j : X \to \mathbb{R}$ be $\mathcal{M}$-measurable for every $j \in \mathbb{Z}^+$. Show that $E \in \mathcal{M}$, where

$$E := \left\{ x \in X : \ \lim_{j \to \infty} f_j(x) \in \mathbb{R} \text{ exists} \right\}.$$

Exercise C.2.18. Let (X, τ) be a topological space, $f_j \in C(X)$ for each $j \in \mathbb{Z}^+$ and $f_j \to f$ uniformly. Show that $f : X \to \mathbb{R}$ is also continuous. This extends Theorem A.9.7.

Remark C.2.19. Let $(X, \mathcal{M}, \mu)$ be as above. By Theorems C.2.13 and C.2.15, if $f_j \to f$ μ-a.e. then $f : X \to [-\infty, +\infty]$ is $\mathcal{M}$-measurable. Moreover, if $f_j \to f$ in measure or $f_j \to f$ almost uniformly then f is $\mathcal{M}$-measurable, and $f(x) \in \mathbb{R}$ for μ-a.e. $x \in X$ (by Theorem C.2.24 and Exercise C.2.20, respectively).

Exercise C.2.20. Let $f_j \to f$ μ-almost uniformly.

  a) Show that $f_j \to f$ in measure μ.

  b) Show that $f_j \to f$ μ-almost everywhere.

These implications cannot be reversed: give examples.

Exercise C.2.21. For each $j \in \mathbb{Z}^+$, let $f_j : X \to \mathbb{R}$ be $\mathcal{M}$-measurable. Let $(f_j)_{j=1}^\infty$ be a *Cauchy sequence in measure* μ, that is

$$\forall \varepsilon > 0 : \quad \mu(\{|f_i - f_j| \geq \varepsilon\}) \xrightarrow[i,j \to \infty]{} 0.$$

Show that there exists $f : X \to [-\infty, +\infty]$ such that $f_j \to f$ in measure μ.

Exercise C.2.22. Let $f_j \to f$ μ-almost everywhere.

  a) Show that $f_j \to f$ in measure μ, if $\mu(X) < \infty$.

  b) Give an example where $\mu(X) = \infty$ and $f_j \not\to f$ in measure μ;

consequently, here also $f_j \not\to f$ μ-almost uniformly, by Exercise C.2.20.

    For **finite** measure spaces, almost everywhere convergence implies almost uniform convergence:

Theorem C.2.23 (Egorov: "finite pointwise is almost uniform"). *Let $(X, \mathcal{M}, \mu)$ be a complete finite measure space. Let $f_j \to f$ μ-almost everywhere. Then $f_j \to f$ almost uniformly.*

Proof. Take $\varepsilon > 0$. We want to find $A_\varepsilon \in \mathcal{M}$ such that $\mu(A_\varepsilon^c) < \varepsilon$ and $(f_j - f)|_{A_\varepsilon} \xrightarrow[j \to \infty]{} 0$ uniformly. Let

$$E := \{|f_j - f| \to 0\}.$$

Now $E \in \mathcal{M}$ and $\mu(E^c) = 0$, because $f_j \to f$ μ-almost everywhere. Moreover,

$$A_{jk} := \bigcap_{i=j}^\infty \left\{ |f_i - f| < \frac{1}{k} \right\} \in \mathcal{M}.$$

We may choose $j_k \in \mathbb{Z}^+$ such that

$$\mu(A_{j_k k}^c) < 2^{-k}\varepsilon, \tag{C.8}$$

because

$$\lim_{j \to \infty} \mu(A_{jk}^c) \overset{\mu(X)<\infty,\ A_{jk}^c \supset A_{(j+1)k}^c \in \mathcal{M}}{=} \mu\left(\bigcap_{j=1}^\infty A_{jk}^c \right) \overset{E \subset \bigcup_{j=1}^\infty A_{jk}}{\leq} \mu(E^c) = 0.$$

Now $A_\varepsilon := \bigcap_{k=1}^{\infty} A_{j_k k} \in \mathcal{M}$ is the desired set:

$$\mu(A_\varepsilon^c) = \mu\left(\bigcup_{k=1}^{\infty} A_{j_k k}^c\right) \leq \sum_{k=1}^{\infty} \mu(A_{j_k k}^c) \overset{(C.8)}{<} \varepsilon,$$

and $(f_i - f)|_{A_\varepsilon} \to 0$ uniformly, because

$$A_\varepsilon = \bigcap_{k=1}^{\infty} \bigcap_{i=j_k}^{\infty} \left\{|f_i - f| < \frac{1}{k}\right\},$$

so that $|f_i(x) - f(x)| < \frac{1}{k}$ for all $x \in A_\varepsilon$ whenever $i \geq j_k$.   □

Theorem C.2.24. *Let $(X, \mathcal{M}, \mu)$ be complete and $f_j \to f$ in measure μ. Then there exists a subsequence $\{f_{j_k}\}_{k=1}^{\infty} \subset \{f_j\}_{j=1}^{\infty}$ such that $f_{j_k} \to f$ μ-almost everywhere.*

Proof. Since $f_j \to f$ in measure, for each $k \in \mathbb{Z}^+$ we may take $j_k \in \mathbb{Z}^+$ such that $j_1 = 1$, $j_{k+1} > j_k$ and

$$\mu^*\left(\left\{|f_j - f| \geq \frac{1}{k}\right\}\right) < 2^k \tag{C.9}$$

whenever $j \geq j_k$. Let $N_k := \left\{|f_{j_k} - f| \geq \frac{1}{k}\right\}$ and

$$N := \limsup_{k\to\infty} N_k = \bigcap_{j=1}^{\infty} \bigcup_{k=j}^{\infty} N_k.$$

Then $\mu^*(N) = 0$ (thus $N \in \mathcal{M}$), because

$$\mu^*(N) \leq \mu^*\left(\bigcup_{k=j}^{\infty} N_k\right) \leq \sum_{k=j}^{\infty} \mu^*(N_k) \xrightarrow[j\to\infty]{(C.9)} 0.$$

Now

$$N^c = \bigcup_{j=1}^{\infty} \bigcap_{k=j}^{\infty} \left\{|f_{j_k} - f| < \frac{1}{k}\right\},$$

so that $|f_{j_k}(x) - f(x)| < \frac{1}{k}$ for all $x \in N^c$ whenever k is large enough. Thus $f_{j_k}(x) \xrightarrow[k\to\infty]{} f(x)$ for all $x \in N^c$.   □

Corollary C.2.25. *If $f_j \to f$ in measure then f is measurable.*   □

Exercise C.2.26. Draw a clear diagram about logical implications between different types of convergences $f_j \to f$.

C.2.3 Approximating measurable functions

Definition C.2.27 (Simple functions). A function $f : X \to \mathbb{R}$ is called *simple* if its range $f(X) = \{f(x) : x \in X\} \subset \mathbb{R}$ is finite. Its *normal form* is then

$$f = \sum_{a \in f(X)} a \, \chi_{f^{-1}(\{a\})}.$$

Definition C.2.28 (Positive and negative parts). The *positive and negative parts* of a function $f : X \to [-\infty, +\infty]$ are $f^+, f^- : X \to [0, \infty]$, respectively, where

$$
\begin{aligned}
f^+(x) &:= \max\{0, f(x)\}, \\
f^- &:= (-f)^+.
\end{aligned}
$$

Theorem C.2.29. *Let* $f : X \to [-\infty, +\infty]$. *Then there exist simple functions* $f_j : X \to \mathbb{R}$ *such that*

$$f_j \to f \text{ pointwise.}$$

Moreover, f_j *can be chosen so that*

$$\text{if } 0 \leq f \text{ then } 0 \leq f_j \leq f_{j+1} \leq f,$$

$$\text{if } f \text{ bounded then } f_j \to f \text{ uniformly,}$$

$$\text{if } f \text{ measurable then } f_j \text{ measurable.}$$

Proof. Since $f = f^+ - f^-$, we may approximate f^+ and f^- separately. Thus assume $f \geq 0$. Define

$$
f_i(x) := \begin{cases}
\frac{k-1}{2^i}, & \text{when } \frac{k-1}{2^i} \leq f(x) < \frac{k}{2^i} \text{ and } 1 \leq k \leq 2^i i, \\
i, & \text{when } f(x) \geq i.
\end{cases}
\tag{C.10}
$$

We leave the further details for the reader. $\qquad\square$

Exercise C.2.30. Check that the functions f_i defined in (C.10) have the desired properties.

Theorem C.2.31 (Luzin: "measurable is almost continuous"). *Let* (X, d) *be a metric space,* $(X, \mathcal{M}, \mu)$ *be a complete finite measure space. Let* $\tau_d \subset \mathcal{M}$ *and* $f : X \to \mathbb{R}$ *be* $\mathcal{M}$-*measurable. Then for every* $\varepsilon > 0$ *there exists a closed set* $F_\varepsilon \subset X$ *such that* $\mu(F_\varepsilon^c) < \varepsilon$ *and* $f|_{F_\varepsilon} : F_\varepsilon \to \mathbb{R}$ *is continuous.*

Remark C.2.32. Notice that if $f : X \to \mathbb{R}$ is continuous and $E \subset X$ then $f|_E : E \to \mathbb{R}$ is continuous; this implication may not be reversible!

Proof. Since $f = f^+ - f^-$, it suffices to assume that $f \geq 0$. For each $i \in \mathbb{Z}^+$, $f(X)$ has a disjoint Borel cover $\{[(j-1)/i, j/i) : j \in \mathbb{Z}^+\}$, and

$$\left\{ A_{ij} := f^{-1}([(j-1)/i, j/i)) \right\}_{j=1}^{\infty} \subset \mathcal{M}$$

is a disjoint cover of X. Take closed sets $F_{ij} \subset X$ such that $F_{ij} \subset A_{ij}$ and

$$\mu(A_{ij} \setminus F_{ij}) < 2^{-(i+j)}\varepsilon.$$

Then

$$\lim_{k\to\infty} \mu(X \setminus \bigcup_{j=1}^{k} F_{ij}) \overset{\mu(X)<\infty}{=} \mu\left(X \setminus \bigcup_{j=1}^{\infty} F_{ij}\right)$$

$$= \mu\left(\bigcup_{j=1}^{\infty}(A_{ij} \setminus F_{ij})\right) = \sum_{j=1}^{\infty} \mu(A_{ij} \setminus F_{ij})$$

$$< 2^{-i}\varepsilon.$$

Thereby let $B_i := \bigcup_{j=1}^{k_i} F_{ij}$, where k_i is so large that

$$\mu(X \setminus B_i) < 2^{-i}\varepsilon.$$

Now $F_\varepsilon := \bigcap_{i=1}^{\infty} B_i \subset X$ is closed, and

$$\mu(F_\varepsilon^c) = \mu\left(\bigcup_{i=1}^{\infty} B_i^c\right) \le \sum_{i=1}^{\infty} \mu(B_i^c) < \varepsilon.$$

Let us define $g_i : B_i \to \mathbb{R}$ such that

$$g_i(x) := j/i, \quad \text{when} \quad x \in F_{ij}.$$

Then $g_i : B_i \to \mathbb{R}$ is continuous, because $g_i|_{F_{ij}}$ is constant and because the closed sets F_{ij} are disjoint. Next, $g_i|_{F_\varepsilon} \to f|_{F_\varepsilon}$ uniformly, because $|f(x) - g_i(x)| < 1/j$ whenever $x \in F_{ij} \subset A_{ij}$; the proof is complete, since continuity is preserved in uniform convergence. □

Exercise C.2.33. Let $(X, \mathcal{M}, \mu)$, where $X = [0,1] \subset \mathbb{R}$ and μ is the restriction of the Lebesgue measure $\lambda_\mathbb{R}$ to X. Consider the characteristic function $f = \chi_{\mathbb{Q}\cap X} : X \to \mathbb{R}$ in the light of Luzin's Theorem C.2.31.

Remark C.2.34 (**Littlewood's principles**). With a pinch of salt, measure theory may be crystallised in *Littlewood's principles*:

1. "A measurable set is almost open"
   (see Topological Approximation Theorem C.1.42).

2. "A measurable function is almost continuous"
   (see Luzin's Theorem C.2.31).

3. "Pointwise convergence is almost uniform convergence"
   (see Egorov's Theorem C.2.23).

C.3 Integration

In this section, let $(X, \mathcal{M}, \mu)$ be a complete measure space. The μ-*integral*

$$\int f \ \mathrm{d}\mu$$

of an $\mathcal{M}$-measurable function $f : X \to [-\infty, +\infty]$ is defined step-by-step:

1. first for a simple non-negative function;
2. then for a non-negative function;
3. finally, the general definition.

Definition C.3.1. Let $s : X \to [0, \infty)$ be an $\mathcal{M}$-measurable simple function. Its *integral* $\int s \ \mathrm{d}\mu \in [0, \infty]$ is defined as

$$\int s \ \mathrm{d}\mu = \int \sum_{a \in s(X)} a \ \chi_{\{s=a\}} \ \mathrm{d}\mu := \sum_{a \in s(X)} a \cdot \mu \left(\{s = a\}\right),$$

with the convention $0 \cdot \infty := 0$. Especially, $\int \chi_E \mathrm{d}\mu = \mu(E)$ for $E \in \mathcal{M}$.

Definition C.3.2. Let $f^+ : X \to [0, \infty]$ be an $\mathcal{M}$-measurable non-negative function. Its *integral* $\int f^+ \mathrm{d}\mu \in [0, \infty]$ is defined as

$$\int f^+ \mathrm{d}\mu := \sup \left\{ \int s \ \mathrm{d}\mu : \ 0 \le s \le f^+, \ s \text{ simple measurable} \right\}.$$

Definition C.3.3 (Integral). Let $f : X \to [-\infty, +\infty]$ be an $\mathcal{M}$-measurable function. Its *integral* $\int f \mathrm{d}\mu$ is defined as

$$\int f \ \mathrm{d}\mu := \int f^+ \ \mathrm{d}\mu - \int f^- \ \mathrm{d}\mu$$

provided that $\int f^+ \mathrm{d}\mu < \infty$ *or* $\int f^- \mathrm{d}\mu < \infty$: we want to avoid a situation $\infty - \infty$ here. If $\int f^+ \mathrm{d}\mu < \infty$ *and* $\int f^- \mathrm{d}\mu < \infty$ then f is called μ-*integrable*. Let $f : X \to \mathbb{C}$ be $\mathcal{M}$-measurable such that $|f| : X \to \mathbb{R}$ is μ-integrable. The μ-integral of f is defined by

$$\int f \ \mathrm{d}\mu := \int \mathrm{Re} f \ \mathrm{d}\mu + \mathrm{i} \int \mathrm{Im} f \ \mathrm{d}\mu,$$

where $\mathrm{Re} f, \mathrm{Im} f : X \to \mathbb{R}$ are the real and imaginary parts of f, respectively. If we want to emphasize the variable in the integration, we may write

$$\int f \ \mathrm{d}\mu = \int f(x) \ \mathrm{d}\mu(x),$$

or even $\int f(x) \ \mathrm{d}x$, if the measure is clear from the context. We shall also use the abbreviation

$$\int_E f \ \mathrm{d}\mu := \int \chi_E f \ \mathrm{d}\mu,$$

where $E \in \mathcal{M}$; this is the *integral of f over* $E \subset X$. The *Lebesgue integral* is the integral with respect to the Lebesgue measure.

C.3.1   Integrating simple non-negative functions

It is simple to integrate simple functions. We leave the details as an exercise for
the reader:

Exercise C.3.4. Let $r, s : X \to [0, \infty)$ be $\mathcal{M}$-measurable simple functions and
$a \in [0, \infty)$. Show that

$$\int ar \, d\mu = a \int r \, d\mu \quad \text{and} \quad \int (r + s) \, d\mu = \int r \, d\mu + \int s \, d\mu.$$

Moreover, if $r \leq s$, show that $\int r \, d\mu \leq \int s \, d\mu$.

C.3.2   Integrating non-negative functions

Let us now concentrate on integrating measurable non-negative functions. Recall
that we are dealing with a complete measure space $(X, \mathcal{M}, \mu)$.

Exercise C.3.5. Let $S \in \mathcal{M}$ and

$$\mu_S(E) := \mu(E \cap S).$$

Show that $(X, \mathcal{M}, \mu_S)$ is complete and that

$$\int f \, d\mu_S = \int_S f \, d\mu$$

for all $\mathcal{M}$-measurable $f \geq 0$.

As an easy consequence of Exercise C.3.4, for $\mathcal{M}$-measurable functions
$f^+, g^+ : X \to [0, \infty]$ and $a \in \mathbb{R}^+$,

$$\int af^+ \, d\mu \quad = \quad a \int f^+ \, d\mu,$$

$$\text{if} \quad f^+ \leq g^+ \quad \text{then} \quad \int f^+ \, d\mu \leq \int g^+ \, d\mu.$$

These observations will be used frequently. However, it is not evident whether

$$\int (f^+ + g^+) \, d\mu = \int f^+ \, d\mu + \int g^+ \, d\mu.$$

This will soon be obtained as a consequence of the following fundamental result:

Theorem C.3.6 (Monotone Convergence Theorem (Lebesgue, Levi)). *For each*
$k \geq 1$, *let* $f_k : X \to [0, \infty]$ *be* $\mathcal{M}$-*measurable such that* $f_k \leq f_{k+1}$. *Then*

$$\lim_{k \to \infty} \int f_k \, d\mu = \int \lim_{k \to \infty} f_k \, d\mu.$$

The case when the limit $f := \lim_{k \to \infty} f_k$ is integrable was proved by Henri-Léon Lebesgue, with the integrability assumption removed by Beppo Levi. In general, the convergence here is meant in $[0, \infty]$, i.e., the limit may be infinite.

Proof of Theorem C.3.6. The function $f := \lim_{k \to \infty} f_k : X :\to [0, \infty]$ is measurable as a limit of measurable functions. Clearly, $f_k \leq f_{k+1} \leq f$, so the increasing sequence of integrals $\int f_k \mathrm{d}\mu \leq \int f \mathrm{d}\mu$ converges to the limit

$$\lim_{k \to \infty} \int f_k \, \mathrm{d}\mu \leq \int f \, \mathrm{d}\mu.$$

Let $0 < \varepsilon < 1$. Take a simple measurable function s such that $s \leq f$ and

$$\int s \, \mathrm{d}\mu \geq (1 - \varepsilon) \int f \, \mathrm{d}\mu.$$

Let $E_k := \{f_k > (1 - \varepsilon)s\}$. Since f_k and s are measurable, $E_k \in \mathcal{M}$. Furthermore,

$$
\begin{aligned}
\int f_k \, \mathrm{d}\mu \quad &\geq \quad \int (1 - \varepsilon)s \, \chi_{E_k} \, \mathrm{d}\mu \\
&= \quad \sum_{a \in s(X)} (1 - \varepsilon)a \cdot \mu \left(E_k \cap \{s = a\} \right) \\
&\xrightarrow[k \to \infty]{} \quad (1 - \varepsilon) \sum_{a \in s(X)} a \cdot \mu \left(\{s = a\} \right) \\
&= \quad (1 - \varepsilon) \int s \, \mathrm{d}\mu \\
&\geq \quad (1 - \varepsilon)^2 \int f \, \mathrm{d}\mu,
\end{aligned}
$$

where the limit is due to $X = \bigcup_{k=1}^{\infty} E_k$, where $E_k \subset E_{k+1} \in \mathcal{M}$. Thus

$$\lim_{k \to \infty} \int f_k \, \mathrm{d}\mu \geq (1 - \varepsilon)^2 \int f \, \mathrm{d}\mu.$$

Taking $\varepsilon \to 0$, the proof is complete. $\square$

Corollary C.3.7. *Let $f, g : X \to [0, \infty]$ be $\mathcal{M}$-measurable. Then*

$$\int (f + g) \, \mathrm{d}\mu = \int f \, \mathrm{d}\mu + \int g \, \mathrm{d}\mu.$$

Proof. Take measurable simple functions $f_k, g_k : X \to [0, \infty)$ such that $f_k \leq f_{k+1}$ and $g_k \leq g_{k+1}$ for each $k \in \mathbb{Z}^+$, and $f_k \to f$ and $g_k \to g$ pointwise. Then $f_k + g_k : X \to [0, \infty)$ is measurable and simple, such that

$$f_k + g_k \leq f_{k+1} + g_{k+1} \xrightarrow[k \to \infty]{} f + g,$$

so that by the Monotone Convergence Theorem C.3.6,

$$\int (f+g)\ \mathrm{d}\mu \quad = \quad \lim_{k\to\infty} \int (f_k + g_k)\ \mathrm{d}\mu$$

$$\overset{\text{Exercise C.3.4}}{=} \quad \lim_{k\to\infty} \left(\int f_k\ \mathrm{d}\mu + \int g_k\ \mathrm{d}\mu \right)$$

$$= \quad \int f\ \mathrm{d}\mu + \int g\ \mathrm{d}\mu,$$

establishing the result.   □

Corollary C.3.8. *Let* $g_j : X \to [0,\infty]$ *be* $\mathcal{M}$*-measurable for each* $j \in \mathbb{Z}^+$*. Then*

$$\int \sum_{j=1}^{\infty} g_j\ \mathrm{d}\mu = \sum_{j=1}^{\infty} \int g_j\ \mathrm{d}\mu.$$

Proof. For each $k \in \mathbb{Z}^+$, let us define functions $f_k, f : X \to [0,\infty]$ by

$$f_k := \sum_{j=1}^{k} g_j \quad \text{and} \quad f := \lim_{k\to\infty} f_k = \sum_{j=1}^{\infty} g_j.$$

These functions are measurable and $f_k \le f_{k+1} \le f$, so

$$\int \lim_{k\to\infty} \sum_{j=1}^{k} g_j\ \mathrm{d}\mu \quad \overset{\substack{\text{Monotone} \\ \text{Convergence}}}{=} \quad \lim_{k\to\infty} \int \sum_{j=1}^{k} g_j\ \mathrm{d}\mu$$

$$\overset{\text{Corollary C.3.7}}{=} \quad \lim_{k\to\infty} \sum_{j=1}^{k} \int g_j\ \mathrm{d}\mu,$$

completing the proof.   □

Exercise C.3.9. Let $f \ge 0$ be $\mathcal{M}$-measurable and $\int f\ \mathrm{d}\mu < \infty$. Prove that

$$\forall \varepsilon > 0\ \exists \delta > 0\ \forall A \in \mathcal{M} : \quad \mu(A) < \delta \Rightarrow \int_A f\ \mathrm{d}\mu < \varepsilon.$$

Theorem C.3.10 (Fatou's lemma). *Let* $g_k : X \to [0,\infty]$ *be* $\mathcal{M}$*-measurable for each* $k \in \mathbb{Z}^+$*. Then*

$$\int \liminf_{k\to\infty} g_k\ \mathrm{d}\mu \le \liminf_{k\to\infty} \int g_k\ \mathrm{d}\mu.$$

Proof. Notice that

$$\liminf_{k\to\infty} g_k = \sup_{k \ge 1} \inf_{j \ge k} g_j.$$

Define $f_k := \inf_{j \geq k} g_j$ for each $k \geq 1$. Now $f_k : X \to [0, \infty]$ is measurable and $f_k \leq f_{k+1}$, so that $\sup_{k \geq 1} f_k = \lim_{k \to \infty} f_k$, and

$$
\begin{aligned}
\int \liminf_{k \to \infty} g_k \, d\mu &= \int \sup_{k \geq 1} f_k \, d\mu \\
&= \int \lim_{k \to \infty} f_k \, d\mu \\
\overset{\text{Monotone Convergence}}{=} \ & \lim_{k \to \infty} \int f_k \, d\mu \\
&= \liminf_{k \to \infty} \int f_k \, d\mu \\
&\leq \liminf_{k \to \infty} \int g_k \, d\mu.
\end{aligned}
$$

The proof is complete. $\qquad\square$

Exercise C.3.11. Sometimes $\int \liminf_{k \to \infty} g_k \, d\mu < \liminf_{k \to \infty} \int g_k \, d\mu$ happens in Fatou's Lemma C.3.10. Find an example.

Exercise C.3.12. Actually, the Monotone Convergence Theorem C.3.6 and Fatou's Lemma C.3.10 are logically equivalent: prove this.

Exercise C.3.13 (Reverse Fatou's lemma). Prove the following *reverse Fatou's lemma*. Let $g_k : X \to [0, \infty]$ be $\mathcal{M}$-measurable for each $k \in \mathbb{Z}^+$. Assume that $g_k \leq g$ for every k, where g is μ-integrable. Then

$$
\int \limsup_{k \to \infty} g_k \, d\mu \geq \limsup_{k \to \infty} \int g_k \, d\mu.
$$

C.3.3 Integration in general

Let $f : X \to [-\infty, +\infty]$ be an $\mathcal{M}$-measurable function. Recall that if

$$
I^+ = \int f^+ \, d\mu < \infty \qquad or \qquad I^- = \int f^- \, d\mu < \infty
$$

then the μ-integral f is $\int f d\mu = I^+ - I^-$. Moreover, if both I^+ and I^- are finite, f is called μ-integrable. We shall be interested mainly in μ-integrable functions.

Theorem C.3.14. *Let $a \in \mathbb{R}$ and $f : X \to [-\infty, +\infty]$ be μ-integrable. Then*

$$
\int af \, d\mu = a \int f \, d\mu.
$$

Moreover, if $g : X \to [-\infty, +\infty]$ is μ-integrable such that $f \leq g$, then

$$\int f \, \mathrm{d}\mu \leq \int g \, \mathrm{d}\mu.$$

Especially, $\left| \int f \, \mathrm{d}\mu \right| \leq \int |f| \, \mathrm{d}\mu.$

Exercise C.3.15. Prove Theorem C.3.14.

Exercise C.3.16. Let $E \in \mathcal{M}$ and $|f| \leq g$, where f is $\mathcal{M}$-measurable and g is μ-integrable. Show that f and $f\chi_E$ are μ-integrable.

Exercise C.3.17 (Chebyshev's inequality). Let $0 < a < \infty$, and let $f : X \to [-\infty, +\infty]$ be $\mathcal{M}$-measurable. Prove *Chebyshev's inequality*

$$\mu(\{|f| > a\}) \leq a^{-1} \int |f| \, \mathrm{d}\mu. \tag{C.11}$$

We continue by noticing the short-sightedness of integrals:

Lemma C.3.18. *Let $f, g : X \to [-\infty, +\infty]$ be μ-integrable. Then*

1. *Let $E \in \mathcal{M}$ such that $\mu(E) = 0$. Then $\displaystyle\int_E f \, \mathrm{d}\mu = 0$.*

2. *Let $f = g$ μ-almost everywhere. Then $\displaystyle\int f \, \mathrm{d}\mu = \int g \, \mathrm{d}\mu$.*

3. *Let $\displaystyle\int |f| \, \mathrm{d}\mu = 0$. Then $f = 0$ μ-almost everywhere.*

Proof. First,

$$
\begin{aligned}
\int_E f^+ \, \mathrm{d}\mu \quad &= \quad \int f^+ \chi_E \, \mathrm{d}\mu \\
&= \quad \sup\left\{ \int s \, \mathrm{d}\mu : s \leq f^+ \chi_E \text{ simple measurable} \right\} \\
&\stackrel{\mu(E)=0}{=} \quad 0,
\end{aligned}
$$

proving the first result. Next, let us suppose $f = g$ μ-almost everywhere. Then

$$
\begin{aligned}
\int f^+ \, \mathrm{d}\mu \quad &= \quad \int \left(f^+ \chi_{\{f=g\}} + f^+ \chi_{\{f \neq g\}} \right) \, \mathrm{d}\mu \\
&\stackrel{\text{Corollary C.3.7}}{=} \quad \int_{\{f=g\}} f^+ \, \mathrm{d}\mu + \int_{\{f \neq g\}} f^+ \, \mathrm{d}\mu \\
&\stackrel{\mu(\{f \neq g\})=0}{=} \quad \int_{\{f=g\}} f^+ \, \mathrm{d}\mu,
\end{aligned}
$$

showing that $\int f^+ \mathrm{d}\mu = \int g^+ \mathrm{d}\mu$, establishing the second result. Finally,

$$
\begin{aligned}
\mu(\{f \neq 0\}) &= \mu\left(\bigcup_{k=1}^{\infty}\{|f| > 1/k\}\right) \\
&\leq \sum_{k=1}^{\infty} \mu(\{|f| > 1/k\}) \\
&= \sum_{k=1}^{\infty} \int \chi_{\{|f| > 1/k\}} \,\mathrm{d}\mu \\
&\leq \sum_{k=1}^{\infty} \int k|f| \,\mathrm{d}\mu \\
&= \sum_{k=1}^{\infty} k \int |f| \,\mathrm{d}\mu
\end{aligned}
$$

so that if $\int |f| \mathrm{d}\mu = 0$, then $\mu(\{f \neq 0\}) = 0$. $\qquad\square$

Proposition C.3.19. *Let $f : X \to [-\infty, +\infty]$ be μ-integrable. Then $f(x) \in \mathbb{R}$ for μ-almost every $x \in X$.*

Proof. First, $\{f^+ = \infty\} = \bigcap_{k=1}^{\infty}\{f^+ > k\} \in \mathcal{M}$, because f^+ is $\mathcal{M}$-measurable. Thereby

$$
\begin{aligned}
\mu(\{f^+ = \infty\}) &= \frac{1}{k}\int k \cdot \chi_{\{f^+ = \infty\}} \,\mathrm{d}\mu \\
&\leq \frac{1}{k}\int f^+ \,\mathrm{d}\mu \xrightarrow[k\to\infty]{} 0,
\end{aligned}
$$

so that $\mu(\{f^+ = \infty\}) = 0$. Similarly, $\mu(\{f^- = \infty\}) = 0$. $\qquad\square$

Remark C.3.20. By Lemma C.3.18 and Proposition C.3.19, when it comes to integration, we may identify a μ-integrable function $f : X \to [-\infty, +\infty]$ with the function $\tilde{f} : X \to \mathbb{R}$ defined by

$$
\tilde{f}(x) = \begin{cases} f(x), & \text{when } f(x) \in \mathbb{R}, \\ 0, & \text{when } |f(x)| = \infty. \end{cases}
$$

We shall establish this identification without any further notice.

Theorem C.3.21 (Sum is integrable). *Let $f, g : X \to [-\infty, +\infty]$ be μ-integrable. Then $f + g$ is μ-integrable and*

$$
\int (f + g) \,\mathrm{d}\mu = \int f \,\mathrm{d}\mu + \int g \,\mathrm{d}\mu.
$$

Proof. For integrable $f, g : X \to \mathbb{R}$, the function $f + g : X \to \mathbb{R}$ is measurable. Notice that

$$f + g = \begin{cases} (f^+ - f^-) + (g^+ - g^-), \\ (f + g)^+ - (f + g)^-. \end{cases}$$

Since $(f + g)^+ \leq f^+ + g^+$, and $(f + g)^- \leq f^- + g^-$, the integrability of $f + g$ follows. Moreover, $(f + g)^+ + f^- + g^- = (f + g)^- + f^+ + g^+$. By Corollary C.3.7,

$$\int (f + g)^+ \mathrm{d}\mu + \int f^- \mathrm{d}\mu + \int g^- \mathrm{d}\mu = \int (f + g)^- \mathrm{d}\mu + \int f^+ \mathrm{d}\mu + \int g^+ \mathrm{d}\mu,$$

implying

$$\begin{aligned} \int (f + g) \, \mathrm{d}\mu &= \int (f + g)^+ \, \mathrm{d}\mu - \int (f + g)^- \, \mathrm{d}\mu \\ &= \int f^+ \, \mathrm{d}\mu - \int f^- \, \mathrm{d}\mu + \int g^+ \, \mathrm{d}\mu - \int g^- \, \mathrm{d}\mu \\ &= \int f \, \mathrm{d}\mu + \int g \, \mathrm{d}\mu. \end{aligned}$$

The proof for the summation is thus complete. $\qquad\qquad\qquad\qquad\qquad\qquad$ $\square$

Theorem C.3.22 (Lebesgue's Dominated Convergence Theorem). *For each $k \geq 1$, let $f_k : X \to [-\infty, +\infty]$ be measurable and $f_k \xrightarrow[k \to \infty]{} f$ pointwise. Assume that $|f_k| \leq g$ for every $k \geq 1$, where g is μ-integrable. Then*

$$\int |f_k - f| \, \mathrm{d}\mu \xrightarrow[k \to \infty]{} 0,$$

$$\int f_k \, \mathrm{d}\mu \xrightarrow[k \to \infty]{} \int f \, \mathrm{d}\mu.$$

Proof. The functions $f_k, f, |f_k - f|$ are μ-integrable, because they are measurable, g is μ-integrable, $|f_k|, |f| \leq g$ and $|f_k - f| \leq 2g$. For each $k \geq 1$, we define function $g_k := 2g - |f_k - f|$. Then the functions $g_k \geq 0$ satisfy the assumptions of Fatou's Lemma C.3.10, yielding

$$\begin{aligned} \int 2g \, \mathrm{d}\mu &= \int \liminf_{k \to \infty} g_k \, \mathrm{d}\mu \\ &\overset{\text{Fatou}}{\leq} \liminf_{k \to \infty} \int g_k \, \mathrm{d}\mu \\ &= \liminf_{k \to \infty} \left(\int 2g \, \mathrm{d}\mu - \int |f_k - f| \, \mathrm{d}\mu \right) \\ &= \int 2g \, \mathrm{d}\mu - \limsup_{k \to \infty} \int |f_k - f| \, \mathrm{d}\mu. \end{aligned}$$

Here we may cancel $\int 2g \, \mathrm{d}\mu \in \mathbb{R}$, getting

$$\limsup_{k \to \infty} \int |f_k - f| \, \mathrm{d}\mu \leq 0,$$

so that $\int |f_k - f| \, \mathrm{d}\mu \xrightarrow[k \to \infty]{} 0$. Finally,

$$\left| \int f_k \, \mathrm{d}\mu - \int f \, \mathrm{d}\mu \right| = \left| \int (f_k - f) \, \mathrm{d}\mu \right| \leq \int |f_k - f| \, \mathrm{d}\mu \xrightarrow[k \to \infty]{} 0,$$

which completes the proof.   □

Remark C.3.23. It is easy to slightly generalise Lebesgue's Dominated Convergence Theorem C.3.22: the same conclusions hold even if we assume only that $f_k \to f$ *almost everywhere*, and that $|f_k| \leq g$ *almost everywhere*, where g is integrable. This is because integrals are not affected if we change values of functions in a set of measure zero.

Exercise C.3.24 (Indispensability of an integrable dominating function). Show that in Theorem C.3.22 it is indispensable to require the μ-integrability of a dominating function g. For this, consider $X = [0, 1]$, μ the Lebesgue measure, and the sequence $(f_k)_{k=1}^\infty$ with $f_k(x) = k$ for $x \in (0, 1/k]$, and $f_k(x) = 0$ for $x \in (1/k, 1]$. Show that the function $h := \sup_k f_k \geq 0$ is not Lebesgue-integrable on $[0, 1]$ (hence no dominating function here can be Lebesgue-integrable). Finally, show that the conclusion of Theorem C.3.22 fails for this sequence $(f_k)_{k=1}^\infty$.

Exercise C.3.25 (Fatou–Lebesgue Theorem). Prove the following **Fatou–Lebesgue Theorem**: *Let $(f_k)_{k=1}^\infty$ be a sequence of $\mathcal{M}$-measurable functions $f_k : X \to \mathbb{R}$ on a measure space $(X, \mathcal{M}, \mu)$. Assume that $|f_k| \leq g$ for every $k \geq 1$, where g is μ-integrable. Then $\liminf_{k \to \infty} f_k$ and $\limsup_{k \to \infty} f_k$ are μ-integrable and we have*

$$\int \liminf_{k \to \infty} f_k \, \mathrm{d}\mu \leq \liminf_{k \to \infty} \int f_k \, \mathrm{d}\mu \leq \limsup_{k \to \infty} \int f_k \, \mathrm{d}\mu \leq \int \limsup_{k \to \infty} f_k \, \mathrm{d}\mu.$$

Proposition C.3.26 (Riemann vs Lebesgue). *Let $f : \mathbb{R} \to \mathbb{R}$ be Riemann-integrable on the closed interval $[a, b] \subset \mathbb{R}$. Then $f\chi_{[a,b]}$ is Lebesgue-integrable and the Riemann- and Lebesgue-integrals coincide:*

$$\int_a^b f(x) \, \mathrm{d}x = \int_{[a,b]} f \, \mathrm{d}\lambda_{\mathbb{R}}.$$

Exercise C.3.27 (Riemann integration). Prove Proposition C.3.26. Recall the definition of the Riemann-integral: Let $g : [a, b] \to \mathbb{R}$ be bounded. A finite sequence $P_n = (x_0, \ldots, x_n)$ is called a *partition of $[a, b]$* if

$$a = x_0 < x_1 < x_2 < \cdots < x_{n-1} < x_n = b,$$

for which the *lower and upper Riemann sums* $L(g, P_n), U(g, P_n)$ are defined by

$$U(g, P_n) = \sum_{k=1}^{n} \left(\sup_{x_{k-1} \leq x < x_k} g(x) \right) (x_k - x_{k-1}),$$

$$L(g, P_n) = \sum_{k=1}^{n} \left(\inf_{x_{k-1} \leq x < x_k} g(x) \right) (x_k - x_{k-1}).$$

Now $L(g) \leq U(g)$, where

$$\begin{cases} U(g) := \inf \{U(g, P) : \ P \text{ is a partition of } [a, b]\}, \\ L(g) := \sup \{L(g, P) : \ P \text{ is a partition of } [a, b]\}. \end{cases}$$

If $L(g) = U(g)$, we say that g is *Riemann-integrable* with Riemann integral

$$\int_a^b g(x) \, dx = L(g).$$

Exercise C.3.28. Prove the following ϵ-*criterion for Riemann integrability*: if for any $\varepsilon > 0$ there is a partition P of $[a, b]$ such that $U(g, P) - L(g, P) < \varepsilon$, then g is Riemann-integrable over $[a, b]$.

Consequently, prove that if g is monotonic on $[a, b]$ or if g is continuous on $[a, b]$, then g is Riemann-integrable over $[a, b]$.

C.4   Integral as a functional

C.4.1   Lebesgue spaces $L^p(\mu)$

In the sequel, $(X, \mathcal{M}, \mu)$ is a complete measure space. For instance, we may have $\mathcal{M} = \mathcal{M}(\mu^*)$.

Definition C.4.1 ($L^p(\mu)$-**norms**). For $1 \leq p < \infty$, the $L^p(\mu)$-*norm* of an $\mathcal{M}$-measurable function $f : X \to [-\infty, +\infty]$ is

$$\|f\|_{L^p(\mu)} := \left(\int |f|^p \, d\mu \right)^{1/p},$$

and let

$$\|f\|_{L^\infty(\mu)} := \inf \{M \in [0, \infty] : \ |f| \leq M \ \mu\text{-a.e.}\},$$

Here L^p is read "L-p" or "Lebesgue-p". If μ is known from the context, notations $L^p = L^p(X) = L^p(\mu)$ are used: e.g., $L^p(\mathbb{R}^n) = L^p(\lambda_{\mathbb{R}^n})$.

Remark C.4.2. The quantities $\|f\|_{L^p(\mu)}$ are not the norms because the non-degeneracy fails: $\|f\|_{L^p(\mu)} = 0$ only implies that $f = 0$ μ-a.e. In fact, clearly, $\|f\|_{L^p} \in$

$[0, \infty]$, $\|f\|_{L^p} = 0$ if and only if $f = 0$ μ-almost everywhere, $\|\lambda f\|_{L^p} = |\lambda| \|f\|_{L^p}$, and $\|f\|_{L^1(\mu)} = \int |f| \, d\mu$. Also, in Theorem C.4.5 we will see that $\|f + g\|_{L^p} \leq \|f\|_{L^p} + \|g\|_{L^p}$ so that the triangle inequality would be satisfied. Thus, we will modify the construction slightly (identifying functions equal μ-a.e.) in Definition C.4.6 to make them norms, so that the terminology "norm" will be justified.

Definition C.4.3 (Lebesgue conjugate). The *Lebesgue conjugate* of $p \in [1, \infty]$ is the number $p' \in [1, \infty]$ defined by

$$\frac{1}{p} + \frac{1}{p'} = 1$$

with the usual convention $1/\infty = 0$.

The converse to the following theorem (the converse of Hölder's inequality) will be shown in Theorem C.4.56.

Theorem C.4.4 (Hölder's inequality). *Let* $1 \leq p \leq \infty$ *and* $q = p'$. *Let* $f, g : X \to [-\infty, +\infty]$ *be* $\mathcal{M}$-*measurable. Then*

$$\|fg\|_{L^1} \leq \|f\|_{L^p} \|g\|_{L^q}.$$

Proof. For $p = 1$,

$$
\begin{aligned}
\|fg\|_{L^1} &= \int |f||g| \, d\mu \\
&\leq \int |f| \, d\mu \; \|g\|_{L^\infty} \\
&= \|f\|_{L^1} \|g\|_{L^\infty};
\end{aligned}
$$

the proof for $p = \infty$ is symmetric. Finally, let us assume that $1 < p < \infty$. We may assume the non-trivial case $0 < \|f\|_{L^p} < \infty$ and $0 < \|g\|_{L^q} < \infty$. Then

$$\|fg\|_{L^1} = \|f\|_{L^p} \|g\|_{L^q} \int ab \, d\mu,$$

where $a = |f|/\|f\|_{L^p}$ and $b = |g|/\|g\|_{L^q}$. The concavity of the logarithm gives

$$
\begin{aligned}
\ln(ab) &= \ln(a^p)/p + \ln(b^q)/q \\
&\overset{1/q = 1 - 1/p}{\leq} \ln\left(a^p/p + b^q/q\right),
\end{aligned}
$$

so $\int ab \, d\mu \leq \int (a^p/p + b^q/q) \, d\mu = 1/p + 1/q = 1$. $\qquad\square$

Theorem C.4.5 (Minkowski's inequality). *Let* $1 \leq p \leq \infty$. *Let* $f, g : X \to [-\infty, +\infty]$ *be* $\mathcal{M}$-*measurable. Then*

$$\|f + g\|_{L^p} \leq \|f\|_{L^p} + \|f\|_{L^p}. \tag{C.12}$$

Proof. First,

$$\|f+g\|_{L^1} = \int |f+g| \, \mathrm{d}\mu$$

$$\leq \int |f| \, \mathrm{d}\mu + \int |g| \, \mathrm{d}\mu$$

$$= \|f\|_{L^1} + \|g\|_{L^1}.$$

Also $|f+g| \leq |f| + |g|$, so that $|f+g| \leq \|f\|_{L^\infty} + \|g\|_{L^\infty}$ almost everywhere, yielding

$$\|f+g\|_{L^\infty} \leq \|f\|_{L^\infty} + \|g\|_{L^\infty}.$$

Finally, assume that $1 < p < \infty$. Then

$$\|f+g\|_{L^p}^p = \int |f+g|^p \, \mathrm{d}\mu$$

$$\leq \int (|f|+|g|) \, |f+g|^{p-1} \, \mathrm{d}\mu$$

$$= \left\| f \, |f+g|^{p-1} \right\|_{L^1} + \left\| g \, |f+g|^{p-1} \right\|_{L^1}$$

$$\overset{\text{Hölder}}{\leq} (\|f\|_{L^p} + \|g\|_{L^p}) \, \left\| |f+g|^{p-1} \right\|_{L^q},$$

where

$$\left\| |f+g|^{p-1} \right\|_{L^q} = \left(\int |f+g|^{(p-1)q} \, \mathrm{d}\mu \right)^{1/q}$$

$$= \left(\int |f+g|^p \, \mathrm{d}\mu \right)^{(p-1)/p}$$

$$= \|f+g\|_{L^p}^{p-1},$$

concluding the proof.   □

Definition C.4.6 ($L^p(\mu)$-spaces). Let $1 \leq p \leq \infty$ and

$$V^p := \{ f : \|f\|_{L^p} < \infty \}.$$

Noticing that $\|\lambda f\|_{L^p} = |\lambda| \, \|f\|_{L^p}$ for any scalar λ, and recalling Minkowski's inequality (C.12), we see that

$$(f \mapsto \|f\|_{L^p}) : V^p \to [0, \infty)$$

is a seminorm on the vector space V^p. Let us define an equivalence relation $\sim$ on V^p by

$$f \sim g \iff \|f-g\|_{L^p} = 0;$$

i.e., $f \sim g \iff f = g$ μ-almost everywhere. Let us denote the equivalence classes by

$$[f] := \{ g \in V^p : f \sim g \}.$$

We obtain the quotient vector space

$$L^p(\mu) \quad := \quad V^p/\sim \quad = \quad \{ [f] : f \in V^p \}$$

with the usual vector space operations. Moreover,

$$([f] \mapsto \|f\|_{L^p}) : L^p(\mu) \to [0, \infty)$$

is a norm on vector space $L^p(\mu)$. Customarily, $\ell^p(X) := L^p(\mu)$, where μ is the *counting measure*, i.e., $\mu(E)$ is the number of points in the set E.

Remark C.4.7. $f \in [f]$ is a function $X \to [-\infty, +\infty]$, but $[f] \in L^p(\mu)$ is not a function, but an equivalence class of functions. However in practice, to avoid cumbersome notation, one often identifies f and $[f]$, e.g., writing briefly $f \in L^p(\mu)$.

Definition C.4.8 (Convergence in $L^p(\mu)$). Let $f \in L^p$ and $\{f_j\}_{j=1}^{\infty} \subset L^p$. We say that $f_j \to f$ *in* L^p if

$$\|f_j - f\|_{L^p} \xrightarrow[j \to \infty]{} 0.$$

Theorem C.4.9. $L^p(\mu)$ *is a Banach space.*

Proof. The case $p = \infty$ is left as Exercise C.4.10; let us consider the case $1 \leq p < \infty$. We already know that $L^p(\mu)$ is a normed space. Given a Cauchy sequence $(f_j)_{j=1}^{\infty}$ in $L^p(\mu)$, we need a candidate f for the limit of this sequence. Now $(f_j)_{j=1}^{\infty}$ is a Cauchy sequence in measure μ, because

$$
\begin{aligned}
\mu\left(\{|f_i - f_j| \geq \varepsilon\}\right) \quad &= \quad \mu\left(\{|f_i - f_j|^p \geq \varepsilon^p\}\right) \\
&\overset{\text{Chebyshev (C.11)}}{\leq} \quad \varepsilon^{-p} \int |f_i - f_j|^p \, d\mu \\
&= \quad \varepsilon^{-p} \|f_i - f_j\|_{L^p}^p \\
&\xrightarrow[i,j \to \infty]{} \quad 0.
\end{aligned}
$$

Hence by Exercise C.2.21, $f_j \to f$ in measure μ for an $\mathcal{M}$-measurable function f. By Theorem C.2.24, $f_{j_k} \to f$ μ-almost everywhere for a subsequence $(f_{j_k})_{k=1}^{\infty}$ of $(f_j)_{j=1}^{\infty}$. Here $f \in L^p(\mu)$, because

$$
\begin{aligned}
\|f\|_{L^p}^p \quad &= \quad \int |f|^p \, d\mu \\
&= \quad \int \liminf_{k \to \infty} |f_{j_k}|^p \, d\mu \\
&\overset{\text{Fatou}}{\leq} \quad \liminf_{k \to \infty} \int |f_{j_k}|^p \, d\mu \\
&\leq \quad \text{constant} < \infty,
\end{aligned}
$$

because Cauchy sequences in a normed space are bounded. Finally, $f_i \to f$ in $L^p(\mu)$, because

$$
\begin{aligned}
\|f_i - f\|_{L^p}^p &= \int |f_i - f|^p \, d\mu \\
&= \int \liminf_{k \to \infty} |f_i - f_{j_k}|^p \, d\mu \\
&\overset{\text{Fatou}}{\leq} \liminf_{k \to \infty} \|f_i - f_{j_k}\|_{L^p}^p \\
&\xrightarrow[\substack{i \to \infty}]{(f_i)_{i=1}^\infty \text{ Cauchy}} 0.
\end{aligned}
$$

Thus $L^p(\mu)$ is a Banach space for $1 \leq p < \infty$. $\qquad\square$

Exercise C.4.10. Complete the previous proof by showing that $L^\infty(\mu)$ is a Banach space.

Exercise C.4.11. Let $1 \leq p < \infty$ and $\|f_j - f\|_{L^p} \to 0$, where $f \in L^p$ and $\{f_j\}_{j=1}^\infty \subset L^p$. Show that

$$
\forall \varepsilon > 0 \; \exists \delta > 0 \; \forall j \in \mathbb{Z}^+ : \quad \mu(E) < \delta \Longrightarrow \int_E |f_j|^p \, d\mu < \varepsilon.
$$

Why is $p \neq \infty$ here?

Lemma C.4.12. Let $g \in L^p(\mu)$, where $1 \leq p < \infty$. Show that

$$
\forall \varepsilon > 0 \; \exists E_g \in \mathcal{M} : \quad \mu(E_g) < \infty \quad \text{and} \quad \int_{E_g^c} |g|^p \, d\mu < \varepsilon.
$$

Exercise C.4.13. Prove Lemma C.4.12.

Theorem C.4.14 (Vitali's Convergence Theorem). Let $1 \leq p < \infty$. Let $f, f_j \in L^p(\mu)$ for each $j \in \mathbb{Z}^+$. Then properties (1,2,3) imply (0), and (0) implies properties (2,3):

(0) $f_j \to f$ in L^p.
(1) $f_j \to f$ μ-almost everywhere.
(2) $\forall \varepsilon > 0 \; \exists E \in \mathcal{M} \; \forall j \in \mathbb{Z}^+ : \mu(E) < \infty, \; \int_{E^c} |f_j|^p \, d\mu < \varepsilon$.
(3) $\forall \varepsilon > 0 \; \exists \delta > 0 \; \forall j \in \mathbb{Z}^+ \; \forall A \in \mathcal{M} : \mu(A) < \delta \Rightarrow \int_A |f_j|^p \, d\mu < \varepsilon$.

Proof. First, let us show that $(1,2,3)$ implies (0). Take $\varepsilon > 0$. Take $\delta > 0$ as in (3). Take $E \in \mathcal{M}$ as in (2). Exploiting (1), Egorov's Theorem C.2.23 says that $(f_j - f)|_E \to 0$ μ-almost uniformly. Hence there exists $B \in \mathcal{M}$ such that

$$
\begin{cases}
B \subset E, \\
\mu(E \setminus B) < \delta, \\
(f_j - f)|_B \to 0 \quad \text{uniformly.}
\end{cases}
\tag{C.13}
$$

We want to show that $\|f_j \to f\|_{L^p} \to 0$:

$$\|f_j - f\|_{L^p}^p \ =\ \int |f_j - f|^p \, d\mu$$

$$=\ \int_B |f_j - f|^p \, d\mu + \int_{B^c} |f_j - f|^p \, d\mu,$$

and here the integral over B tends to 0 as $j \to \infty$, by (C.13). What about the integral over B^c? Since $(t \mapsto t^p) : \mathbb{R}^+ \to \mathbb{R}$ is a convex function, we have $(a/2 + b/2)^p \le a^p/2 + b^p/2$, so that

$$\int_{B^c} |f_j - f|^p \, d\mu$$

$$\le\ \int_{B^c} 2^{p-1} \left(|f_j|^p + |f|^p \right) \, d\mu$$

$$=\ 2^{p-1} \left(\int_{E^c} |f_j|^p \, d\mu + \int_{E\backslash B} |f|^p \, d\mu + \int_{B^c} \liminf_{j\to\infty} |f_j|^p \, d\mu \right)$$

$$\overset{(2),\,(3),\,\text{Fatou}}{<}\ 2^{p-1} \left(\varepsilon + \varepsilon + 2\varepsilon \right);$$

thus $\|f_j - f\|_{L^p} \to 0$: we have proven that (0) follows from (1, 2, 3).

Implication (0) $\Rightarrow$ (3) is left as Exercise C.4.15.

Let us show that (0) $\Rightarrow$ (2). Let $f_j \to f$ in $L^p(\mu)$. Take $\varepsilon > 0$. Take $j_\varepsilon \in \mathbb{Z}^+$ such that $\|f_j - f\|_{L^p} < \varepsilon^{1/p}$ whenever $j > j_\varepsilon$. Take $E_f, E_{f_j} \in \mathcal{M}$ as in Lemma C.4.12. Let

$$E := E_f \cup \bigcup_{j=1}^{j_\varepsilon} E_{f_j}.$$

Then $E \in \mathcal{M}$ and $\mu(E) < \infty$. If $j \le j_\varepsilon$ then

$$\int_{E^c} |f_j|^p \, d\mu \le \int_{E_{f_j}^c} |f_j|^p \, d\mu < \varepsilon.$$

If $j > j_\varepsilon$ then

$$\int_{E^c} |f_j|^p \, d\mu \overset{\text{Minkowski}}{\le} \left(\|\chi_{E^c}(f_j - f)\|_{L^p} + \|\chi_{E^c} f\|_{L^p} \right)^p$$

$$\le\ \left(\varepsilon^{1/p} + \varepsilon^{1/p} \right)^p,$$

so that $\int_{E^c} |f_j|^p \, d\mu \le 2^p \varepsilon$ for every $j \in \mathbb{Z}^+$. We have shown that (0) $\Rightarrow$ (2).

Finally, let us prove that $(0) \Rightarrow (1)$. We have

$$
\begin{aligned}
\mu(\{|f_j - f| \geq \varepsilon\}) &= \mu(\{|f_j - f|^p \geq \varepsilon^p\}) \\
&\underset{\text{Chebyshev}}{\leq} \varepsilon^{-p} \int |f_j - f|^p \, d\mu \\
&\xrightarrow[j\to\infty]{(0)} 0,
\end{aligned}
$$

so that $f_j \to f$ in measure μ. By Theorem C.2.24, there is a subsequence $(f_{j_k})_{k=1}^{\infty}$ such that $f_{j_k} \to f$ μ-almost everywhere. We have shown that $(0) \Rightarrow (1)$. $\qquad\square$

Exercise C.4.15. Complete the proof of Vitali's Convergence Theorem C.4.14 by showing that $(0) \Rightarrow (3)$.

Exercise C.4.16. Let $1 \leq p \leq \infty$ and $f_j \to f$ μ-a.e., where $\{f_j\}_{j=1}^{\infty} \subset L^p$.

(a) Let $f_j \to g$ in L^p. Show that $f = g$ μ-a.e.

(b) Give an example where $f \in L^p$, but $f_j \not\to f$ in L^p.

Finally, we give without proof a very useful interpolation theorem. But first we introduce

Definition C.4.17 (Semifinite measures). A measure μ is called *semifinite* if for every $E \in \mathcal{M}$ with $\mu(E) = \infty$ there exists $F \in \mathcal{M}$ such that $F \subset E$ and $0 < \mu(F) < \infty$.

Theorem C.4.18 (M. Riesz–Thorin interpolation theorem). *Let μ, ν be semifinite measures and let $1 \leq p_0, p_1, q_0, q_1 \leq \infty$. For every $0 < t < 1$ define p_t and q_t by*

$$
\frac{1}{p_t} = \frac{1-t}{p_0} + \frac{t}{p_1}, \qquad \frac{1}{q_t} = \frac{1-t}{q_0} + \frac{t}{q_1}.
$$

Assume that A is a linear operator such that

$$
\|Af\|_{L^{q_0}(\nu)} \leq C_0 \|f\|_{L^{p_0}(\mu)}, \qquad \|Af\|_{L^{q_1}(\nu)} \leq C_1 \|f\|_{L^{p_1}(\mu)},
$$

for all $f \in L^{p_0}(\mu)$ and $f \in L^{p_1}(\mu)$, respectively. Then for all $0 < t < 1$, the operator A extends to a bounded linear operator from $L^{p_t}(\mu)$ to $L^{q_t}(\nu)$ and we have

$$
\|Af\|_{L^{q_t}(\nu)} \leq C_0^{1-t} C_1^t \|f\|_{L^{p_t}(\mu)}
$$

for all $f \in L^{p_t}(\mu)$.

C.4.2   Signed measures

Definition C.4.19 (Signed measures). Let $\mathcal{M}$ be a σ-algebra on X. A mapping $\nu : \mathcal{M} \to \mathbb{R}$ is called a *signed measure* on X if

$$
\nu\left(\bigcup_{j=1}^{\infty} E_j\right) = \sum_{j=1}^{\infty} \nu(E_j)
$$

for any disjoint countable family $\{E_j\}_{j=1}^{\infty} \subset \mathcal{M}$.

Example. Let $\mu, \nu : \mathcal{M} \to [0, \infty]$ be *finite* measures on X, that is $\mu(X) < \infty$ and $\nu(X) < \infty$. Then

$$\mu - \nu : \mathcal{M} \to \mathbb{R}$$

is a signed measure. It will turn out that there are no other types of signed measures on X, see the Jordan decomposition result in Corollary C.4.26

Remark C.4.20. For simplicity and in view of the planned applications of this notion we restrict the exposition to what may be called finite signed measures. In principle, one can allow $\nu : \mathcal{M} \to [-\infty, +\infty]$ assuming that only one of infinities may be achieved. The statements and the proofs remain largely similar, so we may leave this case as an exercise for an interested reader. For example, only one of the measures in Theorem C.4.25 would be finite, etc.

Exercise C.4.21. Let $(X, \mathcal{M}, \mu)$ be a measure space and let $f : X \to [-\infty, +\infty]$ be μ-integrable. Define $\nu : \mathcal{M} \to \mathbb{R}$ by

$$\nu(E) := \int_E f \, d\mu. \tag{C.14}$$

Show that ν is a signed measure. Moreover, prove that ν is a (finite) measure if and only if $f \geq 0$ μ-almost everywhere.

Definition C.4.22 (Variations of measures). Let $\nu : \mathcal{M} \to \mathbb{R}$ be a signed measure. Define mappings $\nu^+, \nu^-, |\nu| : \mathcal{M} \to [0, \infty]$ by

$$
\begin{aligned}
\nu^+(E) &:= \sup_{A \in \mathcal{M}:\ A \subset E} \nu(A), \\
\nu^- &:= (-\nu)^+, \\
|\nu| &:= \nu^+ + \nu^-.
\end{aligned}
$$

The mappings ν^+, ν^- are called the *positive and negative variations* (respectively) of ν, and the pair (ν^+, ν^-) is the *Jordan decomposition* of ν. The mapping $|\nu|$ is the *total variation* of ν.

Exercise C.4.23. Show that $\nu^+, \nu^-, |\nu| : \mathcal{M} \to [0, \infty]$ are measures.

Exercise C.4.24. Let $\nu(E) = \int_E f \, d\mu$ as in (C.14). Show that

$$\nu^+(E) = \int_E f^+ \, d\mu \quad \text{and} \quad \nu^-(E) = \int_E f^- \, d\mu.$$

Hence here $\nu = \nu^+ - \nu^-$, but this happens even generally:

Theorem C.4.25. *Let $\nu : \mathcal{M} \to \mathbb{R}$ be a signed measure. Then the measures $\nu^+, \nu^- : \mathcal{M} \to [0, \infty]$ are finite.*

Proof. By Exercise C.4.23, ν^+ and ν^- are measures. Let us show that ν^+ (and similarly ν^-) is finite. To get a contradiction, assume that $\nu^+(X) = \infty$. Take $E_0 \in \mathcal{M}$ such that $\nu(E_0) \geq 0$. Take $A_0 \in \{E_0, X \setminus E_0\}$ such that $\nu^+(A_0) = \infty$. For $k \in \mathbb{Z}^+$, suppose $E_k, A_k \in \mathcal{M}$ have been chosen so that $\nu^+(A_k) = \infty$. Take $E_{k+1} \in \mathcal{M}$ such that

$$E_{k+1} \subset A_k \quad \text{and} \quad \nu(E_{k+1}) \geq 1 + \nu(E_k).$$

Take $A_{k+1} \in \{E_{k+1}, A_k \setminus E_{k+1}\}$ such that $\nu^+(A_{k+1}) = \infty$. Then

1. either $\quad \exists k_0 \ \forall k \geq k_0 : A_{k+1} = E_{k+1}$
2. or $\quad \forall k_0 \ \exists k \geq k_0 : A_{k+1} = A_k \setminus E_{k+1}$.

Here in the first case, $E \supset E_k \supset E_{k+1}$ for every $k \geq k_0$, and

$$
\begin{aligned}
\nu(E_{k_0}) &= \nu\left(\bigcap_{k=k_0}^{\infty} E_k\right) + \sum_{k=k_0}^{\infty} \nu(E_k \setminus E_{k+1}) \\
&= \nu\left(\bigcap_{k=k_0}^{\infty} E_k\right) + \sum_{k=k_0}^{\infty} (\nu(E_k) - \nu(E_{k+1})) \\
&= -\infty;
\end{aligned}
$$

of course, this is a contradiction, excluding the first case. In the second case, take a disjoint family $\{E_{k_j}\}_{j=1}^{\infty}$ where $k_{j+1} > k_j \in \mathbb{Z}^+$, so that

$$\nu\left(\bigcup_{j=1}^{\infty} E_{k_j}\right) = \sum_{j=1}^{\infty} \nu(E_{k_j}) = +\infty,$$

again a contradiction; therefore ν^+ and ν^- must be finite measures. $\qquad \square$

Corollary C.4.26 (Jordan Decomposition). *Let $\nu : \mathcal{M} \to \mathbb{R}$ be a signed measure. Then*

$$\nu = \nu^+ - \nu^-.$$

Proof. Let $E \in \mathcal{M}$. For any $A \in \mathcal{M}$ we have

$$
\begin{aligned}
\nu(E) &= \nu(A \cap E) + \nu(A^c \cap E) \\
&\leq \nu^+(E) - (-\nu)(A^c \cap E),
\end{aligned}
$$

yielding $\nu(E) \leq \nu^+(E) - \nu^-(E)$. Similarly,

$$(-\nu)(E) \leq (-\nu)^+(E) - (-\nu)^-(E) = \nu^-(E) - \nu^+(E),$$

so that $\nu(E) \geq \nu^+(E) - \nu^-(E)$. $\qquad \square$

Exercise C.4.27. Let $\mathcal{M}$ be a σ-algebra on X. Let $M(\mathcal{M})$ be the real vector space of all signed measures $\nu : \mathcal{M} \to \mathbb{R}$. For $\nu \in M(\mathcal{M})$, let $\|\nu\| = |\nu|(X)$; show that this gives a Banach space norm on $M(\mathcal{M})$.

Definition C.4.28 (Hahn decomposition). A pair (P, P^c) is called a *Hahn decomposition* of a signed measure $\nu : \mathcal{M} \to \mathbb{R}$ if $P \in \mathcal{M}$ and

$$\forall E \in \mathcal{M} : \quad \nu(P \cap E) \geq 0 \geq \nu(P^c \cap E).$$

Then P is called a *ν-positive set* and P^c is a *ν-negative set*.

Example. Let $\nu(E) = \displaystyle\int_E f \, d\mu$ as in (C.14). Then (P, P^c) and (Q, Q^c) are Hahn decompositions of $\nu : \mathcal{M} \to \mathbb{R}$, where

$$P := \{f \geq 0\}, \qquad Q := \{f > 0\}.$$

Definition C.4.29 (Mutually singular measures). The measures $\mu, \lambda : \mathcal{M} \to [0, \infty]$ are *mutually singular*, denoted by $\mu \perp \lambda$, if there exists $P \in \mathcal{M}$ such that

$$\mu(P) = 0 = \lambda(P^c).$$

Here, the zero-measure condition $\mu(P) = 0$ can be interpreted so that the measure μ *does not see* the set $P \in \mathcal{M}$.

Theorem C.4.30 (Hahn Decomposition). *Let $\nu : \mathcal{M} \to \mathbb{R}$ be a signed measure. Then ν has a Hahn decomposition (P, P^c). More precisely,*

$$\begin{cases} \nu^+(E) = +\nu(P \cap E), \\ \nu^-(E) = -\nu(P^c \cap E) \end{cases}$$

for each $E \in \mathcal{M}$. Especially, $\nu^- \perp \nu^+$ such that $\nu^-(P) = 0 = \nu^+(P^c)$.

Proof. For each $k \in \mathbb{Z}^+$, take $A_k \in \mathcal{M}$ such that $\nu^+(X) - \nu(A_k) < 2^{-k}$. Then

$$P := \limsup_{k \to \infty} A_k = \bigcap_{j=1}^{\infty} \bigcup_{k=j}^{\infty} A_k \in \mathcal{M}.$$

Moreover,

$$\begin{aligned} \nu^-(P) \quad &\leq \quad \sum_{k=j}^{\infty} \nu^-(A_k) \\[2mm] &\overset{\text{Corollary C.4.26}}{=} \quad \sum_{k=j}^{\infty} \left(\nu^+(A_k) - \nu(A_k)\right) \\[2mm] &\leq \quad \sum_{k=j}^{\infty} 2^{-k} \quad = \quad 2^{1-j}, \end{aligned}$$

so that $\nu^-(P) = 0$. On the other hand,

$$
\begin{aligned}
\nu^+(P^c) &= \nu^+(\liminf_{k\to\infty} A_k^c) \\
&= \lim_{j\to\infty} \nu^+\left(\bigcap_{k=j}^{\infty} A_k^c\right) \\
&\leq \lim_{j\to\infty} \nu^+(A_j^c) \\
&\leq \lim_{j\to\infty} 2^{-j} = 0,
\end{aligned}
$$

so that $\nu^+(P^c) = 0$. Thereby

$$
\begin{aligned}
\nu(P \cap E) \overset{\text{Jordan}}{=}\; & \nu^+(P \cap E) - \nu^-(P \cap E) \\
=\; & \nu^+(P \cap E) + \nu^+(P^c \cap E) \\
=\; & \nu^+(E),
\end{aligned}
$$

and similarly $\nu(P^c \cap E) = -\nu^-(E)$.   □

Exercise C.4.31. Let (P, P^c) and (Q, Q^c) be two Hahn decompositions of a signed measure ν. Show that $|\nu|(P \setminus Q) = 0$. The moral here is that all the Hahn decompositions are "essentially the same".

Exercise C.4.32. Let $\nu = \alpha - \beta$, where $\alpha, \beta : \mathcal{M} \to [0, \infty]$ are finite measures and $\alpha \perp \beta$. Show that

$$
\alpha = \nu^+ \quad \text{and} \quad \beta = \nu^-.
$$

In this respect the Jordan decomposition is the most natural decomposition of ν as a difference of two measures.

C.4.3   Derivatives of signed measures

In this section we study which signed measures $\nu : \mathcal{M} \to \mathbb{R}$ can be written in the integral form as in (C.15). The key property is the *absolute continuity of ν with respect to μ*, and the key result is the Radon–Nikodym Theorem C.4.38.

Definition C.4.33 (Radon–Nikodym derivative). Let $(X, \mathcal{M}, \mu)$ be a measure space and $f : X \to [-\infty, +\infty]$ be μ-integrable. Let a signed measure $\nu : \mathcal{M} \to \mathbb{R}$ be defined by

$$
\nu(E) := \int_E f \, d\mu. \tag{C.15}
$$

Then $\dfrac{d\nu}{d\mu} := f$ is the *Radon–Nikodym derivative* of ν with respect to μ.

Remark C.4.34. Actually, the Radon–Nikodym derivative $d\nu/d\mu = f$ is not an integrable function $X \to [-\infty, +\infty]$ but is the equivalence class

$$\{g : X \to [-\infty, +\infty] \mid f \sim g\},$$

where $f \sim g \iff f = g$ μ-almost everywhere. The classical derivative of a function (a limit of a difference quotient) is connected to the Radon–Nikodym derivative in the case of the Lebesgue measure $\mu = \lambda_{\mathbb{R}}$, but this shall not be investigated here.

Definition C.4.35 (Absolutely continuous measures). A signed measure $\nu : \mathcal{M} \to \mathbb{R}$ is *absolutely continuous* with respect to a measure $\mu : \mathcal{M} \to [0, \infty]$, denoted by $\nu \ll \mu$, if

$$\forall E \subset \mathcal{M}: \quad \mu(E) = 0 \Rightarrow \nu(E) = 0.$$

Example. If $\nu(E) = \displaystyle\int_E f \, d\mu$ as in (C.15) then $\nu \ll \mu$.

The following (ε, δ)-result justifies the term *absolute continuity* here:

Theorem C.4.36. *Let $\nu : \mathcal{M} \to \mathbb{R}$ be a signed measure and $\mu : \mathcal{M} \to [0, \infty]$ a measure. Then the following conditions are equivalent:*

(a) $\nu \ll \mu$.

(b) $\forall \varepsilon > 0 \; \exists \delta > 0 \; \forall E \in \mathcal{M}: \; \mu(E) < \delta \Rightarrow |\nu(E)| < \varepsilon$.

Proof. The (ε, δ)-condition trivially implies $\nu \ll \mu$. On the other hand, let us show that $\nu \not\ll \mu$, when we assume

$$\exists \varepsilon > 0 \; \forall \delta > 0 \; \exists E_\delta \in \mathcal{M}: \quad \mu(E_\delta) < \delta \quad \text{and} \quad |\nu(E_\delta)| \geq \varepsilon.$$

Then

$$E := \limsup_{k \to \infty} E_{2^{-k}} = \bigcap_{j=1}^{\infty} \bigcup_{k=j}^{\infty} E_{2^{-k}} \in \mathcal{M},$$

and $\mu(E) = 0$, because

$$\mu(E) \leq \mu\left(\bigcup_{k=j}^{\infty} E_{2^{-k}}\right) \leq \sum_{k=j}^{\infty} 2^{-k} = 2^{1-j}.$$

Now $|\nu|(E) > 0$, because

$$|\nu|(E) = \lim_{j \to \infty} |\nu|\left(\bigcup_{k=j}^{\infty} E_{2^{-k}}\right) \geq \varepsilon.$$

Hence $\nu^+(E) > 0$ or $\nu^-(E) > 0$, so that $|\nu(A)| > 0$ for some $A \subset E$, where $A \in \mathcal{M}$. Here $\mu(A) = 0$, so that $\nu \not\ll \mu$. $\qquad\qquad\square$

Exercise C.4.37. Show that the following conditions are equivalent:

1. $\nu \ll \mu$.
2. $|\nu| \ll \mu$.
3. $\nu^+ \ll \mu$ and $\nu^- \ll \mu$.

Theorem C.4.38 (Radon–Nikodym). *Let $\mu : \mathcal{M} \to [0,\infty]$ be a finite measure and $\nu \ll \mu$. Then there exists a Radon–Nikodym derivative $\mathrm{d}\nu/\mathrm{d}\mu$, i.e.,*

$$\nu(E) = \int_E \frac{\mathrm{d}\nu}{\mathrm{d}\mu} \, \mathrm{d}\mu$$

for every $E \in \mathcal{M}$.

Exercise C.4.39 (σ-finite Radon–Nikodym). A measure space is called σ-*finite* if it is a countable union of sets of finite measure. Generalise the Radon–Nikodym Theorem to σ-finite measure spaces. For example, if μ is σ-finite, we can find a sequence $E_j \nearrow X$ with $\mu(E_j) < \infty$ and define

$$\frac{\mathrm{d}\nu}{\mathrm{d}\mu} := \sup_j \frac{\mathrm{d}\nu|_{E_j}}{\mathrm{d}\mu|_{E_j}}.$$

Exercise C.4.40. Let $\nu = \lambda_{\mathbb{R}^n}$ be the Lebesgue measure, and let $(X, \mathcal{M}, \mu) = (\mathbb{R}^n, \mathcal{M}(\lambda^*_{\mathbb{R}^n}), \mu)$, where μ is the counting measure; this measure space is not σ-finite, but $\nu \ll \mu$. Show that ν cannot be of the form $\nu(E) = \int_E f \, \mathrm{d}\mu$. Thus there is no anologue to the Radon–Nikodym Theorem in this case.

Before proving the Radon–Nikodym Theorem C.4.38, let us deal with the essential special case of the result:

Lemma C.4.41. *Let $\mu, \nu : \mathcal{M} \to [0,\infty]$ be finite measures such that $\nu \leq \mu$. Then there exists a Radon–Nikodym derivative $\mathrm{d}\nu/\mathrm{d}\mu$, i.e.,*

$$\nu(E) = \int_E \frac{\mathrm{d}\nu}{\mathrm{d}\mu} \, \mathrm{d}\mu$$

for every $E \in \mathcal{M}$. Moreover,

$$\int g^+ \, \mathrm{d}\nu = \int g^+ \frac{\mathrm{d}\nu}{\mathrm{d}\mu} \, \mathrm{d}\mu \tag{C.16}$$

when $g^+ : X \to [0,\infty]$ is $\mathcal{M}$-measurable.

Proof. An $\mathcal{M}$-*partition* of a set X is a finite disjoint collection $\mathcal{P} \subset \mathcal{M}$, for which $X = \bigcup \mathcal{P}$. Let us define a partial order $\leq$ on the family the $\mathcal{M}$-partitions by $\mathcal{P} \leq \mathcal{Q}$ if and only if for every $Q \in \mathcal{Q}$ there exists $P \in \mathcal{P}$ such that $Q \subset P$. The *common refinement* of $\mathcal{M}$-partitions $\mathcal{P}, \mathcal{Q}$ is the $\mathcal{M}$-partition

$$\uparrow \{\mathcal{P}, \mathcal{Q}\} = \{P \cap Q : P \in \mathcal{P}, \ Q \in \mathcal{Q}\}.$$

For an $\mathcal{M}$-partition $\mathcal{P}$, let us define $d_{\mathcal{P}} : \mathcal{M} \to \mathbb{R}$ by

$$d_{\mathcal{P}}(E) := \begin{cases} \frac{\nu(P)}{\mu(P)}, & \text{if } x \in P \in \mathcal{P} \text{ and } \mu(P) > 0, \\ 0, & \text{otherwise.} \end{cases}$$

Then $0 \le d_{\mathcal{P}} \le 1$, $d_{\mathcal{P}}$ is simple and μ-integrable, and

$$d_{\mathcal{P}} = \sum_{P \in \mathcal{P}} \frac{\nu(P)}{\mu(P)} \, \chi_P.$$

The idea in the following is that the Radon–Nikodym derivative $d\nu/d\mu$ will be approximated by functions $d_{\mathcal{P}}$ is the $L^2(\mu)$-sense. If $\mathcal{P} \le \mathcal{Q}$ and $E \in \mathcal{P}$ then

$$\nu(E) = \int_E d_{\mathcal{P}} \, d\mu = \int_E d_{\mathcal{Q}} \, d\mu, \tag{C.17}$$

because

$$
\begin{aligned}
\int_E d_{\mathcal{Q}} \, d\mu &= \int_E \sum_{Q \in \mathcal{Q}} \frac{\nu(Q)}{\mu(Q)} \chi_Q \, d\mu \\
&= \sum_{Q \in \mathcal{Q}} \frac{\nu(Q)}{\mu(Q)} \int \chi_{Q \cap E} \, d\mu \\
&\overset{E \in \mathcal{P} \le \mathcal{Q}}{=\!=} \sum_{Q \in \mathcal{Q}: \, Q \subseteq E} \frac{\nu(Q)}{\mu(Q)} \mu(Q) \\
&\overset{E \in \mathcal{P} \le \mathcal{Q}}{=\!=} \nu(E).
\end{aligned}
$$

Moreover, here

$$\|d_{\mathcal{P}}\|^2_{L^2(\mu_E)} \le \|d_{\mathcal{Q}}\|^2_{L^2(\mu_E)} = \|d_{\mathcal{P}}\|^2_{L^2(\mu_E)} + \|d_{\mathcal{Q}} - d_{\mathcal{P}}\|^2_{L^2(\mu_E)}, \tag{C.18}$$

because

$$
\begin{aligned}
\|d_{\mathcal{P}}\|^2_{L^2(\mu_E)} &\le \|d_{\mathcal{P}}\|^2_{L^2(\mu_E)} + \|d_{\mathcal{Q}} - d_{\mathcal{P}}\|^2_{L^2(\mu_E)} \\
&= \int_E d_{\mathcal{P}}^2 \, d\mu + \int_E (d_{\mathcal{Q}} - d_{\mathcal{P}})^2 \, d\mu \\
&= \int_E d_{\mathcal{Q}}^2 \, d\mu + 2 \int_E d_{\mathcal{P}} \, (d_{\mathcal{P}} - d_{\mathcal{Q}}) \, d\mu \\
&\overset{E \in \mathcal{P}}{=\!=} \|d_{\mathcal{Q}}\|^2_{L^2(\mu_E)} + 2 \int_E \frac{\nu(E)}{\mu(E)} \left(\frac{\nu(E)}{\mu(E)} - d_{\mathcal{Q}} \right) d\mu \\
&\overset{(C.17)}{=\!=} \|d_{\mathcal{Q}}\|^2_{L^2(\mu_E)} + 2 \frac{\nu(E)}{\mu(E)} \left(\frac{\nu(E)}{\mu(E)} - \nu(E) \right) \\
&= \|d_{\mathcal{Q}}\|^2_{L^2(\mu_E)}.
\end{aligned}
$$

Now

$$M \quad := \quad \sup_{0 \leq d_{\mathcal{P}} \leq 1} \left\{ \|d_{\mathcal{P}}\|^2_{L^2(\mu)} \mid \mathcal{P} \text{ is an } \mathcal{M}\text{-partition} \right\}$$

$$\leq \quad \mu(X) \quad < \quad \infty.$$

Take a sequence of $\mathcal{M}$-partitions $\mathcal{P}_k$ such that

$$\|d_{\mathcal{P}_k}\|^2_{L^2(\mu)} \xrightarrow[k \to \infty]{} M.$$

We obtain an increasing sequence of partitions $\mathcal{Q}_k$ by common refinements:

$$\begin{cases} \mathcal{Q}_1 := \mathcal{P}_1, \\ \mathcal{Q}_{k+1} := \uparrow \{\mathcal{P}_{k+1}, \mathcal{Q}_k\}. \end{cases}$$

Let us show that the sequence of functions $f_k := d_{\mathcal{Q}_k}$ converges to the Radon–Nikodym derivative $d\nu/d\mu$ in $L^2(\mu)$. First, $f_k \in L^2(\mu)$, because $\mu(X) < \infty$. Moreover, these functions form a Cauchy sequence, because

$$\|f_j - f_k\|^2_{L^2(\mu)} \stackrel{\text{(C.18)}}{=} \left| \|f_j\|^2_{L^2(\mu)} - \|f_k\|^2_{L^2(\mu)} \right|$$

$$\xrightarrow[j,k \to \infty]{} 0,$$

as $M \geq \|f_k\|^2_{L^2(\mu)} \geq \|d_{\mathcal{P}_k}\|^2_{L^2(\mu)} \to M$. Since $L^2(\mu)$ is a Banach space, there exists $f \in L^2(\mu)$ for which $\|f - f_k\|_{L^2(\mu)} \to 0$. Let us show that $f = d\nu/d\mu$. Take $E \in \mathcal{M}$. Let $d_k := d_{\mathcal{R}_k}$, where

$$\mathcal{R}_k := \uparrow \{\mathcal{Q}_k, \{E, X \setminus E\}\}.$$

Then

$$\nu(E) \stackrel{\text{(C.17)}}{=} \int_E d_k \, d\mu$$

$$= \int_E (d_k - f_k) \, d\mu + \int_E f_k \, d\mu$$

$$\xrightarrow[k \to \infty]{} \int_E f \, d\mu,$$

because $\int_E f_k \, d\mu \to \int E f \, d\mu$ by the Monotone Convergence Theorem C.3.6 and by Vitali's Convergence Theorem C.4.14, and because

$$\left| \int_E (d_k - f_k) \, d\mu \right| \leq \int_E |d_k - f_k| \, d\mu$$

$$\stackrel{\text{Hölder}}{\leq} \left(\int_E |d_k - f_k|^2 \, d\mu \right)^{1/2} \left(\int_E d\mu \right)^{1/2}$$

$$\leq \|d_k - f_k\|^2_{L^2(\mu)} \, \mu(X)^{1/2}$$

$$\stackrel{\text{(C.18)}}{\leq} \left| \|d_k\|^2_{L^2(\mu)} - \|f_k\|^2_{L^2(\mu)} \right| \mu(X)^{1/2}$$

$$\xrightarrow[k \to \infty]{} 0.$$

Thus $f = d\nu/d\mu$. Finally, let $g^+ \geq 0$ be $\mathcal{M}$-measurable. Take simple $\mathcal{M}$-measurable functions s_k for which

$$0 \leq s_k(x) \leq s_{k+1}(x) \xrightarrow[k \to \infty]{} g^+(x).$$

Then

$$
\begin{aligned}
\int g^+ \, d\nu &= \int \lim_{k \to \infty} s_k \, d\mu \\
&\overset{\text{Mon. conv.}}{=} \lim_{k \to \infty} \int s_k \, d\mu \\
&\overset{(*)}{=} \lim_{k \to \infty} \int s_k \, \frac{d\nu}{d\mu} \, d\mu \\
&\overset{\text{Mon. conv.}}{=} \int \lim_{k \to \infty} s_k \, \frac{d\nu}{d\mu} \, d\mu \\
&= \int g^+ \frac{d\nu}{d\mu} \, d\mu,
\end{aligned}
$$

where equality $(*)$ easily follows from $\int \chi_E \, d\nu = \int \chi_E \, \frac{d\nu}{d\mu} \, d\mu$. $\qquad \square$

Proof of the Radon–Nikodym Theorem C.4.38. Since $\nu = (\nu^+ - \nu^-) \ll \mu$, we have also $\nu^+, \nu^- \ll \mu$. If the Radon–Nikodym derivatives $d\nu^+/d\mu$ and $d\nu^-/d\mu$ exist, then by the linearity of the integral we have

$$\frac{d\nu}{d\mu} = \frac{d\nu^+}{d\mu} - \frac{d\nu^-}{d\mu}.$$

Thus we may assume that ν, μ are finite measures, where $\nu \ll \mu$. Then also $\mu + \nu : \mathcal{M} \to [0, \infty]$ is a finite measure. By Lemma C.4.41, the Radon–Nikodym derivatives $d\mu/d(\mu + \nu)$ and $d\nu/d(\mu + \nu)$ exist. Let

$$A := \left\{ \frac{d\mu}{d(\mu + \nu)} > 0 \right\};$$

let us show that $d\nu/d\mu = g^+$, where $g^+ : X \to [0, \infty]$ is defined by

$$g^+(x) := \begin{cases} \frac{d\nu}{d(\mu+\nu)} \Big/ \frac{d\mu}{d(\mu+\nu)}, & \text{when } x \in A, \\ 0, & \text{when } x \in A^c. \end{cases}$$

Here

$$\mu(A^c) \overset{\text{(C.16)}}{=} \int_{A^c} \frac{d\mu}{d(\mu + \nu)} \, d(\mu + \nu) = 0,$$

and if $E \in \mathcal{M}$ then

$$
\begin{aligned}
\nu(E) &= \nu(A \cap E) + \nu(A^c \cap E) \\
&\stackrel{\mu(A^c \cap E)=0,\ \nu \ll \mu}{=} \nu(A \cap E) \\
&\stackrel{\text{(C.16)}}{=} \int_{A \cap E} \frac{\mathrm{d}\nu}{\mathrm{d}(\mu + \nu)} \, \mathrm{d}(\mu + \nu) \\
&= \int_{A \cap E} g^+ \frac{\mathrm{d}\mu}{\mathrm{d}(\mu + \nu)} \, \mathrm{d}(\mu + \nu) \\
&\stackrel{\text{(C.16)}}{=} \int_{A \cap E} g^+ \, \mathrm{d}\mu \\
&\stackrel{\mu(A^c)=0}{=} \int_{E} g^+ \, \mathrm{d}\mu.
\end{aligned}
$$

Thus $g^+ = \mathrm{d}\nu/\mathrm{d}\mu$, and the Radon–Nikodym Theorem C.4.38 is proven. $\qquad\square$

Exercise C.4.42. Let $\lambda, \mu, \nu : \mathcal{M} \to [0, \infty]$ be σ-finite measures. Prove:

(a) If $\lambda \ll \mu$, $E \in \mathcal{M}$ and g is $\mathcal{M}$-measurable, then

$$
\int_E g \, \mathrm{d}\lambda = \int_E g \frac{\mathrm{d}\lambda}{\mathrm{d}\mu} \, \mathrm{d}\mu.
$$

(b) If $\lambda \ll \nu$ and $\mu \ll \nu$, then $\frac{\mathrm{d}(\lambda + \mu)}{\mathrm{d}\nu} = \frac{\mathrm{d}\lambda}{\mathrm{d}\nu} + \frac{\mathrm{d}\mu}{\mathrm{d}\nu}$.

(c) If $\lambda \ll \mu$ and $\mu \ll \nu$, then $\frac{\mathrm{d}\lambda}{\mathrm{d}\nu} = \frac{\mathrm{d}\lambda}{\mathrm{d}\mu} \frac{\mathrm{d}\mu}{\mathrm{d}\nu}$.

(d) If $\lambda \ll \mu$ and $\mu \ll \lambda$, then $\frac{\mathrm{d}\lambda}{\mathrm{d}\mu} = \left(\frac{\mathrm{d}\mu}{\mathrm{d}\lambda} \right)^{-1}$.

Definition C.4.43 (Lebesgue decomposition). Let $\mu, \nu : \mathcal{M} \to [0, \infty]$ be measures. A *Lebesgue decomposition* of ν with respect to μ is a pair (ν_0, ν_1) of measures $\nu_0, \nu_1 : \mathcal{M} \to [0, \infty]$ satisfying

$$
\nu = \nu_0 + \nu_1, \quad
\begin{cases}
\nu_0 \perp \mu, \\
\nu_1 \ll \mu.
\end{cases}
$$

Theorem C.4.44 (Existence of Lebesgue decomposition). *Let $\mu, \nu : \mathcal{M} \to [0, \infty]$ be σ-finite measures. Then there exists a unique Lebesgue decomposition of ν with respect to μ.*

Proof. The Radon–Nikodym Theorem C.4.38 was formulated for a finite measure, but it can be easily generalised to σ-finite spaces: showing this was left as Exercise C.4.39. Let

$$
A := \left\{ \frac{\mathrm{d}\mu}{\mathrm{d}(\mu + \nu)} > 0 \right\}.
$$

Define measures $\nu_0, \nu_1 : \mathcal{M} \to [0, \infty]$ by

$$\begin{cases} \nu_0(E) := \nu(A^c \cap E), \\ \nu_1(E) := \nu(A \cap E). \end{cases}$$

Clearly $\nu = \nu_0 + \nu_1$, and $\nu_0 \perp \mu$ because

$$\begin{cases} \nu_0(A) = \nu(A^c \cap A) = \nu(\emptyset) = 0, \\ \mu(A^c) \overset{(C.16)}{=} \int_{A^c} \frac{d\mu}{d(\mu + \nu)} \, d(\mu + \nu) = 0. \end{cases}$$

We will now prove that $\nu_1 \ll \mu$. Let $A_k := \left\{ \dfrac{d\mu}{d(\mu + \nu)} \geq 1/k \right\}$. Take $E \in \mathcal{M}$ such that $\mu(E) = 0$. Now $\nu_1(E) = 0$, because

$$
\begin{aligned}
\nu_1(E) \quad &= \quad \nu(A \cap E) \\
&\leq \quad (\mu + \nu)(A \cap E) \\
&= \quad \lim_{k \to \infty} (\mu + \nu)(A_k \cap E) \\
&\leq \quad \int_{A_k \cap E} k \, \frac{d\mu}{d(\mu + \nu)} \, d(\mu + \nu) \\
\overset{\text{Radon-Nikodym}}{=} \quad & k \, \mu(A_k \cap E) \\
&\leq \quad k \, \mu(E) \quad = \quad 0.
\end{aligned}
$$

Proving the uniqueness part is left as Exercise C.4.45.   □

Exercise C.4.45. Show that the Lebesgue decomposition in Theorem C.4.44 is unique.

C.4.4 Integration as functional on function spaces

Assume $(X, \mathcal{M}, \mu)$ is a measure space, possibly with topology. On function spaces like $L^p(\mu)$ or $C(X) = C(X, \mathbb{R})$ (when X is, e.g., a compact Hausdorff space), integration acts as a bounded linear functional by

$$f \mapsto \int fg \, d\mu,$$

when g is a suitable weight function on X. It is natural to study necessary and sufficient conditions for g, and ask whether all the bounded linear functionals are of this form. The general functional analytic outline is as follows: Let V be a real Banach space, e.g., $L^p(\mu)$ or $C(X) = C(X, \mathbb{R})$. The dual of V is the Banach space

$$V' = \mathcal{L}(V, \mathbb{R}) := \{\phi : V \to \mathbb{R} \mid \phi \text{ bounded and linear}\},$$

endowed with the (operator) norm

$$\phi \mapsto \|\phi\| := \sup_{f \in V:\ \|f\|_V \le 1} |\phi(f)|,$$

see Definition B.4.15 and Exercise B.4.16. Given a "concrete" space V, we would like to discover an intuitive representation of the dual.

C.4.5 Integration as functional on $L^p(\mu)$

Now we are going to find a concrete presentation for the dual of $V = L^p(\mu)$, where $(X, \mathcal{M}, \mu)$ is a measure space. We shall assume that $\mu(X) < \infty$, though often this technical assumption can be removed, since everything works for σ-finite measures just as well.

Lemma C.4.46. *Let μ be a finite measure. Let $1 \le p \le \infty$, and let $q = p'$ be its Lebesgue conjugate, i.e., $1/p + 1/q = 1$. Let $g \in L^q(\mu)$. Then $\phi_g \in L^p(\mu)'$, where*

$$\phi_g(f) := \int fg \, d\mu,$$

and $\|\phi_g\| = \|g\|_{L^q}$.

Exercise C.4.47. Prove Lemma C.4.46.

Remark C.4.48. You may generalise Lemma C.4.46 as follows: the conclusion holds for a general measure μ if $1 < p \le \infty$, and for a σ-finite measure μ if $1 \le p \le \infty$.

Remark C.4.49. The next Theorem C.4.50 roughly says that the dual of L^p "is" L^q, under some technical assumptions. The result holds for a general measure μ if $1 < p < \infty$, and for a σ-finite measure if $1 \le p < \infty$.

Theorem C.4.50 (Dual of $L^p(\mu)$). *Let μ be a finite measure. Let $1 \le p < \infty$, and let $q = p'$ be its Lebesgue conjugate. Then the mapping*

$$(g \mapsto \phi_g) : L^q(\mu) \to L^p(\mu)'$$

is an isometric isomorphism, i.e., $L^p(\mu)' \cong L^q(\mu)$.

Proof. By the previous Lemma C.4.46, it suffices to show that $\psi \in L^p(\mu)'$ is of the form $\psi = \phi_g$ for some $g \in L^q(\mu)$. Let us define $\nu : \mathcal{M} \to \mathbb{R}$ such that

$$\nu(E) := \psi(\chi_E),$$

where $\chi_E \in L^p(\mu)$ because $\mu(X) < \infty$.

The idea in the proof is to show that $d\nu/d\mu \in L^q(\mu)$ and that

$$\psi(f) = \int f \frac{d\nu}{d\mu} \, d\mu. \tag{C.19}$$

The first step is to show that ν is a signed measure: Let $\{E_j\}_{j=1}^\infty \subset \mathcal{M}$ be a disjoint collection. Then $\nu(\bigcup_{j=1}^\infty E_j) = \sum_{j=1}^\infty \nu(E_j)$, because

$$
\left| \nu\left(\bigcup_{j=1}^\infty E_j\right) - \sum_{j=1}^k \nu(E_j) \right| \quad = \quad \left| \psi(\chi_{\bigcup_{j=1}^\infty E_j}) - \sum_{j=1}^k \psi(E_j) \right|
$$

$$
= \quad \left| \psi\left(\sum_{j=k+1}^\infty \chi_{E_j}\right) \right|
$$

$$
\leq \quad \|\psi\| \sum_{j=k+1}^\infty \|\chi_{E_j}\|_{L^p(\mu)}
$$

$$
\xrightarrow[k\to\infty]{\mu(X)<\infty} \quad 0,
$$

where we used the linearity and boundedness of ψ, and the disjointness of $\{E_j\}_{j=1}^\infty$. Thus ν is a signed measure. Moreover $\nu \ll \mu$, because if $\mu(E) = 0$ then $\chi_E = 0$ μ-almost everywhere, implying

$$\nu(E) = \psi(\chi_E) = 0$$

as $\psi \in L^p(\mu)'$. Thus $d\nu/d\mu \in L^1(\mu)$ exists by the Radon–Nikodym Theorem C.4.38. We have to show that $d\nu/d\mu \in L^q(\mu)$ and that (C.19) holds for every $f \in L^p(\mu)$. At least

$$\psi(\chi_E) = \nu(E) \overset{\text{Radon–Nikodym}}{=} \int \chi_E \frac{d\nu}{d\mu} \, d\mu$$

for every $E \in \mathcal{M}$; by the linearity of ψ, this extends to

$$\psi(s) = \int s \frac{d\nu}{d\mu} \, d\mu \tag{C.20}$$

for every simple $\mathcal{M}$-measurable $s : X \to \mathbb{R}$. Next we show that $d\nu/d\mu \in L^q(\mu)$.

For a moment, let $p = 1$ (so that $q = \infty$). We shall soon see that $\|d\nu/d\mu\|_{L^q} \leq \|\psi\|$. Take any $M > \|\psi\|$ and let

$$A_M := \left\{ \left| \frac{d\nu}{d\mu} \right| > M \right\}.$$

Then

$$
\begin{aligned}
M\,\mu(A_M) \;&=\; \int_{A_M} M \; \mathrm{d}\mu \\[2mm]
&\le\; \int_{A_M} \left|\frac{\mathrm{d}\nu}{\mathrm{d}\mu}\right| \, \mathrm{d}\mu \\[2mm]
&=\; \chi_{A_M} \, \mathrm{sgn}\left(\frac{\mathrm{d}\nu}{\mathrm{d}\mu}\right) \frac{\mathrm{d}\nu}{\mathrm{d}\mu} \, \mathrm{d}\mu \\[2mm]
&\overset{(C.20)}{=}\; \psi\left(\chi_{A_M} \, \mathrm{sgn}\left(\frac{\mathrm{d}\nu}{\mathrm{d}\mu}\right)\right) \\[2mm]
&\le\; \|\psi\| \, \left\|\chi_{A_M} \, \mathrm{sgn}\left(\frac{\mathrm{d}\nu}{\mathrm{d}\mu}\right)\right\|_{L^p(\mu)} \\[2mm]
&\overset{p=1}{\le}\; \|\psi\| \, \mu(A_M).
\end{aligned}
$$

Since $M > \|\psi\|$ and $\mu(A_M) < \infty$, we must have $\mu(A_M) = 0$, so $\|\mathrm{d}\nu/\mathrm{d}\mu\|_{L^\infty(\mu)} \le \|\psi\|$.

Now let $1 < p < \infty$ (so that $\infty > q > 1$). Take simple $\mathcal{M}$-measurable functions $h_k : X \to \mathbb{R}$ such that

$$
0 \le h_k(x) \le h_{k+1}(x) \xrightarrow[k\to\infty]{} \left|\frac{\mathrm{d}\nu}{\mathrm{d}\mu}(x)\right|.
$$

Then

$$
\left\|\frac{\mathrm{d}\nu}{\mathrm{d}\mu}\right\|_{L^q(\mu)}^q = \int \left|\frac{\mathrm{d}\nu}{\mathrm{d}\mu}\right| \, \mathrm{d}\mu \overset{\text{Fatou}}{\le} \liminf_{k\to\infty} \int h_k^q \, \mathrm{d}\mu,
$$

so that $\mathrm{d}\nu/\mathrm{d}\mu \in L^q(\mu)$ follows if we show that $\|h_k\|_{L^q(\mu)} \le constant < \infty$ for every $k \in \mathbb{Z}^+$:

$$
\begin{aligned}
\|h_k\|_{L^q(\mu)}^q \;&=\; \int h_k^q \, \mathrm{d}\mu \\[2mm]
&\le\; \int h_k^{q-1} \left|\frac{\mathrm{d}\nu}{\mathrm{d}\mu}\right| \, \mathrm{d}\mu \\[2mm]
&=\; \int h_k^{q-1} \, \mathrm{sgn}\left(\frac{\mathrm{d}\nu}{\mathrm{d}\mu}\right) \frac{\mathrm{d}\nu}{\mathrm{d}\mu} \, \mathrm{d}\mu \\[2mm]
&\overset{(C.20)}{=}\; \psi\left(h_k^{q-1} \, \mathrm{sgn}\left(\frac{\mathrm{d}\nu}{\mathrm{d}\mu}\right)\right) \\[2mm]
&\le\; \|\psi\| \, \left\|h_k^{q-1}\right\|_{L^p(\mu)} \\[2mm]
&=\; \|\psi\| \, \|h_k\|_{L^{q/p}(\mu)},
\end{aligned}
$$

because $p(q-1) = q$. Hence $\|h_k\|_{L^q(\mu)} = \|h_k\|_{L^q(\mu)}^{q(1-1/p)} \le \|\psi\|$.

Finally, we have to show that (C.19) holds for $f \in L^p(\mu)$. Take simple $\mathcal{M}$-measurable functions $f_k : X \to [-\infty, +\infty]$ such that $f_k \to f$ in $L^p(\mu)$. Then

$$
\left| \psi(f) - \int f \, \frac{\mathrm{d}\nu}{\mathrm{d}\mu} \, \mathrm{d}\mu \right|
$$

$$
\stackrel{(\mathrm{C.20})}{=} \left| \psi(f - f_k) + \int (f_k - f) \, \frac{\mathrm{d}\nu}{\mathrm{d}\mu} \, \mathrm{d}\mu \right|
$$

$$
\leq \quad |\psi(f - f_k)| + \int |f_k - f| \left| \frac{\mathrm{d}\nu}{\mathrm{d}\mu} \right| \mathrm{d}\mu
$$

$$
\stackrel{\text{Hölder}}{\leq} \quad \|\psi\| \, \|f - f_k\|_{L^p(\mu)} + \|f_k - f\|_{L^p(\mu)} \left\| \frac{\mathrm{d}\nu}{\mathrm{d}\mu} \right\|_{L^q(\mu)}
$$

$$
\xrightarrow[k \to \infty]{} 0.
$$

Thus the proof is complete. □

Exercise C.4.51. Generalise Theorem C.4.50 to the case where μ is σ-finite (and $1 \leq p < \infty$).

Exercise C.4.52. Generalise Theorem C.4.50 to the case where μ is any measure and $1 < p < \infty$ (so that $1 < q < \infty$ also). (Hint: apply the result of Exercise C.4.51.)

Remark C.4.53. We have not dealt with the dual of $L^\infty(\mu)$. This case actually resembles the other L^p-cases, but is slightly different, see details, e.g., in [153]. Often, however, $L^\infty(\mu)' \not\cong L^1(\mu)$.

Exercise C.4.54. Let $X = [0, 1]$ and $\mu = (\lambda_{\mathbb{R}})_X$. Show that there exists $\psi \in L^\infty(\mu)'$ which is not of the form $f \mapsto \int fg \, \mathrm{d}\mu$ for any $g \in L^1(\mu)$.
(Hint: Define a suitable bounded linear functional $f \mapsto \varphi(f)$ for continuous functions f, and extend it to ψ using the Hahn–Banach Theorem, see Theorem B.4.25.)

Exercise C.4.55. Let $(X, \mathcal{M}, \mu)$ be a measure space, where X is uncountable, $\mathcal{M} = \{E \subset X : E \text{ or } E^c \text{ is countable}\}$ and μ is the counting measure. Show that there exists $\psi \in L^1(\mu)'$ which is not of the form $f \mapsto \int fg \, \mathrm{d}\mu$ for any $g \in L^\infty(\mu)$.
(Hint: You may use that there exists $S \in \mathcal{P}(X) \setminus \mathcal{M}$, which follows by using the Hausdorff Maximal Principle or other equivalents to the Axiom of Choice.)

Theorem C.4.56 (Converse of Hölder's inequality). *Let μ be a σ-finite measure, $1 \leq p \leq \infty$, and $\frac{1}{p} + \frac{1}{q} = 1$. Let S be the space of all simple functions that vanish outside a set of finite measure. Let g be $\mathcal{M}$-measurable such that $fg \in L^1(\mu)$ for all $f \in S$, and such that*

$$
M_q(g) := \sup \left\{ \left| \int fg \, \mathrm{d}\mu \right| : f \in S, \, \|f\|_{L^p(\mu)} = 1 \right\}
$$

is finite. Then $g \in L^q(\mu)$ and $M_q(g) = \|g\|_{L^q(\mu)}$.

Proof. From Hölder's inequality (Theorem C.4.4) we have the inequality $M_q(g) \leq \|g\|_{L^q(\mu)}$. For the proof of $\|g\|_{L^q(\mu)} \leq M_q(g)$ we follow [35]. Assume first that $q < \infty$. Let $E_n \subset X$ be an increasing sequence of sets such that $0 < \mu(E_n) < \infty$ for all n, and such that $\bigcup_{n=1}^{\infty} E_n = X$. Let φ_n be a sequence of simple functions such that $\varphi_n \to g$ pointwise and $|\varphi_n| \leq g$, and let $g_n := \varphi_n \chi_{E_n}$, where χ_{E_n} is the characteristic function of the set E_n. Then $g_n \to g$ pointwise, $|g_n| \leq |g|$ and $g_n \in S$. Define $f_n := \|g_n\|_{L^q(\mu)}^{1-q} |g_n|^{q-1} \frac{g}{|g|}$ when $g \neq 0$ and $f_n := 0$ when $g = 0$. The relation $\frac{1}{p} + \frac{1}{q} = 1$ implies $(q-1)p = q$, so that $\|f_n\|_{L^p(\mu)} = 1$, and by Fatou's lemma C.3.10 we have:

$$\|g\|_{L^q(\mu)} \leq \liminf \|g_n\|_{L^q(\mu)} \;=\; \liminf \int |f_n g_n| \, d\mu$$

$$\leq \liminf \int |f_n g| \, d\mu = \liminf \int f_n g \, d\mu \leq M_q(g).$$

The case $q = \infty$ is slightly different. Take $\epsilon > 0$ and denote $A := \{x \in X : |g(x)| \geq M_\infty(g) + \epsilon$. We need to show that $\mu(A) = 0$. If $\mu(A) > 0$, there exists some $B \subset A$ such that $0 < \mu(B) < \infty$. Let us define $f := \frac{1}{\mu(B)} \frac{g}{|g|} \chi_B$ when $g \neq 0$ and $f := 0$ when $g = 0$. Then $\|f\|_{L^1(\mu)} = 1$ and $\int fg \, d\mu \geq M_\infty(g) + \epsilon$, a contradiction. $\square$

C.4.6   Integration as functional on $C(X)$

Measure theory and topology have fundamental connections, as exemplified in this passage. For our purposes, it is enough to study compact Hausdorff spaces, though analogies hold for locally compact Hausdorff spaces. Let (X, τ) be a compact Hausdorff space and let $C(X) = C(X, \mathbb{R})$ denote the Banach space of continuous functions $f : X \to \mathbb{R}$, endowed with the supremum norm:

$$\|f\| = \|f\|_{C(X)} := \sup_{x \in X} |f(x)|.$$

Appealing to the "geometry" of X, we are going to characterise the dual $C(X)' = \mathcal{L}(C(X), \mathbb{R})$.

Exercise C.4.57. Let (X, τ) be a compact Hausdorff space. Actually, $C(X)$ contains all the information about (X, τ): a set $S \subset X$ is closed if and only if $S = \{f = 0\}$ for some $f \in C(X)$. Prove this.

Remark C.4.58. Let (X, τ) be a topological space. Recall that the vector space of signed (Borel) measures

$$M(X) = M(\Sigma(\tau)) := \{\nu : \Sigma(\tau) \to \mathbb{R} \mid \nu \text{ is a signed measure}\}$$

is a Banach space with the norm

$$\|\nu\| := |\nu|(X).$$

Lemma C.4.59. *Let* $\nu : \mathcal{M} \to \mathbb{R}$ *be a signed measure on* X, *where* $\tau \subset \mathcal{M}$. *For* $f \in C(X)$, *let*

$$T_\nu(f) \;:=\; \int f \, d\nu$$

$$= \int f \, d\nu^+ - \int f \, d\nu^-.$$

Then $T_\nu \in C(X)'$ *and* $\|T_\nu\| \leq \|\nu\|$.

Proof. Each $f \in C(X)$ is $\mathcal{M}$-measurable, as $\tau \subset \mathcal{M}$. Furthermore, $\nu^+, \nu^- : \mathcal{M} \to [0, \infty]$ are finite measures. Consequently, $f \in C(X)$ is $\nu^\pm$-integrable, so $T_\nu(f) \in \mathbb{R}$ is well defined, and

$$|T_\nu(f)| \;\leq\; \int |f| \, d\nu^+ + \int |f| \, d\nu^-$$

$$\leq\; \|f\| \left(\nu^+(X) + \nu^-(X) \right) \;=\; \|f\| \, \|\nu\|.$$

The operator $T_\nu : C(X) \to \mathbb{R}$ is linear since integration is linear. $\qquad\square$

Theorem C.4.60 (F. Riesz's Topological Representation Theorem). *Let* (X, τ) *be a compact Hausdorff space. Let* $M(X)$ *and* $T_\nu \in C(X)'$ *be as above. Then*

$$(\nu \mapsto T_\nu) : M(X) \to C(X)'$$

is an isometric isomorphism.

In other words, bounded linear functionals on $C(X)$ are exactly integrations with respect to signed measures, with the natural norms coinciding. We shall soon prove the Riesz Representation Theorem C.4.60 step-wise.

Definition C.4.61 (Positive functionals). Let (X, τ) be a compact Hausdorff space. A functional $T : C(X) \to \mathbb{R}$ is called *positive* if $T(f) \geq 0$ whenever $f \geq 0$.

Exercise C.4.62. Show that a positive linear functional $T \in C(X)'$ is bounded and that

$$\|T\| = T1,$$

where $1 \in C(X)$ is the constant function $x \mapsto 1$.

Lemma C.4.63. *Let* $T \in C(X)'$, *where* (X, τ) *is a compact Hausdorff space. Then there exist positive* $T^+, T^- \in C(X)'$ *such that*

$$T \;=\; T^+ - T^-,$$
$$\|T\| \;=\; \|T^+\| + \|T^-\|.$$

Proof. For $f = f^+ - f^- \in C(X)$, let us define

$$T^+(f) := T^+(f^+) - T^+(f^-),$$

where

$$T^+(g^+) := \sup\left\{T(h^+) \mid h^+ \in C(X),\ 0 \le h^+ \le g^+\right\}.$$

Obviously, $0 = T(0) \le T^+(g^+) \le T(\|g^+\|1) = \|T\|\,\|g^+\|$. Thereby the functional $T^+ : C(X) \to \mathbb{R}$ is well defined and positive. Let us show that T^+ is linear. If $0 < \lambda^+ \in \mathbb{R}$ then

$$
\begin{aligned}
T^+(\lambda^+ f^+) &= \sup\left\{Th \mid h \in C(X),\ 0 \le h \le \lambda^+ f^+\right\} \\
&= \sup\left\{T(\lambda^+ h) \mid h \in C(X),\ 0 \le h \le f^+\right\} \\
&\overset{T\ \text{linear}}{=} \lambda^+\, T^+(f^+);
\end{aligned}
$$

from this we easily see that

$$T^+(\lambda f) = \lambda\, T^+(f)$$

for every $\lambda \in \mathbb{R}$ and $f \in C(X)$. Next,

$$T^+(f^+ + g^+) = T^+(f^+) + T^-(g^+)$$

whenever $0 \le f^+, g^+ \in C(X)$, because

$$
\text{if}\quad
\begin{cases}
0 \le h \le f^+ + g^+, \\
0 \le h_1 \le f^+, \\
0 \le h_2 \le g^+
\end{cases}
\quad\text{then}\quad
\begin{cases}
0 \le h_1 + h_2 \le f^+ + g^+, \\
0 \le \min(f^+, h) \le f^+, \\
0 \le h - \min(f^+, h) \le g^+.
\end{cases}
$$

Since

$$(f+g)^+ + f^- + g^- = (f+g)^- + f^+ + g^+,$$

we get

$$T^+((f+g)^+) + T^+(f^-) + T^+(g^-) = T^+((f+g)^-) + T^+(f^+) + T^+(g^+),$$

so that

$$
\begin{aligned}
T^+(f+g) &= T^+((f+g)^+) - T^+((f+g)^-) \\
&= \left(T^+(f^+) - T^+(f^-)\right) + \left(T^+(g^+) - T^+(g^-)\right) \\
&= T^+(f) + T^+(g).
\end{aligned}
$$

Hence we have seen that $T^+ : C(X) \to \mathbb{R}$ is linear and positive, and that $\|T^+\| \le \|T\|$. Next, let us define $T^- := T^+ - T \in C(X)'$. Then T^- is positive, because

$$T^-(f^+) = \sup\left\{Th - T(f^+) \mid h \in C(X) : 0 \le h \le f^+\right\}.$$

Finally,

$$
\begin{aligned}
\|T\| &= \|T^+ - T^-\| \\
&\leq \|T^+\| + \|T^-\| \\
&= T^+(1) + T^-(1) \\
&= 2\,T^+(1) - T(1) \\
&= \sup\{T(2h-1) \mid h \in C(X) : 0 \leq h \leq 1\} \\
&= \sup\{T(g) \mid g \in C(X) : -1 \leq g \leq 1\} \\
&= \|T\|,
\end{aligned}
$$

so $\|T\| = \|T^+\| + \|T^-\|$. $\qquad\square$

Remark C.4.64. Recall that the *support* $\mathrm{supp}(f) \subset X$ of a function $f \in C(X)$ is the closure of the set $\{f \neq 0\}$. Moreover, abbreviations

$$
K \prec f, \quad f \prec U
$$

mean that $0 \leq f \leq 1$, $K \subset X$ is compact such that $\chi_K \leq f$, and $U \subset X$ is open such that $\mathrm{supp}(f) \subset U$.

Theorem C.4.65. *Let $T^+ \in C(X)'$ be positive, where (X, τ) is a compact Hausdorff space. Then there exists a finite Borel measure $\mu : \Sigma(\tau) \to [0, \infty]$ such that*

$$
Tf = \int f \, d\mu
$$

for every $f \in C(X)$.

Proof. Let us define a measurelet $m : \tau \to [0, \infty]$ such that

$$
m(U) := \sup\{Tf \mid f \prec U\}.
$$

Indeed, $m(\emptyset) = T(0) = 0$. Thus m generates an outer measure $m^* : \mathcal{P}(X) \to [0, \infty]$ by

$$
m^*(E) = \inf\left\{\sum_{j=1}^{\infty} m(U_j) : \quad E \subset \bigcup_{j=1}^{\infty} U_j \in \tau, U_j \in \tau\right\}.
$$

We have to show $\mu := m^*|_{\Sigma(\tau)}$ is the desired measure. First,

$$
m^*(E) = \inf\{m(U) : \quad E \subset U \in \tau\} \tag{C.21}
$$

follows, if we show that

$$
m\left(\bigcup_{j=1}^{\infty} U_j\right) \leq \sum_{j=1}^{\infty} m(U_j).
$$

So let $f \prec \bigcup_{j=1}^{\infty} U_j$. Now $\mathrm{supp}(f) \subset X$ is compact, so $\mathrm{supp}(f) \subset \bigcup_{j=1}^{n} U_j$ for some $n \in \mathbb{Z}^+$. Let $\{g_j\}_{j=1}^{n}$ be a partition of unity for which

$$K \prec \sum_{j=1}^{n} g, \quad g_j \prec U_j.$$

Thereby

$$
\begin{aligned}
Tf \quad &= \quad T\Big(f \sum_{j=1}^{n} g_j\Big) \\[2mm]
&\overset{T \ \mathrm{linear}}{=} \quad \sum_{j=1}^{n} T(fg_j) \\[2mm]
&\overset{fg_j \prec U_j}{\leq} \quad \sum_{j=1}^{n} m(U_j) \\[2mm]
&\leq \quad \sum_{j=1}^{\infty} m(U_j),
\end{aligned}
$$

proving (C.21). Next, we show that $\tau \subset \mathcal{M}(m^*)$ by proving that $m^*(A \cup B) = m^*(A) + m^*(B)$ whenever $A \subset U \in \tau$ and $B \subset U^c$; let us assume the non-trivial case $m^*(A), m^*(B) < \infty$. Given $\varepsilon > 0$, there exists $V \in \tau$ such that $A \cup B \subset V$ and $m^*(A \cup B) + \varepsilon > m(V)$. Moreover, let

$$
\begin{cases}
f \prec U \cap V : & m(U \cap V) < Tf + \varepsilon, \\
g \prec \mathrm{supp}(f)^c \cap V : & m(\mathrm{supp}(f)^c \cap V) < Tg + \varepsilon.
\end{cases}
$$

We notice that $U \in \mathcal{M}(m^*)$, because

$$
\begin{aligned}
m^*(A \cup B) + \varepsilon \quad &> \quad m(V) \\[2mm]
&\overset{f+g \prec V}{\geq} \quad T(f + g) \\[2mm]
&\overset{T \ \mathrm{linear}}{=} \quad Tf + Tg \\[2mm]
&> \quad m(U \cap V) + m(\mathrm{supp}(f)^c \cap V) - 2\varepsilon \\[2mm]
&\geq \quad m^*(A) + m^*(B) - 2\varepsilon \\[2mm]
&\geq \quad m^*(A \cup B) - 2\varepsilon.
\end{aligned}
$$

Thus we can define the Borel measure $\mu := m^*|_{\Sigma(\tau)}$. Notice that $m(U) = \mu(U)$, $\mu(X) = T1 < \infty$ and that m^* is Borel-regular. If $\chi_E \leq g \leq \chi_F$, where $g \in C(X)$ and $E, F \in \Sigma(\tau)$, then

$$\int \chi_E \, d\mu \leq \int g \, d\mu \leq \int \chi_F \, d\mu; \tag{C.22}$$

moreover,

$$\mu(E) \le Tg \le \mu(F),\qquad\qquad(\text{C.23})$$

because

$$\mu(E)\ \overset{E\subset\{g\ge1\}}{\le}\ \mu(\{g\ge1\})$$
$$\overset{\delta>0}{\le}\ \mu(\{g>1-\delta\})$$
$$\overset{\{g>1-\delta\}\in\tau}{=}\ \sup\{Tf:\ f\prec\{g>1-\delta\}\}$$
$$\overset{T\ \text{positive},\ 0<\delta<1}{\le}\ T(g/(1-\delta))$$
$$\xrightarrow[\delta\to0]{T\ \text{linear}}\ Tg$$

implies $\mu(E) \le Tg$, which implies

$$\mu(X\setminus F)\ \overset{\chi_{F^c}\le1-g}{T}\ (1-g)=\mu(X)-Tg,$$

so $Tg \le \mu(F)$. Let us show that $T(f^+) = \int f^+ \, d\mu$, when $0 \le f^+ \in C(X)$. Take $\varepsilon > 0$. Let us define $f_\varepsilon \in C(X)$ by

$$f_\varepsilon(x) := \min\{f^+(x), \varepsilon\}.$$

Then

$$f^+ = \sum_{k=0}^{\infty}\left(f_{(k+1)\varepsilon} - f_{k\varepsilon}\right),\qquad\qquad(\text{C.24})$$

where the sum is actually finite, as f^+ is bounded. Combining

$$\varepsilon\chi_{\{f^+\ge(k+1)\varepsilon\}} \le f_{(k+1)\varepsilon} - f_{k\varepsilon} \le \varepsilon\chi_{\{f^+\ge k\varepsilon\}}$$

with inequalities (C.22,C.23), we get

$$\begin{cases}\varepsilon\mu(\{f^+\ge(k+1)\varepsilon\}) \le \int(f_{(k+1)\varepsilon}-f_{k\varepsilon})\,d\mu \le \varepsilon\mu(\{f^+\ge k\varepsilon\}),\\ \varepsilon\mu(\{f^+\ge(k+1)\varepsilon\}) \le T(f_{(k+1)\varepsilon}-f_{k\varepsilon}) \le \varepsilon\mu(\{f^+\ge k\varepsilon\}).\end{cases}\qquad(\text{C.25})$$

We obtain

$$\left|\int f^+\,d\mu - T(f^+)\right|\ \overset{(\text{C.24})}{\le}\ \sum_{k=0}^{\infty}\left|\int(f_{(k+1)\varepsilon}-f_{k\varepsilon})\,d\mu - T(f_{(k+1)\varepsilon}-f_{k\varepsilon})\right|$$
$$\overset{(\text{C.25})}{\le}\ \varepsilon\sum_{k=0}^{\infty}\left(\mu(\{f^+\ge k\varepsilon\}) - \mu(\{f^+\ge(k+1)\varepsilon\})\right)$$
$$=\ \varepsilon\mu(X)$$
$$\xrightarrow[\varepsilon\to0]{\mu(X)<\infty}\ 0,$$

i.e., $T(f^+) = \int f^+ \, d\mu$. Consequently, $Tf = \int f \, d\mu$ for every $f \in C(X)$. $\square$

Proof of the Riesz Representation Theorem C.4.60. Let $T \in C(X)'$. Then $T = T^+ - T^-$ by Lemma C.4.63, and Theorem C.4.65 provides finite Borel measures $\mu, \lambda : \Sigma(\tau) \to [0, \infty]$ such that

$$\begin{cases} T^+(f) = \int f \, d\mu, \\ T^-(f) = \int f \, d\lambda \end{cases}$$

for all $f \in C(X)$. Thus $\mu - \lambda : \Sigma(\tau) \to \mathbb{R}$ is a signed measure for which

$$Tf = \int f \, d(\mu - \lambda)$$

for every $f \in C(X)$. Moreover,

$$\begin{aligned} \|T\| &= \|T_{\mu-\lambda}\| \\ &\overset{\text{Lemma C.4.59}}{\leq} \|\mu - \lambda\| \\ &\leq \|\mu\| + \|\lambda\| \\ &= \mu(X) + \lambda(X) \quad = \quad T^+ 1 + T^- 1 \\ &= \|T^+\| + \|T^-\| \\ &= \|T\|, \end{aligned}$$

so $\|T\| = \|\mu - \nu\|$.   □

Remark C.4.66. The Riesz Representation Theorem C.4.60 can be generalised. For instance, let (X, τ) be a locally compact Hausdorff space, which is also second countable (i.e., τ has a countable base). Endow the vector spaces

$$\begin{aligned} C_0(X) &:= \{f \in C(X) \mid \{f \geq \varepsilon\} \subset X \text{ is compact for every } \varepsilon > 0\}, \\ M(X) &:= \{\nu : \Sigma(\tau) \to \mathbb{R} \mid \nu \text{ is a signed measure}\} \end{aligned}$$

with respective complete norms $f \mapsto \|f\| = \sup\{|f(x)| : x \in X\}$ and $\|\nu\| := |\nu|(X)$; the rough idea is that a function $f \in C_0(X)$ *"vanishes at infinity"*. Then

$$(\nu \mapsto T_\nu) : M(X) \to C_0(X)'$$

is an isometric isomorphism, where

$$T_\nu f := \int f \, d\nu = \int f \, d\nu^+ - \int f \, d\nu^-.$$

We shall not prove this claim here.

Exercise C.4.67. Let μ be a finite Borel measure on a compact metric space (X, d). Prove that $C(X)$ is dense in $L^p(\mu)$ (with the natural embedding) if $1 \leq p < \infty$. What could be a problem with $p = \infty$?

Exercise C.4.68. Let us define $(\tau_h f)(x) := f(x+h)$, when $f : \mathbb{R} \to [-\infty, \infty]$. Show that $\|\tau_h f - f\|_{L^p(\mathbb{R}^n)} \to_{h \to 0} 0$ if $f \in L^p(\mathbb{R}^n)$ and $1 \leq p < \infty$. Why is $p \neq \infty$?

C.5 Product measure and integral

Let $(X, \mathcal{M}_\mu, \mu)$ and $(Y, \mathcal{M}_\nu, \nu)$ be measure spaces. We are going to study the possibility of changing the order of iterated integrals, i.e., whether

$$\int_X \int_Y f \; d\nu \; d\mu = \int_Y \int_X f \; d\mu \; d\nu$$

under some reasonable assumptions on μ, ν and $f : X \times Y \to [-\infty, +\infty]$. Of course, there are many technical issues involved: for instance,

$$\int_X \int_Y f \; d\nu \; d\mu = \int_X \left(x \mapsto \int_Y (y \mapsto f(x, y)) \; d\nu \right) d\nu,$$

where $y \mapsto f(x, y)$ is $\mathcal{M}_\nu$-measurable for μ-almost every $x \in X$, and so on. We start by defining the product measure $\mu \times \nu$ on $X \times Y$.

Definition C.5.1 (Product measures). Let $\mathcal{A} := \{A \times B : A \in \mathcal{M}_\mu, \; B \in \mathcal{M}_\nu\}$. Define a measurelet $m : \mathcal{A} \to [0, \infty]$ on $X \times Y$ by

$$m(A \times B) := \mu(A) \; \nu(B).$$

The outer measure $m^* : \mathcal{P}(X \times Y) \to [0, \infty]$ generated by the measurelet m is called the *product outer measure* of the measures μ and ν. The *product measure* of μ and ν is the restricted measure $\mu \times \nu := m^*|_{\mathcal{M}_{\mu \times \nu}}$, where $\mathcal{M}_{\mu \times \nu} := \mathcal{M}(m^*)$.

Remark C.5.2. Recall that m generates m^* by

$$m^*(S) = \inf \left\{ \sum_{j=1}^\infty m(A_j \times B_j) : S \subset \bigcup_{j=1}^\infty A_j \times B_j, \; \{A_j \times B_j\}_{j=1}^\infty \subset \mathcal{A} \right\}.$$

Actually, we can do better here: we may demand that $\{A_j \times B_j\}_{j=1}^\infty \subset \mathcal{A}$ is disjoint. Why is that? For any two families $\{A_j\}_{j=1}^\infty \subset \mathcal{M}_\mu$ and $\{B_j\}_{j=1}^\infty \subset \mathcal{M}_\nu$, let

$$\begin{cases} F_1 := A_1, \\ F_{k+1} := A_{k+1} \setminus \bigcup_{j=1}^k A_j, \end{cases} \qquad \begin{cases} G_1 := B_1, \\ G_{k+1} := B_{k+1} \setminus \bigcup_{j=1}^k B_j. \end{cases}$$

Clearly, $\{F_j \times G_k \mid j, k \in \mathbb{Z}^+\} \subset \mathcal{A}$ is a disjoint family, and it is easy to check that

$$E = \bigcup \{F_j \times G_k \mid j, k \in \mathbb{Z}^+ : F_j \times G_k \subset E\},$$

where $E = \bigcup_{i=1}^\infty A_i$. Moreover,

$$\sum_{j=1}^\infty m(A_j \times B_j) \geq \sum_{j, k \leq n: \; F_j \times G_k \subset E} m(F_j \times G_k)$$

for each $n \in \mathbb{Z}^+$. Letting $n \to \infty$ yields the claim.

Proposition C.5.3. *Let $A \in \mathcal{M}_\mu$ and $B \in \mathcal{M}_\nu$. Then $A \times B \in \mathcal{M}_{\mu \times \nu}$, and hence the $\mathcal{A}$-generated σ-algebra $\Sigma(\mathcal{A}) \subset \mathcal{M}_{\mu \times \nu}$. Moreover,*

$$(\mu \times \nu)(A \times B) = \mu(A)\, \nu(B).$$

Remark C.5.4. Of course $\int_X \int_Y \chi_{A \times B}\, d\nu\, d\mu = \int_Y \int_X \chi_{A \times B}\, d\mu\, d\nu = \mu(A)\nu(B)$ here, but this certainly does not prove the claims above.

Proof. To prove that $A \times B \in \mathcal{M}_{\mu \times \nu}$, it suffices to show that

$$m^*(S \cup T) \geq m^*(S) + m^*(T)$$

whenever $S \subset A \times B$ and $T \subset (A \times B)^c$ such that $m^*(S), m^*(T) < \infty$. Take $\varepsilon > 0$. Let $\{A_j \times B_j\}_{j=1}^\infty \subset \mathcal{A}$ be disjoint such that

$$S \cup T \quad \subset \quad \bigcup_{j=1}^\infty A_j \times B_j,$$

$$\varepsilon + m^*(S \cup T) \quad > \quad \sum_{j=1}^\infty m(A_j \times B_j).$$

Let us define $S_j, T_j, U_j \subset X \times Y$ by

$$\begin{aligned}
S_j &:= (A_j \times B_j) \cap (A \times B), \\
T_j &:= (A_j \times B_j) \cap (A \times (Y \setminus B)), \\
U_j &:= (A_j \times B_j) \cap ((X \setminus A) \times Y).
\end{aligned}$$

Then $\{S_j, T_j, U_j\}_{j=1}^\infty \subset \mathcal{A}$ is disjoint, and $A_j \times B_j = S_j \cup T_j \cup U_j$. Moreover,

$$S \subset \bigcup_{j=1}^\infty S_j, \quad T \subset \bigcup_{j=1}^\infty (T_j \cup U_j),$$

so that

$$\begin{aligned}
\varepsilon + m^*(S \cup T) \quad &> \quad \sum_{j=1}^\infty m(A_j \times B_j) \\
&= \quad \sum_{j=1}^\infty m(S_j) + \sum_{j=1}^\infty (m(T_j) + m(U_j)) \\
&\geq \quad m^*(S) + m^*(T).
\end{aligned}$$

Thus we have shown that $A \times B \in \mathcal{M}_{\mu \times \nu}$. Finally, if $\{A_j \times B_j\}_{j=1}^{\infty} \subset \mathcal{A}$ is a cover of $A \times B$ then

$$
\begin{aligned}
m^*(A \times B) &\overset{\text{trivial}}{\leq} m(A \times B) \\
&= \mu(A) \, \nu(B) \\
&= \int_X \int_Y \chi_{A \times B} \, d\nu \, d\mu \\
&\leq \int_X \int_Y \sum_{j=1}^{\infty} \chi_{A_j \times B_j} \, d\nu \, d\mu \\
&\overset{\text{Mon. Conv.}}{=} \sum_{j=1}^{\infty} m(A_j \times B_j).
\end{aligned}
$$

Therefore $m^*(A \times B) = \mu(A) \, \nu(B)$. $\square$

Exercise C.5.5. Show that for the Lebesgue measures, $\lambda_{\mathbb{R}^m} \times \lambda_{\mathbb{R}^n} = \lambda_{\mathbb{R}^{m+n}}$.

Definition C.5.6. For $x \in X$, the x-slice $S^x \subset Y$ of a set $S \subset X \times Y$ is

$$
S^x := \{y \in Y \mid (x, y) \in S\}.
$$

Remark C.5.7. Let

$$
\mathcal{B} = \{R \in \Sigma(\mathcal{A}) : \ R^x \in \mathcal{M}_\nu \text{ for all } x \in X\}.
$$

Clearly $X \times Y \in \mathcal{A} \subset \mathcal{B}$. If $R \in \mathcal{B}$ then also $R^c = (X \times Y) \setminus R \in \mathcal{B}$, because $(R^c)^x = (R^x)^c$. Similarly, if $\{R_j\}_{j=1}^{\infty} \subset \mathcal{B}$ then $\bigcup_{j=1}^{\infty} R_j \in \mathcal{B}$, because $(\bigcup_{j=1}^{\infty} R_j)^x = \bigcup_{j=1}^{\infty}(R_j)^x$. Thus $\Sigma(\mathcal{A}) \subset \mathcal{B}$.

Lemma C.5.8. *The product outer measure m^* is $\Sigma(\mathcal{A})$-regular: for any $S \subset X \times Y$ there exists $R \in \Sigma(\mathcal{A})$ such that $S \subset R$ and $m^*(S) = m^*(R)$. Moreover, if μ, ν are finite then the x-slice $R^x \in \mathcal{M}_\nu$ for every $x \in X$, $x \mapsto \nu(R^x)$ is $\mathcal{M}_\mu$-measurable, and*

$$
m^*(R) = \int_X \int_Y \chi_R \, d\nu \, d\mu.
$$

Proof. For each $k \in \mathbb{Z}^+$, take a disjoint family $\{A_{kj} \times B_{kj}\}_{j=1}^{\infty} \subset \mathcal{A}$ such that

$$
S \subset \bigcup_{j=1}^{\infty} A_{kj} \times B_{kj},
$$

$$
m^*(S) + \frac{1}{k} \geq \sum_{j=1}^{\infty} m(A_{kj} \times B_{kj}).
$$

Let $R_n := \bigcap_{k=1}^{n} \bigcup_{j=1}^{\infty} (A_{kj} \times B_{kj})$ and $R := \bigcap_{n=1}^{\infty} R_n$. Then $S \subset R \in \Sigma(\mathcal{A})$. Moreover, we have $m^*(S) = m^*(R)$, because

$$
\begin{aligned}
m^*(S) &\leq m^*(R) \\
&\leq m^*(\bigcup_{j=1}^{\infty} A_{nj} \times B_{nj}) \\
&= \sum_{j=1}^{\infty} m(A_{nj} \times B_{nj}) \\
&\leq m^*(S) + \frac{1}{n}.
\end{aligned}
$$

The set R_n is the union of a disjoint family $\{C_{nj} \times D_{nj}\}_{j=1}^{\infty} \subset \mathcal{A}$, and

$$
\chi_{R_n^x}(y) = \sum_{j=1}^{\infty} \chi_{C_{nj}}(x) \, \chi_{D_{nj}}(y).
$$

Consequently, $\chi_{R_n^x} : Y \to \mathbb{R}$ is $\mathcal{M}_\nu$-measurable for all $x \in X$,

$$
1 \leq \chi_{R_n^x}(y) \leq \chi_{R_{n+1}^x}(y) \xrightarrow[n\to\infty]{} \chi_{R^x}(y) \geq 0.
$$

Lebesgue's Dominated Convergence Theorem C.3.22 yields

$$
h_n(x) := \int_Y \chi_{R_n^x} \, d\nu \xrightarrow[n\to\infty]{\nu(Y)<\infty} \int_Y \chi_{R^x} \, d\nu =: h(x).
$$

Then $h_n : X \to [0, \infty)$ is $\mathcal{M}_\mu$-measurable and

$$
h_n(x) = \sum_{j=1}^{\infty} \nu(D_{nj}) \, \chi_{C_{nj}}(x).
$$

Moreover,

$$
\infty > \nu(Y) \geq h_n(x) \geq h_{n+1}(x) \xrightarrow[n\to\infty]{} h(x) \geq 0,
$$

so also $h : X \to [0, \infty)$ is $\mathcal{M}_\mu$-measurable and

$$
\int_X \int_Y \chi_{R_n} \, d\nu \, d\mu = \int_X h_n \, d\mu \xrightarrow[n\to\infty]{\mu(X)<\infty} \int_X h \, d\mu = \int_X \int_Y \chi_R \, d\nu \, d\mu,
$$

where Lebesgue's Dominated Convergence Theorem was used again. $\qquad\square$

The following result will neatly justify calling m^* *the* product outer measure.

Corollary C.5.9. *Let μ, ν be finite. Then $(\mu \times \nu)^* = m^*$.*

Proof. If $E \subset X \times Y$ then

$$
\begin{aligned}
(\mu \times \nu)^*(E) \quad &= \quad \inf\{(\mu \times \nu)(S) \mid E \subset S \in \mathcal{M}_{\mu \times \nu}\} \\
&\overset{\text{Lemma C.5.8}}{=} \quad \inf\{m^*(R) \mid E \subset R \in \Sigma(\mathcal{A})\} \\
&\geq \quad m^*(E) \\
&\geq \quad (\mu \times \nu)^*(E),
\end{aligned}
$$

that is $(\mu \times \nu)^*(E) = m^*(E)$. $\qquad\square$

Exercise C.5.10. Generalise Corollary C.5.9 for σ-finite measures.

Exercise C.5.11. Let $\mu_0 = \mu^*|_{\mathcal{M}(\mu^*)}$ and $\nu_0 = \nu^*|_{\mathcal{M}(\nu^*)}$. Is $\mu \times \nu = \mu_0 \times \nu_0$?

Proposition C.5.12. *Let μ, ν be finite and complete, $S \subset \mathcal{M}_{\mu \times \nu}$. Then the x-slice $S^x \in \mathcal{M}_\nu$ for μ-almost every $x \in X$, and $x \mapsto \nu(S^x)$ is $\mathcal{M}_\mu$-measurable. Moreover,*

$$
(\mu \times \nu)(S) = \int_X \int_Y \chi_S \, \mathrm{d}\nu \, \mathrm{d}\mu.
$$

Proof. By Lemma C.5.8, there exists a set $R \in \Sigma(\mathcal{A}) \subset \mathcal{M}_{\mu \times \nu}$ such that $S \subset R$ and

$$
\begin{aligned}
(\mu \times \nu)(S) \quad &= \quad m^*(S) \\
&= \quad m^*(R) \\
&= \quad \int_X \int_Y \chi_R \, \mathrm{d}\nu \, \mathrm{d}\mu.
\end{aligned}
$$

Thus the result follows, if we can show that $\nu(R^x) = \nu(S^x)$ for μ-almost every $x \in X$. Here $R^x = S^x \cup (R \setminus S)^x$. By Lemma C.5.8, there exists a set $Q \in \Sigma(\mathcal{A})$ such that $R \setminus S \subset Q$ and

$$
m^*(R \setminus S) = \int_X \int_Y \chi_Q \, \mathrm{d}\nu \, \mathrm{d}\mu.
$$

Now $m^*(R \setminus S) = m^*(R) - m^*(S) = 0$, because the measures are finite. Therefore $\nu(Q^x) = 0$ for μ-almost every $x \in X$. Because $R \setminus S \subset Q$ and ν is complete, we see that

$$
(R \setminus S)^x \in \mathcal{M}_\nu, \quad \nu((R \setminus S)^x) = 0
$$

for μ-almost every $x \in X$. This shows that $\nu(R^x) = \nu(S^x)$ for μ-almost every $x \in X$. $\qquad\square$

Theorem C.5.13 (Fubini–Tonelli Theorem). *Let μ, ν be finite complete measures, and let $f \geq 0$ be $\mathcal{M}_{\mu \times \nu}$-measurable. Then*

$$x \mapsto f(x,y) \quad is \quad \mathcal{M}_\mu\text{-measurable } \nu\text{-a.e.,}$$

$$y \mapsto \int_X (x \mapsto f(x,y)) \, \mathrm{d}\mu \quad is \quad \mathcal{M}_\nu\text{-measurable, and}$$

$$\int_{X \times Y} f \, \mathrm{d}(\mu \times \nu) \; = \; \int_Y \int_X f \, \mathrm{d}\mu \, \mathrm{d}\nu$$

$$= \int_X \int_Y f \, \mathrm{d}\nu \, \mathrm{d}\mu.$$

Proof. Take $\mathcal{M}_{\mu \times \nu}$-measurable simple functions $f_k : X \times Y \to [0, \infty)$ for which

$$f_k \leq f_{k+1} \xrightarrow[k \to \infty]{\text{pointwise}} f.$$

By Proposition C.5.12, take $N_k \in \mathcal{M}_\mu$ such that $\mu(N_k) = 0$ and

$$(y \mapsto f_k(x,y)) : Y \to [0, \infty)$$

is $\mathcal{M}_\nu$-measurable for all $x \in X \setminus N_k$. Then $N := \bigcup_{k=1}^\infty N_k \in \mathcal{M}_\mu$, $\mu(N) = 0$ and

$$(y \mapsto f(x,y)) : Y \to [0, \infty]$$

is $\mathcal{M}_\nu$-measurable for all $x \in X \setminus N$. Let us define $g_k, g : X \to [0, \infty]$ such that $g_k(x) := 0 =: g(x)$ if $x \in N$, and otherwise

$$g_k(x) := \int_Y f_k(x,y) \, \mathrm{d}\nu(y) \xrightarrow[k \to \infty]{\text{Mon. Conv.}} \int_Y f(x,y) \, \mathrm{d}\nu(y) =: g(x).$$

By Proposition C.5.12, $g_k : X \to [0, \infty]$ is $\mathcal{M}_\mu$-measurable, and clearly

$$g_k \leq g_{k+1} \xrightarrow[k \to \infty]{\text{pointwise}} g.$$

Thus $g : X \to [0, \infty]$ is $\mathcal{M}_\mu$-measurable, and

$$\int_X \int_Y f \, \mathrm{d}\nu \, \mathrm{d}\mu \quad = \quad \int_X g \, \mathrm{d}\mu$$

$$\overset{\text{Mon. Conv.}}{=} \quad \lim_{k \to \infty} \int g_k \, \mathrm{d}\mu$$

$$= \quad \lim_{k \to \infty} \int_X \int_Y f_k \, \mathrm{d}\nu \, \mathrm{d}\mu$$

$$\overset{\text{Prop. C.5.12}}{=} \quad \lim_{k \to \infty} \int_{X \times Y} f_k \, \mathrm{d}(\mu \times \nu)$$

$$\overset{\text{Mon. Conv.}}{=} \quad \int_{X \times Y} f \, \mathrm{d}(\mu \times \nu).$$

This concludes the proof. $\qquad\qquad\qquad\qquad\qquad\qquad\qquad\qquad\qquad\qquad\qquad\quad\square$

Corollary C.5.14 (Fubini Theorem). *Let $f \in L^1(\mu \times \nu)$. Then*

$$\int_{X \times Y} f \, d(\mu \times \nu) = \int_X \int_Y f \, d\nu \, d\mu = \int_Y \int_X f \, d\mu \, d\nu. \qquad \square$$

Theorem C.5.15 (σ-finite Fubini–Tonelli and Fubini Theorems). *The results in Proposition C.5.12, the Fubini–Tonelli Theorem C.5.13 and its Corollary C.5.14 hold also for σ-finite complete measures μ, ν.*

Exercise C.5.16. Prove Theorem C.5.15 by taking an increasing sequence of measurable sets E_n of finite measure such that $\bigcup_n E_n = X \times Y$.

Exercise C.5.17. Let $X = [0,1] = Y$, $\mu := (\lambda_{\mathbb{R}})_X$ and $\nu := (E \mapsto \#E) : \mathcal{P}(Y) \to [0,\infty]$. Let $f = \chi_S$, where $S := \{(x,y) \in X \times Y : x = y\}$. Calculate

(a) $\int_X \int_Y f \, d\nu \, d\mu$,
(b) $\int_Y \int_X f \, d\mu \, d\nu$,
(c) $\int_{X \times Y} f \, d(\mu \times \nu)$.

Remark C.5.18. In Exercise C.5.17, the measure ν is not σ-finite. Thus there is no contradiction with Theorem C.5.15.

Exercise C.5.19. Let $(X, \mathcal{M}, \mu)$ be a complete σ-finite measure space. Let $f : X \to [0,\infty]$ be $\mathcal{M}$-measurable. Show that

$$\int f \, d\mu = \int_{[0,\infty)} \mu(\{f > t\}) \, d\lambda_{\mathbb{R}}(t).$$

Corollary C.5.20 (Young's inequality). *Let μ, ν be σ-finite and $1 < p < \infty$. Assume that $K : X \times Y \to \mathbb{C}$ is a $\mathcal{M}_{\mu \times \nu}$-measurable function satisfying*

$$C_1 := \sup_{y \in Y} \int_X |K(x,y)| \, d\mu(x),$$

$$C_2 := \sup_{x \in X} \int_Y |K(x,y)| \, d\nu(y),$$

where $C_1, C_2 < \infty$. For any $u \in L^p(\nu)$ define $Au : X \to \mathbb{C}$ by

$$Au(x) = \int_Y K(x,y) \, u(y) \, d\nu(y).$$

Then

$$\|Au\|_{L^p(\mu)} \leq C_1^{1/p} \, C_2^{1/q} \, \|u\|_{L^p(\nu)},$$

where q is the conjugate exponent of p.

Remark C.5.21. Notice that this defines a unique bounded linear operator $A : L^p(\nu) \to L^p(\mu)$.

Remark C.5.22. It is clear that we can replace sup by the esssup in the definition of C_1, C_2, where the esssup would be taken with respect to ν and μ in C_1 and C_2, respectively:

$$C_1 \; := \; \nu - \mathrm{esssup}_{y \in Y} \int_X |K(x,y)| \; \mathrm{d}\mu(x),$$

$$C_2 \; := \; \mu - \mathrm{esssup}_{x \in X} \int_Y |K(x,y)| \; \mathrm{d}\nu(y),$$

with the same proof.

Proof of Corollary C.5.20. First, $y \mapsto K(x,y) \, u(y)$ is $\mathcal{M}_\nu$-measurable, and

$$
\begin{aligned}
|Au(x)| \quad &\leq \quad \int_Y \left(|K(x,y)|^{1/p} |u(y)| \right) \left(|K(x,y)|^{1/q} \right) \; \mathrm{d}\nu(y) \\
&\overset{\text{Hölder}}{\leq} \quad \left(\int_Y |K(x,y)| |u(y)|^p \; \mathrm{d}\nu(y) \right)^{1/p} \left(\int_Y |K(x,y)| \; \mathrm{d}\nu(y) \right)^{1/q} \\
&\leq \quad \left(\int_Y |K(x,y)| |u(y)|^p \; \mathrm{d}\nu(y) \right)^{1/p} C_2^{1/q}.
\end{aligned}
$$

Using this we get

$$
\begin{aligned}
\|Au\|_{L^p(\mu)}^p \quad &= \quad \int_X |Au(x)|^p \; \mathrm{d}\mu(x) \\
&\leq \quad C_2^{p/q} \int_X \int_Y |K(x,y)| |u(y)|^p \; \mathrm{d}\nu(y) \; \mathrm{d}\mu(x) \\
&\overset{\text{Fubini}}{=} \quad C_2^{p/q} \int_Y |u(y)|^p \int_X |K(x,y)| \; \mathrm{d}\mu(x) \; \mathrm{d}\nu(y) \\
&\leq \quad C_1 C_2^{p/q} \int_Y |u(y)|^p \; \mathrm{d}\nu(y) \\
&= \quad \left(C_1^{1/p} C_2^{1/q} \right)^p \|u\|_{L^p(\mu)}^p,
\end{aligned}
$$

which gives the result. $\qquad\qquad\qquad\qquad\qquad\qquad\qquad\qquad\qquad\qquad$ $\square$

Theorem C.5.23 (Minkowski's inequality for integrals). *Let μ, ν be σ-finite and let $f : X \times Y \to \mathbb{C}$ be a $\mathcal{M}_{\mu \times \nu}$-measurable function. Let $1 \leq p < \infty$. Then*

$$\left\{ \int \left(\int |f(x,y)| \; \mathrm{d}\nu(y) \right)^p \mathrm{d}\mu(x) \right\}^{1/p} \leq \int \left(\int |f(x,y)|^p \; \mathrm{d}\mu(x) \right)^{1/p} \mathrm{d}\nu(y).$$

Proof. If $p = 1$ the result follows from Theorem C.5.15 exchanging the order of the integration. For $1 < p < \infty$, taking $g \in L^q(\mu)$, $\frac{1}{p} + \frac{1}{q} = 1$, we have

$$\int \left(\int |f(x,y)| \, \mathrm{d}\nu(y) \right) |g(x)| \, \mathrm{d}\mu(x)$$

$$\overset{\text{Fubini}}{=} \int \left(\int |f(x,y)||g(x)| \, \mathrm{d}\mu(x) \right) \mathrm{d}\nu(y)$$

$$\overset{\text{Hölder}}{=} \int \left(\int |f(x,y)|^p \, \mathrm{d}\mu(x) \right)^{1/p} \mathrm{d}\nu(y) \, \|g\|_{L^q(\mu)}.$$

Now the statement follows from the converse of Hölder's inequality (Theorem C.4.56). $\qquad\square$

As a consequence, we obtain the second part of Minkowski's inequality for integrals:

Corollary C.5.24 (Monotonicity of L^p-norm). *Let μ, ν be σ-finite and let $f : X \times Y \to \mathbb{C}$ be a $\mathcal{M}_{\mu \times \nu}$-measurable function. Let $1 \le p \le \infty$. Assume that $f(\cdot, y) \in L^p(\mu)$ for ν-a.e. y, and assume that the function $y \mapsto \|f(\cdot, y)\|_{L^p(\mu)}$ is in $L^1(\mu)$. Then $f(x, \cdot) \in L^1(\nu)$ for μ-a.e. x, the function $x \mapsto \int f(x,y) \, \mathrm{d}\nu(y)$ is in $L^p(\mu)$, and*

$$\left\| \int f(\cdot, y) \, \mathrm{d}\nu(y) \right\|_{L^p(\mu)} \le \int \|f(\cdot, y)\|_{L^p(\mu)} \, \mathrm{d}\nu(y).$$

Proof. For $p = \infty$ the statement follows from Theorem C.3.14. For $1 \le p < \infty$ it follows from Theorem C.5.23 and Fubini's Theorem C.5.15. $\qquad\square$

Chapter D

Algebras

An algebra is a vector space endowed with a multiplication, satisfying some compatibility conditions. In the sequel, we are going to deal with spectral properties of algebras under various additional assumptions.

D.1   Algebras

Definition D.1.1 (Algebra). A vector space $\mathcal{A}$ over the field $\mathbb{C}$ is an *algebra* if there exists an element $\mathbf{1}_\mathcal{A} \in \mathcal{A} \setminus \{0\}$ and a mapping $\mathcal{A} \times \mathcal{A} \to \mathcal{A}$, $(x, y) \mapsto xy$, satisfying

$$x(yz) = (xy)z,$$
$$x(y + z) = xy + xz, \quad (x + y)z = xz + yz,$$
$$\lambda(xy) = (\lambda x)y = x(\lambda y),$$
$$\mathbf{1}_\mathcal{A} x = x = x\mathbf{1}_\mathcal{A},$$

for all $x, y, z \in \mathcal{A}$ and $\lambda \in \mathbb{C}$. We briefly write $xyz := x(yz)$. The element $\mathbf{1} := \mathbf{1}_\mathcal{A}$ is called the *unit of* $\mathcal{A}$, and an element $x \in \mathcal{A}$ is called *invertible* (with the unique *inverse* x^{-1}) if there exists $x^{-1} \in \mathcal{A}$ such that

$$x^{-1}x = \mathbf{1} = xx^{-1}.$$

If $xy = yx$ for every $x, y \in \mathcal{A}$ then $\mathcal{A}$ is called *commutative*.

Remark D.1.2. Warnings: in some books the algebra axioms allow $\mathbf{1}_\mathcal{A}$ to be 0, but then the resulting algebra is simply $\{0\}$; we have omitted such a triviality. In some books the existence of a unit is omitted from the algebra axioms; what we have called an algebra is there called a *unital algebra*.

Example. Let us give some examples of algebras:

1. $\mathbb{C}$ is the most important algebra. The operations are the usual ones for complex numbers, and the unit element is $\mathbf{1}_\mathbb{C} = 1 \in \mathbb{C}$. Clearly $\mathbb{C}$ is a commutative algebra.

2. The algebra $\mathcal{F}(X) := \{f \mid f : X \to \mathbb{C}\}$ of complex-valued functions on a (finite or infinite) set X is endowed with the usual algebra structure (pointwise operations). Function algebras are commutative, because $\mathbb{C}$ is commutative.

3. The algebra $L(V) := \{A : V \to V \mid A \text{ is linear}\}$ of linear operators on a vector space $V \neq \{0\}$ over $\mathbb{C}$ is endowed with the usual vector space structure and with the multiplication $(A, B) \mapsto AB$ (composition of operators); the unit element is $1_{L(V)} = (v \mapsto v) : V \to V$, the identity operator on V. This algebra is non-commutative if V is at least two-dimensional.

Exercise D.1.3. Let $\mathcal{A}$ be an algebra and $x, y \in \mathcal{A}$. Prove the following claims:

(a) If x, xy are invertible then y is invertible.
(b) If xy, yx are invertible then x, y are invertible.

Exercise D.1.4. Give an example of an algebra $\mathcal{A}$ and elements $x, y \in \mathcal{A}$ such that $xy = 1_\mathcal{A} \neq yx$. Prove that then $(yx)^2 = yx \neq 0$. (Hint: Such an algebra is necessarily infinite-dimensional.)

Exercise D.1.5 (Commutators). In an algebra $\mathcal{A}$, let $[A, B] = AB - BA$. If λ is a scalar and A, B, C are elements of the algebra $\mathcal{A}$ show that

$$
\begin{aligned}
[B, A] &= -[A, B], \\
[\lambda A, B] &= \lambda[A, B], \\
[A + B, C] &= [A, C] + [B, C], \\
[AB, C] &= A[B, C] + [A, C]B, \\
C[A, B]C^{-1} &= [CAC^{-1}, CBC^{-1}].
\end{aligned}
$$

Definition D.1.6 (Spectrum). Let $\mathcal{A}$ be an algebra. The *spectrum* $\sigma(x)$ of an element $x \in \mathcal{A}$ is the set

$$
\sigma_\mathcal{A}(x) = \sigma(x) = \{\lambda \in \mathbb{C} : \lambda 1 - x \text{ is not invertible}\}.
$$

Example. Let us give some examples of invertibility and spectra:

1. An element $\lambda \in \mathbb{C}$ is invertible if and only if $\lambda \neq 0$; the inverse of an invertible λ is the usual $\lambda^{-1} = 1/\lambda$. Generally, $\sigma_\mathbb{C}(\lambda) = \{\lambda\}$.

2. An element $f \in \mathcal{F}(X)$ is invertible if and only if $f(x) \neq 0$ for every $x \in X$. The inverse of an invertible f is g with $g(x) = f(x)^{-1}$. Generally, $\sigma_{\mathcal{F}(X)}(f) = f(X) := \{f(x) \mid x \in X\}$.

3. An element $A \in L(V)$ is invertible if and only if it is a bijection (if and only if $0 \notin \sigma_{L(V)}(A)$).

Exercise D.1.7. Let $\mathcal{A}$ be an algebra and $x, y \in \mathcal{A}$. Prove the following claims:

(a) $1 - yx$ is invertible if and only if $1 - xy$ is invertible.
(b) $\sigma(yx) \subset \sigma(xy) \cup \{0\}$.
(c) If x is invertible then $\sigma(xy) = \sigma(yx)$.

Definition D.1.8 (Ideals). Let $\mathcal{A}$ be an algebra. An *ideal* $\mathcal{J} \subset \mathcal{A}$ (more precisely, a *two-sided ideal*) is a vector subspace $\mathcal{J} \neq \mathcal{A}$ satisfying

$$\forall x \in \mathcal{A} \; \forall y \in \mathcal{J} : \; xy, yx \in \mathcal{J},$$

i.e., $x\mathcal{J}, \mathcal{J}x \subset \mathcal{J}$ for every $x \in \mathcal{A}$. A *maximal ideal* is an ideal not contained in any other ideal.

Remark D.1.9. In some books our ideals are called *proper ideals*, and there *ideal* is either a proper ideal or the whole algebra. In the case of (proper) ideals, the vector space $\mathcal{A}/\mathcal{J} := \{x + \mathcal{J} \mid x \in \mathcal{A}\}$ becomes an algebra with the operation $(x + \mathcal{J}, y + \mathcal{J}) \mapsto xy + \mathcal{J}$ and the unit element $\mathbf{1}_{\mathcal{A}/\mathcal{J}} := \mathbf{1}_{\mathcal{A}} + \mathcal{J}$. It is evident that no proper ideal contains any invertible elements. We will drop the word "proper" since it is incorporated in Definition D.1.8.

Remark D.1.10. Let $\mathcal{J} \subset \mathcal{A}$ be an ideal. Because $x\mathbf{1} = x$ for every $x \in \mathcal{A}$, we notice that $\mathbf{1} \notin \mathcal{J}$. Therefore an invertible element $x \in \mathcal{A}$ cannot belong to an ideal (since $x^{-1}x = \mathbf{1} \notin \mathcal{J}$).

Example. Let us give examples of ideals. Intuitively, an ideal of an algebra is a subspace resembling a multiplicative zero; consider equations $x0 = 0 = 0x$.

1. Let $\mathcal{A}$ be an algebra. Then $\{0\} \subset \mathcal{A}$ is an ideal.

2. The only ideal of $\mathbb{C}$ is $\{0\} \subset \mathbb{C}$.

3. Let X be a set, and $\emptyset \neq S \subset X$. Now

$$\mathcal{I}(S) := \{f \in \mathcal{F}(X) \mid \forall x \in S : \; f(x) = 0\}$$

   is an ideal of the function algebra $\mathcal{F}(X)$. If $x \in X$ then $\mathcal{I}(\{x\})$ is a maximal ideal of $\mathcal{F}(X)$, because it is of co-dimension 1 in $\mathcal{F}(X)$. Notice that $\mathcal{I}(S) \subset \mathcal{I}(\{x\})$ for every $x \in S$; an ideal may be contained in many different maximal ideals (cf. Krull's Theorem D.1.13 in the sequel).

4. Let X be an infinite-dimensional Banach space. The set

$$\mathcal{LC}(X) := \{A \in \mathcal{L}(X) \mid A \text{ is compact}\}$$

   of compact linear operators $X \to X$ is an ideal of the algebra $\mathcal{L}(X)$ of bounded linear operators $X \to X$.

Definition D.1.11 (Semisimple algebras). The *radical* $\mathrm{Rad}(\mathcal{A})$ of an algebra $\mathcal{A}$ is the intersection of all the maximal ideals of $\mathcal{A}$; $\mathcal{A}$ is called *semisimple* if $\mathrm{Rad}(\mathcal{A}) = \{0\}$.

Exercise D.1.12 (Ideals spanned by sets). Show that any intersection of ideals is an ideal. Hence for any set $S \subset \mathcal{A}$ in an algebra $\mathcal{A}$ there exists a smallest possible ideal $\mathcal{J} \subset \mathcal{A}$ such that $S \subset \mathcal{J}$; this $\mathcal{J}$ is called the *ideal spanned by the set S*.

Theorem D.1.13 (W. Krull). *Every ideal is contained in a maximal ideal.*

Proof. Let $\mathcal{J}$ be an ideal of an algebra $\mathcal{A}$. Let P be the set of those ideals of $\mathcal{A}$ that contain $\mathcal{J}$. The inclusion relation is the natural partial order on P; the **Hausdorff Maximal Principle** (Theorem A.4.9) says that there is a maximal chain $C \subset P$. Let $\mathcal{M} := \bigcup C$. Clearly $\mathcal{J} \subset \mathcal{M}$. Let $\lambda \in \mathbb{C}$, $x, y \in \mathcal{M}$ and $z \in \mathcal{A}$. Then there exists $\mathcal{I} \in C$ such that $x, y \in \mathcal{I}$, so that

$$\lambda x \in \mathcal{I} \subset \mathcal{M}, \quad x + y \in \mathcal{I} \subset \mathcal{M}, \quad xz, zx \in \mathcal{I} \subset \mathcal{M};$$

moreover,

$$1 \in \bigcap_{\mathcal{I} \in C} (\mathcal{A} \setminus \mathcal{I}) = \mathcal{A} \setminus \bigcup_{\mathcal{I} \in C} \mathcal{I} = \mathcal{A} \setminus \mathcal{M},$$

so that $\mathcal{M} \neq \mathcal{A}$. We have proven that $\mathcal{M}$ is an ideal. The maximality of the chain C implies that $\mathcal{M}$ is maximal. $\qquad\qquad\square$

Lemma D.1.14. *Let $\mathcal{A}$ be a commutative algebra and let $\mathcal{M}$ be an ideal. Then $\mathcal{M}$ is maximal if and only if $[0]$ is the only non-invertible element of $\mathcal{A}/\mathcal{M}$.*

Proof. Of course, here $[x]$ means $x + \mathcal{M}$, where $x \in \mathcal{A}$. Assume that $\mathcal{M}$ is a maximal ideal. Take $[x] \neq [0]$, so that $x \notin \mathcal{M}$. Define

$$\mathcal{J} := \mathcal{A}x + \mathcal{M} = \{ax + m \mid a \in \mathcal{A}, \ m \in \mathcal{M}\}.$$

Then clearly $\mathcal{J} \neq \mathcal{M} \subset \mathcal{J}$, and $\mathcal{J}$ is a vector subspace of $\mathcal{A}$. If $y \in \mathcal{A}$ then

$$\mathcal{J}y = y\mathcal{J} = y\mathcal{A}x + y\mathcal{M} \subset \mathcal{A}x + \mathcal{M} = \mathcal{J},$$

so that either $\mathcal{J}$ is an ideal or $\mathcal{J} = \mathcal{A}$. But since $\mathcal{M}$ is a maximal ideal contained properly in $\mathcal{J}$, we must have $\mathcal{J} = \mathcal{A}$. Thus there exist $a \in \mathcal{A}$ and $m \in \mathcal{M}$ such that $ax + m = \mathbf{1}_{\mathcal{A}}$. Then

$$[a][x] = \mathbf{1}_{\mathcal{A}/\mathcal{M}} = [x][a],$$

$[x]$ is invertible in $\mathcal{A}/\mathcal{M}$.

Conversely, assume that all the non-zero elements are invertible in $\mathcal{A}/\mathcal{M}$. Assume that $\mathcal{J} \subset \mathcal{A}$ is an ideal containing $\mathcal{M}$. **Suppose** $\mathcal{J} \neq \mathcal{M}$, and pick $x \in \mathcal{J} \setminus \mathcal{M}$. Now $[x] \neq [0]$, so that for some $y \in \mathcal{A}$ we have $[x][y] = [\mathbf{1}_{\mathcal{A}}]$. Thereby

$$\mathbf{1}_{\mathcal{A}} \in xy + \mathcal{M} \overset{x \in \mathcal{J}}{\subset} \mathcal{J} + \mathcal{M} \subset \mathcal{J} + \mathcal{J} = \mathcal{J},$$

which is a **contradiction**, since no ideal can contain invertible elements. Therefore we must have $\mathcal{J} = \mathcal{M}$, meaning that $\mathcal{M}$ is maximal. $\qquad\qquad\square$

Definition D.1.15 (Quotient algebra). Let $\mathcal{A}$ be an algebra with an ideal $\mathcal{J}$. For $x \in \mathcal{A}$, let us denote

$$[x] := x + \mathcal{J} = \{x + j \mid j \in \mathcal{J}\}.$$

Then the set $\mathcal{A}/\mathcal{J} := \{[x] \mid x \in \mathcal{A}\}$ can be endowed with a natural algebra structure. Indeed, let us define

$$\lambda[x] := [\lambda x], \quad [x] + [y] := [x + y], \quad [x][y] := [xy], \quad \mathbf{1}_{\mathcal{A}/\mathcal{J}} := [\mathbf{1}_{\mathcal{A}}];$$

first of all, these operations are well defined, since if $\lambda \in \mathbb{C}$ and $j, j_1, j_2 \in \mathcal{J}$ then

$$
\begin{aligned}
\lambda(x + j) &= \lambda x + \lambda j \in [\lambda x], \\
(x + j_1) + (y + j_2) &= (x + y) + (j_1 + j_2) \in [x + y], \\
(x + j_1)(y + j_2) &= xy + j_1 y + x j_2 + j_1 j_2 \in [xy].
\end{aligned}
$$

Secondly, $[\mathbf{1}_{\mathcal{A}}] = \mathbf{1}_{\mathcal{A}} + \mathcal{J} \neq \mathcal{J} = [0]$, because $\mathbf{1}_{\mathcal{A}} \notin \mathcal{J}$. Moreover,

$$
\begin{aligned}
(x + j_1)(\mathbf{1}_{\mathcal{A}} + j_2) &= x + j_1 + x j_2 + j_1 j_2 \in [x], \\
(\mathbf{1}_{\mathcal{A}} + j_2)(x + j_1) &= x + j_1 + j_2 x + j_2 j_1 \in [x].
\end{aligned}
$$

Now the reader may verify that $\mathcal{A}/\mathcal{J}$ is really an algebra; it is called the *quotient algebra of $\mathcal{A}$ modulo $\mathcal{J}$*.

Remark D.1.16. Notice that $\mathcal{A}/\mathcal{J}$ is commutative if $\mathcal{A}$ is commutative. Also notice that $[0] = \mathcal{J}$ is the zero element in the quotient algebra.

Definition D.1.17 (Algebra homomorphism). Let $\mathcal{A}$ and $\mathcal{B}$ be algebras. A mapping $\phi : \mathcal{A} \to \mathcal{B}$ is called an *algebra homomorphism* (or simply a *homomorphism*) if it is a linear mapping satisfying

$$\phi(xy) = \phi(x)\phi(y)$$

for every $x, y \in \mathcal{A}$ (multiplicativity) and

$$\phi(\mathbf{1}_{\mathcal{A}}) = \mathbf{1}_{\mathcal{B}}.$$

The set of all homomorphisms $\mathcal{A} \to \mathcal{B}$ is denoted by

$$\mathrm{Hom}(\mathcal{A}, \mathcal{B}).$$

A bijective homomorphism $\phi : \mathcal{A} \to \mathcal{B}$ is called an *isomorphism*, denoted by $\phi : \mathcal{A} \cong \mathcal{B}$.

Example. Let us give examples of algebra homomorphisms:

1. The only homomorphism $\mathbb{C} \to \mathbb{C}$ is the identity mapping, i.e., $\mathrm{Hom}(\mathbb{C}, \mathbb{C}) = \{x \mapsto x\}$.
2. Let $x \in X$. Let us define the evaluation mapping $\phi_x : \mathcal{F}(X) \to \mathbb{C}$ by $f \mapsto f(x)$. Then $\phi_x \in \mathrm{Hom}(\mathcal{F}(X), \mathbb{C})$.
3. Let $\mathcal{J}$ be an ideal of an algebra $\mathcal{A}$, and denote $[x] = x + \mathcal{J}$. Then $(x \mapsto [x]) \in \mathrm{Hom}(\mathcal{A}, \mathcal{A}/\mathcal{J})$.

Exercise D.1.18. Let $\phi \in \mathrm{Hom}(\mathcal{A}, \mathcal{B})$. If $x \in \mathcal{A}$ is invertible then $\phi(x) \in \mathcal{B}$ is invertible. For any $x \in \mathcal{A}$, $\sigma_{\mathcal{B}}(\phi(x)) \subset \sigma_{\mathcal{A}}(x)$.

Exercise D.1.19. Let $\mathcal{A}$ be the set of matrices

$$\begin{pmatrix} \alpha & \beta \\ 0 & \alpha \end{pmatrix} \qquad (\alpha, \beta \in \mathbb{C}).$$

Show that $\mathcal{A}$ is a commutative algebra. Classify (up to an isomorphism) all the two-dimensional algebras. (Hint: Prove that in a two-dimensional algebra either $\exists x \neq 0: \ x^2 = 0$ or $\exists x \notin \{\mathbf{1}, -\mathbf{1}\}: \ x^2 = \mathbf{1}$.)

Proposition D.1.20. *Let $\mathcal{A}$ and $\mathcal{B}$ be algebras, and $\phi \in \mathrm{Hom}(\mathcal{A}, \mathcal{B})$. Then $\phi(\mathcal{A}) \subset \mathcal{B}$ is a subalgebra, $\mathrm{Ker}(\phi) := \{x \in \mathcal{A} \mid \phi(x) = 0\}$ is an ideal of $\mathcal{A}$, and $\mathcal{A}/\mathrm{Ker}(\phi) \cong \phi(\mathcal{A})$.*

Exercise D.1.21. Prove Proposition D.1.20.

Definition D.1.22 (Tensor product algebra). The *tensor product algebra* of a $\mathbb{K}$-vector space V is the $\mathbb{K}$-vector space

$$T := \bigoplus_{m=0}^{\infty} \otimes^m V,$$

where $\otimes^0 V := \mathbb{K}$, $\otimes^{m+1} V := (\otimes^m V) \otimes V$; the multiplication of this algebra is given by

$$(x, y) \mapsto xy := x \otimes y$$

with the identifications $W \otimes \mathbb{K} \cong W \cong \mathbb{K} \otimes W$ for a $\mathbb{K}$-vector space W, so that the unit element $\mathbf{1}_T \in T$ is the unit element $1 \in \mathbb{K}$.

D.2 Topological algebras

Definition D.2.1 (Topological algebra). A topological space $\mathcal{A}$ with the structure of an algebra is called a *topological algebra* if

1. $\{0\} \subset \mathcal{A}$ is a closed subset, and
2. the algebraic operations are continuous, i.e., the mappings

$$\begin{aligned} ((\lambda, x) \mapsto \lambda x): \quad & \mathbb{C} \times \mathcal{A} \to \mathcal{A}, \\ ((x, y) \mapsto x + y): \quad & \mathcal{A} \times \mathcal{A} \to \mathcal{A}, \\ ((x, y) \mapsto xy): \quad & \mathcal{A} \times \mathcal{A} \to \mathcal{A} \end{aligned}$$

are continuous.

Remark D.2.2. Similarly, a *topological vector space* is a topological space and a vector space, in which $\{0\}$ is a closed subset and the vector space operations $(\lambda, x) \mapsto \lambda x$ and $(x, y) \mapsto x + y$ are continuous.

Remark D.2.3. Some books omit the assumption that $\{0\}$ should be a closed set; then, e.g., any algebra $\mathcal{A}$ with a topology $\tau = \{\emptyset, \mathcal{A}\}$ would become a topological algebra. However, such generalisations are seldom useful. And it will turn out soon, that actually our topological algebras are indeed Hausdorff spaces! $\{0\}$ being a closed set puts emphasis on closed ideals and continuous homomorphisms, as we shall see later.

Example. Let us give examples of topological algebras:

1. The commutative algebra $\mathbb{C}$ endowed with its usual topology (given by the absolute value norm $x \mapsto |x|$) is a topological algebra.

2. If $(X, x \mapsto \|x\|)$ is a normed space, $X \neq \{0\}$, then $\mathcal{L}(X)$ is a topological algebra with the norm

$$A \mapsto \|A\| := \sup_{x \in X : \|x\| \leq 1} \|Ax\|.$$

   Notice that $\mathcal{L}(\mathbb{C}) \cong \mathbb{C}$, and $\mathcal{L}(X)$ is non-commutative if $\dim(X) \geq 2$.

3. Let X be a set. Then

$$\mathcal{F}_b(X) := \{f \in \mathcal{F}(X) \mid f \text{ is bounded}\}$$

   is a commutative topological algebra with the supremum norm

$$f \mapsto \|f\| := \sup_{x \in X} |f(x)|.$$

   Similarly, if X is a topological space then the algebra

$$C_b(X) := \{f \in C(X) \mid f \text{ is bounded}\}$$

   of bounded continuous functions on X is a commutative topological algebra when endowed with the supremum norm.

4. If (X, d) is a metric space then the algebra

$$\mathrm{Lip}(X) := \{f : X \to \mathbb{C} \mid f \text{ is Lipschitz continuous and bounded}\}$$

   is a commutative topological algebra with the norm

$$f \mapsto \|f\| := \max \left\{ \sup_{x \in X} |f(x)|, \ \sup_{x \neq y} \frac{|f(x) - f(y)|}{d(x, y)} \right\}.$$

5. $\mathcal{E}(\mathbb{R}) := C^\infty(\mathbb{R})$ is a commutative topological algebra with the metric

$$(f, g) \mapsto \sum_{m=1}^\infty 2^{-m} \frac{p_m(f - g)}{1 + p_m(f - g)}, \quad \text{where } p_m(f) := \max_{|x| \leq m, k \leq m} |f^{(k)}(x)|.$$

   This algebra is not normable (can not be endowed with a norm).

6. The topological dual $\mathcal{E}'(\mathbb{R})$ of $\mathcal{E}(\mathbb{R})$ is the space of compactly supported distributions (see Definition 1.4.8). There the multiplication is the convolution, which is defined for nice enough f, g by

$$(f, g) \mapsto f * g, \quad (f * g)(x) := \int_{-\infty}^{\infty} f(x - y)\, g(y)\, \mathrm{d}y.$$

The unit element of $\mathcal{E}(\mathbb{R})$ is the Dirac delta distribution δ_0 at the origin $0 \in \mathbb{R}$. This is a commutative topological algebra with the weak*-topology, but it is not metrisable.

7. Convolution algebras of compactly supported distributions on Lie groups are non-metrisable topological algebras; such an algebra is commutative if and only if the group is commutative.

Remark D.2.4. Let $\mathcal{A}$ be a topological algebra, $U \subset \mathcal{A}$ open, and $S \subset \mathcal{A}$. Due to the continuity of $((\lambda, x) \mapsto \lambda x) : \mathbb{C} \times \mathcal{A} \to \mathcal{A}$ the set $\lambda U = \{\lambda u \mid u \in U\}$ is open if $\lambda \neq 0$. Due to the continuity of $((x, y) \mapsto x + y) : \mathcal{A} \times \mathcal{A} \to \mathcal{A}$ the set $U + S = \{u + s \mid u \in U,\ s \in S\}$ is open.

Exercise D.2.5. Show that topological algebras are Hausdorff spaces.

Remark D.2.6. Notice that in the previous exercise you actually need only the continuities of the mappings $(x, y) \mapsto x + y$ and $x \mapsto -x$, and the fact that $\{0\}$ is a closed set. Indeed, the commutativity of the addition operation is not needed, so that you can actually prove a proposition "Topological groups are Hausdorff spaces"!

Exercise D.2.7. Let $\{\mathcal{A}_j \mid j \in J\}$ be a family of topological algebras. Endow $\mathcal{A} := \prod_{j \in J} \mathcal{A}_j$ with a structure of a topological algebra.

Exercise D.2.8. Let $\mathcal{A}$ be an algebra and a normed space. Prove that it is a topological algebra if and only if there exists a constant $C < \infty$ such that

$$\|xy\| \leq C \|x\| \|y\|$$

for every $x, y \in \mathcal{A}$.

Exercise D.2.9. Let $\mathcal{A}$ be an algebra. The *commutant* of a subset $S \subset \mathcal{A}$ is

$$\Gamma(S) := \{x \in \mathcal{A} \mid \forall y \in S : xy = yx\}.$$

Prove the following claims:

(a) $\Gamma(S) \subset \mathcal{A}$ is a subalgebra; $\Gamma(S)$ is closed if $\mathcal{A}$ is a topological algebra.

(b) $S \subset \Gamma(\Gamma(S))$.

(c) If $xy = yx$ for every $x, y \in S$ then $\Gamma(\Gamma(S)) \subset \mathcal{A}$ is a commutative subalgebra, where $\sigma_{\Gamma(\Gamma(S))}(z) = \sigma_{\mathcal{A}}(z)$ for every $z \in \Gamma(\Gamma(S))$.

Closed ideals

In topological algebras, the good ideals are the closed ones.

Example. Let $\mathcal{A}$ be a topological algebra; then $\{0\} \subset \mathcal{A}$ is a closed ideal. Let $\mathcal{B}$ be another topological algebra, and $\phi \in \mathrm{Hom}(\mathcal{A}, \mathcal{B})$ be **continuous**. Then it is easy to see that $\mathrm{Ker}(\phi) = \phi^{-1}(\{0\}) \subset \mathcal{A}$ is a **closed** ideal; this is actually a canonical example of closed ideals.

Proposition D.2.10. *Let $\mathcal{A}$ be a topological algebra and $\mathcal{J}$ its ideal. Then either* $\overline{\mathcal{J}} = \mathcal{A}$ *or* $\overline{\mathcal{J}} \subset \mathcal{A}$ *is a closed ideal.*

Proof. Let $\lambda \in \mathbb{C}$, $x, y \in \overline{\mathcal{J}}$, and $z \in \mathcal{A}$. Take $V \in \mathcal{V}(\lambda x)$. Then there exists $U \in \mathcal{V}(x)$ such that $\lambda U \subset V$ (due to the continuity of the multiplication by a scalar). Since $x \in \overline{\mathcal{J}}$, we may pick $x_0 \in \mathcal{J} \cap U$. Now

$$\lambda x_0 \in \mathcal{J} \cap (\lambda U) \subset \mathcal{J} \cap V,$$

which proves that $\lambda x \in \overline{\mathcal{J}}$. Next take $W \in \mathcal{V}(x + y)$. Then for some $U \in \mathcal{V} \longrightarrow(x)$ and $V \in \mathcal{V}(y)$ we have $U + V \subset W$ (due to the continuity of the mapping $(x, y) \mapsto x + y$). Since $x, y \in \overline{\mathcal{J}}$, we may pick $x_0 \in \mathcal{J} \cap U$ and $y_0 \in \mathcal{J} \cap V$. Now

$$x + y \in \mathcal{J} \cap (U + V) \subset \mathcal{J} \cap W,$$

which proves that $x + y \in \overline{\mathcal{J}}$. Finally, we should show that $xz, zx \in \overline{\mathcal{J}}$, but this proof is so similar to the previous steps that it is left for the reader as an easy task. $\qquad\square$

Definition D.2.11 (Topology for quotient algebra). Let $\mathcal{J}$ be an ideal of a topological algebra $\mathcal{A}$. Let τ be the topology of $\mathcal{A}$. For $x \in \mathcal{A}$, define $[x] = x + \mathcal{J}$, and let $[S] = \{[x] \mid x \in S\}$. Then it is easy to check that $\{[U] \mid U \in \tau\}$ is a topology of the quotient algebra $\mathcal{A}/\mathcal{J}$; it is called the *quotient topology.*

Remark D.2.12. Let $\mathcal{A}$ be a topological algebra and $\mathcal{J}$ and ideal in $\mathcal{A}$. The quotient map $(x \mapsto [x]) \in \mathrm{Hom}(\mathcal{A}, \mathcal{A}/\mathcal{J})$ is continuous: namely, if $x \in \mathcal{A}$ and $[V] \in \mathcal{V}_{\mathcal{A}/\mathcal{J}}([x])$ for some $V \in \tau$ then $U := V + \mathcal{J} \in \mathcal{V}(x)$ and $[U] = [V]$.

Lemma D.2.13. *Let $\mathcal{J}$ be an ideal of a topological algebra $\mathcal{A}$. Then the algebra operations on the quotient algebra $\mathcal{A}/\mathcal{J}$ are continuous.*

Proof. Let us check the continuity of the multiplication in the quotient algebra: Suppose $[x][y] = [xy] \in [W]$, where $W \subset \mathcal{A}$ is an open set (recall that every open set in the quotient algebra is of the form $[W]$). Then

$$xy \in W + \mathcal{J}.$$

Since $\mathcal{A}$ is a topological algebra, there are open sets $U \in \mathcal{V}_{\mathcal{A}}(x)$ and $V \in \mathcal{V}_{\mathcal{A}}(y)$ satisfying

$$UV \subset W + \mathcal{J}.$$

Now $[U] \in \mathcal{V}_{A/\mathcal{J}}([x])$ and $[V] \in \mathcal{V}_{A/\mathcal{J}}([y])$. Furthermore, $[U][V] \subset [W]$ because

$$(U + \mathcal{J})(V + \mathcal{J}) \subset UV + \mathcal{J} \subset W + \mathcal{J};$$

we have proven the continuity of the multiplication $([x], [y]) \mapsto [x][y]$. As an easy exercise, we leave it for the reader to verify the continuity of the mappings $(\lambda, [x]) \mapsto \lambda[x]$ and $([x], [y]) \mapsto [x] + [y]$.   □

Exercise D.2.14. Complete the previous proof by showing the continuity of the mappings $(\lambda, [x]) \mapsto \lambda[x]$ and $([x], [y]) \mapsto [x] + [y]$.

With Lemma D.2.13, we conclude:

Proposition D.2.15. *Let $\mathcal{J}$ be an ideal of a topological algebra A. Then $A/\mathcal{J}$ is a topological algebra if and only if $\mathcal{J}$ is closed.*

Proof. If the quotient algebra is a topological algebra then $\{[0]\} = \{\mathcal{J}\}$ is a closed subset of $A/\mathcal{J}$; since the quotient homomorphism is a continuous mapping, $\mathcal{J} = \mathrm{Ker}(x \mapsto [x]) \subset A$ must be a closed set.

Conversely, suppose $\mathcal{J}$ is a closed ideal of a topological algebra A. Then we deduce that

$$(A/\mathcal{J}) \setminus \{[0]\} = [A \setminus \mathcal{J}]$$

is an open subset of the quotient algebra, so that $\{[0]\} \subset A/\mathcal{J}$ is closed.   □

Remark D.2.16. Let X be a topological vector space and M be its subspace. The reader should be able to define the *quotient topology* for the quotient vector space $X/M = \{[x] := x + M \mid x \in X\}$. Now X/M is a topological vector space if and only if M is a closed subspace.

Let $M \subset X$ be a closed subspace. If d is a metric on X then there is a natural metric for X/M:

$$([x], [y]) \mapsto d([x], [y]) := \inf_{z \in M} d(x - y, z),$$

and if X is a complete metric space then X/M is also complete. Moreover, if $x \mapsto \|x\|$ is a norm on X then there is a natural norm for X/M:

$$[x] \mapsto \|[x]\| := \inf_{z \in M} \|x - z\|.$$

D.3   Banach algebras

Definition D.3.1 (Banach algebra). An algebra $\mathcal{A}$ is called a *Banach algebra* if it is a Banach space satisfying

$$\|xy\| \le \|x\| \, \|y\|$$

for all $x, y \in \mathcal{A}$ and

$$\|\mathbf{1}\| = 1.$$

Exercise D.3.2. Let K be a compact space. Show that $C(K)$ is a Banach algebra with the norm $f \mapsto \|f\| = \max_{x \in K} |f(x)|$.

Example. Let X be a Banach space. Then the Banach space $\mathcal{L}(X)$ of bounded linear operators $X \to X$ is a Banach algebra if the multiplication is the composition of operators, since

$$\|AB\| \leq \|A\| \, \|B\|$$

for every $A, B \in \mathcal{L}(X)$; the unit is the identity operator $I : X \to X$, $x \mapsto x$. Actually, this is not far away from characterising all the Banach algebras:

Theorem D.3.3 (Characterisation of Banach algebras). *A Banach algebra $\mathcal{A}$ is isometrically isomorphic to a norm closed subalgebra of $\mathcal{L}(X)$ for a Banach space X.*

Proof. Here $X := \mathcal{A}$. For $x \in \mathcal{A}$, let us define

$$m(x) : \mathcal{A} \to \mathcal{A} \quad \text{by} \quad m(x)y := xy.$$

Obviously $m(x)$ is a linear mapping, $m(xy) = m(x)m(y)$, $m(1_\mathcal{A}) = 1_{\mathcal{L}(\mathcal{A})}$, and

$$
\begin{aligned}
\|m(x)\| &= \sup_{y \in \mathcal{A}: \, \|y\| \leq 1} \|xy\| \\
&\leq \sup_{y \in \mathcal{A}: \, \|y\| \leq 1} (\|x\| \, \|y\|) = \|x\| = \|m(x)1_\mathcal{A}\| \\
&\leq \|m(x)\| \, \|1_\mathcal{A}\| = \|m(x)\|;
\end{aligned}
$$

briefly, $m = (x \mapsto m(x)) \in \mathrm{Hom}(\mathcal{A}, \mathcal{L}(\mathcal{A}))$ is isometric. Thereby $m(\mathcal{A}) \subset \mathcal{L}(\mathcal{A})$ is a closed subspace and hence a Banach algebra. $\qquad\square$

Proposition D.3.4. *A maximal ideal in a Banach algebra is closed.*

Proof. In a topological algebra, the closure of an ideal is either an ideal or the whole algebra. Let $\mathcal{M}$ be a maximal ideal of a Banach algebra $\mathcal{A}$. The set $G(\mathcal{A}) \subset \mathcal{A}$ of all invertible elements is open, and $\mathcal{M} \cap G(\mathcal{A}) = \emptyset$ (because no ideal contains invertible elements). Thus $\mathcal{M} \subset \overline{\mathcal{M}} \subset \mathcal{A} \setminus G(\mathcal{A})$, so that $\overline{\mathcal{M}}$ is an ideal containing a maximal ideal $\mathcal{M}$; thus $\overline{\mathcal{M}} = \mathcal{M}$. $\qquad\square$

Proposition D.3.5. *Let $\mathcal{J}$ be a closed ideal of a Banach algebra $\mathcal{A}$. Then the quotient vector space $\mathcal{A}/\mathcal{J}$ is a Banach algebra; moreover, $\mathcal{A}/\mathcal{J}$ is commutative if $\mathcal{A}$ is commutative.*

Proof. Let us denote $[x] := x + \mathcal{J}$ for $x \in \mathcal{A}$. Since $\mathcal{J}$ is a closed vector subspace, the quotient space $\mathcal{A}/\mathcal{J}$ is a Banach space with norm

$$[x] \mapsto \|[x]\| = \inf_{j \in \mathcal{J}} \|x + j\|.$$

Let $x, y \in \mathcal{A}$ and $\varepsilon > 0$. Then there exist $i, j \in \mathcal{J}$ such that

$$\|x + i\| \leq \|[x]\| + \varepsilon, \quad \|y + j\| \leq \|[y]\| + \varepsilon.$$

Now $(x+i)(y+j) \in [xy]$, so that

$$
\begin{aligned}
\|[xy]\| &\leq \|(x+i)(y+j)\| \\
&\leq \|x+i\| \, \|y+j\| \\
&\leq (\|[x]\| + \varepsilon) \, (\|[y]\| + \varepsilon) \\
&= \|[x]\| \, \|[y]\| + \varepsilon(\|[x]\| + \|[y]\| + \varepsilon);
\end{aligned}
$$

since $\varepsilon > 0$ is arbitrary, we have

$$
\|[x][y]\| \leq \|[x]\| \, \|[y]\|.
$$

Finally, $\|[\mathbf{1}]\| \leq \|\mathbf{1}\| = 1$ and $\|[x]\| = \|[x][\mathbf{1}]\| \leq \|[x]\| \, \|[\mathbf{1}]\|$, so that we have $\|[\mathbf{1}]\| = 1$. $\qquad\square$

Exercise D.3.6. Let $\mathcal{A}$ be a Banach algebra, and let $x, y \in \mathcal{A}$ satisfy

$$
x^2 = x, \quad y^2 = y, \quad xy = yx.
$$

Show that either $x = y$ or $\|x - y\| \geq 1$. Give an example of a Banach algebra $\mathcal{A}$ with elements $x, y \in \mathcal{A}$ such that $x^2 = x \neq y = y^2$ and $\|x - y\| < 1$.

Proposition D.3.7. *Let $\mathcal{A}$ be a Banach algebra. Then $\mathrm{Hom}(\mathcal{A}, \mathbb{C}) \subset \mathcal{A}'$ and $\|\phi\| = 1$ for every $\phi \in \mathrm{Hom}(\mathcal{A}, \mathbb{C})$.*

Proof. Let $x \in \mathcal{A}$, $\|x\| < 1$. Let

$$
y_n := \sum_{j=0}^{n} x^j,
$$

where $x^0 := \mathbf{1}$. If $n > m$ then

$$
\begin{aligned}
\|y_n - y_m\| &= \|x^m + x^{m+1} + \cdots + x^n\| \\
&\leq \|x\|^m \left(1 + \|x\| + \cdots + \|x\|^{n-m}\right) \\
&= \|x\|^m \frac{1 - \|x\|^{n-m+1}}{1 - \|x\|} \to_{n > m \to \infty} 0;
\end{aligned}
$$

thus $(y_n)_{n=1}^{\infty} \subset \mathcal{A}$ is a Cauchy sequence. There exists $y = \lim_{n \to \infty} y_n \in \mathcal{A}$, because $\mathcal{A}$ is complete. Since $x^n \to 0$ and

$$
y_n(\mathbf{1} - x) = \mathbf{1} - x^{n+1} = (\mathbf{1} - x)y_n,
$$

we deduce $y = (\mathbf{1} - x)^{-1}$. Suppose $\lambda = \phi(x)$, $|\lambda| > \|x\|$; now $\|\lambda^{-1}x\| = |\lambda|^{-1} \|x\| < 1$, so that $\mathbf{1} - \lambda^{-1}x$ is invertible. Then

$$
\begin{aligned}
1 &= \phi(\mathbf{1}) = \phi\left((\mathbf{1} - \lambda^{-1}x)(\mathbf{1} - \lambda^{-1}x)^{-1}\right) \\
&= \phi\left(\mathbf{1} - \lambda^{-1}x\right) \phi\left((\mathbf{1} - \lambda^{-1}x)^{-1}\right) \\
&= (1 - \lambda^{-1}\phi(x)) \, \phi\left((\mathbf{1} - \lambda^{-1}x)^{-1}\right) = 0,
\end{aligned}
$$

a contradiction; hence

$$\forall x \in \mathcal{A} : \ |\phi(x)| \leq \|x\|,$$

that is $\|\phi\| \leq 1$. Finally, $\phi(\mathbf{1}) = 1$, so that $\|\phi\| = 1$. $\qquad\square$

Lemma D.3.8. *Let $\mathcal{A}$ be a Banach algebra. The set $G(\mathcal{A}) \subset \mathcal{A}$ of its invertible elements is open. The mapping $(x \mapsto x^{-1}) : G(\mathcal{A}) \to G(\mathcal{A})$ is a homeomorphism.*

Proof. Take $x \in G(\mathcal{A})$ and $h \in \mathcal{A}$. As in the proof of the previous Proposition, we see that

$$x - h = x(\mathbf{1} - x^{-1}h)$$

is invertible if $\|x^{-1}\| \, \|h\| < 1$, that is $\|h\| < \|x^{-1}\|^{-1}$; thus $G(\mathcal{A}) \subset \mathcal{A}$ is open.

The mapping $x \mapsto x^{-1}$ is clearly its own inverse. Moreover

$$
\begin{aligned}
\|(x - h)^{-1} - x^{-1}\| &= \ \|(\mathbf{1} - x^{-1}h)^{-1}x^{-1} - x^{-1}\| \\
&\leq \ \|(\mathbf{1} - x^{-1}h)^{-1} - \mathbf{1}\| \, \|x^{-1}\| = \|\sum_{n=1}^{\infty}(x^{-1}h)^n\| \, \|x^{-1}\| \\
&\leq \ \|h\| \left(\sum_{n=1}^{\infty} \|x^{-1}\|^{n+1} \, \|h\|^{n-1} \right) \to_{h \to 0} 0;
\end{aligned}
$$

hence $x \mapsto x^{-1}$ is a homeomorphism. $\qquad\square$

Exercise D.3.9 (Topological zero divisors). Let $\mathcal{A}$ be a Banach algebra. We say that $x \in \mathcal{A}$ is a *topological zero divisor* if there exists a sequence $(y_n)_{n=1}^{\infty} \subset \mathcal{A}$ such that $\|y_n\| = 1$ for all n and

$$\lim_{n \to \infty} xy_n = 0 = \lim_{n \to \infty} y_n x.$$

(a) Show that if $(x_n)_{n=1}^{\infty} \subset G(\mathcal{A})$ satisfies $x_n \to x \in \partial G(\mathcal{A})$ then $\|x_n^{-1}\| \to \infty$.
(b) Using this result, show that the boundary points of $G(\mathcal{A})$ are topological zero divisors.
(c) In what kind of Banach algebras 0 is the only topological zero divisor?

Theorem D.3.10 (Gelfand, 1939). *Let $\mathcal{A}$ be a Banach algebra and $x \in \mathcal{A}$. Then the spectrum $\sigma(x) \subset \mathbb{C}$ is a non-empty compact set.*

Proof. Let $x \in \mathcal{A}$. Then $\sigma(x)$ belongs to a 0-centered disc of radius $\|x\|$ in the complex plane: for if $\lambda \in \mathbb{C}$, $|\lambda| > \|x\|$ then $\mathbf{1} - \lambda^{-1}x$ is invertible, equivalently $\lambda\mathbf{1} - x$ is invertible.

The mapping $g : \mathbb{C} \to \mathcal{A}$, $\lambda \mapsto \lambda\mathbf{1} - x$, is continuous; the set $G(\mathcal{A}) \subset \mathcal{A}$ of invertible elements is open, so that

$$\mathbb{C} \setminus \sigma(x) = g^{-1}(G(\mathcal{A}))$$

is open. Thus $\sigma(x) \in \mathbb{C}$ is closed and bounded, i.e., compact by the Heine–Borel Theorem (Corollary A.13.7).

The hard part is to prove the non-emptiness of the spectrum. Let us define the *resolvent mapping* $R : \mathbb{C} \setminus \sigma(x) \to G(\mathcal{A})$ by

$$R(\lambda) = (\lambda \mathbf{1} - x)^{-1}.$$

We know that this mapping is continuous, because it is composed of continuous mappings

$$(\lambda \mapsto \lambda \mathbf{1} - x) : \mathbb{C} \setminus \sigma(x) \to G(\mathcal{A}) \quad \text{and} \quad (y \mapsto y^{-1}) : G(\mathcal{A}) \to G(\mathcal{A}).$$

We want to show that R is *weakly holomorphic*, that is $f \circ R \in H(\mathbb{C} \setminus \sigma(x))$ for every $f \in \mathcal{A}' = \mathcal{L}(\mathcal{A}, \mathbb{C})$. Let $z \in \mathbb{C} \setminus \sigma(x)$, $f \in \mathcal{A}'$. Then we calculate

$$\frac{(f \circ R)(z + h) - (f \circ R)(z)}{h}$$

$$= \quad f\left(\frac{R(z + h) - R(z)}{h}\right)$$

$$= \quad f\left(\frac{R(z + h)R(z)^{-1} - \mathbf{1}}{h} R(z)\right)$$

$$= \quad f\left(\frac{R(z + h)(R(z + h)^{-1} - h\mathbf{1}) - \mathbf{1}}{h} R(z)\right)$$

$$= \quad f(-R(z + h)R(z))$$

$$\to_{h \to 0} \quad f(-R(z)^2),$$

because f and R are continuous; thus R is weakly holomorphic.

Suppose $|\lambda| > \|x\|$. Then

$$\begin{aligned}
\|R(\lambda)\| \quad &= \quad \|(\lambda \mathbf{1} - x)^{-1}\| \\
&= \quad |\lambda|^{-1} \|(1 - x/\lambda)^{-1}\| \\
&= \quad |\lambda|^{-1} \left\|\sum_{j=0}^{\infty} (x/\lambda)^j\right\| \\
&\leq \quad |\lambda|^{-1} \sum_{j=0}^{\infty} \|x/\lambda\|^{-j} \\
&= \quad |\lambda|^{-1} \frac{1}{1 - \|x/\lambda\|} \\
&= \quad \frac{1}{|\lambda| - \|x\|} \\
&\to_{|\lambda| \to \infty} \quad 0.
\end{aligned}$$

Thereby

$$(f \circ R)(\lambda) \to_{|\lambda| \to \infty} 0$$

for every $f \in \mathcal{A}'$. To get a contradiction, suppose $\sigma(x) = \emptyset$. Then $f \circ R \in H(\mathbb{C})$ is 0 by Liouville's Theorem D.6.2 for every $f \in \mathcal{A}'$; the Hahn–Banach Theorem B.4.25 says that then $R(\lambda) = 0$ for every $\lambda \in \mathbb{C}$; this is a contradiction, since $0 \notin G(\mathcal{A})$. Thus $\sigma(x) \neq \emptyset$. $\qquad\square$

Exercise D.3.11. Let $\mathcal{A}$ be a Banach algebra, $x \in \mathcal{A}$, $\Omega \subset \mathbb{C}$ an open set, and $\sigma(x) \subset \Omega$. Then

$$\exists \delta > 0 \; \forall y \in \mathcal{A} : \; \|y\| < \delta \Rightarrow \sigma(x+y) \subset \Omega.$$

Exercise D.3.12. Alternatively, in the proof of Theorem D.3.10 one could use the Neumann series

$$R(\lambda) = R(\lambda_0) \sum_{k=0}^{\infty} \left((\lambda_0 - \lambda) R(\lambda_0) \right)^k,$$

for all $\lambda_0 \in \mathbb{C} \backslash \sigma(x)$ and $|\lambda - \lambda_0| \|R(\lambda_0)\| < 1$. Then $R(\lambda)$ is analytic in $\mathbb{C} \backslash \sigma(x)$ and satisfies $R(\lambda) \to 0$ as $\lambda \to \infty$. Consequently, use Liouville's theorem (Theorem D.6.2) to conclude the statement.

Corollary D.3.13 (Gelfand–Mazur Theorem). *Let $\mathcal{A}$ be a Banach algebra where $0 \in \mathcal{A}$ is the only non-invertible element. Then $\mathcal{A}$ is isometrically isomorphic to $\mathbb{C}$.*

Proof. Take $x \in \mathcal{A}$, $x \neq 0$. Since $\sigma(x) \neq \emptyset$, pick $\lambda(x) \in \sigma(x)$. Then $\lambda(x)\mathbf{1} - x$ is non-invertible, so that it must be 0; $x = \lambda(x)\mathbf{1}$. By defining $\lambda(0) = 0$, we have an algebra isomorphism

$$\lambda : \mathcal{A} \to \mathbb{C}.$$

Moreover, $|\lambda(x)| = \|\lambda(x)\mathbf{1}\| = \|x\|$. $\qquad\square$

Exercise D.3.14. Let $\mathcal{A}$ be a Banach algebra, and suppose that there exists $C < \infty$ such that

$$\|x\| \, \|y\| \leq C \, \|xy\|$$

for every $x, y \in \mathcal{A}$. Show that $\mathcal{A} \cong \mathbb{C}$ isometrically.

Definition D.3.15 (Spectral radius). Let $\mathcal{A}$ be a Banach algebra. The *spectral radius* of $x \in \mathcal{A}$ is

$$\rho(x) := \sup_{\lambda \in \sigma(x)} |\lambda|;$$

this is well defined, because due to Gelfand's Theorem D.3.10 the spectrum is non-empty. In other words, $\overline{\mathbb{D}(0, \rho(x))} = \{\lambda \in \mathbb{C} : |\lambda| \leq \rho(x)\}$ is the smallest 0-centered closed disk containing $\sigma(x) \subset \mathbb{C}$. Notice that $\rho(x) \leq \|x\|$, since $\lambda\mathbf{1} - x = \lambda(\mathbf{1} - x/\lambda)$ is invertible if $|\lambda| > \|x\|$.

Theorem D.3.16 (Spectral Radius Formula (Beurling, 1938; Gelfand, 1939)). *Let $\mathcal{A}$ be a Banach algebra, $x \in \mathcal{A}$. Then*

$$\rho(x) = \lim_{n \to \infty} \|x^n\|^{1/n}.$$

Proof. For $x = 0$ the claim is trivial, so let us assume that $x \neq 0$. By Gelfand's Theorem D.3.10, $\sigma(x) \neq \emptyset$. Let $\lambda \in \sigma(x)$ and $n \geq 1$. Notice that in an algebra, if both ab and ba are invertible then the elements a, b are invertible. Therefore

$$\lambda^n \mathbf{1} - x^n = (\lambda \mathbf{1} - x) \left(\sum_{k=0}^{n-1} \lambda^{n-1-k} x^k \right) = \left(\sum_{k=0}^{n-1} \lambda^{n-1-k} x^k \right) (\lambda \mathbf{1} - x)$$

implies that $\lambda^n \in \sigma(x^n)$. Thus $|\lambda^n| \leq \|x^n\|$, so that

$$\rho(x) = \sup_{\lambda \in \sigma(x)} |\lambda| \leq \liminf_{n \to \infty} \|x^n\|^{1/n}.$$

Let $f \in \mathcal{A}'$ and $\lambda \in \mathbb{C}$, $|\lambda| > \|x\|$. Then

$$\begin{aligned} f(R(\lambda)) &= f\left((\lambda \mathbf{1} - x)^{-1} \right) = f\left(\lambda^{-1} (1 - \lambda^{-1} x)^{-1} \right) \\ &= f\left(\lambda^{-1} \sum_{n=0}^{\infty} \lambda^{-n} x^n \right) \\ &= \lambda^{-1} \sum_{n=0}^{\infty} f(\lambda^{-n} x^n). \end{aligned}$$

This formula is true also when $|\lambda| > \rho(x)$, because $f \circ R$ is holomorphic in $\mathbb{C} \setminus \sigma(x) \supset \mathbb{C} \setminus \overline{\mathbb{D}(0, \rho(x))}$. Hence if we define $T_{\lambda,x,n} \in \mathcal{A}'' = \mathcal{L}(\mathcal{A}', \mathbb{C})$ by $T_{\lambda,x,n}(f) := f(\lambda^{-n} x^n)$, we obtain

$$\sup_{n \in \mathbb{N}} |T_{\lambda,x,n}(f)| = \sup_{n \in \mathbb{N}} |f(\lambda^{-n} x^n)| < \infty \qquad \text{(when } |\lambda| > \rho(x))$$

for every $f \in \mathcal{A}'$; the Banach–Steinhaus Theorem B.4.29 applied on the family $\{T_{\lambda,x,n}\}_{n \in \mathbb{N}}$ shows that

$$M_{\lambda,x} := \sup_{n \in \mathbb{N}} \|T_{\lambda,x,n}\| < \infty,$$

so that we have

$$\begin{aligned} \|\lambda^{-n} x^n\| \;\overset{\text{Hahn–Banach}}{=}\; & \sup_{f \in \mathcal{A}' : \|f\| \leq 1} |f(\lambda^{-n} x^n)| \\ = \; & \sup_{f \in \mathcal{A}' : \|f\| \leq 1} |T_{\lambda,x,n}(f)| \\ = \; & \|T_{\lambda,x,n}\| \\ \leq \; & M_{\lambda,x}. \end{aligned}$$

Hence

$$\|x^n\|^{1/n} \leq M_{\lambda,x}^{1/n} |\lambda| \to_{n \to \infty} |\lambda|,$$

when $|\lambda| > \rho(x)$. Thus

$$\limsup_{n \to \infty} \|x^n\|^{1/n} \leq \rho(x);$$

collecting the results, the Spectral Radius Formula is verified. $\qquad \square$

Remark D.3.17. The Spectral Radius Formula contains startling information: the spectral radius $\rho(x)$ is purely an algebraic property (though related to a topological algebra), but the quantity $\lim \|x^n\|^{1/n}$ relies on both algebraic and metric properties! Yet the results are equal!

Remark D.3.18. $\rho(x)^{-1}$ is the radius of convergence of the $\mathcal{A}$-valued power series

$$\lambda \mapsto \sum_{n=0}^{\infty} \lambda^n x^n.$$

Remark D.3.19. Let $\mathcal{A}$ be a Banach algebra and $\mathcal{B}$ a Banach subalgebra. If $x \in \mathcal{B}$ then

$$\sigma_{\mathcal{A}}(x) \subset \sigma_{\mathcal{B}}(x)$$

and the inclusion can be proper, but the spectral radii for both Banach algebras are the same, since

$$\rho_{\mathcal{A}}(x) = \lim_{n \to \infty} \|x^n\|^{1/n} = \rho_{\mathcal{B}}(x).$$

Exercise D.3.20. Let $\mathcal{A}$ be a Banach algebra, $x, y \in \mathcal{A}$. Show that $\rho(xy) = \rho(yx)$. Show that if $x \in \mathcal{A}$ is *nilpotent* (i.e., $x^k = 0$ for some $k \in \mathbb{N}$) then $\sigma(x) = \{0\}$.

Exercise D.3.21. Let $\mathcal{A}$ be a Banach algebra and $x, y \in \mathcal{A}$ such that $xy = yx$. Prove that $\rho(xy) \leq \rho(x)\rho(y)$.

Exercise D.3.22. In the proof of Theorem D.3.16 argue as follows. For $\lambda > \|x\|$ note that the resolvent satisfies

$$R(\lambda) = \lambda^{-1} \sum_{k=0}^{\infty} (\lambda^{-1} x)^k,$$

and this Laurent series converges for all $|\lambda| > \rho(x)$. Consequently, its (Hadamard) radius of convergence satisfies

$$\rho(x) \leq \liminf_{n \to \infty} \|x^n\|^{1/n}.$$

At the same time, the convergence for $|\lambda| > \rho(x)$ implies $\lim_{n \to \infty} \lambda^{-n} x^n = 0$, which means that $\|x^n\| \leq |\lambda|^n$ for large enough n.

D.4 Commutative Banach algebras

In this section we are interested in maximal ideals of commutative Banach algebras. We shall learn that such algebras are closely related to algebras of continuous functions on compact Hausdorff spaces: there is a natural, far from trivial, homomorphism from a commutative Banach algebra $\mathcal{A}$ to an algebra of functions on the set $\mathrm{Hom}(\mathcal{A}, \mathbb{C})$, which can be endowed with a canonical topology – related mathematics is called the **Gelfand theory**. In the sequel, one should ponder this dilemma: which is more fundamental, a space or algebras of functions on it?

Example. Let us give examples of commutative Banach algebras:

1. Our familiar $C(K)$, when K is a compact space.

2. $L^\infty([0,1])$, when $[0,1]$ is endowed with the Lebesgue measure.

3. $A(\Omega) := C(\overline{\Omega}) \cap H(\Omega)$, when $\Omega \subset \mathbb{C}$ is open, $H(\Omega)$ are holomorphic functions in Ω, and $\overline{\Omega} \subset \mathbb{C}$ is compact.

4. $M(\mathbb{R}^n)$, the convolution algebra of complex Borel measures on $\mathbb{R}^n$, with the Dirac delta distribution at $0 \in \mathbb{R}^n$ as the unit element, and endowed with the total variation norm.

5. The algebra of matrices $\begin{pmatrix} \alpha & \beta \\ 0 & \alpha \end{pmatrix}$, where $\alpha, \beta \in \mathbb{C}$; notice that this algebra contains nilpotent elements!

Definition D.4.1 (Spectrum and characters of an algebra). The *spectrum of an algebra* $\mathcal{A}$ is

$$\mathrm{Spec}(\mathcal{A}) := \mathrm{Hom}(\mathcal{A}, \mathbb{C}),$$

i.e., the set of homomorphisms $\mathcal{A} \to \mathbb{C}$; such a homomorphism is called a *character of* $\mathcal{A}$.

Remark D.4.2. The concept of spectrum is best suited for **commutative** algebras, as $\mathbb{C}$ is a commutative algebra; here a character $\mathcal{A} \to \mathbb{C}$ should actually be considered as an algebra representation $\mathcal{A} \to \mathcal{L}(\mathbb{C})$. In order to fully capture the structure of a **non-commutative** algebra, we should study representations of type $\mathcal{A} \to \mathcal{L}(X)$, where the vector spaces X are multi-dimensional; for instance, if $\mathcal{H}$ is a Hilbert space of dimension 2 or greater then $\mathrm{Spec}(\mathcal{L}(\mathcal{H})) = \emptyset$. However, the spectrum of a commutative Banach algebra is rich, as there is a bijective correspondence between characters and maximal ideals. Moreover, the spectrum of the algebra is akin to the spectra of its elements:

Theorem D.4.3 (Gelfand, 1940). *Let $\mathcal{A}$ be a commutative Banach algebra. Then:*

(a) *Every maximal ideal of $\mathcal{A}$ is of the form $\mathrm{Ker}(h)$ for some $h \in \mathrm{Spec}(\mathcal{A})$;*

(b) *$\mathrm{Ker}(h)$ is a maximal ideal for every $h \in \mathrm{Spec}(\mathcal{A})$;*

(c) *$x \in \mathcal{A}$ is invertible if and only if $\forall h \in \mathrm{Spec}(\mathcal{A}): h(x) \neq 0$;*

(d) *$x \in \mathcal{A}$ is invertible if and only if it is not in any ideal of $\mathcal{A}$;*

(e) *$\sigma(x) = \{h(x) \mid h \in \mathrm{Spec}(\mathcal{A})\}$.*

Proof. (a) Let $\mathcal{M} \subset \mathcal{A}$ be a maximal ideal; let $[x] := x + \mathcal{M}$ for $x \in \mathcal{A}$. Since $\mathcal{A}$ is commutative and $\mathcal{M}$ is maximal, every non-zero element in the quotient algebra $\mathcal{A}/\mathcal{M}$ is invertible. We know that $\mathcal{M}$ is closed, so that $\mathcal{A}/\mathcal{M}$ is a Banach algebra. Due to the Gelfand–Mazur Theorem (Corollary D.3.13), there exists an isometric isomorphism $\lambda \in \mathrm{Hom}(\mathcal{A}/\mathcal{M}, \mathbb{C})$. Then

$$h = (x \mapsto \lambda([x])) : \mathcal{A} \to \mathbb{C}$$

is a character, and

$$\mathrm{Ker}(h) = \mathrm{Ker}((x \mapsto [x]) : \mathcal{A} \to \mathcal{A}/\mathcal{M}) = \mathcal{M}.$$

(b) Let $h : \mathcal{A} \to \mathbb{C}$ be a character. Now h is a linear mapping, so that the co-dimension of $\mathrm{Ker}(h)$ in $\mathcal{A}$ equals the dimension of $h(\mathcal{A}) \subset \mathbb{C}$, which clearly is 1. Any ideal of co-dimension 1 in an algebra must be maximal, so that $\mathrm{Ker}(h)$ is maximal.

(c) If $x \in \mathcal{A}$ is invertible and $h \in \mathrm{Spec}(\mathcal{A})$ then $h(x) \in \mathbb{C}$ is invertible, that is $h(x) \neq 0$. For the converse, assume that $x \in \mathcal{A}$ is non-invertible. Then

$$\mathcal{A}x = \{ax \mid a \in \mathcal{A}\}$$

is an ideal of $\mathcal{A}$ (notice that $\mathbf{1} = ax = xa$ would mean that $a = x^{-1}$). Hence by Krull's Theorem D.1.13, there is a maximal ideal $\mathcal{M} \subset \mathcal{A}$ such that $\mathcal{A}x \subset \mathcal{M}$. Then (a) provides a character $h \in \mathrm{Spec}(\mathcal{A})$ for which $\mathrm{Ker}(h) = \mathcal{M}$. Especially, $h(x) = 0$.

(d) This follows from (a,b,c) directly.

(e) (c) is equivalent to

"$x \in \mathcal{A}$ is non-invertible if and only if $\exists h \in \mathrm{Spec}(\mathcal{A}) : h(x) = 0$",

which is equivalent to

"$\lambda \mathbf{1} - x$ is non-invertible if and only if $\exists h \in \mathrm{Spec}(\mathcal{A}) : h(x) = \lambda$". $\qquad \square$

Exercise D.4.4. Let $\mathcal{A}$ be a Banach algebra and $x, y \in \mathcal{A}$ such that $xy = yx$. Prove that $\sigma(x + y) \subset \sigma(x) + \sigma(y)$ and $\sigma(xy) \subset \sigma(x)\sigma(y)$.

Exercise D.4.5. Let $\mathcal{A}$ be the algebra of those functions $f : \mathbb{R} \to \mathbb{C}$ for which

$$f(x) = \sum_{n \in \mathbb{Z}} f_n \, e^{ix \cdot n}, \quad \|f\| = \sum_{n \in \mathbb{Z}} |f_n| < \infty.$$

Show that $\mathcal{A}$ is a commutative Banach algebra. Show that if $f \in \mathcal{A}$ and $\forall x \in \mathbb{R} : f(x) \neq 0$ then $1/f \in \mathcal{A}$.

Definition D.4.6 (Gelfand transform). Let $\mathcal{A}$ be a commutative Banach algebra. The *Gelfand transform* $\widehat{x}$ *of an element* $x \in \mathcal{A}$ is the function

$$\widehat{x} : \mathrm{Spec}(\mathcal{A}) \to \mathbb{C}, \quad \widehat{x}(\phi) := \phi(x).$$

Let $\widehat{\mathcal{A}} := \{\widehat{x} : \mathrm{Spec}(\mathcal{A}) \to \mathbb{C} \mid x \in \mathcal{A}\}$. The mapping

$$\mathcal{A} \to \widehat{\mathcal{A}}, \quad x \mapsto \widehat{x},$$

is called the *Gelfand transform of* $\mathcal{A}$. We endow the set $\mathrm{Spec}(\mathcal{A})$ with the $\widehat{\mathcal{A}}$-induced topology, called the *Gelfand topology*; this topological space is called the *maximal ideal space of* $\mathcal{A}$ (for a good reason, in the light of the previous theorem). In other words, the Gelfand topology is the weakest topology on $\mathrm{Spec}(\mathcal{A})$ making every $\widehat{x}$ a continuous function, i.e., the weakest topology on $\mathrm{Spec}(\mathcal{A})$ for which $\widehat{\mathcal{A}} \subset C(\mathrm{Spec}(\mathcal{A}))$.

Theorem D.4.7 (Gelfand, 1940). *Let $\mathcal{A}$ be a commutative Banach algebra. Then $K = \mathrm{Spec}(\mathcal{A})$ is a compact Hausdorff space in the Gelfand topology, the Gelfand transform is a continuous homomorphism $\mathcal{A} \to C(K)$, and $\|\widehat{x}\| = \sup\limits_{\phi \in K} |\widehat{x}(\phi)| = \rho(x)$ for every $x \in \mathcal{A}$.*

Proof. The Gelfand transform is a homomorphism, since

$$\widehat{\lambda x}(\phi) = \phi(\lambda x) = \lambda\phi(x) = \lambda\widehat{x}(\phi) = (\lambda\widehat{x})(\phi),$$
$$\widehat{x+y}(\phi) = \phi(x+y) = \phi(x) + \phi(y) = \widehat{x}(\phi) + \widehat{y}(\phi) = (\widehat{x}+\widehat{y})(\phi),$$
$$\widehat{xy}(\phi) = \phi(xy) = \phi(x)\phi(y) = \widehat{x}(\phi)\widehat{y}(\phi) = (\widehat{x}\widehat{y})(\phi),$$
$$\widehat{1_{\mathcal{A}}}(\phi) = \phi(1_{\mathcal{A}}) = 1 = 1_{C(K)}(\phi),$$

for every $\lambda \in \mathbb{C}$, $x, y \in \mathcal{A}$ and $\phi \in K$. Moreover,

$$\widehat{x}(K) = \{\widehat{x}(\phi) \mid \phi \in K\} = \{\phi(x) \mid \phi \in \mathrm{Spec}(\mathcal{A})\} = \sigma(x),$$

implying

$$\|\widehat{x}\| = \rho(x) \leq \|x\|.$$

Clearly K is a Hausdorff space. What about compactness?

Now $K = \mathrm{Hom}(\mathcal{A}, \mathbb{C})$ is a subset of the closed unit ball of the dual Banach space $\mathcal{A}'$; by the Banach–Alaoglu Theorem B.4.36, this unit ball is compact in the weak*-topology. Recall that the weak*-topology $\tau_{\mathcal{A}'}$ of $\mathcal{A}'$ is the $\mathcal{A}$-induced topology, with the interpretation $\mathcal{A} \subset \mathcal{A}''$; thus the Gelfand topology τ_K is the relative weak*-topology, i.e.,

$$\tau_K = \tau_{\mathcal{A}'}|_K.$$

To prove that τ_K is compact, it is sufficient to show that $K \subset \mathcal{A}'$ is closed in the weak*-topology.

Let $f \in \mathcal{A}'$ be in the weak*-closure of K. We have to prove that $f \in K$, i.e.,

$$f(xy) = f(x)f(y) \quad \text{and} \quad f(1) = 1.$$

Let $x, y \in \mathcal{A}$, $\varepsilon > 0$. Let $S := \{1, x, y, xy\}$. Using the notation of the proof of the Banach–Alaoglu Theorem B.4.36,

$$U(f, S, \varepsilon) = \{\psi \in \mathcal{A}' : z \in S \Rightarrow |\psi z - fz| < \varepsilon\}$$

is a weak*-neighbourhood of f. Thus choose $h_\varepsilon \in K \cap U(f, S, \varepsilon)$. Then

$$|1 - f(1)| = |h_\varepsilon(1) - f(1)| < \varepsilon;$$

$\varepsilon > 0$ being arbitrary, we have $f(\mathbf{1}) = 1$. Noticing that $|h_\varepsilon(x)| \le \|x\|$, we get

$$
\begin{aligned}
&|f(xy) - f(x)f(y)| \\
\le\ & |f(xy) - h_\varepsilon(xy)| + |h_\varepsilon(xy) - h_\varepsilon(x)f(y)| + |h_\varepsilon(x)f(y) - f(x)f(y)| \\
=\ & |f(xy) - h_\varepsilon(xy)| + |h_\varepsilon(x)| \cdot |h_\varepsilon(y) - f(y)| + |h_\varepsilon(x) - f(x)| \cdot |f(y)| \\
\le\ & \varepsilon \left(1 + \|x\| + |f(y)|\right).
\end{aligned}
$$

This holds for every $\varepsilon > 0$, so that actually

$$
f(xy) = f(x)f(y);
$$

we have proven that f is a homomorphism, $f \in K$. $\qquad\qquad\square$

Exercise D.4.8 (Radicals). Let $\mathcal{A}$ be a commutative Banach algebra. Its *radical* $\mathrm{Rad}(\mathcal{A})$ is the intersection of all the maximal ideals of $\mathcal{A}$. Show that

$$
\mathrm{Rad}(\mathcal{A}) = \mathrm{Ker}(x \mapsto \widehat{x}) = \{x \in \mathcal{A} \mid \rho(x) = 0\},
$$

where $x \mapsto \widehat{x}$ is the Gelfand transform. Show that nilpotent elements of $\mathcal{A}$ belong to the radical.

Exercise D.4.9. Let X be a finite set. Describe the Gelfand transform of $\mathcal{F}(X)$.

Exercise D.4.10. Describe the Gelfand transform of the algebra of matrices $\left(\begin{smallmatrix} \alpha & \beta \\ 0 & \alpha \end{smallmatrix}\right)$, where $\alpha, \beta \in \mathbb{C}$.

Theorem D.4.11 (When is $\mathrm{Spec}(C(X))$ homeomorphic to X?). *Let X be a compact Hausdorff space. Then $\mathrm{Spec}(C(X))$ is homeomorphic to X.*

Proof. For $x \in X$, let us define the function

$$
h_x : C(X) \to \mathbb{C}, \quad f \mapsto f(x) \quad \text{(evaluation at } x \in X).
$$

This is clearly a homomorphism, and hence we may define the mapping

$$
\phi : X \to \mathrm{Spec}(C(X)), \quad x \mapsto h_x.
$$

Let us prove that ϕ is a homeomorphism.

If $x, y \in X$, $x \ne y$, then Urysohn's Lemma (Theorem A.12.11) provides $f \in C(X)$ such that $f(x) \ne f(y)$. Thereby $h_x(f) \ne h_y(f)$, yielding $\phi(x) = h_x \ne h_y = \phi(y)$; thus ϕ is injective. It is also surjective: namely, let us **assume** that $h \in \mathrm{Spec}(C(X)) \setminus \phi(X)$. Now $\mathrm{Ker}(h) \subset C(X)$ is a maximal ideal, and for every $x \in X$ we may choose

$$
f_x \in \mathrm{Ker}(h) \setminus \mathrm{Ker}(h_x) \subset C(X).
$$

Then $U_x := f_x^{-1}(\mathbb{C} \setminus \{0\}) \in \mathcal{V}(x)$, so that

$$
\mathcal{U} = \{U_x \mid x \in X\}
$$

is an open cover of X, which due to the compactness has a finite subcover $\{U_{x_j}\}_{j=1}^n \subset \mathcal{U}$. Since $f_{x_j} \in \mathrm{Ker}(h)$, the function

$$f := \sum_{j=1}^n |f_{x_j}|^2 = \sum_{j=1}^n f_{x_j} \overline{f_{x_j}}$$

belongs to $\mathrm{Ker}(h)$. Clearly $f(x) \neq 0$ for every $x \in X$. Therefore $g \in C(X)$ with $g(x) = 1/f(x)$ is the inverse element of f; this is a **contradiction**, since no invertible element belongs to an ideal. Thus ϕ must be surjective.

We have proven that $\phi : X \to \mathrm{Spec}(C(X))$ is a bijection. Thereby X and $\mathrm{Spec}(C(X))$ can be identified as sets. The Gelfand-topology of $\mathrm{Spec}(C(X))$ is then identified with the $C(X)$-induced topology σ of X, which is weaker than the original topology τ of X. Hence $\phi : (X, \tau) \to \mathrm{Spec}(C(X))$ is continuous. Actually, $\sigma = \tau$, because a continuous bijection from a compact space to a Hausdorff space is a homeomorphism, see Proposition A.12.7. $\qquad\square$

Corollary D.4.12. *Let X and Y be compact Hausdorff spaces. Then the Banach algebras $C(X)$ and $C(Y)$ are isomorphic if and only if X is homeomorphic to Y.*

Proof. By Theorem D.4.11, $X \cong \mathrm{Spec}(C(X))$ and $Y \cong \mathrm{Spec}(C(Y))$. If $C(X)$ and $C(Y)$ are isomorphic Banach algebras then

$$X \cong \mathrm{Spec}(C(X)) \overset{C(X) \cong C(Y)}{\cong} \mathrm{Spec}(C(Y)) \cong Y.$$

Conversely, a homeomorphism $\phi : X \to Y$ begets a Banach algebra isomorphism

$$\Phi : C(Y) \to C(X), \quad (\Phi f)(x) := f(\phi(x)),$$

as the reader easily verifies. $\qquad\square$

Exercise D.4.13. Let K be a compact Hausdorff space, $\emptyset \neq S \subset K$, and $\mathcal{J} \subset C(K)$ be an ideal. Let us define

$$\mathcal{I}(S) := \{f \in C(K) \mid \forall x \in S : f(x) = 0\},$$
$$V(\mathcal{J}) := \{x \in K \mid \forall f \in \mathcal{J} : f(x) = 0\}.$$

Prove that

(a) $\mathcal{I}(S) \subset C(K)$ a closed ideal,

(b) $V(\mathcal{J}) \subset K$ is a closed non-empty subset,

(c) $V(\mathcal{I}(S)) = \overline{S}$ (hint: Urysohn), and

(d) $\mathcal{I}(V(\mathcal{J})) = \overline{\mathcal{J}}$.

Lesson to be learned: topology of K goes hand in hand with the (closed) ideal structure of $C(K)$.

D.5 C*-algebras

Now we are finally in the position to abstractly characterise algebras $C(X)$ among Banach algebras: according to Gelfand and Naimark, the category of compact Hausdorff spaces is equivalent to the category of commutative C*-algebras. The class of C*-algebras behaves nicely, and the related functional analysis adequately deserves the name "non-commutative topology".

Definition D.5.1 (Involutive algebra). An algebra $\mathcal{A}$ is a *-algebra ("star-algebra" or an *involutive algebra*) if there is a mapping $(x \mapsto x^*) : \mathcal{A} \to \mathcal{A}$ satisfying

$$(\lambda x)^* = \bar{\lambda} x^*, \quad (x + y)^* = x^* + y^*, \quad (xy)^* = y^* x^*, \quad (x^*)^* = x,$$

for all $x, y \in \mathcal{A}$ and $\lambda \in \mathbb{C}$; such a mapping is called an *involution*. In other words, an involution is a conjugate-linear anti-multiplicative self-invertible mapping $\mathcal{A} \to \mathcal{A}$.

A *-*homomorphism* $\phi : \mathcal{A} \to \mathcal{B}$ between involutive algebras $\mathcal{A}$ and $\mathcal{B}$ is an algebra homomorphism satisfying

$$\phi(x^*) = \phi(x)^*$$

for every $x \in \mathcal{A}$. The set of all *-homomorphisms between *-algebras $\mathcal{A}$ and $\mathcal{B}$ is denoted by $\mathrm{Hom}^*(\mathcal{A}, \mathcal{B})$.

Definition D.5.2 (C*-algebra). A *C*-algebra* $\mathcal{A}$ is an involutive Banach algebra such that

$$\|x^* x\| = \|x\|^2$$

for every $x \in \mathcal{A}$.

Example. Let us consider some involutive algebras:

1. The Banach algebra $\mathbb{C}$ is a C*-algebra with the involution $\lambda \mapsto \lambda^* = \bar{\lambda}$, i.e., the complex conjugation.

2. If K is a compact space then $C(K)$ is a commutative C*-algebra with the involution $f \mapsto f^*$ by complex conjugation, $f^*(x) := \overline{f(x)}$.

3. $L^\infty([0, 1])$ is a C*-algebra, when the involution is as above.

4. $A(\mathbb{D}(0, 1)) = C\left(\overline{\mathbb{D}(0, 1)}\right) \cap H(\mathbb{D}(0, 1))$ is an involutive Banach algebra with $f^*(z) := \overline{f(\bar{z})}$, but it is not a C*-algebra. Here $H(\mathbb{D}(0, 1))$ are functions holomorphic in the unit disc.

5. The radical of a commutative C*-algebra is always the trivial ideal $\{0\}$, and thus 0 is the only nilpotent element. Hence for instance the algebra of matrices $\begin{pmatrix} \alpha & \beta \\ 0 & \alpha \end{pmatrix}$ (where $\alpha, \beta \in \mathbb{C}$) cannot be a C*-algebra.

6. If $\mathcal{H}$ is a Hilbert space then $\mathcal{L}(\mathcal{H})$ is a C*-algebra when the involution is the usual adjunction $A \mapsto A^*$, and clearly any norm-closed involutive subalgebra of $\mathcal{L}(\mathcal{H})$ is also a C*-algebra. Actually, there are no others, but in the sequel we shall not prove the related Gelfand–Naimark Theorem D.5.3:

Theorem D.5.3 (Gelfand–Naimark Theorem (1943)). *If $\mathcal{A}$ is a C*-algebra then there exists a Hilbert space $\mathcal{H}$ and an isometric *-homomorphism onto a closed involutive subalgebra of $\mathcal{L}(\mathcal{H})$.*

However, we shall characterise the commutative case: the Gelfand transform of a commutative C*-algebra $\mathcal{A}$ will turn out to be an isometric isomorphism $\mathcal{A} \to C(\mathrm{Spec}(\mathcal{A}))$, so that $\mathcal{A}$ "is" the function algebra $C(K)$ for the compact Hausdorff space $K = \mathrm{Spec}(\mathcal{A})$! Before going into this, we prove some related results.

Proposition D.5.4. *Let $\mathcal{A}$ be a *-algebra. Then $\mathbf{1}^* = \mathbf{1}$, $x \in \mathcal{A}$ is invertible if and only if $x^* \in \mathcal{A}$ is invertible, and $\sigma(x^*) = \overline{\sigma(x)} := \{\bar{\lambda} \mid \lambda \in \sigma(x)\}$.*

Proof. First,
$$\mathbf{1}^* = \mathbf{1}^*\mathbf{1} = \mathbf{1}^*(\mathbf{1}^*)^* = (\mathbf{1}^*\mathbf{1})^* = (\mathbf{1}^*)^* = \mathbf{1};$$
second,
$$(x^{-1})^* x^* = (x x^{-1})^* = \mathbf{1}^* = \mathbf{1} = \mathbf{1}^* = (x^{-1}x)^* = x^*(x^{-1})^*;$$
third,
$$\bar{\lambda}\mathbf{1} - x^* = (\lambda\mathbf{1}^*)^* - x^* = (\lambda\mathbf{1})^* - x^* = (\lambda\mathbf{1} - x)^*,$$
which concludes the proof. $\square$

Proposition D.5.5. *Let $\mathcal{A}$ be a C*-algebra, and $x = x^* \in \mathcal{A}$. Then $\sigma(x) \subset \mathbb{R}$.*

Proof. **Assume** that $\lambda \in \sigma(x) \setminus \mathbb{R}$, i.e., $\lambda = \lambda_1 + i\lambda_2$ for some $\lambda_j \in \mathbb{R}$ with $\lambda_2 \neq 0$. Hence we may define $y := (x - \lambda_1\mathbf{1})/\lambda_2 \in \mathcal{A}$. Now $y^* = y$. Moreover, $i \in \sigma(y)$, because
$$i\mathbf{1} - y = \frac{\lambda\mathbf{1} - x}{\lambda_2}.$$
Take $t \in \mathbb{R}$. Then $t + 1 \in \sigma(t\mathbf{1} - iy)$, because
$$(t+1)\mathbf{1} - (t\mathbf{1} - iy) = -i(i\mathbf{1} - y).$$
Thereby
$$
\begin{aligned}
(t+1)^2 \quad &\leq \quad \rho(t\mathbf{1} - iy)^2 \\
&\leq \quad \|t\mathbf{1} - iy\|^2 \\
&\overset{\mathrm{C}^*}{=} \quad \|(t\mathbf{1} - iy)^*(t\mathbf{1} - iy)\| \\
&\overset{t\in\mathbb{R},\ y^*=y}{=} \quad \|(t\mathbf{1} + iy)(t\mathbf{1} - iy)\| = \|t^2\mathbf{1} + y^2\| \\
&\leq \quad t^2 + \|y\|,
\end{aligned}
$$
so that $2t + 1 \leq \|y\|$ for every $t \in \mathbb{R}$; a **contradiction**. $\square$

Corollary D.5.6. *Let $\mathcal{A}$ be a C^*-algebra, $\phi : \mathcal{A} \to \mathbb{C}$ a homomorphism, and $x \in \mathcal{A}$. Then $\phi(x^*) = \overline{\phi(x)}$, i.e., ϕ is a $*$-homomorphism.*

Proof. Define the "real part" and the "imaginary part" of x by

$$u := \frac{x + x^*}{2}, \quad v := \frac{x - x^*}{2i}.$$

Then $x = u + iv$, $u^* = u$, $v^* = v$, and $x^* = u - iv$. Since a homomorphism maps invertibles to invertibles, we have $\phi(u) \in \sigma(u)$; we know that $\sigma(u) \subset \mathbb{R}$, because $u^* = u$. Similarly we obtain $\phi(v) \in \mathbb{R}$. Thereby

$$\phi(x^*) = \phi(u - iv) = \phi(u) - i\phi(v) = \overline{\phi(u) + i\phi(v)} = \overline{\phi(u + iv)} = \overline{\phi(x)};$$

this means that $\mathrm{Hom}^*(\mathcal{A}, \mathbb{C}) = \mathrm{Hom}(\mathcal{A}, \mathbb{C})$. □

Exercise D.5.7. Let $\mathcal{A}$ be a Banach algebra, $\mathcal{B}$ a closed subalgebra, and $x \in \mathcal{B}$. Prove the following facts:

(a) $G(\mathcal{B})$ is open and closed in $G(\mathcal{A}) \cap \mathcal{B}$.
(b) $\sigma_{\mathcal{A}}(x) \subset \sigma_{\mathcal{B}}(x)$ and $\partial \sigma_{\mathcal{B}}(x) \subset \partial \sigma_{\mathcal{A}}(x)$.
(c) If $\mathbb{C} \setminus \sigma_{\mathcal{A}}(x)$ is connected then $\sigma_{\mathcal{A}}(x) = \sigma_{\mathcal{B}}(x)$.

Using the results of the exercise above, the reader can prove the following important fact on the invariance of the spectrum in C^*-algebras:

Exercise D.5.8. Let $\mathcal{A}$ be a C^*-algebra and $\mathcal{B}$ a C^*-subalgebra. Show that $\sigma_{\mathcal{B}}(x) = \sigma_{\mathcal{A}}(x)$ for every $x \in \mathcal{B}$.

Lemma D.5.9. *Let $\mathcal{A}$ be a C^*-algebra. Then $\|x\|^2 = \rho(x^*x)$ for every $x \in \mathcal{A}$.*

Proof. Now

$$\|(x^*x)^2\| = \|(x^*x)(x^*x)\| = \|(x^*x)^*(x^*x)\| \stackrel{C^*}{=} \|x^*x\|^2,$$

so that by induction

$$\|(x^*x)^{2^n}\| = \|x^*x\|^{2^n}$$

for every $n \in \mathbb{N}$. Therefore applying the Spectral Radius Formula, we get

$$\rho(x^*x) = \lim_{n \to \infty} \|(x^*x)^{2^n}\|^{1/2^n} = \lim_{n \to \infty} \|x^*x\|^{2^n/2^n} = \|x^*x\|,$$

the result we wanted. □

Exercise D.5.10. Let $\mathcal{A}$ be a C^*-algebra. Show that there can be at most one C^*-algebra norm on an involutive Banach algebra. Moreover, prove that if $\mathcal{A}, \mathcal{B}$ are C^*-algebras then $\phi \in \mathrm{Hom}^*(\mathcal{A}, \mathcal{B})$ is continuous and has norm $\|\phi\| = 1$.

Theorem D.5.11 (Commutative Gelfand–Naimark). *Let $\mathcal{A}$ be a commutative C^*-algebra. Then the Gelfand transform $(x \mapsto \widehat{x}) : \mathcal{A} \to C(\mathrm{Spec}(\mathcal{A}))$ is an isometric $*$-isomorphism.*

Proof. Let $K = \mathrm{Spec}(\mathcal{A})$. We already know that the Gelfand transform is a Banach algebra homomorphism $\mathcal{A} \to C(K)$. Let $x \in \mathcal{A}$ and $\phi \in K$. Since ϕ is actually a $*$-homomorphism, we get

$$\widehat{x^*}(\phi) = \phi(x^*) = \overline{\phi(x)} = \overline{\widehat{x}(\phi)} = \widehat{x}^*(\phi);$$

the Gelfand transform is a $*$-homomorphism.

Now we have proven that $\widehat{\mathcal{A}} \subset C(K)$ is an involutive subalgebra separating the points of K. Stone–Weierstrass Theorem A.14.4 thus says that $\widehat{\mathcal{A}}$ is dense in $C(K)$. If we can show that the Gelfand transform $\mathcal{A} \to \widehat{\mathcal{A}}$ is an isometry then we must have $\widehat{\mathcal{A}} = C(K)$: Take $x \in \mathcal{A}$. Then

$$\|\widehat{x}\|^2 = \|\widehat{x}^*\widehat{x}\| = \|\widehat{x^*x}\| \overset{\mathrm{Gelfand}}{=} \rho(x^*x) \overset{\mathrm{Lemma}}{=} \|x\|^2,$$

i.e., $\|\widehat{x}\| = \|x\|$. $\square$

Exercise D.5.12. Show that an injective $*$-homomorphism between C*-algebras is an isometry. (Hint: Gelfand transform.)

Exercise D.5.13. A linear functional f on a C*-algebra $\mathcal{A}$ is called *positive* if $f(x^*x) \geq 0$ for every $x \in \mathcal{A}$. Show that the positive functionals separate the points of $\mathcal{A}$.

Exercise D.5.14. Prove that the involution of a C*-algebra cannot be altered without destroying the C*-property $\|x^*x\| = \|x\|^2$.

Definition D.5.15 (Normal element). An element x of a C*-algebra is called *normal* if $x^*x = xx^*$.

We use the commutative Gelfand–Naimark Theorem to create the so-called continuous functional calculus at a normal element – a non-commutative C*-algebra admits some commutative studies:

Theorem D.5.16 (Functional calculus at the normal element). *Let $\mathcal{A}$ be a C*-algebra, and $x \in \mathcal{A}$ be a normal element. Let $\iota = (\lambda \mapsto \lambda) : \sigma(x) \to \mathbb{C}$. Then there exists a unique isometric $*$-homomorphism $\phi : C(\sigma(x)) \to \mathcal{A}$ such that $\phi(\iota) = x$ and $\phi(C(\sigma(x)))$ is the C*-algebra generated by x, i.e., the smallest C*-algebra containing $\{x\}$.*

Proof. Let $\mathcal{B}$ be the C*-algebra generated by x. Since x is normal, $\mathcal{B}$ is commutative. Let $\mathrm{Gel} = (y \mapsto \widehat{y}) : \mathcal{B} \to C(\mathrm{Spec}(\mathcal{B}))$ be the Gelfand transform of $\mathcal{B}$. The reader may easily verify that

$$\widehat{x} : \mathrm{Spec}(\mathcal{B}) \to \sigma(x)$$

is a continuous bijection from a compact space to a Hausdorff space; hence it is a homeomorphism. Let us define the mapping

$$C_{\widehat{x}} : C(\sigma(x)) \to C(\mathrm{Spec}(\mathcal{B})), \quad (C_{\widehat{x}}f)(h) := f(\widehat{x}(h)) = f(h(x));$$

$C_{\widehat{x}}$ can be thought as a "transpose" of $\widehat{x}$. Let us define

$$\phi = \mathrm{Gel}^{-1} \circ C_{\widehat{x}} : C(\sigma(x)) \to \mathcal{B} \subset \mathcal{A}.$$

Then $\phi : C(\sigma(x)) \to \mathcal{A}$ is obviously an isometric $*$-homomorphism. Furthermore,

$$\phi(\iota) = \mathrm{Gel}^{-1}(C_{\widehat{x}}(\iota)) = \mathrm{Gel}^{-1}(\widehat{x}) = \mathrm{Gel}^{-1}(\mathrm{Gel}(x)) = x.$$

Due to the Stone–Weierstrass Theorem A.14.4, the $*$-algebra generated by $\iota \in C(\sigma(x))$ is dense in $C(\sigma(x))$; since the $*$-homomorphism ϕ maps the generator ι to the generator x, the uniqueness of ϕ follows $\qquad\square$

Remark D.5.17. The $*$-homomorphism $\phi : C(\sigma(x)) \to \mathcal{A}$ above is called the *(continuous) functional calculus at the normal element* $\phi(\iota) = x \in \mathcal{A}$. If $p = (z \mapsto \sum_{j=1}^{n} a_j z^j) : \mathbb{C} \to \mathbb{C}$ is a polynomial then it is natural to define $p(x) := \sum_{j=1}^{n} a_j x^j$. Then actually

$$p(x) = \phi(p);$$

hence it is natural to define $f(x) := \phi(f)$ for every $f \in C(\sigma(x))$. It is easy to check that if $f \in C(\sigma(x))$ and $h \in \mathrm{Spec}(\mathcal{B})$ then $f(h(x)) = h(f(x))$.

Exercise D.5.18. Let $\mathcal{A}$ be a C*-algebra, $x \in \mathcal{A}$ normal, $f \in C(\sigma(x))$, and $g \in C(f(\sigma(x)))$. Show that $\sigma(f(x)) = f(\sigma(x))$ and that $(g \circ f)(x) = g(f(x))$.

D.6 Appendix: Liouville's Theorem

Here we prove Liouville's Theorem D.6.2 from complex analysis which was used in the proof of Gelfand's Theorem D.3.10.

Definition D.6.1 (Holomorphic function). Let $\Omega \subset \mathbb{C}$ be open. A function $f : \Omega \to \mathbb{C}$ is called *holomorphic in* Ω, denoted by $f \in \mathbb{H}(\Omega)$, if the limit

$$f'(z) := \lim_{h \to 0} \frac{f(z+h) - f(z)}{h}$$

exists for every $z \in \Omega$. Then Cauchy's integral formula provides a power series representation

$$f(z) = \sum_{n=0}^{\infty} c_n (z-a)^n$$

converging uniformly on the compact subsets of the disk

$$\mathbb{D}(a, r) = \{z \in \mathbb{C} : |z - a| < r\} \subset \Omega;$$

here $c_n = f^{(n)}(a)/n!$, where $f^{(0)} = f$ and $f^{(n+1)} = f^{(n)\prime}$.

Theorem D.6.2 (Liouville's Theorem). *Let* $f \in \mathbb{H}(\mathbb{C})$ *such that* $|f|$ *is bounded. Then* f *is constant, i.e.,* $f(z) \equiv f(0)$ *for every* $z \in \mathbb{C}$.

Proof. Since $f \in \mathbb{H}(\mathbb{C})$, we have a power series representation

$$f(z) = \sum_{n=0}^{\infty} c_n z^n$$

converging uniformly on the compact sets in the complex plane. Thereby

$$\frac{1}{2\pi} \int_0^{2\pi} |f(re^{i\phi})|^2 \, d\phi \;=\; \frac{1}{2\pi} \int_0^{2\pi} \sum_{n,m} c_n \, \overline{c_m} \, r^{n+m} \, e^{i(n-m)\phi} \, d\phi$$

$$=\; \sum_{n,m} c_n \, \overline{c_m} \, r^{n+m} \, \frac{1}{2\pi} \int_0^{2\pi} e^{i(n-m)\phi} \, d\phi$$

$$=\; \sum_{n=0}^{\infty} |c_n|^2 r^{2n}$$

for every $r > 0$. Hence the fact

$$\sum_{n=0}^{\infty} |c_n|^2 r^{2n} = \frac{1}{2\pi} \int_0^{2\pi} |f(re^{i\phi})|^2 \, d\phi \leq \sup_{z \in \mathbb{C}} |f(z)|^2 < \infty$$

implies $c_n = 0$ for every $n \geq 1$; thus $f(z) \equiv c_0 = f(0)$.   □

A more general Liouville's theorem for harmonic functions will be given in Theorem 2.6.14, with a proof relying on the Fourier analysis instead.

Part II

Commutative Symmetries

In Part II we present the theory of pseudo-differential operators on commutative groups.

The first commutative case is the Euclidean space $\mathbb{R}^n$ where the theory of pseudo-differential operators is developed most and many things may be considered well-known, so here we only review basics which are useful to contrast it with constructions on other spaces. We start by introducing elements of Fourier analysis in Chapter 1, trying to make an independent exposition of the theory, reducing references to general measure theory to a minimum. In Chapter 2 we develop the most important elements of the theory of pseudo-differential operators on $\mathbb{R}^n$. There we do not aim at developing a comprehensive treatment since there are several excellent monographs already available. Instead, we focus in Chapter 4 on aspects of the theory that have analogues on the torus, and on more general (compact) Lie groups in Part IV. From this point of view Chapters 1 and 2 can be regarded as an introduction to the subject and that is why we have taken special care to accommodate a possibly less experienced reader there.

The second commutative case is the case of the torus $\mathbb{T}^n = \mathbb{R}^n/\mathbb{Z}^n$ considered in Chapter 4. On one hand, pseudo-differential operators on $\mathbb{T}^n$ can be viewed as a special case of periodic pseudo-differential operators on $\mathbb{R}^n$, with all the consequences. However, in this way one may lose many important features of the underlying torus. On the other hand, carrying out the analysis in the intrinsic language of the underlying space is usually a more natural point of view that also has chances of extension to other Lie groups that are not so intimately related to the Euclidean space.

Here the literature on the general theory of periodic pseudo-differential operators in the "toroidal language" is rather non-existent and only a few results seem to be available. This fact is quite surprising given that Fourier analysis on $\mathbb{T}^n$ is nothing else but the periodic Fourier transform on $\mathbb{R}^n$ and, as such, constitutes a starting point of applications of Fourier analysis to numerous problems in applied mathematics and engineering. In particular, such applications (and especially real life or computational applications) do often rely on the toroidal language of the Fourier coefficients and the Fourier series rather than on the Euclidean language of the Fourier transform.

Since every connected commutative Lie group G can be identified with the product $G \cong \mathbb{T}^n \times \mathbb{R}^m$, the combination of these two settings essentially exhausts all compact commutative Lie groups. Indeed, every compact (disconnected) commutative group is isomorphic to the product of a torus and a finite commutative group, so that being connected is not really a restriction and thus it is sufficient to study the case of the torus again.

In Chapter 5 we discuss commutator characterisations of pseudo-differential operators on $\mathbb{R}^n$ and $\mathbb{T}^n$, as well as on closed manifolds which becomes useful in the sequel. In particular, Section 5.2 contains a short introduction to pseudo-differential operators on manifolds.

Chapter 1

Fourier Analysis on $\mathbb{R}^n$

In this chapter we review basic elements of Fourier analysis on $\mathbb{R}^n$. Consequently, we introduce spaces of distributions, putting emphasis on the space of tempered distributions $\mathcal{S}'(\mathbb{R}^n)$. Finally, we discuss Sobolev spaces and approximation of functions and distributions by smooth functions. Throughout, we fix the measure on $\mathbb{R}^n$ to be Lebesgue measure. For convenience, we may repeat a few definitions in the context of $\mathbb{R}^n$ although they may have already appeared in Chapter C on measure theory. From this point of view, the present chapter can be read essentially independently. The notation used in this chapter and also in Chapter 2 is $\langle \xi \rangle = (1 + |\xi|^2)^{1/2}$ where $|\xi| = (\xi_1^2 + \cdots + \xi_n^2)^{1/2}$, $\xi \in \mathbb{R}^n$.

1.1 Basic properties of the Fourier transform

Let $\Omega \subset \mathbb{R}^n$ be a measurable subset of $\mathbb{R}^n$. For simplicity, we may always think of Ω being open or closed in $\mathbb{R}^n$. In this section we will mostly have $\Omega = \mathbb{R}^n$.

Definition 1.1.1 (L^p-spaces)**.** Let $1 \leq p < \infty$. A function $f : \Omega \to \mathbb{C}$ is said to be in $L^p(\Omega)$ if it is measurable and its norm

$$||f||_{L^p(\Omega)} := \left(\int_\Omega |f(x)|^p \; dx \right)^{1/p}$$

is finite. In the case $p = \infty$, f is said to be in $L^\infty(\Omega)$ if it is measurable and essentially bounded, i.e., if

$$||f||_{L^\infty(\Omega)} := \operatorname{esssup}_{x \in \Omega} |f(x)| < \infty.$$

Here $\operatorname{esssup}_{x \in \Omega} |f(x)|$ is defined as the smallest M such that $|f(x)| \leq M$ for almost all $x \in \Omega$.

In particular, $L^1(\Omega)$ is the space of absolutely integrable functions on Ω with $||f||_{L^1(\Omega)} = \int_\Omega |f(x)| \; dx$. We will often abbreviate the $||f||_{L^p(\Omega)}$ norm by $||f||_{L^p}$, or by $||f||_p$, if the choice of Ω is clear from the context.

We note that it is customary to abuse the notation slightly by talking about functions in $L^p(\mathbb{R}^n)$ while in reality elements in $L^p(\mathbb{R}^n)$ are equivalence classes of functions that are equal almost everywhere. However, this is a minor technical issue, see Definition C.4.6 for details.

Definition 1.1.2 (Fourier transform in $L^1(\mathbb{R}^n)$). For $f \in L^1(\mathbb{R}^n)$ we define its Fourier transform by

$$(\mathcal{F}_{\mathbb{R}^n} f)(\xi) = (\mathcal{F}f)(\xi) = \widehat{f}(\xi) := \int_{\mathbb{R}^n} e^{-2\pi i x \cdot \xi} f(x) \, \mathrm{d}x.$$

Remark 1.1.3. Other similar definitions are often encountered in the literature. For example, one can use $e^{-i x \cdot \xi}$ instead of $e^{-2\pi i x \cdot \xi}$, multiply the integral by the constant $(2\pi)^{-n/2}$, etc. Changes in definitions may lead to changes in constants in formulae. It may also seem that our notation for the Fourier transform is a bit excessive. However, $\widehat{f}$ is a useful shorthand notation, while $\mathcal{F}_{\mathbb{R}^n} f$ is useful in the sequel when we want to explicitly distinguish it from the Fourier transform $\mathcal{F}_{\mathbb{T}^n} f$ for functions on the torus $\mathbb{T}^n$ considered in Chapters 3 and 4. However, in this chapter as well as in Chapter 2 we may omit the subscript and write simply $\mathcal{F}$ since there should be no confusion.

It is easy to check that $\mathcal{F} : L^1(\mathbb{R}^n) \to L^\infty(\mathbb{R}^n)$ is a bounded linear operator with norm one:

$$||\widehat{f}||_\infty \le ||f||_1. \tag{1.1}$$

Moreover, if $f \in L^1(\mathbb{R}^n)$, its Fourier transform $\widehat{f}$ is continuous, which follows from Lebesgue's dominated convergence theorem. For Lebesgue's dominated convergence theorem on general measure spaces we refer to Theorem C.3.22, but for completeness, we also state it here in a form useful to us:

Theorem 1.1.4 (Lebesgue's dominated convergence theorem). *Let $(f_k)_{k=1}^\infty$ be a sequence of measurable functions on Ω such that $f_k \to f$ pointwise almost everywhere on Ω as $k \to \infty$. Suppose there is an integrable function $g \in L^1(\Omega)$ such that $|f_k| \le g$ for all k. Then f is integrable and $\int_\Omega f \, \mathrm{d}x = \lim_{k \to \infty} \int_\Omega f_k \, \mathrm{d}x$.*

Exercise 1.1.5. Prove that if $f \in L^1(\mathbb{R}^n)$ then $\widehat{f}$ is continuous everywhere.

Exercise 1.1.6. Let $u, f \in L^1(\mathbb{R}^n)$ satisfy $\mathcal{L}u = f$, where $\mathcal{L} = \frac{\partial^2}{\partial x_1^2} + \cdots + \frac{\partial^2}{\partial x_n^2}$ is the Laplace operator. Prove that $\int_{\mathbb{R}^n} f(x) \, \mathrm{d}x = 0$.

Exercise 1.1.7. Let $u, f \in L^1(\mathbb{R}^n)$ satisfy $(1 - \mathcal{L})u = f$. Suppose that f satisfies

$$|\widehat{f}(\xi)| \le \frac{C}{(1 + |\xi|)^{n-1}}, \quad \text{for all } \xi \in \mathbb{R}^n.$$

Prove that u is a bounded continuous function on $\mathbb{R}^n$.

It is quite difficult to characterise the image of the space $L^1(\mathbb{R}^n)$ under the Fourier transform. But we have the following theorem:

Theorem 1.1.8 (Riemann–Lebesgue lemma). *Let $f \in L^1(\mathbb{R}^n)$. Then its Fourier transform $\hat{f}$ is a continuous function on $\mathbb{R}^n$ vanishing at infinity, i.e., $\hat{f}(\xi) \to 0$ as $\xi \to \infty$.*

Proof. It is enough to make an explicit calculation for f being a characteristic function of a cube and then use a standard limiting argument. Thus, let f be a characteristic function of the unit cube, i.e., $f(x) = 1$ for $x \in [-1, 1]^n$ and $f(x) = 0$ otherwise. Then

$$
\begin{aligned}
\hat{f}(\xi) &= \int_{[-1,1]^n} e^{-2\pi i x \cdot \xi} \, dx \\
&= \prod_{j=1}^{n} \int_{-1}^{1} e^{-2\pi i x_j \xi_j} \, dx_j \\
&= \prod_{j=1}^{n} \frac{1}{-2\pi i \xi_j} e^{-2\pi i x_j \xi_j} \Big|_{-1}^{1} \\
&= \left(\frac{i}{2\pi} \right)^n \frac{1}{\xi_1 \cdots \xi_n} \prod_{j=1}^{n} \left[e^{-2\pi i \xi_j} - e^{2\pi i \xi_j} \right] \\
&= \prod_{j=1}^{n} \frac{\sin(2\pi \xi_j)}{\pi \xi_j}.
\end{aligned}
$$

The product of exponents is bounded, so the whole expression tends to zero as $\xi \to \infty$ away from the coordinate axis. In case some of the ξ_j's are zero, an obvious modification of this argument yields the same result. $\qquad \square$

Exercise 1.1.9. Complete the proof of Theorem 1.1.8 in the case when some of the ξ_j's are zero.

Definition 1.1.10 (Multi-index notation). When working in $\mathbb{R}^n$, the following notation is extremely useful. For multi-indices $\alpha = (\alpha_1, \ldots, \alpha_n)$ and $\beta = (\beta_1, \ldots, \beta_n)$ with integer entries $\alpha_j, \beta_j \geq 0$, we define

$$
\partial^\alpha \varphi(x) = \frac{\partial^{\alpha_1}}{\partial x_1^{\alpha_1}} \cdots \frac{\partial^{\alpha_n}}{\partial x_n^{\alpha_n}} \varphi(x)
$$

and $x^\beta = x_1^{\beta_1} \cdots x_n^{\beta_n}$. For such multi-indices we will write $\alpha, \beta \geq 0$. For multi-indices α and β, $\alpha \leq \beta$ means $\alpha_j \leq \beta_j$ for all $j \in \{1, \ldots, n\}$. The length of the multi-index α will be denoted by $|\alpha| = \alpha_1 + \cdots + \alpha_n$, and $\alpha! = \alpha_1! \cdots \alpha_n!$.

Space $L^1(\mathbb{R}^n)$ has its limitations for the Fourier analysis because its elements may be quite irregular. The following space is an excellent alternative because its elements are smooth and have strong decay properties, thus allowing us not to worry about the convergence of integrals as well as allowing the use of analytic techniques such as integration by parts, etc.

Definition 1.1.11 (Schwartz space $\mathcal{S}(\mathbb{R}^n)$). We define the *Schwartz space* $\mathcal{S}(\mathbb{R}^n)$ of rapidly decreasing functions as follows. We say that $\varphi \in \mathcal{S}(\mathbb{R}^n)$ if φ is smooth on $\mathbb{R}^n$ and if

$$\sup_{x \in \mathbb{R}^n} \left| x^\beta \partial^\alpha \varphi(x) \right| < \infty$$

for all multi-indices $\alpha, \beta \geq 0$.

Exercise 1.1.12. Show that a smooth function f is in the Schwartz space if and only if for all $\alpha \geq 0$ and $N \geq 0$ there is a constant $C_{\alpha,N}$ such that $|\partial^\alpha \varphi(x)| \leq C_{\alpha,N}(1+|x|)^{-N}$ for all $x \in \mathbb{R}^n$.

The space $\mathcal{S}(\mathbb{R}^n)$ is a topological space. Let us now introduce the *convergence of functions in $\mathcal{S}(\mathbb{R}^n)$*.

Definition 1.1.13 (Convergence in $\mathcal{S}(\mathbb{R}^n)$). We will say that $\varphi_j \to \varphi$ in $\mathcal{S}(\mathbb{R}^n)$ as $j \to \infty$, if $\varphi_j, \varphi \in \mathcal{S}(\mathbb{R}^n)$ and if

$$\sup_{x \in \mathbb{R}^n} |x^\beta \partial^\alpha (\varphi_j - \varphi)(x)| \to 0 \text{ as } j \to \infty, \tag{1.2}$$

for all multi-indices $\alpha, \beta \geq 0$.

Remark 1.1.14. The Schwartz space $\mathcal{S}(\mathbb{R}^n)$ contains C^∞-smooth functions on $\mathbb{R}^n$ that decay rapidly at infinity, i.e.,

$$\mathcal{S}(\mathbb{R}^n) := \left\{ f \in C^\infty(\mathbb{R}^n) : \ p_{\alpha\beta}(\varphi) := \sup_{x \in \mathbb{R}^n} |x^\beta \partial^\alpha \varphi(x)| < \infty \ (\alpha, \beta \in \mathbb{N}_0^n) \right\}.$$

If one is familiar with functional analysis, one can take the expressions $p_{\alpha\beta}(\varphi)$ as seminorms on the space $\mathcal{S}(\mathbb{R}^n)$, see Definition B.4.1. This collection turns $\mathcal{S}(\mathbb{R}^n)$ into a locally convex linear topological space. Moreover, it is also a Fréchet space with the natural topology induced by the seminorms $p_{\alpha\beta}$ (see Exercise B.3.9), and it is a nuclear Montel space (see Exercises B.3.37 and B.3.51).

Definition 1.1.15 (Useful notation D^α). Since the definition of the Fourier transform contains the complex exponential, it is often convenient to use the notation $D_j = \frac{1}{2\pi i}\frac{\partial}{\partial x_j}$ and $D^\alpha = D_1^{\alpha_1} \cdots D_n^{\alpha_n}$. If D_j is applied to a function of ξ it will obviously mean $\frac{1}{2\pi i}\frac{\partial}{\partial \xi_j}$. However, there should be no confusion with this convention. If we want to additionally emphasize the variable for differentiation, we will write D_x^α or D_ξ^α.

The following theorem relates multiplication with differentiation, with respect to the Fourier transform.

Theorem 1.1.16. *Let $\varphi \in \mathcal{S}(\mathbb{R}^n)$. Then $\widehat{D_j\varphi}(\xi) = \xi_j \widehat{\varphi}(\xi)$ and $\widehat{x_j\varphi}(\xi) = -D_j\widehat{\varphi}(\xi)$.*

Proof. From the definition of the Fourier transform we readily see that

$$D_j\widehat{\varphi}(\xi) = \int_{\mathbb{R}^n} e^{-2\pi i x \cdot \xi}(-x_j)\varphi(x) \, dx.$$

This gives the second formula. Since the integrals converge absolutely, we can integrate by parts with respect to x_j in the following expression to get

$$\xi_j \widehat{\varphi}(\xi) = \int_{\mathbb{R}^n} \left(-D_j \, e^{-2\pi i x \cdot \xi} \right) \varphi(x) \, dx = \int_{\mathbb{R}^n} e^{-2\pi i x \cdot \xi} D_j \varphi(x) \, dx.$$

This implies the first formula. Note that we do not get boundary terms when integrating by parts because function φ vanishes at infinity. □

Remark 1.1.17. This theorem allows one to tackle some differential equations already. For example, let us look at the equation $\mathcal{L}u = f$ with the Laplace operator $\mathcal{L} = \frac{\partial^2}{\partial x_1^2} + \cdots + \frac{\partial^2}{\partial x_n^2}$. Taking the Fourier transform and using the theorem we arrive at the equation $-4\pi^2 |\xi|^2 \widehat{u} = \widehat{f}$. If we knew how to invert the Fourier transform we could find the solution $u = -\mathcal{F}^{-1} \left(\frac{1}{4\pi^2 |\xi|^2} \widehat{f} \right)$.

Corollary 1.1.18. *Let $\varphi \in \mathcal{S}(\mathbb{R}^n)$. Then*

$$\xi^\beta D_\xi^\alpha \widehat{\varphi}(\xi) = \int_{\mathbb{R}^n} e^{-2\pi i x \cdot \xi} D_x^\beta ((-x)^\alpha \varphi(x)) \, dx.$$

Hence also

$$\sup_{\xi \in \mathbb{R}^n} \left| \xi^\beta D_\xi^\alpha \widehat{\varphi}(\xi) \right| \leq C \sup_{x \in \mathbb{R}^n} \left((1 + |x|)^{n+1} \left| D_x^\beta (x^\alpha \varphi(x)) \right| \right),$$

with $C = \int_{\mathbb{R}^n} (1 + |x|)^{-n-1} \, dx < \infty$.

Here we used the following useful criterion:

Exercise 1.1.19 (Integrability criterion). Show that we have $\int_{\mathbb{R}^n} \frac{dx}{(1+|x|)^\rho} < \infty$ if and only if $\rho > n$. We also have $\int_{|x| \leq 1} \frac{dx}{|x|^\rho} < \infty$ if and only if $\rho < n$. Both of these criteria can be easily checked by passing to polar coordinates.

Remark 1.1.20 (**Fourier transform in $\mathcal{S}(\mathbb{R}^n)$**). Corollary 1.1.18 implies that the Fourier transform $\mathcal{F}$ maps $\mathcal{S}(\mathbb{R}^n)$ to itself. In fact, later we will show that much more is true. Let us note for now that Corollary 1.1.18 together with Lebesgue's dominated convergence theorem imply that the Fourier transform $\mathcal{F} : \mathcal{S}(\mathbb{R}^n) \to \mathcal{S}(\mathbb{R}^n)$ is continuous, i.e., $\varphi_j \to \varphi$ in $\mathcal{S}(\mathbb{R}^n)$ implies $\widehat{\varphi_j} \to \widehat{\varphi}$ in $\mathcal{S}(\mathbb{R}^n)$.

Theorem 1.1.21 (Fourier inversion formula). *The Fourier transform $\mathcal{F} : \varphi \mapsto \widehat{\varphi}$ is an isomorphism of $\mathcal{S}(\mathbb{R}^n)$ into $\mathcal{S}(\mathbb{R}^n)$, whose inverse is given by*

$$\varphi(x) = \int_{\mathbb{R}^n} e^{2\pi i x \cdot \xi} \, \widehat{\varphi}(\xi) \, d\xi. \tag{1.3}$$

This formula is called the Fourier inversion formula and the inverse Fourier transform is denoted by

$$(\mathcal{F}_{\mathbb{R}^n}^{-1} f)(x) \equiv (\mathcal{F}^{-1} f)(x) := \int_{\mathbb{R}^n} e^{2\pi i x \cdot \xi} \, f(\xi) \, d\xi.$$

Thus, we can say that

$$\mathcal{F} \circ \mathcal{F}^{-1} = \mathcal{F}^{-1} \circ \mathcal{F} = identity \quad on \quad \mathcal{S}(\mathbb{R}^n).$$

The proof of this theorem will rely on several lemmas which have a signifi-cance on their own.

Lemma 1.1.22 (Multiplication formula for the Fourier transform). *Let* $f, g \in L^1(\mathbb{R}^n)$. *Then* $\int_{\mathbb{R}^n} \widehat{f} g \, dx = \int_{\mathbb{R}^n} f \widehat{g} \, dx$.

Proof. We will apply Fubini's theorem. Thus,

$$
\int_{\mathbb{R}^n} \widehat{f} g \, dx = \int_{\mathbb{R}^n} \left[\int_{\mathbb{R}^n} e^{-2\pi i x \cdot y} f(y) \, dy \right] g(x) \, dx
$$

$$
= \int_{\mathbb{R}^n} \left[\int_{\mathbb{R}^n} e^{-2\pi i x \cdot y} g(x) \, dx \right] f(y) \, dy
$$

$$
= \int_{\mathbb{R}^n} \widehat{g} f \, dy,
$$

proving the lemma.   □

Lemma 1.1.23 (Fourier transform of Gaussian distributions). *We have the equality*

$$
\int_{\mathbb{R}^n} e^{-2\pi i x \cdot \xi} \, e^{-\epsilon \pi^2 |x|^2} \, dx = (\pi \epsilon)^{-n/2} \, e^{-|\xi|^2/\epsilon},
$$

for every $\epsilon > 0$. *By the change of* $2\pi x \to x$ *and* $\epsilon \to 2\epsilon$ *it is equivalent to*

$$
\int_{\mathbb{R}^n} e^{-i x \cdot \xi} \, e^{-\epsilon |x|^2/2} \, dx = \left(\frac{2\pi}{\epsilon} \right)^{n/2} e^{-|\xi|^2/(2\epsilon)}. \tag{1.4}
$$

Proof. We will use the standard identities

$$
\int_{-\infty}^{\infty} e^{-t^2/2} \, dt = \sqrt{2\pi} \quad \text{and} \quad \int_{\mathbb{R}^n} e^{-|x|^2/2} \, dx = (2\pi)^{n/2}. \tag{1.5}
$$

In fact, (1.4) will follow from the one-dimensional case, when we have

$$
\int_{-\infty}^{\infty} e^{-it\tau} \, e^{-t^2/2} \, dt = e^{-\tau^2/2} \int_{-\infty}^{\infty} e^{-(t+i\tau)^2/2} \, dt
$$

$$
= e^{-\tau^2/2} \int_{-\infty}^{\infty} e^{-t^2/2} \, dt
$$

$$
= \sqrt{2\pi} \, e^{-\tau^2/2},
$$

where we used the Cauchy theorem about changing the contour of integration for analytic functions and formula (1.5). Changing $t \to \sqrt{\epsilon} t$ and $\tau \to \tau/\sqrt{\epsilon}$ gives

$$
\sqrt{\epsilon} \int_{-\infty}^{\infty} e^{-it\tau} \, e^{-\epsilon t^2/2} \, dt = \sqrt{2\pi} \, e^{-\tau^2/(2\epsilon)}.
$$

Extending this to n dimensions yields (1.4).   □

Proof of Theorem 1.1.21. For $\varphi \in \mathcal{S}(\mathbb{R}^n)$, we want to prove (1.3), i.e., that

$$\varphi(x) = \int_{\mathbb{R}^n} e^{2\pi i x \cdot \xi} \, \widehat{\varphi}(\xi) \, d\xi.$$

By the Lebesgue dominated convergence theorem (Theorem 1.1.4) we can replace the right-hand side of this formula by

$$\int_{\mathbb{R}^n} e^{2\pi i x \cdot \xi} \widehat{\varphi}(\xi) \, d\xi$$

$$= \lim_{\epsilon \to 0} \int_{\mathbb{R}^n} e^{2\pi i x \cdot \xi} \widehat{\varphi}(\xi) \, e^{-2\epsilon \pi^2 |\xi|^2} \, d\xi$$

$$= \lim_{\epsilon \to 0} \int_{\mathbb{R}^n} \int_{\mathbb{R}^n} e^{2\pi i (x-y) \cdot \xi} \varphi(y) \, e^{-2\epsilon \pi^2 |\xi|^2} \, dy \, d\xi \quad \text{(change } y \to y + x)$$

$$= \lim_{\epsilon \to 0} \int_{\mathbb{R}^n} \int_{\mathbb{R}^n} e^{-2\pi i y \cdot \xi} \varphi(y + x) \, e^{-2\epsilon \pi^2 |\xi|^2} \, dy \, d\xi \quad \text{(Fubini's theorem)}$$

$$= \lim_{\epsilon \to 0} \int_{\mathbb{R}^n} \varphi(y + x) \, dy \left(\int_{\mathbb{R}^n} e^{-2\pi i y \cdot \xi} \, e^{-2\epsilon \pi^2 |\xi|^2} \, d\xi \right) \quad \text{(F.T. of Gaussian)}$$

$$= \lim_{\epsilon \to 0} \int_{\mathbb{R}^n} \varphi(y + x)(2\pi \epsilon)^{-n/2} \, e^{-|y|^2/(2\epsilon)} \, dy \quad \text{(change } y = \sqrt{\epsilon} z)$$

$$= (2\pi)^{-n/2} \lim_{\epsilon \to 0} \int_{\mathbb{R}^n} \varphi(\sqrt{\epsilon} z + x) \, e^{-|z|^2/2} \, dz \quad \text{(use (1.5))}$$

$$= \varphi(x).$$

This finishes the proof. □

Remark 1.1.24. In fact, in the proof we implicitly established the following useful relation between Fourier transforms and translations of functions. Let $h \in \mathbb{R}^n$ and define $(\tau_h f)(x) = f(x - h)$. Then we also see that

$$(\widehat{\tau_h f})(\xi) = \int_{\mathbb{R}^n} e^{-2\pi i x \cdot \xi} (\tau_h f)(x) \, dx$$

$$= \int_{\mathbb{R}^n} e^{-2\pi i x \cdot \xi} f(x - h) \, dx \quad \text{(change } y = x - h)$$

$$= \int_{\mathbb{R}^n} e^{-2\pi i (y+h) \cdot \xi} f(y) \, dy$$

$$= e^{-2\pi i h \cdot \xi} \widehat{f}(\xi).$$

Exercise 1.1.25 (Fourier transform and linear transformations). Let $f \in L^1(\mathbb{R}^n)$. Show that if $A \in \mathbb{R}^{n \times n}$ satisfies $\det A \neq 0$, and $B = (A^T)^{-1}$, then

$$\widehat{f \circ A} = |\det A|^{-1} \widehat{f} \circ B.$$

In particular, conclude that the *Fourier transform commutes with rotations*: if A is orthogonal (i.e., $A^T = A^* = A^{-1}$, so that A defines a rotation of $\mathbb{R}^n$), then

$\widehat{f \circ A} = \widehat{f} \circ A$. Consequently, conclude also that the *Fourier transform of a radial function is radial*: if $f(x) = h_1(|x|)$ for some h_1, then $\widehat{f}(\xi) = h_2(|\xi|)$ for some h_2.

Definition 1.1.26 (Convolutions). For functions $f, g \in L^1(\mathbb{R}^n)$, we define their *convolution* by

$$(f * g)(x) := \int_{\mathbb{R}^n} f(x - y)\, g(y)\, dy. \tag{1.6}$$

It is easy to see that $f * g \in L^1(\mathbb{R}^n)$ with norm inequality

$$\|f * g\|_{L^1(\mathbb{R}^n)} \le \|f\|_{L^1(\mathbb{R}^n)} \|g\|_{L^1(\mathbb{R}^n)} \tag{1.7}$$

and that $f * g = g * f$. Also, in particular for $f, g \in \mathcal{S}(\mathbb{R}^n)$, integrals are absolutely convergent and we can differentiate under the integral sign to get

$$\partial^\alpha (f * g) = \partial^\alpha f * g = f * \partial^\alpha g. \tag{1.8}$$

Remark 1.1.27. We can note that a more rigorous way of defining the convolution would be first defining (1.6) for $f, g \in \mathcal{S}(\mathbb{R}^n)$ and then extending it to a mapping $* : L^1(\mathbb{R}^n) \times L^1(\mathbb{R}^n) \to L^1(\mathbb{R}^n)$ by (1.7) avoiding the convergence question of the integral in (1.6) for functions from $L^1(\mathbb{R}^n)$.

Exercise 1.1.28. Prove the commutativity of convolution: if $f, g \in L^1(\mathbb{R}^n)$, then $f * g = g * f$. If $f, g, \in \mathcal{S}(\mathbb{R}^n)$, prove formula (1.8).

Exercise 1.1.29. Prove the associativity of convolution: if $f, g, h \in L^1(\mathbb{R}^n)$, prove that

$$(f * g) * h = f * (g * h).$$

The following properties relate convolutions with Fourier transforms.

Theorem 1.1.30. *Let* $\varphi, \psi \in \mathcal{S}(\mathbb{R}^n)$. *Then we have*

(i) $\int_{\mathbb{R}^n} \varphi\, \overline{\psi}\, dx = \int_{\mathbb{R}^n} \widehat{\varphi}\, \overline{\widehat{\psi}}\, d\xi$;

(ii) $\widehat{\varphi * \psi}(\xi) = \widehat{\varphi}(\xi)\widehat{\psi}(\xi)$;

(iii) $\widehat{\varphi\, \psi}(\xi) = (\widehat{\varphi} * \widehat{\psi})(\xi)$.

Proof. (i) Let us denote

$$\chi(\xi) = \overline{\widehat{\psi}}(\xi) = \int_{\mathbb{R}^n} e^{2\pi i x \cdot \xi} \overline{\psi}(x)\, dx = \mathcal{F}^{-1}(\overline{\psi})(\xi),$$

so that $\widehat{\chi} = \overline{\psi}$. It follows now that

$$\int_{\mathbb{R}^n} \varphi\overline{\psi} = \int_{\mathbb{R}^n} \varphi\widehat{\chi} = \int_{\mathbb{R}^n} \widehat{\varphi}\chi = \int_{\mathbb{R}^n} \widehat{\varphi}\overline{\widehat{\psi}},$$

where we used the multiplication formula for the Fourier transform in Lemma 1.1.22.

(ii) We can easily calculate

$$\widehat{\varphi * \psi}(\xi) = \int_{\mathbb{R}^n} e^{-2\pi i x \cdot \xi} (\varphi * \psi)(x) \, dx = \int_{\mathbb{R}^n} \int_{\mathbb{R}^n} e^{-2\pi i x \cdot \xi} \varphi(x - y)\psi(y) \, dy \, dx$$

$$= \int_{\mathbb{R}^n} \int_{\mathbb{R}^n} e^{-2\pi i (x-y) \cdot \xi} \varphi(x - y) \, e^{-2\pi i y \cdot \xi} \psi(y) \, dy \, dx$$

$$= \int_{\mathbb{R}^n} \int_{\mathbb{R}^n} e^{-2\pi i z \cdot \xi} \varphi(z) \, e^{-2\pi i y \cdot \xi} \psi(y) \, dy \, dz = \widehat{\varphi}(\xi)\widehat{\psi}(\xi),$$

where we used the substitution $z = x - y$. We leave (iii) as Exercise 1.1.31. $\quad\square$

Exercise 1.1.31. Prove part (iii) of Theorem 1.1.30.

1.2 Useful inequalities

This section will be devoted to several important inequalities which are very useful in Fourier analysis and in many types of analysis involving spaces of functions.

Proposition 1.2.1 (Cauchy's inequality). *For all $a, b \in \mathbb{R}$ we have $ab \leq \frac{a^2}{2} + \frac{b^2}{2}$. Moreover, for any $\epsilon > 0$, we also have $ab \leq \epsilon a^2 + \frac{b^2}{4\epsilon}$. As a consequence, we immediately obtain Cauchy's inequality for functions:*

$$\int_\Omega |f(x)g(x)| \, dx \leq \frac{1}{2} \int_\Omega (|f(x)|^2 + |g(x)|^2) \, dx,$$

which is

$$\|fg\|_{L^1(\Omega)} \leq \frac{1}{2}\Big(\|f\|^2_{L^2(\Omega)} + \|g\|^2_{L^2(\Omega)}\Big).$$

Proof. The first inequality follows from $0 \leq (a - b)^2 = a^2 - 2ab + b^2$. The second one follows if we apply the first one to $ab = (\sqrt{2\epsilon}a)(b/\sqrt{2\epsilon})$. $\quad\square$

Proposition 1.2.2 (Cauchy–Schwarz inequality). *Let $x, y \in \mathbb{R}^n$. Then we have $|x \cdot y| \leq |x||y|$.*

Proof. For $\epsilon > 0$, we have $0 \leq |x \pm \epsilon y|^2 = |x|^2 \pm 2\epsilon x \cdot y + \epsilon^2 |y|^2$. This implies $\pm x \cdot y \leq \frac{1}{2\epsilon}|x|^2 + \frac{\epsilon}{2}|y|^2$. Setting $\epsilon = \frac{|x|}{|y|}$, we obtain the required inequality, provided $y \neq 0$ (if $x = 0$ or $y = 0$ it is trivial).

An alternative proof may be given as follows. We can observe that the inequality $0 \leq |x + \epsilon y|^2 = |x|^2 + 2\epsilon x \cdot y + \epsilon^2 |y|^2$ implies that the discriminant of the quadratic (in ϵ) polynomial on the right-hand side must be non-positive, which means $|x \cdot y|^2 - |x|^2|y|^2 \leq 0$. $\quad\square$

Proposition 1.2.3 (Young's inequality). *Let $1 < p, q < \infty$ be such that $\frac{1}{p} + \frac{1}{q} = 1$. Then*

$$ab \leq \frac{a^p}{p} + \frac{b^q}{q} \quad \text{for all} \quad a, b > 0.$$

Moreover, if $\epsilon > 0$, we have $ab \leq \epsilon a^p + C(\epsilon)b^q$ for all $a, b > 0$, where $C(\epsilon) = (\epsilon p)^{-q/p}q^{-1}$.

As a consequence, we immediately obtain that if $f \in L^p(\Omega)$ and $g \in L^q(\Omega)$, then $fg \in L^1(\Omega)$ with

$$||fg||_{L^1} \leq \frac{1}{p}||f||_{L^p}^p + \frac{1}{q}||g||_{L^q}^q.$$

Proof. To prove the first inequality, we will use the fact that the exponential function $x \mapsto e^x$ is convex (a function $f : \mathbb{R} \to \mathbb{R}$ is called convex if $f(\tau x + (1 - \tau)y) \leq \tau f(x) + (1 - \tau)f(y)$, for all $x, y \in \mathbb{R}$ and all $0 \leq \tau \leq 1$). This implies

$$ab = e^{\ln a + \ln b} = e^{\frac{1}{p}\ln a^p + \frac{1}{q}\ln b^q} \leq \frac{1}{p}e^{\ln a^p} + \frac{1}{q}e^{\ln b^q} = \frac{a^p}{p} + \frac{b^q}{q}.$$

The second inequality with ϵ follows if we apply the first one to the product $ab = \big((\epsilon p)^{1/p}a\big)\big(b/(\epsilon p)^{1/p}\big)$. $\square$

Proposition 1.2.4 (Hölder's inequality). *Let $1 \leq p, q \leq \infty$ with $\frac{1}{p} + \frac{1}{q} = 1$. Let $f \in L^p(\Omega)$ and $g \in L^q(\Omega)$. Then $fg \in L^1(\Omega)$ and*

$$||fg||_{L^1(\Omega)} \leq ||f||_{L^p(\Omega)}||g||_{L^q(\Omega)}.$$

In the formulation we use the standard convention of setting $1/\infty = 0$. In the case of $p = q = 2$ Hölder's inequality is often called the Cauchy–Schwarz inequality.

Hölder's inequality in the setting of general measures was given in Theorem C.4.4, but here we give a short proof also in $\mathbb{R}^n$ for transparency.

Proof. In the case $p = 1$ or $p = \infty$ the inequality is obvious, so let us assume $1 < p < \infty$. Let us first consider the case when $||f||_{L^p} = ||g||_{L^q} = 1$. Then by Young's inequality with $1 < p, q < \infty$, we have

$$||fg||_{L^1} \leq \frac{1}{p}||f||_{L^p}^p + \frac{1}{q}||g||_{L^q}^q = \frac{1}{p} + \frac{1}{q} = 1 = ||f||_{L^p}||g||_{L^q},$$

which is the desired inequality. Now, let us consider general f, g. We observe that we may assume that $||f||_{L^p} \neq 0$ and $||g||_{L^q} \neq 0$, since otherwise one of the functions is zero almost everywhere in Ω and Hölder's inequality becomes trivial. It follows from the considered case that

$$\int_\Omega \left|\frac{f}{||f||_p}\frac{g}{||g||_q}\right| \, \mathrm{d}x \leq 1,$$

which implies the general case by the linearity of the integral. $\square$

Proposition 1.2.5 (General Hölder's inequality). *Let* $1 \leq p_1, \ldots, p_m \leq \infty$ *be such that* $\frac{1}{p_1} + \cdots + \frac{1}{p_m} = 1$. *Let* $f_k \in L^{p_k}(\Omega)$ *for all* $k = 1, \ldots, m$. *Then the product* $f_1 \cdots f_m \in L^1(\Omega)$ *and*

$$\|f_1 \cdots f_m\|_{L^1(\Omega)} \leq \prod_{k=1}^{m} \|f_k\|_{L^{p_k}(\Omega)}.$$

This inequality readily follows from Hölder's inequality by induction on the number of functions.

Exercise 1.2.6. Prove Proposition 1.2.5. Formulate and prove the corresponding general version of Theorem C.4.4.

Proposition 1.2.7 (Minkowski's inequality). *Let* $1 \leq p \leq \infty$. *Let* $f, g \in L^p(\Omega)$. *Then*

$$\|f + g\|_{L^p(\Omega)} \leq \|f\|_{L^p(\Omega)} + \|g\|_{L^p(\Omega)}.$$

In particular, this means that $\|\cdot\|_{L^p}$ *satisfies the triangle inequality and is a norm, so* $L^p(\Omega)$ *is a normed space.*

Minkowski's inequality in the setting of general measures was given in Theorem C.4.5.

Proof. The cases of $p = 1$ or $p = \infty$ follow from the triangle inequality for complex numbers and are, therefore, trivial. So we may assume $1 < p < \infty$. Then we have

$$\|f + g\|_{L^p(\Omega)}^p = \int_\Omega |f + g|^p \, dx \leq \int_\Omega |f + g|^{p-1}(|f| + |g|) \, dx$$

$$= \int_\Omega |f + g|^{p-1}|f| \, dx + \int_\Omega |f + g|^{p-1}|g| \, dx$$

$$\left(\text{use Hölder's inequality with } p = p, q = \frac{p}{p-1}\right)$$

$$\leq \left(\int_\Omega |f + g|^p \, dx\right)^{\frac{p-1}{p}} \left[\left(\int_\Omega |f|^p \, dx\right)^{\frac{1}{p}} + \left(\int_\Omega |g|^p \, dx\right)^{\frac{1}{p}}\right]$$

$$= \|f + g\|_{L^p(\Omega)}^{p-1} \left(\|f\|_{L^p(\Omega)} + \|g\|_{L^p(\Omega)}\right),$$

which implies the desired inequality. $\qquad\square$

Proposition 1.2.8 (Young's inequality for convolutions). *Let* $1 \leq p \leq \infty$, $f \in L^1(\mathbb{R}^n)$ *and* $g \in L^p(\mathbb{R}^n)$. *Then* $f * g \in L^p(\mathbb{R}^n)$ *and*

$$\|f * g\|_{L^p} \leq \|f\|_{L^1} \|g\|_{L^p}.$$

Proof. We will not prove it from the beginning because the proof is much shorter if we use Minkowski's inequality for integrals in Theorem C.5.23 or the monotonicity of the L^p-norm in Corollary C.5.24. Indeed, we can write

$$\|f * g\|_{L^p} = \left\| \int_{\mathbb{R}^n} f(y)\, g(\cdot - y)\, \mathrm{d}y \right\|_{L^p}$$

$$\leq \int |f(y)|\, \|g(\cdot - y)\|_{L^p}\, \mathrm{d}y = \|f\|_{L^1}\, \|g\|_{L^p}. \qquad \square$$

Exercise 1.2.9. Let $f \in L^1(\mathbb{R}^n)$ and $g \in C^k(\mathbb{R}^n)$ be such that $\partial^\alpha g \in L^\infty(\mathbb{R}^n)$ for all $|\alpha| \leq k$. Prove that $f * g \in C^k$. Consequently, show that $\partial^\alpha(f * g) = f * \partial^\alpha g$ at all points.

Proposition 1.2.10 (General Young's inequality for convolutions). *Let $1 \leq p, q, r \leq \infty$ be such that $\frac{1}{p} + \frac{1}{q} = 1 + \frac{1}{r}$. Let $f \in L^p(\mathbb{R}^n)$ and $g \in L^q(\mathbb{R}^n)$. Then $f * g \in L^r(\mathbb{R}^n)$ and*

$$\|f * g\|_{L^r} \leq \|f\|_{L^p}\, \|g\|_{L^q}.$$

Proof. The proof follows by the Riesz–Thorin interpolation theorem C.4.18 from Proposition 1.2.8 and the estimate $\|f * g\|_{L^\infty} \leq \|f\|_{L^p} \|g\|_{L^q}$ in the case of $\frac{1}{p} + \frac{1}{q} = 1$. $\qquad \square$

Exercise 1.2.11. If $\frac{1}{p} + \frac{1}{q} = 1$, $f \in L^p(\mathbb{R}^n)$ and $g \in L^q(\mathbb{R}^n)$, prove the estimate

$$\|f * g\|_{L^\infty} \leq \|f\|_{L^p} \|g\|_{L^q}.$$

(Hint: Hölder's inequality.)

Remark 1.2.12. If $\frac{1}{p} + \frac{1}{q} = 1$, $f \in L^p(\mathbb{R}^n)$ and $g \in L^q(\mathbb{R}^n)$, one can actually show that $f * g$ is not only bounded, but also uniformly continuous. Consequently, if $1 < p, q < \infty$, then $f * g(x) \to 0$ as $x \to \infty$.

Exercise 1.2.13. Prove this remark. (Hint: for the uniform continuity use Hölder's inequality. For the second part check that the statement is obviously true for compactly supported f and g, and then pass to the limit as supports of f and g grow; this is possible in view of the uniform continuity.)

Proposition 1.2.14 (Interpolation for L^p-norms). *Let $1 \leq s \leq r \leq t \leq \infty$ be such that $\frac{1}{r} = \frac{\theta}{s} + \frac{1-\theta}{t}$ for some $0 \leq \theta \leq 1$. Let $f \in L^s(\Omega) \bigcap L^t(\Omega)$. Then $f \in L^r(\Omega)$ and*

$$\|f\|_{L^r(\Omega)} \leq \|f\|_{L^s(\Omega)}^\theta \|f\|_{L^t(\Omega)}^{1-\theta}.$$

Proof. To prove this, we use that $\frac{\theta r}{s} + \frac{(1-\theta)r}{t} = 1$ and so we can apply Hölder's inequality in the following way:

$$\int_\Omega |f|^r\, \mathrm{d}x = \int_\Omega |f|^{\theta r} |f|^{(1-\theta)r}\, \mathrm{d}x$$

$$\leq \left(\int_\Omega |f|^{\theta r \cdot \frac{s}{\theta r}}\, \mathrm{d}x \right)^{\frac{\theta r}{s}} \left(\int_\Omega |f|^{(1-\theta)r \cdot \frac{t}{(1-\theta)r}}\, \mathrm{d}x \right)^{\frac{(1-\theta)r}{t}},$$

which is the desired inequality. $\qquad \square$

1.3 Tempered distributions

In this section we will introduce several spaces of distributions and will extend the Fourier transform to more general spaces of functions than $\mathcal{S}(\mathbb{R}^n)$ or $L^1(\mathbb{R}^n)$ considered in Section 1.1. The main problem with the immediate extension is that the integral in the definition of the Fourier transform in Definition 1.1.2 may no longer converge if we go beyond the space $L^1(\mathbb{R}^n)$ of integrable functions. We give preference to tempered distributions over general distributions since our main focus in this chapter is Fourier analysis.

Definition 1.3.1 (Tempered distributions $\mathcal{S}'(\mathbb{R}^n)$). We define the *space of tempered distributions* $\mathcal{S}'(\mathbb{R}^n)$ as the space of all continuous linear functionals on $\mathcal{S}(\mathbb{R}^n)$. This means that $u \in \mathcal{S}'(\mathbb{R}^n)$ if it is a functional $u : \mathcal{S}(\mathbb{R}^n) \to \mathbb{C}$ such that:

1. u is linear, i.e., $u(\alpha\varphi + \beta\psi) = \alpha u(\varphi) + \beta u(\psi)$ for all $\alpha, \beta \in \mathbb{C}$ and all $\varphi, \psi \in \mathcal{S}(\mathbb{R}^n)$;

2. u is continuous, i.e., $u(\varphi_j) \to u(\varphi)$ in $\mathbb{C}$ whenever $\varphi_j \to \varphi$ in $\mathcal{S}(\mathbb{R}^n)$.

We can also define the *convergence in the space* $\mathcal{S}'(\mathbb{R}^n)$ of tempered distributions.[1] Let $u_j, u \in \mathcal{S}'(\mathbb{R}^n)$. We will say that $u_j \to u$ in $\mathcal{S}'(\mathbb{R}^n)$ as $j \to \infty$ if $u_j(\varphi) \to u(\varphi)$ in $\mathbb{C}$ as $j \to \infty$, for all $\varphi \in \mathcal{S}(\mathbb{R}^n)$.

Functions in $\mathcal{S}(\mathbb{R}^n)$ are called the *test functions* for tempered distributions in $\mathcal{S}'(\mathbb{R}^n)$. Another notation for $u(\varphi)$ will be $\langle u, \varphi \rangle$.

Here one can also recall the definition of the convergence $\varphi_j \to \varphi$ in $\mathcal{S}(\mathbb{R}^n)$ from (1.2), which said that $\varphi_j \to \varphi$ in $\mathcal{S}(\mathbb{R}^n)$ as $j \to \infty$, if $\varphi_j, \varphi \in \mathcal{S}(\mathbb{R}^n)$ and if $\sup_{x \in \mathbb{R}^n} |x^\beta \partial^\alpha (\varphi_j - \varphi)(x)| \to 0$ as $j \to \infty$, for all multi-indices $\alpha, \beta \geq 0$.

1.3.1 Fourier transform of tempered distributions

Here we show that the Fourier transform can be extended from $\mathcal{S}(\mathbb{R}^n)$ to $\mathcal{S}'(\mathbb{R}^n)$ by duality. We also establish Plancherel's and Parseval's equalities on the space $L^2(\mathbb{R}^n)$.

Definition 1.3.2 (Fourier transform of tempered distributions). If $u \in \mathcal{S}'(\mathbb{R}^n)$, we can define its *(generalised) Fourier transform* by setting

$$\widehat{u}(\varphi) := u(\widehat{\varphi}),$$

for all $\varphi \in \mathcal{S}(\mathbb{R}^n)$.

Proposition 1.3.3 (Fourier transform on $\mathcal{S}'(\mathbb{R}^n)$). *The Fourier transform from Definition 1.3.2 is well defined and continuous from $\mathcal{S}'(\mathbb{R}^n)$ to $\mathcal{S}'(\mathbb{R}^n)$.*

---

[1] We will not discuss here topological properties of spaces of distributions. See Remark A.19.3 for some properties as well as Section B.3 and Section 10.12.

Proof. First, we can readily see that if $u \in \mathcal{S}'(\mathbb{R}^n)$ then also $\widehat{u} \in \mathcal{S}'(\mathbb{R}^n)$. Indeed, since $\varphi \in \mathcal{S}(\mathbb{R}^n)$, it follows that $\widehat{\varphi} \in \mathcal{S}(\mathbb{R}^n)$ and so $u(\widehat{\varphi})$ is a well-defined complex number. Moreover, $\widehat{u}$ is linear since both u and the Fourier transform $\mathcal{F}$ are linear. Finally, $\widehat{u}$ is continuous because $\varphi_j \to \varphi$ in $\mathcal{S}(\mathbb{R}^n)$ implies $\widehat{\varphi}_j \to \widehat{\varphi}$ in $\mathcal{S}(\mathbb{R}^n)$ by Remark 1.1.20, and hence

$$\widehat{u}(\varphi_j) = u(\widehat{\varphi}_j) \to u(\widehat{\varphi}) = \widehat{u}(\varphi)$$

by the continuity of both u from $\mathcal{S}(\mathbb{R}^n)$ to $\mathbb{C}$ and of the Fourier transform $\mathcal{F}$ as a mapping from $\mathcal{S}(\mathbb{R}^n)$ to $\mathcal{S}(\mathbb{R}^n)$ (see Corollary 1.1.18).

Now, it follows that it is also continuous as a mapping from $\mathcal{S}'(\mathbb{R}^n)$ to $\mathcal{S}'(\mathbb{R}^n)$, i.e., $u_j \to u$ in $\mathcal{S}'(\mathbb{R}^n)$ implies that $\widehat{u}_j \to \widehat{u}$ in $\mathcal{S}'(\mathbb{R}^n)$. Indeed, if $u_j \to u$ in $\mathcal{S}'(\mathbb{R}^n)$, we have $\widehat{u}_j(\varphi) = u_j(\widehat{\varphi}) \to u(\widehat{\varphi}) = \widehat{u}(\varphi)$ for all $\varphi \in \mathcal{S}(\mathbb{R}^n)$, which means that $\widehat{u}_j \to \widehat{u}$ in $\mathcal{S}'(\mathbb{R}^n)$. $\qquad\square$

Now we give two immediate but important principles for distributions.

Proposition 1.3.4 (Convergence principle). *Let X be a topological subspace in $\mathcal{S}'(\mathbb{R}^n)$ (i.e., convergence in X implies convergence in $\mathcal{S}'(\mathbb{R}^n)$). Suppose that $u_j \to u$ in $\mathcal{S}'(\mathbb{R}^n)$ and that $u_j \to v$ in X. Then $u \in X$ and $u = v$.*

This statement is simply a consequence of the fact that the space $\mathcal{S}'(\mathbb{R}^n)$ is Hausdorff, hence it has the uniqueness of limits property (recall that a topological space is called Hausdorff if any two points have open disjoint neighbourhoods, i.e., open disjoint sets containing them). The convergence principle is also related to another principle which we call

Proposition 1.3.5 (Uniqueness principle for distributions). *Let $u, v \in \mathcal{S}'(\mathbb{R}^n)$ and suppose that $u(\varphi) = v(\varphi)$ for all $\varphi \in \mathcal{S}(\mathbb{R}^n)$. Then $u = v$.*

This can be reformulated by saying that if an element $o \in \mathcal{S}'(\mathbb{R}^n)$ satisfies $o(\varphi) = 0$ for all $\varphi \in \mathcal{S}(\mathbb{R}^n)$, then o is the zero element in $\mathcal{S}'(\mathbb{R}^n)$.

Exercise 1.3.6. Let $f \in L^p(\mathbb{R}^n)$, $1 \le p \le \infty$, and assume that we have

$$\int_{\mathbb{R}^n} f(x)\, \varphi(x)\, \mathrm{d}x = 0$$

for all $\varphi \in C^\infty(\mathbb{R}^n)$ for which the integral makes sense. Prove that $f = 0$ almost everywhere. Do also a local version of this statement in Exercise 1.4.20.

Remark 1.3.7 **(Functions as distributions).** We can interpret functions in $L^p(\mathbb{R}^n)$, $1 \le p \le \infty$, as tempered distributions. If $f \in L^p(\mathbb{R}^n)$, we define the functional u_f by

$$u_f(\varphi) := \int_{\mathbb{R}^n} f(x)\, \varphi(x)\, \mathrm{d}x, \tag{1.9}$$

for all $\varphi \in \mathcal{S}(\mathbb{R}^n)$. By Hölder's inequality, we observe that $|u_f(\varphi)| \le ||f||_{L^p} ||\varphi||_{L^q}$, for $\frac{1}{p} + \frac{1}{q} = 1$, and hence $u_f(\varphi)$ is well defined in view of the simple inclusion

$\mathcal{S}(\mathbb{R}^n) \subset L^q(\mathbb{R}^n)$, for all $1 \leq q \leq \infty$. It needs to be verified that u_f is a linear continuous functional on $\mathcal{S}(\mathbb{R}^n)$. It is clearly linear in φ, while its continuity follows by Hölder's inequality (Proposition 1.2.4) from

$$|u_f(\varphi_j) - u_f(\varphi)| \leq ||f||_{L^p} ||\varphi_j - \varphi||_{L^q}$$

and the following lemma:

Lemma 1.3.8. *We have* $\mathcal{S}(\mathbb{R}^n) \subset L^q(\mathbb{R}^n)$ *with continuous embedding, i.e.,* $\varphi_j \to \varphi$ *in* $\mathcal{S}(\mathbb{R}^n)$ *implies that* $\varphi_j \to \varphi$ *in* $L^q(\mathbb{R}^n)$.

Exercise 1.3.9. Prove this lemma.

To summarise, any function $f \in L^p(\mathbb{R}^n)$ leads to a tempered distribution $u_f \in \mathcal{S}'(\mathbb{R}^n)$ in the *canonical way* given by (1.9). In this way we will view functions in $L^p(\mathbb{R}^n)$ as tempered distributions and continue to simply write f instead of u_f. There should be no confusion with this notation since writing $f(x)$ suggests that f is a function while $f(\varphi)$ suggests that it is applied to test functions and so it is viewed as a distribution u_f.

Remark 1.3.10 (**Consistency of all definitions**). With this identification, Definition 1.1.2 of the Fourier transform for functions in $L^1(\mathbb{R}^n)$ agrees with Definition 1.3.2 of the Fourier transforms of tempered distributions. Indeed, let $f \in L^1(\mathbb{R}^n)$. Then we have two ways of looking at its Fourier transforms:

1. We can use the first definition $\widehat{f}(\xi) = \int_{\mathbb{R}^n} e^{-2\pi i x \cdot \xi} f(x) \, dx$, and then we know that $\widehat{f} \in L^\infty(\mathbb{R}^n)$. In this way we get $u_{\widehat{f}} \in \mathcal{S}'(\mathbb{R}^n)$.

2. We can immediately think of $f \in L^1(\mathbb{R}^n)$ as of tempered distribution $u_f \in \mathcal{S}'(\mathbb{R}^n)$, and the second definition then produces its Fourier transform $\widehat{u_f} \in \mathcal{S}'(\mathbb{R}^n)$.

Fortunately, these two approaches are consistent and produce the same tempered distribution $u_{\widehat{f}} = \widehat{u_f} \in \mathcal{S}'(\mathbb{R}^n)$. Indeed, we have

$$u_{\widehat{f}}(\varphi) = \int_{\mathbb{R}^n} \widehat{f} \, \varphi \, dx = \int_{\mathbb{R}^n} f \, \widehat{\varphi} \, dx = u_f(\widehat{\varphi}).$$

Here we used the multiplication formula for the Fourier transform in Lemma 1.1.22 and the fact that both $u \in L^1(\mathbb{R}^n)$ and $\widehat{u} \in L^\infty(\mathbb{R}^n)$ can be viewed as tempered distributions in the canonical way (see Remark 1.3.7). It follows that we have $\widehat{f}(\varphi) = f(\widehat{\varphi})$ which justifies Definition 1.3.2.

Remark 1.3.11. We note that if $u \in L^1(\mathbb{R}^n)$ and also $\widehat{u} \in L^1(\mathbb{R}^n)$, then the Fourier inversion formula in Theorem 1.1.21 holds for almost all $x \in \mathbb{R}^n$. A more general Fourier inversion formula for tempered distributions will be given in Theorem 1.3.25.

Exercise 1.3.12. Let $1 \leq p \leq \infty$. Show that if $f_k \to f$ in $L^p(\mathbb{R}^n)$ then $f_k \to f$ in $\mathcal{S}'(\mathbb{R}^n)$.

It turns out that the Fourier transform acts especially nicely on one of the spaces $L^p(\mathbb{R}^n)$, namely on the space $L^2(\mathbb{R}^n)$, which is also a Hilbert space. These two facts lead to a very rich Fourier analysis on $L^2(\mathbb{R}^n)$ which we will deal with only briefly.

Theorem 1.3.13 (Plancherel's and Parseval's formulae). *Let $u \in L^2(\mathbb{R}^n)$. Then $\widehat{u} \in L^2(\mathbb{R}^n)$ and*

$$||\widehat{u}||_{L^2(\mathbb{R}^n)} = ||u||_{L^2(\mathbb{R}^n)} \qquad \text{(Plancherel's identity)}.$$

Moreover, for all $u, v \in L^2(\mathbb{R}^n)$ we have

$$\int_{\mathbb{R}^n} u\, \overline{v}\, \mathrm{d}x = \int_{\mathbb{R}^n} \widehat{u}\, \overline{\widehat{v}}\, \mathrm{d}\xi \qquad \text{(Parseval's identity)}.$$

Proof. We will use the fact (a special case of this fact follows from Theorem 1.3.31 to be proved later) that $\mathcal{S}(\mathbb{R}^n)$ is sequentially dense in $L^2(\mathbb{R}^n)$, i.e., that for every $u \in L^2(\mathbb{R}^n)$ there exists a sequence $u_j \in \mathcal{S}(\mathbb{R}^n)$ such that $u_j \to u$ in $L^2(\mathbb{R}^n)$. Then Theorem 1.1.30, (i), with $\varphi = \psi = u_j - u_k$, implies that

$$||\widehat{u_j} - \widehat{u_k}||_{L^2}^2 = ||u_j - u_k||_{L^2}^2 \to 0,$$

since u_j is a convergent sequence in $L^2(\mathbb{R}^n)$. Thus, $\widehat{u_j}$ is a Cauchy sequence in the complete (Banach, see Theorem C.4.9) space $L^2(\mathbb{R}^n)$. It follows that it must converge to some $v \in L^2(\mathbb{R}^n)$. By the continuity of the Fourier transform in $\mathcal{S}'(\mathbb{R}^n)$ (see Proposition 1.3.3) we must also have $\widehat{u_j} \to \widehat{u}$ in $\mathcal{S}'(\mathbb{R}^n)$. By the convergence principle for distributions in Proposition 1.3.4, we get that $\widehat{u} = v \in L^2(\mathbb{R}^n)$. Applying Theorem 1.1.30, (i), again, to $\varphi = \psi = u_j$, we get $||\widehat{u_j}||_{L^2}^2 = ||u_j||_{L^2}^2$. Passing to the limit, we get $||\widehat{u}||_{L^2}^2 = ||u||_{L^2}^2$, which is Plancherel's formula.

Finally, for $u, v \in L^2(\mathbb{R}^n)$, let $u_j, v_j \in \mathcal{S}(\mathbb{R}^n)$ be such that $u_j \to u$ and $v_j \to v$ in $L^2(\mathbb{R}^n)$. Applying Theorem 1.1.30, (i), to $\varphi = u_j, \psi = v_j$, and passing to the limit, we obtain Parseval's identity. $\qquad \square$

Corollary 1.3.14 (Hausdorff–Young inequality). *Let $1 \leq p \leq 2$ and $\frac{1}{p} + \frac{1}{q} = 1$. If $u \in L^p(\mathbb{R}^n)$ then $\widehat{u} \in L^q(\mathbb{R}^n)$ and*

$$||\widehat{u}||_{L^q(\mathbb{R}^n)} \leq ||u||_{L^p(\mathbb{R}^n)}.$$

Proof. The statement follows by the Riesz–Thorin Interpolation Theorem C.4.18 from estimates $||\widehat{u}||_{L^\infty(\mathbb{R}^n)} \leq ||u||_{L^1(\mathbb{R}^n)}$ in (1.1) and Plancherel's identity $||\widehat{u}||_{L^2(\mathbb{R}^n)} = ||u||_{L^2(\mathbb{R}^n)}$ in Theorem 1.3.13. $\qquad \square$

1.3.2 Operations with distributions

Besides the Fourier transform, there are several other operations that can be extended from functions in $\mathcal{S}(\mathbb{R}^n)$ to tempered distributions in $\mathcal{S}'(\mathbb{R}^n)$.

For example, partial differentiation $\frac{\partial}{\partial x_j}$ can be extended to a continuous operator $\frac{\partial}{\partial x_j} : \mathcal{S}'(\mathbb{R}^n) \to \mathcal{S}'(\mathbb{R}^n)$. Indeed, for $u \in \mathcal{S}'(\mathbb{R}^n)$ and $\varphi \in \mathcal{S}(\mathbb{R}^n)$, let us define

$$\left(\frac{\partial}{\partial x_j}u\right)(\varphi) := -u\left(\frac{\partial \varphi}{\partial x_j}\right).$$

It is necessary to include the negative sign in this definition. Indeed, if $u \in \mathcal{S}(\mathbb{R}^n)$, then the integration by parts formula and the identification of functions with distributions in Remark 1.3.7 yield

$$\left(\frac{\partial}{\partial x_j}u\right)(\varphi) = \int_{\mathbb{R}^n} \left(\frac{\partial u}{\partial x_j}\right)(x)\varphi(x)\ \mathrm{d}x$$

$$= -\int_{\mathbb{R}^n} u(x)\left(\frac{\partial \varphi}{\partial x_j}\right)(x)\ \mathrm{d}x = -u\left(\frac{\partial \varphi}{\partial x_j}\right),$$

which explains the sign. This also shows the consistency of this definition of the derivative with the usual definition for differentiable functions.

Definition 1.3.15 (Distributional derivatives). More generally, for any multi-index α, one can define

$$(\partial^\alpha u)(\varphi) = (-1)^{|\alpha|} u(\partial^\alpha \varphi),$$

for $\varphi \in \mathcal{S}(\mathbb{R}^n)$.

Proposition 1.3.16. *If $u \in \mathcal{S}'(\mathbb{R}^n)$, then $\partial^\alpha u \in \mathcal{S}'(\mathbb{R}^n)$ and operator $\partial^\alpha : \mathcal{S}'(\mathbb{R}^n) \to \mathcal{S}'(\mathbb{R}^n)$ is continuous.*

Proof. Indeed, if $\varphi_k \to \varphi$ in $\mathcal{S}(\mathbb{R}^n)$, then clearly also $\partial^\alpha \varphi_k \to \partial^\alpha \varphi$ in $\mathcal{S}(\mathbb{R}^n)$, and, therefore,

$$(\partial^\alpha u)(\varphi_k) = (-1)^{|\alpha|} u(\partial^\alpha \varphi_k) \to (-1)^{|\alpha|} u(\partial^\alpha \varphi) = (\partial^\alpha u)(\varphi),$$

which means that $\partial^\alpha u \in \mathcal{S}'(\mathbb{R}^n)$. Moreover, let $u_k \to u \in \mathcal{S}'(\mathbb{R}^n)$. Then $\partial^\alpha u_k(\varphi) = (-1)^{|\alpha|} u_k(\partial^\alpha \varphi) \to (-1)^{|\alpha|} u(\partial^\alpha \varphi) = \partial^\alpha u(\varphi)$, for all $\varphi \in \mathcal{S}(\mathbb{R}^n)$, i.e., ∂^α is continuous on $\mathcal{S}'(\mathbb{R}^n)$. $\square$

Exercise 1.3.17. Show that if $u \in \mathcal{S}'(\mathbb{R}^n)$, then $\partial^\alpha \partial^\beta u = \partial^\beta \partial^\alpha u = \partial^{\alpha+\beta} u$.

Exercise 1.3.18. Let $\chi : \mathbb{R} \to \mathbb{R}$ be the characteristic function of the interval $[-1, 1]$, i.e., $\chi(y) = 1$ for $-1 \leq y \leq 1$ and $\chi(y) = 0$ for $|y| > 1$. Calculate the distributional derivative χ'. Define operator T by $Tf(x) = \frac{d}{dx}(\chi * f)(x)$, $x \in \mathbb{R}$, $f \in \mathcal{S}(\mathbb{R}^n)$. Prove that $Tf(x) = f(x+1) - f(x-1)$.

Remark 1.3.19 **(Multiplication by functions).** If a smooth function $f \in C^\infty(\mathbb{R}^n)$ and all of its derivatives are bounded by some polynomial functions, we can define the multiplication of a tempered distribution u by f by setting $(fu)(\varphi) := u(f\varphi)$. This is well defined since $\varphi \in \mathcal{S}(\mathbb{R}^n)$ implies $f\varphi \in \mathcal{S}(\mathbb{R}^n)$.

Exercise 1.3.20 (Hadamard's principal value). Show that $\log|x|$ is a tempered distribution on $\mathbb{R}$. Let $u = \frac{d}{dx}\log|x|$. Show that

$$u(\varphi) = \lim_{\epsilon \searrow 0} \int_{\mathbb{R}\setminus[-\epsilon,\epsilon]} \frac{1}{x}\,\varphi(x)\,dx$$

for all $\varphi \in C^1(\mathbb{R})$ vanishing outside a bounded set. The distribution u is called the *principal value* of $\frac{1}{x}$ and is denoted by $p.v.\frac{1}{x}$.

Remark 1.3.21 **(Schwartz' impossibility result).** One has to be careful when multiplying distributions as the following example shows:

$$0 = \frac{1}{x}\cdot(x\cdot\delta) = \left(\frac{1}{x}\cdot x\right)\cdot\delta = \delta,$$

where $\frac{1}{x}$ may be any inverse of x, for example $p.v.\frac{1}{x}$. In general, distributions can not be multiplied, as was noted by Laurent Schwartz in [104], and which is called the *Schwartz' impossibility result*. Still, some multiplication is possible, as is demonstrated by Remark 1.3.19.

Exercise 1.3.22. Define the distribution $\frac{1}{x\pm i0}$ by

$$\left(\frac{1}{x\pm i0}\right)(\varphi) := \lim_{\epsilon\to 0\pm}\int_{\mathbb{R}}\frac{1}{x+i\epsilon}\,\varphi(x)\,dx,$$

for $\varphi \in S(\mathbb{R}^n)$. Prove that

$$\frac{1}{x\pm i0} = p.v.\frac{1}{x}\mp i\pi\delta.$$

However, as we have seen, statements on $S(\mathbb{R}^n)$ can usually be extended to corresponding statements on $S'(\mathbb{R}^n)$. This applies to the Fourier inversion formula as well.

Definition 1.3.23 (Inverse Fourier transform). Define $\mathcal{F}^{-1}$ on $S'(\mathbb{R}^n)$ by

$$(\mathcal{F}^{-1}u)(\varphi) := u(\mathcal{F}^{-1}\varphi),$$

for $u \in S'(\mathbb{R}^n)$ and $\varphi \in S(\mathbb{R}^n)$.

Exercise 1.3.24. Show that $\mathcal{F}^{-1} : S'(\mathbb{R}^n) \to S'(\mathbb{R}^n)$ is well defined and continuous.

Theorem 1.3.25 (Fourier inversion formula for tempered distributions). *Operators $\mathcal{F}$ and $\mathcal{F}^{-1}$ are inverse to each other on $S'(\mathbb{R}^n)$, i.e.,*

$$\mathcal{F}\mathcal{F}^{-1} = \mathcal{F}^{-1}\mathcal{F} = identity \quad on \quad S'(\mathbb{R}^n).$$

Proof. To prove this, let $u \in S'(\mathbb{R}^n)$ and $\varphi \in S(\mathbb{R}^n)$. Then by Theorem 1.1.21 and Definitions 1.3.2 and 1.3.23, we get

$$(\mathcal{F}\mathcal{F}^{-1}u)(\varphi) = (\mathcal{F}^{-1}u)(\mathcal{F}\varphi) = u(\mathcal{F}^{-1}\mathcal{F}\varphi) = u(\varphi),$$

so $\mathcal{F}\mathcal{F}^{-1}u = u$ by the uniqueness principle for distributions in Proposition 1.3.5. A similar argument applies to show that $\mathcal{F}^{-1}\mathcal{F} = id$. $\square$

Remark 1.3.26. To give an example of these operations, let us define the *Heaviside function H* on $\mathbb{R}$ by setting

$$H(x) = \begin{cases} 0, & \text{if } x < 0, \\ 1, & \text{if } x \geq 0. \end{cases}$$

Clearly $H \in L^\infty(\mathbb{R})$, so in particular, it is a tempered distribution in $\mathcal{S}'(\mathbb{R}^n)$. Let us also define the *Dirac δ-distribution* by setting $\delta(\varphi) = \varphi(0)$ for all $\varphi \in \mathcal{S}(\mathbb{R})$. It is easy to see that $\delta \in \mathcal{S}'(\mathbb{R}^n)$.

We claim first that $H' = \delta$. Indeed, we have

$$H'(\varphi) = -H(\varphi') = -\int_0^\infty \varphi'(x)\, dx = \varphi(0) = \delta(\varphi),$$

hence $H' = \delta$ by the uniqueness principle for distributions.

Let us now calculate the Fourier transform of δ. According to the definitions, we have

$$\widehat{\delta}(\varphi) = \delta(\widehat{\varphi}) = \widehat{\varphi}(0) = \int_\mathbb{R} \varphi(x)\, dx = 1(\varphi),$$

hence $\widehat{\delta} = 1$. Here we used the fact that the constant 1 is in $L^\infty(\mathbb{R}^n)$, hence also a tempered distribution.

Exercise 1.3.27. Check that we also have $\widehat{1} = \delta$.

1.3.3 Approximating by smooth functions

It turns out that although elements of $\mathcal{S}'(\mathbb{R}^n)$ can be very irregular and the space is quite large, tempered distributions can still be approximated by smooth compactly supported functions.

Definition 1.3.28 (Space $C_0^\infty(\Omega)$). For an open set $\Omega \subset \mathbb{R}^n$, the space $C_0^\infty(\Omega)$ of *smooth compactly supported functions* is defined as the space of smooth functions $\varphi : \Omega \to \mathbb{C}$ with compact support. Here the support of φ is defined as the closure of the set where φ is non-zero, i.e., by

$$\operatorname{supp} \varphi = \overline{\{x \in \Omega : \varphi(x) \neq 0\}}.$$

Remark 1.3.29 (**How large is $C_0^\infty(\Omega)$?**). We can see that this space is non-empty. For example, if we define function $\chi(t)$ by $\chi(t) = e^{-1/t^2}$ for $t > 0$ and by $\chi(t) = 0$ for $t \leq 0$, then $f(t) = \chi(t)\chi(1-t)$ is a smooth compactly supported function on $\mathbb{R}$. Consequently, $\varphi(x) = f(x_1) \cdots f(x_n)$ is a function in $C_0^\infty(\mathbb{R}^n)$, with $\operatorname{supp} \varphi = [0,1]^n$.

Another example is the function ψ defined by $\psi(x) = e^{1/(|x|^2-1)}$ for $|x| < 1$ and by $\psi(x) = 0$ for $|x| \geq 1$. We have $\psi \in C_0^\infty(\mathbb{R}^n)$ with $\operatorname{supp} \psi = \{|x| \leq 1\}$.

Remark 1.3.30. For the functional analytic description of the topology of the space $C_0^\infty(\Omega)$ we refer to Exercise B.3.12. It is also a nuclear Montel space, see Exercises B.3.35 and B.3.51.

Although these examples are quite special, products of these functions with any other smooth function as well as their derivatives are all in $C_0^\infty(\mathbb{R}^n)$. On the other hand, $C_0^\infty(\mathbb{R}^n)$ can not contain analytic functions, thus making it relatively small. Still, it is dense in very large spaces of functions/distributions in their respective topologies.

Theorem 1.3.31 (Sequential density of $C_0^\infty(\Omega)$ in $\mathcal{S}'(\mathbb{R}^n)$). *The space $C_0^\infty(\mathbb{R}^n)$ is sequentially dense in $\mathcal{S}'(\mathbb{R}^n)$, i.e., for every $u \in \mathcal{S}'(\mathbb{R}^n)$ there exists a sequence $u_k \in C_0^\infty(\mathbb{R}^n)$ such that $u_k \to u$ in $\mathcal{S}'(\mathbb{R}^n)$ as $k \to \infty$.*

Lemma 1.3.32. *The space $C_0^\infty(\mathbb{R}^n)$ is sequentially dense in $\mathcal{S}(\mathbb{R}^n)$, i.e., for every $\varphi \in \mathcal{S}(\mathbb{R}^n)$ there exists a sequence $\varphi_k \in C_0^\infty(\mathbb{R}^n)$ such that $\varphi_k \to \varphi$ in $\mathcal{S}(\mathbb{R}^n)$ as $k \to \infty$.*

Proof. Let $\varphi \in \mathcal{S}(\mathbb{R}^n)$. Let us fix some $\psi \in C_0^\infty(\mathbb{R}^n)$ such that $\psi = 1$ in a neighbourhood of the origin and let us define $\psi_k(x) = \psi(x/k)$. Then it can be easily checked that $\varphi_k = \psi_k\varphi \to \varphi$ in $\mathcal{S}(\mathbb{R}^n)$, as $k \to \infty$. $\qquad\square$

Proof of Theorem 1.3.31. Let $u \in \mathcal{S}'(\mathbb{R}^n)$ and let ψ and ψ_k be as in the proof of Lemma 1.3.32. Then $\psi u \in \mathcal{S}'(\mathbb{R}^n)$ is well defined by $(\psi u)(\varphi) = u(\psi\varphi)$, for all $\varphi \in \mathcal{S}(\mathbb{R}^n)$. We have that $\psi_k u \to u$ in $\mathcal{S}'(\mathbb{R}^n)$. Indeed, we have that $(\psi_k u)(\varphi) = u(\psi_k\varphi) \to u(\varphi)$ by Lemma 1.3.32. Similarly, we have that $\psi_k\widehat{u} \to \widehat{u}$ in $\mathcal{S}'(\mathbb{R}^n)$, and hence also $\mathcal{F}^{-1}(\psi_k\widehat{u}) \to u$ in $\mathcal{S}'(\mathbb{R}^n)$ because of the continuity of the Fourier transform in $\mathcal{S}'(\mathbb{R}^n)$, see Proposition 1.3.3. Consequently, we have

$$u_{kj} = \psi_j(\mathcal{F}^{-1}(\psi_k\widehat{u})) \to u \quad \text{in } \mathcal{S}'(\mathbb{R}^n) \quad \text{as } k, j \to \infty.$$

It remains to show that $u_{kj} \in C_0^\infty(\mathbb{R}^n)$. In general, let $\chi \in C_0^\infty(\mathbb{R}^n)$ and let $w = \chi\widehat{u}$. We claim that $\mathcal{F}^{-1}w \in C^\infty(\mathbb{R}^n)$. Indeed, we have

$$(\mathcal{F}^{-1}w)(\varphi) = w(\mathcal{F}^{-1}\varphi) = w_\xi\left(\int_{\mathbb{R}^n} e^{2\pi i x \cdot \xi}\varphi(x)\,\mathrm{d}x\right)$$

$$= \int_{\mathbb{R}^n} w_\xi(\,e^{2\pi i x \cdot \xi})\varphi(x)\,\mathrm{d}x,$$

where we write w_ξ to emphasize that w acts on the test function as the function of ξ-variable, and where we used the continuity of w and the fact that $w_\xi(\,e^{2\pi i x \cdot \xi}) = \widehat{u}(\chi\,e^{2\pi i x \cdot \xi})$ is well defined. Now, it follows that $\mathcal{F}^{-1}w$ can be identified with the function $(\mathcal{F}^{-1}w)(x) = \widehat{u}_\xi(\chi(\xi)\,e^{2\pi i x \cdot \xi})$, which is smooth with respect to x. Indeed, we can note first that the right-hand side depends continuously on x because of the continuity of $\widehat{u}$ on $\mathcal{S}(\mathbb{R}^n)$. Here we also use that everything is well defined since $\chi \in C_0^\infty(\mathbb{R}^n)$. Moreover, since the function $\chi(\xi)\,e^{2\pi i x \cdot \xi}$ is compactly supported in ξ, so are its derivatives with respect to x, and hence all the derivatives of $(\mathcal{F}^{-1}w)(x)$ are also continuous in x, proving the claim and the theorem. $\qquad\square$

Exercise 1.3.33. Prove that $\mathcal{S}(\mathbb{R}^n)$ is sequentially dense in $L^2(\mathbb{R}^n)$, i.e., that for every $u \in L^2(\mathbb{R}^n)$ there exists a sequence $u_j \in \mathcal{S}(\mathbb{R}^n)$ such that $u_j \to u$ in $L^2(\mathbb{R}^n)$. Prove that this is also true for $L^p(\mathbb{R}^n)$, for all $1 \le p < \infty$.

Exercise 1.3.34 (Uncertainty principle). Prove that $C_0^\infty(\mathbb{R}^n) \cap \mathcal{F}C_0^\infty(\mathbb{R}^n) = \{0\}$. (Hint: it is enough to know that polynomials are dense in $L^2(K)$ for any compact K.)

Exercise 1.3.35 (Scaling operators). For $\lambda \in \mathbb{R}$, $\lambda \ne 0$, define the mapping $m_\lambda : \mathbb{R}^n \to \mathbb{R}^n$ by $m_\lambda(x) = \lambda x$.

(i) Let $\varphi \in \mathcal{S}(\mathbb{R}^n)$. Prove that $\widehat{\varphi \circ m_\lambda}(\xi) = \lambda^{-n} (\widehat{\varphi} \circ m_{\lambda^{-1}})(\xi)$ for all $\xi \in \mathbb{R}^n$.

(ii) Let $u \in \mathcal{S}'(\mathbb{R}^n)$. Define the distribution $u \circ m_\lambda$ by

$$(u \circ m_\lambda)(\varphi) := \lambda^{-n} u(\varphi \circ m_{\lambda^{-1}}),$$

for all $\varphi \in \mathcal{S}(\mathbb{R}^n)$. Prove that this definition is consistent with $\mathcal{S}(\mathbb{R}^n)$, i.e., show that if $u \in \mathcal{S}(\mathbb{R}^n)$, $(u \circ m_\lambda)(x) = u(\lambda x)$, and if we identify u with its canonical distribution, then we have

$$(u \circ m_\lambda)(\varphi) = \lambda^{-n} u(\varphi \circ m_{\lambda^{-1}})$$

for all $\varphi \in \mathcal{S}(\mathbb{R}^n)$.

(iii) Let $u \in \mathcal{S}'(\mathbb{R}^n)$. Prove that $\widehat{u \circ m_\lambda} = \lambda^{-n} \widehat{u} \circ m_{\lambda^{-1}}$.

1.4 Distributions

Since our main interest is in Fourier analysis, we started with the space $\mathcal{S}'(\mathbb{R}^n)$ of tempered distributions which allows the definition and use of the Fourier transform. However, there is a bigger space of distributions which we will sketch here. It will contain some important classes of functions that $\mathcal{S}'(\mathbb{R}^n)$ does not contain. For much more comprehensive treatments of spaces of distributions and their properties we refer the reader to monographs [8, 10, 39, 106, 105].

1.4.1 Localisation of L^p-spaces and distributions

Definition 1.4.1 (Localisations of L^p-spaces). We define local versions of the spaces $L^p(\Omega)$ as follows. We will say that $f \in L^p_{\text{loc}}(\Omega)$ if $\varphi f \in L^p(\Omega)$ for all $\varphi \in C_0^\infty(\Omega)$. We note that the spaces $L^p_{\text{loc}}(\mathbb{R}^n)$ are not subspaces of $\mathcal{S}'(\mathbb{R}^n)$ since they do not encode any information on the global behaviour of functions. For example, $e^{|x|^2}$ is smooth, and hence belongs to all $L^p_{\text{loc}}(\mathbb{R}^n)$, $1 \le p \le \infty$, but it is not in $\mathcal{S}'(\mathbb{R}^n)$. There is a natural notion of *convergence* in the localised spaces $L^p_{\text{loc}}(\Omega)$. Thus, we will write $f_m \to f$ in $L^p_{\text{loc}}(\Omega)$ as $m \to \infty$, if f and f_m belong to $L^p(\Omega)_{\text{loc}}$ for all m, and if $\varphi f_m \to \varphi f$ in $L^p(\Omega)$ as $m \to \infty$, for all $\varphi \in C_0^\infty(\Omega)$.

The difference between the space of distributions $\mathcal{D}'(\mathbb{R}^n)$ that we are going to introduce now, and the space of tempered distributions $\mathcal{S}'(\mathbb{R}^n)$ is the choice of the set $C_0^\infty(\mathbb{R}^n)$ rather than $\mathcal{S}(\mathbb{R}^n)$ as the space of test functions. At the same time, choosing $C_0^\infty(\Omega)$ as test functions allows one to obtain the space $\mathcal{D}'(\Omega)$ of distributions in Ω, rather than on the whole space $\mathbb{R}^n$.

The definition and facts below are sketched only as they are similar to Definition 1.3.1.

Definition 1.4.2 (Distributions $\mathcal{D}'(\Omega)$). We say that $\varphi_k \to \varphi$ in $C_0^\infty(\Omega)$ if $\varphi_k, \varphi \in C_0^\infty(\Omega)$, if there is a compact set $K \subset \Omega$ such that $\operatorname{supp}\varphi_k \subset K$ for all k, and if $\sup_{x\in\Omega} |\partial^\alpha(\varphi_k - \varphi)(x)| \to 0$ for all multi-indices α. Then $\mathcal{D}'(\Omega)$ is defined as the set of all linear continuous functionals $u : C_0^\infty(\Omega) \to \mathbb{C}$, i.e., all functionals $u : C_0^\infty(\Omega) \to \mathbb{C}$ such that:

1. u is linear, i.e., $u(\alpha\varphi + \beta\psi) = \alpha u(\varphi) + \beta u(\psi)$ for all $\alpha, \beta \in \mathbb{C}$ and all $\varphi, \psi \in C_0^\infty(\Omega)$;

2. u is continuous, i.e., $u(\varphi_j) \to u(\varphi)$ in $\mathbb{C}$ whenever $\varphi_j \to \varphi$ in $C_0^\infty(\Omega)$.

Exercise 1.4.3 (Order of a distribution). Show that a linear operator $u : C_0^\infty(\Omega) \to \mathbb{C}$ belongs to $\mathcal{D}'(\Omega)$ if and only if for every compact set $K \subset \Omega$ there exist constants C and m such that

$$|u(\varphi)| \le C \max_{|\alpha|\le m} \sup_{x\in\Omega} |\partial^\alpha\varphi(x)|, \tag{1.10}$$

for all $\varphi \in C_0^\infty(\Omega)$ with $\operatorname{supp}\varphi \subset K$. The smallest m for which (1.10) holds is called the *order of u in K*. The smallest m which works for all compact sets if called the *order of the distribution u*. Show that δ-distribution has order 1. Find examples of distributions of infinite order.

Remark 1.4.4 (**Distributions of zero order as measures**). If $u \in \mathcal{D}'(\Omega)$ is a distribution of order zero, by (1.10) it defines a continuous functional on $C(\Omega)$. Then it follows from Theorem C.4.60 that u is a measure (at least when Ω is compact, or when u acts on continuous compactly supported functions).

Remark 1.4.5 (**Continuous inclusion $\mathcal{S}'(\mathbb{R}^n) \subset \mathcal{D}'(\mathbb{R}^n)$**). It is easy to see that $C_0^\infty(\mathbb{R}^n) \subset \mathcal{S}(\mathbb{R}^n)$ and that if $\varphi_k \to \varphi$ in $C_0^\infty(\mathbb{R}^n)$, then $\varphi_k \to \varphi$ in $\mathcal{S}(\mathbb{R}^n)$. Thus, if $u \in \mathcal{S}'(\mathbb{R}^n)$ and if $\varphi_k \to \varphi$ in $C_0^\infty(\mathbb{R}^n)$, we have $u(\varphi_k) \to u(\varphi)$, which means that $u \in \mathcal{D}'(\mathbb{R}^n)$. Thus, we showed that $\mathcal{S}'(\mathbb{R}^n) \subset \mathcal{D}'(\mathbb{R}^n)$. We say that $u_k \to u \in \mathcal{D}'(\Omega)$ if $u_k, u \in \mathcal{D}'(\Omega)$ and if $u_k(\varphi) \to u(\varphi)$ for all $\varphi \in C_0^\infty(\Omega)$.

Exercise 1.4.6. Show that $u_k \to u$ in $\mathcal{S}'(\mathbb{R}^n)$ implies $u_k \to u$ in $\mathcal{D}'(\mathbb{R}^n)$, i.e., the inclusion $\mathcal{S}'(\mathbb{R}^n) \subset \mathcal{D}'(\mathbb{R}^n)$ is continuous.

Exercise 1.4.7. Prove that the canonical identification in Remark 1.3.7 yields the inclusions $L_{\mathrm{loc}}^p(\Omega) \subset \mathcal{D}'(\Omega)$ for all $1 \le p \le \infty$.

Definition 1.4.8 (Compactly supported distributions $\mathcal{E}'(\Omega)$). We say that $\varphi_k \to \varphi$ in $C^\infty(\Omega)$ if $\varphi_k, \varphi \in C^\infty(\Omega)$ and if $\sup_{x\in K} |\partial^\alpha(\varphi_k - \varphi)(x)| \to 0$ for all multi-indices α and all compact subsets K of Ω. Then $\mathcal{E}'(\Omega)$ is defined as the set of all linear continuous functionals $u : C^\infty(\Omega) \to \mathbb{C}$, i.e., all functionals $u : C^\infty(\Omega) \to \mathbb{C}$ such that:

1. u is linear, i.e., $u(\alpha\varphi + \beta\psi) = \alpha u(\varphi) + \beta u(\psi)$ for all $\alpha, \beta \in \mathbb{C}$ and all $\varphi, \psi \in C^\infty(\Omega)$;

2. u is continuous, i.e., $u(\varphi_j) \to u(\varphi)$ in $\mathbb{C}$ whenever $\varphi_j \to \varphi$ in $C^\infty(\Omega)$.

Exercise 1.4.9. Show that the restriction of $u \in \mathcal{E}'(\Omega)$ to $C_0^\infty(\Omega)$ is an injective linear mapping from $\mathcal{E}'(\Omega)$ to $\mathcal{D}'(\Omega)$.

Exercise 1.4.10. Show that $\mathcal{E}'(\Omega) \subset \mathcal{D}'(\Omega)$ and that $\mathcal{E}'(\mathbb{R}^n) \subset \mathcal{S}'(\mathbb{R}^n) \subset \mathcal{D}'(\mathbb{R}^n)$. Show also that all these inclusions are continuous.

Definition 1.4.11 (Support of a distribution). We say that $u \in \mathcal{D}'(\Omega)$ is *supported* in the set $K \subset \Omega$ if $u(\varphi) = 0$ for all $\varphi \in C^\infty(\Omega)$ such that $\varphi = 0$ on K. The smallest closed set in which u is supported is called the *support* of u and is denoted by supp u.

Exercise 1.4.12. Formulate and prove the analogue of the criterion in Exercise 1.4.3 for compactly supported distributions in $\mathcal{E}'(\Omega)$.

Exercise 1.4.13. Show that distributions in $\mathcal{E}'(\Omega)$ have compact support (justifying the name of "compactly supported" distributions in Definition 1.4.8).

Exercise 1.4.14 (Distributions with compact support). Prove that if the support of $u \in \mathcal{D}'(\mathbb{R}^n)$ is compact then u is of finite order. Prove that all compactly supported distributions belong to $\mathcal{E}'(\Omega)$.

Exercise 1.4.15 (Distributions with point support). Prove that if a distribution $u \in \mathcal{D}'(\mathbb{R}^n)$ of order m has support supp $u = \{0\}$, then there exist constants $a_\alpha \in \mathbb{C}$ such that $u = \sum_{|\alpha| \le m} a_\alpha \partial^\alpha \delta$.

Definition 1.4.16 (Singular support). The *singular support* of $u \in \mathcal{D}'(\Omega)$ is defined as the complement of the set where u is smooth. Namely, $x \notin$ sing supp u if there is an open neighbourhood U of x and a smooth function $f \in C^\infty(U)$ such that $u(\varphi) = f(\varphi)$ for all $\varphi \in C_0^\infty(U)$.

Exercise 1.4.17. Show that if $u \in \mathcal{D}'(\Omega)$ then its singular support is closed.

Exercise 1.4.18. Show that sing supp $|x| = \{0\}$ and that sing supp $\delta = \{0\}$.

Exercise 1.4.19. Show that if $u \in \mathcal{E}'(\mathbb{R}^n)$, then $\hat{u}$ is a smooth function of the so-called *slow growth* (i.e., $\hat{u}(\xi)$ and all of its derivatives are of at most polynomial growth). Hint: the slow growth follows from testing u on the exponential functions $e_\xi(x) = e^{2\pi i x \cdot \xi}$. Indeed, show first that $u(\xi) = \langle u, \overline{e_\xi} \rangle$ and thus

$$\partial^\alpha \hat{u}(\xi) = \langle u, \partial_\xi^\alpha \overline{e_\xi} \rangle = (-2\pi i)^{|\alpha|} \langle u, x^\alpha \overline{e_\xi} \rangle.$$

Consequently, by an analogue of (1.10) in Exercise 1.4.3 we conclude that

$$|\partial^\alpha \hat{u}(\xi)| \le C \sum_{|\beta| \le m} \sup_{|x| \le R} |\partial_x^\beta (x^\alpha \overline{e_\xi}(x))| \le \tilde{C}(1 + R)^{|\alpha|}(1 + |\xi|)^m.$$

Exercise 1.4.20. Prove the following stronger version of Exercise 1.3.6. Let $f \in L^1_{\mathrm{loc}}(\mathbb{R}^n)$ and assume that $\int_{\mathbb{R}^n} f(x)\,\varphi(x)\,\mathrm{d}x = 0$ for all $\varphi \in C_0^\infty(\mathbb{R}^n)$. Prove that $f = 0$ almost everywhere.

1.4.2 Convolution of distributions

We can write the convolution of two functions $f, g \in \mathcal{S}(\mathbb{R}^n)$ in the following way:

$$(f * g)(x) = \int_{\mathbb{R}^n} f(z) g(x - z) \, dz = \int_{\mathbb{R}^n} f(z) (\tau_x R g)(z) \, dz,$$

where $(Rg)(x) = g(-x)$ and $(\tau_h g)(x) = g(x - h)$, so that

$$(\tau_x R g)(z) = (Rg)(z - x) = g(x - z).$$

Recalling our identification of functions with distributions in Remark 1.3.7, we can write $(f * g)(x) = f(\tau_x R g)$. This can now be extended to distributions.

Definition 1.4.21 (Convolution with a distribution). For $u \in \mathcal{S}'(\mathbb{R}^n)$ and $\varphi \in \mathcal{S}(\mathbb{R}^n)$, define

$$(u * \varphi)(x) := u(\tau_x R \varphi).$$

The definition makes sense since $\tau_x R \varphi \in \mathcal{S}(\mathbb{R}^n)$ and since $\tau_x, R : \mathcal{S}(\mathbb{R}^n) \to \mathcal{S}(\mathbb{R}^n)$ are continuous.

Corollary 1.4.22. *For example, for $\psi \in \mathcal{S}(\mathbb{R}^n)$ we have $\delta * \psi = \psi$ since for every $x \in \mathbb{R}^n$ we have*

$$(\delta * \psi)(x) = \delta(\tau_x R \psi) = \psi(x - z)|_{z=0} = \psi(x).$$

Lemma 1.4.23. *Let $u \in \mathcal{S}'(\mathbb{R}^n)$ and $\varphi \in \mathcal{S}(\mathbb{R}^n)$. Then $u * \varphi \in C^\infty(\mathbb{R}^n)$.*

Proof. We can observe that $(u * \varphi)(x) = u(\tau_x R \varphi)$ is continuous in x since $\tau_x : \mathcal{S}(\mathbb{R}^n) \to \mathcal{S}(\mathbb{R}^n)$ and $u : \mathcal{S}(\mathbb{R}^n) \to \mathbb{C}$ are continuous. The same applies when we look at derivatives in x, implying that $u * \varphi$ is smooth. Here we note that we are allowed to pass the limit through u since it is a continuous functional. $\qquad\square$

Exercise 1.4.24. Prove that if $u, v, \varphi \in \mathcal{S}(\mathbb{R}^n)$ then $(u * v)(\varphi) = u(Rv * \varphi)$.

Exercise 1.4.25 (Reflection of a distribution). For $v \in \mathcal{S}'(\mathbb{R}^n)$, define its *reflection* Rv by

$$(Rv)(\varphi) := v(R\varphi),$$

for $\varphi \in \mathcal{S}(\mathbb{R}^n)$. Prove that $Rv \in \mathcal{S}'(\mathbb{R}^n)$. Prove also that this definition is consistent with the definition of $(Rg)(x) = g(-x)$ for $g \in C^\infty(\mathbb{R}^n)$.

Exercise 1.4.26. Show that if $v \in \mathcal{S}'(\mathbb{R}^n)$, then the mapping $\varphi \mapsto Rv * \varphi$ is continuous from $C^\infty(\mathbb{R}^n)$ to $C^\infty(\mathbb{R}^n)$.

Consequently, if $v \in \mathcal{S}'(\mathbb{R}^n)$ and $\varphi \in \mathcal{S}(\mathbb{R}^n)$, we have $Rv * \varphi \in C^\infty(\mathbb{R}^n)$ by Lemma 1.4.23. This motivates the following:

Definition 1.4.27 (Convolution of distributions). Let $u \in \mathcal{E}'(\mathbb{R}^n)$ and $v \in \mathcal{S}'(\mathbb{R}^n)$. Define the convolution $u * v$ of u and v by

$$(u * v)(\varphi) := u(Rv * \varphi),$$

for all $\varphi \in \mathcal{S}(\mathbb{R}^n)$.

Exercise 1.4.28. We see from Exercise 1.4.24 that this definition is consistent with $\mathcal{S}(\mathbb{R}^n)$. Prove that if $u \in \mathcal{E}'(\mathbb{R}^n)$ and $v \in \mathcal{S}'(\mathbb{R}^n)$ then $u * v \in \mathcal{S}'(\mathbb{R}^n)$.

Exercise 1.4.29. Prove that if $u \in \mathcal{E}'(\mathbb{R}^n)$ and $v \in \mathcal{S}'(\mathbb{R}^n)$ then

$$\text{sing supp } (u * v) \subset \text{sing supp } u + \text{sing supp } v.$$

Exercise 1.4.30. Extend the notion of convolution to two distributions $u, v \in \mathcal{D}'(\mathbb{R}^n)$ when at least one of them has compact support.

Exercise 1.4.31 (Diagonal property). Show that the convolution $u * v$ of two distributions exists if for every compact set K the intersection $(\text{supp } u \times \text{supp } v) \cap \{(x, y) : x + y \in K\}$ is compact. This allows, for example, to take a convolution of two Heaviside functions H, yielding $H * H = xH$.

Remark 1.4.32. Let us show an example of the calculation with distributions. Let $v \in \mathcal{S}'(\mathbb{R}^n)$. We will show that $v * \delta = \delta * v = v$. Indeed, on one hand we have

$$(v * \delta)(\varphi) \overset{1.4.27}{=} v(R\delta * \varphi) = v(\delta * \varphi) \overset{1.4.22}{=} v(\varphi)$$

in view of $R\delta = \delta$:

$$\langle R\delta, \psi \rangle \overset{1.4.25}{=} \langle \delta, R\psi \rangle = R\psi(0) = \psi(0) = \langle \delta, \psi \rangle.$$

On the other hand, we have

$$(\delta * v)(\varphi) \overset{1.4.27}{=} \delta(Rv * \varphi) = (Rv * \varphi)(0) \overset{1.4.21}{=} Rv(\tau_0 R\varphi) = Rv(R\varphi) \overset{1.4.25}{=} v(\varphi).$$

Note that in view of Exercise 1.4.30 we could have taken $v \in \mathcal{D}'(\mathbb{R}^n)$ here.

Exercise 1.4.33. Let $u \in \mathcal{S}'(\mathbb{R}^n)$ and $v \in \mathcal{E}'(\mathbb{R}^n)$. Define $u * v$ as $v * u$, i.e., $u * v := v * u$. Prove that this coincides with Definition 1.4.27 when $u, v \in \mathcal{E}'(\mathbb{R}^n)$.

Exercise 1.4.34. Prove that the extension of (1.8) holds, i.e., that

$$\partial^\alpha (f * g) = \partial^\alpha f * g = f * \partial^\alpha g.$$

Remark 1.4.35 (Non-associativity of convolution). In Exercise 1.1.29 we formulated the associativity of a convolution. However, for distributions one has to be careful. Indeed, recalling the relation $H' = \delta$ from Remark 1.3.26 and assuming the associativity one could prove

$$
\begin{aligned}
1 \quad &\overset{1.4.32}{=} \quad \delta * 1 \\
&\overset{1.4.32}{=} \quad (\delta * \delta) * 1 \\
&\overset{1.4.34}{=} \quad (H * \delta') * 1 \\
&\overset{\text{"}=\text{"}}{} \quad H * (\delta' * 1) \\
&\overset{1.4.34}{=} \quad H * (\delta * 1') \\
&= \quad H * 0 \\
&= \quad 0.
\end{aligned}
$$

Exercise 1.4.36. Why does the associativity in Remark 1.4.35 fail? How could we restrict the spaces of distributions for the convolution to be still associative?

Exercise 1.4.37. Show that if $u \in \mathcal{E}'(\mathbb{R}^n)$ and $v \in \mathcal{S}'(\mathbb{R}^n)$, then $\widehat{u * v} = \widehat{u}\widehat{v}$, where the product on the right-hand side makes sense in view of Remark 1.3.19 and Exercise 1.4.19.

We now formulate a couple of useful properties of translations:

Exercise 1.4.38 (Translation is continuous in $L^p(\mathbb{R}^n)$). Prove that translation is continuous in $L^p(\mathbb{R}^n)$, namely that translations $\tau_x : L^p(\mathbb{R}^n) \to L^p(\mathbb{R}^n)$, $(\tau_x f)(y) = f(y - x)$, satisfy $\|\tau_x f - f\|_{L^p} \to 0$ as $x \to 0$, for every $f \in L^p(\mathbb{R}^n)$.

Exercise 1.4.39 (Translations of convolutions). For $f, g \in L^1(\mathbb{R}^n)$, show that the convolution of f and g satisfies

$$\tau_x(f * g) = (\tau_x f) * g = f * (\tau_x g).$$

Can you extend this to some classes of distributions?

1.5   Sobolev spaces

In this section we discuss Sobolev spaces L_k^p with integer orders $k \in \mathbb{N}$. After introducing the necessary elements of the theory of pseudo-differential operators we will come back to this topic in Section 2.6.3 to also discuss Sobolev spaces L_s^p for all real $s \in \mathbb{R}$.

1.5.1   Weak derivatives and Sobolev spaces

There is a notion of a weak derivative which is a special case of the distributional derivative from Definition 1.3.15. However, it allows a realisation in an integral form and we mention it here briefly.

Definition 1.5.1 (Weak derivative). Let Ω be an open subset of $\mathbb{R}^n$ and let $u, v \in L^1_{\text{loc}}(\Omega)$. We say that v is the αth-*weak partial derivative of* u if

$$\int_\Omega u \, \partial^\alpha \varphi \, dx = (-1)^{|\alpha|} \int_\Omega v \, \varphi \, dx, \quad \text{for all} \quad \varphi \in C_0^\infty(\Omega).$$

In this case we also write $v = \partial^\alpha u$.

The constant $(-1)^{|\alpha|}$ stands for the consistency with the corresponding definition for smooth functions when using integration by parts in Ω. It is the same reason as to include the constant $(-1)^{|\alpha|}$ in Definition 1.3.15. The weak derivative defined in this way is uniquely determined:

Lemma 1.5.2. *Let* $u \in L^1_{\text{loc}}(\Omega)$. *If a weak* α^{th} *derivative of* u *exists, it is uniquely defined up to a set of measure zero.*

Proof. Indeed, assume that there are two functions $v, w \in L^1_{loc}(\Omega)$ such that

$$\int_\Omega u \, \partial^\alpha \varphi \, dx = (-1)^{|\alpha|} \int_\Omega v \, \varphi \, dx = (-1)^{|\alpha|} \int_\Omega w \, \varphi \, dx,$$

for all $\varphi \in C_0^\infty(\Omega)$. Then $\int_\Omega (v - w)\varphi \, dx = 0$ for all $\varphi \in C_0^\infty(\Omega)$. A standard result from measure theory (e.g., Theorem C.4.60) now implies that $v = w$ almost everywhere in Ω. $\square$

Exercise 1.5.3. Let us define $u, v : \mathbb{R} \to \mathbb{R}$ by

$$u(x) = \begin{cases} x, & \text{if } x \leq 1, \\ 1, & \text{if } x > 1, \end{cases} \qquad v(x) = \begin{cases} 1, & \text{if } x \leq 1, \\ 0, & \text{if } x > 1, \end{cases}$$

Prove that $u' = v$ weakly.

Exercise 1.5.4. Define $u : \mathbb{R} \to \mathbb{R}$ by

$$u(x) = \begin{cases} x, & \text{if } x \leq 1, \\ 2, & \text{if } x > 1. \end{cases}$$

Prove that u has no weak derivative. Calculate the distributional derivative of u.

Exercise 1.5.5. Prove that the Dirac δ-distribution is not an element of $L^1_{loc}(\mathbb{R}^n)$.

There are different ways to define Sobolev spaces[2]. Here we choose the one using weak or distributional derivatives.

Definition 1.5.6 (Sobolev spaces). Let $1 \leq p \leq \infty$ and let $k \in \mathbb{N} \cup \{0\}$. The *Sobolev space* $L^p_k(\Omega)$ (or $W^{p,k}(\Omega)$) consists of all $u \in L^1_{loc}(\Omega)$ such that for all multi-indices α with $|\alpha| \leq k$, $\partial^\alpha u$ exists weakly (or distributionally) and $\partial^\alpha u \in L^p(\Omega)$. For $u \in L^p_k(\Omega)$, we define

$$\|u\|_{L^p_k(\Omega)} := \left(\sum_{|\alpha| \leq k} \|\partial^\alpha u\|_{L^p}^p \right)^{1/p} = \left(\sum_{|\alpha| \leq k} \int_\Omega |\partial^\alpha u|^p \, dx \right)^{1/p},$$

for $1 \leq p < \infty$, and for $p = \infty$ we define

$$\|u\|_{L^\infty_k(\Omega)} := \max_{|\alpha| \leq k} \operatorname{esssup}_\Omega |\partial^\alpha u|.$$

Since $p \geq 1$, we know that $L^p_{loc}(\Omega) \subset L^1_{loc}(\Omega)$, e.g., by Hölder's inequality (Proposition 1.2.4), so we note that it does not matter whether we take a weak or a distributional derivative.

In the case $p = 2$, one often uses the notation $H^k(\Omega)$ for $L^2_k(\Omega)$, and in the case $p = 2$ and $k = 0$, we get $H^0(\Omega) = L^2(\Omega)$. As usual, we identify functions in $L^p_k(\Omega)$ which are equal almost everywhere (see Definition C.4.6).

---

[2]We come back to this subject in Section 2.6.3.

Proposition 1.5.7. *The functions* $||\cdot||_{L_k^p(\Omega)}$ *in Definition 1.5.6 are norms on* $L_k^p(\Omega)$.

Proof. Indeed, we clearly have $||\lambda u||_{L_k^p} = |\lambda|||u||_{L_k^p}$ and $||u||_{L_k^p} = 0$ if and only if $u = 0$ almost everywhere. For the triangle inequality, the case $p = \infty$ is straightforward. For $1 \leq p < \infty$ and for $u, v \in L_k^p(\Omega)$, Minkowski's inequality (Proposition 1.2.7) implies

$$||u + v||_{L_k^p} = \left(\sum_{|\alpha| \leq k} ||\partial^\alpha u + \partial^\alpha v||_{L^p}^p \right)^{1/p} \leq \left(\sum_{|\alpha| \leq k} (||\partial^\alpha u||_{L^p} + ||\partial^\alpha v||_{L^p})^p \right)^{1/p}$$

$$\leq \left(\sum_{|\alpha| \leq k} ||\partial^\alpha u||_{L^p}^p \right)^{1/p} + \left(\sum_{|\alpha| \leq k} ||\partial^\alpha v||_{L^p}^p \right)^{1/p}$$

$$= ||u||_{L_k^p} + ||v||_{L_k^p},$$

completing the proof. $\qquad\qquad\qquad\qquad\qquad\qquad\qquad\qquad\qquad\qquad\qquad\quad$ $\square$

We define local versions of spaces $L_k^p(\Omega)$ similarly to local versions of L^p-spaces.

Definition 1.5.8 (Localisations of Sobolev spaces). We will say that $f \in L_k^p(\Omega)_{\text{loc}}$ if $\varphi f \in L_k^p(\Omega)$ for all $\varphi \in C_0^\infty(\Omega)$. We will write $f_m \to f$ in $L_k^p(\Omega)_{\text{loc}}$ as $m \to \infty$, if f and f_m belong to $L_k^p(\Omega)_{\text{loc}}$ for all m, and if $\varphi f_m \to \varphi f$ in $L_k^p(\Omega)$ as $m \to \infty$, for all $\varphi \in C_0^\infty(\Omega)$.

Example (**Example of a point singularity**). An often encountered example of a function with a point singularity is $u(x) = |x|^{-a}$ defined for $x \in \Omega = B(0,1) \subset \mathbb{R}^n$, $x \neq 0$. We may ask a question: for which $a > 0$ do we have $u \in L_1^p(\Omega)$?

First we observe that away from the origin, u is a smooth function and can be differentiated pointwise with $\partial_{x_j} u = -a x_j |x|^{-a-2}$ and hence also $|\nabla u(x)| = |a|||x|^{-a-1}$, $x \neq 0$. In particular, $|\nabla u| \in L^1(\Omega)$ for $a+1 < n$ (here Exercise 1.1.19 is of use). We also have $|\nabla u| \in L^p(\Omega)$ for $(a+1)p < n$. So we must assume $a+1 < n$ and $(a+1)p < n$. Let us now calculate the weak (distributional) derivative of u in Ω. Let $\varphi \in C_0^\infty(\Omega)$. Let $\epsilon > 0$. On $\Omega \backslash B(0, \epsilon)$ we can integrate by parts to get

$$\int_{\Omega \backslash B(0,\epsilon)} u \partial_{x_j} \varphi \, dx = - \int_{\Omega \backslash B(0,\epsilon)} \partial_{x_j} u \varphi \, dx + \int_{\partial B(0,\epsilon)} u \varphi \nu^j \, d\sigma, \qquad (1.11)$$

where $d\sigma$ is the surface measure on the sphere $\partial B(0, \epsilon)$ and $\nu = (\nu^1, \ldots, \nu^n)$ is the inward pointing normal on $\partial B(0, \epsilon)$. Now, since $u = |\epsilon|^{-a}$ on $\partial B(0, \epsilon)$, we can estimate

$$\left| \int_{\partial B(0,\epsilon)} u \varphi \nu^j \, d\sigma \right| \leq ||\varphi||_{L^\infty} \int_{\partial B(0,\epsilon)} \epsilon^{-a} \, d\sigma \leq C \epsilon^{n-1-a} \to 0$$

as $\epsilon \to 0$, since $a + 1 < n$. Passing to the limit in the integration by parts formula (1.11), we get $\int_\Omega u \partial_{x_j} \varphi \, dx = -\int_\Omega \partial_{x_j} u \varphi \, dx$, which means that $\partial_{x_j} u$ is also the weak derivative of u. So, $u \in L_1^p(\Omega)$ if $u, |\nabla u| \in L^p(\Omega)$, which holds for $(a+1)p < n$, i.e., for $a < (n - p)/p$.

Exercise 1.5.9. Find conditions on a in the above example for which $u \in L_k^p(\Omega)$.

1.5.2 Some properties of Sobolev spaces

Since $L^p(\Omega) \subset \mathcal{D}'(\Omega)$, we can work with $u \in L^p(\Omega)$ as with functions or as with distributions. In particular, we can differentiate them distributionally, etc. Moreover, as we have already seen, the equality of objects (be it functions, functionals, distributions, etc.) depends on the spaces in which the equality is considered. In Sobolev spaces we can use tools from measure theory so we work with functions defined almost everywhere. Thus, an equality in Sobolev spaces (as in the following theorem) means pointwise equality almost everywhere.

Theorem 1.5.10 (Properties of Sobolev spaces). *Let* $u, v \in L_k^p(\Omega)$, *and let* α *be a multi-index with* $|\alpha| \leq k$. *Then:*

(i) $\partial^\alpha u \in L_{k-|\alpha|}^p(\Omega)$, *and* $\partial^\alpha(\partial^\beta u) = \partial^\beta(\partial^\alpha u) = \partial^{\alpha+\beta} u$, *for all multi-indices* α, β *such that* $|\alpha| + |\beta| \leq k$.

(ii) *For all* $\lambda, \mu \in \mathbb{C}$ *we have* $\lambda u + \mu v \in L_k^p(\Omega)$ *and* $\partial^\alpha(\lambda u + \mu v) = \lambda \partial^\alpha u + \mu \partial^\alpha v$.

(iii) *If* $\widetilde{\Omega}$ *is an open subset of* Ω, *then* $u \in L_k^p(\widetilde{\Omega})$.

(iv) *If* $\chi \in C_0^\infty(\Omega)$, *then* $\chi u \in L_k^p(\Omega)$ *and we have the Leibniz formula*

$$\partial^\alpha(\chi u) = \sum_{\beta \leq \alpha} \binom{\alpha}{\beta} (\partial^\beta \chi)(\partial^{\alpha-\beta} u),$$

where $\binom{\alpha}{\beta} = \frac{\alpha!}{\beta!(\alpha-\beta)!}$ *is the binomial coefficient.*

(v) $L_k^p(\Omega)$ *is a Banach space.*

Proof. Statements (i), (ii), and (iii) are easy. For example, if $\varphi \in C_0^\infty(\Omega)$ then also $\partial^\beta \varphi \in C_0^\infty(\Omega)$, and (i) follows from

$$\int_\Omega \partial^\alpha u \, \partial^\beta \varphi \, dx = (-1)^{|\alpha|} \int_\Omega u \, \partial^{\alpha+\beta} \varphi \, dx = (-1)^{|\alpha|+|\alpha+\beta|} \int_\Omega \partial^{\alpha+\beta} u \, \varphi \, dx,$$

since $(-1)^{|\alpha|+|\alpha+\beta|} = (-1)^{|\beta|}$.

Let us now show (iv). The proof will be carried out by induction on $|\alpha|$. For $|\alpha| = 1$, writing $\langle u, \varphi \rangle$ for $u(\varphi) = \int_\Omega u \varphi \, dx$, we get

$$\langle \partial^\alpha(\chi u), \varphi \rangle = (-1)^{|\alpha|} \langle u, \chi \partial^\alpha \varphi \rangle$$
$$= -\langle u, \partial^\alpha(\chi \varphi) - (\partial^\alpha \chi)\varphi \rangle = \langle \chi \partial^\alpha u, \varphi \rangle + \langle (\partial^\alpha \chi)u, \varphi \rangle,$$

which is what was required. Now, suppose that the Leibniz formula is valid for all $|\beta| \leq l$, and let us take α with $|\alpha| = l+1$. Then we can write $\alpha = \beta + \gamma$ with some

$|\beta| = l$ and $|\gamma| = 1$. We get

$$
\begin{aligned}
\langle \chi u, \partial^\alpha \varphi \rangle &= \langle \chi u, \partial^\beta (\partial^\gamma \varphi) \rangle \\
&= (-1)^{|\beta|} \langle \partial^\beta (\chi u), \partial^\gamma \varphi \rangle \quad \text{(by induction hypothesis)} \\
&= (-1)^{|\beta|} \left\langle \sum_{\sigma \leq \beta} \binom{\beta}{\sigma} \partial^\sigma \chi \, \partial^{\beta-\sigma} u, \partial^\gamma \varphi \right\rangle \quad \text{(by definition)} \\
&= (-1)^{|\beta|+|\gamma|} \left\langle \sum_{\sigma \leq \beta} \binom{\beta}{\sigma} \partial^\gamma (\partial^\sigma \chi \, \partial^{\beta-\sigma} u), \varphi \right\rangle \quad (\text{set } \rho = \sigma + \gamma) \\
&= (-1)^{|\alpha|} \left\langle \sum_{\sigma \leq \beta} \binom{\beta}{\sigma} (\partial^\rho \chi \, \partial^{\alpha-\rho} u + \partial^\sigma \chi \, \partial^{\alpha-\sigma} u), \varphi \right\rangle \\
&= (-1)^{|\alpha|} \left\langle \sum_{\rho \leq \alpha} \binom{\alpha}{\rho} \partial^\rho \chi \, \partial^{\alpha-\rho} u, \varphi \right\rangle,
\end{aligned}
$$

where we used that $\binom{\beta}{\sigma} + \binom{\beta}{\sigma} = \binom{\alpha-\gamma}{\rho-\gamma} + \binom{\alpha-\gamma}{\rho} = \binom{\alpha}{\rho}$.

Now let us prove (v). We have already shown in Proposition 1.5.7 that $L_k^p(\Omega)$ is a normed space. Let us show now that the completeness of $L^p(\Omega)$ (Theorem C.4.9) implies the completeness of $L_k^p(\Omega)$. Let u_m be a Cauchy sequence in $L_k^p(\Omega)$. Then $\partial^\alpha u_m$ is a Cauchy sequence in $L^p(\Omega)$ for any $|\alpha| \leq k$. Since $L^p(\Omega)$ is complete, there exists some $u_\alpha \in L^p(\Omega)$ such that $\partial^\alpha u_m \to u_\alpha$ in $L^p(\Omega)$. Let $u = u_{(0,\ldots,0)}$, so in particular, we have $u_m \to u$ in $L^p(\Omega)$. Let us now show that in fact $u \in L_k^p(\Omega)$ and $\partial^\alpha u = u_\alpha$ for all $|\alpha| \leq k$. Let $\varphi \in C_0^\infty(\Omega)$. Then

$$
\begin{aligned}
\langle \partial^\alpha u, \varphi \rangle &= (-1)^{|\alpha|} \langle u, \partial^\alpha \varphi \rangle \\
&= (-1)^{|\alpha|} \lim_{m \to \infty} \langle u_m, \partial^\alpha \varphi \rangle \\
&= \lim_{m \to \infty} \langle \partial^\alpha u_m, \varphi \rangle \\
&= \langle u_\alpha, \varphi \rangle,
\end{aligned}
$$

which implies $u \in L_k^p(\Omega)$ and $\partial^\alpha u = u_\alpha$. Moreover, we have $\partial^\alpha u_m \to \partial^\alpha u$ in $L^p(\Omega)$ for all $|\alpha| \leq k$, which means that $u_m \to u$ in $L_k^p(\Omega)$ and hence $L_k^p(\Omega)$ is complete. $\square$

Exercise 1.5.11 (An embedding theorem). Prove that if $s > k + n/2$ and $s \in \mathbb{N}$ then $H^s(\mathbb{R}^n) \subset C^k(\mathbb{R}^n)$ and the inclusion is continuous. Do also Exercise 2.6.17 for a sharper version of this embedding.

1.5.3 Mollifiers

In Theorem 1.3.31 we saw that we can approximate quite irregular functions or (tempered) distributions by much more regular functions. The argument relied on the use of Fourier analysis and worked well on $\mathbb{R}^n$. Such a technique is very powerful, as could have been seen from the proof of Plancherel's formula in Theorem 1.3.13. On the other hand, when working in subsets of $\mathbb{R}^n$ we may be unable to use the Fourier transform (since for its definition we used the whole space $\mathbb{R}^n$). Thus, we want to be able to approximate functions (or distributions) by smooth functions without using Fourier techniques. This turns out to be possible using the so-called *mollification* of functions.

Assume for a moment that we are in $\mathbb{R}^n$ again and let us first argue very informally. Let us first look at the Fourier transform of the convolution with a δ-distribution. Thus, for a function f we must have $\widehat{\delta * f} = \widehat{\delta}\widehat{f} = \widehat{f}$, if we use that $\widehat{\delta} = 1$. Taking the inverse Fourier transform we obtain the important identity

$$\delta * f = f,$$

which will be justified formally later. Now, if we take a sequence of smooth functions η_ϵ approximating the δ-distribution, i.e., if $\eta_\epsilon \to \delta$ in some sense as $\epsilon \to 0$, and if this convergence is preserved by the convolution, we should get

$$\eta_\epsilon * f \to \delta * f = f \quad \text{as } \epsilon \to 0.$$

Now, the convolution $\eta_\epsilon * f$ may be defined locally in $\mathbb{R}^n$, and functions $\eta_\epsilon * f$ will be smooth if η_ϵ are, thus giving us a way to approximate f. We will now make this argument precise. For this, we will deal in a straightforward manner by looking at the limit of $\eta_\epsilon * f$ for a suitably chosen sequence of functions η_ϵ, referring neither to δ-distribution nor to the Fourier transform.

Definition 1.5.12 (Mollifiers). For an open set $\Omega \subset \mathbb{R}^n$ and $\epsilon > 0$ we define $\Omega_\epsilon = \{x \in \Omega : \text{dist}(x, \partial\Omega) > \epsilon\}$. Let us define $\eta \in C_0^\infty(\mathbb{R}^n)$ by

$$\eta(x) = \begin{cases} C\, e^{\frac{1}{|x|^2 - 1}}, & \text{if } |x| < 1, \\ 0, & \text{if } |x| \geq 1, \end{cases}$$

where the constant C is chosen so that $\int_{\mathbb{R}^n} \eta \, dx = 1$. Such a function η is called a (Friedrichs) *mollifier*. For $\epsilon > 0$, we define

$$\eta_\epsilon(x) = \frac{1}{\epsilon^n}\eta\left(\frac{x}{\epsilon}\right),$$

so that $\text{supp}\, \eta_\epsilon \subset B(0, \epsilon)$ and $\int_{\mathbb{R}^n} \eta_\epsilon \, dx = 1$.

Let $f \in L^1_{\text{loc}}(\Omega)$. A mollification of f corresponding to η is a family $f^\epsilon = \eta_\epsilon * f$ in Ω_ϵ, i.e.,

$$f^\epsilon(x) = \int_\Omega \eta_\epsilon(x - y)f(y)\, dy = \int_{B(0,\epsilon)} \eta_\epsilon(y)f(x - y)\, dy, \quad \text{for } x \in \Omega_\epsilon.$$

Theorem 1.5.13 (Properties of mollifications). *Let $f \in L^1_{loc}(\Omega)$. Then we have the following properties.*

(i) $f^\epsilon \in C^\infty(\Omega_\epsilon)$.

(ii) $f^\epsilon \to f$ *almost everywhere as* $\epsilon \to 0$.

(iii) *If $f \in C(\Omega)$, then $f^\epsilon \to f$ uniformly on compact subsets of Ω.*

(iv) $f^\epsilon \to f$ *in* $L^p_{loc}(\Omega)$ *for all* $1 \le p < \infty$.

Proof. To show (i), we can differentiate $f^\epsilon(x) = \int_\Omega \eta_\epsilon(x - y) f(y) \, dy$ under the integral sign and use the fact that $f \in L^1_{loc}(\Omega)$. The proof of (ii) will rely on the following

Theorem 1.5.14 (Lebesgue's differentiation theorem). *Let $f \in L^1_{loc}(\Omega)$. Then*

$$\lim_{r \to 0} \frac{1}{|B(x,r)|} \int_{B(x,r)} |f(y) - f(x)| \, dy = 0 \quad \text{for a.e. } x \in \Omega.$$

Now, for all x for which the statement of Lebesgue's differentiation theorem is true, we can estimate

$$|f^\epsilon(x) - f(x)| = \left| \int_{B(x,\epsilon)} \eta_\epsilon(x - y)(f(y) - f(x)) \, dy \right|$$

$$\le \epsilon^{-n} \int_{B(x,\epsilon)} \eta\left(\frac{x - y}{\epsilon} \right) |f(y) - f(x)| \, dy$$

$$\le C \frac{1}{|B(x,\epsilon)|} \int_{B(x,\epsilon)} |f(y) - f(x)| \, dy,$$

where the last expression goes to zero as $\epsilon \to 0$, by the choice of x. For (iii), let K be a compact subset of Ω. Let $K_0 \subset \Omega$ be another compact set such that K is contained in the interior of K_0. Then f is uniformly continuous on K_0 and the limit in the Lebesgue differentiation theorem holds uniformly for $x \in K$. The same argument as in (ii) then shows that $f^\epsilon \to f$ uniformly on K.

Finally, to show (iv), let us choose open sets $U \subset V \subset \Omega$ such that $U \subset V_\delta$ and $V \subset \Omega_\delta$ for some small $\delta > 0$. Let us show first that $\|f^\epsilon\|_{L^p(U)} \le \|f\|_{L^p(V)}$ for all sufficiently small $\epsilon > 0$. Indeed, for all $x \in U$, we can estimate

$$|f^\epsilon(x)| = \left| \int_{B(x,\epsilon)} \eta_\epsilon(x - y) f(y) \, dy \right|$$

$$\le \int_{B(x,\epsilon)} \eta_\epsilon^{1-1/p}(x - y) \eta_\epsilon^{1/p}(x - y) |f(y)| \, dy \quad \text{(Hölder's inequality)}$$

$$\le \left(\int_{B(x,\epsilon)} \eta_\epsilon(x - y) \, dy \right)^{1-1/p} \left(\int_{B(x,\epsilon)} \eta_\epsilon(x - y) |f(y)|^p \, dy \right)^{1/p}.$$

Since $\int_{B(x,\epsilon)} \eta_\epsilon(x-y) \, dy = 1$, we get

$$
\int_U |f^\epsilon(x)|^p \, dx \leq \int_U \left(\int_{B(x,\epsilon)} \eta_\epsilon(x-y)|f(y)|^p \, dy \right) dx
$$

$$
\leq \int_V \left(\int_{B(y,\epsilon)} \eta_\epsilon(x-y) \, dx \right) |f(y)|^p \, dy
$$

$$
= \int_V |f(y)|^p \, dy.
$$

Now, let $\delta > 0$ and let us choose $g \in C(V)$ such that $||f - g||_{L^p(V)} < \delta$ (here we use the fact that $C(V)$ is (sequentially) dense in $L^p(V)$). Then

$$
||f^\epsilon - f||_{L^p(U)} \leq ||f^\epsilon - g^\epsilon||_{L^p(U)} + ||g^\epsilon - g||_{L^p(U)} + ||g - f||_{L^p(U)}
$$

$$
\leq 2||f - g||_{L^p(V)} + ||g^\epsilon - g||_{L^p(U)}
$$

$$
< 2\delta + ||g^\epsilon - g||_{L^p(U)}.
$$

Since $g^\epsilon \to g$ uniformly on the closure of V by (iii), it follows that $||f^\epsilon - f||_{L^p(U)} \leq 3\delta$ for small enough $\epsilon > 0$, completing the proof of (iv). $\qquad \square$

As a consequence of Theorem 1.5.13 we obtain

Corollary 1.5.15. *The space $C^\infty(\Omega)$ is sequentially dense in the space $C_0(\Omega)$ of all continuous functions with compact support in Ω. Also, $C^\infty(\Omega)$ is sequentially dense in $L^p_{\text{loc}}(\Omega)$ for all $1 \leq p < \infty$.*

Exercise 1.5.16. Prove a simple but useful corollary of the Lebesgue differentiation theorem, partly explaining its name:

Corollary 1.5.17 (Corollary of the Lebesgue differentiation theorem). *Let $f \in L^1_{\text{loc}}(\Omega)$. Then*

$$
\lim_{r \to 0} \frac{1}{|B(x,r)|} \int_{B(x,r)} f(y) \, dy = f(x) \quad \text{for a.e. } x \in \Omega.
$$

1.5.4 Approximation of Sobolev space functions

With the use of mollifications we can approximate functions in Sobolev spaces by smooth functions. We have a local approximation in localised Sobolev spaces $L^p_k(\Omega)_{\text{loc}}$, a global approximation in $L^p_k(\Omega)$, and further approximations dependent on the regularity of the boundary of Ω. Although the set Ω is bounded, we still say that an approximation in $L^p_k(\Omega)$ is global if it works *up to the boundary*.

Proposition 1.5.18 (Local approximation by smooth functions). *Assume that $\Omega \subset \mathbb{R}^n$ is open. Let $f \in L^p_k(\Omega)$ for $1 \leq p < \infty$ and $k \in \mathbb{N} \cup \{0\}$. Let $f^\epsilon = \eta_\epsilon * f$ in Ω_ϵ be the mollification of f, $\epsilon > 0$. Then $f^\epsilon \in C^\infty(\Omega_\epsilon)$ and $f^\epsilon \to f$ in $L^p_k(\Omega)_{\text{loc}}$ as $\epsilon \to 0$, i.e., $f^\epsilon \to f$ in $L^p_k(K)$ as $\epsilon \to 0$ for all compact $K \subset \Omega$.*

Proof. It was already proved in Theorem 1.5.3, (i), that $f^\epsilon \in C^\infty(\Omega_\epsilon)$. Since f is locally integrable, we can differentiate the convolution under the integral sign to get $\partial^\alpha f^\epsilon = \eta_\epsilon * \partial^\alpha u$ in Ω_ϵ. Now, let U be an open and bounded subset of Ω containing K. Then by Theorem 1.5.3, (iv), we get $\partial^\alpha f^\epsilon \to \partial^\alpha f$ in $L^p(U)$ as $\epsilon \to 0$, for all $|\alpha| \le k$. Hence

$$||f^\epsilon - f||^p_{L^p_k(U)} = \sum_{|\alpha| \le k} ||\partial^\alpha f^\epsilon - \partial^\alpha f||^p_{L^p(U)} \to 0$$

as $\epsilon \to 0$, proving the statement. $\qquad\qquad\qquad\qquad\qquad\qquad\qquad\qquad\qquad\square$

Proposition 1.5.19 (Global approximation by smooth functions). *Assume that $\Omega \subset \mathbb{R}^n$ is open and bounded. Let $f \in L^p_k(\Omega)$ for $1 \le p < \infty$ and $k \in \mathbb{N} \cup \{0\}$. Then there is a sequence $f_m \in C^\infty(\Omega) \cap L^p_k(\Omega)$ such that $f_m \to f$ in $L^p_k(\Omega)$.*

Proof. Let us write $\Omega = \bigcup_{j=1}^\infty \Omega_j$, where $\Omega_j = \{x \in \Omega : \text{dist}(x, \partial\Omega) > 1/j\}$. Let $V_j = \Omega_{j+3} \backslash \overline{\Omega}_{j+1}$ (this definition will be very important). Take also any open V_0 with $\overline{V_0} \subset \Omega$ so that $\Omega = \bigcup_{j=0}^\infty V_j$. Let χ_j be a partition of unity subordinate to V_j, i.e., a family $\chi_j \in C_0^\infty(V_j)$ such that $0 \le \chi_j \le 1$ and $\sum_{j=0}^\infty \chi_j = 1$ in Ω. Then $\chi_j f \in L^p_k(\Omega)$ and $\text{supp}(\chi_j f) \subset V_j$. Let us fix some $\delta > 0$ and choose $\epsilon_j > 0$ so small that the function $f^j = \eta_{\epsilon_j} * (\chi_j f)$ is supported in $W_j = \Omega_{j+4} \backslash \overline{\Omega}_j$ and satisfies $||f^j - \chi_j f||_{L^p_k(\Omega)} \le \delta 2^{-j-1}$ for all j. Let now $g = \sum_{j=0}^\infty f^j$. Then $g \in C^\infty(\Omega)$ since in any open set U in Ω there are only finitely many non-zero terms in the sum. Moreover, since $f = \sum_{j=0}^\infty \chi_j f$, for each such U we have

$$||g - f||_{L^p_k(U)} \le \sum_{j=0}^\infty ||f^j - \chi_j f||_{L^p_k(\Omega)} \le \delta \sum_{j=0}^\infty \frac{1}{2^{j+1}} = \delta.$$

Taking the supremum over all open subsets U of Ω, we obtain $||g - f||_{L^p_k(\Omega)} \le \delta$, completing the proof. $\qquad\qquad\qquad\qquad\qquad\qquad\qquad\qquad\qquad\qquad\qquad\square$

In general, there are many versions of these results depending on the set Ω, in particular on the regularity of its boundary. For example, we give here without proof the following

Further result. *Let Ω be a bounded subset of $\mathbb{R}^n$ with C^1 boundary. Let $f \in L^p_k(\Omega)$ for $1 \le p < \infty$ and $k \in \mathbb{N} \cup \{0\}$. Then there is a sequence $f_m \in C^\infty(\overline{\Omega})$ such that $f_m \to f$ in $L^p_k(\Omega)$.*

Finally, we use mollifiers to establish a smooth version of Urysohn's lemma in Theorem A.12.11.

Theorem 1.5.20 (Smooth Urysohn's lemma). *Let $K \subset \mathbb{R}^n$ be compact and $U \subset \mathbb{R}^n$ be open such that $K \subset U$. Then there exists $f \in C_0^\infty(U)$ such that $0 \le f \le 1$, $f = 1$ on K and $\text{supp } f \subset U$.*

Proof. First we observe that the distance $\delta := \mathrm{dist}(K, \mathbb{R}^n \backslash U) > 0$ because K is compact and $\mathbb{R}^n \backslash U$ is closed. Let $V := \{x \in \mathbb{R}^n : \mathrm{dist}(x, K) < \delta/3\}$. If η is the Friedrichs mollifier from Definition 1.5.12, then $\chi := \eta_{\delta/3}$ satisfies $\mathrm{supp}\,\chi \subset \{x \in \mathbb{R}^n : |x| \leq \delta/3\}$ and $\int_{\mathbb{R}^n} \chi(x)\,\mathrm{d}x = 1$. The desired function f can then be obtained as $f := \mathbb{I}_V * \chi$, where $\mathbb{I}_V$ is the characteristic function of the set V. We have that $f \in C^\infty$ by Theorem 1.5.13 and $\mathrm{supp}\,f \subset U$ by Exercise[3] 1.4.29. We have $0 \leq f \leq 1$ from its definition, and $f = 1$ on K follows by a direct verification. $\qquad\square$

1.6 Interpolation

The Riesz–Thorin interpolation theorem C.4.18 was already useful in establishing various inequalities in L^p (for example, it was used to prove the general Young's inequality for convolutions in Proposition 1.2.10, or the Hausdorff–Young inequality in Corollary 1.3.14).

The aim of this section is to prove another very useful interpolation result: the Marcinkiewicz interpolation theorem. Here μ will stand for the Lebesgue measure on $\mathbb{R}^n$.

Definition 1.6.1 (Distribution functions). For a function $f : \mathbb{R}^n \to \mathbb{C}$ we define its *distribution function* $\mu_f(\lambda)$ by

$$\mu_f(\lambda) = \mu\{x \in \mathbb{R}^n : |f(x)| \geq \lambda\}.$$

We have the following useful relation between the L^p-norm and the distribution of a function.

Theorem 1.6.2. *Let $f \in L^p(\mathbb{R}^n)$. Then we have the identity*

$$\int_{\mathbb{R}^n} |f(x)|^p\,\mathrm{d}x = p \int_0^\infty \mu_f(\lambda)\lambda^{p-1}\,\mathrm{d}\lambda.$$

Proof. Let us define a measure on $\mathbb{R}$ by setting

$$
\begin{aligned}
\nu((a, b]) &:= \mu_f(b) - \mu_f(a) \\
&= -\mu\{x \in \mathbb{R}^n : a < |f(x)| \leq b\} \\
&= -\mu(|f|^{-1}((a, b])).
\end{aligned}
$$

By the standard extension property of measures we can then extend ν to all Borel sets $E \subset (0, \infty)$ by setting $\nu(E) = -\mu(|f|^{-1}(E))$. We note that this definition is well defined since $|f|$ is measurable if f is measurable (Theorem C.2.9). Then we claim that we have the following property for, say, integrable functions $\varphi :$ $[0, \infty) \to \mathbb{R}$:

$$\int_{\mathbb{R}^n} \varphi \circ |f|\,\mathrm{d}\mu = -\int_0^\infty \varphi(\alpha)\,\mathrm{d}\nu(\alpha). \tag{1.12}$$

---

[3]In fact, in this case the property $\mathrm{supp}\,f \subset \mathrm{supp}\,\mathbb{I}_V + \mathrm{supp}\,\chi \subset V + B_{\delta/3}(0) = \{x \in \mathbb{R}^n : d(x, K) \leq 2\delta/3\} \subset U$ can be easily checked directly.

Indeed, if $\varphi = \chi_{[a,b]}$ is a characteristic function of a set $[a,b]$, i.e., equal to one on $[a,b]$ and zero on its complement, then the definition of ν implies

$$\int_{\mathbb{R}^n} \chi_{[a,b]} \circ |f| \, \mathrm{d}\mu = \int_{a<|f(x)|\leq b} \mathrm{d}\mu = -\int_a^b \mathrm{d}\nu = -\int_0^\infty \chi_{[a,b]} \, \mathrm{d}\nu,$$

which verifies (1.12) for characteristic functions. By the linearity of integrals, we then have (1.12) for finite linear combinations of characteristic functions and, consequently, for all integrable functions by the monotone convergence theorem (Theorem C.3.6). Now, taking $\varphi(\alpha) = \alpha^p$ in (1.12), we get

$$\int_{\mathbb{R}^n} |f|^p \, \mathrm{d}\mu = -\int_0^\infty \alpha^p \, \mathrm{d}\nu(\alpha) = p \int_0^\infty \alpha^{p-1} \mu_f(\alpha) \, \mathrm{d}\alpha,$$

where we integrated by parts in the last equality. The proof is complete. $\qquad\square$

Definition 1.6.3 (Weak type (p,p)). We say that operator T is of *weak type (p,p)* if there is a constant $C > 0$ such that for every $\lambda > 0$ we have

$$\mu\{x \in \mathbb{R}^n : |Tu(x)| > \lambda\} \leq C \frac{||u||_{L^p}^p}{\lambda^p}.$$

Proposition 1.6.4. *If T is bounded from $L^p(\mathbb{R}^n)$ to $L^p(\mathbb{R}^n)$ then T is also of weak type (p,p).*

Proof. If $v \in L^1(\mathbb{R}^n)$ then for all $\rho > 0$ we have a simple estimate

$$\rho\mu\{x \in \mathbb{R}^n : |v(x)| > \rho\} \leq \int_{|v(x)|>\rho} |v(x)| \, \mathrm{d}\mu(x) \leq ||v||_{L^1}.$$

Now, if we take $v(x) = |Tu(x)|^p$ and $\rho = \lambda^p$, this readily implies that T is of weak type (p,p). $\qquad\square$

The following theorem is extremely valuable in proving L^p-continuity of operators since it reduces the analysis to a weaker type continuity only for two values of indices.

Theorem 1.6.5 (Marcinkiewicz' interpolation theorem). *Let $r < q$ and assume that operator T is of weak types (r,r) and (q,q). Then T is bounded from $L^p(\mathbb{R}^n)$ to $L^p(\mathbb{R}^n)$ for all $r < p < q$.*

Proof. Let $u \in L^p(\mathbb{R}^n)$. For each $\lambda > 0$ we can define functions u_1 and u_2 by $u_1(x) = u(x)$ for $|u(x)| > \lambda$ and by $u_2(x) = u(x)$ for $|u(x)| \leq \lambda$, and to be zero otherwise. Then we have the identity $u = u_1 + u_2$ and estimates $|u_1|, |u_2| \leq |u|$. It follows that

$$\mu_{Tu}(2\lambda) \leq \mu_{Tu_1}(\lambda) + \mu_{Tu_2}(\lambda) \leq C_1 \frac{||u_1||_{L^r}^r}{\lambda^r} + C_2 \frac{||u_2||_{L^q}^q}{\lambda^q},$$

since T is of weak types (r, r) and (q, q). Therefore, we can estimate

$$\int_{\mathbb{R}^n} |Tu(x)|^p \, dx = p \int_0^\infty \lambda^{p-1} \mu_{Tu}(\lambda) \, d\lambda$$

$$\leq C_1 p \int_0^\infty \lambda^{p-1-r} \left(\int_{|u|>\lambda} |u(x)|^r \, dx \right) d\lambda$$

$$+ C_2 p \int_0^\infty \lambda^{p-1-q} \left(\int_{|u|\leq\lambda} |u(x)|^q \, dx \right) d\lambda.$$

Using Fubini's theorem, the first term on the right-hand side can be rewritten as

$$\int_0^\infty \lambda^{p-1-r} \left(\int_{|u|>\lambda} |u(x)|^r \, dx \right) d\lambda$$

$$= \int_0^\infty \lambda^{p-1-r} \left(\int_{\mathbb{R}^n} \chi_{|u|>\lambda} |u(x)|^r \, dx \right) d\lambda$$

$$= \int_{\mathbb{R}^n} |u(x)|^r \left(\int_0^\infty \lambda^{p-1-r} \chi_{|u|>\lambda} \, d\lambda \right) dx$$

$$= \int_{\mathbb{R}^n} |u(x)|^r \left(\int_0^{|u(x)|} \lambda^{p-1-r} \, d\lambda \right) dx$$

$$= \frac{1}{p-r} \int_{\mathbb{R}^n} |u(x)|^r |u(x)|^{p-r} \, dx$$

$$= \frac{1}{p-r} \int_{\mathbb{R}^n} |u(x)|^p \, dx,$$

where $\chi_{|u|>\lambda}$ is the characteristic function of the set $\{x \in \mathbb{R}^n : |u(x)| > \lambda\}$. Similarly, we have

$$\int_0^\infty \lambda^{p-1-q} \left(\int_{|u|\leq\lambda} |u(x)|^q \, dx \right) d\lambda = \frac{1}{q-p} \int_{\mathbb{R}^n} |u(x)|^p \, dx,$$

completing the proof. $\square$

As an important tool (which will not be used here so it is given just for the information) for proving various results of boundedness in $L^1(\mathbb{R}^n)$ or of weak type $(1, 1)$, we have the following fundamental decomposition of integrable functions.

Theorem 1.6.6 (Calderón–Zygmund covering lemma). *Let $u \in L^1(\mathbb{R}^n)$ and $\lambda > 0$. Then there exist $v, w_k \in L^1(\mathbb{R}^n)$ and there exists a collection of disjoint cubes Q_k,*

$k \in \mathbb{N}$, *centred at some points* x_k, *such that the following properties are satisfied:*

$$u = v + \sum_{k=1}^{\infty} w_k, \quad ||v||_{L^1} + \sum_{k=1}^{\infty} ||w_k||_{L^1} \le 3||u||_{L^1},$$

$$\operatorname{supp} w_k \subset Q_k, \quad \int_{Q_k} w_k(x) \, \mathrm{d}x = 0,$$

$$\sum_{k=1}^{\infty} \mu(Q_k) \le \lambda^{-1} ||u||_{L^1}, \quad |v(x)| \le 2^n \lambda.$$

This theorem is one of the starting points of the harmonic analysis of operators on $L^p(\mathbb{R}^n)$, but we will not pursue this topic here, and can refer to, e.g., [118] or [132] for many further aspects.

Chapter 2

Pseudo-differential Operators on $\mathbb{R}^n$

The subject of pseudo-differential operators on $\mathbb{R}^n$ is well studied and there are many excellent monographs on the subject, see, e.g., [27, 33, 55, 71, 101, 112, 130, 135, 152], as well as on the more general subject of Fourier integral operators, microlocal analysis, and related topics in, e.g., [30, 56, 45, 81, 113]. Therefore, here we only sketch main elements of the theory. In this chapter, we use the notation $\langle \xi \rangle = (1 + |\xi|^2)^{1/2}$.

2.1 Motivation and definition

We will start with an informal observation that if T is a translation invariant linear operator on some space of functions on $\mathbb{R}^n$, then we can write

$$T(e^{2\pi i x \cdot \xi}) = a(\xi)\, e^{2\pi i x \cdot \xi} \quad \text{for all} \quad \xi \in \mathbb{R}^n. \tag{2.1}$$

Indeed, more explicitly, if T acts on functions of the variable y, we can write $f(x, \xi) = T(e^{2\pi i y \cdot \xi})(x) = (T e_\xi)(x)$, where $e_\xi(x) = e^{2\pi i x \cdot \xi}$. Let $(\tau_h f)(x) = f(x - h)$ be the translation operator by $h \in \mathbb{R}^n$. We say that T is *translation invariant* if $T\tau_h = \tau_h T$ for all h. By our assumptions on T we get

$$f(x + h, \xi) = T(e^{2\pi i (y+h) \cdot \xi})(x) = e^{2\pi i h \cdot \xi} T(e^{2\pi i y \cdot \xi})(x) = e^{2\pi i h \cdot \xi} f(x, \xi).$$

Now, setting $x = 0$, we get $f(h, \xi) = e^{2\pi i h \cdot \xi} f(0, \xi)$, so we obtain formula (2.1) with $a(\xi) = f(0, \xi)$. In turn, this $a(\xi)$ can be found from formula (2.1), yielding

$$a(\xi) = e^{-2\pi i x \cdot \xi} T(e^{2\pi i y \cdot \xi})(x).$$

If we now formally apply T to the Fourier inversion formula

$$f(x) = \int_{\mathbb{R}^n} e^{2\pi i x \cdot \xi} \widehat{f}(\xi)\, d\xi$$

and use the linearity of T, we obtain

$$Tf(x) = \int_{\mathbb{R}^n} T(\,\mathrm{e}^{2\pi i x \cdot \xi})\widehat{f}(\xi) \,\mathrm{d}\xi = \int_{\mathbb{R}^n} \mathrm{e}^{2\pi i x \cdot \xi}a(\xi)\widehat{f}(\xi) \,\mathrm{d}\xi.$$

This formula allows one to reduce certain properties of the operator T to properties of the multiplication by the corresponding function $a(\xi)$, called the *symbol* of T. For example, continuity of T on L^2 would reduce to the boundedness of $a(\xi)$, composition of two operators $T_1 \circ T_2$ would reduce to the multiplication of their symbols $a_1(\xi)a_2(\xi)$, etc. Pseudo-differential operators extend this construction to functions which are not necessarily translation invariant. In fact, as we saw above we can always write $a(x,\xi) := \mathrm{e}^{-2\pi i x \cdot \xi}(Te_\xi)(x)$, so that we would have $T(\,\mathrm{e}^{2\pi i x \cdot \xi}) = \mathrm{e}^{2\pi i x \cdot \xi}a(x,\xi)$. Consequently, reasoning as above, we could analogously arrive at the formula

$$Tf(x) = \int_{\mathbb{R}^n} \mathrm{e}^{2\pi i x \cdot \xi} \, a(x,\xi) \, \widehat{f}(\xi) \,\mathrm{d}\xi. \tag{2.2}$$

Now, in order to avoid several rather informal conclusions in the arguments above, one usually takes the opposite route and adopts formula (2.2) as the definition of the *pseudo-differential operator* with symbol $a(x,\xi)$. Such operators are then often denoted by $\mathrm{Op}(a)$, by $a(X,D)$, or by T_a.

The simplest and perhaps most useful class of symbols allowing this approach to work well is the following class denoted by $S_{1,0}^m(\mathbb{R}^n \times \mathbb{R}^n)$, or simply by $S^m(\mathbb{R}^n \times \mathbb{R}^n)$.

Definition 2.1.1 (Symbol classes $S^m(\mathbb{R}^n \times \mathbb{R}^n)$). We will say that $a \in S^m(\mathbb{R}^n \times \mathbb{R}^n)$ if $a = a(x,\xi)$ is smooth on $\mathbb{R}^n \times \mathbb{R}^n$ and if the estimates

$$|\partial_x^\beta \partial_\xi^\alpha a(x,\xi)| \le A_{\alpha\beta}(1 + |\xi|)^{m-|\alpha|} \tag{2.3}$$

hold for all α, β and all $x, \xi \in \mathbb{R}^n$. Constants $A_{\alpha\beta}$ may depend on a, α, β but not on x, ξ. The operator T defined by (2.2) is called the *pseudo-differential operator* with symbol a. The class of operators of the form (2.2) with symbols from $S^m(\mathbb{R}^n \times \mathbb{R}^n)$ is denoted by $\Psi^m(\mathbb{R}^n \times \mathbb{R}^n)$ or by $\mathrm{Op}\,S^m(\mathbb{R}^n \times \mathbb{R}^n)$.

Remark 2.1.2. We will insist on writing $S^m(\mathbb{R}^n \times \mathbb{R}^n)$ and not abbreviating it to S^m or even to $S^m(\mathbb{R}^n)$. The reason is that in Chapter 4 we will want to distinguish between symbol class $S^m(\mathbb{T}^n \times \mathbb{R}^n)$ which will be 1-periodic symbols from $S^m(\mathbb{R}^n \times \mathbb{R}^n)$ and symbol class $S^m(\mathbb{T}^n \times \mathbb{Z}^n)$ which will be the class of toroidal symbols.

Remark 2.1.3 **(Symbols of differential operators).** Note that for partial differential operators symbols are just the characteristic polynomials. One can readily see that the symbol of the differential operator $L = \sum_{|\alpha| \le m} a_\alpha(x)\partial_x^\alpha$ is $a(x,\xi) = \sum_{|\alpha| \le m} a_\alpha(x)(2\pi i \xi)^\alpha$ and $a \in S^m(\mathbb{R}^n \times \mathbb{R}^n)$ if the coefficients a_α and all of their derivatives are smooth and bounded on $\mathbb{R}^n$.

Remark 2.1.4 (**Powers of the Laplacian**). For example, the symbol of the Laplacian $\mathcal{L} = \frac{\partial^2}{\partial x_1^2} + \cdots + \frac{\partial^2}{\partial x_n^2}$ is $-4\pi^2|\xi|^2$ and it is an element of $S^2(\mathbb{R}^n \times \mathbb{R}^n)$. Consequently, for any $\mu \in \mathbb{R}$, we can define the operators $(1-\mathcal{L})^\mu$ as pseudo-differential operators with symbol $(1 + 4\pi^2|\xi|^2)^{\mu/2} \in S^\mu(\mathbb{R}^n \times \mathbb{R}^n)$.

Exercise 2.1.5. Let $u \in C(\mathbb{R}^n)$ satisfy $|u(x)| \leq C\langle x \rangle^N$ for some constants C, N, where $\langle x \rangle = (1 + |x|^2)^{1/2}$. Let $k > N + n$. Let us define

$$v_k(\phi) = \int_{\mathbb{R}^n} \int_{\mathbb{R}^n} e^{-2\pi i x \cdot \xi} u(x) \langle x \rangle^{-k} (1 - \mathcal{L})^{k/2} \phi(\xi) \, dx \, d\xi,$$

where $\phi \in \mathcal{S}(\mathbb{R}^n)$. Prove that $v_k \in \mathcal{S}'(\mathbb{R}^n)$. Prove that there is $v \in \mathcal{S}'(\mathbb{R}^n)$ such that $v = v_k$ for all $k > N + n$. Show that $v = \widehat{u}$.

We now proceed in establishing basic properties of pseudo-differential operators.

Theorem 2.1.6 (Pseudo-differential operators on $\mathcal{S}(\mathbb{R}^n)$). *Let $a \in S^m(\mathbb{R}^n \times \mathbb{R}^n)$ and $f \in \mathcal{S}(\mathbb{R}^n)$. We define the pseudo-differential operator with symbol a by*

$$a(X, D)f(x) := \int_{\mathbb{R}^n} e^{2\pi i x \cdot \xi} a(x, \xi) \widehat{f}(\xi) \, d\xi. \tag{2.4}$$

Then $a(X, D)f \in \mathcal{S}(\mathbb{R}^n)$.

Proof. First we observe that the integral in (2.4) converges absolutely. The same is true for all of its derivatives with respect to x by Lebesgue's dominated convergence theorem (Theorem 1.1.4), which implies that $a(X, D)f \in C^\infty(\mathbb{R}^n)$. Let us show now that in fact $a(X, D)f \in \mathcal{S}(\mathbb{R}^n)$. Introducing the operator

$$L_\xi = (1 + 4\pi^2|x|^2)^{-1}(I - \mathcal{L}_\xi)$$

(where $\mathcal{L}_\xi$ is the Laplace operator with respect to ξ-variables) with the property $L_\xi \, e^{2\pi i x \cdot \xi} = e^{2\pi i x \cdot \xi}$, integrating (2.2) by parts N times yields

$$a(X, D)f(x) = \int_{\mathbb{R}^n} e^{2\pi i x \cdot \xi} (L_\xi)^N [a(x, \xi) \widehat{f}(\xi)] \, d\xi.$$

From this we get $|a(X, D)f(x)| \leq C_N(1 + |x|)^{-2N}$ for all N, so $a(X, D)f$ is rapidly decreasing. The same argument applies to derivatives of $a(X, D)f$ to show that $a(X, D)f \in \mathcal{S}(\mathbb{R}^n)$. $\qquad \square$

The following generalisation of symbol class $S^m(\mathbb{R}^n \times \mathbb{R}^n)$ is often useful:

Definition 2.1.7 (Symbol classes $S^m_{\rho,\delta}(\mathbb{R}^n \times \mathbb{R}^n)$). Let $0 \leq \rho, \delta \leq 1$. We will say that $a \in S^m_{\rho,\delta}(\mathbb{R}^n \times \mathbb{R}^n)$ if $a = a(x, \xi)$ is smooth on $\mathbb{R}^n \times \mathbb{R}^n$ and if

$$|\partial_x^\beta \partial_\xi^\alpha a(x, \xi)| \leq A_{\alpha\beta}(1 + |\xi|)^{m - \rho|\alpha| + \delta|\beta|} \tag{2.5}$$

for all α, β and all $x, \xi \in \mathbb{R}^n$. Constants $A_{\alpha\beta}$ may depend on a, α, β but not on x, ξ. The operator T defined by (2.2) is called the *pseudo-differential operator with symbol a of order m and type (ρ, δ)*. The class of operators of the form (2.2) with symbols from $S_{\rho,\delta}^m(\mathbb{R}^n \times \mathbb{R}^n)$ is denoted by $\Psi_{\rho,\delta}^m(\mathbb{R}^n \times \mathbb{R}^n)$ or by $\mathrm{Op}\, S_{\rho,\delta}^m(\mathbb{R}^n \times \mathbb{R}^n)$.

Definition 2.1.8 (Symbol σ_A of operator A). If $A \in \Psi_{\rho,\delta}^m(\mathbb{R}^n \times \mathbb{R}^n)$ we denote its symbol by $\sigma_A = \sigma_A(x, \xi)$. It is well defined in view of Theorem 2.5.6 later on, which also gives a formula for $\sigma_A \in S_{\rho,\delta}^m(\mathbb{R}^n \times \mathbb{R}^n)$.

Exercise 2.1.9. Extend the statement of Theorem 2.1.6 to operators of type (ρ, δ). Namely, let $0 \le \rho, \delta \le 1$, and let $a \in S_{\rho,\delta}^m(\mathbb{R}^n \times \mathbb{R}^n)$ and $f \in \mathcal{S}(\mathbb{R}^n)$. Show that $a(X, D)f \in \mathcal{S}(\mathbb{R}^n)$.

The following convergence criterion will be useful in the sequel. It follows directly from the Lebesgue dominated convergence Theorem 1.1.4.

Proposition 2.1.10 (Convergence criterion for pseudo-differential operators). *Suppose we have a sequence of symbols $a_k \in S^m(\mathbb{R}^n \times \mathbb{R}^n)$ which satisfies the uniform symbolic estimates*

$$|\partial_x^\beta \partial_\xi^\alpha a_k(x, \xi)| \le A_{\alpha\beta}(1 + |\xi|)^{m-|\alpha|},$$

for all α, β, all $x, \xi \in \mathbb{R}^n$, and all k, with constants $A_{\alpha\beta}$ independent of x, ξ and k. Suppose that $a \in S^m(\mathbb{R}^n \times \mathbb{R}^n)$ is such that $a_k(x, \xi)$ and all of its derivatives converge to $a(x, \xi)$ and its derivatives, respectively, pointwise as $k \to \infty$. Then $a_k(X, D)f \to a(X, D)f$ in $\mathcal{S}(\mathbb{R}^n)$ for any $f \in \mathcal{S}(\mathbb{R}^n)$.

Exercise 2.1.11. Verify the details of the proof of Proposition 2.1.10.

Remark 2.1.12. More general families of pseudo-differential operators are introduced in, e.g., [55] and [130]. Yet $S^m(\mathbb{R}^n) = S_{1,0}^m(\mathbb{R}^n)$ contained in the Hörmander classes $S_{\rho,\delta}^m(\mathbb{R}^n)$ is definitely the most important case, and [135], [130], [55], and [118] concentrate on it. Compressed information about pseudo-differential operators and nonlinear partial differential equations can be found in [131]. The spectral properties of pseudo-differential operators are considered in [112], and we have also left out the matrix-valued pseudo-differential operators.

Remark 2.1.13. The relation between operators and symbols can be also viewed as follows. Let $u \in \mathcal{S}(\mathbb{R}^n)$ and fix $s < -n/2$. The function $\psi : \mathbb{R}^n \to H^s(\mathbb{R}^n)$, $\psi(\xi) = e_\xi$, where $e_\xi(x) = e^{2\pi i x \cdot \xi}$, is Bochner-integrable (see [53]) with respect to $\hat{u}(\xi)\, d\xi$, and therefore

$$(Au)(x) = \int_{\mathbb{R}^n} e^{2\pi i x \cdot \xi}\, \sigma_A(x, \xi)\, \hat{u}(\xi)\, d\xi$$

for symbols of order zero. The distribution Au can be viewed as a σ_A-weighted inverse Fourier transform of $\hat{u}$. Unfortunately, the algebra of the finite order operators on the Sobolev scale is too large to admit fruitful symbol analysis, while the non-trivial restrictions by the symbol inequalities (2.3) yield a well-behaving subalgebra.

2.2 Amplitude representation of pseudo-differential operators

If we write out the Fourier transform in (2.4) as an integral, we obtain

$$a(X, D)f(x) = \int_{\mathbb{R}^n} \int_{\mathbb{R}^n} e^{2\pi i(x-y)\cdot\xi} \, a(x, \xi) \, f(y) \, dy \, d\xi. \qquad (2.6)$$

However, a problem with this formula is that the ξ-integral does not converge absolutely even for $f \in \mathcal{S}(\mathbb{R}^n)$. To overcome this difficulty, one uses the idea to approximate $a(x, \xi)$ by symbols with compact support. To this end, let us fix some $\gamma \in C_0^\infty(\mathbb{R}^n \times \mathbb{R}^n)$ such that $\gamma = 1$ near the origin. Let us now define $a_\epsilon(x, \xi) = a(x, \xi)\gamma(\epsilon x, \epsilon \xi)$. Then one can readily check that $a_\epsilon \in C_0^\infty(\mathbb{R}^n \times \mathbb{R}^n)$ and that the following holds:

- if $a \in S^m(\mathbb{R}^n \times \mathbb{R}^n)$, then $a_\epsilon \in S^m(\mathbb{R}^n \times \mathbb{R}^n)$ uniformly in $0 < \epsilon \leq 1$ (this means that constants $A_{\alpha\beta}$ in symbolic inequalities in Definition 2.1.1 may be chosen independent of $0 < \epsilon \leq 1$);

- $a_\epsilon \to a$ pointwise as $\epsilon \to 0$, uniformly in $0 < \epsilon \leq 1$. The same is true for derivatives of a_ϵ and a.

It follows now from the convergence criterion Proposition 2.1.10 that

$$a_\epsilon(X, D)f \to a(X, D)f \text{ in } \mathcal{S}(\mathbb{R}^n) \text{ as } \epsilon \to 0,$$

for all $f \in \mathcal{S}(\mathbb{R}^n)$. Here $a(X, D)f$ is defined as in (2.4). Now, formula (2.6) does make sense for $a_\epsilon \in C_0^\infty$, so we may define the double integral in (2.6) as the limit in $\mathcal{S}(\mathbb{R}^n)$ of $a_\epsilon(X, D)f$, i.e., take

$$a(X, D)f(x) := \lim_{\epsilon \to 0} \int_{\mathbb{R}^n} \int_{\mathbb{R}^n} e^{2\pi i(x-y)\cdot\xi} a_\epsilon(x, \xi)f(y) \, dy \, d\xi, \quad f \in \mathcal{S}(\mathbb{R}^n).$$

Pseudo-differential operators on $\mathcal{S}'(\mathbb{R}^n)$. Recall that we can define the L^2-adjoint $a(X, D)^*$ of an operator $a(X, D)$ by the formula

$$(a(X, D)f, g)_{L^2} = (f, a(X, D)^*g)_{L^2},$$

$f, g \in \mathcal{S}(\mathbb{R}^n)$, where

$$(u, v)_{L^2} = \int_{\mathbb{R}^n} u(x)\overline{v(x)} \, dx$$

is the usual L^2-inner product. From (2.6) and this formula we can readily calculate that

$$a(X, D)^*g(y) = \lim_{\epsilon \to 0} \int_{\mathbb{R}^n} \int_{\mathbb{R}^n} e^{2\pi i(y-x)\cdot\xi} \, \overline{a_\epsilon(x, \xi)} \, g(x) \, dx \, d\xi, \quad g \in \mathcal{S}(\mathbb{R}^n).$$

With the same understanding of non-convergent integrals as in (2.6) and replacing x by z to eliminate any confusion, we can write

$$a(X, D)^*g(y) = \int_{\mathbb{R}^n} \int_{\mathbb{R}^n} e^{2\pi i(y-z)\cdot\xi} \, \overline{a(z, \xi)} \, g(z) \, dz \, d\xi, \quad g \in \mathcal{S}(\mathbb{R}^n). \qquad (2.7)$$

Exercise 2.2.1. As before, by integration by parts, check that $a(X, D)^* : \mathcal{S}(\mathbb{R}^n) \to \mathcal{S}(\mathbb{R}^n)$ is continuous.

Definition 2.2.2 (Pseudo-differential operators on $\mathcal{S}'(\mathbb{R}^n)$). Let $u \in \mathcal{S}'(\mathbb{R}^n)$. We define $a(X, D)u$ by the formula

$$(a(X, D)u)(\varphi) := u(\overline{a(X, D)^* \overline{\varphi}}) \quad \text{for all } \varphi \in \mathcal{S}(\mathbb{R}^n).$$

Remark 2.2.3 (**Consistency**). We clearly have

$$\overline{a(X, D)^* \overline{\varphi}}(y) = \int_{\mathbb{R}^n} \int_{\mathbb{R}^n} e^{2\pi i (z-y) \cdot \xi} \, a(z, \xi) \, \varphi(z) \, \mathrm{d}z \, \mathrm{d}\xi,$$

so if $u, \varphi \in \mathcal{S}(\mathbb{R}^n)$, we have the consistency in

$$(a(X, D)u)(\varphi) = \int_{\mathbb{R}^n} a(X, D)u(x)\varphi(x) \, \mathrm{d}x = (a(X, D)u, \overline{\varphi})_{L^2}$$

$$= (u, a(X, D)^* \overline{\varphi})_{L^2} = \int_{\mathbb{R}^n} u(x)\overline{a(X, D)^* \overline{\varphi}(x)} \, \mathrm{d}x = u(\overline{a(X, D)^* \overline{\varphi}}).$$

Proposition 2.2.4. *If $a \in S^m(\mathbb{R}^n \times \mathbb{R}^n)$ and $u \in \mathcal{S}'(\mathbb{R}^n)$ then $a(X, D)u \in \mathcal{S}'(\mathbb{R}^n)$. Moreover, operator $a(X, D) : \mathcal{S}'(\mathbb{R}^n) \to \mathcal{S}'(\mathbb{R}^n)$ is continuous.*

Proof. Indeed, let $u_k \to u$ in $\mathcal{S}'(\mathbb{R}^n)$. Then we have

$$(a(X, D)u_k)(\varphi) = u_k(\overline{a(X, D)^* \overline{\varphi}}) \to u(\overline{a(X, D)^* \overline{\varphi}}) = (a(X, D)u)(\varphi),$$

so $a(X, D)u_k \to a(X, D)u$ in $\mathcal{S}'(\mathbb{R}^n)$ and, therefore, $a(X, D) : \mathcal{S}'(\mathbb{R}^n) \to \mathcal{S}'(\mathbb{R}^n)$ is continuous. $\qquad\square$

Exercise 2.2.5. Let $0 \le \rho \le 1$ and $0 \le \delta < 1$. Show that if $a \in S^m_{\rho,\delta}(\mathbb{R}^n \times \mathbb{R}^n)$ and $u \in \mathcal{S}'(\mathbb{R}^n)$ then $a(X, D)u \in \mathcal{S}'(\mathbb{R}^n)$, and that the operator $a(X, D) : \mathcal{S}'(\mathbb{R}^n) \to \mathcal{S}'(\mathbb{R}^n)$ is continuous.

2.3 Kernel representation of pseudo-differential operators

Summarising Sections 2.1 and 2.2, we can write pseudo-differential operators in different ways:

$$a(X, D)f(x) = \int_{\mathbb{R}^n} e^{2\pi i x \cdot \xi} \, a(x, \xi) \, \widehat{f}(\xi) \, \mathrm{d}\xi = \int_{\mathbb{R}^n} \int_{\mathbb{R}^n} e^{2\pi i (x-y) \cdot \xi} \, a(x, \xi) \, f(y) \, \mathrm{d}y \, \mathrm{d}\xi$$

$$= \int_{\mathbb{R}^n} \int_{\mathbb{R}^n} e^{2\pi i z \cdot \xi} \, a(x, \xi) \, f(x - z) \, \mathrm{d}z \, \mathrm{d}\xi = \int_{\mathbb{R}^n} k(x, z) \, f(x - z) \, \mathrm{d}z$$

$$= \int_{\mathbb{R}^n} K(x, y) \, f(y) \, \mathrm{d}y,$$

with kernels

$$K(x,y) = k(x, x - y), \quad k(x,z) = \int_{\mathbb{R}^n} e^{2\pi i z \cdot \xi} \, a(x,\xi) \, d\xi.$$

Theorem 2.3.1 (Kernel of a pseudo-differential operator). *Let* $a \in S^m(\mathbb{R}^n \times \mathbb{R}^n)$. *Then the kernel* $K(x,y)$ *of pseudo-differential operator* $a(X, D)$ *satisfies*

$$|\partial_{x,y}^\beta K(x,y)| \leq C_{N\beta} |x - y|^{-N}$$

for $N > m + n + |\beta|$ *and* $x \neq y$. *Thus, for* $x \neq y$, *the kernel* $K(x,y)$ *is a smooth function, rapidly decreasing as* $|x - y| \to \infty$.

Proof. We notice that $k(x, \cdot)$ is the inverse Fourier transform of $a(x, \cdot)$. It follows then that $(-2\pi i z)^\alpha \partial_z^\beta k(x, z)$ is the inverse Fourier transform with respect to ξ of the derivative $\partial_\xi^\alpha [(2\pi i \xi)^\beta a(x, \xi)]$, i.e.,

$$(-2\pi i z)^\alpha \partial_z^\beta k(x,z) = \mathcal{F}_\xi^{-1} \left(\partial_\xi^\alpha [(2\pi i \xi)^\beta a(x,\xi)] \right)(z).$$

Since $(2\pi i \xi)^\beta a(x, \xi) \in S^{m+|\beta|}(\mathbb{R}^n \times \mathbb{R}^n)$ is a symbol of order $m + |\beta|$, we have that

$$\left| \partial_\xi^\alpha [(2\pi i \xi)^\beta a(x,\xi)] \right| \leq C_{\alpha\beta} \langle \xi \rangle^{m+|\beta|-|\alpha|}.$$

Therefore, $\partial_\xi^\alpha [(2\pi i \xi)^\beta a(x, \xi)]$ is in $L^1(\mathbb{R}_\xi^n)$ with respect to ξ, if $|\alpha| > m + n + |\beta|$. Consequently, its inverse Fourier transform is bounded:

$$(-2\pi i z)^\alpha \partial_z^\beta k(x,z) \in L^\infty(\mathbb{R}_z^n)$$

for $|\alpha| > m + n + |\beta|$. Since taking derivatives of $k(x, z)$ with respect to x does not change the argument, this implies the statement of the theorem. □

As an immediate consequence of Theorem 2.3.1 we obtain the information on how the singular support is mapped by a pseudo-differential operator (see Remark 4.10.8 for more details):

Corollary 2.3.2 (Singular supports). *Let* $T \in \Psi^m(\mathbb{R}^n \times \mathbb{R}^n)$. *Then for every* $u \in \mathcal{S}'(\mathbb{R}^n)$ *we have*

$$\text{sing supp } Au \subset \text{sing supp } u. \tag{2.8}$$

Definition 2.3.3 (Local and pseudolocal operators). An operator A is called *pseudolocal* if the property (2.8) holds for all u. This is in analogy to the term "local" where an operator A is called *local* if

$$\text{supp } Au \subset \text{supp } u$$

for all u. By Corollary 2.3.2 every operator in $\Psi^m(\mathbb{R}^n \times \mathbb{R}^n)$ is pseudolocal. The converse is not true, as stated in Exercise 2.3.6.

Exercise 2.3.4 (Partial differential operators are local). Let A be a linear differential operator

$$Af(x) = \sum_{|\alpha| \leq m} a_\alpha(x) \partial_x^\alpha f(x)$$

with coefficients $a_\alpha \in C^\infty(\mathbb{R}^n)$, $|\alpha| \leq m$. Prove that supp $Af \subset$ supp f, for all $f \in C^\infty(\mathbb{R}^n)$.

Exercise 2.3.5 (Peetre's theorem). Prove the converse to Exercise 2.3.4 which is known as *Peetre's theorem:* if $A : C^\infty(\Omega) \to C^\infty(\Omega)$ is a continuous linear operator which is local, then A is a partial differential operator with smooth coefficients.

Exercise 2.3.6 (Pseudo–Peetre's theorem?). Prove that we can not add the word "pseudo" to Peetre's theorem. Namely, a pseudolocal linear continuous operator on $C^\infty(\mathbb{R}^n)$ does not have to be a pseudo-differential operator.

We refer to Section 4.10 for a further exploration of these properties.

We will now discuss an important class of operators which are usually taken to be negligible when one works with pseudo-differential operators. One of the reasons is that whenever they are applied to distributions they produce smooth functions, and so such operators can be neglected from the point of view of the analysis of singularities. However, it is important to understand these operators in order to know exactly what we are allowed to neglect.

Definition 2.3.7 (Smoothing operators). We can define *symbols of order* $-\infty$ by setting $S^{-\infty}(\mathbb{R}^n \times \mathbb{R}^n) := \bigcap_{m \in \mathbb{R}} S^m(\mathbb{R}^n \times \mathbb{R}^n)$, so that $a \in S^{-\infty}(\mathbb{R}^n \times \mathbb{R}^n)$ if $a \in C^\infty$ and if

$$|\partial_x^\beta \partial_\xi^\alpha a(x,\xi)| \leq A_{\alpha\beta N}(1 + |\xi|)^{-N}$$

holds for all N, and all $x, \xi \in \mathbb{R}^n$. The constants $A_{\alpha\beta N}$ may depend on a, α, β, N but not on x, ξ. Pseudo-differential operators with symbols in $S^{-\infty}$ are called *smoothing pseudo-differential operators.*

Exercise 2.3.8. Show that the class $S^{-\infty}(\mathbb{R}^n \times \mathbb{R}^n)$ is independent of ρ and δ in the sense that $S^{-\infty}(\mathbb{R}^n \times \mathbb{R}^n) = \bigcap_{m \in \mathbb{R}} S_{\rho,\delta}^m(\mathbb{R}^n \times \mathbb{R}^n)$ for all ρ and δ.

Proposition 2.3.9. *Let* $a \in S^{-\infty}(\mathbb{R}^n \times \mathbb{R}^n)$. *Then the integral kernel K of $a(X, D)$ is smooth on* $\mathbb{R}^n \times \mathbb{R}^n$.

Proof. Since $a(x, \cdot) \in L^1(\mathbb{R}^n)$, we immediately get $k \in L^\infty(\mathbb{R}^n)$. Moreover,

$$\partial_x^\beta \partial_z^\alpha k(x, z) = \int_{\mathbb{R}^n} e^{2\pi i z \cdot \xi} (2\pi i \xi)^\alpha \partial_x^\beta a(x, \xi) \, d\xi.$$

Since $(2\pi i \xi)^\alpha \partial_x^\beta a(x, \xi)$ is absolutely integrable, it follows that from the Lebesgue dominated convergence theorem (Theorem 1.1.4) that $\partial_x^\beta \partial_z^\alpha k$ is continuous. This is true for all α, β, hence k, and then also K, are smooth. $\qquad\square$

Let us write $k_x(\cdot) = k(x, \cdot)$.

Corollary 2.3.10. *Let $a \in S^{-\infty}(\mathbb{R}^n \times \mathbb{R}^n)$. Then $k_x \in \mathcal{S}(\mathbb{R}^n)$. We have*

$$a(X, D)f(x) = (k_x * f)(x)$$

and, consequently, $a(X, D)f \in C^\infty(\mathbb{R}^n)$ for all $f \in \mathcal{S}'(\mathbb{R}^n)$.

We note that the convolution in the corollary is understood in the sense of distributions, see Section 1.4.2.

Proof of Corollary 2.3.10. Now Corollary 2.3.10 follows from the fact that for $a \in S^{-\infty}$ we can write

$$a(X, D)f(x) = (f * k_x)(x)$$

with $k_x(\cdot) = k(x, \cdot) \in \mathcal{S}(\mathbb{R}^n)$. So

$$(a(X, D)f)(x) = f(\tau_x R k_x).$$

If now $f \in \mathcal{S}'(\mathbb{R}^n)$, it follows that $a(X, D)f \in C^\infty$ because of the continuity of $f(\tau_x R k_x)$ and all of its derivatives with respect to x. $\qquad\square$

Exercise 2.3.11 (Non-locality). Let T be an operator defined by

$$Tf(x) = \int_{\mathbb{R}^n} K(x, y) \, f(y) \, dy,$$

with $K \in C_0^\infty(\mathbb{R}^n \times \mathbb{R}^n)$. Prove that T defines a continuous operator from $\mathcal{S}(\mathbb{R}^n)$ to $\mathcal{S}(\mathbb{R}^n)$ and from $\mathcal{S}'(\mathbb{R}^n)$ to $\mathcal{S}'(\mathbb{R}^n)$. For operators T as above with $K \not\equiv 0$, show that we can never have the property supp $Tf \subset$ supp f for all $f \in C^\infty(\mathbb{R}^n)$.

2.4 Boundedness on $L^2(\mathbb{R}^n)$

In this section we prove that pseudo-differential operators with symbols in $S^0(\mathbb{R}^n \times \mathbb{R}^n)$ are bounded on $L^2(\mathbb{R}^n)$. The corresponding result in Sobolev spaces will be given in Theorem 2.6.11. First we prepare the following general result that shows that in many similar situations we only have to verify the estimate for the operator on a smaller space:

Proposition 2.4.1. *Let $A : \mathcal{S}'(\mathbb{R}^n) \to \mathcal{S}'(\mathbb{R}^n)$ be a continuous linear operator such that $A(\mathcal{S}(\mathbb{R}^n)) \subset L^2(\mathbb{R}^n)$ and such that there exists C for which the estimate*

$$\|Af\|_{L^2(\mathbb{R}^n)} \leq C\|f\|_{L^2(\mathbb{R}^n)} \tag{2.9}$$

holds for all $f \in \mathcal{S}(\mathbb{R}^n)$. Then A extends to a bounded linear operator from $L^2(\mathbb{R}^n)$ to $L^2(\mathbb{R}^n)$, and estimate (2.9) holds for all $f \in L^2(\mathbb{R}^n)$, with the same constant C.

Proof. Indeed, let $f \in L^2(\mathbb{R}^n)$ and let $f_k \in \mathcal{S}(\mathbb{R}^n)$ be a sequence of rapidly decreasing functions such that $f_k \to f$ in $L^2(\mathbb{R}^n)$. Such a sequence exists because $\mathcal{S}(\mathbb{R}^n)$ is dense in $L^2(\mathbb{R}^n)$ (Exercise 1.3.33). Then by (2.9) applied to $f_k - f_m$ we have

$$\|A(f_k - f_m)\|_{L^2(\mathbb{R}^n)} \leq C\|f_k - f_m\|_{L^2(\mathbb{R}^n)},$$

so Af_k is a Cauchy sequence in $L^2(\mathbb{R}^n)$. By the completeness of $L^2(\mathbb{R}^n)$ (Theorem C.4.9) there is some $g \in L^2(\mathbb{R}^n)$ such that $Af_k \to g$ in $L^2(\mathbb{R}^n)$. On the other hand $Af_k \to Af$ in $\mathcal{S}'(\mathbb{R}^n)$ because $f_k \to f$ in $L^2(\mathbb{R}^n)$ implies that $f_k \to f$ in $\mathcal{S}(\mathbb{R}^n)$ (Exercise 1.3.12). By the uniqueness principle in Proposition 1.3.5 we have $Af = g \in L^2(\mathbb{R}^n)$. Passing to the limit in (2.9) applied to f_k, we get $||Af||_{L^2(\mathbb{R}^n)} \leq C||f||_{L^2(\mathbb{R}^n)}$, with the same constant C, completing the proof. $\square$

There are different proofs of the L^2-result. For the proof of Theorem 2.4.2 below we follow [118] but an alternative proof based on the calculus will be also given later in Section 2.5.4.

Theorem 2.4.2 (L^2-boundedness of pseudo-differential operators).   *Let $a \in S^0(\mathbb{R}^n \times \mathbb{R}^n)$. Then $a(X, D)$ extends to a bounded linear operator from $L^2(\mathbb{R}^n)$ to $L^2(\mathbb{R}^n)$.*

Proof. First of all, we note that by a standard functional analytic argument in Proposition 2.4.1 it is sufficient to show the boundedness inequality (2.9) for $A = a(X, D)$ only for $f \in \mathcal{S}(\mathbb{R}^n)$, with constant C independent of the choice of f.

The proof of (2.9) will consist of two parts. In the first part we establish it for compactly supported (with respect to x) symbols and in the second part we will extend it to the general case of $a \in S^0(\mathbb{R}^n \times \mathbb{R}^n)$.

So, let us first assume that $a(x, \xi)$ has compact support with respect to x. This will allow us to use the Fourier transform with respect to x, in particular the formulae

$$a(x, \xi) = \int_{\mathbb{R}^n} e^{2\pi i x \cdot \lambda} \, \widehat{a}(\lambda, \xi) \, \mathrm{d}\lambda, \quad \widehat{a}(\lambda, \xi) = \int_{\mathbb{R}^n} e^{-2\pi i x \cdot \lambda} \, a(x, \xi) \, \mathrm{d}x,$$

with absolutely convergent integrals. We will use the fact that $a(\cdot, \xi) \in C_0^\infty(\mathbb{R}^n) \subset \mathcal{S}(\mathbb{R}^n)$, so that $a(\cdot, \xi)$ is in the Schwartz space in the first variable. Consequently, we have $\widehat{a}(\cdot, \xi) \in \mathcal{S}(\mathbb{R}^n)$ uniformly in ξ. To see the uniformity, we can notice that

$$(2\pi i \lambda)^\alpha \, \widehat{a}(\lambda, \xi) = \int_{\mathbb{R}^n} e^{-2\pi i x \cdot \lambda} \, \partial_x^\alpha a(x, \xi) \, \mathrm{d}x,$$

and hence $|(2\pi i \lambda)^\alpha \widehat{a}(\lambda, \xi)| \leq C_\alpha$ for all $\xi \in \mathbb{R}^n$. It follows that

$$\sup_{\xi \in \mathbb{R}^n} |\widehat{a}(\lambda, \xi)| \leq C_N (1 + |\lambda|)^{-N}$$

for all N. Now we can write

$$a(X, D)f(x) = \int_{\mathbb{R}^n} e^{2\pi i x \cdot \xi} \, a(x, \xi) \, \widehat{f}(\xi) \, \mathrm{d}\xi$$

$$= \int_{\mathbb{R}^n} \int_{\mathbb{R}^n} e^{2\pi i x \cdot \xi} \, e^{2\pi i x \cdot \lambda} \, \widehat{a}(\lambda, \xi) \, \widehat{f}(\xi) \, \mathrm{d}\lambda \, \mathrm{d}\xi$$

$$= \int_{\mathbb{R}^n} (Sf)(\lambda, x) \, \mathrm{d}\lambda,$$

where

$$(Sf)(\lambda, x) = e^{2\pi i x \cdot \lambda}(\widehat{a}(\lambda, D)f)(x).$$

Here $\widehat{a}(\lambda, D)f$ is a Fourier multiplier with symbol $\widehat{a}(\lambda, \xi)$ independent of x, so by Plancherel's identity Theorem 1.3.13 we get

$$||\widehat{a}(\lambda, D)f||_{L^2} = ||\mathcal{F}(\widehat{a}(\lambda, D)f)||_{L^2} = ||\widehat{a}(\lambda, \cdot)\widehat{f}||_{L^2}$$
$$\leq \sup_{\xi \in \mathbb{R}^n} |\widehat{a}(\lambda, \xi)| \, ||\widehat{f}||_{L^2} \leq C_N(1 + |\lambda|)^{-N}||f||_{L^2},$$

for all $N \geq 0$. Hence we get

$$||a(X, D)f||_{L^2} \leq \int_{\mathbb{R}^n} ||Sf(\lambda, \cdot)||_{L^2} \, d\lambda$$
$$\leq C_N \int_{\mathbb{R}^n} (1 + |\lambda|)^{-N}||f||_{L^2} \, d\lambda \leq C||f||_{L^2},$$

if we take $N > n$.

Now, to pass to symbols which are not necessarily compactly supported with respect to x, we will use the inequality

$$\int_{|x-x_0|\leq 1} |a(X, D)f(x)|^2 \, dx \leq C_N \int_{\mathbb{R}^n} \frac{|f(x)|^2 \, dx}{(1 + |x - x_0|)^N}, \qquad (2.10)$$

which holds for every $x_0 \in \mathbb{R}^n$ and for every $N \geq 0$, with C_N independent of x_0 and dependent only on constants in the symbolic inequalities for a. Let us show first that (2.10) implies (2.9). Writing $\chi_{|x-x_0|\leq 1}$ for the characteristic function of the set $|x - x_0| \leq 1$ and integrating (2.10) with respect to x_0 yields

$$\int_{\mathbb{R}^n} \left(\int_{\mathbb{R}^n} \chi_{|x-x_0|\leq 1}|a(X, D)f(x)|^2 \, dx \right) dx_0$$
$$\leq C_N \int_{\mathbb{R}^n} \left(\int_{\mathbb{R}^n} \frac{|f(x)|^2 \, dx}{(1 + |x - x_0|)^N} \right) dx_0.$$

Changing the order of integration, we arrive at

$$\text{vol}(B(1)) \int_{\mathbb{R}^n} |a(X, D)f(x)|^2 \, dx \leq \widetilde{C}_N \int_{\mathbb{R}^n} |f(x)|^2 \, dx,$$

which is (2.9). Let us now prove (2.10).

Let us prove it for $x_0 = 0$ first. We can write $f = f_1 + f_2$, where f_1 and f_2 are smooth functions such that $|f_1| \leq |f|$, $|f_2| \leq |f|$, and $\text{supp} f_1 \subset \{|x| \leq 3\}$, $\text{supp} f_2 \subset \{|x| \geq 2\}$. We will do the estimate for f_1 first. Let us fix $\eta \in C_0^\infty(\mathbb{R}^n)$ such that $\eta(x) = 1$ for $|x| \leq 1$. Then $\eta(a(X, D)) = (\eta a)(X, D)$ is a pseudo-differential operator with a compactly supported in x symbol $\eta(x)a(x, \xi)$, thus by the first part we have

$$\int_{\{|x|\leq 1\}} |a(X,D)f_1(x)|^2 \, dx = \int_{\mathbb{R}^n} |(\eta a)(X,D)f_1(x)|^2 \, dx$$

$$\leq C \int_{\mathbb{R}^n} |f_1(x)|^2 \, dx$$

$$\leq C \int_{\{|x|\leq 3\}} |f(x)|^2 \, dx,$$

which is the required estimate for f_1. Let us now do the estimate for f_2. If $|x| \leq 1$, then $x \notin \operatorname{supp} f_2$, so we can write

$$a(X,D)f_2(x) = \int_{\{|x|\geq 2\}} k(x, x-y)f_2(y) \, dy,$$

where k is the kernel of $a(X,D)$. Since $|x| \leq 1$ and $|y| \geq 2$, we have $|x - y| \geq 1$ and hence by Theorem 2.3.1 we can estimate

$$|k(x, x-z)| \leq C_1|x-y|^{-N} \leq C_2|y|^{-N}$$

for all $N \geq 0$. Thus we can estimate

$$|a(X,D)f_2(x)| \leq C_1 \int_{\{|y|\geq 2\}} \frac{|f_2(y)|}{|y|^N} \, dy$$

$$\leq C_2 \int_{\mathbb{R}^n} \frac{|f(y)|}{(1+|y|)^N} \, dy$$

$$\leq C_3 \left(\int_{\mathbb{R}^n} \frac{|f(y)|^2}{(1+|y|)^N} \, dy \right)^{1/2},$$

where we used the Cauchy-Schwarz inequality (Proposition 1.2.4) and that

$$C_3 = C_2 \left(\int_{\mathbb{R}^n} \frac{1}{(1+|y|)^N} \, dy \right)^{1/2} < \infty$$

for $N > n$ (Exercise 1.1.19). This in turn implies

$$\int_{\{|x|\leq 1\}} |a(X,D)f_2(x)|^2 \, dx \leq C \left(\int_{\mathbb{R}^n} \frac{|f(y)|^2}{(1+|y|)^N} \, dy \right)^{1/2},$$

which is the required estimate for f_2. These estimates for f_1 and f_2 imply (2.10) with $x_0 = 0$. We note that constant C_0 depends only on the dimension and on the constants in symbolic inequalities for a.

Let us now show (2.10) with an arbitrary $x_0 \in \mathbb{R}^n$. Let us define $a_{x_0}(x,\xi) = a(x - x_0, \xi)$. Then we immediately see that estimate (2.10) for $a(X,D)$ in the ball $\{|x - x_0| \leq 1\}$ is equivalent to the same estimate for $a_{x_0}(X,D)$ in the ball $\{|x| \leq 1\}$. Finally we note that since constants in symbolic inequalities for a and a_{x_0} are the same, we obtain (2.10) with constant C_N independent of x_0. This completes the proof of Theorem 2.4.2. $\square$

2.5 Calculus of pseudo-differential operators

In this section we establish formulae for the composition of pseudo-differential operators, adjoint operators and and discuss the transformation of symbols under changes of variables.

2.5.1 Composition formulae

First we analyse compositions of pseudo-differential operators.

Theorem 2.5.1 (Composition of pseudo-differential operators). *Let*

$$a \in S^{m_1}(\mathbb{R}^n \times \mathbb{R}^n) \quad and \quad b \in S^{m_2}(\mathbb{R}^n \times \mathbb{R}^n).$$

Then there exists some symbol $c \in S^{m_1+m_2}(\mathbb{R}^n \times \mathbb{R}^n)$ such that

$$c(X, D) = a(X, D) \circ b(X, D).$$

Moreover, we have the asymptotic formula

$$c \sim \sum_{\alpha} \frac{(2\pi i)^{-|\alpha|}}{\alpha!} (\partial_\xi^\alpha a)(\partial_x^\alpha b), \qquad (2.11)$$

which means that for all $N > 0$ we have

$$c - \sum_{|\alpha| < N} \frac{(2\pi i)^{-|\alpha|}}{\alpha!} (\partial_\xi^\alpha a)(\partial_x^\alpha b) \in S^{m_1+m_2-N}(\mathbb{R}^n \times \mathbb{R}^n).$$

Proof. Let us assume first that all symbols are compactly supported, so we can change the order of integration freely. Indeed, we can think of, say, symbol $a(x, \xi)$ as of $a_\epsilon(x, \xi) = a(x, \xi)\gamma(\epsilon x, \epsilon \xi)$, make all the calculations uniformly in $0 < \epsilon \leq 1$, use that $a_\epsilon \in S^{m_1}(\mathbb{R}^n \times \mathbb{R}^n)$ uniformly in ϵ, and then take a limit as $\epsilon \to 0$. Let us now plug

$$b(X, D)f(y) = \int_{\mathbb{R}^n} \int_{\mathbb{R}^n} e^{2\pi i(y-z)\cdot\xi} \, b(y, \xi) \, f(z) \, dz \, d\xi$$

into

$$a(X, D)g(x) = \int_{\mathbb{R}^n} \int_{\mathbb{R}^n} e^{2\pi i(x-y)\cdot\eta} \, a(x, \xi) \, g(y) \, dy \, d\eta,$$

yielding

$$a(X, D)(b(X, D)f)(x)$$

$$= \int_{\mathbb{R}^n} \int_{\mathbb{R}^n} \int_{\mathbb{R}^n} \int_{\mathbb{R}^n} e^{2\pi i(x-y)\cdot\eta} \, a(x, \eta) \, e^{2\pi i(y-z)\cdot\xi} \, b(y, \xi) \, f(z) \, dz \, d\xi \, dy \, d\eta$$

$$= \int_{\mathbb{R}^n} \int_{\mathbb{R}^n} e^{2\pi i(x-z)\cdot\xi} \, c(x, \xi) \, f(z) \, dz \, d\xi,$$

with

$$c(x,\xi) = \int_{\mathbb{R}^n} \int_{\mathbb{R}^n} e^{2\pi i (x-y)\cdot(\eta-\xi)}\, a(x,\eta)\, b(y,\xi)\, dy\, d\eta$$

$$= \int_{\mathbb{R}^n} e^{2\pi i x \cdot (\eta-\xi)}\, a(x,\eta)\, \widehat{b}(\eta-\xi,\xi)\, d\eta$$

$$= \int_{\mathbb{R}^n} e^{2\pi i x \cdot \eta}\, a(x,\xi+\eta)\, \widehat{b}(\eta,\xi)\, d\eta. \tag{2.12}$$

Here $\widehat{b}$ is the Fourier transform with respect to the first variable, and we used that

$$(x-y)\cdot\eta + (y-z)\cdot\xi - (x-z)\cdot\xi = (x-y)\cdot(\eta-\xi).$$

Asymptotic formula. Let us assume first that $b(x,\xi)$ has compact support in x (although if we think of b as b_ϵ, it will have a compact support, but the size of the support is not uniform in ϵ, so we need to treat operators with compact support in a uniform way first). Since b is compactly supported in x and $b \in S^{m_2}(\mathbb{R}^n \times \mathbb{R}^n)$, its Fourier transform with respect to x is rapidly decaying uniformly in ξ, so we can estimate

$$|\widehat{b}(\eta,\xi)| \leq C_M (1+|\eta|)^{-M}(1+|\xi|)^{m_2},$$

for all $M \geq 0$. The Taylor's expansion formula for $a(x,\xi+\eta)$ in the second variable gives

$$a(x,\xi+\eta) = \sum_{|\alpha|<N} \frac{1}{\alpha!} \partial_\xi^\alpha a(x,\xi)\eta^\alpha + R_N(x,\xi,\eta),$$

with a remainder R_N that we will analyse later. Plugging this formula into (2.12) and looking at terms in the sum, we get

$$\int_{\mathbb{R}^n} e^{2\pi i x \cdot \eta}\left[\partial_\xi^\alpha a(x,\xi)\eta^\alpha\right]\widehat{b}(\eta,\xi)\, d\eta = (2\pi i)^{-|\alpha|}\partial_\xi^\alpha a(x,\xi)\partial_x^\alpha b(x,\xi),$$

so that we have

$$c(x,\xi) = \frac{(2\pi i)^{-|\alpha|}}{\alpha!} \sum_{|\alpha|<N} (\partial_\xi^\alpha a(x,\xi))\partial_x^\alpha b(x,\xi) + \int_{\mathbb{R}^n} e^{2\pi i x \cdot \eta} R_N(x,\xi,\eta)\widehat{b}(\eta,\xi)\, d\eta.$$

Remainder. For the remainder in the Taylor series we have the estimate

$$|R_N(x,\xi,\eta)|$$
$$\leq C_N|\eta|^N \times \max\left\{|\partial_\xi^\alpha a(x,\zeta)| : |\alpha| = N, \zeta \text{ on the line between } \xi \text{ and } \xi+\eta\right\}.$$

Now, if $|\eta| \leq |\xi|/2$, then points on the line between ξ and $\xi+\eta$ are proportional to ξ, so we can estimate

$$|R_N(x,\xi,\eta)| \leq C_N|\eta|^N (1+|\xi|)^{m_1-N} \quad (|\eta| \leq |\xi|/2).$$

On the other hand, if $N \geq m_1$, we get the following estimate for all ξ and η:

$$|R_N(x,\xi,\eta)| \leq C_N|\eta|^N \quad (N \geq m_1).$$

Using the estimate for $\widehat{b}(\eta, \xi)$ and these two estimates, we get

$$\left| \int_{\mathbb{R}^n} e^{2\pi i x \cdot \eta} \, R_N(x, \xi, \eta) \, \widehat{b}(\eta, \xi) \, d\eta \right|$$

$$\leq C_{N,M}(1 + |\xi|)^{m_1 + m_2 - N} \int_{\mathbb{R}^n} (1 + |\eta|)^{-M} |\eta|^N \, d\eta$$

$$+ (1 + |\xi|)^{m_2} \int_{|\eta| \geq |\xi|/2} (1 + |\eta|)^{-M} |\eta|^N \, d\eta.$$

Taking M large enough, we can estimate both terms on the right-hand side by $C(1 + |\xi|)^{m_1 + m_2 - N}$. Making an estimate for $\partial_x^\alpha \partial_\xi^\beta R_N$ in a similar way, we get

$$\left| \int_{\mathbb{R}^n} e^{2\pi i x \cdot \eta} [\partial_x^\alpha \partial_\xi^\beta R_N(x, \xi, \eta)] \widehat{b}(\eta, \xi) \, d\eta \right| \leq C(1 + |\xi|)^{m_1 + m_2 - N - |\beta|},$$

implying the statement of the theorem for b compactly supported with respect to x.

General symbols. We note that it is sufficient to have the asymptotic formula and an estimate for the remainder for x near some fixed point x_0, uniformly in x_0. Let $\chi \in C_0^\infty(\mathbb{R}^n)$ be such that $\operatorname{supp} \eta \subset \{x : |x - x_0| \leq 2\}$ and such that $\eta(x) = 1$ for $|x - x_0| \leq 1$. Let us decompose

$$b = \chi b + (1 - \eta)b =: b_1 + b_2.$$

Since the symbol $b_1 = \chi b$ is compactly supported with respect to x, the composition formula for $a(X, D) \circ b_1(X, D)$ is given by the theorem, and it is equal to the claimed series for $a(X, D) \circ b(X, D)$ near x_0. We now have to show that the operator $a(X, D) \circ b_2(X, D)$ is smoothing, i.e., its symbol $c_2(x, \xi)$ is of order $-\infty$, and does not change the asymptotic formula for the composition. Indeed, we already know that the symbol of operator $a(X, D) \circ b_2(X, D)$ is given by

$$c_2(x, \xi) = \int_{\mathbb{R}^n} \int_{\mathbb{R}^n} e^{2\pi i (x - y) \cdot (\eta - \xi)} \, a(x, \eta) \, b_2(y, \xi) \, dy \, d\eta,$$

and we claim that

$$|c_2(x, \eta)| \leq C_N(1 + |\xi|)^{m_1 + m_2 - N} \quad \text{for all } N \geq 0, \ |x - x_0| \leq \frac{1}{2}.$$

In the above integral for c_2 we can integrate by parts to derive various properties of its decay. For example, we can integrate by parts with respect to the η-Laplacian Δ_η using the identity

$$\Delta_\eta^{N_1} e^{2\pi i (x - y) \cdot (\eta - \xi)} = (-4\pi^2)^{N_1} |x - y|^{2N_1} e^{2\pi i (x - y) \cdot (\eta - \xi)},$$

to see that in the integral we can replace $a(x, \eta)$ by $\frac{\Delta_\eta^{N_1} a(x, \eta)}{(-4\pi^2)^{N_1} |x - y|^{2N_1}}$, i.e.,

$$c_2(x, \xi) = \int_{\mathbb{R}^n} \int_{\mathbb{R}^n} e^{2\pi i (x - y) \cdot (\eta - \xi)} \frac{\Delta_\eta^{N_1} a(x, \eta)}{(-4\pi^2)^{N_1} |x - y|^{2N_1}} b_2(y, \xi) \, dy \, d\eta.$$

Here $|x - x_0| \leq 1/2$ and $y \in \text{supp}(1 - \chi)$ implies $|x - y| \geq 1/2$, so the integration by parts is well defined. We can also integrate by parts with respect to y-Laplacian Δ_y using the identity

$$(1 - \Delta_y)^{N_2} \, e^{2\pi i(x-y)\cdot(\eta-\xi)} = (1 + 4\pi^2|\xi - \eta|^2)^{N_2} \, e^{2\pi i(x-y)\cdot(\eta-\xi)}.$$

Moreover, we can use that

$$1 + |\xi| \leq 1 + |\xi - \eta| + |\eta| \leq (1 + |\xi - \eta|)(1 + |\eta|),$$

and hence

$$\frac{1}{(1 + |\xi - \eta|)^{2N_2}} \leq \frac{(1 + |\eta|)^{2N_2}}{(1 + |\xi|)^{2N_2}}.$$

Thus, integrating by parts with respect to y and using this estimate together with symbolic estimates for a and b_2, we get

$$|c_2(x, \xi)| \leq C \int_{\mathbb{R}^n} \int_{\mathbb{R}^n} \frac{(1 + |\eta|)^{m_1 - 2N_1}}{(1 + |x - y|)^{2N_1}} \frac{(1 + |\xi|)^{m_2}}{(1 + |\xi - \eta|)^{2N_2}} \, d\eta$$

$$\leq C \int_{\mathbb{R}^n} (1 + |\eta|)^{m_1 - 2N_1 + 2N_2}(1 + |\xi|)^{m_2 - 2N_2} \, dy \, d\eta$$

$$\leq C(1 + |\xi|)^{m_1 + m_2 - N} \int_{\mathbb{R}^n} (1 + |\eta|)^{N - 2N_1} \, d\eta,$$

if we take $N = m_1 - 2N_2$. Now, taking large N_2 and $2N_1 > N + n$, we obtain the desired estimate for $c_2(x, \xi)$. Similar estimates can be done for $\partial_x^\alpha \partial_\xi^\beta c_2(x, \xi)$ simply using symbolic inequalities for a and b_2, so we obtain that $c_2 \in S^{-\infty}$ for $|x - x_0| \leq 1/2$.

Finally, we notice that (in the general case) we have never used any information on the size of the support of symbols, so all constants depend only on the constants in symbolic inequalities. Thus, they remain uniformly bounded in $0 < \epsilon \leq 1$ for the composition $c_\epsilon(X, D) = a_\epsilon(X, D) \circ b_\epsilon(X, D)$, and we have $c_\epsilon \in S^{m_1 + m_2}$ with symbolic constants uniform in ϵ. Moreover, the asymptotic formula is satisfied uniformly in ϵ. Since $c_\epsilon \to c$ pointwise as $\epsilon \to 0$, we can conclude that $c \in S^{m_1 + m_2}$, that $c(X, D) = a(X, D) \circ b(X, D)$, and that the asymptotic formula of the theorem holds. $\qquad\qquad\qquad\qquad\qquad\qquad\qquad\qquad\qquad\qquad\qquad\square$

Exercise 2.5.2. Adopt the proof to get the following (ρ, δ) version of the composition formula in Theorem 2.5.1. Namely, let $0 \leq \delta < \rho \leq 1$ and let

$$a \in S_{\rho,\delta}^{m_1}(\mathbb{R}^n \times \mathbb{R}^n) \quad \text{and} \quad b \in S_{\rho,\delta}^{m_2}(\mathbb{R}^n \times \mathbb{R}^n).$$

Then there exists a symbol $c \in S_{\rho,\delta}^{m_1 + m_2}(\mathbb{R}^n \times \mathbb{R}^n)$ such that $c(X, D) = a(X, D) \circ b(X, D)$. Moreover, we have asymptotic formula (2.11) which now means that for all $N > 0$ we have

$$c - \sum_{|\alpha| < N} \frac{(2\pi i)^{-|\alpha|}}{\alpha!}(\partial_\xi^\alpha a)(\partial_x^\alpha b) \in S_{\rho,\delta}^{m_1 + m_2 - (\rho - \delta)N}(\mathbb{R}^n \times \mathbb{R}^n).$$

Exercise 2.5.3. Let $p(x, \xi) = \sum_{|\alpha| \leq m} a_\alpha(x) \xi^\alpha$, $a_\alpha \in C^\infty(\mathbb{R}^n)$, and let $p(X, D)$ be the differential operator with symbol $p(x, \xi)$.

(i) Prove that

$$p(X, D)(f(x)g(x)) = \sum_\alpha \frac{1}{\alpha!} [p^{(\alpha)}(X, D)f(x)][D^\alpha g(x)]$$

for all $f, g \in \mathcal{S}$, where $p^{(\alpha)}(X, D)$ is a differential operator with symbol $p^{(\alpha)}(x, \xi) = \partial_\xi^\alpha p(x, \xi)$.

(ii) Explain how to deduce the composition formula in Theorem 2.5.1 for differential operators from (i).

(iii) Let $p(x, D) \in \Psi^m(\mathbb{R}^n \times \mathbb{R}^n)$ be now a general pseudo-differential operator of order m. Let $f \in L^2(\mathbb{R}^n)$ and $g \in \mathcal{S}(\mathbb{R}^n)$. Prove that

$$p(X, D)(f(x)g(x)) - \sum_{|\alpha| \leq m} \frac{1}{\alpha!} [p^{(\alpha)}(X, D)f(x)][D^\alpha g(x)] \in H^1(\mathbb{R}^n).$$

We have already seen that when taking the adjoint of a pseudo-differential operator, we get the symbol which depends on the "wrong" set of variables: (y, ξ) instead of (x, ξ). Nevertheless, we want the adjoint to be a pseudo-differential operator as well. For this we need to study operators with symbols depending on all combinations of variables. This leads to the definition of an amplitude.

Definition 2.5.4 (Amplitudes). We will write $c = c(x, y, \xi) \in \mathcal{A}^m(\mathbb{R}^n)$ if $c \in C^\infty(\mathbb{R}^n \times \mathbb{R}^n \times \mathbb{R}^n)$ and if

$$\left| \partial_y^\gamma \partial_x^\beta \partial_\xi^\alpha c(x, y, \xi) \right| \leq C_{\alpha, \beta, \gamma} (1 + |\xi|)^{m - |\alpha|}$$

holds for all $x, y, \xi \in \mathbb{R}^n$ and all multi-indices α, β, γ. Constants $C_{\alpha, \beta, \gamma}$ may depend on c, α, β, γ but not on x, y, ξ. The corresponding operator $c(X, Y, D)$ is defined by

$$c(X, Y, D)f(x) := \int_{\mathbb{R}^n} \int_{\mathbb{R}^n} e^{2\pi i(x - y) \cdot \xi} \, c(x, y, \xi) \, f(y) \, dy \, d\xi. \qquad (2.13)$$

The operator $c(X, Y, D)$ is called an *amplitude operator* with amplitude $c = c(x, y, \xi)$. Analogously, for $0 \leq \rho \leq 1$ and $0 \leq \delta < 1$ we will write $c = c(x, y, \xi) \in \mathcal{A}^m_{\rho, \delta}(\mathbb{R}^n)$ if $c \in C^\infty(\mathbb{R}^n \times \mathbb{R}^n \times \mathbb{R}^n)$ and if

$$\left| \partial_y^\gamma \partial_x^\beta \partial_\xi^\alpha c(x, y, \xi) \right| \leq C_{\alpha, \beta, \gamma} (1 + |\xi|)^{m - \rho|\alpha| + \delta(|\beta| + |\gamma|)}$$

holds for all $x, y, \xi \in \mathbb{R}^n$ and all multi-indices α, β, γ. The class of operators $c(X, Y, D)$ with $c \in \mathcal{A}^m_{\rho, \delta}(\mathbb{R}^n)$ will be denoted by $\mathrm{Op}(\mathcal{A}^m_{\rho, \delta}(\mathbb{R}^n))$.

Remark 2.5.5. As in (2.6), we can justify formula (2.13) by considering

$$c_\epsilon(x, y, \xi) = c(x, y, \xi)\gamma(\epsilon y, \epsilon\xi),$$

with the same γ as was considered there. Then $c_\epsilon \to c$ pointwise (also with the pointwise convergence of derivatives), uniformly in $\mathcal{A}^m(\mathbb{R}^n)$ for $0 < \epsilon \leq 1$, so $c_\epsilon(X, Y, D)f \to c(X, Y, D)f$ in $\mathcal{S}(\mathbb{R}^n)$ for all $f \in \mathcal{S}(\mathbb{R}^n)$, by a suitable extension of Proposition 2.1.10. Thus, $c(X, Y, D)$ is well defined and continuous as an operator from $\mathcal{S}(\mathbb{R}^n)$ to $\mathcal{S}(\mathbb{R}^n)$ if $c \in \mathcal{A}^m(\mathbb{R}^n)$ or if $c \in \mathcal{A}^m_{\rho,\delta}(\mathbb{R}^n)$ for $0 \leq \rho \leq 1$ and $0 \leq \delta < 1$.

We are now in position to justify our starting point formula (2.1).

Theorem 2.5.6 (Quantization of operators). *A continuous linear operator* $T :$ $\mathcal{S}'(\mathbb{R}^n) \to \mathcal{S}'(\mathbb{R}^n)$ *is a pseudo-differential operator with symbol* $a(x, \xi)$ *if and only if*

$$a(x, \xi) = e^{-2\pi i x \cdot \xi} T(e^{2\pi i x \cdot \xi}) \in S^\infty(\mathbb{R}^n \times \mathbb{R}^n). \qquad (2.14)$$

In particular, a pseudo-differential operator $T \in \Psi^m(\mathbb{R}^n \times \mathbb{R}^n)$ *defines its symbol* $a \in S^m(\mathbb{R}^n \times \mathbb{R}^n)$ *uniquely, so that* $T = a(X, D)$. *The symbol* $a(x, \xi)$ *is defined by the formula* (2.14).

The notation used in the statement is a useful abbreviation for

$$
\begin{aligned}
e^{-2\pi i x \cdot \xi} T(e^{2\pi i x \cdot \xi}) &= e^{-2\pi i x \cdot \xi}\left(T(e^{2\pi i y \cdot \xi})\right)(x) \\
&= e^{-2\pi i x \cdot \xi}\left(T(e^{2\pi i \langle \cdot, \xi \rangle})\right)(x) \\
&= e_{-\xi}(x)(T e_\xi)(x),
\end{aligned}
$$

where $e_\xi(x) = e^{2\pi i x \cdot \xi}$.

Formula (2.14) for the symbol can be justified either in $\mathcal{S}'(\mathbb{R}^n)$ or as the limit in the expression

$$e^{-2\pi i x \cdot \xi} T(e^{2\pi i x \cdot \xi}\varphi_\epsilon) \to a(x, \xi),$$

as $\varphi_\epsilon \to 1$ for a family of $\varphi_\epsilon \in C_0^\infty(\mathbb{R}^n)$. In Theorem 4.1.4 we will prove this on the torus and in Theorem 10.4.4 we will give a proof of this statement on general Lie groups so now we leave it as

Exercise 2.5.7. Prove Theorem 2.5.6 and formula (2.14) for the symbol.

Theorem 2.5.8 (Symbols of amplitude operators). *Let* $c \in \mathcal{A}^m(\mathbb{R}^n)$ *be an amplitude. Then there exists a symbol* $a \in S^m(\mathbb{R}^n \times \mathbb{R}^n)$ *such that* $a(X, D) = c(X, Y, D)$. *Moreover, the asymptotic expansion for* a *is given by*

$$a(x, \xi) - \sum_{|\alpha| < N} \frac{(2\pi i)^{-|\alpha|}}{\alpha!} \partial_\xi^\alpha \partial_y^\alpha c(x, y, \xi)|_{y=x} \in S^{m-N}(\mathbb{R}^n \times \mathbb{R}^n), \qquad (2.15)$$

for all $N \geq 0$.

Proof. The proof is similar to the proof of the composition formula in Theorem 2.5.1. As in that proof, we first formally conclude that we must have

$$a(x, \xi) = e^{-2\pi i x \cdot \xi} \, T(\, e^{2\pi i x \cdot \xi})$$

$$= \int_{\mathbb{R}^n} \int_{\mathbb{R}^n} e^{-2\pi i x \cdot \xi} \, e^{2\pi i (x-y) \cdot \eta} \, c(x, y, \eta) \, e^{2\pi i y \cdot \xi} \, dy \, d\eta$$

$$= \int_{\mathbb{R}^n} c(x, y, \eta) \, e^{2\pi i (x-y) \cdot (\eta - \xi)} \, dy \, d\eta$$

$$= \int_{\mathbb{R}^n} \widehat{c}(x, \eta, \xi + \eta) \, e^{2\pi i x \cdot \eta} \, d\eta,$$

where $\widehat{c} = \mathcal{F}_y c$ is the Fourier transform of c with respect to y and where we used the change of variables $\eta \mapsto \eta + \xi$. By the same justification as in Theorem 2.5.1, we may work with amplitudes compactly supported in y, and make sure that all our arguments do not depend on the size of the support of c in y. Taking the Taylor's expansion of $\widehat{c}(x, \eta, \xi + \eta)$ at ξ, we obtain

$$\widehat{c}(x, \eta, \xi + \eta) = \sum_{|\alpha| < N} \frac{1}{\alpha!} \partial_\xi^\alpha \widehat{c}(x, \eta, \xi) \eta^\alpha + R_N(x, \eta, \xi),$$

where, as before, we can show that the remainder $R_N(x, \eta, \xi)$ satisfies estimates

$$|R_N(x, \eta, \xi)| \leq A |\eta|^N (1 + |\eta|)^{-M} (1 + |\xi|)^{m-N}$$

for large M and for $2|\eta| \leq |\xi|$, and

$$|R_N(x, \eta, \xi)| \leq A |\eta|^N (1 + |\eta|)^{-M} \quad \text{for all large } M \text{ and } N.$$

The last estimate is used in the region $2|\eta| \geq |\xi|$ and, similar to the proof of the composition formula in Theorem 2.5.1, we can complete the proof in the case of $c(x, y, \xi)$ compactly supported in y.

To treat the general case where $c(x, y, \xi)$ is not necessarily compactly supported in y, we reduce the analysis to the case when $c(x, y, \xi)$ vanishes for y away from $y = x$. For this, we can use the same argument as in the proof of the composition formula in Theorem 2.5.1, where the estimate for the remainder from the integration by parts argument becomes

$$\int_{\mathbb{R}^n} \int_{\mathbb{R}^n} \frac{(1 + |\eta|)^{m-2N_1}}{1 + |\eta - \xi|)^{2N_2}} \frac{1}{(1 + |x - y|)^{2N_1}} \, dy \, d\eta = O((1 + |\xi|)^{m-N}),$$

for large N_1, N_2. The proof is complete. $\qquad \square$

Exercise 2.5.9. Justify the following alternative argument. We know by Theorem 2.5.6 that the symbol of T can be defined by the formula

$$a(x, \xi) = e^{-2\pi i x \cdot \xi} \left(T(\, e^{2\pi i y \cdot \xi}) \right)(x)$$

but we do not know that $a \in S^m(\mathbb{R}^n \times \mathbb{R}^n)$. Let now $T = c(X, Y, D)$ be the operator with amplitude symbol c. Then Theorem 2.5.8 says that we can write

$$c(X, Y, D)u(x) = \int_{\mathbb{R}^n} \int_{\mathbb{R}^n} e^{2\pi i(x-y)\cdot\xi} \, a(x, \xi) \, u(y) \, dy \, d\xi,$$

for all $u \in C_0^\infty(\mathbb{R}^n)$. To show this, we write the Fourier inversion formula

$$u(y) = \int_{\mathbb{R}^n} e^{2\pi i y \cdot \xi} \, \widehat{u}(\xi) \, d\xi,$$

and, justifying the application of T to it and using the formula for the symbol from Theorem 2.5.6, we get

$$(Tu)(x) = \int_{\mathbb{R}^n} T\left(e^{2\pi i y \cdot \xi} \right)(x) \, \widehat{u}(\xi) \, d\xi = \int_{\mathbb{R}^n} e^{2\pi i x \cdot \xi} \, a(x, \xi) \, \widehat{u}(\xi) \, d\xi.$$

For the asymptotic expansion, we write

$$(T \, e^{2\pi i \langle \cdot, \xi \rangle})(x) = \int_{\mathbb{R}^n} \int_{\mathbb{R}^n} e^{2\pi i(x-y)\cdot\eta} \, c(x, y, \eta) \, e^{2\pi i y \cdot \xi} \, dy \, d\eta$$

$$= e^{2\pi i x \cdot \xi} \int_{\mathbb{R}^n} \int_{\mathbb{R}^n} e^{-2\pi i y \cdot \eta} \, c(x, x + y, \xi + \eta) \, dy \, d\eta,$$

where we changed variables $y \mapsto y + x$ and $\eta \mapsto \eta + \xi$, as well as recalculated the phase in new variables as

$$(x - (y + x)) \cdot (\eta + \xi) + (y + x) \cdot \xi = -y \cdot \eta + x \cdot \xi.$$

Now, using the Taylor expansion and estimates for the remainder, we can obtain Theorem 2.5.8 again.

Exercise 2.5.10. Prove the (ρ, δ) version of Theorem 2.5.8. Namely, if $c \in \mathcal{A}_{\rho,\delta}^m(\mathbb{R}^n)$ is an amplitude and $0 \leq \delta < \rho \leq 1$, then there exists a symbol $a \in S_{\rho,\delta}^m(\mathbb{R}^n \times \mathbb{R}^n)$ such that $a(X, D) = c(X, Y, D)$. Moreover, the asymptotic expansion for a is given by formula (2.15), but with $S^{m-N}(\mathbb{R}^n \times \mathbb{R}^n)$ replaced by $S_{\rho,\delta}^{m-(\rho-\delta)N}(\mathbb{R}^n \times \mathbb{R}^n)$.

Exercise 2.5.11. Let

$$Af(x) = \int_{\mathbb{R}^n} \int_{\mathbb{R}^n} e^{2\pi i(x-y)\cdot\xi} \, c(x, y, \xi) \, f(y) \, dy \, d\xi$$

be an operator with amplitude $c \in \mathcal{A}^m(\mathbb{R}^n)$. Prove that if f is smooth in some neighbourhood of a point $x \in \mathbb{R}^n$, then Au is also smooth in some neighbourhood of the same point x. How can one reformulate this in terms of the singular supports?

Exercise 2.5.12. Let $Bu(x) = \int_{\mathbb{R}^n} K(x, y) \, u(y) \, dy$ be an operator from $C_0^\infty(\mathbb{R}^n)$ to $C^\infty(\mathbb{R}^n)$ with kernel $K = K(x, y) \in \mathcal{S}(\mathbb{R}^n \times \mathbb{R}^n)$. Show that B is an amplitude operator of the form (2.13) with amplitude $c \in \mathcal{A}^{-\infty}(\mathbb{R}^n)$.

We now use Theorem 2.5.8 to obtain the asymptotic formula for the adjoint of a pseudo-differential operator.

Theorem 2.5.13 (Adjoint operator). *Let $a \in S^m(\mathbb{R}^n \times \mathbb{R}^n)$. Then there exists a symbol $a^* \in S^m(\mathbb{R}^n \times \mathbb{R}^n)$ such that $a(X, D)^* = a^*(X, D)$, where $a(X, D)^*$ is the L^2-adjoint operator of $a(X, D)$. Moreover, we have the asymptotic expansion*

$$a^*(x, \xi) - \sum_{|\alpha| < N} \frac{(2\pi i)^{-|\alpha|}}{\alpha!} \partial_\xi^\alpha \partial_x^\alpha \overline{a(x, \xi)} \in S^{m-N}(\mathbb{R}^n \times \mathbb{R}^n),$$

for all $N \geq 0$, where $\overline{a(x, \xi)}$ is the complex conjugate of $a(x, \xi)$.

Proof. As we have already calculated in (2.7), we have $a^*(X, D) = c(X, Y, D)$, which is an operator with amplitude $c(x, y, \xi) = \overline{a(y, \xi)}$. Applying Theorem 2.5.8, we obtain the statement of Theorem 2.5.13. □

Exercise 2.5.14. Prove the (ρ, δ) version of Theorem 2.5.13 for $0 \leq \delta < \rho \leq 1$. Namely, show that if $a(X, D) \in \Psi^m_{\rho, \delta}(\mathbb{R}^n \times \mathbb{R}^n)$ then $a(X, D)^* \in \Psi^m_{\rho, \delta}(\mathbb{R}^n \times \mathbb{R}^n)$ and the asymptotic formula for its symbol is as in Theorem 2.5.13, with $S^{m-N}(\mathbb{R}^n \times \mathbb{R}^n)$ replaced by $S^{m-(\rho-\delta)N}_{\rho,\delta}(\mathbb{R}^n \times \mathbb{R}^n)$.

Definition 2.5.15 (Transpose operators). For $u \in \mathcal{S}'(\mathbb{R}^n)$, define *the transpose* T^t of a linear operator $T : \mathcal{S}(\mathbb{R}^n) \to \mathcal{S}(\mathbb{R}^n)$ by the formula

$$(T^t u)(\varphi) := u(T\varphi)$$

for all $\varphi \in \mathcal{S}(\mathbb{R}^n)$.

Exercise 2.5.16 (Properties of the transpose). Prove that if T is a continuous linear operator from $\mathcal{S}(\mathbb{R}^n) \to \mathcal{S}(\mathbb{R}^n)$, then T^t is a continuous linear operator $\mathcal{S}'(\mathbb{R}^n) \to \mathcal{S}'(\mathbb{R}^n)$. In particular, show that if $u \in \mathcal{S}'(\mathbb{R}^n)$ then $T^t u \in \mathcal{S}'(\mathbb{R}^n)$. Prove that the adjoint operator T^* satisfies

$$(T^* \overline{u})(\psi) = \overline{u(T\overline{\psi})}$$

for all $\psi \in \mathcal{S}(\mathbb{R}^n)$. Then also prove that

$$T^t u = \overline{T^* \overline{u}}.$$

Here $\overline{u} \in \mathcal{S}'(\mathbb{R}^n)$ is the complex conjugate of u, defined by $\overline{u}(\varphi) = \overline{u(\overline{\varphi})}$, for all $\varphi \in \mathcal{S}(\mathbb{R}^n)$, and $\overline{u(\overline{\varphi})}$ is the complex conjugate of $u(\overline{\varphi}) \in \mathbb{C}$.

Remark 2.5.17. Recalling the canonical identification of functions with distributions in Remark 1.3.7 we see that if $u \in \mathcal{S}(\mathbb{R}^n)$, then the transpose T^t of T satisfies the relation

$$\int_{\mathbb{R}^n} (T^t u)(x) \, \varphi(x) \, \mathrm{d}x = \int_{\mathbb{R}^n} u(x) \, (T\varphi)(x) \, \mathrm{d}x.$$

This is the difference with the adjoint operator T^* which would satisfy

$$\int_{\mathbb{R}^n} (T^*u)(x)\, \overline{\varphi(x)}\, \mathrm{d}x = \int_{\mathbb{R}^n} u(x)\, \overline{(T\varphi)(x)}\, \mathrm{d}x.$$

Exercise 2.5.18 (Asymptotic expansion of the transpose). Let now T be a pseudo-differential operator with symbol $a \in S^m(\mathbb{R}^n \times \mathbb{R}^n)$. Prove that T^t is also a pseudo-differential operator of order m, with symbol having an asymptotic expansion

$$\text{symbol of } T_a^t \sim \sum_\alpha \frac{(2\pi i)^{-|\alpha|}}{\alpha!} \partial_\xi^\alpha \partial_x^\alpha \left[a(x, -\xi) \right].$$

Formulate and prove the (ρ, δ)-version of this result.

We finish with a brief discussion of properly supported operators.

Remark 2.5.19 (**Proper mappings**). Let $X \subset \mathbb{R}^n$ and $Y \subset \mathbb{R}^m$ be open sets. We recall that if a mapping $\Phi : X \to Y$ is continuous then the images of compact sets in X under Φ are compact in Y. This is not true in general if instead of "images" we take "preimages". Thus, a continuous mapping $\Phi : X \to Y$ is called *proper* if for every compact set $K \subset Y$ its preimage $\Phi^{-1}(K) = \{x \in X : \Phi(x) \in K\} \subset X$ is compact.

Definition 2.5.20 (Properly supported operators). A pseudo-differential operator T is called *properly supported* if the projections $\pi_1(x, y) := x$, $\pi_2(x, y) := y$ from $\operatorname{supp} K$ to $\mathbb{R}^n$ are proper mappings. Here $\operatorname{supp} K \subset \mathbb{R}^n \times \mathbb{R}^n$ is the support of the integral kernel of T.

Exercise 2.5.21. Prove that T is properly supported if and only if T^* is properly supported.

Proposition 2.5.22. *Let $A \in \Psi^m(\mathbb{R}^n \times \mathbb{R}^n)$. Then there exists a properly supported amplitude operator $B \in \operatorname{Op} \mathcal{A}^m(\mathbb{R}^n)$ such that $A - B \in \Psi^{-\infty}(\mathbb{R}^n \times \mathbb{R}^n)$.*

Proof. Let $A \in \Psi^m(\mathbb{R}^n \times \mathbb{R}^n)$ be an operator with symbol σ_A. By Theorem 2.3.1 the singular support of its integral kernel $K(x, y)$ is contained in the diagonal $\{x = y\}$. Let Ω be a neighbourhood of the diagonal $\{x = y\}$ of $\mathbb{R}^n \times \mathbb{R}^n$ such that $\pi_1 : \overline{\Omega} \to \mathbb{R}^n$ and $\pi_2 : \overline{\Omega} \to \mathbb{R}^n$ are proper. Let $\chi \in C^\infty(\mathbb{R}^n \times \mathbb{R}^n)$ be such that $\chi(x, x) = 1$ for all $x \in \mathbb{R}^n$ and such that $\operatorname{supp} \chi = \overline{\Omega}$. Let B be an operator with kernel χK, i.e., $Bf(x) := \int_{\mathbb{R}^n} \chi(x, y)\, K(x, y)\, f(y)\, \mathrm{d}y$. Then B is properly supported. Moreover, B is an amplitude operator with amplitude $b(x, y, \xi) = \chi(x, y)\sigma_A(x, \xi) \in \mathcal{A}^m(\mathbb{R}^n)$ and $A - B \in \Psi^{-\infty}(\mathbb{R}^n \times \mathbb{R}^n)$ because its integral kernel is smooth (see Exercise 2.5.12). $\qquad\square$

Corollary 2.5.23. *Because every amplitude operator is a pseudo-differential operator by Theorem 2.5.8 we see that a pseudo-differential operator differs from a properly supported operator by a smoothing operator: if $A \in \Psi^m(\mathbb{R}^n \times \mathbb{R}^n)$, then there exists a properly supported pseudo-differential operator $B \in \Psi^m(\mathbb{R}^n \times \mathbb{R}^n)$ such that $A - B \in \Psi^{-\infty}(\mathbb{R}^n \times \mathbb{R}^n)$.*

Remark 2.5.24. The advantage of working with properly supported operators is that they take compactly supported functions to compactly supported functions (or distributions). For example, a linear continuous operator $T : C_0^\infty(\mathbb{R}^n) \to C_0^\infty(\mathbb{R}^n)$ can be extended to an operator $\mathcal{D}'(\mathbb{R}^n) \to \mathcal{D}'(\mathbb{R}^n)$ by duality.

2.5.2 Changes of variables

It is clear from the definition of the symbol class $S^m(\mathbb{R}^n \times \mathbb{R}^n)$ that it is locally invariant under smooth changes of variables, i.e., if we take a local change of the variable x in the symbol from symbol class $S^m(\mathbb{R}^n \times \mathbb{R}^n)$, it will still belong to the same symbol class $S^m(\mathbb{R}^n \times \mathbb{R}^n)$. We will now investigate what happens with pseudo-differential operators when we make a change of variable in space.

Let $U, V \subset \mathbb{R}^n$ be open bounded sets and let $\tau : V \to U$ be a surjective diffeomorphism. For $f \in C_0^\infty(V)$, we define its pullback by the change of variables τ by

$$(\tau f)(x) = f(\tau^{-1}(x)).$$

It easily follows that the new function satisfies $\tau f \in C_0^\infty(U)$. Let now $a \in S^m(\mathbb{R}^n \times \mathbb{R}^n)$ be a symbol of order m with compact support with respect to x, and assume that this support is contained in U. Then we will show that there exists a symbol $b \in S^m(\mathbb{R}^n \times \mathbb{R}^n)$ with compact support with respect to x which is contained in V such that $\tau^{-1}a(X, D)\tau = b(X, D)$. In other words, we have

$$\left[a(X, D)(f \circ \tau^{-1})\right](\tau(x)) = b(X, D)f(x)$$

for all $f \in C_0^\infty(V)$ and all $x \in V$. More precisely, we have the following expression for the "main part" of b:

Proposition 2.5.25. *We have* $b(x, \xi) = a(\tau(x), \left[\frac{\partial \tau}{\partial x}\right]' \xi)$ *modulo* $S^{m-1}(\mathbb{R}^n \times \mathbb{R}^n)$, *where* $\left[\frac{\partial \tau}{\partial x}\right]' = \left(\left(\frac{\partial \tau}{\partial x}\right)^T\right)^{-1}$.

Proof. We can write

$$(\tau^{-1}a(X, D)\tau f)(x)$$

$$= \int_{\mathbb{R}^n} \int_{\mathbb{R}^n} e^{2\pi i(\tau(x)-y)\cdot\xi} \, a(\tau(x), \xi) \, f(\tau^{-1}(y)) \, dy \, d\xi$$

$$= \int_{\mathbb{R}^n} \int_{\mathbb{R}^n} e^{2\pi i(\tau(x)-\tau(y))\cdot\xi} \, a(\tau(x), \xi) \, f(y) \, \left|\det \frac{\partial \tau}{\partial y}\right| \, dy \, d\xi, \qquad (2.16)$$

where we changed variables $y \mapsto \tau(y)$ to get the last equality. Now we will argue that the main contribution in this integral comes from variables y close to x. Indeed, let us insert a cut-off function $\chi(x, y)$ in the integral, where χ is a smooth function supported in the set where $|y - x|$ is small and where $\chi(x, x) = 1$ for all x. The remaining integral

$$\int_{\mathbb{R}^n} \int_{\mathbb{R}^n} e^{2\pi i(\tau(x)-\tau(y))\cdot\xi} \, (1 - \chi(x, y)) \, a(\tau(x), \xi) \, f(y) \, \left|\det \frac{\partial \tau}{\partial y}\right| \, dy \, d\xi$$

has a smooth kernel since we can integrate by parts any number of times with respect to ξ to see that the symbol of this operator belongs to symbol classes $S^{-N}(\mathbb{R}^n \times \mathbb{R}^n)$ for all $N \geq 0$. Let us now analyse what happens for y close to x. By the mean value theorem, for y sufficiently close to x, we have

$$\tau(x) - \tau(y) = L_{x,y}(x - y), \tag{2.17}$$

where $L_{x,y}$ is an invertible linear mapping which is smooth in x and y, and satisfies

$$L_{x,x} = \frac{\partial \tau}{\partial x}(x).$$

Using (2.17) in (2.16), we get

$$\int_{\mathbb{R}^n} \int_{\mathbb{R}^n} e^{2\pi i L_{x,y}(x-y)\cdot\xi} \, \chi(x,y) \, a(\tau(x),\xi) \, \left| \det \frac{\partial \tau}{\partial y} \right| f(y) \, dy \, d\xi$$

$$= \int_{\mathbb{R}^n} \int_{\mathbb{R}^n} e^{2\pi i (x-y)\cdot\xi} \, \chi(x,y) \, a(\tau(x), L'_{x,y}\xi) \, \left| \det \frac{\partial \tau}{\partial y} \right| \, |\det L_{x,y}^{-1}| \, f(y) \, dy \, d\xi,$$

where we changed variables $L_{x,y}^T \xi \mapsto \xi$ and where $L_{x,y}^T$ is the transpose matrix of $L_{x,y}$ and $L'_{x,y} = (L_{x,y}^T)^{-1}$. Thus, we get an operator with the amplitude c defined by

$$c(x,y,\xi) = \chi(x,y) \, a(\tau(x), L'_{x,y}\xi) \, \left| \det \frac{\partial \tau}{\partial y} \right| \, |\det L_{x,y}^{-1}|.$$

Applying the asymptotic expansion in Theorem 2.5.8, we see that the first term of this expansion is given by

$$\begin{aligned}
c(x,y,\xi)|_{y=x} &= \chi(x,x) \, a(\tau(x), L'_{x,x}\xi) \, \left| \det \frac{\partial \tau}{\partial x} \right| \, |\det L_{x,x}^{-1}| \\
&= a(\tau(x), \left[\frac{\partial \tau}{\partial x} \right]' \xi),
\end{aligned}$$

completing the proof. $\qquad\qquad\qquad\qquad\qquad\qquad\qquad\qquad\qquad\qquad\qquad\qquad \square$

2.5.3 Principal symbol and classical symbols

We see in Proposition 2.5.25 that the equivalence class modulo $S^{m-1}(\mathbb{R}^n \times \mathbb{R}^n)$ has some meaning for the changes of variables. In fact, we may notice that the transformation

$$(x,\xi) \mapsto (\tau(x), \left[\frac{\partial \tau}{\partial x} \right]' \xi)$$

is the same as the change of variables in the cotangent bundle $T^*\mathbb{R}^n$ of $\mathbb{R}^n$ which is induced by the change of variables $x \mapsto \tau(x)$ in $\mathbb{R}^n$. This observation allows one to make an invariant geometric interpretation of the class of symbols in $S^m(\mathbb{R}^n \times \mathbb{R}^n)$ modulo terms of order $S^{m-1}(\mathbb{R}^n \times \mathbb{R}^n)$. We note that we can use the asymptotic expansion for amplitudes in the proof of Proposition 2.5.25 to find also lower-order terms in the asymptotic expansion of $b(x, \xi)$, but these terms will not have such nice invariant interpretation. Without going into detail, let us just mention that Proposition 2.5.25 allows one to introduce a notion of a *principal symbol* of a pseudo-differential operator with symbol in $S^m(\mathbb{R}^n \times \mathbb{R}^n)$ as the equivalence class of this symbol modulo the subclass $S^{m-1}(\mathbb{R}^n \times \mathbb{R}^n)$, and this principal symbol is defined on the cotangent bundle of $\mathbb{R}^n$. This construction can be further carried out on manifolds leading to many remarkable applications, in particular to those in geometry and index theory.

We will not pursue this path further in this chapter, but we will clarify the notion of such equivalent classes for the so-called *classical symbols* which form a class of symbols that plays a very important role in applications to partial differential equations. First, we define homogeneous functions/symbols.

Definition 2.5.26 (Homogeneous symbols). We will say that a symbol

$$a_k = a_k(x, \xi) \in S^k(\mathbb{R}^n \times \mathbb{R}^n)$$

is *positively homogeneous* (or simply *homogeneous*) of order k if for all $x \in \mathbb{R}^n$ we have

$$a_k(x, \lambda\xi) = \lambda^k a_k(x, \xi) \quad \text{for all } \lambda > 1 \text{ and all } |\xi| > r,$$

for some $r > 0$ (independent of λ, x and ξ).

We note that we exclude small ξ from this definition because if we assumed the property $a_k(x, \lambda\xi) = \lambda^k a_k(x, \xi)$ for all $\xi \in \mathbb{R}^n$, such a function a_k would not be in general smooth at $\xi = 0$.

Definition 2.5.27 (Classical symbols). We will say that a symbol $a \in S_{cl}^m(\mathbb{R}^n \times \mathbb{R}^n)$ is a *classical symbol of order m* if $a \in S^m(\mathbb{R}^n \times \mathbb{R}^n)$ and if there is an asymptotic expansion $a \sim \sum_{k=0}^{\infty} a_{m-k}$, where each a_{m-k} is positively homogeneous of order $m - k$, and if $a - \sum_{k=0}^{N} a_{m-k} \in S^{m-N-1}(\mathbb{R}^n \times \mathbb{R}^n)$, for all $N \geq 0$.

Now, for a classical symbol $a \in S_{cl}^m(\mathbb{R}^n \times \mathbb{R}^n)$ the principal symbol, i.e., its equivalence class modulo $S^{m-1}(\mathbb{R}^n \times \mathbb{R}^n)$ can be easily found. This is simply the first term a_m in the asymptotic expansion in its definition. We will discuss asymptotic sums in more detail in Section 2.5.5, namely we will show that if we start with the asymptotic sum $\sum_{k=0}^{\infty} a_{m-k}$, we can in turn interpret it as a symbol from $S^m(\mathbb{R}^n \times \mathbb{R}^n)$. In the following exercises, one may take classical symbols to avoid unnecessary technicalities.

Exercise 2.5.28. For operators $A \in \Psi^{m_1}(\mathbb{R}^n \times \mathbb{R}^n)$ and $B \in \Psi^{m_2}(\mathbb{R}^n \times \mathbb{R}^n)$, find the principal symbol of the commutator $[A, B] = AB - BA$.

Exercise 2.5.29. Let $a_m \in S^m(\mathbb{R}^n \times \mathbb{R}^n)$ be the principal symbol of $A \in \Psi^m(\mathbb{R}^n \times \mathbb{R}^n)$. Assume that a_m is real valued, i.e., $a_m(x, \xi) \in \mathbb{R}$ for all $x, \xi \in \mathbb{R}^n$. Show that $[A, A^*] = AA^* - A^*A \in \Psi^{2m-2}(\mathbb{R}^n \times \mathbb{R}^n)$.

Exercise 2.5.30. Assume that $A \in \Psi^m(\mathbb{R}^n \times \mathbb{R}^n)$ is self-adjoint, i.e., that $A = A^*$. Show that the principal symbol of A is real valued.

Exercise 2.5.31 (Euler's identity). Prove that $f \in C^1(\mathbb{R}^n \backslash 0)$ is positively homogeneous of order $k \in \mathbb{R}$ for all $\xi \neq 0$, i.e., f satisfies $f(t\xi) = t^k f(\xi)$ for all $t > 0$ and $\xi \neq 0$, if and only if f satisfies *Euler's identity*:

$$\nabla f(\xi) \cdot \xi = k f(\xi)$$

for all $\xi \neq 0$. Consequently, show that a symbol $a_k = a_k(x, \xi) \in S^k(\mathbb{R}^n \times \mathbb{R}^n)$ is positively homogeneous of order k if and only if function

$$\sum_{j=1}^{n} \xi_j \frac{\partial a_k}{\partial \xi_j} - k a_k$$

has a bounded support with respect to ξ uniformly with respect to x.

2.5.4   Calculus proof of L^2-boundedness

In this section we will give a short proof of the fact that pseudo-differential operators of order zero are bounded on $L^2(\mathbb{R}^n)$ which was given in Theorem 2.4.2. The proof will rely on the following lemma that is a special case of Young's inequality for σ-finite measures in Corollary C.5.20.

Lemma 2.5.32 (Schur's lemma). *Let T be an integral operator of the form*

$$Tu(x) = \int_{\mathbb{R}^n} K(x, y) u(y) \, \mathrm{d}y$$

with kernel $K \in L^1_{\mathrm{loc}}(\mathbb{R}^n \times \mathbb{R}^n)$ satisfying

$$\sup_{x \in \mathbb{R}^n} \int_{\mathbb{R}^n} |K(x, y)| \, \mathrm{d}y \leq C, \qquad \sup_{y \in \mathbb{R}^n} \int_{\mathbb{R}^n} |K(x, y)| \, \mathrm{d}x \leq C.$$

Then T is bounded on $L^2(\mathbb{R}^n)$.

Calculus proof of Theorem 2.4.2. Let us now give an alternative proof of Theorem 2.4.2 based on the calculus of pseudo-differential operators. Let $T \in \Psi^0(\mathbb{R}^n \times \mathbb{R}^n)$ be a pseudo-differential operator of order zero with symbol $a \in S^0(\mathbb{R}^n \times \mathbb{R}^n)$ and principal symbol σ_a (for simplicity one may assume that a is a classical symbol here). Then its adjoint satisfies $T^* \in \Psi^0(\mathbb{R}^n \times \mathbb{R}^n)$ and hence also the composition $T^*T \in \Psi^0(\mathbb{R}^n \times \mathbb{R}^n)$ by Theorem 2.5.1. The operator T^*T has a bounded principal symbol $|\sigma_a(x, \xi)|^2$ by the composition formula, so that $|\sigma_a(x, \xi)|^2 < M$ for some constant $0 < M < \infty$. Then the function $p(x, \xi) = \sqrt{M - |\sigma_a(x, \xi)|^2}$ is well defined and it is easy to check that $p \in S^0(\mathbb{R}^n \times \mathbb{R}^n)$. Let $P \in \Psi^0(\mathbb{R}^n \times \mathbb{R}^n)$ be the pseudo-differential operator with symbol p. By the calculus again we have the identity

$$T^*T = M - P^*P + R,$$

for some pseudo-differential operator $R \in \Psi^{-1}(\mathbb{R}^n \times \mathbb{R}^n)$. Then

$$\begin{aligned}
||Tu||^2_{L^2} = \langle Tu, Tu \rangle &= \langle T^*Tu, u \rangle \\
&= M||u||^2_{L^2} - ||Pu||^2_{L^2} + \langle Ru, u \rangle \le M||u||^2_{L^2} + \langle Ru, u \rangle,
\end{aligned}$$

so that T is bounded on $L^2(\mathbb{R}^n)$ if R is. The boundedness of R on $L^2(\mathbb{R}^n)$ can be proved by induction. Indeed, using the estimate

$$||Ru||^2_{L^2} = \langle Ru, Ru \rangle = \langle R^*Ru, u \rangle \le C||R^*Ru||_{L^2}||u||_{L^2}$$

we see that $R \in \Psi^{-1}(\mathbb{R}^n \times \mathbb{R}^n)$ is bounded on $L^2(\mathbb{R}^n)$ if $R^*R \in \Psi^{-2}(\mathbb{R}^n \times \mathbb{R}^n)$ is bounded on $L^2(\mathbb{R}^n)$. Continuing this argument we can reduce the question of L^2-boundedness to the boundedness of pseudo-differential operators $S \in \Psi^m(\mathbb{R}^n \times \mathbb{R}^n)$ for some sufficiently negative $m < 0$. We can now use Theorem 2.3.1 with $N = 0$ for x close to y to show that the integral kernel $K(x, y)$ of S is bounded for $|y - x| \le 1$ while it decreases fast for large $|x - y|$ if we take sufficiently large N. Therefore, we can use Lemma 2.5.4 to conclude that S must be bounded on $L^2(\mathbb{R}^n)$ thus completing the proof. $\qquad\square$

2.5.5 Asymptotic sums

Our objective in Section 2.6 will be to show how the constructed calculus of pseudo-differential operators can be applied to "easily" solve so-called elliptic partial differential equations. However, in order to carry out this application we will need another very useful construction.

Proposition 2.5.33 (Asymptotic sums). *Let* $a_j \in S^{m_j}(\mathbb{R}^n \times \mathbb{R}^n)$, $j = 0, 1, 2, \ldots$, *where the sequence* m_j *of orders satisfies* $m_0 > m_1 > m_2 > \cdots$ *and* $m_j \to -\infty$ *as* $j \to \infty$. *Then there exists a symbol* $a \in S^{m_0}(\mathbb{R}^n \times \mathbb{R}^n)$ *such that*

$$a \sim \sum_{j=0}^{\infty} a_j = a_0 + a_1 + a_2 + \cdots,$$

which means that we have

$$a - \sum_{j=0}^{k-1} a_j \in S^{m_k}(\mathbb{R}^n \times \mathbb{R}^n),$$

for all $k \in \mathbb{N}$.

Proof. Let us fix a function $\chi \in C^\infty(\mathbb{R}^n)$ such that $\chi(\xi) = 1$ for all $|\xi| \geq 1$ and such that $\chi(\xi) = 0$ for all $|\xi| \leq 1/2$. Then, for some sequence τ_j increasing sufficiently fast and to be chosen later, we define

$$a(x, \xi) = \sum_{j=0}^{\infty} a_j(x, \xi) \chi\left(\frac{\xi}{\tau_j}\right).$$

We note that this sum is well defined pointwise because it is in fact locally finite since $\chi\left(\frac{\xi}{\tau_j}\right) = 0$ for $|\xi| < \tau_j/2$. In order to show that $a \in S^{m_0}(\mathbb{R}^n \times \mathbb{R}^n)$ we first take a sequence τ_j such that the inequality

$$\left|\partial_x^\beta \partial_\xi^\alpha \left[a_j(x, \xi) \chi\left(\frac{\xi}{\tau_j}\right)\right]\right| \leq 2^{-j}(1 + |\xi|)^{m_j + 1 - |\alpha|} \tag{2.18}$$

is satisfied for all $|\alpha|, |\beta| \leq j$. We first show that function $\xi^\alpha \partial_\xi^\alpha \chi\left(\frac{\xi}{\tau_j}\right)$ is uniformly bounded in ξ for each j. Indeed, we have

$$\xi^\alpha \partial_\xi^\alpha \chi\left(\frac{\xi}{\tau_j}\right) = \begin{cases} 0, & |\xi| < \tau_j/2, \\ \text{bounded by } C\left|\left(\frac{\xi}{\tau_j}\right)^\alpha\right|, & \tau_j/2 \leq |\xi| \leq |\tau|, \\ 0, & \tau_j < |\xi|, \end{cases}$$

so that $\left|\xi^\alpha \partial_\xi^\alpha \chi\left(\frac{\xi}{\tau_j}\right)\right| \leq C$ is uniformly bounded for all ξ, for any given j. Using this fact, we can also estimate

$$\left|\partial_x^\beta \partial_\xi^\alpha \left[a_j(x, \xi) \chi\left(\frac{\xi}{\tau_j}\right)\right]\right| = \left|\sum_{\alpha_1 + \alpha_2 = \alpha} c_{\alpha_1 \alpha_2} \partial_x^\beta \partial_\xi^{\alpha_1} a_j(x, \xi) \partial_\xi^{\alpha_2} \chi\left(\frac{\xi}{\tau_j}\right)\right|$$

$$\leq \sum_{\alpha_1 + \alpha_2 = \alpha} |c_{\alpha_1 \alpha_2}| (1 + |\xi|)^{m_j - |\alpha_1|} (1 + |\xi|)^{-|\alpha_2|}$$

$$\leq C(1 + |\xi|)^{m_j - |\alpha|}$$

$$= \left[C(1 + |\xi|)^{-1}\right] (1 + |\xi|)^{m_j + 1 - |\alpha|}.$$

Now, the left-hand side in estimate (2.18) is zero for $|\xi| < \tau_j/2$, so we may assume that $|\xi| \geq \tau_j/2$. Hence we can have

$$C(1 + |\xi|)^{-1} \leq C(1 + |\tau_j/2|)^{-1} < 2^{-j}$$

if we take τ_j sufficiently large. This implies that we can take the sum of $\partial_x^\beta \partial_\xi^\alpha$-derivatives in the definition of $a(x, \xi)$ and (2.18) implies that $a \in S^{m_0}(\mathbb{R}^n \times \mathbb{R}^n)$. Finally, to show the asymptotic formula, we can write

$$a - \sum_{j=0}^{k-1} a_j = \sum_{j=k}^{\infty} a_j(x, \xi) \chi\left(\frac{\xi}{\tau_j}\right),$$

and so

$$\left| \partial_x^\beta \partial_\xi^\alpha \left[a - \sum_{j=0}^{k-1} a_j \right] \right| \leq C(1 + |\xi|)^{m_k - |\alpha|}.$$

In this argument we fix α and β first, and then use the required estimates for all $j \geq |\alpha|, |\beta|$. This shows that $a - \sum_{j=0}^{k-1} a_j \in S^{m_k}(\mathbb{R}^n \times \mathbb{R}^n)$ finishing the proof. $\quad\square$

Exercise 2.5.34. Prove that Proposition 2.5.33 remains valid in (ρ, δ) classes for all ρ and δ.

2.6 Applications to partial differential equations

The main question in the theory of partial differential equations is how to solve the equation $Au = f$ for a given partial differential operator A and a given function f. In other words, how to find the *inverse of A*, i.e., an operator A^{-1} such that

$$A \circ A^{-1} = A^{-1} \circ A = I \tag{2.19}$$

is the identity operator (on some space of functions where everything is well defined). In this case function $u = A^{-1}f$ gives a solution to the partial differential equation $Au = f$.

First of all we can observe that if the operator A is an operator with variable coefficients in most cases it is impossible or very hard to find an explicit formula for its inverse A^{-1} (even when it exists). However, in many questions in the theory of partial differential equations one is actually not so much interested in having a precise explicit formula for A^{-1}. Indeed, in reality one is mostly interested not in knowing the solution u to the equation $Au = f$ explicitly but rather in knowing some fundamental properties of u. One of the most important properties is the position and the strength of singularities of u. Thus, the question becomes whether we can say something about singularities of u knowing singularities of $f = Au$. In this case we do not need to solve equation $Au = f$ exactly but it is sufficient to know its solution modulo the class of smooth functions. Namely, instead of A^{-1} in (2.19) one is interested in finding an "approximate" inverse of A modulo smooth functions, i.e., an operator B such that $u = Bf$ solves the equation $Au = f$ modulo smooth functions, i.e., if $(BA - I)f$ and $(AB - I)f$ are smooth for all functions f from some class. Recalling that operators in $\Psi^{-\infty}(\mathbb{R}^n \times \mathbb{R}^n)$ have such a property, we have the following definition, which applies to all pseudo-differential operators A:

Definition 2.6.1 (Parametrix). Operator B is called the *right parametrix* of A if $AB - I \in \Psi^{-\infty}(\mathbb{R}^n \times \mathbb{R}^n)$. Operator C is called the *left parametrix* of A if $CA - I \in \Psi^{-\infty}(\mathbb{R}^n \times \mathbb{R}^n)$.

Remark 2.6.2 (**Left or right parametrix?**). In fact, the left and right parametrix are closely related. Indeed, by definition we have $AB - I = R_1$ and $CA - I = R_2$ with some $R_1, R_2 \in \Psi^{-\infty}(\mathbb{R}^n \times \mathbb{R}^n)$. Then we have

$$C = C(AB - R_1) = (CA)B - CR_1 = B + R_2B - CR_1.$$

If A, B, C are pseudo-differential operators of finite orders, the composition formula in Theorem 2.5.1 implies that $R_2B, CR_1 \in \Psi^{-\infty}(\mathbb{R}^n \times \mathbb{R}^n)$, i.e., $C - B$ is a smoothing operator. Thus, we will be mainly interested in the right parametrix B because $u = Bf$ immediately solves the equation $Au = f$ modulo smooth functions.

We also note that since we work here modulo smoothing operators (i.e., operators in $\Psi^{-\infty}(\mathbb{R}^n \times \mathbb{R}^n)$), parametrices are obviously not unique – finding one of them is already very good because any two parametrices differ by a smoothing operator.

2.6.1   Freezing principle for PDEs

The following freezing principle provides a good and well-known motivation (see, e.g., [118]) for the use of the symbolic analysis in finding parametrices. Suppose we want to solve the following equation for an unknown function $u = u(x)$:

$$(Au)(x) := \sum_{1 \leq i,j \leq n} a_{ij}(x) \frac{\partial^2 u}{\partial x_i \partial x_j}(x) = f(x),$$

where the matrix $\{a_{ij}(x)\}_{i,j=1}^n$ is real valued, smooth, symmetric and positive definite. If we want to proceed in analogy to the Laplace equation in Remark 1.1.17, we should look for the inverse of the operator A. In the case of an operator with variable coefficients this may turn out to be difficult, so we may look for an approximate inverse B such that $AB = I + E$, where the error E is small in some sense. To be able to argue similar to Remark 1.1.17, we "freeze" the operator A at x_0 to get the constant coefficients operator

$$A_{x_0} = \sum_{1 \leq i,j \leq n} a_{ij}(x_0) \frac{\partial^2}{\partial x_i \partial x_j}.$$

Now, A_{x_0} has the exact inverse which is the operator of multiplication by

$$\left(-4\pi^2 \sum_{1 \leq i,j \leq n} a_{ij}(x_0)\xi_i\xi_j \right)^{-1}$$

on the Fourier transform side. To avoid a singularity at the origin, we introduce a cut-off function $\chi \in C^\infty$ which is 0 near the origin and 1 for large ξ. Then we define

$$(B_{x_0}f)(x) = \int_{\mathbb{R}^n} e^{2\pi i x \cdot \xi} \left(-4\pi^2 \sum_{1\le i,j\le n} a_{ij}(x_0)\xi_i\xi_j \right)^{-1} \chi(\xi) \, \widehat{f}(\xi) \, d\xi.$$

Consequently, we can readily see that

$$\begin{aligned}
(A_{x_0}B_{x_0}f)(x) &= \int_{\mathbb{R}^n} e^{2\pi i x \cdot \xi} \, \chi(\xi) \, \widehat{f}(\xi) \, d\xi \\
&= f(x) + \int_{\mathbb{R}^n} e^{2\pi i x \cdot \xi} \, (\chi(\xi) - 1) \, \widehat{f}(\xi) \, d\xi.
\end{aligned}$$

It follows that $A_{x_0}B_{x_0} = I + E_{x_0}$, where

$$(E_{x_0}f)(x) = \int_{\mathbb{R}^n} e^{2\pi i x \cdot \xi} \, (\chi(\xi) - 1) \, \widehat{f}(\xi) \, d\xi$$

is an operator of multiplication by a compactly supported function on the Fourier transform side. Writing it as a convolution with a smooth test function we can readily see that it is a smoothing operator.

Exercise 2.6.3. Prove this.

Now, we can "unfreeze" the point x_0 expecting that the inverse B will be close to B_{x_0} for x close to x_0, and define

$$\begin{aligned}
(Bf)(x) &= (B_x f)(x) \\
&= \int_{\mathbb{R}^n} e^{2\pi i x \cdot \xi} \left(-4\pi^2 \sum_{1\le i,j\le n} a_{ij}(x)\xi_i\xi_j \right)^{-1} \chi(\xi) \, \widehat{f}(\xi) \, d\xi.
\end{aligned}$$

This does not yield a parametrix yet, but it will be clear from the composition formula that we still have $AB = I + E_1$ with error $E_1 \in \Psi^{-1}(\mathbb{R}^n \times \mathbb{R}^n)$ being "smoothing of order 1". We can then set up an iterative procedure to improve the approximation of the inverse operator relying on the calculus of the appearing operators, and to find a parametrix for A. This will be done in Theorem 2.6.7.

2.6.2 Elliptic operators

We will now show how we can use the calculus to "solve" elliptic partial differential equations. First, we recall the notion of ellipticity.

Definition 2.6.4 (Elliptic symbols). A symbol $a \in S^m(\mathbb{R}^n \times \mathbb{R}^n)$ is called *elliptic* if for some $A > 0$ it satisfies

$$|a(x,\xi)| \ge A|\xi|^m$$

for all $|\xi| \geq n_0$ and all $x \in \mathbb{R}^n$, for some $n_0 > 0$. We also say that the symbol a is *elliptic* in $U \subset \mathbb{R}^n$ if the above estimate holds for all $x \in U$. Pseudo-differential operators with elliptic symbols are also called *elliptic*.

Exercise 2.6.5. Show that the constant n_0 is not essential in this definition. Namely, for an elliptic symbol $a \in S^m(\mathbb{R}^n \times \mathbb{R}^n)$ show that there exists a symbol $b \in S^m(\mathbb{R}^n \times \mathbb{R}^n)$ satisfying

$$|b(x, \xi)| \geq \tilde{c}(1 + |\xi|)^m$$

for all $x, \xi \in \mathbb{R}^n$, such that b differs from a by a symbol in $S^{-\infty}(\mathbb{R}^n \times \mathbb{R}^n)$.

Now, let $L = a(X, D)$ be an elliptic pseudo-differential operator with symbol $a \in S^m(\mathbb{R}^n \times \mathbb{R}^n)$ (which is then also elliptic by definition). Let us introduce a cut-off function $\chi \in C^\infty(\mathbb{R}^n)$ such that $\chi(\xi) = 0$ for small ξ, e.g., for $|\xi| \leq 1$, and such that $\chi(\xi) = 1$ for large ξ, e.g., for $|\xi| > 2$. The ellipticity of $a(x, \xi)$ assures that it can be inverted pointwise for $|\xi| \geq 1$, so we can define the symbol

$$b(x, \xi) = \chi(\xi) \left[a(x, \xi) \right]^{-1}.$$

Since $a \in S^m(\mathbb{R}^n \times \mathbb{R}^n)$ is elliptic, we easily see that $b \in S^{-m}(\mathbb{R}^n \times \mathbb{R}^n)$. If we take $P_0 = b(X, D)$ then by the composition Theorem 2.5.1 we obtain

$$LP_0 = I + E_1, \quad PL = I + E_2,$$

for some $E_1, E_2 \in \Psi^{-1}(\mathbb{R}^n \times \mathbb{R}^n)$. Thus, we may view P_0 as a good first approximation for a parametrix of L. In order to find a parametrix of L, we need to modify P_0 in such a way that E_1 and E_2 would be in $\Psi^{-\infty}(\mathbb{R}^n \times \mathbb{R}^n)$. This construction can be carried out in an iterative way. Indeed, we now show that ellipticity is equivalent to invertibility in the algebra $\Psi^\infty(\mathbb{R}^n \times \mathbb{R}^n)/\Psi^{-\infty}(\mathbb{R}^n \times \mathbb{R}^n)$:

Theorem 2.6.6 (Elliptic $\Longleftrightarrow \exists$ Parametrix). *Operator $A \in \Psi^m(\mathbb{R}^n \times \mathbb{R}^n)$ is elliptic if and only if there exists $B \in \Psi^{-m}(\mathbb{R}^n \times \mathbb{R}^n)$ such that $BA \sim I \sim AB$ modulo $\Psi^{-\infty}(\mathbb{R}^n \times \mathbb{R}^n)$.*

Proof. Let σ_A and σ_B denote symbols of A and B, respectively. Assume first that $A \in \Psi^m(\mathbb{R}^n \times \mathbb{R}^n)$ and $B \in \Psi^{-m}(\mathbb{R}^n \times \mathbb{R}^n)$ satisfy $BA = I - T$ and $AB = I - T'$ with $T, T' \in \Psi^{-\infty}(\mathbb{R}^n \times \mathbb{R}^n)$. Then $1 - \sigma_{BA} = \sigma_T \in S^{-\infty}(\mathbb{R}^n \times \mathbb{R}^n)$ and consequently by Theorem 2.5.1 we have $1 - \sigma_B \sigma_A \in S^{-1}(\mathbb{R}^n \times \mathbb{R}^n)$, so that $|1 - \sigma_B \sigma_A| \leq C\langle \xi \rangle^{-1}$. Hence $1 - |\sigma_B| \cdot |\sigma_A| \leq C\langle \xi \rangle^{-1}$, or equivalently $|\sigma_B| \cdot |\sigma_A| \geq 1 - C\langle \xi \rangle^{-1}$. If we choose $n_0 > C$, then $|\sigma_B(x, \xi)| \cdot |\sigma_A(x, \xi)| \geq 1 - C\langle n_0 \rangle^{-1} > 0$ for any $|\xi| \geq n_0$. Thus, $\sigma_A(x, \xi) \neq 0$ for $|\xi| > n_0$ and

$$\frac{1}{|\sigma_A(x, \xi)|} \leq C|\sigma_B(x, \xi)| \leq C\langle \xi \rangle^{-m}.$$

Hence A is elliptic of order m. This yields the first part of the proof.

Conversely, assume that A and $\sigma_A(x,\xi)$ are elliptic. We will construct the symbol b as an asymptotic sum

$$b \sim b_0 + b_1 + b_2 + \cdots$$

and then use Proposition 2.5.5 to justify this infinite sum. Then we take operators B_j with symbols b_j and the operator B with symbol b will be the parametrix for A. We will also work with $|\xi| \geq n_0$ since small ξ are not relevant for symbolic constructions. Moreover, once we have the left parametrix, we also have the right one in view of Remark 2.6.2.

First, we take $b_0 = 1/\sigma_A$ which is well defined for $|\xi| \geq n_0$ in view of the ellipticity of σ_A. Then we have

$$\sigma_{B_0 A} = 1 - e_0, \quad b_0 \in S^{-m}, \quad \text{with } e_0 = 1 - \sigma_{B_0 A} \in S^{-1}.$$

Then we take $b_1 = e_0/\sigma_A \in S^{-m-1}$ so that we have

$$\sigma_{(B_0+B_1)A} = 1 - e_0 + \sigma_{B_1 A} = 1 - e_1, \quad \text{with } e_1 = e_0 - \sigma_{B_1 A} \in S^{-2}.$$

Inductively, we define $b_j = e_{j-1}/\sigma_A \in S^{-m-j}$ and we have

$$\sigma_{(B_0+B_1+\cdots+B_j)A} = 1 - e_j, \quad \text{with } e_j = e_{j-1} - \sigma_{B_j A} \in S^{-j-1}.$$

Now, Proposition 2.5.5 shows that $b \in S^{-m}$ and it satisfies $\sigma_{BA} = 1 - e$ with $e \in S^{-\infty}$ by its construction, completing the proof. $\qquad\square$

We now give a slightly more general statement which is useful for other purposes as well. It is a consequence of Theorem 2.6.6 and composition Theorem 2.5.1.

Corollary 2.6.7 (Local parametrix). *Let $a \in S^m(\mathbb{R}^n \times \mathbb{R}^n)$ be elliptic on an open set $U \subset \mathbb{R}^n$, i.e., there exists some $A > 0$ such that $|a(x,\xi)| \geq A|\xi|^m$ for all $x \in U$ and all $|\xi| \geq 1$. Let $c \in S^l$ be a symbol of order l whose support with respect to x is a compact subset of U. Then there exists a symbol $b \in S^{l-m}$ such that $b(X,D)a(X,D) = c(X,D) - e(X,D)$ for some symbol $e \in S^{-\infty}$.*

We also have the following local version of this for partial differential operators:

Corollary 2.6.8 (Parametrix for elliptic differential operators). *Let*

$$L = \sum_{|\alpha| \leq m} a_\alpha(x)\partial_x^\alpha$$

be an elliptic partial differential operator in an open set $U \subset \mathbb{R}^n$. Let $\chi_1, \chi_2, \chi_3 \in C_0^\infty(\mathbb{R}^n)$ be such that $\chi_2 = 1$ on the support of χ_1 and $\chi_3 = 1$ on the support of χ_2. Then there is an operator $P \in S^{-m}(\mathbb{R}^n \times \mathbb{R}^n)$ such that

$$P(\chi_2 L) = \chi_1 I + E\chi_3,$$

for some $E \in \Psi^{-\infty}(\mathbb{R}^n \times \mathbb{R}^n)$.

Proof. We take $a(X,D) = \chi_2 L$ and $c(X,D) = \chi_1 I$ in Corollary 2.6.7. Then $a(X,D)$ is elliptic on the support of χ_2 and we can take $P = b(X,D)$ with $b \in S^{-m}$ from Corollary 2.6.7. $\qquad\square$

We will now apply this result to obtain a statement on the regularity of solution to elliptic partial differential equations. We assume that the order m below is an integer which is certainly true when L is a partial differential operator. However, if we take into account the discussion from the next section, we will see that the statements below are still true for any $m \in \mathbb{R}$.

Theorem 2.6.9 (A-priori estimate). *Let $L \in \Psi^m$ be an elliptic pseudo-differential operator in an open set $U \subset \mathbb{R}^n$ and let $Lu = f$ in U. Assume that $f \in (L^2_k(U))_{\mathrm{loc}}$. Then $u \in (L^2_{m+k}(U))_{\mathrm{loc}}$.*

This theorem shows that if u is a solution of an elliptic partial differential equation $Lu = f$ then there is local gain of m derivatives for u compared to f, where m is the order of the operator L.

Proof. Let $\chi_1, \chi_2, \chi_3 \in C_0^\infty(U)$ be non-zero functions such that $\chi_2 = 1$ on the support of χ_1 and $\chi_3 = 1$ on the support of χ_2. Then, similar to the proof of Corollary 2.6.8 we have

$$P(\chi_2 L) = \chi_1 I + E\chi_3,$$

with some $P \in \Psi^{-m}$. Since $f \in (L^2_k)_{\mathrm{loc}}$ we have $P(\chi_2 f) \in (L^2_{m+k})_{\mathrm{loc}}$. Also, $E(\chi_3 u) \in C^\infty$ so that $\|\chi_3 E(\chi_3 u)\|_{L^2_k} \leq \|\chi_3 u\|_{L^2}$ for any k. Summarising and using

$$P(\chi_2 f) = \chi_1 u + E(\chi_3 u),$$

we obtain

$$\|\chi_1 u\|_{L^2_{k+m}} \leq C\left(\|\chi_2 f\|_{L^2_k} + \|\chi_3 u\|_{L^2}\right),$$

which implies that $u \in (L^2_{m+k})_{\mathrm{loc}}$ in U. $\qquad\square$

Remark 2.6.10. We can observe from the proof that properties of solution u by the calculus and the existence of a parametrix are reduced to the fact that pseudo-differential operators in Ψ^{-m} map L^2_k to L^2_{k+m}. In fact, in this way many properties of solutions to partial differential equations are reduced to questions about general pseudo-differential operators. In the following statement for now one can think of m and k being integers or zeros such that $m \leq k$, but if we adopt the definition of Sobolev spaces from Definition 2.6.15, it is valid for all $m, k \in \mathbb{R}$. We will prove it completely in the case of $p = 2$, and in the case $p \neq 2$ we will show how to reduce it to the L^p-boundedness of pseudo-differential operators.

Theorem 2.6.11 (L^p_k-continuity). *Let $T \in S^m(\mathbb{R}^n \times \mathbb{R}^n)$ be a pseudo-differential operator of order $m \in \mathbb{R}$, let $1 < p < \infty$, and let $k \in \mathbb{R}$. Then T extends to a bounded linear operator from the Sobolev space $L^p_k(\mathbb{R}^n)$ to the Sobolev space $L^p_{k-m}(\mathbb{R}^n)$.*

We will prove this statement in the next section. As an immediate consequence, by the same argument as in the proof of this theorem, we also obtain

Corollary 2.6.12 (Local L_k^p-continuity). *Let $L \in \Psi^m$ be an elliptic pseudo-differential operator in an open set $U \subset \mathbb{R}^n$, let $1 < p < \infty$, $m, k \in \mathbb{R}$, and let $Lu = f$ in U. Assume that $f \in (L_k^p(U))_{\mathrm{loc}}$. Then $u \in (L_{m+k}^p(U))_{\mathrm{loc}}$.*

Let us briefly discuss an application of the established a priori estimates.

Definition 2.6.13 (Harmonic functions). A distribution $f \in \mathcal{D}'(\mathbb{R}^n)$ is called *harmonic* if $\mathcal{L}f = 0$, where $\mathcal{L} = \frac{\partial^2}{\partial x_1^2} + \cdots + \frac{\partial^2}{\partial x_n^2}$ is the usual Laplace operator.

Taking real and imaginary parts of holomorphic functions, we see that Liouville's theorem D.6.2 for holomorphic functions follows from

Theorem 2.6.14 (Liouville's theorem for harmonic functions). *Every harmonic function $f \in L^\infty(\mathbb{R}^n)$ is constant.*

Proof. Since $\mathcal{L}$ is elliptic, by Theorem 2.6.9 it follows from the equation $\mathcal{L}f = 0$ that $f \in C^\infty(\mathbb{R}^n)$. Taking the Fourier transform of $\mathcal{L}f = 0$ we obtain $-4\pi^2 |\xi|^2 \widehat{f} = 0$ which means that $\operatorname{supp} \widehat{f} \subset \{0\}$. By Exercise 1.4.15 it follows that $\widehat{f} = \sum_{|\alpha| \leq m} a_\alpha \partial^\alpha \delta$. Taking the inverse Fourier transform we see that $f(x)$ must be a polynomial. Finally, the assumption that f is bounded implies that f must be constant. $\qquad\square$

2.6.3 Sobolev spaces revisited

Up to now we defined Sobolev spaces L_k^p assuming that the index k is an integer. In fact, using the calculus of pseudo-differential operators we can show that these spaces can be defined for all $k \in \mathbb{R}$ thus allowing one to measure the regularity of functions much more precisely. In the following discussion we assume the statement on the L^p-continuity of pseudo-differential operators from Theorem 2.6.22.

We recall from Definition 1.5.6 that for an integer $k \in \mathbb{N}$ we defined the Sobolev space $L_k^p(\mathbb{R}^n)$ as the space of all $f \in L^p(\mathbb{R}^n)$ such that their distributional derivatives satisfy $\partial_x^\alpha f \in L^p(\mathbb{R}^n)$, for all $0 \leq |\alpha| \leq k$. This space is equipped with a norm $\|f\|_{L_k^p} = \sum_{|\alpha| \leq k} \|\partial_x^\alpha f\|_{L^p}$ (or with any equivalent norm) for $1 \leq p < \infty$, with a modification for $p = \infty$.

Let $\mathcal{L} = \frac{\partial^2}{\partial x_1^2} + \cdots + \frac{\partial^2}{\partial x_n^2}$ be the Laplace operator, so that its symbol is equal to $4\pi^2 |\xi|^2$. Let $s \in \mathbb{R}$ be a real number and let us consider operators $(I - \mathcal{L})^{s/2} \in \Psi^s(\mathbb{R}^n \times \mathbb{R}^n)$ which are pseudo-differential operators with symbols $a(x, \xi) = (1 + 4\pi^2 |\xi|^2)^{s/2}$.

Definition 2.6.15 (Sobolev spaces). We will say f is in the Sobolev space $L_s^p(\mathbb{R}^n)$, i.e., $f \in L_s^p(\mathbb{R}^n)$, if $(I - \mathcal{L})^{s/2} f \in L^p(\mathbb{R}^n)$. We equip this space with the norm $\|f\|_{L_s^p} := \|(I - \mathcal{L})^{s/2} f\|_{L^p}$.

Proposition 2.6.16. *If $s \in \mathbb{N}$ is an integer, the space $L_s^p(\mathbb{R}^n)$ coincides with the space $L_k^p(\mathbb{R}^n)$ with $k = s$, with equivalence of norms.*

Proof. We will use the index k for both spaces. Since operator $(I - \mathcal{L})^{k/2}$ is a pseudo-differential operator of order k, by Theorem 2.6.11 we get that it is bounded from L_k^p to L^p, i.e., we have

$$\|(I - \mathcal{L})^{k/2} f\|_{L^p} \le C \sum_{|\alpha| \le k} \|\partial_x^\alpha f\|_{L^p}.$$

Conversely, let P_α be a pseudo-differential operator defined by $P_\alpha = \partial_x^\alpha (I - \mathcal{L})^{-k/2}$, i.e., a pseudo-differential operator with symbol $p_\alpha(x, \xi) = (2\pi i \xi)^\alpha (1 + 4\pi^2 |\xi|^2)^{-k/2}$, independent of x. If $|\alpha| \le k$, we get that $p_\alpha \in S^{|\alpha| - k} \subset S^0$, so that $P_\alpha \in S^0(\mathbb{R}^n \times \mathbb{R}^n)$ for all $|\alpha| \le k$. By Theorem 2.6.22 operators P_α are bounded on $L^p(\mathbb{R}^n)$. Therefore, we obtain

$$\sum_{|\alpha| \le k} \|\partial_x^\alpha f\|_{L^p} = \sum_{|\alpha| \le k} \|P_\alpha (I - \mathcal{L})^{k/2} f\|_{L^p} \le C \|(I - \mathcal{L})^{k/2} f\|_{L^p},$$

completing the proof.   □

Exercise 2.6.17 (Sobolev embedding theorem). Prove that if $s > k + n/2$ then $H^s(\mathbb{R}^n) \subset C^k(\mathbb{R}^n)$ and the inclusion is continuous. This gives a sharper version of Exercise 1.5.11.

Exercise 2.6.18 (Distributions as Sobolev space functions). Recall from Exercise 1.4.14 that if $u \in \mathcal{E}'(\mathbb{R}^n)$ then u is a distribution of some finite order m. Prove that if $s < -m - n/2$ then $u \in H^s$. Contrast this with Exercise 2.6.17.

Exercise 2.6.19. Prove that

$$\mathcal{S}(\mathbb{R}^n) = \bigcap_{s \in \mathbb{R}} \langle x \rangle^{-s} H^s(\mathbb{R}^n) \text{ and } \mathcal{S}'(\mathbb{R}^n) = \bigcup_{s \in \mathbb{R}} \langle x \rangle^s H^s(\mathbb{R}^n).$$

Note that the equalities fail without weights: for example, show that we have $\frac{\sin x}{x} \in \bigcap_{k \in \mathbb{N}_0} H^k(\mathbb{R})$ but $\frac{\sin x}{x} \notin \mathcal{S}(\mathbb{R})$. The situation on the torus will be somewhat simpler, see Corollary 3.2.12.

Finally, let us justify Theorem 2.6.11 However, we will assume without proof that pseudo-differential operators of order zero are bounded on $L^p(\mathbb{R}^n)$ for all $1 < p < \infty$, see Theorem 2.6.22.

Proof of Theorem 2.6.11. Let $f \in L_s^p(\mathbb{R}^n)$. By definition this means that $(I - \mathcal{L})^{s/2} f \in L^p(\mathbb{R}^n)$. Then we can write using the calculus of pseudo-differential operators (composition Theorem 2.5.1):

$$(I - \mathcal{L})^{(s-\mu)/2} T f = (I - \mathcal{L})^{(s-\mu)/2} T (I - \mathcal{L})^{-s/2} (I - \mathcal{L})^{s/2} f \in L^p(\mathbb{R}^n)$$

since operator $(I - \mathcal{L})^{(s-\mu)/2} T (I - \mathcal{L})^{-s/2}$ is a pseudo-differential operator of order zero and is, therefore, bounded on $L^p(\mathbb{R}^n)$ by Theorem 2.6.22 if $p \ne 2$ and by Theorem 2.4.2 if $p = 2$.   □

Remark 2.6.20. It is often very useful to conclude something about properties of functions in one Sobolev space knowing about their properties in another Sobolev space. One instance of such a conclusion will be used in the proof of Theorem 4.2.3 on the Sobolev boundedness of operators on the L^2-space on the torus. A general Banach space setting for such conclusions will be presented in Section 3.5. Here we present without proof another instance of this phenomenon:

Theorem 2.6.21 (Rellich's theorem). *Let* $(f_k)_{k=1}^{\infty} \subset H^s(\mathbb{R}^n)$ *be a uniformly bounded sequence of functions: there exists* C *such that* $\|f_k\|_{H^s(\mathbb{R}^n)} \leq C$ *for all* k. *Assume that all functions* f_k *are supported in a fixed compact set. Then there exists a subsequence of* $(f_k)_{k=1}^{\infty}$ *which converges in* $H^\sigma(\mathbb{R}^n)$ *for all* $\sigma < s$.

Remarks on L^p-continuity of pseudo-differential operators. Let $a \in S^0(\mathbb{R}^n \times \mathbb{R}^n)$. Then by Theorem 2.3.1 the integral kernel $K(x,y)$ of pseudo-differential operator $a(X, D)$ satisfies estimates

$$|\partial_x^\alpha \partial_y^\beta K(x,y)| \leq A_{\alpha\beta}|x - y|^{-n-|\alpha|-|\beta|}$$

for all $x \neq y$. In particular, for $\alpha = \beta = 0$ this gives

$$|K(x,y)| \leq A|x - y|^{-n} \text{ for all } x \neq y. \tag{2.20}$$

Moreover, if we use it for $\alpha = 0$ and $|\beta| = 1$, we get

$$\int_{|x-z|\geq 2\delta} |K(x,y) - K(x,z)| \, dx \leq A \text{ if } |x - z| \leq \delta, \text{ for all } \delta > 0. \tag{2.21}$$

Now, if we take a general integral operator T of the form

$$Tu(x) = \int_{\mathbb{R}^n} K(x,y)u(y) \, dy,$$

properties (2.20) and (2.21) of the kernel are the starting point of the so-called *Calderón–Zygmund theory of singular integral operators*. In particular, one can conclude that such operators are of weak type $(1, 1)$, i.e., they satisfy the estimate

$$\mu\{x \in \mathbb{R}^n : |Tu(x)| > \lambda\} \leq \frac{\|u\|_{L^1}}{\lambda},$$

(see Definition 1.6.3 and the discussion following it for more details). Since we also know from Theorem 2.4.2 that $a(X, D) \in \Psi^0(\mathbb{R}^n \times \mathbb{R}^n)$ are bounded on $L^2(\mathbb{R}^n)$ and since we also know from Proposition 1.6.4 that this implies that $a(X, D)$ is of weak type $(2, 2)$, we get that pseudo-differential operators of order zero are of weak types $(1, 1)$ and $(2, 2)$. Then, by Marcinkiewicz' interpolation Theorem 1.6.5, we conclude that $a(X, D)$ is bounded on $L^p(\mathbb{R}^n)$ for all $1 < p < 2$. By the standard duality argument, this implies that $a(X, D)$ is bounded on $L^p(\mathbb{R}^n)$ also for all $2 < p < \infty$.

Since we also have the boundedness of $L^2(\mathbb{R}^n)$, we obtain

Theorem 2.6.22. *Let $T \in \Psi^0(\mathbb{R}^n \times \mathbb{R}^n)$. Then T extends to a bounded operator from $L^p(\mathbb{R}^n)$ to $L^p(\mathbb{R}^n)$, for all $1 < p < \infty$.*

We note that there exist different proofs of this theorem. On one hand, it follows automatically from the Calderón–Zygmund theory of singular integral operators which include pseudo-differential operators considered here, if we view them as integral operators with singular kernels. There are many other proofs that can be found in monographs on pseudo-differential operators. Another alternative and more direct method is to reduce the L^p-boundedness to the question of uniform boundedness of Fourier multipliers in $L^p(\mathbb{R}^n)$ which then follows from Hörmander's theorem on Fourier multipliers. However, in this monograph we decided not to immerse ourselves in the L^p-world since our aims here are different. We can refer to [91] and to [92] for more information on the L^p-boundedness of general Fourier integral operators in (ρ, δ)-classes with real and complex phase functions, respectively.

Chapter 3

Periodic and Discrete Analysis

In this chapter we will review basics of the periodic and discrete analysis which will be necessary for development of the theory of pseudo-differential operators on the torus in Chapter 4. Our aim is to make these two chapters accessible independently for people who choose periodic pseudo-differential operators as a starting point for learning about pseudo-differential operators on $\mathbb{R}^n$. This may be a fruitful idea in the sense that many technical issues disappear on the torus as opposed to $\mathbb{R}^n$. Among them is the fact that often one does not need to worry about convergence of the integrals in view of the torus being compact. Moreover, the theory of distributions on the torus is much simpler than that on $\mathbb{R}^n$, at least in the form required for us. The main reason is that the periodic Fourier transform takes functions on $\mathbb{T}^n = \mathbb{R}^n/\mathbb{Z}^n$ to functions on $\mathbb{Z}^n$ where, for example, tempered distributions become pointwise defined functions on the lattice $\mathbb{Z}^n$ of polynomial growth at infinity. Also, on the lattice $\mathbb{Z}^n$ there are no questions of regularity since all the objects are defined on a discrete set. However, there are many parallels between Euclidean and toroidal theories of pseudo-differential operators, so looking at proofs of similar results in different chapters may be beneficial. In many cases we tried to avoid overlaps by presenting a different proof or by giving a different explanation.

Therefore, we also try to make the reading self-contained and elementary, avoiding cross-references to other chapters unless they increase the didactic value of the material.

Yet, being written for people working with analysis, this chapter only briefly states the related notations and facts of more general function analysis. Supplementary material is, of course, referred to. The reader should have a basic knowledge of Banach and Hilbert spaces (the necessary background material is provided in Chapter B); some familiarity with distributions and point set topology definitely helps (this material can be found in Chapter A and in Chapter 1 if necessary). A word of warning has to be said: in order to use the theory of periodic pseudo-differential operators as a tool, there is no demand to dwell deeply on these prerequisites. One is rather encouraged to read the appropriate theory only when it is encountered and needed, and that is why we present a summary of necessary things here as well.

We will use the following notation in the sequel. Triangles $\triangle$ and $\overline{\triangle}$ will denote the forward and backward difference operators, respectively. The Laplacian will be denoted by $\mathcal{L}$ to avoid any confusion. The Dirac delta at x will be denoted by δ_x and the Kronecker delta at ξ will be denoted by $\delta_{\xi,\eta}$.

As is common, $\mathbb{R}$ and $\mathbb{C}$ are written for real and complex numbers, respectively, $\mathbb{Z}$ stands for the integers, while $\mathbb{N} = \mathbb{Z}^+ := \{n \in \mathbb{Z} \mid n \geq 1\}$ and $\mathbb{N}_0 := \mathbb{Z}^+ \cup \{0\}$ are the sets of positive integers and nonnegative integers, respectively. We would also like to draw the reader's attention to the notation $|\alpha|$ and $\|\xi\|$ in (3.2) and (3.3), respectively, that we will be using in this chapter as well as in Chapters 4 and 5. This is especially of relevance in these chapters as both multi-indices $\alpha \in \mathbb{N}_0^n$ and frequencies $\xi \in \mathbb{Z}^n$ are integers.

3.1 Distributions and Fourier transforms on $\mathbb{T}^n$ and $\mathbb{Z}^n$

We fix the notation for the torus as $\mathbb{T}^n = (\mathbb{R}/\mathbb{Z})^n = \mathbb{R}^n/\mathbb{Z}^n$. Often we may identify $\mathbb{T}^n$ with the cube $[0,1)^n \subset \mathbb{R}^n$, where we identify the measure on the torus with the restriction of the Euclidean measure on the cube. Functions on $\mathbb{T}^n$ may be thought as those functions on $\mathbb{R}^n$ that are 1-periodic in each of the coordinate directions. We will often simply say that such functions are 1-periodic (instead of $\mathbb{Z}^n$-periodic). More precisely, on the Euclidean space $\mathbb{R}^n$ we define an equivalence relation

$$x \sim y \overset{\text{definition}}{\Longleftrightarrow} x - y \in \mathbb{Z}^n,$$

where the equivalence classes are

$$
\begin{aligned}
[x] &= \{y \in \mathbb{R}^n : x \sim y\} \\
&= \{x + k : k \in \mathbb{Z}^n\}.
\end{aligned}
$$

A point $x \in \mathbb{R}^n$ is naturally mapped to a point $[x] \in \mathbb{T}^n$, and usually there is no harm in writing $x \in \mathbb{T}^n$ instead of the actual $[x] \in \mathbb{T}^n$. We may identify functions on $\mathbb{T}^n$ with $\mathbb{Z}^n$-periodic functions on $\mathbb{R}^n$ in a natural manner, $f : \mathbb{T}^n \to \mathbb{C}$ being identified with $g : \mathbb{R}^n \to \mathbb{C}$ satisfying

$$g(x) = f([x])$$

for all $x \in \mathbb{R}^n$. In such a case we typically even write $g = f$ and $g(x) = f(x)$, and we might casually say things like

- "f is periodic",
- "$g \in C^\infty(\mathbb{T}^n)$" when actually "$g \in C^\infty(\mathbb{R}^n)$ is periodic",
- etc.

The reader has at least been warned. Moreover, the one-dimensional torus $\mathbb{T}^1 = \mathbb{R}^1/\mathbb{Z}^1$ is isomorphic to the circle

$$
\begin{aligned}
\mathbb{S}^1 &= \{z \in \mathbb{R}^2 : \|z\| = 1\} \\
&= \{(\cos(t), \sin(t)) : t \in \mathbb{R}\}
\end{aligned}
$$

by the obvious mapping

$$[t] \mapsto (\cos(2\pi t), \sin(2\pi t)),$$

so we may identify functions on $\mathbb{T}^1$ with functions on $\mathbb{S}^1$.

Remark 3.1.1 (**What makes $\mathbb{T}^1$ and $\mathbb{T}^n$ special?**). At this point, we must emphasize how fundamental the study on the one-dimensional torus $\mathbb{T}^1 = \mathbb{R}^1/\mathbb{Z}^1$ is. First, smooth Jordan curves, especially the one-dimensional sphere $\mathbb{S}^1$, are diffeomorphic to $\mathbb{T}^1$. Secondly, the theory on the n-dimensional torus $\mathbb{T}^n = \mathbb{R}^n/\mathbb{Z}^n$ sometimes reduces to the case of $\mathbb{T}^1$. Furthermore, compared to the theory of pseudo-differential operators on $\mathbb{R}^n$, the case of $\mathbb{T}^n$ is beautifully simple. This is due to the fact that $\mathbb{T}^n$ is a compact Abelian group – whereas $\mathbb{R}^n$ is only locally compact – on which the powerful aid of Fourier series is at our disposal. However, the results on $\mathbb{R}^n$ and $\mathbb{T}^n$ are somewhat alike. Many general results concerning series on the torus and their properties can be found in, e.g., [155].

To make this chapter more self-contained, let us also briefly review the multi-index notation. A vector $\alpha = (\alpha_j)_{j=1}^n \in \mathbb{N}_0^n$ is called a *multi-index*. If $x = (x_j)_{j=1}^n \in \mathbb{R}^n$ and $\alpha \in \mathbb{N}_0^n$, we write $x^\alpha := x_1^{\alpha_1} \cdots x_n^{\alpha_n}$. For multi-indices, $\alpha \le \beta$ means $\alpha_j \le \beta_j$ for all $j \in \{1, \ldots, n\}$. We also write $\beta! := \beta_1! \cdots \beta_n!$ and

$$\binom{\alpha}{\beta} := \frac{\alpha!}{\beta! \, (\alpha - \beta)!} = \binom{\alpha_1}{\beta_1} \cdots \binom{\alpha_n}{\beta_n},$$

so that

$$(x + y)^\alpha = \sum_{\beta \le \alpha} \binom{\alpha}{\beta} x^{\alpha - \beta} y^\beta. \tag{3.1}$$

For $\alpha \in \mathbb{N}_0^n$ and $x \in \mathbb{R}^n$ we shall write

$$|\alpha| := \sum_{j=1}^n \alpha_j, \tag{3.2}$$

$$\|x\| := \left(\sum_{j=1}^n x_j^2 \right)^{1/2}, \tag{3.3}$$

$$\partial_x^\alpha := \partial_{x_1}^{\alpha_1} \cdots \partial_{x_n}^{\alpha_n},$$

where $\partial_{x_j} = \frac{\partial}{\partial x_j}$ etc. We will also use the notation $D_{x_j} = -\mathrm{i}2\pi\partial_{x_j} = -\mathrm{i}2\pi\frac{\partial}{\partial x_j}$, where $\mathrm{i} = \sqrt{-1}$ is the imaginary unit. We have chosen the notation $\|x\|$ for the

Euclidean distance in this chapter, to contrast it with $|\alpha|$ used for multi-indices.
We also denote

$$\langle x \rangle := (1 + \|x\|^2)^{1/2}.$$

Exercise 3.1.2. Prove (3.1).

Exercise 3.1.3. Show that $(\sum_{j=1}^n x_j)^m = \sum_{|\alpha|=m} \frac{m!}{\alpha!} x^\alpha$, where $x \in \mathbb{R}^n$ and $m \in \mathbb{N}_0$.

Definition 3.1.4 (Periodic functions). A function $f : \mathbb{R}^n \to Y$ is *1-periodic* if
$f(x+k) = f(x)$ for every $x \in \mathbb{R}^n$ and $k \in \mathbb{Z}^n$. We shall consider these functions to
be defined on $\mathbb{T}^n = \mathbb{R}^n/\mathbb{Z}^n = \{x + \mathbb{Z}^n | x \in \mathbb{R}^n\}$. The space of 1-periodic m times
continuously differentiable functions is denoted by $C^m(\mathbb{T}^n)$, and the test functions
are the elements of the space $C^\infty(\mathbb{T}^n) := \bigcap_{m \in \mathbb{Z}^+} C^m(\mathbb{T}^n)$.

Remark 3.1.5. The natural inherent topology of $C^\infty(\mathbb{T}^n)$ is induced by the seminorms that one gets by demanding the following convergence: $u_j \to u$ if and only
if $\partial^\alpha u_j \to \partial^\alpha u$ uniformly, for all $\alpha \in \mathbb{N}_0^n$. Thus, e.g., by [89, 1.46] $C^\infty(\mathbb{T}^n)$ is a
Fréchet space, but it is not normable as it has the Heine–Borel property.

Let $\mathcal{S}(\mathbb{R}^n)$ denote the space of the Schwartz test functions from Definition
1.1.11, and let $\mathcal{S}'(\mathbb{R}^n)$ be its dual, i.e., the space of the tempered distributions
from Definition 1.3.1. The integer lattice $\mathbb{Z}^n$ plays an important role in periodic
and discrete analysis.

Definition 3.1.6 (Schwartz space $\mathcal{S}(\mathbb{Z}^n)$). Let $\mathcal{S}(\mathbb{Z}^n)$ denote the space of *rapidly
decaying functions* $\mathbb{Z}^n \to \mathbb{C}$. That is, $\varphi \in \mathcal{S}(\mathbb{Z}^n)$ if for any $M < \infty$ there exists a
constant $C_{\varphi,M}$ such that

$$|\varphi(\xi)| \leq C_{\varphi,M} \langle \xi \rangle^{-M}$$

holds for all $\xi \in \mathbb{Z}^n$. The topology on $\mathcal{S}(\mathbb{Z}^n)$ is given by the seminorms p_k, where
$k \in \mathbb{N}_0$ and $p_k(\varphi) := \sup_{\xi \in \mathbb{Z}^n} \langle \xi \rangle^k |\varphi(\xi)|$.

Exercise 3.1.7 (Tempered distributions $\mathcal{S}'(\mathbb{Z}^n)$). Show that the continuous linear
functionals on $\mathcal{S}(\mathbb{Z}^n)$ are of the form

$$\varphi \mapsto \langle u, \varphi \rangle := \sum_{\xi \in \mathbb{Z}^n} u(\xi)\, \varphi(\xi),$$

where functions $u : \mathbb{Z}^n \to \mathbb{C}$ grow at most polynomially at infinity, i.e., there exist
constants $M < \infty$ and $C_{u,M}$ such that

$$|u(\xi)| \leq C_{u,M} \langle \xi \rangle^M$$

holds for all $\xi \in \mathbb{Z}^n$. Such distributions $u : \mathbb{Z}^n \to \mathbb{C}$ form the space $\mathcal{S}'(\mathbb{Z}^n)$.
Note that compared to $\mathcal{S}'(\mathbb{R}^n)$, distributions in $\mathcal{S}'(\mathbb{Z}^n)$ are pointwise well-defined
functions (!) on the lattice $\mathbb{Z}^n$.

To contrast Euclidean and toroidal Fourier transforms, they will be denoted by $\mathcal{F}_{\mathbb{R}^n}$ and $\mathcal{F}_{\mathbb{T}^n}$, respectively. Let $\mathcal{F}_{\mathbb{R}^n} : S(\mathbb{R}^n) \to S(\mathbb{R}^n)$ be the *Euclidean Fourier transform* defined by

$$(\mathcal{F}_{\mathbb{R}^n})f(\xi) := \int_{\mathbb{R}^n} e^{-2\pi i x \cdot \xi} f(x) \, dx.$$

Mapping $\mathcal{F}_{\mathbb{R}^n} : S(\mathbb{R}^n) \to S(\mathbb{R}^n)$ is a bijection, and its inverse $\mathcal{F}_{\mathbb{R}^n}^{-1}$ is given by

$$f(x) = \int_{\mathbb{R}^n} e^{2\pi i x \cdot \xi} (\mathcal{F}_{\mathbb{R}^n} f)(\xi) \, d\xi,$$

see Theorem 1.1.21. As is well known, this Fourier transform can be uniquely extended to $\mathcal{F}_{\mathbb{R}^n} : S'(\mathbb{R}^n) \to S'(\mathbb{R}^n)$ by duality, see Definition 1.3.2. We refer to Section 1.1 for further details concerning the Euclidean Fourier transform.

Definition 3.1.8 (Toroidal/periodic Fourier transform). Let

$$\mathcal{F}_{\mathbb{T}^n} = (f \mapsto \widehat{f}) : C^\infty(\mathbb{T}^n) \to S(\mathbb{Z}^n)$$

be the *toroidal Fourier transform* defined by

$$\widehat{f}(\xi) := \int_{\mathbb{T}^n} e^{-i2\pi x \cdot \xi} f(x) \, dx. \tag{3.4}$$

Then $\mathcal{F}_{\mathbb{T}^n}$ is a bijection and its inverse $\mathcal{F}_{\mathbb{T}^n}^{-1} : S(\mathbb{Z}^n) \to C^\infty(\mathbb{T}^n)$ is given by

$$f(x) = \sum_{\xi \in \mathbb{Z}^n} e^{i2\pi x \cdot \xi} \widehat{f}(\xi),$$

so that for $h \in S(\mathbb{Z}^n)$ we have

$$(\mathcal{F}_{\mathbb{T}^n}^{-1} h)(x) := \sum_{\xi \in \mathbb{Z}^n} e^{i2\pi x \cdot \xi} h(\xi).$$

Remark 3.1.9 (**Notations** $i2\pi x \cdot \xi$ **vs** $2\pi i x \cdot \xi$). We note that in the case of the toroidal Fourier transform we write $i2\pi x \cdot \xi$ in the exponential with i in front to emphasize that $2\pi x$ is now a periodic variable, and also to distinguish it from the Euclidean Fourier transform in which case we usually write $2\pi i$ in the exponential. We will write $\mathcal{F}_{\mathbb{T}^n}$ instead of $\widehat{f}$ in this and the next chapters only if we want to emphasize that we want to take the periodic Fourier transform.

Exercise 3.1.10 (Two Fourier inversion formulae). Prove that the Fourier transform $\mathcal{F}_{\mathbb{T}^n} : C^\infty(\mathbb{T}^n) \to S(\mathbb{Z}^n)$ is a bijection, that $\mathcal{F}_{\mathbb{T}^n} : C^\infty(\mathbb{T}^n) \to S(\mathbb{Z}^n)$ and $\mathcal{F}_{\mathbb{T}^n}^{-1} : S(\mathbb{Z}^n) \to C^\infty(\mathbb{T}^n)$ are continuous, and that

$$\mathcal{F}_{\mathbb{T}^n} \circ \mathcal{F}_{\mathbb{T}^n}^{-1} : S(\mathbb{Z}^n) \to S(\mathbb{Z}^n) \text{ and } \mathcal{F}_{\mathbb{T}^n}^{-1} \circ \mathcal{F}_{\mathbb{T}^n} : C^\infty(\mathbb{T}^n) \to C^\infty(\mathbb{T}^n)$$

are identity mappings on $S(\mathbb{Z}^n)$ and $C^\infty(\mathbb{T}^n)$, respectively.

Let us study an example of periodic distributions, the space $L^2(\mathbb{T}^n)$.

Definition 3.1.11 (Space $L^2(\mathbb{T}^n)$). Space $L^2(\mathbb{T}^n)$ is a Hilbert space with the inner product

$$(u, v)_{L^2(\mathbb{T}^n)} := \int_{\mathbb{T}^n} u(x)\, \overline{v(x)}\, dx, \qquad (3.5)$$

where $\bar{z}$ is the complex conjugate of $z \in \mathbb{C}$. The Fourier coefficients of $u \in L^2(\mathbb{T}^n)$ are

$$\widehat{u}(\xi) = \int_{\mathbb{T}^n} e^{-i2\pi x \cdot \xi}\, u(x)\, dx \quad (\xi \in \mathbb{Z}^n), \qquad (3.6)$$

and they are well defined for all ξ due to Hölder's inequality (Proposition 1.2.4) and compactness of $\mathbb{T}^n$.

Remark 3.1.12 (**Fourier series on $L^2(\mathbb{T}^n)$**). The family $\{e_\xi : \xi \in \mathbb{Z}^n\}$ defined by

$$e_\xi(x) := e^{i2\pi x \cdot \xi} \qquad (3.7)$$

forms an orthonormal basis on $L^2(\mathbb{T}^n)$, which will be proved in Theorem 3.1.20. Thus the partial sums of the *Fourier series* $\sum_{\xi \in \mathbb{Z}^n} \widehat{u}(\xi)\, e^{i2\pi x \cdot \xi}$ converge to u in the L^2-norm, so that we shall identify u with its Fourier series representation:

$$u(x) = \sum_{\xi \in \mathbb{Z}^n} \widehat{u}(\xi)\, e^{i2\pi x \cdot \xi}.$$

As before, we call $\widehat{u} : \mathbb{Z}^n \to \mathbb{C}$ the *Fourier transform of u*.

As a consequence of the Plancherel identity on general compact topological groups to be proved in Corollary 7.6.7 we obtain:

Remark 3.1.13 (**Plancherel's identity**). If $u \in L^2(\mathbb{T}^n)$ then $\widehat{u} \in \ell^2(\mathbb{Z}^n)$ and

$$||\widehat{u}||_{\ell^2(\mathbb{Z}^n)} = ||u||_{L^2(\mathbb{T}^n)}.$$

Exercise 3.1.14. Give a simple direct proof of Remark 3.1.13. (Hint: it is similar to the proof on $\mathbb{R}^n$ but simpler.)

Exercise 3.1.15. Show that $\mathcal{S}(\mathbb{Z}^n)$ is dense in $\ell^2(\mathbb{Z}^n)$.

Remark 3.1.16 (**Functions e_ξ**). We can observe that the functions $e_\xi(x) = e^{i2\pi x \cdot \xi}$ from (3.7) satisfy $e_\xi(x + y) = e_\xi(x)e_\xi(y)$ and $|e_\xi(x)| = 1$ for all $x \in \mathbb{T}^n$. The converse is also true, namely:

Theorem 3.1.17 (Unitary representations of $\mathbb{T}^n$). *If $f \in L^1(\mathbb{T}^n)$ is such that we have $f(x + y) = f(x)f(y)$ and $|f(x)| = 1$ for all $x, y \in \mathbb{T}^n$, then there exists some $\xi \in \mathbb{Z}^n$ such that $f = e_\xi$.*

Remark 3.1.18. It is a nice exercise to show this directly and we do it below. However, we note that employing a more general terminology of Chapter 7, the conditions on f mean that $f : \mathbb{T}^n \to U(1)$ is a unitary representation of $\mathbb{T}^n$, automatically irreducible since it is one-dimensional. Moreover, these conditions imply that f is continuous, and hence $f \in \widehat{\mathbb{T}^n}$, the unitary dual of f. Since functions e_ξ exhaust the unitary dual by the Peter–Weyl theorem (see, e.g., Remark 7.5.17), we obtain the result.

Proof of Theorem 3.1.17. We will prove the one-dimensional case since the general case of $\mathbb{T}^n$ follows from it if we look at functions $f(\tau e_j)$ where e_j is the jth unit basis vector of $\mathbb{R}^n$.

Thus, $x \in \mathbb{T}^1$, we can think of $\mathbb{T}$ as of periodic $\mathbb{R}$, and we choose $\lambda > 0$ such that $\Lambda = \int_0^\lambda f(\tau)\, d\tau \neq 0$. Such λ exists because otherwise we would have $f = 0$ a.e. by Corollary 1.5.17 of the Lebesgue differentiation theorem, contradicting the assumptions. Consequently we can write

$$f(x) = \Lambda^{-1} \int_0^\lambda f(x)f(\tau)\, d\tau = \Lambda^{-1} \int_0^\lambda f(x+\tau)\, d\tau = \Lambda^{-1} \int_x^{x+\lambda} f(\tau)\, d\tau.$$

From this we can observe that $f \in L^1(\mathbb{R})$ implies that f is continuous at x. Since this is true for all $x \in \mathbb{T}$ we get $f \in C^1(\mathbb{T})$. By induction, we get that actually $f \in C^\infty(\mathbb{T})$. Differentiating the equality above, we see that f satisfies the equation

$$f'(x) = \Lambda^{-1}(f(x+\lambda) - f(x)) = \Lambda^{-1}(f(x)f(\lambda) - f(x)) = C_0 f(x),$$

with $C_0 = \Lambda^{-1}(f(\lambda)-1)$. Solving this equation we find $f(x) = f(0)\, e^{C_0 x}$. Recalling that $|f(0)| = 1$ we get that $|f(x)| = e^{\operatorname{Re} C_0 x}$. Since $|f(x)| = 1$ we see that $\operatorname{Re} C_0 = 0$, and thus $C_0 = i2\pi\xi$ for some $\xi \in \mathbb{R}$. Finally, the fact that f is periodic implies that $\xi \in \mathbb{Z}$. $\qquad\square$

Exercise 3.1.19. Work out the details of the extension of the proof from $\mathbb{T}^1$ to $\mathbb{T}^n$. Also, show that the conclusion of Theorem 3.1.17 remains true if we replace $\mathbb{T}^n$ by $\mathbb{R}^n$ and condition $f \in L^1(\mathbb{T}^n)$ by $f \in L^1_{\text{loc}}(\mathbb{R}^n)$, but in this case $\xi \in \mathbb{R}^n$ does not have to be in the lattice $\mathbb{Z}^n$.

Theorem 3.1.20 (An orthonormal basis of $L^2(\mathbb{T}^n)$). *The collection $\{e_\xi : \xi \in \mathbb{Z}^n\}$ is an orthonormal basis of $L^2(\mathbb{T}^n)$.*

Remark 3.1.21. Let us make some general remarks first. From the general theory of Hilbert spaces we know that $L^2(\mathbb{T}^n)$ has an orthonormal basis, which is countable by Theorem B.5.35 if we can check that $L^2(\mathbb{R}^n)$ is separable. On the other hand, a more precise conclusion is possible from the general theory if we use that $\mathbb{T}^n$ is a group. Indeed, Theorem 3.1.17 (see also Remark 3.1.18) implies that $\widehat{\mathbb{T}^n} \cong \{e_\xi : \xi \in \mathbb{Z}^n\}$. Theorem 3.1.20 is then a special case of the Peter–Weyl theorem (see, e.g., Remark 7.5.17). However, at this point we give a more direct proof:

Proof of Theorem 3.1.20. It is easy to check the orthogonality property

$$(e_\xi, e_\eta)_{L^2(\mathbb{T}^n)} = 0 \quad \text{for} \quad \xi \neq \eta,$$

and the normality

$$(e_\xi, e_\xi)_{L^2(\mathbb{T}^n)} = 1 \quad \text{for all} \ \ \xi \in \mathbb{Z}^n,$$

so the real issue is to show that we have the basis according to Definition B.5.34. We want to apply the Stone–Weierstrass theorem A.14.4 to show that the set $E = \mathrm{span}\{e_\xi : \xi \in \mathbb{Z}^n\}$ is dense in $C(\mathbb{T}^n)$. If we have this, we can use the density of $C(\mathbb{T}^n)$ in $L^2(\mathbb{T}^n)$, so that by Theorem B.5.32 it would be a basis. We note that in fact the density of E in both $C(\mathbb{T}^n)$ and $L^2(\mathbb{T}^n)$ is a special case of Theorem 7.6.2 on general topological groups, but we give a direct short proof here. In view of the Stone–Weierstrass theorem A.14.4 all we have to show is that E is an involutive algebra separating the points of $\mathbb{T}^n$. It is clear that E separates points. Finally, from the identity $e_\xi e_\eta = e_{\xi+\eta}$ it follows that E is an algebra, which is also involutive because of the identity $\overline{e_\xi} = e_{-\xi}$. $\qquad\square$

Exercise 3.1.22. Show explicitly how E separates the points of $\mathbb{T}^n$, as well as verify the orthonormality statement in the proof.

Definition 3.1.23 (Spaces $L^p(\mathbb{T}^n)$). For $1 \le p < \infty$ let $L^p(\mathbb{T}^n)$ be the space of all $u \in L^1(\mathbb{T}^n)$ such that

$$||u||_{L^p(\mathbb{T}^n)} := \left(\int_{\mathbb{T}^n} |u(x)|^p \, dx \right)^{1/p} < \infty.$$

For $p = \infty$, let $L^\infty(\mathbb{T}^n)$ be the space of all $u \in L^1(\mathbb{T}^n)$ such that

$$||u||_{L^\infty(\mathbb{T}^n)} := \mathrm{esssup}_{x \in \mathbb{T}^n} |u(x)| < \infty.$$

These are Banach spaces by Theorem C.4.9.

Corollary 3.1.24 (Hausdorff–Young inequality). *Let $1 \le p \le 2$ and $\frac{1}{p} + \frac{1}{q} = 1$. If $u \in L^p(\mathbb{T}^n)$ then $\widehat{u} \in \ell^q(\mathbb{Z}^n)$ and*

$$||\widehat{u}||_{\ell^q(\mathbb{Z}^n)} \le ||u||_{L^p(\mathbb{T}^n)}.$$

Proof. The statement follows by the Riesz–Thorin interpolation theorem C.4.18 from the simple estimate $||\widehat{u}||_{\ell^\infty(\mathbb{Z}^n)} \le ||u||_{L^1(\mathbb{T}^n)}$ and Plancherel's identity $||\widehat{u}||_{\ell^2(\mathbb{Z}^n)} = ||u||_{L^2(\mathbb{T}^n)}$ in Remark 3.1.13. $\qquad\square$

Definition 3.1.25 (Periodic distribution space $\mathcal{D}'(\mathbb{T}^n)$). The dual space $\mathcal{D}'(\mathbb{T}^n) = \mathcal{L}(C^\infty(\mathbb{T}^n), \mathbb{C})$ is called the space of *periodic distributions*. For $u \in \mathcal{D}'(\mathbb{T}^n)$ and $\varphi \in C^\infty(\mathbb{T}^n)$, we shall write

$$u(\varphi) = \langle u, \varphi \rangle.$$

For any $\psi \in C^\infty(\mathbb{T}^n)$,

$$\varphi \mapsto \int_{\mathbb{T}^n} \varphi(x)\, \psi(x) \, dx$$

is a periodic distribution, which gives the embedding $\psi \in C^\infty(\mathbb{T}^n) \subset \mathcal{D}'(\mathbb{T}^n)$. Note that the same argument also shows the embedding of the spaces $L^p(\mathbb{T}^n)$, $1 \le p \le$

∞, into $\mathcal{D}'(\mathbb{T}^n)$. Due to the test function equality $\langle \partial^\alpha \psi, \varphi \rangle = \langle \psi, (-1)^{|\alpha|} \partial^\alpha \varphi \rangle$, it is natural to define distributional derivatives by

$$\langle \partial^\alpha f, \varphi \rangle := \langle f, (-1)^{|\alpha|} \partial^\alpha \varphi \rangle.$$

The topology of $\mathcal{D}'(\mathbb{T}^n) = \mathcal{L}(C^\infty(\mathbb{T}^n), \mathbb{C})$ is the weak*-topology.

Remark 3.1.26 (**Trigonometric polynomials**). The space $\mathrm{TrigPol}(\mathbb{T}^n)$ of *trigonometric polynomials* on the torus is defined by

$$\mathrm{TrigPol}(\mathbb{T}^n) := \mathrm{span}\{e_\xi : \xi \in \mathbb{Z}^n\}.$$

Thus, $f \in \mathrm{TrigPol}(\mathbb{T}^n)$ is of the form

$$f(x) = \sum_{\xi \in \mathbb{Z}^n} \widehat{f}(\xi) e^{i2\pi x \cdot \xi},$$

where $\widehat{f}(\xi) \neq 0$ for only finitely many $\xi \in \mathbb{Z}^n$. In the proof of Theorem 3.1.20 we showed that $\mathrm{TrigPol}(\mathbb{T}^n)$ is dense in both $C(\mathbb{T}^n)$ and in $L^2(\mathbb{T}^n)$ in the corresponding norms. Now, the set of trigonometric polynomials is actually also dense in $C^\infty(\mathbb{T}^n)$, so that a distribution is characterised by evaluating it at the vectors e_ξ for all $\xi \in \mathbb{Z}^n$. We note that there exist linear mappings $u \in L(\mathrm{span}\{e_\xi \mid \xi \in \mathbb{Z}^n\}, \mathbb{C})$ that do not belong to $\mathcal{L}(C^\infty(\mathbb{T}^n), \mathbb{C})$, but for which the determination of the Fourier coefficients $\widehat{u}(\xi) = u(e_\xi)$ makes sense.

Definition 3.1.27 (Fourier transform on $\mathcal{D}'(\mathbb{T}^n)$). By dualising the inverse $\mathcal{F}_{\mathbb{T}^n}^{-1} : \mathcal{S}(\mathbb{Z}^n) \to C^\infty(\mathbb{T}^n)$, the Fourier transform is extended uniquely to the mapping $\mathcal{F}_{\mathbb{T}^n} : \mathcal{D}'(\mathbb{T}^n) \to \mathcal{S}'(\mathbb{Z}^n)$ by the formula

$$\langle \mathcal{F}_{\mathbb{T}^n} u, \varphi \rangle := \langle u, \iota \circ \mathcal{F}_{\mathbb{T}^n}^{-1} \varphi \rangle, \tag{3.8}$$

where $u \in \mathcal{D}'(\mathbb{T}^n)$, $\varphi \in \mathcal{S}(\mathbb{Z}^n)$, and ι is defined by $(\iota \circ \psi)(x) = \psi(-x)$.

Exercise 3.1.28. Prove that if $u \in \mathcal{D}'(\mathbb{T}^n)$ then $\mathcal{F}_{\mathbb{T}^n} u \in \mathcal{S}'(\mathbb{Z}^n)$. Note that by Exercise 3.1.7 it means in particular that $\mathcal{F}_{\mathbb{T}^n} u$ is defined pointwise on $\mathbb{Z}^n$.

Exercise 3.1.29 (Compatibility). Check that extension (3.8) when restricted to $C^\infty(\mathbb{T}^n)$, is compatible with the definition (3.4). Here, the inclusion $C^\infty(\mathbb{T}^n) \subset \mathcal{D}'(\mathbb{T}^n)$ is interpreted in the standard way by

$$\langle u, \varphi \rangle = u(\varphi) = \int_{\mathbb{T}^n} u(x)\, \varphi(x)\, \mathrm{d}x.$$

Remark 3.1.30 (**Notice: different spaces**). Observe that spaces of functions where the toroidal Fourier transform $\mathcal{F}_{\mathbb{T}^n}$ acts are different: one is the space of functions on the torus $C^\infty(\mathbb{T}^n)$ while the other is the space of functions on the lattice $\mathcal{S}(\mathbb{Z}^n)$. That is why one has to be more careful on the torus, e.g., compared to the Fourier transform for distributions on $\mathbb{R}^n$ in Definition 1.3.2. This difference will be even more apparent in the case of compact Lie groups in Chapter 10.

Remark 3.1.31 (**Bernstein's theorem**). The Fourier transform can be studied on other spaces on the torus. For example, let $\Lambda_s(\mathbb{T})$ be the space of Hölder continuous functions of order $0 < s < 1$ on the one-dimensional torus $\mathbb{T}^1$, defined as

$$\Lambda_s(\mathbb{T}) := \left\{ f \in C(\mathbb{T}) : \sup_{x,h \in \mathbb{T}} \frac{|f(x+h) - f(x)|}{|h|^s} < \infty. \right\}.$$

Then **Bernstein's theorem** holds: if $f \in \Lambda_s(\mathbb{T})$ with $s > \frac{1}{2}$, then $\widehat{f} \in \ell^1(\mathbb{Z})$. We refer to [35] for further details on the Hölder continuity on the torus.

Working on the lattice it is always useful to keep in mind the following:

Definition 3.1.32 (Dirac delta comb). The *Dirac delta comb* $\delta_{\mathbb{Z}^n} : \mathcal{S}(\mathbb{R}^n) \to \mathbb{C}$ is defined by

$$\langle \delta_{\mathbb{Z}^n}, \varphi \rangle := \sum_{x \in \mathbb{Z}^n} \varphi(x),$$

and the sum here is absolutely convergent.

Exercise 3.1.33. Prove that $\delta_{\mathbb{Z}^n} \in \mathcal{S}'(\mathbb{R}^n)$.

We recall that the Dirac delta $\delta_x \in \mathcal{S}'(\mathbb{R}^n)$ at x is defined by $\delta_x(\varphi) = \varphi(x)$ for all $\varphi \in \mathcal{S}(\mathbb{R}^n)$. It may be not surprising that we obtain the Dirac delta comb by summing up Dirac deltas over the integer lattice:

Proposition 3.1.34. *We have the convergence*

$$\sum_{x \in \mathbb{Z}^n:\ |x| \leq j} \delta_x \ \xrightarrow[j \to \infty]{\mathcal{S}'(\mathbb{R}^n)} \ \delta_{\mathbb{Z}^n}.$$

Proof. Let us denote $P_j := \sum_{x \in \mathbb{Z}^n:\ |x| \leq j} \delta_x$. If $\varphi \in \mathcal{S}(\mathbb{R}^n)$ then

$$|\langle P_j - \delta_{\mathbb{Z}^n}, \varphi \rangle| \leq \sum_{x \in \mathbb{Z}^n:\ |x| > j} |\varphi(x)| \leq \sum_{x \in \mathbb{Z}^n:\ |x| > j} c_M \langle x \rangle^{-M} \xrightarrow[j \to \infty]{} 0$$

for M large enough (e.g., $M = n + 1$), proving the claim. $\qquad\qquad\square$

Another sequence converging to the Dirac delta comb will be shown in Proposition 4.6.8.

3.2   Sobolev spaces $H^s(\mathbb{T}^n)$

Fortunately, we have rich structures to work on. The periodic Sobolev spaces $H^s(\mathbb{T}^n)$ that we introduce in Definition 3.2.2 are actually Hilbert spaces (and in Section 3.5 we prove several auxiliary theorems about continuity and extensions in Banach spaces that apply in our situation). Here we shall deal with periodic functions and distributions on $\mathbb{R}^n$ and we shall pursue another more applicable definition of distributions: a Hilbert topology will be given for certain distribution subspaces, which are the Sobolev spaces. It happens that every periodic distribution belongs to some of these spaces.

Thus, we are attempting to create spaces which include $L^2(\mathbb{T}^n)$ as a special case and which would pay attention to smoothness properties of distributions. To give an informal motivation, assume that $u \in L^2(\mathbb{T}^n)$ also satisfies $\partial^\alpha u \in L^2(\mathbb{T}^n)$ for some $\alpha \in \mathbb{N}_0^n$. Then writing $\partial^\alpha u$ in a Fourier series we have

$$\partial^\alpha u = \sum_{\xi \in \mathbb{Z}^n} (\mathrm{i}2\pi\xi)^\alpha \, \widehat{u}(\xi) \, e_\xi,$$

with e_ξ as in (3.7), from which by Parseval's equality we obtain

$$\int_{\mathbb{T}^n} |\partial^\alpha u(x)|^2 \, \mathrm{d}x = (2\pi)^{2|\alpha|} \sum_{\xi \in \mathbb{Z}^n} |\xi^\alpha \widehat{u}(\xi)|^2 \, ;$$

with $\alpha = 0$ this is just the L^2-norm. Let us define

$$\langle \xi \rangle := (1 + \|\xi\|^2)^{1/2},$$

where we recall the notation $\|\xi\|$ for the Euclidean norm in (3.3).

Remark 3.2.1. This function will be used for measuring decay rates, and other possible analogues for $(\xi \mapsto \langle \xi \rangle) : \mathbb{Z}^n \to \mathbb{R}^+$ would be $1 + \|\xi\|$, or a function equal to $\|\xi\|$ for $\xi \neq 0$ and to 1 for $\xi = 0$. The idea here is to get a function $\xi \mapsto \langle \xi \rangle$, which behaves asymptotically like the norm $\xi \mapsto \|\xi\|$ when $\|\xi\| \to \infty$, and which satisfies a form of Peetre's inequality (see Proposition 3.3.31), thus vanishing nowhere.

Definition 3.2.2 (Sobolev spaces $H^s(\mathbb{T}^n)$). For $u \in \mathcal{D}'(\mathbb{T}^n)$ and $s \in \mathbb{R}$ we define the norm $\| \cdot \|_{H^s(\mathbb{T}^n)}$ by

$$\|u\|_{H^s(\mathbb{T}^n)} := \left(\sum_{\xi \in \mathbb{Z}^n} \langle \xi \rangle^{2s} \, |\widehat{u}(\xi)|^2 \right)^{1/2}. \tag{3.9}$$

The *Sobolev space* $H^s(\mathbb{T}^n)$ is then the space of 1-periodic distributions u for which $\|u\|_{H^s(\mathbb{T}^n)} < \infty$. For them, we will formally write their Fourier series representation $\sum_{\xi \in \mathbb{Z}^n} \widehat{u}(\xi) \, \mathrm{e}^{\mathrm{i}2\pi x \cdot \xi}$, and in Remark 3.2.5 we give a justification for this. Thus, such u will be also called 1-*periodic distributions, represented by the formal series* $\sum_{\xi \in \mathbb{Z}^n} \widehat{u}(\xi) \, \mathrm{e}^{\mathrm{i}2\pi x \cdot \xi}$. Note that in the definition (3.9) we again take an advantage of $\mathbb{T}^n$: compared to $\mathbb{R}^n$ the distributions on the lattice $\mathbb{Z}^n$ take pointwise values, see Exercise 3.1.7.

Exercise 3.2.3. For example, the 1-periodic Dirac delta δ is expressed by $\delta(x) = \sum_{\xi \in \mathbb{Z}^n} \mathrm{e}^{\mathrm{i}2\pi x \cdot \xi}$, or by $\left(\widehat{\delta}(\xi) \right)_{\xi \in \mathbb{Z}^n}$, where $\widehat{\delta}(\xi) \equiv 1$. Show that δ belongs to $H^s(\mathbb{T}^n)$ if and only if $s < -n/2$.

Exercise 3.2.4. For the function $e_\xi(x) = e^{i2\pi x \cdot \xi}$ show that $\|e_\xi\|_{H^s(\mathbb{T}^n)} = \langle \xi \rangle^s$.

Remark 3.2.5. One can readily see that the union $\bigcup_{s \in \mathbb{R}} H^s(\mathbb{T}^n)$ is the dual of $C^\infty(\mathbb{T}^n)$ in its uniform topology from Remark 3.1.5 (see Corollary 3.2.12). For the details concerning this duality we refer, e.g., to [11, Theorem 6.1]). Hence our definition of the 1-periodic distributions in Definition 3.2.2 coincides with the "official" one in view of the equality

$$\mathcal{D}'(\mathbb{T}^n) = \mathcal{L}(C^\infty(\mathbb{T}^n), \mathbb{C}) = \bigcup_{s \in \mathbb{R}} H^s(\mathbb{T}^n). \tag{3.10}$$

Proposition 3.2.6 (Sobolev spaces are Hilbert spaces). *For every $s \in \mathbb{R}$, the Sobolev space $H^s(\mathbb{T}^n)$ is a Hilbert space with the inner product*

$$(u, v)_{H^s(\mathbb{T}^n)} := \sum_{\xi \in \mathbb{Z}^n} \langle \xi \rangle^{2s} \, \widehat{u}(\xi) \, \overline{\widehat{v}(\xi)}.$$

Proof. The spaces $H^0(\mathbb{T}^n)$ and $H^s(\mathbb{T}^n)$ are isometrically isomorphic by the *canonical isomorphism* $\varphi_s : H^0(\mathbb{T}^n) \to H^s(\mathbb{T}^n)$, defined by

$$\varphi_s u(x) := \sum_{\xi \in \mathbb{Z}^n} \langle \xi \rangle^{-s} \, \widehat{u}(\xi) \, e^{i2\pi x \cdot \xi}.$$

Indeed, φ_s is a linear isometry between $H^t(\mathbb{T}^n)$ and $H^{t+s}(\mathbb{T}^n)$ for every $t \in \mathbb{R}$, and it is true that $\varphi_{s_1} \varphi_{s_2} = \varphi_{s_1+s_2}$ and $\varphi_s^{-1} = \varphi_{-s}$. Then the completeness of $L^2(\mathbb{T}^n) = H^0(\mathbb{T}^n)$ is transferred to that of $H^s(\mathbb{T}^n)$ for every $s \in \mathbb{R}$. $\square$

Exercise 3.2.7. For $k \in \mathbb{N}_0$ the traditional Sobolev norm $\| \cdot \|'_k$ is defined by

$$\|u\|'_k := \left(\sum_{|\alpha| \leq k} \int_{\mathbb{T}^n} |\partial^\alpha u(x)|^2 \, dx \right)^{1/2}.$$

Show that

$$\|u\|_{H^k(\mathbb{T}^n)} \leq \|u\|'_k \leq C_k \, \|u\|_{H^k(\mathbb{T}^n)},$$

and try to find the best possible constant $C_k < \infty$. This resembles Definitions 1.5.6 and 2.6.15 in the case of $\mathbb{R}^n$, with the equivalence of norms proved in Proposition 2.6.16.

Definition 3.2.8 (Banach and Hilbert dualities). We can define different dualities between Sobolev spaces. The Sobolev space $H^{-s}(\mathbb{T}^n)$ is the dual space of $H^s(\mathbb{T}^n)$ via the *Banach duality product* $\langle \cdot, \cdot \rangle$ defined by

$$\langle u, v \rangle := \sum_{\xi \in \mathbb{Z}^n} \widehat{u}(\xi) \, \widehat{v}(-\xi),$$

where $u \in H^s(\mathbb{T}^n)$ and $v \in H^{-s}(\mathbb{T}^n)$. Note that $\langle u, v \rangle = \int_{\mathbb{T}^n} u(x)\, v(x)\, dx$, when $s = 0$. Accordingly, the L^2- (or H^0-) inner product

$$(u, v)_{H^0(\mathbb{T}^n)} = \int_{\mathbb{T}^n} u(x)\, \overline{v(x)}\, dx$$

is the *Hilbert duality product*, and $H^s(\mathbb{T}^n)$ and $H^{-s}(\mathbb{T}^n)$ are duals of each other with respect to this duality. If A is a linear operator between two Sobolev spaces, we shall denote its Banach and Hilbert adjoints by $A^{(*B)}$ and $A^{(*H)}$, respectively. Often, the Banach adjoint is called the *transpose* of the operator A and is denoted by A^t. Then the Hilbert adjoint is simply called the *adjoint* and denoted by A^*. For the relation between Banach and Hilbert adjoints see Definition 2.5.15, Exercise 2.5.16 and Remark 2.5.15.

Exercise 3.2.9 (Trigonometric polynomials are dense). Prove that the trigonometric polynomials (and hence also $C^\infty(\mathbb{T}^n)$) are dense in every $H^s(\mathbb{T}^n)$.

Exercise 3.2.10 (Embeddings are compact). Prove that the inclusion $\iota : H^t(\mathbb{T}^n) \hookrightarrow H^s(\mathbb{T}^n)$ is compact for $s < t$.

Exercise 3.2.11 (An embedding theorem). Let $m \in \mathbb{N}_0$ and $s > m + n/2$. Prove that $H^s(\mathbb{T}^n) \subset C^m(\mathbb{T}^n)$.

As a corollary, we get

Corollary 3.2.12. *We have the equality $\bigcap_{s \in \mathbb{R}} H^s(\mathbb{T}^n) = C^\infty(\mathbb{T}^n)$. By the duality in Definition 3.2.8 it is related to (3.10) in Remark 3.2.5. Note that the situation on $\mathbb{R}^n$ is somewhat more complicated, see Exercise 2.6.19.*

Definition 3.2.13 (Biperiodic Sobolev spaces). The biperiodic Sobolev space $H^{s,t}(\mathbb{T}^n \times \mathbb{T}^n)$ $(s, t \in \mathbb{R})$ is the subspace of biperiodic distributions having the norm $\| \cdot \|_{s,t}$ defined by

$$\|v\|_{s,t} := \left[\sum_{\eta \in \mathbb{Z}^n} \sum_{\xi \in \mathbb{Z}^n} \langle \eta \rangle^{2s} \langle \xi \rangle^{2t} |\widehat{v}(\eta, \xi)|^2 \right]^{1/2}, \tag{3.11}$$

where

$$\widehat{v}(\eta, \xi) = \int_{\mathbb{T}^n} \int_{\mathbb{T}^n} e_{-\eta}(x)\, e_{-\xi}(y)\, v(x, y)\, dy\, dx \tag{3.12}$$

are the Fourier coefficients. It is true that the family of C^∞-smooth biperiodic functions satisfies $C^\infty(\mathbb{T}^n \times \mathbb{T}^n) = \bigcap_{s,t \in \mathbb{R}} H^{s,t}(\mathbb{T}^n \times \mathbb{T}^n)$. In an obvious manner one relates all these definitions for 1-periodic spaces $\mathbb{T}^n = \mathbb{R}^n / \mathbb{Z}^n$.

3.3 Discrete analysis toolkit

In this section we provide tools for the study of periodic pseudo-differential operators. In fact, some of the discrete results presented date back to the 18th and 19th centuries, but seem to have been forgotten in the advent of modern numerical analysis. Global investigation of periodic functions also requires a special treatment, presented in the last subsection, as well as periodic Taylor series in Section 3.4.

Defining functions on the discrete space $\mathbb{Z}^n$ instead of $\mathbb{R}^n$, we lose the traditional limit concepts of differential calculus. However, it is worth viewing differences and sums as relatives to derivatives and integrals, and what we shall come up with is a theory that quite nicely resembles differential calculus. Therefore it is no wonder that this theory is known as the calculus of finite differences.

3.3.1   Calculus of finite differences

In this section we develop the discrete calculus which will be needed in the sequel. In particular, we will formulate and prove a discrete version of the Taylor expansion formula on the lattice $\mathbb{Z}^n$. Let us first list some conventions that will be spotted in the formulae: a sum over an empty index set is 0 (empty product is 1), $0! = 1$, and heretically $0^0 = 1$. When the index set is known from the context, we may even leave it out.

Definition 3.3.1 (Forward and backward differences $\triangle_\xi^\alpha$ and $\overline{\triangle}_\xi^\alpha$). Let $\sigma : \mathbb{Z}^n \to \mathbb{C}$ and $1 \leq i, j \leq n$. Let $\delta_j \in \mathbb{N}_0^n$ be defined by

$$(\delta_j)_i := \begin{cases} 1, & \text{if} \quad i = j, \\ 0, & \text{if} \quad i \neq j. \end{cases}$$

We define the *forward* and *backward* partial difference operators $\triangle_{\xi_j}$ and $\overline{\triangle}_{\xi_j}$, respectively, by

$$\triangle_{\xi_j} \sigma(\xi) \;\; := \;\; \sigma(\xi + \delta_j) - \sigma(\xi),$$
$$\overline{\triangle}_{\xi_j} \sigma(\xi) \;\; := \;\; \sigma(\xi) - \sigma(\xi - \delta_j),$$

and for $\alpha \in \mathbb{N}_0^n$ define

$$\triangle_\xi^\alpha \;\; := \;\; \triangle_{\xi_1}^{\alpha_1} \cdots \triangle_{\xi_n}^{\alpha_n},$$
$$\overline{\triangle}_\xi^\alpha \;\; := \;\; \overline{\triangle}_{\xi_1}^{\alpha_1} \cdots \overline{\triangle}_{\xi_n}^{\alpha_n}.$$

Remark 3.3.2 (**Classical relatives**). Several familiar formulae from classical analysis have discrete relatives: for instance, it can be easily checked that these difference operators commute, i.e., that

$$\triangle_\xi^\alpha \triangle_\xi^\beta = \triangle_\xi^\beta \triangle_\xi^\alpha = \triangle_\xi^{\alpha+\beta}$$

for all multi-indices $\alpha, \beta \in \mathbb{N}_0^n$. Moreover,

$$\triangle_\xi^\alpha (s\varphi + t\psi)(\xi) = s\triangle_\xi^\alpha \varphi(\xi) + t\triangle_\xi^\alpha \psi(\xi),$$

where s and t are scalars.

Exercise 3.3.3. Prove these formulae.

Proposition 3.3.4 (Formulae for $\triangle_\xi^\alpha$ and $\overline{\triangle}_\xi^\alpha$). *Let $\phi : \mathbb{Z}^n \to \mathbb{C}$. We have*

$$
\triangle_\xi^\alpha \phi(\xi) = \sum_{\beta \leq \alpha} (-1)^{|\alpha-\beta|} \binom{\alpha}{\beta} \phi(\xi+\beta),
$$

$$
\overline{\triangle}_\xi^\alpha \phi(\xi) = \sum_{\beta \leq \alpha} (-1)^{|\beta|} \binom{\alpha}{\beta} \phi(\xi-\beta).
$$

Proof. Let us introduce *translation operators* $E_j := (I + \triangle_{\xi_j})$, acting on functions $\phi : \mathbb{Z}^n \to \mathbb{C}$ by

$$
E_j \phi(\xi) := (I + \triangle_{\xi_j})\phi(\xi) = \phi(\xi + \delta_j).
$$

Let $E^\alpha := E_1^{\alpha_1} \cdots E_n^{\alpha_n}$. An application of the binomial formula is enough:

$$
\begin{aligned}
\triangle_\xi^\alpha \phi(\xi) &= (E - I)^\alpha \phi(\xi) \\
&= \sum_{\beta \leq \alpha} \binom{\alpha}{\beta} (-1)^{|\alpha-\beta|} E^\beta \phi(\xi) \\
&= \sum_{\beta \leq \alpha} (-1)^{|\alpha-\beta|} \binom{\alpha}{\beta} \phi(\xi+\beta).
\end{aligned}
$$

The backward difference equality is left for the reader to prove as Exercise 3.3.5. □

Exercise 3.3.5. Notice that $E_j \overline{\triangle}_{\xi_j} = \triangle_{\xi_j} = \overline{\triangle}_{\xi_j} E_j$. Complete the proof of Proposition 3.3.4.

The discrete Leibniz formula is complicated enough to have a proof of its own, and it can be compared with the Leibniz formula on $\mathbb{R}^n$ in Theorem 1.5.10, (iv).

Lemma 3.3.6 (Discrete Leibniz formula). *Let $\phi, \psi : \mathbb{Z}^n \to \mathbb{C}$. Then*

$$
\triangle_\xi^\alpha (\phi\psi)(\xi) = \sum_{\beta \leq \alpha} \binom{\alpha}{\beta} \left(\triangle_\xi^\beta \phi(\xi)\right) \triangle_\xi^{\alpha-\beta} \psi(\xi+\beta). \tag{3.13}
$$

Proof. (Another proof idea, not using induction, can be found in [117, p. 11] and [52, p. 16].) First, we have an easy check

$$
\begin{aligned}
\triangle_{\xi_j}(\varphi\psi)(\xi) &= (\varphi\psi)(\xi+\delta_j) - (\varphi\psi)(\xi) \\
&= \varphi(\xi)\left(\psi(\xi+\delta_j) - \psi(\xi)\right) + \left(\varphi(\xi+\delta_j) - \varphi(\xi)\right)\psi(\xi+\delta_j) \\
&= \varphi(\xi)\triangle_{\xi_j}\psi(\xi) + \left(\triangle_{\xi_j}\varphi(\xi)\right)\psi(\xi+\delta_j).
\end{aligned}
$$

We use this and induction on $\alpha \in \mathbb{N}_0^n$:

$$\triangle_\xi^{\alpha+\delta_j}(\varphi\psi)(\xi) = \triangle_{\xi_j}\triangle_\xi^\alpha(\varphi\psi)(\xi)$$

$$= \triangle_{\xi_j} \sum_{\beta\leq\alpha} \binom{\alpha}{\beta} \left(\triangle_\xi^\beta\varphi(\xi)\right) \triangle_\xi^{\alpha-\beta}\psi(\xi+\beta)$$

$$= \sum_{\beta\leq\alpha} \binom{\alpha}{\beta} \left[\left(\triangle_\xi^\beta\varphi(\xi)\right) \triangle_\xi^{\alpha+\delta_j-\beta}\psi(\xi+\beta)\right.$$

$$\left. + \left(\triangle_\xi^{\beta+\delta_j}\varphi(\xi)\right) \triangle_\xi^{\alpha-\beta}\psi(\xi+\beta+\delta_j)\right]$$

$$\overset{(\star)}{=} \sum_{\beta\leq\alpha+\delta_j} \left[\binom{\alpha}{\beta} + \binom{\alpha}{\beta-\delta_j}\right] \left(\triangle_\xi^\beta\varphi(\xi)\right) \triangle_\xi^{\alpha+\delta_j-\beta}\psi(\xi+\beta)$$

$$= \sum_{\beta\leq\alpha+\delta_j} \binom{\alpha+\delta_j}{\beta} \left(\triangle_\xi^\beta\varphi(\xi)\right) \triangle_\xi^{\alpha+\delta_j-\beta}\psi(\xi+\beta).$$

In $(\star)$ above, we used the convention that $\binom{\alpha}{\gamma} = 0$ if $\gamma \nleq \alpha$ or if $\gamma \notin \mathbb{N}_0^n$. The proof is complete. $\square$

Exercise 3.3.7. Verify that

$$\binom{\alpha}{\beta} + \binom{\alpha}{\beta-\delta_j} = \binom{\alpha+\delta_j}{\beta}$$

in the proof of (3.13).

Remark 3.3.8 (**Discrete product rule – notice the shifts**). Notice the shift in (3.13) in the argument of ψ. For example, already the product rule becomes

$$\triangle_{\xi_j}(\varphi\psi)(\xi) = \varphi(\xi)\,\triangle_{\xi_j}\psi(\xi) + \left(\triangle_{\xi_j}\varphi(\xi)\right)\,\psi(\xi+\delta_j).$$

The shift is caused by the difference operator $\triangle_\xi$, and it is characteristic to the calculus of finite differences – in classical Euclidean analysis it is not present. This shift will have its consequences for the whole theory of pseudo-differential operators on the torus in Chapter 4, especially for the formulae in the calculus.

Exercise 3.3.9. Prove the following form of the discrete Leibniz formula:

$$\triangle_\xi^\alpha \left(\varphi(\xi)\,\psi(\xi)\right) = \sum_{\beta\leq\alpha} \binom{\alpha}{\beta} \left(\triangle_\xi^\beta\varphi(\xi)\right) \overline{\triangle}_\xi^{\alpha-\beta}\psi(\xi+\alpha).$$

As it is easy to guess, in the calculus of finite differences, sums correspond to integrals of classical analysis, and the theory of series (presented, e.g., in [66]) serves as an integration theory. Assuming convergence of the following series, it holds that

$$\sum_\xi (s\varphi(\xi) + t\psi(\xi)) = s\sum_\xi \varphi(\xi) + t\sum_\xi \psi(\xi),$$

and when $a \leq b$ on $\mathbb{Z}^1$, we have an analogue of the fundamental theorem of calculus:

$$\sum_{\xi=a}^{b} \triangle_\xi \psi(\xi) = \psi(b+1) - \psi(a).$$

Difference and partial difference equations (cf. differential and partial differential) are handled in several books concerning combinatorics or difference methods (e.g., [52]), but various mean value theorems have no straightforward interpretation here, since the functions are usually defined only on a discrete set of points (although one can use some suitable interpolation; we refer to Theorem 3.3.39 and Section 4.5). Integration by parts can be, however, translated for our purposes:

Lemma 3.3.10 (Summation by parts). *Assume that* $\varphi, \psi : \mathbb{Z}^n \to \mathbb{C}$. *Then*

$$\sum_{\xi \in \mathbb{Z}^n} \varphi(\xi)\, \triangle_\xi^\alpha \psi(\xi) = (-1)^{|\alpha|} \sum_{\xi \in \mathbb{Z}^n} \left(\overline{\triangle_\xi}^\alpha \varphi(\xi) \right) \psi(\xi) \qquad (3.14)$$

provided that both series are absolutely convergent.

Proof. Let us check the case $|\alpha| = 1$:

$$\begin{aligned}
\sum_{\xi \in \mathbb{Z}^n} \varphi(\xi)\, \triangle_{\xi_j} \psi(\xi) &= \sum_{\xi \in \mathbb{Z}^n} (\psi(\xi + \delta_j) - \psi(\xi))\, \varphi(\xi) \\
&= \sum_{\xi \in \mathbb{Z}^n} \psi(\xi)\, (-\varphi(\xi) + \varphi(\xi - \delta_j)) \\
&= (-1)^1 \sum_{\xi \in \mathbb{Z}^n} \psi(\xi)\, \overline{\triangle}_{\xi_j} \varphi(\xi).
\end{aligned}$$

For any $\alpha \in \mathbb{N}_0^n$ the result is obtained recursively. $\qquad \square$

Exercise 3.3.11. Complete the proof of (3.14) for $|\alpha| \geq 2$.

3.3.2 Discrete Taylor expansion and polynomials on $\mathbb{Z}^n$

The usual polynomials $\theta \mapsto \theta^\alpha$ do not behave naturally with respect to differences: typically $\triangle_\theta^\gamma \theta^\alpha \neq c_{\alpha\gamma}\, \theta^{\alpha-\gamma}$ for any constant $c_{\alpha\gamma}$. Thus let us introduce polynomials $\theta \mapsto \theta^{(\alpha)}$ to cure this defect:

Definition 3.3.12 (Discrete polynomials). For $\theta \in \mathbb{Z}^n$ and $\alpha \in \mathbb{N}_0^n$, we define $\theta^{(\alpha)} = \theta_1^{(\alpha_1)} \cdots \theta_n^{(\alpha_n)}$, where $\theta_j^{(0)} = 1$ and

$$\theta_j^{(k+1)} = \theta_j^{(k)}(\theta_j - k) = \theta_j(\theta_j - 1) \dots (\theta_j - k). \qquad (3.15)$$

Exercise 3.3.13. Show that

$$\triangle_\theta^\gamma \theta^{(\alpha)} = \alpha^{(\gamma)}\, \theta^{(\alpha-\gamma)},$$

in analogy to the Euclidean case where $\partial_\theta^\gamma \theta^\alpha = \alpha^{(\gamma)}\, \theta^{\alpha-\gamma}$.

Remark 3.3.14. Difference operators lessen the degree of a polynomial by 1. In the literature on numerical analysis the polynomials $\theta \mapsto \theta^{(\alpha)}$ appear sometimes in a concealed form using the binomial coefficients:

$$\theta^{(\alpha)} = \alpha! \binom{\theta}{\alpha}.$$

Next, let us consider "discrete integration".

Definition 3.3.15 (Discrete integration). For $b \geq 0$, let us write

$$I_k^b := \sum_{0 \leq k < b} \quad \text{and} \quad I_k^{-b} := - \sum_{-b \leq k < 0}. \tag{3.16}$$

In the sequel, we adopt the notational conventions

$$I_{k_1}^{\theta} I_{k_2}^{k_1} \cdots I_{k_\alpha}^{k_{\alpha-1}} 1 = \begin{cases} 1, & \text{if } \alpha = 0, \\ I_{k_1}^{\theta} 1, & \text{if } \alpha = 1, \\ I_{k_1}^{\theta} I_{k_2}^{k_1} 1, & \text{if } \alpha = 2, \end{cases}$$

and so on.

Remark 3.3.16. One can think of $I_\xi^\theta \cdots$ as a discrete version of the one-dimensional integral $\int_0^\theta \cdots d\xi$; in this discrete context, the difference $\triangle_\xi$ takes the role of the differential operator $d/d\xi$.

Lemma 3.3.17 (Discrete "fundamental theorem of calculus" 1D). *If $\theta \in \mathbb{Z}$ and $\alpha \in \mathbb{N}_0$ then*

$$I_{k_1}^{\theta} I_{k_2}^{k_1} \cdots I_{k_\alpha}^{k_{\alpha-1}} 1 = \frac{1}{\alpha!} \, \theta^{(\alpha)}. \tag{3.17}$$

Proof. We observe simple equalities $k^{(0)} \equiv 1$, $\triangle_k k^{(i)} = i \, k^{(i-1)}$ and $I_k^b \triangle_k k^{(i)} = b^{(i)}$, from which (3.17) follows by induction. $\qquad \square$

Remark 3.3.18. We note that Lemma 3.3.17 can be viewed as a discrete trivial version of the fundamental theorem of calculus:

$$\int_0^\theta f'(\xi) \, d\xi = f(\theta) - f(0)$$

for smooth enough $f : \mathbb{R} \to \mathbb{C}$ corresponds to

$$I_\xi^\theta \triangle_\xi f(\xi) = f(\theta) - f(0)$$

for $f : \mathbb{Z} \to \mathbb{C}$.

Lemma 3.3.17 immediately implies its multidimensional version:

Corollary 3.3.19 (Discrete "fundamental theorem of calculus"). *If $\theta \in \mathbb{Z}^n$ and $\alpha \in \mathbb{N}_0^n$ then*

$$\prod_{j=1}^{n} I_{k(j,1)}^{\theta_j} I_{k(j,2)}^{k(j,1)} \cdots I_{k(j,\alpha_j)}^{k(j,\alpha_j-1)} 1 = \frac{1}{\alpha!}\, \theta^{(\alpha)}, \tag{3.18}$$

where $\prod_{j=1}^{n} I_j$ means $I_1 I_2 \cdots I_n$, where $I_j := I_{k(j,1)}^{\theta_j} I_{k(j,2)}^{k(j,1)} \cdots I_{k(j,\alpha_j)}^{k(j,\alpha_j-1)}$.

Exercise 3.3.20. Work out the details for the proof of Corollary 3.3.19.

Now we are about to present a combinatorial tool of uttermost importance: a discrete version of the classical Taylor polynomial expansion theorem. The result is exactly what one might expect, as differences correspond to derivatives, and polynomials $\theta \mapsto \theta^{(\alpha)}$ replace $\theta \mapsto \theta^\alpha$; also the error estimate seems familiar. In the sequel, the word "series" will also be used in those cases, where the summation index set is finite.

Theorem 3.3.21 (Discrete Taylor expansion on $\mathbb{Z}^n$). *Let $p : \mathbb{Z}^n \to \mathbb{C}$. Then we can write*

$$p(\xi + \theta) = \sum_{|\alpha| < M} \frac{1}{\alpha!}\, \theta^{(\alpha)} \, \triangle_\xi^\alpha p(\xi) + r_M(\xi, \theta),$$

with the remainder $r_M(\xi, \theta)$ satisfying

$$\left| \triangle_\xi^\omega r_M(\xi, \theta) \right| \leq C_M \max_{|\alpha|=M,\ \nu \in Q(\theta)} \left| \theta^{(\alpha)} \triangle_\xi^{\alpha+\omega} p(\xi + \nu) \right|, \tag{3.19}$$

where $Q(\theta) := \{ \nu \in \mathbb{Z}^n : |\nu_j| \leq |\theta_j| \text{ for all } j = 1, \dots, n \}$.

Remark 3.3.22. Notice that the estimate above resembles closely the Lagrange form of the error in the traditional Taylor expansion theorem for $f \in C^\infty(\mathbb{R}^n)$:

$$f(x + h) = \sum_{|\alpha| < M} \frac{1}{\alpha!}\, \partial^\alpha f(x)\, h^\alpha + R_M(x, h),$$

$$R_M(x, h) = \sum_{|\alpha| = M} \frac{1}{\alpha!}\, R_\alpha(x, h)\, h^\alpha,$$

$$|R_\alpha(x, h)| \leq \sup_\theta |\partial^\alpha f(\theta)|,$$

where $\sup_\theta$ is taken over the segment between x and $x + h$.

Exercise 3.3.23. Let $|\beta| < M = |\alpha|$. Regarding Theorem 3.3.21, show that

$$\begin{cases} \triangle_\theta^\alpha r_M(\xi, \theta) = \triangle_\theta^\alpha p(\xi + \theta), \\ \triangle_\xi^\beta r_M(\xi, \theta) \big|_{\theta=0} = 0. \end{cases}$$

Formally, this involves a discrete Cauchy problem.

The essential ideas for proving Theorem 3.3.21 are most transparent in the case of $n = 1$:

Proof of Theorem 3.3.21 for dimension $n = 1$. We claim that the remainder is

$$r_M(\xi, \theta) = I_{k_1}^{\theta} I_{k_2}^{k_1} \cdots I_{k_M}^{k_{M-1}} \triangle_{\xi}^M p(\xi + k_M). \tag{3.20}$$

This is easily verified for

$$r_1(\xi, \theta) = p(\xi + \theta) - p(\xi) = I_{k_1}^{\theta} \triangle_{\xi} p(\xi + k_1).$$

We proceed by induction. Thus

$$
\begin{aligned}
r_{M+1}(\xi, \theta) \quad &= \quad r_M(\xi, \theta) - \frac{1}{M!}\, \theta^{(M)} \triangle_{\xi}^M p(\xi) \\
&\overset{(3.20),\,(3.17)}{=} \quad I_{k_1}^{\theta} I_{k_2}^{k_1} \cdots I_{k_M}^{k_{M-1}} \triangle_{\xi}^M \left(p(\xi + k_M) - p(\xi)\right) \\
&= \quad I_{k_1}^{\theta} I_{k_2}^{k_1} \cdots I_{k_M}^{k_{M-1}} I_{k_{M+1}}^{k_M} \triangle_{\xi}^{M+1} p(\xi + k_{M+1}).
\end{aligned}
$$

Applying (3.17) to (3.20), we obtain

$$\left|\triangle_{\xi}^{\omega} r_M(\xi, \theta)\right| \leq \frac{1}{M!}\, \left|\theta^{(M)}\right| \max_{\nu \in \{0, \ldots, \theta\}} \left|\triangle_{\xi}^{M+\omega} p(\xi + \nu)\right|. \tag{3.21}$$

Hence the proof of the one-dimensional discrete Taylor theorem is complete. □

Remark 3.3.24. The discrete Taylor formula is not a new invention: it can be obtained from the Newton interpolation formula, but this direct proof is beautifully simple, assumes the function to be defined on a discrete space only, and exposes a fundamental structure of the calculus of finite differences. In fact, Niels Nörlund's ([85, p. 11]) *Newtonsche Interpolationsformel* and Steffensen's ([117, p. 22]) *Newton's interpolation-formula with finite differences* are equivalent to Theorem 3.3.21, but they are disguised beyond recognition. Already George Boole presented the *Remainder in the Generalised form of Taylor's* (sic) *theorem* in [15, p. 146]. However, Boole's R_N does not easily give a nice approximation for $|\triangle_{\eta}^l R_N(\xi, \eta)|$; in any case it reveals the connection to differential calculus better than later books. Charles Jordan's ([60, p. 75, 164]) *generalised Newton series* masks the Taylor-likeness under the binomial coefficients, and the convergence of an infinite series is assumed, thus avoiding the study of the remainder. In the modern literature, the situation is even worse as the Newton interpolation formula is usually spoiled in myriad ways destroying its Taylor appearance. However, *Newton's* (or *Gregory's*) *formula* in [52, p. 9] is a welcome exception, but again it expresses the error in terms of derivatives.

Numerical analysis, especially interpolation theory, may yield useful tools for discrete purposes. Briefly, almost everything that you can do in classical analysis is in some sense legal when working with differences.

Now we prove the multidimensional discrete version of the Taylor expansion formula.

Proof of Theorem 3.3.21 for the n-dimensional Taylor expansion. We encourage the reader to compare this general proof to the earlier deduction of the low-dimensional case $n = 1$. For $0 \neq \alpha \in \mathbb{N}_0^n$, let us denote $m_\alpha := \min\{j : \alpha_j \neq 0\}$. For $\theta \in \mathbb{Z}^n$ and $i \in \{1, \ldots, n\}$, let us define $\nu(\theta, i, k) \in \mathbb{Z}^n$ by

$$\nu(\theta, i, k) := (\theta_1, \ldots, \theta_{i-1}, k, 0, \ldots, 0),$$

i.e.,

$$\nu(\theta, i, k)_j = \begin{cases} \theta_j, & \text{if } 1 \leq j < i, \\ k, & \text{if } j = i, \\ 0, & \text{if } i < j \leq n. \end{cases}$$

We claim that the remainder can be written in the form

$$r_M(\xi, \theta) = \sum_{|\alpha|=M} r_\alpha(\xi, \theta), \tag{3.22}$$

where for each α, we have

$$r_\alpha(\xi, \theta) = \prod_{j=1}^n I_{k(j,1)}^{\theta_j} I_{k(j,2)}^{k(j,1)} \cdots I_{k(j,\alpha_j)}^{k(j,\alpha_j-1)} \, \triangle_\xi^\alpha p(\xi + \nu(\theta, m_\alpha, k(m_\alpha, \alpha_{m_\alpha}))); \tag{3.23}$$

recall (3.16) and (3.18). The proof of (3.23) is by induction. The first remainder term r_1 is of the claimed form, since

$$r_1(\xi, \theta) = p(\xi + \theta) - p(\xi) = \sum_{i=1}^n r_{\delta_i}(\xi, \theta),$$

where

$$r_{\delta_i}(\xi, \theta) = I_k^{\theta_i} \triangle_\xi^{\delta_i} p(\xi + \nu(\theta, i, k));$$

here r_{δ_i} is of the form (3.23) for $\alpha = \delta_i$, $m(\alpha) = i$ and $\alpha_{m_\alpha} = 1$. So suppose that the claim (3.23) is true up to order $|\alpha| = M$. Then

$$\begin{aligned} r_{M+1}(\xi, \theta) &= r_M(\xi, \theta) - \sum_{|\alpha|=M} \frac{1}{\alpha!} \theta^{(\alpha)} \triangle_\xi^\alpha p(\xi) \\ &= \sum_{|\alpha|=M} \left(r_\alpha(\xi, \theta) - \frac{1}{\alpha!} \theta^{(\alpha)} \triangle_\xi^\alpha p(\xi) \right) \\ &= \sum_{|\alpha|=M} \prod_{j=1}^n I_{k(j,1)}^{\theta_j} I_{k(j,2)}^{k(j,1)} \cdots I_{k(j,\alpha_j)}^{k(j,\alpha_j-1)} \\ &\qquad \times \triangle_\xi^\alpha \left[p(\xi + \nu(\theta, m_\alpha, k(m_\alpha, \alpha_{m_\alpha}))) - p(\xi) \right], \end{aligned}$$

where we used (3.23) and (3.18) to obtain the last equality. Combining this with the equality

$$p(\xi + \nu(\theta, m_\alpha, k)) - p(\xi) = \sum_{i=1}^{m_\alpha} I_\ell^{\nu(\theta, m_\alpha, k)_i} \, \triangle_\xi^{\delta_i} p(\xi + \nu(\theta, i, \ell)),$$

we get

$$
\begin{aligned}
r_{M+1}(\xi, \theta) &= \sum_{|\alpha|=M} \prod_{j=1}^{n} I_{k(j,1)}^{\theta_j} I_{k(j,2)}^{k(j,1)} \cdots I_{k(j,\alpha_j)}^{k(j,\alpha_j-1)} \sum_{i=1}^{m_\alpha} I_{\ell(i)}^{\nu(\theta, m_\alpha, k(m_\alpha, \alpha_{m_\alpha}))_i} \\
&\quad \triangle_\xi^{\alpha+\delta_i} p(\xi + \nu(\theta, i, \ell(i))) \\
&= \sum_{|\beta|=M+1} \prod_{j=1}^{n} I_{k(j,1)}^{\theta_j} I_{k(j,2)}^{k(j,1)} \cdots I_{k(j,\beta_j)}^{k(j,\beta_j-1)} \\
&\quad \triangle_\xi^{\beta} p(\xi + \nu(\theta, m_\beta, k(m_\beta, \beta_{m_\beta})));
\end{aligned}
$$

the last step here is just simple tedious book-keeping. Thus the induction proof of (3.23) is complete. Finally, let us prove estimate (3.19). By (3.23), we obtain

$$
\begin{aligned}
&\left| \triangle_\xi^\omega r_M(\xi, \theta) \right| \\
&= \left| \sum_{|\alpha|=M} \triangle_\xi^\omega r_\alpha(\xi, \theta) \right| \\
&= \left| \sum_{|\alpha|=M} \prod_{j=1}^{n} I_{k(j,1)}^{\theta_j} I_{k(j,2)}^{k(j,1)} \cdots I_{k(j,\alpha_j)}^{k(j,\alpha_j-1)} \triangle_\xi^{\alpha+\omega} p(\xi + \nu(\theta, m_\alpha, k(m_\alpha, \alpha_{m_\alpha}))) \right| \\
&\le \sum_{|\alpha|=M} \frac{1}{\alpha!} \left| \theta^{(\alpha)} \right| \max_{\nu \in Q(\theta)} \left| \triangle_\xi^{\alpha+\omega} p(\xi + \nu) \right|,
\end{aligned}
$$

where in the last step we used (3.18). The proof is complete. $\square$

Remark 3.3.25 (**Remainder**). If $n \ge 2$, there are many alternative forms of remainders $r_\alpha(\xi, \theta)$. This is due to the fact that there may be many different shortest discrete step-by-step paths in the space $\mathbb{Z}^n$ from ξ to $\xi + \theta$; in the proof above, we chose just one such path, traveling via the points

$$\xi, \quad \xi + \theta_1 \delta_1, \quad \ldots, \quad \xi + \sum_{i=1}^{j} \theta_i \delta_i, \quad \ldots, \quad \xi + \theta.$$

But if $n = 1$, there is just one shortest discrete path from $\xi \in \mathbb{Z}$ to $\theta \in \mathbb{Z}$, and in that case

$$r_M(\xi, \theta) = I_{k_1}^{\theta} I_{k_2}^{k_1} \cdots I_{k_M}^{k_{M-1}} \triangle_\xi^M p(\xi + k_M).$$

In Theorem 3.3.21 we estimated the remainder over the discrete box $Q(\theta)$, but an estimate over a discrete path would have been enough.

Notice also that the discrete Taylor theorem presented above implies the following smooth Taylor result:

Corollary 3.3.26 (Discrete $\implies$ continuous). *Let $p \in C^\infty(\mathbb{R}^n)$ and*

$$r_M(\xi, \theta) := p(\xi + \theta) - \sum_{|\alpha| < M} \frac{1}{\alpha!} \, \theta^\alpha \left(\frac{\partial}{\partial \xi} \right)^\alpha p(\xi).$$

Then we have

$$\left| \partial_\xi^\omega r_M(\xi, \theta) \right| \leq c_M \max_{|\alpha| = M, \, \nu \in Q_{\mathbb{R}^n}(\theta)} \left| \theta^\alpha \, \partial_\xi^{\alpha + \omega} p(\xi + \nu) \right|, \tag{3.24}$$

where $Q_{\mathbb{R}^n}(\theta) := \{ \nu \in \mathbb{R}^n : \min(0, \theta_j) \leq \nu_j \leq \max(0, \theta_j) \}$.

Remark 3.3.27. It is evident from the proof that, in the remainder estimates above, the cubes $Q(\theta) \subset \mathbb{Z}^n$ and $Q_{\mathbb{R}^n}(\theta) \subset \mathbb{R}^n$ could be replaced by (discrete, respectively continuous) paths from 0 to θ; e.g., on $\mathbb{R}^n$, the cube could be replaced by the straight line from 0 to θ.

3.3.3 Several discrete inequalities

In the symbolic analysis of periodic pseudo-differential operators we frequently need the inequalities of Young and Peetre, and that is why we present them separately here. Here $\ell^p = \ell^p(\mathbb{Z}^n)$, $1 \leq p < \infty$, is the space of those complex "sequences" $f = (f(\xi))_{\xi \in \mathbb{Z}^n}$ (i.e., mappings $f : \mathbb{Z}^n \to \mathbb{C}$), for which the norm

$$\|f\|_{\ell^p} := \left[\sum_{\xi \in \mathbb{Z}^n} |f(\xi)|^p \right]^{1/p}$$

is finite. For $p = \infty$ the expression of the norm is modified into

$$\|f\|_{\ell^\infty} = \sup_{\xi \in \mathbb{Z}^n} |f(\xi)|.$$

We will write ℓ^p for $\ell^p(\mathbb{Z}^n)$ here.

Looking back to the definition of the periodic Sobolev spaces $H^s(\mathbb{T}^n)$, it becomes evident that the exponent $p = 2$ is the significant one, as $\| \cdot \|_{H^0(\mathbb{T}^n)} = \| \cdot \|_{\ell^2}$. Naturally, the Hölder's and Minkowski's inequalities hold true in discrete cases.

Lemma 3.3.28 (Discrete Hölder's inequality). *Let $1 \leq p, q \leq \infty$ be conjugate, i.e., $\frac{1}{p} + \frac{1}{q} = 1$. Let $f \in \ell^p$ and $g \in \ell^q$. Then*

$$\|fg\|_{\ell^1} \leq \|f\|_{\ell^p} \|g\|_{\ell^q}.$$

Of course, the general Hölder inequality in Theorem C.4.4 implies our lemma, but this special case is so much easier to prove that it is noteworthy to study it independently:

Exercise 3.3.29. Find a simple proof of Hölder's inequality in Lemma 3.3.28.

Lemma 3.3.30 (Discrete Young's inequality). *Assume that* $h : \mathbb{Z}^n \times \mathbb{Z}^n \to \mathbb{C}$ *is a function satisfying*

$$C_1 := \sup_{\xi \in \mathbb{Z}^n} \sum_{\eta \in \mathbb{Z}^n} |h(\eta, \xi)| < \infty, \quad C_2 := \sup_{\eta \in \mathbb{Z}^n} \sum_{\xi \in \mathbb{Z}^n} |h(\eta, \xi)| < \infty.$$

Let $1 \le p \le \infty$. *For any sequence* $f \in \ell^p$ *let us define* $g : \mathbb{Z}^n \to \mathbb{C}$ *by* $g(\eta) = \sum_{\xi \in \mathbb{Z}^n} h(\eta, \xi) f(\xi)$. *Then*

$$\|g\|_{\ell^p} \le C_1^{1/p} C_2^{1/q} \|f\|_{\ell^p},$$

where q *is the conjugate exponent of* p, *i.e.,* $\frac{1}{p} + \frac{1}{q} = 1$.

Proof. By the discrete Hölder's inequality (Lemma 3.3.28) we have

$$
\begin{aligned}
\sum_{\xi \in \mathbb{Z}^n} |h(\eta, \xi)|\, |f(\xi)| &= \sum_{\xi \in \mathbb{Z}^n} \left[|h(\eta, \xi)|^{1/p}\, |f(\xi)| \right] \left[|h(\eta, \xi)|^{1/q} \right] \\
&\le \left[\sum_{\xi \in \mathbb{Z}^n} |h(\eta, \xi)|\, |f(\xi)|^p \right]^{1/p} \left[\sum_{\xi \in \mathbb{Z}^n} |h(\eta, \xi)| \right]^{1/q} \\
&\le \left[\sum_{\xi \in \mathbb{Z}^n} |h(\eta, \xi)|\, |f(\xi)|^p \right]^{1/p} C_2^{1/q}.
\end{aligned}
$$

Using this we get

$$
\begin{aligned}
\|g\|_{\ell^p}^p &= \sum_{\eta \in \mathbb{Z}^n} \left| \sum_{\xi \in \mathbb{Z}^n} h(\eta, \xi)\, f(\xi) \right|^p \\
&\le \sum_{\eta \in \mathbb{Z}^n} \left[\sum_{\xi \in \mathbb{Z}^n} |h(\eta, \xi)\, f(\xi)| \right]^p \\
&\le C_2^{p/q} \sum_{\eta \in \mathbb{Z}^n} \sum_{\xi \in \mathbb{Z}^n} |h(\eta, \xi)|\, |f(\xi)|^p \\
&= C_2^{p/q} \sum_{\xi \in \mathbb{Z}^n} |f(\xi)|^p \sum_{\eta \in \mathbb{Z}^n} |h(\eta, \xi)| \\
&\le C_1 C_2^{p/q} \sum_{\xi \in \mathbb{Z}^n} |f(\xi)|^p \\
&= \left(C_1^{1/p} C_2^{1/q} \right)^p \|f\|_{\ell^p}^p. \qquad \square
\end{aligned}
$$

Recall that $\langle \xi \rangle = (1 + \|\xi\|^2)^{1/2}$, with $\|\xi\|$ as in (3.3). Working with pseudo-differential operators, some form of the elementary inequality of Jaak Peetre has to be derived at some stage. The most important variant for us is the following:

Proposition 3.3.31 (Peetre's inequality). *For all* $s \in \mathbb{R}$ *and* $\xi, \eta \in \mathbb{R}^n$, *we have*

$$\langle \xi + \eta \rangle^s \leq 2^{|s|} \langle \xi \rangle^s \langle \eta \rangle^{|s|}.$$

Proof. We have $\langle \xi + \eta \rangle \leq \langle \eta \rangle + \langle \xi \rangle \leq 2 \langle \eta \rangle \langle \xi \rangle$, so that

$$\langle \xi + \eta \rangle^t \leq 2^t \langle \eta \rangle^t \langle \xi \rangle^t,$$

when $t \geq 0$. Thus also

$$\langle \xi \rangle^t \leq 2^t \langle -\eta \rangle^t \langle \xi + \eta \rangle^t$$

implying

$$\langle \xi + \eta \rangle^{-t} \leq 2^t \langle \eta \rangle^t \langle \xi \rangle^{-t}.$$

This completes the proof. $\qquad\qquad\qquad\qquad\qquad\qquad\qquad\qquad\qquad\qquad\square$

Exercise 3.3.32. Let $\xi, \eta \in \mathbb{R}^n$. Prove the following kindred Peetre inequalities:

$$
\begin{aligned}
(1 + \|\xi + \eta\|)^s &\leq 2^{|s|} (1 + \|\xi\|)^{|s|} (1 + \|\eta\|)^s, &\qquad (3.25) \\
\underline{\xi + \eta}^s &\leq 2^{|s|} \underline{\xi}^{|s|} \underline{\eta}^s,
\end{aligned}
$$

where $\underline{\omega} := \max\{1, \|\omega\|\}$. This latter form of the weight is used, e.g., in [102].

3.3.4 Linking differences to derivatives

In the light of contemporary analysis, the words of Charles Jordan ([60]) have not expired: "*The importance of* Stirling's *numbers in Mathematical Calculus has not yet been fully recognised, and they are seldom used. This is especially due to the fact that different authors have reintroduced them under different definitions and notations, often not knowing, or not mentioning, that they deal with the same numbers...*". Many properties of these numbers can be found in [2], and they can be defined in various ways. We shall explain their relation to combinatorics, yet we start with another approach, since our purpose is to establish a connection between difference and differential calculi:

Definition 3.3.33 (Stirling numbers). Let $x \in \mathbb{R}$ and $j, k \in \mathbb{N}_0$ such that $j \leq k$. The *Stirling numbers* $S_k^{(j)}$ *of the first kind* are defined by the formula

$$x^{(k)} = \sum_{j=0}^{k} S_k^{(j)} x^j.$$

The uniqueness of $S_k^{(j)}$ is obvious. *Stirling numbers* $\left\{\begin{matrix} k \\ j \end{matrix}\right\}$ *of the second kind* are a sort of dual for the first kind:

$$x^k = \sum_{j=0}^{k} \left\{\begin{matrix} k \\ j \end{matrix}\right\} x^{(j)}.$$

For $j < 0$ and $j > k$, it is natural to extend these definitions by

$$S_k^{(j)} := 0 \quad \text{and} \quad \left\{\begin{matrix} k \\ j \end{matrix}\right\} := 0.$$

For multi-indices $\alpha, \beta \in \mathbb{Z}^n$, let

$$S_\alpha^{(\beta)} \;\; := \;\; S_{\alpha_1}^{(\beta_1)} \cdots S_{\alpha_n}^{(\beta_n)},$$

$$\left\{\begin{matrix} \alpha \\ \beta \end{matrix}\right\} \;\; := \;\; \left\{\begin{matrix} \alpha_1 \\ \beta_1 \end{matrix}\right\} \cdots \left\{\begin{matrix} \alpha_n \\ \beta_n \end{matrix}\right\}.$$

Lemma 3.3.34. *Let $j, k \in \mathbb{N}_0$. Then*

$$S_k^{(j)} = \frac{1}{j!} \left(\frac{\mathrm{d}}{\mathrm{d}x}\right)^j x^{(k)}\Big|_{x=0}, \qquad \left\{\begin{matrix} k \\ j \end{matrix}\right\} = \frac{1}{j!} \triangle_\xi^j \, \xi^k\Big|_{\xi=0}.$$

Proof. The first formula is a direct consequence of the definition of $S_k^{(j)}$. Of course, so is the second one, but since we do not have as profound an experience of difference calculus, we go through a simple calculation:

$$\triangle_\xi^j \, \xi^k\Big|_{\xi=0} = \triangle_\xi^j \sum_{i=0}^{k} \left\{\begin{matrix} k \\ i \end{matrix}\right\} \xi^{(i)}\Big|_{\xi=0} = \sum_{i=0}^{k} \left\{\begin{matrix} k \\ i \end{matrix}\right\} i^{(j)} \, \delta_{i,j} = j! \left\{\begin{matrix} k \\ j \end{matrix}\right\}. \qquad \square$$

Exercise 3.3.35. Let $x \in \mathbb{R}^n$ and $\alpha, \beta \in \mathbb{N}_0^n$. Show that

$$x^{(\alpha)} = \sum_{\beta \le \alpha} S_\alpha^{(\beta)} x^\beta, \qquad S_\alpha^{(\beta)} = \frac{1}{\beta!} \, \partial_x^\beta x^{(\alpha)}\Big|_{x=0},$$

$$x^\alpha = \sum_{\beta \le \alpha} \left\{\begin{matrix} \alpha \\ \beta \end{matrix}\right\} x^{(\beta)}, \qquad \left\{\begin{matrix} \alpha \\ \beta \end{matrix}\right\} = \frac{1}{\beta!} \, \triangle_\xi^\beta \xi^\alpha\Big|_{\xi=0}.$$

Lemma 3.3.36 (Recursion formulae for Stirling numbers). *Stirling numbers are uniquely determined by the recursions*

$$\left\{ \begin{array}{lll} S_k^{(0)} = \delta_{0,k}, & \left\{\begin{matrix} k \\ 0 \end{matrix}\right\} = \delta_{0,k}, & k \in \mathbb{N}_0, \\[2mm] S_k^{(k)} = 1, & \left\{\begin{matrix} k \\ k \end{matrix}\right\} = 1, & k \in \mathbb{N}_0, \\[2mm] S_k^{(j)} = 0, & \left\{\begin{matrix} k \\ j \end{matrix}\right\} = 0, & j < 0 \text{ or } j > k, \\[2mm] S_{k+1}^{(j)} = S_k^{(j-1)} - k S_k^{(j)}, & \left\{\begin{matrix} k+1 \\ j \end{matrix}\right\} = \left\{\begin{matrix} k \\ j-1 \end{matrix}\right\} + j \left\{\begin{matrix} k \\ j \end{matrix}\right\}, & j \ge 1, \ k \ge 0, \end{array} \right.$$

where $\delta_{p,q}$ is the Kronecker delta, defined by $\delta_{p,p} = 1$ and by $\delta_{p,q} = 0$ for $p \neq q$.

Proof. From $x^{(0)} = 1 = S_0^{(0)}$ and $x^{(k+1)} = x(x-1)\cdots(x-k)$ we see that it has to be $S_k^{(0)} = \delta_{0,k}$ and that $S_k^{(k)} = 1$ for every $k \in \mathbb{N}_0$. The statement $S_k^{(j)} = 0 = \begin{Bmatrix} k \\ j \end{Bmatrix}$ when $j < 0$ or $j > k$ simply rephrases a part of the extended definition for Stirling numbers. Suppose that $x^{(k)} = \sum_{j=0}^{k} S_k^{(j)} x^j$. Then

$$
\begin{aligned}
\sum_{j=0}^{k+1} S_{k+1}^{(j)} x^j &= x^{(k+1)} = (x-k)x^{(k)} \\
&= S_k^{(k)} x^{k+1} - k\, S_k^{(0)} + \sum_{j=1}^{k} \left[S_k^{(j-1)} - k\, S_k^{(j)} \right] x^j \\
&= \sum_{j=0}^{k+1} \left[S_k^{(j-1)} - k\, S_k^{(j)} \right] x^j.
\end{aligned}
$$

The case of the first kind is thus concluded. No doubt $\begin{Bmatrix} k \\ k \end{Bmatrix} = 1$ for every $k \in \mathbb{N}_0$, and clearly $\begin{Bmatrix} k \\ 0 \end{Bmatrix} = \delta_{0,k}$. Assume that $x^k = \sum_{j=0}^{k} \begin{Bmatrix} k \\ j \end{Bmatrix} x^{(j)}$. Then

$$
\begin{aligned}
\sum_{j=0}^{k+1} \begin{Bmatrix} k+1 \\ j \end{Bmatrix} x^{(j)} &= x^{k+1} = x\, x^k = x \sum_{j=0}^{k} \begin{Bmatrix} k \\ j \end{Bmatrix} x^{(j)} \\
&= \sum_{j=0}^{k} (x - j + j) \begin{Bmatrix} k \\ j \end{Bmatrix} x^{(j)} \\
&= \sum_{j=0}^{k} \left[\begin{Bmatrix} k \\ j \end{Bmatrix} x^{(j+1)} + j \begin{Bmatrix} k \\ j \end{Bmatrix} x^{(j)} \right] \\
&= \sum_{j=0}^{k+1} \left[\begin{Bmatrix} k \\ j-1 \end{Bmatrix} + j \begin{Bmatrix} k \\ j \end{Bmatrix} \right] x^{(j)},
\end{aligned}
$$

so that we can calculate $\begin{Bmatrix} k \\ j \end{Bmatrix}$ by recursion. $\qquad\square$

The general solution of the difference equation $S_{k+1}^{(j)} = S_k^{(j-1)} - k\, S_k^{(j)}$ is unknown (see [60, p. 143]), but for the second kind there is a closed form without recursion (it is easily obtained by applying Proposition 3.3.4 on Lemma 3.3.34):

$$
\begin{Bmatrix} k \\ j \end{Bmatrix} = \frac{1}{j!} \sum_{i=0}^{j} (-1)^{j-i} \binom{j}{i} i^k.
$$

There are combinatorial ideas behind the Stirling numbers, as explained, e.g., in [24]. The following exercise collects these ideas, using notations $\binom{k}{j}, \left[\begin{smallmatrix} k \\ j \end{smallmatrix}\right], \left\{\begin{smallmatrix} k \\ j \end{smallmatrix}\right\}$ of [67]:

Exercise 3.3.37 (Combinatorial background). Let $j, k \in \mathbb{N}_0$ such that $j \leq k$, and let S be a set with exactly k elements. Show that S has

$$\binom{k}{j} = \frac{k!}{j! \, (k-j)!}$$

subsets of exactly j elements (as usual, read: "k *choose* j"). Moreover, prove that

$$\left[\begin{matrix} k \\ j \end{matrix}\right] := (-1)^{k-j} S_k^{(j)}$$

is the number of permutations of S with precisely j cycles. Finally, show that

$$\left\{\begin{matrix} k \\ j \end{matrix}\right\}$$

is the number of ways to partition S into j non-empty subsets, i.e., the number of the equivalence relations on S having j equivalence classes (read: "k *quotient* j"). Hint: exploit the recursions in Lemma 3.3.36.

In the following matrices some of the Stirling numbers are presented. The index j is used for rows and k for columns:

$$\left(S_k^{(j)} \right)_{j,k=0}^{5} = \begin{pmatrix} 1 & 0 & 0 & 0 & 0 & 0 \\ 0 & 1 & -1 & 2 & -6 & 24 \\ 0 & 0 & 1 & -3 & 11 & -50 \\ 0 & 0 & 0 & 1 & -6 & 35 \\ 0 & 0 & 0 & 0 & 1 & -10 \\ 0 & 0 & 0 & 0 & 0 & 1 \end{pmatrix},$$

$$\left(\left\{\begin{matrix} k \\ j \end{matrix}\right\} \right)_{j,k=0}^{5} = \begin{pmatrix} 1 & 0 & 0 & 0 & 0 & 0 \\ 0 & 1 & 1 & 1 & 1 & 1 \\ 0 & 0 & 1 & 3 & 7 & 15 \\ 0 & 0 & 0 & 1 & 6 & 25 \\ 0 & 0 & 0 & 0 & 1 & 10 \\ 0 & 0 & 0 & 0 & 0 & 1 \end{pmatrix}.$$

Such matrices are inverses to each other:

Lemma 3.3.38. *Assume that* $i, j, N \in \mathbb{N}_0$ *such that* $0 \leq i, j \leq N$. *Then*

$$\sum_{k=0}^{N} S_k^{(i)} \left\{\begin{matrix} j \\ k \end{matrix}\right\} = \delta_{i,j} = \sum_{k=0}^{N} \left\{\begin{matrix} k \\ i \end{matrix}\right\} S_j^{(k)}.$$

Proof. Due to the symmetry, it suffices to prove only that the first sum equals $\delta_{i,j}$:

$$x^j = \sum_{k=0}^{j} \begin{Bmatrix} j \\ k \end{Bmatrix} x^{(k)} = \sum_{k=0}^{N} \begin{Bmatrix} j \\ k \end{Bmatrix} x^{(k)}$$

$$= \sum_{k=0}^{N} \begin{Bmatrix} j \\ k \end{Bmatrix} \sum_{i=0}^{k} S_k^{(i)} x^i = \sum_{k=0}^{N} \begin{Bmatrix} j \\ k \end{Bmatrix} \sum_{i=0}^{N} S_k^{(i)} x^i$$

$$= \sum_{i=0}^{N} x^i \sum_{k=0}^{N} S_k^{(i)} \begin{Bmatrix} j \\ k \end{Bmatrix}. \qquad \qquad \square$$

In the sequel, we shall present two alternative definitions for periodic pseudo-differential operators. To build a bridge between these approaches (in Lemma 4.7.1), we have to know how to approximate differences by derivatives. This problem is considered, for example, in [60, p. 164–165, 189–192], where the error estimates are neglected, as well as in the treatise of the subject in [15]. Francis Hildebrand ([52, p. 123–125]) makes a notice on the estimates, but does not calculate them, and there the approximation is thoroughly presented only for degrees $j = 1, 2$, and without a connection to the Stirling numbers. The finest account is by Steffensen in [117, p. 60-70], where the presentation of *Markoff's formulae* is general with error terms for any degree, but still it lacks the Stirling numbers.

The following theorem is in one dimension, so here $\varphi^{(k)}$ denotes the usual k^{th} derivative of φ.

Theorem 3.3.39 (Approximating differences by derivatives). *There exist constants $c_{N,j}^{\triangle}, c_{N,j}^{d} > 0$ for any $N \in \mathbb{N}_0$ and $j < N$ such that for every $\varphi \in C^{\infty}(\mathbb{R}^1)$ and $\xi \in \mathbb{R}^1$ the following inequalities hold:*

$$\left| \triangle_\xi^j \varphi(\xi) - \sum_{k=j}^{N-1} \frac{j!}{k!} \begin{Bmatrix} k \\ j \end{Bmatrix} \varphi^{(k)}(\xi) \right| \leq c_{N,j}^{\triangle} \max_{\eta \in [0,j]} \left| \varphi^{(N)}(\xi + \eta) \right|, \qquad (3.26)$$

$$\left| \varphi^{(j)}(\xi) - \sum_{k=j}^{N-1} \frac{j!}{k!} S_k^{(j)} \triangle_\xi^k \varphi(\xi) \right| \leq c_{N,j}^{d} \max_{\eta \in [0,N-1]} \left| \varphi^{(N)}(\xi + \eta) \right|. \qquad (3.27)$$

Proof. Note that in the following we cannot apply the discrete Taylor series, because its remainder is defined only on $\mathbb{Z}^1$ with respect to the variable η. The classical Taylor series does not have this disadvantage:

$$\varphi(\xi + \eta) = \sum_{k=0}^{N-1} \frac{1}{k!} \varphi^{(k)}(\xi) \eta^k + \frac{1}{N!} \varphi^{(N)}(\theta(\eta)) \eta^N.$$

We use the Lagrange form of the error term. Here $\theta(\eta)$ is some point in the segment connecting ξ and $\xi + \eta$. Assume that $N > j$. Applying $\triangle_\eta^j$ at $\eta = 0$, and using

Lemma 3.3.34 we get

$$
\triangle_\xi^j \varphi(\xi) = \triangle_\eta^j \left[\sum_{k=0}^{N-1} \frac{1}{k!} \, \varphi^{(k)}(\xi) \, \eta^k + \frac{1}{N!} \, \varphi^{(N)}(\theta(\eta)) \, \eta^N \right] \Bigg|_{\eta=0}
$$

$$
= \sum_{k=j}^{N-1} \frac{j!}{k!} \begin{Bmatrix} k \\ j \end{Bmatrix} \varphi^{(k)}(\xi) + \frac{1}{N!} \, \triangle_\eta^j \left[\varphi^{(N)}(\theta(\eta)) \, \eta^N \right] \Bigg|_{\eta=0}. \tag{3.28}
$$

Using the Leibniz formula on the remainder term, we see that its absolute value is majorised by $c_{N,j}^\triangle \left| \varphi^{(N)}(\theta_j) \right|$ for some $\theta_j \in [\xi, \xi+j]$, and hence (3.26) is true. For the latter inequality (3.27), the "orthogonality" of Stirling numbers (Lemma 3.3.38), and (3.28) are essential:

$$
\sum_{k=i}^{N-1} \frac{i!}{k!} \, S_k^{(i)} \, \triangle_\xi^k \varphi(\xi) = \sum_{k=i}^{N-1} \frac{i!}{k!} \, S_k^{(i)} \sum_{j=k}^{N-1} \frac{k!}{j!} \begin{Bmatrix} j \\ k \end{Bmatrix} \varphi^{(j)}(\xi)
$$

$$
+ \sum_{k=i}^{N-1} \frac{i!}{k!} \, S_k^{(i)} \, \frac{1}{N!} \, \triangle_\eta^k \left[\varphi^{(N)}(\theta(\eta)) \, \eta^N \right] \Bigg|_{\eta=0}
$$

$$
= \sum_{j=i}^{N-1} \frac{i!}{j!} \, \varphi^{(j)}(\xi) \sum_{k=i}^{j} S_k^{(i)} \begin{Bmatrix} j \\ k \end{Bmatrix}
$$

$$
+ \sum_{k=i}^{N-1} \frac{i!}{k!} \, S_k^{(i)} \, \frac{1}{N!} \, \triangle_\eta^k \left[\varphi^{(N)}(\theta(\eta)) \, \eta^N \right] \Bigg|_{\eta=0}
$$

$$
= \varphi^{(i)}(\xi) + \sum_{k=i}^{N-1} \frac{i!}{k!} \, S_k^{(i)} \, \frac{1}{N!} \, \triangle_\eta^k \left[\varphi^{(N)}(\theta(\eta)) \, \eta^N \right] \Bigg|_{\eta=0},
$$

where the absolute value of the remainder part is estimated above by some $c_{N,j}^d \left| \varphi^{(N)}(\theta_N) \right|$ (cf. the proof of (3.26)). $\qquad\square$

Inequality (3.27) is not actually needed in this work, but as a dual statement to (3.26) it is justified. Note that in (3.26) the maximum of $\left| \varphi^{(N)}(\xi+\eta) \right|$ is taken over the interval $\eta \in [0, j]$, whereas in (3.27) over $\eta \in [0, \xi-1]$.

Exercise 3.3.40. Let $\alpha, \beta \in \mathbb{N}_0^n$, $\xi \in \mathbb{Z}^n$ and $\varphi \in C^\infty(\mathbb{R}^n)$. Estimate

$$
\left| \triangle_\xi^\beta \varphi(\xi) - \sum_{|\alpha|<N} \frac{\beta!}{\alpha!} \begin{Bmatrix} \alpha \\ \beta \end{Bmatrix} \partial^\alpha \varphi(\xi) \right| \quad \text{and} \quad \left| \partial^\beta \varphi(\xi) - \sum_{|\alpha|<N} \frac{\beta!}{\alpha!} S_\alpha^{(\beta)} \triangle_\xi^\alpha \varphi(\xi) \right|
$$

in the manner of Theorem 3.3.39.

3.4 Periodic Taylor expansion

For the global analysis on the torus, the ordinary Taylor series is useless due to the lack of periodicity. We are now going to present a tool to fill in this gap. It must be admitted, though, that this approach is not a necessary one, as we can study the theory of periodic pseudo-differential operators without it. Nevertheless, the periodic Taylor series provides an appropriate alternative, and one never knows when it is gravely needed.

Definition 3.4.1 (Discrete modification of partial derivatives). For $\alpha \in \mathbb{N}_0^n$, let us introduce an abbreviation for partial derivatives,

$$
\begin{aligned}
D_x^\alpha &:= D_{x_1}^{\alpha_1} \cdots D_{x_n}^{\alpha_n}, \\
D_x^{(\alpha)} &:= D_{x_1}^{(\alpha_1)} \cdots D_{x_n}^{(\alpha_n)},
\end{aligned}
$$

where for $k \in \mathbb{N}_0$,

$$
D_{x_l}^k := \left(\frac{1}{\mathrm{i} 2\pi} \frac{\partial}{\partial x_l} \right)^k,
$$

$$
D_{x_l}^{(k)} := \prod_{j=0}^{k-1} \left(\frac{1}{\mathrm{i} 2\pi} \frac{\partial}{\partial x_l} - j \right), \tag{3.29}
$$

in the spirit of Stirling numbers. We interpret $D_x^0 = I = D_x^{(0)}$.

Theorem 3.4.2 (Periodic Taylor expansion for $\mathbb{T}^1$). *Any $a \in C^\infty(\mathbb{T}^1)$ has the periodic Taylor representation*

$$
a(x) = \sum_{j=0}^{N-1} \frac{1}{j!} \left(\mathrm{e}^{\mathrm{i} 2\pi x} - 1 \right)^j \left. D_z^{(j)} a(z) \right|_{z=0} + a_N(x) \left(\mathrm{e}^{\mathrm{i} 2\pi x} - 1 \right)^N,
$$

where $a_N \in C^\infty(\mathbb{T}^1)$.

Proof. For $j \in \mathbb{N}_0$, we define the functions a_j by

$$
a_0(x) := a(x), \quad a_{j+1}(x) := \begin{cases} \frac{a_j(x) - a_j(0)}{\mathrm{e}^{\mathrm{i} 2\pi x} - 1}, & \text{if } x \neq 0, \\ D_x a_j(x), & \text{if } x = 0. \end{cases}
$$

Inductively we obtain that a_{j+1} is in $C^\infty(\mathbb{T}^1)$. Thus

$$
a_j(x) = a_j(0) + a_{j+1}(x) \left(\mathrm{e}^{\mathrm{i} 2\pi x} - 1 \right),
$$

and recursively

$$
a(x) = \sum_{j=0}^{N-1} \left(\mathrm{e}^{\mathrm{i} 2\pi x} - 1 \right)^j a_j(0) + a_N(x) \left(\mathrm{e}^{\mathrm{i} 2\pi x} - 1 \right)^N, \tag{3.30}
$$

so we have to prove that $a_j(0) = \frac{1}{j!} \left. D_x^{(j)} a(x) \right|_{x=0}$.

Clearly expression $D_x^{(j)} \left(e^{i2\pi x} - 1\right)^k \Big|_{x=0}$ vanishes, if $j < k$. It vanishes also when $j > k$, since

$$\left(\frac{1}{i2\pi}\frac{\partial}{\partial x} - k\right)\left(e^{i2\pi x} - 1\right)^k = k\,e^{i2\pi x}\left(e^{i2\pi x} - 1\right)^{k-1} - k\left(e^{i2\pi x} - 1\right)^k$$
$$= k\left(e^{i2\pi x} - 1\right)^{k-1}$$

implies that

$$\left[\prod_{m=1}^{k}\left(\frac{1}{i2\pi}\frac{\partial}{\partial x} - m\right)\right]\left(e^{i2\pi x} - 1\right)^k = k!.$$

We use this instantly:

$$D_x^{(k)}\left(e^{i2\pi x} - 1\right)^k\Big|_{x=0} = \left(\frac{1}{i2\pi}\frac{\partial}{\partial x} - k + k\right)\prod_{j=1}^{k-1}\left(\frac{1}{i2\pi}\frac{\partial}{\partial x} - j\right)\left(e^{i2\pi x} - 1\right)^k\Big|_{x=0}$$
$$= k!.$$

Hence we get $D_x^{(j)}\left(e^{i2\pi x} - 1\right)^k\Big|_{x=0} = j!\,\delta_{j,k}$, so that by applying $D_x^{(j)}$ to both sides of the equality (3.30) we obtain $D_x^{(j)}a(x)\Big|_{x=0} = j!\,a_j(0)$ proving the claim. $\qquad\square$

As an immediate consequence of Theorem 3.4.2 we get the biperiodic Taylor series:

Corollary 3.4.3 (Biperiodic Taylor expansion). *Any function* $a \in C^\infty(\mathbb{T}^1 \times \mathbb{T}^1)$ *has the Taylor representation*

$$a(x,y) = \sum_{j=0}^{N-1}\frac{1}{j!}\left(e^{i2\pi(y-x)} - 1\right)^j D_z^{(j)}a(z,y)\Big|_{z=x}$$
$$+ a_N(x,y)\left(e^{i2\pi(y-x)} - 1\right)^N,$$

where $a_N \in C^\infty(\mathbb{T}^1 \times \mathbb{T}^1)$.

We shall return to the biperiodic expansion in the study of the amplitudes of periodic pseudo-differential operators (see Lemma 4.3.4 and Theorem 4.4.5). Instead of such biperiodic series, we may use also the following multidimensional periodic expansion, another corollary to Theorem 3.4.2:

Theorem 3.4.4 (Periodic Taylor expansion for $\mathbb{T}^n$**).** *Any* $a \in C^\infty(\mathbb{T}^n)$ *has the periodic Taylor representation*

$$a(x) = \sum_{|\alpha|<N}\frac{1}{\alpha!}\left(e^{i2\pi x} - 1\right)^\alpha D_z^{(\alpha)}a(z)\Big|_{z=0} + \sum_{|\alpha|=N}a_\alpha(x)\left(e^{i2\pi x} - 1\right)^\alpha,$$

where $a_\alpha \in C^\infty(\mathbb{T}^n)$ *and*

$$\left(e^{i2\pi x} - 1\right)^\alpha := \left(e^{i2\pi x_1} - 1\right)^{\alpha_1}\cdots\left(e^{i2\pi x_n} - 1\right)^{\alpha_n}.$$

Proof. We define the functions $a_\alpha \in C^\infty(\mathbb{T}^n)$ with the aid of Theorem 3.4.2: let $a_0 := a$, and if $\alpha_k > 0$ for each $k \in \{1, \dots, n\}$ then

$$
\begin{aligned}
a_{(\alpha_1,0,0,\dots,0)}(x) &:= a_{\alpha_1}(x_1, x_2, \dots, x_n), \\
a_{(\alpha_1,\alpha_2,0,\dots,0)}(x) &:= \left(a_{(\alpha_1,0,0,\dots,0)}\right)_{\alpha_2}(0, x_2, \dots, x_n), \\
&\vdots \\
a_{(\alpha_1,\dots,\alpha_k,0,\dots,0)}(x) &:= \left(a_{(\alpha_1,\dots,\alpha_{k-1},0,\dots,0)}\right)_{\alpha_k}(0, \dots, 0, x_k, \dots, x_n), \\
&\vdots \\
a_{(\alpha_1,\dots,\alpha_n)}(x) &:= \left(a_{(\alpha_1,\dots,\alpha_{n-1},0)}\right)_{\alpha_n}(0, \dots, 0, x_n).
\end{aligned}
$$

Then we obtain

$$
\begin{aligned}
a(x) &= \sum_{\alpha_1=0}^{N-1} \left(e^{i2\pi x_1} - 1\right)^{\alpha_1} a_{(\alpha_1,0,\dots,0)}(0, x_2, \dots, x_n) \\
&\quad + a_{(N,0,\dots,0)}(x) \left(e^{i2\pi x_1} - 1\right)^N \\
&= \sum_{\alpha_1+\alpha_2 < N} \left(e^{i2\pi x_1} - 1\right)^{\alpha_1} \left(e^{i2\pi x_2} - 1\right)^{\alpha_2} \\
&\qquad \times a_{(\alpha_1,\alpha_2,0,\dots,0)}(0, 0, x_3, \dots, x_n) \\
&\quad + \sum_{\alpha_1+\alpha_2 = N} a_{(\alpha_1,\alpha_2,0,\dots,0)}(x) \left(e^{i2\pi x_1} - 1\right)^{\alpha_1} \left(e^{i2\pi x_2} - 1\right)^{\alpha_2} \\
&= \dots,
\end{aligned}
$$

which iteratively leads to

$$
a(x) = \sum_{|\alpha|<N} \left(e^{i2\pi x} - 1\right)^\alpha a_\alpha(0) + \sum_{|\alpha|=N} a_\alpha(x) \left(e^{i2\pi x} - 1\right)^\alpha.
$$

Observing that $D_x^{(\beta)} \left(e^{i2\pi x} - 1\right)^\alpha \big|_{x=0} = \beta!\, \delta_{\alpha,\beta}$, this implies that we have

$$
D_x^{(\beta)} a(x) \big|_{x=0} = \beta!\, a_\beta(0),
$$

leading to the claimed expansion. $\qquad\square$

3.5 Appendix: on operators in Banach spaces

Elsewhere, we will need several functional analysis results to be used in the analysis of periodic Sobolev spaces and we present them here. The reader is encouraged to skip this section at the first reading and consult it only when a reference is encountered in some proof.

Let A be an operator defined on a set X, and let $W \subset X$. Then the restriction of A to W is the operator $A|_W$, defined on W and satisfying $A|_W(w) = A(w)$ (where $w \in W$). An evaluation is occasionally written as $A(x)|_{x=x_0} := A(x_0)$. If $\mathcal{U}$ is a family of subsets of X, the restriction of $\mathcal{U}$ to $W \subset X$ means $\mathcal{U}|_W := \{U \cap W | U \in \mathcal{U}\}$.

The topology τ_X of a set X is the family of the open subsets of X. Assume that X is a vector space with $\mathbb{R}$ or $\mathbb{C}$ as the scalar field $\mathbb{K}$, and that $\mathbb{K}$ has the usual topology induced by the absolute value norm. Elements of the theory of topological vector spaces are presented in Chapters A and C, but here it is enough to recognise the outlines concerning Banach and Hilbert spaces, the most important topological vector spaces.

Let X and Y be vector spaces. The convergence of the sequence $(u_j)_{j \in \mathbb{Z}^+} \subset X$ to $u \in X$ in the topology τ_X is marked by $u_j \xrightarrow{\tau_X} u$. The closure of a subset $U \subset X$ in this topology is $cl_X(U)$, or when the topology is known from the context, $cl(U)$ or merely $\overline{U}$. From now on, the notation τ_X is reserved for the vector space topology of X. The set of all linear operators with X as the domain, and the range in Y is denoted by $L(X, Y)$, and if the spaces are equipped with vector space topologies τ_X, τ_Y, then $\mathcal{L}(X, Y)$ is the set of continuous linear mappings with respect to these topologies. The norm of X, if there exists such, is $\| \cdot \|_X$, and the operator norm between normed spaces X and Y is $\| \cdot \|_{\mathcal{L}(X,Y)}$. As before, τ_E refers to the vector space topology of space E.

Theorem 3.5.1. *Let E, F, and G be Banach spaces such that $F \subset G$ and $\tau_G|_F \subset \tau_F$. If $A \in \mathcal{L}(E, G)$ maps E into F, then $A \in \mathcal{L}(E, F)$.*

Proof. (Note that in this case $\mathcal{L}(E, F) \subset \mathcal{L}(E, G)$ is topologically trivial, as the inclusion $F \hookrightarrow G$ is continuous.) Take a sequence $(u_j)_{j \in \mathbb{Z}^+} \subset E$ such that $u_j \xrightarrow{\tau_E} u \in E$ and $Au_j \xrightarrow{\tau_F} v \in F$. Now $Au_j \xrightarrow{\tau_G} v$, since $\tau_G|_F \subset \tau_F$. On the other hand, $Au_j \xrightarrow{\tau_G} Au \in G$, because $A \in \mathcal{L}(E, G)$. Hence it has to be $v = Au$. The proof is now completed by the Closed Graph Theorem (Theorem B.4.34). $\square$

The periodic Sobolev spaces $H^s(\mathbb{T}^n)$ $(s \in \mathbb{R})$ are nested in a chain so that $s < t$ implies $H^s(\mathbb{T}^n) \supset H^t(\mathbb{T}^n)$, with $\tau_{H^t(\mathbb{T}^n)}$ finer than the restriction of $\tau_{H^s(\mathbb{T}^n)}$ to $H^t(\mathbb{T}^n)$. The continuous linear operators are most conveniently defined first on the trigonometric polynomials, which are dense in every $H^s(\mathbb{T}^n)$, and then extended to the Sobolev spaces using the next theorem:

Theorem 3.5.2. *Let E_i, F_j $(i, j = 1, 2)$ be Banach spaces such that $E_1 \subset E_2$ and $F_1 \subset F_2$, $\tau_{E_2}|_{E_1} \subset \tau_{E_1}$ and $\tau_{F_2}|_{F_1} \subset \tau_{F_1}$. Assume that $X \subset E_1$ is dense in E_1 and E_2, and that $A \in L(X, F_1)$ is continuous as a map $E_1 \to F_1$ and $E_2 \to F_2$. Then A can be uniquely extended to a map $A \in \mathcal{L}(E_2, F_2)$ satisfying $A|_{E_1} \in \mathcal{L}(E_1, F_1)$.*

Proof. We first show that A can be extended to a continuous operator $A = A_1 \in L(E_1, F_1)$. Assume that $(u_j)_{j \in \mathbb{Z}^+} \subset X : u_j \xrightarrow{\tau_{E_1}} u \in E_1$. Now $(Au_j)_{j \in \mathbb{Z}^+} \subset F_1$ is a Cauchy sequence, because A is continuous from X to F_1 with respect to τ_{E_1}

and τ_{F_1}. Since F_1 is complete, $(Au_j)_{j\in\mathbb{Z}^+}$ converges to some $v \in F_1$, and we define $Au := v$. Uniqueness and linearity of this extension are trivial. Let us denote $\|A\|_{E_1|X,F_1} := \sup_{u\in X:\|u\|_{E_1}\leq 1}\|Au\|_{F_1}$. Then

$$
\begin{aligned}
\|Au\|_{F_1} &= \lim_{j\to\infty}\|Au_j\|_{F_1} \\
&\leq \|A\|_{E_1|X,F_1}\lim_{k\to\infty}\|u_j\|_{E_1} \\
&= \|A\|_{E_1|X,F_1}\|u\|_{E_1},
\end{aligned}
$$

so that $A \in \mathcal{L}(E_1, F_1)$ with $\|A\|_{\mathcal{L}(E_1,F_1)} \leq \|A\|_{E_1|X,F_1}$. Indeed, $\|A\|_{\mathcal{L}(E_1,F_1)} = \|A\|_{E_1|X,F_1}$. The proof that $A = A_2 \in \mathcal{L}(E_2, F_2)$ follows exactly the same pattern. Note that $A = A_1$ is also in $\mathcal{L}(E_1, F_2)$, and that $u_j \xrightarrow{\ \tau_{E_2}\ } u$, since $\tau_{F_2}|_{F_1} \subset \tau_{F_1}$ and $\tau_{E_2}|_{E_1} \subset \tau_{E_1}$. Thus

$$
\begin{aligned}
&\|(A_1 - A_2)u\|_{F_2} \\
\leq\ &\|A_1(u - u_j)\|_{F_2} + \|(A_1 - A_2)u_j\|_{F_2} + \|A_2(u_j - u)\|_{F_2} \\
\leq\ &\|A_1\|_{\mathcal{L}(E_1,F_2)}\|u - u_j\|_{E_1} + 0 + \|A_2\|_{\mathcal{L}(E_2,F_2)}\|u_j - u\|_{E_2} \\
\xrightarrow[j\to\infty]{}\ &0.
\end{aligned}
$$

Hence $A_1 = A_2|_{E_1}$. $\qquad\qquad\qquad\qquad\qquad\qquad\qquad\qquad\qquad\qquad\qquad\square$

Chapter 4

Pseudo-differential Operators on $\mathbb{T}^n$

Pseudo-differential operators on the torus $\mathbb{T}^n = \mathbb{R}^n/\mathbb{Z}^n$, or the *periodic pseudo-differential operators*, are studied next. The presentation is written in a way for a reader to be able to compare and to contrast it to the general theory of pseudo-differential operators on the Euclidean space from Chapter 2.

However, while in $\mathbb{R}^n$ in Chapter 2 we aimed at avoiding technicalities by restricting ourselves to symbols of type $S_{1,0}^m$, here we will also discuss symbols of types $S_{\rho,\delta}^m$. One reason for this is that no comprehensive treatment of operators on $\mathbb{T}^n$ seems to be available in the literature so we may treat a more general situation also to exhibit the dependence of some results on the values of parameters ρ and δ.

In Chapter 3 a sound basis for the development of the theory of periodic pseudo-differential operator was founded, where the pieces of information have been known practically for decades, and in some cases, for centuries; however, these fragments of wisdom have been scattered widely apart in the field of mathematics.

We will see how closely periodic pseudo-differential operators are tied to the general pseudo-differential operators, the main difference actually being that in the periodic case, the theory appears to be more crystallised. This is definitely good news for those who want to grasp the ideas of any pseudo-differential theory.

In 1979 Agranovich [3] proposed, crediting L.R. Volevich, a global definition of pseudo-differential operators on the unit circle $\mathbb{S}^1$, called the periodic pseudo-differential operators. Of course, the definition was readily generalisable for any torus $\mathbb{T}^n$. Due to the group structure of $\mathbb{T}^n$, by exploiting the Fourier series representation these new operators admitted globally defined symbols instead of mere local analysis. We also note here that a similar representation of operators has been already used by Petrovski in [86] in the analysis of the Cauchy problem for systems of partial differential equations.

It is a non-trivial fact, however, that the definitions of pseudo-differential operators on a torus given by Agranovich and Hörmander are equivalent. Agranovich proved this in [4] in the special case of classical operators, and later without some details in [5] in the case of the Hörmander $(1, 0)$-operators. Another treatise of the classical operators was presented in [103]. A complete proof was provided by

McLean [76] for all the Hörmander (ρ, δ)-classes. McLean proved equivalence of the global and local definitions by directly studying charts of the tori. Another proof of this type was given in [79] for the $(1,0)$-class. In the sequel, we give one more approach, based on extension and periodisation techniques, providing the equality of (ρ, δ)-symbol classes (Corollary 4.6.13), also yielding an explicit relation between operators (Theorem 4.6.12).

Periodic integral operators are a major source of applications for the periodic pseudo-differential operator theory. Unfortunately, there is not much room for discussing periodic integral operators here, except for an application in Section 4.11. Important further results on this subject are fast methods of solving periodic integral equations presented in [142] and [102], which certainly is recommended for further reading on these topic. Other applications to the numerical analysis of periodic equations in mechanics and aerodynamics can be found in, e.g., [144, 145]. Numerical aspects of Fourier transforms on general compact groups can be found in, e.g., [75].

From the point of view of these applications, a theory of pseudo-differential operators expressed in terms of Fourier coefficients and discrete operations is appealing. Periodic pseudo-differential operators were briefly considered, e.g., in [34], and certain aspects studied in [6, 7], [142], [102] and [138]. We note that analysis of vector fields has an obvious embedding into the theory of pseudo-differential operators on the torus, and thus, for example, questions of global hypoellipticity and solvability of vector fields (e.g., [42], [43], [44], [13], etc.) obtain a more fundamental ground. Some aspects of the analysis presented in this chapter appeared in [97, 98].

In this chapter we develop the foundations of the theory of pseudo-differential operators on the torus $\mathbb{T}^n$. This includes toroidal quantization of operators, toroidal symbol classes, toroidal amplitudes, asymptotic expansions, symbolic calculus, boundedness on $L^2(\mathbb{T}^n)$ and on the Sobolev spaces $H^s(\mathbb{T}^n)$, questions of ellipticity and regularity. Section 4.11 gives an application to periodic integral equations.

In Sections 4.12 we consider toroidal wave front sets relating them to the standard Hörmander wave front sets in $\mathbb{R}^n$. In Section 4.13 we introduce Fourier series operators and study their compositions with pseudo-differential operators in terms of the toroidal symbols. In Section 4.14 we establish the boundedness of Fourier series operators in $L^2(\mathbb{T}^n)$ and $H^s(\mathbb{T}^n)$, and in Section 4.15 we discuss an application to the Cauchy problem for hyperbolic partial differential equations.

Fourier series operators considered here are analogues of the Fourier integral operators on the torus and we study them in terms of the toroidal quantization. The main new difficulty here is that while pseudo-differential operators do not move the wave front sets of distributions, this is no longer the case for Fourier series operators. Thus, we are forced to make extensions of functions from the integer lattice to the Euclidean space on the frequency side in Theorem 4.13.11. However, the other composition formula in Theorem 4.13.8 is still expressed entirely in the toroidal language.

4.1 Toroidal symbols

In this section we discuss the quantization of operators that we will call *toroidal quantization* and the corresponding classes of toroidal symbols.

4.1.1 Quantization of operators on $\mathbb{T}^n$

First, we discuss the ideas informally.

Informal discussion. The main informal underlying idea here is that for a given linear operator $A : C^\infty(\mathbb{T}^n) \to \mathcal{D}'(\mathbb{T}^n)$, if we study how it maps the functions $e_\xi = (x \mapsto e^{i2\pi x \cdot \xi})$ for all $\xi \in \mathbb{Z}^n$, then A would be completely determined. To collect this kind of information we define the symbol σ_A of A by testing A on the waves e_ξ yielding $Ae_\xi(x) = e^{i2\pi x \cdot \xi} \sigma_A(x, \xi)$, i.e.,

$$\sigma_A(x, \xi) := e^{-i2\pi x \cdot \xi} Ae_\xi(x). \tag{4.1}$$

Conversely, a function $\sigma : \mathbb{T}^n \times \mathbb{Z}^n \to \mathbb{C}$ defines a linear operator $\mathrm{Op}(\sigma) : C^\infty(\mathbb{T}^n) \to \mathcal{D}'(\mathbb{T}^n)$ by

$$\mathrm{Op}(\sigma)u(x) := \sum_{\xi \in \mathbb{Z}^n} e^{i2\pi x \cdot \xi} \, \sigma(x, \xi) \, \widehat{u}(\xi), \tag{4.2}$$

provided that there are some reasonable restrictions on $\sigma(x, \xi)$. Clearly σ is the symbol of $\mathrm{Op}(\sigma)$.

Example. A partial differential operator $A = \sum_{|\alpha| \leq M} a_j(x) \, (\partial/\partial x)^\alpha$, where $a_\alpha \in C^\infty(\mathbb{T}^n)$, has the symbol

$$\sigma_A(x, \xi) = \sum_{|\alpha| \leq M} a_j(x) \, (i2\pi\xi)^\alpha,$$

a polynomial of degree M in ξ. This observation motivates the concept of periodic pseudo-differential operators.

Thus, given a continuous linear operator $A : C^\infty(\mathbb{T}^n) \to C^\infty(\mathbb{T}^n)$, we can consider its toroidal quantization

$$A\varphi(x) = \sum_{\xi \in \mathbb{Z}^n} e^{i2\pi x \cdot \xi} \, \sigma_A(x, \xi) \, \widehat{f}(\xi) \, \mathrm{d}y,$$

where its *toroidal symbol* $\sigma_A \in C^\infty(\mathbb{T}^n \times \mathbb{Z}^n)$ will be uniquely determined by the formula

$$\sigma_A(x, \xi) = e^{-i2\pi x \cdot \xi} Ae_\xi(x),$$

where $e_\xi(x) := e^{i2\pi x \cdot \xi}$. We note that for periodic pseudo-differential operators on $\mathbb{R}^n$ this could be just another quantization of the same class, as several quantizations on $\mathbb{R}^n$ already exist, including the Kohn-Nirenberg quantization from Chapter 1.

More formal presentation.   We will now make this more precise.

Remark 4.1.1 (**Periodic Schwartz kernel**). For $\psi, \varphi \in C^\infty(\mathbb{T}^n)$, let $\psi \otimes \varphi \in C^\infty(\mathbb{T}^{2n})$ be defined by $(\psi \otimes \varphi)(x,y) := \psi(x)\varphi(y)$. If $A : C^\infty(\mathbb{T}^n) \to \mathcal{D}'(\mathbb{T}^n)$ is a continuous linear operator then one can verify that

$$\langle K_A, \psi \otimes \varphi \rangle := \langle A\varphi, \psi \rangle$$

defines the periodic Schwartz distributional kernel $K_A \in \mathcal{D}'(\mathbb{T}^{2n})$ of the operator $A \in \mathcal{L}(C^\infty(\mathbb{T}^n), \mathcal{D}'(\mathbb{T}^n))$; a common informal notation is

$$A\varphi(x) = \int_{\mathbb{T}^n} K_A(x,y)\, \varphi(y)\, dy.$$

The convolution kernel $k_A \in \mathcal{D}'(\mathbb{T}^{2n})$ of A is related to the Schwartz kernel by

$$K_A(x,y) = k_A(x, x-y),$$

i.e., we have

$$A\varphi(x) = \int_{\mathbb{T}^n} k_A(x, x-y)\, \varphi(y)\, dy$$

in the sense of distributions. We write $k_A(x)(y) := k_A(x,y)$.

We now define symbols of operators on the torus.

Definition 4.1.2 (Toroidal symbols of operators on $\mathbb{T}^n$). Let $e_\xi(x) = e^{i2\pi x \cdot \xi}$. The *toroidal symbol* of a linear continuous operator $A : C^\infty(\mathbb{T}^n) \to C^\infty(\mathbb{T}^n)$ at $x \in \mathbb{T}^n$ and $\xi \in \mathbb{Z}^n$ is defined by

$$\sigma_A(x, \xi) := \widehat{k_A(x)}(\xi) = \mathcal{F}_{\mathbb{T}^n}(k_A(x))(\xi).$$

Hence

$$\sigma_A(x, \xi) = \int_{\mathbb{T}^n} k_A(x,y)\, e^{-i2\pi y \cdot \xi}\, dy = \langle k_A(x), e_{-\xi} \rangle.$$

By the Fourier inversion formula the convolution kernel can be regained from the symbol:

$$k_A(x,y) = \sum_{\xi \in \mathbb{Z}^n} e^{i2\pi y \cdot \xi}\, \sigma_A(x, \xi) \tag{4.3}$$

in the sense of distributions. We now show that an operator A can be represented by its symbol:

Theorem 4.1.3 (Quantization of operators on $\mathbb{T}^n$). *Let σ_A be the toroidal symbol of a continuous linear operator $A : C^\infty(\mathbb{T}^n) \to C^\infty(\mathbb{T}^n)$. Then*

$$Af(x) = \sum_{\xi \in \mathbb{Z}^n} e^{i2\pi x \cdot \xi}\, \sigma_A(x, \xi)\, \widehat{f}(\xi) \tag{4.4}$$

for every $f \in C^\infty(\mathbb{T}^n)$ and $x \in \mathbb{T}^n$.

Proof. Let us define a convolution operator $A_{x_0} \in \mathcal{L}(C^\infty(\mathbb{T}^n))$ by the kernel $k_A(x_0, y) = k_{x_0}(y)$, i.e., by

$$A_{x_0} f(x) := \int_{\mathbb{T}^n} f(y) \, k_{x_0}(x - y) \, \mathrm{d}y = (f * k_{x_0})(x).$$

Thus

$$\sigma_{A_{x_0}}(x, \xi) = \widehat{k_{x_0}}(\xi) = \sigma_A(x_0, \xi),$$

so that we have

$$
\begin{aligned}
A_{x_0} f(x) &= \sum_{\xi \in \mathbb{Z}^n} e^{i2\pi x \cdot \xi} \, \widehat{A_{x_0} f}(\xi) \\
&= \sum_{\xi \in \mathbb{Z}^n} e^{i2\pi x \cdot \xi} \, \sigma_A(x_0, \xi) \, \widehat{f}(\xi),
\end{aligned}
$$

where we used that $\widehat{f * k_{x_0}} = \widehat{f}\widehat{k_{x_0}}$ by the same calculation as in Theorem 1.1.30, (iii). This implies the result, because $Af(x) = A_x f(x)$. $\qquad\square$

Theorem 4.1.4 (Formula for the toroidal symbol). *Let σ_A be the toroidal symbol of a continuous linear operator $A : C^\infty(\mathbb{T}^n) \to C^\infty(\mathbb{T}^n)$. Then for all $x \in \mathbb{T}^n$ and $\xi \in \mathbb{Z}^n$ we have*

$$\sigma_A(x, \xi) = e^{-i2\pi x \cdot \xi}(A e_\xi)(x) = \overline{e_\xi(x)}(A e_\xi)(x). \tag{4.5}$$

Proof. First we observe that

$$\widehat{e_\xi}(\eta) = \int_{\mathbb{T}^n} e^{-i2\pi x \cdot \eta} \, e^{i2\pi x \cdot \xi} \, \mathrm{d}x = \delta_{\xi, \eta}$$

is Kronecker's delta. We can now calculate

$$e^{-i2\pi x \cdot \xi}(A e_\xi)(x) \overset{(4.4)}{=} e^{-i2\pi x \cdot \xi} \sum_{\eta \in \mathbb{Z}^n} e^{i2\pi x \cdot \eta} \, \sigma_A(x, \eta) \, \widehat{e_\xi}(\eta) = \sigma_A(x, \xi),$$

completing the proof. $\qquad\square$

Remark 4.1.5 (Mapping $\sigma = (A \mapsto \sigma_A)$). The mapping $\sigma = (A \mapsto \sigma_A)$ can be viewed as a linear mapping: $\sigma_{sA+tB} = s\sigma_A + t\sigma_B$. However, σ is not multiplicative, as usually $\sigma_{AB} \neq \sigma_A \sigma_B$. It should be emphasized that the symbol of a linear operator is unique ($A = 0 \Leftrightarrow \sigma_A(x, \xi) = 0$), and this fact will be used every now and then.

4.1.2 Toroidal symbols

We will now proceed in the direction opposite to that of the proceeding section and discuss classes of symbols that correspond to the toroidal quantization of operators.

Definition 4.1.6 (Space $C^\infty(\mathbb{T}^n \times \mathbb{Z}^n)$). We will write $a \in C^\infty(\mathbb{T}^n \times \mathbb{Z}^n)$ when function $a(\cdot, \xi)$ is smooth on $\mathbb{T}^n$ for all $\xi \in \mathbb{Z}^n$.

Definition 4.1.7 (Toroidal symbol class $S^m_{\rho,\delta}(\mathbb{T}^n \times \mathbb{Z}^n)$). Let $m \in \mathbb{R}$, $0 \leq \delta, \rho \leq 1$. Then the *toroidal symbol class* $S^m_{\rho,\delta}(\mathbb{T}^n \times \mathbb{Z}^n)$ consists of those functions $a(x, \xi)$ which are smooth in x for all $\xi \in \mathbb{Z}^n$, and which satisfy *toroidal symbol inequalities*

$$\left| \triangle^\alpha_\xi \partial^\beta_x a(x, \xi) \right| \leq C_{a\alpha\beta m} \, \langle \xi \rangle^{m - \rho|\alpha| + \delta|\beta|} \tag{4.6}$$

for every $x \in \mathbb{T}^n$, for every $\alpha, \beta \in \mathbb{N}^n_0$, and for all $\xi \in \mathbb{Z}^n$. Here $\triangle^\alpha_\xi$ are the difference operators introduced in Definition 3.3.1. The constants $C_{a\alpha\beta m}$ are independent of $x \in \mathbb{T}^n$ and $\xi \in \mathbb{Z}^n$. The class $S^m_{1,0}(\mathbb{T}^n \times \mathbb{Z}^n)$ (the smallest of the (ρ, δ) classes) will be often denoted by writing simply $S^m(\mathbb{T}^n \times \mathbb{Z}^n)$. Furthermore, we define

$$S^{-\infty}(\mathbb{T}^n \times \mathbb{Z}^n) \quad := \quad \bigcap_{m \in \mathbb{R}} S^m(\mathbb{T}^n \times \mathbb{Z}^n),$$

$$S^\infty_{\rho,\delta}(\mathbb{T}^n \times \mathbb{Z}^n) \quad := \quad \bigcup_{m \in \mathbb{R}} S^m_{\rho,\delta}(\mathbb{T}^n \times \mathbb{Z}^n).$$

Exercise 4.1.8. Show that for any ρ and δ we have

$$\bigcap_{m \in \mathbb{R}} S^m_{\rho,\delta}(\mathbb{T}^n \times \mathbb{Z}^n) = S^{-\infty}(\mathbb{T}^n \times \mathbb{Z}^n).$$

Definition 4.1.9 (Toroidal pseudo-differential operators). If $a \in S^m_{\rho,\delta}(\mathbb{T}^n \times \mathbb{Z}^n)$, we denote by $a(X, D) = \mathrm{Op}(a)$ the corresponding *toroidal pseudo-differential operator* defined by

$$\mathrm{Op}(a)f(x) = a(X, D)f(x) := \sum_{\xi \in \mathbb{Z}^n} e^{\mathrm{i}2\pi x \cdot \xi} \, a(x, \xi) \, \widehat{f}(\xi). \tag{4.7}$$

The series (4.7) converges if, for example, $f \in C^\infty(\mathbb{T}^n)$. The set of operators $\mathrm{Op}(a)$ of the form (4.7) with $a \in S^m_{\rho,\delta}(\mathbb{T}^n \times \mathbb{Z}^n)$ will be denoted by $\mathrm{Op}(S^m_{\rho,\delta}(\mathbb{T}^n \times \mathbb{Z}^n))$, or by $\Psi^m_{\rho,\delta}(\mathbb{T}^n \times \mathbb{Z}^n)$. If an operator A satisfies $A \in \mathrm{Op}\, S^m_{\rho,\delta}(\mathbb{T}^n \times \mathbb{Z}^n)$, we will denote its *toroidal symbol* by $\sigma_A = \sigma_A(x, \xi)$, $x \in \mathbb{T}^n$, $\xi \in \mathbb{Z}^n$. Naturally, $\sigma_{a(X,D)}(x, \xi) = a(x, \xi)$. We also write

$$\mathrm{Op}(S^{-\infty}(\mathbb{T}^n \times \mathbb{Z}^n)) \quad := \quad \bigcap_{m \in \mathbb{R}} \mathrm{Op}(S^m(\mathbb{T}^n \times \mathbb{Z}^n)),$$

$$\mathrm{Op}(S^\infty_{\rho,\delta}(\mathbb{T}^n \times \mathbb{Z}^n)) \quad := \quad \bigcup_{m \in \mathbb{R}} \mathrm{Op}(S^m_{\rho,\delta}(\mathbb{T}^n \times \mathbb{Z}^n)).$$

Remark 4.1.10 (Toroidal $S^m_{\rho,\delta}(\mathbb{T}^n \times \mathbb{Z}^n)$ vs Euclidean $S^m_{\rho,\delta}(\mathbb{T}^n \times \mathbb{R}^n)$). To contrast this with Euclidean (Hörmander's) symbol classes, we write $b \in S^m_{\rho,\delta}(\mathbb{R}^n \times \mathbb{R}^n)$ if $b \in C^\infty(\mathbb{R}^n \times \mathbb{R}^n)$ and if

$$\left| \partial^\alpha_\xi \partial^\beta_x b(x, \xi) \right| \leq C_{b\alpha\beta m} \, \langle \xi \rangle^{m - \rho|\alpha| + \delta|\beta|}$$

holds for every $x \in \mathbb{R}^n$, for every $\alpha, \beta \in \mathbb{N}_0^n$, and for all $x, \xi \in \mathbb{R}^n$. If in addition $b(\cdot, \xi)$ is 1-periodic for every ξ we will write $b \in S_{\rho,\delta}^m(\mathbb{T}^n \times \mathbb{R}^n)$ and call it a *Euclidean symbol* on the torus. The corresponding (Euclidean) pseudo-differential operator on the torus is then given by

$$b(X, D)f(x) = \int_{\mathbb{R}^n} \int_{\mathbb{T}^n} e^{i2\pi(x-y)\cdot\xi} \, b(x, \xi) \, f(y) \, dy \, d\xi.$$

Remark 4.1.11 (**Topology on** $S_{\rho,\delta}^m(\mathbb{T}^n \times \mathbb{Z}^n)$). The set $S_{\rho,\delta}^m(\mathbb{T}^n \times \mathbb{Z}^n)$ of symbols has a natural topology. Let us consider the functions $p_{\alpha\beta}^m : S_{\rho,\delta}^m(\mathbb{T}^n \times \mathbb{Z}^n) \to \mathbb{R}$ ($\alpha, \beta \in \mathbb{N}_0^n$, $m \in \mathbb{R}$) defined by

$$p_{\alpha\beta}^m(\sigma) := \sup \left\{ \frac{\left| \triangle_\xi^\alpha \partial_x^\beta \sigma(x, \xi) \right|}{\langle \xi \rangle^{m - \rho|\alpha| + \delta|\beta|}} : (x, \xi) \in \mathbb{T}^n \times \mathbb{Z}^n \right\}.$$

Now $\left\{ p_{\alpha\beta}^m : \alpha, \beta \in \mathbb{N}_0^n \right\}$ is a countable family of seminorms, and they define a Fréchet topology on $S_{\rho,\delta}^m(\mathbb{T}^n \times \mathbb{Z}^n)$. Due to the bijective correspondence of $\text{Op}(S_{\rho,\delta}^m(\mathbb{T}^n \times \mathbb{Z}^n))$ and $S_{\rho,\delta}^m(\mathbb{T}^n \times \mathbb{Z}^n)$, this directly topologises the set of operators. These spaces are not normable, and the topologies have but a marginal role.

Remark 4.1.12. On $\mathbb{T}^n$, Hörmander's usual (ρ, δ) class of pseudo-differential operators $\text{Op}\, S_{\rho,\delta}^m(\mathbb{R}^n \times \mathbb{R}^n)$ of order $m \in \mathbb{R}$ which are 1-periodic in x coincides with the class $\text{Op}\, S_{\rho,\delta}^m(\mathbb{T}^n \times \mathbb{Z}^n)$, i.e.,

$$\text{Op}\, S_{\rho,\delta}^m(\mathbb{T}^n \times \mathbb{R}^n) = \text{Op}\, S_{\rho,\delta}^m(\mathbb{T}^n \times \mathbb{Z}^n),$$

see Corollary 4.6.13. This fact was originally proved in [76] by studying local coordinate charts, but the techniques of the extension of symbols and periodisation of operators developed here in addition give a precise relation between actual symbols, see, e.g., Theorem 4.5.3, and yield the relation between the corresponding operators which is then given in Theorem 4.6.12.

Proposition 4.1.13. *Let* $f \in C^\infty(\mathbb{T}^n)$. *Then* $\text{Op}(a)f$ *in* (4.7) *is well defined and* $\text{Op}(a)f \in C^\infty(\mathbb{T}^n)$. *Moreover, operator* $\text{Op}(a) : C^\infty(\mathbb{T}^n) \to C^\infty(\mathbb{T}^n)$ *is continuous.*

Proof. Since $\widehat{f} \in \mathcal{S}(\mathbb{Z}^n)$, the series in (4.7) converges absolutely and $\text{Op}(a)f \in C^\infty(\mathbb{T}^n)$. We can write

$$\text{Op}(a)f(x) = \sum_{\xi \in \mathbb{Z}^n} e^{i2\pi x \cdot \xi} \, a(x, \xi) \, \widehat{f}(\xi)$$

$$= \sum_{\xi \in \mathbb{Z}^n} \int_{\mathbb{T}^n} e^{i2\pi(x-y)\cdot\xi} \, a(x, \xi) \, f(y) \, dy$$

$$= \sum_{\xi \in \mathbb{Z}^n} \langle \xi \rangle^{-2q} \, a(x, \xi) \int_{\mathbb{T}^n} e^{i2\pi(x-y)\cdot\xi} \left(I - \frac{\mathcal{L}_y}{4\pi^2} \right)^q f(y) \, dy,$$

where $\mathcal{L}_y$ is the usual Laplacian with respect to y. Then, if we take $q \in \mathbb{Z}^+$ large enough, the series converges absolutely. Consequently, if we have the convergence $f_j \to f$ in $C^\infty(\mathbb{T}^n)$, we can pass to the limit in the series and in the integral by Lebesgue's dominated convergence theorem (Theorem C.3.22) to see that also $\mathrm{Op}(a)f_j \to \mathrm{Op}(a)f$ in $C^\infty(\mathbb{T}^n)$. $\square$

Exercise 4.1.14. How large a q should be chosen for the sum here to converge absolutely, if $\sigma \in S^m_{\rho,\delta}(\mathbb{T}^n \times \mathbb{Z}^n)$?

Remark 4.1.15 (**Difference formula for symbols**). We now mention a nice formula for differences for symbols which follows immediately from Proposition 3.3.4:

$$\triangle^\alpha_\xi \partial^\beta_x \sigma_A(x,\xi) = \sum_{\gamma \leq \alpha} (-1)^{|\alpha-\gamma|} \binom{\alpha}{\gamma} \partial^\beta_x \sigma_A(x, \xi+\gamma). \qquad (4.8)$$

This formula shows that properties of toroidal symbols automatically imply certain properties for differences. For example, decay properties of the symbol automatically imply similar decay properties for differences applied to the symbol. This is a considerable advantage over the Euclidean quantization in classes $S^m(\mathbb{T}^n \times \mathbb{R}^n)$ where such a property does not hold.

4.1.3   Toroidal amplitudes

Remark 4.1.16 (**Alternative representation**). By choosing appropriate q large enough in the proof of Proposition 4.1.13, we see that $\mathrm{Op}(a)f$ can be differentiated arbitrarily many times. Writing out the Fourier transform as an integral suggests the notation

$$\mathrm{Op}(a)f(x) =: \sum_{\xi \in \mathbb{Z}^n} \int_{\mathbb{T}^n} e^{i2\pi(x-y)\cdot\xi}\, a(x,\xi)\, f(y)\, \mathrm{d}y. \qquad (4.9)$$

It should be noted that here the right-hand side is not meant to be read as an integral operator, but rather as an operator defined by the formal integration by parts, so that we can perform operations like the exchange of summation and integral, etc. The Schwartz kernel $K(x,y)$ of (4.9) defined by the formal sum $K(x,y) = \sum_{\xi \in \mathbb{Z}^n} a(x,\xi)\, e^{i2\pi(x-y)\cdot\xi}$ may be singular only at the diagonal $x = y$; anywhere else it is smooth (see Theorem 4.3.6).

Equation (4.9), however, inspires a possible generalisation: why do we not allow function a to depend on the variable y?

Definition 4.1.17 (Toroidal amplitudes). The class $\mathcal{A}^m_{\rho,\delta}(\mathbb{T}^n)$ (or $S^m_{\rho,\delta}(\mathbb{T}^n \times \mathbb{T}^n \times \mathbb{Z}^n)$) of *toroidal amplitudes* consists of the functions $a(x,y,\xi)$ which are smooth in x and y for all $\xi \in \mathbb{Z}^n$ and which satisfy

$$|\triangle^\alpha_\xi \partial^\beta_x \partial^\gamma_y a(x,y,\xi)| \leq C_{a\alpha\beta\gamma m}\, \langle\xi\rangle^{m-\rho|\alpha|+\delta|\beta+\gamma|} \qquad (4.10)$$

for every $x, y \in \mathbb{T}^n$, for every $\alpha, \beta, \gamma \in \mathbb{N}_0^n$, and for all $\xi \in \mathbb{Z}^n$. The constants $C_{a\alpha\beta\gamma m} < \infty$ are independent of x, y, ξ. Such a function a is called a *toroidal amplitude of order* $m \in \mathbb{R}$ *of type* (ρ, δ). Formally we may also define

$$(\mathrm{Op}(a)f)(x) := \sum_{\xi \in \mathbb{Z}^n} \int_{\mathbb{T}^n} \mathrm{e}^{\mathrm{i}2\pi(x-y)\cdot\xi} \, a(x, y, \xi) \, f(y) \, \mathrm{d}y \qquad (4.11)$$

for $f \in C^\infty(\mathbb{T}^n)$. Sometimes we denote $\mathrm{Op}(a)$ by $a(X, Y, D)$.

Clearly we can regard the symbols of periodic pseudo-differential operators as a special class of amplitudes, namely the ones independent of the middle argument. The family of amplitudes of order m and type (ρ, δ) is denoted by $\mathcal{A}_{\rho,\delta}^m(\mathbb{T}^n)$. We also write $\mathcal{A}^m(\mathbb{T}^n) := \mathcal{A}_{1,0}^m(\mathbb{T}^n)$ as well as

$$\mathcal{A}^{-\infty}(\mathbb{T}^n) := \bigcap_{m \in \mathbb{R}} \mathcal{A}^m(\mathbb{T}^n) \quad \text{and} \quad \mathcal{A}_{\rho,\delta}^\infty(\mathbb{T}^n) := \bigcup_{m \in \mathbb{R}} \mathcal{A}_{\rho,\delta}^m(\mathbb{T}^n).$$

In Elias Stein's language (see [118, p.258]) amplitudes are called *compound symbols*.

Remark 4.1.18 (**Amplitude operators**). Formula (4.11) has to be interpreted as a result of a formal integration by parts, being a short-hand writing for

$$\mathrm{Op}(a)f(x) = \sum_{\xi \in \mathbb{Z}^n} \langle\xi\rangle^{-2q} \int_{\mathbb{T}^n} \mathrm{e}^{\mathrm{i}2\pi(x-y)\cdot\xi} \left(I - \frac{1}{4\pi^2}\mathcal{L}_y\right)^q (a(x, y, \xi) \, f(y)) \, \mathrm{d}y,$$

where $\mathcal{L}_y$ is the Laplacian in y. With this explanation we will take the liberty of changing the order of integration and summation, keeping in mind that it can be justified by the integration by parts. Most of the time, we shall use the less cumbersome notation of (4.11). This operator is called an *amplitude operator of degree* m. We define

$$\mathrm{Op}(\mathcal{A}^{-\infty}(\mathbb{T}^n)) := \bigcap_{m \in \mathbb{R}} \mathrm{Op}(\mathcal{A}^m(\mathbb{T}^n))$$

and

$$\mathrm{Op}(\mathcal{A}_{\rho,\delta}^\infty(\mathbb{T}^n)) := \bigcup_{m \in \mathbb{R}} \mathrm{Op}(\mathcal{A}_{\rho,\delta}^m(\mathbb{T}^n)).$$

Later on, $\mathrm{Op}(a)$ and $\mathrm{Op}(\sigma)$ for amplitudes a and symbols σ, respectively, will be extended to operators acting on the Sobolev spaces $H^s(\mathbb{T}^n)$, $s \in \mathbb{R}$. Symbols have the advantage over amplitudes in showing directly how the extension works. We have considered the amplitudes as a generalisation of symbols, but fortunately (or unfortunately?), it turns out that the family of amplitude operators coincides exactly with the family of periodic pseudo-differential operators. This is easy to believe, since on $C^\infty(\mathbb{T}^n)$ every linear operator is uniquely defined by its symbol. Of course, we do not know yet whether an amplitude operator has a symbol that satisfies inequalities (4.6). The proof of this fact is a long one, and we are going to present it in small parts. Nevertheless, the concept of amplitudes is highly justified as a tool in the symbolic analysis. Moreover, amplitudes literally manifest themselves in certain integral operators, a topic we will briefly discuss later.

In the following, symbols are considered as a subspace of amplitudes independent of y.

Definition 4.1.19 (Equivalence of amplitudes). We say that amplitudes a, a' are $m(\rho, \delta)$-equivalent ($m \in \mathbb{R}$), $a \overset{m,\rho,\delta}{\sim} a'$, if $a - a' \in \mathcal{A}_{\rho,\delta}^m(\mathbb{T}^n)$; they are asymptotically equivalent, $a \sim a'$ (or $a \overset{-\infty}{\sim} a'$ if we need additional clarity), if $a - a' \in \mathcal{A}^{-\infty}(\mathbb{T}^n)$. For the related operators we mark $\mathrm{Op}(a) \overset{m,\rho,\delta}{\sim} \mathrm{Op}(a')$ and $\mathrm{Op}(a) \sim \mathrm{Op}(a')$ (or $\mathrm{Op}(a) \overset{-\infty}{\sim} \mathrm{Op}(a')$ if we need additional clarity), respectively. It is obvious that $\overset{m(\rho,\delta)}{\sim}$ and $\sim$ are equivalence relations; the classes modulo $\sim$ are denoted by $[\cdot]$ for both amplitudes and operators.

Remark 4.1.20. From the algebraic point of view, we could handle the amplitudes, symbols, and operators modulo the equivalence relation $\sim$, because the periodic pseudo-differential operators form a $*$-algebra with $\mathrm{Op}(S^{-\infty}(\mathbb{T}^n \times \mathbb{Z}^n))$ as a subalgebra (see Section 4.7). In fact, in [135] (in a similar setting of $\mathbb{R}^n$) a "symbol" means the class $[\sigma] = \sigma + S^{-\infty}$. However, neither topological considerations nor applications support an extensive use of these equivalence classes.

4.2 Pseudo-differential operators on Sobolev spaces

In this section we show that domains of periodic pseudo-differential operators can be extended to the Sobolev spaces. This will be important for the subsequent analysis. Later, in Corollary 4.8.3 we will present conditions on the symbol that ensure that the corresponding operator is bounded between Sobolev spaces on the torus. The proof will rely on the composition formulae for pseudo-differential operators which will be proved in Theorem 4.7.10. At the moment, also to show another idea we will give a direct proof of the Sobolev boundedness for operators with symbols of type $(0, 0)$.

When studying the orders of periodic pseudo-differential operators, estimates for the differences of the Fourier coefficients of symbols and amplitudes are useful, and therefore we present an auxiliary result on this subject:

Lemma 4.2.1. *Let* $0 \leq \delta$, $0 \leq \rho$. *Assume that* $\sigma \in S_{\rho,\delta}^m(\mathbb{T}^n \times \mathbb{Z}^n)$. *Let* $\widehat{\sigma}$ *be the Fourier transform of the symbol with respect to* x, *i.e., the Fourier transform of the smooth function* $x \mapsto \sigma(x, \cdot)$. *Then for every* $\alpha \in \mathbb{N}_0^n$ *and* $r \in \mathbb{N}_0$ *estimate*

$$\left| \triangle_\xi^\alpha \widehat{\sigma}(\eta, \xi) \right| \leq c_{r,\alpha} \, \langle \eta \rangle^{-r} \langle \xi \rangle^{m - \rho|\alpha| + \delta r} \tag{4.12}$$

holds for all $\eta, \xi \in \mathbb{Z}^n$. *Respectively, let* $a \in \mathcal{A}_{\rho,\delta}^m(\mathbb{T}^n)$ *and let* $\widehat{a}$ *be the Fourier transform of* $(x, y) \mapsto a(x, y, \cdot)$. *Then for every* $\alpha \in \mathbb{N}_0^n$ *and* $q, r \in \mathbb{N}_0$ *estimate*

$$\left| \triangle_\xi^\alpha \widehat{a}(\lambda, \eta, \xi) \right| \leq c_{q,r,\alpha} \, \langle \lambda \rangle^{-q} \langle \eta \rangle^{-r} \langle \xi \rangle^{m - \rho|\alpha| + \delta(q+r)}, \tag{4.13}$$

holds for all λ, η, $\xi \in \mathbb{Z}^n$. *The constants* $c_{r,\alpha}$ *and* $c_{q,r,\alpha}$ *are independent of* λ, η *and* ξ.

Proof. Clearly

$$\left(1 - (2\pi)^{-2}\mathcal{L}_x\right)^q \, e^{-i2\pi x\cdot\eta} = \langle\eta\rangle^{2q} \, e^{-i2\pi x\cdot\eta},$$

where $\mathcal{L}_x$ is the Laplacian in x variables. Integrating by parts, we get

$$
\begin{aligned}
\left|\triangle_\xi^\alpha \widehat{\sigma}(\eta,\xi)\right| &= \left|\triangle_\xi^\alpha \int_{\mathbb{T}^n} e^{-i2\pi x\cdot\eta} \, \sigma(x,\xi) \, dx\right| \\
&= \langle\eta\rangle^{-2q} \left|\int_{\mathbb{T}^n} e^{-i2\pi x\cdot\eta} \left(1 - (2\pi)^{-2}\mathcal{L}_x\right)^q \triangle_\xi^\alpha \sigma(x,\xi) \, dx\right| \\
&\leq \langle\eta\rangle^{-2q} \int_{\mathbb{T}^n} \left|\left(1 - (2\pi)^{-2}\mathcal{L}_x\right)^q \triangle_\xi^\alpha \sigma(x,\xi)\right| dx.
\end{aligned}
$$

Since $\sigma \in S_{\rho,\delta}^m(\mathbb{T}^n \times \mathbb{Z}^n)$ estimate (4.12) follows for even $r = 2q$. By taking a geometric mean we get it also for odd r. The proof for amplitudes is similar. $\square$

Exercise 4.2.2. Prove (4.13).

We use Lemma 4.2.1 for the following:

Proposition 4.2.3 (Sobolev space boundedness for $S_{0,0}^m$). *Let $A = \mathrm{Op}\,\sigma_A$ be a pseudo-differential operator with toroidal symbol $\sigma_A \in S_{0,0}^m(\mathbb{T}^n \times \mathbb{Z}^n)$. Then the operator A extends to a bounded linear operator from $H^s(\mathbb{T}^n)$ to $H^{s-m}(\mathbb{T}^n)$ for every $s \in \mathbb{R}$.*

Proof. We have $A = \mathrm{Op}(\sigma_A)$ with $\sigma_A \in S_{0,0}^m(\mathbb{T}^n \times \mathbb{Z}^n)$, i.e.,

$$\left|\triangle_\xi^\alpha \partial_x^\beta \sigma_A(x,\xi)\right| \leq C_{\alpha\beta} \, \langle\xi\rangle^m.$$

Let $u \in C^\infty(\mathbb{T}^n)$ and let us calculate the Fourier coefficients of Au:

$$
\begin{aligned}
Au(x) &= \sum_{\xi\in\mathbb{Z}^n} e^{i2\pi x\cdot\xi} \, \sigma_A(x,\xi) \, \widehat{u}(\xi) \\
&= \sum_{\xi\in\mathbb{Z}^n} e^{i2\pi x\cdot\xi} \left[\sum_{\eta\in\mathbb{Z}^n} e^{i2\pi x\cdot\eta} \, \widehat{\sigma_A}(\eta,\xi)\right] \widehat{u}(\xi) \\
&= \sum_{\eta\in\mathbb{Z}^n} e^{i2\pi x\cdot\eta} \left[\sum_{\xi\in\mathbb{Z}^n} \widehat{\sigma_A}(\eta-\xi,\xi) \, \widehat{u}(\xi)\right],
\end{aligned}
$$

and we note that both series are absolutely convergent. Therefore, we have

$$\widehat{Au}(\eta) = \sum_{\xi\in\mathbb{Z}^n} \widehat{\sigma_A}(\eta-\xi,\xi) \, \widehat{u}(\xi).$$

Let us now estimate $\|Au\|_{H^{s-m}(\mathbb{T}^n)}$. We have

$$
\|Au\|^2_{H^{s-m}(\mathbb{T}^n)} = \sum_{\eta \in \mathbb{Z}^n} \langle \eta \rangle^{2(s-m)} |\widehat{Au}(\eta)|^2
$$

$$
= \sum_{\eta \in \mathbb{Z}^n} \left| \sum_{\xi \in \mathbb{Z}^n} \langle \eta \rangle^{s-m} \, \widehat{\sigma_A}(\eta - \xi, \xi) \, \widehat{u}(\xi) \right|^2
$$

$$
\leq \sum_{\eta \in \mathbb{Z}^n} \left[\sum_{\xi \in \mathbb{Z}^n} \langle \eta \rangle^{s-m} \, |\widehat{\sigma_A}(\eta - \xi, \xi)| \, |\widehat{u}(\xi)| \right]^2 ,
$$

which, by Peetre's inequality (Proposition 3.3.31), can be estimated by

$$
2^{2|s-m|} \sum_{\eta \in \mathbb{Z}^n} \left[\sum_{\xi \in \mathbb{Z}^n} \langle \eta - \xi \rangle^{|s-m|} \, \langle \xi \rangle^{-m} |\widehat{\sigma_A}(\eta - \xi, \xi)| \langle \xi \rangle^s |\widehat{u}(\xi)| \right]^2 .
$$

Consequently, by the discrete Young's inequality (Lemma 3.3.30), this can be estimated by

$$
2^{2|s-m|} \left[\sup_{\eta \in \mathbb{Z}^n} \sum_{\xi \in \mathbb{Z}^n} \langle \eta - \xi \rangle^{|s-m|} \, \langle \xi \rangle^{-m} |\widehat{\sigma_A}(\eta - \xi, \xi)| \right]
$$

$$
\times \left[\sup_{\xi \in \mathbb{Z}^n} \sum_{\eta \in \mathbb{Z}^n} \langle \eta - \xi \rangle^{|s-m|} \, \langle \xi \rangle^{-m} |\widehat{\sigma_A}(\eta - \xi, \xi)| \right] \left[\sum_{\xi \in \mathbb{Z}^n} \langle \xi \rangle^{2s} |\widehat{u}(\xi)|^2 \right] .
$$

Applying (4.12) with $\alpha = 0$ and $\rho = \delta = 0$ we obtain

$$
\|Au\|^2_{H^{s-m}(\mathbb{T}^n)} \leq 2^{2|s-m|} \left[\sup_{\eta \in \mathbb{Z}^n} \sum_{\xi \in \mathbb{Z}^n} \langle \eta - \xi \rangle^{|s-m|-r} \, C_r \right]
$$

$$
\times \left[\sup_{\xi \in \mathbb{Z}^n} \sum_{\eta \in \mathbb{Z}^n} \langle \eta - \xi \rangle^{|s-m|-r} \, C_r \right] \|u\|^2_{H^s(\mathbb{T}^n)}
$$

$$
= 2^{2|s-m|} \left[C_r \sum_{\eta \in \mathbb{Z}^n} \langle \eta \rangle^{|s-m|-r} \right]^2 \|u\|^2_{H^s(\mathbb{T}^n)} .
$$

If we take r large enough we obtain $\|Au\|_{H^{s-m}(\mathbb{T}^n)} \leq C\|u\|^2_{H^s(\mathbb{T}^n)}$. The desired extension on $H^s(\mathbb{T}^n)$ is then obtained by Theorem 3.5.2 (or by a more direct explanation of the type given in Proposition 2.4.1 on how estimate (2.9) implies the boundedness). $\qquad\square$

In particular, we see from the proof that it is enough to take $r \in \mathbb{N}$ such that $r > |m - s| + n$ for the proof to work. Combining this with Remark 4.2.9 below on the symbol class $S_{0,0}^m(\mathbb{T}^n \times \mathbb{Z}^n)$ we conclude:

Corollary 4.2.4 (How many derivatives are needed?). *Let $m, s \in \mathbb{R}$ and let an integer $r \in \mathbb{N}$ be such that $r > |s - m| + n$. Let $a \in C^\infty(\mathbb{T}^n \times \mathbb{Z}^n)$ satisfy estimates*

$$\left| \partial_x^\beta a(x, \xi) \right| \le C_{a\beta m} \langle \xi \rangle^m$$

for all multi-indices $|\beta| \le r$, all $x \in \mathbb{T}^n$ and all $\xi \in \mathbb{Z}^n$, with constants $C_{a\beta m}$ independent of x and ξ. Then the operator $\mathrm{Op}(a)$ extends to a bounded linear operator from $H^s(\mathbb{T}^n)$ to $H^{s-m}(\mathbb{T}^n)$.

We note a useful formula from the proof that does not require that the symbol of A is in a symbol class.

Corollary 4.2.5. *Let A be a linear continuous operator from $C^\infty(\mathbb{T}^n)$ to $C^\infty(\mathbb{T}^n)$ and let $u \in C^\infty(\mathbb{T}^n)$. Then*

$$\widehat{Au}(\xi) = \sum_{\eta \in \mathbb{Z}^n} \widehat{\sigma_A}(\xi - \eta, \eta) \, \widehat{u}(\eta),$$

where $\xi \in \mathbb{Z}^n$, σ_A is the toroidal symbol of A, and $\widehat{\sigma_A}$ its Fourier transform in the space variable.

Exercise 4.2.6. In the proof of Theorem 4.2.3, how are the constants C_r related to the symbol inequalities for σ_A?

Exercise 4.2.7. Explain how the discrete Young's inequality (Lemma 3.3.30) was exploited in the proof of Theorem 4.2.3.

Exercise 4.2.8. Suppose $\sigma_A \in S_{\rho,\delta}^m(\mathbb{T}^n \times \mathbb{Z}^n)$ does not depend on the x-variable, i.e., $\sigma_A(x, \xi) = \sigma(\xi)$. Find the norm $\|A\|_{\mathcal{L}(H^s(\mathbb{T}^n), H^{s-m}(\mathbb{T}^n))}$.

Remark 4.2.9 (**Characterisation of classes $S_{0,0}^m(\mathbb{T}^n \times \mathbb{Z}^n)$**). In view of Remark 4.1.15 we see that in the case of $S_{0,0}^m(\mathbb{T}^n \times \mathbb{Z}^n)$ we can drop difference conditions from the definition of this class: we have $a \in S_{0,0}^m(\mathbb{T}^n \times \mathbb{Z}^n)$ if and only if $a \in C^\infty(\mathbb{T}^n \times \mathbb{Z}^n)$ satisfies

$$\left| \partial_x^\beta a(x, \xi) \right| \le C_{a\beta m} \langle \xi \rangle^m$$

for all multi-indices β, all $x \in \mathbb{T}^n$ and all $\xi \in \mathbb{Z}^n$, with constants $C_{a\beta m}$ independent of x and ξ. From this point of view Proposition 4.2.3 will be equivalent to Corollary 4.8.3 if we use the argument in the proof of Theorem 2.6.11 to change the orders of Sobolev spaces. However, the proof of Corollary 4.8.3 is different: it will follow by the toroidal calculus from Theorem 4.8.1 which gives an estimate on the number of x-derivatives that ensures the boundedness of operators on $L^2(\mathbb{T}^n)$. In any case, Corollary 4.2.4 provides a result under weaker assumptions.

An analogous proof can be carried out for amplitudes. Let us denote the Fourier transform of an amplitude a with respect to the first (or second) variable by $\widehat{a}_1$ (or $\widehat{a}_2$), respectively. The notation $\widehat{a}$ is then reserved for the Fourier transform with respect to both the first and the second variables.

Theorem 4.2.10. *Let* $a \in \mathcal{A}^m_{0,0}(\mathbb{T}^n)$. *Then* $\mathrm{Op}(a)$ *extends to a bounded linear operator from* $H^s(\mathbb{T}^n)$ *to* $H^{s-m}(\mathbb{T}^n)$, *for any* $s \in \mathbb{R}$.

Proof. Let $A = \mathrm{Op}(a)$, $a \in \mathcal{A}^m_{0,0}(\mathbb{T}^n)$, and $u \in C^\infty(\mathbb{T}^n)$. First, we want to find the Fourier coefficients of Au. For this we have

$$
\begin{aligned}
Au(x) &= \sum_{\xi \in \mathbb{Z}^n} \int_{\mathbb{T}^n} e^{i2\pi(x-y)\cdot\xi} \, a(x,y,\xi) \, u(y) \, dy \\
&= \sum_{\xi \in \mathbb{Z}^n} \int_{\mathbb{T}^n} e^{i2\pi(x-y)\cdot\xi} \\
&\quad \times \left[\sum_{\eta \in \mathbb{Z}^n} e^{i2\pi x\cdot(\eta-\xi)} \, \widehat{a}_1(\eta-\xi,y,\xi) \right] \sum_{\kappa \in \mathbb{Z}^n} e^{i2\pi y\cdot\kappa} \, \widehat{u}(\kappa) \, dy \\
&= \sum_{\eta \in \mathbb{Z}^n} e^{i2\pi x\cdot\eta} \sum_{\kappa \in \mathbb{Z}^n} \sum_{\xi \in \mathbb{Z}^n} \widehat{u}(\kappa) \int_{\mathbb{T}^n} \widehat{a}_1(\eta-\xi,y,\xi) \, e^{-i2\pi y\cdot(\xi-\kappa)} \, dy \\
&= \sum_{\eta \in \mathbb{Z}^n} e^{i2\pi x\cdot\eta} \sum_{\kappa \in \mathbb{Z}^n} \sum_{\xi \in \mathbb{Z}^n} \widehat{u}(\kappa) \, \widehat{a}(\eta-\xi,\xi-\kappa,\xi),
\end{aligned}
$$

from which we obtain

$$
\widehat{Au}(\eta) = \sum_{\kappa \in \mathbb{Z}^n} \sum_{\xi \in \mathbb{Z}^n} \widehat{u}(\kappa) \, \widehat{a}(\eta-\xi,\xi-\kappa,\xi).
$$

Then we estimate

$$
\begin{aligned}
\|Au\|^2_{H^{s-m}(\mathbb{T}^n)} &= \sum_{\eta \in \mathbb{Z}^n} \langle\eta\rangle^{2(s-m)} \, |\widehat{Au}(\eta)|^2 \\
&= \sum_{\eta \in \mathbb{Z}^n} \left| \sum_{\xi \in \mathbb{Z}^n} \sum_{\kappa \in \mathbb{Z}^n} \langle\eta\rangle^{s-m} \, \widehat{a}(\eta-\xi,\xi-\kappa,\xi) \, \widehat{u}(\kappa) \right|^2 \\
&\leq \sum_{\eta \in \mathbb{Z}^n} \left[\sum_{\xi \in \mathbb{Z}^n} \sum_{\kappa \in \mathbb{Z}^n} \langle\eta\rangle^{s-m} \, |\widehat{a}(\eta-\xi,\xi-\kappa,\xi)| \, |\widehat{u}(\kappa)| \right]^2. \quad (4.14)
\end{aligned}
$$

Now, using Peetre's inequality (Proposition 3.3.31) twice, we can estimate

$$
\begin{aligned}
\langle\eta\rangle^{s-m} &\leq 2^{2|s-m|} \langle\eta-\xi\rangle^{|s-m|} \langle\xi\rangle^{s-m} \\
&\leq 2^{4|s-m|} \langle\eta-\xi\rangle^{|s-m|} \langle\xi-\kappa\rangle^{|s|} \langle\kappa\rangle^s \langle\xi\rangle^{-m}.
\end{aligned}
$$

Consequently, by (4.13) with $\rho = \delta = 0$, we have

$$\langle \eta \rangle^{s-m} \, |\widehat{a}(\eta - \xi, \xi - \kappa, \xi)| \leq 2^{4|s-m|} \, c_{q,r,0}^2 \langle \eta - \xi \rangle^{|s-m|-q} \, \langle \xi - \kappa \rangle^{|s|-r} \, \langle \kappa \rangle^s.$$

Plugging this estimate into (4.14) we get

$$\|Au\|_{H^{s-m}(\mathbb{T}^n)}^2 \leq C \sum_{\eta \in \mathbb{Z}^n} \left[\sum_{\xi \in \mathbb{Z}^n} \sum_{\kappa \in \mathbb{Z}^n} \langle \eta - \xi \rangle^{|s-m|-q} \, \langle \xi - \kappa \rangle^{|s-m|-r} \, \langle \kappa \rangle^s \, |\widehat{u}(\kappa)| \right]^2$$

Now, using the discrete Young's inequality (Lemma 3.3.30) twice this can be estimated as

$$\leq C \left[\sup_{\xi \in \mathbb{Z}^n} \sum_{\eta \in \mathbb{Z}^n} \langle \eta - \xi \rangle^{|s-m|-q} \right] \sum_{\xi \in \mathbb{Z}^n} \left[\sum_{\kappa \in \mathbb{Z}^n} \langle \xi - \kappa \rangle^{|s|-r} \, \langle \kappa \rangle^s \, |\widehat{u}(\kappa)| \right]^2$$

$$\leq C \left[\sup_{\xi \in \mathbb{Z}^n} \sum_{\eta \in \mathbb{Z}^n} \langle \eta - \xi \rangle^{|s-m|-q} \right] \left[\sup_{\kappa \in \mathbb{Z}^n} \sum_{\xi \in \mathbb{Z}^n} \langle \xi - \kappa \rangle^{|s|-r} \right]$$

$$\times \left[\sum_{\kappa \in \mathbb{Z}^n} \langle \kappa \rangle^{2s} \, |\widehat{u}(\kappa)|^2 \right]$$

$$= C \left[\sum_{\eta \in \mathbb{Z}^n} \langle \eta \rangle^{|s-m|-q} \right] \left[\sum_{\xi \in \mathbb{Z}^n} \langle \xi \rangle^{|s|-r} \right] \|u\|_{H^s(\mathbb{T}^n)}^2.$$

Thus, if q and r are large enough we obtain $\|Au\|_{H^{s-m}(\mathbb{T}^n)} \leq C\|u\|_{H^s(\mathbb{T}^n)}^2$. Again, A can be extended to $H^s(\mathbb{T}^n)$ by Theorem 3.5.2. $\square$

Exercise 4.2.11. Try to do Exercises 4.2.7 and 4.2.8 in the context of amplitude operators.

4.3 Kernels of periodic pseudo-differential operators

We start by describing $\Psi^{-\infty}(\mathbb{T}^n \times \mathbb{Z}^n) = \mathrm{Op}(S^{-\infty}(\mathbb{T}^n \times \mathbb{Z}^n))$-operators conclusively, showing that there is every right to call them the infinitely smoothing periodic pseudo-differential operators. Recall that by Theorem 3.5.1 a linear mapping

$$A \in L(H^s(\mathbb{T}^n), H^{t_1}(\mathbb{T}^n)) \cap \mathcal{L}(H^s(\mathbb{T}^n), H^{t_2}(\mathbb{T}^n))$$

belongs to $\mathcal{L}(H^s(\mathbb{T}^n), H^{t_1}(\mathbb{T}^n))$, even when $t_1 > t_2$.

Theorem 4.3.1 (Smoothing). *The following conditions are equivalent:*

(i) $A \in \mathcal{L}(H^s(\mathbb{T}^n), H^t(\mathbb{T}^n))$ *for every* $s, t \in \mathbb{R}$.

(ii) $\sigma_A \in S^{-\infty}(\mathbb{T}^n \times \mathbb{Z}^n)$.

(iii) *There exists* $K_A \in C^{\infty}(\mathbb{T}^n \times \mathbb{T}^n)$ *such that for all* $u \in C^{\infty}(\mathbb{T}^n)$ *we have*

$$Au(x) = \int_{\mathbb{T}^n} K_A(x, y)\, u(y)\, \mathrm{d}y.$$

Proof. Assume that A satisfies (i). To obtain (ii), it is enough to prove

$$|\partial_x^{\beta} \sigma_A(x, \xi)| \leq c_{\beta, r} \langle \xi \rangle^{-r} \quad \text{for every} \quad r \in \mathbb{R},$$

because by Proposition 3.3.4 we have formula (4.8) which we recall here:

$$\triangle_{\xi}^{\alpha} \partial_x^{\beta} \sigma_A(x, \xi) = \sum_{\gamma \leq \alpha} (-1)^{|\alpha - \gamma|} \binom{\alpha}{\gamma} \partial_x^{\beta} \sigma_A(x, \xi + \gamma); \qquad (4.15)$$

reasoning why this is enough is left as Exercise 4.3.2. Recall that $e_{\xi}(x) = \mathrm{e}^{\mathrm{i}2\pi x \cdot \xi}$ so that $\|e_{\xi}\|_{H^s(\mathbb{T}^n)} = \langle \xi \rangle^s$. We now prepare another estimate:

$$
\begin{aligned}
\|e_{-\xi} f\|_{H^{|\beta|+t}(\mathbb{T}^n)}^2 &= \sum_{\eta \in \mathbb{Z}^n} \langle \eta \rangle^{2|\beta|+2t} |\widehat{e_{-\xi} f}(\eta)|^2 \\
&\leq 2^{2|\beta|+2|t|} \sum_{\eta \in \mathbb{Z}^n} \langle \eta + \xi \rangle^{2|\beta|+2|t|} \langle \xi \rangle^{2|\beta|+2t} |\widehat{f}(\eta + \xi)|^2 \\
&= 2^{2|\beta|+2|t|} \langle \xi \rangle^{2|\beta|+2t} \|f\|_{H^{|\beta|+t}(\mathbb{T}^n)}^2,
\end{aligned}
$$

where we applied Peetre's inequality (Proposition 3.3.31). Finally, choosing $t > n/2$ and using the Sobolev embedding theorem (see, e.g., Exercise 2.6.17), we get

$$
\begin{aligned}
|\partial_x^{\beta} \sigma_A(x, \xi)| &\leq \sum_{\eta \in \mathbb{Z}^n} (2\pi)^{|\beta|} \langle \eta \rangle^{|\beta|} \, |\widehat{\sigma_A}(\eta, \xi)| \\
&\leq C_{\beta, t} \, \|x \mapsto \sigma_A(x, \xi)\|_{H^{|\beta|+t}(\mathbb{T}^n)} \\
&= C_{\beta, t} \, \|e_{-\xi} A e_{\xi}\|_{H^{|\beta|+t}(\mathbb{T}^n)} \\
&\leq C_{\beta, t} \, \|e_{-\xi} I\|_{\mathcal{L}(H^{|\beta|+t}(\mathbb{T}^n), H^{|\beta|+t}(\mathbb{T}^n))} \\
&\qquad \times \|A\|_{\mathcal{L}(H^s(\mathbb{T}^n), H^{|\beta|+t}(\mathbb{T}^n))} \, \|e_{\xi}\|_{H^s(\mathbb{T}^n)} \\
&\leq 2^{|\beta|+t} C_{\beta, t} \, \|A\|_{\mathcal{L}(H^s(\mathbb{T}^n), H^{|\beta|+t}(\mathbb{T}^n))} \, \langle \xi \rangle^{s+|\beta|+t}
\end{aligned}
$$

where $C_{\beta, t} = (2\pi)^{|\beta|} \left[\sum_{\eta \in \mathbb{Z}^n} \langle \eta \rangle^{-2t} \right]^{1/2}$. Since $s \in \mathbb{R}$ is arbitrary, we get (ii).

Let us now show that (ii) implies (iii). If $\sigma_A \in S^{-\infty}(\mathbb{T}^n \times \mathbb{Z}^n)$, the Schwartz kernel

$$K_A(x, y) := \sum_{\xi \in \mathbb{Z}^n} \sigma_A(x, \xi) \, \mathrm{e}^{\mathrm{i}2\pi(x - y) \cdot \xi}$$

is in $C^\infty(\mathbb{T}^n \times \mathbb{T}^n)$. Indeed, formally we can differentiate K_A to obtain

$$\partial_x^\alpha \partial_y^\beta K_A(x,y) = \sum_{\xi \in \mathbb{Z}^n} (-\mathrm{i}2\pi\xi)^\beta \sum_{\gamma \leq \alpha} \binom{\alpha}{\gamma} \left[\partial_x^\gamma \sigma_A(x,\xi)\right] \partial_x^{\alpha-\gamma} \mathrm{e}^{\mathrm{i}2\pi(x-y)\cdot\xi};$$

this is justified, as the convergence of the resulting series is absolute, because $|\partial_x^\gamma \sigma_A(x,\xi)| \leq c_{\gamma,r} \langle\xi\rangle^{-r}$ for any $r \in \mathbb{R}$. This gives (**iii**).

Finally, assume that (**iii**) holds. Define the amplitude a by $a(x,y,\xi) := \delta_{0,\xi} K_A(x,y)$. Now $a \in \mathcal{A}^{-\infty}(\mathbb{T}^n)$, since

$$\begin{aligned}
\left|\partial_x^\alpha \partial_y^\beta \triangle_\xi^\gamma a(x,y,\xi)\right| &\leq 2^{|\gamma|} \left|\partial_x^\alpha \partial_y^\beta K_A(x,y)\right| \chi_{[-|\gamma|,|\gamma|]^n} \\
&\leq C_{r\alpha\beta\gamma} \langle\xi\rangle^{-r}
\end{aligned}$$

for every $r \in \mathbb{R}$, where $\chi_{[-|\gamma|,|\gamma|]^n}$ is the characteristic function of the cube $[-|\gamma|,|\gamma|]^n \subset \mathbb{Z}^n$. On the other hand,

$$\begin{aligned}
\mathrm{Op}(a)u(x) &= \sum_{\xi \in \mathbb{Z}^n} \int_{\mathbb{T}^n} \mathrm{e}^{\mathrm{i}2\pi(x-y)\cdot\xi} \, a(x,y,\xi) \, u(y) \, \mathrm{d}y \\
&= \int_{\mathbb{T}^n} K_A(x,y) \, u(y) \, \mathrm{d}y = Au(x).
\end{aligned}$$

Property (**i**) now follows by Theorem 4.2.10. $\square$

Exercise 4.3.2. In the proof above, based on (4.15), explain why it sufficed to prove $|\partial_x^\beta \sigma_A(x,\xi)| \leq c_{\beta,r} \langle\xi\rangle^{-r}$.

Because the inclusion of a Sobolev space into a strictly larger one is compact (see Exercise 3.2.10), we also obtain

Corollary 4.3.3. *Operators from* $\mathrm{Op}(S^{-\infty}(\mathbb{T}^n \times \mathbb{Z}^n))$ *are compact between any spaces* $H^s(\mathbb{T}^n), H^t(\mathbb{T}^n)$.

Unlike in the case of symbols, the correspondence of amplitudes and amplitude operators is not bijective: several different amplitudes may define the same operator. As an example we are now going to study how the multiplication of an amplitude by

$$\left(\mathrm{e}^{\mathrm{i}2\pi(y-x)} - 1\right)^\gamma := \prod_{j=1}^n \left(\mathrm{e}^{\mathrm{i}2\pi(y_j-x_j)} - 1\right)^{\gamma_j} \tag{4.16}$$

affects the amplitude operator. Notice that this multiplier was encountered in the biperiodic Taylor series (see Corollary 3.4.3 and Theorem 3.4.4).

Lemma 4.3.4. *Let* $c \in \mathcal{A}_{\rho,\delta}^m(\mathbb{T}^n)$, *and define*

$$b_\gamma(x,y,\xi) := \left(\mathrm{e}^{\mathrm{i}2\pi(y-x)} - 1\right)^\gamma c(x,y,\xi),$$

where $\gamma \in \mathbb{N}_0^n$. *Then* $\mathrm{Op}(b_\gamma) = \mathrm{Op}(\triangle_\xi^\gamma c) \in \mathrm{Op}\left(\mathcal{A}_{\rho,\delta}^{m-\rho|\gamma|}(\mathbb{T}^n)\right)$.

Proof. First we note the identity

$$e^{i2\pi(x-y)\cdot\xi}\left(e^{i2\pi(y-x)}-1\right)^\gamma = (-1)^{|\gamma|}\overline{\triangle}_\xi^\gamma e^{i2\pi(x-y)\cdot\xi},$$

where $\overline{\triangle}_\xi^\gamma$ is the backward difference operator (see Definition 3.3.1) which we leave as Exercise 4.3.5. Consequently, the summation by parts (see Lemma 3.3.10) yields

$$
\begin{aligned}
\mathrm{Op}(b_\gamma)u(x) &= \sum_{\xi\in\mathbb{Z}^n}\int_{\mathbb{T}^n} e^{i2\pi(x-y)\cdot\xi}\left[\left(e^{i2\pi(y-x)}-1\right)^\gamma c(x,y,\xi)\right]u(y)\,dy \\
&= \int_{\mathbb{T}^n}\left[\sum_{\xi\in\mathbb{Z}^n} c(x,y,\xi)\,(-1)^{|\gamma|}\,\overline{\triangle}_\xi^\gamma e^{i2\pi(x-y)\cdot\xi}\right]u(y)\,dy \\
&= \int_{\mathbb{T}^n}\left[\sum_{\xi\in\mathbb{Z}^n} e^{i2\pi(x-y)\cdot\xi}\,\triangle_\xi^\gamma c(x,y,\xi)\right]u(y)\,dy.
\end{aligned}
$$

Thus $\mathrm{Op}(b_\gamma) = \mathrm{Op}(\triangle_\xi^\gamma c)$, and clearly $\triangle_\xi^\gamma c \in \mathcal{A}_{\rho,\delta}^{m-\rho|\gamma|}(\mathbb{T}^n)$, since $c \in \mathcal{A}_{\rho,\delta}^m(\mathbb{T}^n)$. $\square$

Exercise 4.3.5. Prove that for every $\gamma \in \mathbb{N}_0^n$ we have the identity

$$e^{i2\pi(x-y)\cdot\xi}\left(e^{i2\pi(y-x)}-1\right)^\gamma = (-1)^{|\gamma|}\overline{\triangle}_\xi^\gamma e^{i2\pi(x-y)\cdot\xi},$$

where $\overline{\triangle}_\xi^\gamma$ is the backward difference operator from Definition 3.3.1.

Surprising or not, but from the smoothness point of view the essential information content of a periodic pseudo-differential operator is in the behavior of its Schwartz kernel in any neighbourhood of the diagonal $x = y$. We note that this can be also seen from the local theory once we know the equality of operator classes $\mathrm{Op}(\mathcal{A}_{\rho,\delta}^m(\mathbb{T}^n))$ and periodic operators in $\mathrm{Op}(\mathcal{A}_{\rho,\delta}^m(\mathbb{R}^n))$. But here we give a direct proof:

Theorem 4.3.6 (Schwartz kernel). *Let $0 < \rho$ and $\delta < 1$. Let $A = \mathrm{Op}(a) \in \mathrm{Op}(\mathcal{A}_{\rho,\delta}^m(\mathbb{T}^n))$ be expressed in the form*

$$Au(x) = \int_{\mathbb{T}^n} K_A(x,y)\,u(y)\,dy,$$

where $K_A(x,y) = \sum_{\xi\in\mathbb{Z}^n} e^{i2\pi(x-y)\cdot\xi}\,a(x,y,\xi)$. Then the Schwartz kernel K_A is a smooth function outside the diagonal $x = y$.

Proof. Let $j \in \{1,\ldots,n\}$. Take $\psi \in C^\infty(\mathbb{T}^n \times \mathbb{T}^n)$ such that $x_j \neq y_j$ for every $(x,y) \in \mathrm{supp}(\psi)$. We have to prove that $(x,y) \mapsto \psi(x,y)\,K_A(x,y)$ belongs to $C^\infty(\mathbb{T}^n \times \mathbb{T}^n)$. Define

$$b(x,y,\xi) := \psi(x,y)\,a(x,y,\xi).$$

By Lemma 4.3.4, the amplitudes

$$(x, y, \xi) \;\; \mapsto \;\; b(x, y, \xi) \quad \text{and}$$

$$(x, y, \xi) \;\; \mapsto \;\; \frac{\triangle_{\xi_j}^k b(x, y, \xi)}{\left(e^{i2\pi(y_j - x_j)} - 1 \right)^k}$$

give the same periodic pseudo-differential operator $B := \mathrm{Op}(b)$.

Hence $b \in \mathcal{A}_{\rho,\delta}^{m-\rho k}(\mathbb{T}^n)$ for every $k \in \mathbb{N}_0^n$, so that it is in $\mathcal{A}^{-\infty}(\mathbb{T}^n)$. Theorem 4.2.10 states that B is continuous between any Sobolev spaces, and then by Theorem 4.3.1 the kernel $(x, y) \mapsto \psi(x, y) \, K_A(x, y)$ belongs to $C^\infty(\mathbb{T}^n \times \mathbb{T}^n)$. $\qquad \square$

Exercise 4.3.7. Derive the quantitative behavior of the kernel $K_A(x, y)$ near the diagonal $x = y$, similarly to Theorem 2.3.1.

4.4 Asymptotic sums and amplitude operators

The next theorem is a prelude to asymptotic expansions, which are the main tool in the symbolic analysis of periodic pseudo-differential operators.

Theorem 4.4.1 (Asymptotic sums of symbols). *Let $(m_j)_{j=0}^\infty \subset \mathbb{R}$ be a sequence such that $m_j > m_{j+1} \xrightarrow[j \to \infty]{} -\infty$, and $\sigma_j \in S_{\rho,\delta}^{m_j}(\mathbb{T}^n \times \mathbb{Z}^n)$ for all $j \in \mathbb{N}_0$. Then there exists a toroidal symbol $\sigma \in S_{\rho,\delta}^{m_0}(\mathbb{T}^n \times \mathbb{Z}^n)$ such that for all $N \in \mathbb{N}_0$,*

$$\sigma \overset{m_N, \rho, \delta}{\sim} \sum_{j=0}^{N-1} \sigma_j.$$

Proof. Choose a function $\varphi \in C^\infty(\mathbb{R}^n)$ satisfying $\|\xi\| \geq 1 \Rightarrow \varphi(\xi) = 1$ and $\|\xi\| \leq 1/2 \Rightarrow \varphi(\xi) = 0$; otherwise φ can be arbitrary. Take a sequence $(\varepsilon_j)_{j=0}^\infty$ of positive real numbers such that $\varepsilon_j > \varepsilon_{j+1} \to 0$ ($j \in \mathbb{N}_0$), and define $\varphi_j \in C^\infty(\mathbb{R}^n)$ by $\varphi_j(\xi) := \varphi(\varepsilon_j \xi)$. When $|\alpha| \geq 1$, the support set of $\triangle_\xi^\alpha \varphi_j$ is bounded, so that by the discrete Leibniz formula (Lemma 3.3.6) we have

$$\left| \triangle_\xi^\alpha \partial_x^\beta \left[\varphi_j(\xi) \sigma_j(x, \xi) \right] \right| \leq C_{j\alpha\beta} \, \langle \xi \rangle^{m_j - \rho|\alpha| + \delta|\beta|}$$

for some constant $C_{j\alpha\beta}$, since $\sigma_j \in S_{\rho,\delta}^{m_j}(\mathbb{T}^n \times \mathbb{Z}^n)$. This means that $((x, \xi) \mapsto \varphi_j(\xi)\sigma_j(x, \xi)) \in S_{\rho,\delta}^{m_j}(\mathbb{T}^n \times \mathbb{Z}^n)$. Examining the support of $\triangle_\xi^\alpha \varphi_j$, we see that $\triangle_\xi^\alpha \left(\varphi_j(\xi) \sigma_j(x, \xi) \right)$ (where $\alpha \in \mathbb{N}_0^n$) vanishes for any fixed $\xi \in \mathbb{Z}^n$, when j is large enough. This justifies the definition

$$\sigma(x, \xi) := \sum_{j=0}^\infty \varphi_j(\xi) \, \sigma_j(x, \xi), \tag{4.17}$$

and clearly $\sigma \in S^{m_0}_{\rho,\delta}(\mathbb{T}^n)$. Furthermore,

$$
\left| \triangle^\alpha_\xi \partial^\beta_x \left[\sigma(x,\xi) - \sum^{N-1}_{j=0} \sigma_j(x,\xi) \right] \right|
$$

$$
\leq \sum^{N-1}_{j=0} \left| \triangle^\alpha_\xi \partial^\beta_x \{ [\varphi_j(\xi) - 1] \; \sigma_j(x,\xi) \} \right| + \sum^\infty_{j=N} \left| \triangle^\alpha_\xi \partial^\beta_x [\varphi_j(\xi) \; \sigma_j(x,\xi)] \right|.
$$

Recall that $\varepsilon_j > \varepsilon_{j+1} \to 0$, so that the $\sum^{N-1}_{j=0}$ part of the sum vanishes, whenever $\|\xi\|$ is large. Hence this part of the sum is dominated by $C_{rN\alpha\beta} \langle\xi\rangle^{-r}$ for any $r \in \mathbb{R}$. The reader may verify that the $\sum^\infty_{j=N}$ part of the sum is majorised by $C'_{N\alpha\beta} \langle\xi\rangle^{m_N - \rho|\alpha| + \delta|\beta|}$. $\square$

Exercise 4.4.2. In the proof of Theorem 4.4.1. estimate the support of $\xi \mapsto \triangle^\alpha_\xi \varphi_j(\xi)$ in terms of α and j. How large should $\|\xi\|$ be for the $\sum^{N-1}_{j=0}$ part of the sum to vanish? Complete the proof by filling in the details. If necessary, consult the Euclidean version of this result which was proved in Proposition 2.5.33.

Definition 4.4.3 (Asymptotic expansions). The formal series $\sum^\infty_{j=0} \sigma_j$ in Theorem 4.4.1 is called an *asymptotic expansion* of the symbol $\sigma \in S^{m_0}_{\rho,\delta}(\mathbb{T}^n \times \mathbb{Z}^n)$ and it is presented in (4.17). In this case we denote

$$
\sigma \sim \sum^\infty_{j=0} \sigma_j
$$

(cf. $a \sim a'$; a different but related meaning). Respectively, $\sum^\infty_{j=0} \mathrm{Op}(\sigma_j)$ is an asymptotic expansion of the operator $\mathrm{Op}(\sigma) \in \mathrm{Op}(S^{m_0}_{\rho,\delta}(\mathbb{T}^n \times \mathbb{Z}^n))$, denoted $\mathrm{Op}(\sigma) \sim \sum^\infty_{j=0} \mathrm{Op}(\sigma_j)$. By altering $\varphi \in C^\infty(\mathbb{R}^n)$ and $(\varepsilon_j)^\infty_{j=0}$ in the proof of Theorem 4.4.1 we get a (possibly) different symbol τ by (4.17). Nevertheless, $\sigma \sim \tau$, which is enough in the symbol analysis of periodic pseudo-differential operators. We are often faced with asymptotic expansions $\sigma \sim \sum^\infty_{j=0} \sigma_j$, where

$$
\sigma_j = \sum_{\gamma \in \mathbb{N}^n_0: \; |\gamma|=j} \sigma_\gamma.
$$

In such case we shall write

$$
\sigma \sim \sum_{\gamma \geq 0} \sigma_\gamma.
$$

Remark 4.4.4 **(Principal symbol).** Assume that in the asymptotic expansion $\sigma \sim \sum^\infty_{j=0} \sigma_j$ we have $\sigma_0 \in S^{m_0}(\mathbb{T}^n \times \mathbb{Z}^n) \setminus S^{m_1}(\mathbb{T}^n \times \mathbb{Z}^n)$, i.e., σ_0 is the most significant term. It is then called the *principal symbol* of $\sigma \sim \sum^\infty_{j=0} \sigma_j$. (In [130, p. 49] the class $\sigma_A + S^{m-1}(\mathbb{R}^n)$ is by definition the principal symbol of the periodic pseudo-differential operator $A \in \mathrm{Op}(S^m(\mathbb{R}^n))$ when $l < m$ implies that $\sigma_A \notin S^l$; it is important due to its invariance under smooth changes of coordinates.)

Next we present an elementary result stating that amplitude operators are merely periodic pseudo-differential operators, and we provide an effective way to calculate the symbol modulo $S^{-\infty}(\mathbb{T}^n \times \mathbb{Z}^n)$ from an amplitude: this theorem has a fundamental status in the symbolic analysis. We give two alternative proofs for it.

Theorem 4.4.5 (Symbols of amplitude operators). *Let $0 \le \delta < \rho \le 1$. For every toroidal amplitude $a \in \mathcal{A}^m_{\rho,\delta}(\mathbb{T}^n)$ there exists a unique toroidal symbol $\sigma \in S^m_{\rho,\delta}(\mathbb{T}^n \times \mathbb{Z}^n)$ satisfying $\mathrm{Op}(a) = \mathrm{Op}(\sigma)$, and σ has the following asymptotic expansion:*

$$\sigma(x,\xi) \sim \sum_{\gamma \ge 0} \frac{1}{\gamma!} \, \triangle^\gamma_\xi D^{(\gamma)}_y a(x,y,\xi)|_{y=x}. \tag{4.18}$$

Proof. As a linear operator in Sobolev spaces, $\mathrm{Op}(a)$ possesses the unique symbol $\sigma = \sigma_{\mathrm{Op}(a)}$ (or as an operator on $C^\infty(\mathbb{T}^n)$, see Definition 4.1.2), but at the moment we do not yet know whether $\sigma \in S^m_{\rho,\delta}(\mathbb{T}^n \times \mathbb{Z}^n)$. By Theorem 4.1.4 the symbol is computed from

$$
\begin{aligned}
\sigma(x,\xi) &= e^{-i2\pi x \cdot \xi} \, \mathrm{Op}(a) \, e_\xi(x) \\
&= e^{-i2\pi x \cdot \xi} \sum_{\eta \in \mathbb{Z}^n} \int_{\mathbb{T}^n} e^{i2\pi(x-y)\cdot\eta} \, a(x,y,\eta) \, e^{i2\pi y \cdot \xi} \, dy.
\end{aligned}
$$

Now, we apply the discrete Taylor formula from Theorem 3.3.21 to obtain

$$
\begin{aligned}
\sigma(x,\xi) &= \sum_{\eta \in \mathbb{Z}^n} \int_{\mathbb{T}^n} e^{i2\pi(x-y)\cdot(\eta-\xi)} \, a(x,y,\eta) \, dy \\
&= \sum_{\eta \in \mathbb{Z}^n} e^{i2\pi x \cdot (\eta-\xi)} \, \widehat{a}_2(x, \eta - \xi, \eta) \\
&= \sum_{\eta \in \mathbb{Z}^n} e^{i2\pi x \cdot \eta} \, \widehat{a}_2(x, \eta, \xi + \eta) \\
&= \sum_{\eta \in \mathbb{Z}^n} e^{i2\pi x \cdot \eta} \sum_{|\gamma|<N} \frac{1}{\gamma!} \, \triangle^\gamma_\xi \widehat{a}_2(x, \eta, \xi) \, \eta^{(\gamma)} + \\
&\quad + \sum_{\eta \in \mathbb{Z}^n} e^{i2\pi x \cdot \eta} \, R_N(x, \eta, \xi, \eta),
\end{aligned}
$$

where $R_N(x,\eta,\xi,p)$ is the error term of the discrete Taylor expansion (Theorem 3.3.21) of $\widehat{a}_2(x,\eta,\xi+p)$. Let us write

$$E_N(x,\xi) := \sum_{\eta \in \mathbb{Z}^n} e^{i2\pi x \cdot \eta} \, R_N(x,\eta,\xi,\eta).$$

Notice that

$$
\begin{aligned}
D_y^{(\gamma)} a(x,y,\xi) &= D_y^{(\gamma)} \sum_{\eta \in \mathbb{Z}^n} e^{i2\pi y \cdot \eta}\, \widehat{a}_2(x,\eta,\xi) \\
&= \sum_{\eta \in \mathbb{Z}^n} e^{i2\pi y \cdot \eta}\, \eta^{(\gamma)}\, \widehat{a}_2(x,\eta,\xi),
\end{aligned}
$$

which yields

$$
\sigma(x,\xi) = \sum_{|\gamma|<N} \frac{1}{\gamma!}\, \triangle_\xi^\gamma D_y^{(\gamma)} a(x,y,\xi)|_{y=x} + E_N(x,\xi). \tag{4.19}
$$

All we need to show now is that $E_N \in S_{\rho,\delta}^{m-N}(\mathbb{T}^n \times \mathbb{Z}^n)$, and for this we have to study the remainder R_N. Recalling the form of the estimate for R_N (see Theorem 3.3.21), noticing that for $|\gamma|=N$, $|\eta^{(\gamma)}| \le (|\eta|+N)^N \le 2^N \langle \eta \rangle^N N^N$, applying (a close variant of) inequality (4.12), and using Peetre's inequality (Proposition 3.3.31), we get

$$
\begin{aligned}
\left| \triangle_\xi^{\alpha'} \partial_x^{\beta'} R_N(x,\eta,\xi,\eta) \right| &\le \frac{1}{N!} \max_{|\gamma|=N} \left| \eta^{(\gamma)} \right| \max_{|\omega|=N,\ \nu \in Q(\eta)} \left| \triangle_\xi^{\alpha'+\omega} \partial_x^{\beta'} \widehat{a}_2(x,\eta,\xi+\nu) \right| \\
&\le \frac{1}{N!} 2^N \langle \eta \rangle^N N^N \\
&\quad \times \max_{\nu \in Q(\eta)} \left[c_{r,\beta',N+\alpha'}\, \langle \eta \rangle^{-r} \langle \xi+\nu \rangle^{m-\rho N-\rho|\alpha'|+\delta|\beta'|+\delta r} \right].
\end{aligned}
$$

Now, let us consider two cases. First, if $|\eta| \le |\xi|/2$, this can be estimated by

$$
C_N \langle \eta \rangle^{N-r} \langle \xi \rangle^{m-\rho N-\rho|\alpha'|+\delta|\beta'|+\delta r},
$$

and taking $N=r$ this can be estimated by $C_N \langle \xi \rangle^{m-(\rho-\delta)N-\rho|\alpha'|+\delta|\beta'|}$ which we can estimate by any $\langle \xi \rangle^{-N'}$ if we take N large enough. On the other hand, for the region $|\eta| > |\xi|/2$, let us fix some (large) N' and take $r=N+N'$. Then we can estimate $\left| \triangle_\xi^{\alpha'} \partial_x^{\beta'} R_N(x,\eta,\xi,\eta) \right|$ by

$$
\begin{aligned}
C_N \langle \eta \rangle^{N-r} &\max_{\nu \in Q(\eta)} \left[c_{r,\beta',N+\alpha'}\, \langle \xi+\nu \rangle^{m-\rho N-\rho|\alpha'|+\delta|\beta'|+\delta\rho} \right] \\
&= \langle \eta \rangle^{-N'} \max_{\nu \in Q(\eta)} \left[c_{r,\beta',N+\alpha'}\, \langle \xi+\nu \rangle^{m-(\rho-\delta)N-\rho|\alpha'|+\delta|\beta'|+\delta N'} \right].
\end{aligned}
$$

Thus, if N is sufficiently large, the maximum term is bounded by a constant in view of $\rho > \delta$, and the remaining term satisfies $\langle \eta \rangle^{-N'} \le C \langle \xi \rangle^{-N'}$. Therefore, taking enough terms in the asymptotic expansion we can estimate $\triangle_\xi^\alpha \partial_x^\beta E_N(x,\xi)$ by any power of $\langle \xi \rangle^{-1}$ and since all the terms are in the necessary symbol classes the estimate for the remainder is complete. Consequently σ belongs to $S_{\rho,\delta}^m(\mathbb{T}^n \times \mathbb{Z}^n)$ by equation (4.19), and Theorem 4.4.1 provides the asymptotic expansion (4.18). $\square$

Remark 4.4.6. Now we can compare the results above with the biperiodic Taylor series. Applying Theorem 4.4.5 and Lemma 4.3.4, we get

$$a(x,y,\xi) \sim \sum_{\gamma=0}^{\infty} \frac{1}{\gamma!} \triangle_\xi^\gamma D_y^{(\gamma)} a(x,y,\xi)|_{y=x}$$

$$\sim \sum_{\gamma=0}^{\infty} \frac{1}{\gamma!} \left(e^{i2\pi(y-x)} - 1\right)^\gamma D_z^{(\gamma)} a(x,z,\xi)|_{z=x},$$

reminding us of the series representation of Corollary 3.4.3.

Alternative proof for Theorem 4.4.5 on $\mathbb{T}^1$. We invoke the biperiodic Taylor expansion for $a(x,y,\xi)$ (see Corollary 3.4.3):

$$a(x,y,\xi) = \sum_{j=0}^{N-1} \frac{1}{j!} \left(e^{i2\pi(y-x)} - 1\right) D_z^{(j)} a(x,z,\xi)|_{z=x}$$

$$+ a_N(x,y,\xi) \left(e^{i2\pi(y-x)} - 1\right)^N.$$

Then we use Lemma 4.3.4 with $b_j(x,y,\xi) = \left(e^{i2\pi(y-x)} - 1\right)^j c_j(x,y,\xi)$, where $c_j(x,y,\xi) = \frac{1}{j!} D_y^{(j)} a(x,y,\xi)|_{y=x}$, to obtain

$$\mathrm{Op}(b_j) = \mathrm{Op}(\triangle_\xi^j c_j) = \mathrm{Op}\left(\frac{1}{j!} \triangle_\xi^j D_y^{(j)} a(x,y,\xi)|_{y=x}\right).$$

By Lemma 4.3.4, the remainder $a_N(x,y,\xi) \left(e^{i2\pi(y-x)} - 1\right)^N$ hence contributes to the operator $\mathrm{Op}(\triangle_\xi^N a_N)$. Thus, in order to get the asymptotic expansion (4.18), we have to prove that $a_N \in \mathcal{A}_{\rho,\delta}^m(\mathbb{T}^1)$ for every $N \in \mathbb{Z}^+$. From the proof of Theorem 3.4.2 we see that a_N is given by

$$a_N(x,y,\xi) = \frac{a_{N-1}(x,y,\xi) - a_{N-1}(x,x,\xi)}{e^{i2\pi(y-x)} - 1},$$

and that it is in $C^\infty(\mathbb{T}^1 \times \mathbb{T}^1)$ for every $\xi \in \mathbb{Z}^1$. Here a_N has the same order as a_{N-1} does, so that recursively $a_N \in \mathcal{A}_{\rho,\delta}^m(\mathbb{T}^1)$, since $a_0 = a \in \mathcal{A}_{\rho,\delta}^m(\mathbb{T}^1)$. This completes the proof. $\square$

Remark 4.4.7 (**Classical periodic pseudo-differential operators**). The operator $A \in \mathrm{Op}(S^m(\mathbb{T}^n))$ is called a *classical periodic pseudo-differential operator*, if its symbol has an asymptotic expansion

$$\sigma_A(x,\xi) \sim \sum_{j=0}^{\infty} \sigma_j(x,\xi),$$

where the symbols σ_j are positively homogeneous of degree $m - j$: they satisfy $\sigma_j(x, \xi) = \sigma_j(x, \xi/\|\xi\|)\|\xi\|^{m-j}$ for large ξ. In [142] and [102], it is shown that any classical periodic pseudo-differential operator can be expressed as a sum of periodic integral operators of the type (4.44) – other contributions to periodic integral operators and classical operators are made in [34], [62], [142], and [102]. The research on these operators is of interest, but in the sequel we will rather concentrate on questions of the symbolic analysis.

4.5 Extension of toroidal symbols

In the study of periodic pseudo-differential operators some of the applications of the calculus of finite differences, for example the discrete Taylor series, can be eliminated. We are going to explain how this can be done by interpolating a symbol $(x, \xi) \mapsto \sigma(x, \xi)$ in the second argument ξ in a smooth way, so that it becomes defined on $\mathbb{T}^n \times \mathbb{R}^n$ instead of $\mathbb{T}^n \times \mathbb{Z}^n$. This process will be called an *extension of the toroidal symbol*. By using such extensions one can work with the familiar tools of classical analysis yielding the same theory as before, and for some practical examples this may be more convenient than operating with differences. However, this approach is quite alien to the idea of periodic symbols, as the results can be derived using quite simple difference calculus. In addition, difference operations can easily be carried out with computers, whereas program realisations of numerical differentiation are computationally expensive and troublesome. Moreover, such an extension explores the intricate relation between $\mathbb{T}^n$ and $\mathbb{R}^n$ and can not be readily generalised to symbols on other compact Lie groups (thus while very useful on $\mathbb{T}^n$ yet unfortunately not providing an additional intuition for operators in Part IV).

Thus, it is often useful to extend toroidal symbols from $\mathbb{T}^n \times \mathbb{Z}^n$ to $\mathbb{T}^n \times \mathbb{R}^n$, ideally getting symbols in Hörmander's symbol classes. The case of $n = 1$ and $(\rho, \delta) = (1, 0)$ was considered in [141] and [102]. This extension can be done with a suitable convolution that respects the symbol inequalities. In the following, $\delta_{0,\xi}$ is the Kronecker delta at $0 \in \mathbb{Z}^n$, i.e., $\delta_{0,0} = 1$, and $\delta_{0,\xi} = 0$ if $\xi \neq 0$. First we prepare the following useful functions $\theta, \phi_\alpha \in \mathcal{S}(\mathbb{R}^n)$:

Lemma 4.5.1. *There exist functions $\phi_\alpha \in \mathcal{S}(\mathbb{R}^n)$ (for each $\alpha \in \mathbb{N}_0^n$) and a function $\theta \in \mathcal{S}(\mathbb{R}^n)$ such that*

$$\mathcal{P}\theta(x) := \sum_{k \in \mathbb{Z}^n} \theta(x + k) \equiv 1,$$

$$(\mathcal{F}_{\mathbb{R}^n}\theta)|_{\mathbb{Z}^n}(\xi) = \delta_{0,\xi} \quad and \quad \partial_\xi^\alpha (\mathcal{F}_{\mathbb{R}^n}\theta)(\xi) = \overline{\triangle}_\xi^\alpha \phi_\alpha(\xi)$$

for all $\xi \in \mathbb{Z}^n$.

The idea of this lemma may be credited to Yves Meyer [29, p. 4]. It will be used in the interpolation presented in Theorem 4.5.3.

Proof. Let us first consider the one-dimensional case. Let $\theta = \theta_1 \in C^\infty(\mathbb{R}^1)$ such that

$$\text{supp}(\theta_1) \subset (-1, 1), \quad \theta_1(-x) = \theta_1(x), \quad \theta_1(1 - y) + \theta_1(y) = 1$$

for $x \in \mathbb{R}$ and for $0 \leq y \leq 1$; these assumptions for θ_1 are enough for us, and of course the choice is not unique. In any case, $\theta_1 \in \mathcal{S}(\mathbb{R}^1)$, so that also $\mathcal{F}_\mathbb{R}\theta_1 \in \mathcal{S}(\mathbb{R}^1)$. If $\xi \in \mathbb{Z}^1$ then we have

$$
\begin{aligned}
\mathcal{F}_\mathbb{R}\theta_1(\xi) &= \int_{\mathbb{R}^1} \theta_1(x) \, e^{-i2\pi x \cdot \xi} \, dx \\
&= \int_0^1 (\theta_1(x - 1) + \theta_1(x)) \, e^{-i2\pi x \cdot \xi} \, dx \\
&= \delta_{0,\xi}.
\end{aligned}
$$

If a desired $\phi_\alpha \in \mathcal{S}(\mathbb{R}^1)$ exists, it must satisfy

$$
\begin{aligned}
\int_{\mathbb{R}^1} e^{i2\pi x \cdot \xi} \, \partial_\xi^\alpha (\mathcal{F}_\mathbb{R}\theta_1)(\xi) \, d\xi &= \int_{\mathbb{R}^1} e^{i2\pi x \cdot \xi} \, \overline{\triangle}_\xi^\alpha \phi_\alpha(\xi) \, d\xi \\
&= \left(1 - e^{i2\pi x}\right)^\alpha \int_{\mathbb{R}^1} e^{i2\pi x \cdot \xi} \phi_\alpha(\xi) \, d\xi
\end{aligned}
$$

due to the bijectivity of $\mathcal{F}_\mathbb{R} : \mathcal{S}(\mathbb{R}^1) \to \mathcal{S}(\mathbb{R}^1)$. Integration by parts leads to the formula

$$(-i2\pi x)^\alpha \theta_1(x) = (1 - e^{i2\pi x})^\alpha (\mathcal{F}_\mathbb{R}^{-1}\phi_\alpha)(x).$$

Thus

$$
(\mathcal{F}_\mathbb{R}^{-1}\phi_\alpha)(x) =
\begin{cases}
\left(\frac{-i2\pi x}{1 - e^{i2\pi x}}\right)^\alpha \theta_1(x), & \text{if } 0 < |x| < 1, \\
1, & \text{if } x = 0, \\
0, & \text{if } |x| \geq 1.
\end{cases}
$$

The general n-dimensional case is reduced to the one-dimensional case, since mapping $\theta = (x \mapsto \theta_1(x_1)\theta_1(x_2) \cdots \theta_1(x_n)) \in \mathcal{S}(\mathbb{R}^n)$ has the desired properties. $\qquad \square$

Remark 4.5.2 **(Periodic symbols on $\mathbb{R}^n$).** The defining symbol inequalities for the class $S_{\rho,\delta}^m(\mathbb{T}^n \times \mathbb{R}^n)$ of *periodic symbols on $\mathbb{R}^n$* are

$$\forall \alpha, \beta \in \mathbb{N}_0^n \; \exists c_{\alpha\beta} > 0 : \left|\partial_\xi^\alpha \partial_x^\beta \sigma(x, \xi)\right| \leq c_{\alpha\beta} \left(1 + \|\xi\|\right)^{m - \rho|\alpha| + \delta|\beta|}. \tag{4.20}$$

To emphasize the difference with toroidal symbols defined on $\mathbb{T}^n \times \mathbb{Z}^n$ we call them *Euclidean symbols.*

Lemma 4.5.1 provides us the means to interpolate between the discrete points of $\mathbb{Z}^n$ in a manner that is faithful to the symbol (cf. inequalities (4.6) and (4.20)):

Theorem 4.5.3 (Toroidal vs Euclidean symbols). *Let $0 < \rho \leq 1$ and $0 \leq \delta \leq 1$. Symbol $\widetilde{a} \in S_{\rho,\delta}^m(\mathbb{T}^n \times \mathbb{Z}^n)$ is a toroidal symbol if and only if there exists a Euclidean symbol $a \in S_{\rho,\delta}^m(\mathbb{T}^n \times \mathbb{R}^n)$ such that $\widetilde{a} = a|_{\mathbb{T}^n \times \mathbb{Z}^n}$. Moreover, this extended symbol a is unique modulo $S^{-\infty}(\mathbb{T}^n \times \mathbb{R}^n)$.*

The relation between the corresponding pseudo-differential operators will be given in Theorem 4.6.12. For the relation between extensions and ellipticity see Theorem 4.9.15.

Proof. Let us first prove the "if" part. Let $a \in S^m_{\rho,\delta}(\mathbb{T}^n \times \mathbb{R}^n)$, and in this part we can actually allow any ρ and δ, for example $0 \leq \rho, \delta \leq 1$. By the Lagrange Mean Value Theorem, if $|\alpha| = 1$ then

$$
\begin{aligned}
\triangle^\alpha_\xi \partial^\beta_x \tilde{a}(x, \xi) &= \triangle^\alpha_\xi \partial^\beta_x a(x, \xi) \\
&= \partial^\alpha_\xi \partial^\beta_x a(x, \xi)|_{\xi=\eta},
\end{aligned}
$$

where η is on the line from ξ to $\xi + \alpha$. By the Mean Value Theorem, for a general multi-index $\alpha \in \mathbb{N}^n_0$, we also have

$$
\triangle^\alpha_\xi \partial^\beta_x \tilde{a}(x, \xi) = \partial^\alpha_\xi \partial^\beta_x a(x, \xi)|_{\xi=\eta}
$$

for some $\eta \in Q := [\xi_1, \xi_1 + \alpha_1] \times \cdots \times [\xi_n, \xi_n + \alpha_n]$. This can be shown by induction. Indeed, let us write $\alpha = \omega + \gamma$ for some $\omega = \delta_j$. Then we can calculate

$$
\begin{aligned}
\triangle^\alpha_\xi \partial^\beta_x \tilde{a}(x, \xi) &= \triangle^\omega_\xi \left(\triangle^\gamma_\xi \partial^\beta_x \tilde{a} \right)(x, \xi) \\
&= \triangle_{\xi_j} \left(\partial^\gamma_\xi \partial^\beta_x a(x, \xi)|_{\xi=\varsigma} \right) \\
&= \partial^\gamma_\xi \partial^\beta_x a(x, \varsigma + \delta_j) - \partial^\gamma_\xi \partial^\beta_x a(x, \varsigma) \\
&= \partial^\alpha_\xi \partial^\beta_x a(x, \xi)|_{\xi=\eta}
\end{aligned}
$$

for some ς and η, where we used the induction hypothesis in the first line. Therefore, we can estimate

$$
\begin{aligned}
\left| \triangle^\alpha_\xi \partial^\beta_x \tilde{a}(x, \xi) \right| &= \left| \partial^\alpha_\xi \partial^\beta_x a(x, \xi)|_{\xi=\eta\in Q} \right| \\
&\leq C_{\alpha\beta m} \langle \eta \rangle^{m-\rho|\alpha|+\delta|\beta|} \\
&\leq C'_{\alpha\beta m} \langle \xi \rangle^{m-\rho|\alpha|+\delta|\beta|}.
\end{aligned}
$$

Let us now prove the "only if" part. First we show the uniqueness. Let $a, b \in S^m_{\rho,\delta}(\mathbb{T}^n \times \mathbb{R}^n)$ and assume that $a|_{\mathbb{T}^n\times\mathbb{Z}^n} = b|_{\mathbb{T}^n\times\mathbb{Z}^n}$. Let $c = a - b$. Then $c \in S^m_{\rho,\delta}(\mathbb{T}^n \times \mathbb{R}^n)$ and it satisfies $c|_{\mathbb{T}^n\times\mathbb{Z}^n} = 0$. If $\xi \in \mathbb{R}^n \setminus \mathbb{Z}^n$, choose $\eta \in \mathbb{Z}^n$ that is the nearest point (or one of the nearest points) to ξ. Then we have the first-order Taylor expansion

$$
\begin{aligned}
c(x, \xi) &= c(x, \eta) + \sum_{\alpha:\, |\alpha|=1} r_\alpha(x, \xi, \xi - \eta)\, (\xi - \eta)^\alpha \\
&= \sum_{\alpha:\, |\alpha|=1} r_\alpha(x, \xi, \xi - \eta)\, (\xi - \eta)^\alpha,
\end{aligned}
$$

where

$$
r_\alpha(x, \xi, \theta) = \int_0^1 (1-t)\, \partial^\alpha_\xi c(x, \xi + t\theta)\, \mathrm{dt}.
$$

Hence we have $|c(x, \xi)| \leq C \, \langle \xi \rangle^{m-\rho}$. Continuing the argument inductively for c and its derivatives and using that $\rho > 0$, we obtain the uniqueness modulo $S^{-\infty}(\mathbb{T}^n \times \mathbb{R}^n)$.

Let us now show the existence. Let $\theta \in \mathcal{S}(\mathbb{R}^n)$ be as in Lemma 4.5.1. Define $a : \mathbb{T}^n \times \mathbb{R}^n \to \mathbb{C}$ by

$$a(x, \xi) := \sum_{\eta \in \mathbb{Z}^n} (\mathcal{F}_{\mathbb{R}^n} \theta)(\xi - \eta) \, \widetilde{a}(x, \eta). \tag{4.21}$$

It is easy to see that $\widetilde{a} = a|_{\mathbb{T}^n \times \mathbb{Z}^n}$. Furthermore, we have

$$
\left| \partial_\xi^\alpha \partial_x^\beta a(x, \xi) \right| = \left| \sum_{\eta \in \mathbb{Z}^n} \partial_\xi^\alpha (\mathcal{F}_{\mathbb{R}^n} \theta)(\xi - \eta) \, \partial_x^\beta \widetilde{a}(x, \eta) \right|
$$

$$
= \left| \sum_{\eta \in \mathbb{Z}^n} \overline{\triangle}_\xi^\alpha \phi_\alpha(\xi - \eta) \, \partial_x^\beta \widetilde{a}(x, \eta) \right|
$$

$$
\overset{(3.14)}{=} \left| \sum_{\eta \in \mathbb{Z}^n} \phi_\alpha(\xi - \eta) \, \triangle_\eta^\alpha \partial_x^\beta \widetilde{a}(x, \eta) \, (-1)^{|\alpha|} \right|
$$

$$
\leq \sum_{\eta \in \mathbb{Z}^n} |\phi_\alpha(\xi - \eta)| \, C_{\alpha \beta m} \, \langle \eta \rangle^{m - \rho|\alpha| + \delta|\beta|}
$$

$$
\leq C'_{\alpha \beta m} \, \langle \xi \rangle^{m - \rho|\alpha| + \delta|\beta|} \sum_{\eta \in \mathbb{Z}^n} |\phi_\alpha(\eta)| \, \langle \eta \rangle^{|m - \rho|\alpha| + \delta|\beta||}
$$

$$
\leq C''_{\alpha \beta m} \, \langle \xi \rangle^{m - \rho|\alpha| + \delta|\beta|},
$$

where we used the summation by parts formula (3.14). In the last two lines we also used that $\phi_\alpha \in \mathcal{S}(\mathbb{R}^n)$ and also a simple fact that for $p > 0$ we have $\langle \xi + \eta \rangle^p \leq \langle \xi \rangle^p \langle \eta \rangle^p$ and $\langle \xi + \eta \rangle^{-p} \langle \eta \rangle^{-p} \leq \langle \xi \rangle^{-p}$, for all $\xi, \eta \in \mathbb{R}^n$. Thus $a \in S_{\rho,\delta}^m(\mathbb{T}^n \times \mathbb{R}^n)$. $\square$

From now on, we can exploit inequalities (4.20), but it is good to remember that all the information was contained already in the original definition of symbols on $\mathbb{T}^n \times \mathbb{Z}^n$. In a sense, the extension is arbitrary (yet unique up to order $-\infty$), as the demands for the function $\theta \in \mathcal{S}(\mathbb{R}^n)$ were quite modest in the proof of Lemma 4.5.1.

Remark 4.5.4. The extension process can also be modified for amplitudes to get $a(x, y, \xi)$ (continuous $\xi \in \mathbb{R}^n$) from $a(x, y, \xi)$ (discrete $\xi \in \mathbb{Z}^n$).

Remark 4.5.5 (**Extension respects ellipticity.**). Moreover, the extension respects ellipticity, as we will show in Theorem 4.9.15.

Exercise 4.5.6. Work out details of the proof of Remark 4.5.4.

We also observe that the same proof yields the following limited regularity version of Theorem 4.5.3:

Corollary 4.5.7 (Limited regularity extensions). *Let the function* $a : \mathbb{T}^n \times \mathbb{R}^n \to \mathbb{C}$ *satisfy*

$$\left| \partial_\xi^\alpha \partial_x^\beta a(x,\xi) \right| \leq c_{\alpha\beta} \, \langle \xi \rangle^{m-\rho|\alpha|+\delta|\beta|} \text{ for all } x \in \mathbb{T}^n, \xi \in \mathbb{R}^n, \tag{4.22}$$

for all $|\alpha| \leq N_1$ *and* $|\beta| \leq N_2$. *Then its restriction* $\tilde{a} := a|_{\mathbb{T}^n \times \mathbb{Z}^n}$ *satisfies*

$$\left| \triangle_\xi^\alpha \partial_x^\beta \tilde{a}(x,\xi) \right| \leq c_{\alpha\beta} \, \langle \xi \rangle^{m-\rho|\alpha|+\delta|\beta|} \text{ for all } x \in \mathbb{T}^n, \xi \in \mathbb{Z}^n, \tag{4.23}$$

and all $|\alpha| \leq N_1$ *and* $|\beta| \leq N_2$. *Conversely, every function* $\tilde{a} : \mathbb{T}^n \times \mathbb{Z}^n \to \mathbb{C}$ *satisfying* (4.23) *for all* $|\alpha| \leq N_1$ *and* $|\beta| \leq N_2$ *is a restriction* $\tilde{a} = a|_{\mathbb{T}^n \times \mathbb{Z}^n}$ *of some function* $a : \mathbb{T}^n \times \mathbb{R}^n \to \mathbb{C}$ *satisfying* (4.22) *for all* $|\alpha| \leq N_1$ *and* $|\beta| \leq N_2$.

4.6 Periodisation of pseudo-differential operators

In this section we describe the relation between operators with Euclidean and toroidal quantizations and between operators corresponding to symbols $a(x,\xi)$ and $\tilde{a} = a|_{\mathbb{T}^n \times \mathbb{Z}^n}$, given by the operator of the periodisation of functions.

First we state a property of a pseudo-differential operator $a(X, D)$ to map the space $\mathcal{S}(\mathbb{R}^n)$ into itself, which will be of importance. The following class will be sufficient for our purposes, and the proof is straightforward.

Proposition 4.6.1. *Let* $a = a(x,\xi) \in C^\infty(\mathbb{R}^n \times \mathbb{R}^n)$ *and assume that there exist* $\epsilon > 0$ *and* $N \in \mathbb{R}$ *such that for every* α, β *there are constants* $C_{\alpha\beta}$ *and* $M(\alpha,\beta)$ *such that the estimate*

$$\left| \partial_x^\alpha \partial_\xi^\beta a(x,\xi) \right| \leq C_{\alpha\beta} \langle x \rangle^{N+(1-\epsilon)|\beta|} \langle \xi \rangle^{M(\alpha,\beta)}$$

holds for all $x, \xi \in \mathbb{R}^n$. *Then the pseudo-differential operator* $a(X, D)$ *with symbol* $a(x,\xi)$ *is continuous from* $\mathcal{S}(\mathbb{R}^n)$ *to* $\mathcal{S}(\mathbb{R}^n)$.

Exercise 4.6.2. Prove Proposition 4.6.1.

Before analysing the relation between operators with Euclidean and toroidal quantizations, we will describe the periodisation operator that will be of importance for such analysis.

Theorem 4.6.3 (Periodisation). *The periodisation* $\mathcal{P}f : \mathbb{R}^n \to \mathbb{C}$ *of a function* $f \in \mathcal{S}(\mathbb{R}^n)$ *is defined by*

$$\mathcal{P}f(x) := \sum_{k \in \mathbb{Z}^n} f(x+k). \tag{4.24}$$

Then $\mathcal{P} : \mathcal{S}(\mathbb{R}^n) \to C^\infty(\mathbb{T}^n)$ *is surjective and* $\|\mathcal{P}f\|_{L^1(\mathbb{T}^n)} \leq \|f\|_{L^1(\mathbb{R}^n)}$. *Moreover, we have*

$$\mathcal{P}f(x) = \mathcal{F}_{\mathbb{T}^n}^{-1} \left((\mathcal{F}_{\mathbb{R}^n} f)|_{\mathbb{Z}^n} \right)(x) \tag{4.25}$$

and

$$\sum_{k \in \mathbb{Z}^n} f(x+k) = \sum_{\xi \in \mathbb{Z}^n} e^{i2\pi x \cdot \xi} \, (\mathcal{F}_{\mathbb{R}^n} f)(\xi). \tag{4.26}$$

Taking $x = 0$ in (4.26), we obtain

Corollary 4.6.4 (Poisson summation formula). *For $f \in \mathcal{S}(\mathbb{R}^n)$ we have*

$$\sum_{k \in \mathbb{Z}^n} f(k) = \sum_{\xi \in \mathbb{Z}^n} \widehat{f}(\xi).$$

As a consequence of the Poisson summation formula and the Fourier transform of Gaussians in Lemma 1.1.23 we obtain the so-called Jacobi identity:

Exercise 4.6.5 (Jacobi identity for Gaussians). For every $\epsilon > 0$ we have

$$\sum_{j=-\infty}^{+\infty} e^{-\pi j^2/\epsilon} = \sqrt{\epsilon} \sum_{j=-\infty}^{+\infty} e^{-\pi \epsilon j^2}.$$

Remark 4.6.6. We note that by Theorem 4.6.3 we may extend the periodisation operator $\mathcal{P}$ to $L^1(\mathbb{R}^n)$, and also this extension is surjective from $L^1(\mathbb{R}^n)$ to $L^1(\mathbb{T}^n)$. This is actually rather trivial compared to the smooth case of Theorem 4.6.3 because we can find a preimage $f \in L^1(\mathbb{R}^n)$ of $g \in L^1(\mathbb{T}^n)$ under the periodisation mapping $\mathcal{P}$ by for example setting $f = g|_{[0,1]^n}$ and $f = 0$ otherwise.

Exercise 4.6.7. Observe that the periodisation operator $\mathcal{P} : \mathcal{S}(\mathbb{R}^n) \to C^\infty(\mathbb{T}^n)$ is dualised to $\mathcal{P}^t : \mathcal{D}'(\mathbb{T}^n) \to \mathcal{S}'(\mathbb{R}^n)$ by the formula

$$\langle \mathcal{P}^t u, \varphi \rangle := \langle u, \mathcal{P}\varphi \rangle \quad \text{for all } \varphi \in \mathcal{S}(\mathbb{R}^n).$$

Indeed, if $\varphi \in \mathcal{S}(\mathbb{R}^n)$ we have that $\mathcal{P}\varphi \in C^\infty(\mathbb{T}^n)$, so that this definition makes sense for $u \in \mathcal{D}'(\mathbb{T}^n)$. What is the meaning of the operator $\mathcal{P}^t$?

Proof of Theorem 4.6.3. The estimate $\|\mathcal{P}f\|_{L^1(\mathbb{T}^n)} \leq \|f\|_{L^1(\mathbb{R}^n)}$ is straightforward. Next, for $\xi \in \mathbb{Z}^n$, we have

$$\begin{aligned}
\mathcal{F}_{\mathbb{T}^n}(\mathcal{P}f)(\xi) &= \int_{\mathbb{T}^n} e^{-i2\pi x \cdot \xi} \, \mathcal{P}f(x) \, dx \\
&= \int_{\mathbb{T}^n} e^{-i2\pi x \cdot \xi} \sum_{k \in \mathbb{Z}^n} f(x + k) \, dx \\
&= \int_{\mathbb{R}^n} e^{-i2\pi x \cdot \xi} \, f(x) \, dx \\
&= (\mathcal{F}_{\mathbb{R}^n} f)(\xi).
\end{aligned}$$

From this we can see that

$$\begin{aligned}
\sum_{k \in \mathbb{Z}^n} f(x + k) &= \mathcal{P}f(x) \\
&= \sum_{\xi \in \mathbb{Z}^n} e^{i2\pi x \cdot \xi} \, \mathcal{F}_{\mathbb{T}^n}(\mathcal{P}f)(\xi) \\
&= \sum_{\xi \in \mathbb{Z}^n} e^{i2\pi x \cdot \xi} \, (\mathcal{F}_{\mathbb{R}^n} f)(\xi),
\end{aligned}$$

proving (4.26). Let us show the surjectivity of $\mathcal{P} : \mathcal{S}(\mathbb{R}^n) \to C^\infty(\mathbb{T}^n)$. Let $\theta \in \mathcal{S}(\mathbb{R}^n)$ be a function defined in Lemma 4.5.1. Then for any $g \in C^\infty(\mathbb{T}^n)$ it holds that

$$\mathcal{P}(g\theta)(x) = \sum_{k \in \mathbb{Z}^n} g(x + k)\, \theta(x + k) = g(x) \sum_{k \in \mathbb{Z}^n} \theta(x + k) = g(x),$$

where $g\theta$ is the product of θ with $\mathbb{Z}^n$-periodic function g on $\mathbb{R}^n$. We omit the straightforward proofs of the other claims. $\qquad\square$

We saw in Proposition 3.1.34 that Dirac delta comb $\delta_{\mathbb{Z}^n}$ can be viewed as a sum of Dirac deltas. We can also relate it to the partial sums of Fourier coefficients:

Proposition 4.6.8. *Let us define $Q_j \in \mathcal{S}'(\mathbb{R}^n)$ by*

$$\langle Q_j, \varphi \rangle := \sum_{k \in \mathbb{Z}^n:\ |k| \leq j} \int_{\mathbb{R}^n} e^{i2\pi k \cdot \xi}\, \varphi(\xi)\, \mathrm{d}\xi \tag{4.27}$$

for $\varphi \in \mathcal{S}(\mathbb{R}^n)$. Then we have the convergence $Q_j \to \delta_{\mathbb{Z}^n}$ in $\mathcal{S}'(\mathbb{R}^n)$ to the Dirac delta comb.

Proof. Indeed, we have

$$
\begin{aligned}
\langle Q_j, \varphi \rangle
&= \sum_{k \in \mathbb{Z}^n:\ |k| \leq j} \int_{\mathbb{R}^n} e^{i2\pi k \cdot \xi}\, \varphi(\xi)\, \mathrm{d}\xi \\
&= \sum_{k \in \mathbb{Z}^n:\ |k| \leq j} (\mathcal{F}_{\mathbb{R}^n}\varphi)(k) \\
&\xrightarrow[j \to \infty]{} \sum_{k \in \mathbb{Z}^n} (\mathcal{F}_{\mathbb{R}^n}\varphi)(k) \\
&\overset{\text{Poisson}}{=} \sum_{\xi \in \mathbb{Z}^n} \varphi(\xi) \\
&= \langle \delta_{\mathbb{Z}^n}, \varphi \rangle,
\end{aligned}
$$

for all $\varphi \in \mathcal{S}(\mathbb{R}^n)$. $\qquad\square$

Remark 4.6.9 (**Inflated torus**). For $N \in \mathbb{N}$ we write $N\mathbb{T}^n = (\mathbb{R}/N\mathbb{Z})^n$ which we call the *N-inflated torus*, or simply an *inflated torus* if the value of N is not of importance. We note that in the case of the N-inflated torus $N\mathbb{T}^n$ we can use the periodisation operator $\mathcal{P}_N$ instead of $\mathcal{P}$, where $\mathcal{P}_N : \mathcal{S}(\mathbb{R}^n) \to C^\infty(N\mathbb{T}^n)$ can be defined by

$$(\mathcal{P}_N f)(x) = \mathcal{F}_{N\mathbb{T}^n}^{-1}\left(\mathcal{F}_{\mathbb{R}^n} f \big|_{\frac{1}{N}\mathbb{Z}^n} \right)(x), \quad x \in N\mathbb{T}^n. \tag{4.28}$$

Exercise 4.6.10. Generalise Theorem 4.6.3 to the N-inflated torus using operator $\mathcal{P}_N$.

Let us now establish some basic properties of pseudo-differential operators with respect to periodisation.

Definition 4.6.11. We will say that a function $a : \mathbb{R}^n \times \mathbb{R}^n \to \mathbb{C}$ is *1-periodic* (we will always mean that it is periodic with respect to the first variable $x \in \mathbb{R}^n$) if the function $x \mapsto a(x, \xi)$ is $\mathbb{Z}^n$-periodic for all ξ.

As in Theorem 4.5.3, we use tilde to denote the restriction of $a \in C^\infty(\mathbb{R}^n \times \mathbb{R}^n)$ to $\mathbb{R}^n \times \mathbb{Z}^n$. If $a(x, \xi)$ is 1-periodic, we can view it as a function on $\mathbb{T}^n \times \mathbb{Z}^n$, and we write $\tilde{a} = a|_{\mathbb{T}^n \times \mathbb{Z}^n}$. For such $\tilde{a}$ the corresponding operator $\mathrm{Op}(\tilde{a}) = \tilde{a}(X, D)$ is defined by (4.7) in Definition 4.1.9.

Theorem 4.6.12 (Periodisation of operators). *Let* $a = a(x, \xi) \in C^\infty(\mathbb{R}^n \times \mathbb{R}^n)$ *be 1-periodic with respect to* x *for every* $\xi \in \mathbb{R}^n$. *Assume that for every* $\alpha, \beta \in \mathbb{N}_0^n$ *there are constants* $C_{\alpha\beta}$ *and* $M(\alpha, \beta)$ *such that the estimate*

$$\left| \partial_x^\alpha \partial_\xi^\beta a(x, \xi) \right| \leq C_{\alpha\beta} \langle \xi \rangle^{M(\alpha,\beta)}$$

holds for all $x, \xi \in \mathbb{R}^n$. *Let* $\tilde{a} = a|_{\mathbb{T}^n \times \mathbb{Z}^n}$. *Then*

$$\mathcal{P} \circ a(X, D) f = \tilde{a}(X, D) \circ \mathcal{P} f \tag{4.29}$$

for all $f \in \mathcal{S}(\mathbb{R}^n)$.

Note that it is not important in this theorem that a is in any of the symbol classes $S_{\rho,\delta}^m(\mathbb{R}^n \times \mathbb{R}^n)$.

Combining Theorems 4.5.3 and 4.6.12 we get

Corollary 4.6.13 (Equality of operator classes). *For* $0 \leq \delta \leq 1$ *and* $0 < \rho \leq 1$ *we have*

$$\mathrm{Op}(S_{\rho,\delta}^m(\mathbb{T}^n \times \mathbb{R}^n)) = \mathrm{Op}(S_{\rho,\delta}^m(\mathbb{T}^n \times \mathbb{Z}^n)),$$

i.e., classes of 1-periodic pseudo-differential operators with Euclidean (Hörmander's) symbols in $S_{\rho,\delta}^m(\mathbb{T}^n \times \mathbb{R}^n)$ *and toroidal symbols in* $S_{\rho,\delta}^m(\mathbb{T}^n \times \mathbb{Z}^n)$ *coincide.*

Remark 4.6.14. Note that by Proposition 4.6.1 both sides of (4.29) are well-defined functions in $C^\infty(\mathbb{T}^n)$. Moreover, equality (4.29) can be justified for f in larger classes of functions. For example, (4.29) remains true pointwise if $f \in C_0^k(\mathbb{R}^n)$ is a C^k compactly supported function for k sufficiently large. In any case, an equality on $\mathcal{S}(\mathbb{R}^n)$ allows an extension to $\mathcal{S}'(\mathbb{R}^n)$ by duality.

Proof of Theorem 4.6.12. Let $f \in \mathcal{S}(\mathbb{R}^n)$. Then we have

$$
\begin{aligned}
\mathcal{P}(a(X, D)f)(x) &= \sum_{k \in \mathbb{Z}^n} (a(X, D)f)(x + k) \\
&= \sum_{k \in \mathbb{Z}^n} \int_{\mathbb{R}^n} e^{i2\pi(x+k)\cdot\xi} \, a(x + k, \xi) \, (\mathcal{F}_{\mathbb{R}^n} f)(\xi) \, d\xi \\
&= \int_{\mathbb{R}^n} \left(\sum_{k \in \mathbb{Z}^n} e^{i2\pi k \cdot \xi} \right) e^{i2\pi x \cdot \xi} \, a(x, \xi) \, (\mathcal{F}_{\mathbb{R}^n} f)(\xi) \, d\xi
\end{aligned}
$$

$$= \int_{\mathbb{R}^n} \delta_{\mathbb{Z}^n}(\xi)\, e^{i2\pi x \cdot \xi}\, a(x,\xi)\, (\mathcal{F}_{\mathbb{R}^n} f)(\xi)\, d\xi$$

$$= \sum_{\xi \in \mathbb{Z}^n} e^{i2\pi x \cdot \xi}\, a(x,\xi)\, (\mathcal{F}_{\mathbb{R}^n} f)(\xi)$$

$$= \sum_{\xi \in \mathbb{Z}^n} e^{i2\pi x \cdot \xi}\, a(x,\xi)\, \mathcal{F}_{\mathbb{T}^n}(\mathcal{P}f)(\xi) \quad = \quad \widetilde{a}(X,D)(\mathcal{P}f)(x),$$

where $\delta_{\mathbb{Z}^n}$ is the Dirac δ comb from Definition 3.1.32. As usual, these calculations can be justified in the sense of distributions (see Remark 4.6.15). $\square$

Remark 4.6.15 (**Distributional justification**). We now give the distributional interpretation of the calculations in Theorem 4.6.12. Let us define some useful variants of the Dirac delta comb from Definition 3.1.32: for $x \in \mathbb{R}$ and $j \in \mathbb{Z}^+$, let $\mathcal{P}^x, \mathcal{P}^x_j \in \mathcal{S}'(\mathbb{R}^n)$ be such that

$$\langle \mathcal{P}^x, \varphi \rangle := \sum_{k \in \mathbb{Z}^n} \varphi(x+k), \qquad \langle \mathcal{P}^x_j, \varphi \rangle := \sum_{k \in \mathbb{Z}^n : \, |k| \leq j} \varphi(x+k),$$

for $\varphi \in \mathcal{S}(\mathbb{R}^n)$. We can easily observe that $\mathcal{P}^x_j \to \mathcal{P}^x$ in $\mathcal{S}'(\mathbb{R}^n)$. Then we can calculate:

$$\mathcal{P}(a(X,D)f)(x)$$

$$= \quad \langle \mathcal{P}^x, a(X,D)f \rangle$$

$$\overset{\mathcal{P}^x_j \to \mathcal{P}^x}{=} \quad \lim_{j \to \infty} \langle \mathcal{P}^x_j, a(X,D)f \rangle$$

$$= \quad \lim_{j \to \infty} \sum_{k \in \mathbb{Z}^n : \, |k| \leq j} (a(X,D)f)(x+k)$$

$$= \quad \lim_{j \to \infty} \sum_{k \in \mathbb{Z}^n : \, |k| \leq j} \int_{\mathbb{R}^n} e^{2\pi i (x+k) \cdot \xi}\, a(x+k,\xi)\, (\mathcal{F}_{\mathbb{R}^n} f)(\xi)\, d\xi$$

$$= \quad \lim_{j \to \infty} \sum_{k \in \mathbb{Z}^n : \, |k| \leq j} \int_{\mathbb{R}^n} e^{i2\pi k \cdot \xi}\, e^{2\pi i x \cdot \xi}\, a(x,\xi)\, (\mathcal{F}_{\mathbb{R}^n} f)(\xi)\, d\xi$$

$$\overset{Q_j \text{ from } (4.27)}{=} \quad \lim_{j \to \infty} \langle Q_j, \xi \mapsto e^{2\pi i x \cdot \xi}\, a(x,\xi)\, (\mathcal{F}_{\mathbb{R}^n} f)(\xi) \rangle$$

$$\overset{Q_j \to \delta_{\mathbb{Z}^n}}{=} \quad \langle \delta_{\mathbb{Z}^n}, \xi \mapsto e^{2\pi i x \cdot \xi}\, a(x,\xi)\, (\mathcal{F}_{\mathbb{R}^n} f)(\xi) \rangle$$

$$\overset{(4.25)}{=} \quad \sum_{\xi \in \mathbb{Z}^n} e^{2\pi i x \cdot \xi}\, a(x,\xi)\, (\mathcal{F}_{\mathbb{R}^n} f)(\xi)$$

$$= \quad \sum_{\xi \in \mathbb{Z}^n} e^{i2\pi x \cdot \xi}\, a(x,\xi)\, \mathcal{F}_{\mathbb{T}^n}(\mathcal{P}f)(\xi)$$

$$= \quad \widetilde{a}(X,D)(\mathcal{P}f)(x).$$

As we can see, the distributional justifications are quite natural, in the end.

Let us now formulate a useful corollary of Theorem 4.6.12 that will be of importance later, in particular in composing a pseudo-differential operator with a Fourier series operator in the proof of Theorem 4.13.11. If in Theorem 4.6.12 we take function f such that $f = g|_{[0,1]^n}$ for some $g \in C^\infty(\mathbb{T}^n)$, and $f = 0$ otherwise, it follows immediately that $g = \mathcal{P}f$. Adjusting this argument by shifting the cube $[0,1]^n$ if necessary and shrinking the support of g to make f smooth, we obtain

Corollary 4.6.16. *Let $a = a(x,\xi)$ be as in Theorem 4.6.12, let $g \in C^\infty(\mathbb{T}^n)$, and let V be an open cube in $\mathbb{R}^n$ with side length equal to 1. Assume that the support of $g|_{\overline{V}}$ is contained in V. Then we have the equality*

$$\widetilde{a}(X,D)g = \mathcal{P} \circ a(X,D)(g|_V),$$

where $g|_V : \mathbb{R}^n \to \mathbb{C}$ is defined as the restriction of g to V and equal to zero outside of V.

Exercise 4.6.17. Work out the details of the proof of Corollary 4.6.16. Especially, the fact that a is 1-periodic plays an important role.

Since we do not always have periodic symbols on $\mathbb{R}^n$ it may be convenient to periodise them.

Definition 4.6.18 (Periodisation of symbols). If $a(X,D)$ is a pseudo-differential operator with symbol $a(x,\xi)$, by $(\mathcal{P}a)(X,D)$ we will denote the pseudo-differential operator with symbol

$$(\mathcal{P}a)(x,\xi) := \sum_{k \in \mathbb{Z}^n} a(x+k,\xi).$$

This procedure makes sense if, for example, a is in $L^1(\mathbb{R}^n)$ with respect to the variable x.

In the following proposition we will assume that supports of symbols and functions are contained in the cube $[-1/2, 1/2]^n$. We note that this is not restrictive if these functions are already compactly supported. Indeed, if supports of $a(\cdot, \xi)$ and f are contained in some compact set independent of ξ, we can find some $N \in \mathbb{N}$ such that they are contained in $[-N/2, N/2]^n$, and then use the analysis on the N-inflated torus, with periodisation operator $\mathcal{P}_N$ instead of $\mathcal{P}$, defined in (4.28).

Proposition 4.6.19 (Operator with periodised symbol). *Let $a = a(x,\xi) \in C^\infty(\mathbb{R}^n \times \mathbb{R}^n)$ satisfy $\operatorname{supp} a \subset [-1/2, 1/2]^n \times \mathbb{R}^n$ and be such that for every $\alpha, \beta \in \mathbb{N}_0^n$ there are constants $C_{\alpha\beta}$ and $M(\alpha, \beta)$ such that the estimate*

$$\left| \partial_x^\alpha \partial_\xi^\beta a(x,\xi) \right| \le C_{\alpha\beta} \langle \xi \rangle^{M(\alpha,\beta)}$$

holds for all $x, \xi \in \mathbb{R}^n$. Then we have

$$a(X,D)f = (\mathcal{P}a)(X,D)f + Rf,$$

*for all $f \in C^\infty(\mathbb{R}^n)$ with supp $f \subset [-1/2, 1/2]^n$. If moreover $a \in S^m_{\rho,\delta}(\mathbb{R}^n \times \mathbb{R}^n)$
with $\rho > 0$, then the operator R extends to a smoothing pseudo-differential operator
$R : \mathcal{D}'(\mathbb{R}^n) \to \mathcal{S}(\mathbb{R}^n)$.*

Proof. By the definition we can write

$$(\mathcal{P}a)(X, D)f(x) = \sum_{k \in \mathbb{Z}^n} \int_{\mathbb{R}^n} e^{i2\pi x \cdot \xi} \, a(x + k, \xi) \, \mathcal{F}_{\mathbb{R}^n} f(\xi) \, d\xi,$$

and let us define $Rf := a(X, D)f - (\mathcal{P}a)(X, D)f$. The assumption on the support
of a implies that for every x there is only one $k \in \mathbb{Z}^n$ for which $a(x + k, \xi) \neq 0$, so
the sum consists of only one term. It follows that $Rf(x) = 0$ for $x \in [-1/2, 1/2]^n$,
because for such x the non-zero term corresponds to $k = 0$. Let now $x \in \mathbb{R}^n \setminus
[-1/2, 1/2]^n$. Since $a(x, \xi) = 0$ for all $\xi \in \mathbb{Z}^n$, we have that the sum

$$Rf(x) = - \sum_{k \in \mathbb{Z}^n, k \neq 0} \int_{\mathbb{R}^n} \int_{\mathbb{R}^n} e^{i2\pi(x-y) \cdot \xi} \, a(x + k, \xi) \, f(y) \, dy \, d\xi$$

is just a single term and $|x - y| > 0$ on supp f, so we can integrate by parts with
respect to ξ any number of times. This implies that $R \in \Psi^{-\infty}(\mathbb{R}^n \times \mathbb{R}^n)$ because
$\rho > 0$ and that Rf decays at infinity faster than any power. The proof is complete
since the same argument can be applied to the derivatives of Rf as well.   □

Exercise 4.6.20. Work out the details for the derivatives of Rf.

Proposition 4.6.19 allows us to extend the formula of Theorem 4.6.12 to
compact perturbations of periodic symbols. We will use it later when $a(X, D)$ is
a sum of a constant coefficient operator and an operator with compactly (in x)
supported symbol.

Corollary 4.6.21 (Periodisation and compactly supported perturbations).
Let $a(X, D)$ be an operator with symbol

$$a(x, \xi) = a_1(x, \xi) + a_0(x, \xi),$$

*where a_1 is as in Theorem 4.6.12, a_1 is 1-periodic in x for every $\xi \in \mathbb{R}^n$, and a_0
is as in Proposition 4.6.19, supported in $[-1/2, 1/2]^n \times \mathbb{R}^n$. Define*

$$\widetilde{b}(x, \xi) := \widetilde{a_1}(x, \xi) + \widetilde{\mathcal{P}a_0}(x, \xi), \quad x \in \mathbb{T}^n, \xi \in \mathbb{Z}^n.$$

Then we have

$$\mathcal{P} \circ a(X, D)f = \widetilde{b}(X, D) \circ \mathcal{P}f + \mathcal{P} \circ Rf \tag{4.30}$$

*for all $f \in \mathcal{S}(\mathbb{R}^n)$, and operator R extends to $R : \mathcal{D}'(\mathbb{R}^n) \to \mathcal{S}(\mathbb{R}^n)$, so that $\mathcal{P} \circ R :
\mathcal{D}'(\mathbb{R}^n) \to C^\infty(\mathbb{T}^n)$. Moreover, if $a_1, a_0 \in S^m_{\rho,\delta}(\mathbb{R}^n \times \mathbb{R}^n)$, then $\widetilde{b} \in S^m_{\rho,\delta}(\mathbb{T}^n \times \mathbb{Z}^n)$.*

Remark 4.6.22. Recalling Remark 4.6.14, (4.30) can be justified for larger function
classes, e.g., for $f \in C^k_0(\mathbb{R}^n)$ for k sufficiently large (which will be of use in Section
4.12).

Proof of Corollary 4.6.21. By Proposition 4.6.19 we can write

$$a(X, D) = a_1(X, D) + (\mathcal{P}a_0)(X, D) + R,$$

with $R : \mathcal{D}'(\mathbb{R}^n) \to \mathcal{S}(\mathbb{R}^n)$. Let us define

$$b(x, \xi) := a_1(x, \xi) + (\mathcal{P}a_0)(x, \xi),$$

so that $\widetilde{b} = b|_{\mathbb{T}^n \times \mathbb{Z}^n}$. The symbol b is 1-periodic, hence for the operator $b(X, D) = a_1(X, D) + (\mathcal{P}a_0)(X, D)$ by Theorem 4.6.12 we have

$$
\begin{aligned}
\mathcal{P} \circ b(X, D) &= \widetilde{b}(X, D) \circ \mathcal{P} \\
&= \widetilde{a_1}(X, D) \circ \mathcal{P} + \widetilde{\mathcal{P}a_0}(X, D) \circ \mathcal{P}.
\end{aligned}
$$

Since $R : \mathcal{D}'(\mathbb{R}^n) \to \mathcal{S}(\mathbb{R}^n)$, we also have $\mathcal{P} \circ R : \mathcal{D}'(\mathbb{R}^n) \to C^\infty(\mathbb{T}^n)$. Finally, if $a_1, a_0 \in S_{\rho,\delta}^m(\mathbb{R}^n \times \mathbb{R}^n)$, then $\widetilde{b} \in S_{\rho,\delta}^m(\mathbb{T}^n \times \mathbb{Z}^n)$ by Theorem 4.5.3. The proof is complete. $\qquad\square$

4.7 Symbolic calculus

In this section we show that (for suitable ρ, δ) the family of periodic pseudo-differential operators is a $*$-algebra, i.e., it is closed under sums (trivially $\sigma_j \in S_{\rho,\delta}^{m_j}(\mathbb{T}^n \times \mathbb{Z}^n) \Rightarrow \sigma_1 + \sigma_2 \in S_{\rho,\delta}^{\max\{m_1, m_2\}}(\mathbb{T}^n \times \mathbb{Z}^n))$, products, and taking adjoints; this algebraic structure is the key property to the applicability of periodic pseudo-differential operators. Furthermore, under these operations the degrees of operators behave as one would expect. In the proofs the symbol analysis techniques are used leaving us with asymptotic expansions, so that there is a point to study periodic pseudo-differential operators that are invertible modulo $\mathrm{Op}(S^{-\infty}(\mathbb{T}^n \times \mathbb{Z}^n))$; that is, the elliptic operators, which are discussed in Section 4.9.

Recall that now there are two types of symbols, toroidal and Euclidean, in (4.6) and (4.20), yielding two alternative (toroidal and Euclidean) quantizations for operators, respectively, see Corollary 4.6.13. As usual, we emphasize this difference by writing $\sigma \in S_{\rho,\delta}^m(\mathbb{T}^n \times \mathbb{Z}^n)$ and $\sigma \in S_{\rho,\delta}^m(\mathbb{T}^n \times \mathbb{R}^n)$, respectively.

Now we will discuss the calculus of pseudo-differential operators with toroidal symbols. For this, let us fix the notation first and recall discrete versions of derivatives from Definition 3.4.1:

$$D_y^{(\alpha)} = D_{y_1}^{(\alpha_1)} \cdots D_{y_n}^{(\alpha_n)}, \qquad (4.31)$$

where $D_{y_j}^{(0)} = I$ and

$$
\begin{aligned}
D_{y_j}^{(k+1)} &= D_{y_j}^{(k)} \left(\frac{\partial}{\mathrm{i}2\pi \partial y_j} - kI \right) \\
&= \frac{\partial}{\mathrm{i}2\pi \partial y_j} \left(\frac{\partial}{\mathrm{i}2\pi \partial y_j} - I \right) \cdots \left(\frac{\partial}{\mathrm{i}2\pi \partial y_j} - kI \right).
\end{aligned}
$$

Also, in this section the equivalence of asymptotic sums in Definition 4.1.19 will be of use. We now observe how the difference operator affects expansions.

Lemma 4.7.1. *Let* $0 \leq \delta < \rho \leq 1$. *Assume that* $\sigma \in S^m_{\rho,\delta}(\mathbb{T}^n \times \mathbb{R}^n)$. *Then*

$$\sum_{\gamma=0}^{\infty} \frac{1}{\gamma!} \, \triangle^\gamma_\xi D^{(\gamma)}_x \sigma(x,\xi) \overset{-\infty}{\sim} \sum_{\gamma=0}^{\infty} \frac{1}{\gamma!} \partial^\gamma_\xi D^\gamma_x \sigma(x,\xi) \tag{4.32}$$

$$= \exp\left(\partial_\xi D_x\right) \sigma(x,\xi),$$

where exp *is used in abbreviating the right-hand side of* (4.32).

In the sequel we will drop the infinity sign from $\overset{-\infty}{\sim}$ and will simply write $\sim$.

Proof. We apply Theorem 3.3.39 in order to translate differences into derivatives, and use the definition of the Stirling numbers[1] of the second kind:

$$\sum_{\alpha \geq 0} \frac{1}{\alpha!} \, \triangle^\alpha_\xi D^{(\alpha)}_x \sigma(x,\xi)$$

$$\overset{m-(\rho-\delta)N, \rho, \delta}{\sim} \sum_{|\alpha|<N} \frac{1}{\alpha!} \left[\sum_{|\gamma|<N} \frac{\alpha!}{\gamma!} \left\{ \begin{matrix} \gamma \\ \alpha \end{matrix} \right\} \partial^\gamma_\xi \right] D^{(\alpha)}_x \sigma(x,\xi)$$

$$= \sum_{|\gamma|<N} \frac{1}{\gamma!} \partial^\gamma_\xi \left[\sum_{|\alpha|<N} \left\{ \begin{matrix} \gamma \\ \alpha \end{matrix} \right\} D^{(\alpha)}_x \right] \sigma(x,\xi)$$

$$= \sum_{|\gamma|<N} \frac{1}{\gamma!} \partial^\gamma_\xi D^\gamma_x \sigma(x,\xi).$$

Since we can do this for any N, the proof is complete. $\qquad\qquad\square$

According to Remark 4.5.4, we can do the same for amplitudes:

Proposition 4.7.2. *Let* $0 \leq \delta < \rho \leq 1$. *If* $a \in \mathcal{A}^m_{\rho,\delta}(\mathbb{T}^n)$ *is an extended defined on* $\mathbb{T}^n \times \mathbb{T}^n \times \mathbb{R}^n$, *then*

$$\sigma_{\mathrm{Op}(a)}(x,\xi) \sim \sum_{\gamma=0}^{\infty} \frac{1}{\gamma!} \partial^\gamma_\xi D^\gamma_y a(x,y,\xi)|_{y=x} \tag{4.33}$$

$$= \exp\left(\partial_\xi D_y\right) a(x,y,\xi)|_{y=x}.$$

Proof. Replace the symbol $\sigma(x,\xi)$ and the operators $D^{(\gamma)}_x$, D^γ_x of (4.32) by $a(x,y,\xi)$ and $D^{(\gamma)}_y$, D^γ_y, respectively. Then the expansion (4.18) yields the claim. $\qquad\square$

---

[1] For Stirling numbers see Definition 3.3.33 and for their properties see Section 3.3.4.

It is time to prove that the adjoints of periodic pseudo-differential operators are periodic pseudo-differential operators by constructing an asymptotic expansion. As a secondary result we notice that the adjoint operation $A \mapsto A^*$ does not change the order of the periodic pseudo-differential operator. A direct proof of this by-product, not using the expansion method, is presented in Remark 4.9.10. In $\mathbb{R}^n$ we proved the adjoint case in Theorem 2.5.13 leaving the transpose as Exercise 2.5.18, and here we will do the opposite. Thus, we will prove this result for transpose operators first, for their definition see Definition 2.5.15 as well as Exercise 2.5.16 and Remark 2.5.17.

Theorem 4.7.3 (Transpose operators). *Let* $0 \le \delta < \rho \le 1$. *Let* $A \in \Psi^m_{\rho,\delta}(\mathbb{T}^n \times \mathbb{Z}^n)$ *be a pseudo-differential operator with toroidal symbol* $\sigma_A \in S^m_{\rho,\delta}(\mathbb{T}^n \times \mathbb{Z}^n)$. *Then the transpose* A^t *of* A *is in* $\Psi^m_{\rho,\delta}(\mathbb{T}^n \times \mathbb{Z}^n)$ *and its symbol* $\sigma_{A^t} \in S^m_{\rho,\delta}(\mathbb{T}^n \times \mathbb{Z}^n)$ *has the following asymptotic expansion:*

$$\sigma_{A^t}(x, \xi) \sim \sum_{\gamma \ge 0} \frac{1}{\gamma!} \, \triangle^\gamma_\xi D^{(\gamma)}_x \sigma_A(x, -\xi). \tag{4.34}$$

Remark 4.7.4. We note that the constant $(2\pi i)^{-|\alpha|}$ which is present in the asymptotic formula in Exercise 2.5.18 does not appear in (4.34). This is due to the fact that discrete modifications of derivatives in (4.31) do not have homogeneous symbols and hence the constant remains incorporated in their definition.

Proof of Theorem 4.7.3. Assume that $u, v \in C^\infty(\mathbb{T}^n)$. We make use of the integral representation of the duality product and the definition of amplitude operators. In general, if the operator A has the amplitude $a(x, y, \xi) = \sigma_A(x, \xi)$, we have

$$\int_{\mathbb{T}^n} v(x) \, A^t u(x) \, dx$$

$$= \int_{\mathbb{T}^n} u(y) \, Av(y) \, dy$$

$$= \int_{\mathbb{T}^n} u(y) \left(\sum_{\xi \in \mathbb{Z}^n} \int_{\mathbb{T}^n} e^{i2\pi(y-x)\cdot\xi} \, a(y, x, \xi) \, v(x) \, dx \right) dy$$

$$= \int_{\mathbb{T}^n} v(x) \left(\sum_{\xi \in \mathbb{Z}^n} \int_{\mathbb{T}^n} e^{i2\pi(y-x)\cdot\xi} \, a(y, x, \xi) \, u(y) \, dy \right) dx.$$

Thus $A^t = \mathrm{Op}(a^t)$ with $a^t(x, y, \xi) = a(y, x, -\xi)$. Especially,

$$(x, y, \xi) \mapsto \sigma_A(y, -\xi)$$

is an amplitude of A^t, so that by (4.18) the asymptotic expansion in the discrete case follows. $\qquad\square$

Remark 4.7.5. Notice that in the proof of Theorem 4.7.3, the sum over ξ converges absolutely if $a \in \mathcal{A}^m_{\rho,\delta}(\mathbb{T}^n)$ with $m < -n$. Otherwise, this can be understood by a suitable integration by parts, as usual.

In fact, we also proved the following:

Corollary 4.7.6. *Let $0 \le \delta < \rho \le 1$. If A is an amplitude operator with toroidal amplitude $a \in \mathcal{A}^m_{\rho,\delta}(\mathbb{T}^n)$, then A^t is also an amplitude operator in $\mathrm{Op}(\mathcal{A}^m_{\rho,\delta}(\mathbb{T}^n))$ with amplitude*

$$a^t(x, y, \xi) = a(y, x, -\xi).$$

Theorem 4.7.7 (Adjoint operators). *Let $0 \le \delta < \rho \le 1$. Let $A \in \Psi^m_{\rho,\delta}(\mathbb{T}^n \times \mathbb{Z}^n)$ be a pseudo-differential operator with toroidal symbol $\sigma_A \in S^m_{\rho,\delta}(\mathbb{T}^n \times \mathbb{Z}^n)$. Then the adjoint A^* of A is in $\Psi^m_{\rho,\delta}(\mathbb{T}^n \times \mathbb{Z}^n)$ and its toroidal symbol $\sigma_{A^*} \in S^m_{\rho,\delta}(\mathbb{T}^n \times \mathbb{Z}^n)$ has the following asymptotic expansion:*

$$\sigma_{A^*}(x, \xi) \sim \sum_{\alpha \ge 0} \frac{1}{\alpha!} \, \triangle^\alpha_\xi D^{(\alpha)}_x \overline{\sigma_A(x, \xi)}. \tag{4.35}$$

Exercise 4.7.8. Accordingly, show that if A is an amplitude operator with toroidal amplitude in $\mathcal{A}^m_{\rho,\delta}(\mathbb{T}^n)$, then A^* is also an amplitude operator in $\mathrm{Op}(\mathcal{A}^m_{\rho,\delta}(\mathbb{T}^n))$ with amplitude $a^*(x, y, \xi) = \overline{a(y, x, \xi)}$. Hence prove Theorem 4.7.7.

The next lemma paves the way for a proof that a composition of periodic pseudo-differential operators is a periodic pseudo-differential operator:

Lemma 4.7.9 (Product of toroidal symbols). *The pointwise product of toroidal symbols $\sigma_A \in S^m_{\rho,\delta}(\mathbb{T}^n \times \mathbb{Z}^n)$ and $\sigma_B \in S^l_{\rho,\delta}(\mathbb{T}^n \times \mathbb{Z}^n)$ is a toroidal symbol in $S^{m+l}_{\rho,\delta}(\mathbb{T}^n \times \mathbb{Z}^n)$.*

Proof. This is a simple calculation exploiting both the classical and the discrete Leibniz formulae (Theorem 1.5.10, (iv), and Lemma 3.3.6) the defining inequality of the symbol of a periodic pseudo-differential operator, and Peetre's inequality (Proposition 3.3.31):

$$\left| \triangle^\alpha_\xi \partial^\beta_x \left[\sigma_B(x, \xi) \, \sigma_A(x, \xi) \right] \right|$$

$$= \left| \sum_{\gamma \le \beta} \sum_{\omega \le \alpha} \binom{\beta}{\gamma} \binom{\alpha}{\omega} \left[\partial^\gamma_x \triangle^\omega_\xi \sigma_B(x, \xi) \right] \partial^{\beta-\gamma}_x \triangle^{\alpha-\omega}_\xi \sigma_A(x, \xi + \omega) \right|$$

$$\le 2^{|\alpha+\beta|} \sum_{\gamma \le \beta} \sum_{\omega \le \alpha} C^B_{\gamma,\omega} \, \langle \xi \rangle^{l - \rho|\omega| + \delta|\gamma|} \, C^A_{\beta-\gamma,\alpha-\omega} \, \langle \xi + \omega \rangle^{m - \rho|\alpha-\omega| + \delta|\beta-\gamma|}$$

$$\le C_{\alpha\beta} \, \langle \xi \rangle^{m + l - \rho|\alpha| + \delta|\beta|}. \qquad \square$$

Curiously enough, $\sigma_A \sigma_B$ is the principal symbol of both AB and BA, and therefore the commutator of periodic pseudo-differential operators is smoother than it may seem at the first look – this has nice consequences.

Theorem 4.7.10 (Toroidal composition formula). *Let* $0 \leq \delta < \rho \leq 1$. *The composition* BA *of two operators* $B \in \mathrm{Op}(S_{\rho,\delta}^{l}(\mathbb{T}^n \times \mathbb{Z}^n))$ *and* $A \in \mathrm{Op}(S_{\rho,\delta}^{m}(\mathbb{T}^n \times \mathbb{Z}^n))$ *is in* $\mathrm{Op}(S_{\rho,\delta}^{m+l}(\mathbb{T}^n \times \mathbb{Z}^n))$, *and its toroidal symbol has the following asymptotic expansion:*

$$\sigma_{BA}(x,\xi) \sim \sum_{\gamma \geq 0} \frac{1}{\gamma!} \left[\triangle_{\xi}^{\gamma} \sigma_B(x,\xi) \right] D_x^{(\gamma)} \sigma_A(x,\xi). \tag{4.36}$$

Thereby the commutator $[A, B] = AB - BA$ *belongs to the operator class*

$$\mathrm{Op}(S_{\rho,\delta}^{m+l-(\rho-\delta)}(\mathbb{T}^n \times \mathbb{Z}^n)).$$

Remark 4.7.11. Formula (4.36) can be compared with the Euclidean formula that we can get by Theorem 4.5.3 by extending toroidal symbols to $\mathbb{T}^n \times \mathbb{R}^n$:

$$\begin{aligned}
\sigma_{BA}(x,\xi) &\sim \sum_{\gamma \geq 0} \frac{1}{\gamma!} \left[\partial_{\xi}^{\gamma} \sigma_B(x,\xi) \right] D_x^{\gamma} \sigma_A(x,\xi) \tag{4.37} \\
&= \exp\left(\partial_{\xi} D_x \right) \left[\sigma_B(y,\xi)\sigma_A(x,\eta) \right]|_{(y,\eta)=(x,\xi)}.
\end{aligned}$$

Remark 4.7.12. In Remark 4.13.9 we will derive another formula for the composition of two pseudo-differential operators where the resulting operator BA is in the form of an amplitude operator with amplitude $c(x, y, \xi)$ having the asymptotic expansion:

$$c(x, z, \xi) \sim \sum_{\alpha \geq 0} \frac{1}{\alpha!} \sigma_B(x,\xi) (-D_z)^{(\alpha)} \triangle_{\xi}^{\alpha} \sigma_A(z,\xi).$$

Consequently, applying Theorem 4.4.5 we can obtain another representation for the symbol σ_{BA}.

Proof of Theorem 4.7.10. Of course, by going through the procedure

$$\begin{aligned}
\sigma_{BA}(x,\xi) &= \mathrm{e}^{-\mathrm{i}2\pi x \cdot \xi} \left[BA e_\xi(x) \right] \\
&= \mathrm{e}^{-\mathrm{i}2\pi x \cdot \xi} \sum_{\eta \in \mathbb{Z}^n} \int_{\mathbb{T}^n} \mathrm{e}^{\mathrm{i}2\pi(x-y)\cdot\eta} \; \sigma_B(x,\eta) \; A e_\xi(y) \; \mathrm{d}y \\
&= \sum_{\eta \in \mathbb{Z}^n} \int_{\mathbb{T}^n} \mathrm{e}^{\mathrm{i}2\pi(x-y)\cdot(\eta-\xi)} \; \sigma_B(x,\eta) \; \sigma_A(y,\xi) \; \mathrm{d}y \\
&= \sum_{\eta \in \mathbb{Z}^n} \mathrm{e}^{\mathrm{i}2\pi x \cdot(\eta-\xi)} \; \sigma_B(x,\eta) \; \widehat{\sigma_A}(\eta - \xi,\xi) \\
&= \sum_{\eta \in \mathbb{Z}^n} \mathrm{e}^{\mathrm{i}2\pi x \cdot \eta} \; \sigma_B(x,\eta + \xi) \; \widehat{\sigma_A}(\eta,\xi),
\end{aligned}$$

one gets the exact symbol of the composition of the periodic pseudo-differential operators. However, as this representation cannot be used to effectively approximate BA, it is of minor importance: this is why we need an asymptotic expansion.

Now we can use the discrete Taylor series of Theorem 3.3.21 for $\sigma_B(x, \eta + \xi)$ to obtain

$$\sigma_{BA}(x, \xi) = \sum_{\eta \in \mathbb{Z}^n} e^{i2\pi x \cdot \eta} \, \sigma_B(x, \eta + \xi) \, \widehat{\sigma_A}(\eta, \xi)$$

$$= \sum_{\eta \in \mathbb{Z}^n} e^{i2\pi x \cdot \eta} \left[\sum_{|\gamma| < N} \frac{1}{\gamma!} \triangle_\xi^\gamma \sigma_B(x, \xi) \, \eta^{(\gamma)} + R_N(x, \xi, \eta) \right] \widehat{\sigma_A}(\eta, \xi)$$

$$= \sum_{|\gamma| < N} \frac{1}{\gamma!} \left[\triangle_\xi^\gamma \sigma_B(x, \xi) \right] \sum_{\eta \in \mathbb{Z}^n} e^{i2\pi x \cdot \eta} \, \widehat{\sigma_A}(\eta, \xi) \, \eta^{(\gamma)}$$

$$+ \sum_{\eta \in \mathbb{Z}^n} e^{i2\pi x \cdot \eta} \, R_N(x, \xi, \eta) \, \widehat{\sigma_A}(\eta, \xi)$$

$$= \sum_{|\gamma| < N} \frac{1}{\gamma!} \left[\triangle_\xi^\gamma \sigma_B(x, \xi) \right] D_x^{(\gamma)} \sigma_A(x, \xi) + E_N(x, \xi),$$

where $E_N(x, \xi) = \sum_{\eta \in \mathbb{Z}^n} e^{i2\pi x \cdot \eta} R_N(x, \xi, \eta) \, \widehat{\sigma_A}(\eta, \xi)$. By Lemma 4.7.9, the first finite sum here is in $S_{\rho,\delta}^{m+l}(\mathbb{T}^n \times \mathbb{Z}^n)$. We only need to prove that the error term $E_N(x, \xi)$ satisfies $E_N \in S_{\rho,\delta}^{m+l-\rho N}(\mathbb{T}^n \times \mathbb{Z}^n)$, or

$$\left| \triangle_\xi^\alpha \partial_x^\beta E_N(x, \xi) \right| = \left| \triangle_\xi^\alpha \partial_x^\beta \sum_{\eta \in \mathbb{Z}^n} e^{i2\pi x \cdot \eta} \, R_N(x, \xi, \eta) \, \widehat{\sigma_A}(\eta, \xi) \right|$$

$$\leq c_{\alpha\beta N} \, \langle \xi \rangle^{m+l-\rho(N+|\alpha|)+\delta|\beta|}.$$

In fact, we will show even more. Indeed, by inequality (4.12) of Lemma 4.2.1 we have, with any $r \in \mathbb{R}$,

$$\left| \triangle_\xi^{\alpha_A} \partial_x^{\beta_A} \left[e^{i2\pi x \cdot \eta} \, \widehat{\sigma_A}(\eta, \xi) \right] \right| \leq c_{\alpha_A, \beta_A, r}^A \, \langle \eta \rangle^{|\beta_A| - r} \langle \xi \rangle^{m - \rho|\alpha_A| + \delta r}. \tag{4.38}$$

The error term of the discrete Taylor series (see Theorem 3.3.21) and $\sigma_B \in S_{\rho,\delta}^l(\mathbb{T}^n \times \mathbb{Z}^n)$ give

$$\left| \triangle_\xi^{\alpha_R} \partial_x^{\beta_R} R_N(x, \xi, \eta) \right| \leq c_{\beta_R, N}^R \, \langle \eta \rangle^N \max_{|\omega| = N, \, \nu \in Q(\eta)} \left| \triangle_\xi^{\omega + \alpha_R} \partial_x^{\beta_R} \sigma_B(x, \xi + \nu) \right|$$

$$\leq c_{\beta_R, N}^R \, \langle \eta \rangle^N \max_{\nu \in Q(\eta)} \left[C_{\beta_R, \omega, \alpha_R}^B \langle \xi + \nu \rangle^{l - \rho(N + |\alpha_R|) + \delta|\beta_R|} \right].$$

Multiplying this estimate with (4.38) and taking $\alpha_A + \alpha_R = \alpha$ and $\beta_A + \beta_R = \beta$, by the discrete Leibniz formula (Lemma 3.3.6) we get

$$\left| \triangle_\xi^\alpha \partial_x^\beta E_N(x, \xi) \right| \leq C \langle \xi \rangle^{m - \rho|\alpha_A| + \delta r} \tag{4.39}$$

$$\times \sum_{\eta \in \mathbb{Z}^n} \langle \eta \rangle^{N + |\beta_A| - r} \max_{\nu \in Q(\eta)} \left[\langle \xi + \nu \rangle^{l - \rho(N + |\alpha_R|) + \delta|\beta_R|} \right].$$

We split the sum in (4.39) into sums $\sum_{|\eta| \le |\xi|/2}$ and $\sum_{|\eta| > |\xi|/2}$. We can estimate

$$\langle \xi \rangle^{m-\rho|\alpha_A|+\delta r} \sum_{|\eta| \le |\xi|/2} \langle \eta \rangle^{N+|\beta_A|-r} \max_{\nu \in Q(\eta)} \left[\langle \xi + \nu \rangle^{l-\rho(N+|\alpha_R|)+\delta|\beta_R|} \right]$$

$$\le C \langle \xi \rangle^{m-\rho|\alpha_A|+\delta r+l-\rho(N+|\alpha_R|)+\delta|\beta_R|+n} \max_{|\eta| \le |\xi|/2} \langle \eta \rangle^{N+|\beta_A|-r}.$$

Taking $r = N + |\beta_A|$ and N large enough we can estimate it by any $\langle \xi \rangle^{-N'}$ in view of $\delta < \rho$. Taking N large enough, the other sum in (4.39) can be estimated as

$$\langle \xi \rangle^{m-\rho|\alpha_A|+\delta r} \sum_{|\eta| > |\xi|/2} \langle \eta \rangle^{N+|\beta_A|-r} \max_{\nu \in Q(\eta)} \left[\langle \xi + \nu \rangle^{l-\rho(N+|\alpha_R|)+\delta|\beta_R|} \right]$$

$$\le C \langle \xi \rangle^{m-\rho|\alpha_A|+\delta r} \sum_{|\eta| > |\xi|/2} \langle \eta \rangle^{N+|\beta_A|-r}$$

$$\le C \langle \xi \rangle^{m-\rho|\alpha_A|+\delta r+N+|\beta_A|-r+1}$$

if r is large compared to N. Taking r large enough we can estimate the final expression by any $\langle \xi \rangle^{-N'}$ in view of $\delta < 1$. Hence $E_N \in S_{\rho,\delta}^{m+l-\rho N}(\mathbb{T}^n \times \mathbb{Z}^n)$, so that by Theorem 4.4.1 formula (4.36) is valid.

Now the study of the commutator remains. Exploiting the first terms of the asymptotic expansions, we see that $\sigma_{BA} = \sigma_B \sigma_A + E_{BA}$ and $\sigma_{AB} = \sigma_A \sigma_B + E_{AB}$, where $E_{BA}, E_{AB} \in S_{\rho,\delta}^{m+l-(\rho-\delta)}(\mathbb{T}^n \times \mathbb{Z}^n)$. Thus

$$\begin{aligned} \sigma_{BA} - \sigma_{AB} &= (\sigma_{BA} - \sigma_B \sigma_A) - (\sigma_{AB} - \sigma_A \sigma_B) \\ &= E_{BA} - E_{AB} \\ &\in S_{\rho,\delta}^{m+l-(\rho-\delta)}(\mathbb{T}^n \times \mathbb{Z}^n), \end{aligned}$$

and consequently $[B, A] \in \mathrm{Op}(S_{\rho,\delta}^{m+l-(\rho-\delta)}(\mathbb{T}^n \times \mathbb{Z}^n))$.

The asymptotic expansion (4.37) in Remark 4.7.11 of extended symbols is obtained by applying (4.32) on (4.36). $\qquad \square$

Remark 4.7.13. Note that one could use the extended symbols in order to be able to apply the classical analysis (derivatives and traditional Taylor series) – then one would use equivalence (4.32) to return to differences. But as it has been expressed, this is not a natural way of handling periodic pseudo-differential operators, as the discrete Taylor series of Theorem 3.3.21 for $\sigma_B(x, \eta + \xi)$ can be used.

For classical periodic pseudo-differential operators, formula (4.37) has been proved in another way by Johannes Elschner in [34, p. 129-130]. Without proofs formulae (4.34), (4.36), and (4.37) have been announced in [142].

4.8   Operators on $L^2(\mathbb{T}^n)$ and Sobolev spaces

In this section we will discuss what conditions on the toroidal symbol a guarantee the L^2-boundedness of the corresponding pseudo-differential operator $\mathrm{Op}(a)$: $C^\infty(\mathbb{T}^n) \to \mathcal{D}'(\mathbb{T}^n)$. We also discuss Sobolev spaces. In Proposition 4.2.3 we showed in particular that operators with toroidal symbols in the class $S^0_{0,0}(\mathbb{T}^n \times \mathbb{Z}^n)$ are bounded on all Sobolev spaces. However, here we show that for the boundedness of an operator on $L^2(\mathbb{T}^n)$ and on $H^s(\mathbb{T}^n)$ it is not necessary that symbols belong to the toroidal symbol class $S^0_{0,0}(\mathbb{T}^n \times \mathbb{Z}^n)$. In fact, already for the Sobolev boundedness in Corollary 4.2.4 we showed that if $m, s \in \mathbb{R}$ and if an integer $r \in \mathbb{N}$ is such that $r > |s - m| + n$, then for $a \in C^\infty(\mathbb{T}^n \times \mathbb{Z}^n)$ satisfying estimates

$$\left|\partial_x^\beta a(x,\xi)\right| \le C_{a\beta m} \, \langle\xi\rangle^m$$

for all multi-indices $|\beta| \le r$, all $x \in \mathbb{T}^n$ and all $\xi \in \mathbb{Z}^n$, the operator $\mathrm{Op}(a)$ extends to a bounded linear operator from $H^s(\mathbb{T}^n)$ to $H^{s-m}(\mathbb{T}^n)$. In particular, if we take $s = m = 0$, this would require $r = n + 1$ derivatives of a with respect to x to be bounded. We now show that this number can be improved:

Theorem 4.8.1 (Boundedness on $L^2(\mathbb{T}^n)$). *Let $k \in \mathbb{N}$ and $k > n/2$. Let $a : \mathbb{T}^n \times \mathbb{Z}^n \to \mathbb{C}$ be such that*

$$\left|\partial_x^\beta a(x,\xi)\right| \le C \quad \text{for all } (x,\xi) \in \mathbb{T}^n \times \mathbb{Z}^n, \tag{4.40}$$

and all $|\beta| \le k$. Then the operator $\mathrm{Op}(a)$ extends to a bounded linear operator on $L^2(\mathbb{T}^n)$.

Remark 4.8.2. We note that compared with several well-known theorems on the L^2-boundedness of pseudo-differential operators (see, e.g., Calderon and Vaillancourt [22], Coifman and Meyer [23], Cordes [25]), Theorem 4.8.1 does not require any regularity with respect to the ξ-variable. In fact, the boundedness of all partial differences of all orders ≥ 1 with respect to ξ follows automatically from (4.40) in view of the formulae for difference operators in Proposition 3.3.4.

Proof of Theorem 4.8.1. Let us define

$$A_y f(x) := \sum_{\xi \in \mathbb{Z}^n} \int_{\mathbb{T}^n} e^{i2\pi(x-z)\cdot\xi} \, a(y,\xi) \, f(z) \, \mathrm{d}z,$$

so that $A_x f(x) = \mathrm{Op}(a) f(x)$. Then

$$\| \mathrm{Op}(a) f \|^2_{L^2(\mathbb{T}^n)} = \int_{\mathbb{T}^n} |A_x f(x)|^2 \, \mathrm{d}x \le \int_{\mathbb{T}^n} \sup_{y \in \mathbb{T}^n} |A_y f(x)|^2 \, \mathrm{d}x,$$

and by an application of the Sobolev embedding theorem we get

$$\sup_{y \in \mathbb{T}^n} |A_y f(x)|^2 \le C \sum_{|\alpha| \le k} \int_{\mathbb{T}^n} |\partial_y^\alpha A_y f(x)|^2 \, \mathrm{d}y.$$

Therefore, using Fubini's theorem to change the order of integration, we obtain

$$
\begin{aligned}
\|\operatorname{Op}(a)f\|_{L^2(\mathbb{T}^n)}^2 \;&\leq\; C\sum_{|\alpha|\leq k}\int_{\mathbb{T}^n}\int_{\mathbb{T}^n}|\partial_y^\alpha A_y f(x)|^2\ \mathrm{d}x\ \mathrm{d}y \\[4pt]
&\leq\; C\sum_{|\alpha|\leq k}\sup_{y\in\mathbb{T}^n}\int_{\mathbb{T}^n}|\partial_y^\alpha A_y f(x)|^2\ \mathrm{d}x \\[4pt]
&=\; C_k\sum_{|\alpha|\leq k}\sup_{y\in\mathbb{T}^n}\|\partial_y^\alpha A_y f\|_{L^2(\mathbb{T}^n)}^2 \\[4pt]
&\leq\; C\sum_{|\alpha|\leq k}\sup_{y\in\mathbb{T}^n}\sup_{\xi\in\mathbb{Z}^n}\|\partial_y^\alpha a(y,\xi)\|_{L^2(\mathbb{T}^n)}^2\|f\|_{L^2(\mathbb{T}^n)}^2,
\end{aligned}
$$

completing the proof. □

From a suitable adaptation of the composition Theorem 4.7.10 we immediately obtain the result in Sobolev spaces:

Corollary 4.8.3 (Boundedness on $H^s(\mathbb{T}^n)$). *Let $m \in \mathbb{R}$ and let $a : \mathbb{T}^n \times \mathbb{Z}^n \to \mathbb{C}$ be such that*

$$
|\partial_x^\beta a(x,\xi)| \leq C\langle\xi\rangle^m \qquad \text{for all } (x,\xi) \in \mathbb{T}^n \times \mathbb{Z}^n, \tag{4.41}
$$

and all multi-indices β. Then the operator $\operatorname{Op}(a)$ extends to a bounded linear operator from $H^s(\mathbb{T}^n)$ to $H^{s-m}(\mathbb{T}^n)$, for any $s \in \mathbb{R}$.

We refer to Remark 4.2.9 for a discussion of this corollary. In particular, we note that Corollary 4.2.4 still yields a better result because it gives an estimate on the number of derivatives sufficient for the Sobolev boundedness. We note that Corollary 4.8.3 can be also obtained by a functional analytic argument without calculus, see Remark 10.8.3.

Now, let us briefly discuss the case of the $L^p(\mathbb{T}^n)$-boundedness. First we discuss Sobolev spaces over $L^p(\mathbb{T}^n)$.

Remark 4.8.4 **(Pseudo-differential description of $H^s(\mathbb{T}^n)$).** Observe that in the pseudo-differential language the definition of Sobolev spaces in Definition 3.2.2 can be reformulated. Indeed, since the toroidal symbol of the Laplacian $\mathcal{L}$ is $\sigma_{\mathcal{L}}(x,\xi) = -4\pi^2|\xi|^2$, it is immediate that $f \in H^s(\mathbb{T}^n)$ if and only if $(1 - (4\pi^2)^{-1}\mathcal{L})^{s/2}f \in L^2(\mathbb{T}^n)$. Of course, constant $(4\pi^2)^{-1}$ here is not essential.

Definition 4.8.5 (Sobolev spaces $L_s^p(\mathbb{T}^n)$). Let $1 \leq p \leq \infty$ and $s \in \mathbb{R}$. We say that $f \in L_s^p(\mathbb{T}^n)$ if $(1 - \mathcal{L})^{s/2}f \in L^p(\mathbb{T}^n)$.

By Theorem C.4.9 we know that $L^p(\mathbb{T}^n)$ is a Banach space, and by a similar proof to that of Theorem 1.5.10, (v), we can conclude that the spaces $L_s^p(\mathbb{T}^n)$ are Banach spaces for $s \in \mathbb{N}$.

Exercise 4.8.6. Prove that $L_s^p(\mathbb{T}^n)$ are Banach spaces for all $s \in \mathbb{R}$.

In [82], Molahajloo and Wong showed that if $k > n/2$, $k \in \mathbb{N}$, and if the symbol $a(x, \xi)$ satisfies estimates

$$\left|\partial_x^\beta \Delta_\xi^\alpha a(x, \xi)\right| \leq C\langle\xi\rangle^{-|\alpha|} \quad \text{for all } (x, \xi) \in \mathbb{T}^n \times \mathbb{Z}^n,$$

and all multi-indices α, β such that $|\alpha| \leq k$ and $|\beta| \leq k$, then the operator $a(X, D)$ is bounded on $L^p(\mathbb{T}^n)$, $1 < p < \infty$. As a consequence of a suitably adapted Theorem 4.7.10 we obtain the boundedness on Sobolev space over $L^p(\mathbb{T}^n)$:

Corollary 4.8.7 (Boundedness on Sobolev spaces $L_s^p(\mathbb{T}^n)$). *Let $m \in \mathbb{R}$, $k \in \mathbb{N}$, $k > n/2$, and let $a : \mathbb{T}^n \times \mathbb{Z}^n \to \mathbb{C}$ be such that*

$$\left|\partial_x^\beta \Delta_\xi^\alpha a(x, \xi)\right| \leq C\langle\xi\rangle^{m-|\alpha|} \quad \text{for all } (x, \xi) \in \mathbb{T}^n \times \mathbb{Z}^n, \tag{4.42}$$

and all multi-indices α such that $|\alpha| \leq k$ and all multi-indices β. Then the operator $\mathrm{Op}(a)$ extends to a bounded linear operator from $L_s^p(\mathbb{T}^n)$ to $L_{s-m}^p(\mathbb{T}^n)$, for any $1 < p < \infty$ and any $s \in \mathbb{R}$.

Exercise 4.8.8. Work out the details of the proof.

4.9 Elliptic pseudo-differential operators on $\mathbb{T}^n$

Clearly, elliptic pseudo-differential operators on compact manifolds have been studied in great depth in view of numerous applications, so we can only touch upon the general theory of this subject. However, here we can analyse them in terms of the developed toroidal quantizations. Thus, next we study elliptic operators on $\mathbb{T}^n$, that is operators that can be described as operators "invertible with respect to smoothness".

Definition 4.9.1 (Elliptic pseudo-differential operators on $\mathbb{T}^n$). A toroidal symbol σ_A and the corresponding periodic pseudo-differential operator $A = \mathrm{Op}(\sigma_A) \in \mathrm{Op}(S^m(\mathbb{T}^n \times \mathbb{Z}^n))$ are called *elliptic of order $m \in \mathbb{R}$*, if σ_A satisfies

$$\forall (x, \xi) \in \mathbb{T}^n \times \mathbb{Z}^n : \quad \|\xi\| \geq n_0 \Rightarrow |\sigma_A(x, \xi)| \geq c_0 \langle\xi\rangle^m \tag{4.43}$$

for some constants $n_0, c_0 > 0$.

Remark 4.9.2. Note that if A is elliptic, it cannot belong to any $\mathrm{Op}(S^l(\mathbb{T}^n \times \mathbb{Z}^n))$ with $l < m$. That is to say, elliptic periodic pseudo-differential operators have sharp minimal degrees. In the definition n_0 does not play any significant role: A differs only by an $\mathrm{Op}(S^{-\infty}(\mathbb{T}^n \times \mathbb{Z}^n))$-operator from a periodic pseudo-differential operator A' that satisfies inequality (4.43) for all $\xi \in \mathbb{Z}^n$. Therefore we may assume without any further comment that $|\sigma_A(x, \xi)| \geq c_0 \langle\xi\rangle^m$ holds for all $\xi \in \mathbb{Z}^n$ for an elliptic operator A.

Remark 4.9.3 (Elliptic $\Longrightarrow$ Fredholm). Elliptic periodic pseudo-differential operators are Fredholm operators, as it will be shown in Theorem 4.9.17. Therefore the Fredholm theory is a closely associated subject in further studies but it falls outside the scope of this treatise. Nevertheless, later we touch upon it briefly.

In concordance with the pointwise product of symbols (Lemma 4.7.9), the toroidal symbol τ of an elliptic operator yields an elliptic toroidal symbol $1/\tau$:

Lemma 4.9.4. *Let σ be a toroidal symbol in $S^m(\mathbb{T}^n \times \mathbb{Z}^n)$, and let $\tau \in S^\ell(\mathbb{T}^n \times \mathbb{Z}^n)$ be an elliptic toroidal symbol such that $|\tau(x,\xi)| \geq c_0 \langle\xi\rangle^\ell$. Then $\sigma/\tau \in S^{m-\ell}(\mathbb{T}^n \times \mathbb{Z}^n)$, and $1/\tau$ is elliptic of order $-\ell$.*

Proof. We begin by observing how derivatives and differences act on σ/τ. The first case is familiar from calculus:

$$\partial_{x_j} \frac{\sigma(x,\xi)}{\tau(x,\xi)} = \frac{\tau(x,\xi)\partial_{x_j}\sigma(x,\xi) - \sigma(x,\xi)\partial_{x_j}\tau(x,\xi)}{\tau(x,\xi)^2}.$$

The action of the difference is quite similar:

$$\triangle_{\xi_k} \frac{\sigma(x,\xi)}{\tau(x,\xi)} = \frac{\tau(x,\xi)\triangle_{\xi_k}\sigma(x,\xi) - \sigma(x,\xi)\triangle_{\xi_k}\tau(x,\xi)}{\tau(x,\xi)\tau(x,\xi+v_k)}.$$

Thus we are motivated to define $\sigma_{0,0} := \sigma$, $\tau_{0,0} := \tau$, and recursively

$$\begin{cases} \sigma_{\alpha,\beta+v_j}(x,\xi) = \tau_{\alpha,\beta}(x,\xi)\partial_{x_j}\sigma_{\alpha,\beta}(x,\xi) - \sigma_{\alpha,\beta}(x,\xi)\partial_{x_j}\tau_{\alpha,\beta}(x,\xi), \\ \tau_{\alpha,\beta+v_j}(x,\xi) = \tau_{\alpha,\beta}(x,\xi)^2, \\ \sigma_{\alpha+v_k,\beta}(x,\xi) = \tau_{\alpha,\beta}(x,\xi)\triangle_{\xi_k}\sigma_{\alpha,\beta}(x,\xi) - \sigma_{\alpha,\beta}(x,\xi)\triangle_{\xi_k}\tau_{\alpha,\beta}(x,\xi), \\ \tau_{\alpha+v_k,\beta}(x,\xi) = \tau_{\alpha,\beta}(x,\xi)\,\tau_{\alpha,\beta}(x,\xi+v_k). \end{cases}$$

By Lemma 4.7.9 we notice that $\sigma_{\alpha,\beta}$ and $\tau_{\alpha,\beta}$ are symbols of periodic pseudo-differential operators, and it holds that

$$\triangle_\xi^\alpha \partial_x^\beta \frac{\sigma(x,\xi)}{\tau(x,\xi)} = \frac{\sigma_{\alpha,\beta}(x,\xi)}{\tau_{\alpha,\beta}(x,\xi)}.$$

Hence induction on both α and β gives

$$\begin{cases} |\sigma_{\alpha,\beta}(x,\xi)| \leq c_{\alpha,\beta} \langle\xi\rangle^{(2^{|\alpha+\beta|}-1)\ell+m-|\alpha|}, \\ d_{\alpha,\beta} \langle\xi\rangle^{2^{|\alpha+\beta|}\ell} \leq |\tau_{\alpha,\beta}(x,\xi)| \leq e_{\alpha,\beta} \langle\xi\rangle^{2^{|\alpha+\beta|}\ell}, \end{cases}$$

where $c_{\alpha,\beta}, d_{\alpha,\beta}, e_{\alpha,\beta} > 0$ are appropriate constants. These estimates complete the proof. $\qquad\square$

Exercise 4.9.5. Let us examine the proof of Lemma 4.9.4 once more. Show that

$$\tau_{\alpha,\beta}(x,\xi) = \prod_{\gamma\leq\alpha} \tau(x,\xi+\gamma)^{2^{|\beta|}}.$$

Moreover, verify the induction in the proof of Lemma 4.9.4.

Now it is time to declare that ellipticity is equivalent to invertibility in the algebra $\mathrm{Op}(S^\infty(\mathbb{T}^n \times \mathbb{Z}^n))/\mathrm{Op}(S^{-\infty}(\mathbb{T}^n \times \mathbb{Z}^n))$. For a detailed discussion and motivation behind the notion of the parametrix in the Euclidean setting of $\mathbb{R}^n$ we refer to Section 2.6.

Theorem 4.9.6 (Elliptic $\iff \exists$ Parametrix). *Operator $A \in \mathrm{Op}(S^m(\mathbb{T}^n \times \mathbb{Z}^n))$ is elliptic if and only if there exists $B \in \mathrm{Op}(S^{-m}(\mathbb{T}^n \times \mathbb{Z}^n))$ such that $BA \sim I \sim AB$. (We call such B a parametrix of A.)*

Proof. The proof of the "if" part is similar to that of Theorem 2.6.6 but we give another argument to prove the "only if" part. Thus, assume first that $A \in \mathrm{Op}(S^m(\mathbb{T}^n \times \mathbb{Z}^n))$ and $B \in \mathrm{Op}(S^{-m}(\mathbb{T}^n \times \mathbb{Z}^n))$ satisfy $BA = I - T$ and $AB = I - T'$ with $T, T' \in \mathrm{Op}(S^{-\infty}(\mathbb{T}^n \times \mathbb{Z}^n))$. Then $1 - \sigma_{BA} = \sigma_T \in S^{-\infty}(\mathbb{T}^n \times \mathbb{Z}^n)$ and consequently by Theorem 4.7.10 we have $1 - \sigma_B \sigma_A \in S^{-1}(\mathbb{T}^n \times \mathbb{Z}^n)$, so that $|1 - \sigma_B \sigma_A| \leq C\langle \xi \rangle^{-1}$. Hence $1 - |\sigma_B| \cdot |\sigma_A| \leq C\langle \xi \rangle^{-1}$, or equivalently $|\sigma_B| \cdot |\sigma_A| \geq 1 - C\langle \xi \rangle^{-1}$. If we choose $n_0 > C$, then $|\sigma_B(x, \xi)| \cdot |\sigma_A(x, \xi)| \geq 1 - C\langle n_0 \rangle^{-1} > 0$ for any $\|\xi\| \geq n_0$. Hence A is elliptic of order m. This yields the "if" part of the proof.

Now assume that $A \in \mathrm{Op}(S^m(\mathbb{T}^n \times \mathbb{Z}^n))$ is elliptic, and define a periodic pseudo-differential operator B_0 by $\sigma_{B_0}(x, \xi) := 1/\sigma_A(x, \xi)$, so that σ_{B_0} is in $S^{-m}(\mathbb{T}^n \times \mathbb{Z}^n)$ by Lemma 4.9.4, and $\sigma_{B_0} \sigma_A \sim 1 \in S^0(\mathbb{T}^n \times \mathbb{Z}^n)$. Then by (4.36) it is true that $\sigma_{B_0 A} = \sigma_{B_0} \sigma_A - \sigma_T \sim 1 - \sigma_T$ for some $\sigma_T \in S^{-1}(\mathbb{T}^n \times \mathbb{Z}^n)$; equivalently $B_0 A \sim I - T$. Notice that

$$\sum_{j=0}^{N-1} T^j \, (I - T) = I - T^N,$$

where $T^N \in S^{-N}(\mathbb{T}^n \times \mathbb{Z}^n)$, so that $R \sim \sum_{j=0}^{\infty} T^j$ is a formal Neumann series of the inverse of $I - T$ in the algebra $\mathrm{Op}(S^\infty(\mathbb{T}^n \times \mathbb{Z}^n))/\mathrm{Op}(S^{-\infty}(\mathbb{T}^n \times \mathbb{Z}^n))$. Then $R B_0 A \sim R(I - T) \sim I$, so that $R B_0$ is a candidate for a parametrix B. Indeed, this technique produces also $B' \in \mathrm{Op}(S^{-m}(\mathbb{T}^n \times \mathbb{Z}^n))$ satisfying $AB' \sim I$. Then

$$B' \sim (BA)B' = B(AB') \sim B$$

completes the proof.   □

From the proof above it can be immediately seen that the principal symbol of a parametrix of an elliptic periodic pseudo-differential operator A is given by the symbol $\sigma_{B_0} = 1/\sigma_A$ (provided that σ_A does not vanish anywhere, which we may assume).

Remark 4.9.7. Notice that the ellipticity of A is really equivalent to the existence of a periodic pseudo-differential operator B satisfying $BA \overset{-\varepsilon}{\sim} I \overset{-\varepsilon}{\sim} AB$ for some $\varepsilon > 0$. In general, there is no restriction in defining "functions" of periodic pseudo-differential operators of negative order by $\sum_{j=0}^{\infty} c_j \, T^j$: the coefficients c_j may even grow arbitrarily fast, as it is only required that $T \in \mathrm{Op}(S^{-\varepsilon}(\mathbb{T}^n \times \mathbb{Z}^n))$ for some $\varepsilon > 0$. For example, one may define sin and cos of these operators, yielding $\sin(T)^2 + \cos(T)^2 \sim I$ etc., or determine fractional powers of elliptic operators (discussed in, e.g., [135, p. 42–44]).

Corollary 4.9.8. *Assume that $A \in \mathrm{Op}(S^m(\mathbb{T}^n \times \mathbb{Z}^n))$. Then its adjoint $A^* \in \mathrm{Op}(S^m(\mathbb{T}^n \times \mathbb{Z}^n))$ and A is elliptic if and only if A^* is.*

Proof. The fact that $A^* \in \mathrm{Op}(S^m(\mathbb{T}^n \times \mathbb{Z}^n))$ was stated in Theorem 4.7.3. Now, suppose A is elliptic with parametrix B such that $BA = I - T$ and $AB = I - T'$, where $T, T' \in \mathrm{Op}(S^{-\infty}(\mathbb{T}^n \times \mathbb{Z}^n))$. Then A^* has B^* as its parametrix, because here $B^*A^* = I - T'^*$ and $A^*B^* = I - T^*$, where $T^*, T'^* \in \mathrm{Op}(S^{-\infty}(\mathbb{T}^n \times \mathbb{Z}^n))$.

$\square$

Exercise 4.9.9. Prove the "transpose" version of Corollary 4.9.8. Namely, show that A is elliptic if and only if A^t is.

Remark 4.9.10. For Corollary 4.9.8, we can give a direct proof that

$$A \in \mathrm{Op}(S^m(\mathbb{T}^n \times \mathbb{Z}^n)) \quad \text{implies} \quad A^* \in \mathrm{Op}(S^m(\mathbb{T}^n \times \mathbb{Z}^n))$$

via the distributional duality by applying (4.12). The same is true for the transpose operator A^t. Since we gave a more detailed proof for the ellipticity of the adjoint in Corollary 4.9.8 and left the transpose case as an exercise, we now do the opposite and give a direct proof of $A^t \in \mathrm{Op}(S^m(\mathbb{T}^n \times \mathbb{Z}^n))$. Indeed, using Theorem 4.1.4, the Fourier coefficients of the symbol of A^t in the first variable can be found as

$$
\begin{aligned}
\widehat{\sigma_{A^t}}(\eta, \xi) &= \langle x \mapsto \sigma_{A^t}(x, \xi), e_{-\eta} \rangle \\
&= \langle e_{-\xi} A^t e_\xi, e_{-\eta} \rangle \\
&= \langle A^t e_\xi, e_{-\eta-\xi} \rangle \\
&= \langle e_\xi, A e_{-\eta-\xi} \rangle \\
&= \langle e_{-\eta}, e_{\eta+\xi} A e_{-\eta-\xi} \rangle \\
&= \langle e_{-\eta}, x \mapsto \sigma_A(x, -\eta-\xi) \rangle \\
&= \widehat{\sigma_A}(\eta, -\eta-\xi).
\end{aligned}
$$

Using this and Peetre's inequality (Proposition 3.3.31) we show that

$$A^t \in \mathrm{Op}(S^m(\mathbb{T}^n \times \mathbb{Z}^n)):$$

$$
\begin{aligned}
\left| \triangle_\xi^\alpha \partial_x^\beta \sigma_{A^t}(x, \xi) \right| &= \left| \triangle_\xi^\alpha \partial_x^\beta \sum_{\eta \in \mathbb{Z}^n} e^{\mathrm{i}2\pi x \cdot \eta} \, \widehat{\sigma_A}(\eta, -\eta-\xi) \right| \\
&\overset{(4.12)}{\leq} \sum_{\eta \in \mathbb{Z}^n} c_{r,\alpha} \langle \eta \rangle^{|\beta|} \langle \eta \rangle^{-r} \langle \xi + \eta \rangle^{m-|\alpha|} \\
&\overset{\text{Peetre}}{\leq} 2^{|m-|\alpha||} \, c_{r,\alpha,\beta} \langle \xi \rangle^{m-|\alpha|} \sum_{\eta \in \mathbb{Z}^n} \langle \eta \rangle^{|m-|\alpha||+|\beta|-r} \\
&\leq C_{\alpha,\beta} \langle \xi \rangle^{m-|\alpha|},
\end{aligned}
$$

if we take r large enough.

$\square$

Exercise 4.9.11. Give a similar direct proof (without using asymptotic expansions) of the fact that $A \in \mathrm{Op}(S^m(\mathbb{T}^n \times \mathbb{Z}^n))$ implies that $A^* \in \mathrm{Op}(S^m(\mathbb{T}^n \times \mathbb{Z}^n))$.

Remark 4.9.12 (∗-**algebra of periodic pseudo-differential operators**). Let us collect the obtained results concerning the ∗-algebra

$$\mathrm{Op}(S^\infty(\mathbb{T}^n \times \mathbb{Z}^n))/\mathrm{Op}(S^{-\infty}(\mathbb{T}^n \times \mathbb{Z}^n)).$$

Here the classes are $[A] = A + \mathrm{Op}(S^{-\infty}(\mathbb{T}^n))$. Now $[A] + [B] = [A + B]$, $[A][B] = [AB]$, and $[A]^* = [A^*]$. If A is elliptic with a parametrix C, it makes sense to write $[A]^{-1} = [C]$. The zero and identity elements are $[0] = \mathrm{Op}(S^{-\infty}(\mathbb{T}^n \times \mathbb{Z}^n))$ and $[I]$, respectively. One could then define the action of $[A]$ in $[u] := u + C^\infty(\mathbb{T}^n)$ ($u \in H^s(\mathbb{T}^n)$) by $[A][u] := [Au]$. We do not develop this line of thought any further.

The proof of Theorem 4.9.6 did not use the asymptotic expansion of the product of periodic pseudo-differential operators. But applying the composition formula (4.36), we can obtain explicitly an asymptotic expansion of the parametrix:

Theorem 4.9.13 (Formula for the parametrix). *Assume that periodic pseudo-differential operators $A \in \mathrm{Op}(S^m(\mathbb{T}^n \times \mathbb{Z}^n))$ and $B \in \mathrm{Op}(S^{-m}(\mathbb{T}^n \times \mathbb{Z}^n))$ are parametrices to each other, that is $AB \sim I \sim BA$. Let $A \sim \sum_{j=0}^\infty A_j$ be an asymptotic expansion, where $A_j \in \mathrm{Op}(S^{m-j}(\mathbb{T}^n \times \mathbb{Z}^n))$. Then an asymptotic expansion $B \sim \sum_{k=0}^\infty B_k$ is obtained by $\sigma_{B_0} = 1/\sigma_{A_0}$, and then recursively*

$$\sigma_{B_N}(x, \xi) = \frac{-1}{\sigma_{A_0}(x, \xi)} \sum_{k=0}^{N-1} \sum_{j=0}^{N-k} \sum_{|\gamma| = N-j-k} \frac{1}{\gamma!} \left[\triangle_\xi^\gamma \sigma_{B_k}(x, \xi) \right] D_x^{(\gamma)} \sigma_{A_j}(x, \xi).$$

Proof. Now $I \sim BA$, so that by (4.36)

$$
\begin{aligned}
1 &\sim \sigma_{BA}(x, \xi) \\
&\sim \sum_{\gamma \geq 0} \frac{1}{\gamma!} \left[\triangle_\xi^\gamma \sigma_B(x, \xi) \right] D_x^{(\gamma)} \sigma_A(x, \xi) \\
&\sim \sum_{\gamma \geq 0} \frac{1}{\gamma!} \left[\triangle_\xi^\gamma \sum_{k=0}^\infty \sigma_{B_k}(x, \xi) \right] D_x^{(\gamma)} \sum_{j=0}^\infty \sigma_{A_j}(x, \xi),
\end{aligned}
$$

where we want to solve σ_{B_k}. Notice that A_0 is elliptic if and only if A is elliptic (and now A is elliptic by Theorem 4.9.6). Moreover, without a loss of generality we may assume that σ_{A_0} does not vanish anywhere. Obviously, we can demand that $1 = \sigma_{B_0}(x, \xi) \, \sigma_{A_0}(x, \xi)$, and that

$$0 = \sum_{j+k+|\gamma|=N} \frac{1}{\gamma!} \left[\triangle_\xi^\gamma \sigma_{B_k}(x, \xi) \right] D_x^{(\gamma)} \sigma_{A_j}(x, \xi).$$

Then the trivial solution of these equations is the recursion of the theorem. The reader may verify that $\sigma_{B_N} \in S^{-m-N}(\mathbb{T}^n \times \mathbb{Z}^n)$. Thus $B \sim \sum_{k=0}^\infty B_k$.   □

Exercise 4.9.14. Show that $\sigma_{B_N} \in S^{-m-N}(\mathbb{T}^n \times \mathbb{Z}^n)$ in Theorem 4.9.13.

Of course, when the extended symbols are used, one directly obtains another asymptotic expansion for parametrices like in the cases of adjoints and products.

In the theorem above the expansions for A and B may sometimes contain only a finite number of terms. The simplest example demonstrating this is the case when $A = I = B$: the expansions are simply $A_0 = I = B_0$.

We now show that the extension of symbols from Theorem 4.5.3 respects ellipticity:

Theorem 4.9.15 (Extensions and ellipticity). *Let the toroidal and Euclidean symbols $\tilde{a} \in S^m(\mathbb{T}^n \times \mathbb{Z}^n)$ and $a \in S^m(\mathbb{T}^n \times \mathbb{R}^n)$ be such that $\tilde{a} = a|_{\mathbb{T}^n \times \mathbb{Z}^n}$. Then $\tilde{a}$ is elliptic if and only if a is elliptic.*

Proof. The "if" part is straightforward because it is simply the restriction of the condition

$$\forall (x, \xi) \in \mathbb{T}^n \times \mathbb{R}^n : \quad \|\xi\| \geq n_0 \Rightarrow |\sigma_A(x, \xi)| \geq c_0 \langle \xi \rangle^m$$

to $\mathbb{T}^n \times \mathbb{Z}^n$. Conversely, assume that $\tilde{a}$ is elliptic. For $\xi \in \mathbb{R}^n$ let $\eta \in \mathbb{Z}^n$ be (one of the) closest integers to ξ. We can assume that ξ and η are sufficiently large. Taking the Taylor expansion of a at η we get

$$a(x, \xi) = a(x, \eta) + \sum_{|\alpha|=1} (\eta - \xi)^\alpha \int_0^t (1-t) \partial_\xi^\alpha a(x, \xi + t(\xi - \eta)) \, \mathrm{d}t.$$

Now, we have $|a(x, \eta)| = |\tilde{a}(x, \eta)| \geq C \langle \eta \rangle^m \geq \tilde{C} \langle \xi \rangle^m$ since $\tilde{a}$ is elliptic. On the other hand,

$$\left| \sum_{|\alpha|=1} (\eta - \xi)^\alpha \int_0^t (1-t) \partial_\xi^\alpha a(x, \xi + t(\xi - \eta)) \, \mathrm{d}t \right| \leq C \langle \xi \rangle^{m-1},$$

implying that a is elliptic. $\qquad\square$

Finally, we discuss the Fredholmness of elliptic periodic pseudo-differential operators.

Definition 4.9.16 (Fredholm operators and index). Let X and Y be Banach spaces. A continuous linear operator $A \in \mathcal{L}(X, Y)$ is a *Fredholm operator* if its kernel $\mathrm{Ker}(A) = \{x \in X : Ax = 0\}$ is finite-dimensional and the range (or image) $\mathrm{Im}(A) = \{Ax : x \in X\}$ is closed and of finite codimension. Then its index is

$$\mathrm{Ind}(A) := \dim \mathrm{Ker}(A) - \mathrm{codim}\, \mathrm{Im}(A),$$

or equivalently

$$\mathrm{Ind}(A) := \dim \mathrm{Ker}(A) - \dim \mathrm{Ker}(A^*),$$

where A^* is the (Banach or Hilbert) adjoint of A. Fredholm operators can be generalised to locally convex topological vector spaces, see, e.g., [135].

Theorem 4.9.17 (Elliptic $\Longrightarrow$ Fredholm). *Let $A \in \mathrm{Op}(S^m(\mathbb{T}^n \times \mathbb{Z}^n))$ be elliptic. Then it is a Fredholm operator $A \in \mathcal{L}(H^s(\mathbb{T}^n), H^{s-m}(\mathbb{T}^n))$ for every $s \in \mathbb{R}$. Moreover, the kernels $\mathrm{Ker}(A), \mathrm{Ker}(A^*)$ are in $C^\infty(\mathbb{T}^n)$, so that $\mathrm{Ker}(A)$, $\mathrm{Ker}(A^*)$, and $\mathrm{Ind}(A)$ are independent of s.*

Proof. Let $B \in \mathrm{Op}(S^{-m}(\mathbb{T}^n \times \mathbb{Z}^n))$ be a parametrix of A such that $BA = I - T$ and $AB = I - T'$, $T, T' \in \mathrm{Op}(S^{-\infty}(\mathbb{T}^n \times \mathbb{Z}^n))$, and let $\mathrm{Ker}_s(A) := \{u \in H^s(\mathbb{T}^n) | Au = 0\}$. Recall that $C^\infty(\mathbb{T}^n) = \bigcap_{t \in \mathbb{R}} H^t(\mathbb{T}^n)$ by Corollary 3.2.12. Now, T and T' are compact (on the basis of Corollary 4.3.3), and

$$\mathrm{Ker}_s(A) \subset \mathrm{Ker}_s(BA) = \mathrm{Ker}_s(I - T) \; \subset \; C^\infty(\mathbb{T}^n),$$
$$A\, H^s(\mathbb{T}^n) \; \supset \; AB\, H^{s-m}(\mathbb{T}^n) = (I - T')H^{s-m}(\mathbb{T}^n).$$

By the Fredholm theory (see, e.g., [55, Chapter XIX] or [69, Chapter 4]), $I-T$ and $I-T'$ are Fredholm operators. Hence $A\, H^s(\mathbb{T}^n)$ is closed and of finite codimension, and $\mathrm{Ker}_s(A) \subset C^\infty(\mathbb{T}^n)$ is finite-dimensional. Similarly $\mathrm{Ker}_{m-s}(A^*) \subset C^\infty(\mathbb{T}^n)$ is finite dimensional. Since these kernels are in $C^\infty(\mathbb{T}^n)$, they do not depend on s, and thus the Fredholm index is independent of s. $\square$

Proposition 4.9.18 (All indices are attainable). *For every $j \in \mathbb{Z}$ there are periodic pseudo-differential operators of index j.*

Proof. For instance, consider $A_1 \in \mathcal{L}(L^2(\mathbb{T}^1))$ defined by

$$A_1 u(x) := [P^+ + \mathrm{e}^{\mathrm{i}2\pi x}P^-]u(x),$$

where $P^+ u(x) := \sum_{\xi \geq 0} \widehat{u}(\xi)$, and $P^- := I - P^+$. Clearly A_1 is an elliptic periodic pseudo-differential operator with one-dimensional kernel, and it maps every $H^s(\mathbb{T}^1)$ onto itself, being thus a Fredholm operator with $\mathrm{Ind}(A_1) = 1$. Generally $\mathrm{Ind}(AB) = \mathrm{Ind}(A) + \mathrm{Ind}(B)$ and $\mathrm{Ind}(A^*) = -\mathrm{Ind}(A)$, so that $\mathrm{Ind}(A_1^j) = j$ and $\mathrm{Ind}(A_1^{*j}) = -j$ for every $j \in \mathbb{N}_0$. $\square$

4.10   Smoothness properties

In Section 3.2 a close connection of Sobolev spaces and differentiability of distributions was put forward. This section handles these smoothness properties and their relation to the continuity properties of periodic pseudo-differential operators.

As it is well known, linear differential operators are local: such an operator A decreases the support of any distribution u: $\mathrm{supp}(Au) \subset \mathrm{supp}(u)$ (see Definition 2.3.3 and exercises thereafter), i.e., for the calculation of $Au(x)$ only the information of the behavior of u in an arbitrary small neighbourhood of x is needed. Periodic pseudo-differential operators do not satisfy this in general, but there is a reasonable version of this property for them, called pseudolocality.

Definition 4.10.1 (Pseudolocality and singular support). The *singular support* sing supp$(u) \subset \mathbb{T}^n$ of a distribution $u \in \mathcal{D}'(\mathbb{T}^n)$ is the complement of the maximal open set where u is C^∞-smooth. An operator A is called *pseudolocal*, if

$$\text{sing supp}(Au) \subset \text{sing supp}(u)$$

for every distribution u in the domain of A. This is to say that pseudolocality means locality with respect to smoothness.

These definitions give rise to further smoothness characterisations:

Definition 4.10.2 (H^t-smoothness). Let $u \in \mathcal{D}'(\mathbb{T}^n)$. Let $U \subset \mathbb{T}^n$ be an open set. We say that u is $H^t(\mathbb{T}^n)$-*smooth in* U, denoted $u \in H^t(U)$, if $\psi(x)u(x) \in H^t(\mathbb{T}^n)$ for every $\psi \in C^\infty(\mathbb{T}^n)$ supported in U. The H^t-*singular support* of u, sing supp$_t(u)$, is the set $\mathbb{T}^n \setminus U$, where U is the maximal open set for which $u \in H^t(U)$. We say that $u \in \mathcal{D}'(\mathbb{T}^n)$ is *locally* H^t-*smooth at* $x \in \mathbb{T}^n$, denoted $u \in H^t(x)$, if $u \in H^t(U)$ for some open set $U \ni x$.

Exercise 4.10.3. How is $C^\infty(U)$ related to the spaces $H^t(U)$?

Exercise 4.10.4. How is sing supp(u) related to $\bigcap_{t \in \mathbb{R}} H^t(x)$?

Theorem 4.10.5 (Sobolev pseudolocality). *Let* $0 \le \delta < \rho \le 1$. *Let*

$$A \in \text{Op}(S^m_{\rho,\delta}(\mathbb{T}^n \times \mathbb{Z}^n)), \quad \text{and} \quad u \in H^s(\mathbb{T}^n).$$

Then $u \in H^t(x)$ *only if* $Au \in H^{t-m}(x)$, *so that*

$$\text{sing supp}_{t-m}(Au) \subset \text{sing supp}_t(u).$$

Proof. (This approach was proposed by Gennadi Vainikko [143].) Assume that $u \in H^t(U)$, where U is an neighbourhood of x, and let $\psi \in C^\infty(\mathbb{T}^n)$ be such that supp$(\psi) \subset U$. We notice that $((x, \xi) \mapsto \psi(x)) \in S^0(\mathbb{T}^n \times \mathbb{Z}^n)$ and supp$(\psi^k) \subset U$ for all $k \in \mathbb{N}_0$. Let us define $A_0 := A$, $u_0 := 0$, and recursively

$$\begin{cases} A_{k+1} := [\psi I, A_k], \\ u_{k+1} := A_k(\psi \cdot u) + \psi \cdot u_k. \end{cases}$$

By induction on k one proves that $\psi^k \cdot Au = A_k u + u_k$, where $u_k \in H^{t-m}(\mathbb{T}^n)$ and $A_k \in \text{Op}(S^{m-k(\rho-\delta)}_{\rho,\delta}(\mathbb{T}^n \times \mathbb{Z}^n))$ (by Theorems 4.2.3 and 4.7.10). Hence we get $\psi^k \cdot Au \in H^{t-m}(U)$, so that $Au \in H^{t-m}(x)$. $\square$

Exercise 4.10.6. Verify the induction in the proof of Theorem 4.10.5.

Immediately we can state:

Corollary 4.10.7. *For* $0 \le \delta < \rho \le 1$, *periodic pseudo-differential operators are pseudolocal.*

Remark 4.10.8. We can present an alternative proof for the pseudolocality of periodic pseudo-differential operators. Indeed, let K_A be the Schwartz kernel of a periodic pseudo-differential operator A: the approach involves an application of smoothness of the kernel outside the diagonal $x = y$ from Theorem 4.3.6. Assume that $u \in \mathcal{D}'(\mathbb{T}^n)$ such that $u|_U \in C^\infty(U)$ for some open set $U \subset \mathbb{T}^n$. Let $(V_j)_{j=1}^\infty$ be a sequence of open sets such that

$$\overline{V_j} \subset V_{j+1} \subset U = \bigcup_{j=1}^\infty V_j,$$

and let $\psi_j \in C^\infty(\mathbb{T}^n)$ be associated functions satisfying $\mathrm{supp}(\psi_j) \subset U$ and $\psi_j|_{\overline{V_j}} = 1$. Then

$$\begin{aligned} Au(x) &= A\left[\psi_j \cdot u + (1 - \psi_j) \cdot u\right](x) \\ &= A(\psi_j \cdot u)(x) + \int_{\mathbb{T}^n} u(y)\,(1 - \psi_j(y))\,K_A(x,y)\,\mathrm{d}y. \end{aligned}$$

As a periodic pseudo-differential operator, A maps $\psi_j \cdot u \in C^\infty(\mathbb{T}^n)$ into $C^\infty(\mathbb{T}^n)$. Now $(1 - \psi_j(y))\,K_A(x,y)$ vanishes when $x = y \in V_j$, so that according to Theorem 4.3.6 it is a C^∞-smooth kernel when $x \in V_j$. Hence Theorem 4.3.1 implies that $Au|_{V_j} \in C^\infty(V_j)$, and thereby $Au|_U \in C^\infty(U)$, as $U = \bigcup_{j=1}^\infty V_j$.   □

   A general periodic pseudo-differential operator can locally (and even globally) smoothen a distribution more than its order indicates. This is not the case with elliptic operators, which faithfully transform the information encoded in smoothness:

Proposition 4.10.9 (Sobolev hypoellipticity). *Let $A \in \mathrm{Op}(S^m(\mathbb{T}^n \times \mathbb{Z}^n))$ be an elliptic periodic pseudo-differential operator, and let $u \in \mathcal{D}'(\mathbb{T}^n)$. Then*

$$\mathrm{sing\ supp}_{t-m}(Au) = \mathrm{sing\ supp}_t(u)$$

for every $t \in \mathbb{R}$, and $\mathrm{sing\ supp}(Au) = \mathrm{sing\ supp}(u)$.

Proof. Let B be a parametrix of A, which exists by Theorem 4.9.6, so that $BA = I - T$. Because $T \in \mathrm{Op}(S^{-\infty}(\mathbb{T}^n \times \mathbb{Z}^n))$, we get by Theorem 4.10.5,

$$\begin{aligned} \mathrm{sing\ supp}_t(u) &= \mathrm{sing\ supp}_t(u - Tu) \\ &= \mathrm{sing\ supp}_t(BAu) \\ &\subset \mathrm{sing\ supp}_{t-m}(Au) \\ &\subset \mathrm{sing\ supp}_t(u), \end{aligned}$$

implying $\mathrm{sing\ supp}_{t-m}(Au) = \mathrm{sing\ supp}_t(u)$.

   Similarly one proves the other equality $\mathrm{sing\ supp}(Au) = \mathrm{sing\ supp}(u)$. Alternatively, one can use the equality $\mathrm{sing\ supp}(u) = \bigcup_t \mathrm{sing\ supp}_t(u)$ to complete the proof.   □

Definition 4.10.10 (Hypoellipticity). Those periodic pseudo-differential operators A which satisfy

$$\text{sing supp}(Au) = \text{sing supp}(u)$$

for every $u \in \mathcal{D}'(\mathbb{T}^n)$ are called *hypoelliptic*. For further information on these operators, we refer to [130] ([135] has a slightly different concept of hypoellipticity).

Recall that by Theorem 3.5.1 a periodic pseudo-differential operator mapping some $H^s(\mathbb{T}^n)$ into $H^{s-l}(\mathbb{T}^n)$ is continuous between these two spaces regardless of its order. Actually, such continuity properties are more widespread.

Theorem 4.10.11. *Let $0 \leq \delta < \rho \leq 1$ and $s \in \mathbb{R}$. Let $\sigma_A \in S^m_{\rho,\delta}(\mathbb{T}^n \times \mathbb{Z}^n)$ and assume that $A\, H^t(\mathbb{T}^n) \subset H^{t-l}(\mathbb{T}^n)$ for some $t, l \in \mathbb{R}$ where $l < m$. Then*

$$\forall \varepsilon > 0 : \quad A \in \mathcal{L}(H^s(\mathbb{T}^n), H^{s-l-\varepsilon}(\mathbb{T}^n)).$$

Furthermore, if $m > l \geq m - (\rho - \delta)$, we can take $\varepsilon = 0$ above.

Proof. Notice that the requirement $l < m$ is not really restricting, since by Theorem 4.2.3 we already know that $A \in \mathcal{L}(H^q(\mathbb{T}^n), H^{q-m}(\mathbb{T}^n))$ for every $q \in \mathbb{R}$. Fix $\varepsilon > 0$ and assume for clarity that $s < t$ (the case $s > t$ is totally symmetric). Then, by choosing $q < s$ small enough, the interpolation theorems

$$\mathcal{L}(H^{t_1}, H^{t_2}) \cap \mathcal{L}(H^{q_1}, H^{q_2}) \quad \subset \quad \mathcal{L}([H^{t_1}, H^{q_1}]_\theta, [H^{t_2}, H^{q_2}]_\theta),$$
$$[H^{t_j}, H^{q_j}]_\theta \quad = \quad H^{\theta t_j + (1-\theta)q_j}$$

(here $0 < \theta < 1$; see [72, Theorems 5.1 and 7.7]) imply that

$$A \in \mathcal{L}(H^s(\mathbb{T}^n), H^{s-l-\varepsilon}(\mathbb{T}^n)).$$

Now suppose $l \geq m - (\rho - \delta)$. With the aid of the canonical Sobolev space isomorphisms φ_γ and Theorem 4.7.10, we get $\varphi_{s-t} A \varphi_{t-s} - A \in Op(S^{m-(\rho-\delta)}_{\rho,\delta}(\mathbb{T}^n \times \mathbb{Z}^n))$. On the other hand,

$$\varphi_{s-t} A \varphi_{t-s}\, H^s(\mathbb{T}^n) \quad = \quad \varphi_{s-t} A\, H^t(\mathbb{T}^n)$$
$$\subset \quad \varphi_{s-t}\, H^{t-l}(\mathbb{T}^n)$$
$$= \quad H^{s-l}(\mathbb{T}^n).$$

This completes the proof. $\square$

The interpolation theorems [72, Theorems 5.1 and 7.7] of the preceding proof enhanced with norm estimates are significant also in the proofs of [142, Lemma 4.3] and [62, Lemma 4.1], which are important results in the analysis of periodic integral operators.

Finally, we study the connection of orders and continuity in the elementary cases when a periodic pseudo-differential operator is either a multiplier or a multiplication. The next theorem, Abel–Dini, dwells in the theory of series. We present only the proof of the divergence part, which Niels Henrik Abel solved in the 1820s.

Theorem 4.10.12 (Abel–Dini). *Let d_j be positive numbers and let $D_N := \sum_{j=1}^N d_j$. Assume that $(D_N)_{N=1}^\infty$ is divergent. Then $\sum_{j=1}^\infty d_j/D_j^r$ diverges exactly when $r \leq 1$.*

Proof (Abel part). (The whole proof is in [66, p. 290-291].) We assume that $r \leq 1$. Since $(D_N)_{N=1}^\infty$ diverges, it is true that for every $i \in \mathbb{Z}^+$ there exists $k_i \in \mathbb{Z}^+$ such that $D_i/D_{i+k_i} \leq 1/2$. Hence

$$\sum_{j=i+1}^{i+k_i} \frac{d_j}{D_j^r} \geq \sum_{j=i+1}^{i+k_i} \frac{d_j}{D_j} \geq \frac{1}{D_{i+k_i}} \sum_{j=i+1}^{i+k_i} d_j = 1 - \frac{D_i}{D_{i+k_i}} \geq \frac{1}{2}.$$

Due to this, $\sum_{j=1}^\infty d_j/D_j^r$ diverges. $\square$

If we say that a sequence *converges to infinity*, $p_j \to \infty$, it is meant that for every $C < \infty$ there exists $j_C \in \mathbb{Z}^+$ such that $p_j > C$ if $j > j_C$.

Corollary 4.10.13. *If $(p_j)_{j\in\mathbb{Z}^+}$ is a monotone sequence of positive real numbers converging to infinity, then there is a convergent series $\sum_{j=1}^\infty c_j$ such that $\sum_{j=1}^\infty p_j\, c_j$ diverges.*

Proof. (A modification of [66, p. 302].) Define

$$\begin{cases} d_1 := p_1, \\ d_{j+1} := p_{j+1} - p_j. \end{cases}$$

Then, in the notation of the Abel–Dini theorem, $D_N = \sum_{j=1}^N d_j = p_N \to \infty$, and $\sum_{j=1}^\infty d_j/D_j = 1 + \sum_{j=1}^\infty (p_{j+1} - p_j)/p_{j+1}$ diverges. Let us define

$$c_j := (p_{j+1} - p_j)/(p_{j+1}\, p_j).$$

Then $\sum_{j=1}^\infty c_j$ converges, because $1/p_j \to 0$:

$$\sum_{j=1}^\infty c_j = \sum_{j=1}^\infty \left(\frac{1}{p_j} - \frac{1}{p_{j+1}} \right) = \frac{1}{p_1}.$$

Clearly, $\sum_{j=1}^\infty p_j c_j = \sum_{j=1}^\infty (p_{j+1} - p_j)/p_{j+1}$ diverges. $\square$

We apply this to obtain the following result concerning multipliers:

Proposition 4.10.14 (Sobolev unboundedness of multipliers).
Assume that $\sigma_A(x,\xi) = \widehat{k}(\xi)$, where for every $C < \infty$ there exists $\xi \in \mathbb{Z}^n$ such that $|\widehat{k}(\xi)| > C\langle\xi\rangle^l$. Then $A\, H^s(\mathbb{T}^n) \not\subset H^{s-l}(\mathbb{T}^n)$ for any $s \in \mathbb{R}$.

Proof. Now there is a subsequence of $((\langle\xi\rangle^{-2l}|\widehat{k}(\xi)|^2)_{\xi\in\mathbb{Z}^n}$ that converges to ∞ as $\|\xi\| \to \infty$. Corollary 4.10.13 then provides the existence of a sequence $(\widehat{u}(\xi))_{\xi\in\mathbb{Z}^n}$ for which $\sum_{\xi\in\mathbb{Z}^n} \langle\xi\rangle^{2s} |\widehat{u}(\xi)|^2$ converges, but for which $\sum_{\xi\in\mathbb{Z}^n} \langle\xi\rangle^{2(s-l)}|\widehat{k}(\xi)\, \widehat{u}(\xi)|^2$ diverges. Thus $u \in H^s(\mathbb{T}^n)$, and it is mapped to $Au \notin H^{s-l}(\mathbb{T}^n)$. $\square$

Example. Proposition 4.2.3 showed that the order of an operator determines its boundedness properties on Sobolev spaces. The converse is not true. Indeed, there is no straightforward way of concluding the order of a symbol from observations about between which spaces the mapping acts. A simple demonstration of this kind of phenomenon is $\sigma(x, \xi) := \sin(\ln(|\xi|)^2)$ (when $|\xi| \geq 1$, $\xi \in \mathbb{R}^1$; the definition of σ for $|\xi| < 1$ is not interesting). This symbol is independent of x, and it is bounded, resulting in that $\mathrm{Op}(\sigma)$ maps $H^s(\mathbb{T}^1)$ into itself for every $s \in \mathbb{R}$. On the other hand, σ defines a periodic pseudo-differential operator of degree ε for any $\varepsilon > 0$, as it is easily verified – however, $\sigma \notin S^0(\mathbb{T}^1)$, because

$$\partial_\xi \sigma(x, \xi) = 2 \frac{\ln(|\xi|)}{\xi} \cos(\ln(|\xi|)^2), \quad (|\xi| > 1),$$

which certainly is not in $\mathcal{O}((1 + |\xi|)^{-1})$.

The case of pure multiplications can be more easily and thoroughly handled:

Proposition 4.10.15. *Any Sobolev space $H^s(\mathbb{T}^n)$ is the intersection of the local Sobolev spaces of the same order, i.e., $H^s(\mathbb{T}^n) = \bigcap_{x \in \mathbb{T}^n} H^s(x)$ for every $s \in \mathbb{R}$. Moreover, if $\varphi \in C^\infty(\mathbb{T}^n)$ such that $\varphi(x) \neq 0$, then φ defines an automorphism of $H^s(x)$ by multiplication.*

Proof. By Theorem 4.2.3, $v \in H^s(\mathbb{T}^n)$ implies $\psi v \in H^s(\mathbb{T}^n)$ for any $\psi \in C^\infty(\mathbb{T}^n)$. Then assume that $v \in H^s(x)$ for every $x \in \mathbb{T}^n$, so that there exist neighbourhoods U_x of points x where $v \in H^s(U_x)$. Since $\mathbb{T}^n$ is compact, there is a finite subcover $\mathcal{U} = \{U_{x(j)}\}_{j=1}^N$. Since there exists a smooth partition of unity subordinate to $\mathcal{U}$ (see Corollary A.12.15 for a continuous partition, and then make it smooth, e.g., by mollification), and $\mathcal{U}$ is finite, it is true that $v \in H^s(\mathbb{T}^n)$ – the first claim is proved.

Let us then show that $u \mapsto \varphi u$ defines an automorphism. As above, $\varphi u \in H^s(x)$. By the continuity of φ on $\mathbb{T}^n$ there exists a neighbourhood U of x such that $\varphi(y) \neq 0$ whenever $y \in \overline{U}$, and furthermore U can be chosen so small that $u \in H^s(U)$. Then take such $\psi \in C^\infty(\mathbb{T}^n)$ that $\psi|_U = 1/\varphi$. Since $\psi \varphi u|_U = u|_U$, and the result is obtained. $\square$

4.11 An application to periodic integral operators

As an example of the symbolic analysis techniques, here we study periodic integral operators. Let A be a linear operator defined on $C^\infty(\mathbb{T}^n)$ by

$$Au(x) := \int_{\mathbb{T}^n} \mathsf{a}(x, y) \, \mathsf{k}(x - y) \, u(y) \, dy, \tag{4.44}$$

where a is a C^∞-smooth biperiodic function, and k is a 1-periodic distribution. Note that when a is a function of a single variable, A is simply a convolution operator composed with multiplication: either $Au(x) = f(x) \int_{\mathbb{T}^n} \mathsf{k}(x - y) \, u(y) \, dy$ if $\mathsf{a}(x, y) = f(x)$, or $Au(x) = \int_{\mathbb{T}^n} g(y) \, \mathsf{k}(x - y) \, u(y) \, dy$ if $\mathsf{a}(x, y) = g(y)$.

We are going to show that whenever A of the type (4.44) is a periodic pseudo-differential operator, it is really something like a convolution operator with multiplication, or on the Fourier side, almost a multiplier:

Theorem 4.11.1. *Let $\rho > 0$. The operator A defined by (4.44) is a periodic pseudo-differential operator of order m if and only if the Fourier coefficients of the distribution k satisfy*

$$\forall \alpha \in \mathbb{N}_0^n \; \exists C_\alpha \in \mathbb{R} \; \forall \xi \in \mathbb{Z}^n : \quad \left| \triangle_\xi^\alpha \widehat{\mathsf{k}}(\xi) \right| \leq C_\alpha \langle \xi \rangle^{m - \rho |\alpha|}. \tag{4.45}$$

In this case $A \in \mathrm{Op}(S_{\rho,0}^m(\mathbb{T}^n \times \mathbb{Z}^n))$ and the symbol of A has the following asymptotic expansion:

$$\sigma_A(x, \xi) \sim \sum_{|\gamma| \geq 0} \frac{1}{\gamma!} \triangle_\xi^\gamma \widehat{\mathsf{k}}(\xi) \; D_y^{(\gamma)} \mathsf{a}(x, y)|_{y=x}.$$

Proof. An amplitude of A is right in front of our eyes:

$$\begin{aligned}
Au(x) &= \int_{\mathbb{T}^n} u(y) \, \mathsf{a}(x, y) \, \mathsf{k}(x - y) \, dy \\
&= \int_{\mathbb{T}^n} u(y) \, \mathsf{a}(x, y) \sum_{\xi \in \mathbb{Z}^n} \widehat{\mathsf{k}}(\xi) \, e^{i2\pi(x - y) \cdot \xi} \, dy \\
&= \mathrm{Op}(a)u(x),
\end{aligned}$$

where $a(x, y, \xi) = \mathsf{a}(x, y)\widehat{\mathsf{k}}(\xi)$. Certainly $\widehat{\mathsf{k}}$ satisfies the estimates (4.45) if and only if $a \in \mathcal{A}_{\rho,0}^m(\mathbb{T}^n)$. Accordingly, a yields the asymptotic expansion in view of (4.18) in Theorem 4.4.5. $\square$

Remark 4.11.2. By Theorem 4.11.1 it is readily seen that a periodic pseudo-differential operator A of the periodic integral operator form (4.44), that is

$$Au(x) = \int_{\mathbb{T}^n} \mathsf{a}(x, y) \, \mathsf{k}(x - y) \, u(y) \, dy, \tag{4.46}$$

is elliptic if and only if $\widehat{\mathsf{k}}(\xi)$ is an elliptic symbol and $\mathsf{a}(x, x) \neq 0$ for all $x \in \mathbb{T}^n$. Consequently in this case by Theorem 4.9.17 it is a Fredholm operator. The index is invariant under compact perturbations (see [55, Corollary 19.1.8], or [135, p. 99]), so that we can add to A any periodic pseudo-differential operator of strictly lower degree and still get an operator with the same index.

Exercise 4.11.3. Let A in (4.46) be elliptic. Show that index $\mathrm{Ind}(A) = 0$.

Theorem 4.11.1 implies that the principal symbol of the periodic integral operator in (4.44) viewed as a periodic pseudo-differential operator is $\mathsf{a}(x, x)\widehat{\mathsf{k}}(\xi)$. By combining Propositions 4.10.14 and 4.10.15 with this observation, we obtain another application to periodic integral operators:

Proposition 4.11.4. *If a periodic pseudo-differential operator A is of the periodic integral operator form* (4.44)

$$Au(x) = \int_{\mathbb{T}^n} u(y) \, \mathbf{a}(x, y) \, \mathbf{k}(x - y) \, dy,$$

where $\mathbf{a}(x, x) \neq 0$ for all $x \in \mathbb{T}^n$ and

$$\forall C \in \mathbb{R} \; \exists \xi \in \mathbb{Z}^n : |\widehat{\mathbf{k}}(\xi)| > C\langle\xi\rangle^l,$$

then $A \, H^s(\mathbb{T}^n) \not\subset H^{s-l}(\mathbb{T}^n)$ for all $s \in \mathbb{R}$.

Remark 4.11.5. In [142] and [102], it is shown that any classical periodic pseudo-differential operator can be expressed as a sum of periodic integral operators of the type (4.44), see Remark 4.4.7. Other contributions to periodic integral operators and classical operators are made, e.g., in [34], [62], [142], and [102].

4.12 Toroidal wave front sets

Here we shall briefly study microlocal analysis not on the cotangent bundle of the torus but on $\mathbb{T}^n \times \mathbb{Z}^n$, which is better suited for the Fourier series representations. Let us define mappings

$$\pi_{\mathbb{R}^n} : \mathbb{R}^n \setminus \{0\} \to \mathbb{S}^{n-1}, \qquad \pi_{\mathbb{R}^n}(\xi) := \xi/\|\xi\|,$$
$$\pi_{\mathbb{T}^n \times \mathbb{R}^n} : \mathbb{T}^n \times (\mathbb{R}^n \setminus \{0\}) \to \mathbb{T}^n \times \mathbb{S}^{n-1}, \qquad \pi_{\mathbb{T}^n \times \mathbb{R}^n}(x, \xi) := (x, \xi/\|\xi\|).$$

We set $\pi_{\mathbb{Z}^n} = \pi_{\mathbb{R}^n}|_{\mathbb{Z}^n} : \mathbb{Z}^n \setminus \{0\} \to \mathbb{S}^{n-1}$.

Definition 4.12.1 (Discrete cones). We say that $K \subset \mathbb{R}^n \setminus \{0\}$ is a cone in $\mathbb{R}^n$ if $\xi \in K$ and $\lambda > 0$ imply $\lambda\xi \in K$. We say that $\Gamma \subset \mathbb{Z}^n \setminus \{0\}$ is a *discrete cone* if $\Gamma = \mathbb{Z}^n \cap K$ for some cone K in $\mathbb{R}^n$; moreover, if this K is open then Γ is called an *open discrete cone*. The set $S := \pi_{\mathbb{R}^n}(\mathbb{Z}^n \setminus \{0\})$ is the set of points with rational directions on the unit sphere.

Proposition 4.12.2. $\Gamma \subset \mathbb{Z}^n \setminus \{0\}$ *is a discrete cone if and only if* $\Gamma = \mathbb{Z}^n \cap \pi_{\mathbb{R}^n}^{-1}(\pi_{\mathbb{R}^n}(\Gamma))$.

Proof. We must show that if K is a cone in $\mathbb{R}^n$ then

$$\mathbb{Z}^n \cap K = \pi_{\mathbb{Z}^n}^{-1} \pi_{\mathbb{Z}^n}(\mathbb{Z}^n \cap K).$$

The inclusion "$\subset$" is obvious. Let us show the inclusion "$\supset$". Let $\xi \in \pi_{\mathbb{Z}^n}^{-1} \pi_{\mathbb{Z}^n}(\mathbb{Z}^n \cap K)$. Then $\xi \in \mathbb{Z}^n$ so we need to show that $\xi \in K$. It follows that $\pi_{\mathbb{Z}^n}(\xi) \in \pi_{\mathbb{Z}^n}(\mathbb{Z}^n \cap K) = S \cap \pi_{\mathbb{R}^n}(K)$, which implies $\xi \in \pi_{\mathbb{Z}^n}^{-1}(S \cap \pi_{\mathbb{R}^n}(K)) \subset K$, completing the proof. $\qquad\square$

Definition 4.12.3 (Toroidal wave front sets). Let $u \in \mathcal{D}'(\mathbb{T}^n)$. The *toroidal wave front set* $\mathrm{WF}_T(u) \subset \mathbb{T}^n \times (\mathbb{Z}^n \setminus \{0\})$ is defined as follows: we say that $(x_0, \xi_0) \in \mathbb{T}^n \times (\mathbb{Z}^n \setminus \{0\})$ does not belong to $\mathrm{WF}_T(u)$ if and only if there exist $\chi \in C^\infty(\mathbb{T}^n)$ and an open discrete cone $\Gamma \subset \mathbb{Z}^n \setminus \{0\}$ such that $\chi(x_0) \neq 0$, $\xi_0 \in \Gamma$ and

$$\forall N > 0 \,\, \exists C_N < \infty \,\, \forall \xi \in \Gamma : \,\, |\mathcal{F}_{\mathbb{T}^n}(\chi u)(\xi)| \leq C_N \langle \xi \rangle^{-N};$$

in such a case we say that $\mathcal{F}_{\mathbb{T}^n}(\chi u)$ *decays rapidly in* Γ.

We say that a pseudo-differential operator $A \in \Psi^m(\mathbb{T}^n \times \mathbb{Z}^n) = \mathrm{Op}\, S^m(\mathbb{T}^n \times \mathbb{Z}^n)$ is *elliptic at the point* $(x_0, \xi_0) \in \mathbb{T}^n \times (\mathbb{Z}^n \setminus \{0\})$ if its toroidal symbol $\sigma_A : \mathbb{T}^n \times \mathbb{Z}^n \to \mathbb{C}$ satisfies

$$|\sigma_A(x_0, \xi)| \geq C \, \langle \xi \rangle^m$$

for some constant $C > 0$ as $\|\xi\| \to \infty$, where $\xi \in \Gamma$ and $\Gamma \subset \mathbb{Z}^n \setminus \{0\}$ is an open discrete cone containing ξ_0. Should $\xi \mapsto \sigma_A(x_0, \xi)$ be rapidly decaying in an open discrete cone containing ξ_0 then A is said to be *smoothing at* (x_0, ξ_0). The *toroidal characteristic set of* $A \in \Psi^m(\mathbb{T}^n \times \mathbb{Z}^n)$ is

$$\mathrm{char}_T(A) := \{(x_0, \xi_0) \in \mathbb{T}^n \times (\mathbb{Z}^n \setminus \{0\}) : \,\, A \text{ is not elliptic at } (x_0, \xi_0)\},$$

and the *toroidal wave front set of* A is

$$\mathrm{WF}_T(A) := \{(x_0, \xi_0) \in \mathbb{T}^n \times (\mathbb{Z}^n \setminus \{0\}) : \,\, A \text{ is not smoothing at } (x_0, \xi_0)\}.$$

Proposition 4.12.4. *We have*

$$\mathrm{WF}_T(A) \cup \mathrm{char}_T(A) = \mathbb{T}^n \times (\mathbb{Z}^n \setminus \{0\}).$$

Proof. The statement follows because if $(x, \xi) \notin \mathrm{char}_T(A)$, it means that A is elliptic at (x, ξ), and hence not smoothing. $\square$

Exercise 4.12.5. Show that $\mathrm{WF}_T(A) = \emptyset$ if and only if A is smoothing, i.e., maps $\mathcal{D}'(\mathbb{T}^n)$ to $C^\infty(\mathbb{T}^n)$ (equivalently, the Schwartz kernel is smooth by Theorem 4.3.6).

Proposition 4.12.6. *Let* $A, B \in \mathrm{Op}\, S^m(\mathbb{T}^n \times \mathbb{Z}^n)$. *Then*

$$\mathrm{WF}_T(AB) \subset \mathrm{WF}_T(A) \cap \mathrm{WF}_T(B).$$

Proof. By Theorem 4.7.10 applied to pseudo-differential operators A and B we notice that the toroidal symbol of $AB \in \mathrm{Op}\, S^{2m}(\mathbb{T}^n \times \mathbb{Z}^n)$ has an asymptotic expansion

$$\sigma_{AB}(x, \xi) \,\, \sim \,\, \sum_{\alpha \geq 0} \frac{1}{\alpha!} \left(\triangle_\xi^\alpha \sigma_A(x, \xi) \right) D_x^{(\alpha)} \sigma_B(x, \xi)$$

$$\sim \,\, \sum_{\alpha \geq 0} \frac{1}{\alpha!} \left(\partial_\xi^\alpha \sigma_A(x, \xi) \right) \partial_x^\alpha \sigma_B(x, \xi),$$

where in the latter expansion we have used smooth extensions of toroidal symbols. This expansion says that AB is smoothing at (x_0, ξ_0) if A or B is smoothing at (x_0, ξ_0). $\square$

The notion of the toroidal wave front set is compatible with the action of pseudo-differential operators:

Proposition 4.12.7 (Transformation of toroidal wave fronts). *Let* $u \in \mathcal{D}'(\mathbb{T}^n)$ *and* $A \in \operatorname{Op} S_{\rho,\delta}^m(\mathbb{T}^n \times \mathbb{Z}^n)$, *where* $0 \leq \rho \leq 1$, $0 \leq \delta < 1$. *Then*

$$\operatorname{WF}_T(Au) \subset \operatorname{WF}_T(u).$$

Especially, if $\varphi \in C^\infty(\mathbb{T}^n)$ *does not vanish, then* $\operatorname{WF}_T(\varphi u) = \operatorname{WF}_T(u)$.

Proof. Let $\mathcal{F}_{\mathbb{T}^n} u$ decay rapidly in an open discrete cone $\Gamma \subset \mathbb{Z}^n$. Let us estimate

$$\mathcal{F}_{\mathbb{T}^n}(Au)(\eta) = \sum_{\xi \in \mathbb{Z}^n} \widehat{\sigma_A}(\eta - \xi, \xi)\, \mathcal{F}_{\mathbb{T}^n} u(\xi),$$

where $\widehat{\sigma_A}(\eta, \xi) = \int_{\mathbb{T}^n} e^{-i2\pi x \cdot \eta}\, \sigma_A(x, \xi)\, dx$. Integration by parts yields

$$|\widehat{\sigma_A}(\eta, \xi)| \leq C_M \, \langle \eta \rangle^{-M} \, \langle \xi \rangle^{m + \delta M},$$

because $\sigma_A \in S_{\rho,\delta}^m(\mathbb{T}^n \times \mathbb{Z}^n)$. Due to the rapid decay of $\mathcal{F}_{\mathbb{T}^n} u$ on Γ, we get

$$
\begin{aligned}
\sum_{\xi \in \Gamma} |\widehat{\sigma_A}(\eta - \xi, \xi)| \, |\mathcal{F}_{\mathbb{T}^n} u(\xi)| &\leq C_{M,N} \sum_{\xi \in \Gamma} \langle \eta - \xi \rangle^{-M} \, \langle \xi \rangle^{m + \delta M} \, \langle \xi \rangle^{-N} \\
&\leq 2^M C_{M,N} \sum_{\xi \in \Gamma} \langle \eta \rangle^{-M} \, \langle \xi \rangle^{m + (1 + \delta) M - N} \\
&\leq C_M \, \langle \eta \rangle^{-M},
\end{aligned}
$$

where we used Peetre's inequality and chose N large enough. Next, take an open discrete cone $\Gamma_1 \subset \Gamma$ such that $\eta \in \Gamma_1$ and that $\langle \omega - \xi \rangle \geq C_1 \max\{\langle \omega \rangle, \langle \xi \rangle\}$ for all $\omega \in \Gamma_1$ and $\xi \in \mathbb{Z}^n \setminus \Gamma$ (where C_1 is a constant). Then $\langle \omega - \xi \rangle \geq C_1 \langle \omega \rangle^{1/k} \langle \xi \rangle^{1 - 1/k}$ for all $k \in \mathbb{N}$. Notice that $|\mathcal{F}_{\mathbb{T}^n} u(\xi)| \leq C_N \langle \xi \rangle^N$ for some positive N. Thereby

$$
\begin{aligned}
\sum_{\xi \in \mathbb{Z}^n \setminus \Gamma} |\widehat{\sigma_A}(\eta - \xi, \xi)| \, |\mathcal{F}_{\mathbb{T}^n} u(\xi)| & \\
\leq C \sum_{\xi \in \mathbb{Z}^n \setminus \Gamma} \langle \eta - \xi \rangle^{-M} \, \langle \xi \rangle^{m + \delta M} \, \langle \xi \rangle^N & \\
\leq C' \sum_{\xi \in \mathbb{Z}^n \setminus \Gamma} \langle \eta \rangle^{-M/k} \, \langle \xi \rangle^{m + (\delta - (k-1)/k) M + N} & \\
\leq C_M \, \langle \eta \rangle^{-M/k}, &
\end{aligned}
$$

where we chose $(k - 1)/k > \delta$ and then M large enough. Thus $\mathcal{F}_{\mathbb{T}^n}(Au)$ decays rapidly in Γ_1. $\qquad \square$

We will not pursue the complete analysis of toroidal wave front sets much further because most of their properties can be obtained from the known properties of the usual wave front sets and the following relation, where $\operatorname{WF}(u)$ stands for the usual Hörmander's wave front set of a distribution u.

Theorem 4.12.8 (Characterisation of toroidal wave front sets). *Let $u \in \mathcal{D}'(\mathbb{T}^n)$. Then*
$$\mathrm{WF}_T(u) = (\mathbb{T}^n \times \mathbb{Z}^n) \cap \mathrm{WF}(u).$$

Proof. Without loss of generality, let $u \in C^k(\mathbb{T}^n)$ for some large k, and let $\chi \in C_0^\infty(\mathbb{R}^n)$ such that $\mathrm{supp}(\chi) \subset (0,1)^n$. If $\mathcal{F}_{\mathbb{R}^n}(\chi u)$ decays rapidly in an open cone $K \subset \mathbb{R}^n$ then $\mathcal{F}_{\mathbb{T}^n}(\mathcal{P}(\chi u)) = \mathcal{F}_{\mathbb{R}^n}(\chi u)|_{\mathbb{Z}^n}$ decays rapidly in the open discrete cone $\mathbb{Z}^n \cap K$. Hence $\mathrm{WF}_T(u) \subset (\mathbb{T}^n \times \mathbb{Z}^n) \cap \mathrm{WF}(u)$.

Next, we need to show that
$$(\mathbb{T}^n \times \mathbb{Z}^n) \setminus \mathrm{WF}_T(u) \subset (\mathbb{T}^n \times \mathbb{Z}^n) \setminus \mathrm{WF}(u).$$

Let $(x_0, \xi_0) \in (\mathbb{T}^n \times \mathbb{Z}^n) \setminus \mathrm{WF}_T(u)$ (where $\xi_0 \neq 0$). We must show that $(x_0, \xi_0) \notin \mathrm{WF}(u)$. There exist $\chi \in C^\infty(\mathbb{T}^n)$ (we may assume that $\mathrm{supp}(\chi) \subset (0,1)^n$ as above) and an open cone $K \subset \mathbb{R}^n$ such that $\chi(x_0) \neq 0$, $\xi_0 \in \mathbb{Z}^n \cap K$ and that $\mathcal{F}_{\mathbb{T}^n}(\mathcal{P}(\chi u))$ decays rapidly in $\mathbb{Z}^n \cap K$.

Let $K_1 \subset \mathbb{R}^n$ be an open cone such that $\xi_0 \in K_1 \subset K$ and that the closure $\overline{K_1} \subset K \cup \{0\}$. Take any function $w \in C^\infty(\mathbb{S}^{n-1})$ such that

$$w(\omega) = \begin{cases} 1, & \text{if } \omega \in \mathbb{S}^{n-1} \cap K_1, \\ 0, & \text{if } \omega \in \mathbb{S}^{n-1} \setminus K. \end{cases}$$

Let $a \in C^\infty(\mathbb{R}^n \times \mathbb{R}^n)$ be independent of x and such that $a(x, \xi) = w(\xi/\|\xi\|)$ whenever $\|\xi\| \geq 1$. Then $a \in S^0(\mathbb{R}^n \times \mathbb{R}^n)$. Let $\tilde{a} = a|_{\mathbb{T}^n \times \mathbb{Z}^n}$, so that $\tilde{a} \in S^0(\mathbb{T}^n \times \mathbb{Z}^n)$ by Theorem 4.5.3.

By Corollary 4.6.21, we have
$$\mathcal{P}(\chi \, a(X, D) f) = \mathcal{P}(\chi) \, \tilde{a}(X, D)(\mathcal{P}f) + \mathcal{P}(Rf)$$

for all Schwartz test functions f, for a smoothing operator $R : \mathcal{D}'(\mathbb{R}^n) \to \mathcal{S}(\mathbb{R}^n)$. By Remark 4.6.22 we also have
$$\mathcal{P}(\chi \, a(X, D)(\chi u)) = \mathcal{P}(\chi) \, \tilde{a}(X, D)(\mathcal{P}(\chi u)) + \mathcal{P}(R(\chi u)),$$

where the right-hand side belongs to $C^\infty(\mathbb{T}^n)$, since its Fourier coefficients decay rapidly on the whole $\mathbb{Z}^n$. Therefore also $\mathcal{P}(\chi \, a(X, D)(\chi u))$ belongs to $C^\infty(\mathbb{T}^n)$.

Thus $\chi \, a(X, D)(\chi u) \in C_0^\infty(\mathbb{R}^n)$. Let $\xi \in K_1$ such that $\|\xi\| \geq 1$. Then we have
$$\mathcal{F}_{\mathbb{R}^n}(a(X, D)(\chi u))(\xi) = w(\xi/\|\xi\|) \, \mathcal{F}_{\mathbb{R}^n}(\chi u)(\xi) = \mathcal{F}_{\mathbb{R}^n}(\chi u)(\xi).$$

Thus $\mathcal{F}_{\mathbb{R}^n}(\chi u)$ decays rapidly on K_1. Therefore (x_0, ξ_0) does not belong to $\mathrm{WF}(u)$. $\qquad\square$

Exercise 4.12.9. Show that for every $u \in \mathcal{D}'(\mathbb{T}^n)$ we have
$$\mathrm{WF}_T(u) = \bigcap_{A \in \Psi^0, \, Au \in C^\infty} \mathrm{char}_T(A).$$

4.13 Fourier series operators

In this section we consider analogues of Fourier integral operators on the torus $\mathbb{T}^n$. We will call such operators Fourier series operators and study their composition formulae with pseudo-differential operators on the torus.

Definition 4.13.1 (Fourier series operators). *Fourier series operators* (FSO) are operators of the form

$$Tu(x) := \sum_{\xi \in \mathbb{Z}^n} \int_{\mathbb{T}^n} e^{2\pi i(\phi(x,\xi) - y \cdot \xi)} \, a(x, y, \xi) \, u(y) \, dy, \qquad (4.47)$$

where $a \in C^\infty(\mathbb{T}^n \times \mathbb{T}^n \times \mathbb{Z}^n)$ is a toroidal amplitude and ϕ is a real-valued phase function such that conditions of the following Remark 4.13.2 are satisfied.

Remark 4.13.2 **(Phase functions).** We note that if $u \in C^\infty(\mathbb{T}^n)$, for the function Tu to be well defined on the torus we need that the integral (4.47) is 1-periodic in x. Therefore, by identifying functions on $\mathbb{T}^n$ with 1-periodic functions on $\mathbb{R}^n$, we will require that the phase function $\phi : \mathbb{R}^n \times \mathbb{Z}^n \to \mathbb{R}$ is such that the function $x \mapsto e^{2\pi i\phi(x,\xi)}$ is 1-periodic for all $\xi \in \mathbb{Z}^n$. Note that here it is not necessary that the function $x \mapsto \phi(x, \xi)$ itself is 1-periodic.

Remark 4.13.3. Assume that the function $\phi : \mathbb{R}^n \times \mathbb{Z}^n \to \mathbb{R}$ is in C^k with respect to x for all $\xi \in \mathbb{Z}^n$. Assume also that the function $x \mapsto e^{2\pi i\phi(x,\xi)}$ is 1-periodic for all $\xi \in \mathbb{Z}^n$. Differentiating it with respect to x we get that the functions $x \mapsto \partial_x^\alpha \phi(x, \xi)$ are 1-periodic for all $\xi \in \mathbb{Z}^n$ and all $\alpha \in \mathbb{N}_0^n$ with $1 \leq |\alpha| \leq k$.

Remark 4.13.4. The operator $T : C^\infty(\mathbb{T}^n) \to \mathcal{D}'(\mathbb{T}^n)$ in (4.47) can be justified in the usual way for oscillatory integrals. If we have more information on the symbol we have better properties, for example:

Proposition 4.13.5. *Let $\phi \in C^\infty(\mathbb{T}^n \times \mathbb{Z}^n)$ be such that the function $x \mapsto e^{2\pi i\phi(x,\xi)}$ is 1-periodic for all $\xi \in \mathbb{Z}^n$, and such that for some $\ell \in \mathbb{R}$ we have*

$$|\partial_x^\alpha \phi(x, \xi)| \leq C_\alpha \langle \xi \rangle^\ell$$

for all multi-indices α, all $x \in \mathbb{T}^n$ and $\xi \in \mathbb{Z}^n$. Let $a \in C^\infty(\mathbb{T}^n \times \mathbb{T}^n \times \mathbb{Z}^n)$ be such that there is $m, \delta_1 \in \mathbb{R}$ and $\delta_2 < 1$ such that for all multi-indices α, β we have

$$|\partial_x^\alpha \partial_y^\beta a(x, y, \xi)| \leq C_{\alpha\beta} \langle \xi \rangle^{m + \delta_1 |\alpha| + \delta_2 |\beta|}$$

for all $x, y \in \mathbb{T}^n$ and $\xi \in \mathbb{Z}^n$. Then the operator T in (4.47) is a well-defined continuous linear operator from $C^\infty(\mathbb{T}^n)$ to $C^\infty(\mathbb{T}^n)$.

Proof. Let $u \in C^\infty(\mathbb{T}^n)$ and let $\mathcal{L}_y$ be the Laplacian with respect to y. Expression (4.47) can be justified by integration by parts with the operator $L_y = 1 - (4\pi^2)^{-1}\mathcal{L}_y$ which satisfies $\langle \xi \rangle^{-2} L_y \, e^{2\pi i y \cdot \xi} = e^{2\pi i y \cdot \xi}$. Consequently, we interpret (4.47) as

$$Tu(x) = \sum_{\xi \in \mathbb{Z}^n} \langle \xi \rangle^{-2N} \int_{\mathbb{T}^n} e^{2\pi i(\phi(x,\xi) - y \cdot \xi)} \, L_y^N \, [a(x, y, \xi) u(y)] \, dy, \qquad (4.48)$$

so that both y-integral and ξ-sum converge absolutely if N is large enough in view of $\delta_2 < 1$. Consequently, Tu is 1-periodic by our assumptions and by Remark 4.13.2, and (4.48) can be differentiated any number of times with respect to x to yield a function $Tu \in C^\infty(\mathbb{T}^n)$ by Remark 4.13.3. Continuity of T on $C^\infty(\mathbb{T}^n)$ follows from Lebesgue's dominated convergence theorem on $\mathbb{T}^n \times \mathbb{Z}^n$ (see Theorems C.3.22 and 1.1.4). $\hfill\square$

Remark 4.13.6. Thus, we will always interpret (4.47) as (4.48). Composition formulae of this section can be compared with those obtained in [94, 96] globally on $\mathbb{R}^n$ under minimal assumptions on phases and amplitudes. However, on the torus, the assumptions on the regularity or boundedness of higher-order derivatives of phases and amplitudes are redundant due to the fact that $\xi \in \mathbb{Z}^n$ takes only discrete values.

We recall the notation for the toroidal version of Taylor polynomials and the corresponding derivatives introduced in (3.15) and (4.31), which will be used in the formulation of the following theorems. However, we need the following:

Definition 4.13.7 (Warning: operators $(-D_y)^{(\alpha)}$). Before we define operators $(-D_y)^{(\alpha)}$ below we warn the reader that one should not formally plug in the minus sign in the definition of the previously defined operators $(D_y)^{(\alpha)}$ in Definition 3.4.1! Please compare these operators with those in (4.31) and observe how the sign changes. With this warning in place, we define

$$(-D_y)^{(\alpha)} = (-D_{y_1})^{(\alpha_1)} \cdots (-D_{y_n})^{(\alpha_n)}, \qquad (4.49)$$

where $-D_{y_j}^{(0)} = I$ and

$$
\begin{aligned}
(-D_{y_j})^{(k+1)} &= (-D_{y_j})^{(k)} \left(-\frac{\partial}{2\pi i \partial y_j} - kI \right) \\
&= -\frac{\partial}{2\pi i \partial y_j} \left(-\frac{\partial}{2\pi i \partial y_j} - I \right) \cdots \left(-\frac{\partial}{2\pi i \partial y_j} - kI \right).
\end{aligned}
$$

We now study composition formulae of Fourier series operators with pseudo-differential operators.

Theorem 4.13.8 (Composition FSO∘ΨDO). *Let $\phi : \mathbb{R}^n \times \mathbb{Z}^n \to \mathbb{R}$ be such that function $x \mapsto e^{2\pi i \phi(x,\xi)}$ is 1-periodic for all $\xi \in \mathbb{Z}^n$. Let $T : C^\infty(\mathbb{T}^n) \to \mathcal{D}'(\mathbb{T}^n)$ be defined by*

$$Tu(x) := \sum_{\xi \in \mathbb{Z}^n} \int_{\mathbb{T}^n} e^{2\pi i(\phi(x,\xi) - y\cdot\xi)} \, a(x,y,\xi) \, u(y) \, dy, \qquad (4.50)$$

where the toroidal amplitude $a \in C^\infty(\mathbb{T}^n \times \mathbb{T}^n \times \mathbb{Z}^n)$ satisfies

$$\left| \partial_x^\alpha \partial_y^\beta a(x,y,\xi) \right| \leq C_{\alpha\beta m} \, \langle \xi \rangle^m$$

for all $x, y \in \mathbb{T}^n$, $\xi \in \mathbb{Z}^n$ *and* $\alpha, \beta \in \mathbb{N}_0^n$. *Let* $p \in S^\ell(\mathbb{T}^n \times \mathbb{Z}^n)$ *be a toroidal symbol and* $P = \mathrm{Op}(p)$ *the corresponding pseudo-differential operator. Then the composition* TP *has the form*

$$TPu(x) = \sum_{\xi \in \mathbb{Z}^n} \int_{\mathbb{T}^n} e^{2\pi i(\phi(x,\xi) - z \cdot \xi)} \, c(x, z, \xi) \, u(z) \, dz,$$

where

$$c(x, z, \xi) = \sum_{\eta \in \mathbb{Z}^n} \int_{\mathbb{T}^n} e^{2\pi i(y-z) \cdot (\eta - \xi)} \, a(x, y, \xi) \, p(y, \eta) \, dy$$

satisfies

$$\left| \partial_x^\alpha \partial_z^\beta c(x, z, \xi) \right| \le C_{\alpha \beta m t} \, \langle \xi \rangle^{m+\ell}$$

for every $x, z \in \mathbb{T}^n$, $\xi \in \mathbb{Z}^n$ *and* $\alpha, \beta \in \mathbb{N}_0^n$. *Moreover, we have an asymptotic expansion*

$$c(x, z, \xi) \sim \sum_{\alpha \ge 0} \frac{1}{\alpha!} \, (-D_z)^{(\alpha)} \left[a(x, z, \xi) \, \triangle_\xi^\alpha p(z, \xi) \right].$$

Furthermore, if $0 \le \delta < \rho \le 1$, $p \in S_{\rho,\delta}^\ell(\mathbb{T}^n \times \mathbb{Z}^n)$ *and* $a \in \mathcal{A}_{\rho,\delta}^m(\mathbb{T}^n)$, *then* $c \in \mathcal{A}_{\rho,\delta}^{m+\ell}(\mathbb{T}^n)$.

Remark 4.13.9. We note that if T in (4.50) is a pseudo-differential operator with phase $\phi(x, \xi) = x \cdot \xi$ and amplitude $a(x, y, \xi) = a(x, \xi)$ independent of y, then the asymptotic expansion formula for the composition of two pseudo-differential operators $T \circ P$ becomes

$$c(x, z, \xi) \sim \sum_{\alpha \ge 0} \frac{1}{\alpha!} \, a(x, \xi) \, (-D_z)^{(\alpha)} \triangle_\xi^\alpha p(z, \xi).$$

This is another representation for the composition compared to Theorem 4.7.10, with an amplitude realisation of the pseudo-differential operator $T \circ P$, see Remark 4.7.12.

Proof of Theorem 4.13.8. Let us calculate the composition TP:

$$
\begin{aligned}
TPu(x) &= \sum_{\xi \in \mathbb{Z}^n} \int_{\mathbb{T}^n} e^{2\pi i(\phi(x,\xi) - y \cdot \xi)} \, a(x, y, \xi) \, Pu(y) \, dy \\
&= \sum_{\xi \in \mathbb{Z}^n} \int_{\mathbb{T}^n} e^{2\pi i(\phi(x,\xi) - y \cdot \xi)} \, a(x, y, \xi) \\
&\qquad \times \sum_{\eta \in \mathbb{Z}^n} \int_{\mathbb{T}^n} e^{2\pi i(y-z) \cdot \eta} \, p(y, \eta) \, u(z) \, dz \, dy \\
&= \sum_{\xi \in \mathbb{Z}^n} \int_{\mathbb{T}^n} e^{2\pi i(\phi(x,\xi) - z \cdot \xi)} \, c(x, z, \xi) \, u(z) \, dz,
\end{aligned}
$$

where

$$c(x, z, \xi) = \sum_{\eta \in \mathbb{Z}^n} \int_{\mathbb{T}^n} e^{2\pi i (y-z)\cdot(\eta-\xi)} \, a(x, y, \xi) \, p(y, \eta) \, dy.$$

Denote $\theta := \eta - \xi$, so that by the discrete Taylor expansion (Theorem 3.3.21), we formally get

$$\begin{aligned}
c(x, z, \xi) \; &\sim \; \sum_{\theta \in \mathbb{Z}^n} \int_{\mathbb{T}^n} e^{2\pi i (y-z)\cdot\theta} \, a(x, y, \xi) \sum_{\alpha \geq 0} \frac{1}{\alpha!} \, \theta^{(\alpha)} \, \triangle_\xi^\alpha p(y, \xi) \, dy \\
&= \; \sum_{\alpha \geq 0} \frac{1}{\alpha!} \sum_{\theta \in \mathbb{Z}^n} \int_{\mathbb{T}^n} \theta^{(\alpha)} \, e^{2\pi i (y-z)\cdot\theta} \, a(x, y, \xi) \, \triangle_\xi^\alpha p(y, \xi) \, dy \\
&= \; \sum_{\alpha \geq 0} \frac{1}{\alpha!} \, (-D_y)^{(\alpha)} \left(a(x, y, \xi) \, \triangle_\xi^\alpha p(y, \xi) \right) |_{y=z}.
\end{aligned}$$

Now we have to justify the asymptotic expansion. First we take a discrete Taylor expansion and using Theorem 3.3.21 again, we obtain

$$p(y, \xi + \theta) = \sum_{|\omega| < M} \frac{1}{\omega!} \, \theta^{(\omega)} \, \triangle_\xi^\omega p(y, \xi) + r_M(y, \xi, \theta).$$

Let $Q(\theta) = \{\nu \in \mathbb{Z}^n : |\nu_j| \leq |\theta_j| \text{ for all } j = 1, \ldots, n\}$ as in Theorem 3.3.21. Then by Peetre's inequality (Proposition 3.3.31) we have

$$\begin{aligned}
\left| \triangle_\xi^\alpha \partial_y^\beta r_M(y, \xi, \theta) \right| \; &\leq \; C \, \langle \theta \rangle^M \max_{\substack{|\omega|=M \\ \nu \in Q(\theta)}} \left| \triangle_\xi^{\alpha+\omega} \partial_y^\beta p(y, \xi + \nu) \right| \\
&\leq \; C \, \langle \theta \rangle^M \max_{\nu \in Q(\theta)} \langle \xi + \nu \rangle^{\ell - |\alpha| - M} \\
&\leq \; C \, \langle \theta \rangle^M \max_{\nu \in Q(\theta)} \langle \nu \rangle^{|\ell - |\alpha| - M|} \, \langle \xi \rangle^{\ell - |\alpha| - M} \\
&\leq \; C \, \langle \theta \rangle^{2M + |\ell| + |\alpha|} \, \langle \xi \rangle^{\ell - |\alpha| - M}.
\end{aligned}$$

The corresponding remainder term in the asymptotic expansion of $c(x, z, \xi)$ is

$$\begin{aligned}
R_M(x, z, \xi) \; &= \; \sum_{\theta \in \mathbb{Z}^n} \int_{\mathbb{T}^n} e^{2\pi i (y-z)\cdot\theta} \, a(x, y, \xi) \, r_M(y, \xi, \theta) \, dy \\
&= \; \sum_{\theta \in \mathbb{Z}^n} \int_{\mathbb{T}^n} e^{2\pi i (y-z)\cdot\theta} \, \langle \theta \rangle^{-2N} \\
&\qquad \times (I - (4\pi^2)^{-1} \mathcal{L}_y)^N \, [a(x, y, \xi) \, r_M(y, \xi, \theta)] \, dy,
\end{aligned}$$

where we integrated by parts exploiting that

$$(I - (4\pi^2)^{-1} \mathcal{L}_y) \, e^{2\pi i (y-z)\cdot\theta} = \langle \theta \rangle^2 \, e^{2\pi i (y-z)\cdot\theta},$$

where $\mathcal{L}_y$ is the Laplacian with respect to y. Thus we get the estimate

$$\left| \triangle_\xi^\alpha \partial_x^\beta \partial_z^\gamma R_M(x,z,\xi) \right| \leq C \sum_{\theta \in \mathbb{Z}^n} \langle \theta \rangle^{|\gamma| - 2N + 2M + |\ell| + |\alpha|} \langle \xi \rangle^{m+\ell-M},$$

and we take $N \in \mathbb{N}$ so large that this sum over θ converges. Hence

$$\left| \triangle_\xi^\alpha \partial_x^\beta \partial_z^\gamma R_M(x,z,\xi) \right| \leq C \langle \xi \rangle^{m+\ell-M}.$$

This completes the proof of the first part of the theorem. Finally, we assume that $a \in \mathcal{A}_{\rho,\delta}^m(\mathbb{T}^n)$. Then also the terms in the asymptotic expansion and the remainder R_M have corresponding decay properties in the ξ-differences, leading to the amplitude $c \in \mathcal{A}_{\rho,\delta}^{m+\ell}(\mathbb{T}^n)$. This completes the proof. $\qquad\square$

Exercise 4.13.10. Work out all the details of the proof in the (ρ, δ)-case.

We now formulate the theorem about compositions of operators in the opposite order.

Theorem 4.13.11 (Composition ΨDO$\circ$FSO). *Let $\phi : \mathbb{R}^n \times \mathbb{Z}^n \to \mathbb{R}$ be 1-periodic for all $\xi \in \mathbb{Z}^n$. Let $T : C^\infty(\mathbb{T}^n) \to \mathcal{D}'(\mathbb{T}^n)$ be such that*

$$Tu(x) := \sum_{\xi \in \mathbb{Z}^n} \int_{\mathbb{T}^n} e^{2\pi i(\phi(x,\xi) - y \cdot \xi)} \, a(x,y,\xi) \, u(y) \, \mathrm{d}y,$$

where $a \in C^\infty(\mathbb{T}^n \times \mathbb{T}^n \times \mathbb{Z}^n)$ satisfies

$$\left| \partial_x^\alpha \partial_y^\beta a(x,y,\xi) \right| \leq C_{\alpha\beta m} \, \langle \xi \rangle^m$$

for all $x, y \in \mathbb{T}^n$, $\xi \in \mathbb{Z}^n$ and $\alpha, \beta \in \mathbb{N}_0^n$. Assume that for some $C > 0$ we have

$$C^{-1} \, \langle \xi \rangle \leq \langle \nabla_x \phi(x,\xi) \rangle \leq C \, \langle \xi \rangle \tag{4.51}$$

for all $x \in \mathbb{T}^n$, $\xi \in \mathbb{Z}^n$, and that

$$\left| \partial_x^\alpha \phi(x,\xi) \right| \leq C_\alpha \, \langle \xi \rangle, \quad \left| \partial_x^\alpha \triangle_\xi^\beta \phi(x,\xi) \right| \leq C_{\alpha\beta} \tag{4.52}$$

for all $x \in \mathbb{T}^n$, $\xi \in \mathbb{Z}^n$ and $\alpha, \beta \in \mathbb{N}_0^n$ with $|\beta| = 1$. Let $\widetilde{p} \in S^\ell(\mathbb{T}^n \times \mathbb{Z}^n)$ be a toroidal symbol let $p(x,\xi)$ denote an extension of $\widetilde{p}(x,\xi)$ to a symbol in $S^\ell(\mathbb{T}^n \times \mathbb{R}^n)$ as given in Theorem 4.5.3. Let $P = \mathrm{Op}(p)$ be the corresponding pseudo-differential operator. Then

$$PTu(x) = \sum_{\xi \in \mathbb{Z}^n} \int_{\mathbb{T}^n} e^{2\pi i(\phi(x,\xi) - z \cdot \xi)} \, c(x,z,\xi) \, u(z) \, \mathrm{d}z,$$

where we have

$$\left| \partial_x^\alpha \partial_z^\beta c(x,z,\xi) \right| \leq C_{\alpha\beta} \, \langle \xi \rangle^{m+\ell}$$

all every $x, z \in \mathbb{T}^n$, $\xi \in \mathbb{Z}^n$ *and* $\alpha, \beta \in \mathbb{N}_0^n$. *Moreover, we have the asymptotic expansion*

$$c(x, z, \xi) \sim \sum_{\alpha \geq 0} \frac{(2\pi i)^{-|\alpha|}}{\alpha!} \, \partial_\eta^\alpha p(x, \eta)|_{\eta = \nabla_x \phi(x, \xi)} \, \partial_y^\alpha \left[e^{2\pi i \Psi(x, y, \xi)} a(y, z, \xi) \right]|_{y=x},$$

$$(4.53)$$

where

$$\Psi(x, y, \xi) := \phi(y, \xi) - \phi(x, \xi) + (x - y) \cdot \nabla_x \phi(x, \xi).$$

Remark 4.13.12. Let us make some remarks about quantities appearing in the asymptotic extension formula (4.53). It is geometrically reasonable to evaluate the symbol $\widetilde{p}(x, \xi)$ at the real Hamiltonian flow generated by the phase function ϕ of the Fourier series operator T. This is the main complication compared with pseudo-differential operators for which we have Proposition 4.12.7. However, although a priori the symbol $\widetilde{p}$ is defined only on $\mathbb{T}^n \times \mathbb{Z}^n$, we can still extend it to a symbol $p(x, \xi)$ on $\mathbb{T}^n \times \mathbb{R}^n$ by Theorem 4.5.3, so that the restriction $\partial_\eta^\alpha p(x, \eta)|_{\eta = \nabla_x \phi(x, \xi)}$ makes sense. We also note that the function $\Psi(x, y, \xi)$ can not be in general considered as a function on $\mathbb{T}^n \times \mathbb{T}^n \times \mathbb{Z}^n$ because it may not be periodic in x and y. However, we can still observe that the derivatives $\partial_y^\alpha \left[e^{2\pi i \Psi(x, y, \xi)} a(y, z, \xi) \right]|_{y=x}$ are periodic in x and z, so all terms in the right-hand side of (4.53) are well-defined functions on $\mathbb{T}^n \times \mathbb{T}^n \times \mathbb{Z}^n$. In any case, for a standard theory of Fourier integral operators on $\mathbb{R}^n$ we refer the reader to [56].

Remark 4.13.13. In Theorem 4.13.11, we note that if $\phi \in S_{\rho,\delta}^1(\mathbb{R}^n \times \mathbb{R}^n)$, $\widetilde{p} \in S_{\rho,\delta}^\ell(\mathbb{T}^n \times \mathbb{Z}^n)$, $a \in \mathcal{A}_{\rho,\delta}^m(\mathbb{T}^n)$, and $0 \leq \delta < \rho \leq 1$, then we also have $c \in \mathcal{A}_{\rho,\delta}^{m+\ell}(\mathbb{T}^n)$.

Exercise 4.13.14. Prove this remark.

Proof of Theorem 4.13.11. To simplify the notation, let us drop writing tilde on p, and denote both symbols $\widetilde{p}$ and p by the same letter p. There should be no confusion since they coincide on $\mathbb{T}^n \times \mathbb{Z}^n$. Let $P = \mathrm{Op}(p)$. We can write

$$\begin{aligned} PTu(x) &= \sum_{\eta \in \mathbb{Z}^n} \int_{\mathbb{T}^n} e^{2\pi i (x-y) \cdot \eta} \, p(x, \eta) \, Tu(y) \, dy \\ &= \sum_{\eta \in \mathbb{Z}^n} \int_{\mathbb{T}^n} e^{2\pi i (x-y) \cdot \eta} \, p(x, \eta) \\ &\qquad \times \sum_{\xi \in \mathbb{Z}^n} \int_{\mathbb{T}^n} e^{2\pi i (\phi(y,\xi) - z \cdot \xi)} \, a(y, z, \xi) \, u(z) \, dz \, dy \\ &= \sum_{\xi \in \mathbb{Z}^n} \int_{\mathbb{T}^n} e^{2\pi i (\phi(x,\xi) - z \cdot \xi)} \, c(x, z, \xi) \, u(z) \, dz, \end{aligned}$$

where

$$c(x, z, \xi) = \sum_{\eta \in \mathbb{Z}^n} p(x, \eta) \int_{\mathbb{T}^n} e^{2\pi i (\phi(y,\xi) - \phi(x,\xi) + (x-y) \cdot \eta)} \, a(y, z, \xi) \, dy.$$

Let us fix some $x \in \mathbb{R}^n$, with corresponding equivalence class $[x] \in \mathbb{T}^n$ which we still denote by x. Let $V \subset \mathbb{R}^n$ be an open cube with side length equal to 1 centred at x. Let $\chi = \chi(x, y) \in C^\infty(\mathbb{T}^n \times \mathbb{T}^n)$ be such that $0 \leq \chi \leq 1$, $\chi(x, y) = 1$ for $\|x - y\| < \kappa$ for some sufficiently small $\kappa > 0$, and such that $\operatorname{supp} \chi(x, \cdot) \cap \overline{V} \subset V$. The last condition means that $\chi(x, \cdot)|_V \in C_0^\infty(V)$ is supported away from the boundaries of the cube V. Let

$$c(x, z, \xi) = c^I(x, z, \xi) + c^{II}(x, z, \xi),$$

where

$$c^I(x, z, \xi) = \sum_{\eta \in \mathbb{Z}^n} \int_{\mathbb{T}^n} e^{2\pi i(\phi(y, \xi) - \phi(x, \xi) + (x - y) \cdot \eta)} \, (1 - \chi(x, y)) \, a(y, z, \xi) \, p(x, \eta) \, dy,$$

and

$$c^{II}(x, z, \xi) = \sum_{\eta \in \mathbb{Z}^n} \int_{\mathbb{T}^n} e^{2\pi i(\phi(y, \xi) - \phi(x, \xi) + (x - y) \cdot \eta)} \, \chi(x, y) \, a(y, z, \xi) \, p(x, \eta) \, dy.$$

1. **Estimate on the support of $1 - \chi$.** By making a decomposition into cones (sectors) centred at x viewed as a point in $\mathbb{R}^n$, it follows that we can assume without loss of generality that the support of $1 - \chi$ is contained in a set where $C < |x_j - y_j|$, for some $1 \leq j \leq n$. In turn, because of the assumption on the support of $\chi(x, \cdot)|_V$ it follows that $C < |x_j - y_j| < 1 - C$, for some $C > 0$. Now we are going to apply the summation by parts formula (3.14) to estimate $c^I(x, z, \xi)$. First we notice that it follows that

$$
\begin{aligned}
\triangle_{\eta_j} e^{2\pi i(x-y) \cdot \eta} &= e^{2\pi i(x-y) \cdot (\eta + e_j)} - e^{2\pi i(x-y) \cdot \eta} \\
&= e^{2\pi i(x-y) \cdot \eta} \left(e^{2\pi i(x_j - y_j)} - 1 \right) \\
&\neq 0
\end{aligned}
$$

on $\operatorname{supp}(1 - \chi)$. Hence by the summation by parts formula (3.14) we get that

$$\sum_{\eta \in \mathbb{Z}^n} e^{2\pi i(x-y) \cdot \eta} \, p(x, \eta) = \left(e^{2\pi i(x_j - y_j)} - 1 \right)^{-N_1} \sum_{\eta \in \mathbb{Z}^n} e^{2\pi i(x-y) \cdot \eta} \, \overline{\triangle}_{\eta_j}^{N_1} p(x, \eta),$$

where the sum on the right-hand side converges absolutely for large enough N_1. On the other hand, we can integrate by parts with the operator

$$^t L_y = \frac{1 - (4\pi^2)^{-1} \mathcal{L}_y}{\langle \nabla_y \phi(y, \xi) \rangle^2 - (2\pi)^{-1} i \, \mathcal{L}_y \phi(y, \xi)},$$

where $\mathcal{L}_y$ is the Laplace operator with respect to y, and for which we have $L_y^{N_2} e^{2\pi i \phi(y, \xi)} = e^{2\pi i \phi(y, \xi)}$. Note that in view of our assumption (4.51) on ϕ, we have

$$|\langle \nabla_y \phi(y, \xi) \rangle^2 - (2\pi)^{-1} i \, \mathcal{L}_y \phi(y, \xi)| \geq |\langle \nabla_y \phi(y, \xi) \rangle|^2 \geq C_1 \langle \xi \rangle^2.$$

Therefore,

$$c^I(x, z, \xi) = \sum_{\eta \in \mathbb{Z}^n} \int_{\mathbb{T}^n} e^{2\pi i(\phi(y,\xi) - \phi(x,\xi) + x \cdot \eta)} L_y^{N_2} \Big\{ e^{-2\pi i y \cdot \eta}$$

$$\times \Big(e^{2\pi i(x_j - y_j)} - 1 \Big)^{-N_1} \overline{\Delta_{\eta_j}}^{N_1} p(x, \eta) \, (1 - \chi(x, y)) \, a(y, z, \xi) \Big\} \, dy.$$

From the properties of amplitudes, we get

$$|c^I(x, z, \xi)| \leq C \sum_{\eta \in \mathbb{Z}^n} \int_{V \cap \{2\pi - c > |x_j - y_j| > c\}} \langle \xi \rangle^{m - 2N_2} \langle \eta \rangle^{2N_2 + \ell - N_1} \, dy$$

$$\leq C \langle \xi \rangle^{-N}$$

for all N, if we choose large enough N_2 and then large enough N_1. We can easily see that similar estimates work for the derivatives of c^I, completing the proof on the support of $1 - \chi$.

2. **Estimate on the support of** χ. Extending $\widetilde{p} \in S^\ell(\mathbb{T}^n \times \mathbb{Z}^n)$ to a symbol in $p \in S^\ell(\mathbb{T}^n \times \mathbb{R}^n)$ as in Theorem 4.5.3, we will make its usual Taylor expansion at $\eta = \nabla_x \phi(x, \xi)$, so that we have

$$p(x, \eta) = \sum_{|\alpha| < M} \frac{(\eta - \nabla_x \phi(x, \xi))^\alpha}{\alpha!} \, \partial_\xi^\alpha p(x, \nabla_x \phi(x, \xi))$$

$$+ \sum_{|\alpha| = M} C_\alpha \, (\eta - \nabla_x \phi(x, \xi))^\alpha \, r_\alpha(x, \xi, \eta - \nabla_x \phi(x, \xi)),$$

$$r_\alpha(x, \xi, \eta - \nabla_x \phi(x, \xi)) = \int_0^1 (1 - t)^{M_1} \, \partial_\xi^\alpha p(x, t\eta + (1 - t)\nabla_x \phi(x, \xi)) \, dt.$$

Then

$$c^{II}(x, z, \xi) = \sum_{|\alpha| < M} \frac{1}{\alpha!} \, c_\alpha(x, z, \xi) + \sum_{|\alpha| = M} C_\alpha R_\alpha(x, z, \xi),$$

where

$$R_\alpha(x, z, \xi) = \sum_{\eta \in \mathbb{Z}^n} \int_{\mathbb{T}^n} e^{2\pi i(\phi(y,\xi) - \phi(x,\xi) + (x - y) \cdot \eta)} (\eta - \nabla_x \phi(x, \xi))^\alpha$$

$$\times \chi(x, y) \, r_\alpha(x, \xi, \eta - \nabla_x \phi(x, \xi)) \, a(y, z, \xi) \, dy,$$

$$c_\alpha(x, z, \xi) = \sum_{\eta \in \mathbb{Z}^n} \int_{\mathbb{T}^n} e^{2\pi i(\phi(y,\xi) - \phi(x,\xi) + (x - y) \cdot \eta)} (\eta - \nabla_x \phi(x, \xi))^\alpha$$

$$\times \chi(x, y) \, a(y, z, \xi) \, \partial_\xi^\alpha p(x, \nabla_x \phi(x, \xi)) \, dy.$$

Now using Corollary 4.6.16 we can calculate

$$
\begin{aligned}
c_\alpha(x, z, \xi) &= \partial_\xi^\alpha p(x, \nabla_x \phi(x, \xi)) \, [D_y - \nabla_x \phi(x, \xi)]^\alpha \\
&\quad \times \left\{ e^{2\pi i (\phi(y,\xi) - \phi(x,\xi))} \, \chi(x, y) \, a(y, z, \xi) \right\}|_{y=x} \\
&= \partial_\xi^\alpha p(x, \nabla_x \phi(x, \xi)) \int_{\mathbb{R}^n} \int_V e^{2\pi i (x - y) \cdot \eta} \, [\eta - \nabla_x \phi(x, \xi)]^\alpha \\
&\quad \times e^{2\pi i (\phi(y,\xi) - \phi(x,\xi))} \, \chi(x, y) \, a(y, z, \xi) \, dy \, d\eta \\
&= \partial_\xi^\alpha p(x, \nabla_x \phi(x, \xi)) \\
&\quad \times D_y^\alpha \left[e^{2\pi i (\phi(y,\xi) - \phi(x,\xi) + (x-y) \cdot \nabla_x \phi(x,\xi))} \chi(x, y) \, a(y, z, \xi) \right]|_{y=x},
\end{aligned}
$$

where we wrote the derivative $[D_y - \nabla_x \phi(x, \xi)]^\alpha$ as a pseudo-differential operator with symbol $[\eta - \nabla_x \phi(x, \xi)]^\alpha$, $x, \xi \in \mathbb{R}^n$, and changed the variables $\eta \mapsto \eta + \nabla_x \phi(x, \xi)$. Since χ is identically equal to one for y near x, we obtain the asymptotic formula (4.53), once the remainders R_α are analysed.

3. **Estimates for the remainder.** Let us first write the remainder in the form

$$
\begin{aligned}
R_\alpha(x, z, \xi) &= \sum_{\eta \in \mathbb{Z}^n} \int_{\mathbb{T}^n} e^{2\pi i (x - y) \cdot \eta} \, r_\alpha(x, \xi, \eta - \nabla_x \phi(x, \xi)) \\
&\quad \times (\eta - \nabla_x \phi(x, \xi))^\alpha \, \chi(x, y) \, g(y) \, dy,
\end{aligned}
\tag{4.54}
$$

with

$$
g(y) = e^{2\pi i (\phi(y,\xi) - \phi(x,\xi))} \, \chi(x, y) \, a(y, z, \xi),
$$

which is a 1-periodic function of y. Now, we can use Corollary 4.6.16 to conclude that $R_\alpha(x, z, \xi)$ in (4.54) is equal to the periodisation with respect to x in the form $R_\alpha(x, z, \xi) = \mathcal{P}_x S_\alpha(x, z, \xi)$, where

$$
\begin{aligned}
S_\alpha(x, z, \xi) &= r_\alpha(x, \xi, D_y - \nabla_x \phi(x, \xi)) \, (D_y - \nabla_x \phi(x, \xi))^\alpha g(y)|_{y=x} \\
&= \int_{\mathbb{R}^n} \int_V e^{2\pi i (x - y) \cdot \eta} \, r_\alpha(x, \xi, \eta - \nabla_x \phi(x, \xi)) \\
&\quad \times (\eta - \nabla_x \phi(x, \xi))^\alpha \, \chi(x, y) \, g(y) \, dy \, d\eta \\
&= \int_{\mathbb{R}^n} \int_V e^{2\pi i (x - y) \cdot \theta} \, e^{2\pi i \Psi(x, y, \xi)} \, \theta^\alpha \, \chi(x, y) \, a(y, z, \xi) \, r_\alpha(x, \xi, \theta) \, dy \, d\theta,
\end{aligned}
$$

where we changed the variables by $\theta = \eta - \nabla_x \phi(x, \xi)$ and where

$$
\Psi(x, y, \xi) := \phi(y, \xi) - \phi(x, \xi) + (x - y) \cdot \nabla_x \phi(x, \xi)
$$

and

$$
r_\alpha(x, \xi, \theta) = \int_0^1 (1 - t)^{M_1} \, \partial_\xi^\alpha p(x, \nabla_x \phi(x, \xi) + t\theta) \, dt.
$$

Since the periodisation $\mathcal{P}_x$ does not change the orders in z and ξ it is enough to derive the required estimates for $S_\alpha(x, z, \xi)$.

Let $\rho \in C_0^\infty(\mathbb{R}^n)$ be such that $\rho(\theta) = 1$ for $\|\theta\| < \epsilon/2$, and $\rho(\theta) = 0$ for $\|\theta\| > \epsilon$, for some small $\epsilon > 0$ to be chosen later. We decompose

$$S_\alpha(x, z, \xi) = S_\alpha^I(x, z, \xi) + S_\alpha^{II}(x, z, \xi),$$

where

$$S_\alpha^I(x, z, \xi) = \int_{\mathbb{R}^n} \int_v e^{2\pi i (x-y)\cdot\theta} \, \rho\left(\frac{\theta}{\langle\xi\rangle}\right)$$
$$\times r_\alpha(x, \xi, \theta) \, D_y^\alpha \left[e^{2\pi i \Psi(x,y,\xi)} \, \chi(x, y) \, a(y, z, \xi) \right] \, dy \, d\theta,$$

$$S_\alpha^{II}(x, z, \xi) = \int_{\mathbb{R}^n} \int_V e^{2\pi i (x-y)\cdot\theta} \left(1 - \rho\left(\frac{\theta}{\langle\xi\rangle}\right)\right)$$
$$\times r_\alpha(x, \xi, \theta) \, D_y^\alpha \left[e^{2\pi i \Psi(x,y,\xi)} \, \chi(x, y) \, a(y, z, \xi) \right] \, dy \, d\theta.$$

3.1. **Estimate for** $\|\theta\| \le \epsilon\langle\xi\rangle$. For sufficiently small $\epsilon > 0$, for any $0 \le t \le 1$, $\langle\nabla_x\phi(x, \xi) + t\theta\rangle$ and $\langle\xi\rangle$ are equivalent. Indeed, if we use the inequalities

$$\langle z \rangle \le 1 + \|z\| \le \sqrt{2}\langle z \rangle,$$

we get

$$\langle\nabla_x\phi(x, \xi) + t\theta\rangle \le (C_2\sqrt{2} + \epsilon)\langle\xi\rangle$$
$$\sqrt{2}\langle\nabla_x\phi(x, \xi) + t\theta\rangle, \ge 1 + \|\nabla_x\phi\| - \|\theta\| \ge \langle\nabla_x\phi\rangle - \|\theta\| \ge (C_1 - \epsilon)\langle\xi\rangle,$$

so we will take $\epsilon < C_1$. This equivalence means that for $\|\theta\| \le \epsilon\langle\xi\rangle$, the function $r_\alpha(x, \xi, \theta)$ is dominated by $\langle\xi\rangle^{\ell-|\alpha|}$ since $p \in S^\ell(\mathbb{T}^n \times \mathbb{R}^n)$. We will need two auxiliary estimates. The first estimate

$$\left| \partial_\theta^\gamma \left(\rho\left(\frac{\theta}{\langle\xi\rangle}\right) r_\alpha(x, \xi, \theta) \right) \right| \le C \sum_{\delta \le \gamma} \left| \partial_\theta^\delta \rho\left(\frac{\theta}{\langle\xi\rangle}\right) \partial_\theta^{\gamma-\delta} r_\alpha(x, \xi, \theta) \right|$$
$$\le C \sum_{\delta \le \gamma} \langle\xi\rangle^{-|\delta|} \langle\xi\rangle^{\ell-|\alpha|-|\gamma-\delta|} \tag{4.55}$$
$$\le C \langle\xi\rangle^{\ell-|\alpha|-|\gamma|}$$

follows from the properties of r_α. Before we state the second estimate, let us analyse the structure of $\partial_y^\alpha e^{2\pi i \Psi(x,y,\xi)}$. It has at most $|\alpha|$ powers of terms $\nabla_y\phi(y, \xi) - \nabla_x\phi(x, \xi)$, possibly also multiplied by at most $|\alpha|$ higher-order derivatives $\partial_y^\delta\phi(y, \xi)$. The product of the terms of the form $\nabla_y\phi(y, \xi) - \nabla_x\phi(x, \xi)$ can be estimated by $C(\|y - x\|\langle\xi\rangle)^{|\alpha|}$. The terms containing no difference $\nabla_y\phi(y, \xi) - \nabla_x\phi(x, \xi)$ are the products of at most $|\alpha|/2$ terms of the type $\partial_y^\delta\phi(y, \xi)$, and the product of all such terms can be estimated by $C\langle\xi\rangle^{|\alpha|/2}$. Altogether, we obtain the estimate

$$|\partial_y^\alpha e^{2\pi i \Psi(x,y,\xi)}| \le C_\alpha (1 + \langle\xi\rangle\|y - x\|)^{|\alpha|} \langle\xi\rangle^{|\alpha|/2}.$$

The second auxiliary estimate now is

$$\left| D_y^\alpha \left[e^{2\pi i \Psi(x,y,\xi)} \, \chi(x,y) \, a(y,z,\xi) \right] \right| \le C_\alpha \, (1 + \langle\xi\rangle \|y - x\|)^{|\alpha|} \, \langle\xi\rangle^{\frac{|\alpha|}{2} + m}. \quad (4.56)$$

Now we are ready to prove a necessary estimate for $S_\alpha^I(x,z,\xi)$. Let

$$L_\theta = \frac{(1 - (4\pi^2)^{-1} \langle\xi\rangle^2 \mathcal{L}_\theta)}{1 + \langle\xi\rangle^2 \|x - y\|^2}, \quad L_\theta^N \, e^{2\pi i (x-y)\cdot\theta} = e^{2\pi i (x-y)\cdot\theta},$$

where $\mathcal{L}_\theta$ is the Laplace operator with respect to θ. Integration by parts with L_θ yields

$$S_\alpha^I(x,z,\xi)$$

$$= \int_{\mathbb{R}^n} \int_V \frac{e^{2\pi i (x-y)\cdot\theta}}{(1 + \langle\xi\rangle^2 \|x - y\|^2)^N} (1 - (4\pi^2)^{-1} \langle\xi\rangle^2 \mathcal{L}_\theta)^N$$

$$\times \left\{ \rho\left(\frac{\theta}{\langle\xi\rangle}\right) r_\alpha(x,\xi,\theta) \, D_y^\alpha \left[e^{2\pi i \Psi(x,y,\xi)} \, \chi(x,y) \, a(y,z,\xi) \right] \right\} \, dy \, d\theta$$

$$= \int_{\mathbb{R}^n} \int_V \frac{e^{2\pi i (x-y)\cdot\theta}}{(1 + \langle\xi\rangle^2 \|x - y\|^2)^N} \sum_{|\gamma| \le 2N} C_\gamma \langle\xi\rangle^{|\gamma|}$$

$$\times \left\{ D_y^\alpha \left[e^{2\pi i \Psi(x,y,\xi)} \, \chi(x,y) \, a(y,z,\xi) \right] \partial_\theta^\gamma \left(\rho\left(\frac{\theta}{\langle\xi\rangle}\right) r_\alpha(x,\xi,\theta) \right) \right\} \, dy \, d\theta.$$

Using estimates (4.55), (4.56) and the fact that the measure of the support of function $\theta \mapsto \rho(\theta/\langle\xi\rangle)$ is estimated by $(\epsilon\langle\xi\rangle)^n$, we obtain the estimate

$$|S_\alpha^I(x,z,\xi)| \le C \sum_{|\gamma| \le 2N} \langle\xi\rangle^{n + |\gamma| + \frac{|\alpha|}{2} + m} \langle\xi\rangle^{\ell - |\alpha| - |\gamma|} \int_V \frac{(1 + \langle\xi\rangle \|y - x\|)^{|\alpha|}}{(1 + \langle\xi\rangle^2 \|x - y\|^2)^N} \, dy$$

$$\le C \langle\xi\rangle^{\ell + m + n - \frac{|\alpha|}{2}},$$

if we choose N large enough, e.g., $N \ge M = |\alpha|$.

Each derivative of $S_\alpha^I(x,z,\xi)$ with respect to x or ξ gives an extra power of θ under the integral. Integrating by parts, this amounts to taking more y-derivatives, giving a higher power of $\langle\xi\rangle$. However, this is not a problem if for the estimate for a given number of derivatives of the remainder $S_\alpha^I(x,z,\xi)$, we choose $M = |\alpha|$ sufficiently large.

3.2. **Estimate for $\|\theta\| > \epsilon\langle\xi\rangle$.** Let us define

$$\omega(x,y,\xi,\theta) := (x - y) \cdot \theta + \Psi(x,y,\xi)$$

$$= (x - y) \cdot (\nabla_x \phi(x,\xi) + \theta) + \phi(y,\xi) - \phi(x,\xi).$$

From (4.51) and (4.52) we have

$$\|\nabla_y \omega\| = \| - \theta + \nabla_y \phi - \nabla_x \phi\| \le 2C_2(\|\theta\| + \langle \xi \rangle),$$
$$\|\nabla_y \omega\| \ge \|\theta\| - \|\nabla_y \phi - \nabla_x \phi\|$$
$$\ge \frac{1}{2}\|\theta\| + \left(\frac{\epsilon}{2} - C_0\|x - y\|\right) \langle \xi \rangle \tag{4.57}$$
$$\ge C(\|\theta\| + \langle \xi \rangle),$$

if we choose $\kappa < \frac{\epsilon}{2C_0}$, since $\|x - y\| < \kappa$ in the support of χ in V (recall that we were free to choose $\kappa > 0$). Let us write

$$\sigma_{\gamma_1}(x, y, \xi) := e^{-2\pi i \Psi(x,y,\xi)} \, D_y^{\gamma_1} \, e^{2\pi i \Psi(x,y,\xi)}.$$

For any ν we have an estimate

$$|\partial_y^\nu \sigma_{\gamma_1}(x, y, \xi)| \le C \langle \xi \rangle^{|\gamma_1|}, \tag{4.58}$$

because of our assumption (4.52) that $|\partial_y^\nu \phi(y, \xi)| \le C_\nu \langle \xi \rangle$. For $M = |\alpha| > \ell$ we also observe that

$$|r_\alpha(x, \xi, \theta)| \le C_\alpha, \ |\partial_y^\nu a(y, z, \xi)| \le C_\beta \langle \xi \rangle^m. \tag{4.59}$$

Let us take ${}^t L_y = i\|\nabla_y \omega\|^{-2} \sum_{j=1}^n (\partial_{y_j} \omega) \partial_{y_j}$. It can be shown by induction that the operator L_y^N has the form

$$L_y^N = \frac{1}{\|\nabla_y \omega\|^{4N}} \sum_{|\nu| \le N} P_{\nu,N} \partial_y^\nu, \quad P_{\nu,N} = \sum_{|\mu|=2N} c_{\nu\mu\delta_j} (\nabla_y \omega)^\mu \partial_y^{\delta_1} \omega \cdots \partial_y^{\delta_N} \omega,$$

where $|\mu| = 2N, |\delta_j| \ge 1, \sum_{j=1}^N |\delta_j| + |\nu| = 2N$. It follows from (4.52) and (4.57) that $|P_{\nu,N}| \le C(\|\theta\| + \langle \xi \rangle)^{3N}$, since for all δ_j we have $|\partial_y^{\delta_j} \omega| \le C(\|\theta\| + \langle \xi \rangle)$. By the Leibniz formula we have

$$S_\alpha^{II}(x, z, \xi)$$
$$= \int_{\mathbb{R}^n} \int_V e^{2\pi i(x-y)\cdot\theta} \left(1 - \rho\left(\frac{\theta}{\langle \xi \rangle}\right)\right) r_\alpha(x, \xi, \theta)$$
$$\times D_y^\alpha \left[e^{2\pi i \Psi(x,y,\xi)} \, \chi(x, y) \, a(y, z, \xi)\right] dy \, d\theta$$
$$= \int_{\mathbb{R}^n} \int_V e^{2\pi i \omega(x,y,\xi,\theta)} \left(1 - \rho\left(\frac{\theta}{\langle \xi \rangle}\right)\right) r_\alpha(x, \xi, \theta)$$
$$\times \sum_{\gamma_1+\gamma_2+\gamma_3=\alpha} \sigma_{\gamma_1}(x, y, \xi) \, D_y^{\gamma_2} \, \chi(x, y) \, D_y^{\gamma_3} a(y, z, \xi) \, dy \, d\theta$$
$$= \int_{\mathbb{R}^n} \int_V e^{2\pi i \omega(x,y,\xi,\theta)} \|\nabla_y \omega\|^{-4N} \sum_{|\nu| \le N} P_{\nu,N}(x, y, \xi, \theta) \left(1 - \rho\left(\frac{\theta}{\langle \xi \rangle}\right)\right)$$
$$\times r_\alpha(x, \xi, \theta) \sum_{\gamma_1+\gamma_2+\gamma_3=\alpha} \partial_y^\nu \left(\sigma_{\gamma_1}(x, y, \xi) \, D_y^{\gamma_2} \, \chi(x, y) \, D_y^{\gamma_3} a(y, z, \xi)\right) \, dy \, d\theta.$$

It follows now from (4.58) and (4.59) that

$$|S_\alpha^{II}(x,z,\xi)| \leq C \int_{\|\theta\|>\epsilon\langle\xi\rangle/2} (\|\theta\| + \langle\xi\rangle)^{-N} \langle\xi\rangle^{|\alpha|} \langle\xi\rangle^m \, d\theta$$
$$\leq C \langle\xi\rangle^{m+|\alpha|+n-N},$$

which yields the desired estimate if we take large enough N. For the derivatives of $S_\alpha^{II}(x,z,\xi)$, similar to Part 3.1 for S_α^I, we can get extra powers of θ, which can be taken care of by choosing large N. The proof of Theorem 4.13.11 is now complete. □

Remark 4.13.15. Note that we could also use the following asymptotic expansion for c based on the discrete Taylor expansion from Theorem 3.3.21:

$$c(x,z,\xi) \sim \sum_{\theta \in \mathbb{Z}^n} \sum_{\alpha \geq 0} \frac{1}{\alpha!} \theta^{(\alpha)} \left[\triangle_\omega^\alpha p(x,\omega)\right]_{\omega=\nabla_x\phi(x,\xi)}$$
$$\times \int_{\mathbb{T}^n} e^{2\pi i(\Psi(x,y,\xi)+(x-y)\cdot\theta)} a(y,z,\xi) \, dy$$
$$= \sum_{\alpha \geq 0} \frac{1}{\alpha!} \left[\triangle_\omega^\alpha p(x,\omega)\right]_{\omega=\nabla_x\phi(x,\xi)}$$
$$\times \sum_{\theta \in \mathbb{Z}^n} \int_{\mathbb{T}^n} \theta^{(\alpha)} e^{2\pi i(x-y)\cdot\theta} e^{2\pi i\Psi(x,y,\xi)} a(y,z,\xi) \, dy$$
$$= \sum_{\alpha \geq 0} \frac{1}{\alpha!} \left[\triangle_\omega^\alpha p(x,\omega)\right]_{\omega=\nabla_x\phi(x,\xi)} D_y^{(\alpha)} \left[e^{2\pi i\Psi(x,y,\xi)} a(y,z,\xi)\right]_{y=x}.$$

Exercise 4.13.16. Justify this expansion to obtain yet another composition formula.

4.14 Boundedness of Fourier series operators on $L^2(\mathbb{T}^n)$

In Theorem 4.8.1 we proved the boundedness of operators on $L^2(\mathbb{T}^n)$ in terms of estimates on their symbols. In particular, in applications it is important to know how many derivatives (or differences in the present toroidal approach) of the symbol must be estimated for the boundedness of the operator. In this section we present the $L^2(\mathbb{T}^n)$-boundedness theorem for Fourier series operators also paying attention to the number of required derivatives for the amplitude.

However, first we need an auxiliary result which is of great importance on its own. The following statement is a modification of the well-known Cotlar's lemma taking into account the fact that operators in our application Theorem 4.14.2, especially the Fourier transform on the torus, act on functions on different Hilbert spaces. The proof below follows [118, p. 280] but there is a difference in how we estimate operator norms because we cannot immediately replace the operator S by S^*S in the estimates since they act on functions on different spaces.

Theorem 4.14.1 (Cotlar's lemma in Hilbert spaces). *Let $\mathcal{H}, \mathcal{G}$ be Hilbert spaces. Assume that a family of bounded linear operators $\{S_j : \mathcal{H} \to \mathcal{G}\}_{j \in \mathbb{Z}^r}$ and positive constants $\{\gamma(j)\}_{j \in \mathbb{Z}^r}$ satisfy*

$$\|S_l^* S_k\|_{\mathcal{H} \to \mathcal{H}} \leq [\gamma(l-k)]^2, \qquad \|S_l S_k^*\|_{\mathcal{G} \to \mathcal{G}} \leq [\gamma(l-k)]^2,$$

and

$$A = \sum_{j \in \mathbb{Z}^r} \gamma(j) < \infty.$$

Then the operator

$$S = \sum_{j \in \mathbb{Z}^r} S_j$$

satisfies

$$\|S\|_{\mathcal{H} \to \mathcal{G}} \leq A.$$

Proof. First let us assume that there are only finitely many (say N) non-zero operators S_j. We want to establish an estimate uniformly in N and then pass to the limit. We observe that we have the estimate $\|S\| \leq \|S^* S\|$ for operator norms because we can estimate

$$\|S\|_{\mathcal{H} \to \mathcal{G}}^2 = \sup_{\|f\|_{\mathcal{H}} \leq 1} (Sf, Sf)_{\mathcal{G}} = \sup_{\|f\|_{\mathcal{H}} \leq 1} (S^* Sf, f)_{\mathcal{H}}$$
$$\leq \|S^* S\|_{\mathcal{H} \to \mathcal{H}}.$$

For any $k \in \mathbb{N}$ and $B \in \mathcal{L}(\mathcal{H})$ we have $\|B\|^{2^k} = \|(B^* B)^{2^{k-1}}\|$, which follows inductively from $\|B\|^2 = \|B^* B\|$. Thus if $m = 2^k$ and $B = S^* S$ then $\|S^* S\|^m = \|(S^* S)^m\|$, so we can conclude

$$\|S\|_{\mathcal{H} \to \mathcal{G}}^{2m} \leq \|S^* S\|_{\mathcal{H} \to \mathcal{H}}^m = \|(S^* S)^m\|_{\mathcal{H} \to \mathcal{H}}$$
$$= \left\| \sum_{i_1, \ldots, i_{2m}} S_{i_1}^* S_{i_2} \cdots S_{i_{2m-1}}^* S_{i_{2m}} \right\|_{\mathcal{H} \to \mathcal{H}}. \tag{4.60}$$

Now, we can group products in the sum in different ways. Grouping the terms in the last product as $(S_{i_1}^* S_{i_2})(S_{i_3}^* S_{i_4}) \cdots (S_{i_{2m-1}}^* S_{i_{2m}})$, we can estimate

$$\left\| S_{i_1}^* S_{i_2} \cdots S_{i_{2m-1}}^* S_{i_{2m}} \right\|_{\mathcal{H} \to \mathcal{H}} \leq \gamma(i_1 - i_2)^2 \, \gamma(i_3 - i_4)^2 \cdots \gamma(i_{2m-1} - i_{2m})^2. \tag{4.61}$$

Alternatively, grouping them as $S_{i_1}^* (S_{i_2} S_{i_3}^*) \cdots (S_{i_{2m-2}} S_{i_{2m-1}}^*) S_{i_{2m}}$, we can estimate

$$\left\| S_{i_1}^* S_{i_2} \cdots S_{i_{2m-1}}^* S_{i_{2m}} \right\|_{\mathcal{H} \to \mathcal{H}} \leq A^2 \, \gamma(i_2 - i_3)^2 \, \gamma(i_4 - i_5)^2 \cdots \gamma(i_{2m-2} - i_{2m-1})^2. \tag{4.62}$$

Taking the geometric mean of (4.61) and (4.62) and using it in (4.60), we get the estimate

$$\|S\|_{\mathcal{H}\to\mathcal{G}}^{2m} \leq \sum_{i_1,\ldots,i_{2m}} A\,\gamma(i_1-i_2)\,\gamma(i_2-i_3)\cdots\gamma(i_{2m-1}-i_{2m}).$$

Now, taking the sum first with respect to i_1 and using that $\sum_{i_1}\gamma(i_1-i_2)\leq A$, then taking the sum with respect to i_2, etc., we can estimate $\|S\|_{\mathcal{H}\to\mathcal{G}}^{2m}\leq A^{2m}\sum_{i_{2m}}1$. Now, if there are only N non-zero S_i's, we obtain the estimate

$$\|S\|_{\mathcal{H}\to\mathcal{G}} \leq A\,N^{\frac{1}{2m}}$$

which proves the statement if we let $m\to\infty$. Since this conclusion is uniform over N, the proof is complete. $\qquad\square$

We recall that in the analysis in this chapter we wrote 2π in the exponential to assure that functions $e^{2\pi i x\cdot\xi}$ are 1-periodic. In this section, the only function that occurs in the exponential is $\phi(x,k)$ and so we do not need to keep writing 2π in the exponential.

Theorem 4.14.2 (Fourier series operators on $L^2(\mathbb{T}^n)$). *Let $T:C^\infty(\mathbb{T}^n)\to\mathcal{D}'(\mathbb{T}^n)$ be defined by*

$$Tu(x) = \sum_{k\in\mathbb{Z}^n} e^{i\phi(x,k)}\,a(x,k)\,(\mathcal{F}_{\mathbb{T}^n}u)(k),$$

where $\phi:\mathbb{R}^n\times\mathbb{Z}^n\to\mathbb{R}$ and $a:\mathbb{T}^n\times\mathbb{Z}^n\to\mathbb{C}$. Assume that the function $x\mapsto e^{i\phi(x,\xi)}$ is 1-periodic for every $\xi\in\mathbb{Z}^n$, and that for all $|\alpha|\leq 2n+1$ and $|\beta|=1$ we have

$$|\partial_x^\alpha a(x,k)|\leq C \quad\text{and}\quad \left|\partial_x^\alpha\triangle_k^\beta\phi(x,k)\right|\leq C \tag{4.63}$$

for all $x\in\mathbb{T}^n$ and $k\in\mathbb{Z}^n$. Assume also that

$$|\nabla_x\phi(x,k) - \nabla_x\phi(x,l)| \geq C|k-l| \tag{4.64}$$

for all $x\in\mathbb{T}^n$ and $k,l\in\mathbb{Z}^n$. Then T extends to a bounded linear operator from $L^2(\mathbb{T}^n)$ to $L^2(\mathbb{T}^n)$.

Remark 4.14.3. Note that condition (4.64) is a discrete version of the usual local graph condition for Fourier integral operators, necessary for the local L^2-boundedness. We also note that conditions on the boundedness of the higher-order differences of phase and amplitude would follow automatically from condition (4.63). Therefore, this theorem relaxes assumptions on the behaviour with respect to the dual variable, compared, for example, with the corresponding global result for Fourier integral operators in [95] in $\mathbb{R}^n$.

Proof of Theorem 4.14.2. Since for $u : \mathbb{T}^n \to \mathbb{C}$ we have $\|u\|_{L^2(\mathbb{T}^n)} = \|\mathcal{F}_{\mathbb{T}^n} u\|_{\ell^2(\mathbb{Z}^n)}$, it is enough to prove that the operator

$$Sw(x) = \sum_{k \in \mathbb{Z}^n} e^{i\phi(x,k)} \, a(x,k) \, w(k)$$

is bounded from $\ell^2(\mathbb{Z}^n)$ to $L^2(\mathbb{T}^n)$. Let us define

$$S_l w(x) := e^{i\phi(x,l)} \, a(x,l) \, w(l),$$

so that $S = \sum_{l \in \mathbb{Z}^n} S_l$. From the identity

$$(w, S^* v)_{\ell^2(\mathbb{Z}^n)} = (Sw, v)_{L^2(\mathbb{T}^n)} = \int_{\mathbb{T}^n} \sum_{k \in \mathbb{Z}^n} e^{i\phi(x,k)} \, a(x,k) \, w(k) \, \overline{v(x)} \, dx$$

we find that the adjoint S^* to S is given by

$$(S^* v)(k) = \int_{\mathbb{T}^n} e^{-i\phi(x,k)} \, \overline{a(x,k)} \, v(x) \, dx$$

and so we also have

$$(S_l^* v)(m) = \delta_{lm} \int_{\mathbb{T}^n} e^{-i\phi(x,m)} \, \overline{a(x,m)} \, v(x) \, dx = \delta_{lm}(S^* v)(l).$$

It follows that

$$
\begin{aligned}
S_k S_l^* v(x) &= e^{i\phi(x,k)} \, a(x,k) \, (S_l^* v)(k) \\
&= \delta_{lk} \int_{\mathbb{T}^n} e^{i\phi(x,k)} \, a(x,k) \, e^{-i\phi(y,k)} \, \overline{a(y,k)} \, v(y) \, dy \\
&= \int_{\mathbb{T}^n} K_{kl}(x,y) \, v(y) \, dy,
\end{aligned}
$$

where

$$K_{kl}(x,y) = \delta_{kl} \, e^{i[\phi(x,k)-\phi(y,k)]} \, a(x,k) \, \overline{a(y,k)}.$$

From (4.63) and compactness of the torus it follows that the kernel K_{kl} is bounded and that $\|S_k S_l^* v\|_{L^2(\mathbb{T}^n)} \leq C\delta_{kl}\|v\|_{L^2(\mathbb{T}^n)}$. In particular, we can trivially conclude that for any $N \geq 0$ we have the estimate

$$\|S_k S_l^*\|_{L^2(\mathbb{T}^n) \to L^2(\mathbb{T}^n)} \leq \frac{C_N}{1 + |k - l|^N}. \tag{4.65}$$

On the other hand, we have

$$
\begin{aligned}
(S_l^* S_k w)(m) &= \delta_{lm} \int_{\mathbb{T}^n} e^{-i\phi(x,l)} \, \overline{a(x,l)} \, (S_k w)(x) \, dx \\
&= \delta_{lm} \int_{\mathbb{T}^n} e^{i[\phi(x,k)-\phi(x,l)]} \, a(x,k) \, \overline{a(x,l)} \, w(k) \, dx \\
&= \sum_{\mu \in \mathbb{Z}^n} \widetilde{K_{lk}}(m,\mu) \, w(\mu),
\end{aligned}
$$

where

$$\widetilde{K_{lk}}(m,\mu) = \delta_{lm}\delta_{k\mu} \int_{\mathbb{T}^n} \mathrm{e}^{\mathrm{i}[\phi(x,k)-\phi(x,l)]}\, a(x,k)\, \overline{a(x,l)}\, \mathrm{d}x.$$

Now, if $k \neq l$, integrating by parts $(2n+1)$-times with operator

$$\frac{1}{\mathrm{i}} \frac{\nabla_x\phi(x,k) - \nabla_x\phi(x,l)}{\|\nabla_x\phi(x,k) - \nabla_x\phi(x,l)\|^2} \cdot \nabla_x$$

and using the periodicity of a and $\nabla_x\phi$ (so there are no boundary terms), we get the estimate

$$|\widetilde{K_{lk}}(m,\mu)| \leq \frac{C\,\delta_{lm}\,\delta_{k\mu}}{1 + |k-l|^{2n+1}}, \tag{4.66}$$

where we also used that by the discrete Taylor expansion (Theorem 3.3.21) the second condition in (4.63) implies that

$$|\nabla_x\phi(x,k) - \nabla_x\phi(x,l)| \leq C|k-l| \quad \text{for all } x \in \mathbb{T}^n,\ k,l \in \mathbb{Z}^n.$$

Estimate (4.66) implies

$$\sup_m \sum_\mu |\widetilde{K_{lk}}(m,\mu)| = |\widetilde{K_{lk}}(l,k)| \leq \frac{C}{1 + |k-l|^{2n+1}},$$

and similarly for $\sup_\mu \sum_m$, so that we have

$$\|S_l^* S_k\|_{\ell^2(\mathbb{Z}^n)\to\ell^2(\mathbb{Z}^n)} \leq \frac{C}{1 + |k-l|^{2n+1}}. \tag{4.67}$$

These estimates for norms $\|S_k S_l^*\|_{L^2(\mathbb{T}^n)\to L^2(\mathbb{T}^n)}$ and $\|S_l^* S_k\|_{\ell^2(\mathbb{Z}^n)\to\ell^2(\mathbb{Z}^n)}$ in (4.65) and (4.67), respectively, imply the theorem by a modification of Cotlar's lemma given in Proposition 4.14.1, which we use with $\mathcal{H} = \ell^2(\mathbb{Z}^n)$ and $\mathcal{G} = L^2(\mathbb{T}^n)$. $\quad\square$

Using Theorems 4.13.8, 4.13.11, and 4.14.2, we obtain the result on the boundedness of Fourier series operators on Sobolev spaces:

Corollary 4.14.4 (Fourier series operators on Sobolev spaces). *Let $T : C^\infty(\mathbb{T}^n) \to \mathcal{D}'(\mathbb{T}^n)$ be defined by*

$$Tu(x) = \sum_{k\in\mathbb{Z}^n} \mathrm{e}^{\mathrm{i}\phi(x,k)}\, a(x,k)\, \widehat{u}(k),$$

where $\phi : \mathbb{T}^n \times \mathbb{Z}^n \to \mathbb{R}$ and $a : \mathbb{T}^n \times \mathbb{Z}^n \to \mathbb{C}$. Assume that for all α and $|\beta| = 1$ we have

$$|\partial_x^\alpha a(x,k)| \leq C_\alpha \langle k\rangle^m,$$

as well as

$$|\partial_x^\alpha \phi(x,k)| \leq C_\alpha \langle k\rangle \quad \text{and} \quad \left|\partial_x^\alpha \triangle_k^\beta \phi(x,k)\right| \leq C_{\alpha\beta}$$

for all $x \in \mathbb{T}^n$ and $k \in \mathbb{Z}^n$. Assume that for some $C > 0$ we have

$$C^{-1} \langle k \rangle \leq \langle \nabla_x \phi(x, k) \rangle \leq C \langle k \rangle$$

for all $x \in \mathbb{T}^n$, $k \in \mathbb{Z}^n$, and that

$$|\nabla_x \phi(x, k) - \nabla_x \phi(x, l)| \geq C|k - l|$$

for all $x \in \mathbb{T}^n$ and $k, l \in \mathbb{Z}^n$. Then T extends to a bounded linear operator from $H^s(\mathbb{T}^n)$ to $H^{s-m}(\mathbb{T}^n)$ for all $s \in \mathbb{R}$.

Exercise 4.14.5. Work out all the details of the proof.

4.15 An application to hyperbolic equations

In this section we briefly discuss how the toroidal analysis can be applied to construct global parametrices for hyperbolic equations on the torus and how to embed certain problems on $\mathbb{R}^n$ into the torus. The finite propagation speed of singularities for solutions to hyperbolic equations allows one to cut-off the equation and the Cauchy data for large x for the local analysis of singularities of solutions for bounded times. Then the problem can be embedded into $\mathbb{T}^n$, or into the inflated torus $N\mathbb{T}^n$ (Remark 4.6.9), in order to apply the periodic analysis developed here. One of the advantages of this procedure is that since phases and amplitudes now are only evaluated at $\xi \in \mathbb{Z}^n$ one can apply this also for problems with low regularity in ξ, in particular to problems for weakly hyperbolic equations or systems with variable multiplicities. For example, if the principal part has constant coefficients then the loss of regularity occurs only in ξ so techniques developed in this chapter can be applied.

Let $a(X, D)$ be a pseudo-differential operator with symbol a satisfying $a = a(x, \xi) \in S^m(\mathbb{R}^n \times \mathbb{R}^n)$ (with some properties to be specified). There is no difference in the subsequent argument if $a = a(t, x, \xi)$ also depends on t. For a function $u = u(t, x)$ of $t \in \mathbb{R}$ and $x \in \mathbb{R}^n$ we write

$$a(X, D)u(t, x) = \int_{\mathbb{R}^n} e^{2\pi i x \cdot \xi}\, a(x, \xi)\, (\mathcal{F}_{\mathbb{R}^n} u)(t, \xi) d\xi$$

$$= \int_{\mathbb{R}^n} \int_{\mathbb{R}^n} e^{2\pi i (x-y) \cdot \xi}\, a(x, \xi)\, u(t, y)\, dy\, d\xi.$$

Let $u(t, \cdot) \in L^1(\mathbb{R}^n)$ $(0 < t < t_0)$ be a solution to the hyperbolic problem

$$\begin{cases} i\frac{\partial}{\partial t} u(t, x) = a(X, D)u(t, x), \\ u(0, x) = f(x), \end{cases} \tag{4.68}$$

where $f \in L^1(\mathbb{R}^n)$ is compactly supported.

Assume now that $a(X, D) = a_1(X, D) + a_0(X, D)$ where $a_1(x, \xi)$ is 1-periodic and $a_0(x, \xi)$ is compactly supported in x (assume even that supp $a_0(\cdot, \xi) \subset [0, 1]^n$). A simple example is a constant coefficient symbol $a_1(x, \xi) = a_1(\xi)$. Let us also assume that supp $f \subset [0, 1]^n$.

We will now describe a way to periodise problem (4.68). According to Proposition 4.6.19, we can replace (4.68) by

$$\begin{cases} i\frac{\partial}{\partial t}u(t, x) = (a_1(x, D) + (\mathcal{P}a_0)(X, D))u(t, x) + Ru(t, x), \\ u(x, 0) = f(x), \end{cases}$$

where the symbol $a_1 + \mathcal{P}a_0$ is periodic and R is a smoothing operator. To study singularities of (4.68), it is sufficient to analyse the Cauchy problem

$$\begin{cases} i\frac{\partial}{\partial t}v(t, x) = (a_1(x, D) + (\mathcal{P}a_0)(X, D))v(t, x), \\ v(x, 0) = f(x) \end{cases}$$

since by Duhamel's formula we have $\mathrm{WF}(u - v) = \emptyset$. This problem can be transferred to the torus. Let $w(t, x) = \mathcal{P}v(\cdot, t)(x)$. By Theorem 4.6.12 it solves the Cauchy problem on the torus $\mathbb{T}^n$, with operator $\mathcal{P}$ from (4.24) in Theorem 4.6.3:

$$\begin{cases} i\frac{\partial}{\partial t}w(t, x) = (\widetilde{a_1}(x, D) + \widetilde{\mathcal{P}a_0}(X, D))w(t, x), \\ w(x, 0) = \mathcal{P}f(x). \end{cases}$$

Now, if $a \in S^1(\mathbb{R}^n \times \mathbb{R}^n)$ is of the first order, the calculus constructed in previous sections yields the solution in the form

$$w(t, x) \equiv T_t f(x) = \sum_{k \in \mathbb{Z}^n} e^{2\pi i \phi(t, x, k)} \, b(t, x, k) \, \mathcal{F}_{\mathbb{T}^n}(\mathcal{P}f)(k),$$

where $\phi(t, x, \xi)$ and $b(t, x, \xi)$ satisfy discrete analogues of the eikonal and transport equations. Here we note that $\mathcal{F}_{\mathbb{T}^n}(\mathcal{P}f)(k) = (\mathcal{F}_{\mathbb{R}^n}f)(k)$. We also note that the phase $\phi(t, x, k)$ is defined for discrete values of $k \in \mathbb{Z}^n$, so there is no issue of regularity, making this representation potentially applicable to low regularity problems and weakly hyperbolic equations.

Example. If the symbol $a_1(x, \xi) = a_1(\xi)$ has constant coefficients and belongs to $S^1(\mathbb{R}^n \times \mathbb{R}^n)$, and a_0 belongs to $S^0(\mathbb{R}^n \times \mathbb{R}^n)$, we can find that the phase is given by $\phi(t, x, k) = x \cdot k + t a_1(k)$. In particular, $\nabla_x \phi(x, k) = k$. Applying $a(X, D)$ to $w(t, x) = T_t f(x)$ and using the composition formula from Theorem 4.13.11 we obtain

$$a(X, D)T_t f(x) = \sum_{k \in \mathbb{Z}^n} \int_{\mathbb{R}^n} e^{2\pi i((x-z) \cdot k + t a_1(k))} \, c(t, x, k) \, f(z) \, \mathrm{d}z,$$

where

$$c(t, x, k) \sim \sum_{\alpha \geq 0} \frac{(2\pi i)^{-|\alpha|}}{\alpha!} \, \partial_\xi^\alpha a(x, \xi)\big|_{\xi=k} \, \partial_x^\alpha b(t, x, k), \tag{4.69}$$

since the function Ψ in Theorem 4.13.11 vanishes. From this we can find amplitude b from the discrete version of the transport equations, details of which we omit here. Finally, we note that we can also have an asymptotic expansion for the amplitude b in (4.69) in terms of the discrete differences $\triangle_\xi^\alpha$ and the corresponding derivatives $\partial_x^{(\alpha)}$ instead of derivatives ∂_ξ^α and ∂_x^α, respectively, if we use Remark 4.13.15 instead of Theorem 4.13.11.

Exercise 4.15.1. Work out the details for the arguments above.

Remark 4.15.2 (**Schrödinger equation**). Let $u(t, x)$, $t \in \mathbb{R}$, $x \in \mathbb{T}^n$, be the solution to a constant coefficients Schrödinger equation on the torus, i.e., u satisfies

$$i\partial_t u + \mathcal{L}u = 0, \quad u|_{t=0} = f,$$

where $\mathcal{L}$ is the Laplace operator. This equation can be solved by taking the Fourier transform, and thus the Fourier series representation of the solution is

$$u(t, x) = e^{it\mathcal{L}}f(x) = \sum_{\xi \in \mathbb{Z}^n} e^{i2\pi(x \cdot \xi - 2\pi t |\xi|^2)}\widehat{f}(\xi).$$

This representation shows, in particular, that the solution is periodic in time. In [16, 17, 18], employing this representation, Bourgain used, for example, the equality $\|u\|_{L^4(\mathbb{T}\times\mathbb{T}^n)}^4 = \|u^2\|_{L^2(\mathbb{T}\times\mathbb{T}^n)}^2$ leading to the corresponding Strichartz estimates and global well-posedness results for nonlinear equations. We can note that since the torus is compact, the usual dispersive estimates fail even locally in time. We will not pursue this topic further, and refer to the aforementioned papers for the details.

Chapter 5

Commutator Characterisation of Pseudo-differential Operators

On a smooth closed manifold the pseudo-differential operators can be characterised by taking commutators with vector fields, i.e., first-order partial derivatives. This approach is due to Beals ([12], 1977), Dunau ([32], 1977), and Coifman and Meyer ([23], 1978); perhaps the first ones to consider these kind of commutator properties were Calderón and his school [21]. For other contributions, see also [26], [133] and [80].

In this chapter we present a Sobolev space version of these characterisations. This will be one of the steps in developing global quantizations of operators on Lie groups in Part IV. Indeed, a commutator characterisation in Sobolev spaces as opposed to only L^2 will have an advantage of allowing us to control the orders of operators.

In particular, the commutators provide us a new, quite simple way of proving the equivalence of local and global definitions of pseudo-differential operators on a torus, and we derive related commutator characterisations for operators of general order on the scale of Sobolev spaces.

The structure of the treatment is the following. First, we review necessary pseudo-differential calculus on $\mathbb{R}^n$, obtaining a commutator characterisation of local pseudo-differential operators (Theorem 5.1.4). After that, the corresponding global characterisation is given on closed manifolds (Theorem 5.3.1). Lastly, we apply this to the global symbolic analysis of periodic pseudo-differential operators on $\mathbb{T}^n$ (Theorem 5.4.1). Section 5.2 is devoted to a brief introduction to the necessary concepts of pseudo-differential operators on manifolds.

5.1 Euclidean commutator characterisation

In this section we discuss the case of the Euclidean space $\mathbb{R}^n$. We will concentrate on the localisation of pseudo-differential operators which is just a local way to look at pseudo-differential operators from Chapter 2 where we dealt with global analysis on $\mathbb{R}^n$. The commutator characterisation of local pseudo-differential operators on $\mathbb{R}^n$ provided by Theorem 5.1.4 is needed in the next section for the commutator characterisation result on closed manifolds.

Definition 5.1.1 (Order of an operator on the Sobolev scale). A linear operator $A : \mathcal{S}(\mathbb{R}^n) \to \mathcal{S}(\mathbb{R}^n)$ is said to be *of order* $m \in \mathbb{R}$ *on the Sobolev scale* $(H^s(\mathbb{R}^n))_{s \in \mathbb{R}}$, if it has bounded extensions $A_{s,s-m} \in \mathcal{L}(H^s(\mathbb{R}^n), H^{s-m}(\mathbb{R}^n))$ for every $s \in \mathbb{R}$. In this case, the extension is unique in the sense that the operator A has the extension $A_{\mathcal{S}'} \in \mathcal{L}(\mathcal{S}'(\mathbb{R}^n))$ satisfying $A_{\mathcal{S}'}|_{H^s(\mathbb{R}^n)} = A_{s,s-m}$. Thereby any of the operators $A_{s,s-m}$ or $A_{\mathcal{S}'}$ is also denoted by A.

By Theorem 2.6.11 a pseudo-differential operator of order m in the class $\Psi^m(\mathbb{R}^n \times \mathbb{R}^n)$ is also of order m on the Sobolev scale.

Definition 5.1.2 (Local pseudo-differential operators). A linear operator

$$A : C_0^\infty(\mathbb{R}^n) \to \mathcal{D}'(\mathbb{R}^n)$$

is called a *local pseudo-differential operator of order* $m \in \mathbb{R}$ *on* $\mathbb{R}^n$, $A \in \mathrm{Op}\,S_{\mathrm{loc}}^m(\mathbb{R}^n)$, if $\phi A \psi \in \mathrm{Op}\,S^m(\mathbb{R}^n)$ for every $\phi, \psi \in C_0^\infty(\mathbb{R}^n)$. Naturally, here

$$((\phi A \psi)u)(x) = \phi(x) A(\psi u)(x).$$

In addition to the symbol inequalities (2.3) in Definition 2.1.1, there is another appealing way of characterising pseudo-differential operators, namely via commutators. This characterisation dates back to [12] by Beals, to [32] by Dunau, and to [23] by Coifman and Meyer. We present a related result, Theorem 5.1.4, about local pseudo-differential operators. First we introduce the following notation:

Definition 5.1.3 (Notation). Let us define the commutators

$$L_j(A) := [\partial_{x_j}, A] \quad \text{and} \quad R_k(A) := [A, M_{x_k}],$$

where M_{x_k} is the multiplication operator $(M_{x_k} f)(x) = x_k f(x)$.

Set $R^\alpha = R_1^{\alpha_1} \cdots R_n^{\alpha_n}$ and accordingly $L^\beta = L_1^{\beta_1} \cdots L_n^{\beta_n}$ for multi-indices α, β, with convention $L_j^0 = I = R_k^0$. Finally, for a *partial differential operator* C on $\mathbb{R}^n$, let $\deg(C)$ denote its order. By Theorem 2.6.11, $\deg(C)$ is also the order of C on the Sobolev scale.

The following theorem characterises local pseudo-differential operators on $\mathbb{R}^n$ in terms of the orders of their commutators on the Sobolev scale:

Theorem 5.1.4 (Commutator characterisation on $\mathbb{R}^n$). *Let $m \in \mathbb{R}$ and let A be a linear operator defined on $C_0^\infty(\mathbb{R}^n)$. Then the following conditions are equivalent:*

(i) $A \in \mathrm{Op}\,S_{\mathrm{loc}}^m(\mathbb{R}^n)$.

(ii) *For any $\phi, \psi \in C_0^\infty(\mathbb{R}^n)$, for any $s \in \mathbb{R}$ and for any sequence $\mathcal{C} = (C_j)_{j=0}^\infty \subset$
   $\mathrm{Op}\, S^1_{\mathrm{loc}}(\mathbb{R}^n)$ of partial differential operators of first order, it holds that*

$$\begin{cases} B_0 = \phi A \psi \in \mathcal{L}(H^s(\mathbb{R}^n), H^{s-m}(\mathbb{R}^n)), \\ B_{k+1} = [B_k, C_k] \in \mathcal{L}(H^s(\mathbb{R}^n), H^{s-m+d_{\mathcal{C},k}}(\mathbb{R}^n)), \end{cases}$$

where $d_{\mathcal{C},k} = \sum_{j=0}^k (1 - \deg(C_j))$.

(iii) *For any $\phi, \psi \in C_0^\infty(\mathbb{R}^n)$, for any $s \in \mathbb{R}$ and for every $\alpha, \beta \in \mathbb{N}_0$, it holds that*

$$R^\alpha L^\beta(\phi A \psi) \in \mathcal{L}(H^s(\mathbb{R}^n), H^{s-(m-|\alpha|)}(\mathbb{R}^n)).$$

Remark 5.1.5. At first sight, condition (ii) in Theorem 5.1.4 may seem awkward, at least when compared to condition (iii). However, this result will be needed in the pseudo-differential analysis on manifolds, and it is crucial in the proof of Theorem 5.3.1. Also notice the similarities in the formulations of Theorems 5.1.4 and 5.3.1, and in the proofs of Theorems 5.1.4 and 5.4.1.

Proof of Theorem 5.1.4. First, let $A \in \mathrm{Op}\, S^m_{\mathrm{loc}}(\mathbb{R}^n)$, and fix $\phi, \psi \in C_0^\infty(\mathbb{R}^n)$. Then $B_0 = \phi A \psi \in \mathrm{Op}\, S^m(\mathbb{R}^n)$. Let $\chi \in C_0^\infty(\mathbb{R}^n)$ be such that $\chi(x) = 1$ in a neighbourhood of the compact set $\mathrm{supp}(\phi) \cup \mathrm{supp}(\psi) \subset \mathbb{R}^n$, so that $B_{k+1} = [B_k, C_k] = [B_k, \chi C_k]$. Notice that $\chi C_k \in \mathrm{Op}\, S^{\deg(C_k)}(\mathbb{R}^n)$. Hence by induction and by the composition Theorem 2.5.1 it follows that $B_{k+1} \in \mathrm{Op}\, S^{m-d_{\mathcal{C},k}}(\mathbb{R}^n)$. This proves the implication $(i) \Rightarrow (ii)$ by Theorem 2.6.11 with $p = 2$.

It is really trivial that (ii) implies (iii).

Finally, let us show that (iii) implies (i). Assume (iii), and fix $\phi, \psi \in C_0^\infty(\mathbb{R}^n)$; we have to prove that $\phi A \psi \in \mathrm{Op}\, S^m(\mathbb{R}^n)$. Let $\chi \in C_0^\infty(\mathbb{R}^n)$ be such that $\chi(x) = 1$ in a neighbourhood of the compact set $\mathrm{supp}(\phi) \cup \mathrm{supp}(\psi) \subset \mathbb{R}^n$. We denote $e_\xi(x) = e^{2\pi i x \cdot \xi}$. Evidently, $\phi A \psi$ is of order m, and

$$\begin{aligned} \partial_\xi^\alpha \partial_x^\beta \sigma_{\phi A \psi}(x, \xi) &= \sigma_{R^\alpha L^\beta(\phi A \psi)}(x, \xi) \\ &= e^{-2\pi i x \cdot \xi}(R^\alpha L^\beta(\phi A \psi)e_\xi)(x) \\ &= e^{-2\pi i x \cdot \xi}(R^\alpha L^\beta(\phi A \psi)(\chi e_\xi))(x). \end{aligned}$$

If $2s > n = \dim(\mathbb{R}^n)$, $s \in \mathbb{N}$, then by the Cauchy-Schwartz inequality for $u \in H^s(\mathbb{R}^n)$ we have:

$$\begin{aligned} |u(x)| &\leq \int_{\mathbb{R}^n} |\hat{u}(\xi)|\, d\xi \\ &\leq \left[\int_{\mathbb{R}^n} (1 + |\xi|)^{-2s}\, d\xi \right]^{1/2} \left[\int_{\mathbb{R}^n} (1 + |\xi|)^{2s} |\hat{u}(\xi)|^2\, d\xi \right]^{1/2} \\ &= C_s \|u\|_{H^s(\mathbb{R}^n)} \leq C_s \left(\sum_{|\gamma| \leq s} \|\partial_x^\gamma u\|^2_{H^0(\mathbb{R}^n)} \right)^{1/2}. \end{aligned}$$

Applied to the symbol $\partial_\xi^\alpha \partial_x^\beta \sigma_{\phi A \psi}$ this implies

$$|\partial_\xi^\alpha \partial_x^\beta \sigma_{\phi A \psi}(x,\xi)| \leq C \left(\sum_{|\gamma| \leq s} \|\partial_\xi^\alpha \partial_x^{\beta+\gamma} \sigma_{\phi A \psi}(\cdot,\xi)\|_{H^0(\mathbb{R}^n)}^2 \right)^{1/2}$$

$$= C \left(\sum_{|\gamma| \leq s} \|e_{-\xi} \left(R^\alpha L^{\beta+\gamma}(\phi A \psi) \right)(\chi e_\xi)\|_{H^0(\mathbb{R}^n)}^2 \right)^{1/2}$$

$$\leq C \left(\sum_{|\gamma| \leq s} \|e_{-\xi}\|_{\mathcal{L}(H^0)}^2 \|R^\alpha L^{\beta+\gamma}(\phi A \psi)\|_{\mathcal{L}(H^{m-|\alpha|}, H^0)}^2 \|\chi e_\xi\|_{H^{m-|\alpha|}}^2 \right)^{1/2}.$$

By a version of Peetre's inequality in (3.25) we have

$$\forall s \in \mathbb{R} \; \forall \eta, \xi \in \mathbb{R}^n : (1+|\eta+\xi|)^s \leq 2^{|s|}(1+|\eta|)^{|s|}(1+|\xi|)^s$$

(where $|\xi| = \|\xi\|$ in the notation of the torus chapters is just the Euclidean norm of the vector ξ), so that we obtain

$$\|\chi e_\xi\|_{H^{m-|\alpha|}(\mathbb{R}^n)} = \left(\int_{\mathbb{R}^n} (1+|\eta|)^{2(m-|\alpha|)} |\widehat{\chi e_\xi}(\eta)|^2 \, d\eta \right)^{1/2}$$

$$= \left(\int_{\mathbb{R}^n} (1+|\eta+\xi|)^{2(m-|\alpha|)} |\widehat{\chi}(\eta)|^2 \, d\eta \right)^{1/2}$$

$$\leq 2^{|m-|\alpha||} \|\chi\|_{H^{|m-|\alpha||}(\mathbb{R}^n)} (1+|\xi|)^{m-|\alpha|}.$$

Hence

$$|\partial_\xi^\alpha \partial_x^\beta \sigma_{\phi A \psi}(x,\xi)| \leq C_{\alpha\beta,\phi,\psi}(1+|\xi|)^{m-|\alpha|},$$

and consequently $A \in \mathrm{Op}\, S_{\mathrm{loc}}^m(\mathbb{R}^n)$. Thus (i) is obtained from (iii). $\qquad \square$

5.2   Pseudo-differential operators on manifolds

Here we briefly provide a background on pseudo-differential operators on manifolds. The differential geometry needed in the study is quite simple, sufficient general reference being any text book in the field, e.g., [54].

Definition 5.2.1 (Atlases on topological spaces). Let X be a topological space. An *atlas on* X is a collection of pairs $\{(U_\alpha, \kappa_\alpha)\}_\alpha$, where all sets $U_\alpha \subset X$ are open in X, $\bigcup_\alpha U_\alpha = X$, and for every α the mapping $\kappa_\alpha : U_\alpha \to \mathbb{R}^n$ is a homeomorphism of U_α onto an open subset of $\mathbb{R}^n$; such n is called the *dimension* of the chart $(U_\alpha, \kappa_\alpha)$, and pairs $(U_\alpha, \kappa_\alpha)$ are called *charts* of the atlas. For every two charts $(U_\alpha, \kappa_\alpha)$ and (U_β, κ_β) with $U_\alpha \cap U_\beta \neq \emptyset$, the functions

$$\kappa_{\alpha\beta} := \kappa_\alpha \circ \kappa_\beta^{-1} : \kappa_\beta(U_\alpha \cap U_\beta) \to \kappa_\alpha(U_\alpha \cap U_\beta)$$

are called *transition maps* of the atlas. We note that each transition map $\kappa_{\alpha\beta}$ is a homeomorphism between open subsets of Euclidean spaces, so that the dimension n is the same for such charts. We will say that a point $x \in X$ belongs to a chart (U, κ) if $x \in U$.

Definition 5.2.2 (Manifolds). Let X be a Hausdorff topological space such that its topology has a countable base[1]. Then X equipped with an atlas $\mathcal{A} = \{(U_\alpha, \kappa_\alpha)\}_\alpha$ of charts of the same dimension n is called a *locally Euclidean* topological space. Since n is the same for all charts, we can set $\dim X := n$ to be the *dimension of X*. A locally Euclidean topological space with atlas $\mathcal{A}$ is called a (*smooth*) *manifold*, or a C^∞ *manifold*, if all the transition maps of the atlas $\mathcal{A}$ are smooth. A manifold M is called *compact* if X is compact.

Example. Simple examples of n-dimensional manifolds include Euclidean spaces $\mathbb{R}^n$, spheres $\mathbb{S}^n$, tori $\mathbb{T}^n$.

Remark 5.2.3. We assume that X has a countable topological base and that it is Hausdorff to ensure that there are not too many open sets and that the topology of compact manifolds is especially nice, respectively. We also note that given two atlases we can look at transition maps in the atlas which is then union. Thus, if the union of two atlases is again an atlas we will call these atlases equivalent. This leads to a notion of *equivalent atlases* and thus a manifold is rather an equivalence class $M = (X, [\mathcal{A}])$, if we do not want to worry about which atlas to fix. However, we will avoid such technicalities because of the limited differential geometry required for our purposes. In the sequel we will often omit writing the atlas at all because on the manifolds that we are dealing with the choice of an atlas will be more or less canonical. However, an important property for us is that if (U, κ) is a chart and $V \subset U$ is open, then (V, κ) is also a chart (in an equivalent atlas, hence a chart in M). We also note that Hausdorff follows from the existence of an atlas, which also implies the existence of a locally countable topological base. Instead of the first countability one may directly assume the existence of a countable atlas.

Definition 5.2.4 (Smooth mappings). Let $f : M \to N$ be a mapping between manifolds $M = (X, \mathcal{A})$ and $N = (Y, \mathcal{B})$. Let $x \in X$, let $(U, \kappa) \in \mathcal{A}$ be a chart in M containing x, and let $(V, \psi) \in \mathcal{B}$ be a chart in N containing $f(x)$. By shrinking the set U if necessary we may assume that $f(U) \subset V$. We will say that f is *smooth at* $x \in X$ if the mapping

$$\psi \circ f \circ \kappa^{-1} : \kappa(U) \to \psi(V) \tag{5.1}$$

is smooth. As usual, f is *smooth* if it is smooth at all points. The space $C^\infty(M)$ is the set of smooth complex-valued functions on M, and $C_0^\infty(U)$ is the set of smooth functions with compact supports in an open set $U \subset M$.

If $k \in \mathbb{N}$ and if all the mappings (5.1) are in $C^k(\kappa(U))$ for all charts, then we will say that $f \in C^k(M)$.

---

[1] For a topological base see Definition A.8.16

Exercise 5.2.5. Check that the definition of "f is smooth at x" does not depend on a particular choice of charts (U, κ) and (V, ψ).

Remark 5.2.6 (**Whitney's embedding theorem**). We will deal only with smooth manifolds. It is a fundamental fact that every compact manifold admits a smooth embedding as a submanifold of $\mathbb{R}^N$ for sufficiently large N. An interesting question is how small can N be. In 1936, in [150], for general (also non-compact) manifolds Whitney showed that one can take $N = 2n + 1$ for this to be true, later also improving it to $N = 2n$. We will not pursue this topic here and can refer to [54] for further details, but we will revisit it in a simpler context of Lie groups in Corollary 8.0.4 as well as use it in Section 10.6.

Remark 5.2.7 (**Orientable manifolds**). The *natural n-form* on $\mathbb{R}^n$ is given by the volume element $\Omega = \mathrm{d}x_1 \wedge \cdots \wedge \mathrm{d}x_n$ which is non-degenerate. For every open $U \subset \mathbb{R}^n$ the restriction $\Omega_U := \Omega|_U$ defines a volume element on U. A diffeomorphism $F : U \to V \subset \mathbb{R}^n$ is called *orientation preserving* if $F^*\Omega_V = f\Omega_U$ for some $f \in C^\infty(U)$ such that $f > 0$ everywhere. A manifold M is called *orientable* if it has an atlas such that all the transition maps are orientation preserving. One can show that orientable manifolds have a non-degenerate volume element, i.e., it is possible to define a smooth n-form on M which is not zero at any point.

Definition 5.2.8 (Localisation of operators). If $A : C^\infty(M) \to C^\infty(M)$ and $\phi, \psi \in C^\infty(M)$, we define the operator $\phi A \psi : C^\infty(M) \to C^\infty(M)$ by $((\phi A \psi)u)(x) = \phi(x) \cdot A(\psi \cdot u)(x)$.

Definition 5.2.9 (κ-transfers). If (U, κ) is a chart on M, the κ-transfer $A_\kappa : C^\infty(\kappa(U)) \to C^\infty(\kappa(U))$ of an operator $A : C^\infty(U) \to C^\infty(U)$ is defined by

$$A_\kappa u := A(u \circ \kappa) \circ \kappa^{-1}.$$

Similarly, the κ-transfer of a function ϕ is $\phi_\kappa = \phi \circ \kappa^{-1}$.

Exercise 5.2.10. Prove that the transfer of a commutator is the commutator of transfers:

$$[A, B]_\kappa = [A_\kappa, B_\kappa]. \tag{5.2}$$

Pseudo-differential operators on the manifold M in the Hörmander sense are defined as follows:

Definition 5.2.11 (Pseudo-differential operators on manifolds). A linear operator $A : C^\infty(M) \to C^\infty(M)$ is a *pseudo-differential operator of order* $m \in \mathbb{R}$ on M, if for every chart (U, κ) and for any $\phi, \psi \in C_0^\infty(U)$, the operator $(\phi A \psi)_\kappa$ is a pseudo-differential operator of order m on $\mathbb{R}^n$. Since the class of pseudo-differential operators of order m on $\mathbb{R}^n$ is diffeomorphism invariant, it follows that the corresponding class on M is well defined. We denote the set of pseudo-differential operators of order m on M by $\Psi^m(M)$.

Exercise 5.2.12. Check that the class of pseudo-differential operators of order m on $\mathbb{R}^n$ is diffeomorphism invariant (and see Section 2.5.2).

Definition 5.2.13 (Diff(M)). Let Diff(M) be the $*$-algebra

$$\text{Diff}(M) = \bigcup_{k=0}^{\infty} \text{Diff}^k(M),$$

where Diff$^k(M)$ is the set of at most k^{th} order partial differential operators on M with smooth coefficients. Here, Diff$^0(M) \cong C^{\infty}(M)$, and Diff$^1(M) \setminus$ Diff$^0(M)$ corresponds to the non-trivial smooth vector fields on M, i.e., the non-trivial smooth sections of the tangent bundle TM.

Definition 5.2.14 (Closed manifolds). A compact manifold without boundary is called *closed*.

Throughout this section and further in this chapter, M will be a closed smooth orientable manifold. Then we can equip it with the volume element from Remark 5.2.7. One can think of it as a suitable pullback of the Euclidean volume n-form (the Lebesgue measure) in local charts.

Remark 5.2.15 **(Spaces $\mathcal{D}(M)$ and $\mathcal{D}'(M)$).** A differential operator $D \in$ Diff(M) defines a seminorm p_D on $C^{\infty}(M)$ by $p_D(u) = \sup_{x \in M} |(Du)(x)|$. The seminorm family $\{p_D : C^{\infty}(M) \to \mathbb{R} \mid D \in \text{Diff}(M)\}$ induces a Fréchet space structure on $C^{\infty}(M)$. This test function space is denoted by $\mathcal{D}(M)$, and the distributions by $\mathcal{D}'(M) = \mathcal{L}(\mathcal{D}(M), \mathbb{C})$. In particular, similar to Remark 1.3.7 in $\mathbb{R}^n$, for $u \in L^p(M)$ and $\varphi \in C^{\infty}(M)$, the duality

$$\langle u, \varphi \rangle := \int_M u(x)\, \varphi(x)\, \mathrm{d}x$$

gives a canonical way to identify $u \in L^p(M)$ with a distribution in $\mathcal{D}'(M)$. Here $\mathrm{d}x$ stands for a volume element on M.

Definition 5.2.16 (Sobolev space $H^s(M)$). The Sobolev space $H^s(M)$ ($s \in \mathbb{R}$) is the set of those distributions $u \in \mathcal{D}'(M)$ such that $(\phi u)_\kappa \in H^s(\mathbb{R}^n)$ for every chart (U, κ) on M and for every $\phi \in C_0^{\infty}(U)$. Let $\mathcal{U} = \{(U_j, \kappa_j)\}$ be a cover of M with charts. Due to the compactness of M, we can require the cover to be finite. Fix a smooth partition of unity $\{(U_j, \phi_j)\}$ with respect to the cover $\mathcal{U}$. We equip the Sobolev space $H^s(M)$ with the norm

$$\|u\|_{H^s(M),\{(U_j,\kappa_j,\phi_j)\}} := \left(\sum_j \|(\phi_j u)_{\kappa_j}\|^2_{H^s(\mathbb{R}^n)} \right)^{1/2}.$$

Exercise 5.2.17. Show that any other choice of U_j, κ_j, ϕ_j would have resulted in an equivalent norm. Prove that $H^s(M)$ is a Hilbert space.

As a consequence of Corollary 1.5.15, as well as Propositions 1.5.18 and 1.5.19, we get:

Corollary 5.2.18 (Density). *Let M be a closed manifold. The space $C^\infty(M)$ is sequentially dense in $L^p(M)$ for all $1 \leq p < \infty$. Also, $C^\infty(M)$ is sequentially dense in $H^s(M)$ for every $s \in \mathbb{R}$.*

Remark 5.2.19. The last statement is true for any $s \in \mathbb{R}$ but requires more of the manifold theory than developed here. Such statements can be easily found in the literature, and in the case of $\mathbb{R}^n$ see an even more general statement in Theorem 1.3.31. However, the reader is encouraged to provide the details for the proof of the density for all $s \in \mathbb{R}$.

Definition 5.2.20 (Order of an operator on the Sobolev scale). A linear operator A on $C^\infty(M)$ is said to be of *order* $m \in \mathbb{R}$ on M, if it extends boundedly between $H^s(M)$ and $H^{s-m}(M)$ for every $s \in \mathbb{R}$. Thereby the operator A has also the continuous extension $A_{\mathcal{D}'} : \mathcal{D}'(M) \to \mathcal{D}'(M)$. As is in the case of $\mathbb{R}^n$ in Definition 5.1.1, any of these extensions coincide in their mutual domains, so that it is meaningful to denote any one of them by A.

Exercise 5.2.21. Prove that

$$C^\infty(M) = \bigcap_{s \in \mathbb{R}} H^s(M) \text{ and } \mathcal{D}'(M) = \bigcup_{s \in \mathbb{R}} H^s(M).$$

Remark 5.2.22 **(All operators are properly supported).** We recall the notion of properly supported operators from Definition 2.5.20. Since the support of the integral kernel is closed, we immediately see that all pseudo-differential operators on a closed manifold are properly supported.

We briefly address the L^p issue on compact manifolds, we formulate

Theorem 5.2.23 (Boundedness on $L^p(M)$). *Let M be a compact manifold and let $A \in \Psi^0(M)$. Then A is bounded from $L^p(M)$ to $L^p(M)$ for any $1 < p < \infty$, and its operator norm is bounded by*

$$\|A\|_{\mathcal{L}(L^p(M))} \leq C \max_{\substack{|\beta| \leq n+1 \\ |\alpha| \leq [n/2]+1}} \left|\partial_x^\beta \partial_\xi^\alpha a(x,\xi)\right| \|\xi\|^{|\alpha|},$$

where $\partial_x^\beta \partial_\xi^\alpha a(x,\xi)$ is defined on one of some finite number of selected coordinate systems covering M.

The proof of this theorem can be carried out by reducing the problem to the corresponding L^p-boundedness statement of pseudo-differential operators in $\Psi^0(\mathbb{R}^n \times \mathbb{R}^n)$ with compactly supported amplitudes which would follow from Theorem 2.6.22. However, the advantage of this theorem is that one also obtains a bound on the number of necessary derivatives (as well as a corresponding result for Theorem 2.6.22) if one reduces the problem to the L^p-multipliers problem. We refer to [130, p. 267] for further details.

We refer to Section 13.1 for a further discussion of these concepts.

5.3   Commutator characterisation on closed manifolds

The main result of this section is Theorem 5.3.1 about the commutator character-isation (cf. Theorems 5.1.4 and 5.4.1), which was stated by Coifman and Meyer [23] in the case of 0-order operators on $L^2(M)$ (see also [32] for a kindred trea-tise). This will be applied in the final part of this chapter concerning periodic pseudo-differential operators (Theorem 5.4.1) and in Part IV (Theorem 10.7.7). Theorem 5.3.1 shows that pseudo-differential operators on closed manifolds can be characterised by the orders of their commutators on the Sobolev scale.

Let M be a closed manifold. Naturally, an operator $D \in \mathrm{Diff}^k(M)$ from Definition 5.2.13 is of order $\deg(D) = k$. Observe that the algebra $\mathrm{Diff}(M)$ has the "almost-commuting property":

$$[\mathrm{Diff}^j(M), \mathrm{Diff}^k(M)] \subset \mathrm{Diff}^{j+k-1}(M),$$

which follows by the Leibniz formula. Actually, more general pseudo-differential operators are also characterised by the "almost-commuting" with differential op-erators:

Theorem 5.3.1 (Commutator characterisation on closed manifolds). *Let* $m \in \mathbb{R}$ *and let* $A : C^\infty(M) \to C^\infty(M)$ *be a linear operator. Then the following conditions are equivalent:*

(i) $A \in \Psi^m(M)$.

(ii) *For any* $s \in \mathbb{R}$ *and for any sequence* $\mathcal{D} = (D_j)_{j=0}^\infty \subset \mathrm{Diff}^1(M)$, *it holds that*

$$\begin{cases} A_0 = A \in \mathcal{L}(H^s(M), H^{s-m}(M)), \\ A_{k+1} = [A_k, D_k] \in \mathcal{L}(H^s(M), H^{s-m+d_{\mathcal{D},k}}(M)), \end{cases}$$

where $d_{\mathcal{D},k} = \sum_{j=0}^k (1 - \deg(D_j))$.

We need the following auxiliary result:

Lemma 5.3.2. *Let* M *be a closed smooth manifold. Then there exists a smooth partition of unity with respect to a cover* $\mathcal{U}$ *on* M *such that* $U \cup V$ *is a chart neighbourhood whenever* $U, V \in \mathcal{U}$.

Proof. Let $\mathcal{V}$ be a cover of M with chart neighbourhoods. Since M is a compact metrisable space by the Whitney embedding theorem (Remark 5.2.6), the cover $\mathcal{V}$ has the Lebesgue number $\lambda > 0$ – i.e., if $S \subset M$ has a small diameter, $\mathrm{diam}(S) < \lambda$, then there exists $V \in \mathcal{V}$ such that $S \subset V$, see Lemma A.13.12. Let $\mathcal{W}$ be a cover of M with chart neighbourhoods of diameter less than $\lambda/2$, and choose a finite subcover $\mathcal{U} \subset \mathcal{W}$. Now there exists a smooth partition of unity on M with respect to $\mathcal{U}$, and if $U, V \in \mathcal{U}$ intersect, then $\mathrm{diam}(U \cup V) < \lambda$. On the other hand, if $U \cap V = \emptyset$, then $U \cup V$ is clearly a chart neighbourhood. $\square$

Proof of Theorem 5.3.1. ((i)$\Rightarrow$(ii)) Assume that $A \in \Psi^m(M)$. Lemma 5.3.2 provides a smooth partition of unity $\{(U_j, \phi_j)\}_{j=1}^N$ such that $U_i \cup U_j$ is always a chart neighbourhood, so that the study can be localised:

$$A = \sum_{i=1}^N \sum_{j=1}^N \phi_i A \phi_j.$$

Let $(U_i \cup U_j, \kappa_{ij})$ be a chart. Now $\phi_i, \phi_j \in C_0^\infty(U_i \cup U_j)$, so that the κ_{ij}-transfer $(\phi_i A \phi_j)_{\kappa_{ij}}$ is a pseudo-differential operator of order m on $\mathbb{R}^n$, hence belonging to $\mathcal{L}(H^s(\mathbb{R}^n), H^{s-m}(\mathbb{R}^n))$ by Theorem 2.6.11 . Thereby $\phi_i A \phi_j = ((\phi_i A \phi_j)_{\kappa_{ij}})_{\kappa_{ij}^{-1}}$ belongs to $\mathcal{L}(H^s(M), H^{s-m}(M))$, and consequently $A \in \mathcal{L}(H^s(M), H^{s-m}(M))$. Thus we have the result $\Psi^m(M) \subset \mathcal{L}(H^s(M), H^{s-m}(M))$.

In order to get (ii), also inclusions

$$[\Psi^m(M), \mathrm{Diff}^1(M)] \subset \Psi^m(M) \quad \text{and} \quad [\Psi^m(M), \mathrm{Diff}^0(M)] \subset \Psi^{m-1}(M)$$

must be proven. Let $A \in \Psi^m(M)$ and $D \in \mathrm{Diff}^1(M)$, and fix an arbitrary chart (U, κ) and arbitrary functions $\phi, \psi \in C_0^\infty(U)$. By a direct calculation,

$$\phi[A, D]\psi = [\phi A\psi, D] - \phi A[\psi, D] - [\phi, D]A\psi,$$

so that

$$(\phi[A, D]\psi)_\kappa = [(\phi A\psi)_\kappa, D_\kappa] - (\phi A[\psi, D])_\kappa - ([\phi, D]A\psi)_\kappa$$

by (5.2). Because $A \in \Psi^m(M)$, Theorem 5.1.4 implies that the operators on the right-hand side of the previous equality are pseudo-differential operators of order $m - (1 - \deg(D))$ on $\mathbb{R}^n$. Therefore $[A, D] \in \Psi^{m-(1-\deg(D))}(M)$, proving implication (i)$\Rightarrow$(ii).

((ii)$\Rightarrow$(i)) Let $A : C^\infty(M) \to C^\infty(M)$ satisfy condition (ii), and fix a chart (U, κ) on M and $\phi, \psi \in C_0^\infty(U)$. To get (i), we have to prove that $(\phi A\psi)_\kappa \in \mathrm{Op}\, S^m(\mathbb{R}^n)$, which by Theorem 5.1.4 follows, if we can prove the following variant of condition (ii):

(ii)$'$ *For any $s \in \mathbb{R}$ and for any sequence $\mathcal{C} = (C_j)_{j=0}^\infty \subset \mathrm{Op}\, S_{\mathrm{loc}}^1(\mathbb{R}^n)$ of partial differential operators, it holds that*

$$\begin{cases} B_0 = (\phi A\psi)_\kappa \in \mathcal{L}(H^s(\mathbb{R}^n), H^{s-m}(\mathbb{R}^n)), \\ B_{k+1} = [B_k, C_k] \in \mathcal{L}(H^s(\mathbb{R}^n), H^{s-m+d_{\mathcal{C},k}}(\mathbb{R}^n)), \end{cases}$$

where $d_{\mathcal{C},k} = \sum_{j=0}^k (1 - \deg(C_j))$.

Indeed, $B_0 = (\phi A\psi)_\kappa \in \mathcal{L}(H^s(\mathbb{R}^n), H^{s-m}(\mathbb{R}^n))$ by (ii). Let $\chi \in C_0^\infty(\kappa(U))$ such that $\chi(x) = 1$ in a neighbourhood of the compact set $\mathrm{supp}(\phi_\kappa) \cup \mathrm{supp}(\psi_\kappa) \subset \mathbb{R}^n$. Then define $\mathcal{D} = (D_j)_{j=0}^\infty \subset \mathrm{Diff}^1(M)$ so that $D_j|_{M \setminus U} = 0$, and $D_j|_U = (\chi C_j)_{\kappa^{-1}}$.

Then $d_{\mathcal{D},k} \geq d_{\mathcal{C},k}$, and due to condition (ii), we get

$$
\begin{aligned}
B_{k+1} &= [B_k, C_k] = [B_k, \chi C_k] = [(B_k)_{\kappa-1}, D_k]_\kappa \\
&\in \mathcal{L}(H^s(\mathbb{R}^n), H^{s-m+d_{\mathcal{D},k}}(\mathbb{R}^n)) \\
&\subset \mathcal{L}(H^s(\mathbb{R}^n), H^{s-m+d_{\mathcal{C},k}}(\mathbb{R}^n)),
\end{aligned}
$$

verifying (ii)'. Hence $A \in \Psi^m(M)$. □

Remark 5.3.3 ($\Psi(M)$ **is a** $*$**-algebra**). The pseudo-differential operators on M form a $*$-algebra

$$
\Psi(M) = \bigcup_{m \in \mathbb{R}} \Psi^m(M),
$$

where $\Psi^m(M) \subset \mathcal{L}(H^s(M), H^{s-m}(M))$. It is true that $\mathrm{Diff}^k(M) \subset \Psi^k(M)$, and $\Psi(M)$ has properties analogous to those of the algebra $\mathrm{Diff}(M)$. Especially, $[\Psi^{m_1}(M), \Psi^{m_2}(M)] \subset \Psi^{m_1+m_2-1}(M)$.

Exercise 5.3.4 (Paracompact manifolds). Generalise the result in Lemma 5.3.2 to smooth paracompact manifolds. Recall that a Hausdorff topological space is called *paracompact* if every open cover admits an open locally finite subcover.

5.4 Toroidal commutator characterisation

On the torus $\mathbb{T}^n = \mathbb{R}^n/\mathbb{Z}^n$ one has a well-defined global symbol analysis of periodic operators from the class $\Psi(\mathbb{T}^n \times \mathbb{Z}^n)$, as developed in Chapter 4. In this section, as one application of the commutator characterisation Theorem 5.3.1, we provide a proof of the equality of operator classes $\Psi(\mathbb{T}^n \times \mathbb{Z}^n) = \Psi(\mathbb{T}^n)$. For the equality of operator classes $\Psi(\mathbb{T}^n \times \mathbb{Z}^n) = \Psi(\mathbb{T}^n \times \mathbb{R}^n)$ see Corollary 4.6.13 that was obtained using the extension and periodisation techniques. However, a similar application of Theorem 5.3.1 will be important on Lie groups (Theorem 10.7.7) where these other techniques are not readily available.

For $1 \leq j, k \leq n$, let us define the operators L_j and R_k acting on periodic pseudo-differential operators by

$$
L_j(A) := [D_{x_j}, A] \quad \text{and} \quad R_k(A) := [A, e^{i2\pi x_k} I].
$$

Moreover, for $\alpha, \beta \in \mathbb{N}_0^n$, let

$$
\begin{aligned}
L^\beta(A) &= L_1^{\beta_1} \cdots L_n^{\beta_n}(A), \\
R^\alpha(A) &= R_1^{\alpha_1} \cdots R_n^{\alpha_n}(A)
\end{aligned}
$$

(here the letters L and R refer to "left" and "right"). By the composition Theorem 4.7.10, if $A \in \mathrm{Op}(S^m(\mathbb{T}^n \times \mathbb{Z}^n))$ then $L_j(A) \in \mathrm{Op}(S^m(\mathbb{T}^n \times \mathbb{Z}^n))$ and $R_k(A) :=$

$\mathrm{Op}(S^{m-1}(\mathbb{T}^n \times \mathbb{Z}^n))$. Let us explain how these commutators arise, and why they are so interesting. First,

$$\begin{aligned}
D_{x_j}\sigma_A(x,\xi) &= D_{x_j}\left(e^{-i2\pi x \cdot \xi}Ae_\xi(x)\right) \\
&= e^{-i2\pi x \cdot \xi}\left(D_{x_j}A - \xi_j A\right)e_\xi(x) \\
&= e^{-i2\pi x \cdot \xi}\left(D_{x_j}A - AD_{x_j}\right)e_\xi(x) \\
&= e^{-i2\pi x \cdot \xi}\left[D_{x_j}, A\right]e_\xi(x) \\
&= \sigma_{L_j(A)}(x,\xi).
\end{aligned}$$

Thus the partial derivative with respect to x_j of the symbol σ_A leads to the symbol of the commutator $[D_{x_j}, A]$. As regards to the difference, the situation is almost similar (where v_k stands for the standard k^{th} unit basis vector of $\mathbb{Z}^n$):

$$\begin{aligned}
\triangle_{\xi_k}\sigma_A(x,\xi) &= \sigma_A(x,\xi + v_k) - \sigma_A(x,\xi) \\
&= e^{-i2\pi x \cdot (\xi+v_k)}Ae_{\xi+v_k}(x) - e^{-i2\pi x \cdot \xi}Ae_\xi(x) \\
&= e^{-i2\pi x_k}e^{-i2\pi x \cdot \xi}\left(A \circ \left(e^{i2\pi x_k}I\right) - \left(e^{i2\pi x_k}I\right) \circ A\right)e_\xi(x) \\
&= e^{-i2\pi x_k}e^{-i2\pi x \cdot \xi}\left[A, e^{i2\pi x_k}I\right]e_\xi(x) \\
&= e^{-i2\pi x_k}\sigma_{R_k(A)}(x,\xi).
\end{aligned}$$

The minor asymmetry in $\triangle_{\xi_k}\sigma_A(x,\xi) = e^{-i2\pi x_k}\sigma_{R_k(A)}(x,\xi)$ caused by $e^{-i2\pi x_k}$ is due to the nature of differences. In [12, p. 46-49] the pseudo-differential operators of certain degree have been characterised using analogues of these commutators representing the differentiations of a symbol. As before, the approach on $\mathbb{T}^n$ is somewhat simpler:

Theorem 5.4.1 (Commutator characterisation on $\mathbb{T}^n$). *Let A be a linear operator on $C^\infty(\mathbb{T}^n)$. Then we have $A \in \mathrm{Op}(S^m(\mathbb{T}^n \times \mathbb{Z}^n))$ if and only if*

$$\forall \alpha, \beta \in \mathbb{N}_0^n : L^\beta R^\alpha(A) \in \mathcal{L}(H^{m-|\alpha|}(\mathbb{T}^n), H^0(\mathbb{T}^n)). \tag{5.3}$$

Thus the classes of periodic pseudo-differential operators and pseudo-differential operators on $\mathbb{T}^n$ coincide. More precisely, for any $m \in \mathbb{R}$ it holds that

$$\mathrm{Op}\, S^m(\mathbb{T}^n \times \mathbb{Z}^n) = \Psi^m(\mathbb{T}^n). \tag{5.4}$$

Proof of Theorem 5.4.1 for the $\mathbb{T}^1$ case. The "only if"-part is trivial by Proposition 4.2.3, since Theorem 4.7.10 implies that $L_1(B) \in \mathrm{Op}(S^l(\mathbb{T}^1 \times \mathbb{Z}^1))$ and $R_1(B) \in \mathrm{Op}(S^{l-1}(\mathbb{T}^1 \times \mathbb{Z}^1))$ for any $B \in \mathrm{Op}(S^l(\mathbb{T}^1 \times \mathbb{Z}^1))$.

For the "if"-part we have to estimate $\triangle_\xi^\alpha \partial_x^\beta \sigma_A(x,\xi)$. Let us define operator R_1' by $R_1'(A) = e^{-i2\pi x}R_1(A)$. Because $u(x) \mapsto e^{-i2\pi x}u(x)$ is a homeomorphism from $H^s(\mathbb{T}^1)$ to $H^s(\mathbb{T}^1)$ for every $s \in \mathbb{R}$, it is true that $L_1^\alpha R_1'^\alpha(A) \in \mathcal{L}(H^{m-\alpha}(\mathbb{T}^1), H^0(\mathbb{T}^1))$. Notice that

$$|u(x)| \leq \sum_{\xi \in \mathbb{Z}^1}|\widehat{u}(\xi)| \leq \left[\sum_{\xi \in \mathbb{Z}^1}\langle\xi\rangle^{-2}\right]^{1/2}\left[\sum_{\xi \in \mathbb{Z}^1}|\xi\widehat{u}(\xi)|^2\right]^{1/2} = C\|\partial_x u\|_{H^0(\mathbb{T}^1)}.$$

Using this we get

$$
\begin{aligned}
|\triangle_\xi^\alpha \partial_x^\beta \sigma_A(x,\xi)| &\leq C \|\triangle_\xi^\alpha \partial_x^{\beta+1} \sigma_A(x,\xi)\|_{H^0(\mathbb{T}^1)} \\
&= C \|e_{-\xi} L_1^{\beta+1} R_1'^\alpha(A) e_\xi\|_{H^0(\mathbb{T}^1)} \\
&\leq C \|e_{-\xi} I\|_{\mathcal{L}(H^0,H^0)} \|L_1^{\beta+1} R_1'^\alpha(A)\|_{\mathcal{L}(H^{m-\alpha},H^0)} \|e_\xi\|_{H^{m-\alpha}} \\
&= C \|L_1^{\beta+1} R_1'^\alpha(A)\|_{\mathcal{L}(H^{m-\alpha}(\mathbb{T}^1),H^0(\mathbb{T}^1))} \langle\xi\rangle^{m-\alpha} \\
&\leq C_{\alpha\beta} \langle\xi\rangle^{m-\alpha}.
\end{aligned}
$$

This completes the proof of the one-dimensional case. $\qquad\square$

General proof of Theorem 5.4.1. Let us first prove the inclusion $\mathrm{Op}\,S^m(\mathbb{T}^n \times \mathbb{Z}^n) \subset \Psi^m(\mathbb{T}^n)$. We know by Proposition 4.2.3 that

$$
\mathrm{Op}\,S^m(\mathbb{T}^n \times \mathbb{Z}^n) \subset \mathcal{L}(H^s(\mathbb{T}^n), H^{s-m}(\mathbb{T}^n)).
$$

Therefore by Theorem 5.3.1, in view of Proposition 4.2.3 it suffices to verify that

$$
[\mathrm{Op}\,S^m(\mathbb{T}^n \times \mathbb{Z}^n), \mathrm{Diff}^1(\mathbb{T}^n)] \subset \mathrm{Op}\,S^m(\mathbb{T}^n \times \mathbb{Z}^n) \tag{5.5}
$$

and that

$$
[\mathrm{Op}\,S^m(\mathbb{T}^n \times \mathbb{Z}^n), \mathrm{Diff}^0(\mathbb{T}^n)] \subset \mathrm{Op}\,S^{m-1}(\mathbb{T}^n \times \mathbb{Z}^n). \tag{5.6}
$$

This is true due to the asymptotic expansion of the composition of two periodic pseudo-differential operators (see Theorem 4.7.10). However, we present a brief independent and instructive proof of the inclusion (5.5). Let $A \in \mathrm{Op}\,S^m(\mathbb{T}^n)$ and let $X \in \mathrm{Diff}^1(\mathbb{T}^n)$, $X_x = \phi(x)\partial_{x_k}$ $(1 \leq k \leq n)$. Now

$$
\begin{aligned}
\sigma_{[A,X]}(x,\xi) &= \mathrm{i}2\pi\xi_k \sum_\eta [\sigma_A(x,\xi+\eta) - \sigma_A(x,\xi)]\hat{\phi}(\eta)e^{\mathrm{i}2\pi x \cdot \eta} \\
&\quad - \phi(x)(\partial_{x_k}\sigma_A)(x,\xi).
\end{aligned}
$$

Notice that

$$
\begin{aligned}
&\sigma_A(x,\xi+\eta) - \sigma_A(x,\xi) \\
&= \sum_{j=1}^n \sum_{\omega_j=(\mathrm{sgn}(\eta_j)-1)/2}^{\eta_j-(\mathrm{sgn}(\eta_j)+1)/2} \mathrm{sgn}(\eta_j)\triangle_{\xi_j}\sigma_A(x,\xi+\eta_1\delta_1+\cdots+\eta_{j-1}\delta_{j-1}+\omega_j\delta_j),
\end{aligned}
$$

where

$$
\mathrm{sgn}(\eta_j) = \begin{cases} 1, & \eta_j > 0, \\ 0, & \eta_j = 0, \\ -1, & \eta_j < 0, \end{cases}
$$

and there are at most $\sum_j |\eta_j| < \sqrt{n}(1+\|\eta\|)$ non-zero terms in the sum. Hence, applying the ordinary Leibniz formula with respect to x, the discrete Leibniz formula with respect to ξ (Lemma 3.3.6), the inequality of Peetre (Proposition

3.3.31) and Lemma 4.2.1, we get

$$|\triangle_\xi^\alpha \partial_x^\beta \sigma_{[A,X]}(x,\xi)|$$

$$\leq C_{\alpha\beta,\phi,r}(1+\|\xi\|)\sum_\eta (1+\|\xi\|)^{m-(|\alpha|+1)}\sqrt{n}(1+\|\eta\|)^{|m-(|\alpha|+1)|+|\beta|+1-r}$$

$$+C_{\alpha\beta,\phi}(1+\|\xi\|)^{m-|\alpha|}.$$

By choosing $r > |m-(|\alpha|+1)|+|\beta|+2$, the series above converges, so that

$$|\triangle_\xi^\alpha \partial_x^\beta \sigma_{[A,X]}(x,\xi)| \leq C_{\alpha\beta}(1+\|\xi\|)^{m-|\alpha|}.$$

Hence $[A, X] \in \operatorname{Op} S^m(\mathbb{T}^n \times \mathbb{Z}^n)$. Similarly, but with less effort, one proves (5.6). Thus $A \in \Psi^m(\mathbb{T}^n)$ by Theorem 5.3.1, and hence also (5.3) by Theorem 5.3.1.

Now assume that $A \in \Psi^m(\mathbb{T}^n)$. We have to prove that σ_A satisfies inequalities defining the toroidal symbol class $S_{1,0}^m(\mathbb{T}^n \times \mathbb{Z}^n)$ in (4.6) from Definition 4.1.7. We also note that $A \in \Psi^m(\mathbb{T}^n)$ implies (5.3) by Theorem 5.3.1. Let us define the transformation $\widetilde{R}_k$ by $\widetilde{R}_k(A) := \mathrm{e}^{-\mathrm{i}2\pi x_k}R_k(A)$, and set $\widetilde{R}^\alpha := \widetilde{R}_1^{\alpha_1}\cdots\widetilde{R}_n^{\alpha_n}$, so that

$$\triangle_\xi^\alpha \partial_x^\beta \sigma_A(x,\xi) = \sigma_{\widetilde{R}^\alpha L^\beta(A)}(x,\xi).$$

By Theorem 5.3.1, we have $\widetilde{R}^\alpha L^\beta(A) \in \mathcal{L}(H^{m-|\alpha|}(\mathbb{T}^n), H^0(\mathbb{T}^n))$. Notice that

$$|u(x)| \;\leq\; \sum_\xi |\widehat{u}(\xi)| \leq \left[\sum_\xi (1+\|\xi\|)^{-2s}\right]^{1/2}\left[\sum_\xi (1+\|\xi\|)^{2s}|\widehat{u}(\xi)|^2\right]^{1/2}$$

$$=\;\; C_s\|u\|_{H^s(\mathbb{T}^n)} \leq C_s\left(\sum_{|\gamma|\leq s}\|\partial_x^\gamma u\|^2_{H^0(\mathbb{T}^n)}\right)^{1/2},$$

where $s \in \mathbb{N}$ satisfies $2s > n = \dim(\mathbb{T}^n)$. Using this we get

$$|\triangle_\xi^\alpha \partial_x^\beta \sigma_A(x,\xi)| \leq C\left(\sum_{|\gamma|\leq s}\|\triangle_\xi^\alpha \partial_x^{\beta+\gamma}\sigma_A(\cdot,\xi)\|^2_{H^0(\mathbb{T}^n)}\right)^{1/2}$$

$$=C\left(\sum_{|\gamma|\leq s}\|e_{-\xi}\widetilde{R}^\alpha L^{\beta+\gamma}(A)e_\xi\|^2_{H^0(\mathbb{T}^n)}\right)^{1/2}$$

$$\leq C\left(\sum_{|\gamma|\leq s}\|e_{-\xi}I\|^2_{\mathcal{L}(H^0(\mathbb{T}^n))}\right.$$

$$\left.\times\|\widetilde{R}^\alpha L^{\beta+\gamma}(A)\|^2_{\mathcal{L}(H^{m-|\alpha|}(\mathbb{T}^n),H^0(\mathbb{T}^n))}\|e_\xi\|^2_{H^{m-|\alpha|}(\mathbb{T}^n)}\right)^{1/2}$$

$$\leq C_{\alpha\beta}(1+\|\xi\|)^{m-|\alpha|},$$

completing the proof. $\square$

Part III

Representation Theory of Compact Groups

We might call the traditional topology and measure theory by the name "commutative geometry", referring to the commutative function algebras; "non-commutative geometry" would then refer to the study of non-commutative algebras. Although the function algebras considered in the sequel are still commutative, the non-commutativity of the corresponding groups is the characteristic feature of Parts III and IV.

Here we present the necessary material on compact groups and their representations. The presentation gradually increases the availability of topological and differentiable structures, thus tracing the development from general compact groups to linear Lie groups. Moreover, we present additional material on the Hopf algebras joining together the material of this part to the analysis of algebras from Chapter D. Nevertheless, we tried to make the exposition self-contained, providing references to Part I when necessary. If the reader wants to gain more profound knowledge of Lie groups, Lie algebras and their representation, there are many excellent monographs available on different aspects of these theories at different levels, for example [9, 19, 20, 31, 36, 37, 38, 47, 48, 49, 50, 51, 58, 61, 64, 65, 73, 74, 88, 123, 127, 147, 148, 149, 154], to mention a few.

Chapter 6

Groups

6.1 Introduction

Loosely speaking, groups encode symmetries of (geometric) objects: if we consider a space X with some specific structure (e.g., a Riemannian manifold), a *symmetry of X* is a bijection $f : X \to X$ preserving the natural involved structure (e.g., the Riemannian metric) – here, the compositions and inversions of symmetries yield new symmetries. In a handful of assumptions, the concept of groups captures the essential properties of wide classes of symmetries, and provides powerful tools for related analysis.

Perhaps the first non-trivial group that mankind encountered was the set $\mathbb{Z}$ of integers; with the usual addition $(x, y) \mapsto x + y$ and "inversion" $x \mapsto -x$ this is a basic example of a group. Intuitively, a group is a set G that has two mappings $G \times G \to G$ and $G \to G$ generalising the properties of the integers in a simple and natural way.

We start by defining the groups, and we study the mappings preserving such structures, i.e., group homomorphisms. Of special interest are representations, that is those group homomorphisms that have values in groups of invertible linear operators on vector spaces. Representation theory is a key ingredient in the theory of groups.

In this framework we study analysis on compact groups, foremost measure theory and Fourier transform. Remarkably, for a compact group G there exists a unique translation-invariant linear functional on $C(G)$ corresponding to a probability measure. We shall construct this so-called Haar measure, closely related to the Lebesgue measure of a Euclidean space. We shall also introduce Fourier series of functions on a group.

Groups having a smooth manifold structure (with smooth group operations) are called Lie groups, and their representation theory is especially interesting. Left-invariant first-order partial differential operators on such a group can be identified with left-invariant vector fields on the group, and the corresponding set called the Lie algebra is studied.

Finally, we introduce Hopf algebras and study the Gelfand theory related to them.

Remark 6.1.1 (**–morphisms**). If X, Y are spaces with the same kind of algebraic structure, the set $\mathrm{Hom}(X, Y)$ of *homomorphisms* consists of mappings $f : X \to Y$ respecting the structure. Bijective homomorphisms are called *isomorphisms*. Homomorphisms $f : X \to X$ are called *endomorphisms* of X, and their set is denoted by $\mathrm{End}(X) := \mathrm{Hom}(X, X)$. Isomorphism-endomorphisms are called *automorphisms*, and their set is $\mathrm{Aut}(X) \subset \mathrm{End}(X)$. If there exist the zero-elements $0_X, 0_Y$ in respective algebraic structures X, Y, the *null space* or the *kernel* of $f \in \mathrm{Hom}(X, Y)$ is

$$\mathrm{Ker}(f) := \{x \in X : f(x) = 0_Y\}.$$

Sometimes algebraic structures might have, say, topology, and then the homomorphisms are typically required to be continuous. Hence, for instance, a homomorphism $f : X \to Y$ between Banach spaces X, Y is usually assumed to be continuous and linear, denoted by $f \in \mathcal{L}(X, Y)$, unless otherwise mentioned; for short, let $\mathcal{L}(X) := \mathcal{L}(X, X)$. The assumptions in theorems etc. will still be explicitly stated.

Conventions

$\mathbb{N}$ is the set of positive integers,

$\mathbb{Z}^+ = \mathbb{N}$,

$\mathbb{N}_0 = \mathbb{N} \cup \{0\}$,

$\mathbb{Z}$ is the set of integers,

$\mathbb{Q}$ the set of rational numbers,

$\mathbb{R}$ the set of real numbers,

$\mathbb{C}$ the set of complex numbers, and

$\mathbb{K} \in \{\mathbb{R}, \mathbb{C}\}$.

6.2 Groups without topology

We start with groups without complications, without assuming supplementary properties. This choice helps in understanding the purely algebraic ideas, and only later we will mingle groups with other structures, e.g., topology.

Definition 6.2.1 (Groups). A *group* consists of a set G having an element $e = e_G \in G$ and endowed with mappings

$$((x, y) \mapsto xy) \;\; : \;\; G \times G \to G,$$
$$(x \mapsto x^{-1}) \;\; : \;\; G \to G$$

satisfying

$$x(yz) = (xy)z,$$
$$ex = x = xe,$$
$$x\, x^{-1} = e = x^{-1}x,$$

for all $x, y, z \in G$. We may freely write $xyz := x(yz) = (xy)z$; the element $e \in G$ is called the *neutral element*, and x^{-1} is the *inverse* of $x \in G$. If the group operations are implicitly known, we may say that G is a *group*. If $xy = yx$ for all $x, y \in G$ then G is called *commutative* (or *Abelian*).

Example. Let us give some examples of groups:

1. (**Symmetric group**). Let $G = \{f : X \to X \mid f \text{ bijection}\}$, where $X \neq \emptyset$; this is a group with operations $(f, g) \mapsto f \circ g$, $f \mapsto f^{-1}$. This group G of bijections on X is called the *symmetric group of X*, and it is non-commutative whenever $|X| \geq 3$, where $|X|$ is the number of elements of X. The neutral element is $\mathrm{id}_X = (x \mapsto x) : X \to X$.

2. The sets $\mathbb{Z}$, $\mathbb{Q}$, $\mathbb{R}$ and $\mathbb{C}$ are commutative groups with operations $(x, y) \mapsto x + y$, $x \mapsto -x$. The neutral element is 0 in each case.

3. Any vector space is a commutative group with operations $(x, y) \mapsto x + y$, $x \mapsto -x$; the neutral element is 0.

4. (**Automorphism group $\mathrm{Aut}(V)$**). Let V be a vector space. The set $\mathrm{Aut}(V)$ of invertible linear operators $V \to V$ forms a group with operations $(A, B) \mapsto AB$, $A \mapsto A^{-1}$; this group is non-commutative when $\dim(V) \geq 2$. The neutral element is $I = (v \mapsto v) : V \to V$.

5. Sets $\mathbb{Q}^\times := \mathbb{Q} \setminus \{0\}$, $\mathbb{R}^\times := \mathbb{R} \setminus \{0\}$, $\mathbb{C}^\times := \mathbb{C} \setminus \{0\}$ (more generally, invertible elements of a unital ring) form multiplicative groups with operations $(x, y) \mapsto xy$ (ordinary multiplication) and $x \mapsto x^{-1}$ (as usual). The neutral element is 1 in each case.

6. (**Affine group $\mathrm{Aff}(V)$**). The set

$$\mathrm{Aff}(V) = \{A_a = (v \mapsto Av + a) : V \to V \mid A \in \mathrm{Aut}(V), \ a \in V\}$$

of affine mappings forms a group with operations defined to be $(A_a, B_b) \mapsto (AB)_{Ab+a}$, $A_a \mapsto (A^{-1})_{A^{-1}a}$; this group is non-commutative when $\dim(V) \geq 1$. The neutral element is I_0.

7. (**Product group**). If G and H are groups then $G \times H$ has a natural group structure:

$$((g_1, h_1), (g_2, h_2)) \mapsto (g_1 h_1, g_2 h_2), \quad (g, h) \mapsto (g^{-1}, h^{-1}).$$

The neutral element is $e_{G \times H} := (e_G, e_H)$.

Exercise 6.2.2. Let G be a group and $x, y \in G$. Prove:

(a) $(x^{-1})^{-1} = x$.

(b) If $xy = e$ then $y = x^{-1}$.

(c) $(xy)^{-1} = y^{-1} x^{-1}$.

Definition 6.2.3 (Finite groups). If a group has finitely many elements it is said to be *finite*.

Example. The symmetry group of a set consisting of n elements is called *the permutation group of n elements*. Such group is a finite group and has $n! = 1 \cdot 2 \cdots n$ elements.

Definition 6.2.4 (Notation). Let G be a group, $x \in A$, $A, B \subset G$ and $n \in \mathbb{Z}^+$. We write

$$
\begin{aligned}
xA &:= \{xa \mid a \in A\}, \\
Ax &:= \{ax \mid a \in A\}, \\
AB &:= \{ab \mid a \in A,\ b \in B\}, \\
A^0 &:= \{e\}, \\
A^{-1} &:= \{a^{-1} \mid a \in A\}, \\
A^{n+1} &:= A^n A, \\
A^{-n} &:= (A^n)^{-1}.
\end{aligned}
$$

Definition 6.2.5 (Subgroups $H < G$, and normal subgroups $H \lhd G$). A set $H \subset G$ is a *subgroup* of a group G, denoted by $H < G$, if

$$ e \in H, \quad xy \in H \quad \text{and} \quad x^{-1} \in H $$

for all $x, y \in H$ (hence H is a group with the inherited operations). A subgroup $H < G$ is called *normal in G* if

$$ xH = Hx $$

for all $x \in G$; then we write $H \lhd G$.

Remark 6.2.6. With the inherited operations, a subgroup is a group. Normal subgroups are the well-behaving ones, as exemplified later in Proposition 6.2.16 and Theorem 6.2.20. In some books normal subgroups of G are called *normal divisors* of G.

Exercise 6.2.7. Let $H < G$. Show that if $H \subset x^{-1}Hx$ for every $x \in G$, then $H \lhd G$.

Exercise 6.2.8. Let $H < G$. Show that $H \lhd G$ if and only if $H = x^{-1}Hx$ for every $x \in G$.

Example. Let us collect some instances and facts about subgroups:

1. (**Trivial subgroups**). We always have normal *trivial subgroups* $\{e\} \lhd G$ and $G \lhd G$. Subgroups of a commutative group are always normal.

2. (**Centre of a group**). The *centre* $Z(G) \lhd G$, where

$$ Z(G) := \{z \in G \mid \forall x \in G : xz = zx\}. $$

Thus, the centre is the collection of elements that commute with all elements of the group.

3. If $F < H$ and $G < H$ then $F \cap G < H$.

4. If $F < H$ and $G \triangleleft H$ then $FG < H$.

5. $\{I_a \mid a \in V\} \triangleleft \mathrm{Aff}(V)$.

6. The following two examples will be of crucial importance later so we formulate them as Remarks 6.2.9 and 6.2.10.

Remark 6.2.9 (**Groups GL$(n, \mathbb{R})$, O(n), SO(n)**). We have

$$\mathrm{SO}(n) < \mathrm{O}(n) < \mathrm{GL}(n, \mathbb{R}) \cong \mathrm{Aut}(\mathbb{R}^n),$$

where the groups consist of real $n \times n$-matrices: $\mathrm{GL}(n, \mathbb{R})$ is the real *general linear* group consisting of invertible real matrices (i.e., determinant non-zero); $\mathrm{O}(n)$ is the *orthogonal* group, where the matrix columns (or rows) form an orthonormal basis for $\mathbb{R}^n$ (so that $A^T = A^{-1}$ for $A \in \mathrm{O}(n)$, $\det(A) \in \{-1, 1\}$); $\mathrm{SO}(n)$ is the *special orthogonal* group, the group of rotation matrices of $\mathbb{R}^n$ around the origin, so that

$$\mathrm{SO}(n) = \{A \in \mathrm{O}(n) : \det(A) = 1\}.$$

Remark 6.2.10 (**Groups GL$(n, \mathbb{C})$, U(n), SU(n)**). We have

$$\mathrm{SU}(n) < \mathrm{U}(n) < \mathrm{GL}(n, \mathbb{C}) \cong \mathrm{Aut}(\mathbb{C}^n),$$

where the groups consist of complex $n \times n$-matrices: $\mathrm{GL}(n, \mathbb{C})$ is the complex *general linear* group consisting of invertible complex matrices (i.e., determinant non-zero); $\mathrm{U}(n)$ is the *unitary* group, where the matrix columns (or rows) form an orthonormal basis for $\mathbb{C}^n$ (so that $A^* = A^{-1}$ for $A \in \mathrm{U}(n)$, $|\det(A)| = 1$); $\mathrm{SU}(n)$ is the *special unitary* group,

$$\mathrm{SU}(n) = \{A \in \mathrm{U}(n) : \det(A) = 1\}.$$

Remark 6.2.11. The mapping

$$(z \mapsto (z)) : \mathbb{C} \to \mathbb{C}^{1 \times 1}$$

identifies complex numbers with complex (1×1)-matrices. Thereby the complex unit circle group $\{z \in \mathbb{C} : |z| = 1\}$ is identified with the group $\mathrm{U}(1)$.

Definition 6.2.12 (Right quotient G/H). Let $H < G$. Then

$$x \sim y \quad \Longleftrightarrow \quad xH = yH$$

defines an equivalence relation on G, as can be easily verified. The (*right*) *quotient of G by H* is the set

$$G/H = \{xH \mid x \in G\}.$$

Notice that $xH = yH$ if and only if $x^{-1}y \in H$.

Similarly, we can define

Definition 6.2.13 (Left quotient $H\backslash G$). Let $H < G$. Then

$$x \sim y \iff Hx = Hy$$

defines an equivalence relation on G. The *(left) quotient of G by H* is the set

$$H\backslash G = \{Hx \mid x \in G\}.$$

Notice that $Hx = Hy$ if and only if $x^{-1}y \in H$.

Remark 6.2.14 (**Right for now**). We will deal mostly with the right quotient G/H in Part III. However, we note that in Part IV we will actually need more the left quotient $H\backslash G$. It should be a simple exercise for the reader to translate all the results from "right" to "left". Indeed, simply replacing the side from which the subgroup acts from right to left, and changing all the words from "right" to "left" should do the job since the situation is completely symmetric. The reason for our change is that once we choose to identify the Lie algebras with the left-invariant vector fields in Part IV it leads to a more natural analysis of pseudo-differential operators on left quotients. However, because our intuition about division may be better suited to the notation G/H we chose to explain the basic ideas for the right quotients, keeping in mind that the situation with the left quotients is completely symmetric.

Remark 6.2.15. It is often useful to identify the points $xH \in G/H$ with the sets $xH \subset G$. Also, for $A \subset G$ we naturally identify the sets

$$AH = \{ah : a \in A, \ h \in H\} \quad \subset \quad G \quad \text{and}$$
$$\{aH : a \in A\} = \{\{ah : h \in H\} : a \in A\} \quad \subset \quad G/H.$$

This provides a nice way to treat the quotient G/H.

Proposition 6.2.16 (When is G/H a group?). *Let $H \triangleleft G$ be normal. Then the quotient G/H can be endowed with the group structure*

$$(xH, yH) \mapsto xyH, \quad xH \mapsto x^{-1}H.$$

Proof. The operations are well-defined mappings $(G/H) \times (G/H) \to G/H$ and $G/H \to G/H$, respectively, since

$$xHyH \overset{H \trianglelefteq G}{=} xyHH \overset{HH=H}{=} xyH,$$

and

$$(xH)^{-1} = H^{-1}x^{-1} \overset{H^{-1}=H}{=} Hx^{-1} \overset{H \trianglelefteq G}{=} x^{-1}H.$$

The group axioms follow, since by simple calculations we have

$$(xH)(yH)(zH) = xyzH,$$
$$(xH)(eH) = xH = (eH)(xH),$$
$$(x^{-1}H)(xH) = H = (xH)(x^{-1}H).$$

Notice that $e_{G/H} = e_G H = H$. $\qquad\qquad\qquad\qquad\qquad\qquad\qquad\qquad\qquad\square$

Definition 6.2.17 (Torus $\mathbb{T}^n$ as a quotient group). The quotient $\mathbb{T}^n := \mathbb{R}^n/\mathbb{Z}^n$ is called the *(flat) n-dimensional torus.*

Definition 6.2.18 (Homomorphisms and isomorphisms). Let G, H be groups. A mapping $\phi : G \to H$ is called a *homomorphism* (or a *group homomorphism*), denoted by $\phi \in \mathrm{Hom}(G, H)$, if

$$\phi(xy) = \phi(x)\phi(y)$$

for all $x, y \in G$. The *kernel* of $\phi \in \mathrm{Hom}(G, H)$ is

$$\mathrm{Ker}(\phi) := \{x \in G \mid \phi(x) = e_H\}.$$

A bijective homomorphism $\phi \in \mathrm{Hom}(G, H)$ is called an *isomorphism*, denoted by $\phi : G \cong H$.

Remark 6.2.19. Group homomorphisms are the natural mappings between groups, preserving the group operations. Notice especially that for a group homomorphism $\phi : G \to H$ it holds that

$$\phi(e_G) = e_H \quad \text{and} \quad \phi(x^{-1}) = \phi(x)^{-1}$$

for all $x \in G$.

Example. Examples of homomorphisms:

1. $(x \mapsto e_H) \in \mathrm{Hom}(G, H)$.
2. For $y \in G$, $(x \mapsto y^{-1}xy) \in \mathrm{Hom}(G, G)$.
3. If $H \triangleleft G$ then $x \mapsto xH$ is a surjective homomorphism $G \to G/H$.
4. For $x \in G$, $(n \mapsto x^n) \in \mathrm{Hom}(\mathbb{Z}, G)$.
5. If $\phi \in \mathrm{Hom}(F, G)$ and $\psi \in \mathrm{Hom}(G, H)$ then $\psi \circ \phi \in \mathrm{Hom}(F, H)$.
6. $\mathbb{T}^1 \cong \mathrm{U}(1) \cong \mathrm{SO}(2)$.

Theorem 6.2.20. *Let $\phi \in \mathrm{Hom}(G, H)$ and $K = \mathrm{Ker}(\phi)$. Then:*

1. $\phi(G) < H$.
2. $K \triangleleft G$.
3. $\psi(xK) := \phi(x)$ *defines a group isomorphism* $\psi : G/K \to \phi(G)$.

Thus we have the commutative diagram

$$
\begin{array}{ccc}
G & \xrightarrow{\;\;\phi\;\;} & H \\
{\scriptstyle x \mapsto xK}\downarrow & & \uparrow{\scriptstyle y \mapsto y} \\
G/K & \xrightarrow[\;\;\psi:G/K\cong\phi(G)\;\;]{} & \phi(G).
\end{array}
$$

Proof. Let $x, y \in G$. Now $\phi(G)$ is a subgroup of H, because

$$
\begin{aligned}
e_H &= \phi(e_G) \in \phi(G), \\
\phi(x)\phi(y) &= \phi(xy) \in \phi(G), \\
\phi(x^{-1})\phi(x) &= \phi(x^{-1}x) = \phi(e_G) \\
&= e_H \\
&= \cdots = \phi(x)\phi(x^{-1});
\end{aligned}
$$

notice that $\phi(x)^{-1} = \phi(x^{-1})$. If $a, b \in \mathrm{Ker}(\phi)$ then

$$
\begin{aligned}
\phi(e_G) &= e_H, \\
\phi(ab) &= \phi(a)\phi(b) = e_H e_H = e_H, \\
\phi(a^{-1}) &= \phi(a)^{-1} = e_H^{-1} = e_H,
\end{aligned}
$$

so that $K = \mathrm{Ker}(\phi) < G$. If moreover $x \in G$ then

$$
\phi(x^{-1}Kx) = \phi(x^{-1})\ \phi(K)\ \phi(x) = \phi(x)^{-1}\ \{e_H\}\ \phi(x) = \{e_H\},
$$

meaning $x^{-1}Kx \subset K$. Thus $K \lhd G$ by Exercise 6.2.8. By Proposition 6.2.16, G/K is a group (with the natural operations). Since $\phi(xa) = \phi(x)$ for every $a \in K$, $\psi = (xK \mapsto \phi(x)) : G/K \to \phi(G)$ is a well-defined surjection. Furthermore,

$$
\psi(xyK) = \phi(xy) = \phi(x)\phi(y) = \psi(xK)\psi(yK),
$$

thus $\psi \in \mathrm{Hom}(G/K, \phi(G))$. Finally,

$$
\psi(xK) = \psi(yK) \iff \phi(x) = \phi(y) \iff x^{-1}y \in K \iff xK = yK,
$$

so that ψ is injective. $\square$

Exercise 6.2.21 (Universality of the permutation groups). Let G be a finite group. Show that there is a set X with finitely many elements such that G is isomorphic to a subgroup of the symmetric group of X.

6.3   Group actions and representations

Spaces can be studied by examining their symmetry groups. On the other hand, it is fruitful to investigate groups when they are acting as symmetries of some nicely structured spaces. Next we study actions of groups on sets. Especially interesting group actions are the linear actions on vector spaces, providing the machinery of linear algebra – this is the fundamental idea in the representation theory of groups.

Definition 6.3.1 (Transitive actions). An *action of a group G on a set $M \neq \emptyset$* is a mapping

$$
((x, p) \mapsto x \cdot p) : G \times M \to M,
$$

for which

$$\begin{cases} x \cdot (y \cdot p) = (xy) \cdot p, \\ e \cdot p = p \end{cases}$$

for all $x, y \in G$ and $p \in M$; the action is *transitive* if

$$\forall p, q \in M \; \exists x \in G : \; x \cdot q = p.$$

If M is a vector space and the mapping $p \mapsto x \cdot p$ is linear for each $x \in G$, the action is called *linear*.

Remark 6.3.2. To be precise, our *action* $G \times M \to M$ in Definition 6.3.1 should be called a *left action*, to make a difference to the *right actions* $M \times G \to M$, which are defined in the obvious way. When G acts on M, it is useful to think of G as a (sub)group of symmetries of M. Transitivity means that M is highly symmetric: there are enough symmetries to move any point to any other point.

Example. Let us present some examples of actions:

1. On a vector space V, the group $\mathrm{Aut}(V)$ acts linearly by $(A, v) \mapsto Av$.
2. If $\phi \in \mathrm{Hom}(G, H)$ then G acts on H by $(x, y) \mapsto \phi(x)y$. Especially, G acts on G transitively by $(x, y) \mapsto xy$.
3. The rotation group $\mathrm{SO}(n)$ acts transitively on the sphere $\mathbb{S}^{n-1} := \{x = (x_j)_{j=1}^n \in \mathbb{R}^n \mid x_1^2 + \cdots + x_n^2 = 1\}$ by $(A, x) \mapsto Ax$.
4. If $H < G$ and $((x, p) \mapsto x \cdot p) : G \times M \to M$ is an action then the restriction $((x, p) \mapsto x \cdot p) : H \times M \to M$ is an action.

Definition 6.3.3 (Isotropy subgroup). Let $((x, p) \mapsto x \cdot p) : G \times M \to M$ be an action. The *isotropy subgroup* of $q \in M$ is

$$G_q := \{x \in G \mid x \cdot q = q\}.$$

That is, $G_q \subset G$ contains those symmetries that fix the point $q \in M$.

Theorem 6.3.4. *Let* $((x, p) \mapsto x \cdot p) : G \times M \to M$ *be a transitive action. Let* $q \in M$. *Then the isotropy subgroup* G_q *is a subgroup for which*

$$f_q := (xG_q \mapsto x \cdot q) : G/G_q \to M$$

is a bijection.

Remark 6.3.5. If $G_q \lhd G$ then G/G_q is a group; otherwise the quotient is just a set. Notice also that the choice of $q \in M$ here is essentially irrelevant.

Example. Let $G = \mathrm{SO}(3)$, $M = \mathbb{S}^2$, and $q \in \mathbb{S}^2$ be the north pole (i.e., $q = (0, 0, 1) \in \mathbb{R}^3$). Then $G_q < \mathrm{SO}(3)$ consists of the rotations around the vertical axis (passing through the north and south poles). Since $\mathrm{SO}(3)$ acts transitively on $\mathbb{S}^2$, we get a bijection $\mathrm{SO}(3)/G_q \to \mathbb{S}^2$. The reader may think how $A \in \mathrm{SO}(3)$ moves the north pole $q \in \mathbb{S}^2$ to $Aq \in \mathbb{S}^2$.

Proof of Theorem 6.3.4. Let $a, b \in G_q$. Then

$$e \cdot q = q,$$
$$(ab) \cdot q = a \cdot (b \cdot q) = a \cdot q = q,$$
$$a^{-1} \cdot q = a^{-1} \cdot (a \cdot q) = (a^{-1}a) \cdot q = e \cdot q = q,$$

so that $G_q < G$. Let $x, y \in G$. Since

$$(xa) \cdot q = x \cdot (a \cdot q) = x \cdot q,$$

$f = (xG_q \mapsto x \cdot q) : G/G_q \to M$ is a well-defined mapping. If $x \cdot q = y \cdot q$ then

$$(x^{-1}y) \cdot q = x^{-1} \cdot (y \cdot q) = x^{-1} \cdot (x \cdot q) = (x^{-1}x) \cdot q = e \cdot q = q,$$

i.e., $x^{-1}y \in G_q$, that is $xG_q = yG_q$; hence f is injective. Take $p \in M$. By transitivity, there exists $x \in G$ such that $x \cdot q = p$. Thereby $f(xG_q) = x \cdot q = p$, i.e., f is surjective. □

Remark 6.3.6. If an action $((x, p) \mapsto x \cdot p) : G \times M \to M$ is not transitive, it is often reasonable to study only the *orbit* of $q \in M$, defined by

$$G \cdot q := \{x \cdot q \mid x \in G\}.$$

Now

$$((x, p) \mapsto x \cdot p) : G \times (G \cdot q) \to (G \cdot q)$$

is transitive, and $(x \cdot q \mapsto xG_q) : G \cdot q \to G/G_q$ is a bijection. Notice that either $G \cdot p = G \cdot q$ or $(G \cdot p) \cap (G \cdot q) = \emptyset$; thus the action of G cuts M into a disjoint union of "slices" (orbits).

Definition 6.3.7 (Unitary groups). Let $(v, w) \mapsto \langle v, w \rangle_{\mathcal{H}}$ be the inner product of a complex vector space $\mathcal{H}$. Recall that the adjoint $A^* \in \text{Aut}(\mathcal{H})$ of $A \in \text{Aut}(\mathcal{H})$ is defined by

$$\langle A^* v, w \rangle_{\mathcal{H}} := \langle v, Aw \rangle_{\mathcal{H}}.$$

The *unitary group* of $\mathcal{H}$ is

$$\mathcal{U}(\mathcal{H}) := \{A \in \text{Aut}(\mathcal{H}) \mid \forall v, w \in \mathcal{H} : \langle Av, Aw \rangle_{\mathcal{H}} = \langle v, w \rangle_{\mathcal{H}}\},$$

i.e., $\mathcal{U}(\mathcal{H})$ contains the unitary linear bijections $\mathcal{H} \to \mathcal{H}$. Clearly $A^* = A^{-1}$ for $A \in \mathcal{U}(\mathcal{H})$. The *unitary matrix group* for $\mathbb{C}^n$ is

$$U(n) := \{A = (a_{ij})_{i,j=1}^n \in \text{GL}(n, \mathbb{C}) \mid A^* = A^{-1}\},$$

see Remark 6.2.10; here $A^* = (\overline{a_{ji}})_{i,j=1}^n = A^{-1}$, i.e.,

$$\sum_{k=1}^n \overline{a_{ki}} a_{kj} = \delta_{ij}.$$

Definition 6.3.8 (Representations). A *representation of a group G on a vector space V* is any $\phi \in \operatorname{Hom}(G, \operatorname{Aut}(V))$; the *dimension* of ϕ is $\dim(\phi) := \dim(V)$. A representation $\psi \in \operatorname{Hom}(G, \mathcal{U}(\mathcal{H}))$ is called a *unitary representation*, and $\psi \in \operatorname{Hom}(G, \operatorname{U}(n))$ is called a *unitary matrix representation*.

Remark 6.3.9. The main idea here is that we can study a group G by using linear algebraic tools via representations $\phi \in \operatorname{Hom}(G, \operatorname{Aut}(V))$.

Remark 6.3.10. There is a bijective correspondence between the representations of G on V and linear actions of G on V. Indeed, if $\phi \in \operatorname{Hom}(G, \operatorname{Aut}(V))$ then

$$((x, v) \mapsto \phi(x)v) : G \times V \to V$$

is an action of G on V. Conversely, if $((x, v) \mapsto x \cdot v) : G \times V \to V$ is a linear action then

$$(x \mapsto (v \mapsto x \cdot v)) \in \operatorname{Hom}(G, \operatorname{Aut}(V))$$

is a representation of G on V.

Example. Let us give some examples of representations:

1. If $G < \operatorname{Aut}(V)$ then $(A \mapsto A) \in \operatorname{Hom}(G, \operatorname{Aut}(V))$.
2. If $G < \mathcal{U}(\mathcal{H})$ then $(A \mapsto A) \in \operatorname{Hom}(G, \mathcal{U}(\mathcal{H}))$.
3. There is always the trivial representation $(x \mapsto I) \in \operatorname{Hom}(G, \operatorname{Aut}(V))$.
4. (**Representations π_L and π_R**). Let $\mathcal{F}(G) = \mathbb{C}^G$, i.e., the complex vector space of functions $f : G \to \mathbb{C}$. Let us define left and right regular representations $\pi_L, \pi_R \in \operatorname{Hom}(G, \operatorname{Aut}(\mathcal{F}(G)))$ by

$$\begin{aligned}(\pi_L(y)f)(x) &:= f(y^{-1}x), \\ (\pi_R(y)f)(x) &:= f(xy)\end{aligned}$$

   for all $x, y \in G$.
5. Let us identify the complex (1×1)-matrices with the complex numbers by the mapping $((z) \mapsto z) : \mathbb{C}^{1\times1} \to \mathbb{C}$. Then $\operatorname{U}(1)$ is identified with the unit circle $\{z \in \mathbb{C} : |z| = 1\}$, and $(x \mapsto e^{ix\cdot\xi}) \in \operatorname{Hom}(\mathbb{R}^n, \operatorname{U}(1))$ for all $\xi \in \mathbb{R}^n$.
6. Analogously, $(x \mapsto e^{i2\pi x\cdot\xi}) \in \operatorname{Hom}(\mathbb{R}^n/\mathbb{Z}^n, \operatorname{U}(1))$ for all $\xi \in \mathbb{Z}^n$.
7. Let $\phi \in \operatorname{Hom}(G, \operatorname{Aut}(V))$ and $\psi \in \operatorname{Hom}(G, \operatorname{Aut}(W))$, where V, W are vector spaces over the same field. Then

$$\phi \oplus \psi = (x \mapsto \phi(x) \oplus \psi(x)) \in \operatorname{Hom}(G, \operatorname{Aut}(V \oplus W)),$$

$$\phi \otimes \psi|_G = (x \mapsto \phi(x) \otimes \psi(x)) \in \operatorname{Hom}(G, \operatorname{Aut}(V \otimes W)),$$

   where $V \oplus W$ is the direct sum and $V \otimes W$ is the tensor product space.
8. If $\phi = (x \mapsto (\phi(x)_{ij})_{i,j=1}^n) \in \operatorname{Hom}(G, \operatorname{GL}(n, \mathbb{C}))$ then the conjugate $\overline{\phi} = (x \mapsto (\overline{\phi(x)_{ij}})_{i,j=1}^n) \in \operatorname{Hom}(G, \operatorname{GL}(n, \mathbb{C}))$.

Definition 6.3.11 (Invariant subspaces and irreducible representations). Let V be a vector space and $A \in \mathrm{End}(V)$. A subspace $W \subset V$ is called A-*invariant* if

$$AW \subset W,$$

where $AW = \{Aw : w \in W\}$. Let $\phi \in \mathrm{Hom}(G, \mathrm{Aut}(V))$. A subspace $W \subset V$ is called ϕ-*invariant* if W is $\phi(x)$-invariant for all $x \in G$ (abbreviated $\phi(G)W \subset W$); moreover, ϕ is *irreducible* if the only ϕ-invariant subspaces are the *trivial subspaces* $\{0\}$ and V.

Remark 6.3.12 (**Restricted representations**). If $W \subset V$ is ϕ-invariant for $\phi \in \mathrm{Hom}(G, \mathrm{Aut}(V))$, we may define the *restricted representation*

$$\phi|_W \in \mathrm{Hom}(G, \mathrm{Aut}(W))$$

by $\phi|_W(x)w := \phi(x)w$. If ϕ is unitary then its restriction is also unitary.

Lemma 6.3.13. *Let $\phi \in \mathrm{Hom}(G, \mathcal{U}(\mathcal{H}))$. Let $W \subset \mathcal{H}$ be a ϕ-invariant subspace. Then its orthocomplement*

$$W^\perp = \{v \in \mathcal{H} \mid \forall w \in W : \langle v, w \rangle_\mathcal{H} = 0\}$$

is also ϕ-invariant.

Proof. If $x \in G$, $v \in W^\perp$ and $w \in W$ then

$$\langle \phi(x)v, w \rangle_\mathcal{H} = \langle v, \phi(x)^* w \rangle_\mathcal{H} = \langle v, \phi(x)^{-1} w \rangle_\mathcal{H} = \langle v, \phi(x^{-1})w \rangle_\mathcal{H} = 0,$$

meaning that $\phi(x)v \in W^\perp$. $\qquad\qquad\qquad\qquad\qquad\qquad\qquad\qquad\qquad\qquad\quad\square$

Definition 6.3.14 (Direct sums). Let V be an inner product space and let $\{V_j\}_{j \in J}$ be some family of its mutually orthogonal subspaces (i.e., $\langle v_i, v_j \rangle_V = 0$ if $v_i \in V_i$, $v_j \in V_j$ and $i \neq j$). The (*algebraic*) *direct sum* of $\{V_j\}_{j \in J}$ is the subspace

$$W = \bigoplus_{j \in J} V_j := \mathrm{span} \bigcup_{j \in J} V_j.$$

If $A_j \in \mathrm{End}(V_j)$ then let us define

$$A = \bigoplus_{j \in J} A_j \in \mathrm{End}(W)$$

by $Av := A_j v$ for all $j \in J$ and $v \in V_j$. If $\phi_j \in \mathrm{Hom}(G, \mathrm{Aut}(V_j))$ then we define

$$\phi = \bigoplus_{j \in J} \phi_j \in \mathrm{Hom}(G, \mathrm{Aut}(W))$$

by $\phi|_{V_j} = \phi_j$ for all $j \in J$, i.e., $\phi(x) := \bigoplus_{j \in J} \phi_j(x)$ for all $x \in G$.

Remark 6.3.15. In a sense, irreducible representations are the building blocks of representations. Given a representation of a group, a fundamental task is to find its invariant subspaces, and describe the representation as a direct sum of irreducible representations. To reach this goal, we often have to assume some extra conditions, e.g., of algebraic or topological nature.

Theorem 6.3.16 (Reducing finite-dimensional representations).
Let $\phi \in \mathrm{Hom}(G, \mathcal{U}(\mathcal{H}))$ be finite-dimensional. Then ϕ is a direct sum of irreducible unitary representations.

Proof (by induction). The claim is true for $\dim(\mathcal{H}) = 1$, since then the only subspaces of $\mathcal{H}$ are the trivial ones. Suppose the claim is true for representations of dimension n or less. Suppose $\dim(\mathcal{H}) = n+1$. If ϕ is irreducible, there is nothing to prove. Hence assume that there exists a non-trivial ϕ-invariant subspace $W \subset \mathcal{H}$. Then also the orthocomplement $W^\perp$ is ϕ-invariant by Lemma 6.3.13. Due to the ϕ-invariance of the subspaces W and $W^\perp$, we may define restricted representations $\phi|_W \in \mathrm{Hom}(G, \mathcal{U}(W))$ and $\phi|_{W^\perp} \in \mathrm{Hom}(G, \mathcal{U}(W^\perp))$. Hence $\mathcal{H} = W \oplus W^\perp$ and $\phi = \phi|_W \oplus \phi|_{W^\perp}$. Moreover, $\dim(W) \leq n$ and $\dim(W^\perp) \leq n$; the proof is complete, since unitary representations up to dimension n are direct sums of irreducible unitary representations by the induction hypothesis. $\square$

Remark 6.3.17. By Theorem 6.3.16, finite-dimensional unitary representations can be decomposed nicely. More precisely, if $\phi \in \mathrm{Hom}(G, \mathcal{U}(\mathcal{H}))$ is finite-dimensional then

$$\mathcal{H} = \bigoplus_{j=1}^{k} W_j, \quad \phi = \bigoplus_{j=1}^{k} \phi|_{W_j},$$

where each $\phi|_{W_j} \in \mathrm{Hom}(G, \mathcal{U}(W_j))$ is irreducible.

Definition 6.3.18 (Equivalent representations). A linear mapping $A : V \to W$ is an *intertwining operator* between representations $\phi \in \mathrm{Hom}(G, \mathrm{Aut}(V))$ and $\psi \in \mathrm{Hom}(G, \mathrm{Aut}(W))$, denoted by $A \in \mathrm{Hom}(\phi, \psi)$, if

$$A\phi(x) = \psi(x)A$$

for all $x \in G$, i.e., if the diagram

$$
\begin{array}{ccc}
V & \xrightarrow{\phi(x)} & V \\
{\scriptstyle A}\downarrow & & \downarrow{\scriptstyle A} \\
W & \xrightarrow{\psi(x)} & W
\end{array}
$$

commutes for every $x \in G$. If $A \in \mathrm{Hom}(\phi, \psi)$ is invertible then ϕ and ψ are said to be *equivalent*, denoted by $\phi \sim \psi$.

Remark 6.3.19. Always $0 \in \mathrm{Hom}(\phi, \psi)$, and $\mathrm{Hom}(\phi, \psi)$ is a vector space. Moreover, if $A \in \mathrm{Hom}(\phi, \psi)$ and $B \in \mathrm{Hom}(\psi, \xi)$ then $BA \in \mathrm{Hom}(\phi, \xi)$.

Exercise 6.3.20. Let G be a finite group and let $\mathcal{F}(G)$ be the vector space of functions $f : G \to \mathbb{C}$. Let

$$\int_G f \, d\mu_G := \frac{1}{|G|} \sum_{x \in G} f(x),$$

when $f \in \mathcal{F}(G)$. Let us endow $\mathcal{F}(G)$ with the inner product

$$\langle f, g \rangle_{L^2(\mu_G)} := \int_G f \, \bar{g} \, d\mu_G.$$

Define $\pi_L, \pi_R : G \to \mathrm{Aut}(\mathcal{F}(G))$ by

$$(\pi_L(y) \, f)(x) := f(y^{-1}x),$$
$$(\pi_R(y) \, f)(x) := f(xy).$$

Show that π_L and π_R are equivalent unitary representations.

Exercise 6.3.21. Let G be non-commutative and $|G| = 6$. Endow $\mathcal{F}(G)$ with the inner product given in Exercise 6.3.20. Find the π_L-invariant subspaces and give orthogonal bases for them.

Exercise 6.3.22 (Torus $\mathbb{T}^n$). Let us endow the n-dimensional torus $\mathbb{T}^n := \mathbb{R}^n/\mathbb{Z}^n$ with the quotient group structure and with the Lebesgue measure. Let $\pi_L, \pi_R :$ $\mathbb{T}^n \to \mathcal{L}(L^2(\mathbb{T}^n))$ be defined by

$$(\pi_L(y) \, f)(x) := f(x - y),$$
$$(\pi_R(y) \, f)(x) := f(x + y)$$

for almost every $x \in \mathbb{T}^n$. Show that π_L and π_R are equivalent reducible unitary representations. Describe the minimal π_L- and π_R-invariant subspaces containing the function $x \mapsto e^{i2\pi x \cdot \xi}$, where $\xi \in \mathbb{Z}^n$.

Remark 6.3.23. One of the main results in the representation theory of groups is Schur's Lemma 6.3.25, according to which the intertwining space $\mathrm{Hom}(\phi, \phi)$ may be rather trivial. The most of the work for such a result is carried out in the proof of the following Proposition 6.3.24:

Proposition 6.3.24. Let $\phi \in \mathrm{Hom}(G, \mathrm{Aut}(V_\phi))$ and $\psi \in \mathrm{Hom}(G, \mathrm{Aut}(V_\psi))$ be irreducible. If $A \in \mathrm{Hom}(\phi, \psi)$ then either $A = 0$ or $A : V_\phi \to V_\psi$ is invertible.

Proof. The image $AV_\phi \subset V_\psi$ of A is ψ-invariant, because

$$\psi(G) \, AV_\phi = A \, \phi(G)V_\phi = AV_\phi,$$

so that either $AV_\phi = \{0\}$ or $AV_\phi = V_\psi$, as ψ is irreducible. Hence either $A = 0$ or A is a surjection.

The kernel $\text{Ker}(A) = \{v \in V_\phi \mid Av = 0\}$ is ϕ-invariant, since

$$A \, \phi(G) \, \text{Ker}(A) = \psi(G) \, A \, \text{Ker}(A) = \psi(G) \, \{0\} = \{0\},$$

so that either $\text{Ker}(A) = \{0\}$ or $\text{Ker}(A) = V_\phi$, as ϕ is irreducible. Hence either A is injective or $A = 0$.

Thus either $A = 0$ or A is bijective. $\qquad\square$

Corollary 6.3.25 (Schur's Lemma (finite-dimensional (1905))).
Let $\phi \in \text{Hom}(G, \text{Aut}(V))$ be irreducible and finite-dimensional. Then $\text{Hom}(\phi, \phi) = \mathbb{C}I = \{\lambda I \mid \lambda \in \mathbb{C}\}$.

Proof. Let $A \in \text{Hom}(\phi, \phi)$. The finite-dimensional linear operator A has an eigenvalue $\lambda \in \mathbb{C}$, i.e., $\lambda I - A : V \to V$ is not invertible. On the other hand, $\lambda I - A \in \text{Hom}(\phi, \phi)$, so that $\lambda I - A = 0$ by Proposition 6.3.24. $\qquad\square$

Corollary 6.3.26 (Representations of commutative groups). *Let G be a commutative group. Irreducible finite-dimensional representations of G are one-dimensional.*

Proof. Let $\phi \in \text{Hom}(G, \text{Aut}(V))$ be irreducible, $\dim(\phi) < \infty$. Due to the commutativity of G,

$$\phi(x)\phi(y) = \phi(xy) = \phi(yx) = \phi(y)\phi(x)$$

for all $x, y \in G$, so that $\phi(G) \subset \text{Hom}(\phi, \phi)$. By Schur's Lemma 6.3.25, $\text{Hom}(\phi, \phi) = \mathbb{C}I$. Hence if $v \in V$ then

$$\phi(G)\text{span}\{v\} = \text{span}\{v\},$$

i.e., $\text{span}\{v\}$ is ϕ-invariant. Therefore either $v = 0$ or $\text{span}\{v\} = V$. $\qquad\square$

Corollary 6.3.27. *Let $\phi \in \text{Hom}(G, \mathcal{U}(\mathcal{H}_\phi))$ and $\psi \in \text{Hom}(G, \mathcal{U}(\mathcal{H}_\psi))$ be finite-dimensional. Then $\phi \sim \psi$ if and only if there exists an isometric isomorphism $B \in \text{Hom}(\phi, \psi)$.*

Remark 6.3.28 (Isometries). An *isometry* $f : M \to N$ between metric spaces (M, d_M) and (N, d_N) satisfies $d_N(f(x), f(y)) = d_M(x, y)$ for all $x, y \in M$.

Proof of Corollary 6.3.27. The "if"-part is trivial. Assume that $\phi \sim \psi$. Recall that there are direct sum decompositions

$$\phi = \bigoplus_{j=1}^{m} \phi_j, \quad \psi = \bigoplus_{k=1}^{n} \psi_k,$$

where ϕ_j, ψ_k are irreducible unitary representations on $\mathcal{H}_{\phi_j}, \mathcal{H}_{\psi_k}$, respectively. Now $n = m$, since $\phi \sim \psi$. Moreover, we may arrange the indices so that $\phi_j \sim \psi_j$ for each j. Choose invertible $A_j \in \text{Hom}(\phi_j, \psi_j)$. Then A_j^* is invertible, and $A_j^* \in$

$\text{Hom}(\psi_j, \phi_j)$: if $x \in G$, $v \in \mathcal{H}_{\phi_j}$ and $w \in \mathcal{H}_{\psi_j}$ then

$$
\begin{aligned}
\langle A_j^* \psi_j(x) w, v \rangle_{\mathcal{H}_\phi} &= \langle w, \psi_j(x)^* A_j v \rangle_{\mathcal{H}_\psi} \\
&= \langle w, \psi_j(x^{-1}) A_j v \rangle_{\mathcal{H}_\psi} \\
&= \langle w, A_j \phi_j(x^{-1}) v \rangle_{\mathcal{H}_\psi} \\
&= \langle \phi_j(x^{-1})^* A_j^* w, v \rangle_{\mathcal{H}_\phi} \\
&= \langle \phi_j(x) A_j^* w, v \rangle_{\mathcal{H}_\phi}.
\end{aligned}
$$

Thereby $A_j^* A_j \in \text{Hom}(\phi_j, \phi_j)$ is invertible. By Schur's Lemma 6.3.25, $A_j^* A_j = \lambda_j I$, where $\lambda_j \neq 0$. Let $v \in \mathcal{H}_{\phi_j}$ such that $\|v\|_{\mathcal{H}_\phi} = 1$. Then

$$
\lambda = \lambda \|v\|_{\mathcal{H}_\phi}^2 = \langle \lambda v, v \rangle_{\mathcal{H}_\phi} = \langle A_j^* A_j v, v \rangle_{\mathcal{H}_\phi} = \langle A_j v, A_j v \rangle_{\mathcal{H}_\psi} = \|A_j v\|_{\mathcal{H}_\psi}^2 > 0,
$$

so that we may define $B_j := \lambda^{-1/2} A_j \in \text{Hom}(\phi_j, \psi_j)$. Then the mapping $B_j : \mathcal{H}_{\phi_j} \to \mathcal{H}_{\psi_j}$ is an isometry, $B_j^* B_j = I$. Finally, define

$$
B := \bigoplus_{j=1}^{m} B_j.
$$

Clearly, $B : \mathcal{H}_\phi \to \mathcal{H}_\psi$ is an isometry, bijection, and $B \in \text{Hom}(\phi, \psi)$. $\square$

We have now dealt with groups in general. In the sequel, by specialising to certain classes of groups, we will obtain fruitful ground for further results in representation theory.

Chapter 7

Topological Groups

A topological group is a natural amalgam of topological spaces and groups: it is a Hausdorff space with continuous group operations. Topology adds a new flavour to representation theory. Especially interesting are compact groups, where group-invariant probability measures exist. Moreover, nice-enough functions on a compact group have Fourier series expansions, which generalise the classical Fourier series of periodic functions.

7.1   Topological groups

Next we marry topology to groups.

Definition 7.1.1 (Topological groups). A group and a topological space G is called a *topological group* if $\{e\} \subset G$ is closed and if the group operations

$$((x, y) \mapsto xy) \quad : \quad G \times G \to G,$$
$$(x \mapsto x^{-1}) \quad : \quad G \to G$$

are continuous.

Remark 7.1.2. The reader may wonder why we assumed that $\{e\} \subset G$ is closed – actually, this condition is left out in some other definitions for a topological group. Notice that the good property brought by this assumption is that the topological groups become even Hausdorff spaces (see Exercise 7.1.3), which appeals to those who work in analysis.

Example. In the following, when not specified, the topologies and the group operations are the usual ones:

1. Any group G endowed with the so-called discrete topology $\mathcal{P}(G) = \{U : U \subset G\}$ is a topological group.

2. $\mathbb{Z}, \mathbb{Q}, \mathbb{R}$ and $\mathbb{C}$ are topological groups when the group operation is the addition and the topology is as usual.

3. $\mathbb{Q}^{\times}$, $\mathbb{R}^{\times}$, $\mathbb{C}^{\times}$ are topological groups when the group operation is the multiplication and the topology is as usual.

4. Topological vector spaces are topological groups with vector addition: such a space is both a vector space and a topological Hausdorff space such that the vector space operations are continuous.

5. Let X be a Banach space. The set $\mathrm{AUT}(X) := \mathrm{Aut}(X) \cap \mathcal{L}(X)$ of invertible bounded linear operators $X \to X$ forms a topological group with respect to the norm topology.

6. Subgroups of topological groups are topological groups.

7. If G and H are topological groups then $G \times H$ is a topological group. Actually, Cartesian products always preserve the topological group structure.

Exercise 7.1.3. Show that a topological group is a Hausdorff space.

Lemma 7.1.4. *Let G be a topological group and $y \in G$. Then*

$$x \mapsto xy, \quad x \mapsto yx, \quad x \mapsto x^{-1}$$

are homeomorphisms $G \to G$.

Proof. The mapping

$$(x \mapsto xy) : G \overset{x \mapsto (x,y)}{\to} G \times G \overset{(a,b) \mapsto ab}{\to} G$$

is continuous as a composition of continuous mappings. Its inverse mapping $(x \mapsto xy^{-1}) : G \to G$ is also continuous; hence this is a homeomorphism. Similarly, $(x \mapsto yx) : G \to G$ is a homeomorphism. By definition, the group operation $(x \mapsto x^{-1}) : G \to G$ is continuous, and it is its own inverse. $\square$

Corollary 7.1.5. *If $U \subset G$ is open and $S \subset G$ then $SU, US, U^{-1} \subset G$ are open.*

Proposition 7.1.6. *Let G be a topological group. If $H < G$ then $\overline{H} < G$. If $H \lhd G$ then $\overline{H} \lhd G$.*

Proof. Let $H < G$. Trivially $e \in H \subset \overline{H}$. Now

$$\overline{H}\,\overline{H} \subset \overline{HH} = \overline{H},$$

where the inclusion is due to the continuity of the mapping $((x, y) \mapsto xy) : G \times G \to G$. The continuity of the inversion $(x \mapsto x^{-1}) : G \to G$ gives

$$\overline{H}^{-1} \subset \overline{H^{-1}} = \overline{H}.$$

Thus $\overline{H} < G$.

Let $H \lhd G$, $y \in G$. Then

$$y\overline{H} = \overline{yH} = \overline{Hy} = \overline{H}y;$$

notice how homeomorphisms $(x \mapsto yx), (x \mapsto xy) : G \to G$ were used. $\square$

Remark 7.1.7. Let $H < G$ and $S \subset G$. For analysis on the quotient space G/H, let us recall Remark 6.2.15: the mapping $(x \mapsto xH) : G \to G/H$ identifies the sets

$$SH = \{sh : s \in S, \ h \in H\} \quad \subset \quad G,$$
$$\{sH : s \in S\} = \{\{sh : h \in H\} : s \in S\} \quad \subset \quad G/H.$$

Definition 7.1.8 (Quotient topology on G/H). Let G be a topological group, $H < G$. The *quotient topology* of G/H is

$$\tau_{G/H} := \{\{uH : u \in U\} : U \subset G \text{ open}\} ;$$

in other words, $\tau_{G/H}$ is the strongest (i.e., largest) topology for which the quotient map $(x \mapsto xH) : G \to G/H$ is continuous. If $U \subset G$ is open, we may identify sets $UH \subset G$ and $\{uH : u \in H\} \subset G/H$.

Proposition 7.1.9. *Let G be a topological group and $H < G$. Then a function $f : G/H \to \mathbb{C}$ is continuous if and only if $(x \mapsto f(xH)) : G \to \mathbb{C}$ is continuous.*

Proof. If $f \in C(G/H)$ then $(x \mapsto f(xH)) \in C(G)$, since it is obtained by composing f and the continuous quotient map $(x \mapsto xH) : G \to G/H$.

Now suppose $(x \mapsto f(xH)) \in C(G)$. Take open $V \subset \mathbb{C}$. Then $U := (x \mapsto f(xH))^{-1}(V) \subset G$ is open, so that $U' := \{uH : u \in U\} \subset G/H$ is open. Trivially, $f(U') = V$. Hence $f \in C(G/H)$. $\qquad\square$

Proposition 7.1.10 (When is G/H Hausdorff?). *Let G be a topological group and $H < G$. Then G/H is a Hausdorff space if and only if H is closed.*

Proof. If G/H is a Hausdorff space then $H = (x \mapsto xH)^{-1}(\{H\}) \subset G$ is closed, because the quotient map is continuous and $\{H\} \subset G/H$ is closed.

Next suppose H is closed. Take $xH, yH \in G/H$ such that $xH \neq yH$. Then $S := ((a,b) \mapsto a^{-1}b)^{-1}(H) \subset G \times G$ is closed, since $H \subset G$ is closed and $((a,b) \mapsto a^{-1}b) : G \times G \to G$ is continuous. Now $(x,y) \notin S$. Take open sets $U \ni x$ and $V \ni y$ such that $(U \times V) \cap S = \emptyset$. Then the sets

$$
\begin{aligned}
U' &:= \{uH : u \in U\} &\subset& \quad G/H, \\
V' &:= \{vH : v \in V\} &\subset& \quad G/H
\end{aligned}
$$

are disjoint and open such that $xH \in U'$ and $yH \in V'$. Thus G/H is Hausdorff. $\qquad\square$

Theorem 7.1.11 (When is G/H a topological group?). *Let G be a topological group and $H \lhd G$. Then*

$$
\begin{aligned}
((xH, yH) \mapsto xyH) &: (G/H) \times (G/H) \to G/H, \\
(xH \mapsto x^{-1}H) &: G/H \to G/H
\end{aligned}
$$

are continuous. Moreover, G/H is a topological group if and only if H is closed.

Proof. We know already that the operations in the theorem are well-defined group operations, because H is normal in G. Recall Remark 7.1.7, how we may identify certain subsets of G with subsets of G/H. Then a neighbourhood of the point $xyH \in G/H$ is of the form UH for some open $U \subset G$, $U \ni xy$. Take open $U_1 \ni x$ and $U_2 \ni y$ such that $U_1 U_2 \subset U$. Then

$$(xH)(yH) \subset (U_1 H)(U_2 H) = U_1 U_2 H \subset UH,$$

so that $((xH, yH) \mapsto xyH) : (G/H) \times (G/H) \to G/H$ is continuous. A neighbourhood of the point $x^{-1}H \in G/H$ is of the form VH for some open $V \subset G$, $V \ni x^{-1}$. But $V^{-1} \ni x$ is open, and $(V^{-1})^{-1} = V$, so that $(xH \mapsto x^{-1}H) : G/H \to G/H$ is continuous.

Notice that $e_{G/H} = H$. If G/H is a topological group, then

$$H = (x \mapsto xH)^{-1}\{e_{G/H}\} \subset G$$

is closed. On the other hand, if $H \lhd G$ is closed then

$$(G/H) \setminus \{e_{G/H}\} \cong (G \setminus H)H \subset G$$

is open, i.e., $\{e_{G/H}\} \subset G/H$ is closed. $\square$

Definition 7.1.12 (Continuous homomorphisms). Let G_1, G_2 be topological groups. Let

$$\mathrm{HOM}(G_1, G_2) := \mathrm{Hom}(G_1, G_2) \cap C(G_1, G_2),$$

i.e., the set of continuous homomorphisms $G_1 \to G_2$.

Remark 7.1.13. By Theorem 7.1.11, closed normal subgroups of G correspond bijectively to continuous surjective homomorphisms from G to some other topological group (up to isomorphism).

Remark 7.1.14. Let us recall some topological concepts: A topological space is *connected* if the only subsets which are both closed and open are the empty set and the whole space. A non-connected space is called *disconnected*. The *component* of a point x in a topological space is the largest connected subset containing x.

Proposition 7.1.15. *Let G be a topological group and $C_e \subset G$ the component of e. Then $C_e \lhd G$ is closed.*

Proof. Components are always closed, and $e \in C_e$ by definition. Since $C_e \subset G$ is connected, also $C_e \times C_e \subset G \times G$ and is connected. By the continuity of the group operations, $C_e C_e \subset G$ and $C_e^{-1} \subset G$ are connected. Since $e = ee \in C_e C_e$, we have $C_e C_e \subset C_e$. And since $e = e^{-1} \in C_e^{-1}$, also $C_e^{-1} \subset C_e$. Take $y \in G$. Then $y^{-1}C_e y \subset G$ is connected, by the continuity of $(x \mapsto y^{-1}xy) : G \to G$. Now $e = y^{-1}ey \in y^{-1}C_e y$, so that $y^{-1}C_e y \subset C_e$; C_e is normal in G. $\square$

Proposition 7.1.16. *Let $((x, p) \mapsto x \cdot p) : G \times M \to M$ be a continuous action of G on M, and let $q \in M$. If G_q and G/G_q are connected then G is connected.*

Proof. Suppose G is disconnected and G_q is connected. Then there are non-empty disjoint open sets $U, V \subset G$ such that $G = U \cup V$. The sets

$$U' := \{uG_q : u \in U\} \subset G/G_q,$$
$$V' := \{vG_q : v \in V\} \subset G/G_q$$

are non-empty and open, and $G/G_q = U' \cup V'$. Take $u \in U$ and $v \in V$. As a continuous image of a connected set, $uG_q = (x \mapsto ux)(G_q) \subset G$ is connected; moreover $u = ue \in uG_q$; thereby $uG_q \subset U$. In the same way we see that $vG_q \subset V$. Hence $U' \cap V' = \emptyset$, so that G/G_q is disconnected. $\qquad\square$

Corollary 7.1.17 (When is a group connected?). *If G is a topological group, $H < G$ is connected and G/H is connected then G is connected.*

Proof. Using the notation of Proposition 7.1.16, let $M = G/H$, $q = H$ and $x \cdot p = xp$, so that $G_q = H$ and $G/G_q = G/H$. $\qquad\square$

Exercise 7.1.18 (Groups SO(n), SU(n) and U(n) are connected). Show that SO(n), SU(n) and U(n) are connected for every $n \in \mathbb{Z}^+$. How about O(n)?

Exercise 7.1.19 (Finiteness of connected components). Prove that a compact topological group can have only finitely many connected components. Consequently, conclude that a discrete compact group is finite.

7.2 Representations of topological groups

Definition 7.2.1 (Strongly continuous representations). Let G be a topological group and $\mathcal{H}$ be a Hilbert space. A representation $\phi \in \mathrm{Hom}(G, \mathcal{U}(\mathcal{H}))$ is *strongly continuous* if

$$(x \mapsto \phi(x)v) : G \to \mathcal{H}$$

is continuous for all $v \in \mathcal{H}$.

Remark 7.2.2. The strong continuity in Definition 7.2.1 means that the mapping $(x \mapsto \phi(x)) : G \to \mathcal{L}(\mathcal{H})$ is continuous, when $\mathcal{L}(\mathcal{H}) \supset \mathcal{U}(\mathcal{H})$ is endowed with the *strong operator topology*:

$$A_j \overset{\text{strongly}}{\to} A \overset{\text{definition}}{\Longleftrightarrow} \forall v \in \mathcal{H} : \|A_j v - Av\|_{\mathcal{H}} \to 0.$$

Why should we not endow $\mathcal{U}(\mathcal{H})$ with the operator norm topology (which is even stronger, i.e., a larger topology)? The reason is that there are interesting unitary representations, which are continuous in the strong operator topology, but not in the operator norm topology. This phenomenon is exemplified by Exercise 7.2.3:

Exercise 7.2.3. Let us define $\pi_L : \mathbb{R}^n \to \mathcal{U}(L^2(\mathbb{R}^n))$ by

$$(\pi_L(y)f)(x) := f(x - y)$$

for almost every $x \in \mathbb{R}^n$. Show that π_L is strongly continuous, but not norm continuous.

Definition 7.2.4 (Topologically irreducible representations). A strongly continuous $\phi \in \mathrm{Hom}(G, \mathcal{U}(\mathcal{H}))$ is called *topologically irreducible* if the only closed ϕ-invariant subspaces are the trivial ones $\{0\}$ and $\mathcal{H}$.

Exercise 7.2.5. Let V be a topological vector space and let $W \subset V$ be an A-invariant subspace, where $A \in \mathrm{Aut}(V)$ is continuous. Show that the closure $\overline{W} \subset V$ is also A-invariant.

Definition 7.2.6 (Cyclic representations and cyclic vectors). A strongly continuous $\phi \in \mathrm{Hom}(G, \mathcal{U}(\mathcal{H}))$ is called a *cyclic representation* if

$$\mathrm{span}\ \phi(G)v \subset \mathcal{H}$$

is dense for some $v \in \mathcal{H}$; such v is called a *cyclic vector*.

Example. If $\phi \in \mathrm{Hom}(G, \mathcal{U}(\mathcal{H}))$ is topologically irreducible then any non-zero $v \in \mathcal{H}$ is cyclic. Indeed, if $V := \mathrm{span}\ \phi(G)v$ then $\phi(G)V \subset V$ and consequently $\phi(G)\overline{V} \subset \overline{V}$, so that $\overline{V}$ is ϕ-invariant. If $v \neq 0$ then $\overline{V} = \mathcal{H}$, because of the topological irreducibility.

Definition 7.2.7 (Representation as a direct sum). A Hilbert space $\mathcal{H}$ is a *direct sum* of closed subspaces $(\mathcal{H}_j)_{j \in J}$, denoted by

$$\mathcal{H} = \bigoplus_{j \in J} \mathcal{H}_j$$

if the subspace family is pairwise orthogonal and the linear span of the set $\cup_{j \in J} \mathcal{H}_j$ is dense in $\mathcal{H}$. Then the vectors in $\mathcal{H}$ have a unique orthogonal series expansion, more precisely

$$\forall x \in \mathcal{H}\ \forall j \in J\ \exists! x_j \in \mathcal{H}_j :\ x = \sum_{j \in J} x_j, \quad \|x\|_{\mathcal{H}}^2 = \sum_{j \in J} \|x_j\|_{\mathcal{H}}^2.$$

If $\phi \in \mathrm{Hom}(G, \mathcal{U}(\mathcal{H}))$ and each $\mathcal{H}_j$ is ϕ-invariant then ϕ is said to be the *direct sum*

$$\phi = \bigoplus_{j \in J} \phi|_{\mathcal{H}_j}$$

where $\phi|_{\mathcal{H}_j} = (x \mapsto \phi(x)v) \in \mathrm{Hom}(G, \mathcal{U}(\mathcal{H}_j))$.

Proposition 7.2.8 (Decomposition of strongly continuous representations). *Let $\phi \in \mathrm{Hom}(G, \mathcal{U}(\mathcal{H}))$ be strongly continuous. Then*

$$\phi = \bigoplus_{j \in J} \phi|_{\mathcal{H}_j},$$

where each $\phi|_{\mathcal{H}_j}$ is cyclic.

Proof. Let $\tilde{J}$ be the family of all closed ϕ-invariant subspaces $V \subset \mathcal{H}$ for which $\phi|_V$ is cyclic. Let

$$S = \left\{ s \subset \tilde{J} \ \middle| \ \forall V, W \in s : \ V = W \text{ or } V \perp W \right\}.$$

It is easy to see that $\{\{0\}\} \in S$, so that $S \neq \emptyset$. Let us introduce a partial order on S by inclusion:

$$s_1 \leq s_2 \overset{\text{definition}}{\iff} s_1 \subset s_2.$$

The chains in S have upper bounds: if $R \subset S$ is a chain then $r \leq \bigcup_{s \in R} s \in S$ for all $r \in R$. Therefore by **Zorn's Lemma**, there exists a maximal element $t \in S$. Let

$$V := \bigoplus_{W \in t} W.$$

To get a contradiction, suppose $V \neq \mathcal{H}$. Then there exists $v \in V^\perp \setminus \{0\}$. Since $\text{span}(\phi(G)v)$ is ϕ-invariant, its closure W_0 is also ϕ-invariant (see Exercise 7.2.5). Clearly $W_0 \subset \overline{V^\perp} = V^\perp$, and $\phi|_{W_0}$ has cyclic vector v, yielding

$$s := t \cup \{W_0\} \in S,$$

where $t \leq s \not\leq t$. This contradicts the maximality of t; thus $V = \mathcal{H}$. $\qquad \square$

Exercise 7.2.9. Fill in the details in the proof of Proposition 7.2.8.

Exercise 7.2.10. Assuming that $\mathcal{H}$ is separable, prove Proposition 7.2.8 by ordinary induction (without resorting to Zorn's Lemma).

7.3 Compact groups

Definition 7.3.1 ((Locally) compact groups). A topological group is a *(locally) compact group* if it is (locally) compact as a topological space.

Remark 7.3.2. We have the following properties:

1. Any group G with the discrete topology is a locally compact group; then G is a compact group if and only if it is finite.

2. $\mathbb{Q}, \mathbb{Q}^\times$ are not locally compact groups;
   $\mathbb{R}, \mathbb{R}^\times, \mathbb{C}, \mathbb{C}^\times$ are locally compact groups, but non-compact.

3. A normed vector space is a locally compact group if and only if it is finite-dimensional.

4. $O(n), SO(n), U(n), SU(n)$ are compact groups.

5. $GL(n)$ is a locally compact group, but non-compact.

6. If G, H are locally compact groups then $G \times H$ is a locally compact group.

7. If $\{G_j\}_{j \in J}$ is a family of compact groups then $\prod_{j \in J} G_j$ is a compact group.

8. If G is a compact group and $H < G$ is closed then H is a compact group.

Proposition 7.3.3. *Let* $((x,p) \mapsto x \cdot p) : G \times M \to M$ *be a continuous action of a compact group* G *on a Hausdorff space* M. *Let* $q \in M$. *Then the mapping*

$$f := (xG_q \mapsto x \cdot q) : G/G_q \to G \cdot q$$

is a homeomorphism.

Proof. We already know that f is a well-defined bijection. We need to show that f is continuous. An open subset of $G \cdot q$ is of the form $V \cap (G \cdot q)$, where $V \subset M$ is open. Since the action is continuous, also $(x \mapsto x \cdot q) : G \to M$ is continuous, so that $U := (x \mapsto x \cdot q)^{-1}(V) \subset G$ is open. Thereby

$$f^{-1}(V \cap (G \cdot q)) = \{xG_q : x \in U\} \subset G/G_q$$

is open. Thus f is continuous. The space G is compact and the quotient map $(x \mapsto xG_q) : G \to G/G_q$ is continuous and surjective, so that G/G_q is compact. From general topology we know that a continuous bijection from a compact space to a Hausdorff space is a homeomorphism (see Proposition A.12.7). □

Corollary 7.3.4. *If* G *is compact,* $\phi \in \mathrm{HOM}(G, H)$ *and* $K = \mathrm{Ker}(\phi)$ *then*

$$\psi := (xK \mapsto \phi(x)) \in \mathrm{HOM}(G/K, \phi(G))$$

is a homeomorphism.

Proof. Using the notation of Proposition 7.3.3, we have $M = H$, $q = e_H$, $x \cdot p = \phi(x)p$, so that $G_q = K$, $G/G_q = G/K$, $G \cdot q = \phi(G)$, $\psi = f$. □

Remark 7.3.5. What could happen if we drop the compactness assumption in Corollary 7.3.4? If G and H are Banach spaces, $\phi \in \mathcal{L}(G, H)$ is compact and $\dim(\phi(G)) = \infty$ then $\psi = (x + \mathrm{Ker}(\phi) \mapsto \phi(x)) : G/\mathrm{Ker}(\phi) \to \phi(G)$ is a bounded linear bijection, but ψ^{-1} is not bounded! But if $\phi \in \mathcal{L}(G, H)$ is a bijection then ϕ^{-1} is bounded by the **Open Mapping Theorem**! (Theorem B.4.31)

Definition 7.3.6 (Uniform continuity on a topological group). Let G be a topological group. A function $f : G \to \mathbb{C}$ is *uniformly continuous* if for every $\varepsilon > 0$ there exists open $U \ni e$ such that

$$\forall x, y \in G : \quad x^{-1}y \in U \implies |f(x) - f(y)| < \varepsilon.$$

Exercise 7.3.7. Under what circumstances is a polynomial $p : \mathbb{R} \to \mathbb{C}$ uniformly continuous? Show that if a continuous function $f : \mathbb{R} \to \mathbb{C}$ is periodic or vanishes outside a bounded set then it is uniformly continuous.

Theorem 7.3.8. *If* G *is a compact group and* $f \in C(G)$ *then* f *is uniformly continuous.*

Proof. Take $\varepsilon > 0$. Define the open disk $\mathbb{D}(z,r) := \{w \in \mathbb{C} : |w - z| < r\}$, where $z \in \mathbb{C}$, $r > 0$. Since f is continuous, the set

$$V_x := f^{-1}(\mathbb{D}(f(x), \varepsilon)) \ni x$$

is open. Then $x^{-1}V_x \ni ee = e$ is open, so that there exist open sets $U_{1,x}, U_{2,x} \ni e$ such that $U_{1,x}U_{2,x} \subset x^{-1}V_x$, by the continuity of the group multiplication. Define $U_x := U_{1,x} \cap U_{2,x}$. Since $\{xU_x : x \in G\}$ is an open cover of the compact space G, there is a finite subcover $\{x_jU_{x_j}\}_{j=1}^n$. Now the set

$$U := \bigcap_{j=1}^n U_{x_j} \ni e$$

is open. Suppose $x, y \in G$ such that $x^{-1}y \in U$. There exists $k \in \{1, \ldots, n\}$ such that $x \in x_kU_{x_k}$, so that

$$x, y \in xU \subset x_kU_{x_k}U_{x_k} \subset x_kx_k^{-1}V_{x_k} = V_{x_k},$$

yielding

$$|f(x) - f(y)| \le |f(x) - f(x_k)| + |f(x_k) - f(y)| < 2\varepsilon. \qquad \square$$

Exercise 7.3.9. Let G be a compact group, $x \in G$ and $A = \{x^n\}_{n=1}^\infty$. Show that $\overline{A} < G$.

7.4　Haar measure and integral

On a group, it would be natural to integrate with respect to measures that are invariant under the group operations: consider, e.g., the Lebesgue integral on $\mathbb{R}^n$. However, it is not obvious whether there exist such invariant integrals in general. Next we will show that on a compact group there exists a unique probability functional, which corresponds to the so-called *Haar measure*.

Definition 7.4.1 (Positive functionals). Let X be a compact Hausdorff space and $\mathbb{K} \in \{\mathbb{R}, \mathbb{C}\}$. Then $C(X, \mathbb{K})$ is a Banach space over $\mathbb{K}$ with the norm

$$f \mapsto \|f\|_{C(X,\mathbb{K})} := \max_{x \in X} |f(x)|.$$

Its dual $C(X, \mathbb{K})' = \mathcal{L}(C(X, \mathbb{K}), \mathbb{K})$ consists of the bounded linear functionals $C(X, \mathbb{K}) \to \mathbb{K}$, and is endowed with the Banach space norm

$$L \mapsto \|L\|_{C(X,\mathbb{K})'} := \sup_{f \in C(X,\mathbb{K}):\ \|f\|_{C(X,\mathbb{K})} \le 1} |Lf|.$$

A functional $L : C(X, \mathbb{K}) \to \mathbb{C}$ is called *positive* if $Lf \ge 0$ whenever $f \ge 0$.

Exercise 7.4.2. Let X be a compact Hausdorff space. Show that a positive linear functional $L : C(X, \mathbb{R}) \to \mathbb{R}$ is bounded.

By the Riesz Representation Theorem (see Theorem C.4.65), if $L \in C(X, \mathbb{K})'$ is positive then there exists a unique positive Borel regular measure μ on X such that

$$Lf = \int_X f \, d\mu$$

for every $f \in C(X, \mathbb{K})$; moreover, $\mu(X) = \|L\|_{C(X,\mathbb{K})'}$. For short, $C(X) := C(X, \mathbb{C})$. Note that this is different from Chapter A (see, e.g., Exercise A.6.6) where we wrote $C(X)$ for $C(X, \mathbb{R})$.

In the sequel, we shall construct a unique positive normalised translation-invariant measure on G. More precisely, we shall prove the following result:

Theorem 7.4.3 (Haar functional). *Let G be a compact group. There exists a unique positive linear functional* $\mathrm{Haar} \in C(G)'$ *such that*

$$\begin{aligned}
\mathrm{Haar}(f) &= \mathrm{Haar}(x \mapsto f(yx)), \\
\mathrm{Haar}(\mathbf{1}) &= 1,
\end{aligned}$$

for all $y \in G$, where $\mathbf{1} = (x \mapsto 1) \in C(G)$. Moreover, this Haar *functional satisfies*

$$\begin{aligned}
\mathrm{Haar}(f) &= \mathrm{Haar}(x \mapsto f(xy)) \\
&= \mathrm{Haar}(x \mapsto f(x^{-1})).
\end{aligned}$$

Remark 7.4.4 **(Haar measure and integral).** By the Riesz Representation Theorem (see Theorem C.4.65), the Haar functional begets a unique Borel regular probability measure μ_G such that

$$\mathrm{Haar}(f) = \int_G f \, d\mu_G = \int_G f(x) \, d\mu_G(x).$$

This μ_G is called the *Haar measure* of G. Often the Haar measure is implicitly assumed, and we may write, e.g.,

$$\int_G f(x) \, dx := \int_G f \, d\mu_G.$$

Obviously,

$$\int_G \mathbf{1} \, d\mu_G = \mu_G(G) = 1,$$

$$\int_G f(x) \, dx = \int_G f(yx) \, dx$$

$$= \int_G f(xy) \, dx = \int_G f(x^{-1}) \, dx.$$

Thus the *Haar integral* $\mathrm{Haar}(f) = \int_G f(x) \, dx$ can be thought of as the most natural average of $f \in C(G)$. In practical applications we can know usually only finitely many values of f, i.e., we are able to take only samples $\{f(x) : x \in S\}$ for a finite set $S \subset G$. Then a natural idea for approximating $\mathrm{Haar}(f)$ would be computing the finite sum

$$\sum_{x \in S} f(x)\, \alpha(x),$$

where sampling weights $\alpha(x) \geq 0$ satisfy $\sum_{x \in S} \alpha(x) = 1$. Of course, such a sum is not usually invariant under the group operations. The problem is to find clever choices for sampling sets and weights, some sort of almost uniformly distributed unit mass on G is needed; for this end we shall introduce convolutions.

Example. If G is finite then

$$\int_G f \, \mathrm{d}\mu_G = \frac{1}{|G|} \sum_{x \in G} f(x).$$

Example (**Haar measure on $\mathbb{T}^n$**). For $\mathbb{T}^n = \mathbb{R}^n / \mathbb{Z}^n$,

$$\int_{\mathbb{T}^n} f \, \mathrm{d}\mu_{\mathbb{T}^n} = \int_{[0,1)^n} f(x + \mathbb{Z}^n) \, \mathrm{d}x,$$

i.e., integration with respect to the Lebesgue measure on $[0,1)^n$.

What follows is preparation for the proof of Theorem 7.4.3.

Definition 7.4.5 (Sampling measures). Let G be a compact group. A function $\alpha : G \to [0,1]$ is a *sampling measure* on G, $\alpha \in \mathcal{SM}_G$, if

$$\mathrm{supp}(\alpha) := \mathrm{cl}\,\{a \in G : \alpha(a) \neq 0\} \quad \text{is finite and} \quad \sum_{a \in G} \alpha(a) = 1.$$

The set $\mathrm{supp}(\alpha) \subset G$ is called the *support* of α. Since $\mathrm{supp}(\alpha)$ is finite we also have $\mathrm{supp}(\alpha) = \{a \in G : \alpha(a) \neq 0\}$ and, therefore, a sampling measure $\alpha \in \mathcal{SM}_G$ can be regarded as a finitely supported probability measure on G, satisfying

$$\int_G f \, \mathrm{d}\alpha = \check{\alpha} * f(e) = f * \check{\alpha}(e),$$

where $\check{\alpha}(a) := \alpha(a^{-1})$.

Remark 7.4.6. A sampling measure is nothing else but

$$\alpha = \sum_j \alpha_j \delta_{a_j},$$

where the sum is finite, $a_j \in G$, δ_{a_j} is the Dirac measure at a_j (i.e., a probability measure supported at a_j), and $\sum_j \alpha_j = 1$.

Definition 7.4.7 (Convolutions). Let $\alpha, \beta \in \mathcal{SM}_G$ and $f \in C(G, \mathbb{K})$. The *convolutions*

$$\alpha * \beta, \alpha * f, f * \beta : G \to \mathbb{K}$$

are defined by

$$\alpha * \beta(b) = \sum_{a \in G} \alpha(a)\beta(a^{-1}b),$$

$$\alpha * f(x) = \sum_{a \in G} \alpha(a)f(a^{-1}x),$$

$$f * \beta(x) = \sum_{b \in G} f(xb^{-1})\beta(b).$$

Notice that these summations are finite, as the sampling measures are supported on finite sets.

Definition 7.4.8 (Semigroups and monoids). A *semigroup* is a non-empty set S with an operation $((r, s) \mapsto rs) : S \times S \to S$ satisfying $r(st) = (rs)t$ for all $r, s, t \in S$. A semigroup is *commutative* if $rs = sr$ for all $r, s \in S$. Moreover, if there exists $e \in S$ such that $es = se = s$ for all $s \in S$ then S is called a *monoid*.

Example. $\mathbb{Z}^+ = \{n \in \mathbb{Z} : n > 0\}$ is a commutative monoid with respect to multiplication, and a commutative semigroup with respect to addition. If V is a vector space then $(\mathrm{End}(V), (A, B) \mapsto AB)$ is a monoid with $e = I$.

Lemma 7.4.9. *The structure $(\mathcal{SM}_G, (\alpha, \beta) \mapsto \alpha * \beta)$ is a monoid.*

Exercise 7.4.10. Prove Lemma 7.4.9. How is $\mathrm{supp}(\alpha * \beta)$ related to $\mathrm{supp}(\alpha)$ and $\mathrm{supp}(\beta)$? In which case is $\mathcal{SM}_G$ is a group? Show that $\mathcal{SM}_G$ is commutative if and only if G is commutative.

Lemma 7.4.11. *If $\alpha \in \mathcal{SM}_G$ then $(f \mapsto \alpha * f), (f \mapsto f * \alpha) \in \mathcal{L}(C(G, \mathbb{K}))$ and*

$$\|\alpha * f\|_{C(G, \mathbb{K})} \leq \|f\|_{C(G, \mathbb{K})},$$
$$\|f * \alpha\|_{C(G, \mathbb{K})} \leq \|f\|_{C(G, \mathbb{K})}.$$

Moreover, $\alpha * 1 = 1 = 1 * \alpha$.

Proof. Trivially, $\alpha * 1 = 1$. Because $(x \mapsto a^{-1}x) : G \to G$ is a homeomorphism and the summation is finite, $\alpha * f \in C(G, \mathbb{K})$. Linearity of $f \mapsto \alpha * f$ is clear. Next,

$$|\alpha * f(x)| \leq \sum_{a \in G} \alpha(a)|f(a^{-1}x)| \leq \sum_{a \in G} \alpha(a)\|f\|_{C(G, \mathbb{K})} = \|f\|_{C(G, \mathbb{K})}.$$

Similar conclusions hold also for $f * \alpha$. □

Definition 7.4.12. Let G be a compact group. Let us define a mapping $p_G : C(G, \mathbb{R}) \to \mathbb{R}$ by

$$p_G(f) := \max(f) - \min(f).$$

Lemma 7.4.13. *If $f \in C(G, \mathbb{R})$ and $\alpha \in \mathcal{SM}_G$ then*

$$\min(f) \leq \min(\alpha * f) \leq \max(\alpha * f) \leq \max(f),$$

$$\min(f) \leq \min(f * \alpha) \leq \max(f * \alpha) \leq \max(f),$$

so that

$$p_G(\alpha * f) \leq p_G(f), \quad p_G(f * \alpha) \leq p_G(f).$$

Proof. Now

$$\min(f) = \sum_{a \in G} \alpha(a) \; \min(f) \leq \min_{x \in G} \sum_{a \in G} \alpha(a) f(a^{-1}x) = \min(\alpha * f),$$

$$\max(\alpha * f) = \max_{x \in G} \sum_{a \in G} \alpha(a) f(a^{-1}x) \leq \sum_{a \in G} \alpha(a) \; \max(f) = \max(f),$$

and clearly $\min(\alpha * f) \leq \max(\alpha * f)$. The proof for $f * \alpha$ is symmetric. $\qquad\square$

Exercise 7.4.14. Show that p_G is a bounded seminorm on $C(G, \mathbb{R})$.

Proposition 7.4.15. *Let $f \in C(G, \mathbb{R})$. For every $\varepsilon > 0$ there exist $\alpha, \beta \in \mathcal{SM}_G$ such that*

$$p_G(\alpha * f) < \varepsilon, \quad p_G(f * \beta) < \varepsilon.$$

Remark 7.4.16. This is the decisive stage in the construction of the Haar measure. The idea is that for a non-constant $f \in C(G)$ we can find sampling measures α, β that tame the oscillations of f so that $\alpha * f$ and $f * \beta$ are almost constant functions. It will turn out that there exists a unique constant function $\mathrm{Haar}(f)\mathbf{1}$ approximated by the convolutions of the type $\alpha * f$ and $f * \beta$. In the sequel, notice how compactness is exploited!

Proof. Let $\varepsilon > 0$. By Theorem 7.3.8, a continuous function on a **compact group** is uniformly continuous. Thus there exists an open set $U \supset e$ such that $|f(x) - f(y)| < \varepsilon$, when $x^{-1}y \in U$. We notice easily that if $\gamma \in \mathcal{SM}_G$ then also $|\gamma * f(x) - \gamma * f(y)| < \varepsilon$, when $x^{-1}y \in U$:

$$
\begin{aligned}
|\gamma * f(x) - \gamma * f(y)| &= \left| \sum_{a \in G} \gamma(a) \left(f(a^{-1}x) - f(a^{-1}y) \right) \right| \\
&\leq \sum_{a \in G} \gamma(a) \left| f(a^{-1}x) - f(a^{-1}y) \right| \\
&< \sum_{a \in G} \gamma(a) \, \varepsilon \\
&= \varepsilon.
\end{aligned}
$$

Now $\{xU : \ x \in G\}$ is an open cover of the **compact** space G, hence having a finite subcover $\{x_jU\}_{j=1}^n$. The set $S := \{x_ix_j^{-1} : \ 1 \leq i, j \leq n\}$ has $|S| \leq n^2$ elements. Define $\gamma_1 \in \mathcal{SM}_G$ by

$$\gamma_1(a) = \begin{cases} |S|^{-1}, & \text{when } a \in S, \\ 0, & \text{otherwise.} \end{cases}$$

Let $\gamma_{k+1} := \gamma_k * \gamma_1 \in \mathcal{SM}_G$. Then

$$
\begin{aligned}
p_G(\gamma_{k+1} * f) &= \max(\gamma_{k+1} * f) - \min(\gamma_{k+1} * f) \\
&\leq \max(\gamma_{k+1} * f) - \min(\gamma_k * f) \\
&= \frac{1}{|S|} \max_{x \in G} \sum_{a \in S} \gamma_k * f(a^{-1}x) - \min(\gamma_k * f) \\
&\overset{(\star)}{<} \frac{1}{|S|} \left[((|S| - 1)\max(\gamma_k * f) + [\min(\gamma_k * f) + \varepsilon] \right] - \min(\gamma_k * f) \\
&= \frac{|S| - 1}{|S|} p_G(\gamma_k * f) + \frac{1}{|S|}\varepsilon,
\end{aligned}
$$

where the last inequality $(\star)$ was obtained by estimating $|S| - 1$ terms in the sum trivially, and finally the remaining term was estimated by recalling the uniform continuity of $\gamma_k * f$. Notice that $(p_G(\gamma_k * f))_{k=1}^\infty \subset \mathbb{R}$ is a non-increasing sequence of non-negative numbers. Thus there exists the limit $\delta := \lim_{k \to \infty} p_G(\gamma_k * f) \geq 0$, and

$$\delta \leq \frac{|S| - 1}{|S|}\delta + \frac{1}{|S|}\varepsilon,$$

so that $\delta \leq \varepsilon$. Hence there exists k_0 such that, say, $p_G(\gamma_k * f) \leq 2\varepsilon$ for every $k \geq k_0$. This proves the claim. $\square$

Exercise 7.4.17. In the proof above, check the validity of inequality $(\star)$.

Definition 7.4.18. The following Corollary 7.4.19 defines the Haar functional Haar : $C(G, \mathbb{R}) \to \mathbb{R}$.

Corollary 7.4.19 (What is the Haar functional Haar(f)?). *For $f \in C(G, \mathbb{R})$ there exists a unique constant function* Haar$(f)\mathbf{1}$ *that belongs to the closure of the set*

$$\{\alpha * f : \ \alpha \in \mathcal{SM}_G\} \subset C(G, \mathbb{R}).$$

Moreover, Haar$(f)\mathbf{1}$ *is the unique constant function that belongs to the closure of the set*

$$\{f * \beta : \ \beta \in \mathcal{SM}_G\} \subset C(G, \mathbb{R}).$$

Proof. Pick any $\alpha_1 \in \mathcal{SM}_G$. Suppose we have chosen $\alpha_k \in \mathcal{SM}_G$. Let $\alpha_{k+1} := \gamma_k * \alpha_k$, where $\gamma_k \in \mathcal{SM}_G$ satisfies

$$p_G(\alpha_{k+1} * f) = p_G(\gamma_k * (\alpha_k * f)) < 2^{-k}.$$

Now

$$\min(\alpha_k * f) \leq \min(\alpha_{k+1} * f) \leq \max(\alpha_{k+1} * f) \leq \max(\alpha_k * f),$$

so that there exists

$$\lim_{k \to \infty} \min(\alpha_k * f) = \lim_{k \to \infty} \max(\alpha_k * f) =: c_1 \in \mathbb{R}.$$

In the same way we may construct a sequence $(\beta_k)_{k=1}^\infty \subset \mathcal{SM}_G$ such that

$$\lim_{k \to \infty} \min(f * \beta_k) = \lim_{k \to \infty} \max(f * \beta_k) =: c_2 \in \mathbb{R}.$$

But

$$
\begin{aligned}
|c_1 - c_2| &= \|c_1\mathbf{1} - c_2\mathbf{1}\|_{C(G,\mathbb{R})} \\
&= \|(c_1\mathbf{1} - \alpha_k * f) * \beta_k + \alpha_k * (f * \beta_k - c_2\mathbf{1})\|_{C(G,\mathbb{R})} \\
&\leq \|(c_1\mathbf{1} - \alpha_k * f) * \beta_k\|_{C(G,\mathbb{R})} + \|\alpha_k * (f * \beta_k - c_2\mathbf{1})\|_{C(G,\mathbb{R})} \\
&\leq \|c_1\mathbf{1} - \alpha_k * f\|_{C(G,\mathbb{R})} + \|f * \beta_k - c_2\mathbf{1}\|_{C(G,\mathbb{R})} \\
&\xrightarrow[k \to \infty]{} 0.
\end{aligned}
$$

Thus $c_1 = c_2 \in \mathbb{R}$ is unique, depending only on the function f. $\qquad\square$

Definition 7.4.20 (Haar functional of $f \in C(G,\mathbb{C})$). The *Haar functional* of $f \in C(G)$ is

$$\mathrm{Haar}(f) := \mathrm{Haar}(\mathrm{Re}(f)) + i\, \mathrm{Haar}(\mathrm{Im}(f)),$$

where $\mathrm{Re}(f), \mathrm{Im}(f)$ are the real and imaginary parts of f, respectively.

Let us now reformulate Theorem 7.4.3:

Theorem 7.4.21 (Haar). *The Haar functional* $\mathrm{Haar} : C(G) \to \mathbb{C}$ *on a compact group G is the unique positive linear functional satisfying*

$$
\begin{aligned}
\mathrm{Haar}(\mathbf{1}) &= 1, \\
\mathrm{Haar}(f) &= \mathrm{Haar}(x \mapsto f(yx)),
\end{aligned}
$$

for all $f \in C(G)$ and $y \in G$. Moreover,

$$\mathrm{Haar}(f) = \mathrm{Haar}(x \mapsto f(xy)) = \mathrm{Haar}(x \mapsto f(x^{-1})).$$

Proof. By Definition 7.4.20 of Haar, it is enough to deal with real-valued functions here. From the construction, it is clear that

$$
\begin{aligned}
f \geq 0 &\Rightarrow \mathrm{Haar}(f) \geq 0, \\
|\mathrm{Haar}(f)| &\leq \|f\|_{C(G)}, \\
\mathrm{Haar}(\lambda f) &= \lambda\, \mathrm{Haar}(f), \\
\mathrm{Haar}(\mathbf{1}) &= 1, \\
\mathrm{Haar}(f) = \mathrm{Haar}(x \mapsto f(yx)) &= \mathrm{Haar}(x \mapsto f(xy)).
\end{aligned}
$$

Choose $\alpha, \beta \in \mathcal{SM}_G$ such that

$$\|\alpha * f - \mathrm{Haar}(f)\mathbf{1}\|_{C(G)} < \varepsilon,$$
$$\|g * \beta - \mathrm{Haar}(g)\mathbf{1}\|_{C(G)} < \varepsilon.$$

Then

$$\|\alpha * (f + g) * \beta - (\mathrm{Haar}(f) + \mathrm{Haar}(g))\mathbf{1}\|_{C(G)}$$
$$= \|(\alpha * f - \mathrm{Haar}(f)\mathbf{1}) * \beta + \alpha * (g * \beta - \mathrm{Haar}(g)\mathbf{1})\|_{C(G)}$$
$$\leq \|(\alpha * f - \mathrm{Haar}(f)\mathbf{1}) * \beta\|_{C(G)} + \|\alpha * (g * \beta - \mathrm{Haar}(g)\mathbf{1})\|_{C(G)}$$
$$\leq \|\alpha * f - \mathrm{Haar}(f)\mathbf{1}\|_{C(G)} + \|g * \beta - \mathrm{Haar}(g)\mathbf{1}\|_{C(G)}$$
$$< 2\varepsilon,$$

so that $\mathrm{Haar}(f + g) = \mathrm{Haar}(f) + \mathrm{Haar}(g)$.

Suppose $L : C(G) \to \mathbb{C}$ is a positive linear functional such that $L(\mathbf{1}) = 1$ and $L(f) = L(x \mapsto f(yx))$ for all $f \in C(G)$ and $y \in G$. Let $f \in C(G)$, $\varepsilon > 0$ and $\alpha \in \mathcal{SM}_G$ be as above. Then

$$|L(f) - \mathrm{Haar}(f)| = |L(\alpha * f - \mathrm{Haar}(f)\mathbf{1})|$$
$$\leq \|L\|_{C(G)'} \|\alpha * f - \mathrm{Haar}(f)\mathbf{1}\|_{C(G)}$$
$$< \|L\|_{C(G)'} \varepsilon$$

yields the uniqueness $L = \mathrm{Haar}$.

Finally, $(f \mapsto \mathrm{Haar}(x \mapsto f(x^{-1}))) : C(G) \to \mathbb{C}$ is a positive linear translation-invariant normalised functional, hence equal to Haar by the uniqueness. $\square$

Exercise 7.4.22. In the previous proof, many properties were declared clear, but the reader is encouraged to verify the claims.

Definition 7.4.23 (Spaces $L^p(\mu_G)$). For $1 \leq p < \infty$, the *Lebesgue-p-space* $L^p(\mu_G)$ on a topological group G is a special case of the Lebesgue-p-space from Definition C.4.6. Because the group is compact, by looking in local coordinates, we see from Exercise 1.3.33 that it is the completion of $C(G)$ with respect to the norm

$$f \mapsto \|f\|_{L^p(\mu_G)} := \left(\int_G |f|^p \, d\mu_G \right)^{1/p}.$$

The space $L^\infty(\mu_G)$ is the usual Banach space of μ_G-essentially bounded functions with the norm $f \mapsto \|f\|_{L^\infty(\mu_G)}$; on the closed subspace $C(G) \subset L^\infty(\mu_G)$ we have $\|f\|_{C(G)} = \|f\|_{L^\infty(\mu_G)}$. Notice that $L^p(\mu_G)$ is a Banach space, but it is a Hilbert space if and only if $p = 2$, having the inner product $(f, g) \mapsto \langle f, g \rangle_{L^2(\mu_G)}$ satisfying

$$\langle f, g \rangle_{L^2(\mu_G)} = \int_G f\bar{g} \, d\mu_G$$

for $f, g \in C(G)$.

Remark 7.4.24. We have now seen that for a compact group G there exists a unique translation-invariant probability functional on $C(G)$, the Haar functional! We also know that it is enough to demand only either left- or right-invariance, since one follows from the other. Moreover, the Haar functional is also inversion-invariant. It must be noted that an inversion-invariant probability functional on $C(G)$ is not necessarily translation-invariant: e.g., let us consider the Dirac point mass δ_e at $e \in G$, for which the functional

$$f \mapsto f(e) = \int_G f(x) \, d\delta_e(x)$$

is inversion-invariant but clearly not translation-invariant (unless $G = \{e\}$). Next we observe that the Haar integral distinguishes continuous functions $f, g \in C(G)$ in the sense that if $\int_G |f - g| \, d\mu_G = 0$ then $f = g$:

Theorem 7.4.25. *Let G be a compact group and $f \in C(G)$. If $\int_G |f| \, d\mu_G = 0$ then $f = 0$.*

Proof. The set $U := f^{-1}(\mathbb{C} \setminus \{0\}) \subset G$ is open, since f is continuous and $\{0\} \subset \mathbb{C}$ is closed. Suppose $f \neq 0$. Then $U \neq \emptyset$, and $\{xU : x \in G\}$ is an open cover for G. By the compactness, there exists a subcover $\{x_j U\}_{j=1}^n$. Define $g \in C(G)$ by

$$g(x) := \sum_{j=1}^n \left| f(x_j^{-1} x) \right|.$$

Now $g(x) > 0$ for all $x \in G$, so that there exists $c := \min_{x \in G} g(x) > 0$ by the compactness. We use the normalisation, positivity and translation-invariance of μ_G to obtain

$$0 < c = \int_G c\mathbf{1} \, d\mu_G \leq \int_G g \, d\mu_G = n \int_G |f| \, d\mu_G,$$

so that $0 < \int_G |f| \, d\mu_G$. $\qquad\square$

Exercise 7.4.26. Let G, H be compact groups. Show that $\mu_{G \times H} = \mu_G \times \mu_H$ (i.e., the Haar measure of the product group is the product of the original Haar measures).

Exercise 7.4.27. Let $\mathcal{M}_G$ denote the σ-algebra of the Haar-measurable sets on the compact group G. Consider mappings $m, p_1, p_2 : G \times G \to G$, where

$$m(x, y) = xy, \quad p_1(x, y) = x, \quad p_2(x, y) = y.$$

Show that they are Haar measurable (that is, $(\mathcal{M}_{G \times G}, \mathcal{M}_G)$-measurable). Moreover, show that

$$\mu_G(E) = \mu_{G \times G}(m^{-1}(E)) = \mu_{G \times G}(p_1^{-1}(E)) = \mu_{G \times G}(p_2^{-1}(E))$$

for all $E \in \mathcal{M}_G$.

7.4.1　Integration on quotient spaces

We have already noticed that the good subgroups of a topological group are the closed ones. Moreover, by now we know that a transitive action of a compact topological group G on a Hausdorff space X begets a homeomorphism $G/H \cong X$ of compact Hausdorff spaces, where H is a closed subgroup of G; effectively, spaces G/H and X are the same. We are now about to show that for X there exists a unique G-action-invariant probability functional on $C(X)$, which might be called the Haar functional of the action; the corresponding measure on G/H will accordingly be denoted by $\mu_{G/H}$. We have seen that continuous functions on G/H (and hence on X) can be interpreted as continuous right-H-translation-invariant functions on G, i.e., $f(xh) = f(x)$ for all $x \in G$ and $h \in H$. Next we show how $f \in C(G)$ casts a shadow $f_{G/H} \in C(G/H)$ in a natural way:

Lemma 7.4.28 (Projection $P_{G/H}$). *Let G be a compact group and $H < G$ closed. If $f \in C(G)$ then $P_{G/H}f \in C(G)$ and $f_{G/H} \in C(G/H)$, where*

$$f_{G/H}(xH) = P_{G/H}f(x) := \int_H f(xh) \; \mathrm{d}\mu_H(h).$$

Furthermore, the projection $P_{G/H} : C(G) \to C(G)$ is bounded, more precisely
$$\|f_{G/H}\|_{C(G/H)} = \|P_{G/H}f\|_{C(G)} \leq \|f\|_{C(G)}.$$

Proof. First, H is a compact group having the Haar measure μ_H. The integration in the definition is legitimate since $f_x := (h \mapsto f(xh)) \in C(H)$ for each $x \in G$. If $x \in G$ and $h_0 \in H$ then

$$P_{G/H}f(xh_0) = \int_H f_x(h_0 h) \; \mathrm{d}\mu_H(h) = \int_H f_x(h) \; \mathrm{d}\mu_H(h) = P_{G/H}f(x),$$

so that $f_{G/H} : G/H \to \mathbb{C}$. Next we prove the continuity. Let $\varepsilon > 0$. A continuous function on a compact group is uniformly continuous, so that for $f \in C(G)$ there exists an open $U \ni e$ such that

$$\forall x, y \in G : \; xy^{-1} \in U \; \Rightarrow \; |f(x) - f(y)| < \varepsilon$$

(apparently, this slightly deviates from our definition of the uniform continuity; however, this is clearly equivalent). Suppose $xy^{-1} \in U$. Then

$$
\begin{aligned}
|P_{G/H}f(x) - P_{G/H}f(y)| &= \left| \int_H f(xh) - f(yh) \; \mathrm{d}\mu_H(h) \right| \\
&\leq \int_H |f(xh) - f(yh)| \; \mathrm{d}\mu_H(h) \\
&< \varepsilon,
\end{aligned}
$$

so that $P_{G/H}f \in C(G)$ and $f_{G/H} \in C(G/H)$. Finally,

$$|P_{G/H}f(x)| \leq \int_H |f(xh)| \; \mathrm{d}\mu_H(h) \leq \int_H \|f\|_{C(G)} \; \mathrm{d}\mu_H(h) = \|f\|_{C(G)}. \qquad \square$$

Exercise 7.4.29. Show that the projection $P_{G/H} \in \mathcal{L}(C(G))$ extends uniquely to an orthogonal projection $P_{G/H} \in \mathcal{L}(L^2(\mu_G))$.

Theorem 7.4.30 (Existence of action-invariant measure on quotient spaces). *Let* $((x,p) \mapsto x \cdot p) : G \times M \to M$ *be a continuous transitive action of a compact group* G *on a Hausdorff space* M. *Then there exists a unique Borel-regular probability measure* μ_M *on* M *which is action-invariant in the sense that*

$$\int_M f_M \; \mathrm{d}\mu_M = \int_M f_M(x \cdot p) \; \mathrm{d}\mu_M(p)$$

for all $f_M \in C(M)$ *and* $x \in G$.

Proof. Given $q \in M$, we know that $M \cong G/G_q$. Hence it is enough to deal with $M = G/H$, where $H < G$ is closed and the action is $((x, yH) \mapsto xyH) : G \times G/H \to G/H$.

We first prove the existence of a G-action-invariant Borel regular probability measure $\mu_{G/H}$ on the compact Hausdorff space G/H. Define $\mathrm{Haar}_{G/H} : C(G/H) \to \mathbb{C}$ by

$$\mathrm{Haar}_{G/H}(f_{G/H}) := \int_G f_{G/H}(xH) \; \mathrm{d}\mu_G(x).$$

Notice that

$$\begin{aligned}
\mathrm{Haar}_{G/H}(f_{G/H}) &= \int_G \int_H f(xh) \; \mathrm{d}\mu_H(h) \; \mathrm{d}\mu_G(x) \\
&\overset{\text{Fubini}}{=} \int_H \int_G f(xh) \; \mathrm{d}\mu_G(x) \; \mathrm{d}\mu_H(h) \\
&= \int_H \mathrm{Haar}_G(f) \; \mathrm{d}\mu_H \\
&= \mathrm{Haar}_G(f).
\end{aligned}$$

It is clear that $\mathrm{Haar}_{G/H}$ is a bounded linear functional, and that

$$\mathrm{Haar}_{G/H}(\mathbf{1}_{G/H}) = \mathrm{Haar}_G(\mathbf{1}_G) = 1.$$

By the Riesz Representation Theorem (see Theorem C.4.65), there exists a unique Borel-regular probability measure $\mu_{G/H}$ on G/H such that

$$\mathrm{Haar}_{G/H}(f_{G/H}) = \int_{G/H} f_{G/H} \; \mathrm{d}\mu_{G/H}$$

for all $f_{G/H} \in C(G/H)$. The action-invariance follows from the left-invariance of the functional Haar_G: if $g(y) := f(xy)$ for all $y \in G$ then $g_{G/H}(yH) = f_{G/H}(xyH)$

and

$$
\begin{aligned}
\mathrm{Haar}_{G/H}(y \mapsto f_{G/H}(xyH)) &= \mathrm{Haar}_{G/H}(g_{G/H}) \\
&= \mathrm{Haar}_G(g) \\
&= \mathrm{Haar}_G(f) \\
&= \mathrm{Haar}_{G/H}(f_{G/H}).
\end{aligned}
$$

Next we shall prove the uniqueness part. Suppose $L : C(G/H) \to \mathbb{C}$ is an action-invariant bounded linear functional for which $L(\mathbf{1}_{G/H}) = 1$. Recall the mapping $(f \mapsto f_{G/H}) : C(G) \to C(G/H)$ in Lemma 7.4.28. Then

$$
\tilde{L}(f) := L(f_{G/H})
$$

defines a bounded linear functional $\tilde{L} : C(G) \to \mathbb{C}$ such that $\tilde{L}(\mathbf{1}_G) = 1$ and

$$
\tilde{L}(y \mapsto f(xy)) = L(y \mapsto f_{G/H}(xyH)) = L(f_{G/H}) = \tilde{L}(f).
$$

Hence $\tilde{L} = \mathrm{Haar}_G$ by Theorem 7.4.21. Consequently,

$$
L(f_{G/H}) = \tilde{L}(f) = \mathrm{Haar}_G(f) = \mathrm{Haar}_{G/H}(f_{G/H}),
$$

yielding $L = \mathrm{Haar}_{G/H}$.   □

Remark 7.4.31. Let G be a compact group and $H < G$ closed. From the proof of Theorem 7.4.30 we see that

$$
\int_G f \; \mathrm{d}\mu_G = \int_{G/H} \int_H f(xh) \; \mathrm{d}\mu_H(h) \; \mathrm{d}\mu_{G/H}(xH)
$$

for all $f \in C(G)$.

Exercise 7.4.32. Let $\omega_j(t) \in \mathrm{SO}(3)$ denote the rotation of $\mathbb{R}^3$ by angle $t \in \mathbb{R}$ around the jth coordinate axis, $j \in \{1, 2, 3\}$. Show that $x \in \mathrm{SO}(3)$ can be represented in the form

$$
x = x(\phi, \theta, \psi) = \omega_3(\phi) \; \omega_2(\theta) \; \omega_3(\psi)
$$

where $0 \le \phi, \psi < 2\pi$ and $0 \le \theta \le \pi$.

Exercise 7.4.33. Let the group $G = \mathrm{SO}(3)$ act on the sphere $M = \mathbb{S}^2$ by rotations. Let $q = (0, 0, 1) \in M$, i.e., q is the north pole. Show that $G_q = \{\omega_3(\psi) : 0 \le \psi < 2\pi\}$. We know that the Lebesgue measure is rotation-invariant. Using the normalised angular part of the Lebesgue measure of $\mathbb{R}^3$, deduce that here

$$
\int_G f \; \mathrm{d}\mu_G = \frac{1}{8\pi^2} \int_0^{2\pi} \int_0^{\pi} \int_0^{2\pi} f(x(\phi, \theta, \psi)) \; \sin(\theta) \; \mathrm{d}\psi \; \mathrm{d}\theta \; \mathrm{d}\phi,
$$

i.e., $\mathrm{d}\mu_{\mathrm{SO}(3)} = \frac{1}{8\pi^2} \sin(\theta) \; \mathrm{d}\psi \; \mathrm{d}\theta \; \mathrm{d}\phi$.

We return to the example of $\mathrm{SO}(3)$ in Chapter 11.

7.5 Peter–Weyl decomposition of representations

In the sequel we apply the Haar integral in studying unitary representations of compact groups. The main result is the Peter–Weyl Theorem 7.5.14, leading to a natural Fourier series representation for functions on a compact group.

Exercise 7.5.1. Let $\phi \in \mathrm{Hom}(G, \mathrm{Aut}(\mathcal{H}))$ be a representation of a compact group G on a finite-dimensional $\mathbb{C}$-vector space $\mathcal{H}$. Construct a G-invariant inner product $((u, v) \mapsto \langle u, v \rangle_\mathcal{H}) : \mathcal{H} \times \mathcal{H} \to \mathbb{C}$, that is

$$\langle \phi(x)u, \phi(x)v \rangle_\mathcal{H} = \langle u, v \rangle_\mathcal{H}$$

for all $x \in G$ and $u, v \in \mathcal{H}$. Notice that now the representation ϕ is unitary with respect to this inner product!

Lemma 7.5.2. *Let G be a compact group and $\mathcal{H}$ be a Hilbert space with the inner product $(u, v) \mapsto \langle u, v \rangle_\mathcal{H}$. Let $\phi \in \mathrm{Hom}(G, \mathcal{U}(\mathcal{H}))$ be cyclic and $w \in \mathcal{H}$ a ϕ-cyclic vector with $\|w\|_\mathcal{H} = 1$. Then*

$$\langle u, v \rangle_\phi := \int_G \langle \phi(x)u, w \rangle_\mathcal{H} \, \langle w, \phi(x)v \rangle_\mathcal{H} \, \mathrm{d}x$$

defines an inner product $(u, v) \mapsto \langle u, v \rangle_\phi$ for $\mathcal{H}$. Moreover, ϕ is unitary also with respect to this new inner product, and $\|u\|_\phi \leq \|u\|_\mathcal{H}$ for all $u \in \mathcal{H}$, where $\|u\|_\phi^2 := \langle u, u \rangle_\phi$.

Proof. Defining $f_u(x) := \langle \phi(x)u, w \rangle_\mathcal{H}$, we notice that $f_u \in C(G)$, because

$$
\begin{aligned}
|f_u(x) - f_u(y)| \quad &= \quad |\langle (\phi(x) - \phi(y))u, w \rangle_\mathcal{H}| \\
&\leq \quad \|(\phi(x) - \phi(y))u\|_\mathcal{H} \, \|w\|_\mathcal{H} \\
&\xrightarrow[x \to y]{} \quad 0 \,,
\end{aligned}
$$

due to the strong continuity of ϕ. Thereby $f_u \overline{f_v}$ is Haar integrable, justifying the definition of $\langle u, v \rangle_\phi$.

Let $\lambda \in \mathbb{C}$ and $t, u, v \in \mathcal{H}$. Then it is easy to verify that

$$
\begin{aligned}
\langle \lambda u, v \rangle_\phi \quad &= \quad \lambda \langle u, v \rangle_\phi, \\
\langle t + u, v \rangle_\phi \quad &= \quad \langle t, v \rangle_\phi + \langle u, v \rangle_\phi, \\
\langle u, v \rangle_\phi \quad &= \quad \overline{\langle v, u \rangle_\phi}, \\
\|u\|_\phi^2 \quad &= \quad \int_G |f_u|^2 \, \mathrm{d}\mu_G \geq 0.
\end{aligned}
$$

What if $0 = \|u\|_\phi^2 = \int_G |f_u|^2 \mathrm{d}\mu_G$? Then $f_u = 0$ by Theorem 7.4.25, i.e.,

$$0 = \langle \phi(x)u, w \rangle_\mathcal{H} = \langle u, \phi(x^{-1})w \rangle_\mathcal{H}$$

for all $x \in G$. Since w is a cyclic vector, $u = 0$ follows. Thus $(u, v) \mapsto \langle u, v \rangle_\phi$ is an inner product on $\mathcal{H}$.

The original norm dominates the ϕ-norm, since

$$
\begin{aligned}
\|u\|_\phi^2 &= \int_G |\langle \phi(x)u, w\rangle_\mathcal{H}|^2 \, dx \\
&\leq \int_G \|\phi(x)u\|_\mathcal{H}^2 \, \|w\|_\mathcal{H}^2 \, dx \\
&= \int_G \|u\|_\mathcal{H}^2 \, dx \\
&= \|u\|_\mathcal{H}^2.
\end{aligned}
$$

The ϕ-unitarity of ϕ follows by

$$
\begin{aligned}
\langle u, \phi(y)^*v\rangle_\phi &= \langle \phi(y)u, v\rangle_\phi \\
&= \int_G \langle \phi(xy)u, w\rangle_\mathcal{H} \, \langle w, \phi(x)v\rangle_\mathcal{H} \, dx \\
&\overset{z=xy}{=} \int_G \langle \phi(z)u, w\rangle_\mathcal{H} \, \langle w, \phi(zy^{-1})v\rangle_\mathcal{H} \, dz \\
&= \langle u, \phi(y)^{-1}v\rangle_\phi,
\end{aligned}
$$

where we applied the translation invariance of the Haar integral. $\square$

Exercise 7.5.3. Check the missing details in the proof of Lemma 7.5.2.

Lemma 7.5.4. *Let $\langle u, v\rangle_\phi$ be as above. Then*

$$
\langle u, Av\rangle_\mathcal{H} := \langle u, v\rangle_\phi
$$

defines a compact self-adjoint operator $A \in \mathcal{L}(\mathcal{H})$. Furthermore, A is positive definite and $A \in \mathrm{Hom}(\phi, \phi)$.

Proof. By Lemma 7.5.2, if $v \in \mathcal{H}$ then $F_v(u) := \langle u, v\rangle_\phi$ defines a linear functional $F_v : \mathcal{H} \to \mathbb{C}$, which is bounded in both norms, since

$$
|F_v(u)| = |\langle u, v\rangle_\phi| \leq \|u\|_\phi \, \|v\|_\phi \leq \|u\|_\mathcal{H} \, \|v\|_\phi.
$$

The Riesz Representation Theorem B.5.19 implies that F_v is represented by a unique vector $A(v) \in \mathcal{H}$, i.e., $F_v(u) = \langle u, A(v)\rangle_\mathcal{H}$ for all $u \in \mathcal{H}$. Thus we have an operator $A : \mathcal{H} \to \mathcal{H}$, which is clearly linear. We obtain a bound $\|A\|_{\mathcal{L}(\mathcal{H})} \leq 1$ from

$$
\|Av\|_\mathcal{H}^2 = \langle Av, Av\rangle_\mathcal{H} = \langle Av, v\rangle_\phi \leq \|Av\|_\phi \, \|v\|_\phi \leq \|Av\|_\mathcal{H} \, \|v\|_\mathcal{H}.
$$

Self-adjointness follows from

$$
\langle u, A^*v\rangle_\mathcal{H} = \langle Au, v\rangle_\mathcal{H} = \overline{\langle v, Au\rangle_\mathcal{H}} = \overline{\langle v, u\rangle_\phi} = \langle u, v\rangle_\phi = \langle u, Av\rangle_\mathcal{H}.
$$

Moreover, A is positive definite, because $\langle u, Au\rangle_\mathcal{H} = \langle u, u\rangle_\phi = \|u\|_\phi^2 \geq 0$, where $\|u\|_\phi = 0$ if and only if $u = 0$.

The property that $A \in \text{Hom}(\phi, \phi)$ is seen from

$$
\begin{aligned}
\langle u, A\phi(y)v \rangle_{\mathcal{H}} &= \langle u, \phi(y)v \rangle_{\phi} \\
&= \langle \phi(y)^{-1}u, v \rangle_{\phi} \\
&= \langle \phi(y)^{-1}u, Av \rangle_{\mathcal{H}} \\
&= \langle u, \phi(y)Av \rangle_{\mathcal{H}}.
\end{aligned}
$$

Let $\mathbb{B} = \{u \in \mathcal{H} : \|u\|_{\mathcal{H}} \leq 1\}$, the closed unit ball of $\mathcal{H}$. To show that $A \in \mathcal{L}(\mathcal{H})$ is compact, we must show that $\overline{A(\mathbb{B})} \subset \mathcal{H}$ is a compact set. So take a sequence $(v_j)_{j=1}^{\infty} \subset A(\mathbb{B})$; we have to find a converging subsequence. Take a sequence $(u_j)_{j=1}^{\infty} \subset \mathbb{B}$ such that $Au_j = v_j$. By the Banach–Alaoglu Theorem B.5.30, the closed ball $\mathbb{B}$ is weakly compact: there exists a subsequence $(u_{j_k})_{k=1}^{\infty}$ such that $u_{j_k} \xrightarrow[k \to \infty]{} u \in \mathbb{B}$ weakly, i.e.,

$$
\langle u_{j_k}, v \rangle_{\mathcal{H}} \xrightarrow[k \to \infty]{} \langle u, v \rangle_{\mathcal{H}}
$$

for all $v \in \mathcal{H}$. Then

$$
\begin{aligned}
\|v_{j_k} - Au\|_{\mathcal{H}}^2 &= \|A(u_{j_k} - u)\|_{\mathcal{H}}^2 \\
&= \langle A(u_{j_k} - u), u_{j_k} - u \rangle_{\phi} \\
&= \int_G g_k \, \mathrm{d}\mu_G
\end{aligned}
$$

where

$$
g_k(x) := \langle \phi(x)A(u_{j_k} - u), w \rangle_{\mathcal{H}} \, \langle w, \phi(x)(u_{j_k} - u) \rangle_{\mathcal{H}}.
$$

Let us show that $\int_G g_k \, \mathrm{d}\mu_G \to 0$ as $k \to \infty$. First, $g_k \in C(G)$ (hence g_k is integrable) and for each $x \in G$,

$$
\begin{aligned}
|g_k(x)| &= \left| \langle u_{j_k} - u, A^*\phi(x^{-1})w \rangle_{\mathcal{H}} \right| \left| \langle \phi(x^{-1})w, u_{j_k} - u \rangle_{\mathcal{H}} \right| \\
&\xrightarrow[k \to \infty]{} 0
\end{aligned}
$$

by the weak convergence. Second,

$$
\begin{aligned}
|g_k(x)| &\leq \|\phi(x)\|_{\mathcal{L}(\mathcal{H})}^2 \, \|A^*\|_{\mathcal{L}(\mathcal{H})} \, \|w\|_{\mathcal{H}}^2 \, \|u_{j_k} - u\|_{\mathcal{H}}^2 \\
&\leq 4,
\end{aligned}
$$

because $\|\phi(x)\|_{\mathcal{L}(\mathcal{H})} = 1$, $\|A\|_{\mathcal{L}(\mathcal{H})} = \|A^*\|_{\mathcal{L}(\mathcal{H})} \leq 1$, $\|w\|_{\mathcal{H}} = 1$ and $u_{j_k}, u \in \mathbb{B}$. Thus $\int_G g_k \, \mathrm{d}\mu_G \xrightarrow[k \to \infty]{} 0$ by the Lebesgue Dominated Convergence Theorem (see Theorem C.3.22). Equivalently, $v_{j_k} \xrightarrow[k \to \infty]{} Au \in A(\mathbb{B})$. We have shown that the set $\overline{A(\mathbb{B})} = A(\mathbb{B}) \subset \mathcal{H}$ is compact. $\qquad\square$

Theorem 7.5.5 (Decomposition in finite-dimensional representations). *Let G be a compact group and $\mathcal{H}$ a Hilbert space. Let $\phi \in \mathrm{Hom}(G, \mathcal{U}(\mathcal{H}))$ be strongly continuous. Then ϕ is a direct sum of finite-dimensional irreducible unitary representations.*

Proof. We know by Proposition 7.2.8 that ϕ is a direct sum of cyclic representations. Therefore it is enough to assume that ϕ itself is cyclic. The operator $A \in \mathrm{Hom}(\phi, \phi)$ in Lemma 7.5.4 is compact and self-adjoint. Hence by the Hilbert–Schmidt Spectral Theorem B.5.26 we have

$$\mathcal{H} = \bigoplus_{\lambda \in \sigma(A)} \mathrm{Ker}(\lambda I - A),$$

where $\dim(\mathrm{Ker}(\lambda I - A)) < \infty$ for each $\lambda \in \sigma(A) \backslash \{0\}$. This can be extended to $\lambda = 0$ as well by Lemma 7.5.4 and the definition of A. Since $A \in \mathrm{Hom}(\phi, \phi)$, the subspace $\mathrm{Ker}(\lambda I - A) \subset \mathcal{H}$ is ϕ-invariant for each λ. Thereby

$$\phi = \bigoplus_{\lambda \in \sigma(A)} \phi|_{\mathrm{Ker}(\lambda I - A)},$$

where $\phi|_{\mathrm{Ker}(\lambda I - A)}$ is finite-dimensional and unitary for all $\lambda \in \sigma(A)$. The proof is concluded, since we know that a finite-dimensional unitary representation is a direct sum of irreducible unitary representations. $\square$

Corollary 7.5.6 (Finite dimensionality of representations!). *Strongly continuous irreducible unitary representations of compact groups are finite-dimensional.*

Definition 7.5.7 (Unitary dual $\widehat{G}$). The *(unitary) dual* $\widehat{G}$ of a locally compact group G is the set consisting of all equivalence classes of strongly continuous irreducible unitary representations of G (for the definition of equivalent representations see Definition 6.3.18).

*Remark 7.5.8 (**Continuity is enough**).* For a compact group G, the set $\widehat{G}$ consists of the equivalence classes of *continuous* irreducible unitary representations (due to the finite-dimensionality), i.e.,

$$\widehat{G} = \{[\phi] \mid \phi \text{ is a continuous irreducible unitary representation of } G\},$$

where $[\phi] = \{\psi \mid \psi \sim \phi\}$ is the equivalence class of ϕ.

*Remark 7.5.9 (**Duals $\widehat{\mathbb{R}^n}$ and $\widehat{\mathbb{T}^n}$**).* It can be proven that

$$\widehat{\mathbb{R}^n} = \left\{ [e_\xi] \mid \xi \in \mathbb{R}^n, \ e_\xi : \mathbb{R}^n \to \mathrm{U}(1), \ e_\xi(x) := e^{i2\pi x \cdot \xi} \right\}.$$

Noticing that $e_\xi e_\eta = e_{\xi+\eta}$ and that $[e_\xi] \neq [e_\eta]$ for $\xi \neq \eta$, we may identify $\widehat{\mathbb{R}^n} \cong \mathbb{R}^n$ as groups. Similarly, and in view of Theorem 3.1.17 and Remark 3.1.18, we have

$$\widehat{\mathbb{T}^n} = \left\{ [e_\xi] \mid \xi \in \mathbb{Z}^n, \ e_\xi : \mathbb{R}^n \to \mathrm{U}(1), \ e_\xi(x) := e^{i2\pi x \cdot \xi} \right\},$$

so that $\widehat{\mathbb{T}^n} \cong \mathbb{Z}^n$ as groups.

Remark 7.5.10 (**Pontryagin duality**). For a commutative locally compact group G the unitary dual $\widehat{G}$ has a natural structure of a commutative locally compact group, and $\widehat{\widehat{G}} \cong G$; this is the so-called *Pontryagin duality*. For a compact non-commutative group G, the unitary dual $\widehat{G}$ is never a group, but still has a sort of weak algebraic structure; we do not consider this in the sequel.

Remark 7.5.11 (**Matrix representations**). Let G be a compact group. For the equivalence class $\xi \in \widehat{G}$ there exists a unitary matrix representation $\phi \in \xi = [\phi]$. That is, we have a homomorphism $\phi = (\phi_{ij})_{i,j=1}^m : G \to U(m)$, where functions $\phi_{ij} : G \to \mathbb{C}$ are continuous. We may find such a representation in the following way: if $\psi \in \xi$, $\psi \in \operatorname{Hom}(G, \mathcal{U}(\mathcal{H}))$ and $\{e_j\}_{j=1}^m \subset \mathcal{H}$ is an orthonormal basis for $\mathcal{H}$, then we can define

$$\phi_{ij}(x) := \langle e_i, \psi(x)e_j \rangle_{\mathcal{H}}.$$

Next we present an L^2-orthogonality result for such functions $\phi_{ij} : G \to \mathbb{C}$.

Lemma 7.5.12 (Orthogonality of representations). *Let G be a compact group. Let $\xi, \eta \in \widehat{G}$, where $\xi \ni \phi = (\phi_{ij})_{i,j=1}^m \in \operatorname{Hom}(G, U(m))$ and $\eta \ni \psi = (\psi_{kl})_{k,l=1}^n \in \operatorname{Hom}(G, U(n))$. Then*

$$\langle \phi_{ij}, \psi_{kl} \rangle_{L^2(\mu_G)} = \begin{cases} 0, & \text{if } \xi \neq \eta, \\ \frac{1}{m}\, \delta_{ik}\delta_{jl}, & \text{if } \phi = \psi. \end{cases}$$

Proof. Fix $1 \leq j \leq m$ and $1 \leq l \leq n$. Define the matrix $E \in \mathbb{C}^{m \times n}$ by $E_{pq} := \delta_{pj}\delta_{lq}$ (i.e., the matrix elements of E are zero except for the (j, l)-element, which is 1.) Define the matrix $A \in \mathbb{C}^{m \times n}$ by

$$A := \int_G \phi(y)\, E\, \psi(y^{-1})\, \mathrm{d}y.$$

Now $A \in \operatorname{Hom}(\psi, \phi)$, since

$$
\begin{aligned}
\phi(x)A &= \int_G \phi(xy)\, E\, \psi(y^{-1})\, \mathrm{d}y \\
&= \int_G \phi(z)\, E\, \psi(z^{-1}x)\, \mathrm{d}z \\
&= A\psi(x).
\end{aligned}
$$

Since ϕ, ψ are finite-dimensional irreducible unitary representations, Schur's Lemma 6.3.25 implies that

$$A = \begin{cases} 0, & \text{if } \phi \not\sim \psi, \\ \lambda I, & \text{if } \phi = \psi \end{cases}$$

for some $\lambda \in \mathbb{C}$. We notice that

$$A_{ik} = \int_G \sum_{p=1}^{m} \sum_{q=1}^{n} \phi_{ip}(y) \; E_{pq} \; \psi_{qk}(y^{-1}) \; dy$$

$$= \int_G \phi_{ij}(y) \; \overline{\psi_{kl}(y)} \; dy = \langle \phi_{ij}, \psi_{kl} \rangle_{L^2(\mu_G)}.$$

Now suppose $\phi = \psi$. Then $m = n$ and

$$\langle \phi_{kj}, \psi_{kl} \rangle_{L^2(\mu_G)} = A_{kk} = \lambda = \frac{1}{m} \; \text{Tr}(A)$$

$$= \frac{1}{m} \int_G \text{Tr}(\phi(y) \; E \; \phi(y^{-1})) \; dy$$

$$= \frac{1}{m} \int_G \text{Tr}(E) \; dy = \frac{1}{m} \delta_{jl},$$

where we used the property $\text{Tr}(BC) = \text{Tr}(CB)$ of the trace functional. The results can be collected from above. $\qquad\square$

Definition 7.5.13 (Left and right regular representations). Let G be a compact group. Its *left* and *right regular representations* $\pi_L, \pi_R : G \to \mathcal{U}(L^2(\mu_G))$ are defined, respectively, by

$$(\pi_L(y) \; f)(x) := f(y^{-1}x),$$

$$(\pi_R(y) \; f)(x) := f(xy)$$

for μ_G-almost every $x \in G$.

The idea here is that G is represented as a natural group of operators on the Hilbert space $L^2(\mu_G)$, enabling the use of functional analytic techniques in studying G. And now for a major result in representation theory:

Theorem 7.5.14 (Peter–Weyl Theorem (1927)). *Let G be a compact group. Then*

$$\mathcal{B} := \left\{ \sqrt{\dim(\phi)} \phi_{ij} \mid \phi = (\phi_{ij})_{i,j=1}^{\dim(\phi)}, \; [\phi] \in \widehat{G} \right\}$$

is an orthonormal basis for $L^2(\mu_G)$. Let $\phi = (\phi_{ij})_{i,j=1}^{m}$, $\phi \in [\phi] \in \widehat{G}$. Then

$$\mathcal{H}_{i,\cdot}^{\phi} := \text{span}\{\phi_{ij} \mid 1 \le j \le m\} \subset L^2(\mu_G)$$

is π_R-invariant and

$$\phi \sim \pi_R|_{\mathcal{H}_{i,\cdot}^{\phi}},$$

$$L^2(\mu_G) = \bigoplus_{[\phi] \in \widehat{G}} \bigoplus_{i=1}^{m} \mathcal{H}_{i,\cdot}^{\phi},$$

$$\pi_R \sim \bigoplus_{[\phi] \in \widehat{G}} \bigoplus_{i=1}^{m} \phi.$$

Remark 7.5.15. Here $\bigoplus_{i=1}^{m} \phi := \phi \oplus \cdots \oplus \phi$, the m-fold direct sum of ϕ; in literature, this is sometimes denoted even by $m\phi$.

Remark 7.5.16 (**Left Peter–Weyl**). We can formulate the Peter–Weyl Theorem 7.5.14 analogously for the left regular representation, as follows: Let $\phi = (\phi_{ij})_{i,j=1}^{m}$, where $\phi \in [\phi] \in \widehat{G}$. Then

$$\mathcal{H}_{\cdot,j}^{\phi} := \text{span}\{\phi_{ij} \mid 1 \le i \le m\} \subset L^2(\mu_G)$$

is π_L-invariant and

$$\phi \sim \pi_L|_{\mathcal{H}_{\cdot,j}^{\phi}},$$

$$L^2(\mu_G) = \bigoplus_{[\phi]\in\widehat{G}} \bigoplus_{j=1}^{m} \mathcal{H}_{\cdot,j}^{\phi},$$

$$\pi_L \sim \bigoplus_{[\phi]\in\widehat{G}} \bigoplus_{j=1}^{m} \phi.$$

Remark 7.5.17 (**Peter–Weyl for $\mathbb{T}^n$**). Let $G = \mathbb{T}^n$. Recall from Remark 7.5.9 that

$$\widehat{\mathbb{T}^n} = \left\{ [e_\xi] \mid \xi \in \mathbb{Z}^n,\ e_\xi(x) = e^{i2\pi x \cdot \xi} \right\}.$$

Now $\mathcal{B} = \{e_\xi \mid \xi \in \mathbb{Z}^n\}$ is an orthonormal basis for $L^2(\mu_{\mathbb{T}^n})$,

$$L^2(\mu_{\mathbb{T}^n}) = \bigoplus_{\xi\in\mathbb{Z}^n} \text{span}\{e_\xi\},$$

$$\pi_L \sim \bigoplus_{\xi\in\mathbb{Z}^n} e_\xi \sim \pi_R.$$

Moreover, for $f \in L^2(\mu_{\mathbb{T}^n})$, we have

$$f = \sum_{\xi\in\mathbb{Z}^n} \widehat{f}(\xi)\, e_\xi,$$

where the Fourier coefficients $\widehat{f}(\xi)$ are calculated by

$$\widehat{f}(\xi) = \int_{\mathbb{T}^n} f\, \overline{e_\xi}\, d\mu_{\mathbb{T}^n} = \langle f, e_\xi \rangle_{L^2(\mu_{\mathbb{T}^n})}.$$

Analogously, the Peter–Weyl Theorem 7.5.14 provides Fourier series expansions for L^2-functions on any compact group. We shall return to the *Fourier series* theme after the proof of the Peter–Weyl theorem.

Proof for the Peter–Weyl Theorem 7.5.14. The π_R-invariance of $\mathcal{H}^\phi_{i,\cdot}$ follows due to

$$\pi_R(y)\phi_{ij}(x) = \phi_{ij}(xy) = \sum_{k=1}^{\dim(\phi)} \phi_{ik}(x)\phi_{kj}(y),$$

i.e., with $\lambda_k(y) = \phi_{kj}(y)$ we have

$$\pi_R(y)\phi_{ij} = \sum_{k=1}^{\dim(\phi)} \lambda_k(y)\ \phi_{ik} \in \operatorname{span}\{\phi_{ik}\}_{k=1}^{\dim(\phi)} = \mathcal{H}^\phi_{i,\cdot}.$$

If $\{e_j\}_{j=1}^{\dim(\phi)} \subset \mathbb{C}^{\dim(\phi)}$ is the standard orthonormal basis then

$$\phi(y)e_j = \sum_{k=1}^{\dim(\phi)} \phi_{kj}(y)e_k,$$

so that the equation

$$A\sum_{j=1}^{\dim(\phi)} \lambda_j e_j := \sum_{j=1}^{\dim(\phi)} \lambda_j \phi_{ij}$$

defines an intertwining isomorphism $A \in \operatorname{Hom}(\phi, \pi_R|_{\mathcal{H}^\phi_{i,\cdot}})$, i.e., $\phi \sim \pi_R|_{\mathcal{H}^\phi_{i,\cdot}}$.

By Lemma 7.5.12, the set $\mathcal{B} \subset L^2(\mu_G)$ is orthonormal. Let

$$\mathcal{H} := \bigoplus_{[\phi]\in\widehat{G}} \bigoplus_{i=1}^{\dim(\phi)} \mathcal{H}^\phi_{i,\cdot}.$$

We assume that $\mathcal{H} \neq L^2(\mu_G)$, and show that this leads to a contradiction (so that $\mathcal{H} = L^2(\mu_G)$ and $\mathcal{B}$ must be a basis). First, clearly $\mathcal{H}$ is π_R-invariant. By our assumption, $\mathcal{H}^\perp$ is a non-trivial π_R-invariant closed subspace. Since $\pi_R|_{\mathcal{H}^\perp}$ is a direct sum of irreducible unitary representations, there exists a non-trivial subspace $E \subset \mathcal{H}^\perp$ and a unitary matrix representation $\phi = (\phi_{ij})_{i,j=1}^m \in \operatorname{HOM}(G, \operatorname{U}(m))$ such that $\phi \sim \pi_R|_E$. The subspace E has an orthonormal basis $\{f_j\}_{j=1}^m$ such that

$$\pi_R(y)f_j = \sum_{i=1}^m \phi_{ij}(y)f_i$$

for all $y \in G$ and $j \in \{1,\ldots,m\}$. Notice that $f_j \in L^2(\mu_G)$ has pointwise values perhaps only μ_G-almost everywhere, so that

$$f_j(xy) = \sum_{i=1}^m \phi_{ij}(y)f_i(x).$$

may hold for only μ_G-almost every $x \in G$. Let us define measurable sets

$$N(y) := \left\{ x \in G : f_j(xy) \neq \sum_{i=1}^{m} \phi_{ij}(y) f_i(x) \right\},$$

$$M(x) := \left\{ y \in G : f_j(xy) \neq \sum_{i=1}^{m} \phi_{ij}(y) f_i(x) \right\},$$

$$K := \left\{ (x,y) \in G \times G : f_j(xy) \neq \sum_{i=1}^{m} \phi_{ij}(y) f_i(x) \right\}.$$

By Exercise 7.4.27, we may utilise the Fubini Theorem to change the order of integration, to get

$$\int_G \mu_G(M(x)) \, d\mu_G(x) = \mu_{G \times G}(K) = \int_G \mu_G(N(y)) \, d\mu_G(y)$$

$$= \int_G 0 \, d\mu_G(y) = 0,$$

meaning that $\mu_G(M(x)) = 0$ for almost every $x \in G$. But it is enough to pick just one $x_0 \in G$ such that $\mu_G(M(x_0)) = 0$. Then

$$f_j(x_0 y) = \sum_{i=1}^{m} \phi_{ij}(y) f_i(x_0)$$

for μ_G-almost every $y \in G$. If we denote $z := x_0 y$ then

$$f_j(z) = \sum_{i=1}^{m} \phi_{ij}(x_0^{-1} z) f_i(x_0)$$

$$= \sum_{i=1}^{m} \sum_{k=1}^{m} \phi_{ik}(x_0^{-1}) \phi_{kj}(z) f_i(x_0)$$

$$= \sum_{k=1}^{m} \phi_{kj}(z) \sum_{i=1}^{m} \phi_{ik}(x_0^{-1}) f_i(x_0)$$

for μ_G-almost every $z \in G$. Hence

$$f_j \in \operatorname{span}\{\phi_{kj}\}_{k=1}^{m} \subset \bigoplus_{k=1}^{m} \mathcal{H}_{k,\cdot}^{\phi} \subset \mathcal{H}$$

for all $j \in \{1, \ldots, m\}$. Thereby

$$E = \operatorname{span}\{f_j\}_{j=1}^{m} \subset \mathcal{H};$$

at the same time $E \subset \mathcal{H}^{\perp}$, yielding $E = \{0\}$. This is a contradiction, since E should be non-trivial. Hence $\mathcal{H} = L^2(\mu_G)$ and $\mathcal{B}$ is a basis. $\qquad \square$

Exercise 7.5.18. Check the details of the proof of the Peter–Weyl theorem. In particular, pay attention to verify the conditions for applying the Fubini Theorem.

7.6   Fourier series and trigonometric polynomials

The classical Fourier series express a periodic function as an infinite sum of elementary waves that behave well under translations. This can be viewed as a special case of a more general phenomenon: a function on a compact group admits an analogous series expansion, thanks to the Peter–Weyl Theorem 7.5.14. We start by discussing the trigonometric polynomials because they play an important role as finite linear combinations of the basis elements of $L^2(\mu_G)$ provided by the Peter–Weyl theorem.

Definition 7.6.1 (Trigonometric polynomials on groups). Let G be a compact group and

$$\mathcal{B} := \left\{ \sqrt{\dim(\phi)}\,\phi_{ij} \mid \phi = (\phi_{ij})_{i,j=1}^{\dim(\phi)},\ [\phi] \in \widehat{G} \right\}$$

as in the Peter–Weyl Theorem 7.5.14. The space of *trigonometric polynomials* on G is

$$\mathrm{TrigPol}(G) = \mathrm{span}(\mathcal{B}).$$

For instance, $f \in \mathrm{TrigPol}(\mathbb{T}^n)$ is of the form

$$f(x) = \sum_{\xi \in \mathbb{Z}^n} \widehat{f}(\xi) e^{i2\pi x \cdot \xi},$$

where $\widehat{f}(\xi) \neq 0$ for only finitely many $\xi \in \mathbb{Z}^n$, see Remark 3.1.26. In the case of the torus the following density statement was already verified in the proof of Theorem 3.1.20.

Theorem 7.6.2 (Density I). $\mathrm{TrigPol}(G)$ *is a dense subalgebra of* $C(G)$.

Proof. It is enough to verify that $\mathrm{TrigPol}(G)$ is an involutive subalgebra of $C(G)$, because the Stone–Weierstrass Theorem A.14.4 provides then the density. We already know that $\mathrm{TrigPol}(G)$ is a subspace of $C(G)$.

First, $\phi = (x \mapsto (1)) \in \mathrm{Hom}(G, \mathrm{U}(1))$ is a continuous irreducible unitary representation, so that $\mathbf{1} = (x \mapsto 1) \in C(G)$ belongs to $\mathcal{B} \subset \mathrm{TrigPol}(G)$.

Let $[\phi] \in \widehat{G}$, $\phi = (\phi_{ij})_{i,j=1}^m$. Then $[\overline{\phi}] \in \widehat{G}$, where $\overline{\phi} = (\overline{\phi_{ij}})_{i,j=1}^m$, as it is easy to verify. Thereby we get the involutivity: $\overline{f} \in \mathrm{TrigPol}(G)$ whenever $f \in \mathrm{TrigPol}(G)$.

Let $[\psi] \in \widehat{G}$, $\psi = (\psi_{kl})_{k,l=1}^n$. Then $\phi \otimes \psi|_G = (x \mapsto \phi(x) \otimes \psi(x)) \in \mathrm{Hom}(G, \mathcal{U}(\mathbb{C}^m \otimes \mathbb{C}^n))$. Let $\{e_i\}_{i=1}^m \subset \mathbb{C}^m$ and $\{f_k\}_{k=1}^n \subset \mathbb{C}^n$ be orthonormal bases. Then $\{e_i \otimes f_k \mid 1 \leq i \leq m,\ 1 \leq k \leq n\}$ is an orthonormal basis for $\mathbb{C}^m \otimes \mathbb{C}^n$, and the $(ik)(jl)$-matrix element of $\phi \otimes \psi|_G$ is calculated by

$$(\phi \otimes \psi|_G)_{(ik)(jl)}(x) = \langle (\phi \otimes \psi|_G)(x)(e_j \otimes f_l), e_i \otimes f_k \rangle_{\mathbb{C}^m \otimes \mathbb{C}^n}$$
$$= \langle \phi(x)e_j, e_i \rangle_{\mathbb{C}^m} \langle \psi(x)f_l, f_k \rangle_{\mathbb{C}^n} = \phi_{ij}(x)\psi_{kl}(x).$$

Hence $\phi_{ij}\psi_{kl}$ is a matrix element of $\phi \otimes \psi|_G$. Representation $\phi \otimes \psi|_G$ can be decomposed as a finite direct sum of irreducible unitary representations. Hence the matrix elements of $\phi \otimes \psi|_G$ can be written as linear combinations of elements of $\mathcal{B}$. Thus $\phi_{ij}\psi_{kl} \in \mathrm{TrigPol}(G)$, so that $fg \in \mathrm{TrigPol}(G)$ for all $f, g \in \mathrm{TrigPol}(G)$. $\square$

Corollary 7.6.3 (Density II). TrigPol(G) *is dense in* $L^2(\mu_G)$.

Remark 7.6.4. Notice that we did not need the Peter–Weyl Theorem 7.5.14 to show that TrigPol(G) $\subset L^2(\mu_G)$ is dense. Therefore this density provides another proof for the Peter–Weyl Theorem 7.5.14.

Remark 7.6.5. By now, we have encountered plenty of translation- and inversion-invariant function spaces on G. For instance, TrigPol(G), $C(G)$ and $L^p(G)$, and more: namely, if $[\phi] \in \widehat{G}$, $\phi = (\phi_{ij})_{i,j=1}^m$, then

$$\pi_L(y)\phi_{i_0 j_0}, \pi_R(y)\phi_{i_0 j_0} \in \operatorname{span}\{\phi_{ij}\}_{i,j=1}^m$$

for all $y \in G$ (and inversion-invariance is clear!).

Exercise 7.6.6. Prove that $f \in C(G)$ is a trigonometric polynomial if and only if

$$\dim\left(\operatorname{span}\{\pi_R(y)f : y \in G\}\right) < \infty.$$

As a direct consequence of knowing the basis of $L^2(\mu_G)$ by the Peter–Weyl theorem, we obtain:

Corollary 7.6.7 (Fourier series and Plancherel (matrix form)). *On a compact group G, a Fourier series presentation of $f \in L^2(\mu_G)$ is given by*

$$f = \sum_{[\phi] \in \widehat{G}} \dim(\phi) \sum_{i,j=1}^{\dim(\phi)} \langle f, \phi_{ij} \rangle_{L^2(\mu_G)} \, \phi_{ij}, \tag{7.1}$$

where we pick just one unitary matrix representation $\phi = (\phi_{ij})_{i,j=1}^{\dim(\phi)}$ from each equivalence class $[\phi] \in \widehat{G}$. Moreover, there is the Plancherel *identity (sometimes called the* Parseval *identity)*

$$\|f\|_{L^2(\mu_G)} = \sum_{[\phi] \in \widehat{G}} \dim(\phi) \sum_{i,j=1}^{\dim(\phi)} |\langle f, \phi_{ij} \rangle_{L^2(\mu_G)}|^2. \tag{7.2}$$

Remark 7.6.8. In $L^2(\mu_G)$, also clearly

$$f = \sum_{[\phi] \in \widehat{G}} \dim(\phi) \sum_{i,j=1}^{\dim(\phi)} \langle f, \overline{\phi_{ij}} \rangle_{L^2(\mu_G)} \, \overline{\phi_{ij}}.$$

A nice thing about the Fourier series is that the basis functions ϕ_{ij} and $\overline{\phi_{ij}}$ are well behaved under translations and inversions.

Definition 7.6.9 (Fourier coefficients and Fourier transform). Let G be a compact group, $f \in L^1(\mu_G)$ and $\phi = (\phi_{ij})_{i,j=1}^m$, $[\phi] \in \widehat{G}$. The ϕ-*Fourier coefficient of f is*

$$\widehat{f}(\phi) := \int_G f(x) \, \phi(x)^* \, dx \in \mathbb{C}^{m \times m},$$

where the integration of the matrix-valued function is element-wise. The matrix-valued function $\widehat{f}$ is called the *Fourier transform* of $f \in L^1(\mu_G)$. We note that this definition immediately extends also to $L^2(\mu_G)$ in view of $L^2(\mu_G) \subset L^1(\mu_G)$, which follows, e.g., by the Hölder's inequality from the compactness of G.

Corollary 7.6.10 (Fourier series and Plancherel). *On a compact group G, a Fourier series presentation of $f \in L^2(\mu_G)$ is given by*

$$f(x) = \sum_{[\phi] \in \widehat{G}} \dim(\phi) \ \mathrm{Tr}\left(\widehat{f}(\phi) \ \phi(x)\right) \tag{7.3}$$

converging for μ_G-almost every $x \in G$, as well as in $L^2(\mu_G)$. The Plancherel identity takes the form

$$\|f\|_{L^2(\mu_G)}^2 = \sum_{[\phi] \in \widehat{G}} \dim(\phi) \ \mathrm{Tr}\left(\widehat{f}(\phi) \ \widehat{f}(\phi)^*\right).$$

If $f, g \in L^2(G)$, then we also have the Parseval identity

$$(f, g)_{L^2(G)} = \sum_{[\xi] \in \widehat{G}} \dim(\phi) \, \mathrm{Tr}\left(\widehat{f}(\phi) \ \widehat{g}(\phi)^*\right) = (\widehat{f}(\phi), \widehat{g}(\phi))_{L^2(\widehat{G})},$$

with $L^2(\widehat{G})$ as in Definition 7.6.11.

Proof. Now

$$\widehat{f}(\phi)_{ij} = \int_G f(x) \ (\phi(x)^*)_{ij} \ \mathrm{d}\mu_G(x) = \int_G f(x) \ \overline{\phi_{ji}(x)} \ \mathrm{d}x = \langle f, \phi_{ji} \rangle_{L^2(\mu_G)},$$

so that

$$
\begin{aligned}
\mathrm{Tr}\left(\widehat{f}(\phi) \ \phi(x)\right) &= \sum_{i=1}^{\dim(\phi)} \left(\widehat{f}(\phi) \ \phi(x)\right)_{ii} \\
&= \sum_{i,j=1}^{\dim(\phi)} \widehat{f}(\phi)_{ij} \ \phi_{ji}(x) \\
&= \sum_{i,j=1}^{\dim(\phi)} \langle f, \phi_{ji} \rangle_{L^2(\mu_G)} \ \phi_{ji}(x).
\end{aligned}
$$

Hence (7.3) follows from (7.1). Finally, if $A = (A_{kl})_{k,l=1}^m \in \mathbb{C}^{m \times m}$ then

$$\|A\|_{HS}^2 := \mathrm{Tr}(A^*A) = \sum_{k,l=1}^m |A_{kl}|^2,$$

completing the proof, if we take $A = \widehat{f}(\phi)$ and use (7.2). The details of the proof of the Parceval identity will be given in Proposition 10.3.17. $\square$

The convergence in $L^2(\mu_G)$ is automatic, see Theorem B.5.32.

Definition 7.6.11 (Hilbert space $L^2(\widehat{G})$). Let G be a compact group. Let $L^2(\widehat{G})$ be the space containing the mappings

$$F : \widehat{G} \to \bigcup_{m=1}^{\infty} \mathbb{C}^{m \times m}$$

satisfying $F([\phi]) \in \mathbb{C}^{\dim(\phi) \times \dim(\phi)}$ such that

$$\sum_{[\phi] \in \widehat{G}} \dim(\phi) \, \|F([\phi])\|_{\mathbb{C}^{\dim(\phi) \times \dim(\phi)}}^2 < \infty.$$

Then $L^2(\widehat{G})$ is a Hilbert space with the inner product

$$\langle E, F \rangle_{L^2(\widehat{G})} := \sum_{[\phi] \in \widehat{G}} \dim(\phi) \, \langle E([\phi]), F([\phi]) \rangle_{\mathbb{C}^{\dim(\phi) \times \dim(\phi)}}.$$

Exercise 7.6.12. Verify that $L^2(\widehat{G})$ is indeed a Hilbert space.

Theorem 7.6.13 (Fourier transform is an isometry $L^2(\mu_G) \to L^2(\widehat{G})$). *Let G be a compact group. The Fourier transform $f \mapsto \widehat{f}$ defines a surjective isometry $L^2(\mu_G) \to L^2(\widehat{G})$.*

Proof. Let us choose one unitary matrix representation ϕ from each $[\phi] \in \widehat{G}$. If we define $F([\phi]) := \widehat{f}(\phi)$ then $F \in L^2(\widehat{G})$, and $f \mapsto F$ is isometric by the Plancherel equality.

Now take any $F \in L^2(\widehat{G})$. We have to show that $F([\phi]) = \widehat{f}(\phi)$ for some $f \in L^2(\mu_G)$, where $\phi \in [\phi] \in \widehat{G}$. Define

$$f(x) := \sum_{[\phi] \in \widehat{G}} \dim(\phi) \, \mathrm{Tr} \left(F([\phi]) \, \phi(x) \right)$$

for μ_G-almost every $x \in G$. This can be done, since

$$f = \sum_{[\phi] \in \widehat{G}} \dim(\phi) \sum_{i,j=1}^{\dim(\phi)} F([\phi])_{ij} \, \phi_{ji}$$

belongs to $L^2(\mu_G)$ by

$$\|f\|_{L^2(\mu_G)}^2 = \|F\|_{L^2(\widehat{G})}^2 < \infty.$$

Clearly $\widehat{f}(\phi) = F([\phi])$, so that the Fourier transform is surjective. $\qquad\square$

We will return to a more detailed analysis of the space $L^2(\widehat{G})$ and of other spaces of functions and distributions on the unitary dual $\widehat{G}$ in Section 10.3.

7.7 Convolutions

For functions f and g on a group, their convolution $f * g$ can be thought as a modulation of one with the other. More precisely, the Fourier coefficients of $f * g$ are the pointwise products of the Fourier coefficients of f and g, as presented in Proposition 7.7.5.

Definition 7.7.1 (Convolutions on compact groups). Let G be a compact group, and let $f \in L^1(\mu_G)$ and $g \in C(G)$ (or $f \in C(G)$ and $g \in L^1(\mu_G)$). The *convolution* $f * g : G \to \mathbb{C}$ is defined by

$$f * g(x) := \int_G f(y)\, g(y^{-1}x)\, dy.$$

Remark 7.7.2. Now $f * g \in C(G)$. Indeed, due to the uniform continuity, for each $\varepsilon > 0$ there exists open $U \ni e$ such that $|g(x) - g(z)| < \varepsilon$ when $z^{-1}x \in U$. Thereby

$$
\begin{aligned}
|f * g(x) - f * g(z)| &\le \int_G |f(y)|\, |g(y^{-1}x) - g(y^{-1}z)|\, dy \\
&\le \|f\|_{L^1(\mu_G)}\, \varepsilon,
\end{aligned}
$$

when $z^{-1}x \in U$. Furthermore, the linear mapping $g \mapsto f * g$ satisfies

$$
\begin{aligned}
\|f * g\|_{C(G)} &\le \|f\|_{L^1(\mu_G)}\, \|g\|_{C(G)}, \\
\|f * g\|_{C(G)} &\le \|f\|_{C(G)}\, \|g\|_{L^1(\mu_G)}, \\
\|f * g\|_{L^1(\mu_G)} &\le \|f\|_{L^1(\mu_G)}\, \|g\|_{L^1(\mu_G)}.
\end{aligned}
$$

Hence we can consider $g \mapsto f * g$ as a bounded operator on $C(G)$ and $L^1(\mu_G)$; of course, we have symmetrical results for $g \mapsto g * f$.

It is also easy to show other L^p-boundedness results, like

$$\|f * g\|_{L^\infty(\mu_G)} \le \|f\|_{L^2(\mu_G)}\, \|g\|_{L^2(\mu_G)}$$

and so on. Notice that the convolution product is commutative if and only if G is commutative.

Proposition 7.7.3. *Let $f, g, h \in L^1(\mu_G)$. Then $f * g \in L^1(\mu_G)$,*

$$\|f * g\|_{L^1(\mu_G)} \le \|f\|_{L^1(\mu_G)}\, \|g\|_{L^1(\mu_G)},$$

and $f * g(x) = \int_G f(y^{-1})\, g(yx)\, dy$ for almost every $x \in G$. Moreover, for μ_G-almost every $x \in G$,

$$
\begin{aligned}
f * g(x) &= \int_G f(xy^{-1})\, g(y)\, dy \\
&= \int_G f(y^{-1})\, g(yx)\, dy \\
&= \int_G f(xy)\, g(y^{-1})\, dy.
\end{aligned}
$$

The convolution product is also associative: $f * (g * h) = (f * g) * h$.

Exercise 7.7.4. Prove Proposition 7.7.3.

Proposition 7.7.5. *For* $f, g \in L^1(\mu_G)$ *it holds that* $\widehat{f * g}(\phi) = \hat{g}(\phi)\, \hat{f}(\phi)$.

Proof. It is enough to assume that $f, g \in C(G)$. Then

$$
\begin{aligned}
\widehat{f * g}(\phi) &= \int_G f * g(x)\, \phi(x)^*\, dx \\
&= \int_G \int_G f(y)\, g(y^{-1}x)\, dy\, \phi(x)^*\, dx \\
&= \int_G \int_G g(y^{-1}x)\, \phi(y^{-1}x)^*\, dx\, f(y)\, \phi(y)^*\, dy \\
&= \int_G g(z)\, \phi(z)^*\, dz \int_G f(y)\, \phi(y)^*\, dy \\
&= \hat{g}(\phi)\, \hat{f}(\phi),
\end{aligned}
$$

completing the proof. $\qquad\square$

Remark 7.7.6. There are plenty of other interesting results concerning the Fourier transform and convolutions on compact groups. For instance, one can study approximate identities for $L^1(\mu_G)$ and prove that the Fourier transform $f \mapsto \hat{f}$ is injective on $L^1(\mu_G)$.

7.8 Characters

Loosely speaking, a character is the trace of a representation, and it contains all the essential information about the corresponding representation.

Definition 7.8.1 (Characters). Let $\phi : G \to \mathrm{Aut}(\mathcal{H})$ be a representation of a group G on a finite-dimensional Hilbert space $\mathcal{H}$. The *character* of ϕ is the function $\chi_\phi : G \to \mathbb{C}$ defined by

$$
\chi_\phi(x) := \mathrm{Tr}\,(\phi(x)).
$$

Remark 7.8.2 **(Purpose of characters).** Notice that here G is just any group, and that the character does not depend on the choice of the basis of $\mathcal{H}$. It turns out that on a compact group, characters provide a way of recognising equivalence of representations: namely, for finite-dimensional unitary representations, $\phi \sim \psi$ if and only if $\chi_\phi = \chi_\psi$, as we shall see.

Proposition 7.8.3 (Properties of characters). *Let* ϕ, ψ *be finite-dimensional representations of a group* G. *Then the following hold:*

(1) *If* $\phi \sim \psi$ *then* $\chi_\phi = \chi_\psi$.
(2) $\chi_\phi(xyx^{-1}) = \chi_\phi(y)$ *for all* $x, y \in G$.
(3) $\chi_{\phi \oplus \psi} = \chi_\phi + \chi_\psi$.
(4) $\chi_{\phi \otimes \psi|_G} = \chi_\phi\, \chi_\psi$.
(5) $\chi_\phi(e) = \dim(\phi)$.

Proof. The results follow from the properties of the trace functionals, see, e.g., Subsection B.5.1.   □

Remark 7.8.4. Since the character depends only on the equivalence class of a representation, we may define $\chi_{[\phi]} := \chi_\phi$, where $[\phi]$ is the equivalence class of ϕ.

Proposition 7.8.5 (Orthonormality of characters). *Let G be a compact group and $\xi, \eta \in \widehat{G}$. Then*

$$\langle \chi_\xi, \chi_\eta \rangle_{L^2(\mu_G)} = \begin{cases} 1 & \text{if } \xi = \eta, \\ 0 & \text{if } \xi \not\sim \eta. \end{cases}$$

Proof. Let $\phi = (\phi_{ij})_{i,j=1}^m \in \xi$ and $\psi = (\psi_{kl})_{k,l=1}^n \in \eta$. Then

$$
\begin{aligned}
\langle \chi_\xi, \chi_\eta \rangle_{L^2(\mu_G)} &= \sum_{j=1}^m \sum_{k=1}^n \langle \phi_{jj}, \psi_{kk} \rangle_{L^2(\mu_G)} \\
&= \begin{cases} 0 & \text{if } \phi \not\sim \psi, \\ 1 & \text{if } \phi = \psi \end{cases}
\end{aligned}
$$

by Lemma 7.5.12.   □

Theorem 7.8.6 (Irreducibility and equivalence characterisations). *Let ϕ, ψ be finite-dimensional continuous unitary representations of a compact group G. Then ϕ is irreducible if and only if $\|\chi_\phi\|_{L^2(\mu_G)} = 1$. Moreover, $\phi \sim \psi$ if and only if $\chi_\phi = \chi_\psi$.*

Proof. We already know the "only if"-parts of the proof. So suppose ϕ is a finite-dimensional unitary representation. Then

$$\phi \sim \bigoplus_{[\xi] \in \widehat{G}} \bigoplus_{j=1}^{m_{[\xi]}} \xi,$$

where $m_{[\xi]} \in \mathbb{N}$ is non-zero for only finitely many $[\xi] \in \widehat{G}$, and with the convention that the empty sum gives zero. Then

$$\chi_\phi = \sum_{[\xi] \in \widehat{G}} m_{[\xi]} \chi_\xi,$$

and if $[\eta] \in \widehat{G}$ then

$$\langle \chi_\phi, \chi_\eta \rangle_{L^2(\mu_G)} = \sum_{[\xi] \in \widehat{G}} m_{[\xi]} \langle \chi_\xi, \chi_\eta \rangle_{L^2(\mu_G)} = m_{[\eta]}.$$

This implies that the multiplicities $m_{[\xi]} \in \mathbb{N}$ can be uniquely obtained by knowing only χ_ϕ; hence if $\chi_\phi = \chi_\psi$ then $\phi \sim \psi$. Moreover,

$$
\begin{aligned}
\|\chi_\phi\|^2_{L^2(\mu_G)} &= \langle \chi_\phi, \chi_\phi \rangle_{L^2(\mu_G)} \\
&= \sum_{[\xi],[\eta] \in \widehat{G}} m_{[\xi]} \, m_{[\eta]} \, \langle \chi_\xi, \chi_\eta \rangle_{L^2(\mu_G)} \\
&= \sum_{[\xi] \in \widehat{G}} m^2_{[\xi]},
\end{aligned}
$$

so that ϕ is irreducible if and only if $\|\chi_\phi\|_{L^2(\mu_G)} = 1$. $\qquad\square$

Exercise 7.8.7. If $f \in L^2(\mu_G)$ then

$$
f = \sum_{[\xi] \in \widehat{G}} \dim(\xi) \, f * \chi_\xi = \sum_{[\xi] \in \widehat{G}} \dim(\xi) \; \mathrm{Tr}\left(\xi(x) \, \widehat{f}(\xi) \right).
$$

Thus, the projection of $f \in L^2(G)$ to $\mathcal{H}^\xi$ is given by $f \mapsto f * \chi_\xi$. The solution of this exercise can be found in Corollary 10.11.6.

We note that the restriction of the representation and the characters to the maximal torus of the group determine them completely:

Theorem 7.8.8 (Cartan's maximal torus theorem). *Let* $\mathbb{T}^n \hookrightarrow G$ *be an injective group homomorphism with the largest possible n. Then two representations* ϕ *and* ψ *of G are equivalent if and only if their restrictions to* $\mathbb{T}^n$ *are equivalent. In particular, the restriction* $\chi_\phi|_{\mathbb{T}^n}$ *of* χ_ϕ *to* $\mathbb{T}^n$ *determines the class* $[\phi]$.

Remark 7.8.9 (**Tensor products of representations**). According to Proposition 7.8.3, (4), we have $\chi_{\phi \otimes \psi|_G} = \chi_\phi \, \chi_\psi$ for any two finite-dimensional representations ϕ and ψ of G. By Theorem 7.5.5 the representation $\phi \otimes \psi|_G = (x \mapsto \phi(x) \otimes \psi(x)) \in \mathrm{Hom}(G, \mathcal{U}(\mathcal{H}_\phi \otimes \mathcal{H}_\psi))$ can be decomposed as a direct sum of irreducible unitary representations:

$$
\phi \otimes \psi|_G = \bigoplus_{[\xi] \in \widehat{G}} \overset{m_{\phi,\psi}([\xi])}{\underset{1}{\bigoplus}} \xi,
$$

where $m_{\phi,\psi}([\xi])$ is the multiplicity of $[\xi]$ in $\phi \otimes \psi|_G$, and only finitely many of $m_{\phi,\psi}([\xi])$ are non-zero in view of the finite dimensionality. We also have

$$
\chi_\phi \, \chi_\psi = \chi_{\phi \otimes \psi|_G} = \sum_{[\xi] \in \widehat{G}} m_{\phi,\psi}([\xi]) \, \chi_\xi
$$

in view of Proposition 7.8.3, (3). The multiplicities $m_{\phi,\psi}([\xi])$ can be analysed using Theorem 7.8.8 because we have, in particular, $\chi_\phi|_{\mathbb{T}^n} \, \chi_\psi|_{\mathbb{T}^n} = \chi_{\phi \otimes \psi|_{\mathbb{T}^n}} = \sum_{[\xi] \in \widehat{G}} m_{\phi,\psi}([\xi]) \, \chi_\xi|_{\mathbb{T}^n}$.

7.9 Induced representations

A group representation trivially gives a representation of its subgroup: if $H < G$ and $\psi \in \mathrm{Hom}(G, \mathrm{Aut}(V))$ then the restriction

$$\mathrm{Res}^G_H \psi := (h \mapsto \psi(h)) \in \mathrm{Hom}(H, \mathrm{Aut}(V)). \tag{7.4}$$

In the sequel, we show how a representation of a subgroup sometimes *induces* a representation for the whole group. This induction process has also plenty of nice properties. Induced representations were defined and studied by Ferdinand Georg Frobenius in 1898 for finite groups, and by George Mackey in 1949 for (most of the) locally compact groups.

    The technical assumptions here are that G is a *compact group*, $H < G$ is *closed* and $\phi \in \mathrm{Hom}(H, \mathcal{U}(\mathcal{H}))$ is *strongly continuous*; then ϕ induces a strongly continuous unitary representation

$$\mathrm{Ind}^G_H \phi \in \mathrm{Hom}\left(G, \mathcal{U}(\mathrm{Ind}^G_\phi \mathcal{H})\right),$$

where the notation will be explained in the sequel. We start by a lengthy definition of the induced representation space $\mathrm{Ind}^G_\phi \mathcal{H}$.

Remark 7.9.1 (**Uniformly continuous Hilbert space-valued mappings**). Since G is a compact group, continuous mappings $G \to \mathcal{H}$ are *uniformly continuous* in the following sense: Let $f \in C(G, \mathcal{H})$ and $\varepsilon > 0$. Then there exists open $U \ni e$ such that $\|f(x) - f(y)\|_{\mathcal{H}} < \varepsilon$ when $xy^{-1} \in U$ (or $x^{-1}y \in U$); the proof of this fact is as in the scalar-valued case. We shall also need to integrate $\mathcal{H}$-valued functions in the weak sense: that is, we need the concept of the Pettis integral, the details of which can be found from exercises related to Definition B.3.28 (see also Remark 7.9.3).

Proposition 7.9.2. *If $f \in C(G, \mathcal{H})$ then $f_\phi \in C(G, \mathcal{H})$, where*

$$f_\phi(x) := \int_H \phi(h) f(xh) \, \mathrm{d}\mu_H(h), \tag{7.5}$$

defined in the weak sense as the Pettis integral. Moreover, we have $f_\phi(xh) = \phi(h)^* f_\phi(x)$ for all $x \in G$ and $h \in H$.

Remark 7.9.3 (**Pettis integral**). The weak (Pettis) integration in (7.5) means that for every $f \in C(G, \mathcal{H})$ there exists a unique $f_\phi \in C(G, \mathcal{H})$ such that for all $u \in \mathcal{H}' = \mathcal{H}$ we have

$$\langle u, f_\phi \rangle_{\mathcal{H}} = \int_H \langle u, \phi(h) f(xh) \rangle_{\mathcal{H}} \, \mathrm{d}\mu_H(h).$$

We denote this f_ϕ as weak integral (7.5). The Riesz Representation Theorem B.5.19 gives the correctness of this integral definition since f_ϕ is clearly a bounded linear functional acting on $u \in \mathcal{H}$. For a more general version of the Pettis integral we refer to Definition B.3.28.

Proof of Proposition 7.9.2. Let $\{e_j\}_{j \in J} \subset \mathcal{H}$ be an orthonormal basis. Then

$$f_\phi(x) = \sum_{j \in J} \langle f_\phi(x), e_j \rangle_\mathcal{H} e_j \in \mathcal{H}$$

is the unique vector defined by inner products

$$\langle f_\phi(x), e_j \rangle_\mathcal{H} = \int_H \langle \phi(h) f(xh), e_j \rangle_\mathcal{H} \, d\mu_H(h).$$

It is easy to prove that the integrals here are sound, since

$$(h \mapsto \langle \phi(h) f(xh), e_j \rangle_\mathcal{H}) \in C(H)$$

because $f \in C(G, \mathcal{H})$ and ϕ is strongly continuous. If $h_0 \in H$ then

$$
\begin{aligned}
f_\phi(xh_0) &= \int_H \phi(h) \, f(xh_0 h) \, d\mu_H(h) \\
&= \int_H \phi(h_0^{-1} h) f(xh) \, d\mu_H(h) \\
&= \phi(h_0)^* f_\phi(x).
\end{aligned}
$$

Take $\varepsilon > 0$. By the uniform continuity of $f \in C(G, \mathcal{H})$ mentioned in Remark 7.9.1, there exists an open set $U \ni e$ such that $\|f(a) - f(b)\|_\mathcal{H} < \varepsilon$ whenever $ab^{-1} \in U$. If $x \in yU$ then

$$
\begin{aligned}
\|f_\phi(x) - f_\phi(y)\|_\mathcal{H}^2 &= \left\| \int_H \phi(h)(f(xh) - f(yh)) \, d\mu_H(h) \right\|_\mathcal{H}^2 \\
&\leq \left(\int_H \|f(xh) - f(yh)\|_\mathcal{H} \, d\mu_H(h) \right)^2 \leq \varepsilon^2,
\end{aligned}
$$

proving the continuity of f_ϕ. $\qquad\square$

Lemma 7.9.4. *If $f, g \in C(G, \mathcal{H})$ then $(xH \mapsto \langle f_\phi(x), g_\phi(x) \rangle_\mathcal{H}) \in C(G/H)$.*

Proof. Let $x \in G$ and $h \in H$. Then

$$
\begin{aligned}
\langle f_\phi(xh), g_\phi(xh) \rangle_\mathcal{H} &= \langle \phi(h)^* f_\phi(x), \phi(h)^* g_\phi(x) \rangle_\mathcal{H} \\
&= \langle f_\phi(x), g_\phi(x) \rangle_\mathcal{H},
\end{aligned}
$$

so that $(xH \mapsto \langle f_\phi(x), g_\phi(x) \rangle_\mathcal{H}) : G/H \to \mathbb{C}$ is well defined. There exists a constant $C < \infty$ such that $\|f_\phi(y)\|_\mathcal{H}, \|g_\phi(x)\|_\mathcal{H} \leq C$ because G is compact and $f_\phi, g_\phi \in C(G, \mathcal{H})$. Thereby

$$
\begin{aligned}
&|\langle f_\phi(x), g_\phi(x) \rangle_\mathcal{H} - \langle f_\phi(y), g_\phi(y) \rangle_\mathcal{H}| \\
&\leq |\langle f_\phi(x) - f_\phi(y), g_\phi(x) \rangle_\mathcal{H}| + |\langle f_\phi(y), g_\phi(x) - g_\phi(y) \rangle_\mathcal{H}| \\
&\leq C \left(\|f_\phi(x) - f_\phi(y)\|_\mathcal{H} + \|g_\phi(x) - g_\phi(y)\|_\mathcal{H} \right) \\
&\xrightarrow[x \to y]{} 0
\end{aligned}
$$

by the continuities of f_ϕ and g_ϕ. $\qquad\square$

Definition 7.9.5 (Induced representation space $\mathrm{Ind}_\phi^G \mathcal{H}$). Let us endow the vector space

$$
\begin{aligned}
C_\phi(G, \mathcal{H}) &:= \{f_\phi \mid f \in C(G, \mathcal{H})\} \\
&= \{e \in C(G, \mathcal{H}) \mid \forall x \in G \ \forall h \in H : \ e(xh) = \phi(h)^* e(x)\}
\end{aligned}
$$

with the inner product defined by

$$
\langle f_\phi, g_\phi \rangle_{\mathrm{Ind}_\phi^G \mathcal{H}} := \int_{G/H} \langle f_\phi(x), g_\phi(x) \rangle_{\mathcal{H}} \ \mathrm{d}\mu_{G/H}(xH).
$$

Let $\mathrm{Ind}_\phi^G \mathcal{H}$ be the completion of $C_\phi(G, \mathcal{H})$ with respect to the corresponding norm

$$
f_\phi \mapsto \|f_\phi\|_{\mathrm{Ind}_\phi^G \mathcal{H}} := \sqrt{\langle f_\phi, f_\phi \rangle_{\mathrm{Ind}_\phi^G \mathcal{H}}};
$$

this Hilbert space is called the *induced representation space*.

Remark 7.9.6. If $\mathcal{H} \neq \{0\}$ then $\{0\} \neq C_\phi(G, \mathcal{H}) \subset \mathrm{Ind}_\phi^G \mathcal{H}$. Why? Let $0 \neq u \in \mathcal{H}$. Due to the strong continuity of ϕ, we can choose open $U \subset G$ such that $e \in U$ and $\|(\phi(h) - \phi(e))u\|_{\mathcal{H}} < \|u\|_{\mathcal{H}}$ for all $h \in H \cap U$. Choose $w \in C(G)$ such that $w \geq 0$, $w|_{G \setminus U} = 0$ and $\int_H w(h) \ \mathrm{d}\mu_H(h) = 1$. Let $f(x) := w(x)u$ for all $x \in G$. Then

$$
\begin{aligned}
\|f_\phi(e) - u\|_{\mathcal{H}} &= \left\| \int_H w(h) \ (\phi(h) - \phi(e))u \ \mathrm{d}\mu_H(h) \right\|_{\mathcal{H}} \\
&= \int_H w(h) \ \|(\phi(h) - \phi(e))u\|_{\mathcal{H}} \ \mathrm{d}\mu_H(h) \\
&< \|u\|_{\mathcal{H}},
\end{aligned}
$$

so that $f_\phi(e) \neq 0$, yielding $f_\phi \neq 0$.

Theorem 7.9.7 (Induced representations). *If $x, y \in G$ and $f_\phi \in C_\phi(G, \mathcal{H})$, let*

$$
\left(\mathrm{Ind}_H^G \phi(y) f_\phi \right)(x) := f_\phi(y^{-1}x).
$$

This begets a unique strongly continuous $\mathrm{Ind}_H^G \phi \in \mathrm{Hom}\left(G, \mathcal{U}(\mathrm{Ind}_\phi^G \mathcal{H}) \right)$, called the representation of G induced by ϕ.

Proof. If $y \in G$ and $f_\phi \in C_\phi(G, \mathcal{H})$ then $\mathrm{Ind}_H^G \phi(y) f_\phi = g_\phi \in C_\phi(G, \mathcal{H})$, where $g \in C(G, \mathcal{H})$ is defined by $g(x) := f(y^{-1}x)$. Thus we have a linear mapping $\mathrm{Ind}_H^G \phi(y) : C_\phi(G, \mathcal{H}) \to C_\phi(G, \mathcal{H})$. Clearly

$$
\mathrm{Ind}_H^G \phi(yz) f_\phi = \mathrm{Ind}_H^G \phi(y) \ \mathrm{Ind}_H^G \phi(z) f_\phi.
$$

Hence $\mathrm{Ind}_H^G \phi \in \mathrm{Hom}\left(G, \mathrm{Aut}(C_\phi(G, \mathcal{H})) \right)$.

If $f, g \in C(G, \mathcal{H})$ then

$$
\begin{aligned}
\left\langle \operatorname{Ind}_H^G \phi(y) f_\phi, g_\phi \right\rangle_{\operatorname{Ind}_\phi^G \mathcal{H}} &= \int_{G/H} \langle f_\phi(y^{-1}x), g_\phi(x) \rangle_{\mathcal{H}} \; \mathrm{d}\mu_{G/H}(xH) \\
&= \int_{G/H} \langle f_\phi(z), g_\phi(yz) \rangle_{\mathcal{H}} \; \mathrm{d}\mu_{G/H}(zH) \\
&= \left\langle f_\phi, \operatorname{Ind}_H^G \phi(y)^{-1} g_\phi \right\rangle_{\operatorname{Ind}_\phi^G \mathcal{H}};
\end{aligned}
$$

hence we have an extension $\operatorname{Ind}_H^G \phi \in \operatorname{Hom}\left(G, \mathcal{U}(\operatorname{Ind}_\phi^G \mathcal{H})\right)$. Next we exploit the uniform continuity of $f \in C(G, \mathcal{H})$: Let $\varepsilon > 0$. Take an open set $U \ni e$ such that $\|f(a) - f(b)\|_{\mathcal{H}} < \varepsilon$ when $ab^{-1} \in U$. Thereby, if $y^{-1}z \in U$ then

$$
\begin{aligned}
\left\| \left(\operatorname{Ind}_H^G \phi(y) - \operatorname{Ind}_H^G \phi(z) \right) f_\phi \right\|_{\operatorname{Ind}_\phi^G \mathcal{H}}^2 & \\
= \int_{G/H} \left\| f_\phi(y^{-1}x) - f_\phi(z^{-1}x) \right\|_{\mathcal{H}}^2 & \; \mathrm{d}\mu_{G/H}(xH) \\
\le \quad \varepsilon^2. &
\end{aligned}
$$

This shows the strong continuity of the induced representation. $\qquad\square$

Remark 7.9.8. In the sequel, some elementary properties of induced representations are deduced. Briefly: induced representations of equivalent representations are equivalent, and the induction process can be taken in stages leading to the same result modulo equivalence.

Proposition 7.9.9. *Let G be a compact group and $H < G$ a closed subgroup. Let $\phi \in \operatorname{Hom}(H, \mathcal{U}(\mathcal{H}_\phi))$ and $\psi \in \operatorname{Hom}(H, \mathcal{U}(\mathcal{H}_\psi))$ be strongly continuous. If $\phi \sim \psi$ then $\operatorname{Ind}_H^G \phi \sim \operatorname{Ind}_H^G \psi$.*

Proof. Since $\phi \sim \psi$, there is an isometric isomorphism $A \in \operatorname{Hom}(\phi, \psi)$. Then

$$
(Bf_\phi)(x) := A(f_\phi(x))
$$

defines a linear mapping $B : C_\phi(G, \mathcal{H}_\phi) \to C_\psi(G, \mathcal{H}_\psi)$, because if $x \in G$ and $h \in H$ then

$$
\begin{aligned}
(Bf_\phi)(xh) &= A(f_\phi(xh)) \\
&= A(\phi(h)^* f_\phi(x)) \\
&= A(\phi(h)^* A^* A(f_\phi(x))) \\
&= A(A^* \psi(h)^* A(f_\phi(x))) \\
&= \psi(h)^* A(f_\phi(x)) \\
&= \psi(h)^* (Bf_\phi)(x).
\end{aligned}
$$

Furthermore, B can be extended to a unique linear isometry $C : \mathrm{Ind}_\phi^G \mathcal{H}_\phi \to \mathrm{Ind}_\psi^G \mathcal{H}_\psi$, since

$$
\begin{aligned}
\|Bf_\phi\|^2_{\mathrm{Ind}_\psi^G \mathcal{H}_\psi} &= \int_{G/H} \|(Bf_\phi)(x)\|^2_{\mathcal{H}_\psi} \ \mathrm{d}\mu_{G/H}(xH) \\
&= \int_{G/H} \|A(f_\phi(x))\|^2_{\mathcal{H}_\psi} \ \mathrm{d}\mu_{G/H}(xH) \\
&= \int_{G/H} \|f_\phi(x)\|^2_{\mathcal{H}_\phi} \ \mathrm{d}\mu_{G/H}(xH) \\
&= \|f_\phi\|^2_{\mathrm{Ind}_\phi^G \mathcal{H}_\phi}.
\end{aligned}
$$

Next, C is a surjection: if $F \in C_\psi(G, \mathcal{H}_\psi)$ then $(y \mapsto A^{-1}(F(y)) \in C_\phi(G, \mathcal{H}_\phi)$ and $\left(C\left(y \mapsto A^{-1}(F(y))\right)\right)(x) = AA^{-1}(F(x)) = F(x)$, and this is enough due to the density of $C_\psi(G, \mathcal{H}_\psi)$ in $\mathrm{Ind}_\phi^G \mathcal{H}$. Finally,

$$
\begin{aligned}
(C \ \mathrm{Ind}_H^G \phi(y) f_\phi)(x) &= A(\mathrm{Ind}_H^G \phi(y) f_\phi(x)) \\
&= A(f_\phi(y^{-1}x)) \\
&= (Cf_\phi)(y^{-1}x) \\
&= (\mathrm{Ind}_H^G \phi(y) Cf_\phi)(x),
\end{aligned}
$$

so that $C \in \mathrm{Hom}\left(\mathrm{Ind}_H^G \phi, \mathrm{Ind}_H^G \psi\right)$ is an isometric isomorphism. $\square$

Corollary 7.9.10. *Let G be a compact group and $H < G$ closed. Let ϕ_1 and ϕ_2 be strongly continuous unitary representations of H. Then $\mathrm{Ind}_H^G(\phi_1 \oplus \phi_2) \sim \left(\mathrm{Ind}_H^G \phi_1\right) \oplus \left(\mathrm{Ind}_H^G \phi_2\right)$.*

Exercise 7.9.11. Prove Corollary 7.9.10.

Corollary 7.9.12. $\mathrm{Ind}_H^G \phi$ *is irreducible only if ϕ is irreducible.*

Exercise 7.9.13. Let G_1, G_2 be compact groups and $H_1 < G_1, H_2 < G_2$ be closed. Let ϕ_1, ϕ_2 be strongly continuous unitary representations of H_1, H_2, respectively. Show that

$$
\mathrm{Ind}_{H_1 \times H_2}^{G_1 \times G_2}(\phi_1 \otimes \phi_2) \sim \left(\mathrm{Ind}_{H_1}^{G_1} \phi_1\right) \otimes \left(\mathrm{Ind}_{H_2}^{G_2} \phi_2\right).
$$

Theorem 7.9.14 (Inducing representations in steps). *Let G be a compact group and $H < K < G$, where H, K are closed. If $\phi \in \mathrm{Hom}(H, \mathcal{U}(\mathcal{H}))$ is strongly continuous then $\mathrm{Ind}_H^G \phi \sim \mathrm{Ind}_K^G \mathrm{Ind}_H^K \phi$.*

Proof. In this proof, $x \in G$, $k, k_0 \in K$ and $h \in H$. Let $\psi := \mathrm{Ind}_H^K \phi$ and $\mathcal{H}_\psi := \mathrm{Ind}_H^K \mathcal{H}$. Let $f_\phi \in C_\phi(G, \mathcal{H})$. Since $(k \mapsto f_\phi(xk)) : K \to \mathcal{H}$ is continuous and $f_\phi(xkh) = \phi(h)^* f_\phi(xk)$, we obtain $(k \mapsto f_\phi(xk)) \in C_\phi(K, \mathcal{H}) \subset \mathcal{H}_\psi$. Let us define $f_\phi^K : G \to \mathcal{H}_\psi$ by

$$
f_\phi^K(x) := (k \mapsto f_\phi(xk)).
$$

If $x \in G$ and $k_0 \in K$ then

$$
\begin{aligned}
f_\phi^K(xk_0)(k) &= f_\phi(xk_0k) \\
&= f_\phi^K(x)(k_0k) \\
&= \left(\psi(k_0)^* f_\phi^K(x)\right)(k),
\end{aligned}
$$

i.e., $f_\phi^K(xk_0) = \psi(k_0)^* f_\phi^K(x)$. Let $\varepsilon > 0$. By the uniform continuity of f_ϕ, take open $U \ni e$ such that $\|f_\phi(a) - f_\phi(b)\|_{\mathcal{H}} < \varepsilon$ if $ab^{-1} \in U$. Thereby if $xy^{-1} \in U$ then

$$
\begin{aligned}
\left\| f_\phi^K(x) - f_\phi^K(y) \right\|_{\mathcal{H}_\psi}^2 &= \int_{K/H} \left\| f_\phi^K(x)(k) - f_\phi^K(y)(k) \right\|_{\mathcal{H}}^2 \, d\mu_{K/H}(kH) \\
&= \int_{K/H} \left\| f_\phi(xk) - f_\phi(yk) \right\|_{\mathcal{H}}^2 \, d\mu_{K/H}(kH) \\
&\leq \varepsilon^2.
\end{aligned}
$$

Hence $f_\phi^K \in C_\psi(G, \mathcal{H}_\psi) \subset \mathrm{Ind}_\psi^G \mathcal{H}_\psi$, so that we indeed have a mapping $(f_\phi \mapsto f_\phi^K) : C_\phi(G, \mathcal{H}) \to C_\psi(G, \mathcal{H}_\psi)$.

Next, we claim that $f_\phi \mapsto f_\phi^K$ defines a surjective linear isometry $\mathrm{Ind}_\phi^G \mathcal{H} \to \mathrm{Ind}_\psi^G \mathcal{H}_\psi$. Isometricity follows by

$$
\begin{aligned}
\left\| f_\phi^K \right\|_{\mathrm{Ind}_\psi^G \mathcal{H}_\psi}^2 &= \int_{G/K} \left\| f_\phi^K(x) \right\|_{\mathcal{H}_\psi}^2 \, d\mu_{G/K}(xK) \\
&= \int_{G/K} \int_{K/H} \left\| f_\phi^K(x)(k) \right\|_{\mathcal{H}}^2 \, d\mu_{K/H}(kH) \, d\mu_{G/K}(xK) \\
&= \int_{G/K} \int_{K/H} \left\| f_\phi(xk) \right\|_{\mathcal{H}}^2 \, d\mu_{K/H}(kH) \, d\mu_{G/K}(xK) \\
&= \int_{G/H} \left\| f_\phi(x) \right\|_{\mathcal{H}}^2 \, d\mu_{G/H}(xH) \\
&= \left\| f_\phi \right\|_{\mathrm{Ind}_\phi^G \mathcal{H}}^2.
\end{aligned}
$$

How about the surjectivity? The representation space $\mathrm{Ind}_\psi^G \mathcal{H}_\psi$ is the closure of $C_\psi(G, \mathcal{H}_\psi)$, and $\mathcal{H}_\psi$ is the closure of $C_\phi(K, \mathcal{H})$. Consequently, $\mathrm{Ind}_\psi^G \mathcal{H}_\psi$ is the closure of the vector space

$$
\begin{aligned}
C_\psi(G, C_\phi(K, \mathcal{H})) := \{ g \in C(G, C(K, \mathcal{H})) \mid \forall x \in G \; \forall k \in K \; \forall h \in H : \\
g(xk) = \psi(k)^* g(x), \; g(x)(kh) = \phi(h)^* g(x)(k) \}.
\end{aligned}
$$

Given $g \in C_\psi(G, C_\phi(K, \mathcal{H}))$, define $f_\phi \in C_\phi(G, \mathcal{H})$ by $f_\phi(x) := g(x)(e)$. Then $f_\phi^K = g$, because

$$
f_\phi^K(x)(k) = f_\phi(xk) = g(xk)(e) = \psi(k)^* g(x)(e) = g(x)(k).
$$

Thus $(f_\phi \mapsto f_\phi^K) : C_\phi(G, \mathcal{H}) \to C_\psi(G, C_\phi(K, \mathcal{H}))$ is a linear isometric bijection. Hence this mapping can be extended uniquely to a linear isometric bijection $A : \mathrm{Ind}_\phi^G \mathcal{H} \to \mathrm{Ind}_\psi^G \mathcal{H}_\psi$.

Finally, $A \in \mathrm{Hom}\left(\mathrm{Ind}_H^G \phi, \mathrm{Ind}_K^G \mathrm{Ind}_H^K \phi\right)$, since

$$
\begin{aligned}
A\left(\mathrm{Ind}_H^G \phi(y) f_\phi(x)\right) &= A f_\phi(y^{-1}x) \\
&= f_\phi^K(y^{-1}x) \\
&= \mathrm{Ind}_K^G \psi(y) f_\phi^K(x) \\
&= \mathrm{Ind}_K^G \psi(y) A f_\phi(x).
\end{aligned}
$$

This completes the proof.   □

Exercise 7.9.15. Let H be a closed subgroup of a compact group G. Let $\phi = (h \mapsto I) \in \mathrm{Hom}(H, \mathcal{U}(\mathcal{H}))$, where $I = (u \mapsto u) : \mathcal{H} \to \mathcal{H}$.
a) Show that $\mathrm{Ind}_\phi^G \mathcal{H} \cong L^2(G/H, \mathcal{H})$, where the $L^2(G/H, \mathcal{H})$ inner product is given by

$$
\langle f_{G/H}, g_{G/H}\rangle_{L^2(G/H, \mathcal{H})} := \int_{G/H} \langle f_{G/H}(xH), g_{G/H}(xH)\rangle_{\mathcal{H}} \, d\mu_{G/H}(xH),
$$

when $f_{G/H}, g_{G/H} \in C(G/H, \mathcal{H})$.
b) Let $K < G$ be closed. Let π_K and π_G be the left regular representations of K and G, respectively. Prove that $\pi_G \sim \mathrm{Ind}_K^G \pi_K$.

Remark 7.9.16 **(Multiplicity of a representation).** A fundamental result for induced representations is the *Frobenius Reciprocity Theorem* 7.9.17, stated below without a proof. Let G be a compact group and $\phi \in \mathrm{Hom}(G, \mathcal{U}(\mathcal{H}))$ be strongly continuous. Let $n\,([\xi], \phi) \in \mathbb{N}$ denote the *multiplicity of* $[\xi] \in \widehat{G}$ in ϕ, defined as follows: if $\phi = \bigoplus_{j=1}^k \phi_j$, where each ϕ_j is a continuous irreducible unitary representation, then
$$
n([\xi], \phi) := |\{j \in \{1, \ldots, k\} : [\phi_j] = [\xi]\}|.
$$

That is, $n([\xi], \phi)$ is the number of times ξ may occur as an irreducible component in a direct sum decomposition of ϕ.

Theorem 7.9.17 (Frobenius Reciprocity Theorem). *Let G be a compact group and $H < G$ be closed. Let ξ, η be continuous such that $[\xi] \in \widehat{G}$ and $[\eta] \in \widehat{H}$. Then*

$$
n\left([\xi], \mathrm{Ind}_H^G \eta\right) = n\left([\eta], \mathrm{Res}_H^G \xi\right),
$$

where $\mathrm{Res}_H^G \xi$ is the restriction[1] of ξ to H.

---

[1]see (7.4) for the definition of $\mathrm{Res}_H^G \xi$.

Example. Let $[\xi] \in \widehat{G}$, $H = \{e\}$ and $\eta = (e \mapsto I) \in \mathrm{Hom}(H, \mathcal{U}(\mathbb{C}))$. Then $\pi_L \sim \mathrm{Ind}_H^G \eta$ by Exercise 7.9.15, and $\widehat{H} = \{[\eta]\}$, so that

$$
\begin{aligned}
n\left([\xi], \mathrm{Ind}_H^G \eta\right) &= n\left([\xi], \pi_L\right) \\
&\overset{\text{Peter--Weyl}}{=} \dim(\xi) \\
&= \dim(\xi)\, n\left([\eta], \eta\right) \\
&= n\left([\eta], \bigoplus_{j=1}^{\dim(\xi)} \eta\right) \\
&= n\left([\eta], \mathrm{Res}_H^G \xi\right).
\end{aligned}
$$

As it should be, this is in accordance with the Frobenius Reciprocity Theorem 7.9.17.

Example. Let $[\xi], [\eta] \in \widehat{G}$. Then by the Frobenius Reciprocity Theorem 7.9.17,

$$
\begin{aligned}
n\left([\xi], \mathrm{Ind}_G^G \eta\right) &= n\left([\eta], \mathrm{Res}_G^G \xi\right) \\
&= n\left([\eta], \xi\right) \\
&= \begin{cases} 1, & \text{when } [\xi] = [\eta], \\ 0, & \text{when } [\xi] \neq [\eta]. \end{cases}
\end{aligned}
$$

Let ϕ be a finite-dimensional continuous unitary representation of G. Then $\phi = \bigoplus_{j=1}^{k} \xi_k$, where each ξ_k is irreducible. Thereby

$$
\mathrm{Ind}_G^G \phi \sim \bigoplus_{j=1}^{k} \mathrm{Ind}_G^G \xi_j \sim \bigoplus_{j=1}^{k} \xi_j \sim \phi;
$$

in other words, induction practically does nothing in this case.

Chapter 8

Linear Lie Groups

In this chapter we study linear Lie groups, i.e., Lie groups which are closed subgroups of $\mathrm{GL}(n, \mathbb{C})$. But first some words about the general Lie groups:

Definition 8.0.1 (Lie groups). A *Lie group* is a C^∞-manifold which is also a group such that the group operations are C^∞-smooth.

We will be mostly interested in the non-commutative Lie groups in view of the following:

Remark 8.0.2 (**Commutative Lie groups**). In the introduction to Part II we mentioned that in the case of commutative groups it is sufficient to study cases of $\mathbb{T}^n$ and $\mathbb{R}^n$. Indeed, we have the following two facts:

- Any compact commutative Lie group is isomorphic to the product of a torus with a finite commutative group.

- Any connected commutative Lie group is isomorphic to the product of a torus and the Euclidean space. In other words, if G is a connected commutative Lie group then $G \cong \mathbb{T}^n \times \mathbb{R}^m$ for some n, m.

We will not prove these facts here but refer to, e.g., [20, p. 25] for further details.

Definition 8.0.3 (Linear Lie groups). A *linear Lie group* is a Lie group which is a closed subgroup of $\mathrm{GL}(n, \mathbb{C})$.

There is a result stating that any compact Lie group is diffeomorphic to a linear Lie group, and thereby the matrix groups are especially interesting. In fact, we have:

Corollary 8.0.4 (Universality of unitary groups). *Let G be a compact Lie group. Then there is some $n \in \mathbb{N}$ such that G is isomorphic to a subgroup of $\mathrm{U}(n)$.*

8.1 Exponential map

The fundamental tool for studying linear Lie groups is the matrix exponential map, treated below.

Let us endow $\mathbb{C}^n$ with the Euclidean inner product

$$(x, y) \mapsto \langle x, y \rangle_{\mathbb{C}^n} := \sum_{j=1}^{n} x_j \overline{y_j}.$$

The corresponding norm is $x \mapsto \|x\|_{\mathbb{C}^n} := \langle x, x \rangle_{\mathbb{C}^n}^{1/2}$. We identify the matrix algebra $\mathbb{C}^{n \times n}$ with $\mathcal{L}(\mathbb{C}^n)$, the algebra of linear operators $\mathbb{C}^n \to \mathbb{C}^n$. Let us endow $\mathbb{C}^{n \times n} \cong \mathcal{L}(\mathbb{C}^n)$ with the operator norm

$$Y \mapsto \|Y\|_{\mathcal{L}(\mathbb{C}^n)} := \sup_{x \in \mathbb{C}^n:\ \|x\|_{\mathbb{C}^n} \leq 1} \|Yx\|_{\mathbb{C}^n}.$$

Notice that $\|XY\|_{\mathcal{L}(\mathbb{C}^n)} \leq \|X\|_{\mathcal{L}(\mathbb{C}^n)} \|Y\|_{\mathcal{L}(\mathbb{C}^n)}$. For a matrix $X \in \mathbb{C}^{n \times n}$, the *exponential* $\exp(X) \in \mathbb{C}^{n \times n}$ is defined by the power series

$$\exp(X) := \sum_{k=0}^{\infty} \frac{1}{k!} X^k,$$

where $X^0 := I$; this series converges in the Banach space $\mathbb{C}^{n \times n} \cong \mathcal{L}(\mathbb{C}^n)$, because

$$\sum_{k=0}^{\infty} \frac{1}{k!} \|X^k\|_{\mathcal{L}(\mathbb{C}^n)} \leq \sum_{k=0}^{\infty} \frac{1}{k!} \|X\|_{\mathcal{L}(\mathbb{C}^n)}^k = e^{\|X\|_{\mathcal{L}(\mathbb{C}^n)}} < \infty.$$

Proposition 8.1.1. *Let* $X, Y \in \mathbb{C}^{n \times n}$. *If* $XY = YX$ *then*

$$\exp(X + Y) = \exp(X) \exp(Y).$$

Therefore $\exp : \mathbb{C}^{n \times n} \to \mathrm{GL}(n, \mathbb{C})$ *satisfies* $\exp(-X) = \exp(X)^{-1}$.

Proof. Now

$$
\begin{aligned}
\exp(X + Y) \quad &= \quad \lim_{l \to \infty} \sum_{k=0}^{2l} \frac{1}{k!} (X + Y)^k \\
&\overset{XY = YX}{=} \quad \lim_{l \to \infty} \sum_{k=0}^{2l} \frac{1}{k!} \sum_{i=0}^{k} \frac{k!}{i!\,(k-i)!} X^i Y^{k-i} \\
&= \quad \lim_{l \to \infty} \left(\sum_{i=0}^{l} \frac{1}{i!} X^i \sum_{j=0}^{l} \frac{1}{j!} Y^j + \sum_{\substack{i,j:\ i+j \leq 2l, \\ \max(i,j) > l}} \frac{1}{i!\,j!} X^i Y^j \right)
\end{aligned}
$$

$$= \lim_{l \to \infty} \left(\sum_{i=0}^{l} \frac{1}{i!} X^i \sum_{j=0}^{l} \frac{1}{j!} Y^j \right)$$

$$= \exp(X) \exp(Y),$$

since the remainder term satisfies

$$\left\| \sum_{\substack{i,j:\ i+j \leq 2l, \\ \max(i,j) > l}} \frac{1}{i!\,j!} X^i Y^j \right\|_{\mathcal{L}(\mathbb{C}^n)} \leq \sum_{\substack{i,j:\ i+j \leq 2l, \\ \max(i,j) > l}} \frac{1}{i!\,j!} \|X\|^i_{\mathcal{L}(\mathbb{C}^n)} \|Y\|^j_{\mathcal{L}(\mathbb{C}^n)}$$

$$\leq l(l+1) \frac{1}{(l+1)!} c^{2l}$$

$$\xrightarrow[l \to \infty]{} 0,$$

where $c := \max\left(1, \|X\|_{\mathcal{L}(\mathbb{C}^n)}, \|Y\|_{\mathcal{L}(\mathbb{C}^n)}\right)$.

Consequently, $I = \exp(0) = \exp(X)\exp(-X) = \exp(-X)\exp(X)$, so that we get $\exp(-X) = \exp(X)^{-1}$. $\qquad \square$

Exercise 8.1.2. Verify the estimates and the ranges of the summation indices in the proof of Proposition 8.1.1.

Lemma 8.1.3. *Let $X \in \mathbb{C}^{n \times n}$ and $P \in \mathrm{GL}(n, \mathbb{C})$. Then*

$$\begin{aligned} \exp\left(X^{\mathrm{T}}\right) &= \exp(X)^{\mathrm{T}}, \\ \exp(X^*) &= \exp(X)^*, \\ \exp(PXP^{-1}) &= P\exp(X)P^{-1}. \end{aligned}$$

Proof. For the adjoint X^*,

$$\exp(X^*) = \sum_{k=0}^{\infty} \frac{1}{k!}(X^*)^k = \sum_{k=0}^{\infty} \frac{1}{k!}(X^k)^* = \left(\sum_{k=0}^{\infty} \frac{1}{k!} X^k \right)^* = \exp(X)^*,$$

and similarly for the transpose X^{T}. Finally,

$$\exp(PXP^{-1}) = \sum_{k=0}^{\infty} \frac{1}{k!}(PXP^{-1})^k = \sum_{k=0}^{\infty} \frac{1}{k!} PX^k P^{-1} = P\exp(X)P^{-1}. \qquad \square$$

Proposition 8.1.4. *If $\lambda \in \mathbb{C}$ is an eigenvalue of $X \in \mathbb{C}^{n \times n}$ then e^λ is an eigenvalue of $\exp(X)$. Consequently*

$$\det(\exp(X)) = \mathrm{e}^{\mathrm{Tr}(X)}.$$

Proof. Choose $P \in \mathrm{GL}(n, \mathbb{C})$ such that $Y := PXP^{-1} \in \mathbb{C}^{n \times n}$ is upper triangular; the eigenvalues of X and Y are the same, and for triangular matrices the eigenvalues are the diagonal elements. Since Y^k is upper triangular for every $k \in \mathbb{N}$,

$\exp(Y)$ is upper triangular. Moreover, $(Y^k)_{jj} = (Y_{jj})^k$, so that $(\exp(Y))_{jj} = e^{Y_{jj}}$. The eigenvalues of $\exp(X)$ and $\exp(Y) = P \exp(X) P^{-1}$ are the same. The determinant of a matrix is the product of its eigenvalues; the trace of a matrix is the sum of its eigenvalues; this implies the last claim. $\qquad\square$

Remark 8.1.5. Recall that $\mathrm{HOM}(G, H)$ is the set of continuous homomorphisms from G to H, see Definition 7.1.12.

Theorem 8.1.6 (The form of $\mathrm{HOM}(\mathbb{R}, \mathrm{GL}(n, \mathbb{C}))$). *We have*

$$\mathrm{HOM}(\mathbb{R}, \mathrm{GL}(n, \mathbb{C})) = \{t \mapsto \exp(tX) \mid X \in \mathbb{C}^{n \times n}\}.$$

Proof. It is clear that $(t \mapsto \exp(tX)) \in \mathrm{HOM}(\mathbb{R}, \mathrm{GL}(n, \mathbb{C}))$, since it is continuous and $\exp(sX) \exp(tX) = \exp((s + t)X)$.

Let $\phi \in \mathrm{HOM}(\mathbb{R}, \mathrm{GL}(n, \mathbb{C}))$. Then $\phi(s + t) = \phi(s)\phi(t)$ implies that

$$\left(\int_0^h \phi(s) \, ds \right) \phi(t) = \int_0^h \phi(s + t) \, ds = \int_t^{t+h} \phi(u) \, du.$$

Recall that if $\|I - A\|_{\mathcal{L}(\mathbb{C}^n)} < 1$ then $A \in \mathbb{C}^{n \times n}$ is invertible; now

$$\left\| I - \frac{1}{h} \int_0^h \phi(s) \, ds \right\|_{\mathcal{L}(\mathbb{C}^n)} = \left\| \frac{1}{h} \int_0^h (I - \phi(s)) \, ds \right\|_{\mathcal{L}(\mathbb{C}^n)}$$

$$\leq \sup_{s: \, |s| \leq |h|} \|I - \phi(s)\|_{\mathcal{L}(\mathbb{C}^n)}$$

$$< 1$$

when $|h|$ is small enough, because $\phi(0) = I$ and ϕ is continuous. Therefore $\int_0^h \phi(s) \, ds$ is invertible for small $|h|$, and we get

$$\phi(t) = \left(\int_0^h \phi(s) \, ds \right)^{-1} \int_t^{t+h} \phi(u) \, du.$$

Since ϕ is continuous, this formula states that ϕ is differentiable. Now

$$\phi'(t) = \lim_{s \to 0} \frac{\phi(s + t) - \phi(t)}{s} = \lim_{s \to 0} \frac{\phi(s) - \phi(0)}{s} \, \phi(t) = X \, \phi(t),$$

where $X := \phi'(0)$. Hence the initial value problem

$$\begin{cases} \psi'(t) = X \, \psi(t), & \psi : \mathbb{R} \to \mathrm{GL}(n, \mathbb{C}), \\ \psi(0) = I \end{cases}$$

has the solutions $\psi = \phi$ and $\psi = \phi_X := (t \mapsto \exp(tX))$. Define $\alpha : \mathbb{R} \to \mathrm{GL}(n, \mathbb{C})$ by $\alpha(t) := \phi(t) \, \phi_X(-t)$. Then $\alpha(0) = \phi(0) \, \phi_X(0) = I$ and

$$\begin{aligned} \alpha'(t) &= \phi'(t) \, \phi_X(-t) - \phi(t) \, \phi'_X(-t) \\ &= X \, \phi(t) \, \phi_X(-t) - \phi(t) \, X \, \phi_X(-t) = 0, \end{aligned}$$

since $X \, \phi(t) = \phi(t) \, X$. Thus $\alpha(t) = I$ for all $t \in \mathbb{R}$, so that $\phi = \phi_X$. $\qquad\square$

Proposition 8.1.7 (Logarithms). *Let $A \in \mathbb{C}^{n \times n}$ be such that $\|I - A\|_{\mathcal{L}(\mathbb{C}^n)} < 1$. The* logarithm

$$\log(A) := -\sum_{k=1}^{\infty} \frac{1}{k}(I - A)^k$$

is well defined, and $\exp(\log(A)) = A$. *Moreover, there exists $r > 0$ such that* $\log(\exp(X)) = X$ *if $\|X\|_{\mathcal{L}(\mathbb{C}^n)} < r$.*

Proof. Let $c := \|I - A\| < 1$ for a matrix $A \in \mathbb{C}^{n \times n}$. Then

$$\sum_{k=1}^{\infty} \frac{1}{k} \left\| (I - A)^k \right\|_{\mathcal{L}(\mathbb{C}^n)} \le \sum_{k=1}^{\infty} \frac{1}{k} \|I - A\|_{\mathcal{L}(\mathbb{C}^n)}^k \le \sum_{k=1}^{\infty} c^k = \frac{c}{1 - c} < \infty,$$

so that $\log(A)$ is well defined. Noticing that I and A commute, we have

$$\exp(\log(A)) = \sum_{k=0}^{\infty} \frac{1}{k!} \left(-\sum_{l=1}^{\infty} \frac{1}{l}(I - A)^l \right)^k = A,$$

because if $|1 - a| < 1$ for a *number* $a \in \mathbb{C}$, then

$$e^{\ln a} = \sum_{k=0}^{\infty} \frac{1}{k!} \left(-\sum_{l=1}^{\infty} \frac{1}{l}(1 - a)^l \right)^k = a. \tag{8.1}$$

Due to the continuity of the exponential function, there exists $r > 0$ such that $|1 - e^x| < 1$ if $x \in \mathbb{C}$ satisfies $|x| < r$, and then

$$\ln(e^x) = -\sum_{l=1}^{\infty} \frac{1}{l}(1 - e^x)^l = -\sum_{l=1}^{\infty} \frac{1}{l} \left(1 - \sum_{k=0}^{\infty} \frac{1}{k!} x^k \right)^l = x, \tag{8.2}$$

so that if $X \in \mathbb{C}^{n \times n}$ satisfies $\|X\|_{\mathcal{L}(\mathbb{C})} < r$ then

$$\log(\exp(X)) = -\sum_{l=1}^{\infty} \frac{1}{l}(I - \exp(X))^l = -\sum_{l=1}^{\infty} \frac{1}{l} \left(I - \sum_{k=0}^{\infty} \frac{1}{k!} X^k \right)^l = X. \quad \square$$

Exercise 8.1.8. Find an estimate for r in Proposition 8.1.7.

Exercise 8.1.9. Justify formulae (8.1) and (8.2) and their matrix forms.

Corollary 8.1.10. *Let r be as above and $\mathbb{B} := \{ X \in \mathbb{C}^{n \times n} : \|X\|_{\mathcal{L}(\mathbb{C}^n)} < r \}$. Then* $(X \mapsto \exp(X)) : \mathbb{B} \to \exp(\mathbb{B})$ *is a diffeomorphism (i.e., a bijective C^{∞}-smooth mapping).*

Proof. As exp and log are defined by power series, they are not just C^{∞}-smooth but also analytic. $\square$

Lemma 8.1.11. *Let $X, Y \in \mathbb{C}^{n \times n}$. Then*

$$\exp(X + Y) = \lim_{m \to \infty} (\exp(X/m) \, \exp(Y/m))^m$$

and

$$\exp([X, Y]) = \lim_{m \to \infty} \{\exp(X/m), \exp(Y/m)\}^{m^2},$$

where $[X, Y] := XY - YX$ and $\{a, b\} := aba^{-1}b^{-1}$.

Proof. As $t \to 0$,

$$\exp(tX) \, \exp(tY) = \left(I + tX + \frac{t^2}{2}X^2 + \mathcal{O}(t^3)\right)\left(I + tY + \frac{t^2}{2}Y^2 + \mathcal{O}(t^3)\right)$$

$$= I + t(X + Y) + \frac{t^2}{2}(X^2 + 2XY + Y^2) + \mathcal{O}(t^3),$$

so that

$$\{\exp(tX), \exp(tY)\} = \left(I + t(X + Y) + \frac{t^2}{2}(X^2 + 2XY + Y^2) + \mathcal{O}(t^3)\right)$$

$$\times \left(I - t(X + Y) + \frac{t^2}{2}(X^2 + 2XY + Y^2) + \mathcal{O}(t^3)\right)$$

$$= I + t^2(XY - YX) + \mathcal{O}(t^3)$$

$$= I + t^2[X, Y] + \mathcal{O}(t^3).$$

Since exp is an injection in a neighbourhood of the origin $0 \in \mathbb{C}^{n \times n}$, we have

$$\exp(tX) \, \exp(tY) = \exp\left(t(X + Y) + \mathcal{O}(t^2)\right),$$
$$\{\exp(tX), \exp(tY)\} = \exp\left(t^2[X, Y] + \mathcal{O}(t^3)\right)$$

as $t \to 0$. Notice that $\exp(X)^m = \exp(mX)$ for all $m \in \mathbb{N}$. Therefore we get

$$\lim_{m \to \infty} (\exp(X/m) \, \exp(Y/m))^m = \lim_{m \to \infty} \exp\left(X + Y + \mathcal{O}(m^{-1})\right)$$

$$= \exp(X + Y),$$

$$\lim_{m \to \infty} \{\exp(X/m), \exp(Y/m)\}^{m^2} = \lim_{m \to \infty} \exp\left([X, Y] + \mathcal{O}(m^{-1})\right)$$

$$= \exp([X, Y]). \qquad \square$$

8.2 No small subgroups for Lie, please

Definition 8.2.1 ("No small subgroups" property). A topological group is said to have the *"no small subgroups"* property if there exists a neighbourhood of the neutral element containing no non-trivial subgroups.

We shall show that this property characterises Lie groups among compact groups.

Example. Let $\{G_j\}_{j\in J}$ be an infinite family of compact groups each having more than one element. Let us consider the compact product group $G := \prod_{j\in J} G_j$. Let

$$H_j := \{x \in G \mid \forall i \in J \setminus \{j\} : x_i = e_{G_i}\}.$$

Then $G_j \cong H_j < G$, and H_j is a non-trivial subgroup of G. If $V \subset G$ is a neighbourhood of $e \in G$ then it contains all but perhaps finitely many H_j, due to the definition of the product topology. Hence in this case G "has small subgroups" (i.e., has not the "no small subgroups" property).

Theorem 8.2.2 (Kernels of representations). *Let G be a compact group and $V \subset G$ open such that $e \in V$. Then there exists $\phi \in \mathrm{HOM}(G, \mathrm{U}(n))$ for some $n \in \mathbb{Z}^+$ such that $\mathrm{Ker}(\phi) \subset V$.*

Proof. First, $\{e\} \subset G$ and $G \setminus V \subset G$ are disjoint closed subsets of a compact Hausdorff space G. By Urysohn's Lemma (Theorem A.12.11), there exists $f \in C(G)$ such that $f(e) = 1$ and $f(G \setminus V) = \{0\}$. Since trigonometric polynomials are dense in $C(G)$ by Theorem 7.6.2, we may take $p \in \mathrm{TrigPol}(G)$ such that $\|p - f\|_{C(G)} < 1/2$. Then

$$\mathcal{H} := \mathrm{span}\,\{\pi_R(x)p \mid x \in G\} \subset L^2(\mu_G)$$

is a finite-dimensional vector space, and $\mathcal{H}$ inherits the inner product from $L^2(\mu_G)$. Let $A : \mathcal{H} \to \mathbb{C}^n$ be a linear isometry, where $n = \dim(\mathcal{H})$. Let us identify $\mathcal{U}(\mathbb{C}^n)$ with $\mathrm{U}(n)$. Define $\phi \in \mathrm{Hom}(G, \mathrm{U}(n))$ by

$$\phi(x) := A\,\pi_R(x)|_{\mathcal{H}}\,A^{-1}.$$

Then ϕ is clearly a continuous unitary representation. For every $x \in G \setminus V$,

$$|p(x) - 0| = |p(x) - f(x)| \le \|p - f\|_{C(G)} < 1/2,$$

so that $p(x) \ne p(e)$, because

$$|p(e) - 1| = |p(e) - f(e)| \le \|p - f\|_{C(G)} < 1/2;$$

consequently $\pi_R(x)p \ne p$. Thus $\mathrm{Ker}(\phi) \subset V$. □

Corollary 8.2.3 (Characterisation of linear Lie groups). *Let G be a compact group. Then G has no small subgroups if and only if it is isomorphic to a linear Lie group.*

Proof. Let G be a compact group without small subgroups. By Theorem 8.2.2, for some $n \in \mathbb{Z}^+$ there exists an injective $\phi \in \mathrm{HOM}(G, \mathrm{U}(n))$. Then $(x \mapsto \phi(x)) : G \to \phi(G)$ is an isomorphism, and a homeomorphism by Proposition A.12.7, because ϕ is continuous, G is compact and $\mathrm{U}(n)$ is Hausdorff. Thus $\phi(G) < \mathrm{U}(n) < \mathrm{GL}(n, \mathbb{C})$ is a compact linear Lie group.

Conversely, suppose $G < \mathrm{GL}(n, \mathbb{C})$ is closed. Recall that the mapping $(X \mapsto \exp(X)) : \mathbb{B} \to \exp(\mathbb{B})$ is a homeomorphism, where

$$\mathbb{B} = \left\{ X \in \mathbb{C}^{n \times n} : \|X\|_{\mathcal{L}(\mathbb{C}^n)} < r \right\}$$

for some small $r > 0$. Hence $V := \exp(\mathbb{B}/2) \cap G$ is a neighbourhood of $I \in G$. In the search for a contradiction, suppose there exists a nontrivial subgroup $H < G$ such that $A \in H \subset V$ and $A \neq I$. Then $0 \neq \log(A) \in \mathbb{B}/2$, so that $m \log(A) \in \mathbb{B} \setminus (\mathbb{B}/2)$ for some $m \in \mathbb{Z}^+$. Thereby

$$\exp(m \log(A)) = \exp(\log(A))^m = A^m \in H \subset V \subset \exp(\mathbb{B}/2),$$

but also

$$\exp(m \log(A)) \in \exp(\mathbb{B} \setminus (\mathbb{B}/2)) = \exp(\mathbb{B}) \setminus \exp(\mathbb{B}/2);$$

this is a contradiction.   □

Remark 8.2.4. Actually, it is shown above that Lie groups have no small subgroups; compactness played no role in this part of the proof.

Exercise 8.2.5. Use the Peter–Weyl Theorem 7.5.14 to provide an alternative proof for Theorem 8.2.2. Hint: For each $x \in G \setminus V$ there exists $\phi_x \in \mathrm{HOM}(G, \mathrm{U}(n_x))$ such that $x \notin \mathrm{Ker}(\phi_x)$, because...

8.3   Lie groups and Lie algebras

Next we deal with representation theory of Lie groups. We introduce Lie algebras, which sometimes still bear the archaic label *"infinitesimal groups"*, quite adequately describing their essence: a Lie algebra is a sort of locally linearised version of a Lie group.

Definition 8.3.1 (Lie algebras). A $\mathbb{K}$-*Lie algebra* is a $\mathbb{K}$-vector space V endowed with a bilinear mapping $((a, b) \mapsto [a, b]_V = [a, b]) : V \times V \to V$ satisfying

$$[a, a] = 0 \quad \text{and} \quad [a, [b, c]] + [b, [c, a]] + [c, [a, b]] = 0$$

for all $a, b, c \in V$; the second identity is called the *Jacobi identity*. Notice that here $[a, b] = -[b, a]$ for all $a, b \in V$. A vector subspace $W \subset V$ of a Lie algebra V is called a *Lie subalgebra* if $[a, b] \in W$ for all $a, b \in W$ (and thus W is a Lie algebra in its own right). A linear mapping $A : V_1 \to V_2$ between Lie algebras V_1, V_2 is called a *Lie algebra homomorphism* if $[Aa, Ab]_{V_2} = A[a, b]_{V_1}$ for all $a, b \in V_1$.

Example.

1. For a $\mathbb{K}$-vector space V, the trivial Lie product $[a, b] := 0$ gives a trivial Lie algebra.

2. A $\mathbb{K}$-algebra $\mathcal{A}$ can be endowed with the canonical Lie product

$$(a, b) \mapsto [a, b] := ab - ba;$$

this Lie algebra is denoted by $\mathrm{Lie}_{\mathbb{K}}(\mathcal{A})$. Important special cases of such Lie algebras are

$$\mathrm{Lie}_{\mathbb{K}}(\mathbb{C}^{n \times n}) \cong \mathrm{Lie}_{\mathbb{K}}(\mathrm{End}(\mathbb{C}^n)), \quad \mathrm{Lie}_{\mathbb{K}}(\mathrm{End}(V)), \quad \mathrm{Lie}_{\mathbb{K}}(\mathcal{L}(X)),$$

where X is a normed space and $\mathrm{End}(V)$ is the algebra of linear operators $V \to V$ on a vector space V. For short, let

$$\mathfrak{gl}(V) := \mathrm{Lie}_{\mathbb{R}}(\mathrm{End}(V)).$$

3. (**Derivations of algebras**). Let $\mathcal{D}(\mathcal{A})$ be the $\mathbb{K}$-vector space of *derivations* of a $\mathbb{K}$-algebra $\mathcal{A}$; that is, $D \in \mathcal{D}(\mathcal{A})$ if it is a linear mapping $\mathcal{A} \to \mathcal{A}$ satisfying the *Leibniz property*

$$D(ab) = D(a)\, b + a\, D(b)$$

for all $a, b \in \mathcal{A}$. Then $\mathcal{D}(\mathcal{A})$ has a Lie algebra structure given by $[D, E] := DE - ED$. An important special case is $\mathcal{A} = C^\infty(M)$, where M is a C^∞-manifold; if $C^\infty(M)$ is endowed with the topology of local uniform convergence for all derivatives, then $D \in \mathcal{D}(C^\infty(M))$ is continuous if and only if it is a linear first-order partial differential operator with smooth coefficients (alternatively, a smooth vector field on M).

Definition 8.3.2. The *Lie algebra* $\mathfrak{Lie}(G) = \mathfrak{g}$ of a *linear Lie group* G is introduced in the following Theorem 8.3.3:

Theorem 8.3.3 (Lie algebras of linear Lie groups). *Let $G < \mathrm{GL}(n, \mathbb{C})$ be closed. The $\mathbb{R}$-vector space*

$$\mathfrak{Lie}(G) = \mathfrak{g} := \{X \in \mathbb{C}^{n \times n} \mid \forall t \in \mathbb{R} : \exp(tX) \in G\}$$

is a Lie subalgebra of the $\mathbb{R}$-Lie algebra $\mathrm{Lie}_{\mathbb{R}}(\mathbb{C}^{n \times n}) \cong \mathfrak{gl}(\mathbb{C}^n)$.

Proof. Let $X, Y \in \mathfrak{g}$ and $\lambda \in \mathbb{R}$. Trivially, $\exp(t\lambda X) \in G$ for all $t \in \mathbb{R}$, yielding $\lambda X \in \mathfrak{g}$. Since G is closed and $\exp$ is continuous,

$$G \ni (\exp(tX/m)\, \exp(tY/m))^m \xrightarrow[m \to \infty]{} \exp(t(X + Y)) \in G$$

$$G \ni \{\exp(tX/m), \exp(Y/m)\}^{m^2} \xrightarrow[m \to \infty]{} \exp(t[X, Y]) \in G$$

by Lemma 8.1.11. Thereby $X + Y, [X, Y] \in \mathfrak{g}$. $\qquad\square$

Exercise 8.3.4. Let $X \in \mathbb{C}^{n \times n}$ be such that $\exp(tX) = I$ for all $t \in \mathbb{R}$. Show that $X = 0$.

Exercise 8.3.5. Let $\mathfrak{g} \subset \mathbb{C}^{n \times n}$ be the Lie algebra of a linear Lie group $G < \mathrm{GL}(n, \mathbb{R})$. Show that $\mathfrak{g} \subset \mathbb{R}^{n \times n}$.

Definition 8.3.6 (Dimension of a linear Lie group). Let G be a linear Lie group and $\mathfrak{g} = \mathfrak{Lie}(G)$. The *dimension of G* is $\dim(G) := \dim(\mathfrak{g}) = k$, hence $\mathfrak{g} \cong \mathbb{R}^k$ as a vector space.

Remark 8.3.7 (**Exponential coordinates**). From Theorem 8.1.6 it follows that

$$\mathrm{HOM}(\mathbb{R}, G) = \{t \mapsto \exp(tX) \mid X \in \mathfrak{g}\}.$$

The mapping $(X \mapsto \exp(X)) : \mathfrak{g} \to G$ is a diffeomorphism in a small neighbourhood of $0 \in \mathfrak{g}$. Hence, given a vector space basis for $\mathfrak{g} \cong \mathbb{R}^k$, a small neighbourhood of $\exp(0) = I \in G$ is endowed with the so-called *exponential coordinates*. If G is compact and connected then $\exp(\mathfrak{g}) = G$, so that the exponential map may "wrap $\mathfrak{g}$ around G"; we shall not prove this.

Remark 8.3.8. Informally speaking, if $X, Y \in \mathfrak{g}$ are near $0 \in \mathfrak{g}$, $x := \exp(X)$ and $y := \exp(Y)$ then $x, y \in G$ are near $I \in G$ and

$$\exp(X + Y) \approx xy, \quad \exp([X, Y]) \approx \{x, y\} = xyx^{-1}y^{-1}.$$

In a sense, the Lie algebra $\mathfrak{g}$ is the infinitesimally linearised G near $I \in G$.

Remark 8.3.9 (**Lie algebra as invariant vector fields**). The Lie algebra $\mathfrak{g}$ can be identified with the tangent space of G at the identity $I \in G$. Using left-translations (resp. right-translations), $\mathfrak{g}$ can be identified with the set of left-invariant (resp. right-invariant) vector fields on G, and vector fields have a natural interpretation as first-order partial differential operators on G: For $x \in G$, $X \in \mathfrak{g}$ and $f \in C^\infty(G)$, define

$$L_X f(x) \quad := \quad \frac{\mathrm{d}}{\mathrm{d}t} f\left(x \, \exp(tX)\right)\Big|_{t=0},$$

$$R_X f(x) \quad := \quad \frac{\mathrm{d}}{\mathrm{d}t} f\left(\exp(tX) \, x\right)\Big|_{t=0}.$$

Then $\pi_L(y)L_X f = L_X \pi_L(y)f$ and $\pi_R(y)R_X f = R_X \pi_R(y)f$ for all $y \in G$, where π_L, π_R are the left and right regular representations of G, respectively.

Definition 8.3.10 (**Abbreviations for Lie algebras**). Some usual abbreviations are

$$
\begin{aligned}
\mathfrak{gl}(n, \mathbb{K}) &= \mathfrak{Lie}(\mathrm{GL}(n, \mathbb{K})), \\
\mathfrak{sl}(n, \mathbb{K}) &= \mathfrak{Lie}(\mathrm{SL}(n, \mathbb{K})), \\
\mathfrak{o}(n) &= \mathfrak{Lie}(\mathrm{O}(n)), \\
\mathfrak{so}(n) &= \mathfrak{Lie}(\mathrm{SO}(n)), \\
\mathfrak{u}(n) &= \mathfrak{Lie}(\mathrm{U}(n)), \\
\mathfrak{su}(n) &= \mathfrak{Lie}(\mathrm{SU}(n)),
\end{aligned}
$$

and so on.

Exercise 8.3.11. Calculate the dimensions of the linear Lie groups mentioned in Definition 8.3.10.

Proposition 8.3.12. *Let G, H be linear Lie groups having the respective Lie algebras $\mathfrak{g}, \mathfrak{h}$. Let $\psi \in \mathrm{HOM}(G, H)$. Then for every $X \in \mathfrak{g}$ there exists a unique $Y \in \mathfrak{h}$ such that $\psi(\exp(tX)) = \exp(tY)$ for all $t \in \mathbb{R}$.*

Proof. Let $X \in \mathfrak{g}$. Then $\phi := (t \mapsto \psi(\exp(tX))) : \mathbb{R} \to H$ is a continuous homomorphism, so that $\phi = (t \mapsto \exp(tY))$, where $Y = \phi'(0) \in \mathfrak{h}$. $\square$

Proposition 8.3.13. *Let F, G, H be closed subgroups of $\mathrm{GL}(n, \mathbb{C})$, with their respective Lie algebras $\mathfrak{f}, \mathfrak{g}, \mathfrak{h}$. Then*

(a) $H < G \Rightarrow \mathfrak{h} \subset \mathfrak{g}$,

(b) *the Lie algebra of $F \cap G$ is $\mathfrak{f} \cap \mathfrak{g}$,*

(c) *the Lie algebra $\mathfrak{c}_I$ of the component $C_I < G$ of the neutral element I is $\mathfrak{g}$.*

Proof. (a) If $H < G$ and $X \in \mathfrak{h}$ then $\exp(tX) \in H \subset G$ for all $t \in \mathbb{R}$, so that $X \in \mathfrak{g}$.

(b) Let $\mathfrak{e}$ be the Lie algebra of $F \cap G$. By (a), $\mathfrak{e} \subset \mathfrak{f} \cap \mathfrak{g}$. If $X \in \mathfrak{f} \cap \mathfrak{g}$ then $\exp(tX) \in F \cap G$ for all $t \in \mathbb{R}$, so that $X \in \mathfrak{e}$. Hence $\mathfrak{e} = \mathfrak{f} \cap \mathfrak{g}$.

(c) By (a), $\mathfrak{c}_I \subset \mathfrak{g}$. Let $X \in \mathfrak{g}$. Now the connectedness of $\mathbb{R}$ (Theorem A.16.9) and the continuity of $t \mapsto \exp(tX)$ by Proposition A.16.3 imply the connectedness of

$$\{\exp(tX) : t \in \mathbb{R}\} \ni \exp(0) = I.$$

Thereby $\{\exp(tX) : t \in \mathbb{R}\} \subset C_I$, so that $X \in \mathfrak{c}_I$. $\square$

Example (**Lie algebra of $\mathrm{SL}(n, \mathbb{K})$**). Let us compute the Lie algebra $\mathfrak{sl}(n, \mathbb{K})$ of the linear Lie group

$$\mathrm{SL}(n, \mathbb{K}) = \{A \in \mathrm{GL}(n, \mathbb{K}) \mid \det(A) = 1\}.$$

Notice that $\mathfrak{sl}(n, \mathbb{K}) \subset \mathbb{K}^{n \times n}$ by Exercise 8.3.5. Hence

$$
\begin{aligned}
\mathfrak{sl}(n, \mathbb{K}) &:= \{X \in \mathbb{K}^{n \times n} \mid \forall t \in \mathbb{R} : \exp(tX) \in \mathrm{SL}(n, \mathbb{K})\} \\
&= \{X \in \mathbb{K}^{n \times n} \mid \forall t \in \mathbb{R} : \exp(tX) \in \mathbb{K}^{n \times n}, \ \det(\exp(tX)) = 1\}.
\end{aligned}
$$

Let $\{\lambda_j\}_{j=1}^n \subset \mathbb{C}$ be the set of eigenvalues of $X \in \mathbb{K}^{n \times n}$. The characteristic polynomial $(z \mapsto \det(zI - X)) : \mathbb{C} \to \mathbb{C}$ of X satisfies

$$
\begin{aligned}
\det(zI - X) &= \prod_{j=1}^n (z - \lambda_j) \\
&= z^n - z^{n-1} \sum_{j=1}^n \lambda_j + \cdots + (-1)^n \prod_{j=1}^n \lambda_j \\
&= z^n - z^{n-1} \mathrm{Tr}(X) + \cdots + (-1)^n \det(X),
\end{aligned}
$$

We know that X is similar to an upper triangular matrix $Y = PXP^{-1}$ for some $P \in \mathrm{GL}(n, \mathbb{K})$. Since

$$
\begin{aligned}
\det(zI - PXP^{-1}) &= \det(P(zI - X)P^{-1}) \\
&= \det(P) \det(zI - X) \det(P^{-1}) \\
&= \det(zI - X),
\end{aligned}
$$

the eigenvalues of X and Y are the same, and they are on the diagonal of Y. Evidently, $\{e^{\lambda_j}\}_{j=1}^n \subset \mathbb{C}$ is the set of the eigenvalues of both $\exp(Y)$ and $\exp(X) = P^{-1}\exp(Y)P$. Since the determinant is the product of the eigenvalues and the trace is the sum of the eigenvalues, we have

$$\det(\exp(X)) = \prod_{j=1}^n e^{\lambda_j} = e^{\sum_{j=1}^n \lambda_j} = e^{\mathrm{Tr}(X)}$$

(see also Proposition 8.1.4). Therefore $X \in \mathfrak{sl}(n, \mathbb{K})$ if and only if $\mathrm{Tr}(X) = 0$ and $\exp(tX) \in \mathbb{K}^{n \times n}$ for all $t \in \mathbb{R}$. Thus

$$\mathfrak{sl}(n, \mathbb{K}) = \left\{ X \in \mathbb{K}^{n \times n} \mid \mathrm{Tr}(X) = 0 \right\}$$

as the reader may check.

Next we ponder the relationship between Lie group and Lie algebra homomorphisms.

Definition 8.3.14 (Differential homomorphisms). Let G, H be linear Lie groups with respective Lie algebras $\mathfrak{g}$, $\mathfrak{h}$. The *differential homomorphism of* $\psi \in \mathrm{HOM}(G, H)$ is the mapping $\psi' = \mathfrak{Lie}(\psi) : \mathfrak{g} \to \mathfrak{h}$ defined by

$$\psi'(X) := \frac{\mathrm{d}}{\mathrm{d}t}\psi(\exp(tX))\Big|_{t=0}.$$

Remark 8.3.15. Above, ψ' is well defined since

$$f := (t \mapsto \psi(\exp(tX))) \in \mathrm{HOM}(\mathbb{R}, H)$$

is of the form $t \mapsto \exp(tY)$ for some $Y \in \mathfrak{h}$, as a consequence of Theorem 8.1.6. Moreover, $Y = f'(0) = \psi'(X)$ holds, so that

$$\psi(\exp(tX)) = \exp(t\psi'(X)).$$

Theorem 8.3.16. *Let F, G, H be linear Lie groups with respective Lie algebras $\mathfrak{f}, \mathfrak{g}, \mathfrak{h}$. Let $\phi \in \mathrm{HOM}(F, G)$ and $\psi \in \mathrm{HOM}(G, H)$. The mapping $\psi' : \mathfrak{g} \to \mathfrak{h}$ defined in Definition 8.3.14 is a Lie algebra homomorphism. Moreover,*

$$(\psi \circ \phi)' = \psi'\phi' \quad \text{and} \quad \mathrm{Id}_G' = \mathrm{Id}_{\mathfrak{g}},$$

where $\mathrm{Id}_G = (x \mapsto x) : G \to G$ *and* $\mathrm{Id}_{\mathfrak{g}} = (X \mapsto X) : \mathfrak{g} \to \mathfrak{g}$.

Proof. Let $X, Y \in \mathfrak{g}$ and $\lambda \in \mathbb{R}$. Then

$$
\begin{aligned}
\psi'(\lambda X) &= \frac{\mathrm{d}}{\mathrm{d}t}\psi(\exp(t\lambda X))|_{t=0} \\
&= \lambda \frac{\mathrm{d}}{\mathrm{d}t}\psi(\exp(tX))|_{t=0} \\
&= \lambda\psi'(X).
\end{aligned}
$$

If $t \in \mathbb{R}$ then

$$
\begin{aligned}
\exp\left(t\psi'(X+Y)\right) &= \psi\left(\exp(tX+tY)\right) \\
&= \psi\left(\lim_{m\to\infty} \left(\exp(tX/m)\,\exp(tY/m)\right)^m\right) \\
&= \lim_{m\to\infty} \left(\psi(\exp(tX/m))\,\psi(\exp(tY/m))\right)^m \\
&= \lim_{m\to\infty} \left(\exp(t\psi'(X)/m)\,\exp(t\psi'(Y)/m)\right)^m \\
&= \exp(t(\psi'(X)+\psi'(Y))),
\end{aligned}
$$

so that $t\psi'(X+Y) = t\left(\psi'(X)+\psi'(Y)\right)$ for small enough $|t|$, as we recall that $\exp$ is injective in a small neighbourhood of $0 \in \mathfrak{g}$. Consequently, $\psi' : \mathfrak{g} \to \mathfrak{h}$ is linear. Next,

$$
\begin{aligned}
\exp\left(t\psi'([X,Y])\right) &= \psi\left(\exp(t[X,Y])\right) \\
&= \psi\left(\lim_{m\to\infty} \{\exp(tX/m),\exp(tY/m)\}^{m^2}\right) \\
&= \lim_{m\to\infty} \{\exp(t\psi'(X)/m),\exp(t\psi'(Y)/m)\}^{m^2} \\
&= \exp\left(t[\psi'(X),\psi'(Y)]\right),
\end{aligned}
$$

so that we get $\psi'([X,Y]) = [\psi'(X),\psi'(Y)]$. Thus $\psi' : \mathfrak{g} \to \mathfrak{h}$ is a Lie algebra homomorphism.

If $Z \in \mathfrak{f}$ then

$$
\begin{aligned}
(\psi \circ \phi)'(Z) &= \frac{\mathrm{d}}{\mathrm{d}t}\psi\left(\phi(\exp(tZ))\right)\big|_{t=0} \\
&= \frac{\mathrm{d}}{\mathrm{d}t}\psi\left(\exp(t\phi'(Z))\right)\big|_{t=0} \\
&= \psi'(\phi'(Z)).
\end{aligned}
$$

Finally, $\frac{\mathrm{d}}{\mathrm{d}t}\exp(tX)|_{t=0} = X$, yielding $\mathrm{Id}'_G = \mathrm{Id}_{\mathfrak{g}}$.   □

Remark 8.3.17. Notice that isomorphic linear Lie groups must have isomorphic Lie algebras. Now we know that a continuous Lie group homomorphism ψ can naturally be linearised to get a Lie algebra homomorphism ψ', so that we have the commutative diagram

$$
\begin{array}{ccc}
G & \xrightarrow{\ \psi\ } & H, \\
{\scriptstyle\exp}\big\uparrow & & \big\uparrow{\scriptstyle\exp} \\
\mathfrak{g} & \xrightarrow{\ \psi'\ } & \mathfrak{h}.
\end{array}
$$

If we are given a Lie algebra homomorphism $f : \mathfrak{g} \to \mathfrak{h}$, does there exist $\phi \in \mathrm{HOM}(G,H)$ such that $\phi' = f$? This problem is studied in the following exercises.

Definition 8.3.18 (Simply connected spaces). A topological space X is said to be *simply connected* if X is path-connected and if every closed curve in X can be shrunken to a point continuously in the set X.

Exercise 8.3.19. Show that the groups $\mathrm{SU}(n)$ and $\mathrm{SL}(n,\mathbb{C})$ are both connected and simply connected.

Exercise 8.3.20. Show that the groups $\mathrm{U}(n)$ and $\mathrm{GL}(n,\mathbb{C})$ are connected but not simply connected.

Exercise 8.3.21. Let G, H be linear Lie groups such that G is simply connected. Let $f : \mathfrak{g} \to \mathfrak{h}$ be a Lie algebra homomorphism. Show that there exists $\phi \in \mathrm{HOM}(G, H)$ such that $\phi' = f$. (This is a rather demanding task unless one knows that $\exp : \mathfrak{g} \to G$ is surjective and uses Lemma 8.1.11. A proof can be found, e.g., in [37].)

Exercise 8.3.22. Related to Exercise 8.3.21, give an example of a non-simply-connected G and a homomorphism $f : \mathfrak{g} \to \mathfrak{h}$ which is not of the form $f = \phi'$.

Lemma 8.3.23. *Let $\mathfrak{g}$ be the Lie algebra of a linear Lie group G, and*

$$S := \left\{ \exp(X_1) \cdots \exp(X_m) \mid m \in \mathbb{Z}^+, \; \{X_j\}_{j=1}^m \subset \mathfrak{g} \right\}.$$

Then $S = C_I$, the component of $I \in G$.

Proof. Now $S < G$ is path-connected, since

$$(t \mapsto \exp(tX_1) \cdots \exp(tX_m)) : [0,1] \to S$$

is continuous, connecting $I \in S$ to the point $\exp(X_1) \cdots \exp(X_m) \in S$. For a small enough neighbourhood $U \subset \mathfrak{g}$ of $0 \in \mathfrak{g}$, we have a homeomorphism $(X \mapsto \exp(X)) : U \to \exp(U)$. Because of

$$\exp(X_1) \cdots \exp(X_m) \in \exp(X_1) \cdots \exp(X_m) \exp(U) \subset S,$$

it follows that $S < G$ is open. But open subgroups are always closed, as the reader can easily verify. Thus $S \ni I$ is connected, closed and open, so that $S = C_I$. $\quad\square$

Corollary 8.3.24. *Let G, H be linear Lie groups and $\phi, \psi \in \mathrm{HOM}(G, H)$. Then:*

(a) *$\mathfrak{Lie}(\mathrm{Ker}(\psi)) = \mathrm{Ker}(\psi')$.*

(b) *If G is connected and $\phi' = \psi'$ then $\phi = \psi$.*

(c) *Let H be connected; then ψ' is surjective if and only if ψ is surjective.*

Proof. (a) $\mathrm{Ker}(\psi) < G < \mathrm{GL}(n,\mathbb{C})$ is a closed subgroup, since ψ is a continuous homomorphism. Thereby

$$
\begin{aligned}
\mathfrak{Lie}(\mathrm{Ker}(\psi)) &= \left\{ X \in \mathbb{C}^{n \times n} \mid \forall t \in \mathbb{R} : \exp(tX) \in \mathrm{Ker}(\psi) \right\} \\
&= \left\{ X \in \mathbb{C}^{n \times n} \mid \forall t \in \mathbb{R} : \exp(t\psi'(X)) = \psi(\exp(tX)) = I \right\} \\
&= \left\{ X \in \mathbb{C}^{n \times n} \mid \psi'(X) = 0 \right\} \\
&= \mathrm{Ker}(\psi').
\end{aligned}
$$

(b) Take $A \in G$. Then $A = \exp(X_1) \cdots \exp(X_m)$ for some $\{X_j\}_{j=1}^m \subset \mathfrak{g}$ by Lemma 8.3.23, so that

$$
\begin{aligned}
\phi(A) &= \exp\left(\phi'(X_1)\right) \cdots \exp\left(\phi'(X_m)\right) \\
&= \exp\left(\psi'(X_1)\right) \cdots \exp\left(\psi'(X_m)\right) \\
&= \psi(A).
\end{aligned}
$$

(c) Suppose $\psi' : \mathfrak{g} \to \mathfrak{h}$ is surjective. Let $B \in H$. Now H is connected, so that Lemma 8.3.23 says that $B = \exp(Y_1) \cdots \exp(Y_m)$ for some $\{Y_j\}_{j=1}^m \subset \mathfrak{h}$. Exploit the surjectivity of ψ' to obtain $X_j \in \mathfrak{g}$ such that $\psi'(X_j) = Y_j$. Then

$$
\begin{aligned}
\psi\left(\exp(X_1) \cdots \exp(X_m)\right) &= \psi\left(\exp(X_1)\right) \cdots \psi\left(\exp(X_m)\right) \\
&= \exp(Y_1) \cdots \exp(Y_m) \\
&= B.
\end{aligned}
$$

Conversely, suppose $\psi : G \to H$ is surjective. Trivially, $\psi'(0) = 0 \in \mathfrak{h}$; let $0 \neq Y \in \mathfrak{h}$. Let $r_0 := r/\|Y\|$, where r is as in Proposition 8.1.7; notice that if $|t| < r_0$ then $\log(\exp(tY)) = tY$. The surjectivity of ψ guarantees that for every $t \in \mathbb{R}$ there exists $A_t \in G$ such that $\psi(A_t) = \exp(tY)$. The set $R := \{A_t : 0 < t < r_0\}$ is uncountable, so that it has an accumulation point $x \in \mathbb{C}^{n \times n}$; and $x \in G$, because $R \subset G$ and $G \subset \mathbb{C}^{n \times n}$ is closed. Let $\varepsilon > 0$. Then there exist $s, t \in]0, r_0[$ such that $s \neq t$ and

$$
\|A_s - x\| < \varepsilon, \quad \|A_t - x\| < \varepsilon, \quad \|A_s^{-1} - x^{-1}\| < \varepsilon.
$$

Thereby

$$
\begin{aligned}
\|A_s^{-1} A_t - I\| &= \|A_s^{-1}(A_t - A_s)\| \\
&\leq \|A_s^{-1}\| \left(\|A_t - x\| + \|x - A_s\|\right) \\
&\leq \left(\|x^{-1}\| + \varepsilon\right) 2\varepsilon.
\end{aligned}
$$

Hence we demand $\|A_s^{-1} A_t - I\| < 1$ and $\|\psi(A_s^{-1} A_t) - I\| < 1$, yielding

$$
\psi(A_s^{-1} A_t) = \psi(A_s)^{-1} \psi(A_t) = \exp((t - s)Y).
$$

Consequently

$$
\psi'\left(\log(A_s^{-1} A_t)\right) = (t - s)Y.
$$

Therefore $\psi'\left(\frac{1}{t-s} \log(A_s^{-1} A_t)\right) = Y$. $\qquad\qquad\square$

Definition 8.3.25 (Adjoint representation of Lie groups). The *adjoint representation of a linear Lie group* G is the mapping $\mathrm{Ad} \in \mathrm{HOM}(G, \mathrm{Aut}(\mathfrak{g}))$ defined by

$$
\mathrm{Ad}(A)X := AXA^{-1},
$$

where $A \in G$ and $G \in \mathfrak{g}$.

Remark 8.3.26. Indeed, $\mathrm{Ad} : G \to \mathrm{Aut}(\mathfrak{g})$, because

$$\exp\left(t\mathrm{Ad}(A)X\right) = \exp\left(tAXA^{-1}\right) = A \exp\left(tX\right) A^{-1}$$

belongs to G if $A \in G$, $X \in \mathfrak{g}$ and $t \in \mathbb{R}$. It is a homomorphism, since

$$\mathrm{Ad}(AB)X = ABXB^{-1}A^{-1} = \mathrm{Ad}(A)(BXB^{-1}) = \mathrm{Ad}(A)\,\mathrm{Ad}(B)\,X,$$

and Ad is trivially continuous.

Exercise 8.3.27. Let $\mathfrak{g}$ be a Lie algebra. Consider $\mathrm{Aut}(\mathfrak{g})$ as a linear Lie group. Show that $\mathfrak{Lie}(\mathrm{Aut}(\mathfrak{g}))$ and $\mathfrak{gl}(\mathfrak{g})$ are isomorphic as Lie algebras.

Definition 8.3.28 (Adjoint representation of Lie algebras). The *adjoint representation of the Lie algebra* $\mathfrak{g}$ *of a linear Lie group* G is the differential representation

$$\mathrm{ad} = \mathrm{Ad}' : \mathfrak{g} \to \mathfrak{Lie}(\mathrm{Aut}(\mathfrak{g})) \cong \mathfrak{gl}(\mathfrak{g}),$$

that is $\mathrm{ad}(X) := \mathrm{Ad}'(X)$, so that

$$
\begin{aligned}
\mathrm{ad}(X)Y &= \frac{\mathrm{d}}{\mathrm{d}t}\left(\exp(tX)Y\exp(-tX)\right)|_{t=0} \\
&= \left(\left(\frac{\mathrm{d}}{\mathrm{d}t}\exp(tX)\right)Y\exp(-tX) + \exp(tX)Y\frac{\mathrm{d}}{\mathrm{d}t}\exp(-tX)\right)|_{t=0} \\
&= XY - YX \\
&= [X,Y].
\end{aligned}
$$

Remark 8.3.29. Notice that the diagram commutes:

$$
\begin{array}{ccc}
G & \xrightarrow{\;\mathrm{Ad}\;} & \mathrm{Aut}(G) \\
{\scriptstyle\exp}\big\uparrow & & \big\uparrow{\scriptstyle\exp} \\
\mathfrak{g} & \xrightarrow{\;\mathrm{Ad}'=\mathrm{ad}\;} & \mathfrak{Lie}(\mathrm{Aut}(\mathfrak{g})).
\end{array}
$$

8.3.1   Universal enveloping algebra

Here we discuss the universal enveloping algebra.

Remark 8.3.30 (**Universal enveloping algebra informally**). We are going to study higher-order partial differential operators on G. Let $\mathfrak{g}$ be the Lie algebra of a linear Lie group G. Next we construct a natural associative algebra $\mathcal{U}(\mathfrak{g})$ generated by $\mathfrak{g}$ modulo an ideal, enabling embedding $\mathfrak{g}$ into $\mathcal{U}(\mathfrak{g})$. Recall that $\mathfrak{g}$ can be interpreted as the vector space of first-order left- (or right-) translation invariant partial differential operators on G. Consequently, $\mathcal{U}(\mathfrak{g})$ can be interpreted as the vector space of finite-order left- (or right-) translation invariant partial differential operators on G.

Definition 8.3.31 (Universal enveloping algebra). Let $\mathfrak{g}$ be a $\mathbb{K}$-Lie algebra. Let

$$\mathcal{T} := \bigoplus_{m=0}^{\infty} \otimes^m \mathfrak{g}$$

be the tensor product algebra of $\mathfrak{g}$, where $\otimes^m \mathfrak{g}$ denotes the m-fold tensor product $\mathfrak{g} \otimes \cdots \otimes \mathfrak{g}$; that is, $\mathcal{T}$ is the linear span of the elements of the form

$$\lambda_{00} \mathbf{1} + \sum_{m=1}^{M} \sum_{k=1}^{K_m} \lambda_{mk}\, X_{mk1} \otimes \cdots \otimes X_{mkm},$$

where $\mathbf{1}$ is the formal unit element of $\mathcal{T}$, $\lambda_{mk} \in \mathbb{K}$, $X_{mkj} \in \mathfrak{g}$ and $M, K_m \in \mathbb{Z}^+$; the product of $\mathcal{T}$ is begotten by the tensor product, i.e.,

$$(X_1 \otimes \cdots \otimes X_p)(Y_1 \otimes \cdots \otimes Y_q) := X_1 \otimes \cdots \otimes X_p \otimes Y_1 \otimes \cdots \otimes Y_q$$

is extended to a unique bilinear mapping $\mathcal{T} \times \mathcal{T} \to \mathcal{T}$. Let $\mathcal{J}$ be the (two-sided) ideal in $\mathcal{T}$ spanned by the set

$$\mathcal{O} := \{ X \otimes Y - Y \otimes X - [X, Y] :\ X, Y \in \mathfrak{g} \};$$

i.e., $\mathcal{J} \subset \mathcal{T}$ is the smallest vector subspace such that $\mathcal{O} \subset \mathcal{J}$ and $DE, ED \in \mathcal{J}$ for every $D \in \mathcal{J}$ and $E \in \mathcal{T}$ (in a sense, $\mathcal{J}$ is a "huge zero" in $\mathcal{T}$). The quotient algebra

$$\mathcal{U}(\mathfrak{g}) := \mathcal{T}/\mathcal{J}$$

is called the *universal enveloping algebra of* $\mathfrak{g}$.

Definition 8.3.32 (Canonical mapping of a Lie algebra). Let $\iota : \mathcal{T} \to \mathcal{U}(\mathfrak{g}) = \mathcal{T}/\mathcal{J}$ be the quotient mapping $t \mapsto t + \mathcal{J}$. A natural interpretation is that $\mathfrak{g} \subset \mathcal{T}$. The restricted mapping $\iota|_{\mathfrak{g}} : \mathfrak{g} \to \mathcal{U}(\mathfrak{g})$ is called the *canonical mapping of* $\mathfrak{g}$.

Remark 8.3.33. Notice that $\iota|_{\mathfrak{g}} : \mathfrak{g} \to \mathrm{Lie}_{\mathbb{K}}(\mathcal{U}(\mathfrak{g}))$ is a Lie algebra homomorphism: it is linear and

$$
\begin{aligned}
\iota|_{\mathfrak{g}}([X, Y]) &= \iota([X, Y]) \\
&= \iota(X \otimes Y - Y \otimes X) \\
&= \iota(X)\iota(Y) - \iota(Y)\iota(X) \\
&= \iota|_{\mathfrak{g}}(X)\iota|_{\mathfrak{g}}(Y) - \iota|_{\mathfrak{g}}(Y)\iota|_{\mathfrak{g}}(X) \\
&= [\iota|_{\mathfrak{g}}(X), \iota|_{\mathfrak{g}}(Y)].
\end{aligned}
$$

Theorem 8.3.34 (Universality of the enveloping algebra). *Let $\mathfrak{g}$ be a $\mathbb{K}$-Lie algebra, $\iota|_{\mathfrak{g}} : \mathfrak{g} \to \mathcal{U}(\mathfrak{g})$ its canonical mapping, $\mathcal{A}$ an associative $\mathbb{K}$-algebra, and*

$$\sigma : \mathfrak{g} \to \mathrm{Lie}_{\mathbb{K}}(\mathcal{A})$$

a *Lie algebra homomorphism. Then there exists a unique algebra homomorphism*

$$\tilde{\sigma} : \mathcal{U}(\mathfrak{g}) \to \mathcal{A}$$

satisfying $\tilde{\sigma}\left(\iota|_{\mathfrak{g}}(X)\right) = \sigma(X)$ *for all* $X \in \mathfrak{g}$, *i.e.,*

$$
\begin{array}{ccc}
\mathcal{U}(\mathfrak{g}) & \xrightarrow{\ \tilde{\sigma}\ } & \mathcal{A} \\
\iota|_{\mathfrak{g}} \uparrow & & \| \\
\mathfrak{g} & \xrightarrow{\ \sigma\ } & \mathrm{Lie}_{\mathbb{K}}(\mathcal{A}).
\end{array}
$$

Proof. Let us define a linear mapping $\sigma_0 : \mathcal{T} \to \mathcal{A}$ by

$$\sigma_0(X_1 \otimes \cdots \otimes X_m) := \sigma(X_1) \cdots \sigma(X_m). \tag{8.3}$$

Then $\sigma_0(\mathcal{J}) = \{0\}$, since

$$
\begin{aligned}
\sigma_0(X \otimes Y - Y \otimes X - [X,Y]) &= \sigma(X)\sigma(Y) - \sigma(Y)\sigma(X) - \sigma([X,Y]) \\
&= \sigma(X)\sigma(Y) - \sigma(Y)\sigma(X) - [\sigma(X), \sigma(Y)] \\
&= 0.
\end{aligned}
$$

Hence if $t, u \in \mathcal{T}$ and $t - u \in \mathcal{J}$ then $\sigma_0(t) = \sigma_0(u)$. Thereby we may define $\tilde{\sigma} := (t + \mathcal{J} \mapsto \sigma_0(t)) : \mathcal{U}(\mathfrak{g}) \to \mathcal{A}$. Finally, it is clear that $\tilde{\sigma}$ is an algebra homomorphism making the diagram above commute. The uniqueness is clear by construction since (8.3) must hold. $\qquad\square$

Corollary 8.3.35 (Ado–Iwasawa Theorem). *Let $\mathfrak{g}$ be the Lie algebra of a linear Lie group G. Then the canonical mapping $\iota|_{\mathfrak{g}} : \mathfrak{g} \to \mathcal{U}(\mathfrak{g})$ is injective.*

Proof. Let $\sigma = (X \mapsto X) : \mathfrak{g} \to \mathfrak{gl}(n, \mathbb{C})$. Due to the universality of $\mathcal{U}(\mathfrak{g})$ there exists an $\mathbb{R}$-algebra homomorphism $\tilde{\sigma} : \mathcal{U}(\mathfrak{g}) \to \mathbb{C}^{n \times n}$ such that $\sigma(X) = \tilde{\sigma}\left(\iota|_{\mathfrak{g}}(X)\right)$ for all $X \in G$, i.e.,

$$
\begin{array}{ccc}
\mathcal{U}(\mathfrak{g}) & \xrightarrow{\ \tilde{\sigma}\ } & \mathbb{C}^{n \times n} \\
\iota|_{\mathfrak{g}} \uparrow & & \| \\
\mathfrak{g} & \xrightarrow{\ \sigma\ } & \mathfrak{gl}(n, \mathbb{C}).
\end{array}
$$

Then $\iota|_{\mathfrak{g}}$ is injective because σ is injective. $\qquad\square$

Remark 8.3.36. By the Ado–Iwasawa Theorem (Corollary 8.3.35), the Lie algebra $\mathfrak{g}$ of a linear Lie group can be considered as a Lie subalgebra of $\mathrm{Lie}_{\mathbb{R}}(\mathcal{U}(\mathfrak{g}))$.

Definition 8.3.37 (ad). Let $\mathfrak{g}$ be a $\mathbb{K}$-Lie algebra. Let us define the linear mapping $\mathrm{ad} : \mathfrak{g} \to \mathrm{End}(\mathfrak{g})$ by $\mathrm{ad}(X)Z := [X, Z]$.

Remark 8.3.38. Let $\mathfrak{g}$ be a $\mathbb{K}$-Lie algebra and $X, Z \in G$. Since

$$
\begin{aligned}
0 &= [[X,Y],Z] + [[Y,Z],X] + [[Z,X],Y] \\
&= [[X,Y],Z] - ([X,[Y,Z]] - [Y,[X,Z]]) \\
&= \mathrm{ad}([X,Y])Z - [\mathrm{ad}(X),\mathrm{ad}(Y)]Z,
\end{aligned}
$$

we notice that

$$
\mathrm{ad}([X,Y]) = [\mathrm{ad}(X),\mathrm{ad}(Y)],
$$

i.e., ad is a Lie algebra homomorphism $\mathfrak{g} \to \mathfrak{gl}(\mathfrak{g})$.

Definition 8.3.39 (Killing form and semisimple Lie groups). The *Killing form* of the Lie algebra $\mathfrak{g}$ is the bilinear mapping $B : \mathfrak{g} \times \mathfrak{g} \to \mathbb{K}$, defined by

$$
B(X,Y) := \mathrm{Tr}\,(\mathrm{ad}(X)\,\mathrm{ad}(Y))
$$

(recall that by Exercise B.5.41, on a finite-dimensional vector space the trace can be defined independent of any inner product). A ($\mathbb{R}$- or $\mathbb{C}$-)Lie algebra $\mathfrak{g}$ is called *semisimple* if its Killing form is *non-degenerate*, i.e., if

$$
\forall X \in \mathfrak{g} \setminus \{0\}\ \exists Y \in \mathfrak{g} : B(X,Y) \neq 0;
$$

equivalently, B is non-degenerate if the matrix $(B(X_i, X_j))_{i,j=1}^n$ is invertible, where $\{X_j\}_{j=1}^n \subset \mathfrak{g}$ is a vector space basis. A connected linear Lie group is called *semisimple* if its Lie algebra is semisimple.

Example. Linear Lie groups $\mathrm{SL}(n,\mathbb{K})$ and $\mathrm{SO}(n)$ are semisimple, but $\mathrm{GL}(n)$ is not semisimple.

Remark 8.3.40. Since $\mathrm{Tr}(ab) = \mathrm{Tr}(ba)$, we have

$$
B(X,Y) = B(Y,X).
$$

We also have

$$
B(X,[Y,Z]) = B([X,Y],Z),
$$

because

$$
\mathrm{Tr}(a(bc - cb)) = \mathrm{Tr}(abc) - \mathrm{Tr}(acb) = \mathrm{Tr}(abc) - \mathrm{Tr}(bac) = \mathrm{Tr}((ab - ba)c)
$$

yields

$$
\begin{aligned}
B(X,[Y,Z]) &= \mathrm{Tr}\,(\mathrm{ad}(X)\,\mathrm{ad}([Y,Z])) \\
&= \mathrm{Tr}\,(\mathrm{ad}(X)\,[\mathrm{ad}(Y),\mathrm{ad}(Z)]) \\
&= \mathrm{Tr}\,([\mathrm{ad}(X),\mathrm{ad}(Y)]\,\mathrm{ad}(Z)) \\
&= \mathrm{Tr}\,(\mathrm{ad}([X,Y])\,\mathrm{ad}(Z)) \\
&= B([X,Y],Z).
\end{aligned}
$$

It can be proven that the Killing form of the Lie algebra of a compact linear Lie group is negative semi-definite, i.e., $B(X,X) \leq 0$. On the other hand, if the Killing form of a Lie group is negative definite, i.e., $B(X,X) < 0$ whenever $X \neq 0$, then the group is compact.

8.3.2   Casimir element and Laplace operator

Here we discuss some properties of the Casimir element and the corresponding
Laplace operator.

Definition 8.3.41 (Casimir element). Let $\mathfrak{g}$ be a semisimple $\mathbb{K}$-Lie algebra with a
vector space basis $\{X_j\}_{j=1}^n \subset \mathfrak{g}$. Let $B : \mathfrak{g} \times \mathfrak{g} \to \mathbb{K}$ be the Killing form of $\mathfrak{g}$, and
define the matrix $R \in \mathbb{K}^{n \times n}$ by $R_{ij} := B(X_i, X_j)$. Let

$$X^i := \sum_{j=1}^n \left(R^{-1} \right)_{ij} X_j,$$

so that $\{X^i\}_{i=1}^n$ is another vector space basis for $\mathfrak{g}$. Then the *Casimir element*
$\Omega \in \mathcal{U}(\mathfrak{g})$ of $\mathfrak{g}$ is defined by

$$\Omega := \sum_{i=1}^n X_i X^i.$$

Remark 8.3.42. The Casimir element $\Omega \in \mathcal{U}(\mathfrak{g})$ for the Lie algebra $\mathfrak{g}$ of a compact
semisimple linear Lie group G can be considered as an elliptic linear second-order
(left and right) translation invariant partial differential operator. In a sense, the
Casimir operator is an analogy of the Euclidean Laplace operator

$$\mathcal{L} = \sum_{j=1}^n \frac{\partial^2}{\partial x_j^2} : C^\infty(\mathbb{R}^n) \to C^\infty(\mathbb{R}^n).$$

Such a Laplace operator can be constructed for any compact Lie group G, and
with it we may define Sobolev spaces on G nicely, etc.

Theorem 8.3.43 (Properties of Casimir element). *The Casimir element of a finite-
dimensional semisimple $\mathbb{K}$-Lie algebra $\mathfrak{g}$ is independent of the choice of the vector
space basis $\{X_j\}_{j=1}^n \subset \mathfrak{g}$. Moreover,*

$$D\Omega = \Omega D$$

for all $D \in \mathcal{U}(\mathfrak{g})$.

Proof. Let $\{X_j\}_{j=1}^n \subset \mathfrak{g}$, $R_{ij} = B(X_i, X_j)$ and Ω be as in Definition 8.3.41. To
simplify notation, we consider only the case $\mathbb{K} = \mathbb{R}$. Let $\{Y_i\}_{i=1}^n \subset \mathfrak{g}$ be a vector
space basis of $\mathfrak{g}$. Then there exists $A = (A_{ij})_{i,j=1}^n \in \mathrm{GL}(n, \mathbb{R})$ such that

$$\left\{ Y_i := \sum_{j=1}^n A_{ij} X_j \right\}_{i=1}^n.$$

Then

$$
\begin{aligned}
S &:= \left(B(Y_i, Y_j)\right)_{i,j=1}^n \\
&= \left(B\left(\sum_{k=1}^n A_{ik}X_k, \sum_{l=1}^n A_{jl}X_l\right)\right)_{i,j=1}^n \\
&= \left(\sum_{k,l=1}^n A_{ik}\, B(X_k, X_l)\, A_{jl}\right)_{i,j=1}^n \\
&= ARA^{\mathrm{T}};
\end{aligned}
$$

hence

$$
S^{-1} = \left((S^{-1})_{ij}\right)_{i,j=1}^n = (A^{\mathrm{T}})^{-1}R^{-1}A^{-1}.
$$

Let us now compute the Casimir element of $\mathfrak{g}$ with respect to the basis $\{Y_j\}_{j=1}^n$:

$$
\begin{aligned}
\sum_{i,j=1}^n (S^{-1})_{ij}Y_iY_j &= \sum_{i,j=1}^n (S^{-1})_{ij}\sum_{k=1}^n A_{ik}X_k \sum_{l=1}^n A_{jl}X_l \\
&= \sum_{k,l=1}^n X_kX_l \sum_{i,j=1}^n A_{ik}(S^{-1})_{ij}A_{jl} \\
&= \sum_{k,l=1}^n X_kX_l \sum_{i,j=1}^n (A^{\mathrm{T}})_{ki}((A^{\mathrm{T}})^{-1}R^{-1}A^{-1})_{ij}A_{jl} \\
&= \sum_{k,l=1}^n X_kX_l(R^{-1})_{kl}.
\end{aligned}
$$

Thus the definition of the Casimir element does not depend on the choice of a vector space basis.

We still have to prove that Ω commutes with every $D \in \mathcal{U}(\mathfrak{g})$. Since

$$
B(X^i, X_j) = \sum_{k=1}^n (R^{-1})_{ik}B(X_k, X_j) = \sum_{k=1}^n (R^{-1})_{ik}R_{kj} = \delta_{ij},
$$

we can extend $(X_i, X_j) \mapsto \langle X_i, X_j\rangle_{\mathfrak{g}} := B(X^i, X_j)$ uniquely to an inner product

$$
((X, Y) \mapsto \langle X, Y\rangle_{\mathfrak{g}}) : \mathfrak{g} \times \mathfrak{g} \to \mathbb{R},
$$

with respect to which the collection $\{X_i\}_{i=1}^n$ is an orthonormal basis. For the Lie product $(x, y) \mapsto [x, y] := xy - yx$ of $\mathrm{Lie}_{\mathbb{R}}(\mathcal{U}(\mathfrak{g}))$ we have

$$
[x, yz] = [x, y]z + y[x, z],
$$

so that for $D \in \mathfrak{g}$ we get

$$[D, \Omega] = [D, \sum_{i=1}^{n} X_i X^i] = \sum_{i=1}^{n} ([D, X_i]X^i + X_i[D, X^i]) .$$

Let $c_{ij}, d_{ij} \in \mathbb{R}$ be defined by

$$[D, X_i] = \sum_{j=1}^{n} c_{ij} X_j, \quad [D, X^i] = \sum_{j=1}^{n} d_{ij} X^j.$$

Then

$$c_{ij} = \langle X_j, [D, X_i] \rangle_{\mathfrak{g}} = B(X^j, [D, X_i]) = B([X^j, D], X_i) = B(-[D, X^j], X_i)$$

$$= B(-\sum_{k=1}^{n} d_{jk} X^k, X_i) = -\sum_{k=1}^{n} d_{jk} B(X^k, X_i)$$

$$= -\sum_{k=1}^{n} d_{jk} \langle X_k, X_i \rangle_{\mathfrak{g}} = -d_{ji},$$

so that

$$[D, \Omega] \quad = \quad \sum_{i,j=1}^{n} (c_{ij} X_j X^i + d_{ij} X_i X^j)$$

$$= \quad \sum_{i,j=1}^{n} (c_{ij} + d_{ji}) X_j X^i$$

$$= \quad 0,$$

i.e., $D\Omega = \Omega D$ for all $D \in \mathfrak{g}$. By induction, we may prove that

$$[D_1 D_2 \cdots D_m, \Omega] \quad = \quad D_1[D_2 \cdots D_m, \Omega] + [D_1, \Omega]D_2 \cdots D_m \quad = \quad 0$$

for every $\{D_j\}_{j=1}^{m} \subset \mathfrak{g}$, so that $D\Omega = \Omega D$ for all $D \in \mathcal{U}(\mathfrak{g})$. $\square$

Exercise 8.3.44. How should the proof of Theorem 8.3.43 be modified if $\mathbb{K} = \mathbb{C}$ instead of $\mathbb{K} = \mathbb{R}$?

Definition 8.3.45 (Laplace operator on G). The Casimir element from Definition 8.3.41, also denoted by

$$\mathcal{L}_G := \Omega \in \mathcal{U}(\mathfrak{g}),$$

and viewed as a second-order partial differential operator on G is also called the *Laplace* operator on G. Here a vector field $Y \in \mathfrak{g}$ is viewed as a differential operator $Y \equiv D_Y : C^\infty(G) \to C^\infty(G)$, defined by

$$Y f(x) \equiv D_Y f(x) = \frac{\mathrm{d}}{\mathrm{d}t} f(x \, \exp(tY)) \Big|_{t=0} .$$

Remark 8.3.46. The Laplace operator $\mathcal{L}_G$ is a negative definite bi-invariant operator on G, by Theorem 8.3.43. If G is equipped with the unique (up to a constant) bi-invariant Riemannian metric, $\mathcal{L}_G$ is its Laplace–Beltrami operator.

In the notation of the right and left Peter–Weyl theorem in Theorem 7.5.14 and Remark 7.5.16, we write

$$\mathcal{H}^\phi := \bigoplus_{i=1}^{\dim \phi} \mathcal{H}^\phi_{i,\cdot} = \bigoplus_{j=1}^{\dim \phi} \mathcal{H}^\phi_{\cdot,j}.$$

Theorem 8.3.47 (Eigenvalues of the Laplacian on G). *For every $[\phi] \in \widehat{G}$ the space $\mathcal{H}^\phi$ is an eigenspace of $\mathcal{L}_G$ and $-\mathcal{L}_G|_{\mathcal{H}^\phi} = \lambda_\phi I$, for some $\lambda_\phi \geq 0$.*

Proof. We will use the notation of Theorem 7.5.14. Note that by Theorem 8.3.43 the Laplace operator $\mathcal{L}_G$ is bi-invariant, so that it commutes with both $\pi_R(x)$ and $\pi_L(x)$, for all $x \in G$. Therefore, by the Peter–Weyl theorem it commutes with all $\phi \in \widehat{G}$. Thus $\mathcal{L}_G(\mathcal{H}^\phi_{\cdot,j}) \subset \mathcal{H}^\phi_{\cdot,j}$ and $\mathcal{L}_G(\mathcal{H}^\phi_{i,\cdot}) \subset \mathcal{H}^\phi_{i,\cdot}$, for all $1 \leq i,j \leq \dim(\phi)$. It follows that $\mathcal{L}_G \phi_{ij} \in \mathcal{H}^\phi_{i,\cdot} \cap \mathcal{H}^\phi_{\cdot,j} = \operatorname{span}(\phi_{ij})$, so that $\mathcal{L}_G \phi_{ij} = c_{ij} \phi_{ij}$ for some constants c_{ij}. Let us now determine these constants. We have

$$
\begin{aligned}
(\mathcal{L}_G \pi_R(y)\phi_{ij})(x) &= \mathcal{L}_G(\phi_{ij}(xy)) \\
&= \mathcal{L}_G \left(\sum_{k=1}^{\dim(\phi)} \phi_{ik}(x)\phi_{kj}(y) \right) \\
&= \sum_{k=1}^{\dim(\phi)} c_{ik}\phi_{ik}(x)\phi_{kj}(y).
\end{aligned}
$$

On the other hand we have

$$
\begin{aligned}
(\pi_R(y)\mathcal{L}_G \phi_{ij})(x) &= c_{ij}\phi_{ij}(xy) \\
&= \sum_{k=1}^{\dim(\phi)} c_{ij}\phi_{ik}(x)\phi_{kj}(y).
\end{aligned}
$$

It follows now from the orthogonality Lemma 7.5.12 that $c_{ik}\phi_{kj}(y) = c_{ij}\phi_{kj}(y)$, or that $c_{ik} = c_{ij}$ for all $1 \leq i,j,k \leq \dim(\phi)$. A similar calculation with the left regular action $\pi_L(y)$ shows that $c_{kj} = c_{ij}$ for all $1 \leq i,j,k \leq \dim(\phi)$. Hence $\mathcal{L}_G \phi_{ij} = c\phi_{ij}$ for all $1 \leq i,j \leq \dim(\phi)$, and since $\mathcal{L}_G$ is negative definite, we obtain the statement with $\lambda_\phi := -c \geq 0$. $\qquad\square$

Chapter 9

Hopf Algebras

Instead of studying a compact group G, we may consider the algebra $C(G)$ of continuous functions $G \to \mathbb{C}$. The structure of the group is encoded in the function algebra, but we shall see that this approach paves the way for a more general functional analytic theory of Hopf algebras, which possess nice duality properties.

9.1 Commutative C^*-algebras

Let $\mathcal{A} := C(X)$, where X is a compact Hausdorff space. We present[1] some fundamental results:

- All the algebra homomorphisms $\mathcal{A} \to \mathbb{C}$ are of the form

$$f \mapsto f(x),$$

  where $x \in X$.
- All the closed ideals of $\mathcal{A}$ are of the form

$$I(K) := \{f \in \mathcal{A} \mid f(K) = \{0\}\},$$

  where $K \subset X$ (with convention $I(\emptyset) := C(X)$). Moreover, $\overline{K} = V(I(K))$, where

$$V(J) = \bigcap_{f \in J} f^{-1}(\{0\});$$

  these results follow by Urysohn's Lemma (Theorem A.12.11).
- Linear functionals $\mathcal{A} \to \mathbb{C}$ are of the form

$$f \mapsto \int_X f \, d\mu, \tag{9.1}$$

  where μ is a Borel-regular measure on X; this is the Riesz Representation Theorem C.4.60.

---

[1] These statements follow essentially from the results in Part I

- Probability functionals $\mathcal{A} \to \mathbb{C}$ are then of the form (9.1), where μ is a Borel-regular probability measure on X.

All in all, we might say that the topology and measure theory of a compact Hausdorff space X is encoded in the algebra $\mathcal{A} = C(X)$, with a dictionary:

Space X	Algebra $\mathcal{A} = C(X)$
homeomorphism $\phi : X \to X$	isomorphism $(f \mapsto f \circ \phi) : \mathcal{A} \to \mathcal{A}$
point $x \in X$	algebra functional $(f \mapsto f(x)) : \mathcal{A} \to \mathbb{C}$
closed set in X	closed ideal in $\mathcal{A}$
X metrisable	$\mathcal{A}$ separable
Borel-regular measure on X	linear functional
Borel-regular probability measure on X	probability functional
$\vdots$	$\vdots$

Remark 9.1.1. In the light of the dictionary above, we are bound to ask:

1. If X is a group, how is this reflected in $C(X)$?

2. Could we study non-commutative algebras just like the commutative ones?

We might call the traditional topology and measure theory by the name "commutative geometry", referring to the commutative function algebras; "non-commutative geometry" would refer to the study of non-commutative algebras. Let us now try to deal with the two questions posed above.

Answering to question 1. Let G be a compact group. By Urysohn's Lemma (Theorem A.12.11), $C(G)$ separates the points of X, so that the associativity of the group operation $((x,y) \mapsto xy) : G \times G \to G$ is encoded by

$$\forall x, y, z \in G \quad \forall f \in C(G) : \quad f((xy)z) = f(x(yz)).$$

Similarly,

$$\exists e \in G \quad \forall x \in G \quad \forall f \in C(G) : \quad f(xe) = f(x) = f(ex)$$

encodes the neutral element $e \in G$. Finally,

$$\forall x \in G \quad \exists x^{-1} \in G \quad \forall f \in C(G) : \quad f(x^{-1}x) = f(e) = f(xx^{-1})$$

encodes the inversion $(x \mapsto x^{-1}) : G \to G$. Thereby let us define linear operators

$$\tilde{\Delta} : C(G) \to C(G \times G), \qquad \tilde{\Delta}f(x,y) := f(xy),$$
$$\tilde{\varepsilon} : C(G) \to \mathbb{C}, \qquad \tilde{\varepsilon}f := f(e),$$
$$\tilde{S} : C(G) \to C(G), \qquad \tilde{S}f(x) := f(x^{-1});$$

the interactions of these algebra homomorphisms contain all the information about the structure of the underlying group! This is a key ingredient in the Hopf algebra theory.

Answering to question 2. Our algebras always have a unit element 1. An involutive $\mathbb{C}$-algebra $\mathcal{A}$ is a C^*-*algebra* if it has a Banach space norm satisfying

$$\|ab\| \leq \|a\| \, \|b\| \quad \text{and} \quad \|a^*a\| = \|a\|^2$$

for all $a, b \in \mathcal{A}$. By Gelfand and Naimark (1943), see Theorem D.5.3, up to an isometric $*$-isomorphism a C^*-algebra is a closed involutive subalgebra of $\mathcal{L}(\mathcal{H})$, where $\mathcal{H}$ is a Hilbert space; moreover, if $\mathcal{A}$ is a commutative unital C^*-algebra then $\mathcal{A} \cong C(X)$ for a compact Hausdorff space X, as explained below:

The *spectrum of* $\mathcal{A}$ is the set $\mathrm{Spec}(\mathcal{A})$ of the algebra homomorphisms $\mathcal{A} \to \mathbb{C}$ (automatically bounded functionals!), endowed with the *Gelfand topology*, which is the relative weak*-topology of $\mathcal{L}(\mathcal{A}, \mathbb{C})$. It turns out that $\mathrm{Spec}(\mathcal{A})$ is a compact Hausdorff space. For $a \in \mathcal{A}$ we define the *Gelfand transform*

$$\widehat{a} : \mathrm{Spec}(\mathcal{A}) \to \mathbb{C}, \quad \widehat{a}(x) := x(a).$$

It turns out that $\widehat{a}$ is continuous, and that

$$(a \mapsto \widehat{a}) : \mathcal{A} \to C(\mathrm{Spec}(\mathcal{A}))$$

is an isometric $*$-algebra isomorphism!

If $\mathcal{B}$ is a non-commutative C^*-algebra, it still has plenty of interesting commutative C^*-subalgebras so that the Gelfand transform provides the nice tools of classical analysis on compact Hausdorff spaces in the study of the algebra. Namely, if $a \in \mathcal{B}$ is *normal*, i.e., $a^*a = aa^*$, then the closure of the algebraic span (polynomials) of $\{a, a^*\}$ is a commutative C^*-subalgebra. E.g., $b^*b \in \mathcal{B}$ is normal for all $b \in \mathcal{B}$.

Synthesis of questions 1 and 2. By the Gelfand–Naimark Theorem D.5.11, the archetypal commutative C^*-algebra is $C(X)$ for a compact Hausdorff space X. In the sequel, we introduce Hopf algebras. In a sense, they are a not-necessarily-commutative analogy of $C(G)$, where G is a compact group. We begin by formally dualising the category of *algebras*, to obtain the category of *co-algebras*. By marrying these concepts in a subtle way, we obtain the category of *Hopf algebras*.

9.2   Hopf algebras

The definition of a Hopf algebra is a lengthy one, yet quite natural. In the sequel, notice the evident dualities in the commutative diagrams.

For $\mathbb{C}$-vector spaces V, W, we define $\tau_{VW} : V \otimes W \to W \otimes V$ by the linear extension of

$$\tau_{VW}(v \otimes w) := w \otimes v.$$

Moreover, in the sequel the identity operation $(v \mapsto v) : V \to V$ for any vector space V is denoted by I. We constantly identify $\mathbb{C}$-vector spaces V and $\mathbb{C} \otimes V$ (and respectively $V \otimes \mathbb{C}$), since $(\lambda \otimes v) \mapsto \lambda v$ defines a linear isomorphism $\mathbb{C} \otimes V \to V$.

In the usual definition of an algebra, the multiplication is regarded as a bilinear map. In order to use dualisation techniques for algebras, we want to linearise the multiplication. Let us therefore give a new, equivalent definition for an algebra:

Definition 9.2.1 (Reformulation of algebras). The triple

$$(\mathcal{A}, m, \eta)$$

is an *algebra* (more precisely, an *associative unital $\mathbb{C}$-algebra*) if $\mathcal{A}$ is a $\mathbb{C}$-vector space, and

$$
\begin{aligned}
m &: \quad \mathcal{A} \otimes \mathcal{A} \to \mathcal{A}, \\
\eta &: \quad \mathbb{C} \to \mathcal{A}
\end{aligned}
$$

are linear mappings such that the following diagrams commute: the *associativity diagram*

$$
\begin{array}{ccc}
\mathcal{A} \otimes \mathcal{A} \otimes \mathcal{A} & \xrightarrow{I \otimes m} & \mathcal{A} \otimes \mathcal{A} \\
{\scriptstyle m \otimes I} \downarrow & & \downarrow {\scriptstyle m} \\
\mathcal{A} \otimes \mathcal{A} & \xrightarrow{\quad m \quad} & \mathcal{A}
\end{array}
$$

and the *unit diagrams*

$$
\begin{array}{ccccc}
\mathcal{A} \otimes \mathbb{C} & \xrightarrow{I \otimes \eta} & \mathcal{A} \otimes \mathcal{A} & \qquad \mathcal{A} \otimes \mathcal{A} & \xleftarrow{\eta \otimes I} & \mathbb{C} \otimes \mathcal{A} \\
{\scriptstyle a \otimes \lambda \mapsto \lambda a} \downarrow & & \downarrow {\scriptstyle m} & {\scriptstyle m} \downarrow & & \downarrow {\scriptstyle \lambda \otimes a \mapsto \lambda a} \\
\mathcal{A} & =\!\!=\!\!= & \mathcal{A}, & \mathcal{A} & =\!\!=\!\!= & \mathcal{A}.
\end{array}
$$

The mapping m is called the *multiplication* and η the *unit mapping*; the algebra $\mathcal{A}$ is said to be *commutative* if $m \tau_{\mathcal{A}\mathcal{A}} = m$. The *unit* of an algebra $(\mathcal{A}, m, \eta)$ is

$$\mathbf{1}_{\mathcal{A}} := \eta(1),$$

and the usual abbreviation for the multiplication is $ab := m(a \otimes b)$. For algebras $(\mathcal{A}_1, m_1, \eta_1)$ and $(\mathcal{A}_2, m_2, \eta_2)$ the *tensor product algebra* $(\mathcal{A}_1 \otimes \mathcal{A}_2, m, \eta)$ is defined by

$$m := (m_1 \otimes m_2)(I \otimes \tau_{\mathcal{A}_1 \mathcal{A}_2} \otimes I),$$

i.e., $(a_1 \otimes a_2)(b_1 \otimes b_2) = (a_1 b_1) \otimes (a_2 b_2)$, and

$$\eta(1) := \mathbf{1}_{\mathcal{A}_1} \otimes \mathbf{1}_{\mathcal{A}_2}.$$

Remark 9.2.2. If an algebra $\mathcal{A} = (\mathcal{A}, m, \eta)$ is finite-dimensional, we can formally dualise its structural mappings m and η; this inspires the concept of the co-algebra:

Definition 9.2.3 (Co-algebras). The triple

$$(\mathcal{C}, \Delta, \varepsilon)$$

is a *co-algebra* (more precisely, a *co-associative co-unital $\mathbb{C}$-co-algebra*) if $\mathcal{C}$ is a $\mathbb{C}$-vector space and

$$\Delta \ : \ \mathcal{C} \to \mathcal{C} \otimes \mathcal{C},$$
$$\varepsilon \ : \ \mathcal{C} \to \mathbb{C}$$

are linear mappings such that the following diagrams commute: the *co-associativity diagram* (notice the duality to the associativity diagram)

$$
\begin{array}{ccc}
\mathcal{C} \otimes \mathcal{C} \otimes \mathcal{C} & \xleftarrow{\ I \otimes \Delta\ } & \mathcal{C} \otimes \mathcal{C} \\
{\scriptstyle \Delta \otimes I} \big\uparrow & & \big\uparrow {\scriptstyle \Delta} \\
\mathcal{C} \otimes \mathcal{C} & \xleftarrow{\ \Delta\ } & \mathcal{C}
\end{array}
$$

and the *co-unit diagrams* (notice the duality to the unit diagrams)

$$
\begin{array}{ccccc}
\mathcal{C} \otimes \mathbb{C} & \xleftarrow{\ I \otimes \varepsilon\ } & \mathcal{C} \otimes \mathcal{C} & & \mathcal{C} \otimes \mathcal{C} & \xrightarrow{\ \varepsilon \otimes I\ } & \mathbb{C} \otimes \mathcal{C} \\
{\scriptstyle \lambda c \mapsto c \otimes \lambda} \big\uparrow & & \big\uparrow {\scriptstyle \Delta} & & {\scriptstyle \Delta} \big\uparrow & & \big\uparrow {\scriptstyle \lambda c \mapsto \lambda \otimes c} \\
\mathcal{C} & = & \mathcal{C}, & & \mathcal{C} & = & \mathcal{C}.
\end{array}
$$

The mapping Δ is called the *co-multiplication* and ε the *co-unit mapping*; the co-algebra $\mathcal{C}$ is *co-commutative* if $\tau_{\mathcal{C}\mathcal{C}}\Delta = \Delta$. For co-algebras $(\mathcal{C}_1, \Delta_1, \varepsilon_1)$ and $(\mathcal{C}_2, \Delta_2, \varepsilon_2)$ the *tensor product co-algebra* $(\mathcal{C}_1 \otimes \mathcal{C}_2, \Delta, \varepsilon)$ is defined by

$$\Delta := (I \otimes \tau_{\mathcal{C}_1 \mathcal{C}_2} \otimes I)(\Delta_1 \otimes \Delta_2)$$

and

$$\varepsilon(c_1 \otimes c_2) := \varepsilon_1(c_1)\varepsilon_2(c_2).$$

Example. A trivial co-algebra example: if $(\mathcal{A}, m, \eta)$ is a finite-dimensional algebra then the vector space dual $\mathcal{A}' = \mathcal{L}(\mathcal{A}, \mathbb{C})$ has a natural co-algebra structure. Indeed, let us identify $(\mathcal{A} \otimes \mathcal{A})'$ and $\mathcal{A}' \otimes \mathcal{A}'$ naturally, so that $m' : \mathcal{A}' \to \mathcal{A}' \otimes \mathcal{A}'$ is the dual mapping to $m : \mathcal{A} \otimes \mathcal{A} \to \mathcal{A}$. Let us identify $\mathbb{C}'$ and $\mathbb{C}$ naturally, so that $\eta' : \mathcal{A}' \to \mathbb{C}$ is the dual mapping to $\eta : \mathbb{C} \to \mathcal{A}$. Then

$$(\mathcal{A}', m', \eta')$$

is a co-algebra (draw the commutative diagrams!). We shall give more interesting examples of co-algebras after the definition of Hopf algebras.

Definition 9.2.4 (Convolution of linear operators). Let $(\mathcal{B}, m, \eta)$ be an algebra and $(\mathcal{B}, \Delta, \varepsilon)$ be a co-algebra. Let $\mathcal{L}(\mathcal{B})$ denote the vector space of linear operators $\mathcal{B} \to \mathcal{B}$. Let us define the *convolution* $A * B \in \mathcal{L}(\mathcal{B})$ *of linear operators* $A, B \in \mathcal{L}(\mathcal{B})$ by

$$A * B := m(A \otimes B)\Delta.$$

Exercise 9.2.5. Show that $\mathcal{L}(\mathcal{B})$ in Definition 9.2.4 is an algebra, when endowed with the convolution product of operators.

Definition 9.2.6 (Hopf algebras). A structure

$$(\mathcal{H}, m, \eta, \Delta, \varepsilon, S)$$

is a *Hopf algebra* if

- $(\mathcal{H}, m, \eta)$ is an algebra,
- $(\mathcal{H}, \Delta, \varepsilon)$ is a co-algebra,
- $\Delta : \mathcal{H} \to \mathcal{H} \otimes \mathcal{H}$ and $\varepsilon : \mathcal{H} \to \mathbb{C}$ are algebra homomorphisms, i.e.,

$$\Delta(fg) = \Delta(f)\Delta(g), \quad \Delta(1_\mathcal{H}) = 1_{\mathcal{H}\otimes\mathcal{H}},$$
$$\varepsilon(fg) = \varepsilon(f)\varepsilon(g), \quad \varepsilon(1_\mathcal{H}) = 1,$$

- and $S : \mathcal{H} \to \mathcal{H}$ is a linear mapping, called the *antipode*, satisfying

$$S * I = \eta\varepsilon = I * S;$$

i.e., $I \in \mathcal{L}(\mathcal{H})$ and $S \in \mathcal{L}(\mathcal{H})$ are inverses to each other in the convolution algebra $\mathcal{L}(\mathcal{H})$.

For Hopf algebras $(\mathcal{H}_1, m_1, \eta_1, \Delta_1, \varepsilon_1, S_1)$ and $(\mathcal{H}_2, m_2, \eta_2, \Delta_2, \varepsilon_2, S_2)$ we define the *tensor product Hopf algebra* $(\mathcal{H}_1 \otimes \mathcal{H}_2, m, \eta, \Delta, \varepsilon, S)$ such that

$$(\mathcal{H}_1 \otimes \mathcal{H}_2, m, \eta)$$

is the usual tensor product algebra,

$$(\mathcal{H}_1 \otimes \mathcal{H}_2, \Delta, \varepsilon)$$

is the usual tensor product co-algebra, and

$$S := S_{\mathcal{H}_1} \otimes S_{\mathcal{H}_2}.$$

Exercise 9.2.7 (Uniqueness of the antipode). Let $(\mathcal{H}, m, \eta, \Delta, \varepsilon, S_j)$ be Hopf algebras, where $j \in \{1, 2\}$. Show that $S_1 = S_2$.

Remark 9.2.8 (**Commutative diagrams for Hopf algebras**). Notice that we now have the *multiplication and co-multiplication diagram*

$$
\begin{array}{ccc}
\mathcal{H} \otimes \mathcal{H} & \xrightarrow{\ \Delta m\ } & \mathcal{H} \otimes \mathcal{H} \\
{\scriptstyle \Delta \otimes \Delta} \downarrow & & \uparrow {\scriptstyle m \otimes m} \\
\mathcal{H} \otimes \mathcal{H} \otimes \mathcal{H} \otimes \mathcal{H} & \xrightarrow{I \otimes \tau_{\mathcal{H}\mathcal{H}} \otimes I} & \mathcal{H} \otimes \mathcal{H} \otimes \mathcal{H} \otimes \mathcal{H},
\end{array}
$$

the *co-multiplication and unit diagram*

$$
\begin{array}{ccc}
\mathcal{H} & \xleftarrow{\ \eta\ } & \mathbb{C} \\
{\scriptstyle \Delta} \downarrow & & \parallel \\
\mathcal{H} \otimes \mathcal{H} & \xleftarrow{\eta \otimes \eta} & \mathbb{C} \otimes \mathbb{C},
\end{array}
$$

the *multiplication and co-unit diagram*

$$
\begin{array}{ccc}
\mathcal{H} & \xrightarrow{\ \varepsilon\ } & \mathbb{C} \\
{\scriptstyle m} \uparrow & & \parallel \\
\mathcal{H} \otimes \mathcal{H} & \xrightarrow{\varepsilon \otimes \varepsilon} & \mathbb{C} \otimes \mathbb{C}
\end{array}
$$

and the *"everyone with the antipode"* diagrams

$$
\begin{array}{ccc}
\mathcal{H} & \xrightarrow{\ \eta\varepsilon\ } & \mathcal{H} \\
{\scriptstyle \Delta} \downarrow & & \uparrow {\scriptstyle m} \\
\mathcal{H} \otimes \mathcal{H} & \xrightarrow[S \otimes I]{I \otimes S} & \mathcal{H} \otimes \mathcal{H}.
\end{array}
$$

Example (**A monoid co-algebra example**). Let G be a finite group and $\mathcal{F}(G)$ be the $\mathbb{C}$-vector space of functions $G \to \mathbb{C}$. Notice that $\mathcal{F}(G) \otimes \mathcal{F}(G)$ and $\mathcal{F}(G \times G)$ are naturally isomorphic by

$$
\sum_{j=1}^{m}(f_j \otimes g_j)(x, y) := \sum_{j=1}^{m} f_j(x)g_j(y).
$$

Then we can define mappings $\Delta : \mathcal{F}(G) \to \mathcal{F}(G) \otimes \mathcal{F}(G)$ and $\varepsilon : \mathcal{F}(G) \to \mathbb{C}$ by

$$
\Delta f(x, y) := f(xy), \quad \varepsilon f := f(e).
$$

In the next example we show that $(\mathcal{F}(G), \Delta, \varepsilon)$ is a co-algebra. But there is still more structure in the group to exploit: let us define an operator $S : \mathcal{F}(G) \to \mathcal{F}(G)$ by $(Sf)(x) := f(x^{-1})\ldots$

Example (**Hopf algebra for finite group**). Let G be a finite group. Now $\mathcal{F}(G)$ from the previous example has a structure of a commutative Hopf algebra; it is co-commutative if and only if G is a commutative group. The algebra mappings are given by

$$\eta(\lambda)(x) := \lambda, \quad m(f \otimes g)(x) := f(x)g(x)$$

for all $\lambda \in \mathbb{C}$, $x \in G$ and $f, g \in \mathcal{F}(G)$. Notice that the identification $\mathcal{F}(G \times G) \cong \mathcal{F}(G) \otimes \mathcal{F}(G)$ gives the interpretation $(ma)(x) = a(x, x)$ for $a \in \mathcal{F}(G \times G)$. Clearly $(\mathcal{F}(G), m, \eta)$ is a commutative algebra. Let $x, y, z \in G$ and $f, g \in \mathcal{F}(G)$. Then

$$
\begin{aligned}
((\Delta \otimes I)\Delta f)(x, y, z) &= (\Delta f)(xy, z) \\
&= f((xy)z) \\
&= f(x(yz)) \\
&= (\Delta f)(x, yz) \\
&= ((I \otimes \Delta)\Delta f)(x, y, z),
\end{aligned}
$$

so that $(\Delta \otimes I)\Delta = (I \otimes \Delta)\Delta$. Next, $(\varepsilon \otimes I)\Delta \cong I \cong (I \otimes \varepsilon)\Delta$, because

$$
\begin{aligned}
(m(\eta\varepsilon \otimes I)\Delta f)(x) &= ((\eta\varepsilon \otimes I)\Delta f)(x, x) \\
&= \Delta f(e, x) \\
&= f(ex) = f(x) = f(xe) \\
&= \cdots = (m(I \otimes \eta\varepsilon)\Delta f)(x).
\end{aligned}
$$

Thereby $(\mathcal{F}(G), \Delta, \varepsilon)$ is a co-algebra. Moreover,

$$\varepsilon(fg) = (fg)(e) = f(e)g(e) = \varepsilon(f)\varepsilon(g),$$

$$\varepsilon(\mathbf{1}_{\mathcal{F}(G)}) = \mathbf{1}_{\mathcal{F}(G)}(e) = 1,$$

so that $\varepsilon : \mathcal{F}(G) \to \mathbb{C}$ is an algebra homomorphism. The co-multiplication $\Delta : \mathcal{F}(G) \to \mathcal{F}(G) \otimes \mathcal{F}(G) \cong \mathcal{F}(G \times G)$ is an algebra homomorphism, because

$$\Delta(fg)(x, y) = (fg)(xy) = f(xy)\, g(xy) = (\Delta f)(x, y)\, (\Delta g)(x, y),$$

$$\Delta(\mathbf{1}_{\mathcal{F}(G)})(x, y) = \mathbf{1}_{\mathcal{F}(G)}(xy) = 1 = \mathbf{1}_{\mathcal{F}(G \times G)}(x, y) \cong (\mathbf{1}_{\mathcal{F}(G)} \otimes \mathbf{1}_{\mathcal{F}(G)})(x, y).$$

Finally,

$$
\begin{aligned}
((I * S)f)(x) &= (m(I \otimes S)\Delta f)(x) \\
&= ((I \otimes S)\Delta f)(x, x) \\
&= (\Delta f)(x, x^{-1}) \\
&= f(xx^{-1}) = f(e) = \varepsilon f \\
&= \cdots = ((S * I)f)(x),
\end{aligned}
$$

so that $I * S = \eta\varepsilon = S * I$. Thereby $\mathcal{F}(G)$ can be endowed with a Hopf algebra structure.

Example (**Hopf algebra for a compact group**). Let G be a compact group. We shall endow the dense subalgebra $\mathcal{H} := \mathrm{TrigPol}(G) \subset C(G)$ of trigonometric polynomials with a natural structure of a commutative Hopf algebra; $\mathcal{H}$ will be co-commutative if and only if G is commutative. Actually, if G is a finite group then $\mathcal{F}(G) = \mathrm{TrigPol}(G) = C(G)$. For a compact group G, it can be shown that here $\mathcal{H} \otimes \mathcal{H} \cong \mathrm{TrigPol}(G \times G)$, where the isomorphism is given by

$$\sum_{j=1}^{m} (f_j \otimes g_j)(x, y) := \sum_{j=1}^{m} f_j(x) g_j(y).$$

The algebra structure

$$(\mathcal{H}, m, \eta)$$

is the usual one for the trigonometric polynomials, i.e., $m(f \otimes g) := fg$ and $\eta(\lambda) = \lambda \mathbf{1}$, where $\mathbf{1}(x) = 1$ for all $x \in G$. By the Peter–Weyl Theorem 7.5.14, the $\mathbb{C}$-vector space $\mathcal{H}$ is spanned by

$$\left\{ \phi_{ij} : \ \phi = (\phi_{ij})_{i,j}^{\dim(\phi)}, \ [\phi] \in \widehat{G} \right\}.$$

Let us define the co-multiplication $\Delta : \mathcal{H} \to \mathcal{H} \otimes \mathcal{H}$ by

$$\Delta \phi_{ij} := \sum_{k=1}^{\dim(\phi)} \phi_{ik} \otimes \phi_{kj};$$

we see that then

$$
\begin{aligned}
(\Delta \phi_{ij})(x, y) &= \sum_{k=1}^{\dim(\phi)} (\phi_{ik} \otimes \phi_{kj})(x, y) \\
&= \sum_{k=1}^{\dim(\phi)} \phi_{ik}(x) \phi_{kj}(y) \\
&= \phi_{ij}(xy).
\end{aligned}
$$

The co-unit $\varepsilon : \mathcal{H} \to \mathbb{C}$ is defined by

$$\varepsilon f := f(e),$$

and the antipode $S : \mathcal{H} \to \mathcal{H}$ by

$$(Sf)(x) := f(x^{-1}).$$

Exercise 9.2.9. In the *Example* about $\mathcal{H} = \mathrm{TrigPol}(G)$ above, check the validity of the Hopf algebra axioms.

Theorem 9.2.10 (Commutative C^*-algebras and Hopf algebras). *Let $\mathcal{H}$ be a commutative C^*-algebra. If $(\mathcal{H}, m, \eta, \Delta, \varepsilon, S)$ is a finite-dimensional Hopf algebra then there exists a Hopf algebra isomorphism $\mathcal{H} \cong C(G)$, where G is a finite group and $C(G)$ is endowed with the Hopf algebra structure given above.*

Proof. Let $G := \mathrm{Spec}(\mathcal{H}) = \mathrm{HOM}(\mathcal{H}, \mathbb{C})$. As $\mathcal{H}$ is a commutative C^*-algebra, it is isometrically $*$-isomorphic to the C^*-algebra $C(G)$ via the Gelfand transform

$$(f \mapsto \widehat{f}) : \mathcal{H} \to C(G), \quad \widehat{f}(x) := x(f).$$

The space G must be finite, because $\dim(C(G)) = \dim(\mathcal{H}) < \infty$.

Now

$$e := \varepsilon \in G,$$

because $\varepsilon : \mathcal{H} \to \mathbb{C}$ is an algebra homomorphism. This $e \in G$ will turn out to be the neutral element of our group.

Let $x, y \in G$. We identify the spaces $\mathbb{C} \otimes \mathbb{C}$ and $\mathbb{C}$, and get an algebra homomorphism $x \otimes y : \mathcal{H} \otimes \mathcal{H} \to \mathbb{C} \otimes \mathbb{C} \cong \mathbb{C}$. Now $\Delta : \mathcal{H} \to \mathcal{H} \otimes \mathcal{H}$ is an algebra homomorphism, so that $(x \otimes y)\Delta : \mathcal{H} \to \mathbb{C}$ is an algebra homomorphism! Let us denote

$$xy := (x \otimes y)\Delta,$$

so that $xy \in G$. This defines the group operation $((x, y) \mapsto xy) : G \times G \to G$!

Inversion $x \mapsto x^{-1}$ will be defined via the antipode $S : \mathcal{H} \to \mathcal{H}$. We shall show that for a *commutative* Hopf algebra, the antipode is an algebra isomorphism. First we prove that $S(\mathbf{1}_{\mathcal{H}}) = \mathbf{1}_{\mathcal{H}}$:

$$
\begin{aligned}
S\mathbf{1}_{\mathcal{H}} &= m(\mathbf{1}_{\mathcal{H}} \otimes S\mathbf{1}_{\mathcal{H}}) \\
&= m(I \otimes S)(\mathbf{1}_{\mathcal{H}} \otimes \mathbf{1}_{\mathcal{H}}) \\
&= m(I \otimes S)\Delta\mathbf{1}_{\mathcal{H}} \\
&= (I * S)\mathbf{1}_{\mathcal{H}} = \eta\varepsilon\mathbf{1}_{\mathcal{H}} \\
&= \mathbf{1}_{\mathcal{H}}.
\end{aligned}
$$

Then we show that $S(gh) = S(h)S(g)$, where $g, h \in \mathcal{H}$, $gh := m(g \otimes h)$. Let us use the so-called *Sweedler notation*

$$\Delta f =: \sum f_{(1)} \otimes f_{(2)} =: f_{(1)} \otimes f_{(2)};$$

consequently

$$
\begin{aligned}
(\Delta \otimes I)\Delta f &= (\Delta \otimes I)(f_{(1)} \otimes f_{(2)}) = f_{(1)(1)} \otimes f_{(1)(2)} \otimes f_{(2)}, \\
(I \otimes \Delta)\Delta f &= (I \otimes \Delta)(f_{(1)} \otimes f_{(2)}) = f_{(1)} \otimes f_{(2)(1)} \otimes f_{(2)(2)},
\end{aligned}
$$

and due to the *co-associativity* we may re-index as follows:

$$(\Delta \otimes I)\Delta f =: f_{(1)} \otimes f_{(2)} \otimes f_{(3)} := (I \otimes \Delta)\Delta f$$

(notice that, e.g., $f_{(2)}$ appears in different meanings above – this is just a matter of notation!). Then

$$
\begin{aligned}
S(gh) &= S(\varepsilon((gh)_{(1)})(gh)_{(2)}) \\
&= \varepsilon((gh)_{(1)})\, S((gh)_{(2)}) \\
&= \varepsilon(g_{(1)}h_{(1)})\, S(g_{(2)}h_{(2)}) \\
&= \varepsilon(g_{(1)})\, \varepsilon(h_{(1)})\, S(g_{(2)}h_{(2)}) \\
&= \varepsilon(g_{(1)})\, S(h_{(1)(1)})\, h_{(1)(2)}\, S(g_{(2)}h_{(2)}) \\
&= \varepsilon(g_{(1)})\, S(h_{(1)})\, h_{(2)}\, S(g_{(2)}h_{(3)}) \\
&= S(h_{(1)})\, \varepsilon(g_{(1)})\, h_{(2)}\, S(g_{(2)}h_{(3)}) \\
&= S(h_{(1)})\, S(g_{(1)(1)})\, g_{(1)(2)}\, h_{(2)}\, S(g_{(2)}h_{(3)}) \\
&= S(h_{(1)})\, S(g_{(1)})\, g_{(2)}\, h_{(2)}\, S(g_{(3)}h_{(3)}) \\
&= S(h_{(1)})\, S(g_{(1)})\, (gh)_{(2)}\, S((gh)_{(3)}) \\
&= S(h_{(1)})\, S(g_{(1)})\, \varepsilon((gh)_{(2)}) \\
&= S(h_{(1)})\, S(g_{(1)})\, \varepsilon(g_{(2)}h_{(2)}) \\
&= S(h_{(1)})\, S(g_{(1)})\, \varepsilon(g_{(2)})\, \varepsilon(h_{(2)}) \\
&= S(h_{(1)}\varepsilon(h_{(2)}))\, S(g_{(1)}\varepsilon(g_{(2)})) \\
&= S(h)\, S(g);
\end{aligned}
$$

this computation can be compared to

$$
\begin{aligned}
(xy)^{-1} &= e(xy)^{-1} \\
&= y^{-1}y(xy)^{-1} \\
&= y^{-1}ey(xy)^{-1} \\
&= y^{-1}x^{-1}xy(xy)^{-1} \\
&= y^{-1}x^{-1}e \\
&= y^{-1}x^{-1}
\end{aligned}
$$

for $x, y \in G$! Since $\mathcal{H}$ is commutative, we have proven that $S : \mathcal{H} \to \mathcal{H}$ is an algebra homomorphism. Thereby $xS : \mathcal{H} \to \mathbb{C}$ is an algebra homomorphism. Let us denote

$$
x^{-1} := xS \in G,
$$

which is the inverse of $x \in G$!

We leave it for the reader to show that $(G, (x, y) \mapsto xy, x \mapsto x^{-1})$ is indeed a group. $\qquad\square$

Exercise 9.2.11. Finish the proof of Theorem 9.2.10.

Exercise 9.2.12 (Universal enveloping algebra as a Hopf algebra). Let $\mathfrak{g}$ be a Lie algebra and $\mathcal{U}(\mathfrak{g})$ its universal enveloping algebra. Let $X \in \mathfrak{g}$; extend definitions

$$\Delta X := X \otimes 1_{\mathcal{U}(\mathfrak{g})} + 1_{\mathcal{U}(\mathfrak{g})} \otimes X, \quad \varepsilon X := 0, \quad SX := -X$$

so that you obtain a Hopf algebra structure $(\mathcal{U}(\mathfrak{g}), m, \eta, \Delta, \varepsilon, S)$.

Exercise 9.2.13. Let $(\mathcal{H}, m, \eta, \Delta, \varepsilon, S)$ be a finite-dimensional Hopf algebra.

(a) Endow the dual $\mathcal{H}' = \mathcal{L}(\mathcal{H}, \mathbb{C})$ with a natural Hopf algebra structure via the duality

$$(f, \phi) \mapsto \langle f, \phi \rangle_{\mathcal{H}} := \phi(f)$$

where $f \in \mathcal{H}$, $\phi \in \mathcal{H}'$.

(b) If G is a finite group and $\mathcal{H} = \mathcal{F}(G)$, what are the Hopf algebra operations for $\mathcal{H}'$?

(c) With a suitable choice for $\mathcal{H}$, give an example of a non-commutative non-co-commutative Hopf algebra $\mathcal{H} \otimes \mathcal{H}'$.

Exercise 9.2.14 (M.E. Sweedler's example). Let $(\mathcal{H}, m, \eta)$ be the algebra spanned by the set $\{1, g, x, gx\}$, where 1 is the unit element and $g^2 = 1$, $x^2 = 0$ and $xg = -gx$. Let us define algebra homomorphisms $\varepsilon : \mathcal{H} \to \mathbb{C}$ and $\Delta : \mathcal{H} \to \mathcal{H} \otimes \mathcal{H}$ by

$$\Delta(g) := g \otimes g, \quad \Delta(x) := x \otimes 1 + g \otimes x,$$

$$\varepsilon(g) := 1, \quad \varepsilon(x) := 0.$$

Let us define a linear mapping $S : \mathcal{H} \to \mathcal{H}$ by

$$S(1) := 1, \quad S(g) := g, \quad S(x) := -gx, \quad S(gx) := -x.$$

Show that $(\mathcal{H}, m, \eta, \Delta, \varepsilon, S)$ is a non-commutative non-co-commutative Hopf algebra.

Remark 9.2.15. In Exercise 9.2.14, a nice concrete matrix example can be given. Let us define $A \in \mathbb{C}^{2 \times 2}$ by

$$A := \begin{pmatrix} 0 & 1 \\ 1 & 0 \end{pmatrix}.$$

Let $g, x \in \mathbb{C}^{4 \times 4}$ be given by

$$g := \begin{pmatrix} A & 0 \\ 0 & -A \end{pmatrix}, \quad x := \begin{pmatrix} 0 & I_{\mathbb{C}^2} \\ 0 & 0 \end{pmatrix}.$$

Then it is easy to see that $\mathcal{H} = \mathrm{span}\{I_{\mathbb{C}^4}, g, x, gx\}$ is a four-dimensional subalgebra of $\mathbb{C}^{4 \times 4}$ such that $g^2 = I_{\mathbb{C}^4}$, $x^2 = 0$ and $xg = -gx$.

Part IV

Non-commutative Symmetries

In this part, we develop a non-commutative quantization of pseudo-differential operators on compact Lie groups. The idea is that it can be constructed in a way to run more or less parallel to the Kohn–Nirenberg quantization of operators on $\mathbb{R}^n$ that was presented in Chapter 2, and to the toroidal quantization of operators on $\mathbb{T}^n$ that was developed in Chapter 4. The main advantage of such an approach is that once the basic notions and definitions are understood, one can see and enjoy a lot of features which are already familiar from the commutative analysis.

The introduced matrix-valued full symbols turn out to have a number of interesting properties. The main difference with the toroidal quantization here is that, due to the non-commutativity of the group, symbols become matrix-valued with sizes depending on the dimensions of the unitary irreducible representations of the group, which are all finite-dimensional because the group is compact.

Among other things, the introduced approach provides a characterisation of the Hörmander class of pseudo-differential operators on a compact Lie group G using a global quantization of operators, thus relying on the representation theory rather than on the usual expressions in local coordinate charts. This yields a notion of the full symbol of an operator as a mapping defined globally on $G \times \widehat{G}$, where $\widehat{G}$ is the unitary dual of G. As such, this presents an advantage over the local theory where only the notion of the principal symbol can be defined globally. In the case of the torus $G = \mathbb{T}^n$, we naturally have $G \times \widehat{G} \cong \mathbb{T}^n \times \mathbb{Z}^n$, and we recapture the notion of a toroidal symbol introduced in Chapter 4, where symbols are scalar-valued (or 1×1 matrix-valued) because all the unitary irreducible representations of the torus are one-dimensional.

As an important example, the approach developed here will give us quite detailed information on the global quantization of operators on the three-dimensional sphere $\mathbb{S}^3$. More generally, we note that if we have a closed simply-connected three-dimensional manifold M, then by the recently resolved Poincaré conjecture there is a global diffeomorphism $M \simeq \mathbb{S}^3 \simeq \mathrm{SU}(2)$ that turns M into a Lie group with a group structure induced by $\mathbb{S}^3$ (or by SU(2)). Thus, we can use the approach developed for SU(2) to immediately obtain the corresponding global quantization of operators on M with respect to this induced group product. In fact, all the formulae remain completely the same since the unitary dual of SU(2) (or $\mathbb{S}^3$ in the quaternionic $\mathbb{R}^4$) is mapped by this diffeomorphism as well; for an example of this construction in the case of $\mathbb{S}^3 \simeq SU(2)$ see Section 12.5. The choice of the group structure on M may be not unique and is not canonical, but after using the machinery that we develop for SU(2), the corresponding quantization can be described entirely in terms of M; for an example compare Theorem 12.5.3 for $\mathbb{S}^3$ and Theorem 12.4.3 for SU(2). In this sense, as different quantizations of operators exist already on $\mathbb{R}^n$ depending on the choice of the underlying structure (e.g., Kohn–Nirenberg quantization, Weyl quantizations, etc.), the possibility to choose different group products on M resembles this. Due to space limitations, we postpone the detailed analysis of operators on the higher-dimensional spheres $\mathbb{S}^n \simeq SO(n+1)/SO(n)$ viewed as homogeneous spaces. However, we will introduce a general machinery on which to obtain the global quantization on homogeneous spaces using the one on the Lie group that acts on the space. Although we do not have general analogues of the diffeomorphic Poincaré conjecture in higher dimensions, this will cover cases when M is a convex surface or a surface with positive curvature tensor, as well as more general manifolds in terms of their Pontryagin class, etc.

Thus, the cases of the three-dimensional sphere $\mathbb{S}^3$ and Lie group SU(2) are analysed in detail in Chapter 12. There we show that pseudo-differential operators from Hörmander's classes $\Psi^m(SU(2))$ and $\Psi^m(\mathbb{S}^3)$ have matrix-valued symbols with a remarkable rapid off-diagonal decay property.

In Chapter 11 we develop the necessary foundations of this analysis on SU(2) which together with Chapter 12 provides a more detailed example of the quantization from Chapter 10. Finally, in Chapter 13 we give an application of these constructions to analyse pseudo-differential operators on homogeneous spaces.

Chapter 10

Pseudo-differential Operators on Compact Lie Groups

10.1 Introduction

In this chapter we develop a global theory of pseudo-differential operators on general compact Lie groups. As usual, $S_{1,0}^m(\mathbb{R}^n \times \mathbb{R}^n) \subset C^\infty(\mathbb{R}^n \times \mathbb{R}^n)$ refers to the Euclidean space symbol class, defined by the symbol inequalities

$$\left|\partial_\xi^\alpha \partial_x^\beta p(x,\xi)\right| \leq C \left(1+|\xi|\right)^{m-|\alpha|}, \tag{10.1}$$

for all multi-indices $\alpha, \beta \in \mathbb{N}_0^n$, $\mathbb{N}_0 = \{0\} \cup \mathbb{N}$, where the constant C is independent of $x, \xi \in \mathbb{R}^n$ but may depend on α, β, p, m. On a compact Lie group G we define the class $\Psi^m(G)$ to be the usual Hörmander class of pseudo-differential operators of order m. Thus, the operator A belongs to $\Psi^m(G)$ if in (all) local coordinates operator A is a pseudo-differential operator on $\mathbb{R}^n$ with some symbol $p(x,\xi)$ satisfying estimates (10.1), see Definition 5.2.11. Of course, symbol p depends on the local coordinate systems.

It is a natural idea to build pseudo-differential operators out of smooth families of convolution operators on Lie groups. In this work, we strive to develop the convolution approach into a symbolic quantization, which always provides a much more convenient framework for the analysis of operators. For this, our analysis of operators and their symbols is based on the representation theory of Lie groups. This leads to the description of the full symbols of pseudo-differential operators on Lie groups as sequences of matrices of growing size equal to the dimension of the corresponding representation of the group. Moreover, the analysis is global and is not confined to neighbourhoods of the neutral element since it does not rely on the exponential map and its properties. We also characterise, in terms of the introduced quantizations, standard Hörmander classes $\Psi^m(G)$ on Lie groups. One of the advantages of the presented approach is that we obtain a notion of full (global) symbols compared with only principal symbols available in the standard theory via localisations.

In our analysis on a Lie group G, at some point we have to make a choice whether to work with left- or right-convolution kernels. Since left-invariant operators on $C^\infty(G)$ correspond to right-convolutions $f \mapsto f * k$, once we decide to identify the Lie algebra $\mathfrak{g}$ of G with the left-invariant vector fields on G, it becomes most natural to work with right-convolution kernels in the sequel, and to define symbols as we do in Definition 10.4.3.

It is also known that globally defined symbols of pseudo-differential operators can be introduced on manifolds in the presence of a connection which allows one to use a suitable globally defined phase function, see, e.g., [151, 100, 109]. However, on compact Lie groups the use of the group structure allows one to develop a theory parallel to those of $\mathbb{R}^n$ and $\mathbb{T}^n$ owing to the fact that the Fourier analysis is well adopted to the underlying representation theory. Some elements of such a theory were discussed in [128, 129] as well as in the PhD thesis of V. Turunen [139]. However, here we present the finite-dimensional symbols and we do not rely on the exponential mapping thus providing a genuine global analysis in terms of the Lie group itself. We also note that the case of the compact commutative group $\mathbb{T}^n$ is also recovered from this general point of view, however a more advanced analysis is possible in this case because of the close relation between $\mathbb{T}^n$ and $\mathbb{R}^n$.

Unless specified otherwise, in this chapter G will stand for a general compact Lie group, and $\mathrm{d}\mu_G$ will stand for the (normalised) Haar measure on G, i.e., the unique regular Borel probability measure which is left-translation-invariant:

$$\int_G f(x) \, \mathrm{d}\mu_G(x) = \int_G f(yx) \, \mathrm{d}\mu_G(x)$$

for all $f \in C(G)$ and $y \in G$. Then also

$$\int_G f(x) \, \mathrm{d}\mu_G(x) = \int_G f(xy) \, \mathrm{d}\mu_G(x) = \int_G f(x^{-1}) \, \mathrm{d}\mu_G(x),$$

see Remark 7.4.4. Usually we abbreviate $\mathrm{d}\mu_G(x)$ to $\mathrm{d}x$ since this should cause no confusion.

10.2 Fourier series on compact Lie groups

We begin with the Fourier series on a compact group G.

Definition 10.2.1 (Rep(G) and $\widehat{G}$). Let Rep(G) denote the set of all strongly continuous irreducible unitary representations of G. In the sequel, whenever we mention unitary representations (of a compact Lie group G), we always mean strongly continuous irreducible unitary representations, which are then automatically smooth. Let $\widehat{G}$ denote the *unitary dual* of G, i.e., the set of equivalence classes of irreducible unitary representations from Rep(G), see Definitions 6.3.18 and 7.5.7. Let $[\xi] \in \widehat{G}$ denote the equivalence class of an irreducible unitary representation $\xi : G \to \mathcal{U}(\mathcal{H}_\xi)$; the representation space $\mathcal{H}_\xi$ is finite-dimensional since G is compact (see Corollary 7.5.6), and we set $\dim(\xi) = \dim \mathcal{H}_\xi$.

We will always equip a compact Lie group G with the Haar measure μ_G, i.e., the uniquely determined bi-invariant Borel regular probability measure, see Remark 7.4.4. For simplicity, we will write $L^p(G)$ for $L^p(\mu_G)$, $\int_G f \, dx$ for $\int_G f \, d\mu_G$, etc. First we collect several definitions scattered over previous chapters in different forms.

Definition 10.2.2 (Fourier coefficients). Let us define the *Fourier coefficient* $\widehat{f}(\xi) \in \operatorname{End}(\mathcal{H}_\xi)$ of $f \in L^1(G)$ by

$$\widehat{f}(\xi) := \int_G f(x) \, \xi(x)^* \, dx; \tag{10.2}$$

more precisely,

$$(\widehat{f}(\xi)u, v)_{\mathcal{H}_\xi} = \int_G f(x) \, (\xi(x)^* u, v)_{\mathcal{H}_\xi} \, dx = \int_G f(x) \, (u, \xi(x)v)_{\mathcal{H}_\xi} \, dx$$

for all $u, v \in \mathcal{H}_\xi$, where $(\cdot, \cdot)_{\mathcal{H}_\xi}$ is the inner product of $\mathcal{H}_\xi$.

Remark 10.2.3. Notice that $\xi(x)^* = \xi(x)^{-1} = \xi(x^{-1})$.

Remark 10.2.4 (**Fourier coefficients on $\mathbb{T}^n$ as a group**). Let $G = \mathbb{T}^n$. Let us naturally identify $\operatorname{End}(\mathbb{C})$ with $\mathbb{C}$, and $\mathcal{U}(\mathbb{C})$ with $\{z \in \mathbb{C} : |z| = 1\}$. For each $k \in \mathbb{Z}^n$, we define $e_k : G \to \mathcal{U}(\mathbb{C})$ by $e_k(x) := e^{i2\pi x \cdot k}$. Then

$$\widehat{f}(k) := \widehat{f}(e_k) = \int_{\mathbb{T}^n} f(x) \, e^{-i2\pi x \cdot k} \, dx$$

is the usual Fourier coefficient of $f \in L^1(\mathbb{T}^n)$.

Remark 10.2.5 (**Intertwining isomorphisms**). Let $U \in \operatorname{Hom}(\eta, \xi)$ be an *intertwining isomorphism*, i.e., let $U : \mathcal{H}_\eta \to \mathcal{H}_\xi$ be a bijective unitary linear mapping such that $U\eta(x) = \xi(x)U$ for all $x \in G$. Then we have

$$\widehat{f}(\eta) = U^{-1}\widehat{f}(\xi) \, U \in \operatorname{End}(\mathcal{H}_\eta). \tag{10.3}$$

Proposition 10.2.6 (Inner automorphisms). *For $u \in G$, consider the inner automorphisms*

$$\phi_u = (x \mapsto u^{-1}xu) : G \to G.$$

Then for all $\xi \in \operatorname{Rep}(G)$ we have

$$\widehat{f \circ \phi_u}(\xi) = \xi(u) \, \widehat{f}(\xi) \, \xi(u)^*. \tag{10.4}$$

Proof. We can calculate

$$\widehat{f \circ \phi_u}(\xi) = \int_G f(u^{-1}xu) \, \xi(x)^* \, dx = \int_G f(x) \, \xi(uxu^{-1})^* \, dx$$

$$= \xi(u) \int_G f(x) \, \xi(x)^* \, dx \, \xi(u)^* = \xi(u) \, \widehat{f}(\xi) \, \xi(u)^*,$$

which gives (10.4). $\qquad \square$

Proposition 10.2.7 (Convolutions). *If* $f, g \in L^1(G)$ *then*

$$\widehat{f * g} = \widehat{g}\, \widehat{f}.$$

Proof. If $\xi \in \mathrm{Rep}(G)$ then

$$
\begin{aligned}
\widehat{f * g}(\xi) &= \int_G (f * g)(x)\, \xi(x)^*\, dx \\
&= \int_G \int_G f(xy^{-1}) g(y)\, dy\, \xi(x)^*\, dx \\
&= \int_G g(y)\, \xi^*(y) \int_G f(xy^{-1})\, \xi(xy^{-1})^*\, dx\, dy \\
&= \widehat{g}(\xi)\, \widehat{f}(\xi),
\end{aligned}
$$

completing the proof. □

Remark 10.2.8. The product $\widehat{g}\, \widehat{f}$ in Proposition 10.2.7 usually differs from $\widehat{f}\, \widehat{g}$ because $\widehat{f}(\xi), \widehat{g}(\xi) \in \mathrm{End}(\mathcal{H}_\xi)$ are operators unless G is commutative when they are scalars (see Corollary 6.3.26) and hence commute. This order exchange is due to the definition of the Fourier coefficients (10.2), where we chose the integration of the function with respect to $\xi(x)^*$ instead of $\xi(x)$. This choice actually serves us well, as we chose to identify the Lie algebra $\mathfrak{g}$ with left-invariant vector fields on the Lie group G: namely, a *left*-invariant continuous linear operator $A : C^\infty(G) \to C^\infty(G)$ can be presented as a *right*-convolution operator $C_a = (f \mapsto f * a)$, resulting in convenient expressions like

$$\widehat{C_a C_b f} = \widehat{a}\, \widehat{b}\, \widehat{f}.$$

However, in Remark 10.4.13 we will still explain what would happen had we chosen another definition for the Fourier transform.

Proposition 10.2.9 (Differentiating the convolution). *Let* $Y \in \mathfrak{g}$ *and let* $D_Y :$ $C^\infty(G) \to C^\infty(G)$ *be defined by*

$$D_Y f(x) = \frac{d}{dt} f(x\, \exp(tY)) \Big|_{t=0}.$$

Let $f, g \in C^\infty(G)$. *Then* $D_Y(f * g) = f * D_Y g$.

Proof. We have

$$D_Y(f * g)(x) = \int_G f(y)\, \frac{d}{dt} g(y^{-1}x\, \exp(tY)) \Big|_{t=0}\, dy = f * D_Y g(x).$$

□

We now summarise properties of the Fourier series as a corollary to the Peter–Weyl Theorem 7.5.14:

Corollary 10.2.10 (Fourier series). *If $\xi : G \to U(d)$ is a unitary matrix representation then*

$$\widehat{f}(\xi) = \int_G f(x)\,\xi(x)^*\,\mathrm{d}x \in \mathbb{C}^{d \times d}$$

has matrix elements

$$\widehat{f}(\xi)_{mn} = \int_G f(x)\,\overline{\xi(x)_{nm}}\,\mathrm{d}x \in \mathbb{C},\ 1 \le m, n \le d.$$

If here $f \in L^2(G)$ then

$$\widehat{f}(\xi)_{mn} = (f, \xi(x)_{nm})_{L^2(G)}, \tag{10.5}$$

and by the Peter–Weyl Theorem 7.5.14 we have

$$
\begin{aligned}
f(x) &= \sum_{[\xi] \in \widehat{G}} \dim(\xi)\, \mathrm{Tr}\left(\xi(x)\,\widehat{f}(\xi)\right) \\
&= \sum_{[\xi] \in \widehat{G}} \dim(\xi) \sum_{m,n=1}^{d} \xi(x)_{nm}\,\widehat{f}(\xi)_{mn} \tag{10.6}
\end{aligned}
$$

for almost every $x \in G$, where the summation is understood so that from each class $[\xi] \in \widehat{G}$ we pick just (any) one representative $\xi \in [\xi]$. The particular choice of a representation from the representation class is irrelevant due to formula (10.3) and the presence of the trace in (10.6). The convergence in (10.6) is not only pointwise almost everywhere on G but also in the space $L^2(G)$.

Example. For $f \in L^2(\mathbb{T}^n)$, we get

$$f(x) = \sum_{k \in \mathbb{Z}^n} \mathrm{e}^{\mathrm{i}2\pi x \cdot k}\,\widehat{f}(k),$$

where $\widehat{f}(k) = \int_{\mathbb{T}^n} f(x)\,\mathrm{e}^{-\mathrm{i}2\pi x \cdot k}\,\mathrm{d}x$ is as in Remark 10.2.4. Here $\mathbb{C}$ was identified with $\mathbb{C}^{1 \times 1}$.

Finally, we record a useful formula for representations:

Remark 10.2.11. Let $e \in G$ be the neutral element of G and let ξ be a unitary matrix representation of G. The unitarity of the representation ξ implies the identity

$$\delta_{mn} = \xi(e)_{mn} = \xi(x^{-1}x)_{mn} = \sum_k \xi(x^{-1})_{mk}\,\xi(x)_{kn} = \sum_k \overline{\xi(x)_{km}}\,\xi(x)_{kn}.$$

Similarly,

$$\delta_{mn} = \sum_k \xi(x)_{mk}\,\overline{\xi(x)_{nk}}.$$

Here, as usual, δ_{mn} is the Kronecker delta: $\delta_{mn} = 1$ for $m = n$, and $\delta_{mn} = 0$ otherwise.

10.3   Function spaces on the unitary dual

In this section we lay down a functional analytic foundation concerning the function spaces that will be useful in the sequel. In particular, distribution space $\mathcal{S}'(\widehat{G})$ is of importance since it provides a distributional interpretation for series on $\widehat{G}$.

10.3.1   Spaces on the group G

We recall Definition 8.3.45 of the Laplace operator $\mathcal{L}_G$ on a Lie group G:

Remark 10.3.1 (**Laplace operator $\mathcal{L} = \mathcal{L}_G$**). The Laplace operator $\mathcal{L}_G$ is a second-order negative definite bi-invariant partial differential operator on G corresponding to the Casimir element of the universal enveloping algebra $\mathcal{U}(\mathfrak{g})$ of the Lie algebra $\mathfrak{g}$ of G. If G is equipped with the unique (up to a constant) bi-invariant Riemannian metric, $\mathcal{L}_G$ is its Laplace–Beltrami operator. We will often denote the Laplace operator simply by $\mathcal{L}$ if there is no need to emphasize the group G in the notation. We refer to Section 8.3.2 for a discussion of its main properties.

In Definition 5.2.4 we defined C^k mappings on a manifold. These can be also characterised globally:

Exercise 10.3.2. Let $n = \dim G$ and let $\{Y_j\}_{j=1}^n$ be a basis of the Lie algebra $\mathfrak{g}$ of G. Show that $f \in C^k(G)$ if and only if $\partial^\alpha f \in C(G)$ for all $\partial^\alpha = Y_1^{\alpha_1} \cdots Y_n^{\alpha_n}$ for all $|\alpha| \le k$, or if and only if $Lf \in C(G)$ for all $L \in \mathcal{U}(\mathfrak{g})$ of degree $\le k$.

Exercise 10.3.3. Show that $f \in C^\infty(G)$ if and only if $(-\mathcal{L}_G)^k f \in C(G)$ for all $k \in \mathbb{N}$. Show that $f \in C^\infty(G)$ if and only if $Lf \in C(G)$ for all $L \in \mathcal{U}(\mathfrak{g})$. (Hint: use a priori estimates from Theorem 2.6.9.)

We can recall from Remark 5.2.15 the definition of the space $\mathcal{D}'(M)$ of distributions on a compact manifold M. Let us be more precise in the case of a compact Lie group G:

Definition 10.3.4 (Distributions $\mathcal{D}'(G)$). We define the *space of distributions* $\mathcal{D}'(G)$ as the space of all continuous linear functionals on $C^\infty(G)$. This means that $u \in \mathcal{D}'(G)$ if it is a functional $u : C^\infty(G) \to \mathbb{C}$ such that

1. u is linear, i.e., $u(\alpha\varphi + \beta\psi) = \alpha u(\varphi) + \beta u(\psi)$ for all $\alpha, \beta \in \mathbb{C}$ and all $\varphi, \psi \in C^\infty(G)$;

2. u is continuous, i.e., $u(\varphi_j) \to u(\varphi)$ in $\mathbb{C}$ whenever $\varphi_j \to \varphi$ in $C^\infty(G)$, as $j \to \infty$.

Here[1] $\varphi_j \to \varphi$ in $C^\infty(G)$ if, e.g.,[2] $\partial^\alpha \varphi_j \to \partial^\alpha \varphi$ for all $\partial^\alpha \in \mathcal{U}(\mathfrak{g})$, as $j \to \infty$.

We define the *convergence in the space* $\mathcal{D}'(G)$ as follows. Let $u_j, u \in \mathcal{D}'(G)$. We will say that $u_j \to u$ in $\mathcal{D}'(G)$ as $j \to \infty$ if $u_j(\varphi) \to u(\varphi)$ in $\mathbb{C}$ as $j \to \infty$, for all $\varphi \in C^\infty(G)$.

---

[1] For a general setting on compact manifolds see Remark 5.2.15.
[2] Exercise 10.3.2 provides more options.

Definition 10.3.5 (Duality $\langle \cdot, \cdot \rangle_G$). Let $u \in \mathcal{D}'(G)$ and $\varphi \in C^\infty(G)$. We write

$$\langle u, \varphi \rangle_G := u(\varphi).$$

If $u \in L^p(G)$, $1 \le p \le \infty$, we can identify u with a distribution in $\mathcal{D}'(G)$ (which we will continue to denote by u) in a canonical way by

$$\langle u, \varphi \rangle_G = u(\varphi) := \int_G u(x) \, \varphi(x) \, dx,$$

where dx is the Haar measure on G.

Exercise 10.3.6. Let $1 \le p \le \infty$. Show that if $u_j \to u$ in $L^p(G)$ as $j \to \infty$ then $u_j \to u$ in $\mathcal{D}'(G)$ as $j \to \infty$.

Remark 10.3.7 (**Derivations in $\mathcal{D}'(G)$**). Similar to operations on distributions in $\mathbb{R}^n$ described in Section 1.3.2, we can define different operations on distributions on G. For example, for $Y \in \mathfrak{g}$, we can differentiate $u \in \mathcal{D}'(G)$ with respect to the vector field Y by defining

$$(Yu)(\varphi) := -u(Y\varphi),$$

for all $\varphi \in C^\infty(G)$. Here the derivative $Y\varphi = D_Y\varphi$ is as in Proposition 10.2.9. Similarly, if $\partial^\alpha \in \mathcal{U}(\mathfrak{g})$ is a differential operator of order $|\alpha|$, we define

$$(\partial^\alpha u)(\varphi) := (-1)^{|\alpha|} u(\partial^\alpha \varphi),$$

for all $\varphi \in C^\infty(G)$.

Exercise 10.3.8. Show that $\partial^\alpha u \in \mathcal{D}'(G)$ and that $\partial^\alpha : \mathcal{D}'(G) \to \mathcal{D}'(G)$ is continuous.

Definition 10.3.9 (Sobolev space $H^s(G)$). First let us note that the Laplacian $\mathcal{L} = \mathcal{L}_G$ is symmetric and $I - \mathcal{L}$ is positive. Set $\Xi := (I - \mathcal{L})^{1/2}$. Then $\Xi^s \in \mathcal{L}(C^\infty(G))$ and $\Xi^s \in \mathcal{L}(\mathcal{D}'(G))$ for every $s \in \mathbb{R}$. Let us define

$$(f, g)_{H^s(G)} := (\Xi^s f, \Xi^s g)_{L^2(G)} \quad (f, g \in C^\infty(G)).$$

The completion of $C^\infty(G)$ with respect to the norm $f \mapsto \|f\|_{H^s(G)} = (f, f)_{H^s(G)}^{1/2}$ gives us the Sobolev space $H^s(G)$ of order $s \in \mathbb{R}$. This is the same space as that in Definition 5.2.16 on general manifolds, or as the Sobolev space obtained using any smooth partition of unity on the compact manifold G, by Corollary 5.2.18. The operator Ξ^r is a Sobolev space isomorphism $H^s(G) \to H^{s-r}(G)$ for every $r, s \in \mathbb{R}$.

Exercise 10.3.10. Show that if $Y \in \mathfrak{g}$, then the differentiation with respect to Y is a bounded linear operator from $H^k(G)$ to $H^{k-1}(G)$ for all $k \in \mathbb{N}$. Extend this to $k \in \mathbb{R}$.

Remark 10.3.11 (**Sobolev spaces and** $C^\infty(G)$). We have

$$\bigcap_{k\in\mathbb{N}} H^{2k}(G) = C^\infty(G). \tag{10.7}$$

This can be seen locally, since $H^{2k}(G) = \mathrm{domain}((-\mathcal{L}_G)^k)$ and $\mathcal{L}_G$ is elliptic, so that (10.7) follows from the local a priori estimates in Theorem 2.6.9.

Since the analysis of Sobolev spaces is closely intertwined with spaces on $\widehat{G}$, we study them also in the next section, in particular we refer to Remark 10.3.24 for their characterisation on the Fourier transform side.

10.3.2 Spaces on the dual $\widehat{G}$

Since we will be mostly using the "right" Peter–Weyl Theorem (Theorem 7.5.14), we may simplify the notation slightly, also adopting it to the analysis of pseudo-differential operators from the next sections. Thus, the space

$$\left\{ \sqrt{\dim(\xi)}\, \xi_{ij} \ : \ \xi = (\xi_{ij})_{i,j=1}^{\dim(\xi)},\ [\xi]\in\widehat{G} \right\}$$

is an orthonormal basis for $L^2(G)$, and the space

$$\mathcal{H}^\xi := \mathrm{span}\{\xi_{ij} \ : \ 1\leq i,j\leq\dim(\xi)\} \subset L^2(G)$$

is π_R-invariant, $\xi\sim\pi_R|_{\mathcal{H}^\xi}$, and

$$L^2(G) = \bigoplus_{[\xi]\in\widehat{G}} \mathcal{H}^\xi.$$

By choosing a unitary matrix representation from each equivalence class $[\xi]\in\widehat{G}$, we can identify $\mathcal{H}^\xi$ with $\mathbb{C}^{\dim(\xi)\times\dim(\xi)}$ by choosing a basis in the linear space $\mathcal{H}^\xi$.

Remark 10.3.12 (**Spaces** $\mathcal{H}^\xi$ **and** $\mathcal{H}_\xi$). We would like to point out a difference between spaces $\mathcal{H}^\xi$ and $\mathcal{H}_\xi$ to eliminate any confusion. Recall that if $\xi\in\mathrm{Rep}(G)$, then ξ is a mapping $\xi : G \to \mathcal{U}(\mathcal{H}_\xi)$, where $\mathcal{H}_\xi$ is the representation space of ξ, with $\dim\mathcal{H}_\xi = \dim(\xi)$. On the other hand, the space $\mathcal{H}^\xi\subset L^2(G)$ is the span of the matrix elements of ξ and $\dim\mathcal{H}^\xi = (\dim(\xi))^2$. In the notation of the right and left Peter–Weyl Theorems in Theorem 7.5.14 and Remark 7.5.16, we have

$$\mathcal{H}^\xi = \bigoplus_{i=1}^{\dim(\xi)} \mathcal{H}_{i,\cdot}^\xi = \bigoplus_{j=1}^{\dim(\xi)} \mathcal{H}_{\cdot,j}^\xi.$$

Informally, spaces $\mathcal{H}_\xi$ can be viewed as "columns/rows" of $\mathcal{H}^\xi$, because for example $\xi(x)v\in\mathcal{H}_\xi$ for every $v\in\mathcal{H}_\xi$.

We recall the important property of the Laplace operator on spaces $\mathcal{H}^\xi$ from Theorem 8.3.47:

Theorem 10.3.13 (Eigenvalues of the Laplacian on G). *For every $[\xi] \in \widehat{G}$ the space $\mathcal{H}^\xi$ is an eigenspace of $\mathcal{L}_G$ and $-\mathcal{L}_G|_{\mathcal{H}^\xi} = \lambda_\xi I$, for some $\lambda_\xi \geq 0$.*

Exercise 10.3.14. Show that $\mathcal{H}^\xi \subset C^\infty(G)$ for all $\xi \in \mathrm{Rep}(G)$. (Hint: Use Theorem 10.3.13 and the ellipticity of $\mathcal{L}_G$.)

From Definition 7.6.11 we recall the Hilbert space $L^2(\widehat{G})$ which we can now describe as follows. But first, we can look at the space of all mappings on $\widehat{G}$:

Definition 10.3.15 (Space $\mathcal{M}(\widehat{G})$). The space $\mathcal{M}(\widehat{G})$ consists of all mappings

$$F : \widehat{G} \to \bigcup_{[\xi] \in \widehat{G}} \mathcal{L}(\mathcal{H}_\xi) \subset \bigcup_{m=1}^{\infty} \mathbb{C}^{m \times m},$$

satisfying $F([\xi]) \in \mathcal{L}(\mathcal{H}_\xi)$ for every $[\xi] \in \widehat{G}$. In matrix representations, we can view $F([\xi]) \in \mathbb{C}^{\dim(\xi) \times \dim(\xi)}$ as a $\dim(\xi) \times \dim(\xi)$ matrix.

The space $L^2(\widehat{G})$ consists of all mappings $F \in \mathcal{M}(\widehat{G})$ such that

$$\|F\|^2_{L^2(\widehat{G})} := \sum_{[\xi] \in \widehat{G}} \dim(\xi) \, \|F([\xi])\|^2_{HS} < \infty,$$

where

$$\|F([\xi])\|_{HS} = \sqrt{\mathrm{Tr}\left(F([\xi]) \, F([\xi])^*\right)}$$

stands for the Hilbert–Schmidt norm of the linear operator $F([\xi])$, see Definition B.5.43. Thus, the space

$$L^2(\widehat{G}) = \left\{ F : \widehat{G} \to \bigcup_{[\xi] \in \widehat{G}} \mathcal{L}(\mathcal{H}_\xi), \; F : [\xi] \mapsto \mathcal{L}(\mathcal{H}_\xi) : \right.$$

$$\left. \|F\|^2_{L^2(\widehat{G})} := \sum_{[\xi] \in \widehat{G}} \dim(\xi) \, \|F([\xi])\|^2_{HS} < \infty \right\} \quad (10.8)$$

is a Hilbert space with the inner product

$$(E, F)_{L^2(\widehat{G})} := \sum_{[\xi] \in \widehat{G}} \dim(\xi) \, \mathrm{Tr}\left(E([\xi]) \, F([\xi])^*\right).$$

Remark 10.3.16 (**Mappings on $\widehat{G}$ or on $\mathrm{Rep}(G)$?**). We note that because the representations in $\widehat{G}$ are all unitary, and because of the Hilbert–Schmidt norm that we use, we can write $F(\xi)$ instead of $F([\xi])$, so that F is defined on $\mathrm{Rep}(G)$ instead of $\widehat{G}$. Thus, to simplify the notation we can write $F(\xi)$ with a convention that $F \in \mathcal{M}(\widehat{G})$ if $F(\xi) = F(\eta)$ whenever $\xi \sim \eta$ (i.e., whenever $[\xi] = [\eta]$).

Proposition 10.3.17 (Parseval's identity). *Let $f, g \in L^2(G)$. Then we have*

$$(f, g)_{L^2(G)} = \sum_{[\xi] \in \widehat{G}} \dim(\xi) \operatorname{Tr}\left(\widehat{f}(\xi)\, \widehat{g}(\xi)^*\right) = (\widehat{f}(\xi), \widehat{g}(\xi))_{L^2(\widehat{G})}.$$

Consequently,

$$\|f\|_{L^2(G)} = \|\widehat{f}\|_{L^2(\widehat{G})}.$$

Proof. Writing f and g as Fourier series (7.3) in Corollary 7.6.10, we obtain

$$(f, g)_{L^2(G)} = \sum_{[\xi],[\eta] \in \widehat{G}} \dim(\xi) \dim(\eta) \int_G \operatorname{Tr}\left(\xi(x)\, \widehat{f}(\xi)\right) \overline{\operatorname{Tr}\left(\eta(x)\, \widehat{g}(\eta)\right)}\, dx$$

$$= \sum_{[\xi],[\eta] \in \widehat{G}} \dim(\xi) \dim(\eta) \int_G \operatorname{Tr}\left(\xi(x)\, \widehat{f}(\xi)\right) \operatorname{Tr}\left(\eta(x)^*\, \widehat{g}(\eta)^*\right)\, dx$$

$$= \sum_{[\xi],[\eta] \in \widehat{G}} \dim(\xi) \dim(\eta) \int_G \sum_{i,j=1}^{\dim(\xi)} \xi(x)_{ij} \widehat{f}(\xi)_{ji} \sum_{k,l=1}^{\dim(\eta)} \overline{\eta(x)}_{lk}\, \widehat{g}(\eta)_{lk}^*\, dx$$

$$= \sum_{[\xi],[\eta] \in \widehat{G}} \dim(\xi) \dim(\eta) \sum_{i,j=1}^{\dim(\xi)} \sum_{k,l=1}^{\dim(\eta)} (\xi(x)_{ij}, \eta(x)_{lk})_{L^2(G)} \widehat{f}(\xi)_{ji}\, \widehat{g}(\eta)_{lk}^*.$$

From this, by Lemma 7.5.12, we obtain

$$(f, g)_{L^2(G)} = \sum_{[\xi] \in \widehat{G}} \dim(\xi) \sum_{i,j=1}^{\dim(\xi)} \widehat{f}(\xi)_{ji}\, \widehat{g}(\xi)_{ij}^*$$

$$= \sum_{[\xi] \in \widehat{G}} \dim(\xi) \operatorname{Tr}\left(\widehat{f}(\xi)\, \widehat{g}(\xi)^*\right),$$

finishing the proof. $\square$

We now define a scale $\langle \xi \rangle$ on Lie groups that is useful in measuring the growth on $\widehat{G}$ and that is associated to the eigenvalues of the Laplace operator $\mathcal{L} = \mathcal{L}_G$ from Remark 10.3.1 and Theorem 10.3.13.

Definition 10.3.18 (Definition of $\langle \xi \rangle$ on Lie groups). Let $\xi \in \operatorname{Rep}(G)$, so that $\xi : G \to \mathcal{U}(\mathcal{H}_\xi)$. Given $v, w \in \mathcal{H}_\xi$, the function $\xi^{vw} : G \to \mathbb{C}$ defined by

$$\xi^{vw}(x) := \langle \xi(x)v, w \rangle_{\mathcal{H}_\xi}$$

is not only continuous but even C^∞-smooth[3]. Let $\operatorname{span}(\xi)$ denote the linear span of $\{\xi^{vw} : v, w \in \mathcal{H}_\xi\}$. If $\xi \sim \eta$ then $\operatorname{span}(\xi) = \operatorname{span}(\eta)$; consequently[4], we may write

$$\operatorname{span}[\xi] := \operatorname{span}(\xi) \subset C^\infty(G).$$

---

[3]See Exercise 10.3.14.
[4]In matrix representations this is the space $\mathcal{H}^\xi$.

It follows from Theorem 10.3.13 that

$$-\mathcal{L}\xi^{vw}(x) = \lambda_{[\xi]}\xi^{vw}(x),$$

where $\lambda_{[\xi]} \geq 0$, and we write

$$\langle\xi\rangle := (1 + \lambda_{[\xi]})^{1/2}. \tag{10.9}$$

A relation between the dimension of the representation ξ and the weight $\langle\xi\rangle$ is given by

Proposition 10.3.19 (Dimension and eigenvalues). *There exists a constant $C > 0$ such that the inequality*

$$\dim(\xi) \leq C\langle\xi\rangle^{\frac{\dim G}{2}}$$

holds for all $\xi \in \mathrm{Rep}(G)$.

Proof. We note by Theorem 10.3.13 that $\langle\xi\rangle$ is an eigenvalue of the first-order elliptic operator $(1 - \mathcal{L}_G)^{1/2}$. The corresponding eigenspace $\mathcal{H}^\xi$ has the dimension $\dim(\xi)^2$. Denoting by $n = \dim G$, the Weyl formula for the counting function of the eigenvalues of $(1 - \mathcal{L}_G)^{1/2}$ yields

$$\sum_{\langle\xi\rangle\leq\lambda} \dim(\xi)^2 = C_0\lambda^n + O(\lambda^{n-1})$$

as $\lambda \to \infty$. This implies the estimate $\dim(\xi)^2 \leq C\langle\xi\rangle^n$ for large $\langle\xi\rangle$, implying the statement. $\square$

Definition 10.3.20 (Space $\mathcal{S}(\widehat{G})$). The space $\mathcal{S}(\widehat{G})$ consists of all mappings $H \in \mathcal{M}(\widehat{G})$ such that for all $k \in \mathbb{N}$ we have

$$p_k(H) := \sum_{[\xi]\in\widehat{G}} \dim(\xi) \langle\xi\rangle^k \, \|H(\xi)\|_{HS} < \infty. \tag{10.10}$$

We will say that $H_j \in \mathcal{S}(\widehat{G})$ converges to $H \in \mathcal{S}(\widehat{G})$ in $\mathcal{S}(\widehat{G})$, and write $H_j \to H$ in $\mathcal{S}(\widehat{G})$ as $j \to \infty$, if $p_k(H_j - H) \to 0$ as $j \to \infty$ for all $k \in \mathbb{N}$, i.e., if

$$\sum_{[\xi]\in\widehat{G}} \dim(\xi) \langle\xi\rangle^k \, \|H_j(\xi) - H(\xi)\|_{HS} \to 0 \text{ as } j \to \infty,$$

for all $k \in \mathbb{N}$.

We can take different families of seminorms on $\mathcal{S}(\widehat{G})$. In particular, the following equivalence will be of importance since it provides a more direct relation with Sobolev spaces on G:

Proposition 10.3.21 (Seminorms on $\mathcal{S}(\widehat{G})$). *For $H \in \mathcal{M}(\widehat{G})$, let $p_k(H)$ be as in (10.10), and let us define a family $q_k(H)$ by*

$$q_k(H) := \left(\sum_{[\xi] \in \widehat{G}} \dim(\xi) \, \langle \xi \rangle^k \, \|H(\xi)\|_{HS}^2 \right)^{1/2}.$$

Then $p_k(H) < \infty$ for all $k \in \mathbb{N}$ if and only if $q_k(H) < \infty$ for all $k \in N$.

Proof. Let $H \in \mathcal{M}(\widehat{G})$. We claim that $p_k(H) \leq C q_{4k}(H)$ if k is large enough, and that $q_{2k}(H) \leq p_k(H)$. Indeed, by the Cauchy–Schwartz inequality (i.e., Hölder's inequality similar to that in Lemma 3.3.28 where we use the discreteness of $\widehat{G}$) we can estimate

$$
\begin{aligned}
p_k(H) &= \sum_{[\xi] \in \widehat{G}} \left\{ (\dim(\xi))^{1/2} \, \langle \xi \rangle^{-k} \right\} \left\{ (\dim(\xi))^{1/2} \, \langle \xi \rangle^{2k} \, \|H(\xi)\|_{HS} \right\} \\
&\leq \left(\sum_{[\xi] \in \widehat{G}} \dim(\xi) \, \langle \xi \rangle^{-2k} \right)^{1/2} \left(\sum_{[\xi] \in \widehat{G}} \dim(\xi) \, \langle \xi \rangle^{4k} \, \|H(\xi)\|_{HS}^2 \right)^{1/2} \\
&\leq C q_{4k}(H),
\end{aligned}
$$

if we choose k large enough and use Proposition 10.3.19. Conversely, we have

$$
\begin{aligned}
p_k(H)^2 &= \sum_{[\xi]=[\eta] \in \widehat{G}} (\dim(\xi))^2 \, \langle \xi \rangle^{2k} \, \|H(\xi)\|_{HS}^2 \\
&\quad + \sum_{[\xi] \neq [\eta]} \dim(\xi) \, \dim(\eta) \, \langle \xi \rangle^k \, \langle \eta \rangle^k \|H(\xi)\|_{HS} \, \|H(\eta)\|_{HS} \\
&\geq \sum_{[\xi] \in \widehat{G}} \dim(\xi) \, \langle \xi \rangle^{2k} \, \|H(\xi)\|_{HS}^2 \\
&= q_{2k}(H)^2.
\end{aligned}
$$
$\square$

As a corollary of Proposition 10.3.21 we have

Corollary 10.3.22. *For $H \in \mathcal{M}(\widehat{G})$, we have*

$$\sum_{[\xi] \in \widehat{G}} \dim(\xi) \, \langle \xi \rangle^{-k} \, \|H(\xi)\|_{HS} < \infty$$

for some $k \in \mathbb{N}$ if and only if

$$\sum_{[\xi] \in \widehat{G}} \dim(\xi) \, \langle \xi \rangle^{-l} \, \|H(\xi)\|_{HS}^2 < \infty$$

for some $l \in \mathbb{N}$.

Let us now summarise properties of the Fourier transform on $L^2(G)$.

Theorem 10.3.23 (Fourier inversion). *Let G be a compact Lie group. The Fourier transform $f \mapsto \mathcal{F}_G f = \widehat{f}$ defines a surjective isometry $L^2(G) \to L^2(\widehat{G})$. The inverse Fourier transform is given by*

$$(\mathcal{F}_G^{-1} H)(x) = \sum_{[\xi] \in \widehat{G}} \dim(\xi) \, \mathrm{Tr}\, (\xi(x) \, H(\xi)), \qquad (10.11)$$

and we have

$$\mathcal{F}_G^{-1} \circ \mathcal{F}_G = id \quad and \quad \mathcal{F}_G \circ \mathcal{F}_G^{-1} = id$$

on $L^2(G)$ and $L^2(\widehat{G})$, respectively. Moreover, the Fourier transform $\mathcal{F}_G$ is unitary, $\mathcal{F}_G{}^* = \mathcal{F}_G^{-1}$, and for any $H \in \mathcal{S}(\widehat{G})$ we have

$$(\mathcal{F}_G{}^* H)(x) = \overline{\mathcal{F}_G^{-1}(H^*)(x^{-1})} = (\mathcal{F}_G^{-1} H)(x)$$

for all $x \in G$, where $H^*(\xi) := H(\xi)^*$ for all $\xi \in \mathrm{Rep}(G)$.

Proof. In Theorem 7.6.13 we have already shown that the Fourier transform is a surjective isometry $L^2(G) \to L^2(\widehat{G})$. By Corollary 7.6.10 the inverse Fourier transform is given by (10.11). Let us show the last part. Let $f \in C^\infty(G)$ and $H \in \mathcal{S}(\widehat{G})$. Then we have

$$(f, \mathcal{F}_G{}^* H)_{L^2(G)} = (\mathcal{F}_G f, H)_{L^2(\widehat{G})} = \sum_{[\xi] \in \widehat{G}} \dim(\xi) \, \mathrm{Tr}\, \left(\widehat{f}(\xi) \, H(\xi)^* \right)$$

$$= \sum_{[\xi] \in \widehat{G}} \dim(\xi) \, \mathrm{Tr}\, \left(\int_G f(x) \, \xi^*(x) \, dx \, H(\xi)^* \right)$$

$$= \int_G f(x) \left\{ \sum_{[\xi] \in \widehat{G}} \dim(\xi) \, \mathrm{Tr}\, \left(\xi(x^{-1}) \, H(\xi)^* \right) \right\} dx$$

$$= \int_G f(x) \, \mathcal{F}_G^{-1}(H^*)(x^{-1}) \, dx,$$

which implies $\mathcal{F}_G{}^* H(x) = \overline{\mathcal{F}_G^{-1}(H^*)(x^{-1})}$. Finally, the unitarity of the Fourier transform follows from continuing the calculation:

$$(f, \mathcal{F}_G{}^* H)_{L^2(G)} = \sum_{[\xi] \in \widehat{G}} \dim(\xi) \, \mathrm{Tr}\, \left(\int_G f(x) \, \xi^*(x) \, dx \, H(\xi)^* \right)$$

$$= \int_G f(x) \left\{ \sum_{[\xi] \in \widehat{G}} \dim(\xi) \, \mathrm{Tr}\, \left(\xi^*(x) \, H(\xi)^* \right) \right\} dx$$

$$= \int_G f(x) \left\{ \sum_{[\xi] \in \widehat{G}} \dim(\xi) \overline{\mathrm{Tr}\left(\xi(x) \, H(\xi)\right)} \right\} \, dx$$

$$= \int_G f(x) \, \overline{(\mathcal{F}_G^{-1} H)(x)} \, dx. \qquad \qquad \Box$$

Remark 10.3.24 (**Sobolev space** $H^s(G)$). On the Fourier transform side, in view of Theorem 10.3.23, the Sobolev space $H^s(G)$ can be characterised by

$$H^s(G) = \{f \in \mathcal{D}'(G) : \langle \xi \rangle^s \widehat{f}(\xi) \in L^2(\widehat{G})\}.$$

We also note that since G is compact, for $s = 2k$ and $k \in \mathbb{N}$, we have that $f \in H^{2k}(G)$ if $(-\mathcal{L}_G)^k f \in L^2(G)$. By Theorem 10.3.25, the Fourier transform $\mathcal{F}_G$ is a continuous bijection from $H^{2k}(G)$ to the space

$$\left\{ F \in \mathcal{M}(\widehat{G}) : \sum_{[\xi] \in \widehat{G}} \dim(\xi) \, \langle \xi \rangle^{2k} \|F(\xi)\|_{HS}^2 < \infty \right\}.$$

We now analyse the Fourier transforms on $C^\infty(G)$ and $\mathcal{S}(\widehat{G})$ in preparation for their extension to the spaces of distributions.

Theorem 10.3.25 (Fourier transform on $C^\infty(G)$ and $\mathcal{S}(\widehat{G})$). *The Fourier transform* $\mathcal{F}_G : C^\infty(G) \to \mathcal{S}(\widehat{G})$ *and its inverse* $\mathcal{F}_G^{-1} : \mathcal{S}(\widehat{G}) \to C^\infty(G)$ *are continuous, satisfying* $\mathcal{F}_G^{-1} \circ \mathcal{F}_G = \mathrm{id}$ *and* $\mathcal{F}_G \circ \mathcal{F}_G^{-1} = \mathrm{id}$ *on* $C^\infty(G)$ *and* $\mathcal{S}(\widehat{G})$, *respectively.*

Proof. By (10.7) in Remark 10.3.11, writing $C^\infty(G) = \bigcap_{k \in \mathbb{N}} H^{2k}(G)$, the smoothness $f \in C^\infty(G)$ is equivalent to $f \in H^{2k}(G)$ for all $k \in \mathbb{N}$, by Remark 10.3.24. This means that

$$\sum_{[\xi] \in \widehat{G}} \dim(\xi) \, \langle \xi \rangle^{2k} \|\widehat{f}\|_{HS}^2 < \infty$$

for all $k \in \mathbb{N}$. Hence $\mathcal{F}_G f \in \mathcal{S}(\widehat{G})$. Consequently, $f_j \to f$ in $C^\infty(G)$ implies that $f_j \to f$ in $H^{2k}(G)$ for all $k \in \mathbb{N}$. Taking the Fourier transform we see that $q_{2k}(\mathcal{F}_G f_j - \mathcal{F}_G f) \to 0$ as $j \to \infty$ for all $k \in \mathbb{N}$, which implies that $\mathcal{F}_G f_j \to \mathcal{F}_G f$ in $\mathcal{S}(\widehat{G})$. Inverting this argument implies the continuity of the inverse Fourier transform $\mathcal{F}_G^{-1}$ from $\mathcal{S}(\widehat{G})$ to $C^\infty(G)$. The last part of the theorem follows from Theorem 10.3.23. $\qquad \Box$

Corollary 10.3.26 ($\mathcal{S}(\widehat{G})$ is a Montel nuclear space). *The space* $\mathcal{S}(\widehat{G})$ *is a Montel space and a nuclear space.*

Proof. This follows from the same properties of $C^\infty(G)$ because $\mathcal{S}(\widehat{G})$ is homeomorphic to $C^\infty(G)$, a homeomorphism given by the Fourier transform. See Exercises B.3.35 and B.3.51. $\qquad \Box$

Definition 10.3.27 (Space $\mathcal{S}'(\widehat{G})$). The space $\mathcal{S}'(\widehat{G})$ of *slowly increasing* or *tempered* distributions on the unitary dual $\widehat{G}$ is defined as the space of all $H \in \mathcal{M}(\widehat{G})$ for which there exists some $k \in \mathbb{N}$ such that

$$\sum_{[\xi] \in \widehat{G}} \dim(\xi) \, \langle \xi \rangle^{-k} \|H(\xi)\|_{HS} < \infty.$$

The convergence in $\mathcal{S}'(\widehat{G})$ is defined as follows. We will say that $H_j \in \mathcal{S}'(\widehat{G})$ converges to $H \in \mathcal{S}'(\widehat{G})$ in $\mathcal{S}'(\widehat{G})$ as $j \to \infty$, if there exists some $k \in \mathbb{N}$ such that

$$\sum_{[\xi] \in \widehat{G}} \dim(\xi) \, \langle \xi \rangle^{-k} \|H_j(\xi) - H(\xi)\|_{HS} \to 0$$

as $j \to \infty$.

Lemma 10.3.28 (Trace and Hilbert–Schmidt norm). *Let $\mathcal{H}$ be a Hilbert space and let $A, B \in S_2(\mathcal{H})$ be Hilbert–Schmidt operators[5]. Then*

$$|\mathrm{Tr}(AB)| \leq \|A\|_{HS} \, \|B\|_{HS}.$$

Proof. By the Cauchy–Schwartz inequality for the (Hilbert–Schmidt) inner product on $S_2(\mathcal{H})$ we have

$$|\mathrm{Tr}(AB)| = |\langle A, B^* \rangle_{HS}| \leq \|A\|_{HS} \, \|B\|_{HS},$$

proving the required estimate. □

Definition 10.3.29 (Duality $\langle \cdot, \cdot \rangle_{\widehat{G}}$). Let $H \in \mathcal{S}'(\widehat{G})$ and $h \in \mathcal{S}(\widehat{G})$. We write

$$\langle H, h \rangle_{\widehat{G}} := \sum_{[\xi] \in \widehat{G}} \dim(\xi) \, \mathrm{Tr}\,(H(\xi)\, h(\xi)). \tag{10.12}$$

The sum is well defined in view of

$$\langle H, h \rangle_{\widehat{G}} \leq \sum_{[\xi] \in \widehat{G}} \dim(\xi) \, |\mathrm{Tr}\,(H(\xi)\, h(\xi))|$$

$$\leq \sum_{[\xi] \in \widehat{G}} \dim(\xi) \, \|H(\xi)\|_{HS} \, \|h(\xi)\|_{HS}$$

$$\leq \left(\sum_{[\xi] \in \widehat{G}} \dim(\xi) \, \langle \xi \rangle^{-k} \, \|H(\xi)\|_{HS}^2 \right)^{1/2} \left(\sum_{[\xi] \in \widehat{G}} \dim(\xi) \, \langle \xi \rangle^{k} \, \|h(\xi)\|_{HS}^2 \right)^{1/2}$$

$$< \infty,$$

---

[5]Recall Definition B.5.43.

which is finite in view of Proposition 10.3.21 and Corollary 10.3.22. Here we used Lemma 10.3.28 and the Cauchy–Schwarz inequality (see Lemma 3.3.28) in the estimate. The bracket $\langle \cdot, \cdot \rangle_{\widehat{G}}$ in (10.12) introduces the duality between $\mathcal{S}'(\widehat{G})$ and $\mathcal{S}(\widehat{G})$ so that $\mathcal{S}'(\widehat{G})$ is the dual space to $\mathcal{S}(\widehat{G})$.

Proposition 10.3.30 (Sequential density of $\mathcal{S}(\widehat{G})$ in $\mathcal{S}'(\widehat{G})$). *The space $\mathcal{S}(\widehat{G})$ is sequentially dense in $\mathcal{S}'(\widehat{G})$.*

Proof. We use the standard approximation in sequence spaces by cutting the sequence to obtain its approximation. Thus, let $H \in \mathcal{S}'(\widehat{G})$. For $\xi \in \widehat{G}$ we define

$$H_j(\xi) := \begin{cases} H(\xi), & \text{if } \langle \xi \rangle \leq j, \\ 0, & \text{if } \langle \xi \rangle > j. \end{cases}$$

Let $k \in \mathbb{N}$ be such that

$$\sum_{[\xi] \in \widehat{G}} \dim(\xi) \, \langle \xi \rangle^{-k} \, \|H(\xi)\|_{HS} < \infty. \tag{10.13}$$

Clearly $H_j \in \mathcal{S}(\widehat{G})$ for all j, and $H_j \to H$ in $\mathcal{S}'(G)$ as $j \to \infty$, because

$$\sum_{[\xi] \in \widehat{G}} \dim(\xi) \, \langle \xi \rangle^{-k} \|H_j(\xi) - H(\xi)\|_{HS} = \sum_{[\xi] \in \widehat{G}, \langle \xi \rangle > j} \dim(\xi) \, \langle \xi \rangle^{-k} \|H(\xi)\|_{HS} \to 0$$

as $j \to 0$, in view of the convergence of the series in (10.13) and the fact that the eigenvalues of the Laplacian on G are increasing to infinity. $\qquad \square$

We now establish a relation of the duality brackets $\langle \cdot, \cdot \rangle_G$ and $\langle \cdot, \cdot \rangle_{\widehat{G}}$ with the Fourier transform and its inverse. This will allow us to extend the actions of the Fourier transforms to the spaces of distributions on G and $\widehat{G}$.

Proposition 10.3.31 (Dualities and Fourier transforms). *Let $\varphi \in C^\infty(G)$, $h \in \mathcal{S}(\widehat{G})$. Then we have the identities*

$$\langle \mathcal{F}_G \varphi, h \rangle_{\widehat{G}} = \langle \varphi, \iota \circ \mathcal{F}_G^{-1} h \rangle_G$$

and

$$\langle \mathcal{F}_G^{-1} h, \varphi \rangle_G = \langle h, \mathcal{F}_G(\iota \circ \varphi) \rangle_{\widehat{G}},$$

where $(\iota \circ \varphi)(x) = \varphi(x^{-1})$.

Proof. We can calculate

$$\langle \mathcal{F}_G \varphi, h \rangle_{\widehat{G}} = \sum_{[\xi] \in \widehat{G}} \dim(\xi) \, \operatorname{Tr} \left(\widehat{\varphi}(\xi) \, h(\xi) \right)$$

$$= \sum_{[\xi] \in \widehat{G}} \dim(\xi) \, \operatorname{Tr} \left(\int_G \varphi(x) \, \xi^*(x) \, \mathrm{d}x \, h(\xi) \right)$$

$$= \int_G \varphi(x) \left\{ \sum_{[\xi] \in \widehat{G}} \dim(\xi) \operatorname{Tr} \left(\xi(x^{-1}) \, h(\xi) \right) \right\} \mathrm{d}x$$

$$\overset{(10.11)}{=} \int_G \varphi(x) \, (\mathcal{F}_G^{-1} h)(x^{-1}) \, \mathrm{d}x$$

$$= \langle \varphi, \iota \circ \mathcal{F}_G^{-1} h \rangle_G.$$

Similarly, we have

$$\langle \mathcal{F}_G^{-1} h, \varphi \rangle_G = \int_G (\mathcal{F}_G^{-1} h)(x) \, \varphi(x) \, \mathrm{d}x$$

$$= \int_G \left\{ \sum_{[\xi] \in \widehat{G}} \dim(\xi) \operatorname{Tr} \left(\xi(x) \, h(\xi) \right) \right\} \varphi(x) \, \mathrm{d}x$$

$$= \sum_{[\xi] \in \widehat{G}} \dim(\xi) \operatorname{Tr} \left(\left\{ \int_G \xi(x^{-1}) \, \varphi(x^{-1}) \, \mathrm{d}x \right\} h(\xi) \right)$$

$$= \sum_{[\xi] \in \widehat{G}} \dim(\xi) \operatorname{Tr} \left(\left\{ \int_G \xi^*(x) \, (\iota \circ \varphi)(x) \, \mathrm{d}x \right\} h(\xi) \right)$$

$$= \sum_{[\xi] \in \widehat{G}} \dim(\xi) \operatorname{Tr} \left(\widehat{\iota \circ \varphi}(\xi) \, h(\xi) \right)$$

$$= \langle h, \mathcal{F}_G(\iota \circ \varphi) \rangle_{\widehat{G}},$$

completing the proof. □

Proposition 10.3.31 motivates the following definitions:

Definition 10.3.32 (Fourier transforms on $\mathcal{D}'(G)$ and $\mathcal{S}'(\widehat{G})$). For $u \in \mathcal{D}'(G)$, we define $\mathcal{F}_G u \equiv \widehat{u} \in \mathcal{S}'(\widehat{G})$ by

$$\langle \mathcal{F}_G u, h \rangle_{\widehat{G}} := \langle u, \iota \circ \mathcal{F}_G^{-1} h \rangle_G, \tag{10.14}$$

for all $h \in \mathcal{S}(\widehat{G})$. For $H \in \mathcal{S}'(\widehat{G})$, we define $\mathcal{F}_G^{-1} H \in \mathcal{D}'(G)$ by

$$\langle \mathcal{F}_G^{-1} H, \varphi \rangle_G := \langle H, \mathcal{F}_G(\iota \circ \varphi) \rangle_{\widehat{G}}, \tag{10.15}$$

for all $\varphi \in C^\infty(G)$.

Theorem 10.3.33 (Well-defined and continuous). *For $u \in \mathcal{D}'(G)$ and $H \in \mathcal{S}'(\widehat{G})$, their forward and inverse Fourier transforms $\mathcal{F}_G u \in \mathcal{S}'(\widehat{G})$ and $\mathcal{F}_G^{-1} H \in \mathcal{D}'(G)$ are well defined. Moreover, the mappings $\mathcal{F}_G : \mathcal{D}'(G) \to \mathcal{S}'(\widehat{G})$ and $\mathcal{F}_G^{-1} : \mathcal{S}'(\widehat{G}) \to \mathcal{D}'(G)$ are continuous, and*

$$\mathcal{F}_G^{-1} \circ \mathcal{F}_G = id \quad and \quad \mathcal{F}_G \circ \mathcal{F}_G^{-1} = id$$

on $\mathcal{D}'(G)$ and $\mathcal{S}'(\widehat{G})$, respectively.

Proof. For $h \in \mathcal{S}(\widehat{G})$, its inverse Fourier transform satisfies $\mathcal{F}_G^{-1} h \in C^\infty(G)$ by Theorem 10.3.25. Therefore, $\iota \circ \mathcal{F}_G^{-1} h \in C^\infty(G)$, so that the definition in (10.14) makes sense. Moreover, the mappings $\mathcal{F}_G^{-1} : \mathcal{S}(\widehat{G}) \to C^\infty(G)$ and $\iota : C^\infty(G) \to C^\infty(G)$ are continuous, so that $\mathcal{F}_G u$ is a continuous functional on $\mathcal{S}(\widehat{G})$, implying that $\mathcal{F}_G u \in \mathcal{S}'(\widehat{G})$. Let us now show that $\mathcal{F}_G : \mathcal{D}'(G) \to \mathcal{S}'(\widehat{G})$ is continuous. Indeed, if $u_j \to u$ in $\mathcal{D}'(G)$, then $\langle u_j, \iota \circ \mathcal{F}_G^{-1} h \rangle_G \to \langle u, \iota \circ \mathcal{F}_G^{-1} h \rangle_G$ as $j \to \infty$ for every $h \in \mathcal{S}(\widehat{G})$, implying that $\mathcal{F}_G u_j \to \mathcal{F}_G u$ in $\mathcal{S}'(\widehat{G})$.

The proof that $\mathcal{F}_G^{-1} H \in \mathcal{D}'(G)$ for every $H \in \mathcal{S}'(\widehat{G})$ is similar, as well as the continuity of $\mathcal{F}_G^{-1} : \mathcal{S}'(\widehat{G}) \to \mathcal{D}'(G)$, and are left as Exercise 10.3.34.

To show that $\mathcal{F}_G^{-1} \circ \mathcal{F}_G = id$ on $\mathcal{D}'(G)$, we take $u \in \mathcal{D}'(G)$ and $\varphi \in C^\infty(G)$, and calculate

$$\langle (\mathcal{F}_G^{-1} \circ \mathcal{F}_G) u, \varphi \rangle_G \overset{(10.15)}{=} \langle \mathcal{F}_G u, \mathcal{F}_G(\iota \circ \varphi) \rangle_{\widehat{G}}$$

$$\overset{(10.14)}{=} \langle u, \iota \circ \mathcal{F}_G^{-1}(\mathcal{F}_G(\iota \circ \varphi)) \rangle_G$$

$$\overset{\text{Theorem } 10.3.25}{=} \langle u, \varphi \rangle_G.$$

Similarly, for $H \in \mathcal{S}'(\widehat{G})$ and $h \in \mathcal{S}(\widehat{G})$, we have

$$\langle (\mathcal{F}_G \circ \mathcal{F}_G^{-1}) H, h \rangle_{\widehat{G}} \overset{(10.14)}{=} \langle \mathcal{F}_G^{-1} H, \iota \circ \mathcal{F}_G^{-1} h \rangle_G$$

$$\overset{(10.15)}{=} \langle H, \mathcal{F}_G(\iota \circ \iota \circ \mathcal{F}_G^{-1} h) \rangle_{\widehat{G}}$$

$$\overset{\text{Theorem } 10.3.25}{=} \langle H, h \rangle_{\widehat{G}},$$

completing the proof.   □

Exercise 10.3.34. Complete the proof of Theorem 10.3.33 by showing that $\mathcal{F}_G^{-1} H \in \mathcal{D}'(G)$ for every $H \in \mathcal{S}'(\widehat{G})$ and that $\mathcal{F}_G^{-1} : \mathcal{S}'(\widehat{G}) \to \mathcal{D}'(G)$ is continuous.

Corollary 10.3.35 (Sequential density of $C^\infty(G)$ in $\mathcal{D}'(G)$). *The space $C^\infty(G)$ is sequentially dense in $\mathcal{D}'(G)$.*

Proof. The statement follows from Proposition 10.3.30 saying that the space $\mathcal{S}(\widehat{G})$ is sequentially dense in $\mathcal{S}'(\widehat{G})$, and properties of the Fourier transform from Theorems 10.3.25 and 10.3.33.   □

10.3.3   Spaces $L^p(\widehat{G})$

Definition 10.3.36 (Spaces $L^p(\widehat{G})$). For $1 \leq p < \infty$, we will write $L^p(\widehat{G}) \equiv \ell^p\left(\widehat{G}, \dim^{p\left(\frac{2}{p}-\frac{1}{2}\right)}\right)$ for the space of all $H \in \mathcal{S}'(\widehat{G})$ such that

$$\|H\|_{L^p(\widehat{G})} := \left(\sum_{[\xi] \in \widehat{G}} (\dim(\xi))^{p\left(\frac{2}{p}-\frac{1}{2}\right)} \|H(\xi)\|_{HS}^p \right)^{1/p} < \infty.$$

For $p = \infty$, we will write $L^\infty(\widehat{G}) \equiv \ell^\infty\left(\widehat{G}, \dim^{-1/2}\right)$ for the space of all $H \in \mathcal{S}'(\widehat{G})$ such that

$$\|H\|_{L^\infty(\widehat{G})} := \sup_{[\xi] \in \widehat{G}} (\dim(\xi))^{-1/2} \|H(\xi)\|_{HS} < \infty.$$

We will usually write $L^p(\widehat{G})$ but the notation $\ell^p\left(\widehat{G}, \dim^{p\left(\frac{2}{p} - \frac{1}{2}\right)}\right)$ can be also used to emphasize that these spaces have a structure of weighted sequence spaces on the discrete set $\widehat{G}$, with the weights given by the powers of the dimensions of the representations.

Exercise 10.3.37. Prove that spaces $L^p(\widehat{G})$ are Banach spaces for all $1 \leq p \leq \infty$.

Remark 10.3.38. Two important cases of $L^2(\widehat{G}) = \ell^2\left(\widehat{G}, \dim^1\right)$ and $L^1(\widehat{G}) = \ell^1\left(\widehat{G}, \dim^{3/2}\right)$ are defined by the norms

$$\|H\|_{L^2(\widehat{G})} := \left(\sum_{[\xi] \in \widehat{G}} \dim(\xi) \|H(\xi)\|_{HS}^2\right)^{1/2},$$

which is already familiar from (10.8), and by

$$\|H\|_{L^1(\widehat{G})} := \sum_{[\xi] \in \widehat{G}} (\dim(\xi))^{3/2} \|H(\xi)\|_{HS}.$$

We now discuss several properties of spaces $L^p(\widehat{G})$. We first recall a result on the interpolation of weighted spaces from [14, Theorem 5.5.1]:

Theorem 10.3.39 (Interpolation of weighted spaces). *Let us write* $d\mu_0(x) = w_0(x) \, d\mu(x)$, $d\mu_1(x) = w_1(x) \, d\mu(x)$, *and write* $L^p(w) = L^p(w \, d\mu)$ *for the weight* w. *Suppose that* $0 < p_0, p_1 < \infty$. *Then*

$$(L^{p_0}(w_0), L^{p_1}(w_1))_{\theta, p} = L^p(w),$$

where $0 < \theta < 1$, $\frac{1}{p} = \frac{1-\theta}{p_0} + \frac{\theta}{p_1}$, *and* $w = w_0^{p(1-\theta)/p_0} w_1^{p\theta/p_1}$.

From this we obtain:

Proposition 10.3.40 (Interpolation of $L^p(\widehat{G})$ spaces). *Let* $1 \leq p_0, p_1 < \infty$. *Then*

$$\left(L^{p_0}(\widehat{G}), L^{p_1}(\widehat{G})\right)_{\theta, p} = L^p(\widehat{G}),$$

where $0 < \theta < 1$ *and* $\frac{1}{p} = \frac{1-\theta}{p_0} + \frac{\theta}{p_1}$.

Proof. The statement follows from Theorem 10.3.39 if we regard $L^p(\widehat{G}) = \ell^p\left(\widehat{G}, \dim^{p\left(\frac{2}{p} - \frac{1}{2}\right)}\right)$ as a weighted sequence space over $\widehat{G}$ with the weight given by $\dim(\xi)^{p\left(\frac{2}{p} - \frac{1}{2}\right)}$. $\qquad\square$

Lemma 10.3.41 (Hilbert–Schmidt norm of representations). *If $\xi \in \mathrm{Rep}(G)$, then*

$$\|\xi(x)\|_{HS} = \sqrt{\dim(\xi)}$$

for every $x \in G$.

Proof. We have

$$\|\xi(x)\|_{HS} = \left(\mathrm{Tr}\,(\xi(x)\,\xi(x)^*)\right)^{1/2} = \left(\mathrm{Tr}\,I_{\dim(\xi)}\right)^{1/2} = \sqrt{\dim(\xi)},$$

by the unitarity of the representation ξ.　　　　　　　　　　　　　　$\square$

Proposition 10.3.42 (Fourier transforms on $L^1(G)$ and $L^1(\widehat{G})$). *The Fourier trans-form $\mathcal{F}_G$ is a linear bounded operator from $L^1(G)$ to $L^\infty(\widehat{G})$ satisfying*

$$\|\widehat{f}\|_{L^\infty(\widehat{G})} \le \|f\|_{L^1(G)}.$$

The inverse Fourier transform $\mathcal{F}_G^{-1}$ is a linear bounded operator from $L^1(\widehat{G})$ to $L^\infty(G)$ satisfying

$$\|\mathcal{F}_G^{-1}H\|_{L^\infty(G)} \le \|H\|_{L^1(\widehat{G})}.$$

Proof. Using $\widehat{f}(\xi) = \int_G f(x)\,\xi(x)^*\,dx$, by Lemma 10.3.41 we get

$$\|\widehat{f}(\xi)\|_{HS} \le \int_G |f(x)|\,\|\xi(x)^*\|_{HS}\,dx \le (\dim(\xi))^{1/2}\|f\|_{L^1(G)}.$$

Therefore,

$$\|\widehat{f}\|_{L^\infty(\widehat{G})} = \sup_{[\xi]\in\widehat{G}} (\dim(\xi))^{-1/2}\|\widehat{f}(\xi)\|_{HS} \le \|f\|_{L^1(G)}.$$

On the other hand, using $(\mathcal{F}_G^{-1}H)(x) = \sum_{[\xi]\in\widehat{G}} \dim(\xi)\,\mathrm{Tr}\,(\xi(x)\,H(\xi))$, by Lemma 10.3.28 and Lemma 10.3.41 we have

$$
\begin{aligned}
|(\mathcal{F}_G^{-1}H)(x)| &\le \sum_{[\xi]\in\widehat{G}} \dim(\xi)\,\|\xi(x)\|_{HS}\,\|H(\xi)\|_{HS}\\
&= \sum_{[\xi]\in\widehat{G}} (\dim(\xi))^{3/2}\,\|H(\xi)\|_{HS}\\
&= \|H\|_{L^1(\widehat{G})},
\end{aligned}
$$

from which we get $\|\mathcal{F}_G^{-1}H\|_{L^\infty(G)} \le \|H\|_{L^1(\widehat{G})}$.　　　　　　$\square$

Theorem 10.3.43 (Hausdorff–Young inequality). *Let $1 \le p \le 2$ and $\frac{1}{p} + \frac{1}{q} = 1$. Let $f \in L^p(G)$ and $H \in L^p(\widehat{G})$. Then $\|\widehat{f}\|_{L^q(\widehat{G})} \le \|f\|_{L^p(G)}$ and $\|\mathcal{F}_G^{-1}H\|_{L^q(G)} \le \|H\|_{L^p(\widehat{G})}$.*

Theorem 10.3.43 follows from the $L^1 \to L^\infty$ and $L^2 \to L^2$ boundedness in Proposition 10.3.42 and in Proposition 10.3.17, respectively, by the following interpolation theorem in [14, Corollary 5.5.4] (which is also a consequence of Theorem 10.3.39):

Theorem 10.3.44 (Stein–Weiss interpolation). *Let $1 \le p_0, p_1, q_0, q_1 < \infty$ and let*

$$T : L^{p_0}(U, w_0 \; d\mu) \to L^{q_0}(V, \widetilde{w_0} \; d\nu), \; T : L^{p_1}(U, w_1 \; d\mu) \to L^{q_1}(V, \widetilde{w_1} \; d\nu),$$

with norms M_0 and M_1, respectively. Then

$$T : L^p(U, w \; d\mu) \to L^q(V, \widetilde{w} \; d\nu)$$

with norm $M \le M_0^{1-\theta} M_1^\theta$, where $\frac{1}{p} = \frac{1-\theta}{p_0} + \frac{\theta}{p_1}$, $\frac{1}{q} = \frac{1-\theta}{q_0} + \frac{\theta}{q_1}$, $w = w_0^{p(1-\theta)/p_0} w_1^{p\theta/p_1}$ and $\widetilde{w} = \widetilde{w_0}^{p(1-\theta)/p_0} \widetilde{w_1}^{p\theta/p_1}$.

We now turn to the duality between spaces $L^p(\widehat{G})$:

Theorem 10.3.45 (Duality of $L^p(\widehat{G})$). *Let $1 \le p < \infty$ and $\frac{1}{p} + \frac{1}{q} = 1$. Then* $\left(L^p(\widehat{G})\right)' = L^q(\widehat{G})$.

Proof. The duality is given by the bracket $\langle \cdot, \cdot \rangle_{\widehat{G}}$ in Definition 10.3.29:

$$\langle H, h \rangle_{\widehat{G}} := \sum_{[\xi] \in \widehat{G}} \dim(\xi) \, \mathrm{Tr}\left(H(\xi) \, h(\xi)\right).$$

Assume first $1 < p < \infty$. Then, if $H \in L^p(\widehat{G})$ and $h \in L^q(\widehat{G})$, using Lemma 10.3.28 we get

$$\left| \langle H, h \rangle_{\widehat{G}} \right| \le \sum_{[\xi] \in \widehat{G}} \dim(\xi) \, \|H(\xi)\|_{HS} \, \|h(\xi)\|_{HS}$$

$$= \sum_{[\xi] \in \widehat{G}} (\dim(\xi))^{\frac{2}{p} - \frac{1}{2}} \, \|H(\xi)\|_{HS} \, (\dim(\xi))^{\frac{2}{q} - \frac{1}{2}} \, \|h(\xi)\|_{HS}$$

$$\le \left(\sum_{[\xi] \in \widehat{G}} (\dim(\xi))^{p\left(\frac{2}{p} - \frac{1}{2}\right)} \, \|H(\xi)\|_{HS}^p \right)^{1/p}$$

$$\times \left(\sum_{[\xi] \in \widehat{G}} (\dim(\xi))^{q\left(\frac{2}{q} - \frac{1}{2}\right)} \, \|h(\xi)\|_{HS}^q \right)^{1/q}$$

$$= \|H\|_{L^p(\widehat{G})} \, \|h\|_{L^q(\widehat{G})},$$

where we also used the discrete Hölder inequality (Lemma 3.3.28). Let now $p = 1$. In this case we have

$$
\begin{aligned}
|\langle H, h \rangle_{\widehat{G}}| &\leq \sum_{[\xi] \in \widehat{G}} (\dim(\xi))^{3/2} \, \|H(\xi)\|_{HS} \, (\dim(\xi))^{-1/2} \, \|h(\xi)\|_{HS} \\
&\leq \|H\|_{L^1(\widehat{G})} \, \|h\|_{L^\infty(\widehat{G})}.
\end{aligned}
$$

We leave the other part of the proof as an exercise.   □

Remark 10.3.46 (**Sobolev spaces** $L_k^p(\widehat{G})$). If we use difference operators $\triangle^\alpha$ from Definition 10.7.1, we can also define Sobolev spaces $L_k^p(\widehat{G})$, $k \in \mathbb{N}$, on the unitary dual $\widehat{G}$ by

$$
L_k^p(\widehat{G}) = \left\{ H \in L^p(\widehat{G}) : \ \triangle^\alpha H \in L^p(\widehat{G}) \text{ for all } |\alpha| \leq k \right\}.
$$

10.4   Symbols of operators

Let G be a compact Lie group. Let us endow $\mathcal{D}(G) = C^\infty(G)$ with the usual test function topology (which is the uniform topology of $C^\infty(G)$; we refer the reader to Section 10.12 for some additional information on the topics of distributions and Schwartz kernels if more introduction is desirable). For a continuous linear operator $A : C^\infty(G) \to C^\infty(G)$, let $K_A, L_A, R_A \in \mathcal{D}'(G \times G)$ denote respectively the Schwartz, left-convolution and right-convolution kernels, i.e.,

$$
\begin{aligned}
Af(x) &= \int_G K_A(x, y) \, f(y) \, \mathrm{d}y \\
&= \int_G L_A(x, xy^{-1}) \, f(y) \, \mathrm{d}y \\
&= \int_G f(y) \, R_A(x, y^{-1}x) \, \mathrm{d}y \qquad\qquad (10.16)
\end{aligned}
$$

in the sense of distributions. To simplify the notation in the sequel, we will often write integrals in the sense of distributions, with a standard distributional interpretation.

Proposition 10.4.1 (Relations between kernels). *We have*

$$
R_A(x, y) = L_A(x, xyx^{-1}), \qquad\qquad (10.17)
$$

as well as

$$
R_A(x, y) = K_A(x, xy^{-1}) \ \text{and} \ L_A(x, y) = K_A(x, y^{-1}x),
$$

with the standard distributional interpretation.

Proof. Equality (10.17) follows directly from (10.16). The proof of the last two equalities is just a change of variables. Indeed, (10.16) implies that $K_A(x, y) = R_A(x, y^{-1}x)$. Denoting $v = y^{-1}x$, we have $y = xv^{-1}$, so that $K_A(x, xv^{-1}) = R_A(x, v)$. Similarly, $K_A(x, y) = L_A(x, xy^{-1})$ from (10.16) and the change $w = xy^{-1}$ yield $y = w^{-1}x$, and hence $K_A(x, w^{-1}x) = L_A(x, w)$. □

We also note that left-invariant operators on $C^\infty(G)$ correspond to right-convolutions $f \mapsto f * k$. Since we identify the Lie algebra $\mathfrak{g}$ of G with the left-invariant vector fields on G, *it will be most natural to study right-convolution kernels* in the sequel. Let us explain this in more detail:

Remark 10.4.2 (**Left or right?**). For $g \in \mathcal{D}'(G)$, define the respective left-convolution and right-convolution operators $l(f), r(f) : C^\infty(G) \to C^\infty(G)$ by

$$l(f)g := f * g,$$
$$r(f)g := g * f.$$

In this notation, the relation between left- and right-convolution kernels of these convolution operators in the notation of (10.16) becomes $L_{l(f)}(x, y) = f(y) = R_{r(f)}(x, y)$. Also, if $Y \in \mathfrak{g}$, then Proposition 10.2.9 implies that $D_Y l(f) = l(f)D_Y$. Let the respective *left and right regular representations of* G be denoted by $\pi_L, \pi_R : G \to \mathcal{U}(L^2(G))$, i.e.,

$$\pi_L(x)f(y) = f(x^{-1}y),$$
$$\pi_R(x)f(y) = f(yx).$$

Operator A is

$$\begin{array}{rll} \text{left-invariant} & \text{if} & \pi_L(x)\, A = A\, \pi_L(x), \\ \text{right-invariant} & \text{if} & \pi_R(x)\, A = A\, \pi_R(x), \end{array}$$

for every $x \in G$. Notice that A is

$$\begin{array}{rcl} \text{left-invariant} & \Longleftrightarrow & \text{right-convolution,} \\ \text{right-invariant} & \Longleftrightarrow & \text{left-convolution.} \end{array}$$

Indeed, we have, for example,

$$\begin{aligned} [\pi_R(x)l(f)g](z) &= (f * g)(zx) \\ &= \int_G f(y)\, g(y^{-1}zx)\, \mathrm{d}y \\ &= \int_G f(y)\, (\pi_R(x)g)(y^{-1}z)\, \mathrm{d}y \\ &= [f * \pi_R(x)g](z) \\ &= [l(f)\pi_R(x)g](z), \end{aligned}$$

so that $\pi_R(x)l(f) = l(f)\pi_R(x)$, and similarly $\pi_L(x)r(f) = r(f)\pi_L(x)$.

The Lie algebras are often (but not always) identified with **left-invariant** vector fields, which are **right-convolutions**, that is why our starting choice in the sequel are **right-convolution kernels**. We refer to Remark 10.4.13 for a further discussion of left and right.

10.4.1 Full symbols

We now define symbols of operators on G.

Definition 10.4.3 (Symbols of operators on G). Let $\xi : G \to \mathcal{U}(\mathcal{H}_\xi)$ be an irreducible unitary representation. The *symbol* of a linear continuous operator $A : C^\infty(G) \to C^\infty(G)$ at $x \in G$ and $\xi \in \mathrm{Rep}(G)$ is defined as

$$\sigma_A(x,\xi) := \widehat{r_x}(\xi) \in \mathrm{End}(\mathcal{H}_\xi),$$

where $r_x(y) = R_A(x,y)$ is the right-convolution kernel of A as in (10.16). Hence

$$\sigma_A(x,\xi) = \int_G R_A(x,y)\, \xi(y)^*\, \mathrm{d}y$$

in the sense of distributions, and by Corollary 10.2.10 the right-convolution kernel can be regained from the symbol as well:

$$R_A(x,y) = \sum_{[\xi]\in\widehat{G}} \dim(\xi)\, \mathrm{Tr}\left(\xi(y)\, \sigma_A(x,\xi)\right), \tag{10.18}$$

where this equality is interpreted distributionally. We now show that operator A can be represented by its symbol:

Theorem 10.4.4 (Quantization of operators). *Let σ_A be the symbol of a continuous linear operator $A : C^\infty(G) \to C^\infty(G)$. Then*

$$Af(x) = \sum_{[\xi]\in\widehat{G}} \dim(\xi)\, \mathrm{Tr}\left(\xi(x)\, \sigma_A(x,\xi)\, \widehat{f}(\xi)\right) \tag{10.19}$$

for every $f \in C^\infty(G)$ and $x \in G$.

Proof. Let us define a right-convolution operator $A_{x_0} \in \mathcal{L}(C^\infty(G))$ by its kernel $R_A(x_0,y) = r_{x_0}(y)$, i.e., by

$$A_{x_0} f(x) := \int_G f(y)\, r_{x_0}(y^{-1}x)\, \mathrm{d}y = (f * r_{x_0})(x).$$

Thus

$$\sigma_{A_{x_0}}(x,\xi) = \widehat{r_{x_0}}(\xi) = \sigma_A(x_0,\xi),$$

so that by (10.6) we have

$$
\begin{aligned}
A_{x_0} f(x) &= \sum_{[\xi] \in \widehat{G}} \dim(\xi) \, \mathrm{Tr} \left(\xi(x) \, \widehat{A_{x_0} f}(\xi) \right) \\
&= \sum_{[\xi] \in \widehat{G}} \dim(\xi) \, \mathrm{Tr} \left(\xi(x) \, \sigma_A(x_0, \xi) \, \widehat{f}(\xi) \right),
\end{aligned}
$$

where we used that $\widehat{f * r_{x_0}} = \widehat{r_{x_0} f}$ by Proposition 10.2.7. This implies the result, because $Af(x) = A_x f(x)$, for each fixed x. $\square$

Definition 10.4.3 and Theorem 10.4.4 justify the following notation:

Definition 10.4.5 (Pseudo-differential operators). For a symbol σ_A, the corresponding operator A defined by (10.19) will be also denoted by $\mathrm{Op}(\sigma_A)$. The operator defined by formula (10.19) will be called the *pseudo-differential operator with symbol σ_A*.

If we fix representations to be matrix representations we can express all the formulae above in matrix components. Thus, if $\xi : G \to \mathrm{U}(\dim(\xi))$ are irreducible unitary matrix representations then

$$
Af(x) = \sum_{[\xi] \in \widehat{G}} \dim(\xi) \sum_{m,n=1}^{\dim(\xi)} \xi(x)_{nm} \left(\sum_{k=1}^{\dim(\xi)} \sigma_A(x, \xi)_{mk} \, \widehat{f}(\xi)_{kn} \right),
$$

and as a consequence of (10.18) and Corollary 10.2.10 we also have formally:

$$
R_A(x, y) = \sum_{[\xi] \in \widehat{G}} \dim(\xi) \sum_{m,n=1}^{\dim(\xi)} \xi(y)_{nm} \, \sigma_A(x, \xi)_{mn}. \tag{10.20}
$$

Alternatively, setting $A\xi(x)_{mn} := (A(\xi_{mn}))(x)$, we have

$$
\sigma_A(x, \xi)_{mn} = \sum_{k=1}^{\dim(\xi)} \overline{\xi_{km}(x)} \, (A\xi_{kn})(x), \tag{10.21}
$$

$1 \leq m, n \leq \dim(\xi)$, which follows from the following theorem:

Theorem 10.4.6 (Formula for the symbol via representations). *Let σ_A be the symbol of a continuous linear operator $A : C^\infty(G) \to C^\infty(G)$. Then for all $x \in G$ and $\xi \in \mathrm{Rep}(G)$ we have*

$$
\sigma_A(x, \xi) = \xi(x)^* (A\xi)(x). \tag{10.22}
$$

Proof. Working with matrix representations $\xi : G \to \mathrm{U}(\dim(\xi))$, we have

$$\sum_{k=1}^{\dim(\xi)} \overline{\xi_{km}(x)}\, (A\xi_{kn})(x)$$

$$\overset{(10.19)}{=} \sum_k \overline{\xi_{km}(x)} \sum_{[\eta]\in\widehat{G}} \dim(\eta)\, \mathrm{Tr}\left(\eta(x)\, \sigma_A(x,\eta)\, \widehat{\overline{\xi_{kn}}}(\eta)\right)$$

$$= \sum_k \overline{\xi_{km}(x)} \sum_{[\eta]\in\widehat{G}} \dim(\eta) \sum_{i,j,l} \eta(x)_{ij}\, \sigma_A(x,\eta)_{jl}\, \widehat{\overline{\xi_{kn}}}(\eta)_{li}$$

$$= \sum_{k,j} \overline{\xi_{km}(x)}\, \xi(x)_{kj}\, \sigma_A(x,\xi)_{jn}$$

$$\overset{\mathrm{Rem.\ 10.2.4}}{=} \sigma_A(x,\xi)_{mn},$$

where we take $\eta = \xi$ if $\eta \in [\xi]$ in the sum, so that $\widehat{\overline{\xi_{kn}}}(\eta)_{li} = \langle \xi_{kn}, \eta_{il}\rangle_{L^2}$ by (10.5), which equals $\frac{1}{\dim(\xi)}$ if $\xi = \eta$, $k = i$ and $n = l$, and zero otherwise. $\qquad\square$

Remark 10.4.7 (**Formula for symbol on** $\mathbb{T}^n$). Since in the case of the torus $G = \mathbb{T}^n$ by Remark 10.2.4 representations of $\mathbb{T}^n$ are given by $e_k(x) = \mathrm{e}^{\mathrm{i}2\pi x\cdot k}$, $k \in \mathbb{Z}^n$, formula (10.22) gives the formula for the *toroidal* symbol

$$\sigma_A(x,k) := \sigma_A(x,e_k) = \mathrm{e}^{-\mathrm{i}2\pi x\cdot k}(A\, \mathrm{e}^{\mathrm{i}2\pi x\cdot k})(x),$$

$k \in \mathbb{Z}^n$, as in Theorem 4.1.4.

Remark 10.4.8 (**Symbol of the Laplace operator**). We note that by Theorem 10.3.13 the symbol of the Laplace operator $\mathcal{L} = \mathcal{L}_G$ on G is $\sigma_{\mathcal{L}}(x,\xi) = -\lambda_{[\xi]}I_{\dim\xi}$, where $I_{\dim\xi}$ is the identity mapping on $\mathcal{H}_\xi$ and $\lambda_{[\xi]}$ are the eigenvalues of $-\mathcal{L}$.

Remark 10.4.9 (**Symbol as a mapping on** $G \times \widehat{G}$). The symbol of $A \in \mathcal{L}(C^\infty(G))$ is a mapping

$$\sigma_A : G \times \mathrm{Rep}(G) \to \bigcup_{\xi\in\mathrm{Rep}(G)} \mathrm{End}(\mathcal{H}_\xi),$$

where $\sigma_A(x,\xi) \in \mathrm{End}(\mathcal{H}_\xi)$ for every $x \in G$ and $\xi \in \mathrm{Rep}(G)$. However, it can be viewed as a *mapping on the space* $G \times \widehat{G}$. Indeed, let $\xi, \eta \in \mathrm{Rep}(G)$ be equivalent via an intertwining isomorphism $U \in \mathrm{Hom}(\xi,\eta)$: i.e., such that there exists a linear unitary bijection $U : \mathcal{H}_\xi \to \mathcal{H}_\eta$ such that $U\, \eta(x) = \xi(x)\, U$ for every $x \in G$, that is $\eta(x) = U^{-1}\, \xi(x)\, U$. Then by Remark 10.2.5 we have $\widehat{f}(\eta) = U^{-1}\, \widehat{f}(\xi)\, U$, and hence also

$$\sigma_A(x,\eta) = U^{-1}\, \sigma_A(x,\xi)\, U.$$

Therefore, taking any representation from the same class $[\xi] \in \widehat{G}$ leads to the same operator A in view of the trace in formula (10.19). In this sense we may think that symbol σ_A is defined on $G \times \widehat{G}$ instead of $G \times \mathrm{Rep}(G)$.

Remark 10.4.10 (**Symbol of right-convolution**). Notice that if $A = (f \mapsto f * a)$ then $R_A(x, y) = a(y)$ and hence

$$\sigma_A(x, \xi) = \widehat{a}(\xi),$$

and hence $\widehat{Af}(\xi) = \widehat{a}(\xi) \, \widehat{f}(\xi) = \sigma_A(x, \xi) \, \widehat{f}(\xi)$.

Proposition 10.4.11 (Symbol of left-convolution). *If $B = (f \mapsto b * f)$ is the left-convolution operator, then*

$$L_B(x, y) = b(y), \quad R_B(x, y) = L_B(x, xyx^{-1}) = b(xyx^{-1}),$$

and the symbol of B is given by

$$\sigma_B(x, \xi) = \xi(x)^* \, \widehat{b}(\xi) \, \xi(x).$$

Exercise 10.4.12. Prove Proposition 10.4.11. (Hint: use (10.4) and (10.17).)

Remark 10.4.13 (**What if we started with left-convolution kernels?**). What if we had chosen *right*-invariant vector fields and corresponding *left*-convolution operators as the starting point of the Fourier analysis? Let us define another "Fourier transform" by

$$\pi_f(\xi) := \int_G f(x) \, \xi(x) \, \mathrm{d}x.$$

Then $\pi_{f*g} = \pi_f \, \pi_g$, and a continuous linear operator $A : C^\infty(G) \to C^\infty(G)$ can be presented by

$$
\begin{aligned}
Af(x) \quad &= \quad \sum_{[\xi] \in \widehat{G}} \dim(\xi) \, \mathrm{Tr}\left(\xi(x) \, \sigma_A(x, \xi) \, \widehat{f}(\xi) \right) \\
&= \quad \sum_{[\xi] \in \widehat{G}} \dim(\xi) \, \mathrm{Tr}\left(\xi(x)^* \, \widetilde{\sigma_A}(x, \xi) \, \pi_f(\xi) \right),
\end{aligned}
$$

where

$$
\begin{aligned}
\widetilde{\sigma_A}(x, \xi) \quad &= \quad \pi_{y \mapsto L_A(x,y)}(\xi) \\
&= \quad \xi(x) \, (A(\xi^*))(x).
\end{aligned}
$$

In the coming symbol considerations this left–right choice is encountered, e.g., as follows:

$$
\begin{aligned}
\sigma_{AB}(x, \xi) &\sim \sigma_A(x, \xi) \, \sigma_B(x, \xi) + \cdots && \text{if we use right-convolutions,} \\
\widetilde{\sigma_{AB}}(x, \xi) &\sim \widetilde{\sigma_B}(x, \xi) \, \widetilde{\sigma_A}(x, \xi) + \cdots && \text{if we use left-convolutions.}
\end{aligned}
$$

There is an explicit link between the left–right cases. We refer to Section 10.11 for a further discussion of these issues for the operator-valued symbols. At the same time, we note that these choices already determine the need to work with right actions on homogeneous spaces in Chapter 13, so that the homogeneous spaces there are $K \backslash G$ instead of G/K, see Remark 13.2.5 for the discussion of this issue.

Remark 10.4.14. We have now associated a unique full symbol σ_A to each continuous linear operator $A : C^\infty(G) \to C^\infty(G)$. Here $\sigma_A(x, \xi) : \mathcal{H}_\xi \to \mathcal{H}_\xi$ is a linear operator for each $x \in G$ and each irreducible unitary representation $\xi : G \to \mathcal{U}(\mathcal{H}_\xi)$. The correspondence $A \mapsto \sigma_A$ is linear in the sense that

$$\sigma_{A+B}(x, \xi) = \sigma_A(x, \xi) + \sigma_B(x, \xi) \quad \text{and} \quad \sigma_{\lambda A}(x, \xi) = \lambda \sigma_A(x, \xi),$$

where $\lambda \in \mathbb{C}$. However, $\sigma_{AB}(x, \xi)$ is not usually $\sigma_A(x, \xi)\sigma_B(x, \xi)$ (unless B is a right-convolution operator, so that the symbol $\sigma_B(x, \xi) = \widehat{b}(\xi)$ does not depend on the variable $x \in G$). A composition formula will be established in Theorem 10.7.8 below.

10.4.2   Conjugation properties of symbols

In the sequel, we will need conjugation properties of symbols which we will now analyse for this purpose.

Definition 10.4.15 (ϕ-pushforwards). Let $\phi : G \to G$ be a diffeomorphism, $f \in C^\infty(G)$, $A : C^\infty(G) \to C^\infty(G)$ continuous and linear. Then the ϕ-pushforwards $f_\phi \in C^\infty(G)$ and $A_\phi : C^\infty(G) \to C^\infty(G)$ are defined by

$$
\begin{aligned}
f_\phi &:= f \circ \phi^{-1}, \\
A_\phi f &:= \left(A(f_{\phi^{-1}})\right)_\phi = A(f \circ \phi) \circ \phi^{-1}.
\end{aligned}
$$

Notice that

$$A_{\phi \circ \psi} = (A_\psi)_\phi .$$

Exercise 10.4.16. Using the local theory of pseudo-differential operators show that $A \in \Psi^\mu(G)$ if and only if $A_\phi \in \Psi^\mu(G)$.

Definition 10.4.17. For $u \in G$, let $u_L, u_R : G \to G$ be defined by

$$u_L(x) := ux \quad \text{and} \quad u_R(x) := xu.$$

Then $(u_L)^{-1} = (u^{-1})_L$ and $(u_R)^{-1} = (u^{-1})_R$. The inner automorphism $\phi_u : G \to G$ defined in Proposition 10.2.6 by $\phi_u(x) := u^{-1}xu$ satisfies

$$\phi_u = u_L^{-1} \circ u_R = u_R \circ u_L^{-1}.$$

Proposition 10.4.18. *Let $u \in G$, $B = A_{u_L}$, $C = A_{u_R}$ and $F = A_{\phi_u}$. Then we have the following relations between symbols:*

$$
\begin{aligned}
\sigma_B(x, \xi) &= \sigma_A(u^{-1}x, \xi), \\
\sigma_C(x, \xi) &= \xi(u)^* \, \sigma_A(xu^{-1}, \xi) \, \xi(u), \\
\sigma_F(x, \xi) &= \xi(u)^* \, \sigma_A(uxu^{-1}, \xi) \, \xi(u).
\end{aligned}
$$

Especially, if $A = (f \mapsto f * a)$, *i.e.,* $\sigma_A(x, \xi) = \widehat{a}(\xi)$, *then*

$$
\begin{aligned}
\sigma_B(x, \xi) &= \widehat{a}(\xi), \\
\sigma_C(x, \xi) &= \xi(u)^* \, \widehat{a}(\xi) \, \xi(u) \\
&= \sigma_F(x, \xi).
\end{aligned}
$$

Proof. We notice that $F = C_{(u^{-1})_L}$, so it suffices to consider only operators B and C. For the operator $B = A_{u_L}$, we get

$$
\begin{aligned}
\int_G f(z) \, R_B(x, z^{-1}x) \, f(z) \, \mathrm{d}z &= Bf(x) \\
&= A(f \circ u_L)(u_L^{-1}(x)) \\
&= \int_G f(uy) \, R_A(u^{-1}x, y^{-1}u^{-1}x) \, \mathrm{d}y \\
&= \int_G f(z) \, R_A(u^{-1}x, z^{-1}x) \, \mathrm{d}z,
\end{aligned}
$$

so $R_B(x, y) = R_A(u^{-1}x, y)$, yielding

$$
\sigma_B(x, \xi) = \sigma_A(u^{-1}x, \xi).
$$

For the operator $C = A_{u_R}$, we have

$$
\begin{aligned}
\int_G f(z) \, R_C(x, z^{-1}x) \, \mathrm{d}z &= Cf(x) \\
&= A(f \circ u_R)(u_R^{-1}(x)) \\
&= \int_G f(yu) \, R_A(xu^{-1}, y^{-1}xu^{-1}) \, f(yu) \, \mathrm{d}y \\
&= \int_G f(z) \, R_A(xu^{-1}, uz^{-1}xu^{-1}) \, \mathrm{d}z,
\end{aligned}
$$

so that $R_C(x, y) = R_A(xu^{-1}, uyu^{-1})$, yielding

$$
\begin{aligned}
\sigma_C(x, \xi) &= \int_G R_C(x, y) \, \xi(y)^* \, \mathrm{d}y \\
&= \int_G R_A(xu^{-1}, uyu^{-1}) \, \xi(y)^* \, \mathrm{d}y \\
&= \int_G R_A(xu^{-1}, z) \, \xi(u^{-1}zu)^* \, \mathrm{d}z \\
&= \int_G R_A(xu^{-1}, z) \, \xi(u)^* \, \xi(z)^* \, \xi(u) \, \mathrm{d}z \\
&= \xi(u)^* \, \sigma_A(xu^{-1}, \xi) \, \xi(u)
\end{aligned}
$$

and completing the proof. $\qquad\qquad\square$

Let us now record how push-forwards by translation affect vector fields.

Lemma 10.4.19 (Push-forwards of vector fields). *Let $u \in G$, $Y \in \mathfrak{g}$ and let $E = D_Y : C^\infty(G) \to C^\infty(G)$ be defined by*

$$D_Y f(x) = \frac{d}{dt} f(x \, \exp(tY)) \Big|_{t=0}. \tag{10.23}$$

Then

$$E_{u_R} = E_{\phi_u} = D_{u^{-1}Yu},$$

i.e.,

$$D_Y(f \circ u_R)(xu^{-1}) = D_Y(f \circ \phi_u)(uxu^{-1}) = D_{u^{-1}Yu}f(x).$$

Proof. We have

$$
\begin{aligned}
E_{u_R}f(x) &= E(f \circ u_R)(xu^{-1}) \\
&= \frac{d}{dt}(f \circ u_R)(xu^{-1}\exp(tY))\big|_{t=0} \\
&= \frac{d}{dt} f(xu^{-1}\exp(tY)u)\big|_{t=0} \\
&= \frac{d}{dt} f(x\exp(tu^{-1}Yu))\big|_{t=0} \\
&= D_{u^{-1}Yu}f(x).
\end{aligned}
$$

Due to the left-invariance, we have $E_{u_L} = E$, so that

$$E_{\phi_u} = (E_{u_L^{-1}})_{u_R} = E_{u_R} = D_{u^{-1}Yu}.$$

For more transparency, we also calculate directly:

$$
\begin{aligned}
E_{\phi_u}f(x) &= E(f \circ \phi_u)(uxu^{-1}) \\
&= \frac{d}{dt}(f \circ \phi_u)(uxu^{-1}\exp(tY))\big|_{t=0} \\
&= \frac{d}{dt} f(xu^{-1}\exp(tY)u)\big|_{t=0} \\
&= \frac{d}{dt} f(x\exp(tu^{-1}Yu))\big|_{t=0} \\
&= D_{u^{-1}Yu}f(x),
\end{aligned}
$$

yielding the same result. $\qquad\square$

Remark 10.4.20 (Symbol of $\mathrm{i}D_Y$ can be diagonalised). Notice first that the complex vector field $\mathrm{i}D_Y$ is symmetric:

$$
\begin{aligned}
(\mathrm{i}D_Y f, g)_{L^2(G)} &= \int_G (\mathrm{i}D_Y f)(x)\, \overline{g(x)}\, dx \\
&= -\mathrm{i}\int_G f(x)\, \overline{D_Y g(x)}\, dx \\
&= (f, \mathrm{i}D_Y g)_{L^2(G)}.
\end{aligned}
$$

Hence it is always possible to choose a representative $\xi \in \mathrm{Rep}(G)$ from each $[\xi] \in$ $\widehat{G}$ such that $\sigma_{iD_Y}(x, \xi)$ is a diagonal matrix $\begin{pmatrix} \lambda_1 & & \\ & \ddots & \\ & & \lambda_{\dim(\xi)} \end{pmatrix}$, with diagonal entries $\lambda_j \in \mathbb{R}$, which follows because symmetric matrices can be diagonalised by unitary matrices. Notice that then also the commutator of symbols satisfy

$$[\sigma_{iD_Y}, \sigma_A](x, \xi)_{mn} = (\lambda_m - \lambda_n)\, \sigma_A(x, \xi)_{mn}.$$

10.5 Boundedness of operators on $L^2(G)$

In this section we will state some natural conditions on the symbol of an operator $A : C^\infty(G) \to C^\infty(G)$ to guarantee its boundedness on $L^2(G)$. Recall first that the Hilbert–Schmidt inner product of matrices is defined as a special case of Definition B.5.43:

Definition 10.5.1 (Hilbert–Schmidt inner product). The *Hilbert–Schmidt inner product* of $A, B \in \mathbb{C}^{m \times n}$ is

$$\langle A, B \rangle_{HS} := \mathrm{Tr}(B^*A) = \sum_{i=1}^{m} \sum_{j=1}^{n} \overline{B_{ij}} A_{ij},$$

with the corresponding norm $\|A\|_{HS} := \langle A, A \rangle_{HS}^{1/2}$, and the operator norm

$$\|A\|_{op} := \sup\left\{ \|Ax\|_{\ell^2} : x \in \mathbb{C}^{n \times 1},\ \|x\|_{\ell^2} \leq 1 \right\} = \|A\|_{\ell^2 \to \ell^2},$$

where $\|x\|_{\ell^2} = (\sum_{j=1}^{n} |x_j|^2)^{1/2}$ is the usual Euclidean norm.

Let $A, B \in \mathbb{C}^{n \times n}$. Then by Theorem 12.6.1 proved in Section 12.6 we have

$$\|AB\|_{HS} \leq \|A\|_{op} \|B\|_{HS}.$$

Moreover, we also have

$$\|A\|_{op} = \sup\left\{ \|AX\|_{HS} : X \in \mathbb{C}^{n \times n},\ \|X\|_{HS} \leq 1 \right\}.$$

By this, taking the Fourier transform of the convolution and using Plancherel's formula (Corollary 7.6.10), by Proposition 10.2.7 we get

Proposition 10.5.2 (Operator norm of convolutions). *We have*

$$\|g \mapsto f * g\|_{\mathcal{L}(L^2(G))} = \|g \mapsto g * f\|_{\mathcal{L}(L^2(G))} = \sup_{\xi \in \mathrm{Rep}(G)} \|\widehat{f}(\xi)\|_{op}. \qquad (10.24)$$

We also note that $\|\widehat{f}(\xi)\|_{op} = \|\widehat{f}(\eta)\|_{op}$ if $[\xi] = [\eta] \in \widehat{G}$.

We now extend this property to operators that are not necessarily left- or right-invariant. First we introduce derivatives of higher order on the Lie group G:

Definition 10.5.3 (Operators ∂^α on G). Let $\{Y_j\}_{j=1}^{\dim(G)}$ be a basis for the Lie algebra of G, and let ∂_j be the left-invariant vector fields corresponding to Y_j, $\partial_j = D_{Y_j}$, as in (10.23). For $\alpha \in \mathbb{N}_0^n$, let us denote $\partial^\alpha = \partial_1^{\alpha_1} \cdots \partial_n^{\alpha_n}$. Sometimes we denote these operators by ∂_x^α.

Remark 10.5.4 **(Orderings).** We note that unless G is commutative, operators ∂_j do not in general commute. Thus, when we talk about "all operators ∂^α", we mean that we take these operators in all orderings. However, if we fix a certain ordering of Y_j's, the commutator of a general ∂^α with ∂^α taken in this particular ordering is an operator of lower order (this can be easily seen either for the simple properties of commutators in Exercise D.1.5 or from the general composition Theorem 10.7.9). The commutator is again a combination of operators of the form ∂^β with $|\beta| \leq |\alpha| - 1$. Thus, since usually we require some property to hold for example "for all ∂^α with $|\alpha| \leq N$", we can rely iteratively on the fact that the assumption is already satisfied for ∂^β, thus making this ordering issue less important.

Theorem 10.5.5 (Boundedness of operators on $L^2(G)$). *Let G be a compact Lie group of dimension n and let k be an integer such that $k > n/2$. Let σ_A be the symbol of a linear continuous operator $A : C^\infty(G) \to C^\infty(G)$. Assume that there is a constant C such that*

$$\|\partial_x^\alpha \sigma_A(x,\xi)\|_{op} \leq C$$

for all $x \in G$, all $\xi \in \mathrm{Rep}(G)$, and all $|\alpha| \leq k$. Then A extends to a bounded operator from $L^2(G)$ to $L^2(G)$.

Proof. Let $Af(x) = (f * r_A(x))(x)$, where $r_A(x)(y) = R_A(x,y)$ is the right-convolution kernel of A. Let $A_y f(x) = (f * r_A(y))(x)$, so that $A_x f(x) = Af(x)$. Then

$$\|Af\|_{L^2(G)}^2 = \int_G |A_x f(x)|^2 \, dx \leq \int_G \sup_{y \in G} |A_y f(x)|^2 \, dx,$$

and by an application of the Sobolev embedding theorem we get

$$\sup_{y \in G} |A_y f(x)|^2 \leq C \sum_{|\alpha| \leq k} \int_G |\partial_y^\alpha A_y f(x)|^2 \, dy.$$

Therefore, using the Fubini theorem to change the order of integration, we obtain

$$\|Af\|_{L^2(G)}^2 \quad \leq \quad C \sum_{|\alpha| \leq k} \int_G \int_G |\partial_y^\alpha A_y f(x)|^2 \, dx \, dy$$

$$\leq \quad C \sum_{|\alpha| \leq k} \sup_{y \in G} \int_G |\partial_y^\alpha A_y f(x)|^2 \, dx$$

$$
\begin{aligned}
&= && C \sum_{|\alpha| \le k} \sup_{y \in G} \|\partial_y^\alpha A_y f\|_{L^2(G)}^2 \\
&\le && C \sum_{|\alpha| \le k} \sup_{y \in G} \|f \mapsto f * \partial_y^\alpha r_A(y)\|_{\mathcal{L}(L^2(G))}^2 \|f\|_{L^2(G)}^2 \\
&\overset{(10.24)}{\le} && C \sum_{|\alpha| \le k} \sup_{y \in G} \sup_{[\xi] \in \widehat{G}} \|\partial_y^\alpha \sigma_A(y, \xi)\|_{op}^2 \|f\|_{L^2(G)}^2,
\end{aligned}
$$

where the last inequality holds due to (10.24). This completes the proof. □

10.6 Taylor expansion on Lie groups

As Taylor polynomial expansions are useful in obtaining symbolic calculus on $\mathbb{R}^n$, we would like to have analogous expansions on a group G. Here, the Taylor expansion formula on G will be obtained by embedding G into some $\mathbb{R}^m$, using the Taylor expansion formula in $\mathbb{R}^m$, and then restricting it back to G.

Let $U \subset \mathbb{R}^m$ be an open neighbourhood of some point $\vec{e} \in \mathbb{R}^m$. The Nth order Taylor polynomial $P_N f : \mathbb{R}^m \to \mathbb{C}$ of $f \in C^\infty(U)$ at $\vec{e}$ is given by

$$
P_N f(\vec{x}) = \sum_{\alpha \in \mathbb{N}_0^m: \ |\alpha| \le N} \frac{1}{\alpha!} (\vec{x} - \vec{e})^\alpha \ \partial_x^\alpha f(\vec{e}).
$$

Then the remainder $E_N f := f - P_N f$ satisfies

$$
E_N f(\vec{x}) = \sum_{|\alpha| = N+1} (\vec{x} - \vec{e})^\alpha \ f_\alpha(\vec{x})
$$

for some functions $f_\alpha \in C^\infty(U)$. In particular,

$$
E_N f(\vec{x}) = \mathcal{O}(|\vec{x} - \vec{e}|^{N+1}) \quad \text{as} \quad \vec{x} \to \vec{e}.
$$

Let G be a compact Lie group; we would like to approximate a smooth function $u : G \to \mathbb{C}$ using a Taylor polynomial type expansion nearby the neutral element $e \in G$. By Corollary 8.0.4 we may assume that G is a closed subgroup of $\mathrm{GL}(n, \mathbb{R}) \subset \mathbb{R}^{n \times n}$, the group of real invertible $(n \times n)$-matrices, and thus a closed submanifold of the Euclidean space of dimension $m = n^2$. This embedding of G into $\mathbb{R}^m$ will be denoted by $x \mapsto \vec{x}$, and the image of G under this embedding will be still denoted by G. Also, if $x \in G$, we may still write x for $\vec{x}$ to simplify the notation. Let $U \subset \mathbb{R}^m$ be a small enough open neighbourhood of $G \subset \mathbb{R}^m$ such that for each $\vec{x} \in U$ there exists a unique nearest point $p(\vec{x}) \in G$ (with respect to the Euclidean distance). For $u \in C^\infty(G)$ we define $f \in C^\infty(U)$ by

$$
f(\vec{x}) := u(p(\vec{x})).
$$

The effect is that f is constant in the directions perpendicular to G. As above, we may define the Euclidean Taylor polynomial $P_N f : \mathbb{R}^m \to \mathbb{C}$ at $e \in G \subset \mathbb{R}^m$. Let us define $P_N u : G \to \mathbb{C}$ as the restriction,

$$P_N u := P_N f|_G.$$

We call $P_N u \in C^\infty(G)$ a *Taylor polynomial of u of order N at $e \in G$*. Then for $x \in G$, we have

$$u(x) - P_N u(x) = \sum_{|\alpha| = N+1} (x - e)^\alpha \, u_\alpha(x)$$

for some functions $u_\alpha \in C^\infty(G)$, where we set $(x-e)^\alpha := (\vec{x}-\vec{e})^\alpha$. There should be no confusion with this notation because there is no substraction on the group G, so subtracting group elements means subtracting them when they are embedded in a higher-dimensional linear space. Taylor polynomials on G are given by

$$P_N u(x) = \sum_{|\alpha| \leq N} \frac{1}{\alpha!} \, (x - e)^\alpha \, \partial_x^{(\alpha)} u(e),$$

where we set

$$\partial_x^{(\alpha)} u(e) := \partial_x^\alpha f(\vec{e}). \tag{10.25}$$

Remark 10.6.1. We note that in this way we can obtain different forms of the Taylor series. For example, it may depend on the embedding of G into $\mathrm{GL}(n, \mathbb{R})$, on the choice of the coordinates in $\mathbb{R}^n \times \mathbb{R}^n$, etc.

Let us now consider the example of $G = \mathrm{SU}(2)$. Recall the quaternionic identification

$$(x_0 \mathbf{1} + x_1 \mathbf{i} + x_2 \mathbf{j} + x_3 \mathbf{k} \mapsto (x_0, x_1, x_2, x_3)) : \mathbb{H} \to \mathbb{R}^4,$$

to be discussed in detail in Section 11.4. Moreover, there is the identification $\mathbb{H} \supset \mathbb{S}^3 \cong \mathrm{SU}(2)$, given by

$$\vec{x} = (x_0, x_1, x_2, x_3) \mapsto \begin{pmatrix} x_0 + ix_3 & x_1 + ix_2 \\ -x_1 + ix_2 & x_0 - ix_3 \end{pmatrix} = \begin{pmatrix} x_{11} & x_{12} \\ x_{21} & x_{22} \end{pmatrix} = x.$$

Hence we identify $(1, 0, 0, 0) \in \mathbb{R}^4$ with the neutral element of $\mathrm{SU}(2)$.

Remark 10.6.2. Notice that the functions

$$\begin{aligned} q_+(x) &= x_{12} = x_1 + ix_2, \\ q_-(x) &= x_{21} = -x_1 + ix_2, \\ q_0(x) &= x_{11} - x_{22} = 2ix_3 \end{aligned}$$

also vanish at the identity element of $\mathrm{SU}(2)$.

A function $u \in C^\infty(\mathbb{S}^3)$ can be extended to $f \in C^\infty(U) = C^\infty(\mathbb{R}^4 \setminus \{0\})$ by

$$f(\vec{x}) := u(\vec{x}/\|\vec{x}\|).$$

Therefore, we obtain $P_N u \in C^\infty(\mathbb{S}^3)$,

$$P_N u(\vec{x}) := \sum_{|\alpha| \leq N} \frac{1}{\alpha!} \, (\vec{x} - \vec{e})^\alpha \, \partial_x^\alpha f(\vec{e}),$$

where $\vec{e} = (1, 0, 0, 0)$. Expressing this in terms of $x \in \mathrm{SU}(2)$, we obtain Taylor polynomials for $x \in \mathrm{SU}(2)$ in the form

$$P_N u(x) = \sum_{|\alpha| \leq N} \frac{1}{\alpha!} \, (x - e)^\alpha \, \partial_x^{(\alpha)} u(e),$$

where we write $\partial_x^{(\alpha)} u(e) := \partial_x^\alpha f(\vec{e})$, and where

$$
\begin{aligned}
(x - e)^\alpha &:= (\vec{x} - \vec{e})^\alpha \\
&= (x_0 - 1)^{\alpha_1} x_1^{\alpha_2} x_2^{\alpha_3} x_3^{\alpha_4} \\
&= \left(\frac{x_{11} + x_{22}}{2} - 1 \right)^{\alpha_1} \left(\frac{x_{12} - x_{21}}{2} \right)^{\alpha_2} \left(\frac{x_{12} + x_{21}}{2i} \right)^{\alpha_3} \left(\frac{x_{11} - x_{22}}{2i} \right)^{\alpha_4}.
\end{aligned}
$$

This gives an example of possible Taylor monomials on $\mathrm{SU}(2)$.

10.7 Symbolic calculus

In this section, we study global symbols of pseudo-differential operators on compact Lie groups, as defined in Definition 10.4.3. We also derive elements of the calculus in quite general classes of symbols. For this, we first introduce difference operators acting on symbols in the ξ-variable. These are analogues of the ∂_ξ-derivatives in $\mathbb{R}^n$ and of the difference operators $\triangle_\xi$ on $\mathbb{T}^n$, and are obtained by the multiplication by "coordinate functions" on the Fourier transform side.

10.7.1 Difference operators

As explained in Section 10.6, smooth functions on a group G can be approximated by Taylor polynomial type expansions. More precisely, there exist partial differential operators $\partial_x^{(\alpha)}$ of order $|\alpha|$ on G such that for every $u \in C^\infty(G)$ we have

$$
\begin{aligned}
u(x) &= \sum_{|\alpha| \leq N} \frac{1}{\alpha!} \, q_\alpha(x^{-1}) \, \partial_x^{(\alpha)} u(e) + \sum_{|\alpha| = N+1} q_\alpha(x^{-1}) \, u_\alpha(x) \\
&\sim \sum_{\alpha \geq 0} \frac{1}{\alpha!} q_\alpha(x^{-1}) \, \partial_x^{(\alpha)} u(e)
\end{aligned}
\tag{10.26}
$$

in a neighbourhood of $e \in G$, where $u_\alpha \in C^\infty(G)$, and $q_\alpha \in C^\infty(G)$ satisfy $q_{\alpha+\beta} = q_\alpha q_\beta$, and $\partial_x^{(\alpha)}$ are as in (10.25). Moreover, here $q_0 \equiv 1$, and $q_\alpha(e) = 0$ if $|\alpha| \geq 1$.

Definition 10.7.1 (Difference operators $\triangle_\xi^\alpha$). Let us define *difference operators* $\triangle_\xi^\alpha$ acting on Fourier coefficients by $\triangle_\xi^\alpha \widehat{f}(\xi) := \widehat{q_\alpha f}(\xi)$. Notice that $\triangle_\xi^{\alpha+\beta} = \triangle_\xi^\alpha \triangle_\xi^\beta$.

Remark 10.7.2. The technical choice of writing $q_\alpha(x^{-1})$ in (10.26) is dictated by our desire to make the asymptotic formulae in Theorems 10.7.8 and 10.7.10 look similar to the familiar Euclidean formulae in $\mathbb{R}^n$, and by an obvious freedom in selecting among different forms of Taylor polynomials q_α, see Remark 10.6.1. For example, on SU(2), if we work with operators $\triangle_+, \triangle_-, \triangle_0$ defined in (12.14)–(12.16), we can choose the form of the Taylor expansion (10.26) adapted to functions q_+, q_-, q_0. On SU(2), we can observe that $q_+(x^{-1}) = -q_-(x)$, $q_-(x^{-1}) = -q_+(x)$, $q_0(x^{-1}) = -q_0(x)$, so that for $|\alpha| = 1$ the functions $q_\alpha(x)$ and $q_\alpha(x^{-1})$ are linear combinations of q_+, q_-, q_0. In terms of the quaternionic identification, these are functions from Remark 10.6.2. Taylor monomials $(x - e)^\alpha$ from the previous section, when restricted to SU(2), can be expressed in terms of functions q_+, q_-, q_0. For an argument of this type we refer to the proof of Lemma 12.4.5.

Remark 10.7.3 **(Differences reduce the order of symbols).** In Theorem 12.3.6 we will apply the differences on the symbols of specific differential operators on SU(2). In general, on a compact Lie group G, a difference operator of order $|\gamma|$ applied to a symbol of a *partial differential operator* of order N gives a symbol of order $N - |\gamma|$. More precisely:

Proposition 10.7.4 (Differences for symbols of differential operators). *Let*

$$D = \sum_{|\alpha| \leq N} c_\alpha(x)\, \partial_x^\alpha \tag{10.27}$$

be a partial differential operator with coefficients $c_\alpha \in C^\infty(G)$, and ∂_x^α as in Definition 10.5.3. For $q \in C^\infty(G)$ such that $q(e) = 0$, we define difference operator $\triangle_q$ acting on symbols by

$$\triangle_q \widehat{f}(\xi) := \widehat{qf}(\xi).$$

Then we obtain

$$\triangle_q \sigma_D(x,\xi) = \sum_{|\alpha| \leq N} c_\alpha(x) \sum_{\beta \leq \alpha} \binom{\alpha}{\beta} (-1)^{|\beta|} \, (\partial_x^\beta q)(e)\, \sigma_{\partial_x^{\alpha-\beta}}(x,\xi), \tag{10.28}$$

which is a symbol of a partial differential operator of order at most $N - 1$. More precisely, if q has a zero of order M at $e \in G$ then $\mathrm{Op}(\triangle_q \sigma_D)$ is of order $N - M$.

Proof. Let D in (10.27) be be a partial differential operator, where $c_\alpha \in C^\infty(G)$ and $\partial_x^\alpha : \mathcal{D}(G) \to \mathcal{D}(G)$ is left-invariant of order $|\alpha|$. If $|\alpha| = 1$ and $\phi, \psi \in C^\infty(G)$ then we have the Leibniz property

$$\partial_x^\alpha(\phi\psi) = \phi \, (\partial_x^\alpha \psi) + (\partial_x^\alpha \phi) \, \psi,$$

leading to

$$0 = \int_G \phi(x) \, \partial_x^\alpha \psi(x) \, dx + \int_G \partial_x^\alpha \phi(x) \, \psi(x) \, dx.$$

More generally, for $|\alpha| \in \mathbb{N}_0$, $\phi \in C^\infty(G)$ and $f \in \mathcal{D}'(G)$, we have for the distributional derivatives

$$\int_G \phi(x) \, \partial_x^\alpha f(x) \, dx = (-1)^{|\alpha|} \int_G \partial_x^\alpha \phi(x) \, f(x) \, dx,$$

with a standard distributional interpretation. Recall that the right-convolution kernel $R_A \in \mathcal{D}'(G \times G)$ of a continuous linear operator $A : \mathcal{D}(G) \to \mathcal{D}(G)$ satisfies

$$A\phi(x) = \int_G \phi(y) \, R_A(x, y^{-1}x) \, dy.$$

For instance, informally

$$\phi(x) = \int_G \phi(y) \, \delta_e(y^{-1}x) \, dy = \int_G \phi(y) \, \delta_x(y) \, dy,$$

where $\delta_p \in \mathcal{D}'(G)$ is the Dirac delta distribution at $p \in G$. Notice that

$$\begin{aligned}
\partial_x^\alpha \phi(x) &= \int_G (-1)^{|\alpha|} \, (\partial_y^\alpha \phi)(xy^{-1}) \, \delta_e(y) \, dy \\
&= \int_G \phi(xy^{-1}) \, \partial_y^\alpha \delta_e(y) \, dy.
\end{aligned}$$

The right-convolution kernel of the operator D from (10.27) is given by

$$R_D(x, y) = \sum_{|\alpha| \le N} c_\alpha(x) \, \partial_y^\alpha \delta_e(y).$$

Let $D_q : C^\infty(G) \to C^\infty(G)$ be defined by

$$\sigma_{D_q}(x, \xi) := \triangle_q \sigma_D(x, \xi),$$

i.e.,

$$R_{D_q}(x, y) = q(y) \, R_D(x, y).$$

Then $D_q = \mathrm{Op}(\sigma_{D_q})$ is a differential operator:

$$\begin{aligned}
D_q \phi(x) &= \int_G \phi(xy^{-1}) \, q(y) \sum_{|\alpha| \le N} c_\alpha(x) \, \partial_y^\alpha \delta_e(y) \, dy \\
&= \sum_{|\alpha| \le N} (-1)^{|\alpha|} \, c_\alpha(x) \int_G \partial_y^\alpha \left(\phi(xy^{-1}) \, q(y) \right) \, \delta_e(y) \, dy \\
&= \sum_{|\alpha| \le N} (-1)^{|\alpha|} \, c_\alpha(x) \sum_{\beta \le \alpha} \binom{\alpha}{\beta} (-1)^{|\alpha - \beta|} \, (\partial_x^\beta q)(e) \, \partial_x^{\alpha - \beta} \phi(x).
\end{aligned}$$

Thus

$$\triangle_q \sigma_D(x, \xi) = \sum_{|\alpha| \leq N} c_\alpha(x) \sum_{\beta \leq \alpha} \binom{\alpha}{\beta} (-1)^{|\beta|} (\partial_x^\beta q)(e) \ \sigma_{\partial_x^{\alpha-\beta}}(x, \xi).$$

Hence if q has a zero of order M at $e \in G$ then D_q is of order $N - M$. $\qquad \square$

Exercise 10.7.5. Provide the distributional interpretation of all the steps in the proof of Proposition 10.7.4.

10.7.2 Commutator characterisation

Definition 10.7.6 (Operator classes $\mathcal{A}_k^m(M)$). For a compact closed manifold M, let $\mathcal{A}_0^m(M)$ denote the set of those continuous linear operators $A : C^\infty(M) \to C^\infty(M)$ which are bounded from $H^m(M)$ to $L^2(M)$. Recursively define $\mathcal{A}_{k+1}^m(M) \subset \mathcal{A}_k^m(M)$ such that $A \in \mathcal{A}_k^m(M)$ belongs to $\mathcal{A}_{k+1}^m(M)$ if and only if $[A, D] = AD - DA \in \mathcal{A}_k^m(M)$ for every smooth vector field D on M.

We now recall a variant of the commutator characterisation of pseudo-differential operators given in Theorem 5.3.1 which assures that the behaviour of commutators in Sobolev spaces characterises pseudo-differential operators:

Theorem 10.7.7. *A continuous linear operator $A : C^\infty(M) \to C^\infty(M)$ belongs to $\Psi^m(M)$ if and only if $A \in \bigcap_{k=0}^\infty \mathcal{A}_k^m(M)$.*

We note that in such a characterisation on a compact Lie group $M = G$, it suffices to consider vector fields of the form $D = M_\phi \partial_x$, where $M_\phi f := \phi f$ is multiplication by $\phi \in C^\infty(G)$, and ∂_x is left-invariant. Notice that

$$[A, M_\phi \partial_x] = M_\phi [A, \partial_x] + [A, M_\phi] \partial_x,$$

where $[A, M_\phi]f = A(\phi f) - \phi A f$. Hence we need to consider compositions $M_\phi A$, $A M_\phi$, $A \circ \partial_x$ and $\partial_x \circ A$. First, we observe that

$$\sigma_{M_\phi A}(x, \xi) = \phi(x) \sigma_A(x, \xi), \tag{10.29}$$
$$\sigma_{A \circ \partial_x}(x, \xi) = \sigma_A(x, \xi) \sigma_{\partial_x}(x, \xi), \tag{10.30}$$
$$\sigma_{\partial_x \circ A}(x, \xi) = \sigma_{\partial_x}(x, \xi) \sigma_A(x, \xi) + (\partial_x \sigma_A)(x, \xi), \tag{10.31}$$

where $\sigma_{\partial_x}(x, \xi)$ is independent of $x \in G$. Here (10.29) and (10.30) are straightforward and (10.31) follows by the Leibniz formula:

$$\begin{aligned}
\partial_x \circ A f(x) &= \partial_x \sum_{[\xi] \in \widehat{G}} \dim(\xi) \operatorname{Tr}\left(\xi(x) \sigma_A(x, \xi) \widehat{f}(\xi) \right) \\
&= \sum_{[\xi] \in \widehat{G}} \dim(\xi) \operatorname{Tr}\left((\partial_x \xi)(x) \sigma_A(x, \xi) \widehat{f}(\xi) \right) \\
&\quad + \sum_{[\xi] \in \widehat{G}} \dim(\xi) \operatorname{Tr}\left(\xi(x) \partial_x \sigma_A(x, \xi) \widehat{f}(\xi) \right),
\end{aligned}$$

where we used that $\sigma_{\partial_x}(x, \xi) = \xi(x)^*(\partial_x \xi)(x)$ by Theorem 10.4.6 to obtain (10.31). Next we claim that we have the formula

$$\sigma_{AM_\phi}(x, \xi) \sim \sum_{\alpha \geq 0} \frac{1}{\alpha!} \triangle_\xi^\alpha \sigma_A(x, \xi) \, \partial_x^{(\alpha)} \phi(x),$$

where $\partial_x^{(\alpha)}$ are certain partial differential operators of order $|\alpha|$. This follows from the general composition formula in Theorem 10.7.8.

10.7.3 Calculus

Here we discuss elements of the symbolic calculus of operators. First we recall the fundamental quantity $\langle \xi \rangle$ from Definition 10.3.18 that will allow us to introduce the orders of operators. We note that this scale $\langle \xi \rangle$ on G is determined by the eigenvalues of the Laplace operator $\mathcal{L}$ on G. We now formulate the result on compositions:

Theorem 10.7.8 (Composition formula I). *Let $m_1, m_2 \in \mathbb{R}$ and $\rho > \delta \geq 0$. Let $A, B : C^\infty(G) \to C^\infty(G)$ be continuous and linear, their symbols satisfy*

$$\left\| \triangle_\xi^\alpha \sigma_A(x, \xi) \right\|_{op} \leq C_\alpha \, \langle \xi \rangle^{m_1 - \rho|\alpha|},$$
$$\left\| \partial_x^\beta \sigma_B(x, \xi) \right\|_{op} \leq C_\beta \, \langle \xi \rangle^{m_2 + \delta|\beta|},$$

for all multi-indices α and β, uniformly in $x \in G$ and $[\xi] \in \widehat{G}$. Then

$$\sigma_{AB}(x, \xi) \sim \sum_{\alpha \geq 0} \frac{1}{\alpha!} (\triangle_\xi^\alpha \sigma_A)(x, \xi) \, \partial_x^{(\alpha)} \sigma_B(x, \xi), \tag{10.32}$$

where the asymptotic expansion means that for every $N \in \mathbb{N}$ we have

$$\left\| \sigma_{AB}(x, \xi) - \sum_{|\alpha| < N} \frac{1}{\alpha!} (\triangle_\xi^\alpha \sigma_A)(x, \xi) \, \partial_x^{(\alpha)} \sigma_B(x, \xi) \right\|_{op} \leq C_N \langle \xi \rangle^{m_1 + m_2 - (\rho - \delta)N}.$$

Proof. First,

$$\begin{aligned}
ABf(x) &= \int_G (Bf)(xz) \, R_A(x, z^{-1}) \, dz \\
&= \int_G \int_G f(xy^{-1}) \, R_B(xz, yz) \, dy \, R_A(x, z^{-1}) \, dz,
\end{aligned}$$

where we use the standard distributional interpretation of integrals. Hence

$$
\begin{aligned}
& \sigma_{AB}(x,\xi) \\
={}& \int_G R_{AB}(x,y)\,\xi(y)^*\,dy \\
={}& \int_G \int_G R_A(x,z^{-1})\,\xi(z^{-1})^*\,R_B(xz,yz)\,\xi(yz)^*\,dz\,dy \\
={}& \sum_{|\alpha|<N} \frac{1}{\alpha!} \int_G \int_G R_A(x,z^{-1})\,q_\alpha(z^{-1})\,\xi(z^{-1})^* \\
& \qquad\qquad \times \partial_x^{(\alpha)} R_B(x,yz)\,\xi(yz)^*\,dz\,dy \\
& + \sum_{|\alpha|=N} \int_G \int_G R_A(x,z^{-1})\,q_\alpha(z^{-1})\,\xi(z^{-1})^*\,u_\alpha(x,yz)\,\xi^*(yz)\,dz\,dy \\
={}& \sum_{|\alpha|<N} \frac{1}{\alpha!}\,(\triangle_\xi^\alpha \sigma_A)(x,\xi)\,\partial_x^{(\alpha)}\sigma_B(x,\xi) + \sum_{|\alpha|=N} (\triangle_\xi^\alpha \sigma_A)(x,\xi)\,\widehat{u_\alpha}(x,\xi).
\end{aligned}
$$

Now the statement follows because we have

$$
\|\widehat{u_\alpha}(x,\xi)\|_{op} \le C\langle\xi\rangle^{m_1+\delta N}
$$

since $u_\alpha(x,y)$ is the remainder in the Taylor expansion of $R_B(x,y)$ in x only and so it satisfies similar estimates to those of σ_B with respect to ξ. This completes the proof. $\qquad\square$

A similar proof yields another version of the composition formula:

Theorem 10.7.9 (Composition formula II). *Let $m_1, m_2 \in \mathbb{R}$ and $\rho > \delta \ge 0$. Let $A, B : C^\infty(G) \to C^\infty(G)$ be continuous and linear, their symbols satisfying*

$$
\begin{aligned}
\left\| \partial_x^\beta \triangle_\xi^\alpha \sigma_A(x,\xi) \right\|_{op} &\le C_\alpha\,\langle\xi\rangle^{m_1-\rho|\alpha|+\delta|\beta|}, \\
\left\| \partial_x^\beta \triangle_\xi^\alpha \sigma_B(x,\xi) \right\|_{op} &\le C_\beta\,\langle\xi\rangle^{m_2-\rho|\alpha|+\delta|\beta|},
\end{aligned}
$$

for all multi-indices α and β, uniformly in $x \in G$ and $[\xi] \in \widehat{G}$. Then

$$
\sigma_{AB}(x,\xi) \sim \sum_{\alpha \ge 0} \frac{1}{\alpha!}\,(\triangle_\xi^\alpha \sigma_A)(x,\xi)\,\partial_x^{(\alpha)}\sigma_B(x,\xi),
$$

where the asymptotic expansion means that for every $N \in \mathbb{N}$ we have

$$
\left\| \triangle_\xi^\gamma \partial_x^\beta \left(\sigma_{AB}(x,\xi) - \sum_{|\alpha|<N} \frac{1}{\alpha!}\,(\triangle_\xi^\alpha \sigma_A)(x,\xi)\,\partial_x^{(\alpha)}\sigma_B(x,\xi) \right) \right\|_{op}
$$
$$
\le C_N \langle\xi\rangle^{m_1+m_2-(\rho-\delta)N-\rho|\gamma|+\delta|\beta|}.
$$

Let us complement Theorems 10.7.8 and 10.7.9 with a formula for the adjoint operator:

Theorem 10.7.10 (Adjoint). *Let* $m \in \mathbb{R}$ *and* $\rho > \delta \geq 0$. *Let* $A : C^\infty(G) \to C^\infty(G)$ *be continuous and linear, with symbol* σ_A *satisfying*

$$\left\| \triangle_\xi^\alpha \partial_x^\beta \sigma_A(x, \xi) \right\|_{op} \leq C_\alpha \, \langle \xi \rangle^{m - \rho|\alpha| + \delta|\beta|},$$

for all multi-indices α, *uniformly in* $x \in G$ *and* $[\xi] \in \widehat{G}$. *Then the symbol of* A^* *is*

$$\sigma_{A^*}(x, \xi) \sim \sum_{\alpha \geq 0} \frac{1}{\alpha!} \, \triangle_\xi^\alpha \partial_x^{(\alpha)} \sigma_A(x, \xi)^*,$$

where the asymptotic expansion means that for every $N \in \mathbb{N}$ *we have*

$$\left\| \triangle_\xi^\gamma \partial_x^\beta \left(\sigma_A(x, \xi) - \sum_{|\alpha| < N} \frac{1}{\alpha!} \, \triangle_\xi^\alpha \partial_x^{(\alpha)} \sigma_A(x, \xi)^* \right) \right\|_{op} \leq C_N \langle \xi \rangle^{m - (\rho - \delta)N - \rho|\gamma| + \delta|\beta|}.$$

Proof. First we observe that writing

$$A^* g(y) = \int_G g(x) \, R_{A^*}(y, x^{-1}y) \, dx,$$

we get the relation

$$R_{A^*}(y, x^{-1}y) = \overline{R_A(x, y^{-1}x)}$$

between kernels, which means that

$$R_{A^*}(x, v) = \overline{R_A(xv^{-1}, v^{-1})}.$$

From this we find

$$
\begin{aligned}
\sigma_{A^*}(x, \xi) &= \int_G R_{A^*}(x, v) \, \xi(v)^* \, dv \\
&= \int_G \overline{R_A(xv^{-1}, v^{-1})} \, \xi(v)^* \, dv \\
&= \sum_{|\alpha| < N} \frac{1}{\alpha!} \int_G q_\alpha(v) \, \partial_x^{(\alpha)} \overline{R_A(x, v^{-1})} \, \xi(v)^* \, dv + \mathcal{R}_N(x, \xi) \\
&= \sum_{|\alpha| < N} \frac{1}{\alpha!} \, \triangle_\xi^\alpha \partial_x^{(\alpha)} \sigma_A(x, \xi)^* + \mathcal{R}_N(x, \xi),
\end{aligned}
$$

where the last formula for the asymptotic expansion follows in view of

$$\sigma_A(x, \xi)^* = \left(\int_G R_A(x, v) \, \xi^*(v) \, dv \right)^* = \int_G \overline{R_A(x, v^{-1})} \, \xi^*(v) \, dv,$$

and estimate for the remainder $\mathcal{R}_N(x, \xi)$ follows by an argument similar to that in the proof of Theorem 10.7.8. $\qquad \square$

Above we considered the symbol of the adjoint. Since $K_{A^t}(x,y) = \overline{K_{A^*}(x,y)}$, Theorem 10.7.10 provides a follow-up:

Corollary 10.7.11 (Transpose). *Let $A : C^\infty(G) \to C^\infty(G)$ be as in Theorem 10.7.10. Then the symbol of the transpose A^t is*

$$\sigma_{A^t}(x,\xi) \sim \sum_{\alpha \geq 0} \frac{1}{\alpha!} \triangle_\xi^\alpha \partial_x^{(\alpha)} \sigma_{\overline{A}}(x,\xi)^*,$$

where the right-convolution kernel of $\overline{A} : C^\infty(G) \to C^\infty(G)$ is defined by $R_{\overline{A}}(x,y) := \overline{R_A(x,y)}$, and where the asymptotic expansion is interpreted as in Theorem 10.7.10.   □

We postpone the discussion of the asymptotic expansion of the parametrix for elliptic operators in Theorem 10.9.10 to after we introduce symbol classes.

10.7.4   Leibniz formula

For studying products of pseudo-differential symbols, it would be beneficial to have a Leibniz-like formula for the "derivatives with respect to the dual variable $\xi \in \widehat{G}$". Classically for smooth functions $\sigma_A, \sigma_B : \mathbb{R} \to \mathbb{C}$, the Leibniz formula is the familiar

$$(\sigma_A \sigma_B)'(\xi) = \sigma_A'(\xi)\, \sigma_B(\xi) + \sigma_A(\xi)\, \sigma_B'(\xi).$$

In the context of pseudo-differential calculus on the torus $\mathbb{T}$, we have $\widehat{\mathbb{T}} \cong \mathbb{Z}$; for functions $\sigma_A, \sigma_B : \mathbb{Z} \to \mathbb{C}$ a useful Leibniz-like formula reads

$$\triangle_\xi(\sigma_A \sigma_B) = (\triangle_\xi \sigma_A)\,\sigma_B + \sigma_A\,(\triangle_\xi \sigma_B) + (\triangle_\xi \sigma_A)\,(\triangle_\xi \sigma_B),$$

where the difference operator is defined by $\triangle_\xi \sigma(\xi) = \sigma(\xi+1) - \sigma(\xi)$.

Let G be a compact Lie group. Given $q \in C^\infty(G)$ and $f \in \mathcal{D}'(G)$, let

$$\triangle_q \widehat{f}(\xi) := \widehat{qf}(\xi). \tag{10.33}$$

Let $\sigma_A(\xi) = \widehat{a}(\xi)$ and $\sigma_B(\xi) = \widehat{b}(\xi)$. Then

$$
\begin{aligned}
\triangle_q(\sigma_A \sigma_B)(\xi) &= \widehat{q(b * a)}(\xi) \\
&= \widehat{b * (qa)}(\xi) + \widehat{Rb}(\xi) \\
&= (\triangle_q \sigma_A(\xi))\,\sigma_B(\xi) + \widehat{Rb}(\xi),
\end{aligned}
$$

where the remainder operator $R : \mathcal{D}'(G) \to \mathcal{D}'(G)$ is given by

$$Rb = q(b * a) - b * (qa),$$

that is

$$Rb(x) = \int_G b(y)\,\left(q(x) - q(y^{-1}x) \right) a(y^{-1}x)\; \mathrm{d}y.$$

Thus by the Taylor expansion from Section 10.6 we have

$$Rb(x) \sim \int_G b(y) \left(\sum_{|\alpha| \geq 0} c(q, \alpha)(x) \, y^{(\alpha)} \right) a(y^{-1}x) \, dy,$$

where

$$c(q, \alpha)(x) := \frac{1}{\alpha!} \, \partial_z^{(\alpha)} \left(q(x) - q(z^{-1}x) \right) \Big|_{z=e}.$$

As usual, according to (10.33), the difference $\triangle_{c(q,\alpha)}$ is

$$\triangle_{c(q,\alpha)} \widehat{f}(\xi) = \widehat{c(q, \alpha)f}(\xi). \tag{10.34}$$

Notice that $c(q, 0)(x) = q(x) - q(x) = 0$, so that we get an asymptotic Leibniz formula:

Theorem 10.7.12 (Asymptotic Leibniz formula). *For symbols σ_A and σ_B, we have*

$$\triangle_q(\sigma_A \sigma_B)(\xi) \sim (\triangle_q \sigma_A(\xi)) \, \sigma_B(\xi) + \sum_{|\alpha| > 0} \triangle_{c(q,\alpha)} \left(\sigma_A(\xi) \, (\triangle_\xi^\alpha \sigma_B(\xi)) \right).$$

10.8 Boundedness on Sobolev spaces $H^s(G)$

In this section we show conditions on the symbol for operators to be bounded on Sobolev spaces $H^s(G)$. The Sobolev space $H^s(G)$ of order $s \in \mathbb{R}$ can be defined via a smooth partition of unity of the closed manifold G, and there are other definitions as well, in particular in terms of the Laplace operator on G, see Definition 10.3.9. We also recall Definition 10.3.18 of the quantity $\langle \xi \rangle$ on the Lie group G, measuring the order of operators compared to that of the Laplacian $\mathcal{L}$: if $-\mathcal{L}\xi^{vw}(x) = \lambda_{[\xi]}\xi^{vw}(x)$, then

$$\langle \xi \rangle := (1 + \lambda_{[\xi]})^{1/2}.$$

according to (10.9). Now we can formulate the result on the Sobolev space boundedness:

Theorem 10.8.1 (Boundedness of operators on Sobolev spaces). *Let G be a compact Lie group. Let A be a continuous linear operator from $C^\infty(G)$ to $C^\infty(G)$ and let σ_A be its symbol. Assume that there are constants $\mu, C_\alpha \in \mathbb{R}$ such that*

$$\|\partial_x^\alpha \sigma_A(x, \xi)\|_{op} \leq C_\alpha \, \langle \xi \rangle^\mu$$

holds for all $x \in G$, $\xi \in \mathrm{Rep}(G)$ and all multi-indices α. Then A extends to a bounded operator from $H^s(G)$ to $H^{s-\mu}(G)$, for all $s \in \mathbb{R}$.

Remark 10.8.2. Notice that we may easily show a special case of this result with $s = \mu$. Namely, if σ_A is as in Theorem 10.8.1, then

$$\left\| \partial_x^\alpha \left(\sigma_A(x, \xi) \langle \xi \rangle^{-\mu} \right) \right\|_{op} \leq C_\alpha$$

for every multi-index α. Here $\sigma_A(x, \xi) \langle \xi \rangle^{-\mu} = \sigma_{A \circ \Xi^{-\mu}}(x, \xi)$, and thus Theorem 10.5.5 implies that $A \circ \Xi^{-\mu}$ is bounded on $L^2(G)$, so that $A \in \mathcal{L}(H^\mu(G), L^2(G))$.

Proof of Theorem 10.8.1. Observing the continuous mapping $\Xi^s : H^s(G) \to L^2(G)$, we have to prove that operator $\Xi^{s-\mu} \circ A \circ \Xi^{-s}$ is bounded from $L^2(G)$ to $L^2(G)$. Let us denote $B = A \circ \Xi^{-s}$, so that the symbol of B satisfies $\sigma_B(x, \xi) = \langle \xi \rangle^{-s} \sigma_A(x, \xi)$ for all $x \in G$ and $\xi \in \mathrm{Rep}(G)$. Since $\Xi^{s-\mu} \in \Psi^{s-\mu}(G)$, by (10.24) and Lemma 10.9.3 its symbol satisfies

$$\| \triangle_\xi^\alpha \sigma_{\Xi^{s-\mu}}(x, \xi) \|_{op} \leq C_\alpha' \, \langle \xi \rangle^{s-\mu-|\alpha|}. \tag{10.35}$$

Now we can observe that the asymptotic formula in Theorem 10.7.8 works for the composition $\Xi^{s-\mu} \circ B$ in view of (10.35), and we obtain

$$\partial_x^\beta \sigma_{\Xi^{s-\mu} \circ B}(x, \xi) \quad \sim \quad \sum_{\alpha \geq 0} \frac{1}{\alpha!} \left(\triangle_\xi^\alpha \sigma_{\Xi^{s-\mu}}(x, \xi) \right) \, \langle \xi \rangle^{-s} \, \partial_x^{(\alpha)} \partial_x^\beta \sigma_A(x, \xi).$$

It follows that

$$\left\| \partial_x^\beta \sigma_{\Xi^{s-\mu} \circ B}(x, \xi) \right\|_{op} \leq C_\beta'',$$

so that $\Xi^{s-\mu} \circ B$ is bounded on $L^2(G)$ by Theorem 10.5.5. This completes the proof. $\qquad\square$

Remark 10.8.3 (**Functional analytic argument**). The boundedness of operators on the Sobolev spaces follows also by a purely functional analytic argument if G is compact. Indeed, because the space $C^\infty(G)$ is nuclear (by adopting Exercise B.3.51), the tensor product space $C^\infty(G) \otimes_\pi \ell^\infty(\widehat{G})$ can be endowed with the Fréchet topology given by the seminorms $\sup_{x \in G, [\xi] \in \widehat{G}} |\partial_x^\alpha a(x, \xi)|$. Hence, as soon as the operator satisfies $\sup_{x \in G, [\xi] \in \widehat{G}} |\partial_x^\alpha a(x, \xi)| < \infty$ for finitely many α, it is bounded on the corresponding space $H^s(G)$.

Exercise 10.8.4. Work out the details of this argument.

10.9   Symbol classes on compact Lie groups

The goal of this section is to describe the pseudo-differential symbol inequalities on compact Lie groups that yield Hörmander's classes $\Psi^m(G)$. On the way to characterising the usual Hörmander classes $\Psi^m(G)$ in Theorem 10.9.6, we need some properties concerning symbols of pseudo-differential operators that we will establish in the next sections.

10.9.1 Some properties of symbols of $\Psi^m(G)$

Given an operator from Hörmander's class $\Psi^m(G)$, we can derive some information about its full symbol as defined in Definition 10.4.3. A precise characterisation of the class $\Psi^m(G)$ will be given in Theorem 10.9.6.

Lemma 10.9.1. *Let $A \in \Psi^m(G)$. Then there exists a constant $C < \infty$ such that*

$$\|\sigma_A(x,\xi)\|_{op} \leq C\langle \xi \rangle^m$$

for all $x \in G$ and $\xi \in \mathrm{Rep}(G)$. Also, if $u \in G$ and if B is an operator with symbol $\sigma_B(x,\xi) = \sigma_A(u,\xi)$, then $B \in \Psi^m(G)$.

Proof. First, $B \in \Psi^m(G)$ follows from the local theory of pseudo-differential operators, by studying

$$Bf(x) = \int_G K_A(u, ux^{-1}y) \, f(y) \, \mathrm{d}y.$$

Hence the right-convolution operator B is bounded from $H^s(G)$ to $H^{s-m}(G)$, implying $\|\sigma_A(u,\xi)\| \leq C\langle \xi \rangle^m$. $\qquad \square$

Exercise 10.9.2. Provide the details for the proof of Lemma 10.9.1.

Lemma 10.9.3. *Let $A \in \Psi^m(G)$. Then $\mathrm{Op}(\triangle_\xi^\alpha \partial_x^\beta \sigma_A) \in \Psi^{m-|\alpha|}(G)$ for all α, β.*

Proof. First, given $A \in \Psi^m(G)$, let us define $\sigma_B(x,\xi) = \triangle_\xi^\alpha \partial_x^\beta \sigma_A(x,\xi)$. We must show that $B \in \Psi^{m-|\alpha|}(G)$. If here $|\beta| = 0$, we obtain

$$Bf(x) = \int_G f(xy^{-1}) \, q_\alpha(y) \, R_A(x,y) \, \mathrm{d}y = \int_G q_\alpha(y^{-1}x) \, K_A(x,y) \, f(y) \, \mathrm{d}y.$$

Moving to local coordinates, we need to study

$$\tilde{B}f(x) = \int_{\mathbb{R}^n} \phi(x,y) \, K_{\tilde{A}}(x,y) \, f(y) \, \mathrm{d}y,$$

where $\tilde{A} \in \Psi^m(\mathbb{R}^n \times \mathbb{R}^n)$ with $\phi \in C^\infty(\mathbb{R}^n \times \mathbb{R}^n)$, the kernel $K_{\tilde{A}}$ being compactly supported. Let us calculate the symbol of $\tilde{B}$:

$$
\begin{aligned}
\sigma_{\tilde{B}}(x,\xi) &= \int_{\mathbb{R}^n} e^{2\pi i(y-x)\cdot\xi} \, \phi(x,y) \, K_{\tilde{A}}(x,y) \, \mathrm{d}y \\
&\sim \sum_{\gamma \geq 0} \frac{1}{\gamma!} \, \partial_z^\gamma \phi(x,z)|_{z=x} \int_G e^{2\pi i(y-x)\cdot\xi} \, (y-x)^\gamma \, K_{\tilde{A}}(x,y) \, \mathrm{d}y \\
&= \sum_{\gamma \geq 0} \frac{1}{\gamma!} \, \partial_z^\gamma \phi(x,z)|_{y=x} \, D_\xi^\gamma \sigma_{\tilde{A}}(x,\xi).
\end{aligned}
$$

This shows that $\tilde{B} \in \Psi^m(\mathbb{R}^n \times \mathbb{R}^n)$. We obtain $\mathrm{Op}(\triangle_\xi^\alpha \sigma_A) \in \Psi^{m-|\alpha|}(G)$ if $A \in \Psi^m(G)$.

Next we show that $B = \mathrm{Op}(\partial_x^\beta \sigma_A) \in \Psi^m(G)$. We may assume that $|\beta| = 1$. Left-invariant vector field ∂_x^β is a linear combination of terms of the type $c(x)D_x$, where $c \in C^\infty(G)$ and D_x is right-invariant. By the previous considerations on $\tilde{B}$, we may remove $c(x)$ here, and consider only $C = \mathrm{Op}(D_x \sigma_A)$. Since $R_A(x, y) = K_A(x, xy^{-1})$, we get

$$D_x R_A(x, y) = (D_x + D_z)K_A(x, z)|_{z=xy^{-1}},$$

leading to

$$Cf(x) = \int_G f(xy^{-1}) \, D_x R_A(x, y) \, \mathrm{d}y = \int_G f(y) \, (D_x + D_y)K_A(x, y) \, \mathrm{d}y.$$

Thus, we study local operators of the form

$$\tilde{C}f(x) \quad = \quad \int_{\mathbb{R}^n} f(y) \left(\phi(x, y)\partial_x^\beta + \psi(x, y)\partial_y^\beta \right) K_{\tilde{A}}(x, y) \, \mathrm{d}y,$$

where the kernel of $\tilde{A} \in \Psi^m(\mathbb{R}^n \times \mathbb{R}^n)$ has compact support, $\phi, \psi \in C^\infty(\mathbb{R}^n \times \mathbb{R}^n)$, and $\phi(x, x) = \psi(x, x)$ for every $x \in \mathbb{R}^n$. Let $\tilde{C} = \tilde{D} + \tilde{E}$, where

$$\tilde{D}f(x) \quad = \quad \int_{\mathbb{R}^n} f(y) \, \phi(x, y) \left(\partial_x^\beta + \partial_y^\beta \right) K_{\tilde{A}}(x, y) \, \mathrm{d}y,$$

$$\tilde{E}f(x) \quad = \quad \int_{\mathbb{R}^n} f(y) \, (\psi(x, y) - \phi(x, y)) \, \partial_y^\beta K_{\tilde{A}}(x, y) \, \mathrm{d}y.$$

By the above considerations about $\tilde{B}$, we may assume that $\phi(x, y) \equiv 1$ here, and obtain $\sigma_{\tilde{D}}(x, \xi) = \partial_x^\beta \sigma_{\tilde{A}}(x, \xi)$. Thus $\tilde{D} \in \Psi^m(\mathbb{R}^n \times \mathbb{R}^n)$. Moreover,

$$\tilde{E}f(x) \sim \sum_{\gamma \geq 0} \frac{1}{\gamma!} \, \partial_z^\gamma \left(\psi(x, z) - \phi(x, z) \right)|_{z=x} \int_{\mathbb{R}^n} f(y) \, (y - x)^\gamma \, \partial_y^\beta K_{\tilde{A}}(x, y) \, \mathrm{d}y,$$

yielding

$$\sigma_{\tilde{E}}(x, \xi) \quad \sim \quad \sum_{\gamma \geq 0} c_\gamma(x) \, \partial_\xi^\gamma \left(\xi^\beta \, \sigma_{\tilde{A}}(x, \xi) \right)$$

for some functions $c_\gamma \in C^\infty(\mathbb{R}^n)$ for which $c_0(x) \equiv 0$. Since $|\beta| = 1$, this shows that $\tilde{E} \in \Psi^m(\mathbb{R}^n \times \mathbb{R}^n)$. Thus $\mathrm{Op}(\partial_x^\beta \sigma_A) \in \Psi^m(G)$ if $A \in \Psi^m(G)$. $\qquad \square$

Lemma 10.9.4. *Let $A \in \Psi^m(G)$ and let $D : C^\infty(G) \to C^\infty(G)$ be a smooth vector field. Then $\mathrm{Op}(\sigma_A \sigma_D) \in \Psi^{m+1}(G)$ and $\mathrm{Op}([\sigma_A, \sigma_D]) \in \Psi^m(G)$.*

Proof. For simplicity, we may assume that $D = M_\phi \partial_x$, where ∂_x is left-invariant and $\phi \in C^\infty(G)$. Now

$$\sigma_A(x, \xi) \, \sigma_D(x, \xi) = \phi(x) \, \sigma_A(x, \xi) \, \sigma_{\partial_x}(\xi) = \sigma_{M_\phi A \circ \partial_x}(x, \xi),$$

and it follows from the local theory that $M_\phi A \circ \partial_x \in \Psi^{m+1}(G)$. Thus $\mathrm{Op}(\sigma_A \sigma_D) \in \Psi^{m+1}(G)$. Next,

$$
\begin{aligned}
\sigma_D(x,\xi)\,\sigma_A(x,\xi) &= \phi(x)\,\sigma_{\partial_x}(\xi)\,\sigma_A(x,\xi)\\
&\overset{(10.31)}{=} \phi(x)\,(\sigma_{\partial_x \circ A}(x,\xi) - (\partial_x \sigma_A)(x,\xi))\\
&= \sigma_{M_\phi \circ \partial_x \circ A}(x,\xi) - \phi(x)\,(\partial_x \sigma_A)(x,\xi).
\end{aligned}
$$

From this we see that

$$
\begin{aligned}
\mathrm{Op}([\sigma_A,\sigma_D]) &= M_\phi A \circ \partial_x - M_\phi \partial_x \circ A + M_\phi\,\mathrm{Op}(\partial_x \sigma_A)\\
&= M_\phi[A,\partial_x] + M_\phi\,\mathrm{Op}(\partial_x \sigma_A).
\end{aligned}
$$

Here $\mathrm{Op}(\partial_x \sigma_A) \in \Psi^m(G)$ by Lemma 10.9.3. Hence $\mathrm{Op}([\sigma_A,\sigma_D])$ belongs to $\Psi^m(G)$ as a sum of operators from $\Psi^m(G)$. $\qquad\square$

10.9.2 Symbol classes $\Sigma^m(G)$

Combined with asymptotic expansion (10.32) for composing operators, commutator characterisation Theorem 10.7.7 motivates defining the following symbol classes

$$
\Sigma^m(G) = \bigcap_{k=0}^{\infty} \Sigma_k^m(G)
$$

that we will show to characterise Hörmander's classes $\Psi^m(G)$. The classes $\Sigma_k^m(G)$ are defined iteratively in the following way:

Definition 10.9.5 (Symbol classes $\Sigma^m(G)$). Let $m \in \mathbb{R}$. We denote $\sigma_A \in \Sigma_0^m(G)$ if

$$
\mathrm{sing\ supp}\,(y \mapsto R_A(x,y)) \subset \{e\} \tag{10.36}
$$

and if

$$
\|\triangle_\xi^\alpha \partial_x^\beta \sigma_A(x,\xi)\|_{op} \le C_{A\alpha\beta m}\,\langle\xi\rangle^{m-|\alpha|}, \tag{10.37}
$$

for all $x \in G$, all multi-indices α,β, and all $\xi \in \mathrm{Rep}(G)$, where $\langle\xi\rangle$ is defined in (10.9). Then we say that $\sigma_A \in \Sigma_{k+1}^m(G)$ if and only if

$$
\begin{aligned}
\sigma_A &\in \Sigma_k^m(G), & (10.38)\\
\sigma_{\partial_j}\sigma_A - \sigma_A \sigma_{\partial_j} &\in \Sigma_k^m(G), & (10.39)\\
(\triangle_\xi^\gamma \sigma_A)\,\sigma_{\partial_j} &\in \Sigma_k^{m+1-|\gamma|}(G), & (10.40)
\end{aligned}
$$

for all $|\gamma| > 0$ and $1 \le j \le \dim(G)$. Let

$$
\Sigma^m(G) := \bigcap_{k=0}^{\infty} \Sigma_k^m(G).
$$

We write $A \in \mathrm{Op}\,\Sigma^m(G)$ if and only if $\sigma_A \in \Sigma^m(G)$.

Theorem 10.9.6 (Equality $\operatorname{Op}\Sigma^m(G) = \Psi^m(G)$). *Let G be a compact Lie group and let $m \in \mathbb{R}$. Then $A \in \Psi^m(G)$ if and only if $\sigma_A \in \Sigma^m(G)$.*

Proof. First, applying Theorem 10.7.8 to $\sigma_A \in \Sigma^m_{k+1}(G)$, we notice that $[A, D] \in \operatorname{Op}\Sigma^m_k(G)$ for any smooth vector field $D : C^\infty(G) \to C^\infty(G)$. Consequently, if here $A \in \operatorname{Op}\Sigma^m(G)$ then also $[A, D] \in \operatorname{Op}\Sigma^m(G)$. By Remark 10.8.2 we obtain $\operatorname{Op}\Sigma^m(G) \subset \mathcal{L}(H^m(G), L^2(G))$. Consequently, Theorem 10.7.7 implies $\operatorname{Op}\Sigma^m(G) \subset \Psi^m(G)$.

Conversely, we have to show that $\Psi^m(G) \subset \operatorname{Op}\Sigma^m(G)$. This follows by Lemma 10.9.1, and Lemmas 10.9.3 and 10.9.4. More precisely, let $A \in \Psi^m(G)$. Then we have

$$
\begin{aligned}
\operatorname{Op}(\triangle^\alpha_\xi \partial^\beta_x \sigma_A) &\in \Psi^{m-|\alpha|}(G), \\
\operatorname{Op}([\sigma_{\partial_j}, \sigma_A]) &\in \Psi^m(G), \\
\operatorname{Op}\left((\triangle^\gamma_\xi \sigma_A)\sigma_{\partial_j}\right) &\in \Psi^{m+1-|\gamma|}(G).
\end{aligned}
$$

Moreover, $\|\sigma_A(x, \xi)\| \leq C\langle\xi\rangle^m$ by Lemma 10.9.1, and the singular support $y \mapsto R_A(x, y)$ is contained in $\{e\} \subset G$. This completes the proof. $\qquad\square$

Corollary 10.9.7. *The set $\Sigma^m(G)$ is invariant under x-freezings, x-translations and ξ-conjugations. More precisely, if $(x, \xi) \mapsto \sigma_A(x, \xi)$ belongs to $\Sigma^m(G)$ and $u \in G$ then also the following symbols belong to $\Sigma^m(G)$:*

$$
\begin{aligned}
(x, \xi) &\mapsto \sigma_A(u, \xi), &\qquad (10.41)\\
(x, \xi) &\mapsto \sigma_A(ux, \xi), &\qquad (10.42)\\
(x, \xi) &\mapsto \sigma_A(xu, \xi), &\qquad (10.43)\\
(x, \xi) &\mapsto \xi(u)^* \, \sigma_A(x, \xi) \, \xi(u). &\qquad (10.44)
\end{aligned}
$$

Proof. The symbol classes $\Sigma^m(G)$ are defined by conditions (10.36)–(10.40), which are checked for points $x \in G$ fixed (with constants uniform in x). Therefore it follows that $\Sigma^m(G)$ is invariant under the x-freezing (10.41), and under the left and right x-translations (10.42),(10.43). The x-freezing property (10.41) would have followed also from Lemma 10.9.1 and Theorem 10.9.6. From the general local theory of pseudo-differential operators it follows that $A \in \Psi^m(G)$ if and only if the ϕ-pullback A_ϕ belongs to the same class $\Psi^m(G)$, where $A_\phi f = A(f \circ \phi) \circ \phi^{-1}$. This, combined with the x-translation invariance and Proposition 10.4.18, implies the conjugation invariance in (10.44). $\qquad\square$

From Theorem 10.9.6 and Lemma 10.9.3 we also obtain:

Corollary 10.9.8. *If $\sigma_A \in \Sigma^m(G)$ then $\triangle^\alpha_\xi \partial^\beta_x \sigma_A \in \Sigma^{m-|\alpha|}(G)$.*

Finally, we formulate a simple relation between convergence of symbols and operators. It is a straightforward consequence of Theorems 10.5.5 and 10.8.1 for $L^2(G)$ and $H^s(G)$ cases, respectively. In Corollary 12.4.11 we will give an improvement of this result on the group $\mathrm{SU}(2)$.

Corollary 10.9.9 (Convergence of symbols and operators). *Let $\sigma \in \Sigma^0(G)$, and assume that a sequence $\sigma_k \in \Sigma^0(G)$ satisfies inequalities (10.37) uniformly in k. Let $N \in \mathbb{N}$ be such that $N > \dim(G)/2$, and assume that for all $|\beta| \le N$ we have the convergence*

$$\partial_x^\beta \sigma_k(x, \xi) \to \partial_x^\beta \sigma(x, \xi) \text{ as } k \to \infty \tag{10.45}$$

in the operator norm, uniformly over all $x \in G$ and $\xi \in \mathrm{Rep}(G)$. Then $\mathrm{Op}\,\sigma_k \to \mathrm{Op}\,\sigma$ strongly on $L^2(G)$.

Moreover, if the convergence (10.45) holds for all multi-indices β, then $\mathrm{Op}\,\sigma_k \to \mathrm{Op}\,\sigma$ strongly on $H^s(G)$ for any $s \in \mathbb{R}$.

From the asymptotic expansion for the composition of pseudo-differential operators in Section 10.7.3, we get an expansion for a parametrix of an elliptic operator:

Theorem 10.9.10 (Parametrix). *Let $\sigma_{A_j} \in \Sigma^{m-j}(G)$, and let*

$$\sigma_A(x, \xi) \sim \sum_{j=0}^\infty \sigma_{A_j}(x, \xi).$$

Assume that A is elliptic in the sense that $\sigma_{A_0}(x, \xi) = \sigma_{B_0}(x, \xi)^{-1}$ is an invertible matrix for every x and ξ, and that $B_0 = \mathrm{Op}(\sigma_{B_0}) \in \Psi^{-m}(G)$. Then there exists $\sigma_B \in \Sigma^{-m}(G)$ such that $I - BA$ and $I - AB$ are smoothing operators. Moreover,

$$\sigma_B(x, \xi) \sim \sum_{k=0}^\infty \sigma_{B_k}(x, \xi),$$

where the operators $B_k \in \Sigma^{-m-k}(G)$ are defined recursively by

$$\sigma_{B_N}(x, \xi) := -\sigma_{B_0}(x, \xi) \sum_{k=0}^{N-1} \sum_{j=0}^{N-k} \sum_{|\gamma|=N-j-k} \frac{1}{\gamma!} \left[\triangle_\xi^\gamma \sigma_{B_k}(x, \xi) \right] \partial_x^{(\gamma)} \sigma_{A_j}(x, \xi).$$

$$\tag{10.46}$$

Proof. If $\sigma_I \sim \sigma_{BA}$ holds for some $\sigma_B \sim \sum_{k=0}^\infty \sigma_{B_k}$, then by Theorem 10.7.8 we have

$$\begin{aligned}
I_{\dim(\xi)} = \sigma_I(x, \xi) \quad &\sim \quad \sigma_{BA}(x, \xi) \\
&\sim \quad \sum_{\gamma \ge 0} \frac{1}{\gamma!} \left[\triangle_\xi^\gamma \sigma_B(x, \xi) \right] \partial_x^{(\gamma)} \sigma_A(x, \xi) \\
&\sim \quad \sum_{\gamma \ge 0} \frac{1}{\gamma!} \left[\triangle_\xi^\gamma \sum_{k=0}^\infty \sigma_{B_k}(x, \xi) \right] \partial_x^{(\gamma)} \sum_{j=0}^\infty \sigma_{A_j}(x, \xi).
\end{aligned}$$

From this, we want to find σ_{B_k}. Now $I_{\dim(\xi)} = \sigma_{B_0}(x, \xi)\, \sigma_{A_0}(x, \xi)$, and for $|\gamma| \geq 1$ we can demand that

$$0 = \sum_{|\gamma|=N-j-k} \frac{1}{\gamma!} \left[\triangle_\xi^\gamma \sigma_{B_k}(x, \xi) \right] \partial_x^{(\gamma)} \sigma_{A_j}(x, \xi).$$

Then (10.46) provides the solution to these equations, and the reader may verify also that $\sigma_{B_N} \in \Sigma^{-m-N}(G)$. Thus $\sigma_B \sim \sum_{k=0}^{\infty} \sigma_{B_k}$. Finally, notice that $\sigma_I \sim \sigma_{BA}$. $\qquad\qquad\qquad\qquad\qquad\qquad\qquad\qquad\qquad\qquad\qquad\qquad\qquad\qquad\qquad\qquad\square$

Exercise 10.9.11. Show that $\sigma_{B_N} \in \Sigma^{-m-N}(G)$ in Theorem 10.9.10.

Exercise 10.9.12. In the notation of Theorem 10.9.10, let us write

$$\sigma_{C_0}(x, \xi) := \sigma_{B_0}(x, \xi) = \sigma_{A_0}(x, \xi)^{-1},$$

and let $\sigma_C \sim \sum_{N=0}^{\infty} \sigma_{C_N}$, where

$$\sigma_{C_N}(x, \xi) = -\sigma_{C_0}(x, \xi) \sum_{k=0}^{N-1} \sum_{j=0}^{N-k} \sum_{|\gamma|=N-j-k} \frac{1}{\gamma!} \left[\triangle_\xi^\gamma \sigma_{A_j}(x, \xi) \right] \partial_x^{(\gamma)} \sigma_{C_k}(x, \xi).$$

Check that also C is a parametrix of A.

10.10   Full symbols on compact manifolds

In this section we discuss how the introduced constructions are mapped by global diffeomorphisms.

Let $\Phi : G \to M$ be a diffeomorphism from a compact Lie group G to a smooth manifold M. Such diffeomorphisms can be obtained for large classes of compact manifolds by the Poincaré conjecture type results. For example, if $\dim G = \dim M = 3$ it is now known that such Φ exists for any closed simply-connected[6] manifold, and we can take $G \cong \mathbb{S}^3 \cong \mathrm{SU}(2)$. We will now explain how the diffeomorphism Φ induces the quantization of operators on M from that on G.

Let us endow M with the natural Lie group structure induced by Φ, i.e., with the group multiplication $((x, y) \mapsto x \cdot y) : M \times M \to M$ defined by

$$x \cdot y := \Phi\left(\Phi^{-1}(x)\, \Phi^{-1}(y)\right).$$

The spaces $C^\infty(G)$ and $C^\infty(M)$ are isomorphic via mappings

$$\Phi_* : C^\infty(G) \to C^\infty(M), \qquad f \mapsto f_\Phi = f \circ \Phi^{-1},$$
$$\Phi^* : C^\infty(M) \to C^\infty(G), \qquad g \mapsto g_{\Phi^{-1}} = g \circ \Phi,$$

---

[6]For the definition of simply-connectedness see Definition 8.3.18.

and the Haar integral on M is given by

$$\int_M g \, \mathrm{d}\mu_M \equiv \int_M g \, \mathrm{d}x := \int_G g \circ \Phi \, \mathrm{d}\mu_G,$$

where μ_G is the Haar measure on G, because

$$
\begin{aligned}
\int_M g(x \cdot y) \, \mathrm{d}x &= \int_M g(\Phi(\Phi^{-1}(x) \, \Phi^{-1}(y))) \, \mathrm{d}x \\
&= \int_G (g \circ \Phi)(\Phi^{-1}(x) \, \Phi^{-1}(y)) \, \mathrm{d}(\Phi^{-1}(x)) \\
&= \int_G (g \circ \Phi)(z) \, \mathrm{d}z \\
&= \int_M g(x) \, \mathrm{d}x.
\end{aligned}
$$

Moreover, $\Phi_* : C^\infty(G) \to C^\infty(M)$ extends to a linear unitary bijection $\Phi_* : L^2(\mu_G) \to L^2(\mu_M)$:

$$\int_M g(x) \, \overline{h(x)} \, \mathrm{d}x = \int_G (g \circ \Phi) \, (\overline{h \circ \Phi}) \, \mathrm{d}\mu_G.$$

Notice also that there is an isomorphism

$$\Phi_* : \mathrm{Rep}(G) \to \mathrm{Rep}(M), \quad \xi \mapsto \Phi_*(\xi) = \xi \circ \Phi$$

of irreducible unitary representations. Thus $\widehat{G} \cong \widehat{M}$ in this sense. This immediately implies that the whole construction of symbols of pseudo-differential operators on M is equivalent to that on G.

In Section 12.5 we will give an example of this identification in the case of $\mathrm{SU}(2) \cong \mathbb{S}^3$.

10.11 Operator-valued symbols

In this section we discuss another notion of a symbol which we call operator-valued symbols of operators.

We recall left- and right-convolution operators from Remark 10.4.2, which will now play an important role:

Definition 10.11.1 (Convolution operators $l(f)$ and $r(f)$). For $f \in \mathcal{D}'(G)$, we define the respective *left-convolution* and *right-convolution operators* $l(f), r(f) : C^\infty(G) \to C^\infty(G)$ by

$$
\begin{aligned}
l(f)g &:= f * g, \\
r(f)g &:= g * f.
\end{aligned}
$$

Exercise 10.11.2. To check that we indeed have $l(f), r(f) : C^\infty(G) \to C^\infty(G)$ one can use a distributional interpretation similar to the one in Section 1.4.2. Work out the details of this.

Remark 10.11.3. For $f \in L^2(G)$, in the literature, operator $l(f)$ is sometimes called a "global" Fourier transform of f. Such terminology can be found in, e.g., W.F. Stinespring [120], where it is denoted by $\pi(f)$, and for related integration theory of operators by I.E. Segal, see [107] and [108]. However, since we are dealing with the quantization of operators on G that have both left- and right-convolution kernels, we want to keep track of both left- and right-convolution operators. At the same time, it will turn out that the proceeding theory of full symbols is better adapted to the operator $r(f)$ because our starting point was right-convolution kernels.

As usual, the *left and right regular representations of G* are denoted by $\pi_L, \pi_R : G \to \mathcal{U}(L^2(G))$, respectively, i.e.,

$$\begin{aligned}
\pi_L(x)f(y) &= f(x^{-1}y), \\
\pi_R(x)f(y) &= f(yx).
\end{aligned}$$

Exercise 10.11.4. Verify that these representations are indeed unitary. For example, check that $\pi_R(x)^{-1} = \pi_R(x^{-1}) = \pi_R(x)^*$, and similarly for $\pi_L(x)$.

Keeping in mind the right and left Peter–Weyl Theorems (Theorem 7.5.14 and Remark 7.5.14), the Fourier inversion formulae may be viewed in the following form:

Proposition 10.11.5 ("Fourier inversion formulae"). *Let* $f \in C^\infty(G)$. *Then we have*

$$r(f) = \int_G f(y)\ \pi_R(y)^*\ \mathrm{d}y \quad and \quad l(f) = \int_G f(y)\ \pi_L(y)\ \mathrm{d}y. \tag{10.47}$$

Conversely, for every $x \in G$ *we have*

$$f(x) = \mathrm{Tr}\ (r(f)\ \pi_R(x)) \quad and \quad f(x) = \mathrm{Tr}\ (l(f)\ \pi_L(x)^*), \tag{10.48}$$

where Tr *is the trace functional, see Definition B.5.38; notice that* $\mathrm{Tr}(AB) = \mathrm{Tr}(BA)$. *These formulae have an almost everywhere extension to* $L^2(\mu_G)$.

Proof. We will prove the case of right-convolutions since for left-convolutions it is similar. First, we can write

$$\begin{aligned}
(r(f)g)(x) &= (g * f)(x) \\
&= \int_G g(xy^{-1})f(y)\ \mathrm{d}y \\
&= \int_G f(y)(\pi_R(y^{-1})g)(x)\ \mathrm{d}y,
\end{aligned}$$

from which we obtain (10.47) in view of the unitarity $\pi_R(y^{-1}) = \pi_R(y)^*$. The proof of (10.48) is somewhat lengthier if we provide the details. First we observe that (10.47) implies that

$$r(f)\pi_R(x) = \int_G f(y)\, \pi_R(y)^*\, \pi_R(x)\, \mathrm{d}y = \int_G f(y)\, \pi_R(y^{-1}x)\, \mathrm{d}y. \qquad (10.49)$$

By the Peter–Weyl Theorem (Theorem 7.5.14) whose notation we will use in this proof, we know that $\{\dim(\phi)\phi_{ij}\}_{\phi,i,j}$ is an orthonormal basis in $L^2(\mu_G)$, so that by Definition B.5.38 of the trace we have

$$\mathrm{Tr}(r(f)\,\pi_R(x)) \quad = \quad \sum_{[\phi]\in\widehat{G}}\sum_{i,j=1}^{\dim(\phi)} \dim(\phi)(r(f)\pi_R(x)\phi_{ij},\phi_{ij}) \qquad (10.50)$$

$$\overset{(10.49)}{=} \quad \sum_{[\phi]\in\widehat{G}}\sum_{i,j=1}^{\dim(\phi)} \dim(\phi)\int_G\int_G f(y)\phi_{ij}(zy^{-1}x)\overline{\phi_{ij}(z)}\,\mathrm{d}y\,\mathrm{d}z$$

$$= \quad \sum_{[\phi]\in\widehat{G}}\sum_{i,j=1}^{\dim(\phi)} \dim(\phi)\int_G\int_G f(xw^{-1})\phi_{ij}(zw)\overline{\phi_{ij}(z)}\,\mathrm{d}w\,\mathrm{d}z,$$

where in the last equality we changed the variables $w := y^{-1}x$. Using $\phi(zw) = \phi(z)\phi(w)$, we obtain

$$\phi(zw)_{ij} = \sum_{k=1}^{\dim(\phi)} \phi_{ik}(z)\phi_{kj}(w) \qquad (10.51)$$

and

$$\phi(y^{-1}x)_{jj} = \sum_{k=1}^{\dim(\phi)} \phi_{ji}(y^{-1})\phi_{ij}(x) = \sum_{k=1}^{\dim(\phi)} \overline{\phi_{ij}(y)}\phi_{ij}(x), \qquad (10.52)$$

to be used later as well. We also notice that

$$\int_G \phi_{ik}(z)\overline{\phi_{ij}(z)}\,\mathrm{d}z = \langle\phi_{ik},\phi_{ij}\rangle_{L^2} = \frac{1}{\dim(\phi)}\delta_{kj} \qquad (10.53)$$

in view of Lemma 7.5.12. Plugging (10.51) and (10.53) into (10.50) we get

$$\mathrm{Tr}(r(f)\,\pi_R(x)) \quad = \quad \sum_{[\phi]\in\widehat{G}}\sum_{i,j=1}^{\dim(\phi)} \int_G f(xw^{-1})\,\phi_{jj}(w)\,\mathrm{d}w$$

$$= \quad \sum_{[\phi]\in\widehat{G}}\sum_{j=1}^{\dim(\phi)} \dim(\phi)\int_G f(xw^{-1})\,\phi_{jj}(w)\,\mathrm{d}w$$

$$\stackrel{y:=xw^{-1}}{=} \sum_{[\phi]\in\widehat{G}} \sum_{j=1}^{\dim(\phi)} \dim(\phi) \int_G f(y)\, \phi_{jj}(y^{-1}x)\, dy$$

$$\stackrel{(10.52)}{=} \sum_{[\phi]\in\widehat{G}} \sum_{j=1}^{\dim(\phi)} \dim(\phi)\, \langle f, \phi_{ij}\rangle_{L^2}\, \phi_{ij}(x)$$

$$= f(x), \tag{10.54}$$

with the last equality in view of Corollary 7.6.7. ☐

Corollary 10.11.6 (Character description). *For every $f \in L^2(G)$ we have*

$$f = \sum_{[\phi]\in\widehat{G}} \dim(\phi)\, f * \chi_\phi,$$

where χ_ϕ is the character of ϕ. Thus, the projection of $f \in L^2(G)$ to H^ϕ is given by $f \mapsto f * \chi_\phi$.

Proof. Writing the expression in the line of (10.54) as a trace, we see that

$$\sum_{[\phi]\in\widehat{G}} \sum_{j=1}^{\dim(\phi)} \dim(\phi) \int_G f(xw^{-1})\phi_{jj}(w)\, dw$$

$$= \sum_{[\phi]\in\widehat{G}} \dim(\phi)(f * \mathrm{Tr}\phi)(x)$$

$$= \sum_{[\phi]\in\widehat{G}} \dim(\phi)(f * \chi_\phi)(x).$$

Consequently (10.54) implies the statement. ☐

Exercise 10.11.7. Provide details for the proof of Proposition 10.11.5 in the left-convolution case.

Definition 10.11.8 (Right and left kernels $R_A(x), L_A(x)$). Let $A : C^\infty(G) \to C^\infty(G)$ be a linear continuous operator, and let $R_A(x,y), L_A(x,y) \in \mathcal{D}(G)\otimes\mathcal{D}'(G)$ be its right-convolution and left-convolution kernels, respectively[7]. For every $x \in G$ we define $R_A(x), L_A(x) \in \mathcal{D}'(G)$ by

$$[R_A(x)](y) := R_A(x,y) \quad \text{and} \quad [L_A(x)](y) := L_A(x,y).$$

In this notation we can write

$$(Af)(x) = [f * R_A(x)](x) = [r(R_A(x))f](x), \tag{10.55}$$

---

[7] As defined in (10.16) in Section 10.4, and shown in Lemma 10.12.5 in Section 10.12.

and similarly
$$(Af)(x) = [L_A(x) * f](x) = [l(L_A(x))f](x). \tag{10.56}$$
This motivates the following definition.

Definition 10.11.9 (Operator-valued symbols). Let $A : C^\infty(G) \to C^\infty(G)$ be a linear continuous operator. We define its *right operator-valued symbol* $r_A : G \to \mathcal{L}(C^\infty(G))$ by
$$x \mapsto r_A(x) := r(R_A(x)).$$
Similarly, we define its *left operator-valued symbol* $l_A : G \mapsto \mathcal{L}(C^\infty(G))$ by
$$x \mapsto l_A(x) := l(L_A(x)).$$
We observe that (10.55) and (10.56) imply the equality
$$[r_A(x)f](x) = Af(x) = [l_A(x)]f(x). \tag{10.57}$$

Lemma 10.11.10. *For every $f \in C^\infty(G)$ and every $x \in G$ we have*
$$r(r_A(x)f) = r_A(x)r(f) \quad and \quad l(l_A(x)f) = l_A(x)l(f).$$

Proof. We prove only the "right" case since the "left" one is similar. We have a simple calculation
$$
\begin{aligned}
r(r_A(x)f)g \quad &= \quad g * (r_A(x)f) \\
&= \quad g * (f * R_A(x)) \\
&\overset{\text{Prop. 7.7.3}}{=} \quad (g * f) * R_A(x) \\
&= \quad r(R_A(x))(g * f) \\
&= \quad r_A(x)r(f)g,
\end{aligned}
$$
completing the proof. □

Exercise 10.11.11. Prove the left cases of Lemma 10.11.10 as well as of the following theorem:

Theorem 10.11.12 (Operator-valued quantization). *Let $A : C^\infty(G) \to C^\infty(G)$ be a linear continuous operator. Then we have*
$$
\begin{aligned}
Af(x) \quad &= \quad \mathrm{Tr}(r_A(x)\, r(f)\, \pi_R(x)) \\
&= \quad \mathrm{Tr}(l_A(x)\, l(f)\, \pi_L(x)^*), \tag{10.58}
\end{aligned}
$$
for all $f \in C^\infty(G)$ and $x \in G$.

Proof. We can write
$$
\begin{aligned}
Af(x) \quad &\overset{(10.57)}{=} \quad (r_A(x)f)(x) \\
&\overset{(10.48)}{=} \quad \mathrm{Tr}\left(r[r_A(x)f]\, \pi_R(x)\right) \\
&\overset{\text{Lemma 10.11.10}}{=} \quad \mathrm{Tr}\left(r_A(x)\, r(f)\, \pi_R(x)\right),
\end{aligned}
$$
completing the proof for the right case. The left one is similar. □

Remark 10.11.13 (**Symbol as a family of convolution operators**). In view of Theorem 10.11.12 the operator-valued symbols l_A and r_A can be regarded as a family of convolution operators obtained from A by "freezing" it at points $x \in G$. We note that if $A \in \mathcal{L}(\mathcal{D}(G))$ is a left-invariant operator, i.e., $A\pi_L(x) = \pi_L(x)A$ for every $x \in G$, then its right operator-valued symbol is the constant mapping $x \mapsto r_A(x) \equiv A$ (A is a right-convolution operator).

Remark 10.11.14 (**Operator-valued symbols and operators**). The quantization $A \mapsto r_A, l_A$ is an injective linear mapping. Conversely, starting from an operator-valued symbol we can define the corresponding operator. Indeed, let $\rho : G \to \mathcal{L}(\mathcal{D}'(G))$ be a mapping such that $\rho(x) = r(s(x))$, for some $s \in \mathcal{D}(G) \otimes \mathcal{D}'(G)$. Then we can define $\mathrm{Op}(\rho) \in \mathcal{L}(\mathcal{D}(G))$ by

$$(\mathrm{Op}(\rho)f)(x) = (\rho(x)f)(x),$$

for which s is the right-convolution kernel and ρ is the right operator-valued symbol.

Corollary 10.11.15 (Decomposition in matrix blocks). *According to the Peter–Weyl Theorem, all operators in* (10.58) *have direct sum decompositions with corresponding finite-dimensional square matrix blocks* $r_A(x, \xi) := r_A(x)|_{\mathcal{H}^\xi}$, $l_A(x, \xi) := l_A(x)|_{\mathcal{H}^\xi}$, $\pi_R(x, \xi) := \pi_R(x)|_{\mathcal{H}^\xi}$, $\pi_L(x, \xi) := \pi_L(x)|_{\mathcal{H}^\xi}$, *where* $\xi \in \mathrm{Rep}(G)$ *is a representation* $\xi : G \to \mathcal{U}(\mathcal{H}_\xi)$. *In this notation* (10.58) *implies*

$$Af(x) \quad = \quad \sum_{\xi \in \widehat{G}} \mathrm{Tr}\left(\pi_R(x, \xi)\, r_A(x, \xi)\, r(f)|_{\mathcal{H}^\xi}\right) \tag{10.59}$$

$$= \quad \sum_{\xi \in \widehat{G}} \mathrm{Tr}\left(\pi_L(x, \xi)^*\, l_A(x, \xi)\, l(f)|_{\mathcal{H}^\xi}\right).$$

The meaning of operators $r(f)|_{\mathcal{H}^\xi}$ *and* $l(f)|_{\mathcal{H}^\xi}$ *can be clarified as follow: if* $\xi \in \mathrm{Rep}(G)$, *then*

$$l(u)\xi(x) = \widehat{u}(\xi)\xi(x) \quad and \quad r(u)\xi(x) = \xi(x)\widehat{u}(\xi).$$

Let us show the last formula for $l(u)$. Indeed, we have

$$l(u)\xi(x) = \int_G u(y)\xi(y^{-1}x)\, \mathrm{d}y = \left(\int_G u(y)\xi(y)^*\, \mathrm{d}y\right)\xi(x) = \widehat{u}(\xi)\xi(x),$$

and the calculation for $r(u)$ is similar, where we can commute $\xi(x)$ and the integral since $\xi(x)$ is finite-dimensional.

We now establish a relation between different quantizations. Namely, for every $x \in G$ the mapping $r_A(x) : C^\infty(G) \to C^\infty(G)$ from Definition 10.11.9 is linear and continuous, so that we can find its symbol $\sigma_{r_A(x)}(y, \xi)$ according to Definition 10.4.3.

Theorem 10.11.16 (Relation between quantizations). *Let $A : C^\infty(G) \to C^\infty(G)$ be a linear continuous operator and let $r_A, l_A : G \to \mathcal{L}(C^\infty(G))$ be its right and left operator-valued symbols, respectively. Then for all $x, y \in G$ and $\xi \in \mathrm{Rep}(G)$ we have*

$$\sigma_{r_A(x)}(y, \xi) = \sigma_A(x, \xi) \tag{10.60}$$

and

$$\sigma_{l_A(x)}(y, \xi) = \xi(y)^* \xi(x) \sigma_A(x, \xi) \xi(x)^* \xi(y). \tag{10.61}$$

The operator-valued symbols take the form

$$[r_A(x)f](y) = \sum_{[\xi] \in \widehat{G}} \dim(\xi) \ \mathrm{Tr}\left(\xi(y) \ \sigma_A(x, \xi) \ \widehat{f}(\xi)\right) \tag{10.62}$$

and

$$[l_A(x)f](y) = \sum_{[\xi] \in \widehat{G}} \dim(\xi) \ \mathrm{Tr}\left(\xi(x) \ \sigma_A(x, \xi) \ \xi(x)^* \ \xi(y) \ \widehat{f}(\xi)\right). \tag{10.63}$$

Finally, we also have

$$Af(x) = \mathrm{Tr}\left(\mathrm{Op}(\sigma_B) \ r(f) \ \pi_R(x)\right), \tag{10.64}$$

where $\sigma_B(y, \xi) := \sigma_A(x, \xi)$.

Remark 10.11.17. Formulae (10.62) and (10.63) when evaluated at $y = x$ give us $Af(x)$, thus recovering equality (10.57). Formula (10.60) shows that the operator $r_A(x)$ when quantized in terms of symbols from Definition 10.4.3 becomes a multiplier, with its symbol independent of y. Formula (10.61) is slightly more complicated as a consequence of the fact that the symbols from Definition 10.4.3 are better adapted to right-convolutions, see Remark 10.4.2. In the "left" case additional conjugations in (10.61) appear in view of Proposition 10.4.11.

Proof of Theorem 10.11.16. We start with the right operator-valued symbol $r_A(x)$. By Theorem 10.4.6 its symbol can be found as

$$\sigma_{r_A(x)}(y, \xi) = \xi(y)^* [r_A(x)\xi](y).$$

Thus, we calculate

$$
\begin{aligned}
[r_A(x)\xi](y) &= [r(R_A(x))\xi](y) \\
&= [\xi * R_A(x)](y) \\
&= \int_G \xi(yz^{-1}) R_A(x)(z) dz \\
&= \xi(y) \int_G \xi(z)^* R_A(x, z) dz \\
&= \xi(y) \sigma_A(x, \xi),
\end{aligned}
$$

yielding (10.60). Here, as usual, we used the finite dimensionality of representations to commute $\xi(y)$ with the integral. This implies (10.62) by Theorem 10.4.4:

$$
\begin{aligned}
[r_A(x)f](y) &= \sum_{[\xi]\in\widehat{G}} \dim(\xi)\ \mathrm{Tr}\left(\xi(y)\ \sigma_{r_A(x)}(y,\xi)\ \widehat{f}(\xi)\right) \\
&= \sum_{[\xi]\in\widehat{G}} \dim(\xi)\ \mathrm{Tr}\left(\xi(y)\ \sigma_A(x,\xi)\ \widehat{f}(\xi)\right).
\end{aligned}
$$

Moreover, (10.64) follows from (10.60) by Theorem 10.11.12, where $B = r_A(x)$. Similarly, in view of

$$
\sigma_{l_A(x)}(y,\xi) = \xi(y)^*[l_A(x)\xi](y)
$$

we calculate

$$
\begin{aligned}
[l_A(x)\xi](y) &= [l(L_A(x))\xi](y) \\
&= [L_A(x) * \xi](y) \\
&= \int_G L_A(x)(yz^{-1})\xi(z)dz \\
&= \int_G L_A(x,yz^{-1})\xi(z)dz.
\end{aligned}
$$

Using formula (10.17), i.e., $R_A(x,w) = L_A(x,xwx^{-1})$ with $w = x^{-1}yz^{-1}x$, gives $z = xw^{-1}x^{-1}y$ and

$$
L_A(x,yz^{-1}) = R_A(x,x^{-1}yz^{-1}x).
$$

Continuing the calculation for $[l_A(x)\xi](y)$ above, we get

$$
\begin{aligned}
[l_A(x)\xi](y) &= \int_G L_A(x,yz^{-1})\xi(z)dz \\
&= \int_G R_A(x,x^{-1}yz^{-1}x)\xi(z)dz \\
&= \int_G R_A(x,w)\xi(xw^{-1}x^{-1}y)dz \\
&= \xi(x)\left(\int_G R_A(x,w)\xi(w)^*\ dw\right)\xi(x^{-1})\xi(y) \\
&= \xi(x)\sigma_A(x,\xi)\xi(x)^*\xi(y),
\end{aligned}
$$

yielding (10.61). This implies (10.63) by Theorem 10.4.4. $\qquad\square$

In Theorem 10.5.5 we presented conditions on the $L^2(G)$-boundedness of operators in terms of symbols $\sigma(x,\xi)$. We now briefly discuss the same question in terms of the operator-valued symbols. We will do this in terms of the right operator-valued symbols because the left case is the same, see Exercise 10.11.21.

Definition 10.11.18 (Derivations of operator-valued symbols). Let operator $A :$ $C^\infty(G) \to C^\infty(G)$ be linear and continuous, with right operator-valued symbol $r_A : G \to \mathcal{L}(C^\infty(G))$. Let $p(D) \in \mathcal{L}(C^\infty(G))$ be a partial differential operator. For each $x \in G$ we define

$$(p(D)r_A)(x) = p(D)r_A(x) := r(R_B(x)),$$

where $R_B(x)(y) = R_B(x,y)$ and

$$R_B = (p(D) \otimes \mathrm{id})R_A.$$

Operator B defined by $Bf(x) = [r(R_B(x))f](x)$ is then also a linear continuous operator from $C^\infty(G))$ to $C^\infty(G))$ because $p(D) \in \mathcal{L}(C^\infty(G))$, $\mathrm{id} \in \mathcal{L}(\mathcal{D}'(G))$.

Theorem 10.11.19 (Boundedness on $L^2(G)$). *Let G be a compact Lie group of dimension n. Let A be a linear continuous operator from $C^\infty(G)$ to $C^\infty(G)$ and assume that its right operator-valued symbol r_A satisfies $r_A \in C^k(G, \mathcal{L}(L^2(G)))$ with $k > n/2$. Then A extends to a bounded linear operator from $L^2(G)$ to $L^2(G)$.*

Proof. By (10.57) we have

$$
\begin{aligned}
\|Af\|^2_{L^2(G)} &= \int_G |(r_A(x)f)(x)|^2 \, \mathrm{d}x \\
&\leq \int_G \sup_{y \in G} |(r_A(y)f)(x)|^2 \, \mathrm{d}x,
\end{aligned}
$$

and by an application of the Sobolev embedding theorem we get

$$\sup_{y \in G} |(r_A(y)f)(x)|^2 \leq C_k \sum_{|\alpha| \leq k} \int_G |((\partial_y^\alpha r_A)(y)f)(x)|^2 \, \mathrm{d}y.$$

Therefore using the Fubini theorem to change the order of integration, we obtain

$$
\begin{aligned}
\|Af\|^2_{L^2(G)} &\leq C \sum_{|\alpha| \leq k} \int_G \int_G |((\partial_y^\alpha r_A)(y)f)(x)|^2 \, \mathrm{d}x \, \mathrm{d}y \\
&\leq C \sum_{|\alpha| \leq k} \sup_{y \in G} \int_G |((\partial^\alpha r_A)(y)f)(x)|^2 \, \mathrm{d}x \\
&= C \sum_{|\alpha| \leq k} \sup_{y \in G} \|(\partial^\alpha r_A)(y)f\|^2_{L^2(G)} \\
&\leq C \sum_{|\alpha| \leq k} \sup_{y \in G} \|(\partial^\alpha r_A)(y)\|^2_{\mathcal{L}(L^2(G))} \|f\|^2_{L^2(G)}.
\end{aligned}
$$

The proof is complete, because G is compact and $r_A \in C^k(G, \mathcal{L}(L^2(G)))$. $\qquad\square$

Theorem 10.11.19 yields the following Sobolev boundedness result:

Corollary 10.11.20. *Let G be a compact Lie group and let A be a linear continuous operator from $C^\infty(G)$ to $C^\infty(G)$ such that its right operator-valued symbol r_A satisfies $r_A \in C^\infty(G, \mathcal{L}(H^m(G), H^0(G)))$, $m \in \mathbb{R}$. Then A extends to a bounded linear operator from $H^m(G)$ to $H^0(G)$.*

Exercise 10.11.21. Show that we can replace r_A by the left operator-valued symbol l_A in Theorem 10.11.19 and Corollary 10.11.20 for them to remain true. Work out details of the proofs.

Remark 10.11.22 (**M.E. Taylor's characterisation**). Let $\mathfrak{g}$ be the Lie algebra of a compact Lie group G, and let $n = \dim(G) = \dim(\mathfrak{g})$. By the exponential mapping $\exp : \mathfrak{g} \to G$, a neighbourhood of the neutral element $e \in G$ can be identified with a neighbourhood of $0 \in \mathfrak{g}$. Let $\mathcal{X}^m = S^m_{1\#} \subset S^m_{1,0}(\mathbb{R}^n \times \mathbb{R}^n)$ consist of the x-invariant symbols $(x, \xi) \mapsto p(\xi)$ in $S^m_{1,0}(\mathbb{R}^n \times \mathbb{R}^n)$ with the usual Fréchet space topology (given by the optimal constants in symbol estimates as seminorms). A distribution $k \in \mathcal{D}'(G)$ with a sufficiently small support is said to belong to space $\widehat{\mathcal{X}^m}$ if

$$\text{sing supp}(k) \subset \{e\} \quad \text{and} \quad \widehat{k} \in \mathcal{X}^m \subset C^\infty(\mathfrak{g}'),$$

where the Fourier transform $\widehat{k}$ is the usual Fourier transform on $\mathfrak{g} \cong \mathbb{R}^n$, and the dual space satisfies $\mathfrak{g}' \cong \mathbb{R}^n$ (and we are using the exponential coordinates for $k(y)$ when $y \approx e \in G$). If $k \in \widehat{\mathcal{X}^m}$ then the convolution operator

$$u \mapsto k * u, \quad k * u(x) = \int_G k(xy^{-1})\, u(y)\, dy,$$

is said to belong to space $OP\mathcal{X}^m$, which is endowed with the natural Fréchet space structure obtained from $\mathcal{X}^m$. Formally, let $k(x, y) = k_x(y)$ be the left-convolution kernel of a linear operator $\mathcal{K} : C^\infty(G) \to C^\infty(G)$, i.e.,

$$\mathcal{K}u(x) = \int_G k_x(xy^{-1})\, u(y)\, dy.$$

In [128], M.E. Taylor showed that $\mathcal{K} \in \Psi^m(G)$ if and only if the mapping

$$(x \mapsto (u \mapsto k_x * u)) : G \to OP\mathcal{X}^m$$

is smooth; here naturally $u \mapsto k_x * u$ must belong to $OP\mathcal{X}^m$ for each $x \in G$. This approach was pursued in [128] as well as in [136, 139] in the setting of the left-convolution kernels, and relied on the exponential mapping and on the pseudo-differential operator on the Lie algebra (see Remark 10.11.22). In this approach many arguments may be restricted to a suitable neighbourhood of the identity element. However, the notion of the symbol σ_A from Definition 10.4.3 appears to be more practical as it allows a finite-dimension realisation of the symbol and works globally on the group.

Remark 10.11.23 (**Operator-valued calculus**). It is possible to construct the calculus of operator-valued symbols, including compositions, adjoints, transposes, and the inverse. Exponential coordinates of the Lie algebra versions of it can be found in [128, 139]. However, using our calculus in Section 10.7 together with Corollary 10.11.15 one can obtain this directly on the group without referring to the exponential coordinates in the neighbourhood of the origin. We leave this as an exercise for an interested reader.

10.11.1 Example on the torus $\mathbb{T}^n$

We will show here how the construction of this section applies for the operator-valued quantization of operator on the torus $\mathbb{T}^n$. The starting point for this example is the operator-valued quantization formula (10.58) in Theorem 10.11.12:

$$
\begin{aligned}
Af(x) &= \mathrm{Tr}(r_A(x)\, r(f)\, \pi_R(x)) \\
&= \mathrm{Tr}(l_A(x)\, l(f)\, \pi_L(x)^*), \tag{10.65}
\end{aligned}
$$

where $\widehat{G}$ is the unitary dual of G, $\pi_L : G \to \mathcal{L}(L^2(G))$ is the left regular representation of G on $L^2(G)$, $r_A, l_A : G \to \mathcal{L}(\mathcal{D}(G))$ are the right and left operator-valued symbol of A, and $r(f)$ and $l(f)$ are the right and left convolution operators. According to the Peter–Weyl Theorem, these operators also have direct sum decompositions with corresponding finite-dimensional square matrix blocks which appear in (10.59).

Let us inspect and summarise what this means in the special case of the n-torus

$$
G = \mathbb{T}^n.
$$

The 1-torus $\mathbb{T} = \mathbb{R}/\mathbb{Z}$ can be identified with the multiplicative group of the unit circle in the plane. Let $U(1)$ be the group of 1×1 unitary matrices; $U(1)$ can be identified with the unit circle, again. For each $\xi \in \mathbb{Z}^n$, define $e_\xi : \mathbb{T}^n \to U(1)$ by

$$
e_\xi(x) = e^{i2\pi x \cdot \xi};
$$

up to isomorphism, an irreducible unitary representation of $G = \mathbb{T}^n$ is some e_ξ. Thus $\widehat{G}$ can be identified with the integer lattice $\mathbb{Z}^n$, and $\{e_\xi : \xi \in \mathbb{Z}^n\}$ is an orthonormal basis for $L^2(G)$, where the Haar measure on $G = \mathbb{T}^n$ is obtained from the Lebesgue measure of $\mathbb{R}^n$. Let us analyse this in terms of the left regular representation $\pi_L : G \to \mathcal{L}(L^2(G))$ which is now

$$
(\pi_L(x)f)(y) = f(y - x) \quad \text{(for almost every } y \in \mathbb{T}^n).
$$

Especially, we notice

$$
(\pi_L(x)^* e_\xi)(y) = (\pi_L(-x)e_\xi)(y) = e_\xi(y + x) = e^{i2\pi x \cdot \xi}\, e_\xi(y). \tag{10.66}
$$

Let $u \in C^\infty(G)$. Then $l(u) \in \mathcal{L}(L^2(G))$ is the convolution operator given by

$$(l(u)f)(x) = (u * f)(x) = \int_G u(x - y)\, f(y)\, \mathrm{d}y = \int_G u(y)\, f(x - y)\, \mathrm{d}y.$$

Especially, we have

$$l(u)e_\xi = \widehat{u}(\xi)\, e_\xi, \tag{10.67}$$

where $\widehat{u}(\xi) \in \mathbb{C}$ is the usual Fourier coefficient of $u : \mathbb{T}^n \to \mathbb{C}$. Let A be a continuous linear operator from $C^\infty(G)$ to $C^\infty(G)$. Recall that here $G = \mathbb{T}^n$ and $\widehat{G} = \mathbb{Z}^n$; let us define $\sigma_A : G \times \widehat{G} \to \mathbb{C}$ by

$$\sigma_A(x, \xi) = \mathrm{e}^{-\mathrm{i}2\pi x \cdot \xi}\, (\sigma_A e_\xi)(x),$$

which is the toroidal symbol of A as well as the 1×1 matrix symbol, see Remark 10.4.7. The left operator-valued symbol of A is the convolution operator-valued mapping $l_A : G \to \mathcal{L}(C^\infty(G))$ that by (10.57) satisfies

$$(l_A(x)u)(x) = (Au)(x).$$

Thereby and by Theorem 10.11.16 we have

$$(l_A(x)u)(y) = \sum_{\xi \in \mathbb{Z}^n} e_\xi(y)\, \sigma_A(x, \xi)\, \widehat{u}(\xi). \tag{10.68}$$

In particular, (10.66) and (10.68) imply that

$$\pi_L(x)^* l_A(x)e_\xi = \pi_L(x)^* e_\xi \sigma_A(x, \xi)\widehat{e}_\xi(\xi) = \mathrm{e}^{\mathrm{i}2\pi x \cdot \xi}\sigma_A(x, \xi)e_\xi. \tag{10.69}$$

The trace of a linear operator $B : L^2(G) \to L^2(G)$ is

$$\mathrm{Tr}(B) = \sum_{\xi \in \mathbb{Z}^n} \langle Be_\xi, e_\xi \rangle_{L^2(G)},$$

where $\langle f, g \rangle_{L^2(G)} = \int_G f(y)\, \overline{g(y)}\, \mathrm{d}y$ is the inner product of $L^2(G)$, see Definition B.5.38. Let us now explore formula (10.65) in the case of $G = \mathbb{T}^n$:

$$
\begin{aligned}
\mathrm{Tr}\left(\pi_L(x)^*\, l_A(x)\, l(u)\right) \quad &= \quad \sum_{\xi \in \mathbb{Z}^n} \langle \pi_L(x)^*\, l_A(x)\, l(u)\, e_\xi, e_\xi \rangle_{L^2(G)} \\[2mm]
&\overset{(10.67)}{=} \quad \sum_{\xi \in \mathbb{Z}^n} \langle \pi_L(x)^*\, l_A(x)\, \widehat{u}(\xi)\, e_\xi, e_\xi \rangle_{L^2(G)} \\[2mm]
&\overset{(10.69)}{=} \quad \sum_{\xi \in \mathbb{Z}^n} \langle\, \mathrm{e}^{\mathrm{i}2\pi x \cdot \xi}\, \sigma_A(x, \xi)\, \widehat{u}(\xi)\, e_\xi, e_\xi \rangle_{L^2(G)} \\[2mm]
&= \quad \sum_{\xi \in \mathbb{Z}^n} \mathrm{e}^{\mathrm{i}2\pi x \cdot \xi}\, \sigma_A(x, \xi)\, \widehat{u}(\xi)\, \langle e_\xi, e_\xi \rangle_{L^2(G)} \\[2mm]
&= \quad \sum_{\xi \in \mathbb{Z}^n} \mathrm{e}^{\mathrm{i}2\pi x \cdot \xi}\, \sigma_A(x, \xi)\, \widehat{u}(\xi) \\[2mm]
&= \quad (Au)(x),
\end{aligned}
$$

where the last equality is the toroidal quantization of the operator A, given in Theorem 4.1.3. This is the case on the commutative group $G = \mathbb{T}^n$, where spaces $\mathcal{H}^\xi = \mathrm{span}(e_\xi)$ are all one-dimensional. On non-commutative compact Lie groups, even using the quantization (10.59), $\pi_L(x, \xi)$, $\sigma_A(x, \xi)$ and $l(u)|_{\mathcal{H}^\xi}$ are no longer numbers (more precisely: no longer 1×1 matrices), but they are $\dim(\xi) \times \dim(\xi)$ matrices, where the dimension $\dim(\xi) \in \mathbb{Z}^+$ depends on the corresponding representation $\xi \in \widehat{G}$ and is usually greater than 1.

Exercise 10.11.24. Work out the above example for the "right" representation π_R and right operator-valued symbols r_A. In particular show that on the torus "left" and "right" coincide.

10.12 Appendix: integral kernels

Here we provide a short appendix with more technical explanations about integral kernels of operators on the compact Lie group G.

Definition 10.12.1 (Duality between $\mathcal{D}(G)$ and $\mathcal{D}'(G)$). Let $\mathcal{D}(G)$ be the set $C^\infty(G)$ equipped with the usual Fréchet space topology defined by seminorms $p_\alpha(f) = \max_{x \in G} |\partial^\alpha f(x)|$, with ∂^α as in Definition 10.5.3. Thus, the convergence on $\mathcal{D}(G)$ is just the uniform convergence of functions and all their derivatives: $f_k \to f$ in $C^\infty(G)$ (or in $\mathcal{D}(G)$) if $\partial^\alpha f_k(x) \to \partial^\alpha f(x)$ for all $x \in G$, due to the compactness of G.

Let $\mathcal{D}'(G) = \mathcal{L}(\mathcal{D}(G), \mathbb{C})$ be its dual, i.e., the set of distributions with $\mathcal{D}(G)$ as the test function space. We equip the space of distributions with the weak*-topology. The duality $\mathcal{D}'(G) \times \mathcal{D}(G) \to \mathbb{C}$ is denoted by

$$\langle f, \phi \rangle := f(\phi),$$

where $\phi \in \mathcal{D}(G)$ and $f \in \mathcal{D}'(G)$, and an embedding $\mathcal{D}(G) \hookrightarrow \mathcal{D}'(G)$, $\psi \mapsto f_\psi = \psi$ is given by

$$\langle \psi, \phi \rangle = \int_G \psi(x)\, \phi(x)\, \mathrm{d}y$$

(in the same way as on $\mathbb{R}^n$ in Remark 1.3.7).

Definition 10.12.2 (Transpose and adjoint). The *transpose* of $A \in \mathcal{L}(\mathcal{D}(G))$ is $A^t \in \mathcal{L}(\mathcal{D}'(G))$ defined by the equality

$$\langle A^t f, \phi \rangle = \langle f, A\phi \rangle,$$

and the *adjoint* $A^* \in \mathcal{L}(\mathcal{D}'(G))$ is given by the equality

$$(A^* f, \phi) = (f, A\phi),$$

where $(f, \phi) = \langle f, \overline{\phi} \rangle$, $\overline{\phi}(x) := \overline{\phi(x)}$ is the complex conjugate.

Exercise 10.12.3. For $f \in \mathcal{D}'(G)$ and for the left-convolution operator $l(f)\phi :=$ $f * \phi$ show that $l(f)^* = l(\hat{f})$ where $\hat{f}(x) = \overline{f(x^{-1})}$, and $l(f)^t = l(\check{f})$ where $\check{f}(x) = f(x^{-1})$.

Remark 10.12.4 **(Schwartz kernel).** Let $\mathcal{D}(G) \widehat{\otimes} \mathcal{D}'(G)$ denote the complete locally convex tensor product of the nuclear spaces $\mathcal{D}(G)$ and $\mathcal{D}'(G)$ (see Section B.3.1). Then $K \in \mathcal{D}(G) \widehat{\otimes} \mathcal{D}'(G)$ defines a linear operator $A \in \mathcal{L}(\mathcal{D}(G))$ by

$$\langle f, A\phi \rangle := \langle K, \phi \otimes f \rangle. \tag{10.70}$$

In fact, the Schwartz Kernel Theorem states that $\mathcal{L}(\mathcal{D}(G))$ and $\mathcal{D}(G) \widehat{\otimes} \mathcal{D}'(G)$ are isomorphic: for every $A \in \mathcal{L}(\mathcal{D}(G))$ there exists a unique $K_A \in \mathcal{D}(G) \widehat{\otimes} \mathcal{D}'(G)$ such that the duality (10.70) is satisfied with $K = K_A$, which is called the *Schwartz kernel of A*. Duality (10.70) gives us also the interpretation for

$$(A\phi)(x) = \int_G K_A(x, y)\, \phi(y)\, \mathrm{d}y.$$

For a more general Schwartz kernel theorem see Theorem B.3.55 and Definition B.3.56.

Lemma 10.12.5. *Let $A \in \mathcal{L}(\mathcal{D}(G))$, and let*

$$L_A(x, y) := K_A(x, y^{-1}x)$$

in the sense of distributions. Then $L_A \in \mathcal{D}(G) \widehat{\otimes} \mathcal{D}'(G)$.

Proof. Notice that $\mathcal{D}(G) \widehat{\otimes} \mathcal{D}(G) \cong \mathcal{D}(G \times G)$. Let us define the *multiplication*

$$m : \mathcal{D}(G) \otimes \mathcal{D}(G) \to \mathcal{D}(G), \quad m(f \otimes g)(x) := f(x)g(x),$$

the *co-multiplication*

$$\Delta : \mathcal{D}(G) \to \mathcal{D}(G) \widehat{\otimes} \mathcal{D}(G), \quad (\Delta f)(x, y) := f(xy),$$

and the *antipode*

$$S : \mathcal{D}(G) \to \mathcal{D}(G), \quad (Sf)(x) := f(x^{-1}).$$

These mappings are a part of the (nuclear Fréchet) Hopf algebra structure of $\mathcal{D}(G)$, see Chapter 9 (as well as, e.g., [1] or [126]). The mappings are all continuous and linear.

The *convolution of operators* $A, B \in \mathcal{L}(\mathcal{D}(G))$ is said to be the operator

$$A * B := m(A \otimes B)\Delta \in \mathcal{L}(\mathcal{D}(G));$$

it is easy to calculate the Schwartz kernel

$$K_{A*B}(x, y) = \int_G K_A(x, yz^{-1})\, K_B(x, z)\, \mathrm{d}z,$$

or

$$K_{A*B} = (m \otimes \Delta^t)(\mathrm{id} \otimes \tau \otimes \mathrm{id})(K_A \otimes K_B),$$

where

$$\tau : \mathcal{D}'(G) \widehat{\otimes} \mathcal{D}(G) \to \mathcal{D}(G) \widehat{\otimes} \mathcal{D}'(G), \quad \tau(f \otimes \phi) := \phi \otimes f,$$

and

$$\Delta^t : \mathcal{D}'(G) \widehat{\otimes} \mathcal{D}'(G) \to \mathcal{D}'(G)$$

is the transpose of the co-multiplication Δ, $\Delta^t(f \otimes g) = f * g$ (i.e., Δ^t extends the convolution of distributions). Now $(A * S)S \in \mathcal{L}(\mathcal{D}(G))$, hence $K_{(A*S)S} \in \mathcal{D}(G) \widehat{\otimes} \mathcal{D}'(G)$ by the Schwartz kernel theorem, and

$$
\begin{aligned}
K_{(A*S)S}(x,y) &= K_{A*S}(x, y^{-1}) \\
&= \int_G K_A(x, y^{-1}z^{-1}) \, K_S(x, z) \, \mathrm{d}z \\
&= K_A(x, y^{-1}x) \\
&= L_A(x, y).
\end{aligned}
$$

$\square$

Remark 10.12.6. Any distribution $s \in \mathcal{D}(G) \widehat{\otimes} \mathcal{D}'(G)$ can be considered as a mapping

$$s : G \to \mathcal{D}'(G), \quad x \mapsto s(x),$$

where $s(x)(y) := s(x,y)$. If $D \in \mathcal{L}(\mathcal{D}(G))$ and $M \in \mathcal{L}(\mathcal{D}'(G))$ then $(D \otimes M)s \in \mathcal{D}(G) \widehat{\otimes} \mathcal{D}'(G)$. For instance, D could be a partial differential operator, and M a multiplication.

Exercise 10.12.7. Prove Lemma 10.12.5 for the right-convolution kernel $R_A(x,y)$.

Chapter 11

Fourier Analysis on SU(2)

In this chapter we develop elements of Fourier analysis on the group SU(2) in a form suitable for the consequent development of the theory of pseudo-differential operators on SU(2) in Chapter 12. Certain results from this chapter can be found in [148], which, together with [154] we can recommend for further reading, including for some instances of explicitly calculated Clebsch–Gordan coefficients. However, on this occasion, with pseudo-differential operators in mind and the form of the analysis necessary for us and adopted to Chapter 10, we present an independent exposition of SU(2) with considerably more direct proofs and different arguments compared to, e.g., [148].

11.1 Preliminaries: groups U(1), SO(2), and SO(3)

First, we discuss a simpler model of commutative groups $\mathrm{U}(1) \cong \mathrm{SO}(2)$, to treat rotation group SO(3) in a similar manner later. Following that, we study the special unitary group SU(2). For the definitions of $\mathrm{SO}(n)$ and $\mathrm{SU}(n)$ we refer to Remarks 6.2.9 and 6.2.10, respectively. We start by discussing Lie algebras of U(1), SO(2), and SO(3).

Definition 11.1.1 (Group U(1)). Let $\mathfrak{g}$ be the real vector space of 1-by-1-matrices

$$X = X(x) = (\mathrm{i}2\pi x), \quad (x \in \mathbb{R}).$$

Of course, one may treat $X(x)$ as a pure imaginary number, but we insist on the matrix interpretation to anticipate future developments. Then

$$\exp(X(x)) = \left(\mathrm{e}^{\mathrm{i}2\pi x}\right)$$

is a matrix belonging to the Lie group $G = \mathrm{U}(1)$, the unitary matrix group of dimension 1. Of course, $\mathrm{U}(1) \cong \{z \in \mathbb{C} : |z| = 1\} \cong \mathbb{T}^1 = \mathbb{R}/\mathbb{Z}$.

Definition 11.1.2 (Group SO(2)). Let $\mathfrak{g}$ be the real vector space of matrices

$$X = X(x) = \begin{pmatrix} 0 & -x \\ x & 0 \end{pmatrix}, \quad (x \in \mathbb{R}).$$

Now

$$\exp(X(x)) = \begin{pmatrix} \cos(x) & -\sin(x) \\ \sin(x) & \cos(x) \end{pmatrix}$$

is a matrix belonging to the Lie group $G = SO(2)$. If $|x| < \pi/2$ and $g = (g_{ij})_{i,j=1}^2 = \exp(X(x))$ then

$$x = \arcsin(g_{21}) = \arcsin\left((g_{21} - g_{12})/2\right).$$

Definition 11.1.3 (Group SO(3)). Let the real vector space $\mathfrak{g}$ consist of matrices of the form

$$X(x) = \begin{pmatrix} 0 & -x_3 & x_2 \\ x_3 & 0 & -x_1 \\ -x_2 & x_1 & 0 \end{pmatrix},$$

where $x = (x_j)_{j=1}^3 \in \mathbb{R}^3$. Thus $x \mapsto X(x)$ identifies $\mathbb{R}^3$ with $\mathfrak{g}$, and we equip $\mathfrak{g}$ with the Euclidean norm $\|X(x)\|_{\mathfrak{g}} := \|x\|_{\mathbb{R}^3} = \sqrt{x_1^2 + x_2^2 + x_3^2}$. Actually, $\mathfrak{g}$ is a Lie algebra with the Lie product $(A, B) \mapsto [A, B] := AB - BA$, since

$$[X(x), X(y)] = X\left(\begin{pmatrix} x_2 y_3 - x_3 y_2 \\ x_3 y_1 - x_1 y_3 \\ x_1 y_2 - x_2 y_1 \end{pmatrix}\right).$$

It turns out that $\mathfrak{g}$ is the Lie algebra of the group $G = SO(3)$. Let

$$X_1 := X((1,0,0)), \ X_2 := X((0,1,0)), \ X_3 = X((0,0,1)).$$

Then

$$[X_1, X_2] = X_3, [X_2, X_3] = X_1 \text{ and } [X_3, X_1] = X_2.$$

If $x \in \mathbb{R}^3$ and $t := \|x\|_{\mathbb{R}^3}$ then we have the *Rodrigues representation formula*

$$\exp(X(x)) = I + X(x)\,\frac{\sin(t)}{t} + X(x)^2\,\frac{1 - \cos(t)}{t^2}$$

which is equal to

$$\begin{pmatrix} 1 + (x_1^2 - t^2)\frac{1-\cos t}{t^2} & -x_3\frac{\sin t}{t} + x_1 x_2\frac{1-\cos t}{t^2} & x_2\frac{\sin t}{t} + x_1 x_3\frac{1-\cos t}{t^2} \\ x_3\frac{\sin t}{t} + x_1 x_2\frac{1-\cos t}{t^2} & 1 + (x_2^2 - t^2)\frac{1-\cos t}{t^2} & -x_1\frac{\sin t}{t} + x_2 x_3\frac{1-\cos t}{t^2} \\ -x_2\frac{\sin t}{t} + x_1 x_3\frac{1-\cos t}{t^2} & x_1\frac{\sin t}{t} + x_2 x_3\frac{1-\cos t}{t^2} & 1 + (x_3^2 - t^2)\frac{1-\cos t}{t^2} \end{pmatrix}.$$

If here $\|x\|_{\mathbb{R}^3} < \pi$ and $g = (g_{ij})_{i,j=1}^3 = \exp(X(x))$, we obtain the formula

$$x = (x_j)_{j=1}^3 = \frac{t}{2\,\sin(t)}\begin{pmatrix} g_{32} - g_{23} \\ g_{13} - g_{31} \\ g_{21} - g_{12} \end{pmatrix},$$

where $\sin(t) = \sqrt{1 - \cos^2(t)}$ with $\cos(t) = (g_{11} + g_{22} + g_{33} - 1)/2$.

11.1.1   Euler angles on $SO(3)$

Euler angles are useful as local coordinates on $SO(3)$. First we note that rotations of $\mathbb{R}^3$ by an angle $t \in \mathbb{R}$ around the x_j-axis, $j = 1, 2, 3$, respectively, are expressed by the matrices $\omega_j(t) = \exp(tX_j)$ given by

$$\begin{pmatrix} 1 & 0 & 0 \\ 0 & \cos t & -\sin t \\ 0 & \sin t & \cos t \end{pmatrix}, \begin{pmatrix} \cos t & 0 & \sin t \\ 0 & 1 & 0 \\ -\sin t & 0 & \cos t \end{pmatrix}, \begin{pmatrix} \cos t & -\sin t & 0 \\ \sin t & \cos t & 0 \\ 0 & 0 & 1 \end{pmatrix}.$$

We represent rotations by *Euler angles* $\phi, \theta, \psi \in \mathbb{R}$. Any $g \in SO(3)$ is of the form

$$g = \omega(\phi, \theta, \psi) := \omega_3(\phi) \, \omega_2(\theta) \, \omega_3(\psi),$$

where $-\pi < \phi, \psi \leq \pi$ and $0 \leq \theta \leq \pi$. If $0 < \theta_1, \theta_2 < \pi$ then $\omega(\phi_1, \theta_1, \psi_1) = \omega(\phi_2, \theta_2, \psi_2)$ if and only if $\theta_1 = \theta_2$ and $\phi_1 \equiv \phi_2 \pmod{2\pi}$ and $\psi_1 \equiv \psi_2 \pmod{2\pi}$; thus we conclude that the Euler angles provide local coordinates for the manifold $SO(3)$ nearby a point $\omega(\phi, \theta, \psi)$ whenever $\theta \not\equiv 0 \pmod{\pi}$.

Let $g = \omega(\phi, \theta, \psi)$ be the Euler angle representation of $g \in SO(3)$, where $-\pi < \phi, \psi \leq \pi$, $0 \leq \theta \leq \pi$, so that $\omega(\phi, \theta, \psi)$ is

$$\begin{pmatrix} \cos\phi\cos\theta\cos\psi - \sin\phi\sin\psi & -\cos\phi\cos\theta\sin\psi - \sin\phi\cos\psi & \cos\phi\sin\theta \\ \sin\phi\cos\theta\cos\psi + \cos\phi\sin\psi & -\sin\phi\cos\theta\sin\psi + \cos\phi\cos\psi & \sin\phi\sin\theta \\ -\sin\theta\cos\psi & \sin\theta\sin\psi & \cos\theta \end{pmatrix}.$$

The group $SO(3)$ acts transitively on the space $\mathbb{S}^2$, as $\omega(\phi, \theta, \psi)$ moves the north pole $e_3 = (0, 0, 1)^T \in \mathbb{S}^2$ to the point

$$\omega(\phi, \theta, \psi)e_3 = \begin{pmatrix} \cos\phi \, \sin\theta \\ \sin\phi \, \sin\theta \\ \cos\theta \end{pmatrix}.$$

If $0 < \theta < \pi$ and $-\pi/2 < \phi, \psi < \pi/2$, then the Euler angles and the exponential coordinates are related by

$$\omega(\phi, \theta, \psi) = \exp(X(x)),$$

where $x \in \mathbb{R}^3$, $0 < t := \|x\|_{\mathbb{R}^3} < \pi$,

$$\cos t = (\cos(\phi + \psi)(1 + \cos\theta) + \cos\theta - 1)/2,$$

and

$$x = \frac{t}{2\sin t} \begin{pmatrix} \sin\theta(\sin\psi - \sin\phi) \\ \sin\theta(\cos\psi + \cos\phi) \\ (1 + \cos\theta)\sin(\phi + \psi) \end{pmatrix}.$$

11.1.2 Partial derivatives on $\mathrm{SO}(3)$

For an element $g \in \mathrm{SO}(3)$ we now introduce the notation for its components. Thus, if g denotes elements of the group $\mathrm{SO}(3)$, we can view it as the identity mapping from $\mathrm{SO}(3)$ to $\mathrm{SO}(3)$. In its matrix components we can write this as $g = (g_{ij})_{i,j=1}^{3} : \mathrm{SO}(3) \to \mathrm{SO}(3)$. In particular, if $g \in \mathrm{SO}(3)$, we can denote its matrix components by g_{ij}, $1 \leq i,j \leq 3$, so that we obtain functions $(g \mapsto g_{ij}) : \mathrm{SO}(3) \to \mathbb{R}$ which are smooth on G: $g_{ij} \in C^{\infty}(G)$; with this identification and notation, $g = (g_{ij})_{i,j=1}^{3} : \mathrm{SO}(3) \to \mathrm{SO}(3)$ is the identity mapping. Define

$$\partial_1 = \partial^{(1,0,0)}, \quad \partial_2 = \partial^{(0,1,0)}, \quad \partial_3 = \partial^{(0,0,1)}.$$

Notice that

$$(\partial_k f)(g) = \frac{\mathrm{d}}{\mathrm{d}t} f(\omega_k(t)g)|_{t=0}.$$

In particular, we have

$$\partial_k g = \omega_k'(0)g.$$

Here

$$\omega_1'(0) = X_1, \quad \omega_2'(0) = X_2,, \quad \omega_3'(0) = X_3,$$

so that $\partial_1 g$, $\partial_2 g$ and $\partial_3 g$ are respectively

$$\begin{pmatrix} 0 & 0 & 0 \\ -g_{31} & -g_{32} & -g_{33} \\ g_{21} & g_{22} & g_{23} \end{pmatrix}, \quad \begin{pmatrix} g_{31} & g_{32} & g_{33} \\ 0 & 0 & 0 \\ -g_{11} & -g_{12} & -g_{13} \end{pmatrix}, \quad \begin{pmatrix} -g_{21} & -g_{22} & -g_{23} \\ g_{11} & g_{12} & g_{13} \\ 0 & 0 & 0 \end{pmatrix}.$$

11.1.3 Invariant integration on $\mathrm{SO}(3)$

We recall that on a compact group G there exists a unique translation-invariant regular Borel probability measure, called the *Haar measure* μ_G (see Remark 7.4.4); customarily $L^2(G)$ refers to $L^2(G, \mu_G)$. Integrations on G are (unless otherwise mentioned) with respect to μ_G, so we may write

$$\int_G f(x) \, \mathrm{d}x$$

instead of $\int_G f \, \mathrm{d}\mu_G = \int_G f(x) \, \mathrm{d}\mu_G(x)$.

Using the Euler angle coordinates on $G = \mathrm{SO}(3)$, we define an orthogonal projection $P_{\mathbb{S}^2} \in \mathcal{L}(L^2(\mathrm{SO}(3)))$ by

$$(P_{\mathbb{S}^2} f)(\omega(\phi, \theta, \psi)) = \frac{1}{2\pi} \int_{-\pi}^{\pi} f(\omega(\phi, \theta, \tilde{\psi})) \, \mathrm{d}\tilde{\psi}$$

for almost all $g = \omega(\phi, \theta, \psi)$. With the natural interpretation we have $P_{\mathbb{S}^2} f \in L^2(\mathbb{S}^2)$, and if $f \in C^{\infty}(\mathrm{SO}(3))$ then $P_{\mathbb{S}^2} f \in C^{\infty}(\mathbb{S}^2)$. Thereby

$$\int_{\mathrm{SO}(3)} f(x) \, \mathrm{d}x = \int_{\mathbb{S}^2} P_{\mathbb{S}^2} f \, \mathrm{d}\sigma,$$

where the measure $d\sigma$ on the sphere is the normalised angular part of the Lebesgue measure of $\mathbb{R}^3$. This yields the *Haar integral* on SO(3):

$$f \mapsto \int_{\text{SO}(3)} f(x)\, dx = \frac{1}{8\pi^2} \int_{-\pi}^{\pi} \int_0^{\pi} \int_{-\pi}^{\pi} f(\omega(\phi, \theta, \psi))\, \sin(\theta)\, d\phi\, d\theta\, d\psi. \quad (11.1)$$

Exercise 11.1.4. The reader can check rigorously now that (11.1) indeed defines the Haar integral on SO(3).

11.2 General properties of SU(2)

From now on, we shall study the group of two-dimensional special unitary matrices, denoted by SU(2). In other words, we study

$$\text{SU}(2) = \left\{ u \in \mathbb{C}^{2\times 2} : \det(u) = 1 \text{ and } u^*u = I \right\},$$

where $I = \begin{pmatrix} 1 & 0 \\ 0 & 1 \end{pmatrix} \in \mathbb{C}^{2\times 2}$ is the identity matrix of dimension 2. Indeed, SU(2) is a matrix group, since for $u, v \in \text{SU}(2)$, we have

$$\begin{aligned}
\det(uv) &= \det(u)\det(v) &= 1, \\
(uv)^*(uv) &= v^*u^*uv &= I, \\
\det(u^*) &= \overline{\det(u)} &= 1, \\
(u^*)^*u^* &= (uu^*)^* &= I^* &= I, \\
u^{-1} &= u^* &\in \text{SU}(2).
\end{aligned}$$

Lemma 11.2.1 (Elements of SU(2)). *The matrix $u \in \mathbb{C}^{2\times 2}$ belongs to SU(2) if and only if it is of the form*

$$u = \begin{pmatrix} \alpha & \beta \\ -\overline{\beta} & \overline{\alpha} \end{pmatrix},$$

where $\alpha, \beta \in \mathbb{C}$ are such that $|\alpha|^2 + |\beta|^2 = 1$. Moreover, SU(2) is a compact group.

Proof. If $u \in \mathbb{C}^{2\times 2}$ is as above, we have $\det(u) = |\alpha|^2 + |\beta|^2 = 1$ and

$$u^*u = \begin{pmatrix} \overline{\alpha} & -\beta \\ \overline{\beta} & \alpha \end{pmatrix} \begin{pmatrix} \alpha & \beta \\ -\overline{\beta} & \overline{\alpha} \end{pmatrix} = \begin{pmatrix} |\alpha|^2 + |\beta|^2 & 0 \\ 0 & |\beta|^2 + |\alpha|^2 \end{pmatrix} = I$$

so that $u \in \text{SU}(2)$. Now suppose

$$u = \begin{pmatrix} u_{11} & u_{12} \\ u_{21} & u_{22} \end{pmatrix} \in \mathbb{C}^{2\times 2}$$

belongs to SU(2). Then $u^{-1} = u^*$ and $1 = \det(u) = u_{11}u_{22} - u_{12}u_{21}$. Specifically,

$$u^{-1} = \frac{1}{\det(u)} \begin{pmatrix} u_{22} & -u_{12} \\ -u_{21} & u_{11} \end{pmatrix} = \begin{pmatrix} u_{22} & -u_{12} \\ -u_{21} & u_{11} \end{pmatrix}$$

and
$$u^{-1} = u^* = \begin{pmatrix} \overline{u_{11}} & \overline{u_{21}} \\ \overline{u_{12}} & \overline{u_{22}} \end{pmatrix}.$$

This yields the desired form for u, with $\alpha = u_{11}$, $\beta = u_{12}$.

From the proof so far, we see that the mapping $(u \mapsto (u_{11}, u_{12})) : SU(2) \to \mathbb{C}^2$ provides a homeomorphism from $SU(2)$ to the Euclidean unit sphere of $\mathbb{C}^2$. Thus $SU(2)$ is compact.   □

11.3 Euler angle parametrisation of $SU(2)$

The unitary group $U(1) = \{u \in \mathbb{C} : u^*u = 1\}$ is often parametrised by the angle $t \in \mathbb{R}$, i.e., $u(t) = e^{it} \in U(1)$; of course, the parameter range $0 \le t < 2\pi$ is sufficient here. These angles provide convenient expressions for group operations:

$$u(t_0)\, u(t_1) = u(t_0 + t_1) \quad \text{and} \quad u(t)^{-1} = u(-t). \tag{11.2}$$

In analogy to this, elements of $SU(2)$ can be endowed with so-called *Euler angles*, our next topic.

Notice that the Euclidean unit sphere of $\mathbb{C}^2$ is naturally identified with the Euclidean unit sphere $\mathbb{S}^3$ of $\mathbb{R}^4$. An easy (non-unique) way to parametrise the points (u_{11}, u_{12}) of the unit sphere of $\mathbb{C}^2$ by $r, s, t \in \mathbb{R}$ is

$$u_{11} = \cos(r)\, e^{is}, \quad u_{12} = i\, \sin(r)\, e^{it},$$

resulting in
$$u = \begin{pmatrix} \cos(r)\, e^{is} & i\, \sin(r)\, e^{it} \\ i\, \sin(r)\, e^{-it} & \cos(r)\, e^{-is} \end{pmatrix}. \tag{11.3}$$

Putting $r = 0$ in (11.3) we obtain a one-parametric subgroup of matrices
$$\begin{pmatrix} e^{is} & 0 \\ 0 & e^{-is} \end{pmatrix},$$

and with $s = 0 = t$ we get another one-parametric subgroup of matrices
$$\begin{pmatrix} \cos(r) & i\, \sin(r) \\ i\, \sin(r) & \cos(r) \end{pmatrix}.$$

Multiplying elements of these one-parametric subgroups, let us define

$$u(2s, 2r, 2t) := \begin{pmatrix} e^{is} & 0 \\ 0 & e^{-is} \end{pmatrix} \begin{pmatrix} \cos(r) & i\, \sin(r) \\ i\, \sin(r) & \cos(r) \end{pmatrix} \begin{pmatrix} e^{it} & 0 \\ 0 & e^{-it} \end{pmatrix}$$

$$= \begin{pmatrix} \cos(r)\, e^{i(s+t)} & i\, \sin(r)\, e^{i(s-t)} \\ i\, \sin(r)\, e^{-i(s-t)} & \cos(r)\, e^{-i(s+t)} \end{pmatrix}.$$

Clearly, by Lemma 11.2.1 any matrix $u \in SU(2)$ is of this form. This leads us to the *Euler angle parametrisation* of $SU(2)$:

Definition 11.3.1 (Euler's angles on $SU(2)$**).** *Euler's angles* (ϕ, θ, ψ) *from the parameter ranges*

$$0 \leq \phi < 2\pi, \quad 0 \leq \theta \leq \pi, \quad -2\pi \leq \psi < 2\pi \tag{11.4}$$

correspond to the group element

$$u(\phi, \theta, \psi) = \begin{pmatrix} \cos(\frac{\theta}{2}) \ e^{i(\phi+\psi)/2} & i \ \sin(\frac{\theta}{2}) \ e^{i(\phi-\psi)/2} \\ i \ \sin(\frac{\theta}{2}) \ e^{-i(\phi-\psi)/2} & \cos(\frac{\theta}{2}) \ e^{-i(\phi+\psi)/2} \end{pmatrix} \in SU(2).$$

Exercise 11.3.2. Check that the parameter ranges for ϕ, θ, ψ in (11.4) are sufficient. Show that the Euler angle parametrisation is almost injective in the sense that if $(\phi_1, \theta_1, \psi_1) \neq (\phi_2, \theta_2, \psi_2)$ with $0 \leq \phi_j < 2\pi, 0 < \theta_j < \pi, -2\pi \leq \psi < 2\pi$, then $u(\phi_1, \theta_1, \psi_1) \neq u(\phi_2, \theta_2, \psi_2)$.

The angle parametrisation of $U(1)$ behaves well with respect to the group operations in (11.2). Unfortunately, the situation with the Euler angles of $SU(2)$ is complicated. Nevertheless, let us study this problem.

Exercise 11.3.3. Let $u(\phi, \theta, \psi) = \begin{pmatrix} a & b \\ c & d \end{pmatrix}$. Verify that

$$\begin{aligned} 2a\bar{a} &= 1 + \cos(\theta), \\ 2ab &= i \ e^{i\phi} \sin(\theta), \\ -2a\bar{b} &= i \ e^{i\psi} \sin(\theta). \end{aligned}$$

Notice how these formulae allow one to recover Euler angles ϕ, θ, ψ from the matrix $u(\phi, \theta, \psi)$.

Let us examine the multiplication

$$u(\phi, \theta, \psi) = u(\phi_0, \theta_0, \psi_0) \ u(\phi_1, \theta_1, \psi_1).$$

Abbreviating $2r_j := \theta_j$ and $s := \psi_0 + \phi_1$, we have

$$u(\phi, \theta, \psi) = u(\phi_0, 0, 0) \ v \ u(0, 0, \psi_1),$$

where

$$\begin{aligned} v &= u(0, \theta_0, \psi_0) \ u(\phi_1, \theta_1, 0) \\ &= \begin{pmatrix} \cos r_0 \ e^{i\psi_0/2} & i \sin r_0 \ e^{-i\psi_0/2} \\ i \sin r_0 \ e^{i\psi_0/2} & \cos r_0 \ e^{-i\psi_0/2} \end{pmatrix} \begin{pmatrix} \cos r_1 \ e^{i\phi_1/2} & i \sin r_1 \ e^{i\phi_1/2} \\ i \sin r_1 \ e^{-i\phi_1/2} & \cos r_1 \ e^{-i\phi_1/2} \end{pmatrix} \\ &= \begin{pmatrix} \cos(r_0) \cos(r_1)e^{is} - \sin(r_0) \sin(r_1)e^{-is} & \cdots \\ i \sin(r_0) \cos(r_1)e^{is} + i \cos(r_0) \sin(r_1)e^{-is} & \cdots \end{pmatrix}. \end{aligned}$$

Notice that there is no need for calculating the second column of the matrix v.

Applying Exercise 11.3.3 to $u = u(\phi_0, 0, 0)$ v $u(0, 0, \psi_1) = \begin{pmatrix} a & b \\ c & d \end{pmatrix}$, we get

$$
\begin{aligned}
1 + \cos(\theta) \quad &= \quad 2a\bar{a} \\
&= \quad 2\cos(r_0)^2 \cos(r_1)^2 + 2\sin(r_0)^2 \sin(r_1)^2 \\
&\quad -4\cos(2s)\cos(r_0)\sin(r_0)\cos(r_1)\sin(r_1) \\
&= \quad \frac{2}{4}(1 + \cos(\theta_0))(1 + \cos(\theta_1)) + \frac{2}{4}(1 - \cos(\theta_0))(1 - \cos(\theta_1)) \\
&\quad -\cos(2s)\sin(\theta_0)\sin(\theta_1) \\
&= \quad 1 + \cos(\theta_0)\cos(\theta_1) - \cos(\psi_0 + \phi_1)\sin(\theta_0)\sin(\theta_1),
\end{aligned}
$$

$$
\begin{aligned}
&ie^{i\phi}\sin(\theta) \\
&= \quad 2ab \\
&= \quad 2ie^{i\phi_0}\left[e^{i2s}\cos(r_0)^2 \cos(r_1)\sin(r_1) - e^{-i2s}\sin(r_0)^2\sin(r_1)\cos(r_1)\right. \\
&\quad \left. +\cos(r_0)\sin(r_0)\cos(r_1)^2 - \sin(r_0)\cos(r_0)\sin(r_1)^2\right] \\
&= \quad ie^{i\phi_0}\left[\cos(2s)\cos(2r_0)\sin(2r_1) + i\sin(2s)\sin(2r_1)\right. \\
&\quad \left. +\sin(2r_0)\cos(2r_1)\right] \\
&= \quad ie^{i\phi_0}\left[\cos(\psi_0 + \phi_1)\cos(\theta_0)\sin(\theta_1) + i\sin(\psi_0 + \phi_1)\sin(\theta_1)\right. \\
&\quad \left. +\sin(\theta_0)\cos(\theta_1)\right],
\end{aligned}
$$

$$
\begin{aligned}
&ie^{i\psi}\sin(\theta) \\
&= \quad -2a\bar{b} \\
&= \quad 2ie^{i\psi_1}\left[e^{i2s}\cos(r_0)\sin(r_0)\cos(r_1)^2 - e^{-i2s}\sin(r_0)\cos(r_0)\sin(r_1)^2\right. \\
&\quad \left. +\cos(r_0)^2\cos(r_1)\sin(r_1) - \sin(r_0)^2\sin(r_1)\cos(r_1)\right] \\
&= \quad ie^{i\psi_1}\left[\cos(2s)\sin(2r_0)\cos(2r_1) + i\sin(2s)\sin(2r_0)\right. \\
&\quad \left. +\cos(2r_0)\sin(2r_1)\right] \\
&= \quad ie^{i\psi_1}\left[\cos(\psi_0 + \phi_1)\sin(\theta_0)\cos(\theta_1) + i\sin(\psi_0 + \phi_1)\sin(\theta_0)\right. \\
&\quad \left. +\cos(\theta_0)\sin(\theta_1)\right].
\end{aligned}
$$

Let us collect the outcomes:

Proposition 11.3.4. *Let $u(\phi, \theta, \psi) = u_0\, u_1$, where $u_j = u(\phi_j, \theta_j, \psi_j)$. Then*

$$\cos(\theta) = \cos(\theta_0)\cos(\theta_1) - \cos(\psi_0 + \phi_1)\sin(\theta_0)\sin(\theta_1),$$

$$e^{i\phi} = e^{i\phi_0}\frac{\cos(\psi_0 + \phi_1)\cos(\theta_0)\sin(\theta_1) + i\sin(\psi_0 + \phi_1)\sin(\theta_1) + \sin(\theta_0)\cos(\theta_1)}{\sin(\theta)},$$

$$e^{i\psi} = e^{i\psi_1}\frac{\cos(\psi_0 + \phi_1)\sin(\theta_0)\cos(\theta_1) + i\sin(\psi_0 + \phi_1)\sin(\theta_0) + \cos(\theta_0)\sin(\theta_1)}{\sin(\theta)}.$$

Exercise 11.3.5. Let $u(\phi_1, \theta_1, \psi_1) = u(\phi_0, \theta_0, \psi_0)^{-1}$, where

$$0 \le \phi_k < 2\pi, \quad 0 \le \theta_k \le \pi \quad \text{and} \quad -2\pi \le \psi_k < 2\pi.$$

Express $(\phi_1, \theta_1, \psi_1)$ in terms of $(\phi_0, \theta_0, \psi_0)$.

11.4 Quaternions

The *quaternion space* $\mathbb{H}$ is the associative $\mathbb{R}$-algebra with a vector space basis $\{\mathbf{1}, \mathbf{i}, \mathbf{j}, \mathbf{k}\}$, where $\mathbf{1} \in \mathbb{H}$ is the unit and

$$\mathbf{i}^2 = \mathbf{j}^2 = \mathbf{k}^2 = -\mathbf{1} = \mathbf{ijk}.$$

The mapping

$$x = (x_m)_{m=0}^3 \mapsto x_0 \mathbf{1} + x_1 \mathbf{i} + x_2 \mathbf{j} + x_3 \mathbf{k}$$

identifies $\mathbb{R}^4$ with $\mathbb{H}$, and the quaternion inner product is given by

$$(x, y) \mapsto \langle x, y \rangle_{\mathbb{H}} := x_0 y_0 + x_1 y_1 + x_2 y_2 + x_3 y_3,$$

the corresponding norm being

$$x \mapsto \|x\|_{\mathbb{H}} := \langle x, x \rangle_{\mathbb{H}}^{1/2}. \tag{11.5}$$

For all $x, y \in \mathbb{H}$ we have

$$\|xy\|_{\mathbb{H}} = \|x\|_{\mathbb{H}} \|y\|_{\mathbb{H}};$$

if $\|x\|_{\mathbb{H}} = 1$ then both $y \mapsto xy$ and $y \mapsto yx$ are linear isometries $\mathbb{H} \to \mathbb{H}$. In particular, the unit sphere $\mathbb{S}^3 \subset \mathbb{R}^4 \cong \mathbb{H}$ is a multiplicative group.

Exercise 11.4.1. Show that $\mathbf{i}, \mathbf{j}, \mathbf{k}$ satisfy the following multiplication rules:

$$\mathbf{i} = \mathbf{jk} = -\mathbf{kj}, \quad \mathbf{j} = \mathbf{ki} = -\mathbf{ik} \text{ and } \mathbf{k} = \mathbf{ij} = -\mathbf{ji}.$$

11.4.1 Quaternions and SU(2)

A bijective homomorphism $\mathbb{S}^3 \to \mathrm{SU}(2)$ is defined by

$$x \mapsto u(x) = \begin{pmatrix} x_0 + ix_3 & x_1 + ix_2 \\ -x_1 + ix_2 & x_0 - ix_3 \end{pmatrix} =: \begin{pmatrix} \alpha & \beta \\ -\overline{\beta} & \overline{\alpha} \end{pmatrix}, \tag{11.6}$$

with $\det u(x) = \|x\|_{\mathbb{H}}^2 = |x|^2 = 1$. Thus from Lemma 11.2.1 we get:

Proposition 11.4.2 ($\mathbb{S}^3 \cong \mathrm{SU}(2)$). *We have $\mathbb{S}^3 \cong \mathrm{SU}(2)$.*

The reader may ask why we reject a perhaps more obvious candidate for a homomorphism, namely

$$x \mapsto \begin{pmatrix} x_0 + ix_1 & x_2 + ix_3 \\ -x_2 + ix_3 & x_0 - ix_1 \end{pmatrix};$$

the reason is that our choice fits perfectly with the choice of traditional Euler angles, where the x_3-axis is the "fundamental one". On $\mathbb{S}^3 \subset \mathbb{R}^4 \cong \mathbb{H}$, the *Euler angle* representation is

$$x = \begin{pmatrix} x_0 \\ x_1 \\ x_2 \\ x_3 \end{pmatrix} = \begin{pmatrix} \cos\frac{\phi+\psi}{2} \cos\frac{\theta}{2} \\ -\sin\frac{\phi-\psi}{2} \sin\frac{\theta}{2} \\ \cos\frac{\phi-\psi}{2} \sin\frac{\theta}{2} \\ \sin\frac{\phi+\psi}{2} \cos\frac{\theta}{2} \end{pmatrix},$$

where the ranges of parameters ϕ, θ, ψ are the same as in (11.4), i.e.,

$$0 \le \phi < 2\pi, \quad 0 \le \theta \le \pi, \quad -2\pi \le \psi < 2\pi.$$

Notice that the Euler angle representation is unique if and only if $0 < \theta < \pi$, but this is not a true problem for us, since $\theta \in \{0, \pi\}$ corresponds to a set of lower dimension on $\mathbb{S}^3$.

Remark 11.4.3. Recall that for $u = u(\phi, \theta, \psi) \in SU(2)$, the Euler angle representation is given by

$$\begin{aligned} u &= \begin{pmatrix} e^{i(\phi+\psi)/2} \cos\frac{\theta}{2} & e^{i(\phi-\psi)/2} i \sin\frac{\theta}{2} \\ e^{-i(\phi-\psi)/2} i \sin\frac{\theta}{2} & e^{-i(\phi+\psi)/2} \cos\frac{\theta}{2} \end{pmatrix} \\ &= \begin{pmatrix} e^{i\phi/2} & 0 \\ 0 & e^{-i\phi/2} \end{pmatrix} \begin{pmatrix} \cos(\theta/2) & i\sin(\theta/2) \\ i\sin(\theta/2) & \cos(\theta/2) \end{pmatrix} \begin{pmatrix} e^{i\psi/2} & 0 \\ 0 & e^{-i\psi/2} \end{pmatrix}, \end{aligned}$$

where $0 \le \phi < 2\pi, 0 \le \theta \le \pi$ and $-2\pi \le \psi < 2\pi$.

11.4.2   Quaternions and $SO(3)$

Let $P : \mathbb{H} \to \mathbb{H}$ be the orthogonal projection

$$P(x) := x_1 \mathbf{i} + x_2 \mathbf{j} + x_3 \mathbf{k} \cong (x_m)_{m=1}^3.$$

The subspace $P(\mathbb{H}) \subset \mathbb{H}$ is naturally identified with $\mathbb{R}^3$, and the cross product of $\mathbb{R}^3$ is a "shadow" of the quaternion product:

$$P(P(x)P(y)) = \frac{xy - yx}{2} \cong (x_m)_{m=1}^3 \times (y_m)_{m=1}^3.$$

Then $\mathbb{S}^2 \subset \mathbb{R}^3 \cong P(\mathbb{H}) \subset \mathbb{H}$ consists of points x with $\cos((\phi + \psi)/2) = 0$, i.e.,

$$x = \begin{pmatrix} 0 \\ \pm\cos\phi \, \sin\frac{\theta}{2} \\ \pm\sin\phi \, \sin\frac{\theta}{2} \\ \pm\cos\frac{\theta}{2} \end{pmatrix} = \begin{pmatrix} 0 \\ \cos\phi \, \sin\tilde{\theta} \\ \sin\phi \, \sin\tilde{\theta} \\ \cos\tilde{\theta} \end{pmatrix},$$

where $0 \le \phi < 2\pi$ and $0 \le \theta, \tilde{\theta} \le \pi$. of $\mathbb{S}^2$. The *quaternion conjugate* of $x \in \mathbb{H}$ is

$$x^* := x_0 \mathbf{1} - x_1 \mathbf{i} - x_2 \mathbf{j} - x_3 \mathbf{k}.$$

It is easy to see that $u(x^*) = u(x)^*$. Now $(x \mapsto x^*) : \mathbb{H} \to \mathbb{H}$ is a linear isometry such that $(x^*)^* = x$, $(xy)^* = y^* x^*$ and $x^* x = xx^* = \|x\|_{\mathbb{H}}^2 \mathbf{1}$. Consequently, if $\|x\|_{\mathbb{H}} = 1$ then

$$(y \mapsto xyx^*) : \mathbb{H} \to \mathbb{H}$$

is a linear orientation-preserving isometry mapping $P(\mathbb{H}) \to P(\mathbb{H})$ bijectively; hence $y \mapsto xyx^*$ corresponds to a rotation $R_x \in \mathrm{SO}(3)$, and $x \mapsto R_x$ is a group homomorphism $\mathbb{S}^3 \to \mathrm{SO}(3)$. This surjective homomorphism is given by

$$x \cong u(x) = u(\phi, \theta, \psi)$$
$$\mapsto \begin{pmatrix} \mathrm{Re}(\alpha^2 + \beta^2) & -\mathrm{Im}(\alpha^2 - \beta^2) & 2\,\mathrm{Im}(\alpha\beta) \\ \mathrm{Im}(\alpha^2 + \beta^2) & \mathrm{Re}(\alpha^2 - \beta^2) & -2\,\mathrm{Re}(\alpha\beta) \\ 2\,\mathrm{Im}(\alpha\bar{\beta}) & 2\,\mathrm{Re}(\alpha\bar{\beta}) & |\alpha|^2 - |\beta|^2 \end{pmatrix} = w(\phi, \theta, \psi), \quad (11.7)$$

i.e., $u(\phi, \theta, \psi) \mapsto w(\phi, \theta, \psi)$, and its kernel is $\{\mathbf{1}, -\mathbf{1}\} \in \mathbb{S}^3$. Thus we obtain

Theorem 11.4.4 ($\mathrm{SO}(3) \cong \mathbb{S}^3/\{\pm 1\}$). *We have*

$$\mathrm{SO}(3) \cong \mathbb{S}^3/\{\pm 1\},$$

and the space $C^\infty(\mathrm{SO}(3))$ *consists of smooth functions* $f \in C^\infty(\mathbb{S}^3)$ *satisfying* $f(-x) = f(x)$.

11.4.3 Invariant integration on $\mathrm{SU}(2)$

Let us consider the doubly covering homomorphism

$$(u(\phi, \theta, \psi) \mapsto w(\phi, \theta, \psi)) : \mathrm{SU}(2) \to \mathrm{SO}(3),$$

presented in formula (11.7) in Section 11.4.2. From this, recalling the Haar integral (11.1) of $\mathrm{SO}(3)$, we obtain the Haar integral for $\mathrm{SU}(2)$:

$$f \mapsto \int_{\mathrm{SU}(2)} f(x) \, \mathrm{d}x = \frac{1}{16\pi^2} \int_{-2\pi}^{2\pi} \int_0^\pi \int_{-\pi}^\pi f(u(\phi, \theta, \psi)) \, \sin(\theta) \, \mathrm{d}\phi \, \mathrm{d}\theta \, \mathrm{d}\psi.$$

Exercise 11.4.5. Deduce the Haar integral of $\mathrm{SU}(2)$ by considering the group as the unit sphere in $\mathbb{R}^4$ and expressing the Lebesgue measure of $\mathbb{R}^4$ in "polar coordinates".

11.4.4 Symplectic groups

Here we briefly review the notion of symplectic groups and show that the group $\mathrm{SU}(2)$ can be also viewed as the first symplectic group $\mathrm{Sp}(1)$. In general, the subgroups

$$\mathrm{O}(n) < \mathrm{GL}(n, \mathbb{R}), \quad \mathrm{U}(n) < \mathrm{GL}(n, \mathbb{C}) \quad \text{and} \quad \mathrm{Sp}(n) < \mathrm{GL}(n, \mathbb{H})$$

have a lot in common, namely all of them are subgroups of *unitary* elements of the respective general linear groups with respect to their inner products, Euclidean, Hermitian, and symplectic, respectively.

For any $n \in \mathbb{N}$, the symplectic group $\mathrm{Sp}(n)$ is the group of automorphisms of $\mathbb{H}^n$ preserving the norm. More precisely, for $x, y \in \mathbb{H}^n$ we can generalise the inner product (11.5) from $\mathbb{H}$ to $\mathbb{H}^n$ by setting

$$\langle x, y \rangle_{\mathbb{H}^n} := \sum_{j=1}^{n} x_j y_j^*,$$

so that the norm in $\mathbb{H}^n$ becomes $||x||_{\mathbb{H}^n} = \langle x, x \rangle_{\mathbb{H}^n}$. The inner product $\langle x, y \rangle_{\mathbb{H}^n}$ is sometimes called the *symplectic inner product*.

Definition 11.4.6 (Symplectic groups $\mathrm{Sp}(n)$). The symplectic group is defined as the set

$$\mathrm{Sp}(n) = \{A \in \mathrm{GL}(n, \mathbb{H}) : ||Ax||_{\mathbb{H}^n} = ||x||_{\mathbb{H}^n} \text{ for all } x \in \mathbb{H}^n\}.$$

As an immediate corollary we obtain:

Corollary 11.4.7. *We have*

$$\mathrm{Sp}(1) \cong \{x \in \mathbb{H} : ||x||_{\mathbb{H}} = 1\} \cong \mathbb{S}^3 \cong \mathrm{SU}(2).$$

Exercise 11.4.8. Show that if $A \in \mathrm{Sp}(n)$ then $\langle Ax, Ay \rangle_{\mathbb{H}^n} = \langle x, y \rangle_{\mathbb{H}^n}$ for all $x, y \in \mathbb{H}^n$.

Remark 11.4.9 **(Symplectic matrices).** The inner-product preserving identification $\mathbb{H} = \mathbb{C}^2 = \mathbb{R}^4$ extends to the inner-product preserving identification $\mathbb{H}^n = \mathbb{C}^{2n} = \mathbb{R}^{4n}$. In particular, norms on $\mathbb{H}^n$ and $\mathbb{C}^{2n}$ coincide. Therefore, the symplectic group $\mathrm{Sp}(n)$ can be identified with a subgroup of $\mathrm{GL}(2n, \mathbb{C})$ preserving the Euclidean norm with respect to matrix multiplication. Thus, $\mathrm{Sp}(n)$ can be identified with the subgroup of $\mathrm{U}(2n)$ of matrices of the form $\begin{pmatrix} \alpha & \beta \\ -\overline{\beta} & \overline{\alpha} \end{pmatrix}$, where $\alpha, \beta \in \mathrm{End}(\mathbb{C}^n)$. Such matrices are called *symplectic matrices*.

Remark 11.4.10 **(Complex symplectic groups).** The complex symplectic group is defined by

$$\mathrm{Sp}(n, \mathbb{C}) = \{A \in \mathrm{GL}(2n, \mathbb{C}) : A^T J A = J\},$$

where $A^T = \overline{A}^*$ is the transpose matrix of A and $J = \begin{pmatrix} 0 & -I_n \\ I_n & 0 \end{pmatrix}$, where I_n is the identity matrix in $\mathrm{GL}(n, \mathbb{C})$. The matrix J has the property $J^2 = I_{2n}$ so it can be viewed as an extension of the complex identity $i^2 = -1$ to higher dimensions. The product in $\mathrm{GL}(2n, \mathbb{C})$ defined by $a \cdot b := a^T J b$ is invariant under the action of $\mathrm{Sp}(n, \mathbb{C})$.

Exercise 11.4.11. Show that $\mathrm{Sp}(n) = \mathrm{Sp}(n, \mathbb{C}) \cap \mathrm{U}(2n)$. Also show that $\mathrm{Sp}(n, \mathbb{C})$ is not compact but that $\mathrm{Sp}(n)$ is.

11.5   Lie algebra and differential operators on $\mathrm{SU}(2)$

Let $f \in C^\infty(\mathrm{SU}(2))$ and $u : \mathbb{R} \to \mathrm{SU}(2)$ be a smooth function. Let $\phi(t), \theta(t), \psi(t)$ be the Euler angles of $u(t)$. If $0 < \phi(0) < 2\pi$, $0 < \theta(0) < \pi$ and $-2\pi < \psi(0) < 2\pi$ then

$$\left. \frac{d}{dt} f(u) \right|_{t=0} = \left(\phi'(0)\frac{\partial}{\partial\phi} + \theta'(0)\frac{\partial}{\partial\theta} + \psi'(0)\frac{\partial}{\partial\psi} \right) f(u(\phi, \theta, \psi)) \bigg|_{t=0}. \qquad (11.8)$$

This is the derivative of f at $u(0) \in \mathrm{SU}(2)$ in the direction $u'(0)$, as expressed in the Euler angles. Of particular interest are those differential operators that arise naturally from the Lie algebra.

Let us consider one-parametric subgroups $\omega_1, \omega_2, \omega_3 : \mathbb{R} \to \mathrm{SU}(2)$, where

$$\omega_1(t) = \begin{pmatrix} \cos(t/2) & i\sin(t/2) \\ i\sin(t/2) & \cos(t/2) \end{pmatrix},$$

$$\omega_2(t) = \begin{pmatrix} \cos(t/2) & -\sin(t/2) \\ \sin(t/2) & \cos(t/2) \end{pmatrix},$$

$$\omega_3(t) = \begin{pmatrix} e^{it/2} & 0 \\ 0 & e^{-it/2} \end{pmatrix}.$$

Let us also make a special choice $w_j := \omega_j(\pi/2)$, i.e.,

$$w_1 = 2^{-1/2} \begin{pmatrix} 1 & i \\ i & 1 \end{pmatrix},$$

$$w_2 = 2^{-1/2} \begin{pmatrix} 1 & -1 \\ 1 & 1 \end{pmatrix},$$

$$w_3 = 2^{1/2} \begin{pmatrix} 1+i & 0 \\ 0 & 1-i \end{pmatrix}.$$

Exercise 11.5.1. Find the Euler angles of $\omega_j(t)$.

Definition 11.5.2 (Basis of $\mathfrak{su}(2)$). Let $Y_j := \omega_j'(0)$, i.e.,

$$Y_1 = \frac{1}{2}\begin{pmatrix} 0 & i \\ i & 0 \end{pmatrix}, \quad Y_2 = \frac{1}{2}\begin{pmatrix} 0 & -1 \\ 1 & 0 \end{pmatrix}, \quad Y_3 = \frac{1}{2}\begin{pmatrix} i & 0 \\ 0 & -i \end{pmatrix}.$$

Matrices Y_1, Y_2, Y_3 constitute a basis for the real vector space $\mathfrak{su}(2)$, the Lie algebra of $\mathrm{SU}(2)$.

Remark 11.5.3 (**Pauli matrices**). We note that the matrices $\frac{2}{i}Y_j$, $j = 1, 2, 3$, are known as Pauli (spin) matrices in physics. In our case, the coefficient $\frac{i}{2}$ appears in front of these Pauli matrices because we obtained Y_j's as elements of the Lie algebra $\mathfrak{su}(2)$ (leading to the coefficient i), and because of the use of Euler angles (leading to the coefficient $\frac{1}{2}$). It can be also noted that $\mathfrak{k} = \mathrm{span}\{Y_3\}$ and $\mathfrak{p} = \mathrm{span}\{Y_1, Y_2\}$ form a *Cartan pair* of the Lie algebra $\mathfrak{su}(2)$, but we will not pursue this topic here.

Exercise 11.5.4. Check the commutation relations

$$[Y_1, Y_2] = Y_3, \quad [Y_2, Y_3] = Y_1, \quad [Y_3, Y_1] = Y_2.$$

State explicitly a Lie algebra isomorphism $\mathfrak{su}(2) \to \mathfrak{so}(3)$.

Exercise 11.5.5. For $x, y \in \mathbb{R}^3$ let us define $X, Y \in \mathfrak{su}(2)$ by

$$X = \sum_{j=1}^{3} x_j Y_j \quad \text{and} \quad Y = \sum_{j=1}^{3} y_j Y_j.$$

What has $[X, Y]$ to do with the vector cross product $x \times y \in \mathbb{R}^3$?

Exercise 11.5.6. Let $Y(y) = y_1 Y_1 + y_2 Y_2 + y_3 Y_3$. Show that

$$u = \exp(Y(y)) = I \, \cos(\|y\|) + Y(y) \, \frac{\sin(\|y\|)}{\|y\|},$$

and (provided that $\|y\|$ is small enough) that

$$y = \frac{\|y\|}{\sin(\|y\|)} \begin{pmatrix} -i(u_{12} + u_{21}) \\ u_{21} - u_{12} \\ i(u_{22} - u_{11}) \end{pmatrix}.$$

Recall that an inner automorphism of a group G is a group isomorphism of the form $(x \mapsto g^{-1}xg) : G \to G$ for some $g \in G$. Let us study specific inner automorphisms of SU(2) conjugating the one-parametric subgroups $\omega_1, \omega_2, \omega_3$ to each other.

Proposition 11.5.7 (Conjugating one-parametric subgroups to each other). *Let* $w_j = \omega_j(\pi/2)$ *and* $t \in \mathbb{R}$. *Then*

$$\begin{aligned} w_1 \, \omega_2(t) \, w_1^{-1} &= \omega_3(t), \\ w_2 \, \omega_3(t) \, w_2^{-1} &= \omega_1(t), \\ w_3 \, \omega_1(t) \, w_3^{-1} &= \omega_2(t). \end{aligned}$$

The differential versions of these formulae are

$$\begin{aligned} w_1 \, Y_2 \, w_1^{-1} &= Y_3, \\ w_2 \, Y_3 \, w_2^{-1} &= Y_1, \\ w_3 \, Y_1 \, w_3^{-1} &= Y_2. \end{aligned}$$

Proof. We can calculate

$$\begin{aligned} w_1^{-1} \, \omega_3(t) \, w_1 &= \frac{1}{2} \begin{pmatrix} 1 & -i \\ -i & 1 \end{pmatrix} \begin{pmatrix} e^{it/2} & 0 \\ 0 & e^{-it/2} \end{pmatrix} \begin{pmatrix} 1 & i \\ i & 1 \end{pmatrix} \\ &= \frac{1}{2} \begin{pmatrix} e^{it/2} + e^{-it/2} & ie^{it/2} - ie^{-it/2} \\ -ie^{it/2} + ie^{-it/2} & e^{it/2} + e^{-it/2} \end{pmatrix} \\ &= \begin{pmatrix} \cos(t/2) & -\sin(t/2) \\ \sin(t/2) & \cos(t/2) \end{pmatrix} = \omega_2(t). \end{aligned}$$

We also have

$$
\begin{aligned}
w_2\, \omega_3(t)\, w_2^{-1} &= \frac{1}{2}\begin{pmatrix} 1 & -1 \\ 1 & 1 \end{pmatrix}\begin{pmatrix} e^{it/2} & 0 \\ 0 & e^{-it/2} \end{pmatrix}\begin{pmatrix} 1 & 1 \\ -1 & 1 \end{pmatrix} \\
&= \frac{1}{2}\begin{pmatrix} e^{it/2}+e^{-it/2} & e^{it/2}-ie^{-it/2} \\ e^{it/2}-ie^{-it/2} & e^{it/2}+e^{-it/2} \end{pmatrix} \\
&= \begin{pmatrix} \cos(t/2) & i\sin(t/2) \\ i\sin(t/2) & \cos(t/2) \end{pmatrix} \\
&= \omega_1(t).
\end{aligned}
$$

Finally, we have

$$
\begin{aligned}
& w_3\, \omega_1(t)\, w_3^{-1} \\
&= \frac{1}{2}\begin{pmatrix} 1+i & 0 \\ 0 & 1-i \end{pmatrix}\begin{pmatrix} \cos(t/2) & i\sin(t/2) \\ i\sin(t/2) & \cos(t/2) \end{pmatrix}\begin{pmatrix} 1-i & 0 \\ 0 & 1+i \end{pmatrix} \\
&= \frac{1}{2}\begin{pmatrix} 1+i & 0 \\ 0 & 1-i \end{pmatrix}\begin{pmatrix} (1-i)\cos(t/2) & (i-1)i\sin(t/2) \\ (i+1)\sin(t/2) & (1+i)\cos(t/2) \end{pmatrix} \\
&= \begin{pmatrix} \cos(t/2) & -\sin(t/2) \\ \sin(t/2) & \cos(t/2) \end{pmatrix} \\
&= \omega_2(t).
\end{aligned}
$$

The differential version follows immediately by differentiating the above formulae with respect to t at $t=0$. □

Definition 11.5.8 (Left-invariant differential operators). To a vector $Y \in \mathfrak{su}(2)$ we associate the left-invariant differential operator $D_Y : C^\infty(\mathrm{SU}(2)) \to C^\infty(\mathrm{SU}(2))$ defined by

$$
D_Y f(u) = \frac{\mathrm{d}}{\mathrm{d}t} f(u\,\exp(tY))\Big|_{t=0}.
$$

In the sequel, we write $D_j := D_{Y_j}$.

Proposition 11.5.9 (Derivatives D_1, D_2, D_3 in Euler angles). *Expressed in Euler angles,*

$$
\begin{aligned}
D_1 &= \cos(\psi)\,\frac{\partial}{\partial\theta} + \frac{\sin(\psi)}{\sin(\theta)}\,\frac{\partial}{\partial\phi} - \frac{\cos(\theta)}{\sin(\theta)}\,\sin(\psi)\,\frac{\partial}{\partial\psi}, \\
D_2 &= -\sin(\psi)\,\frac{\partial}{\partial\theta} + \frac{\cos(\psi)}{\sin(\theta)}\,\frac{\partial}{\partial\phi} - \frac{\cos(\theta)}{\sin(\theta)}\,\cos(\psi)\,\frac{\partial}{\partial\psi}, \\
D_3 &= \frac{\partial}{\partial\psi},
\end{aligned}
$$

provided that $\sin(\theta) \neq 0$, i.e., $0 < \theta < \pi$.

Proof. The simplest case is

$$D_3 f(u) = \frac{\mathrm{d}}{\mathrm{d}t} f(u(\phi, \theta, \psi) \, \omega_3(t)) \Big|_{t=0} = \frac{\mathrm{d}}{\mathrm{d}t} f(u(\phi, \theta, \psi + t)) \Big|_{t=0} = \frac{\partial}{\partial \psi} f(u).$$

Next we shall deal with D_1. According to (11.8), we need to calculate $\phi'(0)$, $\theta'(0)$, $\psi'(0)$, where $u(\phi, \theta, \psi) = u(\phi_0, \theta_0, \psi_0) \, \omega_1(t)$. Let us exploit Proposition 11.3.4 with $u(\phi_1, \theta_1, \psi_1) = \omega_1(t) = u(0, t, 0)$, yielding

$$\cos(\theta) = \cos(\theta_0) \cos(t) - \cos(\psi_0) \sin(\theta_0) \sin(t),$$

$$e^{i(\phi - \phi_0)} \sin(\theta) = \cos(\psi_0) \cos(\theta_0) \sin(t) + i \sin(\psi_0) \sin(t) + \sin(\theta_0) \cos(t),$$

$$e^{i\psi} \sin(\theta) = \cos(\psi_0) \sin(\theta_0) \cos(t) + i \sin(\psi_0) \sin(\theta_0) + \cos(\theta_0) \sin(t).$$

Differentiating these equalities with respect to t at $t = 0$, we obtain

$$-\theta'(0) \sin(\theta_0) = -\cos(\psi_0) \sin(\theta_0),$$

$$i\phi'(0) \sin(\theta_0) + \theta'(0) \cos(\theta_0) = \cos(\psi_0) \cos(\theta_0) + i \sin(\psi_0),$$

$$e^{i\psi_0} [i\psi'(0) \sin(\theta_0) + \theta'(0) \cos(\theta_0)] = \cos(\theta_0).$$

Thereby $\theta'(0) = \cos(\theta_0)$, $\phi'(0) = \sin(\psi_0)/\sin(\theta_0)$, and

$$i\psi'(0) \sin(\theta_0) = \left[e^{-i\psi_0} - \cos(\psi_0) \right] \cos(\theta_0) = -i \sin(\psi_0) \cos(\theta_0),$$

yielding $\psi'(0) = -\sin(\psi_0) \cos(\theta_0)/\sin(\theta_0)$. This proves the expression for D_1. Lastly, the D_2 formula must be deduced. Again, we use equation (11.8), where this time $u(\phi, \theta, \psi) = u(\phi_0, \theta_0, \psi_0) \, \omega_2(t)$. Thus let

$$u(\phi_1, \theta_1, \psi_1) = \omega_2(t) = u(\pi/2, t, -\pi/2),$$

and apply Proposition 11.3.4: noticing that $\sin(\psi_0 + \pi/2) = \cos(\psi_0)$ and $\cos(\psi_0 + \pi/2) = -\sin(\psi_0)$, we get

$$\cos(\theta) = \cos(\theta_0) \cos(t) + \sin(\psi_0) \sin(\theta_0) \sin(t),$$

$$\begin{aligned} e^{i(\phi - \phi_0)} \sin(\theta) = &-\sin(\psi_0) \cos(\theta_0) \sin(t) + i \cos(\psi_0) \sin(t) \\ &+ \sin(\theta_0) \cos(t), \end{aligned}$$

$$\begin{aligned} e^{i(\psi + \pi/2)} \sin(\theta) = &-\sin(\psi_0) \sin(\theta_0) \cos(t) + i \cos(\psi_0) \sin(\theta_0) \\ &+ \cos(\theta_0) \sin(t). \end{aligned}$$

Differentiating these equalities with respect to t at $t = 0$, we get

$$-\theta'(0) \sin(\theta_0) = \sin(\psi_0) \sin(\theta_0),$$

$$i\phi'(0) \sin(\theta_0) + \theta'(0) \cos(\theta_0) = -\sin(\psi_0) \cos(\theta_0) + i \cos(\psi_0),$$

$$ie^{i\psi_0} [i\psi'(0) \sin(\theta_0) + \theta'(0) \cos(\theta_0)] = \cos(\theta_0).$$

In this case, $\theta'(0) = -\sin(\psi_0)$, $\phi'(0) = \cos(\psi_0)/\sin(\theta_0)$, and

$$i\psi'(0) \sin(\theta_0) = \left[\sin(\psi_0) - ie^{-i\psi_0} \right] \cos(\theta_0) = -i \cos(\psi_0) \cos(\theta_0),$$

so that $\psi'(0) = -\cos(\psi_0) \cos(\theta_0)/\sin(\theta_0)$. This completes the proof. □

Endowing SU(2) with the Riemannian metric inherited from $\mathbb{R}^4 \subset \mathbb{S}^3 \cong$ SU(2), we could show that $\{Y_1, Y_2, Y_3\}$ is orthogonal, so that operators D_1, D_2, D_3 form a good choice for generating the first-order left-invariant partial differential operators. However, in order to simplify notation in the sequel, we will work with slightly different operators.

Definition 11.5.10 (Creation, annihilation, and neutral operators). Let us define left-invariant first-order partial differential operators $\partial_+, \partial_-, \partial_0 : C^\infty(\mathrm{SU}(2)) \to C^\infty(\mathrm{SU}(2))$, called *creation*, *annihilation*, and *neutral* operators, respectively, by

$$\begin{cases} \partial_+ := iD_1 - D_2, \\ \partial_- := iD_1 + D_2, \\ \partial_0 := iD_3, \end{cases} \text{, i.e.,} \quad \begin{cases} D_1 = \frac{-i}{2}(\partial_- + \partial_+), \\ D_2 = \frac{1}{2}(\partial_- - \partial_+), \\ D_3 = -i\partial_0. \end{cases}$$

The reader may ask why we work with operators $\partial_+, \partial_-, \partial_0$ instead of vector fields D_1, D_2, D_3. The reason is that many calculations become considerably shorter. Indeed, already expressions in Euler's angles in Corollary 11.5.13 are quite simpler than those in Proposition 11.5.9, as well as expressions in Theorem 11.9.3 are simpler than those in Proposition 11.9.2. The other reason is that the symbols of $\partial_+, \partial_-, \partial_0$ will turn out to have fewer non-zero elements, see Theorem 12.2.1. The terminology of "creation", "annihilation", and "neutral" operators is explained in Remark 12.2.3.

Remark 11.5.11 **(Laplacian).** The Laplacian $\mathcal{L}$ satisfies $\mathcal{L} = D_1^2 + D_2^2 + D_3^2$ and we have

$$[\mathcal{L}, D_j] = 0$$

for every $j \in \{1, 2, 3\}$. Notice that it can be expressed as

$$\mathcal{L} = -\partial_0^2 - (\partial_+\partial_- + \partial_-\partial_+)/2.$$

Exercise 11.5.12. Show that the operators $\partial_+, \partial_-, \partial_0$ satisfy commutator relations

$$[\partial_0, \partial_+] = \partial_+, \quad [\partial_-, \partial_0] = \partial_-, \quad [\partial_+, \partial_-] = 2\partial_0.$$

The operators $\partial_+, \partial_-, \partial_0$ coincide with operators $\hat{H}_+, \hat{H}_-, \hat{H}_0$ in Vilenkin [148, p. 140].

By Proposition 11.5.9, we also have

Corollary 11.5.13 (Operators $\partial_+, \partial_-, \partial_0$ in Euler angles).

$$\partial_+ = e^{-i\psi}\left(i\frac{\partial}{\partial\theta} - \frac{1}{\sin(\theta)}\frac{\partial}{\partial\phi} + \frac{\cos(\theta)}{\sin(\theta)}\frac{\partial}{\partial\psi}\right),$$

$$\partial_- = e^{i\psi}\left(i\frac{\partial}{\partial\theta} + \frac{1}{\sin(\theta)}\frac{\partial}{\partial\phi} - \frac{\cos(\theta)}{\sin(\theta)}\frac{\partial}{\partial\psi}\right),$$

$$\partial_0 = i\frac{\partial}{\partial\psi}.$$

Exercise 11.5.14. For $u = \begin{pmatrix} a(u) & b(u) \\ c(u) & d(u) \end{pmatrix} \in \mathrm{SU}(2)$, let us write

$$D_Y u := \begin{pmatrix} D_Y a & D_Y b \\ D_Y c & D_Y d \end{pmatrix}.$$

Without resorting to Euler angles, show that

$$D_1 u = \frac{i}{2}\begin{pmatrix} b & a \\ d & c \end{pmatrix}, \quad D_2 u = \frac{1}{2}\begin{pmatrix} b & -a \\ d & -c \end{pmatrix}, \quad D_3 u = \frac{i}{2}\begin{pmatrix} a & -b \\ c & -d \end{pmatrix},$$

$$\partial_+ u = \begin{pmatrix} -b & 0 \\ -d & 0 \end{pmatrix}, \quad \partial_- u = \begin{pmatrix} 0 & -a \\ 0 & -c \end{pmatrix}, \quad \partial_0 u = \frac{1}{2}\begin{pmatrix} -a & b \\ -c & d \end{pmatrix}.$$

11.6   Irreducible unitary representations of $\mathrm{SU}(2)$

In this section we shall determine irreducible unitary representations of $\mathrm{SU}(2)$ up to unitary equivalence. Our first task is to find a natural representation to start with. Matrices act naturally on vectors by matrix multiplication, and this is indeed a good idea: let us identify $z = (z_1, z_2) \in \mathbb{C}^2$ with matrix $z = \begin{pmatrix} z_1 & z_2 \end{pmatrix} \in \mathbb{C}^{1 \times 2}$, and consider

$$T : \mathrm{SU}(2) \to \mathrm{GL}(\mathbb{C}[z_1, z_2]), \quad (T(u)f)(z) := f(zu),$$

where $\mathrm{GL}(\mathbb{C}[z_1, z_2])$ is the space of invertible linear mappings on the complex vector space $\mathbb{C}[z_1, z_2]$ consisting of two-variable polynomials $f : \mathbb{C}^2 \to \mathbb{C}$. Clearly T is a representation of $\mathrm{SU}(2)$ on $\mathbb{C}[z_1, z_2]$, necessarily reducible as $\mathbb{C}[z_1, z_2]$ is infinite-dimensional. So, decomposing T is the next step.

It is clear that the orders of polynomials $f, T(u)f \in \mathbb{C}[z_1, z_2]$ are the same, so that, e.g., polynomials $f \in \mathbb{C}[z_1, z_2]$ of order less than $k \in \mathbb{N}_0$ form a T-invariant subspace. But these subspaces can clearly be decomposed further. For each $l \in \frac{1}{2}\mathbb{N}_0$, let V_l be the subspace of $\mathbb{C}[z_1, z_2]$ containing the homogeneous polynomials of order $2l \in \mathbb{N}_0$, i.e.,

$$V_l = \left\{ f \in \mathbb{C}[z_1, z_2] : f(z_1, z_2) = \sum_{k=0}^{2l} a_k z_1^k z_2^{2l-k}, \quad \{a_k\}_{k=0}^{2l} \subset \mathbb{C} \right\}.$$

Let

$$T_l : \mathrm{SU}(2) \to \mathrm{GL}(V_l), \quad (T_l(u)f)(z) = f(zu),$$

denote the restriction of T on the T-invariant subspace V_l of dimension $(2l+1) \in \mathbb{Z}^+$. Our objective is to show that T_l is irreducible, unitary with respect to a natural inner product of V_l, and that (up to unitary equivalence) there are no other irreducible unitary representations for $\mathrm{SU}(2)$. By regarding $f \in V_l$ as a natural function on $\mathrm{SU}(2)$, we shall endow V_l with the L^2-inner product of the group.

A natural basis for the vector space V_l is $\{p_{lk} : k \in \{0, 1, \ldots, 2l\}\}$, where

$$p_{lk}(z) = z_1^k z_2^{2l-k}. \tag{11.9}$$

Theorem 11.6.1. *As defined above, T_l is irreducible. Moreover, p_{lk} is an eigenfunction of $T_l(u(\phi,0,0))$ with eigenvalue $e^{i\phi(k-l/2)}$.*

Proof. By Schur's Lemma (Corollary 6.3.25), irreducibility of T_l follows, if we can show that its intertwining operators are necessarily scalar. So let $A \in \mathrm{End}(V_l)$ be an intertwining operator for T_l, i.e., such that for every $u \in \mathrm{SU}(2)$ it holds that

$$AT_l(u) = T_l(u)A.$$

Due to the decomposition

$$u(s,r,t) = u(s,0,0)u(0,r,0)u(0,0,t),$$

checking the cases $u = u(0,0,s)$ and $u = u(0,r,0)$ suffices. First,

$$T_l(u(2s,0,0))p_{lk}(z) = p_{lk}(zu(2s,0,0)) = p_{lk}(z_1 e^{is}, z_2 e^{-is}) = e^{is(2k-l)} \, p_{lk}(z),$$

so that p_{lk} is an eigenvector of $T_l(u(2s,0,0))$ with eigenvalue $e^{is(2k-l)}$. The same thing is true for the vector $Ap_{lk} \in V_l$:

$$
\begin{aligned}
T_l(u(2s,0,0))Ap_{lk}(z) &= AT_l(u(2s,0,0))p_{lk}(z) \\
&= A\left(z \mapsto e^{is(2k-l)}p_{lk}(z)\right) = e^{is(2k-l)}Ap_{lk}(z).
\end{aligned}
$$

Let $Ap_{lk} = \sum_j a_j\, p_{lj} \in V_l$. Then

$$
\begin{aligned}
e^{is(2k-l)}Ap_{lk} &= T_l(u(2s,0,0))\sum_j a_j\, p_{lj} \\
&= \sum_j a_j\, T_l(u(2s,0,0))\, p_{lj} = \sum_j a_j\, e^{is(2j-l)}\, p_{lj},
\end{aligned}
$$

which is possible only if $Ap_{lk} = a_k\, p_{lk}$. This yields especially

$$
\begin{aligned}
T_l(u(0,2r,0))Ap_{l0}(z) &= T_l(u(0,2r,0))\, a_0 p_{l0}(z) = a_0\, p_{l0}(zu(0,2r,0)) \\
&= a_0\, (z_1 \sin(r) + \cos(r)z_2)^l \\
&= a_0 \sum_k \binom{l}{k} \sin(r)^k \cos(r)^{l-k} p_{lk}(z).
\end{aligned}
$$

On the other hand, this coincides with

$$
\begin{aligned}
AT_l(u(0,2r,0))p_{l0}(z) &= A\left(z \mapsto p_{l0}(zu(0,2r,0))\right) \\
&= A\left(z \mapsto (z_1 \sin(r) + \cos(r)z_2)^l\right) \\
&= A\sum_k \binom{l}{k} \sin(r)^k \cos(r)^{l-k}\, p_{lk}(z) \\
&= \sum_k \binom{l}{k} \sin(r)^k \cos(r)^{l-k}\, a_k p_{lk}(z).
\end{aligned}
$$

Choosing r so that $\sin(r) \neq 0 \neq \cos(r)$, we see that $a_k = a_0$ for each k. Thus $A = a_0 I$ is scalar. $\qquad\square$

Next we shall find out that, up to equivalence, representations T_l provide a complete set of irreducible representations for SU(2).

Lemma 11.6.2 (Decomposition of elements of SU(2)). *For $g \in$ SU(2) there is a decomposition $g = uhu^{-1}$, where $u, h \in$ SU(2) such that h is diagonal.*

Proof. Let $g = \begin{pmatrix} a & b \\ -\bar{b} & \bar{a} \end{pmatrix} \in$ SU(2). If $b = 0$ then the claim is trivial with the choice $h := g$ and $u = I$, so we assume $b \neq 0$. The characteristic polynomial of g is $P_g : \mathbb{C} \to \mathbb{C}$, where

$$
\begin{aligned}
P_g(z) \quad &= \quad \det(zI - g) \\
&= \quad (z - a)(z - \bar{a}) - b(-\bar{b}) \\
&\overset{|a|^2 + |b|^2 = 1}{=} \quad z^2 - 2\mathrm{Re}(a)z + 1.
\end{aligned}
$$

Now $|a| \neq 1$, because $b \neq 0$, and therefore P_g has two distinct roots, i.e., g has two different eigenvalues $z_1, z_2 \in \mathbb{C}$. Let

$$
h = \begin{pmatrix} z_1 & 0 \\ 0 & z_2 \end{pmatrix},
$$

and let the column vectors of $v \in \mathbb{C}^{2 \times 2}$ be the corresponding normalised eigenvectors of g. Now $g = vhv^{-1}$. The eigenvectors corresponding to the different eigenvalues are orthogonal. Thus $v \in$ U(2), and $u := \lambda v \in$ SU(2) with a suitable choice of $\lambda \in \mathbb{C}$, $|\lambda| = 1$. Now we have $g = uhu^{-1}$. $\square$

Theorem 11.6.3 (Completeness of the representation series). *Let $T_\infty :$ SU(2) $\to$ GL(V) be an irreducible representation. Then T_∞ is equivalent to T_l, where $\dim(V) = 2l + 1$.*

Proof. Let $m \in \frac{1}{2} \mathbb{N}_0 \cup \{\infty\}$. Let $\chi_m :$ SU(2) $\to \mathbb{C}$ be the character of T_m. By Theorem 7.8.6 it is enough to show that $\langle \chi_\infty, \chi_l \rangle_{L^2(\mathrm{SU}(2))} \neq 0$ for some $l \in \frac{1}{2} \mathbb{N}_0$. Notice that

$$
\chi_m(uhu^{-1}) = \chi_m(h).
$$

By Lemma 11.6.2, we may identify χ_m with $f_m : \mathbb{R} \to \mathbb{C}$ defined by

$$
f_m(t) := \chi_m(h(t)), \quad \text{where} \quad h(t) = \begin{pmatrix} e^{it} & 0 \\ 0 & e^{-it} \end{pmatrix}.
$$

Thereby

$$
\begin{aligned}
\langle \chi_\infty, \chi_l \rangle_{L^2(\mathrm{SU}(2))} \quad &= \quad \int_{\mathrm{SU}(2)} \chi_\infty(g) \, \overline{\chi_l(g)} \, \mathrm{d}\mu_{\mathrm{SU}(2)}(g) \\
&= \quad \frac{1}{2\pi} \int_0^{2\pi} f_\infty(t) \, \overline{f_l(t)} \, \mathrm{d}t.
\end{aligned}
$$

Recall that $\{p_{lk} : 0 \le k \le 2l\}$ is a basis for vector space V_l, and

$$T_l(h(t))p_{lk}(z) = p_{lk}(z\,h(t)) = p_{lk}(z_1 e^{it}, z_2 e^{-it}) = e^{it(2k-l)}p_{lk}(z).$$

Thus we get

$$f_l(t) = \mathrm{Tr}\,(T_l(h(t))) = \sum_{k=0}^{2l} e^{it(2k-l)} = \sum_{k=-l}^{l} e^{itk}.$$

Thereby $\mathrm{span}\{f_l : l \in \mathbb{N}_0/2\}$ is dense in the space of 2π-periodic continuous even functions. Because $f_\infty : \mathbb{R} \to \mathbb{C}$ is a 2π-periodic continuous even function,

$$\int_0^{2\pi} f_\infty(t)\,\overline{f_l(t)}\,\mathrm{d}t \neq 0$$

for some $l \in \mathbb{N}_0/2$. This completes the proof.   □

11.6.1   Representations of SO(3)

We now briefly review representations of SO(3) and related spherical harmonics. We will only sketch this topic since it is analogous to constructing representations of SU(2). However, to provide an additional insight into the topic we indicate the relation to spherical harmonics.

The group SO(3) is a subgroup of the group $GL(3, \mathbb{R})$ of three-dimensional matrices acting on $\mathbb{R}^3$ by matrix multiplication. In analogy to the spaces V_l in the case of SU(2), we introduce the space P_l to be the complex subspace of the space $\mathbb{C}[x_1, x_2, x_3]$ containing all the homogeneous polynomials of order $l \in \mathbb{N}_0$ in three variables. The group SO(3) acts on P_l in the same way as in the case of SU(2), namely we have

$$T_l : SO(3) \to GL(P_l), \quad (T_l(u)f)(x) = f(xu).$$

Exercise 11.6.4. Show that $\dim P_l = \frac{1}{2}(l+1)(l+2)$.

The problem with the spaces P_l is that they do not yield irreducible representation spaces for $l \ge 2$. For example, the space generated by the polynomial $x_1^2 + x_2^2 + x_3^2$ is a non-trivial SO(3)-invariant subspace of P_2. Thus, we have to decompose the spaces P_l further which is done in the following exercises.

Exercise 11.6.5 (Spherical harmonics). Let $\mathcal{L} = \frac{\partial^2}{\partial x_1^2} + \frac{\partial^2}{\partial x_2^2} + \frac{\partial^2}{\partial x_3^2}$ be the Laplacian on $\mathbb{R}^3$. A polynomial $f \in P_l$ is called a *harmonic polynomial* of order l if $\mathcal{L}f = 0$, and we set $H_l := \{f \in P_l : \mathcal{L}f = 0\}$. Show that $\dim H_l = 2l + 1$. Restrictions of harmonic polynomials to the sphere $\mathbb{S}^2$ are called *spherical harmonics of order l*. Consequently, since harmonic polynomials are homogeneous, the dimension of the space of spherical harmonics of order l is also $2l + 1$.

Exercise 11.6.6. Show that $\mathcal{L}$ is SO(3)-invariant. Consequently, show that H_l is an SO(3)-invariant subspace of P_l.

Exercise 11.6.7. Show that the mapping

$$T_l : \mathrm{SO}(3) \to \mathrm{GL}(H_l), \quad (T_l(u)f)(x) = f(xu)$$

is an irreducible representation of $\mathrm{SO}(3)$. Moreover, show that an irreducible representation $T_\infty : \mathrm{SO}(3) \to \mathrm{GL}(V)$ is equivalent to T_l, where $\dim(V) = 2l + 1$ for $l \in \mathbb{N}_0$.

11.7   Matrix elements of representations of $\mathrm{SU}(2)$

Recalling polynomials in (11.9), we see that the collection

$$\{q_{lk} : \; k \in \{-l, -l+1, \dots, +l-1, +l\}\}$$

is an orthonormal basis for the representation space V_l, where

$$q_{lk}(z) = \frac{z_1^{l-k} z_2^{l+k}}{\sqrt{(l-k)!(l+k)!}}.$$

Let us compute the matrix elements $t_{mn}^l(\phi, \theta, \psi)$ of $T_l(u(\phi, \theta, \psi))$ with respect to this basis.

Theorem 11.7.1 (Matrix elements of T^l). *Let $u \in \mathrm{SU}(2)$ be given by*

$$u = u(\phi, \theta, \psi) = \begin{pmatrix} a & b \\ c & d \end{pmatrix} = \begin{pmatrix} e^{\mathrm{i}(\phi+\psi)/2} \cos\frac{\theta}{2} & e^{\mathrm{i}(\phi-\psi)/2}\mathrm{i}\sin\frac{\theta}{2} \\ e^{-\mathrm{i}(\phi-\psi)/2}\mathrm{i}\sin\frac{\theta}{2} & e^{-\mathrm{i}(\phi+\psi)/2}\cos\frac{\theta}{2} \end{pmatrix}.$$

Then

$$t_{mn}^l(u) = \left(\frac{\mathrm{d}}{\mathrm{d}z_1}\right)^{l-m} \left(\frac{\mathrm{d}}{\mathrm{d}z_2}\right)^{l+m} \frac{(z_1 a + z_2 c)^{l-n}(z_1 b + z_2 d)^{l+n}}{\sqrt{(l-m)!(l+m)!(l-n)!(l+n)!}}, \quad (11.10)$$

where one can see that the right-hand side does not depend on z. In Euler angles

$$t_{mn}^l(\phi, \theta, \psi) = e^{-\mathrm{i}(m\phi + n\psi)} \; P_{mn}^l(\cos(\theta)),$$

where

$$P_{mn}^l(x) = c_{mn}^l \frac{(1-x)^{(n-m)/2}}{(1+x)^{(m+n)/2}} \left(\frac{\mathrm{d}}{\mathrm{d}x}\right)^{l-m} \left[(1-x)^{l-n}(1+x)^{l+n}\right]$$

with

$$c_{mn}^l = 2^{-l} \frac{(-1)^{l-n} \, \mathrm{i}^{n-m}}{\sqrt{(l-n)! \, (l+n)!}} \sqrt{\frac{(l+m)!}{(l-m)!}}.$$

Moreover, we have

$$
t^l_{mn}(u) = \sqrt{\frac{(l-m)!(l+m)!}{(l-n)!(l+n)!}} \sum_{i=\max\{0,n-m\}}^{\min\{l-n,l-m\}} \tag{11.11}
$$

$$
\times \frac{(l-n)!(l+n)!}{i!(l-n-i)!(l-m-i)!(n+m+i)!} a^i b^{l-m-i} c^{l-n-i} d^{n+m+i}.
$$

Definition 11.7.2 (Representations t^l). The matrix $(t^l_{mn})_{m,n}$ with indices m, n such that $-l \leq m, n \leq l$ and $l - m, l - n \in \mathbb{Z}$ will be denoted by t^l. The usual convention in all the formulae is that $0! = 1$. We note that although l, m, n can be half-integers, the *spacing between them* is such that the differences between any of them are always integers!

Remark 11.7.3. Let $0 \leq \theta \leq \pi$ and $x = \cos(\theta)$. Notice that

$$
\begin{aligned}
(1-x)^{1/2} &= 2^{1/2} \sin(\theta/2), \\
(1+x)^{1/2} &= 2^{1/2} \cos(\theta/2).
\end{aligned}
$$

In [148] the classical orthogonal polynomials of Legendre and Jacobi are connected to functions P^l_{mn}, which consequently could be called, e.g., *generalised Legendre–Jacobi functions.*

Proof of Theorem 11.7.1. First, notice that if vectors q_m form an orthonormal basis of a finite-dimensional inner product space V, then we can write $u = \sum_m (u, q_m)_V \, q_m$ for all $u \in V$. Especially, $Tq_n = \sum_m T_{mn} \, q_m$ for a linear operator $T : V \to V$, where the numbers $T_{mn} = (Tq_n, q_m)_V$ are the matrix elements of T with respect to the chosen basis. Thus for the linear operator $T_l(u) : V_l \to V_l$, the matrix coefficients $t^l_{mn}(u)$ satisfy

$$
\begin{aligned}
T_l(u) \, q_{ln}(z) &= \sum_m t^l_{mn}(u) \, q_{lm}(z) \\
&= \sum_m t^l_{mn}(u) \, \frac{z_1^{l-m} z_2^{l+m}}{\sqrt{(l-m)!(l+m)!}}.
\end{aligned}
$$

Since $T_l(u)q_{ln}(z) = q_{ln}(zu)$, we have

$$
\begin{aligned}
t^l_{mn}(u) &= \frac{\sqrt{(l-m)!(l+m)!}}{(l-m)!(l+m)!} \left(\frac{d}{dz_1}\right)^{l-m} \left(\frac{d}{dz_2}\right)^{l+m} T_l(u)q_{ln}(z) \\
&= \left(\frac{d}{dz_1}\right)^{l-m} \left(\frac{d}{dz_2}\right)^{l+m} \frac{(z_1 a + z_2 c)^{l-n}(z_1 b + z_2 d)^{l+n}}{\sqrt{(l-m)!(l+m)!(l-n)!(l+n)!}},
\end{aligned}
$$

proving (11.10). Here we can use that the total number of derivatives is $2l$ and also that $(z_1 a + z_2 c)^{l-n}(z_1 b + z_2 d)^{l+n}$ is a homogeneous polynomial of degree

$2l$ in (z_1, z_2) so that the value of the derivative corresponds to the coefficient of $z_1^{l-m} z_2^{l+m}$ in this polynomial, and the right-hand side does not depend on z. So we can calculate

$$\left(\frac{d}{dz_1}\right)^{l-m} \left(\frac{d}{dz_2}\right)^{l+m} \left[(z_1 a + z_2 c)^{l-n}(z_1 b + z_2 d)^{l+n}\right]$$

$$= \left(\frac{d}{dz_1}\right)^{l-m} \left(\frac{d}{dz_2}\right)^{l+m} \sum_{i=0}^{l-n}\sum_{j=0}^{l+n} \binom{l-n}{i}\binom{l+n}{j}$$

$$\times a^i b^j c^{l-n-i} d^{l+n-j} z_1^{i+j} z_2^{2l-i-j}$$

$$= \left(\frac{d}{dy}\right)^{l-m} (l+m)! \sum_{i=0}^{l-n}\sum_{j=0}^{l+n} \binom{l-n}{i}\binom{l+n}{j}$$

$$\times a^i b^j c^{l-n-i} d^{l+n-j} y^{i+j}\Big|_{y=0}$$

$$= (l+m)! \left(\frac{d}{dy}\right)^{l-m} \left[(ay + c)^{l-n}(by + d)^{l+n}\right]_{y=0}.$$

Recalling $ad - bc = \det(u) = 1$, we notice that $a(by + d) - b(ay + c) = 1$, inspiring a change of variables: let x be such that $a(by + d) = (x + 1)/2$, i.e., $b(ay + c) = (x - 1)/2$. Hence $dx/dy = 2ab$, so that

$$t_{mn}^l = \frac{1}{\sqrt{(l-m)!(l+m)!(l-n)!(l+n)!}}(l+m)!$$

$$\times \left(\frac{d}{dy}\right)^{l-m} \left[(ay + c)^{l-n}(by + d)^{l+n}\right]_{y=0}$$

$$= \frac{1}{\sqrt{(l-m)!(l+m)!(l-n)!(l+n)!}}(l+m)!$$

$$\times (2ab)^{l-m} \left(\frac{d}{dx}\right)^{l-m} \left[\left(\frac{x-1}{2b}\right)^{l-n}\left(\frac{x+1}{2a}\right)^{l+n}\right]_{x=2ad-1}$$

$$= \frac{1}{\sqrt{(l-m)!(l+m)!(l-n)!(l+n)!}}(l+m)!$$

$$\times 2^{-l-m}\frac{b^{n-m}}{a^{m+n}} \left(\frac{d}{dx}\right)^{l-m} \left[(x-1)^{l-n}(x+1)^{l+n}\right]_{x=2ad-1}.$$

Here $x = 2ad - 1 = 2\cos(\theta/2)^2 - 1 = \cos(\theta)$, i.e., $\cos(\theta/2)^2 = (1 + x)/2$ and $\sin(\theta/2)^2 = (1 - x)/2$, $a = e^{i(\phi+\psi)/2}\cos(\theta/2)$ and $b = e^{i(\phi-\psi)/2}i\sin(\theta/2)$. Thus

$$t_{mn}^l = \sqrt{\frac{(l+m)!}{(l-m)!(l-n)!(l+n)!}}\, 2^{-l-m} \frac{\left(e^{i(\phi-\psi)/2}\right)^{n-m}}{\left(e^{i(\phi+\psi)/2}\right)^{n+m}} i^{n-m}$$

$$\times \frac{\sin(\theta/2)^{n-m}}{\cos(\theta/2)^{m+n}} \left(\frac{d}{dx}\right)^{l-m} \left[(x-1)^{l-n}(x+1)^{l+n}\right]_{x=\cos\theta}$$

$$= \sqrt{\frac{(l+m)!}{(l-m)!(l-n)!(l+n)!}}\, 2^{-l-m}\, e^{-i(m\phi+n\psi)}\, i^{n-m}$$

$$\times \frac{\left(\frac{1-x}{2}\right)^{(n-m)/2}}{\left(\frac{1+x}{2}\right)^{(m+n)/2}} \left(\frac{d}{dx}\right)^{l-m} \left[(x-1)^{l-n}(x+1)^{l+n}\right]_{x=\cos\theta}$$

$$= \sqrt{\frac{(l+m)!}{(l-m)!(l-n)!(l+n)!}}\, 2^{-l}\, e^{-i(m\phi+n\psi)}\, i^{n-m}$$

$$\times \frac{(1-x)^{(n-m)/2}}{(1+x)^{(m+n)/2}} \left(\frac{d}{dx}\right)^{l-m} \left[(x-1)^{l-n}(x+1)^{l+n}\right]_{x=\cos\theta}.$$

Let us finally establish formula (11.11). Again, we note that the total number of derivatives in (11.10) is $2l$ and also that $(z_1 a + z_2 c)^{l-n}(z_1 b + z_2 d)^{l+n}$ is a homogeneous polynomial of degree $2l$ in (z_1, z_2). Therefore,

$$t_{mn}^l(u) = (l-m)!(l+m)!\,\alpha,$$

where α is the coefficient of $z_1^{l-m} z_2^{l+m}$ in the polynomial $(z_1 a + z_2 c)^{l-n}(z_1 b + z_2 d)^{l+n}$. Using the binomial formula, we have

$$\alpha = \sum_{\substack{i=0 \\ i+j=l-m}}^{l-n} \sum_{j=0}^{l+n} \binom{l-n}{i}\binom{l+n}{j} a^i b^j c^{l-n-i} d^{l+n-j}. \tag{11.12}$$

The restriction $j \le l+n$ implies $l-m = i+j \le i+l+n$, so that $i \ge n-m$. On the other hand, the restriction $j \ge 0$ implies $l-m = i+j \ge i$. Thus, substituting $j = l-m-i$ in (11.12) and letting i vary in the range $\max\{0, n-m\} \le i \le \min\{l-n, l-m\}$, we obtain formula (11.11). This completes the proof of Theorem 11.7.1. $\qquad\square$

Exercise 11.7.4. Prove that

$$(-1)^{l+m}(-1)^{l+n} = (-1)^{m-n} \tag{11.13}$$

if $l \in \frac{1}{2}\mathbb{N}_0 = \{k : 2k \in \mathbb{N}_0\}$ and $m, n \in \{-l, -l+1, \dots, +l-1, +l\}$.

Corollary 11.7.5. *For each θ, we have the identities*

$$P_{-m,-n}^l(\cos(\theta)) = P_{mn}^l(\cos(\theta)), \tag{11.14}$$

$$P_{nm}^l(\cos(\theta)) = P_{mn}^l(\cos(\theta)). \tag{11.15}$$

Proof. By Theorem 11.7.1, notice that $P_{mn}^l(\cos(\theta)) = t_{mn}^l(u)$ if

$$u = \begin{pmatrix} a & b \\ b & a \end{pmatrix}, \quad \text{where} \quad \begin{cases} a = \cos(\theta/2), \\ b = i\sin(\theta/2). \end{cases}$$

Then Theorem 11.7.1 yields

$$t^l_{mn}(u) = \left(\frac{\mathrm{d}}{\mathrm{d}z_1}\right)^{l-m} \left(\frac{\mathrm{d}}{\mathrm{d}z_2}\right)^{l+m} \frac{(z_1 a + z_2 b)^{l-n} (z_1 b + z_2 a)^{l+n}}{\sqrt{(l-m)!(l+m)!(l-n)!(l+n)!}},$$

and the $(\pm m, \pm n)$-symmetry in this formula implies (11.14). Notice also that

$$
\begin{aligned}
t^l_{mn}(u^*) &= \left(\frac{\mathrm{d}}{\mathrm{d}z_1}\right)^{l-m} \left(\frac{\mathrm{d}}{\mathrm{d}z_2}\right)^{l+m} \frac{(z_1 a - z_2 b)^{l-n}(-z_1 b + z_2 a)^{l+n}}{\sqrt{(l-m)!(l+m)!(l-n)!(l+n)!}} \\
&= \left(\frac{\mathrm{d}}{\mathrm{d}x_1}\right)^{l-m} \left(\frac{-\mathrm{d}}{\mathrm{d}x_2}\right)^{l+m} \frac{(x_1 a + x_2 b)^{l-n}(-x_1 b - x_2 a)^{l+n}}{\sqrt{(l-m)!(l+m)!(l-n)!(l+n)!}} \\
&= (-1)^{l+m}(-1)^{l+n}\, t^l_{mn}(u) \\
&\overset{(11.13)}{=} (-1)^{m-n}\, t^l_{mn}(u).
\end{aligned}
$$

This leads to

$$
\begin{aligned}
P^l_{nm}(\cos(\theta)) &= t^l_{nm}(u) \\
&= \overline{t^l_{mn}(u^*)} \\
&= (-1)^{m-n}\,\overline{t^l_{mn}(u)} \\
&= (-1)^{m-n}\,\overline{P^l_{mn}(\cos(\theta))} \\
&\overset{\text{Theorem } 11.7.1}{=} P^l_{mn}(\cos(\theta)),
\end{aligned}
$$

proving (11.15).   □

11.8 Multiplication formulae for representations of SU(2)

On a compact group G, a function $f : G \to \mathbb{C}$ is called a *trigonometric polynomial* if its translates span the finite-dimensional vector space

$$\mathrm{span}\left\{(x \mapsto f(y^{-1}x)) : G \to \mathbb{C} \mid y \in G\right\},$$

see Section 7.6 for details. Trigonometric polynomials can be expressed as a linear combination of matrix elements of irreducible unitary representations. Thus a trigonometric polynomial is continuous, and on a Lie group even C^∞-smooth. Moreover, trigonometric polynomials form an algebra when endowed with the usual multiplication. On SU(2), actually,

$$t^{l'}_{m'n'}\, t^l_{mn} = \sum_{k=|l-l'|}^{l+l'} C^{ll'(l+k)}_{m'm(m'+m)}\, C^{ll'(l+k)}_{n'n(n'+n)}\, t^{l+k}_{(m'+m)(n'+n)},$$

where $C^{ll'(l+k)}_{m'm(m'+m)}$ are so-called Clebsch-Gordan coefficients, for which there are explicit formulae [148]. We are going to derive basic multiplication formulae on trigonometric polynomials $t^l_{mn} : SU(2) \to \mathbb{C}$; for general multiplication of trigonometric polynomials, one can use this computation iteratively.

Theorem 11.8.1. *Let*

$$\begin{pmatrix} t_{--} & t_{-+} \\ t_{+-} & t_{++} \end{pmatrix} := t^{1/2} = \begin{pmatrix} t^{1/2}_{-1/2,-1/2} & t^{1/2}_{-1/2,+1/2} \\ t^{1/2}_{+1/2,-1/2} & t^{1/2}_{+1/2,+1/2} \end{pmatrix}$$

and $x^{\pm} := x \pm 1/2$ for $x \in \mathbb{R}$. Then

$$
\begin{aligned}
t^l_{mn} t_{--} &= \frac{\sqrt{(l-m+1)(l-n+1)}}{2l+1}\, t^{l+}_{m^-n^-} + \frac{\sqrt{(l+m)(l+n)}}{2l+1}\, t^{l-}_{m^-n^-}, \\
t^l_{mn} t_{++} &= \frac{\sqrt{(l+m+1)(l+n+1)}}{2l+1}\, t^{l+}_{m^+n^+} + \frac{\sqrt{(l-m)(l-n)}}{2l+1}\, t^{l-}_{m^+n^+}, \\
t^l_{mn} t_{-+} &= \frac{\sqrt{(l-m+1)(l+n+1)}}{2l+1}\, t^{l+}_{m^-n^+} - \frac{\sqrt{(l+m)(l-n)}}{2l+1}\, t^{l-}_{m^-n^+}, \\
t^l_{mn} t_{+-} &= \frac{\sqrt{(l+m+1)(l-n+1)}}{2l+1}\, t^{l+}_{m^+n^-} - \frac{\sqrt{(l-m)(l+n)}}{2l+1}\, t^{l-}_{m^+n^-}.
\end{aligned}
$$

Remark 11.8.2. Notice the pattern of the $\pm$ signs above!

Proof. It is enough to consider multiplication formulae for functions P^l_{mn}, because by Theorem 11.7.1,

$$t^l_{mn}(\phi,\theta,\psi) = e^{-i(\phi m+\psi n)} P^l_{mn}(\cos(\theta)).$$

Moreover, by Corollary 11.7.5, the $t^l_{mn}t_{--}$ formula implies the $t^l_{mn}t_{++}$ formula, and the $t^l_{mn}t_{-+}$ formula implies the $t^l_{mn}t_{+-}$ formula. Indeed, evaluating P below at $x = \cos\theta$, we have

$$
\begin{aligned}
P^l_{mn} P^{1/2}_{1/2,1/2} &= P^l_{-m,-n} P^{1/2}_{-1/2,-1/2} \\
&= \frac{\sqrt{(l+m+1)(l+n+1)}}{2l+1}\, P^{l+}_{-m^+,-n^+} + \frac{\sqrt{(l-m)(l-n)}}{2l+1}\, P^{l-}_{-m^+,-n^+} \\
&= \frac{\sqrt{(l+m+1)(l+n+1)}}{2l+1}\, P^{l+}_{m^+,n^+} + \frac{\sqrt{(l-m)(l-n)}}{2l+1}\, P^{l-}_{m^+,n^+}
\end{aligned}
$$

and

$$
\begin{aligned}
P^l_{mn} P^{1/2}_{1/2,-1/2} &= P^l_{-m,-n} P^{1/2}_{-1/2,1/2} \\
&= \frac{\sqrt{(l+m+1)(l-n+1)}}{2l+1}\, P^{l+}_{-m^+,-n^-} + \frac{\sqrt{(l-m)(l+n)}}{2l+1}\, P^{l-}_{-m^+,-n^-} \\
&= \frac{\sqrt{(l+m+1)(l-n+1)}}{2l+1}\, P^{l+}_{m^+,n^-} + \frac{\sqrt{(l-m)(l+n)}}{2l+1}\, P^{l-}_{m^+,n^-}.
\end{aligned}
$$

Let us now prove the $t^l_{mn}t_{--}$ formula. By Theorem 11.7.1, we have

$$P^{l^+}_{m-n-}(x) = c^{l^+}_{m-n-} \frac{(1-x)^{(n-m)/2}}{(1+x)^{(m+n-1)/2}} \left(\frac{\mathrm{d}}{\mathrm{d}x}\right)^{l-m+1} \left[(1-x)^{l-n+1}(1+x)^{l+n}\right],$$

$$= c^{l^+}_{m-n-} \frac{(1-x)^{(n-m)/2}}{(1+x)^{(m+n-1)/2}} \left(\frac{\mathrm{d}}{\mathrm{d}x}\right)^{l-m}$$

$$\times \frac{\mathrm{d}}{\mathrm{d}x} \left[(1-x)^{l-n+1}(1+x)^{l+n}\right],$$

where

$$\frac{\mathrm{d}}{\mathrm{d}x} \left[(1-x)^{l-n+1}(1+x)^{l+n}\right]$$

$$= (1-x)^{l-n}(1+x)^{l+n-1} \left[-(l-n+1)(1+x) + (l+n)(1-x)\right]$$

$$= (1-x)^{l-n}(1+x)^{l+n-1} \left[-(2l+1)(1+x) + 2(l+n)\right]$$

$$= -(2l+1)(1-x)^{l-n}(1+x)^{l+n} + 2(l+n)(1-x)^{l-n}(1+x)^{l+n-1},$$

yielding

$$P^{l^+}_{m-n-}(x) = (2l+1) c^{l^+}_{m-n-} \frac{P^l_{mn}(x)}{c^l_{mn}} \frac{P^{1/2}_{-1/2,-1/2}(x)}{c^{1/2}_{-1/2,-1/2}}$$

$$+ 2(l+n) c^{l^+}_{m-n-} \frac{P^{l^-}_{m-n-}(x)}{c^{l^-}_{m-n-}}$$

$$= (2l+1) \frac{P^l_{mn}(x) P^{1/2}_{-1/2,-1/2}(x)}{\sqrt{(l-m+1)(l-n+1)}}$$

$$- \sqrt{(l+m)(l+n)} \frac{P^{l^-}_{m-n-}(x)}{\sqrt{(l-m+1)(l-n+1)}}.$$

Thereby

$$(2l+1) P^l_{mn}(x) P^{1/2}_{-1/2,-1/2}(x)$$

$$= \sqrt{(l-m+1)(l-n+1)} P^{l^+}_{m-n-}(x) + \sqrt{(l+m)(l+n)} P^{l^-}_{m-n-}(x),$$

proving the $t^l_{mn}t_{--}$ formula. The $t^l_{mn}t_{-+}$ case is similar. Indeed, by Theorem 11.7.1, we have

$$P^{l^+}_{m-n+}(x) = c^{l^+}_{m-n+} \frac{(1-x)^{(n-m+1)/2}}{(1+x)^{(m+n)/2}} \left(\frac{\mathrm{d}}{\mathrm{d}x}\right)^{l-m+1} \left[(1-x)^{l-n}(1+x)^{l+n+1}\right],$$

$$= c^{l^+}_{m-n+} \frac{(1-x)^{(n-m+1)/2}}{(1+x)^{(m+n)/2}} \left(\frac{\mathrm{d}}{\mathrm{d}x}\right)^{l-m} \frac{\mathrm{d}}{\mathrm{d}x} \left[(1-x)^{l-n}(1+x)^{l+n+1}\right],$$

where

$$\frac{d}{dx}\left[(1-x)^{l-n}(1+x)^{l+n+1}\right]$$
$$= (1-x)^{l-n-1}(1+x)^{l+n}\left[-(l-n)(1+x)+(l+n+1)(1-x)\right]$$
$$= (1-x)^{l-n-1}(1+x)^{l+n}\left[(2l+1)(1-x)-2(l-n)\right]$$
$$= (2l+1)\,(1-x)^{l-n}(1+x)^{l+n} - 2(l-n)\,(1-x)^{l-n-1}(1+x)^{l+n},$$

yielding

$$P^{l+}_{m-n+}(x) = (2l+1)\,c^{l+}_{m-n+}\,\frac{P^l_{mn}(x)}{c^l_{mn}}\,\frac{P^{1/2}_{-1/2,+1/2}(x)}{c^{1/2}_{-1/2,+1/2}}$$

$$-2(l-n)\,c^{l+}_{m-n+}\,\frac{P^{l-}_{m-n+}(x)}{c^{l-}_{m-n+}}$$

$$= (2l+1)\frac{P^l_{mn}(x)\,P^{1/2}_{-1/2,+1/2}(x)}{\sqrt{(l-m+1)(l+n+1)}}$$

$$+\sqrt{(l+m)(l-n)}\,\frac{P^{l-}_{m-n+}(x)}{\sqrt{(l-m+1)(l+n+1)}}.$$

Thereby

$$(2l+1)\,P^l_{mn}(x)\,P^{1/2}_{-1/2,+1/2}(x)$$
$$= \sqrt{(l-m+1)(l+n+1)}P^{l+}_{m-n+}(x) - \sqrt{(l+m)(l-n)}P^{l-}_{m-n+}(x),$$

proving the $t^l_{mn}t_{-+}$ formula. □

Exercise 11.8.3. Calculate formulae for $t^l_{mn}t_{++}$ and $t^l_{mn}t_{+-}$ in Theorem 11.8.1 directly from definitions of t^l_{mn}, t_{++} and t_{+-}.

Exercise 11.8.4. There are other forms of multiplication formulae that can be derived. For example, by using symmetries as in Corollary 11.7.5 derive the following formulae:

$$t^l_{mn}t_{++} = \frac{\sqrt{(l-m+1)(l-n+1)}}{2l+1}\,t^{l+}_{m-n-} + \frac{\sqrt{(l+m)(l+n)}}{2l+1}\,t^{l-}_{m-n-},$$

$$t^l_{mn}t_{+-} = \frac{\sqrt{(l-m+1)(l+n+1)}}{2l+1}\,t^{l+}_{m-n+} - \frac{\sqrt{(l+m)(l-n)}}{2l+1}\,t^{l-}_{m-n+}.$$

11.9 Laplacian and derivatives of representations on $\mathrm{SU}(2)$

Our aim now is to derive the formula for the Laplacian in terms of Euler angles and determine its eigenvalues and eigenfunctions. We also want to find formulae for derivatives of representations. First, recall the invariant differential operators D_1, D_2, D_3 and their expressions in Euler angles in view of Proposition 11.5.9:

$$D_1 := \cos(\psi)\,\frac{\partial}{\partial\theta} + \frac{\sin(\psi)}{\sin(\theta)}\,\frac{\partial}{\partial\phi} - \frac{\cos(\theta)}{\sin(\theta)}\,\sin(\psi)\,\frac{\partial}{\partial\psi},$$

$$D_2 := -\sin(\psi)\,\frac{\partial}{\partial\theta} + \frac{\cos(\psi)}{\sin(\theta)}\,\frac{\partial}{\partial\phi} - \frac{\cos(\theta)}{\sin(\theta)}\,\cos(\psi)\,\frac{\partial}{\partial\psi},$$

$$D_3 := \frac{\partial}{\partial\psi}.$$

The Laplacian $\mathcal{L}$ is given by $\mathcal{L} = D_1^2 + D_2^2 + D_3^2$ and we have $[\mathcal{L}, D_j] = 0$, see Remark 11.5.11. Therefore, we see that $D_3^2 = \partial^2/\partial\psi^2$, as well as calculate

$$
\begin{aligned}
D_1^2 \;=\; & \cos\psi\left(\cos\psi\,\frac{\partial^2}{\partial\theta^2} + \frac{-\cos\theta\sin\psi}{\sin(\theta)^2}\,\frac{\partial}{\partial\phi} + \frac{\sin\psi}{\sin\theta}\,\frac{\partial^2}{\partial\theta\,\partial\phi}\right.\\
& \left. -\frac{-\sin(\theta)^2 - \cos(\theta)^2}{\sin(\theta)^2}\,\sin\psi\,\frac{\partial}{\partial\psi} - \frac{\cos\theta}{\sin\theta}\,\sin\psi\,\frac{\partial^2}{\partial\theta\,\partial\psi}\right)\\
& +\frac{\sin\psi}{\sin\theta}\left(\cos\psi\,\frac{\partial^2}{\partial\theta\,\partial\phi} + \frac{\sin\psi}{\sin\theta}\,\frac{\partial^2}{\partial\phi^2} - \frac{\cos\theta}{\sin\theta}\,\sin\psi\,\frac{\partial^2}{\partial\phi\,\partial\psi}\right)\\
& -\frac{\cos\theta}{\sin\theta}\,\sin\psi\left(-\sin\psi\,\frac{\partial}{\partial\theta} + \cos\psi\,\frac{\partial^2}{\partial\theta\,\partial\psi}\right.\\
& \qquad\qquad +\frac{\cos\psi}{\sin\theta}\,\frac{\partial}{\partial\phi} + \frac{\sin\psi}{\sin\theta}\,\frac{\partial^2}{\partial\phi\,\partial\psi}\\
& \qquad\qquad \left. -\frac{\cos\theta}{\sin\theta}\,\cos\psi\,\frac{\partial}{\partial\psi} - \frac{\cos\theta}{\sin\theta}\,\sin\psi\,\frac{\partial^2}{\partial\psi^2}\right),
\end{aligned}
$$

so that after cancellations we get

$$
\begin{aligned}
D_1^2 \;=\; & \cos(\psi)^2\,\frac{\partial^2}{\partial\theta^2} - 2\frac{\cos\psi\,\cos\theta\,\sin\psi}{\sin(\theta)^2}\,\frac{\partial}{\partial\phi} + \frac{2\cos\psi\sin\psi}{\sin\theta}\,\frac{\partial^2}{\partial\theta\,\partial\phi}\\
& +\frac{(1+\cos(\theta)^2)\cos\psi\sin\psi}{\sin(\theta)^2}\,\frac{\partial}{\partial\psi} - 2\frac{\cos\psi\cos\theta}{\sin\theta}\,\sin\psi\,\frac{\partial^2}{\partial\theta\,\partial\psi}\\
& +\frac{\sin(\psi)^2}{\sin(\theta)^2}\,\frac{\partial^2}{\partial\phi^2} - 2\frac{\cos\theta}{\sin(\theta)^2}\,\sin(\psi)^2\,\frac{\partial^2}{\partial\phi\,\partial\psi}\\
& +\frac{\cos(\theta)}{\sin(\theta)}\,\sin(\psi)^2\,\frac{\partial}{\partial\theta}\\
& +\frac{\cos(\theta)^2}{\sin(\theta)^2}\,\sin(\psi)^2\,\frac{\partial^2}{\partial\psi^2}.
\end{aligned}
$$

Similarly, we have

$$
D_2^2 = -\sin\psi \left(-\sin\psi \frac{\partial^2}{\partial\theta^2} \right.
$$

$$
+ \frac{-\cos\theta\cos\psi}{\sin(\theta)^2} \frac{\partial}{\partial\phi} + \frac{\cos\psi}{\sin\theta} \frac{\partial^2}{\partial\theta\,\partial\phi}
$$

$$
\left. - \frac{-\sin(\theta)^2 - \cos(\theta)^2}{\sin(\theta)^2} \cos\psi \frac{\partial}{\partial\psi} - \frac{\cos\theta}{\sin\theta} \cos\psi \frac{\partial^2}{\partial\theta\,\partial\psi} \right)
$$

$$
+ \frac{\cos\psi}{\sin\theta} \left(-\sin\psi \frac{\partial^2}{\partial\theta\,\partial\phi} + \frac{\cos\psi}{\sin\theta} \frac{\partial^2}{\partial\phi^2} - \frac{\cos\theta}{\sin\theta} \cos\psi \frac{\partial^2}{\partial\phi\,\partial\psi} \right)
$$

$$
- \frac{\cos\theta}{\sin\theta} \cos\psi \left(-\cos\psi \frac{\partial}{\partial\theta} - \sin\psi \frac{\partial^2}{\partial\theta\,\partial\psi} \right.
$$

$$
+ \frac{-\sin\psi}{\sin\theta} \frac{\partial}{\partial\phi} + \frac{\cos\psi}{\sin\theta} \frac{\partial^2}{\partial\phi\,\partial\psi}
$$

$$
\left. + \frac{\cos\theta}{\sin\theta} \sin\psi \frac{\partial}{\partial\psi} - \frac{\cos\theta}{\sin\theta} \cos\psi \frac{\partial^2}{\partial\psi^2} \right),
$$

and after cancellations we get

$$
D_2^2 = \sin(\psi)^2 \frac{\partial^2}{\partial\theta^2} + 2\frac{\cos\theta\cos\psi\sin\psi}{\sin(\theta)^2} \frac{\partial}{\partial\phi} - 2\frac{\cos\psi\sin\psi}{\sin\theta} \frac{\partial^2}{\partial\theta\,\partial\phi}
$$

$$
- \frac{1+\cos(\theta)^2}{\sin(\theta)^2} \cos\psi\sin\psi \frac{\partial}{\partial\psi} + 2\frac{\cos\theta}{\sin\theta} \cos\psi\sin\psi \frac{\partial^2}{\partial\theta\,\partial\psi}
$$

$$
+ \frac{\cos(\psi)^2}{\sin(\theta)^2} \frac{\partial^2}{\partial\phi^2} - 2\frac{\cos\theta}{\sin(\theta)^2} \cos(\psi)^2 \frac{\partial^2}{\partial\phi\,\partial\psi}
$$

$$
+ \frac{\cos\theta}{\sin\theta} \cos(\psi)^2 \frac{\partial}{\partial\theta}
$$

$$
+ \frac{\cos(\theta)^2}{\sin(\theta)^2} \cos(\psi)^2 \frac{\partial^2}{\partial\psi^2}.
$$

Hence we obtain

Proposition 11.9.1 (Laplacian on $SU(2)$). *In terms of Euler angles, the Laplacian on $SU(2)$ is given by*

$$
\mathcal{L} = D_1^2 + D_2^2 + D_3^2
$$

$$
= \frac{\partial^2}{\partial\theta^2} + \frac{1}{\sin(\theta)^2} \frac{\partial^2}{\partial\phi^2} - 2\frac{\cos\theta}{\sin(\theta)^2} \frac{\partial^2}{\partial\phi\,\partial\psi} + \frac{1}{\sin(\theta)^2} \frac{\partial^2}{\partial\psi^2}
$$

$$
+ \frac{\cos\theta}{\sin\theta} \frac{\partial}{\partial\theta}.
$$

We now determine derivatives of representations.

Proposition 11.9.2 (Derivatives D_1, D_2, D_3 of t^l_{mn}). *We have*

$$D_1 t^l_{mn} = \frac{\sqrt{(l-n)(l+n+1)}}{-2i} t^l_{m,n+1} + \frac{\sqrt{(l+n)(l-n+1)}}{-2i} t^l_{m,n-1},$$

$$D_2 t^l_{mn} = \frac{\sqrt{(l-n)(l+n+1)}}{2} t^l_{m,n+1} - \frac{\sqrt{(l+n)(l-n+1)}}{2} t^l_{m,n-1},$$

$$D_3 t^l_{mn} = -in\ t^l_{mn}.$$

Proof. Recall that

$$D_j f(u) = \frac{d}{dt} f(u\ \omega_j(t))|_{t=0},$$

where $f : \mathrm{SU}(2) \to \mathbb{C}$. Recall also that t^l_{mn} is a matrix element of the irreducible unitary representation $T_l : \mathrm{SU}(2) \to \mathrm{End}(V_l)$, acting by

$$T_l(u)v(z) = v(zu),$$

where $v \in V_l$ is a homogeneous polynomial $v : \mathbb{C}^2 \to \mathbb{C}$ of order $2l \in \mathbb{N}_0$. Matrix elements $t^l_{mn} : \mathrm{SU}(2) \to \mathbb{C}$ of T_l with respect to the basis

$$\{q_{lk} \mid k \in \{-l, -l+1, \ldots, +l-1, +l\}\}$$

satisfy

$$q_{ln}(zu) = T_l(u)q_{ln}(z) = \sum_m t^l_{mn}(u)\ q_{lm}(z),$$

where

$$q_{lm}(z) = \frac{z_1^{l-m}\ z_2^{l+m}}{\sqrt{(l-m)!\ (l+m)!}}.$$

Especially,

$$D_j q_{ln}(zu) := D_j\left(u \mapsto q_{ln}(zu)\right) = \sum_m \left(D_j t^l_{mn}\right)(u)\ q_{lm}(z).$$

First,

$$
\begin{aligned}
D_3 q_{ln}(zu) &= \frac{d}{dt}\ q_{ln}(z\ u\ \omega_3(t))\Big|_{t=0} \\
&= \frac{d}{dt}\ \frac{\left((zu)_1\ e^{it/2}\right)^{l-n} \left((zu)_2\ e^{-it/2}\right)^{l+n}}{\sqrt{(l-n)!\ (l+n)!}}\Bigg|_{t=0} \\
&= \frac{d}{dt}\ e^{-itn}\Big|_{t=0}\ \frac{(zu)_1^{l-n}(zu)_2^{l+n}}{\sqrt{(l-n)!\ (l+n)!}} \\
&= -in\ q_{ln}(zu).
\end{aligned}
$$

Thereby

$$D_3 t^l_{mn}(u) = -in\ t^l_{mn}(u).$$

Next,

$$D_2 q_{ln}(zu)$$

$$= \left. \frac{d}{dt} q_{ln}(z \, u \, \omega_2(t)) \right|_{t=0}$$

$$= \left. \frac{d}{dt} \frac{\left((zu)_1 \cos \frac{t}{2} + (zu)_2 \sin \frac{t}{2}\right)^{l-n} \left(-(zu)_1 \sin \frac{t}{2} + (zu)_2 \cos \frac{t}{2}\right)^{l+n}}{\sqrt{(l-n)! \, (l+n)!}} \right|_{t=0}$$

$$= \frac{(l-n)(zu)_2 (zu)_1^{l-n-1} (zu)_2^{l+n} - (zu)_1^{l-n}(l+n)(zu)_1(zu)_2^{l+n-1}}{2\sqrt{(l-n)! \, (l+n)!}}$$

$$= \frac{\sqrt{(l-n)(l+n+1)} \, q_{l,n+1}(zu) - \sqrt{(l+n)(l-n+1)} \, q_{l,n-1}(zu)}{2}.$$

Thus $D_2 t^l_{mn}(u)$ equals

$$\frac{\sqrt{(l-n)(l+n+1)} \, t^l_{m,n+1}(u) - \sqrt{(l+n)(l-n+1)} \, t^l_{m,n-1}(u)}{2}.$$

Finally,

$$D_1 q_{ln}(zu)$$

$$= \left. \frac{d}{dt} q_{ln}(z \, u \, \omega_1(t)) \right|_{t=0}$$

$$= \left. \frac{d}{dt} \frac{\left((zu)_1 \cos \frac{t}{2} + (zu)_2 i \sin \frac{t}{2}\right)^{l-n} \left((zu)_1 i \sin \frac{t}{2} + (zu)_2 \cos \frac{t}{2}\right)^{l+n}}{\sqrt{(l-n)! \, (l+n)!}} \right|_{t=0}$$

$$= \frac{(l-n)(zu)_2 (zu)_1^{l-n-1} (zu)_2^{l+n} + (zu)_1^{l-n}(l+n)(zu)_1(zu)_2^{l+n-1}}{-2i \sqrt{(l-n)! \, (l+n)!}}$$

$$= \frac{\sqrt{(l-n)(l+n+1)} \, q_{l,n+1}(zu) + \sqrt{(l+n)(l-n+1)} \, q_{l,n-1}(zu)}{-2i}.$$

Thus $D_1 t^l_{mn}(u)$ equals

$$\frac{\sqrt{(l-n)(l+n+1)} \, t^l_{m,n+1}(u) + \sqrt{(l+n)(l-n+1)} \, t^l_{m,n-1}(u)}{-2i},$$

completing the proof.   □

In the sequel, we will work with operators $\partial_+, \partial_-, \partial_0$ rather than with D_1, D_2, D_3, and the relation between them was given in Definition 11.5.10, which we recall:

$$\begin{cases} \partial_+ := iD_1 - D_2, \\ \partial_- := iD_1 + D_2, \\ \partial_0 := iD_3, \end{cases} \quad \text{, i.e.,} \quad \begin{cases} D_1 = \frac{-i}{2}\left(\partial_- + \partial_+\right), \\ D_2 = \frac{1}{2}\left(\partial_- - \partial_+\right), \\ D_3 = -i\partial_0. \end{cases} \quad (11.16)$$

Theorem 11.9.3 (Derivatives $\partial_+, \partial_-, \partial_0$ and Laplacian $\mathcal{L}$ of t_{mn}^l). *We have*

$$\begin{aligned}
\partial_+ t_{mn}^l &= -\sqrt{(l-n)(l+n+1)}\, t_{m,n+1}^l, \\
\partial_- t_{mn}^l &= -\sqrt{(l+n)(l-n+1)}\, t_{m,n-1}^l, \\
\partial_0 t_{mn}^l &= n\, t_{mn}^l, \\
\mathcal{L} t_{mn}^l &= -l(l+1)\, t_{mn}^l.
\end{aligned}$$

Proof. Formulae for $\partial_+, \partial_-, \partial_0$ follow from formulae in Proposition 11.9.2 and formulae (11.16). Since by Remark 11.5.11 we have

$$\mathcal{L} = -\partial_0^2 - (\partial_+\partial_- + \partial_-\partial_+)/2,$$

we get

$$\begin{aligned}
&\mathcal{L} t_{mn}^l \\
=\ & -n^2\, t_{mn}^l + \frac{1}{2}\left(\sqrt{(l+n)(l-n+1)}\, \partial_+ t_{m,n-1}^l \right. \\
& \left. + \sqrt{(l-n)(l+n+1)}\, \partial_- t_{m,n+1}^l \right) \\
=\ & \frac{-1}{2}\left(2n^2 + \sqrt{(l+n)(l-n+1)}\sqrt{(l-(n-1))(l+(n-1)+1)} \right. \\
& \left. + \sqrt{(l-n)(l+n+1)}\sqrt{(l+(n+1))(l-(n+1)+1)} \right) t_{mn}^l \\
=\ & \frac{-1}{2}\left(n^2 + (l+n)(l-n+1) + (l-n)(l+n+1) \right) t_{mn}^l \\
=\ & \frac{-1}{2}\left(2n^2 + 2(l^2 - n^2) + (l+n) + (l-n) \right) t_{mn}^l \\
=\ & -l(l+1)\, t_{mn}^l,
\end{aligned}$$

completing the proof. $\square$

Remark 11.9.4. In Proposition 11.9.2 we saw that

$$\begin{aligned}
D_1 t_{mn}^l &= \frac{\sqrt{(l-n)(l+n+1)}\, t_{m,n+1}^l + \sqrt{(l+n)(l-n+1)}\, t_{m,n-1}^l}{-2i}, \\
D_2 t_{mn}^l &= \frac{\sqrt{(l-n)(l+n+1)}\, t_{m,n+1}^l - \sqrt{(l+n)(l-n+1)}\, t_{m,n-1}^l}{2}, \\
D_3 t_{mn}^l &= -in\, t_{mn}^l,
\end{aligned}$$

so that

$$\begin{aligned}
\overline{D_1 t_{mn}^l} &= \frac{\sqrt{(l-n)(l+n+1)}\, \overline{t_{m,n+1}^l} + \sqrt{(l+n)(l-n+1)}\, \overline{t_{m,n-1}^l}}{+2i}, \\
\overline{D_2 t_{mn}^l} &= \frac{\sqrt{(l-n)(l+n+1)}\, \overline{t_{m,n+1}^l} - \sqrt{(l+n)(l-n+1)}\, \overline{t_{m,n-1}^l}}{2}, \\
\overline{D_3 t_{mn}^l} &= +in\, \overline{t_{mn}^l},
\end{aligned}$$

implying

$$
\begin{aligned}
\partial_+ \overline{t^l_{mn}} &= +\sqrt{(l+n)(l-n+1)}\ \overline{t^l_{m,n-1}}, \\
\partial_- \overline{t^l_{mn}} &= +\sqrt{(l-n)(l+n+1)}\ \overline{t^l_{m,n+1}}, \\
\partial_0 \overline{t^l_{mn}} &= -n\ \overline{t^l_{mn}}.
\end{aligned}
$$

11.10 Fourier series on $\mathrm{SU}(2)$ and on $\mathrm{SO}(3)$

By the Peter–Weyl theorem (Theorem 7.5.14), an orthogonal basis for $L^2(\mathrm{SU}(2))$ consists of functions t^l_{nm} for $l \in \mathbb{N}_0/2$ and $-l \le m, n \le l$, where $l - m, l - n \in \mathbb{Z}$ and

$$
t^l_{nm}(\omega(\phi, \theta, \psi)) = e^{-i(n\phi + m\psi)}\ P^l_{nm}(\cos(\theta)),
$$

and where

$$
\begin{aligned}
P^l_{nm}(z) \ = \ & 2^{-l}\, \frac{(-1)^{l-m}\ i^{m-n}}{\sqrt{(l-m)!(l+m)!}}\, \sqrt{\frac{(l+n)!}{(l-n)!}} \\
& \times \frac{(1-z)^{\frac{m-n}{2}}}{(1+z)^{\frac{n+m}{2}}}\, \frac{d^{l-n}}{dz^{l-n}}\left[(1-z)^{l-m}(1+z)^{l+m}\right].
\end{aligned}
$$

Note that here we changed the order of indices (m, n) into (n, m) compared with the formulation of the Peter–Weyl theorem in order to have (m, n) entries for the Fourier coefficients in (11.17) and in the sequel.

Notice that for $\mathrm{SO}(3)$, these same formulae are valid with appropriate Euler's angles, but then $l \in \mathbb{N}_0$ (not $l \in \mathbb{N}_0/2$). Nevertheless, $t^l(u) \in \mathrm{U}(2l+1) \subset \mathbb{C}^{(2l+1)\times(2l+1)}$. For instance,

$$
\begin{aligned}
T_0(u(\phi, \theta, \psi)) &= \ (1), \\
T_{1/2}(u(\phi, \theta, \psi)) &= \begin{pmatrix} e^{i(\phi+\psi)/2}\cos\frac{\theta}{2} & e^{i(\phi-\psi)/2}i\sin\frac{\theta}{2} \\ e^{-i(\phi-\psi)/2}i\sin\frac{\theta}{2} & e^{-i(\phi+\psi)/2}\cos\frac{\theta}{2} \end{pmatrix} \\
&= \ u(\phi, \theta, \psi) = \begin{pmatrix} a & b \\ c & d \end{pmatrix}, \\
T_1(u(\phi, \theta, \psi)) &= \begin{pmatrix} e^{i(\phi+\psi)}\cos^2\frac{\theta}{2} & e^{i\phi}\frac{i\sin\theta}{\sqrt{2}} & -e^{i(\phi-\psi)}\sin^2\frac{\theta}{2} \\ e^{i\psi}\frac{i\sin\theta}{\sqrt{2}} & \cos\theta & e^{-i\psi}\frac{i\sin\theta}{\sqrt{2}} \\ -e^{-i(\phi-\psi)}\sin^2\frac{\theta}{2} & e^{-i\phi}\frac{i\sin\theta}{\sqrt{2}} & e^{-i(\phi+\psi)}\cos^2\frac{\theta}{2} \end{pmatrix} \\
&= \begin{pmatrix} a^2 & \sqrt{2}ab & b^2 \\ \sqrt{2}ac & ad+bc & \sqrt{2}bd \\ c^2 & \sqrt{2}cd & d^2 \end{pmatrix}.
\end{aligned}
$$

Consequently, the collection

$$
\{\sqrt{2l+1}\ t^l_{nm} : l \in \tfrac{1}{2}\mathbb{N}_0, \ -l \le m, n \le l, \ l - m, l - n \in \mathbb{Z}\}
$$

is an orthonormal basis for $L^2(\mathrm{SU}(2))$, and thus by the Peter–Weyl theorem (Theorem 7.5.14) any function $f \in C^\infty(\mathrm{SU}(2))$ has a Fourier series representation

$$f(x) = \sum_{l \in \frac{1}{2}\mathbb{N}_0} (2l + 1) \sum_m \sum_n \widehat{f}(l)_{mn}\, t^l_{nm}(x), \qquad (11.17)$$

where the Fourier coefficients are computed by

$$\widehat{f}(l)_{mn} := \int_{\mathrm{SU}(2)} f(x)\, \overline{t^l_{nm}(x)}\, \mathrm{d}x = \langle f, t^l_{nm} \rangle_{L^2(\mathrm{SU}(2))},$$

so that $\widehat{f}(l) \in \mathbb{C}^{(2l+1)\times(2l+1)}$. The series (11.17) converges almost everywhere on SU(2) as well as in $L^2(\mathrm{SU}(2))$.

Definition 11.10.1 (Quantum numbers, notation $\widehat{f}(l)$, and summation on SU(2)).
In the case of SU(2), we simplify the notation writing $\widehat{f}(l)$ instead of $\widehat{f}(t^l)$, etc.
In sums (11.17), our convention will be that summations $\sum_m \sum_n$ are over m, n such that $-l \leq m, n \leq l$ and $l - m, l - n \in \mathbb{Z}$. The index l is called the *quantum number*.

On SO(3), the collection

$$\{\sqrt{2l + 1}\, t^l_{nm} : l \in \mathbb{N}_0,\ m, n \in \mathbb{Z},\ -l \leq m, n \leq l\}$$

is an orthonormal basis for $L^2(\mathrm{SO}(3))$, and thus $f \in C^\infty(\mathrm{SO}(3))$ has a Fourier series representation

$$f(x) = \sum_{l=0}^{\infty} (2l + 1) \sum_{m=-l}^{l} \sum_{n=-l}^{l} \widehat{f}(l)_{mn}\, t^l_{nm}(x),$$

where the Fourier coefficients are computed by

$$\widehat{f}(l)_{mn} := \langle f, t^l_{nm} \rangle_{L^2(\mathrm{SO}(3))} = \int_{\mathrm{SO}(3)} f(x)\, \overline{t^l_{nm}(x)}\, \mathrm{d}\mu_{\mathrm{SO}(3)}(x).$$

Notice that by Remark 10.2.3 we have

$$\overline{t^l_{nm}(x)} = (t^l(x)^*)_{mn} = t^l_{mn}(x^{-1}).$$

Evidently, the values of $f \in C^\infty(\mathbb{S}^2) \subset C^\infty(\mathrm{SO}(3))$ do not depend on the Euler angle ψ, so that in this case $\widehat{f}(l)_{nm} = 0$ whenever $n \neq 0$.

Chapter 12

Pseudo-differential Operators on SU(2)

In this chapter we carry out the analysis of operators on SU(2) with an application to the operators on the three-dimensional sphere $\mathbb{S}^3$. In particular, we derive a much simpler symbolic characterisation of pseudo-differential operators on SU(2) than the one given in Definition 10.9.5. In turn, this will also yield a characterisation of full symbols of pseudo-differential operators on the 3-sphere $\mathbb{S}^3$. We note that this approach works globally on the whole sphere, since the version of the Fourier analysis that we use is different from the one in, e.g., [110, 122, 111] which covers only a hemisphere, with singularities at the equator. For a general introduction and motivation for the analysis on SU(2) we refer the reader to the introduction in Part IV where the cases of SU(2) and $\mathbb{S}^3$ were put in a perspective.

On SU(2), the conventional abbreviations in summation indices are

$$\sum_{l} = \sum_{l \in \frac{1}{2}\mathbb{N}_0}, \qquad \sum_{l}\sum_{m,n} = \sum_{l \in \frac{1}{2}\mathbb{N}_0} \sum_{|m| \leq l,\ l+m \in \mathbb{Z}} \sum_{|n| \leq l,\ l+n \in \mathbb{Z}},$$

where $\mathbb{N}_0 = \{0\} \cup \mathbb{N} = \{0, 1, 2, \dots\}$. As before, the space of all linear mappings from a finite-dimensional (inner-product) vector space $\mathcal{H}$ to itself is denoted by $\mathrm{End}(\mathcal{H})$, a mapping $U \in \mathcal{L}(\mathcal{H})$ is called unitary if $U^* = U^{-1}$, and the space of all unitary linear mappings on a finite-dimensional inner product space $\mathcal{H}$ is denoted by $\mathcal{U}(\mathcal{H})$.

12.1 Symbols of operators on SU(2)

First we summarise the approach to symbols from Section 10.4 in the case of SU(2), also simplifying the notation in this case. We recall that in the case of SU(2), we simplify the notation writing $\widehat{f}(l)$ instead of $\widehat{f}(t^l)$, etc.

By the Peter–Weyl theorem (Theorem 7.5.14) the collection

$$\{\sqrt{2l+1}\, t^l_{nm} : l \in \frac{1}{2}\mathbb{N}_0,\ -l \le m, n \le l,\ l-m, l-n \in \mathbb{Z}\} \tag{12.1}$$

is an orthonormal basis for $L^2(\mathrm{SU}(2))$, and thus $f \in C^\infty(\mathrm{SU}(2))$ has a Fourier series representation

$$\begin{aligned} f(x) &= \sum_{l \in \frac{1}{2}\mathbb{N}_0} (2l+1) \sum_m \sum_n \widehat{f}(l)_{mn}\, t^l_{nm}(x) \\ &= \sum_{l \in \frac{1}{2}\mathbb{N}_0} (2l+1)\, \mathrm{Tr}\Big(\widehat{f}(l)\, t^l(x)\Big), \end{aligned}$$

where the Fourier coefficients are computed by

$$\widehat{f}(l)_{mn} := \int_{\mathrm{SU}(2)} f(x)\, \overline{t^l_{nm}(x)}\, \mathrm{d}x = \langle f, t^l_{nm}\rangle_{L^2(\mathrm{SU}(2))},$$

so that $\widehat{f}(l) \in \mathbb{C}^{(2l+1)\times(2l+1)}$. We recall that $t^l \in \mathbb{C}^{(2l+1)\times(2l+1)}$ by Definition 11.10.1 is a matrix with components t^l_{nm}, with the convention that indices m and n vary as in (12.1).

Let $A : C^\infty(\mathrm{SU}(2)) \to C^\infty(\mathrm{SU}(2))$ be a continuous linear operator and let $R_A \in \mathcal{D}'(\mathrm{SU}(2) \times \mathrm{SU}(2))$ be its right-convolution kernel, i.e.,

$$Af(x) = \int_{\mathrm{SU}(2)} f(y)\, R_A(x, y^{-1}x)\, \mathrm{d}y = (f * R_A(x, \cdot))(x)$$

in the sense of distributions. According to Definition 10.4.3, by the symbol of A we mean the sequence of matrix-valued mappings

$$(x \mapsto \sigma_A(x, l)) : \mathrm{SU}(2) \to \mathbb{C}^{(2l+1)\times(2l+1)},$$

where $2l \in \mathbb{N}_0$, obtained from

$$\sigma_A(x, l)_{mn} = \int_{\mathrm{SU}(2)} R_A(x, y)\, \overline{t^l_{nm}(y)}\, \mathrm{d}y. \tag{12.2}$$

That is, $\sigma_A(x, l)$ is the lth Fourier coefficient of the function $y \mapsto R_A(x, y)$. Then by Theorem 10.4.4 we have

$$\begin{aligned} Af(x) &= \sum_l (2l+1)\, \mathrm{Tr}\Big(t^l(x)\, \sigma_A(x, l)\, \widehat{f}(l)\Big) \\ &= \sum_l (2l+1) \sum_{m,n} t^l(x)_{nm} \left(\sum_k \sigma_A(x, l)_{mk}\, \widehat{f}(l)_{kn}\right). \end{aligned}$$

Alternatively, by Theorem 10.4.6 we have

$$\sigma_A(x, l) = t^l(x)^* \left(A t^l \right)(x),\tag{12.3}$$

that is

$$\sigma_A(x, l)_{mn} = \sum_k \overline{t^l_{km}(x)} (A t^l_{kn})(x),\tag{12.4}$$

by formula (10.21). Formula (10.20) expressing the right-convolution kernel in terms of the symbol becomes

$$R_A(x, y) = \sum_l (2l+1) \sum_{m,n} t^l_{nm}(y)\, \sigma_A(x, l)_{mn},\tag{12.5}$$

with a similar distributional interpretation for the series. In the case of SU(2), the quantity $\langle t^l \rangle$ in (10.9) for the representation $\xi = t^l$ can be calculated as

$$\langle t^l \rangle = (1 + \lambda_{[t^l]})^{1/2} = (1 + l(l+1))^{1/2},$$

in view of Theorem 11.9.3. Consequently, Definition 10.9.5 of the symbol class $\Sigma^m(\mathrm{SU}(2))$ becomes:

Definition 12.1.1. We write that $\sigma_A \in \Sigma_0^m(\mathrm{SU}(2))$ if

$$\text{sing supp}\,(y \mapsto R_A(x, y)) \subset \{e\}\tag{12.6}$$

and if

$$\left\| \triangle_l^\alpha \partial_x^\beta \sigma_A(x, l) \right\|_{\mathbb{C}^{2l+1} \to \mathbb{C}^{2l+1}} \leq C_{A\alpha\beta m}\, (1+l)^{m-|\alpha|}\tag{12.7}$$

for all $x \in G$, all multi-indices α, β, and $l \in \frac{1}{2}\mathbb{N}_0$. Here

$$\partial_x^\beta = \partial_0^{\beta_1} \partial_+^{\beta_2} \partial_-^{\beta_3} \text{ and } \triangle_l^\alpha = \triangle_0^{\alpha_1} \triangle_+^{\alpha_2} \triangle_-^{\alpha_3}$$

are defined in the general situation in Definition 10.7.1, but in Section 12.3 we discuss the simplification of these difference operators. Moreover, $\sigma_A \in \Sigma_{k+1}^m(\mathrm{SU}(2))$ if and only if

$$\sigma_A \quad \in \quad \Sigma_k^m(\mathrm{SU}(2)),\tag{12.8}$$

$$[\sigma_{\partial_j}, \sigma_A] = \sigma_{\partial_j}\sigma_A - \sigma_A\sigma_{\partial_j} \quad \in \quad \Sigma_k^m(\mathrm{SU}(2)),\tag{12.9}$$

$$(\triangle_l^\gamma \sigma_A)\,\sigma_{\partial_j} \quad \in \quad \Sigma_k^{m+1-|\gamma|}(\mathrm{SU}(2)),\tag{12.10}$$

for all $|\gamma| > 0$ and $j \in \{0, +, -\}$. Let

$$\Sigma^m(\mathrm{SU}(2)) := \bigcap_{k=0}^\infty \Sigma_k^m(\mathrm{SU}(2)),$$

so that by Theorem 10.9.6 we have $\mathrm{Op}\,\Sigma^m(\mathrm{SU}(2)) = \Psi^m(\mathrm{SU}(2))$.

Remark 12.1.2. The ordering of operators $\partial_0, \partial_+, \partial_-$ in $\partial_x^\beta = \partial_0^{\beta_1} \partial_+^{\beta_2} \partial_-^{\beta_3}$ may seem to be of importance because they do not commute (see Exercise 11.5.12). However, this is not really an issue in view of Remark 10.5.4: indeed applying their commutator relations from Exercise 11.5.12 iteratively, we see that we can take any ordering in ∂_x^β in (12.7) to obtain the same class of symbols.

Remark 12.1.3. We would like to provide a more direct definition for the symbol class $\Sigma^m(\mathrm{SU}(2))$, without resorting to classes $\Sigma_k^m(\mathrm{SU}(2))$. Condition (12.7) is just an analogy of the usual symbol inequalities. Conditions (12.6) and (12.8) are straightforward. We may have difficulties with differences $\triangle_l^\alpha$, but derivatives ∂_x^β do not cause problems; if we want, we may assume that the symbols are constant in x. In Section 12.4 we present such a simplification, thus providing a more straightforward characterisation of operators from $\Psi^m(\mathrm{SU}(2))$ in terms of quantizations and full symbols developed here.

Exercise 12.1.4. From the definition of operators $\triangle_l^\alpha$ and ∂_x^β, verify the following properties:

$$\triangle_l^\alpha \partial_x^\beta \sigma_A(x, l) = \partial_x^\beta \triangle_l^\alpha \sigma_A(x, l),$$
$$\partial_j \left(\sigma_A(x, l)\, \sigma_B(x, l) \right) = \left(\partial_j \sigma_A(x, l) \right) \sigma_B(x, l) + \sigma_A(x, l)\, \partial_j \sigma_A(x, l),$$
$$\partial_y^\beta \left(\sigma_A(x, l)\, \sigma_B(y, l)\, \sigma_C(z, l) \right) = \sigma_A(x, l) \left(\partial_y^\beta \sigma_B(y, l) \right) \sigma_C(z, l).$$

12.2 Symbols of $\partial_+, \partial_-, \partial_0$ and Laplacian $\mathcal{L}$

In this section we calculate symbols of the creation, annihilation and neutral operators $\partial_+, \partial_-, \partial_0$, and of the Laplacian $\mathcal{L}$. We will use the fact that the symbol of the operator A is obtained by

$$\sigma_A(x, l) = t^l(x)^* \, (At^l)(x),$$

that is, $\sigma_A(x, l)_{mn} = \sum_k \overline{t_{km}^l(x)}\, (At_{kn}^l)(x)$, see (12.3) and (12.4).

Theorem 12.2.1. *We have*

$$
\begin{aligned}
\sigma_{\partial_+}(x, l)_{mn} &= -\sqrt{(l - n)(l + n + 1)}\; \delta_{m, n+1} \\
&= -\sqrt{(l - m + 1)(l + m)}\; \delta_{m-1, n}, \\
\sigma_{\partial_-}(x, l)_{mn} &= -\sqrt{(l + n)(l - n + 1)}\; \delta_{m, n-1} \\
&= -\sqrt{(l + m + 1)(l - m)}\; \delta_{m+1, n}, \\
\sigma_{\partial_0}(x, l)_{mn} &= n\, \delta_{mn} = m\, \delta_{mn}, \\
\sigma_{\mathcal{L}}(x, l)_{mn} &= -l(l + 1)\, \delta_{mn},
\end{aligned}
$$

where δ_{mn} *is the Kronecker delta:* $\delta_{mn} = 1$ *for* $m = n$ *and,* $\delta_{mn} = 0$ *otherwise.*

Proof. Let $e \in SU(2)$ be the neutral element of $SU(2)$ and let t^l be a unitary matrix representation of $SU(2)$. First we note that

$$\delta_{mn} = t^l(e)_{mn} = t^l(x^{-1}x)_{mn} = \sum_k t^l(x^{-1})_{mk}\, t^l(x)_{kn} = \sum_k \overline{t^l(x)_{km}}\, t^l(x)_{kn},$$

see also Remark 10.2.11 for the general form of this identity. Similarly, we note the identity

$$\delta_{mn} = \sum_k t^l(x)_{mk}\, \overline{t^l(x)_{nk}}.$$

From this, formula (12.4), and Theorem 11.9.3 we get

$$
\begin{aligned}
\sigma_{\partial_+}(x,l)_{mn} &= \sum_k \overline{t^l_{km}(x)}\,(\partial_+ t^l_{kn})\,(x) \\
&= -\sqrt{(l-n)(l+n+1)}\sum_k \overline{t^l_{km}(x)}\, t^l_{k,n+1}(x) \\
&= -\sqrt{(l-n)(l+n+1)}\,\delta_{m,n+1},
\end{aligned}
$$

and the case of $\sigma_{\partial_-}(x,l)$ is analogous:

$$
\begin{aligned}
\sigma_{\partial_-}(x,l)_{mn} &= \sum_k \overline{t^l_{km}(x)}\,(\partial_- t^l_{kn})\,(x) \\
&= -\sqrt{(l+n)(l-n+1)}\sum_k \overline{t^l_{km}(x)}\, t^l_{k,n-1}(x) \\
&= -\sqrt{(l+n)(l-n+1)}\,\delta_{m,n-1},
\end{aligned}
$$

Finally,

$$\sigma_{\partial_0}(x,l)_{mn} = \sum_k \overline{t^l_{km}(x)}\,(\partial_0 t^l_{kn})\,(x) = n\sum_k \overline{t^l_{km}(x)}\, t^l_{kn}(x) = n\,\delta_{m,n},$$

and similarly for $\mathcal{L}$. □

Exercise 12.2.2. Complete the proof of Theorem 12.2.1 for the symbol $\sigma_{\mathcal{L}}$ of the Laplacian $\mathcal{L}$.

Remark 12.2.3. Notice that $\sigma_{\partial_0}(x,l)$ and $\sigma_{\mathcal{L}}(x,l)$ are diagonal matrices. The non-zero elements reside just above the diagonal of $\sigma_{\partial_+}(x,l)$, and just below the diagonal of $\sigma_{\partial_-}(x,l)$. Because of this the operators ∂_0, ∂_+ and ∂_- may be called *neutral, creation* and *annihilation* operators, respectively, and this explains our preference to work with them rather than with D_j's, which have more non-zero entries.

Finally we note that vector fields D_1, D_2, D_3 related to $\partial_+, \partial_-, \partial_0$ by Definition 11.5.10 can be conjugated as follows (recall the definition of conjugations and their properties, e.g., in Proposition 10.4.18):

Proposition 12.2.4 (Conjugation of D_1, D_2, D_3 and their symbols). *We have*

$$(D_3)_{(w_1)_R} = D_2, \quad (D_1)_{(w_2)_R} = D_3, \quad (D_2)_{(w_3)_R} = D_1. \tag{12.11}$$

The symbols of the operators D_1, D_2 can be transformed to that of D_3 by taking suitable conjugations:

$$
\begin{aligned}
\sigma_{D_1}(x, l) &= t^l(w_2) \, \sigma_{D_3}(x, l) \, t^l(w_2)^*, & (12.12) \\
\sigma_{D_2}(x, l) &= t^l(w_1)^* \, \sigma_{D_3}(x, l) \, t^l(w_1). & (12.13)
\end{aligned}
$$

Moreover, if $D \in \mathfrak{su}(2)$ there exists $u \in SU(2)$ such that

$$\sigma_D(l) = t^l(u)^* \, \sigma_{D_3}(l) \, t^l(u).$$

Proof. Combining Lemma 10.4.19 with Proposition 11.5.7, we see that formulae (12.11) hold. Since D_1, D_2, D_3 are left-invariant operators, their symbols $\sigma_{D_j}(x, l)$ do not depend on $x \in G$, and by Proposition 10.4.18 we obtain (12.12) and (12.13). The last statement follows from Proposition 10.4.18 since D is a rotation of D_3. □

12.3   Difference operators for symbols

We are now going to introduce "difference operators" $\triangle_+, \triangle_-, \triangle_0$ acting on symbols of operators on SU(2) that resemble first-order forward and backward differences $\triangle_\xi^\alpha$ acting on symbols $\sigma_A(x, \xi)$ on a torus, where we had $\xi \in \mathbb{Z}^n$ (see Definition 3.3.1). Then we will apply these differences to first-order differential operators, as well as to products of special type.

12.3.1   Difference operators on SU(2)

From Theorem 11.7.1 and Theorem 11.8.1 we recall the notation

$$
\begin{aligned}
t^{1/2} &= \begin{pmatrix} t_{--} & t_{-+} \\ t_{+-} & t_{++} \end{pmatrix} = \begin{pmatrix} t^{1/2}_{-1/2,-1/2} & t^{1/2}_{-1/2,+1/2} \\ t^{1/2}_{+1/2,-1/2} & t^{1/2}_{+1/2,+1/2} \end{pmatrix} \\
&= \begin{pmatrix} \cos(\theta/2) \, e^{i(+\phi+\psi)/2} & i\sin(\theta/2) \, e^{i(+\phi-\psi)/2} \\ i\sin(\theta/2) \, e^{i(-\phi+\psi)/2} & \cos(\theta/2) \, e^{i(-\phi-\psi)/2} \end{pmatrix}.
\end{aligned}
$$

Definition 12.3.1 (Differences $\triangle_q$ for Fourier coefficients). For $q \in C^\infty(SU(2))$ and $f \in \mathcal{D}'(SU(2))$, let

$$\triangle_q \widehat{f}(l) := \widehat{qf}(l).$$

We shall use abbreviations

$$\triangle_+ = \triangle_{q_+}, \triangle_- = \triangle_{q_-} \text{ and } \triangle_0 = \triangle_{q_0},$$

where

$$q_- := t_{-+} = t^{1/2}_{-1/2,+1/2},$$
$$q_+ := t_{+-} = t^{1/2}_{+1/2,-1/2},$$
$$q_0 := t_{--} - t_{++} = t^{1/2}_{-1/2,-1/2} - t^{1/2}_{+1/2,+1/2}.$$

Thus each trigonometric polynomial $q_+, q_-, q_0 \in C^\infty(\mathrm{SU}(2))$ vanishes at the neutral element $e \in \mathrm{SU}(2)$. In this sense trigonometric polynomials $q_- + q_+, q_- - q_+, q_0$ on $\mathrm{SU}(2)$ are analogues of polynomials x_1, x_2, x_3 in the Euclidean space $\mathbb{R}^3$.

Now our aim is to let these difference operators act on symbols. For this purpose we may only look at symbols independent of x corresponding to left-invariant (right-convolution) operators since the following construction is independent of x:

Definition 12.3.2 (Differences $\triangle_+, \triangle_-, \triangle_0$ acting on symbols). Let $a = a(\xi)$ be a symbol as in Definition 10.4.3. It follows that $a = \widehat{s}$ for some right-convolution kernel $s \in \mathcal{D}'(\mathrm{SU}(2))$ so that the operator $\mathrm{Op}(a)$ is given by

$$\mathrm{Op}(a)f = f * s.$$

We define *"difference operators"* $\triangle_+, \triangle_-, \triangle_0$ acting on the symbol a by

$$\triangle_+ a := \widehat{q_+ s}, \tag{12.14}$$
$$\triangle_- a := \widehat{q_- s}, \tag{12.15}$$
$$\triangle_0 a := \widehat{q_0 s}. \tag{12.16}$$

Obviously, Definitions 12.3.1 and 12.3.2 are consistent. We note once more that this construction is analogous to the one producing usual derivatives in $\mathbb{R}^n$ or difference operators on the torus $\mathbb{T}^n$.

To analyse the structure of difference operators on $\mathrm{SU}(2)$, we first need to know how to multiply functions t^l_{mn} by q_+, q_-, q_0, and the necessary formulae are given in Theorem 11.8.1. We recall the notation

$$t^{1/2} = \begin{pmatrix} t_{--} & t_{-+} \\ t_{+-} & t_{++} \end{pmatrix} = \begin{pmatrix} t_{--} & q_- \\ q_+ & t_{++} \end{pmatrix}, \quad q_0 = t_{--} - t_{++},$$

and summarise the multiplication formulae as follows:

Corollary 12.3.3. *For $x \in \mathbb{R}$, let $x^\pm := x \pm 1/2$. Then*

$$(2l+1)q_+ t^l_{mn} = +\sqrt{(l+m+1)(l-n+1)} t^{l^+}_{m^+ n^-} - \sqrt{(l-m)(l+n)} t^{l^-}_{m^+ n^-},$$

$$(2l+1)q_- t^l_{mn} = +\sqrt{(l-m+1)(l+n+1)} t^{l^+}_{m^- n^+} - \sqrt{(l+m)(l-n)} t^{l^-}_{m^- n^+},$$

$$(2l+1)q_0 t^l_{mn} = +\sqrt{(l-m+1)(l-n+1)} t^{l^+}_{m^- n^-} + \sqrt{(l+m)(l+n)} t^{l^-}_{m^- n^-}$$
$$-\sqrt{(l+m+1)(l+n+1)} t^{l^+}_{m^+ n^+} - \sqrt{(l-m)(l-n)} t^{l^-}_{m^+ n^+}.$$

We also recall the relation between symbols and kernels which follows from (12.2) and (12.5), and where we switch the order of m and n to adopt it to the proof of Theorem 12.3.5:

Corollary 12.3.4 (Kernel and symbol). *Symbol $a(x, l) = a(l)$ and kernel $s(x)$ of an operator are related by*

$$a(x, l)_{nm} = a^l_{nm} = \widehat{s}(l)_{nm} = \int_{SU(2)} s(y)\, \overline{t^l_{mn}(y)}\, dy,$$

and

$$s(x) = \sum_l (2l + 1)\, \mathrm{Tr}\left(a(x, l)\, t^l(x)\right) = \sum_l (2l + 1) \sum_{m,n} a^l_{nm}\, t^l_{mn}. \qquad (12.17)$$

Let us now derive explicit expressions for the first-order difference operators $\triangle_+, \triangle_-, \triangle_0$ defined in (12.14)–(12.16). To abbreviate the notation, we will also write $a^l_{nm} = a(x, l)_{nm}$, even if symbol $a(x, l)$ depends on x, keeping in mind that the following theorem holds pointwise in x.

Theorem 12.3.5 (Formulae for difference operators $\triangle_+, \triangle_-, \triangle_0$). *The difference operators are given by*

$$(\triangle_- a)^l_{nm} = \frac{\sqrt{(l-m)(l+n)}}{2l+1} a^{l-}_{n-m+} - \frac{\sqrt{(l+m+1)(l-n+1)}}{2l+1} a^{l+}_{n-m+},$$

$$(\triangle_+ a)^l_{nm} = \frac{\sqrt{(l+m)(l-n)}}{2l+1} a^{l-}_{n+m-} - \frac{\sqrt{(l-m+1)(l+n+1)}}{2l+1} a^{l+}_{n+m-},$$

$$(\triangle_0 a)^l_{nm} = \frac{\sqrt{(l-m)(l-n)}}{2l+1} a^{l-}_{n+m+} + \frac{\sqrt{(l+m+1)(l+n+1)}}{2l+1} a^{l+}_{n+m+}$$

$$- \frac{\sqrt{(l+m)(l+n)}}{2l+1} a^{l-}_{n-m-} - \frac{\sqrt{(l-m+1)(l-n+1)}}{2l+1} a^{l+}_{n-m-},$$

where $k^{\pm} = k \pm \frac{1}{2}$, and satisfy commutator relations

$$[\triangle_0, \triangle_+] = [\triangle_0, \triangle_-] = [\triangle_-, \triangle_+] = 0. \qquad (12.18)$$

Proof. Identities (12.18) follow immediately from (12.14)–(12.16). As discussed before, we can abbreviate $a(x, l)$ by $a(l)$ since none of the arguments in the proof will act on the variable x. In the proof we will heavily rely on the relation between kernels and symbols in Proposition 12.3.4. Also, in the calculation below we will not worry about boundaries of summations, keeping in mind that we can always view finite matrices as infinite ones simply by extending them by zeros. Recalling that $q_- = t_{-+}$ and using Theorem 11.8.1, we can calculate

$$q_-\, s \overset{(12.17)}{=} \sum_l (2l + 1) \sum_{m,n} a^l_{nm}\, q_-\, t^l_{mn}$$

$$\underset{\mathrm{Cor}\ \underline{12.3.3}}{=} \sum_l \sum_{m,n} a^l_{nm} \left[t^{l+}_{m-n+} \sqrt{(l-m+1)(l+n+1)} \right.$$

$$\left. - t^{l-}_{m-n+} \sqrt{(l+m)(l-n)} \right]$$

$$= \sum_l \sum_{m,n} t^l_{mn} \left[a^{l-}_{n-m+} \sqrt{(l-m)(l+n)} \right.$$

$$\left. - a^{l+}_{n-m+} \sqrt{(l+m+1)(l-n+1)} \right].$$

Since $\triangle_- a = \widehat{q_- s}$, we obtain the desired formula for $\triangle_-$:

$$(\triangle_- a)^l_{nm} = \frac{\sqrt{(l-m)(l+n)}}{2l+1} a^{l-}_{n-m+} - \frac{\sqrt{(l+m+1)(l-n+1)}}{2l+1} a^{l+}_{n-m+}.$$

Similarly, for $\triangle_+$, we calculate

$$q_+ s = \sum_l (2l+1) \sum_{m,n} a^l_{nm}\ q_+\ t^l_{mn}$$

$$= \sum_l \sum_{m,n} a^l_{nm} \left[t^{l+}_{m+n-} \sqrt{(l+m+1)(l-n+1)} \right.$$

$$\left. - t^{l-}_{m+n-} \sqrt{(l-m)(l+n)} \right]$$

$$= \sum_l \sum_{m,n} t^l_{mn} \left[a^{l-}_{n+m-} \sqrt{(l+m)(l-n)} \right.$$

$$\left. - a^{l+}_{n+m-} \sqrt{(l-m+1)(l+n+1)} \right].$$

From this we obtain the desired formula for $\triangle_+$:

$$(\triangle_+ a)^l_{nm} = \frac{\sqrt{(l+m)(l-n)}}{2l+1} a^{l-}_{n+m-} - \frac{\sqrt{(l-m+1)(l+n+1)}}{2l+1} a^{l+}_{n+m-}.$$

Finally, for $\triangle_0$, we calculate

$$q_0 s = \sum_l (2l+1) \sum_{m,n} a^l_{nm}\ q_0\ t^l_{mn}$$

$$= \sum_l \sum_{m,n} a^l_{nm} \left[t^{l+}_{m-n-} \sqrt{(l-m+1)(l-n+1)} + t^{l-}_{m-n-} \sqrt{(l+m)(l+n)} \right.$$

$$\left. - t^{l+}_{m+n+} \sqrt{(l+m+1)(l+n+1)} - t^{l-}_{m+n+} \sqrt{(l-m)(l-n)} \right]$$

$$= \sum_l \sum_{m,n} t^l_{mn} \left[a^{l-}_{n+m+} \sqrt{(l-m)(l-n)} + a^{l+}_{n+m+} \sqrt{(l+m+1)(l+n+1)} \right.$$

$$\left. - a^{l-}_{n-m-} \sqrt{(l+m)(l+n)} - a^{l+}_{n-m-} \sqrt{(l-m+1)(l-n+1)} \right].$$

From this we obtain the desired formula for $\triangle_0$:

$$(\triangle_0 a)^l_{nm} = \frac{\sqrt{(l-m)(l-n)}}{2l+1} a^{l-}_{n+m+} + \frac{\sqrt{(l+m+1)(l+n+1)}}{2l+1} a^{l+}_{n+m+}$$
$$- \frac{\sqrt{(l+m)(l+n)}}{2l+1} a^{l-}_{n-m-} - \frac{\sqrt{(l-m+1)(l-n+1)}}{2l+1} a^{l+}_{n-m-},$$

and the proof of Theorem 12.3.5 is complete. $\qquad\square$

12.3.2 Differences for symbols of $\partial_+, \partial_-, \partial_0$ and Laplacian $\mathcal{L}$

Remark 10.7.3 and Proposition 10.7.4 said that application of difference operators reduces the order of the symbol of differential operators. However, for operators $\partial_+, \partial_-, \partial_0$ and for the Laplacian $\mathcal{L}$ we calculate this now more explicitly:

Theorem 12.3.6. *We have*

$$\sigma_I = \triangle_+ \sigma_{\partial_+} = \triangle_- \sigma_{\partial_-} = \triangle_0 \sigma_{\partial_0}. \tag{12.19}$$

If $\mu, \nu \in \{+, -, 0\}$ are such that $\mu \neq \nu$, then

$$\triangle_\mu \sigma_{\partial_\nu} = 0, \tag{12.20}$$

and for every $\nu \in \{+, -, 0\}$, we have

$$\triangle_\nu \sigma_I(x) = 0. \tag{12.21}$$

Moreover, if $\mathcal{L}$ is the bi-invariant Laplacian, then

$$\triangle_+ \sigma_{\mathcal{L}} = -\sigma_{\partial_-}, \quad \triangle_- \sigma_{\mathcal{L}} = -\sigma_{\partial_+}, \quad \triangle_0 \sigma_{\mathcal{L}} = -2\sigma_{\partial_0}. \tag{12.22}$$

The proof of this theorem will depend on explicit calculations. In trying to simplify the presentation, we prove this theorem in the form of several propositions dealing with different parts of the statement. We recall that the symbols of the first-order partial differential operators $\partial_+, \partial_-, \partial_0$ have many zero elements, and altogether they are as in Theorem 12.2.1:

$$\sigma_{\partial_+}(l)_{mn} = \begin{cases} -\sqrt{(l-n)(l+n+1)}, & \text{if } m = n+1, \\ 0, & \text{otherwise.} \end{cases} \tag{12.23}$$

$$\sigma_{\partial_-}(l)_{mn} = \begin{cases} -\sqrt{(l+n)(l-n+1)}, & \text{if } m = n-1, \\ 0, & \text{otherwise.} \end{cases}$$

$$\sigma_{\partial_0}(l)_{mn} = \begin{cases} n, & \text{if } m = n, \\ 0, & \text{otherwise.} \end{cases}$$

Proposition 12.3.7. *We have identities* (12.19), *i.e.,*

$$\sigma_I = \triangle_+\sigma_{\partial_+} = \triangle_-\sigma_{\partial_-} = \triangle_0\sigma_{\partial_0}.$$

Proof. From Corollary 12.3.3 we get an expression for $q_+t_{mn}^l$, which is used in the following calculation together with (12.17) in Corollary 12.3.4 and Theorem 12.2.1:

$$q_+ \, s_{\partial_+} \quad \overset{(12.17)}{=} \quad q_+ \sum_l (2l+1) \sum_{m,n} \sigma_{\partial_+}(l)_{mn} \, t_{nm}^l$$

$$\overset{(12.23)}{=} \quad \sum_l \sum_n \sigma_{\partial_+}(l)_{n+1,n} \, (2l+1) \, q_+ \, t_{n,n+1}^l$$

$$\overset{\text{Cor }12.3.3}{=} \quad -\sum_l \sum_n \left(\sqrt{(l-n)(l+n+1)}\right)^2 \left(t_{n^+n^+}^{l^+} - t_{n^+n^+}^{l^-}\right)$$

$$\overset{(12.24)}{=} \quad \sum_l (2l+1) \sum_k t_{kk}^l$$

$$\overset{(12.17)}{=} \quad \delta_e$$

$$= \quad s_I,$$

where we made a change in indices and used the identity

$$(l^- - n)(l^- + n + 1) - (l^+ - n)(l^+ + n + 1) = -2l - 1. \tag{12.24}$$

Hence $\triangle_+\sigma_{\partial_+} = \sigma_I$. Similarly, $\triangle_-\sigma_{\partial_-} = \sigma_I$, because

$$q_- \, s_{\partial_-} \quad = \quad q_- \sum_l (2l+1) \sum_{m,n} \sigma_{\partial_-}(l)_{mn} \, t_{nm}^l$$

$$= \quad \sum_l \sum_n \sigma_{\partial_-}(l)_{n-1,n} \, (2l+1) \, q_- \, t_{n,n-1}^l$$

$$= \quad \sum_l \sum_n -\left(\sqrt{(l-n+1)(l+n)}\right)^2 \left(t_{n^-n^-}^{l^+} - t_{n^-n^-}^{l^-}\right)$$

$$= \quad \sum_l (2l+1) \sum_k t_{kk}^l$$

$$= \quad \delta_e$$

$$= \quad s_I.$$

Moreover,

$$q_0 \, s_{\partial_0} = q_0 \sum_l (2l+1) \sum_{m,n} \sigma_{\partial_0}(l)_{mn} \, t_{nm}^l$$

$$= \sum_l \sum_n \sigma_{\partial_0}(l)_{nn} \, (2l+1) \, q_0 \, t_{nn}^l$$

$$= \sum_l \sum_n n \left(+(l-n+1)\, t^{l+}_{n-n-} + (l+n)\, t^{l-}_{n-n-} \right.$$
$$\left. -(l+n+1)\, t^{l+}_{n+n+} - (l-n)\, t^{l-}_{n+n+} \right)$$

$$= \sum_l \sum_n \left(t^{l+}_{n+n+} \left((n+1)(l-n) - n(l+n+1)\right) \right.$$
$$\left. + t^{l-}_{n+n+} \left((n+1)(l+n+1) - n(l-n)\right) \right)$$

$$= \sum_l \sum_n \left(t^{l+}_{n+n+} \left(l - 2n^2 - 2n\right) \right.$$
$$\left. + t^{l-}_{n+n+} \left(l + 1 + 2n^2 + 2n\right) \right)$$

$$= \sum_l \sum_n t^{l+}_{n+n+} \left((l - 2n^2 - 2n) + (l + 2 + 2n^2 + 2n)\right)$$

$$= \sum_l (2l+2) \sum_n t^{l+}_{n+n+}$$

$$= \sum_l (2l+1) \sum_n t^l_{nn}$$

$$= \delta_e$$

$$= s_I. \qquad\qquad \square$$

Proposition 12.3.8. *We have identities* (12.20), *i.e.,*

$$\triangle_\mu \sigma_{\partial_\nu} = 0,$$

where $\mu, \nu \in \{+, -, 0\}$ *such that* $\mu \neq \nu$.

Proof. We can calculate

$$q_+ \, s_{\partial_-} \;=\; q_+ \sum_l (2l+1) \sum_{m,n} \sigma_{\partial_-}(l)_{mn}\, t^l_{nm}$$

$$= \sum_l \sum_n \sigma_{\partial_-}(l)_{n-1,n}\, (2l+1)\, q_+\, t^l_{n,n-1}$$

$$= \sum_l \sum_n -\sqrt{(l+n)(l-n+1)}$$
$$\times \left(\sqrt{(l+n+1)(l-n+2)}\, t^{l+1/2}_{n+1/2,n-3/2} \right.$$
$$\left. - \sqrt{(l-n)(l+n-1)}\, t^{l-1/2}_{n+1/2,n-3/2} \right)$$

$$= \sum_l \sum_n t^{l+1/2}_{n+1/2,n-3/2}$$
$$\times \left(-\sqrt{(l+n+1)(l-n+2)}\sqrt{(l-n+1)(l+n)} \right.$$
$$\left. + \sqrt{(l+n)(l-n+1)}\sqrt{(l+n+1)(l-n+2)} \right) \quad = \quad 0.$$

Analogously,

$$
\begin{aligned}
q_- \, s_{\partial_+} &= q_- \sum_l (2l+1) \sum_{m,n} \sigma_{\partial_+}(l)_{mn} \, t^l_{nm} \\
&= \sum_l \sum_n \sigma_{\partial_+}(l)_{n+1,n} \, (2l+1) \, q_- \, t^l_{n,n+1} \\
&= \sum_l \sum_n -\sqrt{(l-n)(l+n+1)} \\
&\qquad \times \left(\sqrt{(l-n+1)(l+n+2)} \, t^{l+1/2}_{n-1/2,n+3/2} \right. \\
&\qquad\qquad \left. - \sqrt{(l+n)(l-n-1)} \, t^{l-1/2}_{n-1/2,n+3/2} \right) \\
&= \sum_l \sum_n t^{l+1/2}_{n-1/2,n+3/2} \\
&\qquad \times \left(-\sqrt{(l-n)(l+n+1)}\sqrt{(l-n+1)(l+n+2)} \right. \\
&\qquad\qquad \left. + \sqrt{(l-n+1)(l+n+2)}\sqrt{(l+n+1)(l-n)} \right) \\
&= 0,
\end{aligned}
$$

and

$$
\begin{aligned}
q_+ \, s_{\partial_0} &= q_+ \sum_l (2l+1) \sum_{m,n} \sigma_{\partial_0}(l)_{mn} \, t^l_{nm} \\
&= \sum_l \sum_n \sigma_{\partial_0}(l)_{nn} \, (2l+1) \, q_+ \, t^l_{nn} \\
&= \sum_l \sum_n n \left(\sqrt{(l+n+1)(l-n+1)} \, t^{l^+}_{n^-,n^+} - \sqrt{(l-n)(l+n)} \, t^{l^-}_{n^-,n^+} \right) \\
&= \sum_l \sum_n n \, t^{l^+}_{n^-,n^+} \left(\sqrt{(l+n+1)(l-n+1)} - \sqrt{(l-n+1)(l+n+1)} \right) \\
&= 0.
\end{aligned}
$$

Analogously,

$$
\begin{aligned}
q_- \, s_{\partial_0} &= q_- \sum_l (2l+1) \sum_{m,n} \sigma_{\partial_0}(l)_{mn} \, t^l_{nm} \\
&= \sum_l \sum_n \sigma_{\partial_0}(l)_{nn} \, (2l+1) \, q_- \, t^l_{nn} \\
&= \sum_l \sum_n n \left(\sqrt{(l+n+1)(l-n+1)} \, t^{l^+}_{n^+,n^-} - \sqrt{(l-n)(l+n)} \, t^{l^-}_{n^+,n^-} \right) \\
&= \sum_l \sum_n n \, t^{l^+}_{n^+,n^-} \left(\sqrt{(l+n+1)(l-n+1)} - \sqrt{(l-n+1)(l+n+1)} \right) \\
&= 0.
\end{aligned}
$$

We also have

$$q_0 \, s_{\partial_-} = q_0 \sum_l (2l+1) \sum_{m,n} \sigma_{\partial_-}(l)_{mn} \, t_{nm}^l$$

$$= \sum_l \sum_n \sigma_{\partial_-}(l)_{n-1,n} \, (2l+1) \, q_0 \, t_{n,n-1}^l$$

$$= \sum_l \sum_n -\sqrt{(l+n)(l-n+1)}$$

$$\times \left(+\sqrt{(l-n+1)(l-n+2)} t_{n-1/2,n-3/2}^{l+1/2} + \sqrt{(l+n)(l+n-1)} t_{n-1/2,n-3/2}^{l-1/2} \right.$$

$$\left. - \sqrt{(l+n+1)(l+n)} t_{n+1/2,n-1/2}^{l+1/2} - \sqrt{(l-n)(l-n+1)} t_{n+1/2,n-1/2}^{l-1/2} \right)$$

$$= \sum_l \sum_n t_{n,n-1}^l$$

$$\times \left(+\sqrt{(l+n)(l-n+1)} \, (l-n) + (l+n+1)\sqrt{(l-n+1)(l+n)} \right.$$

$$\left. - (l+n-1) \, \sqrt{(l-n+1)(l+n)} - \sqrt{(l+n)(l-n+1)} \, (l-n+2) \right)$$

$$= 0.$$

Analogously,

$$q_0 \, s_{\partial_+} = q_0 \sum_l (2l+1) \sum_{m,n} \sigma_{\partial_+}(l)_{mn} \, t_{nm}^l$$

$$= \sum_l \sum_n \sigma_{\partial_+}(l)_{n+1,n} \, (2l+1) \, q_0 \, t_{n,n+1}^l$$

$$= \sum_l \sum_n -\sqrt{(l-n)(l+n+1)}$$

$$\times \left(+\sqrt{(l-n+1)(l-n)} t_{n-1/2,n+1/2}^{l+1/2} + \sqrt{(l+n)(l+n+1)} t_{n-1/2,n+1/2}^{l-1/2} \right.$$

$$\left. - \sqrt{(l+n+1)(l+n+2)} t_{n+1/2,n+3/2}^{l+1/2} - \sqrt{(l-n)(l-n-1)} t_{n+1/2,n+3/2}^{l-1/2} \right)$$

$$= \sum_l \sum_n t_{n,n+1}^l$$

$$\times \left(+(l-n-1) \, \sqrt{(l+n+1)(l-n)} + \sqrt{(l-n)(l+n+1)} \, (l+n+2) \right.$$

$$\left. - \sqrt{(l-n)(l+n+1)} \, (l+n) - (l-n+1) \, \sqrt{(l+n+1)(l-n)} \right)$$

$$= 0. \qquad \qquad \square$$

Proposition 12.3.9. *We have identities* (12.21), *i.e.,*

$$\triangle_\nu \sigma_I(x) = 0,$$

for every $\nu \in \{+, -, 0\}$.

Proof. Similarly to the propositions before, we calculate

$$
\begin{aligned}
q_+\, s_I &= q_+ \sum_l (2l+1) \sum_{m,n} \delta_{mn} t^l_{nm} \\
&= \sum_l \sum_n (2l+1)\, q_+\, t^l_{nn} \\
&= \sum_l \sum_n \left(\sqrt{(l+n+1)(l-n+1)}\, t^{l^+}_{n^+,n^-} - \sqrt{(l-n)(l+n)}\, t^{l^-}_{n^+,n^-} \right) \\
&= \sum_l \sum_n t^{l^-}_{n^+,n^-} \left(\sqrt{(l+n)(l-n)} - \sqrt{(l-n)(l+n)} \right) \\
&= 0.
\end{aligned}
$$

Analogously,

$$
\begin{aligned}
q_-\, s_I &= q_- \sum_l (2l+1) \sum_{m,n} \delta_{mn} t^l_{nm} \\
&= \sum_l \sum_n (2l+1)\, q_-\, t^l_{nn} \\
&= \sum_l \sum_n \left(\sqrt{(l-n+1)(l+n+1)}\, t^{l^+}_{n^-,n^+} - \sqrt{(l+n)(l-n)}\, t^{l^-}_{n^-,n^+} \right) \\
&= \sum_l \sum_n t^{l^-}_{n^-,n^+} \left(\sqrt{(l-n)(l+n)} - \sqrt{(l+n)(l-n)} \right) \\
&= 0.
\end{aligned}
$$

Moreover,

$$
\begin{aligned}
q_0\, s_I &= q_0 \sum_l (2l+1) \sum_{m,n} \delta_{mn} t^l_{nm} \\
&= \sum_l \sum_n (2l+1)\, q_0\, t^l_{nn} \\
&= \sum_l \sum_n \left(+(l-n+1)\, t^{l^+}_{n^-n^-} + (l+n)\, t^{l^-}_{n^-n^-} \right. \\
&\qquad\qquad \left. -(l+n+1)\, t^{l^+}_{n^+n^+} - (l-n)\, t^{l^-}_{n^+n^+} \right) \\
&= \sum_l \sum_n t^l_{nn} \left(+(l-n) + (l+n+1) - (l+n) - (l-n+1) \right) \\
&= 0. \qquad\qquad\qquad\qquad\qquad\qquad\qquad\qquad\qquad\qquad\qquad \square
\end{aligned}
$$

Proposition 12.3.10. *We have identities (12.22), i.e.,*

$$
\triangle_+ \sigma_{\mathcal{L}} = -\sigma_{\partial_-}, \quad \triangle_- \sigma_{\mathcal{L}} = -\sigma_{\partial_+}, \quad \triangle_0 \sigma_{\mathcal{L}} = -2\sigma_{\partial_0}.
$$

Proof. Since

$$\sigma_{\mathcal{L}}(x,l)_{mn} = -l(l+1)\,\delta_{mn},$$

we get

$$
\begin{aligned}
q_+\, s_{-\mathcal{L}} &= q_+ \sum_l (2l+1) \sum_{m,n} \sigma_{-\mathcal{L}}(x,l)_{mn}\, t^l_{nm} \\
&= \sum_l (2l+1) \sum_n l(l+1)\, q_+\, t^l_{nn} \\
&= \sum_l \sum_n l(l+1) \left(+\sqrt{(l+n+1)(l-n+1)}\; t^{l^+}_{n^-,n^+} \right.\\
&\qquad\qquad\qquad \left. -\sqrt{(l-n)(l+n)}\; t^{l^-}_{n^-,n^+} \right) \\
&= \sum_l \sum_n t^{l^+}_{n^-,n^+} \left(+l(l+1)\sqrt{(l+n+1)(l-n+1)} \right.\\
&\qquad\qquad\qquad \left. -(l+1)(l+2)\sqrt{(l-n+1)(l+n+1)} \right) \\
&= \sum_l \sum_n -2(l+1)\sqrt{(l+n+1)(l-n+1)}\; t^{l^+}_{n^-,n^+} \\
&= \sum_l (2l+1) \sum_n -\sqrt{(l+n)(l-n+1)}\; t^l_{n-1,n} \\
&= s_{\partial_-}.
\end{aligned}
$$

Analogously,

$$
\begin{aligned}
q_-\, s_{-\mathcal{L}} &= q_- \sum_l (2l+1) \sum_{m,n} \sigma_{-\mathcal{L}}(x,l)_{mn}\, t^l_{nm} \\
&= \sum_l (2l+1) \sum_n l(l+1)\, q_-\, t^l_{nn} \\
&= \sum_l \sum_n l(l+1) \left(+\sqrt{(l-n+1)(l+n+1)}\; t^{l^+}_{n^-,n^+} \right.\\
&\qquad\qquad\qquad \left. -\sqrt{(l+n)(l-n)}\; t^{l^-}_{n^-,n^+} \right) \\
&= \sum_l \sum_n t^{l^+}_{n^-,n^+} \left(+l(l+1)\sqrt{(l-n+1)(l+n+1)} \right.\\
&\qquad\qquad\qquad \left. -(l+1)(l+2)\sqrt{(l+n+1)(l-n+1)} \right) \\
&= \sum_l \sum_n -2(l+1)\sqrt{(l+n+1)(l-n+1)}\; t^{l^+}_{n^-,n^+} \\
&= \sum_l (2l+1) \sum_n -\sqrt{(l+n)(l-n+1)}\; t^l_{n-1,n} \\
&= s_{\partial_+}.
\end{aligned}
$$

Moreover,

$$
\begin{aligned}
q_0 \, s_{-\mathcal{L}} \;&=\; q_0 \sum_l (2l+1) \sum_{m,n} \sigma_{-\mathcal{L}}(x,l)_{mn} t^l_{nm} \\[4pt]
&=\; \sum_l (2l+1) \sum_n l(l+1) \, q_0 \, t^l_{nn} \\[4pt]
&=\; \sum_l \sum_n l(l+1) \left(+(l-n+1)\, t^{l^+}_{n-\,n-} + (l+n)\, t^{l^-}_{n-\,n-} \right. \\[4pt]
&\qquad\qquad\left. -(l+n+1)\, t^{l^+}_{n+\,n+} - (l-n)\, t^{l^-}_{n+\,n+} \right) \\[4pt]
&=\; \sum_l l(l+1) \sum_n \\[4pt]
&\qquad \left(t^{l^+}_{n+\,n+}\big((l-n)-(l+n+1)\big) + t^{l^-}_{n+\,n+}\big((l+n+1)-(l-n)\big) \right) \\[4pt]
&=\; \sum_l l(l+1) \sum_n -(2n+1)\left(t^{l^+}_{n+\,n+} - t^{l^-}_{n+\,n+} \right) \\[4pt]
&=\; -\sum_l \sum_n t^{l^+}_{n+\,n+}(2n+1)\,\big(l(l+1)-(l+1)(l+2)\big) \\[4pt]
&=\; \sum_l \sum_n t^{l^+}_{n+\,n+}(2n+1)\,(2l+2) \;=\; \sum_l (2l+1) \sum_n 2n\, t^l_{nn} \;=\; 2\, s_{\partial_0},
\end{aligned}
$$

completing the proof. □

Remark 12.3.11. The proof of Theorem 12.3.6 relied on explicit calculations in representations and we decided to include them in detail for didactic purposes as well as in preparation for the proof of Theorem 12.3.12. However, we may also argue using formula (10.28) in Proposition 10.7.4. There, if $q(e) = 0$, we get

$$
\begin{aligned}
\triangle_q \sigma_{\partial_+} \;&=\; -(\partial_+ q)(e)\, \sigma_I, \\
\triangle_q \sigma_{\partial_-}(x,\xi) \;&=\; -(\partial_- q)(e)\, \sigma_I, \\
\triangle_q \sigma_{\partial_0}(x,\xi) \;&=\; -(\partial_0 q)(e)\, \sigma_I.
\end{aligned}
$$

If we write $u \in \mathrm{SU}(2)$ as $u = \begin{pmatrix} a & b \\ c & d \end{pmatrix}$ then

$$
q_+ = c, \quad q_- = b, \quad q_0 = a - d,
$$

and by Exercise 11.5.14 we have

$$
\begin{aligned}
\partial_+ u\big|_{u=e} \;&=\; \begin{pmatrix} -b & 0 \\ -d & 0 \end{pmatrix}_{u=e} \;=\; \begin{pmatrix} 0 & 0 \\ -1 & 0 \end{pmatrix}, \\[6pt]
\partial_- u\big|_{u=e} \;&=\; \begin{pmatrix} 0 & -a \\ 0 & -c \end{pmatrix}_{u=e} \;=\; \begin{pmatrix} 0 & -1 \\ 0 & 0 \end{pmatrix}, \\[6pt]
\partial_0 u\big|_{u=e} \;&=\; \begin{pmatrix} -a/2 & 0 \\ 0 & d/2 \end{pmatrix}_{u=e} \;=\; \begin{pmatrix} -1/2 & 0 \\ 0 & 1/2 \end{pmatrix}.
\end{aligned}
$$

This implies that

$$
\begin{aligned}
\sigma_I &= \triangle_+\sigma_{\partial_+} = \triangle_-\sigma_{\partial_-} = \triangle_0\sigma_{\partial_0}, \\
0 &= \triangle_+\sigma_{\partial_-} = \triangle_+\sigma_{\partial_0} \\
&= \triangle_-\sigma_{\partial_+} = \triangle_-\sigma_{\partial_0} \\
&= \triangle_0\sigma_{\partial_+} = \triangle_0\sigma_{\partial_-}.
\end{aligned}
$$

Let us now apply the difference operators to the symbol of the Laplacian $\mathcal{L}$. First, by Remark 11.5.11 we write

$$
\mathcal{L} = -\partial_0^2 - (\partial_+\partial_- + \partial_-\partial_+)/2.
$$

Now

$$
\begin{aligned}
\partial_0^2 u &= \partial_0 \frac{1}{2} \begin{pmatrix} -a & b \\ -c & d \end{pmatrix} = \frac{1}{4} \begin{pmatrix} a & b \\ c & d \end{pmatrix} \overset{u=e}{=} \frac{1}{4} \begin{pmatrix} 1 & 0 \\ 0 & 1 \end{pmatrix}, \\
\partial_+\partial_- u &= \partial_+ \begin{pmatrix} 0 & -a \\ 0 & -c \end{pmatrix} = \begin{pmatrix} 0 & b \\ 0 & d \end{pmatrix} \overset{u=e}{=} \begin{pmatrix} 0 & 0 \\ 0 & 1 \end{pmatrix}, \\
\partial_-\partial_+ u &= \partial_- \begin{pmatrix} -b & 0 \\ -d & 0 \end{pmatrix} = \begin{pmatrix} a & 0 \\ c & 0 \end{pmatrix} \overset{u=e}{=} \begin{pmatrix} 1 & 0 \\ 0 & 0 \end{pmatrix},
\end{aligned}
$$

so that

$$
\begin{aligned}
\triangle_+\sigma_{\partial_0^2} &= q_+(e)\,\sigma_{\partial_0^2} - 2(\partial_0 q_+)(e)\,\sigma_{\partial_0} + (\partial_0^2 q_+)(e)\,\sigma_I \\
&= 0, \\
\triangle_+\sigma_{\partial_+\partial_-} &= q_+(e)\,\sigma_{\partial_+\partial_-} - [(\partial_+ q_+)(e)\,\sigma_{\partial_-} \\
&\quad + (\partial_- q_-)(e)\,\sigma_{\partial_+}] + (\partial_+\partial_- q_+)(e)\,\sigma_I \\
&= \sigma_{\partial_-}, \\
\triangle_+\sigma_{\partial_-\partial_+} &= q_+(e)\,\sigma_{\partial_-\partial_+} - [(\partial_+ q_+)(e)\,\sigma_{\partial_-} \\
&\quad + (\partial_- q_-)(e)\,\sigma_{\partial_+}] + (\partial_-\partial_+ q_+)(e)\,\sigma_I \\
&= \sigma_{\partial_-}.
\end{aligned}
$$

Therefore

$$
\triangle_+\sigma_{\mathcal{L}} = -\sigma_{\partial_-}.
$$

Analogously,

$$
\triangle_-\sigma_{\mathcal{L}} = -\sigma_{\partial_+}.
$$

Finally,

$$
\triangle_0\sigma_{\mathcal{L}} = -2\sigma_{\partial_0},
$$

because

$$\begin{aligned}
\triangle_0 \sigma_{\partial_0^2} &= q_0(e)\, \sigma_{\partial_0^2} - 2(\partial_0 q_0)(e)\, \sigma_{\partial_0} + (\partial_0^2 q_0)(e)\, \sigma_I \\
&= 2\sigma_{\partial_0}, \\
\triangle_0 \sigma_{\partial_+ \partial_-} &= q_0(e)\, \sigma_{\partial_+ \partial_-} - [(\partial_+ q_0)(e)\, \sigma_{\partial_-} \\
&\quad + (\partial_- q_-)(e)\, \sigma_{\partial_+}] + (\partial_+ \partial_- q_0)(e)\, \sigma_I \\
&= -\sigma_I, \\
\triangle_0 \sigma_{\partial_- \partial_+} &= q_0(e)\, \sigma_{\partial_- \partial_+} - [(\partial_+ q_0)(e)\, \sigma_{\partial_-} \\
&\quad + (\partial_- q_-)(e)\, \sigma_{\partial_+}] + (\partial_- \partial_+ q_0)(e)\, \sigma_I \\
&= +\sigma_I.
\end{aligned}$$

12.3.3 Differences for $a\sigma_{\partial_0}$

Let us now calculate higher-order differences of the symbol $a\sigma_{\partial_0}$ which will be needed in the sequel.

Theorem 12.3.12. *For any $\alpha \in \mathbb{N}_0^3$, we have the formula*

$$\left[\triangle_+^{\alpha_1} \triangle_-^{\alpha_2} \triangle_0^{\alpha_3} (a\sigma_{\partial_0}) \right]_{nm}^l$$
$$= (m - \alpha_1/2 + \alpha_2/2) \left[\triangle_+^{\alpha_1} \triangle_-^{\alpha_2} \triangle_0^{\alpha_3} a \right]_{nm}^l + \alpha_3 \left[\overline{\triangle_0} \triangle_+^{\alpha_1} \triangle_-^{\alpha_2} \triangle_0^{\alpha_3 - 1} a \right]_{nm}^l,$$

where $\overline{\triangle_0}$ is given by

$$\begin{aligned}
(\overline{\triangle_0} a)_{nm}^l = \frac{1}{2} \Bigg[&\frac{\sqrt{(l-m)(l-n)}}{2l+1} a_{n^+ m^+}^{l^-} + \frac{\sqrt{(l+m+1)(l+n+1)}}{2l+1} a_{n^+ m^+}^{l^+} \\
&+ \frac{\sqrt{(l+m)(l+n)}}{2l+1} a_{n^- m^-}^{l^-} + \frac{\sqrt{(l-m+1)(l-n+1)}}{2l+1} a_{n^- m^-}^{l^+} \Bigg],
\end{aligned}$$

and satisfies $[\triangle_0, \overline{\triangle_0}] = 0$.

Proof. First we observe that we have

$$(a\, \sigma_{\partial_0})_{nm}^l = \sum_k a_{nk}^l \, k\, \delta_{km} = m\, a_{nm}^l.$$

Then using Theorem 12.3.5, we get

$$\triangle_-(a\sigma_{\partial_0})_{nm}^l = \frac{\sqrt{(l-m)(l+n)}}{2l+1} m^+ a_{n^- m^+}^{l^-} - \frac{\sqrt{(l+m+1)(l-n+1)}}{2l+1} m^+ a_{n^- m^+}^{l^+}$$
$$= (m^+ \triangle_- a)_{nm}^l,$$

and we can abbreviate this by writing $\triangle_-(a\sigma_{\partial_0}) = m^+\triangle_-a$. Further, we have

$$
\begin{aligned}
\triangle_-(\triangle_-(a\sigma_{\partial_0}))^l_{nm} &= \frac{\sqrt{(l-m)(l+n)}}{2l+1}[\triangle_-(a\sigma_{\partial_0})]^{l^-}_{n-m+} \\
&\quad - \frac{\sqrt{(l+m+1)(l-n+1)}}{2l+1}[\triangle_-(a\sigma_{\partial_0})]^{l^+}_{n-m+} \\
&= \frac{\sqrt{(l-m)(l+n)}}{2l+1}(m+1)(\triangle_-a)^{l^-}_{n-m+} \\
&\quad - \frac{\sqrt{(l+m+1)(l-n+1)}}{2l+1}(m+1)(\triangle_-a)^{l^+}_{n-m+} \\
&= (m+1)(\triangle_-^2 a)^l_{nm}.
\end{aligned}
$$

Continuing this calculation we can obtain

$$
\left[\triangle_-^k(a\sigma_{\partial_0})\right]^l_{nm} = (m+k/2)(\triangle_-^k a)^l_{nm}. \tag{12.25}
$$

By Theorem 12.3.5 we also have

$$
\begin{aligned}
[\triangle_+(\triangle_-(a\sigma_{\partial_0}))]^l_{nm} &= \left[\triangle_+\left(m^+\triangle_-a\right)\right]^l_{nm} \\
&= \frac{\sqrt{(l+m)(l-n)}}{2l+1}\left(m^+\triangle_-a\right)^{l^-}_{n+m-} \\
&\quad - \frac{\sqrt{(l-m+1)(l+n+1)}}{2l+1}\left(m^+\triangle_-a\right)^{l^+}_{n+m-} \\
&= m(\triangle_+\triangle_-a)^l_{nm}.
\end{aligned}
$$

By induction, and using (12.25), we then get

$$
\left[\triangle_+^{k_1}\triangle_-^{k_2}(a\sigma_{\partial_0})\right]^l_{nm} = (m-k_1/2+k_2/2)(\triangle_+^{k_1}\triangle_-^{k_2}a)^l_{nm}. \tag{12.26}
$$

The situation with $\triangle_0$ is more complicated because there are more terms. Using Theorem 12.3.5 we have

$$
\begin{aligned}
&\triangle_0(a\sigma_{\partial_0})^l_{nm} \\
&= \frac{\sqrt{(l-m)(l-n)}}{2l+1}(ma)^{l^-}_{n+m+} + \frac{\sqrt{(l+m+1)(l+n+1)}}{2l+1}(ma)^{l^+}_{n+m+} \\
&\quad - \frac{\sqrt{(l+m)(l+n)}}{2l+1}(ma)^{l^-}_{n-m-} - \frac{\sqrt{(l-m+1)(l-n+1)}}{2l+1}(ma)^{l^+}_{n-m-} \\
&= \frac{\sqrt{(l-m)(l-n)}}{2l+1}m^+a^{l^-}_{n+m+} + \frac{\sqrt{(l+m+1)(l+n+1)}}{2l+1}m^+a^{l^+}_{n+m+} \\
&\quad - \frac{\sqrt{(l+m)(l+n)}}{2l+1}m^-a^{l^-}_{n-m-} - \frac{\sqrt{(l-m+1)(l-n+1)}}{2l+1}m^-a^{l^+}_{n-m-}
\end{aligned}
$$

$$= m(\triangle_0 a)_{nm}^l + \frac{1}{2}\left[\frac{\sqrt{(l-m)(l-n)}}{2l+1}a_{n+m+}^{l-} + \frac{\sqrt{(l+m+1)(l+n+1)}}{2l+1}a_{n+m+}^{l+}\right.$$

$$\left.+\frac{\sqrt{(l+m)(l+n)}}{2l+1}a_{n-m-}^{l-} + \frac{\sqrt{(l-m+1)(l-n+1)}}{2l+1}a_{n-m-}^{l+}\right]$$

$$= m(\triangle_0 a)_{nm}^l + (\overline{\triangle}_0 a)_{nm}^l,$$

i.e.,

$$\triangle_0(a\sigma_{\partial_0})_{nm}^l = m(\triangle_0 a)_{nm}^l + (\overline{\triangle}_0 a)_{nm}^l, \qquad (12.27)$$

where $\overline{\triangle}_0$ is a weighted averaging operator given by

$$(\overline{\triangle}_0 a)_{nm}^l = \frac{1}{2}\left[\frac{\sqrt{(l-m)(l-n)}}{2l+1}a_{n+m+}^{l-} + \frac{\sqrt{(l+m+1)(l+n+1)}}{2l+1}a_{n+m+}^{l+}\right.$$

$$\left.+\frac{\sqrt{(l+m)(l+n)}}{2l+1}a_{n-m-}^{l-} + \frac{\sqrt{(l-m+1)(l-n+1)}}{2l+1}a_{n-m-}^{l+}\right].$$

We want to find a formula for $\triangle_0^k$, and for this we first calculate

$$\left[\triangle_0(\overline{\triangle}_0 a)\right]_{nm}^l$$

$$= \frac{\sqrt{(l-m)(l-n)}}{2l+1}(\overline{\triangle}_0 a)_{n+m+}^{l-} + \frac{\sqrt{(l+m+1)(l+n+1)}}{2l+1}(\overline{\triangle}_0 a)_{n+m+}^{l+}$$

$$- \frac{\sqrt{(l+m)(l+n)}}{2l+1}(\overline{\triangle}_0 a)_{n-m-}^{l-} - \frac{\sqrt{(l-m+1)(l-n+1)}}{2l+1}(\overline{\triangle}_0 a)_{n-m-}^{l+}$$

$$= \frac{\sqrt{(l-m)(l-n)}}{2l+1}\frac{1}{2}\left[\frac{\sqrt{(l^- - m^+)(l^- - n^+)}}{2l^-+1}a_{n++m++}^{l--}\right.$$

$$+ \frac{\sqrt{(l^- + m^+ + 1)(l^- + n^+ + 1)}}{2l^-+1}a_{n++m++}^{l-+}$$

$$+ \frac{\sqrt{(l^- + m^+)(l^- + n^+)}}{2l^-+1}a_{n+-m+-}^{l--}$$

$$\left.+ \frac{\sqrt{(l^- - m^+ + 1)(l^- - n^+ + 1)}}{2l^-+1}a_{n+-m+-}^{l-+}\right]$$

$$+ \frac{\sqrt{(l+m+1)(l+n+1)}}{2l+1}\frac{1}{2}\frac{1}{2l^++1}\left[\sqrt{(l^+ - m^+)(l^+ - n^+)}a_{n++m++}^{l+-}\right.$$

$$+ \sqrt{(l^+ + m^+ + 1)(l^+ + n^+ + 1)}a_{n++m++}^{l++}$$

$$+ \sqrt{(l^+ + m^+)(l^+ + n^+)}a_{n+-m+-}^{l+-}$$

$$\left.+ \sqrt{(l^+ - m^+ + 1)(l^+ - n^+ + 1)}a_{n+-m+-}^{l++}\right]$$

$$- \frac{\sqrt{(l+m)(l+n)}}{2l+1}\frac{1}{2}\frac{1}{2l^-+1}\left[\sqrt{(l^- - m^-)(l^- - n^-)}a_{n-+m-+}^{l--}\right.$$

$$+ \sqrt{(l^- + m^- + 1)(l^- + n^- + 1)} a_{n^-+m^-+}^{l^-+}$$
$$+ \sqrt{(l^- + m^-)(l^- + n^-)} a_{n^--m^--}^{l^--}$$
$$+ \sqrt{(l^- - m^- + 1)(l^- - n^- + 1)} a_{n^--m^--}^{l^-+} \Big]$$
$$- \frac{\sqrt{(l - m + 1)(l - n + 1)}}{2l + 1} \frac{1}{2} \frac{1}{2l^+ + 1} \Big[\sqrt{(l^+ - m^-)(l^+ - n^-)} a_{n^-+m^-+}^{l^+-}$$
$$+ \sqrt{(l^+ + m^- + 1)(l^+ + n^- + 1)} a_{n^-+m^-+}^{l^++}$$
$$+ \sqrt{(l^+ + m^-)(l^+ + n^-)} a_{n^--m^--}^{l^+-}$$
$$+ \sqrt{(l^+ - m^- + 1)(l^+ - n^- + 1)} a_{n^--m^--}^{l^++} \Big].$$

From this we get

$$\left[\triangle_0 (\overline{\triangle_0} a) \right]_{nm}^l = \frac{\sqrt{(l - m)(l - n)}}{2l + 1} \frac{1}{2} \frac{1}{2l} \Big[\sqrt{(l - m - 1)(l - n - 1)} a_{n++m++}^{l--}$$
$$+ \sqrt{(l + m + 1)(l + n + 1)} a_{n++m++}^l$$
$$+ \sqrt{(l + m)(l + n)} a_{nm}^{l--} + \sqrt{(l - m)(l - n)} a_{nm}^l \Big]$$
$$+ \frac{\sqrt{(l + m + 1)(l + n + 1)}}{2l + 1} \frac{1}{2} \frac{1}{2l + 2} \Big[\sqrt{(l - m)(l - n)} a_{n++m++}^l$$
$$+ \sqrt{(l + m + 2)(l + n + 2)} a_{n++m++}^{l++}$$
$$+ \sqrt{(l + m + 1)(l + n + 1)} a_{nm}^l + \sqrt{(l - m + 1)(l - n + 1)} a_{nm}^{l++} \Big]$$
$$- \frac{\sqrt{(l + m)(l + n)}}{2l + 1} \frac{1}{2} \frac{1}{2l} \Big[\sqrt{(l - m)(l - n)} a_{nm}^{l--}$$
$$+ \sqrt{(l + m)(l + n)} a_{nm}^l$$
$$+ \sqrt{(l + m - 1)(l + n - 1)} a_{n--m--}^{l--}$$
$$+ \sqrt{(l - m + 1)(l - n + 1)} a_{n--m--}^l \Big]$$
$$- \frac{\sqrt{(l - m + 1)(l - n + 1)}}{2l + 1} \frac{1}{2} \frac{1}{2l + 2} \Big[\sqrt{(l - m + 1)(l - n + 1)} a_{nm}^l$$
$$+ \sqrt{(l + m + 1)(l + n + 1)} a_{nm}^{l++}$$
$$+ \sqrt{(l + m)(l + n)} a_{n--m--}^l + \sqrt{(l - m + 2)(l - n + 2)} a_{n--m--}^{l++} \Big],$$
$$= \frac{\sqrt{(l - m)(l - n)}}{2l + 1} \frac{1}{2} \frac{1}{2l} \Big[\sqrt{(l - m - 1)(l - n - 1)} a_{n++m++}^{l--} +$$
$$+ \sqrt{(l + m + 1)(l + n + 1)} a_{n++m++}^l \Big]$$
$$+ \frac{\sqrt{(l + m + 1)(l + n + 1)}}{2l + 1} \frac{1}{2} \frac{1}{2l + 2} \Big[\sqrt{(l - m)(l - n)} a_{n++m++}^l$$

$$+ \sqrt{(l+m+2)(l+n+2)}a^{l++}_{n++m++} + \Big]$$

$$- \frac{\sqrt{(l+m)(l+n)}}{2l+1}\frac{1}{2}\frac{1}{2l}\Big[\sqrt{(l+m-1)(l+n-1)}a^{l--}_{n--m--} +$$

$$+ \sqrt{(l-m+1)(l-n+1)}a^{l}_{n--m--}\Big]$$

$$- \frac{\sqrt{(l-m+1)(l-n+1)}}{2l+1}\frac{1}{2}\frac{1}{2l+2}$$

$$\times \Big[\sqrt{(l+m)(l+n)}a^{l}_{n--m--} + \sqrt{(l-m+2)(l-n+2)}a^{l++}_{n--m--}\Big],$$

where we used that pairs of terms with a^{l--}_{nm}, a^{l++}_{nm} canceled, and also four terms with a^{l}_{nm} canceled in view of the identity

$$\frac{(l-m)(l-n)}{(2l+1)(2l)} + \frac{(l+m+1)(l+n+1)}{(2l+1)(2l+2)}$$

$$- \frac{(l+m)(l+n)}{(2l+1)(2l)} - \frac{(l-m+1)(l-n+1)}{(2l+1)(2l+2)}$$

$$= \frac{-2l(m+n)}{(2l+1)(2l)} + \frac{(2l+2)(m+n)}{(2l+1)(2l+2)} = 0.$$

Calculating in the other direction, we get

$$\big[\overline{\triangle}_0(\triangle_0 a)\big]^{l}_{nm}$$

$$= \frac{1}{2}\frac{\sqrt{(l-m)(l-n)}}{2l+1}(\triangle_0 a)^{l-}_{n+m+} + \frac{1}{2}\frac{\sqrt{(l+m+1)(l+n+1)}}{2l+1}(\triangle_0 a)^{l+}_{n+m+}$$

$$+ \frac{1}{2}\frac{\sqrt{(l+m)(l+n)}}{2l+1}(\triangle_0 a)^{l-}_{n-m-} + \frac{1}{2}\frac{\sqrt{(l-m+1)(l-n+1)}}{2l+1}(\triangle_0 a)^{l+}_{n-m-}$$

$$= \frac{\sqrt{(l-m)(l-n)}}{2l+1}\frac{1}{2}\frac{1}{2l^-+1}\Big[\sqrt{(l^--m^+)(l^--n^+)}a^{l--}_{n++m++}$$

$$+ \sqrt{(l^-+m^++1)(l^-+n^++1)}a^{l-+}_{n++m++}$$

$$- \sqrt{(l^-+m^+)(l^-+n^+)}a^{l--}_{n+-m+-}$$

$$- \sqrt{(l^--m^++1)(l^--n^++1)}a^{l-+}_{n+-m+-}\Big]$$

$$+ \frac{\sqrt{(l+m+1)(l+n+1)}}{2l+1}\frac{1}{2}\frac{1}{2l^++1}\Big[\sqrt{(l^+-m^+)(l^+-n^+)}a^{l+-}_{n++m++}$$

$$+ \sqrt{(l^++m^++1)(l^++n^++1)}a^{l++}_{n++m++}$$

$$- \sqrt{(l^++m^+)(l^++n^+)}a^{l+-}_{n+-m+-}$$

$$- \sqrt{(l^+-m^++1)(l^+-n^++1)}a^{l++}_{n+-m+-}\Big]$$

$$+ \frac{\sqrt{(l+m)(l+n)}}{2l+1}\frac{1}{2}\frac{1}{2l^-+1}\Big[\sqrt{(l^--m^-)(l^--n^-)}a^{l--}_{n-+m-+}$$

$$+ \sqrt{(l^- + m^- + 1)(l^- + n^- + 1)}a^{l^{-+}}_{n^-+m^-+}$$
$$- \sqrt{(l^- + m^-)(l^- + n^-)}a^{l^{--}}_{n^--m^--}$$
$$- \sqrt{(l^- - m^- + 1)(l^- - n^- + 1)}a^{l^{-+}}_{n^--m^--}\Big]$$
$$+ \frac{\sqrt{(l-m+1)(l-n+1)}}{2l+1}\frac{1}{2}\frac{1}{2l+1}\Big[\sqrt{(l^+ - m^-)(l^+ - n^-)}a^{l^{+-}}_{n^-+m^-+}$$
$$+ \sqrt{(l^+ + m^- + 1)(l^+ + n^- + 1)}a^{l^{++}}_{n^-+m^-+}$$
$$- \sqrt{(l^+ + m^-)(l^+ + n^-)}a^{l^{+-}}_{n^--m^--}$$
$$- \sqrt{(l^+ - m^- + 1)(l^+ - n^- + 1)}a^{l^{++}}_{n^--m^--}\Big].$$

From this we get

$$\big[\overline{\triangle_0}(\triangle_0 a)\big]^l_{nm}$$
$$= \frac{\sqrt{(l-m)(l-n)}}{2l+1}\frac{1}{2}\frac{1}{2l}\Big[\sqrt{(l-m-1)(l-n-1)}a^{l^{--}}_{n++m++}$$
$$+ \sqrt{(l+m+1)(l+n+1)}a^{l}_{n++m++}$$
$$- \sqrt{(l+m)(l+n)}a^{l^{--}}_{nm} - \sqrt{(l-m)(l-n)}a^{l}_{nm}\Big]$$
$$+ \frac{\sqrt{(l+m+1)(l+n+1)}}{2l+1}\frac{1}{2}\frac{1}{2l+2}\Big[\sqrt{(l-m)(l-n)}a^{l}_{n++m++}$$
$$+ \sqrt{(l+m+2)(l+n+2)}a^{l^{++}}_{n++m++}$$
$$- \sqrt{(l+m+1)(l+n+1)}a^{l}_{nm} - \sqrt{(l-m+1)(l-n+1)}a^{l^{++}}_{nm}\Big]$$
$$+ \frac{\sqrt{(l+m)(l+n)}}{2l+1}\frac{1}{2}\frac{1}{2l}\Big[\sqrt{(l-m)(l-n)}a^{l^{--}}_{nm}$$
$$+ \sqrt{(l+m)(l+n)}a^{l}_{nm}$$
$$- \sqrt{(l+m-1)(l+n-1)}a^{l^{--}}_{n--m--}$$
$$- \sqrt{(l-m+1)(l-n+1)}a^{l}_{n--m--}\Big]$$
$$+ \frac{\sqrt{(l-m+1)(l-n+1)}}{2l+1}\frac{1}{2}\frac{1}{2l+2}\Big[\sqrt{(l-m+1)(l-n+1)}a^{l}_{nm}$$
$$+ \sqrt{(l+m+1)(l+n+1)}a^{l^{++}}_{nm}$$
$$- \sqrt{(l+m)(l+n)}a^{l}_{n--m--} - \sqrt{(l-m+2)(l-n+2)}a^{l^{++}}_{n--m--}\Big]$$
$$= \frac{\sqrt{(l-m)(l-n)}}{2l+1}\frac{1}{2}\frac{1}{2l}\Big[\sqrt{(l-m-1)(l-n-1)}a^{l^{--}}_{n++m++}$$
$$+ \sqrt{(l+m+1)(l+n+1)}a^{l}_{n++m++}\Big]$$

$$+ \frac{\sqrt{(l+m+1)(l+n+1)}}{2l+1} \frac{1}{2} \frac{1}{2l+2} \left[\sqrt{(l-m)(l-n)} a_{n++m++}^{l} \right.$$

$$\left. + \sqrt{(l+m+2)(l+n+2)} a_{n++m++}^{l++} \right]$$

$$+ \frac{\sqrt{(l+m)(l+n)}}{2l+1} \frac{1}{2} \frac{1}{2l} \left[\sqrt{(l+m-1)(l+n-1)} a_{n--m--}^{l--} \right.$$

$$\left. - \sqrt{(l-m+1)(l-n+1)} a_{n--m--}^{l} \right]$$

$$+ \frac{\sqrt{(l-m+1)(l-n+1)}}{2l+1} \frac{1}{2} \frac{1}{2l+2}$$

$$\times \left[-\sqrt{(l+m)(l+n)} a_{n--m--}^{l} - \sqrt{(l-m+2)(l-n+2)} a_{n--m--}^{l++} \right],$$

where we used again the fact that terms a_{nm}^{l}, a_{nm}^{l--} and a_{nm}^{l++} canceled. Comparing these calculations we get the commutativity property

$$\overline{\triangle}_0 \triangle_0 a = \triangle_0 \overline{\triangle}_0 a.$$

From this and (12.27) we obtain

$$\triangle_0^2 (ma) = \triangle_0 (m \triangle_0 a + \overline{\triangle}_0 a) = m \triangle_0^2 a + 2 \overline{\triangle}_0 \triangle_0 a,$$

and, consequently, by induction we get

$$\triangle_0^k (ma) = m \triangle_0^k a + k \overline{\triangle}_0 \triangle_0^{k-1} a.$$

Let us now apply this to (12.26). Using commutativity of $\triangle_0, \triangle_+$ and $\triangle_-$ from Theorem 12.3.5, we get

$$\left[\triangle_+^{k_1} \triangle_-^{k_2} \triangle_0^{k_3} (a\sigma_{\partial_0}) \right]_{nm}^{l}$$

$$= \left[\triangle_0^{k_3} \triangle_+^{k_1} \triangle_-^{k_2} (a\sigma_{\partial_0}) \right]_{nm}^{l}$$

$$= \left[\triangle_0^{k_3} \left((m - k_1/2 + k_2/2) \triangle_+^{k_1} \triangle_-^{k_2} a \right) \right]_{nm}^{l}$$

$$= \left[\triangle_0^{k_3} \left(m \triangle_+^{k_1} \triangle_-^{k_2} a \right) \right]_{nm}^{l} - \left[\triangle_0^{k_3} \left((k_1/2 - k_2/2) \triangle_+^{k_1} \triangle_-^{k_2} a \right) \right]_{nm}^{l}$$

$$= m \left[\triangle_0^{k_3} \triangle_+^{k_1} \triangle_-^{k_2} a \right]_{nm}^{l} + k_3 \left[\overline{\triangle}_0 \triangle_0^{k_3-1} \triangle_+^{k_1} \triangle_-^{k_2} a \right]_{nm}^{l}$$

$$- (k_1/2 - k_2/2) \left[\triangle_0^{k_3} \triangle_+^{k_1} \triangle_-^{k_2} a \right]_{nm}^{l}$$

$$= (m - k_1/2 + k_2/2) \left[\triangle_+^{k_1} \triangle_-^{k_2} \triangle_0^{k_3} a \right]_{nm}^{l} + k_3 \left[\overline{\triangle}_0 \triangle_+^{k_1} \triangle_-^{k_2} \triangle_0^{k_3-1} a \right]_{nm}^{l},$$

completing the proof. $\square$

12.4   Symbol classes on $\mathrm{SU}(2)$

The goal of this section is to simplify the symbol class $\Sigma^m(\mathrm{SU}(2))$ from Definition 12.1.1, yielding Hörmander's class $\Psi^m(\mathrm{SU}(2))$ of pseudo-differential operators on $\mathrm{SU}(2)$. For this purpose, we introduce and investigate the symbol class $S^m(\mathrm{SU}(2))$.

Definition 12.4.1 (Symbol class $S^m(\mathrm{SU}(2))$). For $u \in \mathrm{SU}(2)$, we write

$$A_u f := A(f \circ \phi) \circ \phi^{-1},$$

where $\phi(x) = xu$; note that by Proposition 10.4.18 we have

$$
\begin{aligned}
R_{A_u}(x, y) &= R_A(xu^{-1}, uyu^{-1}), \\
\sigma_{A_u}(x, l) &= t^l(u)^* \, \sigma_A(xu^{-1}, l) \, t^l(u).
\end{aligned}
$$

The symbol class $S^m(\mathrm{SU}(2))$ consists of the symbols σ_A of those operators $A \in \mathcal{L}(C^\infty(\mathrm{SU}(2)))$ for which

$$\mathrm{sing\ supp}\ (y \mapsto R_A(x, y)) \subset \{e\},$$

and for which

$$\left| \triangle_l^\alpha \partial_x^\beta \sigma_{A_u}(x, l)_{ij} \right| \leq C_{A\alpha\beta mN} \, \langle i - j \rangle^{-N} \, (1 + l)^{m - |\alpha|} \qquad (12.28)$$

uniformly in $x, u \in \mathrm{SU}(2)$, for every $N \geq 0$, all $l \in \frac{1}{2}\mathbb{N}_0$, all multi-indices $\alpha, \beta \in \mathbb{N}_0^3$, and for all matrix column/row numbers i, j. Thus, the constant in (12.28) may depend on A, α, β, m and N, but not on x, u, l, i, j.

Remark 12.4.2 **(Rapid off-diagonal decay).** We note that inequality (12.28) contains the *rapid off-diagonal decay property* since we can take N as large as we want.

We now formulate the main theorem of this section:

Theorem 12.4.3 (Equality of classes $\mathrm{Op}\, S^m(\mathrm{SU}(2)) = \Psi^m(\mathrm{SU}(2))$). *We have $A \in \Psi^m(\mathrm{SU}(2))$ if and only if $\sigma_A \in S^m(\mathrm{SU}(2))$. Moreover, we have the equality of symbol classes*

$$S^m(\mathrm{SU}(2)) = \Sigma^m(\mathrm{SU}(2)). \qquad (12.29)$$

In fact, we need to prove only the equality of symbol classes (12.29), from which the first part of the theorem would follow by Theorem 10.9.6. In the process of proving equality (12.29), we establish a number of auxiliary results.

Remark 12.4.4. By Corollary 10.9.8, if $\sigma_A \in \Sigma^m(\mathrm{SU}(2))$ then

$$\triangle_l^\gamma \partial_x^\delta \sigma_A \in \Sigma^{m-|\gamma|}(\mathrm{SU}(2)).$$

We show the analogous result for $S^m(\mathrm{SU}(2))$:

Lemma 12.4.5. *If $\sigma_A \in S^m(\mathrm{SU}(2))$ then $\sigma_B = \triangle_l^\gamma \partial_x^\delta \sigma_A \in S^{m-|\gamma|}(\mathrm{SU}(2))$.*

Proof. First, let $|\gamma| = 1$. Then $\triangle_l^\gamma \widehat{f}(l) = \widehat{qf}(l)$ for some

$$q \in Pol_1(\mathrm{SU}(2)) := \mathrm{span}\left\{t_{ij}^{1/2} : i,j \in \{-1/2, +1/2\}\right\}$$

for which $q(e) = 0$. Let $r(y) := q(uyu^{-1})$. Then $r \in Pol_1(\mathrm{SU}(2))$, because

$$t_{ij}^{1/2}(uyu^{-1}) = \sum_{k,m} t_{ik}^{1/2}(u) \, t_{km}^{1/2}(y) \, t_{mj}^{1/2}(u^{-1}).$$

Moreover, we have $r(e) = 0$. Hence $\widehat{f}(l) \mapsto \widehat{rf}(l)$ is a linear combination of difference operators $\triangle_0, \triangle_+, \triangle_-$ because $\{f \in Pol^1(\mathrm{SU}(2)) : f(e) = 0\}$ is a three-dimensional vector space spanned by q_0, q_+, q_-. Now let $\gamma \in \mathbb{N}_0^3$ and $\sigma_B = \triangle_l^\gamma \partial_x^\delta \sigma_A$. We have

$$
\begin{aligned}
\triangle_l^\alpha \partial_x^\beta \sigma_{B_u}(x,l) &= \triangle_l^\alpha \partial_x^\beta \left(t^l(u)^* \, \sigma_B(xu^{-1},l) \, t^l(u)\right) \\
&= \triangle_l^\alpha \partial_x^\beta \left(t^l(u)^* \left(\triangle_l^\gamma \partial_x^\delta \sigma_A(xu^{-1},l)\right) t^l(u)\right) \\
&= \sum_{|\gamma'|=|\gamma|} \lambda_{u,\gamma'} \, \triangle_l^{\alpha+\gamma'} \partial_x^{\beta+\delta} \, \sigma_{A_u}(x,l),
\end{aligned}
$$

for some scalars $\lambda_{u,\gamma'} \in \mathbb{C}$ depending only on $u \in \mathrm{SU}(2)$ and multi-indices $\gamma' \in \mathbb{N}_0^3$. $\square$

Remark 12.4.6. Let D be a left-invariant vector field on $\mathrm{SU}(2)$. From the very definition of the symbol classes $\Sigma^m(\mathrm{SU}(2)) = \bigcap_{k=0}^\infty \Sigma_k^m(\mathrm{SU}(2))$, it is evident that $[\sigma_D, \sigma_A] \in \Sigma^m(\mathrm{SU}(2))$ if $\sigma_A \in \Sigma^m(\mathrm{SU}(2))$. We now prove the similar invariance for $S^m(\mathrm{SU}(2))$.

Lemma 12.4.7. *Let D be a left-invariant vector field on $\mathrm{SU}(2)$. Let $\sigma_A \in S^m(\mathrm{SU}(2))$. Then $[\sigma_D, \sigma_A] \in S^m(\mathrm{SU}(2))$ and $\sigma_A \, \sigma_D \in S^{m+1}(\mathrm{SU}(2))$.*

Proof. For $D \in \mathfrak{su}(2)$ we write $D = iE$, so that $E \in i\,\mathfrak{su}(2)$. By Proposition 12.2.4 there is some $u \in \mathrm{SU}(2)$ such that $\sigma_E(l) = t^l(u)^* \, \sigma_{\partial_0}(l) \, t^l(u)$. Now, we have

$$
\begin{aligned}
[\sigma_E, \sigma_A](l) &= t^l(u)^* \, \left[\sigma_{\partial_0}(l), t^l(u) \, \sigma_A(x,l) \, t^l(u)^*\right] \, t^l(u) \\
&= \left[\sigma_{\partial_0}, \sigma_{A_{u^{-1}}}\right]_u (l).
\end{aligned}
$$

Next, notice that $S^m(\mathrm{SU}(2))$ is invariant under the mappings $\sigma_B \mapsto \sigma_{B_u}$ and $\sigma_B \mapsto [\sigma_{\partial_0}, \sigma_B]$; here $[\sigma_{\partial_0}, \sigma_B](l)_{ij} = (i-j)\,\sigma_B(l)_{ij}$. Finally,

$$
\begin{aligned}
\sigma_A(x,l) \, \sigma_E(l) &= t^l(u)^* \, t^l(u) \, \sigma_A(x,l) \, t^l(u)^* \, \sigma_{\partial_0}(l) \, t^l(u) \\
&= \left(\sigma_{A_{u^{-1}}}(x,l) \, \sigma_{\partial_0}(l)\right)_u.
\end{aligned}
$$

Just as in the first part of the proof, we see that $\sigma_A \, \sigma_D$ belongs to $S^{m+1}(\mathrm{SU}(2))$ since $\sigma_B \, \sigma_{\partial_0} \in S^{m+1}(\mathrm{SU}(2))$ if $\sigma_B \in S^m(\mathrm{SU}(2))$, by Theorem 12.3.12. $\square$

Proof of Theorem 12.4.3. We have to show that $S^m(\mathrm{SU}(2)) = \Sigma^m(\mathrm{SU}(2))$, so that the theorem would follow from Theorem 10.9.6. Note that both classes $S^m(\mathrm{SU}(2))$ and $\Sigma^m(\mathrm{SU}(2))$ require the singular support condition $(y \mapsto R_A(x,y)) \subset \{e\}$, so we do not have to consider this; moreover, the x-dependence of the symbol is not essential here, and therefore we abbreviate $\sigma_A(l) := \sigma_A(x,l)$. First, let us show that $\Sigma^m(\mathrm{SU}(2)) \subset S^m(\mathrm{SU}(2))$. Take $\sigma_A \in \Sigma^m(\mathrm{SU}(2))$. Then also $\sigma_{A_u} \in \Sigma^m(\mathrm{SU}(2))$ (either by the well-known properties of pseudodifferential operators and Theorem 10.9.6, or by checking directly that the definition of the classes $\Sigma_k^m(\mathrm{SU}(2))$ is conjugation-invariant). Let us define $c_N(B)$ by

$$\sigma_{c_N(B)}(l)_{ij} := (i-j)^N \ \sigma_B(l)_{ij}.$$

Now $\sigma_{c_N(A_u)} \in \Sigma^m(\mathrm{SU}(2))$ for every $N \in \mathbb{Z}^+$, because $\sigma_{A_u} \in \Sigma^m(\mathrm{SU}(2))$ and

$$[\sigma_{\partial_0}, \sigma_B](l)_{ij} = (i-j) \ \sigma_B(l)_{ij}.$$

This implies the "rapid off-diagonal decay" of σ_{A_u}:

$$|\sigma_{A_u}(x,l)_{ij}| \le C_{AmN} \ \langle i-j \rangle^{-N} \ (1+l)^m,$$

implying the norm comparability

$$\|\cdots \sigma_{A_u}(l)\|_{op} \sim \sup_{i,j} |\cdots \sigma_{A_u}(l)_{ij}| \qquad (12.30)$$

in view of Lemma 12.6.5 in Section 12.6. Moreover, $\triangle_l^\alpha \partial_x^\beta \sigma_{A_u} \in \Sigma^{m-|\alpha|}(\mathrm{SU}(2))$ by Corollary 10.9.8, so that we obtain the symbol inequalities (12.28) from (10.37). Thereby $\Sigma^m(\mathrm{SU}(2)) \subset S^m(\mathrm{SU}(2))$.

Now we have to show that $S^m(\mathrm{SU}(2)) \subset \Sigma^m(\mathrm{SU}(2))$. Again, we may exploit the norm comparabilities (12.30): thus clearly $S^m(\mathrm{SU}(2)) \subset \Sigma_0^m(\mathrm{SU}(2))$. Consequently, $S^m(\mathrm{SU}(2)) \subset \Sigma_k^m(\mathrm{SU}(2))$ for all $k \in \mathbb{Z}^+$, due to Lemmas 12.4.5 and 12.4.7. $\qquad \square$

Remark 12.4.8. Notice that in Definition 12.4.1, we demanded inequalities (12.28) uniformly in $u \in \mathrm{SU}(2)$. However, it suffices to assume this for only $u \in \{e, \omega_1(\pi/2)\}$, where

$$e = \begin{pmatrix} 1 & 0 \\ 0 & 1 \end{pmatrix} \quad \text{and} \quad \omega_1\left(\frac{\pi}{2}\right) = \frac{\sqrt{2}}{2} \begin{pmatrix} 1 & i \\ i & 1 \end{pmatrix}.$$

Exercise 12.4.9. Prove the claim of Remark 12.4.8. Hint: Notice that

$$u(\phi, \theta, \psi) = \omega_3(\phi) \ \omega_2(\theta) \ \omega_3(\psi) \quad \text{and that} \quad \omega_2(\theta) = w_1^{-1}\omega_3(\theta)w_1,$$

where $w_1 = \omega_1(\pi/2) = \frac{\sqrt{2}}{2} \begin{pmatrix} 1 & i \\ i & 1 \end{pmatrix}$. Recall also Proposition 10.4.18. Conjugating a symbol $\sigma_A(x,l)$ with $t^l(\omega_3(t))$ does not affect the "rapid off-diagonal decay" property. Let $q \in Q_1 := \mathrm{span}\{q_+, q_-, q_0\} \setminus \{0\}$, and let $\triangle_q$ be the corresponding first-order difference operator, i.e.,

$$\triangle_q \widehat{f}(l) = \widehat{qf}(l).$$

Defining $q_1 \in Q_1$ by $q_1(z) := q(u^{-1}zu)$ and $\sigma_B := \triangle_{q_1}\sigma_A$, we have $\triangle_q \sigma_{A_u}(l) = \sigma_{B_u}(l)$. This shows us that difference operators can essentially be moved through the $t^l(u)$-conjugations; also, such conjugations do not affect the x-derivatives; moreover, these conjugations behave well with respect to taking commutators of symbols.

Remark 12.4.10 (**Topology on** $S^m(\mathrm{SU}(2))$). It is natural to define the topology on $S^m(\mathrm{SU}(2))$ by seminorms

$$p_{\alpha,\beta,m,i,j,N,u}(\sigma_A) := \sup_{x \in \mathrm{SU}(2), l \in \frac{1}{2}\mathbb{N}_0} \left\{ \langle i-j \rangle^N \frac{|\triangle_l^\alpha \partial_x^\beta \sigma_{A_u}(x,l)_{ij}|}{(1+l)^{m-|\alpha|}} \right\}. \tag{12.31}$$

Notice that by Exercise 12.4.9, it is sufficient to consider only the cases $u \in \{e, \omega_2(\pi/2)\}$.

Compared to Corollary 10.9.9, we can replace the convergence in the Hilbert–Schmidt norm by the pointwise ℓ^∞ convergence to relate the convergence of symbols to the convergence of operators:

Corollary 12.4.11 (Convergence of symbols and operators). *Let* $\sigma \in S^0(\mathrm{SU}(2))$, *and assume that a sequence* $\sigma_k \in S^0(\mathrm{SU}(2))$ *satisfies inequalities* (12.28) *uniformly in* k (*i.e., with constants independent of* k). *Assume that for all* $|\beta| \leq 2$ *we have the convergence*

$$\partial_x^\beta \sigma_k(x,l) \to \partial_x^\beta \sigma(x,l) \ \textit{as} \ k \to \infty \tag{12.32}$$

in the ℓ^∞ *norm, uniformly over all* $x \in G$ *and all* $l \in \frac{1}{2}\mathbb{N}_0$. *Then* $\mathrm{Op}\,\sigma_k \to \mathrm{Op}\,\sigma$ *strongly on* $L^2(\mathrm{SU}(2))$.

Moreover, if the convergence (12.32) *holds for all* β, *then* $\mathrm{Op}\,\sigma_k \to \mathrm{Op}\,\sigma$ *strongly on* $H^s(\mathrm{SU}(2))$ *for any* $s \in \mathbb{R}$.

Proof. We observe that Theorem 10.5.5 implies that

$$\| \mathrm{Op}\,\sigma_k - \mathrm{Op}\,\sigma \|_{\mathcal{L}(L^2(\mathrm{SU}(2)))}$$
$$\leq \ C_1 \sup_{x \in \mathrm{SU}(2), l \in \frac{1}{2}\mathbb{N}_0, |\beta| \leq 2} \| \partial_x^\beta \sigma_k(x,l) - \partial_x^\beta \sigma(x,l) \|_{op}$$
$$\leq \ C_2 \sup_{x \in \mathrm{SU}(2), l \in \frac{1}{2}\mathbb{N}_0, |\beta| \leq 2} \| \partial_x^\beta \sigma_k(x,l) - \partial_x^\beta \sigma(x,l) \|_{\ell^\infty},$$

where the last estimate follows from Lemma 12.6.5 in Section 12.6, with constant C_2 independent of k. The strong convergence on $L^2(\mathrm{SU}(2))$ now follows directly from (12.32). The strong convergence on $H^s(\mathrm{SU}(2))$ follows from Theorem 10.8.1 by the same argument. $\square$

12.5   Pseudo-differential operators on $\mathbb{S}^3$

In this section we discuss how the construction of Section 10.10 yields full symbols
of pseudo-differential operators on the 3-sphere $\mathbb{S}^3$. For this, we will use a global
isomorphism $\mathbb{S}^3 \cong SU(2)$ from Proposition 11.4.2. First we recall the quaternion
space $\mathbb{H}$ from Section 11.4 which is the associative $\mathbb{R}$-algebra with a vector space
basis $\{\mathbf{1}, \mathbf{i}, \mathbf{j}, \mathbf{k}\}$, where $\mathbf{1} \in \mathbb{H}$ is the unit and

$$\mathbf{i}^2 = \mathbf{j}^2 = \mathbf{k}^2 = -\mathbf{1} = \mathbf{ijk}.$$

The mapping $x = (x_m)_{m=0}^3 \mapsto x_0 \mathbf{1} + x_1 \mathbf{i} + x_2 \mathbf{j} + x_3 \mathbf{k}$ identifies $\mathbb{R}^4$ with $\mathbb{H}$. In
particular, the unit sphere $\mathbb{S}^3 \subset \mathbb{R}^4 \cong \mathbb{H}$ is a multiplicative group. A bijective
homomorphism $\Phi^{-1} : \mathbb{S}^3 \to SU(2)$ in (11.6) is defined by

$$x \mapsto \Phi^{-1}(x) = \begin{pmatrix} x_0 + ix_3 & x_1 + ix_2 \\ -x_1 + ix_2 & x_0 - ix_3 \end{pmatrix},$$

and its inverse $\Phi : SU(2) \to \mathbb{S}^3$ gives rise to the global quantisation of pseudo-
differential operators on $\mathbb{S}^3$ induced by that on $SU(2)$, as shown in Section 10.10.

The diffeomorphism Φ induces the Fourier analysis on $\mathbb{S}^3$ in terms of the
representations of $SU(2)$. To fix the notation for this in terms of $\mathbb{S}^3$, let

$$t^l : \mathbb{S}^3 \to U(2l+1) \subset \mathbb{C}^{(2l+1)\times(2l+1)},$$

$l \in \frac{1}{2}\mathbb{N}_0$, be a family of group homomorphisms, which are the irreducible continuous (and hence smooth) unitary representations of $\mathbb{S}^3$ when it is endowed with
the $SU(2)$ structure via the quaternionic product, see Section 12.5 for details. The
Fourier coefficient $\widehat{f}(l)$ of $f \in C^\infty(\mathbb{S}^3)$ is defined by

$$\widehat{f}(l) = \int_{\mathbb{S}^3} f(x)\, t^l(x)^* \, \mathrm{d}x,$$

where the integration is performed with respect to the Haar measure, and $\widehat{f}(l) \in
\mathbb{C}^{(2l+1)\times(2l+1)}$. The corresponding Fourier series is given by

$$f(x) \quad = \quad \sum_{l \in \frac{1}{2}\mathbb{N}_0} (2l+1)\, \mathrm{Tr}\left(\widehat{f}(l)\, t^l(x)\right).$$

Now, if $A : C^\infty(\mathbb{S}^3) \to C^\infty(\mathbb{S}^3)$ is a continuous linear operator, we define its full
symbol as a mapping

$$(x, l) \mapsto \sigma_A(x, l), \quad \sigma_A(x, l) = t^l(x)^* (At^l)(x) \in \mathbb{C}^{(2l+1)\times(2l+1)}.$$

Then we have the representation of operator A in the form

$$Af(x) = \sum_{l \in \frac{1}{2}\mathbb{N}_0} (2l+1)\, \mathrm{Tr}\left(t^l(x)\, \sigma_A(x, l)\, \widehat{f}(l)\right),$$

see Theorem 10.4.4. We also note that if

$$Af(x) = \int_{\mathbb{S}^3} K_A(x,y)\, f(y)\, \mathrm{d}y = \int_{\mathbb{S}^3} f(y)\, R_A(x, y^{-1}x)\, \mathrm{d}y,$$

where R_A is the right-convolution kernel of A, then

$$\sigma_A(x,l) = \int_{\mathbb{S}^3} R_A(x,y)\, t^l(y)^*\, \mathrm{d}y$$

by Theorem 10.4.6, where, as usual, the integration is performed with respect to the Haar measure with a standard distributional interpretation.

We now introduce symbol classes $S^m(\mathbb{S}^3)$ which allow us to characterise operators from Hörmander's class $\Psi^m(\mathbb{S}^3)$.

Definition 12.5.1 (Symbol class $S^m(\mathbb{S}^3)$). We write $\sigma_A \in S^m(\mathbb{S}^3)$ if the corresponding kernel $K_A(x,y)$ is smooth outside the diagonal $x = y$ and if we have the estimate

$$\left| \triangle_l^\alpha \partial_x^\beta \sigma_{A_u}(x,l)_{ij} \right| \leq C_{A\alpha\beta m N}\, (1 + |i - j|)^{-N} (1 + l)^{m - |\alpha|}, \tag{12.33}$$

for every $N \geq 0$, every $u \in \mathbb{S}^3$, and all multi-indices α, β, where the symbol σ_{A_u} is the symbol of the operator

$$A_u f = A(f \circ \varphi_u) \circ \varphi_u^{-1},$$

with $\varphi_u(x) = xu$ the quaternionic product. We write $\triangle_l^\alpha = \triangle_+^{\alpha_1} \triangle_-^{\alpha_2} \triangle_0^{\alpha_3}$, where the operators $\triangle_+, \triangle_-, \triangle_0$ are discrete difference operators acting on matrices $\sigma_A(x,l)$ in the variable l, and explicit formulae for them and their properties are given in Definition 12.3.2 and Theorem 12.3.5, with polynomials q_+, q_- and q_0 defined as in Remark 10.6.2. Constants $C_{A\alpha\beta m N}$ in (12.33) may depend on A, α, β, m, N but not on i, j, l.

Remark 12.5.2. As in the case of SU(2), the symbols of A_u and A are related by

$$\sigma_{A_u}(x,l) = t^l(u)^* \sigma_A(xu^{-1}, l)\, t^l(u),$$

see Proposition 10.4.18. We notice that imposing the same conditions on all symbols σ_{A_u} in (12.33) simply refers to the well-known fact that the class $\Psi^m(\mathbb{S}^3)$ should be in particular "translation"-invariant (i.e., invariant under the changes of variables induced by quaternionic products φ_u), namely that $A \in \Psi^m(\mathbb{S}^3)$ if and only if $A_u \in \Psi^m(\mathbb{S}^3)$, for all $u \in \mathbb{S}^3$. Condition (12.33) is the growth condition with respect to the *quantum number l* combined with the condition that matrices $\sigma_A(x,l)$ must have a rapid off-diagonal decay.

With Definition 12.5.1, we have the following characterisation, which follows immediately from Theorem 12.4.3:

Theorem 12.5.3 (Equality $\mathrm{Op}\, S^m(\mathbb{S}^3) = \Psi^m(\mathbb{S}^3)$). *We have $A \in \Psi^m(\mathbb{S}^3)$ if and only if $\sigma_A \in S^m(\mathbb{S}^3)$.*

12.6   Appendix: infinite matrices

In this section we discuss infinite matrices. The main conclusion that we need is
that the operator-norm and the l^∞-norm are equivalent for matrices arising as
full symbols of pseudo-differential operators in $\Psi^m(SU(2))$, used in (12.30) in the
proof of Theorem 12.4.3.

The reader should already know basic linear algebra, but for the sake of
completeness, we review necessary matrix operations. Let $\mathbb{C}^{m \times n}$ denote the complex vector space of matrices with m rows and n columns; the rows are numbered
$1, \ldots, m$ downwards, the columns $1, \ldots, n$ from left to right. Let $A_{ij} \in \mathbb{C}$ denote the element of matrix $A \in \mathbb{C}^{m \times n}$ on row i and column k. Let $\lambda \in \mathbb{C}$ and
$A, B \in \mathbb{C}^{m \times n}$; let matrices $\lambda A, A + B \in \mathbb{C}^{m \times n}$ and the adjoint $A^* \in \mathbb{C}^{n \times m}$ be
defined by

$$(\lambda A)_{ij} := \lambda A_{ij},$$

$$(A + B)_{ij} := A_{ij} + B_{ij},$$

$$(A^*)_{ij} := \overline{A_{ji}}.$$

The product of $A \in \mathbb{C}^{m \times p}$ and $B \in \mathbb{C}^{p \times n}$ is $AB \in \mathbb{C}^{m \times n}$ defined by

$$(AB)_{ij} := \sum_{k=1}^{p} A_{ik} B_{kj}.$$

The trace of $A \in \mathbb{C}^{n \times n}$ is

$$\mathrm{Tr}(A) := \sum_{j=1}^{n} A_{jj}.$$

We may naturally identify vector space $\mathbb{C}^n$ with $\mathbb{C}^{n \times 1}$, and a mapping $(x \mapsto Ax) :$
$\mathbb{C}^{m \times 1} \to \mathbb{C}^{n \times 1}$ can be seen as a linear mapping $\mathbb{C}^m \to \mathbb{C}^n$.

The *Euclidean inner product* (or *Hilbert–Schmidt inner product*) of $A, B \in$
$\mathbb{C}^{m \times n}$ is a special case of that in Subsection B.5.1, and is given by

$$\langle A, B \rangle_{HS} := \mathrm{Tr}(B^* A)^{1/2} = \sum_{i=1}^{m} \sum_{j=1}^{n} \overline{B_{ij}} A_{ij}$$

and the corresponding norm of A is $\|A\|_{HS} := \langle A, A \rangle_{HS}^{1/2}$. The *operator norm* of
$A \in \mathbb{C}^{m \times n}$ is

$$\|A\| = \|A\|_{op} := \sup \left\{ \|Ax\|_{\ell^2} : \ x \in \mathbb{C}^{n \times 1}, \ \|x\|_{\ell^2} \leq 1 \right\},$$

where $\|x\|_{\ell^2} = (\sum_{j=1}^{n} |x_j|^2)^{1/2}$ is the usual Euclidean norm. Of course, due to the
finite dimensionality here, supremum could be replaced by maximum.

Theorem 12.6.1. *Let* $A, B \in \mathbb{C}^{n \times n}$. *Then*

$$\|AB\|_{HS} \le \|A\| \, \|B\|_{HS}.$$

Moreover, $\|A\| = \sup \left\{ \|AX\|_{HS} : X \in \mathbb{C}^{n \times n}, \, \|X\|_{HS} \le 1 \right\}.$

Proof. Let $b_j \in \mathbb{C}^{n \times 1}$ denote the jth column vector of the matrix B. Then $Ab_j \in \mathbb{C}^{n \times 1}$ is the jth column vector of the matrix AB, so that

$$\|AB\|_{HS}^2 = \sum_{j=1}^{n} \|Ab_j\|_{HS}^2 \le \|A\|^2 \sum_{j=1}^{n} \|b_j\|_{HS}^2 = \|A\|^2 \, \|B\|_{HS}^2.$$

Let $x \in \mathbb{C}^{n \times 1}$ be such that $\|x\|_{HS} = 1$ and $\|Ax\|_{HS} = \|A\|$. Let each column vector of $X \in \mathbb{C}^{n \times n}$ be $x/\sqrt{n}$. Then $\|X\|_{HS} = 1$ and

$$\|AX\|_{HS}^2 = \sum_{j=1}^{n} \|Ax\|_{HS}^2 / n = \|A\|^2.$$

Taking square roots completes the proof. □

Definition 12.6.2 (Operator as an infinite matrix). Let $\mathbb{C}^{\mathbb{Z}}$ denote the space of complex sequences $\hat{x} : \mathbb{Z} \to \mathbb{C}$. We will write $\hat{x} = (x_j)_{j \in \mathbb{Z}}$, where $x_j = \hat{x}(j)$. Let

$$V = \left\{ \hat{x} \in \mathbb{C}^{\mathbb{Z}} : \{ j \in \mathbb{Z} : x_j \ne 0 \} \text{ finite} \right\}.$$

A *matrix* $A \in \mathbb{C}^{\mathbb{Z} \times \mathbb{Z}}$ is a function $A : \mathbb{Z} \times \mathbb{Z} \to \mathbb{C}$, sometimes presented as an infinite table

$$A = (A_{ij})_{i,j \in \mathbb{Z}} = \begin{pmatrix} \ddots & \vdots & \vdots & \vdots & \vdots & \vdots & \\ \cdots & A_{-1,-1} & A_{-1,0} & A_{-1,1} & A_{-1,2} & A_{-1,3} & \cdots \\ \cdots & A_{0,-1} & A_{00} & A_{01} & A_{02} & A_{03} & \cdots \\ \cdots & A_{1,-1} & A_{10} & A_{11} & A_{12} & A_{13} & \cdots \\ \cdots & A_{2,-1} & A_{20} & A_{21} & A_{22} & A_{23} & \cdots \\ \cdots & A_{3,-1} & A_{30} & A_{31} & A_{32} & A_{33} & \cdots \\ & \vdots & \vdots & \vdots & \vdots & \vdots & \ddots \end{pmatrix},$$

where $A_{ij} := A(i,j)$. The *(standard)* *matrix* of a linear operator $A : V \to \mathbb{C}^{\mathbb{Z}}$ is the matrix $(A_{ij})_{i,j \in \mathbb{Z}}$ where the numbers $A_{ij} \in \mathbb{C}$ are obtained from

$$(A\hat{x})_i = \sum_{j \in \mathbb{Z}} A_{ij} x_j;$$

there should be no confusion in denoting the linear operator and the corresponding standard matrix with the same letter.

Remark 12.6.3. Let $\delta_i = (\delta_{ij})_{j \in \mathbb{Z}} \in \ell^2 = \ell^2(\mathbb{Z})$, where $\delta_{ii} = 1$ and $\delta_{ij} = 0$ if $i \neq j$. Then for a linear operator $A : \ell^2 \to \ell^2$, the standard matrix is

$$A_{ij} = \langle A\delta_j, \delta_i \rangle_{\ell^2}, \tag{12.34}$$

i.e., the standard matrix is the matrix of the operator A with respect to the basis $\{\delta_i : i \in \mathbb{Z}\}$.

Definition 12.6.4. Let $A : V \to \mathbb{C}^{\mathbb{Z}}$ be a linear operator. For each $k \in \mathbb{Z}$, let us define a linear operator $A(k) : V \to \mathbb{C}^{\mathbb{Z}}$ by

$$A(k)_{ij} = \begin{cases} A_{ij}, & \text{if } i - j = k, \\ 0, & \text{if } i - j \neq k \end{cases}.$$

Notice that now $A = \sum_{k \in \mathbb{Z}} A(k)$ and

$$\|A(k)\|_{\ell^2 \to \ell^2} = \sup_j |A(k)_{j+k,j}|. \tag{12.35}$$

Let

$$\|A\|_{\ell^\infty} := \sup_{i,j \in \mathbb{Z}} |A_{ij}|,$$

and recall the notation $\langle i - j \rangle = (1 + |i - j|^2)^{1/2}$.

Lemma 12.6.5. *Let $A : V \to \mathbb{C}^{\mathbb{Z}}$ be a linear operator. Then*

$$\|A\|_{\ell^\infty} \leq \|A\|_{\ell^2 \to \ell^2}.$$

Moreover, if $|A_{ij}| \leq c_r \langle i - j \rangle^{-r} \|A\|_{\ell^\infty}$ for a constant $c_r < \infty$ where $r > 1$, then

$$\|A\|_{\ell^2 \to \ell^2} \leq c'_r \|A\|_{\ell^\infty}$$

for the constant $c'_r = c_r \sum_{k \in \mathbb{Z}} \langle k \rangle^{-r} < \infty$; hence in this case the norms $\| \cdot \|_{\ell^2 \to \ell^2}$ and $\| \cdot \|_{\ell^\infty}$ are equivalent.

Proof. The first claim follows from the Cauchy–Schwarz inequality:

$$|A_{ij}| \overset{(12.34)}{=} |(A\delta_j, \delta_i)_{\ell^2}| \leq \|A\|_{\ell^2 \to \ell^2}.$$

Next, we can assume that $\|A\|_{\ell^\infty} < \infty$ since otherwise there is nothing to prove. Since $A = \sum_{k \in \mathbb{Z}} A(k)$, we get

$$
\begin{aligned}
\|A\|_{\ell^2 \to \ell^2} \quad &\leq \quad \sum_{k \in \mathbb{Z}} \|A(k)\|_{\ell^2 \to \ell^2} \\
&\overset{(12.35)}{=} \quad \sum_{k \in \mathbb{Z}} \sup_j |A(k)_{j+k,j}| \\
&\leq \quad \sum_{k \in \mathbb{Z}} c_r \langle k \rangle^{-r} \|A\|_{\ell^\infty} \quad =: \quad c'_r \|A\|_{\ell^\infty},
\end{aligned}
$$

this last sum converging since $r > 1$. This concludes the proof. $\qquad\square$

Definition 12.6.6 (Matrices with rapid off-diagonal decay). A matrix $A \in \mathbb{C}^{\mathbb{Z} \times \mathbb{Z}}$ is said to *decay (rapidly) off-diagonal* if

$$|A_{ij}| \leq c_{Ar} \langle i - j \rangle^{-r}$$

for every $i, j \in \mathbb{Z}$ and $r \in \mathbb{N}$, where constants $c_{Ar} < \infty$ depend on r, A, but not on i, j. The set of off-diagonally decaying matrices is denoted by $\mathcal{D}$.

Proposition 12.6.7. *Let $A, B \in \mathcal{D}$. Then $AB \in \mathcal{D}$.*

Proof. In principle, we must be cautious here, since linear operators $A, B : V \to \mathbb{C}^{\mathbb{Z}}$ in general cannot be composed to get $AB : V \to \mathbb{C}^{\mathbb{Z}}$. Here, however, there is no problem as $A, B \in \mathcal{D}$, so that

$$\left(\sum_{k} (AB)_{ik} x_k \right)_{i \in \mathbb{Z}} = (AB)\widehat{x} = A(B\widehat{x}) = \left(\sum_{j} A_{ij} \sum_{k} B_{jk} x_k \right)_{i \in \mathbb{Z}},$$

where

$$
\begin{aligned}
|(AB)_{ik}| &= \left| \sum_{j} A_{ij} B_{jk} \right| \leq \sum_{j} |A_{ij}| |B_{jk}| \\
&\leq c_{Ar} c_{Bs} \sum_{j} \langle i - j \rangle^{-r} \langle j - k \rangle^{s} \\
&\overset{\text{Peetre 3.3.31}}{\leq} 2^{|r|} c_{Ar} c_{Bs} \sum_{j} \langle i - k \rangle^{-r} \langle k - j \rangle^{|r|} \langle j - k \rangle^{s} \\
&= 2^{|r|} c_{Ar} c_{Bs} \langle i - k \rangle^{-r} \sum_{j} \langle j \rangle^{|r|+s},
\end{aligned}
$$

which converges if $|r| + s < -1$. This shows that $AB \in \mathcal{D}$. $\qquad\square$

Remark 12.6.8. Proposition 12.6.7 dealt with matrix multiplication in $\mathcal{D}$. For matrices $A, B \in \mathbb{C}^{\mathbb{Z} \times \mathbb{Z}}$ in general, notice that

$$
\begin{cases}
(A + B)_{ij} &= A_{ij} + B_{ij}, \\
(\lambda A)_{ij} &= \lambda A_{ij}.
\end{cases}
$$

Moreover, we may define *involution* $A \mapsto A^*$ by

$$(A^*)_{ij} := \overline{A_{ji}}.$$

Of course, on the algebra $\mathcal{L}(\mathcal{H})$ this corresponds to the usual adjoint operation $A \mapsto A^*$, where $\langle A^* x, y \rangle_{\mathcal{H}} = \langle x, Ay \rangle_{\mathcal{H}}$ for every $x, y \in \mathcal{H}$. We may collect these observations:

Theorem 12.6.9. *$\mathcal{D} \subset \mathcal{L}(\ell^2)$ is a unital involutive algebra. Moreover, for $A \in \mathcal{D}$, the norms $\|A\|_{op}$ and $\|A\|_{\ell^\infty}$ are equivalent.*

Chapter 13

Pseudo-differential Operators on Homogeneous Spaces

In this chapter we discuss pseudo-differential operators on homogeneous spaces. The main question addressed here is how operators on such a space are related to pseudo-differential operators on the group that acts on the space. Once such a correspondence is established, one can use it to map the whole construction developed earlier from the group to the homogeneous space. We also note that among other things, this chapter provides an application to the characterisation of pseudo-differential operators in terms of Σ^m-classes in Theorem 10.9.6. An important class of examples to keep in mind here are the spheres $\mathbb{S}^n \cong \mathrm{SO}(n)\backslash\mathrm{SO}(n+1) \cong \mathrm{SO}(n+1)/\mathrm{SO}(n)$.

13.1 Analysis on closed manifolds

We start with closed manifolds. Let M be a C^∞-smooth, closed (i.e., compact, without a boundary) orientable manifold. We refer to Section 5.2 for the basic constructions on M, so we now review only a few of them. The test function space $\mathcal{D}(M)$ is the space of $C^\infty(M)$ endowed with the usual Fréchet space topology. Its dual $\mathcal{D}'(M) = \mathcal{L}(\mathcal{D}(M), \mathbb{C})$ is the space of distributions, endowed with the weak-$*$-topology, see Remark 5.2.15. The duality is expressed by the bracket $\langle f, \varphi \rangle = f(\varphi)$ $(\varphi \in \mathcal{D}(M), f \in \mathcal{D}'(M))$. The embedding $\mathcal{D}(M) \hookrightarrow \mathcal{D}'(M)$ is interpreted by

$$\langle \psi, \varphi \rangle := \int_M \psi(x)\, \varphi(x)\, \mathrm{d}x.$$

The Schwartz kernel theorem states that $\mathcal{L}(\mathcal{D}(M))$ is isomorphic to $\mathcal{D}(M) \widehat{\otimes} \mathcal{D}'(M)$; the isomorphism is given by

$$\langle A\varphi, f \rangle = \langle K_A, \varphi \otimes f \rangle,$$

where $A \in \mathcal{L}(\mathcal{D}(M))$, $\varphi \in \mathcal{D}(M)$, $f \in \mathcal{D}'(M)$, and distribution $K_A \in \mathcal{D}(M) \widehat{\otimes} \mathcal{D}'(M)$ is called the *Schwartz kernel* of A. Then A can be uniquely extended (by duality) to $A \in \mathcal{L}(\mathcal{D}'(M))$, and it is customary to write informally

$$(Af)(x) = \int_M K_A(x, y) \, f(y) \, dy$$

instead of $\varphi \mapsto \langle Af, \varphi \rangle$ ($\varphi \in \mathcal{D}(M)$). Recall that $L^2(M) = H^0(M)$, $\mathcal{D}'(M) = \cup_{s \in \mathbb{R}} H^s(M)$ and $\mathcal{D}(M) = \cap_{s \in \mathbb{R}} H^s(M)$, where $H^s(M)$ is the (L^2-type) Sobolev space of order $s \in \mathbb{R}$, see Definition 5.2.16. There are different spaces of distributions available more specifically on homogeneous spaces, see, e.g., [93] for spaces $\mathcal{D}'_{L^1}(M)$ of summable distributions.

An operator $A \in \mathcal{L}(\mathcal{D}(M))$ is *a pseudo-differential operator of order $m \in \mathbb{R}$ on M*, $A \in \Psi^m(M)$, if $(M_\phi A M_\psi)_\kappa \in \Psi^m(\mathbb{R}^{\dim(M)})$ for every chart (U, κ) of M and for every $\phi, \psi \in C_0^\infty(U)$, where M_ϕ is the multiplication operator $f \mapsto \phi f$, and

$$(M_\phi A M_\psi)_\kappa f := (M_\phi A M_\psi (f \circ \kappa)) \circ \kappa^{-1} \quad (f \in C^\infty(\kappa U)).$$

We sometimes write $M_\phi A M_\psi \in \Psi^m(\mathbb{R}^{\dim(M)})$, thus omitting the subscript κ and leaving the chart mapping implicit. Equivalently, pseudo-differential operators can be characterised by commutators (see Theorem 5.3.1): $A \in \mathcal{L}(\mathcal{D}(M))$ belongs to $\Psi^m(M)$ if and only if $(A_k)_{k=0}^\infty \subset \mathcal{L}(H^m(M), H^0(M))$ for every sequence of smooth vector fields $(D_k)_{k=1}^\infty$ on M, where $A_0 := A$ and $A_{k+1} := [D_{k+1}, A_k]$.

Definition 13.1.1 (Right transformation group). A smooth *right transformation group* is

$$(G, M, m),$$

where G is a Lie group, M is a C^∞-manifold and $m : M \times G \to M$ is a C^∞-mapping called a right *action*, satisfying $m(p, e) = p$ and $m(m(p, y), x) = m(p, yx)$ for all $x, y \in G$ and $p \in M$, where $e \in G$ is the neutral element of the group. The action is *free*, if $m(p, x) = p$ implies $x = e$. It is evident how one defines a *left* transformation group (G, M, m) with a *left* action $m : G \times M \to M$.

Definition 13.1.2 (Fiber bundles). A smooth *fiber bundle* is

$$(E, B, F, p_{E \to B}),$$

where E, B, F are C^∞-manifolds and $p_{E \to B} \in C^\infty(E, B)$ is a surjective mapping such that there exists an open cover $\mathcal{U} = \{U_j \mid j \in J\}$ of B and diffeomorphisms $\phi_j : p^{-1}(U_j) \to U_j \times F$ satisfying $\phi_j(x) = (p_{E \to B}(x), \psi_j(x))$ for every $x \in p_{E \to B}^{-1}(U_j)$. The spaces E, B, F are called the *total space*, the *base space*, and the *fiber* of the bundle, respectively. The cover $\mathcal{U}$ is called a *locally trivialising cover* of the bundle. Sometimes only the mapping $p_{E \to B}$ is called the fiber bundle.

Definition 13.1.3 (Principal fiber bundles). A *principal fiber bundle* is

$$(E, B, F, p_{E \to B}, m),$$

where $(E, B, F, p_{E \to B})$ is a smooth fiber bundle with cover $\mathcal{U}$ and mappings ϕ_j, ψ_j as above and (F, E, m) is a smooth right transformation group with a free action satisfying $p_{E \to B}(m(x, y)) = p_{E \to B}(x)$ for every $(x, y) \in E \times F$ and $\psi_j(m(x, y)) = \psi_j(x)y$ for every $(x, y) \in p_{E \to B}^{-1}(U_j) \times F$.

13.2   Analysis on compact homogeneous spaces

Here we review some elements of the analysis on homogeneous spaces. The group will be acting on the right to adopt the construction to the previously constructed symbolic calculus on groups.

Definition 13.2.1 (Homogeneous spaces I). Let (G, M, m) be a smooth right transformation group. The manifold M is called a *homogeneous space* if the action $m : M \times G \to M$ is *transitive*, i.e., if for every $p, q \in M$ there exists $x \in G$ such that $m(p, x) = q$.

For this line of thought we can refer to, e.g., [149]. However, let us also give another, equivalent definition for a homogeneous space:

Definition 13.2.2 (Homogeneous spaces II). Let G be a Lie group with a closed subgroup K. The *homogeneous space* $K \backslash G$ is the set of classes $Kx = \{kx \mid k \in K\}$ $(x \in G)$ endowed with the topology co-induced by $x \mapsto Kx$, and equipped with the unique C^∞-manifold structure such that the mapping $(x, Ky) \mapsto Kyx$ belongs to $C^\infty(G \times (K \backslash G), K \backslash G)$, and such that there is a neighbourhood $U \subset K \backslash G$ of $Ke \in K \backslash G$ and a mapping $\psi \in C^\infty(U, G)$ satisfying $K\psi(Kx) = Kx$. The group G acts smoothly from the right on the manifold $K \backslash G$ by $(Ky, x) \mapsto Kyx$.

Exercise 13.2.3. Actually a smooth homogeneous space M is diffeomorphic to $G_p \backslash G$, where $G_p = \{x \in G \mid m(p, x) = p\}$ is the isotropy subgroup (see Definition 6.3.3 and Theorem 6.3.4). Thus, show that two definitions are equivalent.

Exercise 13.2.4. Show that $(G, K \backslash G, K, x \mapsto Kx, (x, k) \mapsto kx)$ has a structure of a principal fiber bundle. For a further development of this point of view see, e.g., [20].

Remark 13.2.5 (**Homogeneous spaces $K \backslash G$ vs G/K**). Clearly one can consider homogeneous spaces G/K with the action of the left transformation group. As it turns out, once we have chosen to identify the Lie algebra of a Lie group with the left invariant vector fields, the further analysis is fixed from the point of view of "right"/"left", see Remark 10.4.2 for the starting point of this choice. Moreover, since on the group we wanted to have the composition formulae for pseudo-differential operators in the usual form $\sigma_{A \circ B} = \sigma_A \sigma_B + \cdots$ and not in the form $\sigma_{A \circ B} = \sigma_B \sigma_A + \cdots$, also the choice of the definition of the Fourier transform was fixed, see Remark 10.4.13. However, the right/left constructions are very symmetric, and since the notation G/K recalling the division of numbers may be more familiar, we chose to introduce homogeneous spaces in this setting in the definition of the right quotient in Definition 6.2.12. Consequently, this led to the corresponding definition of the quotient topology in Definition 7.1.8, as well as the corresponding discussion of the group actions in Section 6.3 and the invariant integration in Section 7.4.1. However, as we pointed out in Remark 6.2.14 the choice between "right" and "left" is completely symmetric, so the reader should have no difficulty in translating those results to the setting of the right actions considered here.

From now on we assume the Lie group G to be compact, and we observe that by Remark 7.3.2, (8), and by Proposition 7.1.10 the space $K\backslash G$ is a compact Hausdorff space.

We can regard functions (or distributions) constant on the cosets Kx $(x \in G)$ as functions (or distributions) on $K\backslash G$; it is obvious how one embeds the spaces $C^\infty(K\backslash G)$ and $\mathcal{D}'(K\backslash G)$ into the spaces $C^\infty(G)$ and $\mathcal{D}'(G)$, respectively. Let us define $P_{K\backslash G} \in \mathcal{L}(C^\infty(G))$ by

$$(P_{K\backslash G}f)(x) := \int_K f(kx)\, \mathrm{d}\mu_K(k), \tag{13.1}$$

where $\mathrm{d}\mu_K$ is the Haar measure on the compact Lie group K. Hence $P_{K\backslash G}f \in C^\infty(K\backslash G)$, and $P_{K\backslash G}$ extends uniquely to the orthogonal projection of $L^2(G)$ onto the subspace $L^2(K\backslash G)$. Let us consider an operator $A \in \mathcal{L}(C^\infty(G))$ with the symbol satisfying

$$\sigma_A(kx,\xi) = \sigma_A(x,\xi) \quad (x \in G,\ k \in K,\ \xi \in \mathrm{Rep}(G)); \tag{13.2}$$

this condition is equivalent to

$$R_A(kx,y) = R_A(x,y) \tag{13.3}$$

in the sense of distributions for the right-convolution kernels, in view of (10.18) in Section 10.4.1. Consequently, the Schwartz integral kernel K_A of A satisfies

$$K_A(kx, kxy^{-1}) = K_A(x, xy^{-1})$$

in view of Proposition 10.4.1. Replacing xy^{-1} by y we have

$$K_A(kx, ky) = K_A(x, y).$$

This means that A maps the space $C^\infty(K\backslash G)$ into itself. Of course, for a general $A \in \mathcal{L}(C^\infty(G))$ the equality (13.2) does not have to be true, but then we can define an operator $A_{K\backslash G} \in \mathcal{L}(\mathcal{D}(G))$ by the right convolution kernel

$$R_{A_{K\backslash G}} := (P_{K\backslash G} \otimes \mathrm{id})R_A,$$

with $P_{K\backslash G}$ as in (13.1). We note that for $A \in \Psi^m(G)$ its right operator-valued symbol r_A in Definition 10.11.9 satisfies the property that

$$r_A \in C^\infty(G, \mathcal{L}(H^m(G), H^0(G))),$$

so that the right operator-valued symbol

$$r_{A_{K\backslash G}}(x) = \int_K r_A(kx)\, \mathrm{d}\mu_K(k)$$

of $A_{K\backslash G}$ exists as a weak integral (Pettis integral), with the interpretation as in Remark 7.9.3. Consequently, by (10.60) in Theorem 10.11.16, or directly by Definition 10.4.3, the symbol of $A_{K\backslash G}$ satisfies

$$\sigma_{A_{K\backslash G}}(x,\xi) = \int_K \sigma_A(kx,\xi)\, \mathrm{d}\mu_K(k)$$

for all $x \in G$ and $\xi \in \mathrm{Rep}(G)$.

Remark 13.2.6 (**Calculus of K-invariant operators**). Suppose we are given symbols of pseudodifferential operators A_1, A_2 on G satisfying the K-invariance (13.2). If we look at the asymptotic expansion formulae for $\sigma_{A_1 A_2}$, $\sigma_{A_1^*}$ and $\sigma_{A_1^t}$ in Section 10.7.3, we see that all the terms there are K-invariant in the same sense. Moreover, for an elliptic K-invariant symbol the terms in the asymptotic expansion for a parametrix in Theorem 10.9.10 are also K-invariant. In this way the calculus of the K-invariant operators is immediately obtained from the corresponding calculus of operators on the group G.

Theorem 13.2.7 and Corollary 13.2.8 below show how to "project" pseudo-differential operators on G to pseudo-differential operators on $K\backslash G$. The history of such averaging processes for pseudo-differential operators can be traced at least back to the work of M.F. Atiyah and I.M. Singer in the 1960s, and H. Stetkær studied related topics for classical pseudo-differential operators in [119].

Theorem 13.2.7 (Averaging of operators). *Let G be a compact Lie group with a closed Lie subgroup K. If $A \in \Psi^m(G)$, then $A_{K\backslash G} \in \Psi^m(G)$.*

Proof. We will use Theorem 10.9.6 characterising symbols of operators from $\Psi^m(G)$. First, notice that $P_{K\backslash G}$ is right-invariant, and hence

$$(\partial_x^\beta \otimes \triangle_\xi^\alpha)(P_{K\backslash G} \otimes \mathrm{id})\sigma_A = (P_{K\backslash G} \otimes \mathrm{id})(\partial_x^\beta \otimes \triangle_\xi^\alpha)\sigma_A$$

for a left-invariant partial differential operator ∂_x^β and a difference operator $\triangle_\xi^\alpha$, for all $\alpha, \beta \in \mathbb{N}_0^{\dim(G)}$. Therefore

$$\mathrm{Op}(\triangle_\xi^\alpha \partial_x^\beta \sigma_{A_{K\backslash G}}) = \big(\mathrm{Op}(\triangle_\xi^\alpha \partial_x^\beta \sigma_A)\big)_{K\backslash G}.$$

Since $A \in \Psi^m(G)$, by Theorem 10.9.6 we have

$$\|\triangle_\xi^\alpha \partial_x^\beta \sigma_A(x,\xi)\|_{op} \le C_{A\alpha\beta m}\langle\xi\rangle^{m-|\alpha|},$$

and hence we can estimate

$$\|\Delta_\xi^\alpha \partial_x^\beta \sigma_{A_{K\backslash G}}(x,\xi)\|_{op} = \left\|\int_K \Delta_\xi^\alpha \partial_x^\beta \sigma_A(kx,\xi)\; \mathrm{d}\mu_K(k)\right\|_{op}$$

$$\le \int_K \|\Delta_\xi^\alpha \partial_x^\beta \sigma_A(kx,\xi)\|_{op}\; \mathrm{d}\mu_K(k)$$

$$\le C_{A\alpha\beta m}\langle\xi\rangle^{m-|\alpha|}.$$

At the same time, formula (13.3) implies that the right-convolution kernel of $\mathrm{Op}(\sigma_{A_{K\backslash G}})$ has singularities only at $y = e$. This proves that $\sigma_{A_{K\backslash G}} \in \Sigma_0^m(G)$.

Let now $B \in \mathcal{L}(C^\infty(G))$ be a left-invariant (right-convolution) pseudo-differential operator. Then $\sigma_B(x,\xi) = \sigma_B(\xi)$ is independent of $x \in G$ in view of Remark 10.4.10, and hence $B = B_{K\backslash G}$. Consequently, we have

$$(\mathrm{Op}(\sigma_A\sigma_B))_{K\backslash G} = \mathrm{Op}(\sigma_{A_{K\backslash G}}\sigma_B)$$

and

$$(\mathrm{Op}(\sigma_B\sigma_A))_{K\backslash G} = \mathrm{Op}(\sigma_B\sigma_{A_{K\backslash G}}).$$

To argue by induction, assume now that for some $k \in \mathbb{N}_0$ we have proven that $\sigma_{C_{K\backslash G}} \in \Sigma_k^r(G)$ for every $C \in \Psi^r(G)$, for every $r \in \mathbb{R}$. By Remark 13.2.6 we hence get

$$[\sigma_{\partial_j}, \sigma_{A_{K\backslash G}}] = [\sigma_{\partial_j}, \sigma_A]_{K\backslash G} \in \Sigma_k^m(G),$$

$$(\Delta_\xi^\gamma \sigma_{\partial_j})\sigma_{A_{K\backslash G}} = ((\Delta_\xi^\gamma\sigma_{\partial_j})\sigma_A)_{K\backslash G} \in \Sigma_k^{m+1-|\gamma|}(G)$$

and

$$(\Delta_\xi^\gamma\sigma_{A_{K\backslash G}})\sigma_{\partial_j} = ((\Delta_\xi^\gamma\sigma_A)\sigma_{\partial_j})_{K\backslash G} \in \Sigma_k^{m+1-|\gamma|}(G);$$

this means that $\sigma_{A_{K\backslash G}} \in \Sigma_{k+1}^m(G)$, so that by induction we get $\sigma_{A_{K\backslash G}} \in \Sigma^m(G) = \bigcap_{k=0}^\infty \Sigma_k^m(G)$. $\qquad\square$

Once we get a pseudo-differential operator of the form $A_{K\backslash G}$, it can be projected to the homogeneous space $K\backslash G$:

Corollary 13.2.8 (Projection of operators). *Let $K\backslash G$ be orientable. Then*

$$A_{K\backslash G}|_{C^\infty(K\backslash G)} \in \Psi^m(K\backslash G)$$

for every $A \in \Psi^m(G)$.

Proof. Let us write

$$\Psi^m(G)_{K\backslash G} := \{A_{K\backslash G} \mid A \in \Psi^m(G)\}$$

and

$$\Psi^m(G)_{K\backslash G}|_{C^\infty(K\backslash G)} = \{A_{K\backslash G}|_{C^\infty(K\backslash G)} : A \in \Psi^m(G)\}.$$

By Theorem 13.2.7 we know that $\Psi^m(G)_{K\backslash G} \subset \Psi^m(G)$. Let D be a smooth vector field on $K\backslash G$. Since by Exercise 13.2.4, $(G, K\backslash G, K, x \mapsto Kx, (x, k) \mapsto kx)$ is a principal fiber bundle, there exists a smooth vector field $X = X_{K\backslash G}$ on G such that $X|_{C^\infty(K\backslash G)} = D$ (see [115]). Then we have

$$[D, \Psi^m(G)_{K\backslash G}|_{C^\infty(K\backslash G)}] = [X, \Psi^m(G)_{K\backslash G}]|_{C^\infty(K\backslash G)} \subset \Psi^m(G)_{K\backslash G}|_{C^\infty(K\backslash G)},$$

and this combined with $\Psi^m(G)_{K\backslash G}|_{C^\infty(K\backslash G)} \subset \mathcal{L}(H^m(K\backslash G), H^0(K\backslash G))$ yields the conclusion due to the commutator characterisation of pseudo-differential operators on closed manifolds in Theorem 5.3.1. □

Definition 13.2.9 (Lifting of operators). We will say that the operator $A \in \Psi^m(G)$ is a *lifting* of the operator $B \in \Psi^m(K\backslash G)$ if $A = A_{K\backslash G}$ and if $A|_{C^\infty(K\backslash G)} = B$.

Remark 13.2.10 **(Calculus of liftings).** It already follows from Corollary 13.2.8 that at least sometimes a pseudo-differential operator on $K\backslash G$ can be (possibly non-uniquely) lifted to a pseudo-differential operator on G. If $B_j \in \Psi^{m_j}(K\backslash G)$ can be lifted to $C_j = (C_j)_{K\backslash G} \in \Psi^{m_j}(G)$ (i.e., $C_j|_{C^\infty(K\backslash G)} = B_j$), then $C_j^* \in \Psi^{m_j}(G)$ is a lifting of the adjoint operator $B_j^* \in \Psi^{m_j}(K\backslash G)$, and $B_1 B_2 \in \Psi^{m_1+m_2}(K\backslash G)$ is lifted to $C_1 C_2 \in \Psi^{m_1+m_2}(G)$. Moreover, if C_1 is elliptic with a parametrix $D \in \Psi^{-m_1}(G)$ as in Theorem 10.9.10, then $D = D_{K\backslash G}$ and $B_1 \in \Psi^{m_1}(K\backslash G)$ is elliptic with a parametrix $D|_{C^\infty(K\backslash G)} \in \Psi^{-m_1}(K\backslash G)$.

13.3 Analysis on $K\backslash G$, K a torus

In this section we assume that the subgroup K of G is a torus, $K \cong \mathbb{T}^q$. For example, K may be the maximal torus which has an additional importance in the representation theory of G in view of Cartan's maximal torus theorem (Theorem 7.8.8). However, it may be a lower-dimensional torus as well.

Remark 13.3.1 **(Sphere $\mathbb{S}^2$).** Let $\mathbb{B}^n$ be the unit ball of the Euclidean space $\mathbb{R}^n$, and $\mathbb{S}^{n-1}$ its boundary, the $(n-1)$-sphere. The two-sphere $\mathbb{S}^2$ can be considered as the base space of the Hopf fibration $\mathbb{S}^3 \to \mathbb{S}^2$, where the fibers are diffeomorphic to the unit circle $\mathbb{S}^1 \subset \mathbb{R}^2$. In the context of harmonic analysis, $\mathbb{S}^3$ is diffeomorphic[1] to the compact non-commutative Lie group $G = \mathrm{SU}(2)$, having a maximal torus $K \cong \mathbb{S}^1 \cong \mathbb{T}^1$. Then the homogeneous space $K\backslash G$ is diffeomorphic to $\mathbb{S}^2$, so that the canonical projection $p_{G \to K\backslash G} : x \mapsto Kx$ is interpreted as the Hopf fiber bundle $G \to K\backslash G$; in the sequel we treat the two-sphere $\mathbb{S}^2$ always as the homogeneous space $K\backslash G$. Notice that also $\mathbb{S}^2 \cong \mathbb{T}^1 \backslash \mathrm{SO}(3)$. For a sketch of operators there see [140].

Remark 13.3.2 **(Spherical symbols).** In [125] a subalgebra of $\Psi^m(\mathbb{S}^2)$ was described in terms of the so-called spherical symbols. Functions $f \in C^\infty(\mathbb{S}^2)$ can be expanded

---

[1]See Proposition 11.4.2.

in series

$$f(\phi, \theta) = \sum_{l=0}^{\infty} \sum_{m=-l}^{l} \widehat{f}(l)_m \, Y_l^m(\phi, \theta),$$

where $(\phi, \theta) \in [0, 2\pi] \times [0, \pi]$ are the spherical coordinates, and the functions Y_l^m are the spherical harmonics with "spherical" Fourier coefficients

$$\widehat{f}(l)_m := \int_0^{\pi} \int_0^{2\pi} f(\phi, \theta) \, \overline{Y_l^m(\phi, \theta)} \sin(\theta) \, d\phi \, d\theta.$$

Let us define

$$(Af)(\phi, \theta) := \sum_{l=0}^{\infty} \sum_{m=-l}^{l} a(l) \, \widehat{f}(l)_m \, Y_l^m(\phi, \theta),$$

where $a : \mathbb{N}_0 \to \mathbb{C}$ is a rational function; in [125], Svensson states that $A \in \Psi^m(\mathbb{S}^2)$ if and only if

$$|a(l)| \le C_{A,m}(l+1)^m.$$

Let us now present another proof for a special case of Theorem 13.2.7 and Corollary 13.2.8 for the torus subgroup K; this method of proof turns out to be useful when we develop an analogous method for showing that the mapping $(A \mapsto A_{K\backslash G}|_{C^\infty(K\backslash G)}) : \Psi^m(G) \to \Psi^m(K\backslash G)$ is surjective if K is a torus subgroup (see the proof of Theorem 13.3.5).

Theorem 13.3.3. *Let G be a compact Lie group with a torus subgroup K. If $A \in \Psi^m(G)$, then $A_{K\backslash G} \in \Psi^m(G)$ and the restriction $A_{K\backslash G}|_{C^\infty(K\backslash G)} \in \Psi^m(K\backslash G)$.*

Proof. Let $\dim(G) = p + q$, where $K \cong \mathbb{T}^q$. Let $\mathcal{V} = \{V_i \mid i \in \mathcal{I}\}$ be a locally trivialising open cover of the base space $K\backslash G$ for the principal fiber bundle $(G, K\backslash G, K, x \mapsto Kx, (x, k) \mapsto kx)$. Let $\mathcal{U} = \{U_j \mid 1 \le j \le N\}$ be an open cover of $K\backslash G$ such that for every $j_1, j_2 \in \{1, \ldots, N\}$ there exists $V_i \in \mathcal{V}$ containing $U_{j_1} \cup U_{j_2}$ whenever $U_{j_1} \cap U_{j_2} \ne \emptyset$; notice that we can always refine any open cover on a finite-dimensional manifold to get a new cover satisfying this additional requirement. Then each $U_i \cup U_j$ $(1 \le i, j \le N)$ is a chart neighbourhood on $K\backslash G$, and furthermore there exist diffeomorphisms $\phi_{ij} : (U_i \cup U_j) \times K \to p_{G \to K\backslash G}^{-1}(U_i \cup U_j)$ such that $p_{G \to K\backslash G}(\phi_{ij}(x, k)) = x$ for every $x \in U_i \cup U_j$ and $k \in K$. To simplify the notation, we treat the neighbourhood $U_i \cup U_j \subset K\backslash G$ as a set $U_i \cup U_j \subset \mathbb{R}^p$, and $p_{G \to G/K}^{-1}(U_i \cup U_j) \subset G$ as a set $(U_i \cup U_j) \times \mathbb{T}^q \subset \mathbb{R}^p \times \mathbb{T}^q$.

Let $\{(U_j, \psi_j) \mid 1 \le j \le N\}$ be a partition of unity subordinate to $\mathcal{U}$, and let $A_{ij} = M_{\psi_i} A M_{\psi_j} \in \Psi^m(G)$. With the localised notation we consider $A_{ij} \in \Psi^m(\mathbb{R}^p \times \mathbb{T}^q; \mathbb{R}^p \times \mathbb{Z}^q)$, so that it has the symbol $\sigma_{A_{ij}} \in S^m(\mathbb{R}^p \times \mathbb{T}^q; \mathbb{R}^p \times \mathbb{Z}^q)$. We note that the notation we use for symbols here is slightly different from before: $\mathbb{R}^p \times \mathbb{T}^q$ stands for the space variables, and $\mathbb{R}^p \times \mathbb{Z}^q$ is for dual frequencies. Then

$$\sigma_{(A_{K\backslash G})_{ij}}(x, \xi) = \sigma_{(A_{ij})_{K\backslash G}}(x, \xi)$$

$$= \int_{\mathbb{T}^q} \sigma_{A_{ij}}(x_1, \ldots, x_p, x_{p+1} + z_1, \ldots, x_{p+q} + z_q; \xi) \, dz_1 \cdots dz_q,$$

and it is now easy to check that $\sigma_{(A_{K\backslash G})_{ij}} \in S^m(\mathbb{R}^p \times \mathbb{T}^q; \mathbb{R}^p \times \mathbb{Z}^q)$. This yields $(A_{K\backslash G})_{ij} \in \Psi^m(G)$, and hence

$$A_{K\backslash G} = \sum_{i,j}(A_{K\backslash G})_{ij} \in \Psi^m(G),$$

completing the proof. $\qquad\qquad\qquad\qquad\qquad\qquad\qquad\qquad\qquad\qquad\qquad\qquad\square$

Theorem 13.3.3 has the inverse which will be given in Theorem 13.3.5. But first, we prepare a lemma on the extension of symbols in the Euclidean space. Because of the commutator characterisations in Chapter 5 (especially the equality (5.4) in Theorem 5.4.1), and in view of Corollary 4.6.13, all of the symbol classes on $\mathbb{T}^n$ in both the Euclidean and toroidal quantizations coincide. That is why, to simplify the notation, we will skip writing the space for the frequency variable and will only write the space which will usually be $\mathbb{R}^p \times \mathbb{T}^q$. Thus, the class $\Psi^m(\mathbb{R}^p \times \mathbb{T}^q)$ will stand for either $\Psi^m(\mathbb{R}^p \times \mathbb{T}^q; \mathbb{R}^p \times \mathbb{Z}^q)$ or for $\Psi^m(\mathbb{R}^p \times \mathbb{T}^q; \mathbb{R}^p \times \mathbb{R}^q)$, which we know to be equal, with the correspondence between the Euclidean and toroidal symbols given in Theorem 4.5.3. The same will apply for symbols, with the quantization clear from the context.

Lemma 13.3.4 (Extension of symbols). *Let $\chi \in C^\infty(\mathbb{R}^{p+q})$ be homogeneous of order 0 in $\mathbb{R}^{p+q} \setminus \mathbb{B}(0,1)$, i.e.,[2] $\chi(\xi) = \chi(\xi/\|\xi\|)$ when $\|\xi\| \geq 1$. Furthermore, assume that χ satisfies $\chi|_{(U\times\mathbb{R}^q)\setminus\mathbb{B}(0,1)} \equiv 0$ and $\chi|_{\mathbb{R}^p\times V} \equiv 1$, where $U \subset \mathbb{R}^p$ and $V \subset \mathbb{R}^q$ are neighbourhoods of zeros. Let $\sigma_B \in S^m(\mathbb{R}^p)$ and write*

$$\sigma_A(x,\xi) := \chi(\xi)\,\sigma_B(Px, P\xi),$$

where $P : \mathbb{R}^{p+q} \to \mathbb{R}^p$ is defined by $P(x_1,\ldots,x_{p+q}) = (x_1,\ldots,x_p)$. Then $\sigma_A \in S^m(\mathbb{R}^{p+q})$ and $\sigma_A|_{(\mathbb{R}^p\times\mathbb{R}^q)\times(\mathbb{R}^p\times\mathbb{Z}^q)} \in S^m(\mathbb{R}^p \times \mathbb{T}^q)$.

Proof. We shall first prove that

$$|(\partial_\xi^\gamma \chi)(\xi)| \leq C_{\gamma r} \langle P\xi \rangle^{-r} \langle \xi \rangle^{r-|\gamma|} \qquad\qquad (13.4)$$

for every $r \in \mathbb{R}$ and for every $\gamma \in \mathbb{N}_0^{p+q}$. It is trivial that $(x,\xi) \mapsto \chi(\xi)$ belongs to $S^0(\mathbb{R}^{p+q})$. If $r \geq 0$ then obviously (13.4) is true. Since we are not interested in the behaviour of the symbols when $\|\xi\|$ is small, we assume that $\|\xi\| > 1$ from here on. There exists $r_0 \in (0,1)$ such that $\chi(\xi) = 0$ when $\|P\xi\| < r_0$. Let $r < 0$ and $\xi \in \text{supp}(\chi)$. Then $\|P\xi\| \geq r_0\|\xi\|$, and thus

$$
\begin{aligned}
|(\partial_\xi^\gamma \chi)(\xi)| &\leq C_\gamma \langle \xi \rangle^{-|\gamma|} \\
&= C_\gamma \langle P\xi \rangle^{-r} \langle P\xi \rangle^r \langle \xi \rangle^{-|\gamma|} \\
&\leq C_\gamma \langle P\xi \rangle^{-r} \langle r_0\xi \rangle^r \langle \xi \rangle^{-|\gamma|} \\
&\leq C_\gamma r_0^r \langle P\xi \rangle^{-r} \langle \xi \rangle^{r-|\gamma|}.
\end{aligned}
$$

---

[2]Here $\mathbb{B}(0,1)$ stands for the unit ball in $\mathbb{R}^{p+q}$ centred at the origin and of radius 1.

Hence the inequality (13.4) is proven. Now

$$|\partial_\xi^\alpha \partial_x^\beta \sigma_A(x,\xi)| \leq \sum_{\gamma \leq \alpha} \binom{\alpha}{\gamma} |(\partial_\xi^\gamma \chi)(\xi)| \, |(\partial_\xi^{\alpha-\gamma} \partial_x^\beta \sigma_B)(Px, P\xi)|$$

$$\leq \sum_{\gamma \leq \alpha} \binom{\alpha}{\gamma} C_{\gamma r_\gamma} \langle P\xi \rangle^{-r_\gamma} \langle \xi \rangle^{r_\gamma - |\gamma|} C_{B(\alpha-\gamma)\beta m} \langle P\xi \rangle^{m-|\alpha-\gamma|}$$

$$\leq C_{B\alpha\beta m\chi} \langle \xi \rangle^{m-|\alpha|},$$

if we choose $r_\gamma = m - |\alpha - \gamma|$. Thereby $\sigma_A \in S^m(\mathbb{R}^{p+q})$. Clearly we can regard this symbol as a function $\sigma_A : (\mathbb{R}^p \times \mathbb{T}^q) \times (\mathbb{R}^p \times \mathbb{R}^q) \to \mathbb{C}$ and study its restriction $\sigma_A|_{(\mathbb{R}^p \times \mathbb{T}^q) \times (\mathbb{R}^p \times \mathbb{Z}^q)}$. We claim that this restriction belongs to $S^m(\mathbb{R}^p \times \mathbb{T}^q)$. Indeed, the Taylor expansion of a function $\sigma \in C^\infty(\mathbb{R}^q)$ yields

$$\triangle_\xi^\gamma \sigma(\xi) = \sum_{\delta \leq \gamma} \binom{\gamma}{\delta} (-1)^{|\gamma-\delta|} \, \sigma(\xi + \delta)$$

$$= \sum_{\delta \leq \gamma} \binom{\gamma}{\delta} (-1)^{|\gamma-\delta|} \left(\sum_{|\rho| < |\gamma|} \frac{1}{\rho!} \delta^\rho \, (\partial_\xi^\rho \sigma)(\xi) + \sum_{|\rho| = |\gamma|} \frac{1}{\rho!} \delta^\rho \, (\partial_\xi^\rho \sigma)(\xi + \theta_\delta \delta) \right)$$

$$= \sum_{|\rho| < |\gamma|} \frac{1}{\rho!} \, (\partial_\xi^\rho \sigma)(\xi) \sum_{\delta \leq \gamma} \binom{\gamma}{\delta} (-1)^{|\gamma-\delta|} \delta^\rho + \sum_{\delta \leq \gamma} \sum_{|\rho| = |\gamma|} \frac{1}{\rho!} \delta^\rho \, (\partial_\xi^\rho \sigma)(\xi + \theta_\delta \delta)$$

$$= \sum_{\delta \leq \gamma} \sum_{|\rho| = |\gamma|} \frac{1}{\rho!} \delta^\rho \, (\partial_\xi^\rho \sigma)(\xi + \theta_\delta \delta),$$

because

$$\sum_{\delta \leq \gamma} \binom{\gamma}{\delta} (-1)^{|\gamma-\delta|} \delta^\rho = \triangle_\xi^\gamma \xi^\rho |_{\xi=0} = 0$$

whenever $|\rho| < |\gamma|$. Therefore

$$|\triangle_\xi^\gamma \sigma(\xi)| \leq \sum_{\delta \leq \gamma} \sum_{|\rho| = |\gamma|} \frac{1}{\rho!} \delta^\rho \, |(\partial_\xi^\rho \sigma)(\xi + \theta_\delta \delta)| \leq c_\gamma \sup_{\eta \in S_\gamma, |\rho| = |\gamma|} |(\partial_\xi^\rho \sigma)(\xi + \eta)|,$$

where S_γ is the rectangle $\prod_{j=1}^q [0, \gamma_j]$. Let $\alpha' = (P\alpha, 0, \ldots, 0)$ and let $\alpha'' = \alpha - \alpha'$; then

$$|\partial_\xi^{\alpha'} \triangle_\xi^{\alpha''} \partial_x^\beta \sigma_A(x,\xi)| \leq C_\alpha \sup_{\eta \in S_{\alpha''}, |\rho| = |\alpha''|} |\partial_\xi^{\alpha'+\rho} \partial_x^\beta \sigma_A(x, \xi + \eta)|$$

$$\leq C_\alpha C_{A\alpha\beta m} \sup_{\eta \in S_\alpha} \langle \xi + \eta \rangle^{m-|\alpha|}$$

$$\leq C_\alpha C_{A\alpha\beta m} 2^{|m-|\alpha||} \sup_{\eta \in S_\alpha} \langle \eta \rangle^{|m-|\alpha||} \langle \xi \rangle^{m-|\alpha|}$$

$$\leq C_\alpha C_{A\alpha\beta m} 2^{|m-|\alpha||} \langle \alpha \rangle^{|m-|\alpha||} \langle \xi \rangle^{m-|\alpha|}$$

$$= C'_{A\alpha\beta m} \langle \xi \rangle^{m-|\alpha|};$$

notice the application of the Peetre inequality (Proposition 3.3.31):

$$\langle \xi + \eta \rangle^s \leq 2^{|s|} \langle \xi \rangle^s \langle \eta \rangle^{|s|}.$$

Hence $\sigma_A|_{(\mathbb{R}^p \times \mathbb{T}^q) \times (\mathbb{R}^p \times \mathbb{Z}^q)} \in S^m(\mathbb{R}^p \times \mathbb{T}^q)$.   □

Now we are ready to give the converse to Theorem 13.3.3:

Theorem 13.3.5 (Lifting of operators). *Let G be a compact Lie group with a torus subgroup K. Let $B \in \Psi^m(K\backslash G)$. Then there exists an operator $A = A_{K\backslash G} \in \Psi^m(G)$ such that $A|_{C^\infty(K\backslash G)} = B$.*

Proof. Let $K \cong \mathbb{T}^q$, $\dim(G) = p + q$, and let $\{(U_j, \psi_j) \mid 1 \leq j \leq N\}$ be the same partition of unity as in the proof of Theorem 13.3.3. Let $B_{ij} = M_{\psi_i} B M_{\psi_j} \in \Psi^m(K\backslash G)$. With the localised notation we consider $B_{ij} \in \Psi^m(\mathbb{R}^p; \mathbb{R}^p)$, so that it has the symbol $\sigma_{B_{ij}} : \mathbb{R}^p \times \mathbb{R}^p \to \mathbb{C}$, and the mapping $(x, \xi) \mapsto \sigma_{B_{ij}}(x, \xi)$ is zero when $x \in \mathbb{R}^p \setminus (U_i \cup U_j)$. By Lemma 13.3.4 there exists a pseudo-differential operator $A_{ij} \in \Psi^m(\mathbb{R}^p \times \mathbb{T}^q; \mathbb{R}^p \times \mathbb{Z}^q)$ such that $\sigma_{A_{ij}} : (\mathbb{R}^p \times \mathbb{T}^q) \times (\mathbb{R}^p \times \mathbb{Z}^q) \to \mathbb{C}$ satisfies

$$\sigma_{A_{ij}}(x; P\xi, 0, \ldots, 0) = \sigma_{B_{ij}}(Px; P\xi),$$

where $Py = (y_1, \ldots, y_p)$ $(y \in \mathbb{R}^{p+q})$. Because A_{ij} are independent of the K-variables, we have $A = A_{K\backslash G} = \sum_{i,j} A_{ij} \in \Psi^m(G)$ and $A|_{C^\infty(K\backslash G)} \in \Psi^m(K\backslash G)$. Let $f = \sum_k f_k \in C^\infty(K\backslash G) \subset C^\infty(G)$, $f_k = f\psi_k$. Then we have

$$(Af)(x) = \sum_{i,j,k} (A_{ij} f_k)(x)$$

$$= \sum_{i,j,k} \int_{\mathbb{R}^p} \sum_{\xi_{p+1}, \ldots, \xi_{p+q} \in \mathbb{Z}} \sigma_{A_{ij}}(x, \xi) \, \widehat{f_k}(\xi) \, \mathrm{e}^{\mathrm{i}2\pi x \cdot \xi} \, \mathrm{d}\xi_1 \cdots \mathrm{d}\xi_p$$

$$= \sum_{i,j,k} \int_{\mathbb{R}^p} \sigma_{A_{ij}}(x; P\xi, 0, \ldots, 0) \, \widehat{f_k}(P\xi, 0, \ldots, 0) \, \mathrm{e}^{\mathrm{i}2\pi (Px) \cdot (P\xi)} \, \mathrm{d}\xi_1 \cdots \mathrm{d}\xi_p$$

$$= \sum_{i,j,k} \int_{\mathbb{R}^p} \sigma_{B_{ij}}(Px; P\xi) \, \widehat{f_k}(P\xi, 0, \ldots, 0) \, \mathrm{e}^{\mathrm{i}2\pi (Px) \cdot (P\xi)} \, \mathrm{d}\xi_1 \cdots \mathrm{d}\xi_p$$

$$= \sum_{i,j,k} (B_{ij} f_k)(Px)$$

$$= (Bf)(Kx),$$

completing the proof.   □

Remark 13.3.6. Theorem 13.3.5 provides just one way of lifting operators in $\Psi^m(K\backslash G)$ to operators in $\Psi^m(G)$, unfortunately destroying ellipticity: this is due to the apparent non-ellipticity of the symbol χ in Lemma 13.3.4. We now discuss this problem and a possibility of other liftings.

Remark 13.3.7 (**Lifting the identity**). Let us lift the identity operator $I \in \Psi^0(\mathbb{R}^p)$ using the process suggested by Lemma 13.3.4. Of course, it would be desirable if $I \in \Psi^0(\mathbb{R}^p)$ could be extended to the identity in $\Psi^0(\mathbb{R}^{p+q})$, but now $\sigma_I(x, \xi) \equiv 1$, and thereby its lifting $A \in \Psi^0(\mathbb{R}^{p+q})$ has the non-elliptic homogeneous symbol $\sigma_A = \chi \in S^0(\mathbb{R}^{p+q})$.

Given an elliptic symbol $\sigma_B \in S^m(\mathbb{R}^p)$ one can occasionally modify the construction in Lemma 13.3.4 to get an extended elliptic symbol in $S^m(\mathbb{R}^{p+q})$. Sometimes the following trick helps: Let $\sigma_{A_1} \in S^m(\mathbb{R}^{p+q})$ be an extension of σ_{B_1} as in Lemma 13.3.4,

$$\sigma_{A_1}(x, \xi) = \chi_1(\xi)\, \sigma_{B_1}(x_1, \ldots, x_p; \xi_1, \ldots, \xi_p),$$

where $\chi_1 \in S^0(\mathbb{R}^{p+q})$ is a homogeneous symbol such that $\chi_1|_{(U \times \mathbb{R}^q) \backslash \mathbb{B}(0,1)} \equiv 0$, $\chi_1|_{\mathbb{R}^p \times V} \equiv 1$, where $U \subset \mathbb{R}^p$ and $V \subset \mathbb{R}^q$ are neighbourhoods of zeros. Take any elliptic symbol $\sigma_{B_2} \in S^m(\mathbb{R}^q)$, and modify Lemma 13.3.4 to construct an extension $\sigma_{A_2} \in S^m(\mathbb{R}^{p+q})$ such that

$$\sigma_{A_2}(x, \xi) = \chi_2(\xi)\, \sigma_{B_2}(x_{p+1}, \ldots, x_{p+q}; \xi_{p+1}, \ldots, \xi_{p+q})$$

for a homogeneous symbol $\chi_2 \in S^0(\mathbb{R}^{p+q})$ satisfying $\chi_2|_{(U \times \mathbb{R}^q) \backslash \mathbb{B}(0,1)} \equiv 1$ and $\chi_2|_{(\mathbb{R}^p \times V) \backslash \mathbb{B}(0,1)} \equiv 0$. Then $\sigma_{A_1} + \sigma_{A_2} \in S^m(\mathbb{R}^{p+q})$ is an extension for σ_{B_1} (modulo infinitely smoothing operators). For instance, if $B_1 = I \in \Psi^0(\mathbb{R}^p)$, let $B_2 = I \in \Psi^0(\mathbb{R}^q)$ and $\chi_2(\xi) = 1 - \chi_1(\xi)$ (for $|\xi| > 1$), then $A_1 + A_2 = I \in \Psi^0(\mathbb{R}^{p+q})$ (modulo infinitely smoothing operators).

Remark 13.3.8 (**No elliptic liftings**). It may happen that any lifting process for an elliptic symbol $\sigma_B \in S^m(\mathbb{R}^p)$ yields a non-elliptic symbol in $S^m(\mathbb{R}^{p+q})$. Consider, for instance, a case where $B \in \Psi^m(\mathbb{R}^2)$ is an elliptic convolution operator and $\xi \mapsto f(\xi) \equiv \sigma_B(x, \xi)$ is homogeneous outside the unit ball $\mathbb{B}(0,1) \subset \mathbb{R}^2$. If the restricted mapping $f|_{\mathbb{S}^1} : \mathbb{S}^1 \to \mathbb{C} \backslash \{0\}$ is not homotopic to a constant mapping (i.e., $f|_{\mathbb{S}^1}$ has a non-zero winding number) then no lifting $\sigma_A \in S^m(\mathbb{R}^3)$ of σ_B can be elliptic.

Multiplications on $K \backslash G$ have already been lifted to multiplications on G via $x \mapsto Kx$, and $A = A_{K \backslash G}$ for any right-convolution operator (multiplier) $A \in \mathcal{L}(C^\infty(G))$ (in fact, then $\sigma_A(x, \xi) = \sigma_A(\xi)$ for every $x \in G$). Sometimes on $K \backslash G$ we have operators that resemble convolution operators. Suppose we are given a right-convolution operator $A \in \Psi^m(SU(2))$. Then the restriction $B = A|_{C^\infty(\mathbb{S}^2)} \in \Psi^m(\mathbb{S}^2)$ is of the form

$$(Bf)(\phi, \theta) = \sum_{l=0}^{\infty} \sum_{m=-l}^{l} \left(\sum_{n=-l}^{l} a(l)_{mn}\, \widehat{f}(l)_n \right) Y_l^m(\phi, \theta), \tag{13.5}$$

where the coefficients $a(l)_{mn} \in \mathbb{C}$ can be calculated from the data

$$\{BY_l^m \mid l \in \mathbb{N}_0,\ m \in \{-l, -l+1, \ldots, l-1, l\}\}.$$

It is even true that the original operator A can be retrieved from the coefficients $a(l)_{mn}$. In fact, any operator $B \in \mathcal{L}(C^{\infty}(\mathbb{S}^2))$ of the form (13.5) can be lifted to a unique right-convolution operator belonging to $\mathcal{L}(C^{\infty}(\mathrm{SU}(2)))$. An interesting special case is

$$(Bf)(x) = \int_{\mathbb{S}^2} \kappa(x \cdot y) \ f(y) \ \mathrm{d}y,$$

where $\kappa \in \mathcal{D}'(\mathbb{S}^2)$, $(x, y) \mapsto x \cdot y$ is the scalar product of $\mathbb{R}^3$, and the integration is with respect to the angular part of the Lebesgue measure of $\mathbb{R}^3$. Then

$$(Bf)(\phi, \theta) = \sum_{l=0}^{\infty} \sum_{m=-l}^{l} c_l \ \widehat{\kappa}(l)_0 \ \widehat{f}(l)_m \ Y_l^m(\phi, \theta)$$

for some normalising constants c_l depending only on $l \in \mathbb{N}_0$.

13.4 Lifting of operators

We now describe the lifting of operators from $K \backslash G$ to G for general closed subgroup K, which can be done similar to the proof of Theorem 13.3.5:

Theorem 13.4.1 (Lifting of operators). *Let G be a compact Lie group with a closed subgroup K. Let $B \in \Psi^m(K \backslash G)$. Then there exists an operator $A = A_{K \backslash G} \in \Psi^m(G)$ such that $A|_{C^{\infty}(K \backslash G)} = B$.*

The rest of this section is devoted to the proof of this theorem. Since we proved Theorem 13.3.5 and Lemma 13.3.4 in detail, we sketch the construction here. We start with the following

Lemma 13.4.2 (Some properties of representations). *Let $\phi_0 \in \mathrm{Rep}(G)$ be the trivial one-dimensional representation given by $\phi_0(x) = 1$ for all $x \in G$. Then for every $\phi \in \mathrm{Rep}(G)$ we have*

$$\widehat{1}(\phi) = \delta_{\phi, \phi_0} \ I_{\dim \phi} = \begin{cases} 1, & \text{if } \phi = \phi_0, \\ 0, & \text{if } \phi \neq \phi_0, \end{cases}$$

where 0 on the right-hand side is the zero operator $0 \in \mathcal{L}(\mathcal{H}_\phi)$. Moreover, for every non-trivial $\phi \in \mathrm{Rep}(G)$, i.e., $\phi \notin [\phi_0]$, we have $\int_G \phi(x) \ \mathrm{d}x = 0$.

Proof. If $\phi \in \mathrm{Rep}(G)$ is such that $\phi \notin [\phi_0]$, then $\int_G \phi(x) \ \mathrm{d}x = 0$ follows from the orthogonality of ϕ_{ij}, $1 \leq i, j \leq \dim \phi$, and ϕ_0, given in Lemma 7.5.12. Consequently, $\widehat{1}(\phi) = \int_G \phi(x)^* \ \mathrm{d}x = 0 \in \mathcal{L}(\mathcal{H}_\phi)$ if $\phi \notin [\phi_0]$, and $\widehat{1}(\phi_0) = \int_G \phi_0(x)^* \ \mathrm{d}x = 1$. $\square$

Exercise 13.4.3. Show that for every $\phi \in \mathrm{Rep}(G)$ we have $\widehat{\phi}(\phi) = I_{\dim \phi}$, the identity operator on $\mathcal{L}(\mathcal{H}_\phi)$.

By the argument similar to that in the proof of Theorem 13.3.5 we can use the partition of unity to reduce the question to the extension of symbols from $\mathbb{R}^p$ to $\mathbb{R}^p \times K$. Let $x = (x', x'') \in \mathbb{R}^p \times K$, let $\xi \in \mathbb{R}^p$ and $\phi \in \text{Rep}(K)$. Assume that $\sigma_B = \sigma(x', \xi) \in S^m(\mathbb{R}^p)$ has an extension to $\mathbb{R}^p \times K$, i.e., that there exists a symbol $\sigma_A = \sigma_A(x', x'', \xi, \phi) \in S^m(\mathbb{R}^p) \otimes \Sigma^m(K)$ such that

$$\sigma_A(x', x'', \xi, \phi_0) = \sigma_B(x', \xi), \tag{13.6}$$

where $\phi_0 \in \text{Rep}(K)$ is the trivial representation. Then by the argument of Theorem 13.3.5, and using Lemma 13.4.2, we have

$$
\begin{aligned}
(Af)(x) &= \sum_{i,j,k} (A_{ij} f_k)(x) \\
&= \sum_{i,j,k} \int_{\mathbb{R}^p} \sum_{[\phi] \in \widehat{K}} \sigma_{A_{ij}}(x', x'', \xi, \phi) \, \widehat{f_k \otimes 1}(\xi, \phi) \, e^{i2\pi x' \cdot \xi} \phi(x'') \, d\xi \\
&= \sum_{i,j,k} \int_{\mathbb{R}^p} \sigma_{A_{ij}}(x', x'', \xi, \phi_0) \, \widehat{f_k}(\xi) \, e^{i2\pi x' \cdot \xi} \, d\xi \\
&= \sum_{i,j,k} (B_{ij} f_k)(x') \\
&= (Bf)(Kx).
\end{aligned}
$$

Thus, we need to construct an extension σ_A of σ_B from $\mathbb{R}^p$ to $\mathbb{R}^p \times K$ satisfying property (13.6). We note that it is enough to do it for symbols σ_B with compact support in x'. First, let us define operator $C \in \Psi^0(\mathbb{R}^p \times K)$ by

$$C := (I - 2\mathcal{L}_K)(I - \mathcal{L}_{\mathbb{R}^p \times K})^{-1},$$

where $\mathcal{L}_K$ is the Laplace operator on K and $\mathcal{L}_{\mathbb{R}^p \times K} = \mathcal{L}_{\mathbb{R}^p} + \mathcal{L}_K$ is the Laplace operator on $\mathbb{R}^p \times K$. For each $\phi \in \text{Rep}(K)$, $\phi : K \to \mathcal{U}(\mathcal{H}_\phi)$, let $\lambda_\phi \geq 0$ be the eigenvalue of $-\mathcal{L}_K$ corresponding to the eigenspace $\mathcal{H}_\phi \subset L^2(K)$. For the details of this construction on compact groups we refer to Theorem 8.3.47. Consequently, we have $\sigma_{I-2\mathcal{L}_K}(\xi, \phi) = (1 + 2\lambda_\phi) I_{\dim \phi}$ and $\sigma_{I-\mathcal{L}_{\mathbb{R}^p \times K}}(\xi, \phi) = (1 + |\xi|^2 + \lambda_\phi) I_{\dim \phi}$, so that

$$\sigma_C(x', x'', \xi, \phi) = \sigma_C(\xi, \phi) = (1 + 2\lambda_\phi)(1 + |\xi|^2 + \lambda_\phi)^{-1} I_{\dim \phi}.$$

Now, let $\chi \in C_0^\infty(\mathbb{R})$ be such that $\chi(t) = 1$ for $|t| \leq 1/2$ and $\chi(t) = 0$ for $|t| \geq 1$. Let us denote

$$A_0 := \chi(C) = \chi\big((I - 2\mathcal{L}_K)(I - \mathcal{L}_{\mathbb{R}^p \times K})^{-1}\big).$$

By writing $\chi(t) = \int_{\mathbb{R}} e^{2\pi i \tau t} \, \widehat{\chi}(\tau) \, d\tau$, we have $A_0 = \int_{\mathbb{R}} e^{2\pi i \tau C} \, \widehat{\chi}(\tau) \, d\tau \in \Psi^0(\mathbb{R}^p \times K)$. This follows from the fact that the operator $u(\tau) := e^{2\pi i \tau C}$ is the solution of the Cauchy problem

$$i\partial_\tau u(\tau) + 2\pi C u(\tau) = 0, \quad u(0) = I,$$

and hence can be constructed by solving the transport equations. We note that such an argument is a special case of solving the hyperbolic Cauchy problems in terms of Fourier integral operators but the situation at hand is simpler because operator C is of order zero. Thus, $A_0 \in \Psi^0(\mathbb{R}^p \times K)$ follows, e.g., from [130, Section XII.1]. For the representation of $\chi(C)$ on the Fourier transform side one can also see that

$$\sigma_{A_0}(x', x'', \xi, \phi) = \sigma_{A_0}(\xi, \phi) = \chi\big((1 + 2\lambda_\phi)(1 + |\xi|^2 + \lambda_\phi)^{-1}\big) I_{\dim \phi}$$

is a diagonal symbol. We claim that the operator $\mathrm{Op}(\sigma_A)$ with $\sigma_A := \sigma_{A_0}\sigma_B$ satisfies the required properties. Let us first show that $A \in \Psi^m(G)$. For this we can apply Theorem 10.9.6 and use the characterisation given in Definition 10.9.5. On one hand we have

$$\|\triangle_\phi^\alpha \sigma_A\|_{op} = \|(\triangle_\phi^\alpha \sigma_{A_0})\sigma_B\|_{op} \leq C_\alpha(\langle\phi\rangle + \langle\xi\rangle)^{-|\alpha|}\langle\xi\rangle^m$$

which implies $\|\triangle_\phi^\alpha \sigma_A\|_{op} \leq C_\alpha'(\langle\phi\rangle + \langle\xi\rangle)^{m-|\alpha|}$, if we use Theorem 10.9.6 for the operator $A_0 \in \Psi^0(G)$ as well as the fact that $\lambda_\phi \leq |\xi|^2$, and hence $\langle\phi\rangle = (1 + \lambda_\phi)^{1/2} \leq \langle\xi\rangle$ on the support of $\sigma_{A_0}(\xi, \phi)$. On the other hand we have

$$\begin{aligned}
\|\partial_{\xi_j}\sigma_A\|_{op} &\leq \|\partial_{\xi_j}\sigma_{A_0}\|_{op}\,|\sigma_B| \\
&+ |\chi((1 + 2\lambda_\phi)(1 + |\xi|^2 + \lambda_\phi)^{-1})|\,|\partial_{\xi_j}\sigma_B| \\
&\leq C_1(\langle\phi\rangle + \langle\xi\rangle)^{-1}\langle\xi\rangle^m + C_2\langle\xi\rangle^{m-1},
\end{aligned}$$

and again this implies $\|\partial_{\xi_j}\sigma_A\|_{op} \leq C_3(\langle\phi\rangle + \langle\xi\rangle)^{m-1}$ if we use that $\langle\phi\rangle \leq \langle\xi\rangle$ on the support of $\sigma_{A_0}(\xi, \phi)$. Similarly, one can extend this to the higher-order derivatives ∂_ξ^α. Let us show (10.36). Using the usual Euclidean formula together with (10.18), the right-convolution kernel of A can be written as

$$R_A(x', x'', y', y'') = \int_{\mathbb{R}^p} e^{2\pi i y' \cdot \xi} \sum_{[\phi] \in \widehat{K}} \dim(\phi)\,\mathrm{Tr}\big(\phi(y')\,\sigma_{A_0}(\xi, \phi)\big)\,\sigma_B(x', \xi)\,\mathrm{d}\xi.$$

Now, if $y' \neq 0$ we can integrate by parts in ξ any number of times. At the same time, viewing σ_{A_0} as a ξ-dependent symbol on K implies that R_A is smooth for $y'' \neq e$ because the same property holds for σ_{A_0}. Hence the singular support of $y \mapsto R_A(x, y)$ is contained in $(0, e)$. The other properties in Definition 10.9.5 follow from the diagonality of the symbol σ_{A_0}. It remains to show (13.6). For this we note that $\phi_0 \equiv 1$ is the eigenfunction of $-\mathcal{L}_K$ with the eigenvalue $\lambda_{\phi_0} = 0$. Consequently, we have

$$\sigma_{A_0}(x', x'', \xi, \phi_0) = \chi\big((1 + |\xi|^2)^{-1}\big) = 1$$

for $|\xi| \geq 1$. Finally we note that we can modify $\sigma_A(x', x'', \xi, \phi_0)$ arbitrarily for small ξ, in particular setting it to be zero for $|\xi| \leq 1$, similarly to Lemma 13.3.4, thus completing the proof. $\square$

Bibliography

[1] E. Abe, *Hopf algebras*. Cambridge University Press, 1980.

[2] M. Abramowitz and I.A. Stegun, *Handbook of mathematical functions*. United States Department of Commerce, 1965.

[3] M.S. Agranovich, *Spectral properties of elliptic pseudodifferential operators on a closed curve*. (Russian) Funktsional. Anal. i Prilozhen. **13** (1979), no. 4, 54–56. (English translation in Functional Analysis and Its Applications. 13, p. 279–281.)

[4] M.S. Agranovich, *Elliptic pseudodifferential operators on a closed curve*. (Russian) Trudy Moskov. Mat. Obshch. **47** (1984), 22–67, 246. (English translation in Transactions of Moscow Mathematical Society. 47, p. 23–74.)

[5] M.S. Agranovich, *Elliptic operators on closed manifolds* (in Russian). Itogi Nauki i Tehniki, Ser. Sovrem. Probl. Mat. Fund. Napravl. 63 (1990), 5–129. (English translation in Encyclopaedia Math. Sci. 63 (1994), 1–130.)

[6] B.A. Amosov, *On the theory of pseudodifferential operators on the circle*. (Russian) Uspekhi Mat. Nauk **43** (1988), 169–170; translation in Russian Math. Surveys **43** (1988), 197–198.

[7] B.A. Amosov, *Approximate solution of elliptic pseudodifferential equations on a smooth closed curve*. (Russian) Z. Anal. Anwendungen **9** (1990), 545–563.

[8] P. Antosik, J. Mikusiński and R. Sikorski, *Theory of distributions. The sequential approach*. Warszawa. PWN – Polish Scientific Publishers, 1973.

[9] A. Baker, *Matrix Groups. An Introduction to Lie Group Theory*. Springer-Verlag, 2002.

[10] J. Barros-Neto, *An introduction to the theory of distributions*. Marcel Dekker, Inc., 1973.

[11] R. Beals, *Advanced mathematical analysis*. Springer-Verlag, 1973.

[12] R. Beals, *Characterization of pseudodifferential operators and applications*. Duke Mathematical Journal. **44** (1977), 45–57.

[13] A.P. Bergamasco and P.L. da Silva, *Solvability in the large for a class of vector fields on the torus*. J. Math. Pures Appl. **86** (2006), 427–447.

[14] J. Bergh and J. Löfström, *Interpolation spaces. An introduction*. Springer-Verlag, 1976.

[15] G. Boole, *Finite differences*. 4th edition, Library of Congress catalogue card no. 57-8495. (1st edition 1860).

[16] J. Bourgain, *Exponential sums and nonlinear Schrödinger equations*. Geom. Funct. Anal. **3** (1993), 157–178.

[17] J. Bourgain, *Fourier transform restriction phenomena for certain lattice subsets and applications to nonlinear evolution equations. I. Schrödinger equations*. Geom. Funct. Anal. **3** (1993), 107–156.

[18] J. Bourgain, *Global solutions of nonlinear Schrödinger equations*. American Mathematical Society Colloquium Publications, 1999.

[19] G.E. Bredon: *Introduction to Compact Transformation Groups*. Academic Press, 1972.

[20] T. Bröcker and T. tom Dieck, *Representations of Compact Lie Groups*. Springer-Verlag, 1985.

[21] A.P. Calderón, *Commutators of singular integral operators*. Proc. Nat. Acad. Sci. USA 53 (1965), 1092–1099.

[22] A.P. Calderón and R. Vaillancourt, *On the boundedness of pseudo-differential operators*, J. Math. Soc. Japan **23** (1971), 374–378.

[23] R.R. Coifman and Y. Meyer, *Au-delà des opérateurs pseudo-différentiels*. Astérisque 57, Société Math. de France. 1978.

[24] L. Comtet, *Advanced combinatorics*. Dordrecht. D.Reidel Publishing Company, 1974.

[25] H.O. Cordes, *On compactness of commutators of multiplications and convolutions, and boundedness of pseudo-differential operators*, J. Funct. Anal. **18** (1975), 115–131.

[26] H.O. Cordes, *On pseudodifferential operators and smoothness of special Lie group representations*. Manuscripta Math. 28 (1979), 51–69.

[27] H.O. Cordes, *The technique of pseudodifferential operators*. Cambridge University Press, 1995.

[28] D. Crespin, *Hahn–Banach Theorem Implies Riesz Theorem*. Portugaliae Mathematica **51** (1994), 217–218.

[29] G. David, *Wavelets and singular integrals on curves and surfaces*. Springer-Verlag, 1992.

[30] J.J. Duistermaat, *Fourier integral operators*. Birkhäuser, 1996.

[31] J.J. Duistermaat and J.A. Kolk, *Lie groups.* Springer-Verlag, 2000.

[32] J. Dunau, *Fonctions d'un operateur elliptique sur une variete compacte.* J. Math. Pures et Appl. **56** (1977), 367–391.

[33] Y.V. Egorov, B.-W. Schulze, *Pseudo-differential operators, singularities, applications.* Operator Theory: Advances and Applications, 93, Birkhäuser, 1997.

[34] J. Elschner, *Singular ordinary differential operators and pseudodifferential equations.* Springer-Verlag, 1985.

[35] G.B. Folland. *Real Analysis. Modern techniques and their applications.* Second edition. A Wiley-Interscience Publication. John Wiley and Sons, Inc., 1999.

[36] G.B. Folland, *Harmonic analysis in phase space.* Princeton Univ. Press, 1989.

[37] H. Freudenthal and H. de Vries, *Linear Lie Groups.* Academic Press, 1969.

[38] S.A. Gaal, *Linear Analysis and Representation Theory.* Springer-Verlag, 1973.

[39] I.M. Gelfand and G.E. Shilov, *Generalized functions.* Vols. 1–3, Academic Press, 1968

[40] F. Geshwind and N.H. Katz, *Pseudodifferential operators on* SU(2), J. Fourier Anal. Appl. **3** (1997), 193–205.

[41] P. Glowacki, *A symbolic calculus and L^2-boundedness on nilpotent Lie groups*, J. Funct. Anal. **206** (2004), 233–251.

[42] T. Gramchev, P. Popivanov and M. Yoshino, *Global solvability and hypoellipticity on the torus for a class of differential operators with variable coefficients.* Proc. Japan Acad. Ser. A Math. Sci. **68** (1992), 53–57.

[43] T. Gramchev, P. Popivanov and M. Yoshino, *Global properties in spaces of generalized functions on the torus for second order differential operators with variable coefficients.* Rend. Sem. Mat. Univ. Politec. Torino **51** (1993), 145–172.

[44] T. Gramchev, *Simultaneous normal forms of perturbations of vector fields on tori with zero order pseudodifferential operators.* Symmetry and perturbation theory (Rome, 1998), 187–195, World Sci. Publ., River Edge, NJ, 1999.

[45] A. Grigis and J. Sjöstrand, *Microlocal analysis for differential operators. An introduction.* Cambridge University Press, 1994.

[46] P.R. Halmos, *Naive Set Theory.* Springer-Verlag. 1974.

[47] S. Helgason, *Differential geometry and symmetric spaces.* Academic Press, 1962.

[48] S. Helgason, *Topics in harmonic analysis on homogeneous spaces*. Birkhäuser, 1981.

[49] S. Helgason, *Groups and geometric analysis*, Academic Press, 1984.

[50] E. Hewitt and K.A. Ross, *Abstract Harmonic Analysis I*. Springer-Verlag, 1963.

[51] E. Hewitt and K.A. Ross, *Abstract Harmonic Analysis II*. Springer-Verlag, 1963.

[52] F.B. Hildebrand, *Finite-difference equations and simulations*. Prentice-Hall, Inc., 1968.

[53] E. Hille and R.S. Phillips, *Functional Analysis and Semi-Groups*. American Mathematical Society, 1981.

[54] M.W. Hirsch, *Differential Topology*. Springer-Verlag, 1976.

[55] L. Hörmander, *The analysis of linear partial differential operators III*. Springer-Verlag, 1985.

[56] L. Hörmander, *The Analysis of Linear Partial Differential Operators IV*. Springer-Verlag, 1985.

[57] R. Howe, *A symbolic calculus for nilpotent groups*, Operator algebras and group representations, Vol. I (Neptun, 1980), 254–277, Monogr. Stud. Math., 17, Pitman, Boston, MA, 1984.

[58] T. Husain, *Introduction to Topological Groups*. W.B. Saunders Company, 1966.

[59] V. Hutson and J.S. Pym, *Applications of functional analysis and operator theory*. Academic Press, 1980.

[60] Ch. Jordan, *Calculus of finite differences*. New York. Chelsea Publishing Company, 1950.

[61] Y. Katznelson, *An Introduction to Harmonic Analysis*. Dover, 1976.

[62] O. Kelle and G. Vainikko, *A fully discrete Galerkin method of integral and pseudodifferential equations on closed curves*. Journal for Analysis and its Applications. **14** (1995), 593–622.

[63] J.L. Kelley, I. Namioka et al.: *Linear Topological Spaces*. D. Van Nostrand Company, Inc. Princeton, New Jersey, 1963.

[64] A.A. Kirillov, *Elements of the Theory of Representations*. Springer-Verlag, 1976.

[65] A. Klimyk and K. Schmüdgen, *Quantum Groups and Their Representations*. Springer-Verlag, 1997.

[66] K. Knopp, *Theory and application of infinite series*. Glasgow. Blackie & Son Limited, 1948.

[67] D.E. Knuth, *Two notes on notation.* Amer. Math. Monthly **99** (1992), 403–422.

[68] J.J. Kohn and L. Nirenberg, *On the algebra of pseudo-differential operators.* Comm. Pure Appl. Math. **18** (1965), 269–305.

[69] R. Kress, *Linear integral equations.* Springer-Verlag, 1989.

[70] E. Kreyszig, *Introductory Functional Analysis with Applications.* John Wiley & Sons 1989.

[71] H. Kumano-go, *Pseudodifferential operators.* MIT Press, Cambridge, Mass.-London, 1981.

[72] J.-L. Lions and E. Magenes, *Non-homogeneous boundary value problems and applications.* Vol. I. Springer-Verlag. 1972.

[73] S. Majid, *Foundations of Quantum Group Theory.* Cambridge University Press, 1995.

[74] S. Majid, *A Quantum Groups Primer.* Cambridge University Press, 2002.

[75] D.K. Maslen, *Efficient computation of Fourier transforms on compact groups.* J. Fourier Anal. Appl. 4 (1998), 19–52.

[76] W. McLean, *Local and global description of periodic pseudodifferential operators.* Math. Nachr. **150** (1991), 151–161.

[77] G.A. Meladze and M.A. Shubin, *A functional calculus of pseudodifferential operators on unimodular Lie groups.* J. Soviet Math. **47** (1989), 2607–2638.

[78] A. Melin, *Parametrix constructions for right invariant differential operators on nilpotent groups*, Ann. Global Anal. Geom. **1** (1983), 79–130.

[79] S.T. Melo, *Characterizations of pseudodifferential operators on the circle.* Proc. Amer. Math. Soc. **125** (1997), 1407–1412.

[80] S.T. Melo, *Smooth operators for the regular representation on homogeneous spaces.* Stud. Math. **142** (2000), 149–157.

[81] R.B. Melrose, *Geometric scattering theory.* Cambridge Univ. Press, 1995.

[82] S. Molahajloo and M.W. Wong, *Pseudo-differential operators on* $\mathbb{S}^1$, in "New Developments in Pseudo-Differential Operators", Operator Theory: Advances and Applications, Vol. 189, 2009, 297–306.

[83] S. Molahajloo, *Pseudo-differential operators on* $\mathbb{Z}$, in " Pseudo-Differential Operators: Complex Analysis and Partial Differential Equations", Operator Theory: Advances and Applications, to appear.

[84] A. Mohammed and M.W. Wong, *Sampling and pseudo-differential operators*, in "New Developments in Pseudo-Differential Operators", Operator Theory: Advances and Applications, Vol. 189, 2009, 323–332.

[85] N.E. Nörlund, *Vorlesungen über Differenzenrechnung.* Berlin. Verlag von Julius Springer, 1924.

[86] I.G. Petrovski, *Über das Cauchysche Problem für Systeme von partiellen Differentialgleichungen.* Mat. Sb. **5** (1937), 815–870.

[87] A. Pietsch, *Nuclear locally convex spaces.* Springer-Verlag, 1972.

[88] D.W. Robinson, *Elliptic operators and Lie groups.* Oxford University Press, 1991.

[89] W. Rudin, *Functional Analysis.* Tata McGraw-Hill, 1974.

[90] W. Rudin, *Real and Complex Analysis.* McGraw-Hill 1987.

[91] M.V. Ruzhansky, *Singularities of affine fibrations in the regularity theory of Fourier integral operators,* Russian Math. Surveys, **55** (2000), 93–161.

[92] M. Ruzhansky, *Regularity theory of Fourier integral operators with complex phases and singularities of affine fibrations.* CWI Tracts, vol. 131, 2001.

[93] M. Ruzhansky, *L^p-distributions on symmetric spaces.* Results Math. 44 (2003), 159–168.

[94] M. Ruzhansky and M. Sugimoto, *Global calculus of Fourier integral operators, weighted estimates, and applications to global analysis of hyperbolic equations,* in Advances in pseudo-differential operators, 65–78, Oper. Theory Adv. Appl., 164, Birkhäuser, 2006.

[95] M. Ruzhansky and M. Sugimoto, Global L^2 boundedness theorems for a class of Fourier integral operators, *Comm. Partial Differential Equations,* **31** (2006), 547–569.

[96] M. Ruzhansky and M. Sugimoto, *Weighted L^2 estimates for a class of Fourier integral operators,* arXiv:0711.2868v1

[97] M. Ruzhansky and V. Turunen, *On the Fourier analysis of operators on the torus,* Modern trends in pseudo-differential operators, 87-105, Oper. Theory Adv. Appl., 172, Birkhäuser, Basel, 2007.

[98] M. Ruzhansky and V. Turunen, *Quantization of pseudo-differential operators on the torus,* arXiv:0805.2892v1

[99] M. Ruzhansky and V. Turunen, *Global quantization of pseudo-differential operators on compact Lie groups, $SU(2)$ and $\mathbb{S}^3$,* arXiv:0812.3961v1.

[100] Yu. Safarov, *Pseudodifferential operators and linear connections.* Proc. London Math. Soc. **74** (1997), 379–416.

[101] X. Saint Raymond, *Elementary introduction to the theory of pseudodifferential operators.* Studies in Advanced Mathematics. CRC Press, Boca Raton, FL, 1991.

[102] J. Saranen and G. Vainikko, *Periodic integral and pseudodifferential equations with numerical approximation.* Springer-Verlag, 2002.

[103] J. Saranen, W.L. Wendland, *The Fourier series representation of pseudodifferential operators on closed curves.* Complex Variables Theory Appl. **8** (1987), 55–64.

[104] L. Schwartz, *Sur l'impossibilité de la multiplication des distributions,* C. R. Acad. Sci. Paris **239** (1954), 847–848.

[105] L. Schwartz, *Théorie des distributions, I, II.* 2nd ed., Hermann, Paris, 1957.

[106] L. Schwartz, *Méthodes mathématiques pour les sciences physiques.* (French) Hermann, Paris 1961

[107] I.E. Segal, *An extension of Plancherel's formula to separable unimodular groups.* Ann. Math. **52** (1950), 272–292.

[108] I.E. Segal, *A non-commutative extension of abstract integration.* Ann. Math. **57** (1953), 401–457.

[109] V.A. Sharafutdinov, *Geometric symbol calculus for pseudodifferential operators. I* [Translation of Mat. Tr. **7** (2004), 159–206]. Siberian Adv. Math. **15** (2005), 81–125.

[110] T. Sherman, *Fourier analysis on the sphere,* Trans. Amer. Math. Soc. **209** (1975), 1–31.

[111] T. Sherman, *The Helgason Fourier transform for compact Riemannian symmetric spaces of rank one,* Acta Math. **164** (1990), 73–144.

[112] M.A. Shubin, *Pseudodifferential operators and spectral theory.* Springer-Verlag, 1987.

[113] J. Sjöstrand, *Microlocal analysis.* In *Development of mathematics 1950–2000,* 967–991, Birkhäuser, Basel, 2000.

[114] R.M. Solovay, *A model of set-theory in which every set of reals is Lebesgue measurable.* Ann. of Math. **92** (1970), 1–56.

[115] M. Spivak, *A comprehensive introduction to differential geometry.* Publish or Perish, Inc., Boston, Mass., 1975.

[116] L.A. Steen and J.A. Seebach, Jr.: *Counterexamples in Topology.* Dover Publications, Inc., 1995.

[117] J.F. Steffensen, *Interpolation.* New York. Chelsea Publishing Company, 1950.

[118] E.M. Stein, *Harmonic analysis: real-variable methods, orthogonality, and oscillatory integrals.* Princeton University Press, 1993.

[119] H. Stetkaer, *Invariant pseudo-differential operators.* Math. Scand. **28** (1971), 105–123.

[120] W.F. Stinespring, *Integration theorems for gages and duality for unimodular groups*. Trans. Amer. Math. Soc. 90 (1959), 15–56.

[121] R. Strichartz, *Invariant pseudo-differential operators on a Lie group*. Ann. Scuola Norm. Sup. Pisa **26** (1972), 587–611.

[122] R. Strichartz, *Local harmonic analysis on spheres*. J. Funct. Anal. **77** (1988), 403–433.

[123] M. Sugiura, *Unitary Representations and Harmonic Analysis – an Introduction*. Kodansha Ltd., 1975.

[124] P. Suppes, *Axiomatic Set Theory*. Dover Publications, Inc., 1972.

[125] S.L. Svensson, *Pseudodifferential operators – a new approach to the boundary problems of physical geodesy*. Manuscripta Geodaetica **8** (1983), 1–40.

[126] M.E. Sweedler, *Hopf Algebras*. W.A. Benjamin, Inc. 1969.

[127] K. Tapp, *Matrix Groups for Undergraduates*. American Mathematical Society, 2005.

[128] M.E. Taylor, *Noncommutative microlocal analysis*. Memoirs AMS **52** (1984), No. 313.

[129] M.E. Taylor, *Noncommutative harmonic analysis*. Mathematical Surveys and Monographs, Vol. 22, Amer. Math. Soc., 1986.

[130] M.E. Taylor, *Pseudodifferential operators*. Princeton University Press, 1981.

[131] M.E. Taylor, *Pseudodifferential operators and nonlinear PDE*. Birkhäuser, 1991.

[132] M.E. Taylor, *Partial differential equations*. Vol. III. Nonlinear equations. Springer-Verlag, 1997.

[133] M.E. Taylor, *Beals–Cordes-type characterizations of pseudodifferential operators*. Proc. Amer. Math. Soc. **125** (1997), 1711–1716.

[134] F. Treves: *Topological Vector Spaces, Distributions and Kernels*. Adademic Press. New York, 1967.

[135] F. Treves, *Introduction to pseudodifferential and Fourier integral operators*. Plenum Press. Vol. 1, Vol. 2., 1980.

[136] V. Turunen, *Commutator characterization of periodic pseudodifferential operators*. Z. Anal. Anw. **19** (2000), 95–108.

[137] V. Turunen, *Periodic convolution integral operators*. An undergraduate essay at Helsinki University of Technology. 19 p., 1996.

[138] V. Turunen, *Symbol analysis of periodic pseudodifferential operators*. Master's Thesis. Helsinki University of Technology, 1997.

[139] V. Turunen, *Pseudodifferential calculus on compact Lie groups and homogeneous spaces.* Helsinki University of Technology, PhD Thesis, 2001.

[140] V. Turunen, *Pseudodifferential calculus on the 2-sphere.* Proc. Estonian Acad. Sci. Phys. Math. **53** (2004), 156–164.

[141] V. Turunen and G. Vainikko, *On symbol analysis of periodic pseudodifferential operators.* Z. Anal. Anw. **17** (1998), 9–22.

[142] G. Vainikko, *Periodic integral and pseudodifferential equations.* Espoo. Helsinki University of Technology, 1996.

[143] G. Vainikko, Personal communication, 1997.

[144] G.M. Vainikko and I.K. Lifanov, *Generalization and use of the theory of pseudodifferential operators in the modeling of some problems in mechanics.* (Russian) Dokl. Akad. Nauk **373** (2000), 157–160.

[145] G.M. Vainikko and I.K. Lifanov, *The modeling of problems in aerodynamics and wave diffraction and the extension of Cauchy-type integral operators on closed and open curves.* (Russian) Differ. Uravn. **36** (2000), 1184–1195, 1293; translation in Differ. Equ. **36** (2000), 1310–1322.

[146] J. Väisälä, *Topologia II.* Limes, 1987.

[147] N. Th. Varopoulos, *Analysis on Lie groups.* Journal of Functional Analysis **76** (1988), 346–410.

[148] N.J. Vilenkin, *Special Functions and the Theory of Group Representations.* American Mathematical Society, 1968.

[149] N.R. Wallach, *Harmonic Analysis on Homogeneous spaces.* Marcel Dekker Inc., 1973.

[150] H. Whitney, *Differentiable manifolds,* Ann. Math. **37** (1936), 645–680.

[151] H. Widom, *A complete symbolic calculus for pseudodifferential operators,* Bull. Sci. Math. (2) **104** (1980), no. 1, 19–63.

[152] M.W. Wong, *An introduction to pseudo-differential operators.* Second edition. World Scientific Publishing Co., Inc., 1999.

[153] K. Yosida, *Functional Analysis.* Reprint of the sixth (1980) edition. Springer-Verlag, Berlin, 1995.

[154] D.P. Zelobenko, *Compact Lie Groups and Their Representations.* American Mathematical Society, 1973.

[155] A. Zygmund, *Trigonometric series.* Cambridge University Press, 1959.

Notation

$\mathcal{S}(\widehat{G})$, p_k, 539

$\mathcal{S}'(\widehat{G})$, 543

$\langle\cdot,\cdot\rangle_{\widehat{G}}$, 543

$L^p(\widehat{G})$, $\ell^p\left(\widehat{G},\dim^{p\left(\frac{2}{p}-\frac{1}{2}\right)}\right)$, 546

$L^p_k(\widehat{G})$, 550

$K_A(x,y)$, $L_A(x,y)$, $R_A(x,y)$, 550

$l(f)$, $r(f)$, 551, 579

∂^α, 534, 560

$\sigma_A(x,\xi)$, 552

f_ϕ, A_ϕ, 556

u_L, u_R, 556

$\langle A,B\rangle_{HS}$, $||A||_{HS}$, $||A||_{op}$, 559

$\triangle^\alpha_\xi$, 564

$\triangle_q$, 564

$\mathcal{A}^m_k(M)$, 566

$\Sigma^m(G)$, $\Sigma^m_k(G)$, 575

π_L, π_R, 580

$R_A(x)$, $L_A(x)$, 582

l_a, $l_a(x)$, r_A, $r_A(x)$, 583

$\mathcal{D}(G)$, 591

$\mathrm{SU}(2)$, 599

$\mathbb{H}$, 603

$\mathrm{Sp}(n)$, 606

$\mathrm{Sp}(n,\mathbb{C})$, 606

w_1, w_2, w_3, 607

Y_1, Y_2, Y_3, 607

D_1, D_2, D_3, 609

∂_+, ∂_-, ∂_0, 611

V_l, T_l, 612

t^l, t^l_{mn}, P^l_{mn}, 617

t_{--}, t_{-+}, t_{+-}, t_{++}, 621

$\widehat{f}(l)_{mn}$, 632

$\sigma_A(x,l)$, $\sigma_A(x,l)_{mn}$, 632

σ_{∂_+}, σ_{∂_-}, σ_{∂_0}, 634

$S^m(\mathrm{SU}(2))$, 656

$S^m(\mathbb{S}^3)$, 661

$\mathcal{D}'_{L^1}(M)$, 668

$p_{E\to B}$, 668

$K\backslash G$, 669

Index

Pseudo-Differential Operators (PDO)
Theorie and Applications

Edited by
M.W. Wong, York University, Canada

In cooperation with an international editorial board

Pseudo-Differential Operators: Theory and Applications is a series of moderately priced graduate-level textbooks and monographs appealing to students and experts alike. Pseudo-differential operators are understood in a very broad sense and include such topics as harmonic analysis, PDE, geometry, mathematical physics, microlocal analysis, time-frequency analysis, imaging and computations. Modern trends and novel applications in mathematics, natural sciences, medicine, scientific computing, and engineering are highlighted.

Forthcoming

Nicola, F. / Rodino, L.
Global Pseudo-Differential Calculus on
Euclidean Spaces
ISBN 978-3-7643-8511-8

Lerner, N.
Metrics on the Phase Space and Non-Selfadjoint
Pseudo-Differential Operators
ISBN 978-3-7643-8509-5

de Gosson, M.A.
Bopp Pseudo-Differential Operators and
Deformation Quantization
ISBN 978-3-7643-9991-7

Available

PDO 2: Ruzhansky, M. / Turunen, V.
Pseudo-Differential Operators and Symmetries.
Background Analysis and Advanced Topics
(2009)
ISBN 978-3-7643-8513-2

This monograph develops a global quantization theory of pseudo-differential operators on compact Lie groups.

Traditionally, the theory of pseudo-differential operators was introduced in the Euclidean setting with the aim of tackling a number of important problems in analysis and in the theory of partial differential equations. This also yields a local theory of pseudo-differential operators on manifolds. The present book takes a different approach by using global symmetries of the space which are often available. First, a particular attention is paid to the theory of periodic operators, which are realized in the form of pseudo-differential and Fourier integral operators on the torus. Then, the cases of the unitary group SU(2) and the 3-sphere are analyzed in extensive detail. Finally, the monograph also develops elements of the theory of pseudo-differential operators on general compact Lie groups and homogeneous spaces.

The exposition of the book is self-contained and provides the reader with the background material surrounding the theory and needed for working with pseudo-differential operators in different settings. The background section of the book may be used for independent learning of different aspects of analysis and is complemented by numerous examples and exercises.

PDO 1: Unterberger, A.
Quantization and Arithmetic (2008)
ISBN 978-3-7643-8790-7

The primary aim of this book is to create situations in which the zeta function, or other L-functions, will appear in spectral-theoretic questions. A secondary aim is to connect pseudo-differential analysis, or quantization theory, to analytic number theory. Both are attained through the analysis of operators on functions on the line by means of their diagonal matrix elements against families of arithmetic coherent states: these are families of discretely supported measures on the line, transforming in specific ways under the part of the metaplectic representation or, more generally, representations from the discrete series of SL(2,R), lying above an arithmetic group such as SL(2,Z).

Printed in the United States
By Bookmasters